SMART ENGINEERING SYSTEM DESIGN:

Neural Networks, Fuzzy Logic, Evolutionary Programming, Data Mining, and Complex Systems

VOLUME 11

ASME PRESS SERIES ON INTELLIGENT ENGINEERING SYSTEMS THROUGH ARTIFICIAL NEURAL NETWORKS

EDITOR
C.H. Dagli, University of Missouri-Rolla, Rolla, Missouri, USA

Intelligent Engineering Systems Through Artificial Neural Networks, Volume 1, edited by Cihan H. Dagli, Soundar R.T. Kumara, and Yung C. Shin, 1991

Intelligent Engineering Systems Through Artificial Neural Networks, Volume 2, edited by Cihan H. Dagli, Laura I. Burke, and Yung C. Shin, 1992

Intelligent Engineering Systems Through Artificial Neural Networks, Volume 3, edited by Cihan H. Dagli, Laura I. Burke, Benito Fernandez, and Joydeep Ghosh, 1993

Intelligent Engineering Systems Through Artificial Neural Networks, Volume 4, edited by Cihan H. Dagli, Benito Fernandez, Joydeep Ghosh, and R.T. Soundar Kumara, 1994

Intelligent Engineering Systems Through Artificial Neural Networks, Volume 5, edited by Cihan H. Dagli, Metin Akay, C.L. Phillip Chen, Benito Fernandez, and Joydeep Ghosh, 1995

Intelligent Engineering Systems Through Artificial Neural Networks, Volume 6, edited by Cihan H. Dagli, Metin Akay, C.L. Phillip Chen, Benito Fernandez, and Joydeep Ghosh, 1996

Intelligent Engineering Systems Through Artificial Neural Networks: Smart Engineering Systems: Neural Networks, Fuzzy Logic, Data Mining and Evolutionary Programming, Volume 7, edited by Cihan H. Dagli, Metin Akay, Okan Ersoy, Benito Fernandez, and Alice Smith, 1997

Intelligent Engineering Systems Through Artificial Neural Networks: Smart Engineering Systems: Neural Networks, Fuzzy Logic, Evolutionary Programming, Data Mining and Rough Sets, Volume 8, edited by Cihan H. Dagli, Metin Akay, Anna L. Buczak, Okan Ersoy, and Benito Fernandez, 1998

Intelligent Engineering Systems Through Artificial Neural Networks: Smart Engineering System Design: Neural Networks, Fuzzy Logic, Evolutionary Programming, Data Mining and Complex Systems, Volume 9, edited by Cihan H. Dagli, Anna L. Buczak, Joydeep Ghosh, Mark Embrechts and Okan Ersoy, 1999

Intelligent Engineering Systems Through Artificial Neural Networks: Smart Engineering System Design: Neural Networks, Fuzzy Logic, Evolutionary Programming, Data Mining and Complex Systems, Volume 10, edited by Cihan H. Dagli, Anna L. Buczak, Joydeep Ghosh, Mark Embrechts, Okan Ersoy and Stephen Kercel, 2000

Intelligent Engineering Systems Through Artificial Neural Networks: Smart Engineering System Design: Neural Networks, Fuzzy Logic, Evolutionary Programming, Data Mining and Complex Systems, Volume 11, edited by Cihan H. Dagli, Anna L. Buczak, Joydeep Ghosh, Mark Embrechts, Okan Ersoy and Stephen Kercel, 2001

SMART ENGINEERING SYSTEM DESIGN:

Neural Networks, Fuzzy Logic, Evolutionary Programming, Data Mining and Complex Systems

VOLUME 11

Proceedings of the Artificial Neural Networks in Engineering Conference (ANNIE 2001), held November 4-7, 2001, in St. Louis, Missouri, U.S.A.

EDITORS
Cihan H. Dagli
Department of Engineering Management
University of Missouri-Rolla

Anna L. Buczak
Philips Research
Briarcliff Manor, New York

Joydeep Ghosh
University of Texas at Austin
Austin, Texas

Mark Embrechts
Rensselaer Polytechnic Institute, RPI
Troy, New York

Okan Ersoy
Purdue University
West Lafayette, Indiana

Stephen Kercel
Oak Ridge National Laboratory
Oak Ridge, Tennessee

ASME PRESS   NEW YORK   2001

CONTENTS

Preface xix

Part 1: Neural Networks 1

Parallel, Self-Organizing, Hierarchical Support Vector Machines
    Hoi-Ming Chi, Okan Ersoy 3

Sample Selection Through Class Probability and Possibility in Support Vector Machines
    Sunghwan Sohn, Cihan Dagli 9

A Pre-Selection Method for Training of Support Vector Machines
    W. Hu, Qing Song 15

Controlled Learning of GREN-Networks
    Iveta Mrazova 21

Critic-Driven Ensemble Classification Via A Learning Method Akin to Boosting
    Lian Yan, David Miller 27

A Virtually Convex Error Criterion for Training Neural Networks
    James Ting-Ho Lo 33

Learning Structures with Flexible Basis Functions
    Chun-Shin Lin, K. Al-Hindi 41

Parallel Neural Networks with Accurate Correctness Decision of Outputs and a Largely Improved Generalization
    Yohtaro Yatsuzuka, Tetsuya Sugiyama 47

Real-Time Reinforcement Learning for Target Tracking
    Ross Lamm, Donald Specht 57

Adaptive Sensor Integration Using Reliability Map
    Dai Naito, Takayuki Suzuki, Hiroshi Yokoi, Yukinori Kakazu 63

The Reference Neuron Model: Game Playing via Trial and Error Learning
    George Schleis, Mateen Rizki, Farshad Fotouhi 69

Punish/Reward: Learning with a Critic in Multilayer Neural Networks
    Gregory Plett 75

Strictly Self-Adaptive Training with XOR ANN
    Shamsuddin Ahmed, Jim Cross, Abdesselam Bouzerdoum   81

Neural-Complexity Measures of Machine Learning Algorithms
    Khalid A Al-Mashouq, Zaygham Nawaz   89

Rule Extraction Based on Hidden Neural Activation Interval Projection on Each Dimensional Axis
    Wiphada Wettayaprasit, Chidchanok Lursinsap   95

Old AI- New AI Old ANN- New ANN New Biological Trends
    John Alexander, Jr.   101

Artificial Parietal Cortex Neurons for 3D Reconstruction
    Jeremiah Neubert, Y.H. Hu, Nicola Ferrier   107

Artificial Neural Cortex
    Alexei Mikhailov, Yang-Ming Pok   113

Part 2: Evolutionary Programming   121

Tackling Multimodal Problems in Hybrid Genetic Algorithms
    Prasanna Parthasarathy, David Goldberg, Scott Burns   123

Modeling Tournament Selection With Replacement Using Apparent Added Noise
    Kumara Sastry, David Goldberg   129

Symbiotic Model for Solutions and Operators
    Moeko Nerome, Satoshi Endo, Koji Yamada, Hayao Miyagi   135

Use of Emulations of the Immune System to Handle Constraints in Evolutionary Algorithms
    Carlos Coello, Nareli Cruz-Cortes   141

Exploring the Use of Entropy in Understanding the Behavior of the Enhanced Genetic Algorithm
    L. X. Yang, Deborah Stacey   147

DNA Local Sequence Alignment with Genetic Algorithms
    Mark Embrechts, Robert Bress   153

Multi-Recombinant Evolution Strategies for Real-Valued Function Optimization
    Yoshiyuki Matsumura, Kazuhiro Ohkura, Kanji Ueda   159

Efficient Evaluation Relaxation Under Integrated Fitness Functions
    Laura Albert, David Goldberg   165

Nonlinear Modeling: Genetic Programming vs. Fast Evolutionary Programming
    Minglei Duan, Richard Povinelli   171

Use of Dominance-based Tournament Selection to Handle Constraints in Genetic Algorithms
    Carlos Coello, Efren Mezura-Montes   177

Yield Modeling in Electronics Manufacturing using a Hybrid System of Neural Networks and Genetic Algorithms
    Swaminathan Vaithianathasamy, Sarah Lam, Krishnaswami Srihari   183

Recovering Tabular Information in ASCII Documents Using Evolutionary Programming
    Michael Zmuda   189

Evolutionary Synthesis of Microelectromechanical Systems (MEMS) Design
    Ningning Zhou, Bo Zhu, Alice Agogino, Kris Pister   197

Detection of Partial Tool Damage by using Genetic Algorithm based Tool Monitoring System
    L. Li, W. Y. Bao, Ibrahim Tansel, N. Reen, C. Kropas-Hughes   203

Intelligent Control of the Electrical Tuning Process in Televisions using Soft Computing Techniques
    Oscar Castillo, Patricia Melin, Felipe Duenas   209

Estimating Time Series Predictability Using Genetic Programming
    Minglei Duan, Richard Povinelli   215

Part 3: Fuzzy Systems   221

Adaptive Fuzzy Controllers Based on Variable Universes Theory
    Hongxing Li, C.L. Philip Chen   223

A New Type of CMAC Learning Controller with Fuzzy Learning Gain
    Yuncan Xue, Jixin Qian   229

Model Reference Adaptive Fuzzy Voltage Control in GTAW
    P. Smithmaitrie, P. Koseeyaporn, George Cook, A. Strauss   235

Nonlinear System Identification Using Fuzzy NARMA Model
    Chokchai Wiwattanakantang, Worapoj Kreesuradej   241

Adaptive Fuzzy Logic Controller for Robots Manipulators
    Youcef Touati, Karim Djouani, Yacine Amirat 247

Control of Highly Coupled Structures Using a Multivariable Sliding Mode Fuzzy Controller
    Hamid Allamehzadeh, John Y. Cheung 255

Intelligent Control of a Robot Arm to Avoid an Obstacle
    Rajab Challoo, Ligong Wang, Robert McLauchlan, S. Iqbal Omar, Shuhui Li 261

Fuzzy Based Drug Dosage Controller
    Divya Kesavan, Ganesh Vaidyanathan, S. Sankaravadivoo 267

Parallel Genetic Algorithms for Tuning A Fuzzy Data Mining System
    Qitao Liu, Susan Bridges, Ioana Banicescu 273

The Containment Problem for Fuzzy Class Algebra
    Daniel J. Buehrer, Lo Tse-Win, Hsieh Chih-Ming, Maxwell Hou 279

Hierarchical Rule-Base Reduction for a Fuzzy Logic Based Human Operator Performance Model
    W. Norris, R. Sreenivas, Qin Zhang, Jose Lopez-Dominguez 285

An Intelligent Fuzzy-Based System for Paroxysmal Atrial Tachychardia Identification
    Warawat Assawasantakul, Rawin Raviwongse, Wattana Watanapa 295

The Effect of Noise on the Dynamic Behavior of a Fuzzy Closed Loop System
    Eyad Elqaq, Roland Priemer 301

Fuzzy Logic for Predicting Soil Hydraulic Conductivity Using CT Images
    Z. Cheng, Steve Anderson, C. Gantzer, Y. Chu 307

Intelligent Technical Stock Analysis Using Fuzzy Logic and Trading Heuristics
    Vamsi Bogullu, David Enke, Cihan Dagli 313

Fuzzy Correlation Used in Text Multi-Categorization Problem
    Hao-En Chueh, Nancy P. Lin 319

Adapting Fuzzy Set-based Trade-off Strategy in Engineering Design
    Jiachuan Wang, Janis Terpenny 325

Part 4: Data Mining — 331

A Modified Version of Parallel, Self-Organizing, Hierarchical Neural Networks for Detecting Rare Events
    Hoi-Ming Chi, Okan Ersoy — 333

Experiments on Rough Set Based Data Mining
    Sarah Coppock, Aijing He, Lawrence J. Mazlack, Yaoyao Zhu — 339

Data Mining Using 2-D Neural Network Sensitivity Analysis for Molecules
    Mark Embrechts, Fabio Arciniegas, Muhsin Ozdemir, Curt Breneman, Kristin Bennett — 345

Visualization of High Dimensional Image Data Using a 3D SOFM
    Archana Sangole, George Knopf — 351

Modelling the Relationship Between Problem Characteristics and Data Mining Algorithm Performance Using Neural Networks
    Kate Smith, Frederick Woo, Vic Ciesielski, Remzi Ibrahim — 357

Visualization of Any Data with Elastic Map Method
    Alexander N. Gorban, A. Pitenko, A. Zinovyev, Donald Wunsch — 363

Robust Multi-resolution Web Usage Mining with Genetic Niche Clustering
    Olfa Nasraoui, Raghu Krishnapuram — 369

Neuro-Fuzzy-Genetic Architecture for Data Mining
    Korakot Hemsathapat, Cihan Dagli, David Enke — 375

Knowledge Discovery from Multispectarl Satellite Images
    Arun Kulkarni, Zhiwei Mo — 381

Data Mining of Mine Equipment Databases
    Tad Golosinski, Hui Hu — 387

Part 5: Complex Systems — 397

Interaction between Agents with Emotional Model
    Sayaka Nakamura, Takashi Ishida, Hiroshi Yokoi, Yukinori Kakazu — 399

Evolving Cooperative Partial Functions for Data Summary and Interpolation
    Peter Johnson, Kenneth Bryden, Dan Ashlock, Edmundo Vasquez — 405

A Development of an Interpreter for Cellular Automata
    Yuhei Akamine, Satoshi Endo, Koji Yamada — 411

New Scientific Borders For Artificial Neural Networks Based in Polynomial Powers of Sigmoid Theory
    Joao Fernando Marar   417

A Proposal of an Immune Optimization Inspired by Cell-Cooperation
    Naruaki Toma, Satoshi Endo, Koji Yamada, Hayao Miyagi   423

The Naming of Things and the Confusion of Tongues
    Florence Reeder   429

Evaluation of Endogenous Systems
    H. John Caulfield, Florence Reeder   435

Hebbian Brain Cell-Assemblies: Nonsynaptic Neurotransmission, Space and Energy Considerations
    Gaetano Aiello, Paul Bach-y-Rita   441

Coding Learning Strategies
    Stefania Brown-VanHoozer, W. VanHoozer   449

Theoretical Biology: Organisms and Mechanisms
    Christopher Landauer   455

Does Incomputable Mean Not Engineerable?
    Stephen Kercel   461

Part 6: Adaptive Control   469

Architectures for Autonomous Computing Systems
    Christopher Landauer, Kirstie Bellman   471

Real-Time Applications on a CMAC Neural Network
    Gilles Mercier, Dana Vilcu, Laurent George   477

Off-Line Adaption in an Iterative Learning Control Scheme with Plant Non-Linearities
    Laszlo Hideg   485

Flexible Structure Multi-cell Robot -Obstacle Avoidance using Amoeboid Self-organization Model
    Nobuyuki Takahashi, Takashi Nagai, Hiroshi Yokoi, Yukinori Kakazu   491

Hessien Algorithm used in Multilayered Neural Networks Controllers in Robot Control
    Joseph Constantin, Chaiban Nasr, Denis Hamad   497

Reinforcement Learning Control of Cooperative Arm Robots
    Kazuaki Yamada, Fumihiro Kojima, Kazuhiro Ohkura, Kanji Ueda   503

LSPB Trajectories used in Multilayered Neural Networks Controllers in Robot Control
    Chaiban Nasr, Joseph Constantin, Denis Hamad   509

Position Control Using a Programmable E/H Valve
    Haibo Hu, Qin Zhang, Xiangdong Kong, Andrew Alleyne   515

Artificial Neural Network for Tool Condition Monitoring in Drilling
    Issam Abu-Mahfouz   521

Tuning of PI Controller Coefficients Using Genetic Algorithms and Artificial Neural Network
    Seydi Vakkas Ustun, Galip Cansever, Mehmet Bulut   527

Monitoring of Drilling Processes
    Herman Leep   533

A Hybrid Neuro- Fuzzy- Fractal Approach for Automated Quality Control in the Manufacturing of Sound Speakers
    Patricia Melin, Oscar Castillo, Fernando Sotelo   539

Monitoring and Knowledge Extraction in Real Time Multivariable Dynamic Processes
    Aaron Ranson, Karen Hernandez, Justo Matheus, Angel Vivas   545

Dynamic Modelling Using Neural Networks. Case Study-Pilot Scale VSA Process for Oxygen Production
    Chris Beh, Kate Smith, Paul Webley   551

Detection and Separation of Extracellular Neuronal Discharges
    Tetyana Aksenova, Igor Tetko, Oleksandr Dryga, Olga Chibirova, Alessandro E.P. Villa   557

Nonlinear Dynamic Systems Modeling and Pattern Recognition
    Jianping Xiang, N. Jones, F. Schlindwein   563

A New Approach of Intelligent Control System Design
    Dorian Aur, Teodora Ghioca   569

Part 7: Pattern Recognition   575

Spatial-Temporal Feature Screening
    Kelly Greene, Kenneth W. Bauer Jr.   577

Scattered Data Interpolation Using Self-Organizing Feature Maps
    George Knopf, Archana Sangole 583

Pilot Mental Workload Calibration
    Jeremy Noel, Kenneth W. Bauer Jr., Jeffrey Lanning 589

Extended Minimal Resource Allocating Neural Networks for Aircraft SFDIA
    Giampiero Campa, M. Fravolini, M Napolitano 595

A Generalized Unstructured Artificial Neural Network Architecture: A First Study
    Robert Woodley, Levent Acar 601

Evaluation of Feature Vectors for ANNs used in Automatic Modulation Recognition
    Sudip Biswas, David Calvert, Orlando Cicchello, Stefan C. Kremer 607

Intelligent Pattern-based Techniques to Monitor the Operation of Nondestructive Analysis Equipment and Contribute to Multivariate Feedback Control
    Gail Cordes, Leo Van Ausdeln, James Jones, Kevin Haskell 613

An Analysis of the Human Vision Process as a Decentralized Continuous-Time Dynamical System
    Robert Woodley, Levent Acar 619

Detection of Defects in Mechanical Equipment from Vibration Signatures Using Neural Networks
    Suresh Pallerla, Robert McLauchlan 625

Active Model-Based Object Recognition Employing Foveal Imagery and Multiresolution Feature Sets
    Qiang Lu, Peter Scott 633

Ball Recognition in Images for Detecting Goal in Football
    Nicola Ancona, Grazia Cicirelli, A. Branca, Arcangelo Distante 639

Classification of Summarized Videos by Using Hidden Markov Models on Compressed Chromaticity Signatures
    Cheng Lu, Mark S. Drew, James Au 645

Printed Thai Character Recognition Using Neural Networks
    Phongthep Ruxpakawong, Arit Thammano 651

Information Content of the Frequency Dictionaries, Reconstruction, Transformation and Classification of Dictionaries and Genetic Texts
    Alexander N. Gorban, T. Popova, M. Sadovsky, Donald Wunsch   657

An Optical Fibre Multipoint Sensor-Utilising U-Bend Sensors-Based on Artificial Neural Network Pattern Recognition
    W. Lyons, Hartmut Ewald, C. Flanagan, E. Lewis   665

Validating Digital Terrain Elevation Data with Neural Networks
    Robert Gray, Thomas Hemminger   673

Recognition of Musical Rhythm Patterns Based on a Neuro-Fuzzy-System
    Tillman Weyde, Klaus Dalinghaus   679

Automotive Design Driven by Pattern Recognition
    Sawsan Aboul-Hassan, Djamel Bouchaffra   685

Characterization of Fractal Morphology of Flocs Formed in Oil-Water Emulsions by Microscopic Image Analysis
    Jiranun Hempoonsert, Berrin Tansel, Ibrahim Tansel   695

Part 8: Prediction   701

Development of Intelligent Systems for Improved WSR-88D Rainfall Estimation
    Theodore Trafalis, Michael Richman, Anderson White, Budi Santosa   703

Using Neural Networks to Predict Rainfall Patterns in South Texas
    Gary Weckman, Robert McLauchlan   709

Comparison of Hourly and Daily Neural Network Models for Forecasting Hourly Electric Load
    Swaminathan Vaitianathasamy, David Enke   715

A Compound Technique for Estimating Missing Data of Wind Speeds
    Punnee Siripitayananon, Hui-Chuan Chen, Kang-Ren Jin   721

Detecting Regimes in Temperature Time Series
    Patrick Clemins, Richard Povinelli   727

Forecasting Warranty Claims: A Comparison of SVMs with Statistical Methods and Neural Networks
    Ratna Babu Chinnam, Vinay Kumar, Gary S. Wasserman   733

Using Neural Networks and Technical Analysis Indicators for Predicting Stock Trends
 Suraphan Thanwornwong, David Enke, Cihan Dagli   739

Predicting Monthly Flour Prices through Neural Networks, RBFs and SVR
 Theodore Trafalis, Budi Santosa   745

Estimation of the Temperature Field in a Steel Plate Through Solution of the Inverse Heat Transfer Problem
 Maria Petkova   751

Using Support Vector Machines for Recognizing Shifts in Correlated and Other Manufacturing Processes
 Ratna Babu Chinnam, Vinay Kumar   757

Improving Accuracy of Nearest Neighbor Algorithm in Highway Accident Prediction
 Chakarida Nukoolkit, Hui-Chuan Chen, David Brown   763

Selection of Key Data Variables Using Sensitivity Analysis to Predict Jet Engine Maintenance Removals with Neural Networks
 Gary Weckman, Robert McLauchlan   769

Hierarchical Neural Networks for Reducing Systematic Prediction Errors
 S. Holl, Andrew Flitman, Kate Smith   775

Part 9: Biology and Medicine   781

Neural Network Classification of Malaria Protein Sequences
 Kate Smith, Terry Spithill, Ross Coppel, Peter Smooker   783

Neural PCA Network for Lung Outline Reconstruction in VQ Scan Images
 Gursel Serpen, R. Iyer, H. Elsamaloty, E. I. Parsai   789

A Dynamic Model for Nuclear Membrane of a Nanoscale Bio-Molecular Motor
 A Sinha, R.M. Pidaparti, P. Sarma, G. Vemuri, A. Gacy   797

Architectures for Equine Gait Analysis
 S. Ahmad, David Calvert, Deborah Stacey, J Thomason   803

Feasibility of Using Recurrent Neural Networks to Classify Mental Workload
 Kelly Greene   809

Feasibility of Using Time Delay Neural Networks to Classify Mental Workload
 Kelly Greene   815

Modeling of Heart Rate Variability
    Rupa Balan, Sudhir Rai, Stephen Hull, Jr., John Y. Cheung ... 821

Dynamic Modeling of Behavior of Central Olfactory System: Simulation Analysis and Information Processing
    Natacha Gueorguieva, Iren Valova, George Georgiev ... 827

Variance Analysis of a Vestibulo-Oculomotor System State in a Sinusoidal Stimulation on the Basis of a Reconstructed Component Method
    Anatoly Boriskevich, Vadim Kudryavtsev ... 833

Application of Support Vector Machines to Breast Cancer Classification Using Mammogram and History Data
    Anab Akanda, Walker Land Jr, Joseph Lo ... 839

Part 10: Smart Engineering Systems ... 845

D3D-A Software for Graphical Representation of Objective Functions
    Petru-Aurelian Simionescu, D. Beale ... 847

A New Approach to Update Probability Distributions Estimates of Air Travel Demand
    Ioana Bilegan, W. El Moudani, K. Achaibou, F. Mora-Camino ... 853

An Heuristic Approach for Throughput Optimization of SMT Placement Machines
    Sreekrishna Palaparthi, Sarah Lam, Krishnaswami Srihari, Dennis Warheit ... 859

A New Self-Organizing Neural Network for Solving the Travelling Salesman Problem
    Fernando Guerrero, Sebastian Lozano, D. Canca, J. Garcia, Kate Smith ... 865

Optimization of the Mold Orientation on an Investment Casting Centrifuge
    Petru-Aurelian Simionescu, D. Beale, R. Overfelt ... 871

Traditional Calculus and Modern Fuzzy Approach to Optimization
    Grace Woo, Peng-Yung Woo ... 877

A Recursive Asynchronous Neural Network That Finds the Largest Input
    Thomas Dranger, Roland Priemer ... 883

An Improved Parallel Algorithm for Finding the Maximum Clique in An Arbitrary Graph
    Songnian Yu, Xiaofeng Qian, Weimin Xu ... 889

A Modified Hopfield Model for Solving Several Types of Optimization Problems
    Ivan Nunes da Silva, Andre Nunes de Souza, Jose Ulson   897

Segmentation of Plant from Background Using Neural Network Approach
    Dev Shrestha, Brian Steward, Eric Bartlett   903

Hybrid Neurochip with Learning On-chip
    Frank Stupmann, Steffen Rode, Norbert Schmidt, Wolfgang Fredrich   909

Automated Separation of Clods from Agricultural Produce Using Mechanical Separator With Integrated Neural Network Machine Vision
    Demian Morquin, Mounir Ghalia, Subhash Bose   915

Neural Network Based Speed Boat Emulator
    Ibrahim Tansel, R. Seltzer, W. Yuen   921

Synchronization of Hyperchaos in Systems with Delay
    Yechiel Crispin   927

Virtual Sensors for Preventive Health Maintenance of Aircraft Engines
    Anna Buczak, O. Uluyol, Emmanuel Nwadiogbu   933

Neural Network Model of Hardening Coefficient of Airengine Details
    Valeriy Dubrovin, Sergey Subbotin, Viktor Yatzenko   939

Intelligent Triangulations which Approximate 3D Points for Production Purposes
    Stefan Pittner, Kishore Pochampally, Srikanth Vadde, Sagar Kamarthi   945

Abstraction Based Software System Design
    Christopher Landauer, Kirstie Bellman   951

A Search Engine for Two-Dimensional Vector Graphics
    Mike Fisher, Dan St. Clair   957

Multi-Agent Transactional Negotiation For E-Commerce
    V.K. Murthy   963

Flexible Design Support System
    Christopher Landauer, Kirstie Bellman   969

A New Decision Making Strategy for Distributed Control of Cellular Manufacturing Systems
    P. Renna, A. Amico, Giovanni Perrone, M Bruccoleri   975

Neural Network System for White Color Balance Adjustment in Television Receivers
    Francisco Del Puerto, Mounir Ghalia   981

Smart Systems, Metric Entropy, Phase Transitions and Time-Arrow
    E. V. Krishnamurthy, Vikram Krishnamurthy   987

Natural Language Translation System Using Neural Networks
    Saroj Kaushik, Manoj Kumar   993

Comparative Performance Analysis of Structured Neural Networks and Observers for CNC Feed Drive Disturbance Force Estimation
    Melik Dolen, Ekrem Kayikci, Robert Lorenz   1001

Load Torque Estimation for Spindle Drives Using Structured Neural Network Topologies
    Melik Dolen, Ekrem Kayikci, Robert Lorenz   1007

Model Identification for Restructurable Control System Considering Fault Tolerance
    Yuji Kuwashima, Hiroshi Yokoi, Yukinori Kakazu   1013

Deepsia- From Supply Chains to Supply Webs
    Pedro Sousa, Joao Pimentao, Adolfo Steiger-Garcao   1019

Computerized Arrangement of Vocal Music
    Matt Johnson, Ralph Wilkerson   1025

Best Value Procurement Using Artificial Intelligent 'Modified Displaced Ideal Model"
    Dean Kashiwagi   1031

Document Control Simulation for Customs Clearance in a Sea Port
    Shamsuddin Ahmed   1037

Subject Index   1043

Author Index   1049

PREFACE

As a follow up to the previous ten volumes of Intelligent Engineering Systems Through Artificial Neural Networks, edited by Dagli, C.H. et. al., this volume contains the edited versions of the technical presentations of ANNIE 2001. The eleventh international gathering of researchers interested in Smart Engineering System Design using neural networks, fuzzy logic, evolutionary programming, data mining and complex systems, was held on November 4-7, 2001 in St. Louis, Missouri, USA. The conference covered the theory of neural networks, fuzzy logic, evolutionary programming, data mining, and complex systems; and their applications in the engineering domain. The papers included in this volume provide a forum for researchers in the field to exchange ideas on smart engineering systems design.

An extended version of each paper selected for inclusion was reviewed by two referees, then revised, edited, and condensed to the format herein. The papers in this edited book are grouped into ten categories:
- Neural Networks
- Evolutionary Programming
- Fuzzy Systems
- Data Mining
- Complex Systems
- Adaptive Control
- Pattern Recognition
- Prediction
- Biology and Medicine
- Smart Engineering Systems

There were nine plenary sessions scheduled for ANNIE 2001. Dr. Lofti Zadeh presented "Toward an Enlargement of the Role of Natural Languages in Information Processing, Decision and Control" at the Monday morning plenary session. At noon on Monday, Dr. Metin Akay offered "Respiratory Related Evoked Responses in Obstructive Sleep Apnea Syndrome." Monday afternoon opened with "Connectionism: What's Wrong With It and How Do We Get Out of It," a presentation by Dr. Asim Roy. In Tuesday's morning plenary session, "Interactive Evolutionary Computation – Addressing Poor-Definition and Uncertainty through Optimal Information-gathering and Problem Re-formulation" was presented by Dr. Ian Parmee. Tuesday's luncheon plenary "Computational Intelligence – Fulfilling Visions for Automation and Control" was offered by Dr. Tariq Samad. The Tuesday afternoon plenary was "Language Learning for Machine Translation – A Study in MT Evaluation and Vocabulary Acquisition" and was presented by Dr. Florence Reeder. Dr. Hans Zimmerman offered "Neural Networks in Econometrics: Principles, Applications" for the Wednesday morning plenary and "Identification and Forecasting of Dynamical Systems by Recurrent Neural Networks" for the luncheon plenary.

Our Banquet Plenary Speaker for this year's ANNIE was Dr. Bruce Wheeler and his talk was entitled "Designable Neuronal Networks in Vitro." Dr. Bruce Wheeler is a Professor of Electrical and Computer Engineering and of the Beckman Institute at the University of Illinois at Urbana-Champaign. He has served as Chair of the Bioengineering and Neuroscience Programs and as Associate Head of the ECE Department. His research interests lie in the application of electrical engineering to neuroscience, especially microlithography for influencing neuronal growth, the topic of the talk, but also biomedical signal processing, including advanced algorithms for hearing aids.

The editors would like to once again thank the plenary speakers, and the authors for their contributions. We would also like to recognize the organizing committee members of ANNIE 2001 for their excellent support in promoting the conference internationally. Further, we wish to express our gratitude to all the referees for their excellent and timely review efforts, which made this edited book possible within a short period of time.

Space will not permit for the recognition of all contributors, but special mention must be made of the IEEE Neural Networks Council for giving the "In-Cooperation Status" to the conference and to the session chairs, whom contributed to the quality of the discussions in the presentations. We would also like to mention our appreciation to the staff of the Engineering Management Department and to Davae Collins for their timely and continuous support and particularly to Patti Peterson. Her ownership of ANNIE conferences and her excellent ability to communicate with authors all over the world and organizing abilities along with putting this volume together made her an integral part of the ANNIE conferences.

Lastly, but most importantly, we would like to thank all of our families for their patience and sustenance during the many long hours that it took to create yet another ANNIE, ANNIE 2001, and subsequently, this book.

Cihan H. Dagli
University of Missouri-Rolla

Anna Buczak
Philips Research

Joydeep Ghosh
University of Texas at Austin

Mark Embrechts
Rensselaer Polytechnic Institute, RPI

Okan Ersoy
Purdue University

Stephen Kercel
Oak Ridge National Laboratory

PART I: NEURAL NETWORKS

PARALLEL, SELF-ORGANIZING, HIERARCHICAL SUPPORT VECTOR MACHINES

HOI-MING CHI
School of Electrical and Computer Engineering, Purdue University, West Lafayette, Indiana, USA

OKAN K. ERSOY
School of Electrical and Computer Engineering, Purdue University, West Lafayette, Indiana, USA

ABSTRACT
This paper proposes a new support vector learning algorithm to be referred to as parallel, self-organizing, hierarchical support vector machine (PSHSVM). It consists of a binary tree structure with a linear support vector machine in each node and the class labels in all leaves. During training, multiple linear hyperplanes are constructed while traversing down the tree. PSHSVM inherits all the advantages from linear SVMs, and is capable of separating both linearly and non-linearly separable data. The only parameter that needs to be chosen by the user is the regularization parameter C, thus eliminating the model selection problem. Moreover, it is computationally efficient, and testing can be done in parallel. Experiments with different data sets show that PSHSVM achieves comparable performance with decision tree, polynomial and Gaussian SVMs.

INTRODUCTION

Support vector learning is a machine learning algorithm recently developed by Vapnik (1995). Its idea originates from the statistical learning theory, and it can be formulated as a quadratic programming problem with equality and inequality constraints. Research in the past few years has shown that support vector machines (SVMs) have many attractive features and promising empirical performance over other conventional learning algorithms in many real-world classification problems. However, SVMs do suffer a few drawbacks, one of which is the model selection problem. Prior to SVM training, a few parameters and the kernel function have to be chosen optimally by the user to ensure good generalization performance. Existing approaches use cross-validation and leave-one-out types of estimators to select the best kernel function and its parameters (Lee, and Lin 2000). However, these approaches can be very time-consuming with large data sets. In this paper, a new and efficient scheme to be referred to as parallel, self-organizing, hierarchical support vector machine (PSHSVM) is proposed to address this problem.

ORIGINAL SV LEARNING ALGORITHM

The original SVM attempts to find the best separating hyperplane in terms of the largest margin and smallest number of training errors (Burges, 1998). It can be formulated as a convex quadratic programming problem as shown below:

$$\max L_D \equiv \sum_i \alpha_i - \frac{1}{2} \sum_{i,j} \alpha_i \alpha_j y_i y_j \mathbf{x}_i \cdot \mathbf{x}_j \qquad (1)$$

subject to the constraints

$$0 \leq \alpha_i \leq C$$

$$\sum_i \alpha_i y_i = 0$$

where x_i is the i-th data vector, y_i is the binary (-1 or 1) class label of the i-th data vector, α_i is the Lagrangian multiplier, ξ_i is the slack variable, C is the regularization parameter, and b is the bias. The above formulation corresponds to linear SV learning. In non-linear SV learning, the dot product is replaced by a kernel function $K(x_i, x_j)$, which is simply the dot product of the data vectors after mapping to a high dimensional Euclidean space. It is hoped that the data vectors become linearly separable in this high dimensional space, so that a linear decision hyperplane can be drawn. The kernel function can take on a few different forms, such as polynomial, Gaussian, and hyperbolic tangent functions. Model selection refers to the optimal choice of the kernel function and the parameters associated with it. Since only linear SVMs are used in the PSHSVM, the only parameter that needs to be chosen prior to training is the regularization parameter C.

PSHSVM ALGORITHM

The idea of PSHSVM originates from the parallel, self-organizing, hierarchical neural networks (PSHNN) developed by Ersoy and Hong (1990), (1993). It consists of a binary tree structure with a linear SVM at each tree node, as shown in Fig. 1. During training, the linear SVM at each node generates a linear decision hyperplane in the input space and separates the input space into two regions. All data vectors in the negative (output of the linear SVM ≤ 0) decision region are further fed into another linear SVM in the left child of the current tree node for training, while all data vectors in the positive (output of the linear SVM > 0) decision region are further fed into another linear SVM in the right child of the current tree node for training. This process continues recursively until all vectors are classified correctly, and a leaf node is reached. Here the class label is assigned to the leaf node.

Before describing the recursive training procedure, we first define a few terms. Common class is defined as the class that has more data vectors in a two-class classification problem. Similarly, rare class is defined as the class that has fewer data vectors. Class ratio is defined as the ratio of the number of data vectors in the common class to that in the rare class.

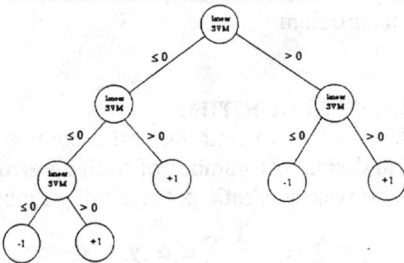

Figure 1. An example of a binary tree of the PSHSVM. The output of the linear SVM decides which child to go to next. The tree leaves hold the final class labels.

Training Procedure

Initialize: Set C to a predefined value, for example 10. Create a tree node and set its parent pointer to NIL (i.e., this node is the root node). Create a pointer, and set it to point to this tree node.

1) Train a linear SVM. After training is done, check to see the following:
   a) if all the data vectors lie on only one side of the decision hyperplane, then randomly select vectors from the rare class and add them to the data set until the class ratio reduces to a value close to 1. Now we have a new data set that has almost equal numbers of data vectors in both classes. Then train another linear SVM using this new data set. After training is done, check to see the following:
      i) if all the data vectors still lie on only one side of the decision hyperplane, make the current node a leaf node, and assign to this leaf node the class label according to the majority of the class labels of the original data set before randomly adding vectors from the rare class. Break out of the current recursion.
      ii) else, store the linear SVM in the node currently pointed to by the pointer, and go to step 2.
   b) else, store the linear SVM in the node currently pointed to by the pointer, and go to step 2.
2) Separate all the vectors in the data set into two subsets by the linear decision hyperplane generated in step 1.
3) For all vectors lying in the negative side of the hyperplane (output of the linear SVM ≤ 0), check to see the following:
   a) if they all belong to the -1 class, assign the class label -1 to this node and make it a leaf node. Break out of the current recursion.
   b) if not, create a new left child of the current node, set the pointer to point to this left child, and recursively go to step 1.
4) For all vectors lying in the positive side of the hyperplane (output of the SVM > 0), check to see the following:
   a) if they all belong to the +1 class, assign the class label $+1$ to this node and make it a leaf node. Break out of the current recursion.
   b) if not, create a new right child of the current tree node, set the pointer to point to this right child, and recursively go to step 1.

If all the data vectors are linearly separable, the PSHSVM classifies them correctly in the first node and stops immediately. In this case, the PSHSVM simply reduces to a linear SVM. On the other hand, if the data vectors are linearly nonseparable, the PSHSVM will try to classify the vectors as best as it can at the current node, and separate all the vectors into two subsets by the linear decision hyperplane (step 2). Each subset is then classified recursively by other linear SVMs in later nodes (steps 3 and 4). Furthermore, if the data set is linearly nonseparable and has under- or un-proportionally represented classes, after training at the current node, all the data vectors will lie on only one side of the hyperplane. The linear decision hyperplane thus fails to divide the data vectors into two subsets, and the PSHSVM cannot further classify the data vectors. In order to overcome this problem, vectors from the rare class are randomly added to the original data set until the class ratio reduces to a value close to 1 (step 1a). This method has been successfully applied to neural networks to increase rare event detection probability (Chi, 2001). However, caution has to be taken here, as excessive use of step 1a will lead to overfitting. This is especially true as the PSHSVM progresses deep in the tree. One way to address this problem is to add fewer and fewer vectors as progressing down the tree.

To classify a test vector, we simply traverse down the tree until a leaf node is reached. If the output of the SVM at the current node is less than zero, we traverse to the left child of the node, and vice versa. The class label associated with the leaf node reached is assigned to the test vector. The testing procedure is described in detail below.

Testing Procedure
Initialize: Set the pointer to the tree root node.
1) Input the test vector to the linear SVM pointed to by the pointer.
2) Check the output of the SVM:
   a) if the output value is less than or equal to 0, set the pointer to point to the left child of the current tree node.
   b) if the output value is greater than 0, set the pointer to point to the right child of the current tree node.
3) If the pointer points to a leaf node, assign the class label associated with this leaf node to the test vector. If not, go to step 1.

The PSHSVM algorithm can also be applied to problems with more than two classes. In k-class classification problems, k PSHSVMs are constructed independently by training vectors from each of the k classes against vectors from the remaining k-1 classes, essentially reducing to k 2-class classfication problems.

EXPERIMENTAL RESULTS

The first part of the experiment involves classifying two sets of linearly nonseparable two-dimensional data using four algorithms: the proposed PSHSVM, nonlinear SVMs with polynomial and Gaussian kernels, and a decision tree. Each linear SVM in the PSHSVM is implemented by using OSU SVM Classifier Matlab Toolbox (Ma, and Ahalt, 2001). All figures are generated in Matlab 5.3.1, and those figures for SVMs with polynomial and Gaussian kernels are created by Steve Gunn's Matlab SVM toolbox (1997). The figures for decision trees are generated by using Craig W. Codrington's Matlab Decision Tree toolbox (1995).

Figures 2 and 3 show the decision regions generated by various algorithms for the first and second data sets, respectively. The numbers in Fig. 2a and 3a dictate the order of the decision boundaries drawn. As seen from the figures, the PSHSVM succcessfully classifies linearly nonseparable data sets, and it generates decision regions similar to those by the polynomial and Gaussian SVMs, and the decision tree.

The second part of the experiment investigates the PSHSVM's ability to classify higher dimensional data. This real-world data set is obtained from John Platt's web site (2001). It contains the training data set for UCI adult benchmark, which is to predict whether a household has an income greater than $50K. There are a total of 123 sparse binary features. Table 1 shows the overall testing accuracy for the adult income prediction using the PSHSVM, polynomial and Gaussian SVMs. The PSHSVM achieves a slightly lower testing accuracy.

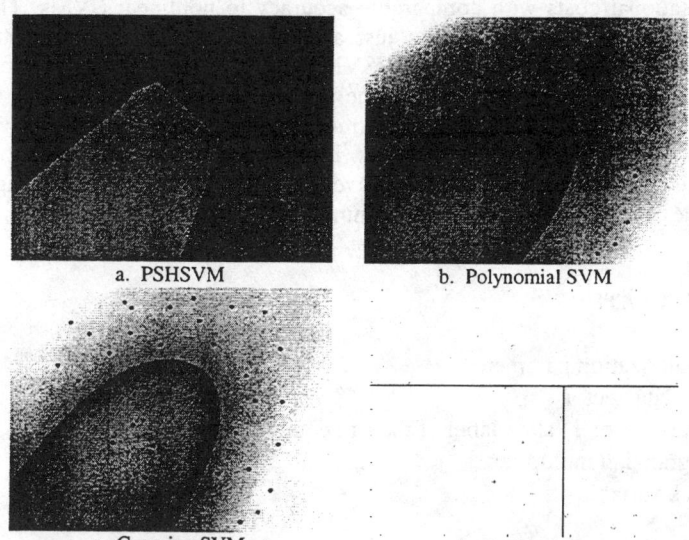

Figure 2. Decision regions generated by various algorithms for the first data set

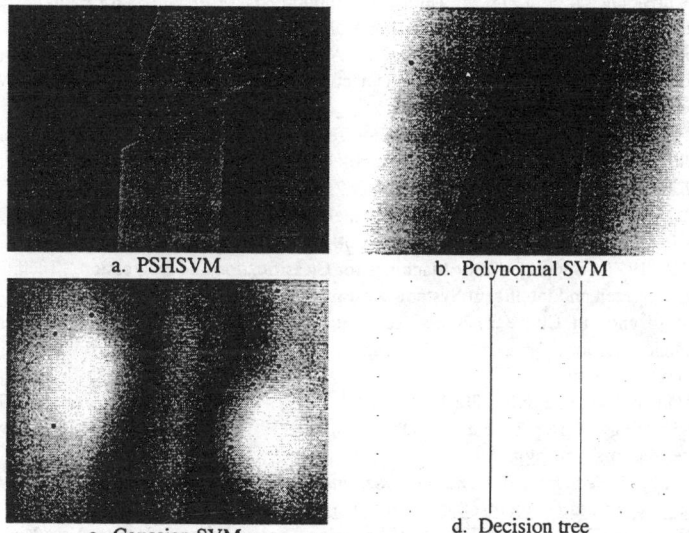

Figure 3. Decision regions generated by various algorithms for the second data set

Table 1. Performance comparison among PSHSVM, polynomial and Gaussian SVMs

	Training Accuracy	Testing Accuracy
PSHSVM with C = 0.001	0.8150	0.8276
Polynomial SVM with degree = 2 and C = 0.01	0.8885	0.8421
Gaussian SVM with σ^2 = 10 and C = 1	0.8866	0.8420

CONCLUSION

The PSHSVM classifies data vectors by generating multiple linear hyperplanes using a binary tree with linear SVMs. It greatly reduces computational costs with comparable accuracy to nonlinear SVMs. The model selection problem is reduced to just a single choice of the parameter C. Its recursive nature enables the PSHSVM to learn linearly nonseparable data vectors. Moreover, testing can be done in parallel. Because of the way it is built in terms of a tree, it is easy to extract decision rules from it, an appealing property for data mining applications. Further research is necessary to develop methods that are more effective in preventing all data vectors from lying on only one side of the hyperplane during training.

NOMENCLATURE

b : bias
C : regularization parameter
x_i : i-th data vector
y_i : binary (-1 or 1) class label of the i-th data vector
α_i : Lagrangian multiplier
ξ_i : slack variable

REFERENCES

Burges, Christopher J. C., 1998, "A Tutorial on Support Vector Machines for Pattern Recognition," *Data Mining and Knowledge Discovery*, Vol. 2, No. 2, pp. 121-167.

Chi, Hoi-Ming, and Ersoy, Okan K., 2001, "A Modified Version of Parallel, Self-organizing, Hierarchical Neural Networks for Detecting Rare Events," Submitted to *ANNIE 2001 Conference Proceedings*.

Codrington, Craig, 1995, *Matlab Decision Tree Toolbox*, Private Communication.

Ersoy, Okan K., and Hong, Daesik, 1990, "Parallel, Self-organizing, Hierarchical Neural Networks," *IEEE Trans. Neural Networks*, Vol. 1, No. 2, pp. 167-178.

Ersoy, Okan K., and Hong, Daesik, 1993, "Parallel, Self-organizing, Hierarchical Neural Networks – II," *IEEE Trans. Industrial Electronics*, Vol. 40, No. 2, pp. 218-227.

Gunn, S. R., 1997, "Support Vector Machines for Classification and Regression," Technical Report, Image Speech and Intelligent Systems Research Group, University of Southampton.

Lee, Jen-Hao, and Lin, Chih-Jen, 2000, "Automatic Model Selection for Support Vector Machines," Technical Report, Department of Computer Science and Information Engineering, National Taiwan University.

Ma, Junshui, and Ahalt, Stanley, 2001, *OSU SVM Classifier Matlab Toolbox Version 2.00*. Electrical Engineering Department, Ohio State University, http://eewww.eng.ohio-state.edu/~maj/osu_svm.

Platt, John, 2001, *Sequential Minimal Optimization*. Signal Processing Group of Microsoft Research http://www.research.microsoft.com/~jplatt/smo.html.

Vapnik, V., 1995, *The Nature of Statistical Learning Theory*, Springer-Verlag, New York.

SAMPLE SELECTION THROUGH CLASS PROBABILITY AND POSSIBILITY IN SUPPORT VECTOR MACHINES

SUNGHWAN SOHN
Smart Engineering Systems Lab
Dept. of Engineering Management
University of Missouri-Rolla
Rolla, MO 65409
ssohn@umr.edu

CIHAN H. DAGLI
Smart Engineering Systems Lab
Dept. of Engineering Management
University of Missouri-Rolla
Rolla, MO 65409
dagli@umr.edu

ABSTRACT

Support vector machines (SVM) have shown attractive potential in classification. However, they have the limitation of size in training large data set and also sensitivity to outliers. In this paper, we used the class membership using the probability of each sample through K-nearest neighbors to reduce the training set. In SVM, if the training set contains outliers, support vectors might not be properly chosen and degrade the classification performance for unseen samples. In this case, the class membership assignment using probability is not appropriate to effectively eliminate outliers. To overcome this problem, we could check the possibility of the sample belonging to the class and eliminate the samples having weak possibility.

1. INTRODUCTION

Support vector machines are basically implented by the application of structual risk minimization using Vapnik-Chervonenkis (VC) dimension (Vapnik, 1982, 1995) in that the error of a learning machine on test data is bounded by the sum of the training error. Unlike the conventional approach, support vector machines provide a method of designing a learning machine by controlling model complexity independently of dimensionality (Vapnik, 1998), (Haykin, 1999). However, support vector machines have the difficulty of the training set having a large size and containing noise or outliers.

The design of the support vector machines depends on a subset of the training data, such as support vectors that lie closest to the decision boundary, and knowing that removing any training samples that are not support vectors might have no effect on building decision hyperplane (Burges, 1998). Therefore, we might choose only samples that are relevant to support vectors as the taining set and so reduce the size of training samples. This can be performed by selecting samples having non-crisp class membership – i.e., not complete class membership – based on the fact that support vectors lie closest to the decision boundary.

In implementing support vector machines, we assume that all necessary information for classification could be represented by support vectors. However, if the training set contains outliers, support vectors might not be properly chosen and degrade the classification performance for unseen samples. Therefore, we need to differentiate outliers from training samples. In this case, traditional class membership assignments are not appropriate to solve this problem since they use the probabilistic method in that the class memberships of a sample sum to one, even though this sample is an outlier. In this reason, we can apply the possibilistic measure to separate non-typical samples from typical training samples.

In the following section the method of sample selection is discussed. In section 3 we presents experimental results and analysis. Finally, in section 4, we conclude the paper.

2. SAMPLE SELECTION IN SVM

2.1 PROBABILISTIC APPROACH

As we discussed in the previous section, support vector machines build the decision function with only the part of training samples that are near the decision boundary region. Intuitively, the samples near the decision boundary region have not complete class membership based on fuzzy concepts (Zadeh, 1965). From this idea, we can simply check each sample x_j's typicalness of the particular class i, $u_i(x_j)$ by counting K nearest neighbors.

$$u_i(x_j) = u_{ij} = n_i / K \qquad (1)$$

where n_i is the number of the neighbors of x_j belonging to the i_{th} class. Then, we can select the samples having non-crisp class membership – i.e., not 1 or 0 – in eq. (1) based on the fact that those samples are near the decision boundary region and are closely relavant to support vectors.

Here the size of the training set could be controlled by adjusting K neighbors when assigning class memberships into training samples. The smaller K used, the smaller the number of training sets we have. In fact, eq. (1) does not gaurantee to keep the true class label if all of the K neighbors are in different classes. Since we merely need to check whether the sample is typical or not, this problem can be ignored.

2.2 POSSIBILISTIC APPROACH

Vapnik provided a bound on the actual risk of support vector machine (Vapnik, 1995), (Burges, 1998):

$$E[P(error)] \leq \frac{E[\text{Number of support vectors}]}{\text{Number of training samples}}$$

where *P(error)* is the actual risk for a machine trained on *l*-1 samples, *E[P(error)]* is the expectation of the actual risk over all choices of training set of size *l*-1, and *E*[Number of support vectors] is the expectation of the number of support vectors over all choices of training sets of size *l*. Support vectors are the most difficult to be classified and the basis for constructing the decision function. If there exist outliers in training samples, they would be considered as support vectors and disturb to build a proper decision function, and consequently increase the risk of support vector machine.

The probabilistic class membership assignment might not eliminate a noise or outlier because there is no possibilistic meaning to the class. In other words, an outlier can have high degree of class membership. For example, in the two class problem, an outlier can have 0.5 class membership for each class if $n_1 = n_2$ in eq. (1). Therefore, we need another measure to check class possibility. In this reason, we can apply vague concepts or classes (Zimmerman and Zysno, 1985) for the class membership.

$$u_{ij} = \frac{1}{1+d_{ij}} \quad (2)$$

where d_{ij} is the distance of sample x_j to the prototype of the i_{th} class. To have the proper possibilistic membership in any classes, the distance has to be normalized in the following manner (Krishnapuram and Keller, 1993):

$$u_{ij} = \frac{1}{1+\left(\frac{d_{ij}^2}{\eta_i}\right)^{\frac{1}{m-1}}} \quad (3)$$

When $m \to 1$, the membership function becomes hard and when $m \to \infty$, the membership function becomes fuzzier. A typical value of m is 2. Where the value of η_i is determined by

$$\eta_i = \frac{\sum_{j=1}^{N} u_{ij}^m d_{ij}^2}{\sum_{j=1}^{N} u_{ij}^m} \quad (4)$$

The η_i relates to the overall size and shape of class i. The samples having a small possibilistic class membership might be considered as an outlier and could be eliminated from the training set. This elimination of outliers could reduce the risk of support vector machine.

3. EXPERIMENT

Two basic experiments were performed to show the performance of sample selection – i.e., sample selection using class probability and class possibility. For class probability we used K-nearest neighbors to calculate the class membership in eq. (1) and for class possibility we used the vague concept to calculate the class membership in eq. (3).

3.1. SAMPLE SELECTION USING CLASS PROBABILITY

We used an artificial 2-class data set (Fig 1 (a)) with 500 training samples and another 500 test samples. The samples having non-crisp membership from eq. (1) were selected as the training set (Fig 1 (b)). The results are in table 1.

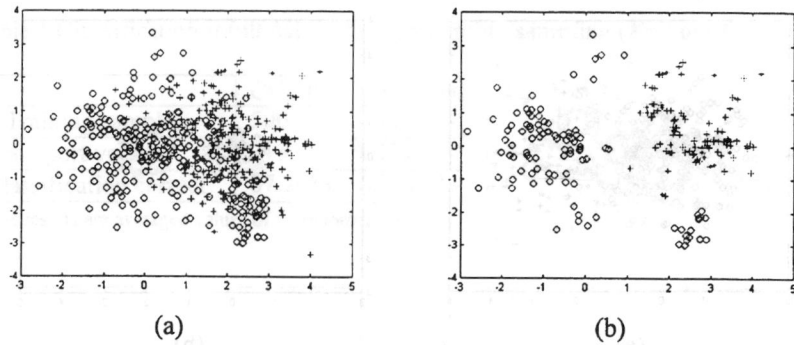

Fig 3 (a) Noisy data (b) Selected data having complete class probability

4. CONCLUSIONS

We proposed two approaches to select proper training samples in SVM classifier. Theses methods are relatively simple and easy to be implemented. With these methods, we can reduce training samples as well as support vectors without loss of classification performance (or even better classification performance) and speed up the training of SVM having a large training set. Those subsets of training data seemed to be enough or better to construct the proper decision boundary.

When using class probability, we can use only samples near the decision boundary region, but if there exist some variations and noise, a class-possibility sampling method seems to be better. In class-possibility sampling, we need to consider the distance measure because selected samples depend on the distance measure chosen. Also, we might partition training samples into some proper groups and then apply a class possibility concept into the individual group.

REFERENCES

- Burges, C. J.C., 1998, "A tutorial on support vector machines for pattern recognition," *Data Mining and Knowledge Discovery* 2, pp. 121-167.
- Haykin, Simon, 1999, *Neural Networks: A Comprehensive Foundation*, Ch 9, Second Edition, Prentice Hall.
- Krishnapuram R. and Keller, J. M., May 1993, "A possibilistic approach to clustering," IEEE Trans. On Fuzzy Systems, vol. 1, No. 2.
- Vapnik, V., 1982, *Estimation of Dependences Based on Empirical Data*. New York: Springer Verlag.
- Vapnik, V. N., 1995, *The Nature of Statistical Leaning Theory*, New York: Springer-Verlag.
- Vapnik, V. N., 1998, *Statistical Learning Theory*, New York: Wiley.
- Zadeh, L. A., 1965, "Fuzzy Sets," *Inf. Control*, vol. 8, pp. 338-353.
- Zimmerman, H. J. and Zysno, P., 1985, "Quantifying vagueness in decision models," *European J. Operational Res.*, vol. 22, pp. 148-158.

A PRE-SELECTION METHOD FOR TRAINING OF SUPPORT VECTOR MACHINES

W.J. HU
School of Electrical and Electronic Engineering,
Nanyang Technological University,
Singapore

Q. SONG
School of Electrical and Electronic Engineering,
Nanyang Technological University,
Singapore, Email: eqsong@ntu.edu.sg

ABSTRACT
The support vector machine (SVM) is a new and very promising classification technique developed by Vapnik and his group. Training of a SVM to obtain the maximum margin classifier requires solving a large-scale quadratic programming problem with linear and box constraints. Osuna proposed a theorem which proves that the large QP problem can be broken down into a series of smaller QP sub-problems and convergence is guaranteed. But fast progress depends heavily on whether the algorithm can select good working sets. This paper proposes a method for pre-selection of the working sets based on Osuna's theorem, which make the training process more efficient. As an implementation of this method, simulation results on the data obtained by the computerized auto-scoring system and UCI dataset are presented.

INTRODUCTION

Support Vector Machine can be seen as an alternative training technique for Polynomial, Radial Basis Function and Multi-layer Perceptron classifiers. The main idea behind the technique is to separate the classes with a surface that maximizes the margin between them. An interesting property of this approach is that it is an approximate implementation of the Structure Risk Minimization (SRM) induction principle [1].

In the *linear case*, we have a data set $\{(\mathbf{x}_i, y_i)\}_{i=1}^{\ell}$ of labeled samples, where $y_i \in \{-1, 1\}$. We wish to find a hyperplane which separates the data and minimizes the generalization error, or an upper bound on it (Structure Risk Minimization). The margin is defined by the distance of the hyperplane to the nearest of the positive and negative samples. Maximizing the margin can be expressed via the following optimization problem [2].

$$\min_{\mathbf{w}, b} \frac{1}{2}\|\mathbf{w}\|^2 \tag{1}$$

subject to $y_i(\mathbf{w} \cdot \mathbf{x} - b) \geq 1, \quad \forall i$

For the *linearly non-separable case*, Cortes and Vapnik [2] suggested a modification to the original optimization statement (1) and the approach for doing this is to allow violations in the constrains in (1), and penalize such violations linearly in the objective function.

The discussion above has only treated linear decision surface, which are definitely not appropriate for many tasks even for the simple XOR problem. The extension to more complex decision surfaces is conceptually quite simple, and is done by mapping the input variable $\mathbf{x}$ into a higher dimensional feature space and by working with linear classification in that space with map $\mathbf{z} = \Phi(\mathbf{x})$ associated with Mercer's Kernel K. The *non-linear* SVM is explicitly computed from the Lagrange multipliers $\alpha = (\alpha_1, \cdots \alpha_\ell)^T$ [2]. $u = \sum y_i \alpha_i K(\mathbf{x}_i, \mathbf{x}) - b$, where $K(\mathbf{x}_i, \mathbf{x})$ is a kernel function that represents the dot product $\Phi(\mathbf{x}_i) \cdot \Phi(\mathbf{x})$ in the feature space, and the transformed dual objective function is still quadratic as following

$$\min_\alpha \Psi(\alpha) = \min_\alpha \frac{1}{2} \sum_1^\ell \sum_1^\ell y_i y_j K(\mathbf{x}_i, \mathbf{x}_j) \alpha_i \alpha_j - \sum_1^\ell \alpha_i$$

$$0 \leq \alpha_i \leq C, \forall i, \qquad (2)$$

$$\sum_1^\ell y_i \alpha_i = 0$$

This implies we do not need to know explicitly the form of map Φ which maps data from input space into the feature space. The only thing we need to do is to decide the "admissibility" of the kernel function.

OSUNA'S DECOMPOSITION ALGORITHM

One important characteristic of Solving the quadratic programming problem (2) is that the quadratic form matrix that appears in the objective function is completely dense and with size square in the number of data vectors. This fact implies that due to memory and computational constraints, problem with large data sets cannot be solved without some kind of problem decomposition. The QP problem that we have to solve in order to train a SVM in compact form is following

$$\min \quad W(\alpha) = -\alpha^T \cdot \mathbf{1} + \frac{1}{2} \alpha^T \mathbf{D} \alpha$$

$$\text{subject to} \quad \alpha^T \mathbf{y} = 0 \qquad (3)$$

$$\alpha - C\mathbf{1} \leq \mathbf{0}$$

$$-\alpha \leq \mathbf{0}$$

Where $(\mathbf{1})_i = 1$, $D_{ij} = y_i y_j K(\mathbf{x}_i, \mathbf{x}_j)$, $\alpha = (\alpha_1, \cdots, \alpha_j)^T$ The choice of the kernel K is left to the user, and it depends on the decision surfaces one expects to work best. Since $\mathbf{D}$ is a positive semi-definite matrix (the kernel function K is positive definite), and the constraints in (3) are linear, so the Kuhn-Tucker (KT) conditions are necessary and sufficient for optimality. The KT conditions are as follows [4],

$$\nabla W(\alpha) + \Gamma - \Pi + \mu \mathbf{y} = \mathbf{0}$$

$$\Gamma^T (\alpha - C\mathbf{1}) = 0$$

$$\Pi^T \alpha = 0 \qquad (4)$$

$$\Gamma \geq \mathbf{0}, \quad \Pi \geq \mathbf{0}$$

$$\alpha^T \mathbf{y} = 0, \quad \alpha - C\mathbf{1} \leq \mathbf{0}, \quad -\alpha \leq \mathbf{0}$$

where μ, $\Gamma = (v_1, \cdots, v_\ell)^T$, $\Pi = (\pi_1, \cdots, \pi_\ell)^T$ are the associated Kuhn-Tucker multipliers

A decomposition algorithm was approached by Osuna [4]. The main idea behind the algorithm is the iteration solution of sub-problems and the evaluation of the stop criteria for the algorithm. The optimality conditions derived above are essential in order for a decomposition strategy that guarantees that at every iteration the objective function is improved. In order to accomplish this goal the index set is partitioned into two sets B and N, where the set B is called the *working set*. Then α is decomposed into two sub-vectors α_B and α_N, keeping fixed α_N and allowing changes only in α_B, thus defining the following subproblem:

$$\min W(\alpha_B) = -\alpha_B^T \mathbf{1} + \frac{1}{2}[\alpha_B^T \mathbf{D}_{BB}\alpha_B + \alpha_B^T \mathbf{D}_{BN}\alpha_N +$$
$$\alpha_N^T \mathbf{D}_{NB}\alpha_B + \alpha_N^T \mathbf{D}_{NN}\alpha_N] - \alpha_N^T \mathbf{1}$$
$$\text{subject to } \alpha_B^T \mathbf{y}_B + \alpha_N^T \mathbf{y}_N = 0 \tag{5}$$
$$\alpha_B - C\mathbf{1} \le 0$$
$$-\alpha_B = 0$$

where $(\mathbf{1})_i = 1$, $D_{ij} = y_i y_j K(\mathbf{x}_i, \mathbf{x}_j)$ and C is a positive constant. Using this decomposition we have the following propositions by Osunna [4].

Proposition 1 *Moving a variable from B to N leaves the cost function unchanged, and the solution is feasible in the subproblem.*

Proposition 2 *Moving variable that violate the optimality condition from N to B gives a strict improvement in the cost function when the subproblem is re-optimized.*

PROPOSED METHOD FOR SELECTING A GOOD WORKING SET

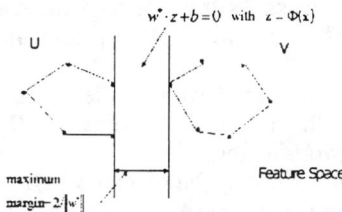

Figure 1: The hyperplane that separates the two classes

The decomposition algorithm discussed above breaks large QP problem into a series of smaller QP subproblems, as long as at least one example that violates the KT conditions is added to the examples for the previous subproblem, each step will reduce the overall objective function and maintain a feasible point that obeys all of the constraints. But Osuna did not point out how to select the working set and just got these points arbitrarily. Evidently, Fast progress depends heavily on whether the algorithm can select good working set.

When selecting the working set, it is desirable to select a set of variables such that these points are Support Vectors. From Fig. 1., it is clear that the support vectors are those points laying on the optimal hyperplanes and they are closer to each other. So if we can evaluate the distance of two points between the two classes in the feature space and select those points with smaller distance, we should get a good working set. Fortunately, we can calculate the distance between two points in the Feature space only in the terms of Kernel as follows $D_{ij} = \left\| \Phi(\mathbf{x}_i) - \Phi(\mathbf{x}_j) \right\|^2 = K(\mathbf{x}_i, \mathbf{x}_i) - 2K(\mathbf{x}_i, \mathbf{x}_j) + K(\mathbf{x}_j, \mathbf{x}_j)$.

Assume that the size of working set is ℓ which is big enough to contain all the support vectors but small enough such that the computer can handle it and optimizing it using some standard QP solver (this assumption is reasonable because in many application case, support vectors are just a small part of the training set.). We can decide working set by computing the distance of data between two classes and select the nearest points as the working set we wanted e.g. $(\mathbf{z}_1, \cdots, \mathbf{z}_\ell)$, where $\mathbf{z}_i = \Phi(\mathbf{x}_i)$. But suppose that there are n points in class 1 and m points in class 2, according the method described above, we should evaluate $n \times m$ distances and arrange them in ascend order. Clearly it is not an efficient way to deal with the pre-selection problem. So we ask for less, rather than evaluating the distance between the training data, we compute the distance between the training data in one class and the center vector of the other class as follows

$$D_{ij} = \left\| \Phi(\mathbf{x}_i) - \frac{1}{m}[\Phi(\mathbf{x}_1) + \ldots + \Phi(\mathbf{x}_\ell)] \right\|^2$$

$$= K(\mathbf{x}_i, \mathbf{x}_i) - \frac{1}{m}\sum_{j=1}^{m} K(\mathbf{x}_i, \mathbf{x}_j) + \quad (6)$$

$$\frac{1}{m}[\Phi(\mathbf{x}_1) + \ldots + \Phi(\mathbf{x}_m)] \cdot \frac{1}{m}[\Phi(\mathbf{x}_1) + \ldots + \Phi(\mathbf{x}_m)]$$

where $\mathbf{x}_i$, $i = 1, \cdots, n$ is the training data in class 1 and $\mathbf{x}_j$, $j = 1, \cdots, m$ is the training data in class 2, For class 1's training data, the third term of (6) is a constant and can be dropped, so $D_{ij} = K(\mathbf{x}_i, \mathbf{x}_j) - \frac{2}{m}\sum_{j=1}^{m} K(\mathbf{x}_i, \mathbf{x}_j)$. The same simplification is added to class 2's training data for class 2's training data as $D_{ji} = K(\mathbf{x}_j, \mathbf{x}_j) - \frac{2}{n}\sum_{i=1}^{n} K(\mathbf{x}_i, \mathbf{x}_j)$. Using the results above we are now ready to formulate the algorithm

1. Choose $\ell/2$ points from class 1 with smallest distance values D_{ij} and $\ell/2$ points from class 2 with smallest distance values D_{ji} as the working set B (ℓ is the size of the working set)
2. Solve the subproblem defined by the variables in B.
3. While there exists some $\alpha_j, j \in N$, such that the KT conditions are violated replace $\alpha_i, i \in B$, with $\alpha_j, j \in N$ and solve the new subproblem.

Evidently, this method is more efficient and easy to be implemented, which is used in the following section.

APPLICATION AND RESULT

In this section, we have implemented the proposed pre-selection method and decomposition algorithm using the standard QP toolbox of MATLAB as the solver for the subproblem. Firstly we tested our technique on the data obtained by the Computerized Auto Scoring System [7]. The focus of this system is to classify the bullet hole images into proper classes. Normally, certain bullet hole images contain two or more bullet holes with different degrees of overlapping. Therefore, it is critical to classify bullet hole images into multi-class sets for the auto-scoring system. The statistical information of the bullet hole images based on 35 samples of target papers is shown in Tab. 1. And Fig. 2. illustrates some representative examples of bullet hole image used in the experiments. To simplify the situation, only one, and two-hole images are considered in this paper.

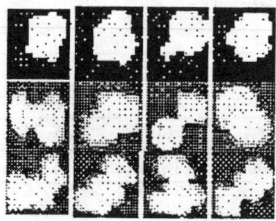

Figure 2: Some respective examples of bullet hole images used in the experiments. The first row: 1-bullet holes; the second row: 2-bullet holes: The third row; 3-bullet holes.

Holes	1-bullet	2-bullet	3-bullet	4-bullet	5-bullet	6-bullet
Number	1496	108	30	4	4	0
Percentage	0.919	0.066	0.009	0.002	0.002	0

Table 1: The statistics of the Bullet Holes

We considered the training data set of 300 data (120 for one-hole image, 180 for two-hole image) and another 200 data (100 for one-hole image, 100 for two-hole image) as the test samples. Each point has 20 features [9]. The kernel function we used here is RBF kernel: $K(\mathbf{x},\mathbf{y}) = e^{-\|x-y\|^2/2\sigma^2}$ and the working set size is 100 points. Tab. 2 shows the result of three methods which are used for training the Support Vector Machine. For Standard QP method and Osunna's method, the total time needed includes the training and testing time, and for Pre-selection method the total time needed includes pre-selection, training and the testing time.

Another simulation result is from the UCI "adult" data set [6]. The goal is to predict whether a household has an income greater than $50000. There are 6 continuous attributes and 8 nominal attributes. After discretization of the continuous attributes, there are 123 binary features, of which 14 features are true. Tab. 3. shows the simulation results of the Income Predict problem with RBF kernel $\sigma = 10, C = 1$. When the sample size becomes larger, using the MATLAB standard QP solver is not enough to meet the computation requirement because of the memory thrashing. So we can use the SMO [5] or SVM light [6] as tools to solve the subproblem defined by the pre-selection method. It clearly shows that if we have some *a priori* knowledge to determine a pre-selection working set, it would be very efficient to begin training SVMs just on those examples and still get almost the same result. This significantly reduces the training time and the number of steps needed to reach to the optimal solution.

	Standard QP	Osuna	Pre-selection
Total Time (Sec)	338.6	184.9	103.6
No. of SVs	65	64	62
Iterative Steps	1	9	4
No. of Errors	0	0	0

Table 2: Experiment result of three methods (Standard QP method, Osuna's method and Pre-selection method)

Samples	200		400		800	
Methods	Osuna	Pre-selection	Osuna	Pre-selection	Osuna	Pre-selection
Time	87.6	62.1	621.9	458.1	9632.9	6324.1
SVs	71	70	170	170	327	328
Steps	5	3	7	5	8	5
Wsize	100	100	180	180	360	360

Table 3: Experiment results of Income Prediction on 200/400/800 samples by Osuna's method and pre-selection method. (Wsize denotes working size of the data set)

CONCLUSIONS

In this paper we have presented a method of pre-selection the working set, which can be combined with the Osuna's decomposition algorithm to make the training of SVM faster and more efficient. The current version of the algorithm has been tested with two real-world problems. one is the data set from Autoscoring system and another is from UCI "adult"data. All simulations are test on PII 400 with 64Mb of RAM. No attempts to optimize and speed up the algorithm have been made yet (such as caching kernel evaluations).

REFERENCES

1. Vapnik, *An Overview of Statistical Learning Theory*, IEEE. Trans. Neural Networks, Vol. 10. No. 5. Sept. 1999.
2. Burges, C.J.C., *A Tutorial on Support Vector Machines for Pattern Recognition*, Submitted to Data mining and Knowledge Discovery, http://svm.research.bell-labs.com/SVMdoc.html, 1998.
3. Cortes, C., Vapnik, V., *Support Vector Networks*, Machine Learning, 20:273-297, 1995.
4. Edgar Osuna, Robert Freund, Federio Girosi, *An Improved Training Algorithm for Support Vector Machines.*, Proc. of IEEE NNSP's 97, Amelia Island, 1997.
5. John C. Platt, *Sequential Minimal Optimization: A Fast Algorithm for training Support Vector Machines*, Tech report. Microsoft Research. 1998.
6. Thorsten Joachims, *Making Large-Scale SVM Learning Practical*, Research report. ISSN 0943-4135. University of Dortmund. 1998
7. W.F.Xie, D.J. Hou and Q.Song, *Bullet-hole Image Classification With Support Vector Machines*, IEEE International Workshop on Neural Networks for Signal Processing11-13 December 2000.

CONTROLLED LEARNING OF GREN-NETWORKS

IVETA MRÁZOVÁ
Department of Software Engineering
Charles University
Prague, Czech Republic

ABSTRACT
Within the framework of our previous research, we have introduced a modular system for training BP-networks without the knowledge of their desired output values. The necessary error terms are determined by the so-called GREN-networks. Anyway, some GREN-networks lack in their ability to train BP-networks. This inability is characterized mainly by their low sensitivity to BP-network outputs. In this paper, we propose a strategy for increasing this kind of sensitivity.

INTRODUCTION

The standard Back-Propagation algorithm proposed by Rumelhart et al. (1986) and its various modifications belong to the most often used algorithms applied for training multi-layer feed-forward neural networks. Further, by a BP-network we mean a feed-forward neural network B with a set of neurons N, a set of oriented inter-connections among neurons C, a set of input neurons I, a set of output neurons O, a weight function $w : C \to R$ and a threshold function $t : N \to R$. An input vector $\vec{x} \in R^n$ being processed by B with $|I| = n$ will be called an input pattern. A vector $\vec{d} \in R^m$ formed by desired outputs of neurons lying in the output layer of B with $|O| = m$ will be called an output pattern. And a vector $\vec{y} \in R^m$ formed by actual outputs of neurons lying in the output layer of B with $m = |O|$ output neurons will be called an actual output of B.

For the standard BP-algorithm, the training set $\mathcal{T}$ is a finite non-empty set of P ordered pairs of input/output patterns: $\mathcal{T} = \{[\vec{x}_1, \vec{d}_1], \ldots, [\vec{x}_P, \vec{d}_P]\}$. Here, we will consider networks with the standard sigmoidal transfer function $f(\xi) = 1/(1 + e^{-\xi})$, where ξ denotes the neuron potential. For a neuron having the weights $(w_1, \ldots, w_n)$, the threshold ϑ and the input vector $\vec{z} = (z_1, \ldots, z_n)$, the potential ξ is to be computed as $\xi = \sum_{i=1}^{n} z_i w_i + \vartheta$. In this context, the thresholds can be represented by weights coming from a fictive neuron with a constant output value equal 1. The objective function E represents the total error between the desired and actual outputs of all the output neurons in the BP-network taken for all training patterns from $\mathcal{T}$.

GREN-NETWORKS

Mrázová (2000) introduced a new learning strategy for training BP-networks based also on steepest gradient descent. However, the error terms necessary for training the "actual" BP-network are determined by the so-called generalized relief error network (GREN-network) and without an explicit knowledge of the desired BP-network outputs. The scheme of such a system is shown in Figure 1.

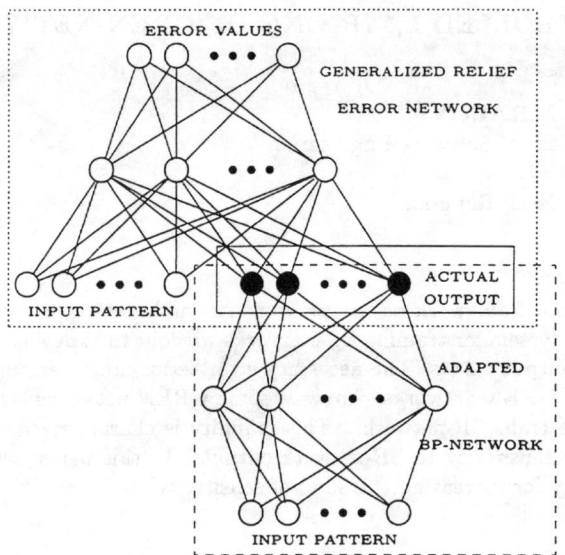

Figure 1: A modular system for training BP-networks with GREN-networks

For such a GREN-network GR_B, an input pattern $\vec{v}$ has the form: $\vec{v} = [\vec{x}, \vec{y}] = (x_1, \ldots, x_n, y_1, \ldots, y_m)$, where $\vec{x} = (x_1, \ldots, x_n)$ corresponds to the input pattern of B and $\vec{y} = (y_1, \ldots, y_m)$ stands for its actual output. An output pattern $\vec{f}$ of GR_B will be denoted as $\vec{f} = (f_1, \ldots, f_t)$ and its actual output is a vector $\vec{e} = (e_1, \ldots, e_t)$. The training set $\mathcal{T}_{GR_B}$ of the GREN-network GR_B consists of Q ($Q \geq 1$) ordered pairs of input/output patterns: $\mathcal{T}_{GR_B} = \{[\vec{v}_1, \vec{f}_1], \ldots, [\vec{v}_Q, \vec{f}_Q]\} = \left\{ [[\vec{x}_1, \vec{y}_1], \vec{f}_1], \ldots, [[\vec{x}_Q, \vec{y}_Q], \vec{f}_Q] \right\}$. The GREN-network GR_B can be trained e.g. by the standard BP-training algorithm. Then, for a BP-network B with n input and m output neurons trained by the GREN-network GR_B, a training set $\mathcal{T}_B$ will be a finite set of P ($P \geq 1$) input patterns: $\mathcal{T}_B = \{ \vec{x}_1, \ldots, \vec{x}_P \}$. The output of GR_B should be w.l.o.g. close to the zero-vector (from R^t) for those inputs $[\vec{x}, \vec{y}]$ where $\vec{y}$ is "close" to the "desired" output of B.

For training a BP-network B by means of GR_B, we can apply the basic idea of Back-Propagation as well. For each presented input pattern $\vec{x}$, we first compute the actual output $\vec{y}$ of the BP-network B. This actual output represents together with the presented input pattern an input pattern $\vec{v}$ for the applied GREN-network GR_B. For this input pattern $\vec{v} = [\vec{x}, \vec{y}]$, we compute the actual output $\vec{e}$ of GR_B. The elements of the vector $\vec{e}$ represent the error values assigned to $\vec{v}$. The total network error E corresponds to:

$$E = \sum_{p=1}^{P} \sum_{e=1}^{t} e_{e,p}^{GR_B} \qquad (1)$$

The values $e_{e,p}^{GR_B}$; ($0 < e_{e,p} < 1$; $\forall p = 1, \ldots, P$; $\forall e = 1, \ldots, t$) stand each one for the respective output neuron and input pattern. With regard to simplicity, we will further omit the index p. The weights w_{ij}^B of B will be adjusted with regard to the error terms $\partial E / \partial y^B$ back-propagated through GR_B to the output neurons of B, for details, see Mrázová (2000). Unfortunately, some GREN-networks have a low sensitivity to their inputs, and in particular to those inputs representing the outputs of the "actual" BP-network. Insensitive GREN-networks are not able to train "actual" BP-networks because the error terms propagated by the GREN-networks are inadequately small even for large errors computed at their outputs. Therefore, the aim of our paper will be to design a method for increasing the sensitivity of GREN-networks already during their training.

INFLUENCING THE SENSITIVITY OF BP-NETWORKS

In the literature, many articles approach the problem of handling the sensitivity of BP-networks with the aim to increase their robustness. An improved generalization and network robustness correspond in this context to a lower sensitivity of network outputs to its inputs. Often, these methods are based on an alternative formulation of the objective function to be minimized during the training process. Additional requirements specify the "desired smoothness" of the network function by means of a regularization term E^{REG} penalizing non-smooth mappings. The "smoothness requirements" are expressed by means of the elements of the Jacobi matrix $J_{ij} = \partial y_j / \partial x_i$, where y_j corresponds to the j-th output value and x_i stands for the i-th input value.

Over-fitted networks are generally characterized by mappings which have a lot of structure and a relatively high curvature. This provides some indirect motivation for weight-decay regularizers as a way of reducing the curvature of the network function. Bishop (1993) and Drucker and Le Cun (1992) consider regularizers which penalizes the curvature explicitly by means of second-order derivatives of the network function. These techniques attempt to improve the generalization abilities of trained networks by favouring smoother network functions. Smoother means in this context functions with smaller values of squared second-order derivatives of network outputs with respect to network inputs. But minimizing extensively the network sensitivity can cause an "artificial" insensitivity of the network to any inputs.

CONTROLLED LEARNING OF GREN-NETWORKS

We require on the contrary GREN-networks that are more "sensitive" to their inputs. More sensitive networks are expected to allow a reliable and quicker error back-propagation – and thus also a quicker learning of the "actual" BP-network. Hence, for sensitive "GREN-experts" the partial derivatives of GREN-network outputs with respect to its inputs should be non-zero. The only reasonable exception represent those cases with correct BP-outputs for the presented input patterns. Therefore, we require non-zero first-order derivatives of network outputs y_s to network inputs y_r instead of minimizing the second-order derivatives. Moreover, we will favour networks with larger values of these derivatives. The regularization term to be minimized during training will have the following form:

$$E^{REG} = \sum_{q=1}^{Q} E_q^{REG} = -\sum_{q=1}^{Q} \sum_{s} \sum_{r>n} \left(\frac{\partial y_{q,s}}{\partial y_{q,r}} \right)^2 \qquad (2)$$

where $y_{q,s}$ denotes the value of the s-th GREN-network output for the q-th pattern and $y_{q,r}$ stands for the value of the r-th BP-network output. Anyway, with regard to simplicity, we will further omit the pattern index q. The additional error function can be expressed for each training pattern q in GREN-networks without hidden layers as it follows:

$$E^{REG} = -\sum_s \sum_{r>n} \left(\frac{\partial y_s}{\partial y_r}\right)^2 = -\sum_s \sum_{r>n} \left[(\delta_s^s)^2 w_{rs}^2\right] \qquad (3)$$

where $\qquad \delta_s^s = \dfrac{\partial y_s}{\partial \xi_s} = y_s(1-y_s) \qquad (4)$

and for GREN-networks with hidden layers, E^{REG} corresponds to:

$$E^{REG} = -\sum_s \sum_{r>n} \left(\frac{\partial y_s}{\partial y_r}\right)^2 = -\sum_s \sum_{r>n} \left(\sum_k \delta_k^s w_{rk}\right)^2 \qquad (5)$$

where $\qquad \delta_k^s = y_k(1-y_k) \sum_l \delta_l^s w_{kl} \qquad (6)$

and l denotes neurons from the layer above neuron k. These expressions correspond to the evaluation of the Jacobian matrix described by Bishop (1996). For each presented input pattern q, the weights w_{ij} will be adjusted additionally also by the term $\Delta_{E^{REG}} w_{ij} = -\partial E^{REG}/\partial w_{ij}$:

$$\Delta_{E^{REG}} w_{ij} = -\frac{\partial E^{REG}}{\partial w_{ij}} = -\frac{\partial}{\partial w_{ij}} \left(-\sum_s \sum_{r>n} \left(\frac{\partial y_s}{\partial y_r}\right)^2\right) \qquad (7)$$

Thus, the particular weights w_{ij} of the GREN-network have to be adjusted by:

$$w_{ij}(T+1) = w_{ij}(T) + \alpha \, \delta_j^E \, y_i + \alpha_c \, \sigma_{ij} + \alpha_m \, (w_{ij}(T) - w_{ij}(T-1)) \qquad (8)$$

where $\delta_j^E = \begin{cases} (d_j - y_j) \, y_j \, (1-y_j) & \text{for an output neuron} \\ y_j \, (1-y_j) \sum_k \delta_k^E w_{jk} & \text{for a hidden neuron} \end{cases} \qquad (9)$

$$\sigma_{ij} = \begin{cases} 2[y_j(1-y_j)]^2 \left[w_{ij} + \left(\sum_{i'>n_1} w_{i'j}^2\right) y_i(1-2y_j)\right] \\ \qquad\qquad\qquad\qquad\qquad\qquad\qquad\qquad \text{for an output neuron} \\[1em] \sum_s \left(\delta_j^s [1-2y_j] y_i\right) \left[\sum_{i'>n_1} w_{i'j} \left(\sum_{k'} \delta_{k'}^s w_{i'k'}\right)\right] + \\ \qquad + \sum_s \delta_j^s \left[\sum_{k'} \delta_{k'}^s w_{ik'} + \delta_j^s w_{ij}\right] \quad \text{for a hidden neuron} \end{cases} \qquad (10)$$

$$\delta_j^s = \begin{cases} y_j(1-y_j) & \text{for the output neuron } j=s \\ 0 & \text{for other output neurons} \\ y_j(1-y_j) \sum_k \delta_k^s w_{jk} & \text{for a hidden neuron} \end{cases} \qquad (11)$$

In the above expressions, k and k' index neurons in the layer above the neuron j, i and i' correspond to neurons from the layer below the neuron j. s stands for neurons from the output layer. d_j denotes the desired and y_j the actual output value of the neuron j. $T+1$, T and $T-1$ index next, present and previous weights, respectively. α and α_c are the constants representing the "standard" and "control" learning rates, resp. α_m denotes the momentum rates. $n_1 = n$ for the input layer of the GREN-network and $n_1 = 1$ for its other layers. Very similar adjustment rules could be derived also for the thresholds and weights from other layers, for details see Mrázová (2001). Moreover, it is possible to force the network sensitivity in the above-described way also for general BP-networks and for arbitrary chosen input neurons.

The Analysis of the Proposed Method

The main idea of controlled learning consists in forcing the sensitivity of GREN-network outputs to its inputs. Larger absolute values of partial derivatives of the outputs of the GREN-network with respect to its inputs are expected to transfer better the errors from the GREN-network and thus also to train "safer" and "quicker" the "actual" BP-network. However, due to the "linear" nature of the error function (1) to be minimized the training process often oscillates and is not able to fit narrow local minima of the error function. Therefore, we have slightly modified the training process of the "actual" BP-network by using "quadratic" error terms $\widehat{E}$ instead of the "linear" ones defined in (1)):

$$\widehat{E} = \sum_{p=1}^{P} \sum_{e=1}^{t} \left(e_{e,p}^{GR_B}\right)^2 \qquad (12)$$

As a consequence, there are considered also the errors computed as the output of GR_B during training the "actual" BP-network. Taking into account both the GREN-network outputs and the sensitivity terms can be crucial for the result of the training process especially in the case of erroneous training patterns with a low sensitivity.

SUPPORTING EXPERIMENTS

In order to compare the capabilities of the discussed training and pattern adjustment strategy with the standard BP-training algorithm, we tested it on the following problem. The training set $\mathcal{T}_B$ for the BP-network B to be trained by means of the standard BP-training algorithm consisted of the following 9 training patterns: $\mathcal{T}_B = \{[(0,0), 1], [(0.5, 0), 0], [(1, 0), 1], [(0, 0.5), 0], [(0.5, 0.5), 1], [(1, 0.5), 0], [(0, 1), 1], [(0.5, 1), 0], [(1, 1), 1]\}$. The same input patterns formed the training set $\mathcal{T}_{B'}$ for the BP-network B' to be trained by means of a GREN-network $GR_{B'}$. The training set $\mathcal{T}_{GR_{B'}}$ for the GREN-network $GR_{B'}$ contained 41 training patterns represented by the above input patterns together with various possible actual outputs of B' and the error value defined as the absolute value of the difference between the respective actual and desired output values.

We have reached the best results for the BP-network B' with 8 neurons in one hidden layer $(2-8-1)$ and with random initial weights from the interval $\langle -1, 1 \rangle$. Each training pattern from $\mathcal{T}_{B'}$ was presented once in every training cycle using parameter values $\alpha = 0.4$, $\alpha_m = 0.6$ and the error function of (12). Within cca 1500 cycles, the error measured as the sum of GREN-network outputs dropped to 1.19. The corresponding summed squared error between actual outputs of B' and output patterns for B (SSE) reached 0.05. The GREN-network $GR_{B'}$ applied for training the BP-network B' had 11 neurons in one hidden layer (3 − 11 − 1) and random initial weights also from the interval $\langle -1, 1 \rangle$. With the same parameters as in the previous case and $\alpha_c = 2000$, we reached after 1300 cycles SSE 0.05 and a reasonable network sensitivity to the outputs of B'.

For the BP-network B with the same architecture, initial weights and training parameters as B', the SSE dropped within 1500 cycles to 0.5. Unfortunately, further training did not converge to lower values of the SSE. Applying other parameter values, the training was possible, anyway, a considerably larger number of cycles was necessary. Our further experiments have shown that using the "squared" error terms (12) for training "actual" BP-networks leads to a bit "slower" but more successful training process comparing with the "linear" case (1). We expect that this effect is caused by a wider and flatter error surface in the areas close to local minima.

CONCLUSIONS

In this paper, we discussed an approach for increasing the ability of GREN-networks to train "actual" BP-networks. The main results achieved so far indicate that it is possible:
- to increase the sensitivity of trained GREN-networks to their inputs,
- to detect "overtraining" in GREN-networks characterized by steeply decreasing sensitivity coefficients,
- to train "actual" BP-networks more efficiently by minimizing the squared GREN-network outputs instead of the "linear" ones. This apparently insignificant difference can be crucial for the success of the training process.

Within the framework of our further research, we plan to study how to improve and to preserve the generalization abilities of trained GREN-networks. We expect that such a "simplified sensitivity control" could be based on punishing the trained network for developing non-efficient large negative weights.

ACKNOWLEDGEMENTS

This research was supported by the grant No. 157/1999/A INF/MFF of the GA UK and by the grant No. 201/99/0236 of the GA ČR.

REFERENCES

Bishop, C. M., 1993, "Curvature-driven smoothing: a learning algorithm for feedforward networks," *IEEE Transactions on Neural Networks*, 4, pp. 882-884.

Bishop, C. M., 1996, *Neural Networks for Pattern Recognition*, Oxford University Press.

Drucker, H. and Le Cun, Y., 1992, "Improving generalization performance using double back-propagation," *IEEE Transactions on Neural Networks*, 3, pp. 991-997.

Mrázová, I., 2000, "Generalized Relief Error Networks," *Smart Engineering System Design: Neural Networks, Fuzzy Logic, Evolutionary Programming, Data Mining, and Complex Systems*, C.H. Dagli, A.L. Buczak, J. Ghosh, M.J. Embrechts, O. Ersoy and S. Kercel (eds), ASME Press Series, pp. 21-26.

Mrázová, I., 2001, "Controlled Learning of GREN-networks," TR No 2001/3, KSI MFF UK, Charles University, Prague, Czech Republic, 16 pp.

Rumelhart, D. E., Hinton, G. E., and Williams, R. J., 1986, "Learning representations by back-propagating errors," *Nature*, Vol. 323, pp. 533-536.

CRITIC-DRIVEN ENSEMBLE CLASSIFICATION VIA A LEARNING METHOD AKIN TO BOOSTING[*]

LIAN YAN
Athene Software, Inc.
Boulder, CO 80302
liany@athenesoft.com

DAVID J. MILLER
The Pennsylvania State University
University Park, PA 16802
miller@perseus.ee.psu.edu

ABSTRACT

A well-known limitation of most ensemble techniques is the requirement that individual classifiers must achieve accuracy greater than 50%. Recently, the authors proposed *critic-driven ensemble classification*, which can overcome the 50% requirement, as proved under an independence assumption, and which achieves improved performance over Bagging. Here we propose an algorithm akin to *boosting* for jointly training critic-driven ensembles. In addition to overcoming the 50% requirement, one advantage of critic-driven boosting lies in the rule for aggregation, which uses *input-dependent weights*, provided by the critics. The advantage of this approach is shown experimentally, on a number of benchmark sets, in comparison with Adaboost.

INTRODUCTION

A well-known limitation of ensemble classification techniques is the prevailing requirement that the individual classifiers must achieve a correct decision rate greater than 50%. Theoretically, ensemble decision accuracy will actually degrade if this requirement is not met. More generally, in practice ensemble techniques will not be effective if the individual classifiers are not sufficiently reliable. Thus, the 50% requirement is pervasive in ensemble work. Recently, the authors proposed a new paradigm, dubbed *critic-driven ensembles*, expressly aiming to practically mitigate (as well as to theoretically remove) this requirement on individual classifier accuracy (Miller and Yan, 1999). In this framework, specific to each classifier (expert) is another classifier called a *critic*. This classifier tackles the problem of deciding, for each example, whether or not its expert's decision is correct. Since the expert is assumed to address a multiclass (> 2 class) problem while the critic only tackles a two-class one, the critic can typically be made more accurate than its expert. Accordingly, critics can be used within the combining rule to help achieve more accurate ensemble decisions. In a hard voting context

---
[*]David J. Miller was supported by National Science Foundation grant IIS-0082214.

critics can suppress expert opinions, thus adding a (judicious) *abstention* mechanism to standard voting procedures. In a soft averaging context, critics provide unequal, feature-dependent weights of confidence for expert aggregation.

In our earlier work (Miller and Yan, 1999), we i) proved that critic-driven schemes can overcome the 50% requirement under an independence assumption; ii) proposed several critic-driven hard voting schemes and also soft averaging methods grounded in information-theoretic and probabilistic criteria; and iii) demonstrated improvement over Bagging (Breiman, 1996) on several real-world examples. In this work, we assumed that there are separate training sets for the learning of each expert-critic pair. This was motivated by applications where a common pool of training data is unavailable, and hence where joint training methods (including boosting) are precluded. Such applications include i) multisensor classification/detection (based on separate, incommensurate sensor "training sets") and 2) the pooling of existing or proprietary classifiers (where even the *feature space* of each classifier may be proprietary or unknown). However, when a common training set is available, separate training of the individual classifiers is suboptimal and will limit performance.

In this work we assume a common pool of data does exist and thus propose an algorithm akin to boosting for the joint training of critic-driven ensembles. In addition to mitigating the 50% accuracy requirement, one significant advantage of critic-driven boosting over standard boosting lies in the method's rule for performing classifier aggregation. In standard boosting, the weight given to each expert's decision is not a function of the input feature vector (Freund and Schapire, 1997), (Schapire, 1999). By contrast, critic-driven combining uses *input-dependent* weights, provided by the critics, to assimilate expert opinions. The advantage of this approach is borne out in our experiments.

BOOSTING CRITIC-DRIVEN ENSEMBLES

In (Miller and Yan, 1999) it was assumed that joint training of the ensemble is either impossible (a lack of common training data) or impractical. Here we develop a critic-driven joint training method inspired by the boosting paradigm. Boosting generates an ensemble by a sequence of training stages, each producing an individual classifier, with the training set reweighted at each stage to focus on the examples that are most difficult to classify by the existing ensemble. Key features in boosting are: 1) a weak learner that generates an expert, 2) a confidence measure for the individual experts, 3) a rule for reweighting the data for the next training stage, and 4) a decision function based on the aggregation of the individual experts and their confidence weights. Boosting decision rules generally take the form $k^*(\underline{x}_i) = \arg\max_k \sum_{j=1}^{N} w_j P_e^{(j)}[k|\underline{x}_i]$, $k = 1, \ldots, C$, $i = 1, \ldots, M$, where $P_e^{(j)}[k|\underline{x}]$ is the discriminant function output for class k from the jth expert. Here, without loss of generality, we will assume $P_e^{(j)}[k|\underline{x}]$ is an *a posteriori* class probability, with $\underline{x}$ an input feature vector, M the total number of samples, C the number of classes, and N the number of experts. Note that the weights w_j, $j = 1, \ldots, N$, are *constants* in standard boosting algorithms.

The pseudocode for a critic-driven variant of boosting is shown in Table 1. As for Adaboost, individual experts (but now with their associated critics) are sequentially trained, based on a reweighted distribution over the training set. If arithmetic averaging is used as the critic-driven combination rule (Miller and Yan, 1999), the ensemble's *a posteriori* class probabilities at round t, $t = 1, \ldots, N$, are in the form

$$P_t(k|\underline{x}_i) = \sum_{j=1}^{t} w_j(\underline{x}_i) P_e^{(j)}[k|\underline{x}_i], k = 1, \ldots, C, \ i = 1, \ldots, M. \quad (1)$$

Here,
$$w_j(\underline{x}_i) = \frac{P_c^{(j)}[1|\underline{x}_i]}{\sum_{l=1}^{t} P_c^{(l)}[1|\underline{x}_i]}, \quad (2)$$

with $P_c^{(j)}[1|\underline{x}_i]$ the probability of critic j accepting its expert's decision. To measure the existing ensemble's performance on each training sample, the cross entropy between $\{P_t[k|\underline{x}_i]\}$ and $\{P_o[k|\underline{x}_i]\}$, the object distribution, can be used. If $P_o[k|\underline{x}_i] \in \{0,1\}$, the cross entropy can be simplified as $-\log(P_t[c_i|\underline{x}_i])$, where c_i is the supervising label for $\underline{x}_i$. When $-\log(P_t[c_i|\underline{x}_i])$ is close to zero, the existing ensemble has accurately classified sample i. Otherwise, the next expert needs to concentrate more on this sample. In particular, the weight $D_{t+1}(i) = -\log(P_t[c_i|\underline{x}_i])/Z$, $Z = \sum_{j=1}^{M} -\log(P_t[c_j|\underline{x}_j])$, is assigned to sample i while training the $t+1$st expert. At each stage, after training expert t, its corresponding critic is also trained in order to generate the expert's *input-dependent* confidence weights. We emphasize that the critics are trained over the original *unweighted* training data, not based on the weights $D_t(i)$.

RELATIONSHIP TO ADABOOST

Critic-driven boosting differs from standard boosting in several main respects. First, note that each learning stage now generates an (expert,critic) pair, rather than just an expert. Second, note that we reweight the data based, explicitly, on the cross entropy disagreement between the true (0-1) distribution and the current *ensemble a posteriori* probabilities. This is only qualitatively similar to the reweighting done in standard boosting. Finally, in our method the confidence weights for each expert are given by our (input-dependent) critic probabilities, while in boosting these weights are constants that depend on each individual classifier's *overall* error rate over the reweighted training set. This is the most significant difference, and one

Given: a training set $S = \{(\underline{x}_i, c_i), i = 1, \ldots, M\}$, where $c_i \in \{1, \ldots, C\}$.
Initialize: $\{D_1(i) = \frac{1}{M}\}$.
For $t = 1, \ldots, N$ {
  Train expert t using distribution $\{D_t(i)\}$.
  Test expert t over training set S, and generate training set S_c for critic t.
  Train critic t over training set S_c.
  Update critic weights $\{w_j(\underline{x}_i)\}$ by (2).
  Calculate a posteriori class probability $\{P_t(k|\underline{x}_i)\}$, e.g., by (1).
  Update: $D_{t+1}(i) = -\log(P_t[c_i|\underline{x}_i])/Z$.
}
Output the classifier: $k^*(\underline{x}) = P_N(k|\underline{x}) = \arg\max_k \sum_{j=1}^{N} w_j(x) P_e^{(j)}[k|\underline{x}]$.

Table 1: Pseudo-code for critic-driven boosting.

which we believe should be the source of performance improvement for our method. Specifically, we believe that, in both the critic-driven and standard boosting cases, boosting will tend to focus the training of individual classifiers on localized regions of the feature space. Yet, in the actual decision rule, the confidence given to each expert in standard boosting is based on the expert's *overall* accuracy (measured with respect to its reweighted distribution), without any dependence on the input feature. Thus, an expert may strongly participate in decisions for localized feature regions that were not regions of emphasis during its training. By contrast, *input-dependent* confidence weights, if accurately learned, should effectively "select" for participation the subset of experts that are reliable for the given input feature vector. This is precisely the role of the critics in the critic-driven combining rule.

Most boosting algorithms, especially the multi-class extensions, explicitly require that all individual experts have an error rate (p) less than 0.5 over the weighted training set, in order to train the ensemble correctly (Freund and Schapire, 1997). In our experiments, we often observed $p > 0.5$. When this happened, we had to bootstrap the original training set and restart the boosting process. Moreover, even if the training set's error rate is less than 0.5, the error rate over the test set may still be larger than 0.5, which will cause the ensemble to not work effectively. However, critic-driven boosting mitigates the $p < 0.5$ requirement (Miller and Yan, 1999), which makes critic-driven boosting more robust than Adaboost. We demonstrate this in Figure 1 for the 10-class *Vowel* data set from the UC Irvine repository. We can see that Adaboost fails when a one hidden unit multilayer perceptron (MLP) structure is used as the base classifier(This one unit MLP has an error rate $p > 0.5$ over the training set.). However, critic-driven boosting, with the same 1 hidden unit MLP structure for the experts, generates very reasonable results. The improvement is completely due to the auxiliary critics (also one hidden unit MLP classifiers).

EXPERIMENTAL RESULTS

We have experimentally evaluated critic-driven boosting and Adaboost (Schapire, 1999), using MLP base classifiers trained by minimizing the cross entropy cost. The *improved combination rule* in (Miller and Yan, 1999) was used for critic-driven boosting. Four data sets were taken from the UCI machine learning repository: the six-class *glass* set (214 samples), the three-class *Waveform* set (5,000 samples), the ten-class *Vowel* set (1,520 samples), and the ten-class *Yeast* set (1,484 samples). We randomly split the data sets into equally sized training and test sets. All the results were obtained by averaging over three random splits.

For each of the data sets, we compared the error rates over the test set between critic-driven boosting and Adaboost for two cases: 1) where the complexity of Adaboost (in terms of total number of parameters to optimize) was matched to that of critic-driven boosting; 2) where the complexity was chosen to give the best performance for Adaboost. Figure 2 shows that critic-driven boosting consistently outperforms Adaboost in both cases, for all four data sets. Interestingly, one can note that, for 10 hidden units, Adaboost severely overfits on *Yeast*, but critic-driven boosting (at the same complexity) still achieves improving results for an increasing number of classifiers.

CONCLUSIONS

We proposed a new boosting algorithm based on the framework of critic-driven ensemble classification. Unlike standard boosting algorithms, which have constant confidence weights, the critic-driven boosting algorithm has input-dependent confidence weights, provided by critics. It was demonstrated over several benchmark data sets that critic-driven boosting consistently outperformed Adaboost. Moreover, while standard boosting algorithms generally require the individual experts to have accuracy greater than 50%, critic-driven boosting eliminates this requirement. Future work may consider a more integrated critic-driven boosting approach that optimizes an overall object function over the training set.

REFERENCES

Breiman, L., 1996, "Bagging Predictors", *Machine Learning*, Vol. 24, pp. 123-140.

Freund, Y., and Schapire, R., 1997, "A Decision-Theoretic Generalization of On-Line Learning and An Application to Boosting", *Journal of Computer and System Sciences*, Vol. 55, pp. 119-139.

Miller, D. J., and Yan, L., 1999, "Critic-Driven Ensemble Classification", *IEEE Transactions on Signal Processing*, Vol. 47, pp. 2833-2844.

Schapire, R., 1999, "A Brief Introduction to Boosting", *Proceedings of the Sixteen International Joint Conference on Artificial Intelligence*.

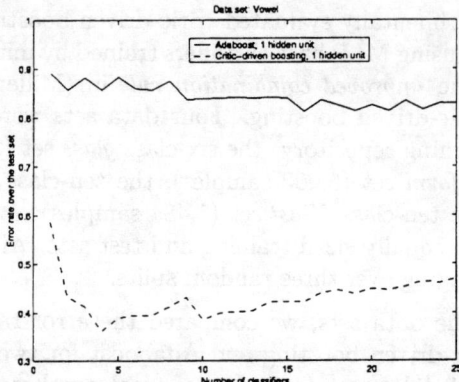

Figure 1: Error rates of critic-driven boosting and Adaboost for the *Vowel* data set, with $p > 0.5$.

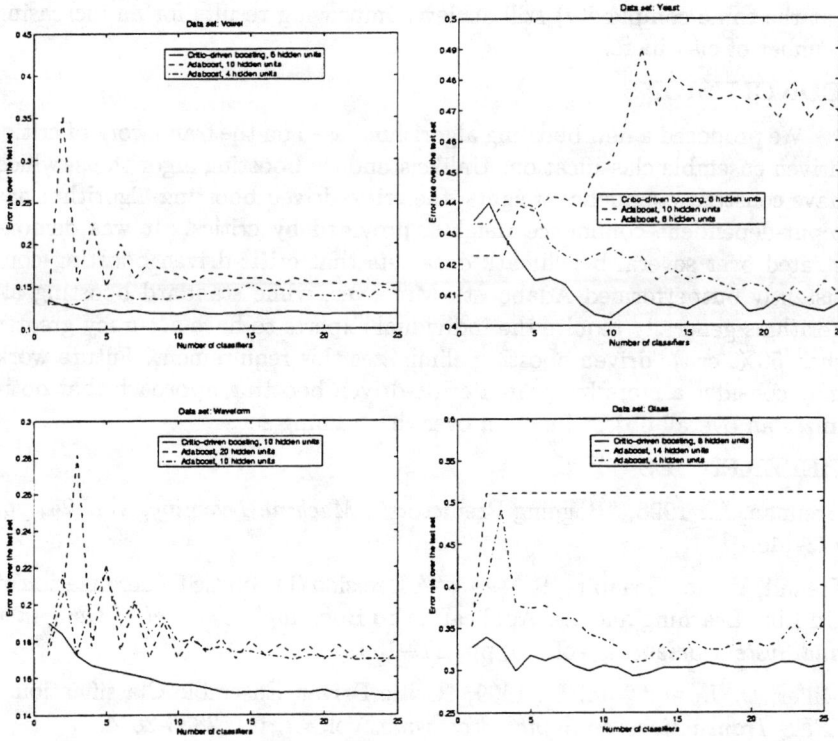

Figure 2: Error rates of critic-driven boosting and Adaboost for the four data sets.

A VIRTUALLY CONVEX ERROR CRITERION FOR TRAINING NEURAL NETWORKS[1]

JAMES T. LO
Department of Mathematics and Statistics
University of Maryland Baltimore County
Baltimore, MD 21228
e-mail: jameslo@math.umbc.edu

ABSTRACT

This paper shows that as the risk-sensitivity index of a risk-averting error criterion increases, the region on which the criterion is convex increases monotonically, creating "worm holes" for a local-search optimization procedure to travel through to a lower local minimum of the selected risk-averting error criterion. This paper also provides a minimax interpretation of the risk-averting error criterion as its risk-sensitivity index increases.

INTRODUCTION

In [3, 4], the novel risk-averting error criteria of order (λ, p) were proposed for training neural networks for function approximation and dynamical system identification. In a risk-averting error criterion, the risk-sensitivity index λ is greater than zero, and the parameter p is greater than or equal to 1. This paper shows that if the risk-sensitivity index λ is sufficiently large, the risk-averting error criterion approaches a minimax error criterion and is "virtually convex" in the sense that the region on which the criterion is convex expands to the entire weight or parameter space R^N except the union of a finite number of manifolds with a dimension less than N. We now describe two versions of the order-(λ, p) risk-averting error criterion in the following.

For approximating an m-vector-valued function $f(x)$, let us denote the output of the feedforward neural network with weight vector w that has just received and processed the vector-valued input x by $\hat{f}(x, w)$. Then the risk-averting error criterion of order (λ, p) is defined as

$$J_{\lambda,p}(w) = \sum_{k=1}^{K} \exp\left[\lambda \left\|y_k - \hat{f}(x_k, w)\right\|_p^p\right] \tag{1}$$

where $\|\cdot\|_p$ is the L_p norm in R^m, and $\{(x_k, y_k) : k = 1, \ldots K\}$ is the training data set, in which y_k denotes the output measurement corresponding to the input x_k and may or may not contain measurement noise.

For approximating a dynamical system,

$$z_k = f(z_{k-1}, \ldots, z_{k-p}, x_k, \ldots, x_{k-q}, \xi_k)$$

---
[1]This work was supported in part by NSF and ARO, but does not necessarily reflect the position or policy of the Government.

where x_k and z_k are the m-vector-valued input and output processes of the dynamical system; and ξ_k denotes a random driver process, we denote the output of the recurrent neural network with weight vector w, that has been initialized (i.e. with the initial dynamical state of the network set properly) at time 0 and has received the input sequence, $x^i := \left[x_i^T, \ldots, x_{i-r}^T\right]^T, i = 1,$ ..., k, one at a time, by $\hat{f}\left(x^k, w\right)$. Here r denotes a preselected integer. If r is greater than 0, tapped delay lines are used to hold x^i as the input vector to the recurrent neural network at time i. Then the risk-averting error criterion of order (λ, p) is defined as

$$J_{\lambda,p}(w) = \sum_{\omega \in S} \sum_{k=1}^{K} \exp\left[\lambda \left\|y_k(\omega) - \hat{f}\left(x^k(\omega), w\right)\right\|_p^p\right] \quad (2)$$

where $\left\{\left(x^k(\omega), y_k(\omega)\right) : k = 1, \ldots, K, \omega \in S\right\}$ denotes the training data, in which $y_k(\omega)$ denotes the output measurement of the network corresponding to the input $x^k(\omega)$ at time k in the sample sequence (or sampled realization) ω. Note that $y_k(\omega)$ is either equal to $z_k(\omega)$ or a noisy measurement of $z_k(\omega)$.

The risk-averting error criteria, (1) and (2), are different from the standard risk-sensitive error criteria used in robust control and signal processing [2, 5, 1] in two ways: First, (1) and (2) are not an expectation. Second, the squares are exponentiated before being summed in (1) and (2), whereas the squares are summed before being exponentiated in the standard risk-sensitive criteria.

A new adaptive risk-averting training method based on (1) and (2) is reported in IJCNN'01 to have the ability to avoid poor local minima of the risk-averting error criterion, that is adaptively selected by the method in the process of training neural networks for function approximation or system identification.

THE HESSIAN MATRIX

Without loss of generality, the risk-averting error criteria, (1) and (2), with $p = 2$ are rewritten as follows

$$J_\lambda(w) = \sum_{k=1}^{K} \exp\left[\lambda \left\|y_k - \hat{y}_k(w)\right\|^2\right] \quad (3)$$

where $\|\cdot\|$ is the L_2 norm $\|\cdot\|_2$ in R^m. Note that in rewriting (2), the training data $\left\{\left(x^k(\omega), y_k(\omega)\right) : k = 1, \ldots, K, \omega \in S\right\}$ are renumbered, the number K in (3) is equal to $K|S|$ in (2), and the dependence of $\hat{y}_k(w)$ on $x^k(\omega)$ is not explicitly indicated. No confusion is expected to be caused by the rewriting. In the following discussion, we consider a useful general risk-averting error criterion:

$$J_\lambda(w) = \sum_{k=1}^{K} \alpha_k(w)$$

where $\alpha_k(w) := \exp\left[\lambda e_k^T(w) Q e_k(w)\right]$, $e_k(w) := y_k - \hat{y}_k(w)$, and Q is a positive definite matrix.

Assume that the the neural network or the regression model $\hat{y}_k(w)$ with the weight vector w is twice continuously differentiable. Then the partial derivative of $J_\lambda(w)$ with respect to the jth entry w_j of w is

$$\frac{\partial J_\lambda(w)}{\partial w_j} = -2\lambda \sum_{k=1}^{K} \alpha_k(w) e_k^T(w) Q \frac{\partial \hat{y}_k(w)}{\partial w_j}$$

The second order partial derivative of $J_\lambda(w)$ with respect to w_j and w_i is

$$\frac{\partial^2 J_\lambda(w)}{\partial w_i \partial w_j} = 2\lambda \sum_{k=1}^{K} \alpha_k(w) \left\{ 2\lambda e_k^T(w) Q \frac{\partial \hat{y}_k(w)}{\partial w_i} \frac{\partial \hat{y}_k^T(w)}{\partial w_j} Q e_k(w) \right.$$
$$\left. + \frac{\partial \hat{y}_k^T(w)}{\partial w_i} Q \frac{\partial \hat{y}_k(w)}{\partial w_j} - e_k^T(w) Q \frac{\partial^2 \hat{y}_k(w)}{\partial w_i \partial w_j} \right\}$$

Denoting the matrices $\left[e_k^T(w) Q (\partial \hat{y}_k(w)/\partial w_i)(\partial \hat{y}_k^T(w)/\partial w_j) Q e_k(w)\right]$, $\left[(\partial \hat{y}_k(w)/\partial w_i) Q (\partial \hat{y}_k^T(w)/\partial w_j)\right]$ and $\left[e_k^T(w) Q (\partial^2 \hat{y}_k(w)/\partial w_i \partial w_j)\right]$ by $A_k(w)$, $B_k(w)$ and $C_k(w)$ respectively, the Hessian matrix of $J_\lambda(w)$ (i.e. $\left[\partial^2 J_\lambda(w)/\partial w_i \partial w_j\right]$) is

$$H_\lambda(w) = 2\lambda \sum_{k=1}^{K} \alpha_k(w) \left\{ 2\lambda A_k(w) + B_k(w) - C_k(w) \right\} \quad (4)$$

VIRTUAL CONVEXITY

We note first that

$$B_k(w) = F_k(w) Q F_k^T(w)$$
$$F_k^T(w) := \left[\partial \hat{y}_k(w)/\partial w_1 \quad \cdots \quad \partial \hat{y}_k(w)/\partial w_N \right]$$
$$A_k(w) = D_k(w) D_k^T(w)$$
$$D_k^T(w) := \left[(\partial \hat{y}_k^T(w)/\partial w_1) Q e_k(w) \quad \cdots \quad (\partial \hat{y}_k^T(w)/\partial w_N) Q e_k(w) \right]$$

Hence, $B_k(w)$ and $A_k(w)$ are positive semi-definite for all w. However, $C_k(w)$ is indefinite.

Straightforward matrix algebra yields

$$\sum_{k=1}^{K} \alpha_k B_k(w) = \mathcal{F}_K(w) \mathcal{A}_K(w) \mathcal{F}_K^T(w) \quad (5)$$
$$\mathcal{F}_K(w) := \left[F_1(w) \quad \cdots \quad F_K(w) \right] \quad (6)$$
$$\mathcal{A}_K(w) := \text{diag}\left[\alpha_1(w) Q \quad \cdots \quad \alpha_K(w) Q \right]$$

and

$$\sum_{k=1}^{K} \alpha_k A_k(w) = \mathcal{D}_K(w) \Phi_K(w) \mathcal{D}_K^T(w) \tag{7}$$

$$\mathcal{D}_K(w) := [\ D_1(w)\ \cdots\ D_K(w)\ ]$$

$$\Phi_K(w) := diag[\ \alpha_1(w)\ \cdots\ \alpha_K(w)\ ]$$

We further note that

$$\mathcal{D}_K(w) = \mathcal{F}_K(w) \mathcal{Q} diag[\ e_1(w)\ \cdots\ e_K(w)\ ] \tag{8}$$

$$\mathcal{Q} := diag[\ Q\ \cdots\ Q\ ]$$

The matrix $\mathcal{F}_K(w)$ is an $N \times Km$ matrix, where we recall that N is the number of weights of a neural network or the number of parameters of a regression model, Km is the product of the dimension m of the output vector of the neural network or the regression model and the number K of input-output pairs in the training data. The rank of $\mathcal{F}_K(w)$ depends on the structure of the neural network or regression model as well as the inputs $\{x^k(\omega) : k = 1, \ldots, K, \omega \in S\}$ in the training data.

LEMMA 1. Assume that the risk-averting error criterion $J_\lambda(w)$ in (3) is twice continuously differentiable at all $w \in R^N$. For each weight vector $w \in R^N$, if $\Gamma(w) := \{k : e_k(w) = 0\} \neq \phi$ with cardinality $|\Gamma(w)|$, we use such a numbering k' of $y_{k'} - \hat{y}_{k'}(w)$ in (3) that if $k_1' \notin \Gamma(w)$ and $k_2' \in \Gamma(w)$, then $k_1' < k_2'$. Denoting $K - |\Gamma(w)|$ by $K'(w)$, let $\mathcal{F}'_{K'(w)}(w)$ be the matrix defined by (6) for this numbering k'. If, at a weight vector $w \in R^N$, the rank of the $N \times K'(w)m$ matrix $\mathcal{F}'_{K'(w)}(w)$ is N, then there is a positive number $\Lambda(w)$ such that $H_\lambda(w) > 0$ for $\lambda \geq \Lambda(w)$.

Proof: Since at the given w, the rank of $\mathcal{F}'_{K'(w)}(w)$ is N, the rank of $\mathcal{F}_K(w)$, of which $\mathcal{F}'_{K'(w)}(w)$ is a submatrix, is obviously N. Since Q and $A_K(w)$ are positive definite for all $w \in R^N$, it is easy to see from (5) that $\sum_{k=1}^{K} \alpha_k B_k(w)$ is a positive definite matrix at the given $w \in R^N$.

For the given $w \in R^N$, rewrite (4) as

$$H_\lambda(w) = 2\lambda \sum_{k'=1}^{K'(w)} \alpha_{k'}(w) \{2\lambda A_{k'}(w) + B_{k'}(w) - C_{k'}(w)\}$$

$$+ 2\lambda \sum_{k'=K'(w)+1}^{K} \alpha_{k'}(w) B_{k'}(w)$$

where the numbering k' is that for the w under consideration. Under the same assumption $\mathcal{F}'_{K'(w)}(w)$ has full rank N, the $N \times K'(w)m$ matrix $\mathcal{F}'_{K'(w)}(w) \mathcal{Q}$ is of full rank N. It follows from (8) that $\mathcal{D}_{K'(w)}(w)$ for the numbering k' is also of full rank N, since the deviation vectors,

$e_{k'}(w)$, $k' = 1, \ldots, K'(w)$, are nonzero vectors for the same w. Hence, $\sum_{k'=1}^{K'} \alpha_{k'} A_{k'}(w) = \sum_{k'=1}^{K'} \alpha_{k'} D_{k'}(w) D_{k'}^T(w)$ is positive definite for the arbitrary w under consideration.

Because of the positive definiteness of $\sum_{k'=1}^{K'} \alpha_{k'} A_{k'}(w)$, there is a number $\Lambda(w) < \infty$ such that $\sum_{k'=1}^{K'(w)} \alpha_{k'}(w) \{2\lambda A_{k'}(w) + B_{k'}(w) - C_{k'}(w)\} > 0$ for $\lambda > \Lambda(w)$ at the same w under consideration. Hence, $H_\lambda(w) > 0$ for $\lambda \geq \Lambda(w)$. This completes the proof of the lemma.

Letting Γ be a subset of $\{1, \ldots, K\}$ with cardinality $|\Gamma|$, we use such a numbering k' of $y_{k'} - \hat{y}_{k'}(w)$ in (3) that if $k_1' \notin \Gamma$ and $k_2' \in \Gamma$, then $k_1' < k_2'$. Denoting $K - |\Gamma|$ by $K'(\Gamma)$, let $\mathcal{F}'_{K'}$ be the $N \times K'(\Gamma) m$ matrix defined by (6) for this numbering k'. Consider the set $M(\Gamma)$ of all such weight vectors $w \in R^N$ that $e_{k'}(w) = 0$ for $k' = K'(\Gamma) + 1, \ldots, K$; and that the rank of $\mathcal{F}'_{K'(\Gamma)}$ is less than N. That $\text{rank} \mathcal{F}'_{K'(\Gamma)} < N$ means that there is a vector $c \in R^N$ such that $\mathcal{F}'^T_{K'(\Gamma)} c = 0$. Hence $M(\Gamma)$ is the solution set of the simultaneous $K - K'(\Gamma) + K'(\Gamma) m$ equations $e_{k'}(w) = 0$ for $k' = K'(\Gamma) + 1, \ldots, K$; and $\sum_{i=1}^{N} c_i \partial \hat{y}_j(w)/\partial w_i = 0$ for $j = 1, \ldots, K'(\Gamma) m$. Each equation defines an $N - 1$ dimensional manifold in the weight vector space R^N. The solution set $M(\Gamma)$ is the intersection of these $K - K'(\Gamma) + K'(\Gamma) m$ manifolds in R^N. Note that $|\Gamma| < K$ and $K - K'(\Gamma) + K'(\Gamma) m = |\Gamma| + (K - |\Gamma|) m \geq K$. Hence, if $K \geq N$, then "under some normality conditions," $M(\Gamma)$ is expected to be either empty or a singleton. The greater K and m are, the more likely this is the case.

Note that there are 2^K different subsets Γ of $\{1, \ldots, K\}$, each corresponding to a solution set $M(\Gamma)$ as described above. Let the union of these 2^K manifolds be denoted by M. This union M is expected to be a finite subset of R^N "under some normality conditions." Exactly what these normality conditions are is still under study and will soon be reported. We are now ready to state our main theorem in the following:

THEOREM 2. Assume that the risk-averting error criterion $J_\lambda(w)$ in (3) is twice continuously differentiable. The sequence of sets $P_\lambda := \{w \in R^N : H_\lambda(w) > 0\}$ is monotone increasing as λ increases. Moreover, $M^c \subset \cup_{\lambda>0} P_\lambda$, where M is the union of solution sets of the 2^K systems of simultaneous K algebraic equations discribed above and M^c denotes the complement of M relative to R^N.

PROOF. Pick an arbitrary w from M^c. By the definition of M given above, the rank of the $N \times K'(w) m$ matrix $\mathcal{F}'_{K'(w)}(w)$ is N. By Lemma 1, there is a positive number $\Lambda(w)$ such that $H_\lambda(w) > 0$ for $\lambda \geq \Lambda(w)$. Hence, $w \in P_\lambda \subset \cup_{\lambda>0} P_\lambda$ for $\lambda \geq \Lambda(w)$, completing the proof.

MINIMAX INTERPRETATION

Let $J_{\lambda,p}(w) := \sum_{k=1}^{K} \exp\left[\lambda \|y_k - \hat{y}_k(w)\|_p^p\right]$ be defined in the same way as $J_\lambda(w)$ in (3), and a minimax criterion be $\eta_p(w) := \max_k \{\|e_k(w)\|_p^p : k = 1, \ldots, K\}$.

THEOREM 3. Let $\{u_\lambda \in R^N, \lambda > 0\}$ be a sequence of weight vectors such that

$$\lim_{\lambda \to \infty} \max_k \|e_k(u_\lambda)\|_p = \inf_w \max_k \|e_k(w)\|_p \qquad (9)$$

If a sequence $\{w_\lambda \in R^N, \lambda > 0\}$ satisfies $J_{\lambda,p}(w_\lambda) \leq J_{\lambda,p}(u_\lambda)$ for all $\lambda \geq \Lambda$ for some $\Lambda > 0$, then

$$\lim_{\lambda \to \infty} \max_k \|e_k(w_\lambda)\|_p = \inf_w \max_k \|e_k(w)\|_p \qquad (10)$$

where $\max_k$ means the maximum over $k \in \{1, \ldots K\}$ and $\inf_w$ means the infimum over $w \in R^N$.

PROOF. We note first that the existence of $\{u_\lambda \in R^N, \lambda > 0\}$ is implied by the definition of $\inf_w \max_k \|e_k(w)\|_p$. Define the notations, $\Phi(w) := \arg\max_k \|e_k(w)\|_p^p$, $b := \inf_w \max_k \|e_k(w)\|_p$ and $b_\lambda := \inf_w J_{\lambda,p}(w)$. Note that $\Phi(w)$ may be a set.

Rewirte the two sides of $J_{\lambda,p}(w_\lambda) \leq J_{\lambda,p}(u_\lambda)$ as follows:

$$\exp\left[\lambda \left\|e_{\phi(w_\lambda)}(w_\lambda)\right\|_p^p\right] + \sum_{k \neq \phi(w_\lambda)} \exp\left[\lambda \|e_k(w_\lambda)\|_p^p\right]$$
$$\leq \exp\left[\lambda \left\|e_{\phi(u_\lambda)}(u_\lambda)\right\|_p^p\right] + \sum_{k \neq \phi(u_\lambda)} \exp\left[\lambda \|e_k(u_\lambda)\|_p^p\right] \qquad (11)$$

where $\phi(u_\lambda) \in \Phi(u_\lambda)$, $\phi(w_\lambda) \in \Phi(w_\lambda)$, the summations $\sum$ are taken over $k = 1, \ldots, K$ with the exception indicated below the summation signs. Because the first term on the right side is a dominant term, the following strict inequality follows from (11),

$$\exp\left[\lambda \left\|e_{\phi(w_\lambda)}(w_\lambda)\right\|_p^p\right] < (K+1) \exp\left[\lambda \left\|e_{\phi(u_\lambda)}(u_\lambda)\right\|_p^p\right] \qquad (12)$$

By (9), the sequence $d_\lambda := \max_k \|e_k(u_\lambda)\|_p^p - b^p$ converges to 0 as λ approaches ∞. Rewrite (12) as

$$\exp\left[\lambda \left\|e_{\phi(w_\lambda)}(w_\lambda)\right\|_p^p\right] < (K+1) \exp[\lambda(b^p + d_\lambda)] \qquad (13)$$

Note that this inequality holds for all $\lambda \geq \Lambda$.

Assume that (10) fails to hold for this sequence $\{w_\lambda, \lambda > 0\}$. Then $\lim_{\lambda \to \infty} \max_k \|e_k(w_\lambda)\|_p > b$ or $\lim_{\lambda \to \infty} \max_k \|e_k(w_\lambda)\|_p$ does not exist. In either case, there is $c > 0$ such that for every $L > 0$, there exists $\lambda > L$ such that $\left\|e_{\phi(w_\lambda)}(w_\lambda)\right\|_p^p \geq b^p + c$. Since $\lim_{\lambda \to \infty} d_\lambda = 0$, there is $L_1 > 0$ such that $d_\lambda < c/2$ for all $\lambda \geq L_1$. It follows from the above inequality (13) that for all $\lambda \geq \max\{\Lambda, L_1\}$,

$$\exp\left[\lambda \left\|e_{\phi(w_\lambda)}(w_\lambda)\right\|_p^p\right] < (K+1) \exp[\lambda(b^p + c/2)] \qquad (14)$$

Then for $L_2 = \max\{\Lambda, L_1, 2\ln(K+2)/c\}$, there exists a $\lambda > L_2$ such that $\|e_{\phi(w_\lambda)}(w_\lambda)\|_p^p \geq b^p + c$, which implies that for this λ, $\exp\left[\lambda \|e_{\phi(w_\lambda)}(w_\lambda)\|_p^p\right]$ $\geq \exp[\lambda b^p + \lambda c/2 + \ln(K+2)] > (K+2)\exp[\lambda(b^p + c/2)]$, which contradicts (13). Therefore, the above assumption is false, and $\{w_\lambda, \lambda > 0\}$ satisfies (10), completing the proof of the theorem.

The following corollary is an immediate consequence of Theorem 3.

COROLLARY 4. If $w_\lambda \in \arg\min_w J_{\lambda,p}(w)$ for all $\lambda > \Lambda$ for some $\Lambda > 0$, then $\lim_{\lambda \to \infty} \max_k \|e_k(w_\lambda)\|_p = \inf_w \max_k \|e_k(w)\|_p$.

Note that Theorem 3 and its proof with the pth power $\|\cdot\|_p^p$ of the L_p norm $\|\cdot\|_p$ replaced with any monotone increasing function of any norm are still valid.

CONCLUSION

Theorem 1 shows that increasing the risk-sensitivity index during training a neural network or estimating the parameters of a regression model reduces the chance for a local-search optimization algorithm to get trapped into a poor local minmum of the error criterion. Theorem 2 gives an interpretation of the risk-averting error criterion with a large risk-sensitivity index, and shows that increasing the risk-sensitivity index during training a neural network or estimating the parameters of a regression model reduces the maximum absolute deviations of the outputs of the neural network or the regression model from the corresponding outputs. Both of these observations have been confirmed in numerical experiments reported in IJCNN'01.

References

[1] T. Basar and P. Bernhard. *H-Infinity-Optimal Control and Related Minimax, Design Problems: A Dynamic Game Approach, 2nd edition.* Birkhauser, Boston, Massachussetts, 1995.

[2] D. H. Jacobson. Optimal stochastic linear systems with exponential performance criteria and their relation to deterministic games. *IEEE Transactions on Automatic Control*, AC-18-2:124–131, 1973.

[3] J. T. Lo. Risk-sensitive approximation of functions by neural networks. In *Proceedings of the 34th Allerton Conference on Communication, Control, and Computing*, Monticello, Illinois, 1996.

[4] J. T. Lo. Risk-sensitive identification of dynamic systems by neural networks. In *Proceedings of the 34th Allerton Conference on Communication, Control, and Computing*, Monticello, Illinois, 1996.

[5] P. Whittle. *Risk Sensitive Optimal Control.* Wiley, New York, New York, 1990.

LEARNING STRUCTURES WITH FLEXIBLE BASIS FUNCTIONS

CHUN-SHIN LIN AND KHALID AL-HINDI
Dept. of Electrical Engineering
University of Missouri-Columbia

ABSTRACT

Radial Basis Function Network is one well-known structure used for implementing static mapping in learning systems. It uses a set of radial basis functions, each dominating a local input area. One drawback is that the number of basis function components could increase abruptly when the input dimension increases. This problem could be relaxed if the used basis function is not restricted to only one specific form such as the Gaussian function. The flexibility can be achieved using the Sum-of-Products Neural Network (SOPNN) structure, which we developed earlier. This paper presents two structures modified from the SOPNN for possible efficient hardware realization; both intend to eliminate the high cost multiplication. The first structure, named the Minimum-Sum Network, uses the MINIMUM function to approximate a multiplication. The second structure, named the Sum-Exponential-Sum Network, uses the sum of logarithmic values for computing a product (*i.e.*, a logarithmic multiplier). A logarithmic multiplier is an adder, which can efficiently handle the multiplication of many input values at one time.

INTRODUCTION

Multilayer Feedforward Neural Networks (MNN) [1, 2] and Radial Basis Function Networks (RBFN) [3, 4] are among the most well-known types of neural networks used for static mapping with applications in classification, control, prediction, signal processing, etc. One weakness of the multilayer feedforward neural network is the possible difficulty in training. On the other hand, a radial basis function network usually has its basis functions dominate local areas. Thus its training involves more local change and is, generally speaking, simpler. However, one drawback is that the number of basis function components is likely to increase abruptly (possibly in an exponential rate) when the input dimension increases. This problem could be significantly relaxed if the basis function components are given more flexibility and not restricted to one specific form such as the Gaussian function. The desired flexibility can be achieved with the use of *memory* for description of basis functions. The Sum-of-Product Neural Network (SOPNN) [5, 6] that we developed previously is capable of implementing one type of flexible basis function neural networks, of which the Gaussian Function Network [3] (one type of RBFN) is a special case. The SOPNN structure consists of several *submodules* for computing the products, which are then added up to generate the network output. Each submodule generates a flexible basis function.

SOPNN involves heavy multiplication computation. Hardware realization of a multiplier is costly and unfavourable. This paper presents two modified structures that are free of traditional multipliers. The first structure, named the

Minimum-Sum Network (MSN) is an approximation of the SOPNN. The MSN uses the MINIMUM function to approximate a multiplication. One advantage of the MSN is that the MINIMUM function can be easily implemented using simple diodes or a "MINIMUM circuit"; its circuit is much simpler than multipliers. The second structure, named the Sum-Exponential-Sum Network (SESN), is a different form of the SOPNN. The SESN uses the sum of logarithmic values for computing a product (*i.e.*, a logarithmic multiplier). The result is then converted to the actual product using an exponential function. Thus in the SESN structure, multipliers used in an SOPNN submodule are replaced by a single logarithmic multiplier, which is actually an adder, plus an exponential amplifier [7]. The SESN and SOPNN are indeed equivalent. Note that both the circuit and the analog logarithmic multiplier can handle many input values at once while a traditional multiplier can only generate a product of two values. This means that with the same amount of resources, a larger size of MSN or SESN can be realized.

In the next section, we will first introduce SOPNN. After that we will present the first structure, MSN, including the ideas in hardware realization and the performance evaluation. The introduction of the second structure, SESN, will then follow.

SOPNN

The SOPNN utilizes small-size memory arrays for description of flexible basis functions. Figure 1 illustrates the structure. Each memory array takes a problem input (x_j) as its input, which is converted to an address for data retrieving. Several memory arrays, one for each input, and a multiplier form a *submodule*. A group of submodules and an adder form an SOPNN. The output of a submodule is the product of the outputs from the memory arrays in the submodule. The output of the SOPNN, which is the sum of all submodule outputs, has the sum-of-product form $\sum_{i=1}^{N_p} \prod_{j=1}^{N_v} f_{ij}(x_j)$, where x_j's are inputs, N_v is the number of inputs, $f_{ij}()$ is a function generated through network training, and N_p is the number of submodules. Note that $f_{ij}()$ represents a stored lookup table. If all $f_{ij}()$'s are Gaussian functions, the neural network degenerates to a Gaussian Function Network [3].

The arrangement creates some attractive advantages. One is that each submodule implements a self-generated basis function, which is developed during the learning. The flexible self-generated basis functions can better fit the desired modeling and can significantly increase the modeling efficiency.

The gradient-descent learning rule can be derived for updating parameters in the memory arrays. To achieve efficient learning, for each training pattern, several memory cells that affect the neighborhood of the training sample are updated. Detailed arrangement for efficient learning can be found in reference 6.

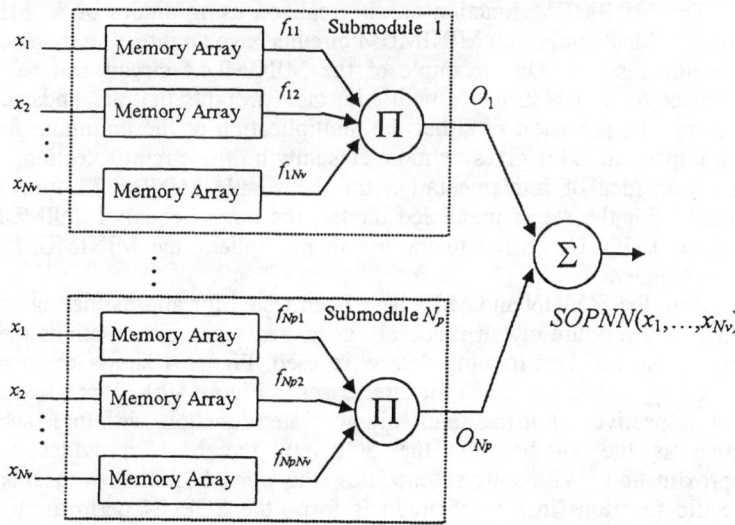

Figure 1. The sum-of-product neural network structure

THE MINIMUM-SUM NETWORK

The SOPNN could involve heavy multiplication computation. Hardware realization of a multiplier is considered costly and undesirable. In addition, multiplication can be done for only two numbers at a time. For the product of N_v numbers, $N_v - 1$ times of multiplication are needed. Thus hardware realization of a large SOPNN is less favorable.

In this section, a new structure, named the Minimum-Sum Network (MSN), is presented. The MSN has the multiplication in SOPNN replaced by a MINIMUM function and is an approximator of the SOPNN. If all the numbers to be multiplied together are within (0, 1), the MINIMUM will provide a reasonable approximation to the product. Theoretically, the limitation of the range to (0,1) can be achieved if a bias as a shifter and a weight as a scalar are included in the output summing unit to scale the result to the desired target range. Figure 2 shows the diagram of a submodule for the MSN.

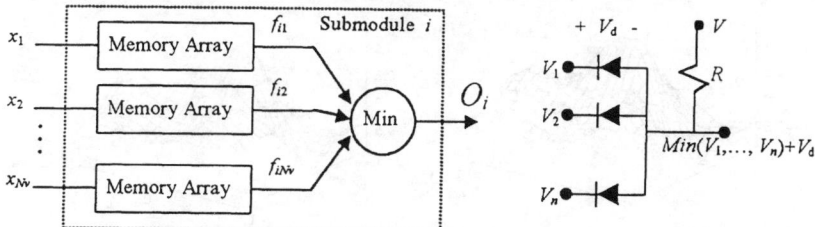

Figure 2. A submodule for the Minimum-Sum Network

Figure 3. Possible generation of a minimum using diodes

The MINIMUM function can be realized using diodes or a "MINIMUM circuit." Most studies on MINIMUM circuits have an intention to generate high precision results. One example of the MINIMUM circuit can be found in reference 8. For a structure with adjustable weights that are updated through learning, the precision of either the multiplication or the minimum function is less important. This gives us more elasticity on the circuit selection. Figure 3 shows an idea of implementation for the simple MINIMUM function using diodes. For the set of paralleled diodes, the output equals MINIMUM(v_1, v_2, ...v_n) + 0.7V. Compared to the use of multipliers, the MINIMUM circuit is much simpler.

Both the SOPNN and MSN have been tested for approximating a function, which is a mixture of ten randomly generated Gaussian functions (see Figure 4(a)). Four hundred training data were used. Figure 4 shows the results. The sum of squared errors (SSE) for three cases in Figure 4(b)-(d) are 1.33, 1.88 and 1.14, respectively. For this randomly generated function, with three submodules (same as the number for the SOPNN), the MSN provides acceptable approximation. With four submodules, the error becomes even smaller. For specific functions in sum-of-products form, the SOPNN performs better than MSN. A few additional submodules will be needed for the MSN to reduce the error level. Since the circuit is much simpler than the traditional multiplier and only one MINIMUM circuit is required for one submodule, more submodules can be built onto the same size of chip. The detail tradeoff can be evaluated only when the hardware design is carried out.

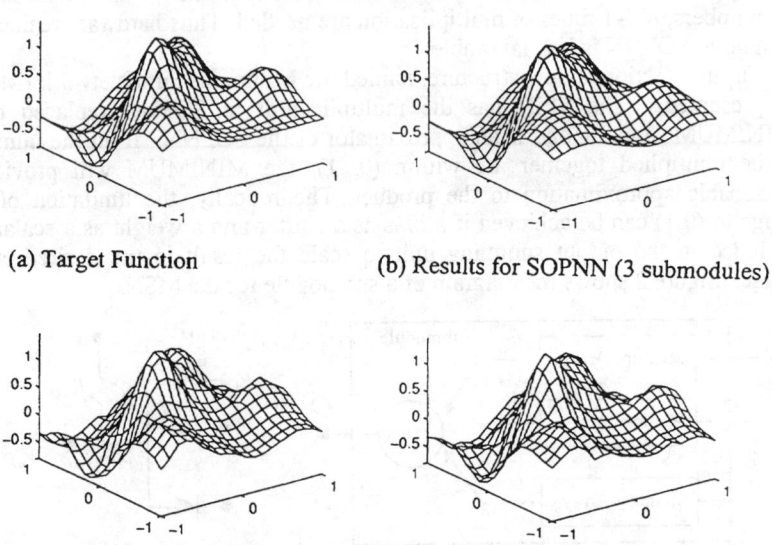

(a) Target Function (b) Results for SOPNN (3 submodules)

(c) Results for MSN (3 submodules) (d) Results for MSN (4 submodules)

Figure 4. Simulation results for SOPNN and MSN

SUM-EXPONENTIAL-SUM NETWORK

As stated in the previous section, multiplication represents the major cost in the SOPNN structure. To reduce the cost, one may use the logarithmic multiplier, which uses the logarithmic values and an adder to implement the multiplication. For instance, $ln\ (a*b*c)$ can be computed as $ln\ a + ln\ b + ln\ c$, and thus $a*b*c$ can be computed as $exp(ln\ a + ln\ b + ln\ c)$. Instead of having the values a, b, c stored in the memory arrays, their logarithmic values will be stored. Figure 5 shows the submodule for the SESN with the use of logarithmic multipliers.

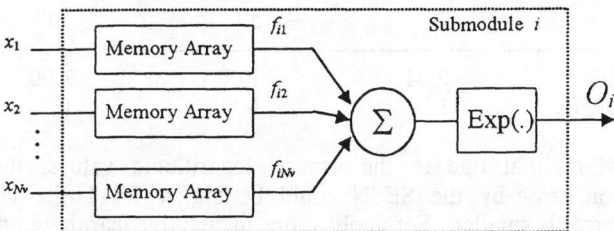

Figure 5. A submodule for the Sum-Exponential-Sum Network

With the change of multiplication to summation, the analog hardware realization becomes much easier. One single logarithmic multiplier can generate the product of N_v values. Figure 6 shows an example of the analog logarithmic multiplier (2-input). An analog adder together with an exponential amplifier performs the multiplication. This can significantly reduce the circuit complexity.

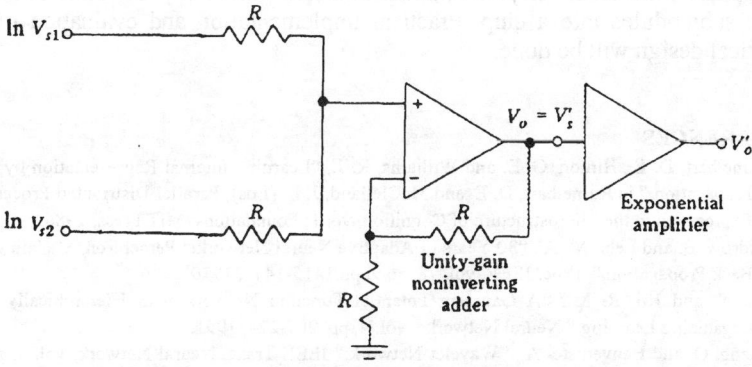

Figure 6. Diagram for a logarithmic multiplier (changed from [7])

Theoretically speaking, the SESN is equivalent to the SOPNN; they differ only on the way of implementation for multiplication. One practical concern is the possible error caused by data imprecision and noise. To illustrate the effect on precision, we can assume that the data to be stored in an SOPNN has a range between a and b. Thus the corresponding values to be stored using an SESN will be between $ln\ a$ and $ln\ b$. Let the range of the analog voltage for data representation be $(0, V)$; voltages 0 and V will be corresponding to a and b for the SOPNN, and $\ln a$ and $\ln b$ for the SESN. One can then calculate the error caused by an step-size error of V/N in the SESN when the input is $\ln y$ by

$$error = e^{\ln y} - e^{(\ln y - \frac{\ln b - \ln a}{N})}$$

On the other hand for SOPNN, the error is simply $(b-a)/N$.
Evaluating the errors using $a = 1$, $b = 100$ and $N = 99$ with $y = 2$, 50 and 100 gives

y	2	50	100
SESN	0.04	1.00	2.00
SOPNN	1	1	1

The data show that due to the use of logarithmic values, the absolute representation error by the SESN could be bigger for larger y while the percentage error is smaller. For applications in adaptive/learning systems, there should not be much difference between these two structures based on their precision.

CONCLUSIONS

This paper presents two structures modified from the Sum-of-Products Neural Network. One uses the MINIMUM circuit to approximate the multiplication and the other uses the logarithmic multiplier. Eliminating multipliers reduces the implementation size and makes it possible to fabricate more submodules into a chip. Practical implementation and evaluation of the practical design will be done.

REFERENCES

1. Rumelhart, D. E., Hinton, G. E. and Williams, R. J., "Learning Internal Representation by Error Propagation," in Rumelhart, D. E. and McClelland, J. L. (Eds), Parallel Distributed Processing: Exploration in the Microstructure of Cognition, vol.1: Foundations. MIT Press, 1986.
2. Widrow, B. and Lehr, M. A., "30 Years of Adaptive Neural Networks: Perceptron, Madaline, and Back Propagation," Proc. IEEE, vol. 78, no.9, pp.1415-1441, 1990.
3. Lee, S. and Kil, R. M., "A Gaussian Potential Function Network with Hierarchically Self-Organizing Learning," Neural Networks, vol. 4, pp. 207-224, 1991.
4. Zhang, Q. and Benveniste, A., "Wavelet Network," IEEE Trans. Neural Network, vol. 3, no. 6, pp. 889-898, November 1992.
5. Lin, C.S. and C. K. Li, "A Memory-based Self-Generated Basis Function Neural Network," *International Journal of Neural Systems*, Vol. 9, No. 1, January 1999, 41-59.
6. Lin, C. S. and Li, C. K., "A Sum-of-Product Form Neural Network (SOPNN)," *Neurocomputing* 30 (2000), 273-291.
7. Millman, J, Microelectronics: Digital and Analog Circuits and Systems, McGraw-Hill, 1979.
8. Huang, C. Y., Wang C. J. and Liu, B. D., "Modular Current-mode Multiple Input Minimum Circuit for Fuzzy Logic Controllers," Electronics Letters, Vol.32, No.12, pp. 1067-1069, June 1996.

PARALLEL NEURAL NETWORKS WITH ACCURATE CORRECTNESS DECISION OF OUTPUTS AND A LARGELY IMPROVED GENERALIZATION

YOHTARO YATSUZUKA
KDDI R & D Labs.
Kamifukuoka Saitama 356, Japan
yohtaro@kddilabs.co.jp

TETSUYA SUGIYAMA
Ecology Simulation Labs.
Kawagoe Saitama 350, Japan

ABSTRACT

A combination of well or weakly trained neural networks is a very attractive approach for improving the generalization performance of a single network. We propose a new parallel neural network not only for improving the generalization performance, but also for providing a correctness decision capability of outputs for unseen inputs with very high accuracy. Each of the 3-layered neural networks are trained with an error-perturbed back-propagation learning algorithm for an anchor training set having mutually distinct target allocations for labels. The selection and correctness decision of an ultimate output of the parallel network are conducted by utilizing both a coincident state among neural network outputs and the estimated correctness of the outputs. The proposed parallel neural networks achieved a very high generalization performance and also a very accurate correctness decision of the ultimate output for unseen input samples.

INTRODUCTION

Correctness decision ability with respect to the output for unseen input samples in supervised learning systems is very attractive for monitoring in applications such as the diagnosis of alarm correlation in telecommunication networks (Jakobson and Weissman, 1993) to be in conformity with operational circumstances. Therefore not only complete convergence in binary space within a small number of training cycles and a high generalization ability, but also adaptability by rapid re-learning, are required for reliable systems.

In multi-layer neural networks, the achievement of a quick and stable convergence and high generalization without overfitting or overtraining are major issues for a wide range of real applications (Baum and Haussler., 1989, Reed, 1993, Plutowski and White, 1993). A 3-layered neural network with an error-perturbed back propagation learning was proposed as a global-minimization approach using a deterministic method for the avoidance of fatal and tenacious local minima, resulting in a quicker and more stable convergence in binary space. The neural network also provided an insensitivity to sets of initial weights and higher generalization performance than conventional networks with a back-propagation algorithm (Yatsuzuka, 1995).

Two main approaches were proposed to improve generalization performance over the single neural network by combining or ensembling several neural networks in parallel. One is to use well-trained neural networks (Sharkey and Sharkey, 1996, 1997), and the other is to use weakly trained ones (Hansen and Salamon, 1990, Sarkar, 1996, Ma and Ji, 1999). Providing neural networks with diverse error patterns is crucial to effectively improve performance through combinations of networks (Sharkey and Sharkey, 1996, Ma and Ji, 1997, 1999).

We propose a parallel network that has a combination of diverse neural networks well-trained for anchor training sets having mutually distinct target allocations for labels. The proposed parallel network also has a selection and correctness decision ability of the ultimate output by using the discretized node output patterns of a hidden layer in the neural networks.

A BASIC SCENARIO OF A COMBINATION OF NEURAL NETWORKS IN PARALLEL

To tackle the problem of generalization performance of a single neural network, both approaches of the well-trained and weakly trained neural networks achieve similar generalization (Ma and Ji, 1999, Wolpert and Macready, 1997). We propose a parallel network combining well-trained neural networks from the following viewpoints:

(1) clarity of the generalization performance of neural networks in advance in the design of the combination,
(2) realization of a neural network providing complete correct outputs for a training set with very quick convergence avoiding tenacious local minima,
(3) capability of the accurate inference of the output correctness of a neural network for unseen input samples,
(4) repetitive learning with an anchor and incremental training sets accounting for very small parts of a huge number of test input samples, providing whole correct outputs for the test set and reasonably high generalization performance,
(5) ease of pluralization of repetitively learned parallel neural networks.

For the learning and test procedures, the input samples are systematically and deliberately grouped into a set of anchor input samples which basically presents a main feature of an input space as a domain knowledge, and several hierarchical test sets some distance away from the anchor training input samples. The test sets largely correlative to the anchor training set were prepared to reliably form the classification boundaries.

In the repetitive learning procedure (Yatsuzuka and Enomoto,1996), a high generalization performance was successfully achieved by using input samples, which were automatically selected from the test sets close to the anchor training set by its own generalization ability of the neural network. The only about 3% volume of the test sets results in a drastic reduction of computations. Although it may not be surprising that the choice of good examples allows one to learn with fewer examples. However, the generalization performance slightly degraded in comparison to direct learning with the whole test input samples. Therefore, the combination of neural networks in parallel is very attractive for the repetitive learning to largely improve performance with a reasonable computation amount.

CONFIGURATION OF PROPOSED PARALLEL NETWORKS

A basic configuration of the proposed parallel network is given in Fig. 1. The case of two neural networks is only shown to illustrate for explanations. The neural networks individually learned anchor training sets as feature vectors having a mutually distinct target allocation for a label, which is assigned to the same anchor training input for category classification. An Output Code Converter converts a discretized output into the second neural network output having the same target allocation as in the first neural network. In a Coincidence Detector a coincident state between the outputs of the first and second neural networks is detected. In a Correctness Estimator, each neural network output is estimated to be correct or incorrect, depending on a Hamming distance obtained using a Hamming Distance Calculator and a threshold control signal. The control signal is obtained from a Test Range Detector that discriminates whether the output of an output layer for an unseen input sample meets a tested output range or not. In a Correctness Decision and Selection of Ultimate Output unit, an ultimate output is selected from the neural network outputs and is decided as either correct or incorrect, according to the estimated correctness and the coincident.

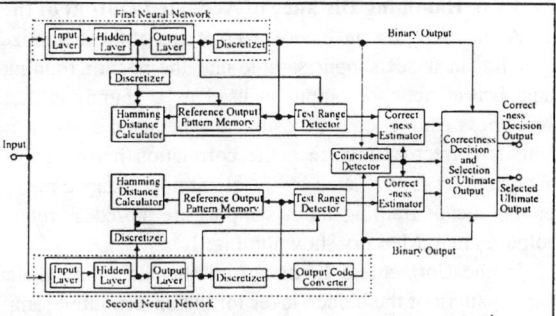

Figure 1 Block diagram of a proposed parallel network with a combination of two neural networks.

(1) Learning of Neural Networks for an Anchor Training Set

In an anchor training set, at lease one anchor input sample should be assigned to each target pattern with a distributed representation in order to prevent the degradation of the generalization performance due to overtraining or overfitting induced by the redundant weights in a multi-layered neural network (Yatsuzuka, 1995).

The clusters of anchor input samples allocated to each target can also basically form the classification boundaries around balanced center positions in the feature space in updating weights having tightly mutual interactions by the distributed representation of the outputs and the error-perturbed back propagation algorithm.

(2) Mutually Distinct Target Allocations for Labels

Combining neural networks with diverse classification domains is crucial to effectively improving the generalization performance. By setting randomized initial weight vectors in the learning procedure, the neural networks can have slightly diverse generalization domains for the same training input samples, as described in ensembled neural networks (Hansen and Salamon, 1990, Sarkar, 1996, Ma and Ji, 1999). However, it is very difficult to stably control the amount of the diversity. The diversity also becomes very small especially under a converged condition having global minima providing all correct outputs. There will also be very little gain from any combination due to the lack of diversity among the well-trained neural networks having almost identical classifications.

Therefore, well-trained neural networks are conventionally designed by using distinct training sets having different input samples or training sets with input samples converted from original input samples (Sharkey, 1996). Both approaches however have difficulty in systematically obtaining a certain diversity without trying out the generalization performance evaluation after training and combining.

On the other hand, in the proposed parallel network the mutually distinct target allocations for labels are applied to the neural networks. The trained neural networks in parallel provide similar generalization performance to the same input samples, in accompany with complete convergence. Further, the classification domains formed by the neural networks are clearly and stably diversified from each other in the input space.

(3) Correctness Estimation of Neural Network Output
i) Hamming Distance of Activated Output of the Hidden Layer

A Hamming distance between patterns of the discretized node outputs of the hidden layer for an unseen input sample and the anchor training input sample providing the same neural network output is used as a Hamming distance metrics for the output correctness estimation. To estimate whether the neural network output of the unseen input is correct or incorrect, the correlation between the unseen and anchor training input samples is evaluated by the Hamming distance metrics. The node output patterns for the anchor training input samples are stored as reference patterns in a Reference Output Pattern Memory shown in Fig. 1.

In the Correctness Estimator, if the Hamming distance of the discretized node output pattern of the hidden layer for the unseen input sample exceeds a given threshold, the neural network output is estimated to be correct. In other case, the neural network output is estimated to be incorrect. The threshold consists of the average and standard deviation values of the Hamming distances and a constant value. The values of the average and standard deviation of the Hamming distances are only calculated in advance for node output patterns, which send the same correct neural network output for a test set. The test set close to the anchor training set was prepared, consisting of test input samples having a Hamming distance of 1 from the anchor input samples. The constant value, Thcont, is optimally selected depending on the coincident state among the neural network outputs in parallel and a state of the margins of the node outputs of the output layer, as given by Eq. 1. The $O(p, i)$ is the p-th node output of the output layer for the unseen input sample in the i-th neural network.

A larger constant value, Thcont1, is set if all of the margins of the node outputs of the output layer for the unseen input sample take values inside a tested range of margin under the condition of coincidence among neural network outputs, observing that the neural networks might have a higher possibility of providing correct neural network outputs. The tested range of margin, TRM [MinOM, MaxOM], of the correct neural network output is calculated for a range from a minimum value, MinOM, to a maximum value, MaxOM, in margins, which are the differences between the binary threshold of 0.5 for discretizing and the node outputs of the output layer sending the same correct output for the

For $p \in [1, P]$,

$\quad$ If $O(p, i) \in$ TRM [MinOM, MaxOM], $\quad$ (1)
$\quad\quad$ then Thcont = Thcont1,
$\quad\quad$ else Thcont = Thcont2,

where P is the number of the output node, and

$$MinOM = \underset{p \in [1,P]}{Min}\{ABS|\ 0.5 - O(p,i)\ |\}.$$

$$MaxOM = \underset{p \in [1,P]}{Max}\{ABS|\ 0.5 - O(p,i)\ |\}. \quad (2)$$

Thcont2 $\leq$ Thcont1.

test set, as given by Eq. 2. Thcont is selected from either Thcont1 or Thcont2, whether or not all of the node outputs of the output layer meet the tested range of margin, as given by Eq 1.

ii) Hamming distance metrics

The Hamming distance metric for the unseen input sample, HDhl(i), in the i-th neural network is given by the Hamming distance calculated by Eqs. 3 and 4, where $i \in [1,2,3]$, M is the number of the hidden nodes of the i-th neural networks, the BOhlinput(m,i) is the binary m-th node output pattern of the hidden layer for the unseen input sample, the BOhlanchor(m,i) is for the anchor input sample providing the same discretized output of the i-th neural network as that for the unseen input sample, and HDmin is a minimum allowable value of the Hamming distance. The distance threshold is also denoted by THdis(i) which consists of the average value of Thave, the standard deviated value of Thstd and the constant value of Thcont. The CE(i) is the correctness estimation output of the i-th neural network in parallel, and is given by Eq. 5.

$$HDhl(i) = \Sigma \; BOhlinput(m,i) \; .ExOR. \; BOhlanchor(m,i), \quad (3)$$
$$m \in [1, M]$$
$$\text{and} \quad \text{if } HDhl(i) \; .LT. \; HDmin, \quad (4)$$
$$\text{then } HDhl(i) = HDmin.$$

$$\text{If } HDhl(i) \leq THdis(i), \quad (5)$$
$$\text{then CE(i) is set to be correct,}$$
$$\text{else CE(i) is set to be incorrect,}$$

TABLE 1 Correctness Decision and the Selection Algorithm of the Ultimate Output of the 3-parallel Neural Network.

Coincident States of Neural Net Outputs	Correct-ness Estimation of Neural Net Outputs	Ultimate Net Output		Coincident States of Neural Net Outputs	Correct-ness Estimation of Neural Net Outputs	Ultimate Net Output	
	O1 O2 O3	Decision	Selection		O1 O2 O3	Decision	Selection
Whole Coincidence among Neural Net Outputs (O1=O2=O3)	C C C C C I C I C C I I I C C I C I I I C I I I	C C C C C C C I	O1 O1 O1 O1 O1 O1 O1 O1	Coincidence between Two Neural Net Outputs (O1=O2)	C C C C C I C I C C I I I C C I C I I I C I I I	C C C C C C C I	O1 O1 O1 O1 O1 O1 O3 O1
Coincidence between Two Neural Net Outputs (O1=O3)	C C C C C I C I C C I I I C C I C I I I C I I I	C C C C C C C I	O1 O1 O1 O1 O1 O2 O1 O1	Coincidence between Two Neural Net Outputs (O2=O3)	C C C C C I C I C C I I I C C I C I I I C I I I	C C C C C C C I	O2 O2 O2 O1 O2 O2 O2 O2
Non-Coincidence among Neural Net Outputs (O1.NE.O2 and O1.NE.O3)	C C C C C I C I C C I I I C C I C I I I C I I I	C C C C C C C I	O1 O1 O1 O1 O2 O2 O3 O1				

(4) Correctness Decision and the Selection of the Ultimate Network Output

The selection and correctness decision of the ultimate output of the parallel network are determined by the coincident state among outputs and the correctness estimations of outputs of the neural networks in parallel, as listed in TABLE 1. In Table 1, the bolded characters of **C** and **I** denote respectively estimations as correct and incorrect outputs having the coincident states. Very Similar scheme is also applied to the 2-parallel network. A 3-majority voting system uses only the coincident states among the neural network outputs.

SIMULATION MODELS AND CONDITIONS

The 3-layered neural network has 75 input nodes, 140 hidden nodes and 7 output nodes. A basic anchor training set composed of feature vectors having 128 anchor input samples, and two other anchor training sets consisting of the same 128 anchor training input samples with mutually distinct target allocations for 128 labels were prepared. A test set, HD1, with the test input samples having a Hamming distance of 1 from the basic anchor training input samples, and validation sets of HDn, where n is an integer of 2 to 10, having a Hamming distance n, were also prepared to validate the performance of the parallel networks. The sizes of the test and validation sets are 9,598 in HD1, 355,040 in HD2 and 12,800 in HD3 to HD10, respectively. The validation sets of HD3 to HD10 were randomly generated.

A unique initial weight vector was randomly chosen in the range of -0.5 to 0.5 under a uniform distribution for use in training individual neural networks by the error-perturbed back propagation algorithm. Different random initial weight vectors were also prepared for each neural network to evaluate the effect on the diversity of the generalization domain.

Neural networks cease learning when a minimum margin among the node outputs of the output layer for the anchor training set exceeds a threshold of 0.42 under the condition of complete correct outputs in the binary space. The margin is given by an absolute value of the difference between the values of the discretizing threshold of 0.5 and the node output sending a correct binary output (Yatsuzuka and Enomoto, 1996).

Simulations were conducted for 2- and 3-parallel networks combining the well-trained neural networks. For comparisons, The 3-majority voting system combining the three neural networks was also simulated. Performance of the parallel networks and the 3-majority voting system concerning a confusion matrix (Weiss and Kulikowsk, 1991) were evaluated for the test set of HD1 and the validation sets of HD2 to HD10 through these simulations. In the confusion matrix listed in Table 2, "Positives" denote physically correct outputs, whereas "Negatives" denote physically incorrect outputs. A decision result of Positives, D_+, means deciding the ultimate output as a correct output, and the result of Negatives, D_- means an incorrect output.

TABLE 2 Confusion Matrix of Decision on correctness of the ultimate output.

Decision Results	Class Positives C_+	Class Negatives C_-
Decision as Positives (D_+) (Correct ones)	True Positives : TP (True decision on correct outputs)	False Negatives : FN (False decision on incorrect outputs)
Decision as Negatives (D_-) (Incorrect ones)	False Positives : FP (False decision on correct outputs)	True Negatives : TN (True decision on incorrect outputs)

DISCUSSIONS

(1) Convergence of Neural Networks

The neural networks simply converged without trapping tenacious local minima, completely achieving correct outputs for the anchor training set, and the training cycles were 85, 84 and 78 for the first, second and third neural networks in parallel, respectively. Conventional back propagation algorithms did not provide all the correct outputs, resulting in trapping local minima.

(2) Effects of the Different Target Allocations for Labels and the Initial Weight Sets on Diversity

In the evaluation of the diversity among classification domains of the neural networks in parallel, the deviations of the generalization domains, which provide different outputs for the same input samples, between two of the three correct neural network outputs were measured. The rate of deviations were about 7.6% and 24.2% respectively for HD1 test set and HD2 validation set between the first and second neural networks, and 7.3% and 23.9% between the first and third ones. The rates of deviations between any two neural network outputs were very similar even for the HD5 validation set. Simulations also indicate that there are no significant differences in these rates whether or not the unique or individually randomized initial weight vector was set in the neural networks. A stable and certainly appropriate diversity of the trained neural networks can be successfully formed by the mutually distinct target allocations for labels, resulting in systematic displacement among the classification domains.

(3) Non-coincident States of Neural network Outputs

The coincident states of neural networks in parallel is utilized in the output correctness estimation. The HD1 test and HD2 validation sets close to the anchor training set give very low percentages of the occurrences of non-coincidence among three neural network outputs such as 1.47% and 9.11%, respectively, due to the complete convergence. The occurrence however increases to a high rate of 48.30% for the HD5 validation set away from the anchor training set. This is a reason why the proposed 3-parallel network is superior to the 3-majority voting system in generalization performance and the accuracy of correctness decision, as shown in Figs. 2 to 5.

(4) Generalization Performance

Generalization performance was evaluated by the HD1 test set and the validation sets of HD2 to HD10. Figure 2 shows the generalization performance in terms of HD input data. The generalization performance gradually deteriorates for the validation sets with an increase in the Hamming distance from the anchor training inputs. This tendency depends on the decrease in the correlation among the input samples in training and validation sets. The generalization performance was similar among the single neural networks. The variance of the generalization performance also became about 5% at maximum even for the test and validation sets, denoting the stable diversity. The higher generalization performance of the first neural network than that of other neural networks is due to the ordered target assignment to labels, i.e., not random. Namely, the binary patterns of the target having 7 bits were assigned to labels in decimal order. Therefore, many adjacent patterns in binary have only one in Hamming distance indicating a close correlation, resulting in the highest generalization ability as shown in Fig. 2. On the other hand, the randomized target allocation resulted in a very similar generalization performance, though having a slightly smaller value.

The proposed 3-parallel network with the correctness decision and the ultimate output selection can successfully handle the generalization performance over the single neural network, and also fairly improve over the 3-majority voting system. In the 3-parallel network the generalization performance was dramatically improved from 96.83% in the single 3-layered neural network to 99.64% for the

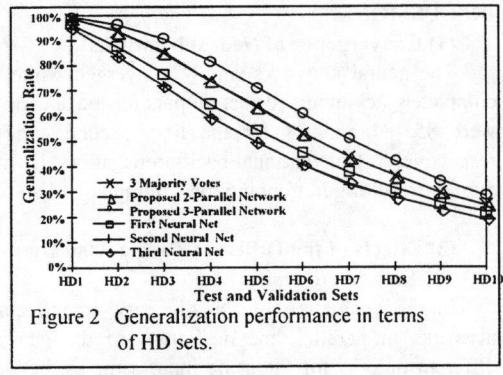

Figure 2 Generalization performance in terms of HD sets.

HD1 test set, from 88.08% to 96.88% for the HD2 validation set, and from 54.52% to 71.19% even for the HD5, respectively, and was superior to those of the 3-majority voting system. However, there is still some room to improve the generalization performance in comparing with the physical limits of 99.83%, 98.48% and 83.12% for HD1, HD2 and HD5, respectively. Further accurate correctness estimation is required to resolve this problem.

(5) Correctness Decision of Outputs

Three kinds of accuracy [15] of correctness decisions were evaluated for the test and validation sets. Namely, the accuracy of decisions on the ultimate outputs was calculated by $(TP + TN) / (C_+ + C_-)$ based on the confusion matrix in terms of HD input samples, and is shown in Fig. 3. The accuracy of decisions as correct ultimate outputs and as incorrect ultimate outputs was also calculated, respectively by TP / D_+ and by TN / D_- as shown in Fig. 4 and Fig. 5. A very high accuracy of correctness decisions of 99.9% on the ultimate outputs for HD1 and 87.9% even for HD10 in the 3-parallel networks, was achieved. The accuracy for both 2- and 3-parallel networks is superior to the 3-majority voting system, as shown in Figs.3, 4 and 5. The Hamming distance metrics is very effective for making an accurate correctness estimation.

The sensitivity of true decisions on correct outputs to the correct ultimate outputs was calculated by TP/C_+ and is shown in Fig. 6. The 3-and 2-parallel networks are also superior to the 3-majority vote. The rate of false decisions on correct outputs to the ultimate outputs was calculated by $FP/(C_+ + C_-)$. The 3-parallel network provides a small rate of false decisions less than 11.5% even for validation sets far away from the anchor training set such as HD8 and HD10.

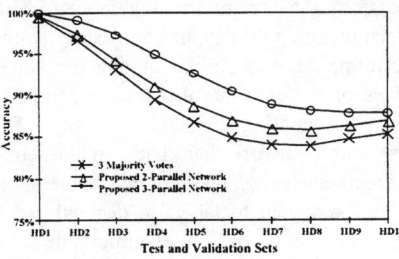

Figure 3 Accuracy of decision on the ultimate outputs in terms of HD sets.

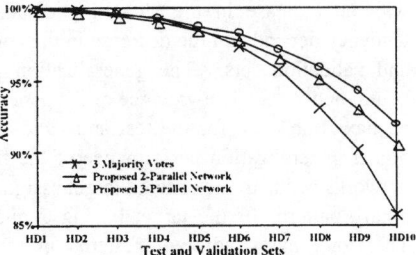

Figure 4 Accuracy of decision as correct ultimate outputs in terms of HD sets.

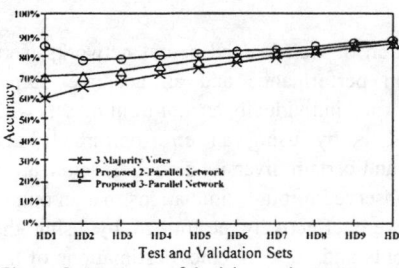 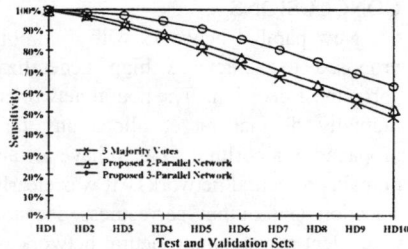

Figure 5 Accuracy of decision as incorrect ultimate outputs in terms of HD sets.

Figure 6 Sensitivity of decision on correct outputs to the correct ultimate outputs in terms of HD sets.

The 3-parallel network is also superior to the 3-majority voting system.

The rates of false decisions on incorrect outputs to the ultimate outputs was calculated by $FN/(C_+ + C_-)$. This factor is very important in some applications, in which mixing up the incorrect ultimate outputs with the correct ultimate outputs of the parallel network becomes serious such as in diagnosis systems. The rates of false decisions are extremely small for test and validation sets in the vicinity of anchor training inputs for both the 3-parallel network and the 3-majority voting system, and are less than 2% even for the HD10 validation set. The 3-Majority voting system is slightly better than the 3-parallel network due to the many occurrences of coincidence among neural network outputs. However, for the validation sets gradually further away from the anchor training set, the parallel networks become superior to the 3-majority voting system.

FURTHER IMPROVEMENT OF THE GENERALIZATION ABILITY BY INCREMENTAL APPROACHES

Generalization performance of the parallel network comprising a combination of well-trained neural networks can be further improved by associating with two incremental approaches one (Yatsuzuka and Sugiyama, 2001). One is a repetitive learning of the neural networks in parallel for incremental training sets, and the other is a pluralization of parallel networks.

The repetitive learning procedure actively detects incremental training sets from test input samples located in the vicinity of category classification boundaries by its own generalization ability of the neural networks. The progress of the repetitive learning procedure can asymptotically adjust the precise classification boundaries without excess overfitting and overtraining. This feature is very attractive for well-trained neural networks in parallel to have a very high generalization capability for unseen input samples away from the anchor training samples by localizing the parallel network.

On the other hand, in the pluralization of parallel networks a new anchor training set quite distant from the anchor training set for the preceding parallel networks is introduced to largely diversify the next parallel network so as to learn different parts from those of an input feature space in the preceding.

Accordingly, the pluralization of parallel networks, which has output selection and correctness decision capabilities, associated with the repetitive learning of neural networks might dramatically improve the generalization performance for an extremely wide range of unseen input samples.

CONCLUSIONS

New parallel networks with a combination of well-learned neural networks were proposed to achieve a high generalization performance and an accurate output correctness decision. The neural networks learned individually anchor training sets with mutually distinct target allocations for labels by using an error-perturbed back propagation algorithm so as to have a stable and certain diversity among generalization domains of neural networks. It was clearly observed through simulations that an output selection and output correctness decision were effectively performed by using the coincident states among neural network outputs and the correctness estimations of the outputs, based on a Hamming distance metrics.

It was concluded that the proposed parallel networks were able to provide a dramatic improvement of the generalization performance over the single neural networks, and also to feature a very accurate correctness decision capability in a binary input space. We believe that mutually distinct target allocations for labels will prove an efficient method as well as output correctness decision for floating-type input samples. We should evaluate these methods by empirical floating-type data to reach broader conclusions (Wolpert and Macready, 1997).

REFERENCES

Jakobson G. and Weissman M.D., 1993, " Alarm Correlation," IEEE Network, pp.52-59, November.

Baum E. and Haussler D., 1989 "What Size Net Gives Valid Generalization?" Neural Computation, Vol.1, pp.151-160.

Reed R., "Pruning Algorithms, 1993, " IEEE Trans. Neural Networks. Vol.4, No.5, pp.740-747, Sept.

Plutowski M. and White H., 1993 "Selecting Concise Training Sets from Clean Data, IEEE Trans. on Neural Network," Vol.4, No.2, March, pp.305-318.

Yatsuzuka Y., 1995, "An Error Perturbation for Learning and Detection of Local Minima in Binary 3-layered Neural Networks," Proc. of the International Conference on Neural Networks, Vol.1, pp.52-59, November.

Yatsuzuka Y. and Enomoto M., 1996 "A Binary Three Layered Neural Network with Switched Error Perturbation and Reiterative Learning Utilizing the Generalization Property," Proc. of the Artificial Neural Networks in Engineering Conference, Vol.6, pp.35-44, ASME Press Series.

Ji Chuanyi and Ma Sheng, 1997, "Combinations of Weak Classifiers," IEEE Trans. on Neural Networks, Vol.8, No.1, pp. 32- 42. January.

Ma Sheng and Ji Chuanyi, 1999, "Performance and Efficiency: Recent Advances in Supervised Learning," Proc. of the IEEE, Vol.87, No.9, pp. 1519- 1535. September 1999.

Sharkey Amanda J. C., 1996, "On Combining Artificial Neural Nets," Connection Science, Vol.8, Nos.3 & 4, pp.299-313.

Sharkey Amanda J.C. and Sharkey Noel E., 1997, "Combining diverse neural Nets," The Knowledge Engineering Review, Vol.12:3, pp.231247.

Sarkar D., 1996, "Randomness in Generalization Ability:A Source to Improve it, " IEEE Trans. Neural Networks, Vol.7, No.3, pp.676-685, May.

Hansen Lars Kai and Salamon Peter, 1990, "Neural Network Ensembles," IEEE Trans. Pattern Analysis and Machine Intelligence, Vol.12, No.10, pp.993-1001, October.

Weiss S.M. and Kulikowsk C.A., 1991, Computer Systems that learn, Morgan Kaufmann Publishers, Inc..

Yatsuzuka Y., Sugiyama T., 2001, "Parallel Neural Networks Repetitively Learned with Incremental Training Data Derived from Classification Boundaries and Pluralization of Parallel Networks, " to appear.

Wolpert David H. and Macready William G., 1997,"No Free Lunch Theorems for Optimization, " IEEE Trans. on Evolutionary Computation, Vol.1, No.1, pp.67-82, April.

REAL-TIME REINFORCEMENT LEARNING
FOR TARGET TRACKING

ROSS D. LAMM, PH.D.
Silicon Recognition, Inc.
Petaluma, California
rosslamm@yahoo.com

DONALD F. SPECHT, PH.D.
Consultant
Los Altos, California
donaldspecht@cs.com

ABSTRACT

Most conventional target tracking systems can track an airplane against a plain blue sky or a boat on the water, but fail miserably if they try to track a car on the freeway that can change in relative size or orientation or pass other vehicles of similar size and shape or go under a freeway overpass or tree and be obscured from view temporarily. The microsecond learning capability of the ZISC chipset allows us to develop an adaptive target tracking system that can track objects even as they change in orientation, scale or shape.

This same technique can be used for a variety of tracking problems including tracking people in crowds, tracking eye movements for human-computer interfaces, smart bombs, scene-guided artillery shells, stabilizing handheld video cameras by tracking objects of interest, and tracking landing sites for autonomous vehicles.

A video demonstrating real-time tracking will be shown.

INTRODUCTION

An adaptive target tracking system that tracks objects even as they change in orientation, scale or shape is presented. The system uses the Zero Instruction Set Computer chip (ZISC®, IBM) which is a silicon based Neural Network. The ZISC chip employs a Radial Basis Function (RBF) Neural Network model for nonlinear feature space mapping and recognition. Due to the fact that the ZISC chip is a hardware implementation of the RBF model, training or recognition of pattern vectors can be accomplished in 4 microseconds. The target of interest must be identified to the tracking system in an initial video frame. The system learns the textural features of the target and searches for these patterns in subsequent video frames. If the target texture pattern starts to change between consecutive images, the system automatically takes a new learning sample and adds this to its knowledge base. The target tracking system is able to quantify the certainty by which it is recognizing the object of interest by successively measuring the difference between the input feature vectors and the known pattern vectors and is able to re-learn a pattern of the tracked object if necessary. The real-time reinforcement capability allows the system to track objects even as they become partially occluded or change in scale, shape or orientation. The true benefit of this system is the ability to first realize when the system is tracking with decreasing certainty and to secondly re-learn the changed target pattern in real-time. This paper describes the ZISC Neural Network chip and its implementation in an adaptive real-time target tracking system. Our test results from a car tracking experiment on a crowded freeway are presented and demonstrate that the method is dynamic and robust.

Many previous techniques for neural target tracking have been reported. Good examples are given by Kruger, et al. (2000) and Rovetta and Zunino (1999). These require training either on a database of similar objects or on multiple poses of the object to be tracked before tracking can begin. Our system requires only one designation of the target and will learn as the target changes position and pose. Parallel processing is often required to achieve real-time speeds (Rovetta and Zunino, 1999). With ZISC, either 36 or 78 parallel processors are implemented on a single chip.

THE ZISC CHIPSET

Real-time relearning was possible because of the availability of the ZISC neural network chipset. It is a parallel processing chip that provides ultra-fast computing power for applications requiring both fast pattern classification and fast training. Each neuron is an independent processor, with integrated memory and learning capability.

The name "ZISC" is a registered trademark of IBM Corporation and an acronym for Zero Instruction Set Computing. The chips themselves are custom logic and are not programmable. This feature means that the core pattern recognition algorithm does not have to be programmed. In operation, instructions do not have to be fetched and interpreted.

The present ZISC chips are capable of performing an evaluation in less than four microseconds, making them ideal for real time applications. ZISC network extension is unlimited, due to a parallel architecture, which enables applications to run at a speed independent of the number of neurons.

A detailed description of the functions of the ZISC hardware is available in the ZISC reference manual (IBM, 1998). An overview follows.

Radial Basis Function (RBF)

The ZISC chipset handles 2 approaches to feature space mapping: Radial Basis Function and K-Nearest Neighbor. The RBF (Radial Basis Function) model is based on mapping an N-dimensional space by prototypes. Each prototype is associated with a category and an influence field representing a region of the N-dimensional space around the prototype, where generalization is possible. A prototype is a vector defining the co-ordinates of a point within the N-dimensional space. The classification task consists of determining if an N-dimension input vector lies within the influence field of any prototype stored in the network. This is done by computing the distance between the input vector and all stored prototypes, and comparing it to the influence field associated with the prototype. Within the network, several prototypes may be associated with one given class, and influence fields may partially overlap.

If the input vector does not lie within any influence fields, it is not recognized.

If the input vector lies within the influence field of one or more prototypes associated with one class, it is declared as belonging to that class. If the input vector lies within the influence field of two or more prototypes associated with different classes, it is declared as recognized but not formally identified.

The learning task consists of mapping the space by prototypes and adjusting the influence fields according to all neighbors. Learning, as opposed to classification, requires that the correct category associated with the prototype be given to the network. The learning mechanism rearranges the network content as a function of the proximity of the new input to the map already recorded in the network. The result of presenting a prototype and the corresponding category to the network can result in:

- No change in the network content,
- One or more influence fields being modified (reduced), or
- The new prototype being stored in the network.

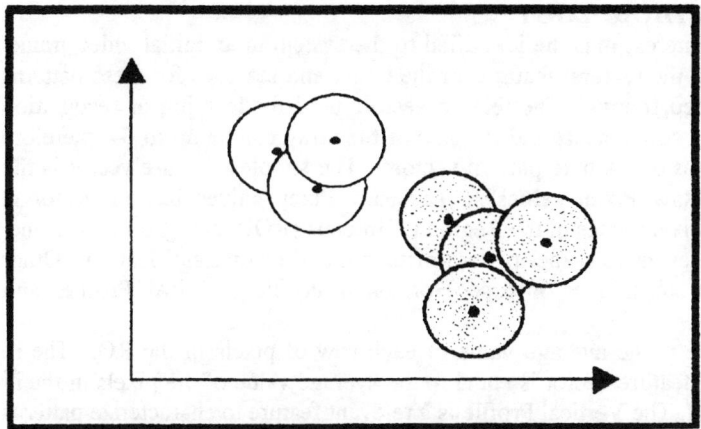

Figure 1. Example of an RBF approach mapping a two-dimensional space.

An RBF neural network topology is a three-layer network in which each neuron is connected to all inputs and to a given category output. The learning process dynamically establishes neuron connectivity to category nodes. An RBF neural network can be considered an expert system, which recognizes and classifies objects or situations and makes instantaneous decisions, based on accumulated knowledge. It accumulates its knowledge 'by example' from data samples and corresponding categories. Its generalization capability allows it to react correctly to objects or situations that were not part of the learning examples. The learning capability of an RBF neural network model is not limited in time, as opposed to some other models. It is capable of additional learning while performing classification tasks.

In software implementations, RBF models usually work with Euclidean distance, and so influence fields represent hyper-spheres. Within ZISC, they represent hyper-polygons whose shape depends on the norm selected by the user. ZISC supports two norms, L1 (city-block distance) and LSUP (absolute value of the largest difference in the components of the vector). The Tracking system utilizes the L1 norm distance evaluation.

The ZISC hardware also has a Nearest Neighbor function that returns the distances between a vector feature and stored prototypes, in ascending order of distance, at a rate of one value every two microseconds. This is often of value in other applications, but was not used for the target tracking application.

To perform recognition, input vector components are fed in sequence. ZISC performs distance calculations for all prototypes as the components are entered. Only a few additional cycles are required to achieve the evaluation.

The ZISC hardware also offers the capability to build networks of several devices to constitute larger networks with no theoretical limit or reduction in performance. Network extension does not require external logic, only resistors and eventually re-powering devices.

TRACKING METHODOLOGY

The target of interest must be identified to the system in an initial video frame. The system learns the textural features of the target and searches for these patterns in subsequent video frames. The feature vectors used for learning or recognition are arrays of 8-bit components and the size of the array can be up to 64, therefore the system operates on 64 byte pattern vectors. The simplest feature vector is the retinal data or "Raw Pixel Values". The Raw Pixel Values feature vector is composed of the pixel values in the Region Of Interest (ROI) arranged line per line. It is especially relevant for exact pattern matching and color classification. Other feature extraction techniques include Horizontal Profile, Vertical Profile, and Composite Profile.

Vertical Profile is the average value of each row of pixels in the ROI. The ith component of the feature vector is equal to the average value of the pixels in the ith column of the ROI. The Vertical Profile is a relevant feature to characterize patterns oriented horizontally.

Similarly, Horizontal Profile is the average value of each column of pixels in the ROI, and is a relevant feature to characterize patterns oriented vertically.

The textural feature used in the target tracking system is Composite Profile. Composite Profile is the combination of Horizontal Profile appended to Vertical profile. The Composite Profile is composed of the average value of the columns in the ROI followed by the average value of rows in the ROI.

Example	Feature	Vector
1 1 1 1 1 1 9 9 9 1 1 9 7 8 1 6 7 9 1 1 1 8 7 1 1	Raw pixels = Horizontal = Vertical = Composite =	1 1 1 1 1 1 9 9 9 1 1 9 7 8 1 6 7 9 1 1 1 8 7 1 1 2 7 7 4 1 1 6 5 5 4 1 6 5 5 4 2 7 7 4 1

Figure 2. Examples of different Feature Extraction Techniques.

If for a given ROI size, the feature vector exceeds the maximum vector length of 64 bytes supported by the ZISC chip, the feature vector has to be mapped to an array of 64 bytes.

If the target texture pattern starts to change significantly between consecutive images, the system automatically takes a new learning sample and adds this to its knowledge base. The tracking system is able to quantify the certainty by which it is recognizing the object of interest by successively measuring the difference between the input feature vectors and the known pattern vectors and is able to re-learn a pattern of the tracked object if the distance exceeds a predetermined threshold value. The real-time reinforcement capability allows the system to track objects even as they become partially occluded or change in scale, shape or orientation.

While in "Normal Track" mode, the system captures each video frame and scans a Region of Search (ROS) in the image. The ROS size is determined by the user and has default values of 20x20 pixels. During each image search, the average x and y location of all the positively found object locations is calculated and the center of the ROS is updated to this location for the next image capture iteration.

In the case that the system suddenly losses track of an object, the system will force itself into "Auto-Reacquire" mode of operation. This mode of operation will assume that the target of interest has been obscured by an object such as an overpass or tree. In the "Auto-Reacquire" mode, the system will successively enlarge its region of search in order to reacquire the object after it has passed under the obscuration. Once the object is reacquired, the ROS is reduced back to its default size.

RESULTS

One important application of this technique involves tracking vehicles on a freeway using a video camera mounted on a moving helicopter. The system was tested on many videos recorded from the helicopter. As an example, Figure 3 shows the system tracking the first bus on the right. Figure 4 shows that even as the second and third bus pass the first, the system is "locked on" and able to continue tracking only the first bus.

For an image 320 x 240 and a Region of Scan 20 x 20 the tracking system can update the offset object location at the full video frame rate. In the event the object changes in size, shape, or orientation, and the system is required to relearn on the fly, the system still operates in real-time. The system has an offset error resolution of 1x1(pixel) on the 320x240 pixel image. The number of neurons required for this kind of application is entirely dependent on the situation and object to be tracked. However, a total of 72 neurons is usually sufficient to track a car on the freeway.

REFERENCES

IBM, 1998, *ZISC036 Neurons User's Manual*, version 1.2 available
  from web page www.fr.ibm.com/france/cdlab/zisc.htm. Further information is available on the Silicon Recognition, Inc. web page, www.silirec.com, and U.S. Patents 5,740,326 and 5,621,863.
Kruger, V., Happe, A., and Sommer, G., 2000, "Affine Real-Time Face Tracking

using Gabor Wavelet Networks," *Proc. 15th Int. Conf. On Pattern Recognition 2000*, Vol. 1, 2000, pp. 127-130.

Rovetta, S., and Zunino, R., 1999, "A Multiprocessor Oriented Visual Tracking System," *IEEE transactions on Industrial Electronics*, Vol. 46, No. 4, August 1999, pp. 842-850.

Figure 3. Tracking the first bus on the right.

Figure 4. Still tracking "only" the original object!

ADAPTIVE SENSOR INTEGRATION USING RELIABILITY MAP

DAI NAITO
Dept. of Complex Systems Engineering, Hokkaido University, Sapporo, Japan

HIROSHI YOKOI
Dept. of Complex Systems Engineering, Hokkaido University, Sapporo, Japan

TAKAYUKI SUZUKI
Mitsubishi Electric Corporation, Nagoya, Japan

YUKINORI KAKAZU
Dept. of Complex Systems Engineering, Hokkaido University, Sapporo, Japan

ABSTRACT
Although there is no doubt that a single sensor is not enough for an autonomous robot nowadays, using many sensors causes the problem how to integrate them. We introduce the concept of the reliability between sensors and we propose the technique of making the reliability as the index of the integration. To deal with a fine change of situations, the reliability should be prepared as the combination of the state of the sensors, and we call it as the reliability map. The reliability is updated at any time when the robot acts and is evaluated. It can also keep up with a large change of the environment including itself.

INTRODUCTION

Nowadays there is no doubt that a single sensor is not enough for an autonomous robot to act intelligently. Adopting multi-sensor gives a big benefit to a robot (Luo, 1994). However in order to receive the clear benefit, it is necessary to consider what to integrate and how to integrate sensors (Bruder,1999; Weckesser, 1998).

Man has various sense organs, such as an eye, an ear, and a tactile sense. However, not all feeling information is used for recognition of an environment or himself, we select a sense organ and/or fuse some feeling information according to a situation (Akamatsu, 1994). If it is applied to an autonomous mobile robot, it can be autonomously determined which sensor should be trusted and how much it should be trusted. Moreover, although the method of the integration should be changed according to situations, the conventional studies had the problems in re-training when the environment changes a lot.

Based on such a background, we introduced the concept of the reliability between sensors and we propose the technique of making the reliability as the index of the integration. To deal with a fine change of situations, the reliability should be prepared as the combination of the state of the sensors, and we call it as the reliability map. The reliability is updated at any time when the robot acts and is evaluated. It can also keep up with a large change of the environment including itself.

In this paper, the sensor integration is represented by the integration of the action policy defined for every sensor, and the reliability is used as the rate of combining them. By learning a suitable reliability map, the robot can perform the sensor integration for an autonomous action. Moreover, the whole amount of learning can be reduced by making learning into two stages, that is a big merit in learning of an autonomous mobile robot. In order to prove the validity of the proposal technique, we shows that suitable sensor integration is performed in the case of sensor trouble by experiment using a real autonomous robot.

DEFINITION OF RELIABILITY

In this paper, we deal with two kinds of sensors: a vision sensor and an ultrasonic distance sensor. The two sensors are provided the policy of action Pv and Pd, respectively. Pv and Pd are the probability of selecting an action from prepared actions when a state is Sv and Sd, respectively. And the two action policies are integrated as a weighted sum of them. The integrated policy of the action is expressed as the following formula.

$$P(S) \bullet (1 \bullet k) \bullet Pv(Sv) \bullet k \bullet Pd(Sd), (0 \bullet k \bullet 1) \qquad (1)$$

By Eq.(1), a robot chooses action stochastically(see Fig.1).
The coefficient k reflects which policy should be more reliable in the state. In this case, the larger the value of k is, the more the ultrasonic sensor is relied.

Pv and Pd are gained by Q-Learning which is one of Reinforcement Learning (Sutton, 1998), respectively. Q-Learning is the learning method which is focused on the combination of a state and an action, and it can gain a sequence of actions which make an expected long-term reward the maximum by giving reward and penalty. This is a learning method suitable for an autonomous robot (Asada, 1995). Before applying Q-Learning, the state and the reward should be specified. Q-Learning is performed before integrating with the reliability. That is, there are two learning phases such as learning of policies and learning of reliabilities.

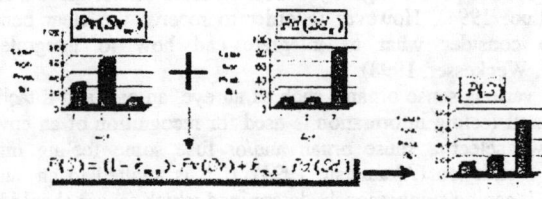

Fig. 1: Integration using the reliability

RELIABILITY MAP

The k defined above is prepared as many as the combination of the state of the vision sensor and the ultrasonic distance sensor, and we call it the reliability map (Fig.2). If the vision sensor has m states and the ultrasonic sensor has n states, the number of k is $m \bullet n$. An environmental change appears as the change of k.

	Sd_1	Sd_2	$\cdots$	Sd_n
Sv_1	$k_{1,1}$	$k_{1,2}$	$\cdots$	$k_{1,n}$
Sv_2	$k_{2,1}$	$k_{2,2}$	$\cdots$	$k_{2,n}$
$\vdots$	$\vdots$	$\vdots$	$\ddots$	$\vdots$
Sv_m	$k_{m,1}$	$k_{m,2}$	$\cdots$	$k_{m,n}$

Fig. 2: Reliability Map (*Sv*: state of vision sensor, *Sd*: state of ultrasonic distance sensor)

UPDATING THE RELIABILITIES

The value of k is changed adaptively by rewards and penalties when a robot moves in a certain environment. If the robot acts a_i and is given reward R, the way of updating the reliabilities is as follow:

(1) when $R>0$, raise the reliability of the sensor (either vision or ultrasonic distance sensor) which has higher action selection probability of a_i than the other.

(2) when $R<0$, lower the reliability of the sensor (either vision or ultrasonic distance sensor) which has higher action selection probability of a_i than the other.

(3) when $R=0$, do not change the reliability.

The amount of change of reliability is fixed to a certain amount.

In this experiment, when a penalty is received, a penalty is also given to the previous state, and reliability k is changed. The amount of change of the reliability at that time is made fewer than usual.

CONSIDERATION ABOUT THE AMOUNT OF LEARNING

The proposed technique in this paper is able to reduce the amount (or the number of times) of learning.** The amount of learning is determined in general Reinforcement Learning by the number of states of a sensor, and the number of actions. Therefore, the amount of learning in the case there are two sensors could be expressed as follows:

$$S_1 \cdot S_2 \cdot A \cdot O(n^3)$$

Since it is learning in two stages, the amount of learning using the proposal technique is as follows:

$$S_1 \cdot A \cdot O(n^2)$$
$$S_2 \cdot A \cdot O(n^2)$$
$$\underline{S_1 \cdot S_2 \cdot O(n^2)}$$
$$\cdot O(n^2)$$

Since one dimension of n has fallen, if the number of the states increases, it can learn efficiently. Moreover, because the action policies acquired once can usually be used in any environments, a robot can be moved only by study of the reliability map.

SPECIFICATION OF THE AUTONOMOUS MOBILE ROBOT

The autonomous mobile robot used for this experiment has a drive system that two wheels run independently like Khepera, as shown in Fig.3. She can take four actions, advance, retreat, left revolution, and right revolution. And she has a vision sensor, an ultrasonic distance sensor and 16 infrared touch sensors. The vision sensor and the ultrasonic sensor can sense in the range of 180 degree by controlling a mounted motor.

Fig. 3: View of robot

BASIC SETUP OF THE EXPERIMENT

The task assigned to the robot is "avoid obstacles and go to the light". The experiment is conducted in a passage in which some obstacles is put on. The position of the light source is judged by the image of the camera.

If the robot moves and catches the light in the front, she is given a value of estimation +1 as a reward. If she collides with an obstacle, she is given -1 as a penalty. There are three kinds of action which the robot can choose: 1) step forward, 2) left revolution and step forward, 3) right revolution and step forward.

The vision sensor and the ultrasonic sensor sense three directions: front, right-front and left-front. The state of the vision sensor is divided into four because of the position of the light is: 1)center, 2)right, 3)left, 4)missing. The state of the ultrasonic sensor is divided into 27 because of the position of an obstacle and distance to an obstacle.

LEARNING EACH SENSOR POLICIES

Before the experiment with the sensor integration, an action policy for every sensor is acquired by Q-Learning by basic setup. The reward values given for the ultrasonic sensor and the vision sensor are set separately.

Although detailed explanation is excluded here, as a result, the vision sensor system acquired an action policy which goes toward the goal, and the ultrasonic sensor system acquired an action policy which avoids an obstacle.

EXPERIMENTS: CHANGE OF A RELIABILITY MAP

·We carry out experiment with the sensor integration based on the proposal method. Figure 5 shows the transition of the evaluation value at the time of moving the robot in the experiment environment as shown in Fig.4. The evaluation value is considered as the total of reward value and penalty value. About the 100th step of the start are considered to be the condition where suitable reliability map is learned, and it turns out that the evaluation value has been stabilized and gone up after that.

A part of the change of the reliability map is shown in Table 1. All the initial value of k is set to 0.5. Each k is changed by the environment and it causes a change of action selection probabilities.

Looking at second from the top, for example, the state is that the light is in left and right-front obstacle is in middle distance. At this time, vision sensor is trusted more because k is changed from 0.5 to 0.

Next, we make the experiment which verifies that it can correspond to change of unpredictable environment. After the experiment of Fig.5 that a suitable reliability map was gained, the ultrasonic sensor is assumed to be broken and can sense in only left-front since 400th step. Because of that, front and right-front states of the ultrasonic sensor are the same one as left-front. The result in this case is shown by the solid line of Fig.6. After the 400th step, the robot could not take good action immediately, but she has adapted to the situation after a while. A part of the change of the reliability map at this time is shown in Table 2. On the whole, it turns out that reliability is changing so that the vision sensor may be trusted.

In order to investigate whether learning of the reliability map after sensor failure is effective, the broken line of Fig.6 shows the case where renewal of reliability is ended by 400 steps as a comparison experiment.

This experiment showed that effective sensor integration could be performed to a big change of a situation like sensor failure by changing the reliability map into a suitable thing.

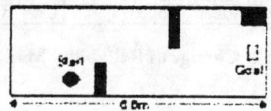

Fig. 4: Surroundings of the experiment

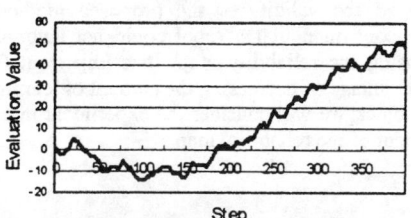

Fig. 5: Evaluation of the sensor integration (ex.1)

State			act selection prob.(%)				act selection prob.(%)		
Sv	Sd	k	Left	Center	Right	k	Left	Center	Right
Left	n n n	0.5	52.05	3.32	44.63	0.46	55.06	3.49	41.45
Left	f f m	0.5	52.59	35.11	12.30	0	89.67	5.51	4.82
Center	f n n	0.5	47.10	52.04	0.85	0.78	72.37	26.94	0.70
Center	f f n	0.5	31.37	66.45	2.19	0.08	6.68	92.01	1.31
Center	f f f	0.5	13.06	58.06	28.88	0	1.98	96.88	1.14
Right	f f n	0.5	31.54	19.08	49.38	0.78	47.90	28.56	23.54
Right	n f f	0.5	16.12	7.20	76.68	0.18	7.30	3.97	88.74
Miss	m m n	0.5	42.67	20.21	37.12	0.78	59.10	24.07	16.83
Miss	n n f	0.5	8.15	8.64	83.20	0.26	10.64	10.89	78.47

Table 1: Change of Reliability Map (ex.1). Explanation of Sd, n:near, m:middle, f:far in distance to an obstacle

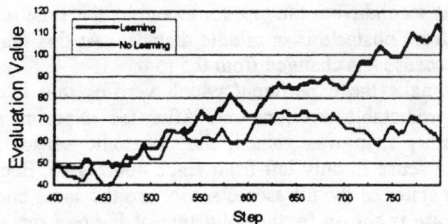

Fig. 6: Evaluation of the sensor integration (ex.2)

State		k	act selection prob.(%)			k	act selection prob.(%)		
Sv	Sd		Left	Center	Right		Left	Center	Right
Left	n n n	0.46	55.06	3.49	41.45	0.7	37.00	2.44	60.56
Left	m m m	0.5	59.24	23.23	17.53	0.28	72.63	15.43	11.94
Center	f f f	0.38	64.76	10.73	24.50	0.08	84.42	6.61	8.97
Center	n n n	0.76	11.44	24.10	64.45	0.5	8.21	49.00	42.79
Center	m m m	0	1.98	96.88	1.14	0.08	4.13	92.41	3.47
Right	f f f	0	1.98	96.88	1.14	0	1.98	96.88	1.14
Right	n n n	0.58	9.35	1.55	89.10	0.16	4.27	1.98	93.75
Miss	m m m	0.38	12.40	16.89	70.71	0	2.33	2.15	95.52
Miss	f f f	0.38	10.62	8.65	80.74	0.42	11.49	9.33	79.18

Table 2: Change of Reliability Map (ex.2)

CONCLUSIONS

We proposed the sensor integration which uses the reliability among sensors, and showed the validity of the proposed method by experiments. Consequently, we confirmed that a robot could act autonomously in strange environment by using the reliability map. It is important for an autonomous robot to have an advantage of decreasing the amount of learning.

As a future subject, we will consider the experiment in a more complicated task and development of the reliability map.

REFERENCES

Ren C.Luo, 1994, "A Perspective on Multisensor Fusion and Integration," Journal of the Robotics Society of Japan Vol.12, No.5, pp.646-649.

Stephen B.H. Bruder, 1999, "An information centric approach to heterogeneous multi-sensor integration for robotic applications," Robotics and Autonomous Systems 26, pp.255-280.

P.Weckesser and R.Dillmann, 1998, "Modeling unknown environments with a mobile robot," Robotics and Autonomous Systems 23, pp.293-300.

M.Akamatsu and T.Kasai, 1994, "Sensor Fusion in Human," Journal of the Robotics Society of Japan Vol.12, No.5, pp.656-663.

Richard S.Sutton and Andrew G.Barto, 1998, "Reinforcement Learning," The MIT Press.

M.Asada, S.Noda, S.Tawaratsumida and K.Hosoda, 1995, "Purspposive Behavior Acquisition for a Robot by Vision-Based Reinforcement Learning", Journal of the Robotics Society of Japan Vol.13, No.1, pp.68-74.

The Reference Neuron Model:
Game Playing via Trial and Error Learning

GEORGE SCHLEIS
Wayne State University
Computer Science Department, Detroit, MI 48202
gms@cs.wayne.edu

MATEEN RIZKI
Wright State University
Computer Science Department, Dayton, Ohio
45435-0001
mrizki@cs.wright.edu

FARSHAD FOTOUHI
Wayne State University
Computer Science Department, Detroit, MI
48202
fotouhi@cs.wayne.edu

In memory of Dr. Michael Conrad without whose vision, advice, and friendship this project would not have been possible.

ABSTRACT
The project investigates a hierarchical neuron model of memory storage and retrieval known as the reference neuron model (RNM). The model is based on the principle of superposition-free memory, i.e., the requirement that the acquisition of new memories by a neural network does not degrade (or hybridize) the previously acquired memories [2]. The memory manipulation mechanisms inherent in the model facilitate the development of associative memory structures and trial and error learning. The architectural and mechanistic features are motivated by biological and psychological theories of brain function. The simulation is used to examine ideas about the sensory input, perception processing, memory and learning abilities exhibited by the brain. Results suggest that the RNM model performs quite well in a traditional Artificial Intelligence game playing scenario (3X3 and 5X5 Tic-Tac-Toe) when compared to traditional search approaches (Minimax search) [16]. Neural network systems perform well on traditional pattern classification problems, but fall short in the area of representing rules and functions. Traditional "procedural" systems have the capability of supplying an explanation as to "why" certain decisions were made, but generally have a rather narrow domain. A consequence is that such systems become rather frail as the limits of the domain are reached. The RNM has both the features found in procedural systems as well as the pattern classification capabilities of neural network approaches.

Keywords: Associative Memory; Discovery Systems; Hybrid Artificial Neural System; Memory-Based Neural Network; Pattern Recognition; Memory & Learning; Hierarchical Networks;

INTRODUCTION
The reference neuron model (RNM) [2,3,4] is a synaptic orchestration model reminiscent of the Hebbian memory model [6], but with the difference that hierarchical control plays a critical role. In physiological terms the RNM acts as a perception-action mediator [1]. The RNM consists of at least two levels of neurons. The two levels include primaries, which respond to external input and secondaries (reference neurons), which are the focal point for later firing of primary neurons. Reference neurons can be activated by other reference neurons for time-ordered memories or by primary neurons for content ordered memories. The mechanisms described in this model allow for associative memory structures to develop via rememorization. The rememorization process provides

the basis for classical conditioning and trial and error learning. The problem domain for the RNM is the game of Tic-Tac-Toe [15]. We used Tic-Tac-Toe as a testbed to elucidate the capabilities of the RNM model.

The early work in the area of neuron models was developed out of the disciplines of biology and psychology. The early work by McCulloch and Pitts, and Hebb, has motivated a significant amount of research using a biological framework [9,6]. In both cases these models explored the information processing ability of neurons in which the input was either averaged, or summed. Once either of these mathematical steps are completed the resultant output is compared to a threshold value to determine whether or not a neuron will fire. If the neuron fires an output will be generated. Both of these models have been used to explain the process by which information is stored and recalled as well as how learning might occur within the brain.

The first detailed computer simulation of a neural model, the Perceptron, was shown to have several limitations [11]. These limitations were eliminated with developments in hardware and software and led to the construction of multi-layered networks that could overcome many of these original limitations [13]. The key questions that these early researchers were attempting to address are still relevant today: 1) How do we remember and store information? 2) What is the process of learning ? 3) What impact does memory have on learning ?

Production systems became the focus of research during the 1960's and early 1970's [8]. In the late-1970's advancements in hardware and software again allowed researchers to focus their attention on biologically oriented models. One of the first models during this era was the RNM which is the focus of this discussion [2,3,4]. Later models, the K-lines model [12] the hippocampal indexing model of memory [14], the hippocampal CA3 network architecture model [10] and multimodal memory networks [7] are similar to the RNM. An earlier physiological model, memory functions in the vertebrate nervous system [5], contains similar features as well.

BIOLOGICAL SUPPORT

A biological correlate for the RNM can be found in the hippocampus and the adjacent cortical regions. Physiological studies of memory have identified a circuit contained within the sheets of cells oriented transverse to the longitudinal (septo-temporal) axis of the hippocampus. This is known as the trisynaptic circuit - strips of cells that form a functional circuit and posses a lamellar organization [14]. In figure 1, the entorhinal projections onto dentate granule cells, the granule cell projections onto CA3 pyramidal cells(the mossy fiber system), and the efferents of CA1 pyramids into the alveus demonstrate a transverse, lamellar, functional orientation (Trisynaptic Circuit).

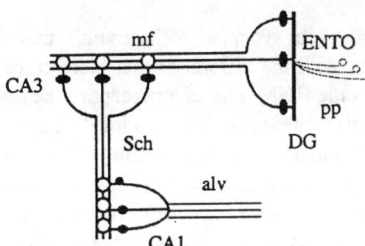

Figure1: Tri-synaptic circuit
Diagramtic representation of the transverse view of the trisynaptic circuit of the hippocampus. Alv, alveus; DG, dentate gyrus; mf, mossy fibres; pp, perforant path; Sch, Schaeffer collaterals. (adapted from Teylor, 1991)

Long Term Potentiation (LTP) is a potential process for long-term information storage displayed by the hippocampus, a structure that has been repeatedly implicated in learning and memory. LTP, is a dramatic long term increase in the postsynaptic response following a short and physiologically realistic period of afferent stimulation. LTP, is seen at excitatory synapses in the hippocampus and elsewhere and is expressed as a 200-300% increase in postsynaptic response, which can last for weeks. Teylor examined the evidence supporting the hypothesis that LTP is a viable candidate for a mnemonic device in the vertebrate nervous system. According to Teylor, there are several arguments supporting this hypothesis [14].

EXPERIMENTAL MOTIVATION

The difference between this model and other approaches is that the architecture and operations allow it to develop both procedural and associative memories as well as learn continuously. In the case of a typical production system, information about a problem is converted into relationship expressions (rules) which are then incorporated into the system. In the RNM such rules are not programmed into the system they are developed by the model itself via Trial and Error Learning over time. Typical neural networks, adjust network connection strengths via numerical computations in order to transform input patterns to the desired output patterns. Consequently the knowledge in such systems is located in the strength of the connections between the nodes rather than the nodes themselves. The RNM does not use connection strengths, simply the connectivity alone as a basis for memories. The system operates in one of three modes: 1). pattern recognition mode - pattern recognition is based upon the summed activation energy of several neurons firing . 2). both pattern recognition and strategic (temporal) mode – previously mentioned pattern recognition processing as well as the development of temporal memories or strategies. 3). Artificial Imagination (A.I.)mode - this mode utilizes both pattern recognition , temporal memories as well as the RNM player playing an imaginary game in order to assist in determining the next move.

On the basis of the previously stated issues we thought that it would be good to use a well understood problem such as Tic-Tac-Toe which also has the potential to scale as needed. One could say that since this problem is a game of complete information and finite in size (certainly in the case of the 3X3 game) learning therefore really isn't important. Obviously this is true, however the criteria for success based upon this game is also very clear.

THE RNM MODEL

The system begins by initializing all of the neurons and synapses for each player (see Table 1) :

Number of neurons	Type of neuron
27	Input receptor primaries
9	Output effector primaries(available)
5	Short-term reference
Generated as needed	Long-term reference

Table 1:3X3 Tic-Tac-Toe Game RNM neurons

In the case of the primary neurons: a particular neuron is associated with a board location and a piece. Ex. "3X" = there is an X in board location 3. Once initialization is completed, the RNM player decides on his first move. A move on the board is generated in one of two ways, either a recall of an old or the creation of a new memory. In both cases the process begins with the system scanning the board. The scan function causes several primary neurons to fire in response to a particular board configuration. The firing of the primary neurons allows them to be loaded by a reference neuron which contacts them for a brief period of time. We arbitrarily selected 40 system time units, however additional experimentation is planned which will vary this time duration in order to better understand the impact of time on this model. In the case of a new memory, the firing of a reference neuron is determined by a supervisory system. The supervisory system selects a reference neuron to fire in an arbitrary fashion we have selected round robin firing. The firing of the reference neuron will facilitate (open) synaptic connections between the reference neuron and the primary neurons in the backward direction.

The development of such a connectivity between a reference neuron and the primaries generates a memory circuit. When that reference neuron re-fires in the future the original primaries will be recalled (re-fired). Similarly, the firing of the reference neuron allows it to be loaded (potentiated) by primary neurons which contact it for a brief period of time. The firing of the primary neurons will facilitate (open) synaptic connections between the primary neurons and the reference neuron in the forward direction This generates a fully connected memory circuit between primaries and references. These memory circuits remain available for recall by the model for a short term period of time (i.e. period of one game).

In the case of an existing memory, the firing of several primary neurons during the scan function generates an activation energy which has a cumulative effect [7] upon a reference neuron which contacts these primaries. If the activation energy exceeds the threshold level, it causes a previously synaptically facilitated reference neuron to re-fire and as such recall the original memory circuit. As the board configurations become more complex, the previous memories will become less and less similar and therefore more reference neurons will become activated simultaneously however the most activated reference neuron will be the most similar. Such a grouping of neurons defines a new short-term memory circuit. If at the end of the game a short-term memory circuit is not stored (transferred) to a long term memory circuit the connections expire and return to an unconnected state. Over the period of a game several memory circuits develop(see RNM Algorithmic Description and RNM Architecture Diagram).

At the conclusion of a game, those short term memory circuits which contributed to the 'maximum number of pieces in a row' are transferred to long-term memory circuits for potential reuse during future games. The process of transferring a short-term memory circuit to a long-term memory circuit is as follows. The short-term memory circuit reference neuron is re-fired which in turn re-fires the associated primary neurons. During this same time period a long-term reference neuron fires which then associates those same primary neurons with it. The only difference between long and short term reference neurons is that long term reference neuron associations last for a longer period of time (several games). Similar to the short term memory circuits, if over a period of games these long term memory circuits are not reused they will expire and return to an unconnected state (are forgotten).

RNM ALGORITHMIC DESCRIPTION

While(Another move is possible)
    Fire input receptor primaries based upon current board configuration

If RNM memory is sufficiently distinct acquire a new memory
    Backward process is the accepting memory process
Else
    Forward process is the recall of a memory process
    Re-fire the old memory circuit and refire original output effector primary

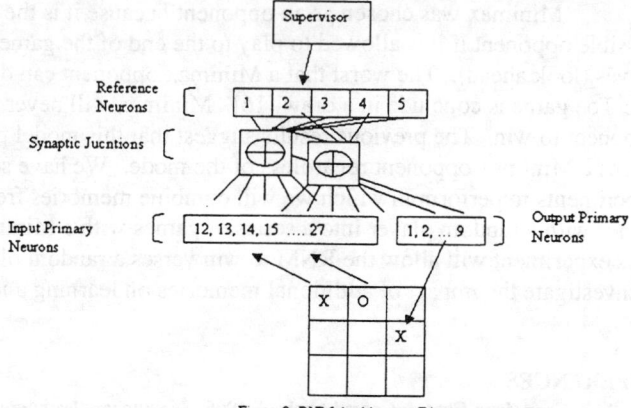

Figure 2: RNM Architecture Diagram

RESULTS

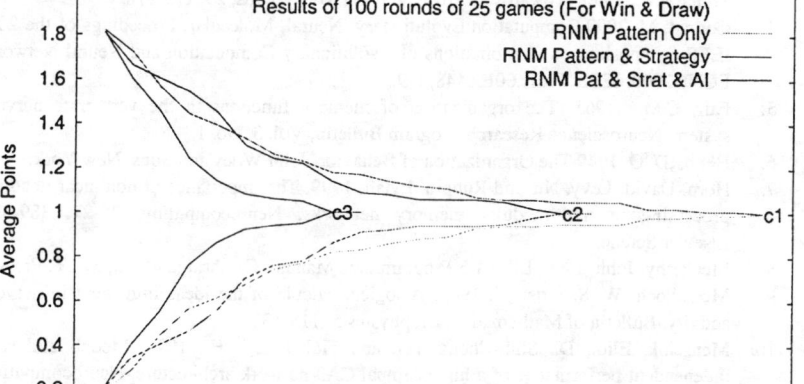

Figure 3: Convergence rate RNM vs Minimax for 100 Rounds
1 Round = 25 Games, 2 points for a win, 1 point for a Draw
c1= 24 games, c2= 16 games, c3 =8 games convergence point

    The previous graph represents the results of the system playing in each of its three basic modes for 100 rounds of 25 games (1 round = 25 games). In this case the RNM is the first player and the second player is a Minimax

Search [16] implementation. The upper curve represents the Minimax player and the lower curve represents the RNM player.

CONCLUSIONS & FUTURE WORK

Minimax was chosen as an opponent because it is the strongest possible opponent if it is allowed to play to the end of the game at each of its moves (look ahead). The worst that a Minimax opponent can do in a 3X3 Tic-Tac-Toe game is conclude in a draw [16]. Minimax will never allow an opponent to win. The previous results suggest that this model performs well verses a Minimax opponent regardless of the mode. We have several additional experiments to perform in which we will combine memories from a series of games with a random player into a series of games with a Minimax opponent. This experiment will allow the RNM to win verses a random player and allow us to investigate the impact of additional memories on learning and capability.

REFERENCES

1. Chen, Jong-Chen & Conrad Michael, 1996, Evolutionary learning with a neuromolecular architecture: A biologically motivated approach to computational adaptability.
2. Conrad, M., 1977, Principles of superposition-free memory, J. of theor. Biol. 1977, vol. 67 213-219.
3. Conrad, M., Kampfner, R.R., Kirby, K.G., Rizki, E.N., Schleis, G.M., Smalz, R., and Trenary, R., 1990, Towards an artificial brain, Biosystems, 23, 175-218.
4. Conrad, M.,2000,Computation:Evolutionary, Neural, Molecular, Procedings of the 2000 IEEE Symposium on Combinations of Evolutionary Computation and Neural Networks ECNN2000, IEEE Press 00EX448, 1-9.
5. Fair, C.M. 1965, The organization of memory functions in the vertebrate nervous system. Neuroscience Research Program Bulletin, Vol. 3, No. 1.
6. Hebb, ,D. O. 1949 The Organization of Behavior. John Wiley and Sons, New York.
7. Horn, David, Levy, Nir and Ruppin, Eytan, 1999, The importance of nonlinear dendritic processing in multimodular memory networks, Neurocomputing, 26-27, 389-394, Elsevier Scienc.
8. McCarthy, John, 1960, LISP 1.5 Programmers Manual, Cambridge Mass., M.I.T. Press.
9. McCulloch, W. S. Pitts, W. 1943, A logical calcula of the ideas immanent in nervous activity. Bulletin of Mathematical Biophysics 5, 115-133.
10. Menschik, Eliot, D., Shih-Cheng Yen and Finkel, Leif H., 1999, Model- and scal-independent performance of a hippocampal CA3 network architecture, Neurocomputing, 26-27, 443-453, Elsevier Science.
11. Minsky, M, Pappert, S. 1969, Perceptrons. MIT Press, Cambridge, Mass..
12. Minsky, M.L., 1980,K-lines: a theory of memory. Cognitive Science 4, 117-133.
13. Rummelhart,D.E.,McClelland, J.L., & the PDP Research Group (Eds.) 1986, Parallel Distributed Processing:Vol.1, Foundations. Cambridge, Mass.:M.I.T. Press.
14. Teylor, T.J. 1991, Learning & memory, Second Ed., Academic Press, 299-327.
15. Thrun, S. 1995, Learning to play the game of chess, Neural Information Processing Systems, vol. 7. 1069-1076.
16. Knuth, Donald E. and Moore, W. Ronald, 1975,, An analysis of alpha-beta pruning, Artificial Inelligence, 6(4), 293-326

PUNISH/REWARD: LEARNING WITH A CRITIC IN MULTILAYER NEURAL NETWORKS

GREGORY L. PLETT
Department of Electrical and Computer Engineering
University of Colorado at Colorado Springs
Colorado Springs, Colorado

ABSTRACT
One of the earliest machine-learning algorithms in the class we now refer to as "reinforcement learning" was the learning-with-a-critic method by Widrow, Gupta, and Maitra. In their work, a punish and reward scheme is used to train a single adaptive neuron. The novel feature of this method is that an error signal is not used as is required with supervised-learning algorithms; neither is the algorithm entirely unsupervised. Instead, an external observer called the critic evaluates the neuron's performance in a qualitative way and decides to either "reward" or "punish" its behavior. The neuron is "rewarded" if its performance is good, and is "punished" if its performance is bad.

In this paper, the punish/reward mechanism is generalized in order to train multilayer neural networks. An example is presented where a neural network is trained to play the childrens' game "tic-tac-toe."

INTRODUCTION

In the early 1960s, the LMS algorithm (Widrow and Hoff, 1960; Widrow and Stearns, 1985) was developed for training a single neuron. This is a supervised learning algorithm which requires a "teacher" to supply a *desired response* for every input training pattern. The difference between the neuron's output and the desired response is used to adapt the neuron's internal parameter values to cause learning. The backpropagation algorithm (Werbos, 1974; Rumelhart et al., 1986) was later developed for supervised learning with multilayer neural networks.

An alternative to supervised learning is unsupervised learning. An example is Kohonen's self-organizing feature map (Kohonen, 1982). Without a teacher, the neurons are only able to identify patterns in the data set and perform data clustering.

A final paradigm is the class of reinforcement learning algorithms (Haykin, 1994). Within this class, Widrow, Gupta, and Maitra (Widrow et al., 1973) invented the concept of learning-with-a-critic for training a single neuron. Critic learning does not require a precise desired response for each input training pattern. An external observer, called the critic, decides whether the neuron's decisions are good or bad, and *rewards* or *punishes* accordingly. Rewarded decisions are more likely to be repeated, and punished decisions are less likely to be repeated.

In this paper, the original critic algorithm is first discussed, and is then extended to be able to train a larger neural network.

ADAPTING A SINGLE NEURON WITH CRITIC LEARNING

We consider a neuron with a signum activation function (see Fig. 1(a)). The neuron computes $y = \text{sgn}(s) = \text{sgn}(W^T X)$, where y is the output of the neuron for input vector X,

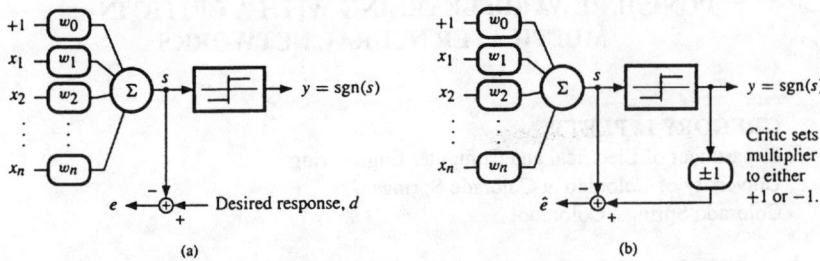

Figure 1. Single-neuron neural network: Adapted (a) using LMS; (b) using critic.

and W are the weight values of the neuron. (X is augmented with a bias value of 1, and W is likewise augmented with a weight to multiply the bias value.) When a desired response is available, an error signal may be computed and used with the LMS algorithm to update the neuron's weight values: $\Delta W = 2\mu e X$, where $e = d - s$ is the difference between the desired response and the neuron's internal sum, and μ is a small positive constant that controls the learning rate.

If no desired response is available, we cannot compute the error e and so cannot use LMS to adapt the weights. Critic learning estimates the desired response. The neuron's output is thresholded and interpreted to be a binary decision. An external observer called the critic judges the the result of the decision to be "good" or "bad."

If the critic decides "good," then the neuron's decision is assumed correct and the weight values are reinforced. This is done by estimating the desired response to be equal to the neuron's output: $\Delta W = 2\mu \hat{e} X$, where $\hat{e} = \text{sgn}(s) - s$. *If the critic decides "bad,"* then the neuron's decision is assumed to be incorrect and the weight values are punished. The desired response is estimated to be the negative of the neuron's output. The same weight update formula is used, but $\hat{e} = -\text{sgn}(s) - s$. Figure 1(b) shows a neuron with error $\hat{e}$. An internal desired response signal is computed with the aid of the external critic which sets the ± 1 switch.

Critic learning was devised for a single neuron, so has limited usefulness. Here, we extend the method so that it is able to train multilayered neural networks and therefore solve more difficult problems. In keeping with the original work by Widrow and colleagues, the resulting algorithm does not require a desired response—a simple reward/punish signal will adapt the entire network.

ADAPTING A NEURAL NETWORK WITH CRITIC LEARNING

A multi-layer neural network is shown in Fig. 2(a). Each neuron in the network computes a nonlinear function of a weighted sum of its inputs. The nonlinear function is usually chosen to be either the sigmoid or the tanh(·) function. Layers of neurons are able to compute very general nonlinear functions, and the network may be trained (with a teacher) using the backpropagation algorithm.

To extend critic learning to train multi-layered neural networks, it is important to notice that Widrow and colleagues' algorithm can be divided into two parts. The first part estimates a desired response for the neuron based on whether the neuron is being rewarded or punished. The error $\hat{e}$ is then computed to be the estimated desired response minus the actual output. The second part of the algorithm uses LMS to adapt the neurons' weights using this error.

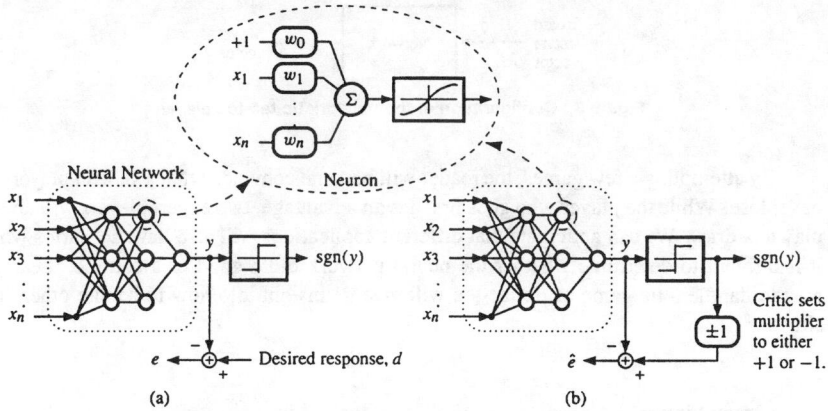

Figure 2. Multilayer neural network: Adapted (a) using backpropagation; (b) using critic.

From this point of view, it is easy to extend critic learning to more general neural-network structures. First, as before, an external observer called the critic supplies a reward/punish signal which is used to estimate a desired response. The network's error is estimated to be $\hat{e} = [\text{sgn}(y) - y]$ for rewarding or $\hat{e} = [-\text{sgn}(y) - y]$ for punishing. This error is used with the backpropagation algorithm to adapt the weights in the neural network. The resulting scheme is shown in Fig. 2(b).

When dealing with multi-output neural networks the reward/punish mechanism needs to be chosen to conform to the interpretation of the network outputs. If the outputs compete in a one-of-many scheme (where the output with the highest value is the selected output), then the error for all neurons in the output layer should be zero except for the neuron that wins the competition. That specific neuron should have its desired response set to either $+1$ or -1 by the critic. If the neurons in the output layer operate independently, then each should either be rewarded or punished based on individual performance.

One final scenario needs to be considered. It is possible that the critic is unable to judge whether the network's performance is good or bad. The performance may be "fair" or "tie." One possible method to generate a desired response when the critic sees no clear winner is described below.

APPLICATION OF CRITIC LEARNING TO PLAYING TIC-TAC-TOE
Problem Specification

Multiple-player games are problems which have an optimal strategy which may not be known a priori. Using a supervised-learning algorithm to train the network is impractical since a desired response cannot be efficiently generated at the end of each turn to function as a training signal. Knowledge of the system's performance is only available at the end of the game, after the players take a number of turns.

"Tic-tac-toe" is a simple two-player game which is played on a grid of three-by-three squares. At the start of the game, the grid is empty, and two players alternately place their mark in an empty position in the grid. One player uses an "×" to mark his or her position, and the other player uses an "o." The first player to get three of his or her marks in a row (horizontally, vertically or diagonally) wins the game. If neither player has three adjacent marks when the grid is full, the game is considered to be tied.

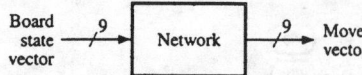

Figure 3. Configuration of the network tic-tac-toe player.

By attempting a few games, the reader will become convinced that a skillful player will never lose. While the player who goes first has an advantage, two expert players will always play to a draw. We can anticipate that different applications will also have scenarios where it is difficult to determine whether the neural network did a good or a bad job. Learning how to handle a tie game of tic-tac-toe will give us insight into how to handle other "tie" situations.

The Linear Player

The network configuration used to play tic-tac-toe is shown in Fig. 3. The nine board squares are represented in the computer as a "board state vector" of nine elements. An element is 1, 0, or −1, when the corresponding board position is "×", empty or "o", respectively. The board state vector is the network input. The output of the network consists of nine floating-point numbers. The index (into the output vector) of the maximum output is chosen to be the network's move.

Our first attempt to teach a network to play tic-tac-toe was to train a single-layer network of nine *linear* neurons—ones without sigmoidal activation functions. Each linear neuron computes a weighted sum of the input board state vector, plus an adaptable bias value: $y_i = W_i^T X$. The opponent's strategy is to automatically make a move by selecting at random one of the remaining available positions. One might envision a different scenario where two networks are trained to play against each other (or, where a single network is trained by playing against itself). We found that the network quickly learned a single winning or tying scenario which was always repeated. We wish to learn a more general strategy. This being the case, a perfectly-trained network competing against a random player will generally win (as opposed to tie), and will never loose. Either the network or the random player may make the first move of the game. Since the strategies for the odd-turn player and the even-turn player are different, two different networks were trained.

Notice that it is possible for the network to make an invalid move: the square it chooses for its move may already be marked. *If the network makes an invalid move,* then the game is terminated and that single invalid move is punished. *If the network makes no invalid moves and wins the game,* then each move made by the network is rewarded: each move of the game is replayed and the board configuration at the beginning of the turn is again used as input to the network. The output neuron corresponding to the move made by the network is rewarded, and all other neurons are unchanged. The amounts of the weight updates for all neurons in the network are computed (using backpropagation) and stored. This procedure is repeated for all of the moves that the network made in the game. Finally, the weights of the network are updated by the amounts calculated in each of the individual sub-steps. *If the network makes no invalid moves and looses the game,* all of the above steps are performed, with the exception that the output neurons corresponding to the moves made by the network are punished. In general, the adaptation constant μ is may be one value when rewarding, and a different value when punishing. Since punish events happen either when the network looses a game or when it makes an invalid move, and reward events happen only when the network wins a game, punish events tend to occur more frequently, especially at the

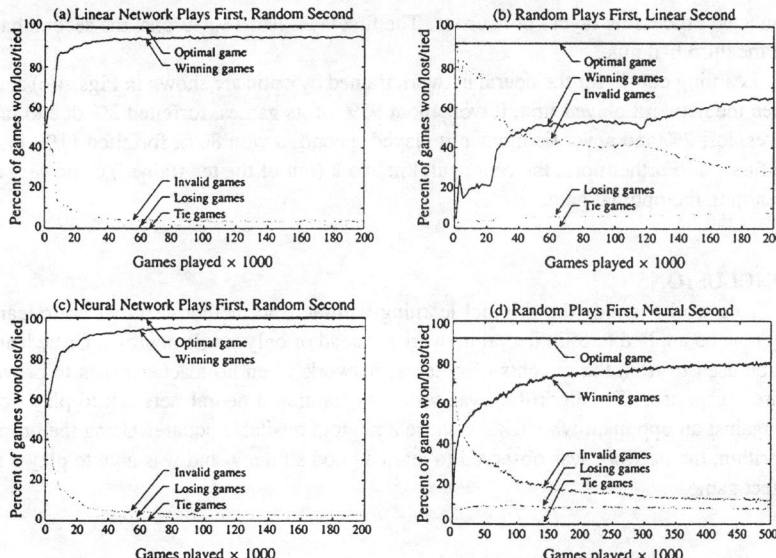

Figure 4. Learning curves: (a) and (b) Linear network; (c) and (d) Neural network.

beginning of training. To compensate, we found that the reward constant should be slightly larger than the punish constant. All that remains is to determine what should be done if the game results in a tie. Because the network played against a random player, we found that tie games were very rare. They were neither rewarded nor punished.

Learning curves resulting from simulations are shown in Figs. 4(a) and (b). Each curve portrays an ensemble average of 10 simulation runs, plotted against games played, where the initial network weights were different in each run. The solid horizontal line in each plot corresponds to the theoretical maximum average performance (too close to 100% in Fig. 4(a) to be clearly seen). In Fig. 4(a), the network was given the advantage of taking the first move of the game. In Fig. 4(b), the network played second. The curves are labeled and represent the ensemble-average percentages of games won, invalid, lost and tied.

When the network played first, it performed better than if it played second, as would be expected. When playing first, it wins about 94% of its games with the random player, forfeits about 4% due to invalid moves, loses about 2% and never ties; when playing second, it wins about 64% of its games, forfeits about 28%; loses about 8% and never ties. To give these results some meaning, simulations were done with two random players playing against each other. The random first-turn player won 58% of the games, the random second-turn player won 29%, and the players tied 13%. Clearly, the network is doing *much* better than random.

Playing against a random opponent, the optimal strategy wins 99.5% of its games and ties the rest, if it takes the *first* turn. If it takes the *second* turn, the optimal strategy wins 91.6% of its games, and to ties the rest.

The Neural-Network Player

While the linear system just described performed well, we expect that a multi-layer neural network can do better. The neural-network trained to play tic-tac-toe had three fully-connected processing layers of neurons. The first layer had 18 neurons; the second had 12; and the third had nine.

Learning curves for the neural network trained by critic are shown in Figs. 4(c) and (d). When the network played first, it won about 95% of its games, forfeited 3% due to invalid moves, lost 2% and never tied; when it played second, it won 80%, forfeited 11%, lost 8% and tied 1%. Furthermore, the best neural network (out of the ten trained) came *very* close to learning the optimal game.

CONCLUSIONS

This work extends the original learning-with-a-critic methods so that critic learning may now be applied to train neural networks instead of only single neurons. Critic learning can be used to train the weights of a neural network when no teacher exists to provide a desired response. The algorithm was tested by training a neural network to play tic-tac-toe against an opponent who always chose a random available square. Using the proposed algorithm, the network was observed to learn a good strategy, and was able to play a near-perfect game.

NOMENCLATURE

W Neuron weight vector.
ΔW Change in weight vector due to adaptation.
X Input vector to neuron.
e Network or neuron output error.
$\hat{e}$ Network or neuron output error, estimated by critic.
s Neuron internal sum (before nonlinearity).
y Neuron or network output.
μ Small positive learning constant.

ACKNOWLEDGMENTS

This work was supported in part by funding from the National Science Foundation under contract ECS-9522085, and the Electric Power Research Institute under contract WO8016-17. Thanks to Dr. Raymond T. Shen for discussions relating to similar work, and to Dr. Bernard Widrow for providing the inspiration for this work, and for reviewing initial drafts of this paper.

REFERENCES

Haykin, S. 1994. *Neural Networks: A Comprehensive Foundation*, MacMillan Publishing Company, New York.

Kohonen, T. 1982. Self-organized formation of topologically correct feature maps, *Biological Cybernetics* 43: 59–69.

Rumelhart, D. E., Hinton, G. E. and Williams, R. J. 1986. Learning internal representations by error propagation, in D. E. Rumelhart and J. L. McClelland (eds), *Parallel Distributed Processing*, Vol. 1, The MIT Press, Cambridge, MA, chapter 8.

Werbos, P. 1974. *Beyond Regression: New Tools for Prediction and Analysis in the Behavioral Sciences*, PhD thesis, Harvard University, Cambridge, MA.

Widrow, B., Gupta, N. K. and Maitra, S. 1973. Punish/reward: Learning with a critic in adaptive threshold systems, *IEEE Transactions on Systems, Man, and Cybernetics* SMC-3(5): 455–65.

Widrow, B. and Hoff, M. E. 1960. Adaptive switching circuits, *1960 IRE WESCON Convention Record*, Vol. 4, IRE, New York, pp. 96–104.

Widrow, B. and Stearns, S. D. 1985. *Adaptive Signal Processing*, Prentice-Hall, Englewood Cliffs, NJ.

STRICTLY SELF-ADAPTIVE TRAINING WITH XOR ANN

SHAMSUDDIN AHMED	JIM CROSS	S. BOUZERDOUM
DBA, CBE	School of Engineering and	School of Engineering and
U. A. E University	Mathematics	Mathematics
POB 17555, Al Ain	Edith Cowan University	Edith Cowan University
U.A.E	Perth, WA, Australia	Perth, WA, Australia
Email:Dr_SUA@Yahoo.Com		

ABSTRACT

In ANN computation the direction along which the error function converges; is conceptualized by a vector of direction defined as directional derivative. An algorithmic map is defined over the ANN error function possessing descent properties. As the training continues, the error function is reduced into a lower dimension and mapped accurately to identify exact learning rate for faster training in ANN. The algorithmic map operates over the controlled error space guided by the directional training map and determination of self-adaptive learning rate that reduces the error function at each epoch. The training as a result converges to the minimum trajectory of the error function at 76% faster rate than that of the standard back propagation method in XOR problem. The oscillation during training is avoided due to the fact that the learning rate is dynamically changing every epoch.

Key Words: Self-adaptive training, Directional derivative, ANN, variable learning rate, XOR.

INTRODUCTION

A self-adaptive training method that computes variable learning rates to train an ANN is discussed in this study. Consider an ANN error function containing m training weights. The magnitudes of weights are incremented appropriately each epoch by a self-adaptive training method. The actual error function in higher dimension is decomposed into several error functions in lower dimension (Ahmed et al., 2000). The transformed error function retains the true convex characteristics of the original error function. The training method selects learning rates for all the weight parameter say w_j, $j=1,2,.....m$, by a factor such that improvement in training is noticed. Each epoch identify m different learning rates. Once all the learning rates are computed, the ANN network weights are then updated and the error function is evaluated. The consequences of this update are noted namely: training improvements, learning behavior of the error function and the convergence rate of the error surface. The training scheme uses least square criteria (Jang et al., 1997).

To demonstrate the concept behind the self-adaptive back propagation-training scheme a XOR ANN with 2-2-1 configuration is considered. Figure 1 shows the variable learning rates per epoch. As the training converges the variable learning rates also converges to lower magnitude. As a result the training converges rapidly. The training reduces the error function value of the order 10^{-3} at the early stage of training and trained at 10^{-6} tolerance limit for accuracy. Figure 1 shows the training performance. Notice that the function convergence and the self-adaptive training parameters gradually reduce to small magnitude without oscillation. The terminal value of the vector is [0.00098, 1.00001, 0.99995, 0.0006] against the desired value [0,1,1,0].

The training method generates different learning rates during training as seen in Figure 1. It is evident from Figure 1 that the considerable reduction in error function is achieved within 20 epochs to train the XOR problem. The standard back propagation (BP) takes on average 464 epochs to train the XOR problem, while the proposed method takes on average 152 epochs.

The standard BP training method and its variant are not self-adaptive but ad-hoc in training method (Fahlman, 1988; Hinton, 1987; Jacobs, 1988; Wier, 1991; Vogl, 1986 and Kamarthi et al., 1999). The proposed method generates self-adaptive variable learning rates during the training and monotonically reduces the error function.

General references on BP neural networks include Rumelhart, McClelland (1989); Muller and Reinhardt (1990); and Hertz, Krogh and Palmer (1991). More discussions on neural networks as statistical model can be found in Cheng and Titterington (1994), Werbos (1994), Kuan and White (1994), Bishop (1995).

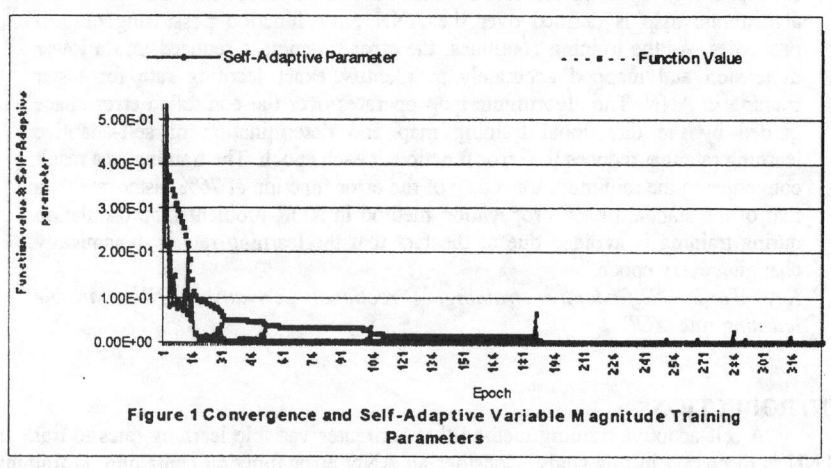

Figure 1 Convergence and Self-Adaptive Variable Magnitude Training Parameters

IDENTIFYING TRAINING DIRECTIONS

In order to describe the self-adaptive iterative training algorithm, the convergence properties of the equivalent XOR error function $f(w)$ in a 2-2-1 ANN framework with log activation function in hidden layer is defined. When an iterative algorithm is applied to an error function with an initial arbitrary weight vector w_k, at the beginning of iteration k, the algorithm generates a sequence of vectors w_{k+1}, w_{k+2},... during epoch $k+1$, $k+2$,....,..., and so on. The iterative algorithm is *globally convergent* if the sequence of vectors converges to a solution set Ω. Consider for example the following training problem, where w is defined over the dimension E^m:

minimize $f(w)$ (1)

subject to: $w \in E^m$.

Let, $\Omega \in E^m$ be the solution set, and the application of an algorithmic map, $\mathcal{B}$, generates the sequence $w_{k+1}, w_{k+2}, \ldots$, starting with weight vector w_k such that $(w_k, w_{k+1}, w_{k+2}, \ldots) \in \Omega$, then the algorithm converges globally and the *algorithmic map is closed* over Ω.

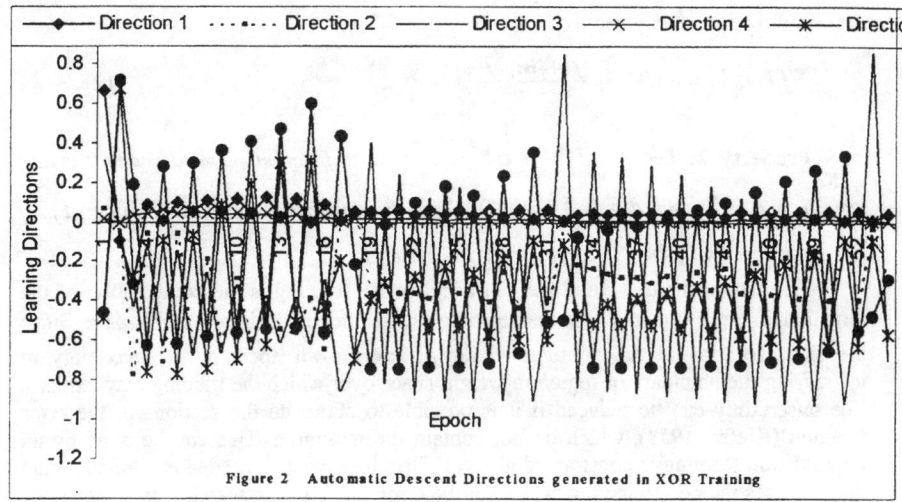

Figure 2 Automatic Descent Directions generated in XOR Training

Consider further that the XOR error function is a minimization problem, since it is possible to train such function (Haykin, 1994). Let Ω be a non-empty compact subset of E^m, and if the algorithmic map $\mathcal{B}$ generates a sequence: $\{w_k\} \in \Omega$ such that $f(w)$ decreases at each iteration while satisfying $f(w_k) > f(w_{k+1}) > f(w_{k+2}), \ldots$, then the error function $f(w)$ is a *descent function*. In ANN computation $f(w)$ is assumed to posses descent properties. It implies that it is convex in nature (Hykin, 1994). Therefore, it is possible to define a descent direction along which the error function can be trained. Following properties demonstrate the concept as to how the self-adaptive variable learning rates are generated.

Property 1: *Suppose that $f : E^m \to E^1$ and the gradient of the error function, $\nabla f(w)$, is defined then there is a directional vector d such that $\nabla f(w)^T d < 0$, and $f(w + \eta d) < f(w) : \{\eta \in (0, \delta), \delta > 0\}$, then the vector d is a descent direction of $f(w)$, where δ is assumed arbitrary positive scalar.*

A vector defined as *directional derivative* as clarified next, conceptualizes the direction along which the error function converges to minimum. Also, assume that error function is smooth and continuous. Subsequently the directional derivative is defined using the following property.

Let, $f : E^m \to E^1$, $w \in E^m$ and d is a non-zero vector satisfying $(w + \eta d) \in E^m$, $\eta > 0$ and $\eta \to 0^+$. The *directional derivative* at w along the descent direction d is given by:

$$\nabla f(w; d) = \lim_{\eta \to 0^+} \frac{f(w+\eta d) - f(w)}{\eta}. \qquad (2)$$

Property 2: *Let $f : E^m \to E^1$ is a descent function. Consider any training weight $w \in E^m$ and $d \in E^m$: $d \neq 0$. Then the directional derivative $\nabla f(w; d)$ of the error function $f(w)$ in direction d always exists.*

The central theme behind the self-adaptive BP training is the computation of the directional search vector d and the learning rate parameter η in m weight space. Since the exact location of the minimum valley is not known, there is an uncertainty in identifying the boundary or region in weight space over which the training may explore. The uncertainty can be reduced if it is possible to eliminate the sections of the error function (Kiefer, 1953), which do not contain the minimum. This can be done by an interpolation learning in constrained interval. Therefore, what is needed is a training map that explores the constrained region, L, of the error surface. The search map samples the error surface with discrete length in a given direction. The following definitions are needed to describe the interpolation-training map. For convenience define the following expression to update the ANN connection weights:

$$u_k \equiv w_{k+1} = w_k + \eta_k d_k \text{ or } u = w + \eta d. \qquad (3)$$

Now consider a training problem with ANN error function defined as:

$$\eta_k \equiv \arg\{\min f(w + \eta d)\} \qquad (4)$$

subject to : $\eta_k \in L$

in a closed interval $L = \{\eta : \eta \in E^1\}$.

Property 3: The interpolation map in restricted error space is defined as:

$A : E^m \times E^m \to E^m$ such that:

$$A(w, d) = \{u : u = (w + \eta \ d) \,|\, \eta\} \ \in L \qquad (5)$$

$$f(u) = \min f(w + \eta \ d) \qquad (6)$$

Subject to : $\eta^* \equiv \eta \in L$.

Suppose that the mapping algorithm $\mathcal{B}$ operating on $f(w)$ produces descent directions. The map is closed as set value mapping (Luenberger, 1984) such that correct learning rate is obtained during the training phase.

It is therefore now possible by any standard technique to identify η_k for each epoch of training cycle k while descending along the minimum trajectory and exploring all the m dimensions. The error function as a result monotonically converges to a minimum value. Finally the algorithm gradually converges to the acceptable minimum limit in few epochs.

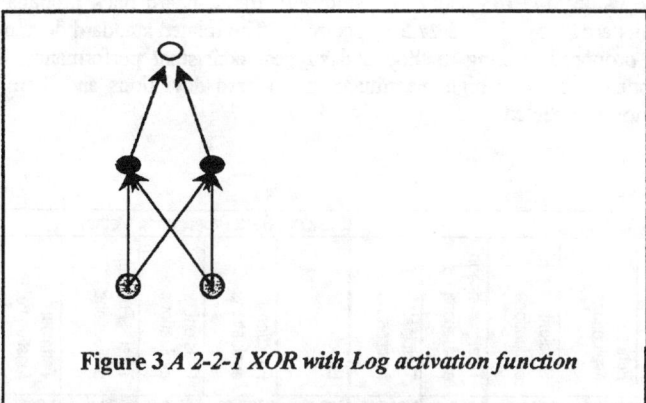

Figure 3 *A 2-2-1 XOR with Log activation function*

TRAINING

One of the most difficult computational steps in identifying the trained weights in ANN is that the ANN error function contains several local minimums within the reasonable range of w_j. The reduced ANN error function can be considered a continuous function of m training weights w_j, $j=1,2,3,...., m$, describing a hyper surface in m dimensional space (Moller, 1997). By suitable reduction of the error surface an approximate error function can be constructed with parabolic hyper surface, so that the learning weights can be found from the approximate function with initial estimate made at the beginning of training. The training initiates with random starting weight vector and few epochs are needed to converge to the minimum trajectory.

DISCUSSIONS ON THE TRAINING RESULTS

Figure 1 shows the self-adaptive parameter and the function value as the training progresses. Observe that within 30 epochs the XOR error function converged to a low magnitude rapidly. Figure 1 also indicates that the convergence is monotonic. The self-adaptive dynamic learning rates are relatively larger at the early stage of the training and gradually reduce to smaller magnitude as the training progress.

The training is continued further to observe the convergence pattern. It is seen that as the training continues the reduction in error function is relatively insignificant. The learning rate at this stage in general is very small in magnitude. Some where between epoch 181 and 191 there is a higher magnitude in learning rate compared to the previous one. It indicates that for some time there is no significant improvement in error function, but at that particular point further reduction in error function has been achieved (Ahmed

& Cross, 2000). As the training continues the improvement is small and in training cycle between 281 and 291 again there is relatively large magnitude of learning rate detected. Similarly at this training epoch the training gains considerable reduction in error function.

Ten simulation experiments are carried out to test the proposed training method against the standard BP algorithm (Rumelhart et al., 1989) with small random starting weights. The corresponding simulation results are shown in Table 1.

The average epoch, function evaluations and gradient evaluations are 151.8, 916.8 and 2601.1 respectively. The corresponding values with the standard back propagation training are 463.6, 2787.6 and 465.6 respectively. The median performance of the proposed algorithm with epoch size, function evaluations and gradient evaluations corresponds to the values 17, 108 and 206, while with the standard back propagation training these counts are 25.5, 159 and 27.5 respectively. The related standard deviations are low with the proposed training method and suggest consistent performance. The standard back propagation show high magnitude in standard deviations and therefore inconsistent behavior is expected.

Statistical Measure	Proposed Method					Standard Back Propagation Method				
	Epoch	Function evaluations	Gradient evaluations	Total Function evaluations	Terminal function value	Epoch	Function evaluations	Gradient evaluations	Total Function evaluations	Terminal function value
Mean	151.8	916.8	2601.1	3518.4	0.00859	463.6	2787.6	465.6	3253.2	0.00557
Median	17	108	206	310.5	1.8E-08	25.5	159	27.5	186.5	0.00478
Standard Deviation	420.828	2524.97	7449.30	9974.04	0.02713	951.121	5706.73	951.121	6657.85	0.00489
Range	1343	8058	23718	31770	0.0858	2667	16002	2667	18669	0.01487
Minimum	6	42	78	126	1.3E-10	20	126	22	148	0.00013
Maximum	1349	8100	23796	31896	0.0858	2687	16128	2689	18817	0.015

Table 1 Comparison with standard back propagation method (2-2-1:ANN XOR)

CONCLUSIONS

It is demonstrated in this study how an ANN error function could be trained dynamically with variable learning rates. A 2-2-1 XOR ANN training function is used to demonstrate the training results. A closed algorithmic map operating over a descent error function determines the variable learning rate. The ANN training explores in all dimensions of error space. These are important features of this training method. The results indicate that as the training progress the descent direction also changes. The learning rates do not have fixed values. The closed algorithmic map identifies the descent directions as well the variable learning rate. The shape and properties of the error function are also changing, as the error function is gradually reduced each iteration. The proposed training algorithm account the constant changes in the geometry of the error function. Some times the error function is flat and the reduction in error function is insignificant, but the proposed algorithm does not prematurely terminate the training. At some training epoch near a flat surface the algorithm identifies a suitable learning rate so that the further reduction in error function is possible, and the training escapes the local minimum.

The proposed self-adaptive training method trained the XOR problem within 20 epochs and produces a vector [0.00098, 1.00001, 0.99995, 0.0006] against the true value [0, 1, 1 0].

REFERENCES

Ahmed, S., and Cross, J., (2000)."Convergence in Artificial Neural Network without Learning Parameter", Second *International Computer Science Conventions on Neural Computations, 2000*, Berlin, Germany

Ahmed, S., Cross, J., and Bouzerdoum, A. (2000). "Performance Analysis of a new multi-directional training algorithm for Feed-Forward Neural Network", *World Neural Network Journal*, 2000.

Bishop, Christopher M. (1995). *"Neural Networks for Pattern Recognition"*. Oxford, UK: Clarendon.

Cheng, Bing and D. M. Titterington. (1994). "Neural Networks: A Review From a Statistical Per-spective." *Statistical Science* 9 (1): 2-54.

Fahlman, S.E., (1988). "Faster-learning variations on back-propagation: an empirical study", in Touretzky, D., Hinton, G. and Sejnowski, T. (Eds), *Proceedings of the Connectionist Models Summer School*, Morgan Kaufmann, San Mateo, 38-51.

Haykin, Simon., (1994). *"Neural Networks: A Comprehensive Foundation".*, Macmillan College Publishing, 1994.

Hertz, John, Anders Krogh, and Richard G. Palmer. (1991). *"Introduction to the Theory of Neural Computation"*. Reading, MA: Addison-Wesley

Hinton, G. E., (1987). "Learning Translation Invariant Recognition in Massively Parallel Networks". In De Bakker, J.W., Nijman, A.J., and Treleaven, P.C. (Eds.), *Proceedings PARLE Conference on Parallel Architectures and Languages Europe*, 1-13. Berlin: Springer-Verlag.

Jacobs, R.A. (1988). "Increased Rate of Convergence Through Learning Rate Adaptation". *Neural Networks, 1*, 295-307.

Jang, J.-S.R., Sun, C.-T., Mizutani, E., (1997). *"Neuro-Fuzzy and Soft Computing"*, Prentice Hall International, Inc, NJ, USA.

Kamarthi, S.V., and Pittner, S., (1999). "Accelerating Neural Network Training Using Weight Extrapolations", *Neural Networks, 12*, 1285-1299.

Kiefer, J., (1953). "Sequential Minimax search for a Maximum". *Proceedings of the American mathematical Society, 4*, 502-506.

Kuan, Chung-Ming and Halbert White. (1994). "Artificial Neural Networks: An Econometric Per-spective." *Econometric Reviews* 13 (1): 1-91.

Luenberger, D. G., (1984). *"Introduction to Linear and Nonlinear Programming".*, 2nd ed. Addison-Wesley, Reading, Mass., USA.

Moller, Martin., (1997). "Efficient training of feed forward neural networks", in Brown Anthony, editor, *Neural Network: Analysis, Architectures and Applications*,: Institute of Physics Publishing, Bristol and Philadelphia.

Muller, Berndt and Joachim Reinhardt. (1990). *"Neural Networks": An Introduction*. Berlin: Springer.

Rumelhart, D.E. and McClelland, J.L. (1989), *"Parallel Distributed Processing: Explorations in the Microstructure of Cognition"*, Vol. I, MIT Press, Cambridge.

Vogl, T. P., Mangis, J.K., Rigler, A.K., Zink, W.T., and Alkon, D.L., (1988). "Accelerating the Convergence of the Back-Propagation Method." *Biological Cybernetics, 59*, 257-263.

Weir, M.K., (1991). "A Method for Self-Determination of Adaptive Learning Rates in Back Propagation". *Neural Networks, 4*, 371-379.

Werbos, Paul. (1994). *"The Roots of Back propagation: From Ordered Derivatives to Neural Net-works and Political Forecasting"*. New York: John Wiley.

NEURAL-COMPLEXITY MEASURES OF MACHINE LEARNING ALGORITHMS

KHALID AL-MASHOUQ
Electrical Engineering Department
King Saud University
Riyadh, Saudi Arabia

ZAYGHAM NAWAZ
Advanced Communications and
Electronics Co., Ltd.
Riyadh, Saudi Arabia

ABSTRACT
In this paper, we consider methods to characterize data sets available for the benchmarking of machine learning algorithms. We propose five characterization parameters, which will be useful if incorporated in the description of the data sets. Some of these parameters are obtained from known neural network algorithms such as the minimum size of a multilayer neural network that attains a prescribed correct classification rate. Other parameters are new, such as the "boundary-crossing rate" and needed branches of a recursive branching network. We evaluate the proposed parameters for five commonly used data sets. In addition, we point out the relation between these apparently different parameters and their use in directing the choice of a neural network model.

INTRODUCTION
In recent years, various research centers around the world have compiled benchmarking data sets for the use in machine learning and in neural networks [1]. These data sets are collected from real world problems. To evaluate a new neural network model, it is a common practice to select a few of these data sets for benchmarking tests. It is well known that there is no single model can solve efficiently all class of problems. Therefore, we need to specify classes. All "similar" problems are grouped into one class. Unfortunately, there is no systematic classification procedure to classify the benchmarking data sets. The existing data sets only have simple characterization parameters such as number of instances, mean, and variance.

In this paper, we propose five characterization parameters to be used in classifying data sets. These measures are designed to meet the needs of neural network model selection. Some of these parameters are directly related to the single layer, multilayer and radial basis networks. The other parameters are related to the clustering of data points and to their "recursive" linear separability. Then we obtain these parameters for several data sets from UCI repository [1]. We show how these parameters can be used to have better insight into the data sets. Moreover, they can be used to guide the process of model selection of neural networks.

RECURSIVE BRANCHING NETWORK

The recursive branching network (RBN) [2] will be used to extract one of the parameters. Now we review the basics of RBN. Consider a perceptron with N input nodes and a single output node. This perceptron is required to classify M patterns $a_1, a_2, \ldots, a_M$, where $a_i \in R^N$. Each pattern is labeled by either +1 or -1. These patterns can be represented by a matrix of size $M \times N$, let this matrix be denoted by A. The labels can also be arranged as vector, D, of dimension $M \times 1$. The ith component of D is the corresponding label of a_i.

The problem is then to find a weight vector w, which yields an approximate solution to the equation *sign (Aw)=D*. For randomly selected patterns, it is shown in [3] that when $M<2N$ and N is large, the network can classify all the M patterns correctly with high probability. However, when the number of patterns exceeds $2N$ then with high probability the 1-layer net will fail to classify all the patterns. Note that if the patterns are clustered (not random), then the 1-layer net will absorb more that $2N$ patterns.

The RBN is used to reduce the number of patterns to be absorbed by the 1-layer net in a systemic way. Let us assume a 2-class problem, where A is the training matrix and the D is the label vector. A single perceptron is used to classify all the patterns. If all the patterns are correctly classified, then the goal is met. However, if classification errors occur then it means that the number of patterns exceeds the capability of the network. To reduce the number of patterns that need to be absorbed by the single layer net, we add two more perceptron branches in parallel. Thus, we have a 1-layer net with three output nodes. The training matrix, which contains all the patterns, is split into three smaller matrices, A_a, A_b and A_c, with (almost) equal number of patterns. Using these matrices we form three matrices A_1, A_2 and A_3 such that any input pattern occurs in at least two of these matrices, where

$$A_1=[A_a;A_b]$$
$$A_2=[A_a;A_c]$$
$$A_3=[A_b;A_c]$$

The semicolon means row augmentation. We form three label vectors D_1, D_2 and D_3, which contain the labels corresponding to the patterns in A_1, A_2 and A_3 respectively.

Now we have three branches with three weight vectors w_1, w_2 and w_3 and three pattern/label sets (A_i, D_i), $i=1,2,3$. Each set is to be absorbed in the corresponding weight vectors. We may use an algorithm, such as the perceptron algorithm, LMS algorithm or even a closed-form solution [4], to find the weights. Majority voting is then used to decide the output label. If two nodes match the correct label, the decision is correct. This method can increase the capacity of a 1-layer net by one third for randomly selected patterns. This is because now each branch is asked to absorb two-third of the total number of patterns.

If any of the new 'root' nodes, say the ith node, still fails to classify all or of its $2/3$ M patterns, we repeat the same procedure described above; we add two more nodes to form a group of three branches which share the same input. Then, we generate in the same way, three pattern matrices $A_{i,1}, A_{i,2}$ and $A_{i,3}$ each of which contains $(2/3)^2 M$ patterns. The selected algorithm is used to find the weight vectors for each branch. The majority voting is again used for each node to decide the output of this group. This output replaces the output of the root node. During training, we repeat this recursively for any node, which cannot absorb its pattern matrix, until we reach satisfactory performance or a prescribed complexity limit.

CHARACTERIZATION PARAMETERS

Here we restrict our attention to binary classification problems. Nonetheless, we can modify our parameters to accept multiclass problems. We propose the following parameters.

PCC_{LC}: Percentage of correct classification using a linear classifier (LC).
MNH_{MLP}: Minimum number of hidden nodes in a 2-layer multilayer perceptron (MLP) network.
MNH_{RB}: Minimum number of nodes in a radial basis (RB) network.
MNB_{RBN}: Minimum number of branches in a recursive branching network.
BXR: Boundary-crossing rate.

All these parameters are intended for the training data. The generalization issue is not within the scope of this work. The first parameter, PCC, measures the linear separability of the data set. If a data set is found to be linearly separable, there is no need for a multilayer network or recursive branching network. For the second parameter, MNH_{MLP}, we find the smallest size 2-layer MLP network that can achieve a prescribed correct classification rate, say 85%. The third parameter, MNH_{RB}, is similar to the second one but for a radial basis network. The fourth parameter, MNB_{RBN}, uses the recursive branching algorithm as described above. We "recursively" repeat the same procedure for any node that fails to correctly classify its instances. Once we reach a prescribed correct classification rate, say 85%, we terminated the branching process. We count the total number of branches. The fifth parameter, BXR, is computed by moving from one instance in the data set to the nearest one (in terms of Euclidean distance). If we therefore also move from one class to the other one, we call that a boundary-crossing event. We count the average number of times we cross boundaries. We call that boundary-crossing rate. This parameter measures the clustering of points and the ease of their separation.

We compute these parameters for some data sets from UCI repository [1] and correlate them with each others. Then we point out the use of these measures in any learning algorithm.

DATA SETS

Simulation results for five data sets taken from [1] are presented here. Multiclass data sets were converted to two-class data sets. Summary of these data sets is shown in Table 1. In the following, we briefly describe these data sets.

Table 1: Summary of the five data sets

Data Set	Number of features	Number of instances
Liver	6	345
Sonar	30	208
Pima-Indian Diabetes	8	576
Ionosphere	34	200
Vowel-2	10	528

LIVER

This is a two-class data set. The purpose of this data is to predict whether an individual will have liver disorder or not, taking into view individuals alcohol consumption. There are six features, which are related to blood measurements. The number of instances is 345.

SONAR

This data set contains 111 patterns obtained by bouncing sonar signals off a metal cylinder at various angles and under various conditions. It also contains 97 patterns obtained from rocks under similar conditions. The transmitted sonar signal is a frequency-modulated chirp, rising in frequency. The data set contains signals obtained from a variety of different aspect angles, spanning 90 degrees for the cylinder and 180 degrees for the rock. The label associated with each record contains the letter "R" if the object is a rock and "M" if it is a mine (metal cylinder).

PIMA-INDIAN DIABETES (PID)

This is a two-class data set. The purpose of this data set is to predict whether an individual is diabetes positive or not based upon the measurements of the plasma glucose level of the individual. The number of features is eight, and there are 576 instances.

IONOSPHERE

This is a two-class problem. The measurements in the data set are the radar returns from the ionosphere. The purpose is to show the presence or absences of some type of 'structure in the ionosphere'. The numbers of instances used for training were 200.

VOWEL-2

This data set contains features extracted from speech segments of several speakers. It is required to determine the uttered vowel among 10 possible vowels, based on these features. We modified this problem to a 2-class problem by deciding only on the gender of the speaker. This data set contains 10 features and the number of instances is 528.

RESULTS

Here we describe our procedure in obtaining the characterization parameters for the five data sets. The first parameter, PCC_{LC}, is obtained by finding the weight vector, w, of a 1-layer net that minimizes the error, e, where

$$e^2 = \|Aw - D\|^2 \quad (1)$$

If $M > N$ and the columns of A are linearly independent, we have the following solution [2,4]

$$w = (A^T A)^{-1} A^T D \quad (2)$$

Equation (2) is used to find w for each of the five data sets. Then we computed the percentage of correct classification, PCC_{LC}, for the training data, which is listed in Table 2. For the Sonar and Ionosphere data sets, PCC_{LC} exceeds the prescribed level, which is 85%. Thus, we don't need to use a multilayer net or use recursive branching. The Liver data set has the lowest PCC_{LC}, while Vowel-2 is very close to our target of 85%. The next parameter, MNH_{MLP}, is obtained training a 2-layer net. We start with a small number of hidden nodes and increase it until we meet the prescribed performance level. As mentioned above, we need to find MNH_{MLP} only for Liver, PID and Vowel-2 data sets. For the other two sets we put $MNH_{MLP} = 0$ since a 1-layer will attain the required performance.

The third parameter is MNH_{RB}, which is obtained by trying the minimum number of nodes in a radial basis network that attains 85% correct classification rate. The Liver and PID data sets required 62 and 59 nodes, respectively. The other three data sets required significantly less number of nodes. The recursive branching algorithm is then applied to Liver, PID and Vowel-2 to obtain MNB_{RBN}. We use (2) to find the weight vector of each branch. Again, the Liver and PID required a large number of branching, which is 133 and 231, respectively. The Vowel-2 needs only 9 branches. The last parameter BXR is also obtained for the five data sets.

All the five parameters are shown in Table 2. From this table, one can see that there are "easy" data sets such as Sonar and Ionosphere. The PCC_{LC} has a strong correlation with MNH_{MLP} and MNH_{RB}. However, it shows less correlation with BXR. We see that BXR as a complement to PCC_{LC}. The BXR parameter gives and indication about the clustering of points. Clustered points (small BXR) do not necessarily mean linear separability. However, It indicates the ease of separation with an appropriate surface such as the MLP. This viewpoint is clearly demonstrated in Vowel-2 where BXR is very low. Nonetheless, Vowel-2 in not linearly separable. However, the nonlinear networks (MLP and RB) were able to reach good performance with small number of nodes.

Table 2: Summary of characterization parameters.

Data Set	PCC$_{LC}$	MNH$_{MLP}$	MNH$_{RB}$	MNB$_{RBN}$	BXR
Liver	68.4%	10	62	133	0.37
Sonar	90.4%	0	10	1	0.23
PID	74.8%	10	59	231	0.33
Ionosphere	87.5%	0	3	1	0.1
Vowel-2	83.7%	1	9	9	0.04

CONCLUSIONS

The characterization of benchmarking data sets is very essential in interpreting the good performance of a neural network with some data sets while failing with others. The currently available descriptions are based on simple statistics, which are of limited use in analyzing results. In this paper, we suggested five parameters to characterize the data sets. These parameters range from simple ones like the correct classification percentage of a linear classifier to more elaborate ones like the minimum number of branching in RBN. Although simple and compact, these parameters extract meaningful information about the data set. This information can be utilized in assessing the choice of a neural network model and its complexity. We demonstrated this by experimental work with five common data sets. We found that the boundary-crossing rate is highly correlated with the network size. If the CCP is large, the classification can be improved further by RBN. The number of branches in RBN has little correlation with the size of multilayer network. One may suggest other parameters, which when used along with the proposed parameters may give even a clearer picture of the data set and the appropriate network structure for an efficient solution.

REFERENCES

[1] P. Murphy and D. Aha, "UCI repository of machine learning databases [machine-readable data] repository", Tech. rep., University of California, Irvine, 1994.
[2] Khalid Al-Mashouq, "Recursive branching net", Proceedings of the IEEE Conference on Systems, Man and Cybernetics, Orlando, Florida, October 1997.
[3] T. M. Cover, "Geometrical and statistical properties of linear threshold devices", Ph.D. thesis, Tech. Rep. 6107-1, Stanford Electron. Labs, Stanford, CA.
[4] G. Strang, *Linear Algebra and Its Applications*, Academic Press, Inc., 1976.

RULE EXTRACTION BASED ON HIDDEN NEURAL ACTIVATION INTERVAL PROJECTION ON EACH DIMENSIONAL AXIS

WIPHADA WETTAYAPRASIT
Department of Computer Science
Faculty of Science
Prince of Songkla University
Songkla Thailand
E-mail: wwettayaprasit@yahoo.com

CHIDCHANOK LURSINSAP
Department of Mathematics
Faculty of Science
Chulalongkorn University
Bangkok Thailand
E-mail: lchidcha@chula.ac.th

ABSTRACT
Generally and accurately, the rules extracted from a trained MLP neural network must be meaningful, explainable, and, yet, understandable. Most current rule extraction techniques capture rules in forms of mathematical functions derived directly from the activation value of each neuron. This makes it difficult to explain the roles of each considered factor. In this paper, we propose a new extraction technique for numeric data based on the projection of the hidden activation intervals on the principle component axis. A MLP neural network is trained to separate each class one by one. Then, the rule-based intervals are derived from the graph between the activation value from each hidden neuron and the value of each factor of each class. Finally, the rule-based interval of the factor is extracted by using the overlapped activation interval. We tested this approach with Iris data and obtained the same accuracy as that from the neural network. In addition, by carefully considering the activation projection graph, the ambiguity or noisy data can be obviously detected and those redundant hidden neurons are possibly pruned. In data mining application, the activation projection can reveal which factor is essential or non-essential. The number of rules extracted by our approach is always equal to the number of classes.

INTRODUCTION

Neural Network is an autonomous machine learning-based knowledge discovery paradigm for data mining. This paradigm consists of a massive number of neurons with a very high degree of interconnectivity and provides a general as well as practical method for learning real-valued and discrete-valued from examples. The disadvantage of using a neural network in Knowledge Discovery in Database (KDD) is incomprehensible for user because of the distributed, low level representation of interconnection weights. That is a neural network returns the predicted class but it cannot provide a comprehensible explanation about why that class was chosen.

Knowledge Acquisition using neural networks, called rule extraction, is important because it is a key to the solution to one of the bottlenecks in artificial intelligence. Feedforward Neural Networks (FNN) have been successfully used as a tool for classification in a variety of real-world applications. The development in algorithm which extracts rules from trained neural networks is a very useful feature to the network classification process. These algorithms can produce symbolic rules and preserve the high accuracy of the network from which they are extracted. The set of "If-Then" rules provides a new knowledge for particular domain problem such as medical application.

In this paper, we propose a new *Rule Extraction Activation Projection (REAP)* technique for continuous/numeric database on each dimensional axis. A Feedforward Multi Layer Perceptron (MLP) neural network is trained to separate each class one by one. Then, the rule-based intervals are derived from the graph between the activation value from each hidden neuron and the value of each factor of each class. Finally, the rule-based interval of the factor is extracted by using the overlapped activation interval. We tested this approach with Iris data and obtained the same accuracy as that from the neural network. The number of rules extracted by our approach is always equal to the number of classes. In addition, by considering the activation projection graph, the ambiguity data can be obviously detected and those redundant hidden neurons are possibly pruned. In data mining application, attribute selection is one of the approaches to speed up data mining. The activation projection algorithm presented in this paper can reveal which factor is essential or non-essential.

This paper is organized into five sections. Section 2 reviews the rule extraction techniques. Section 3 describes the Rule Extraction Activation Projection (REAP) approach. The experimental results are given in Section 4. The last section concludes the papers.

RULE EXTRACTION REVIEWS

Extracting explicit rules using neural networks is quite a fascinating task, because it has a potential applicability in knowledge acquisition in artificial intelligence. This is, however, very hard due to the difficulty in interpreting hidden neurons. We have seen more and more research works of rule extraction algorithms in the literature. Gupta, Park and Lam (1999) have reviewed a classification framework for rule extraction algorithms. There are two general approaches to extract symbolic knowledge from a trained neural network. The first approach is the analytical approach which extracts rules by directly interpreting the strengths of connection weights in the trained network (Setiono, 2000). This approach is a nonsearch based or an *open-box approach*. The second approach is *the generate-and-test* approach which only the input/output behaviors are observed. It implies that the rule extraction requires a high computation complexity. The searching is a search based or *black-box approach*. Taha and Ghosh (1999) have discussed some representative rule extraction techniques on Link Rule Extraction (LRE), Black-box Rule Extraction (BRE), extracting fuzzy rules from artificial neural networks and extracting rule from recurrent network.

RULE EXTRACTION ACTIVATION PROJECTION (REAP) APPROACH

In this paper we propose an approach to data mining in which explicit rules are extracted from trained feedforward neural networks for the purpose of knowledge discovering in a manner of understandable by decision makers. The neural network module has three layers (number of input units : number of hidden units : number of output units) fully connected architecture with a continuous/numeric input data. The MLP network is trained on a set of training data using the backpropagation learning rule. Then a single hidden layer of neurons partitions the feature space into convex intersections of halfspace determined by neuron hyperplanes, when an output layer of neuron is required to join combinations of those convex region into nonconvex classes in the feature space. The analytical approach which extracts rules by directly interpreting the strengths of connection weights in the trained network or an *open-box approach* is used as a rule extraction module. The rule extraction module extracts the "If-Then" rule from the weights of the trained neural network.

How A Feedforward Neural Network Classifies

A pattern is assigned as the quantitative description of an object. The goal of pattern classification is to specify a physical object to one of the prespecified categories. Let $X = \{p_1, p_2, ..., p_n\}$, where $p_i = [x_1 \ x_2 \ ... \ x_m]^T$ be a set of given patterns to be classified into C category. The category of each pattern p_i is predetermined prior to the classification process. Generally, a neuron linearly classifies a set of patterns by computing the activation values and passes these values to a threshold function. The activation value is defined as the dot product of the input pattern p_i and its weight vector w_j. For example, the activation value h_j with respect to pattern p_1 of neuron j is computed by

$$h_j(x) = w_{j1}x_1 + w_{j2}x_2 + b_j \qquad (1)$$

where w_{ji} is weight link i of neuron j, b_j is the bias of neuron j.

It is obvious that equation (1) acts as a separating line in a two dimensional space. Similarly, a neuron can non-linearly separate a set of patterns by using a nonlinear activation value. The use of either linear or nonlinear activation depends upon the applications.

Analytic Geometry in Eucledean Space in Cartesian Coordinates

The rule extraction concept of neuron j is based on the observation of the activation value of input vector p_i and the separating hyperplane H. This activation values is computed from dot products between the input vector p_i and the weight vector w_j of neuron j. Hyperplane H dichotomously separates the input vectors into class A and class B. Without loss of generality, we define class A as a class containing input vectors whose dot products with w_j are greater than or equal to zero and class B as a class whose dot products of input vectors with w_j are less than zero. An example of this is illustrated in Figure 1. In this figure, all vectors denoted by asterik are in class A since they are above the separating line. But all vectors denoted by boxes are in class B. To meaningfully relate the interval of each element x_k with each class, we individually consider the activation value of each input vector p_i and the value of each x_k. This can be easily achieved by plotting the activation value of each input vector p_i against the values of x_k as shown in Figure 2 and Figure 3. Here, there are five input vectors in class A and five input vectors in class B. Let p_1, p_2, p_3, p_4, p_5 be input vectors in class A and $p_6, p_7, p_8, p_9, p_{10}$ be input vectors in class B with the following coordinates: $p_1 = [1\ 4]^T$, $p_2 = [2\ 6]^T$, $p_3 = [4\ 5]^T$, $p_4 = [4\ 8]^T$, $p_5 = [5\ 7]^T$, $p_6 = [2\ 1]^T$, $p_7 = [3\ 2]^T$, $p_8 = [4\ 2]^T$, $p_9 = [5\ 1]^T$ and $p_{10} = [6\ 3]^T$. The weight vector w_j is equal to $[-1\ 1]^T$. The activation values of all input vectors, p_1 to p_{10}, are as follows $\{3, 4, 1, 4, 2, -1, -1, -2, -4, -3\}$. Suppose we are interested in extracting the condition defined by the second element of an input vector, namely x_2. The activation values are plotted against the values of x_2. All the positive activation values (class A) are laid above the zero line and all the negative activation values (class B) are laid beneath the zero line. There is no mixture of both positive and negative activation values for a particular class as shown in Figure 2. This implies that there exist intervals of x_2 that absolutely define input vectors in class A and those in class B. The intervals, in this case, are [4,8] for class A and [1,3] for class B. In case of x_1, we can see that there are mixtures of positive and negative activation values in both classes A and B as illustrated in Figure 3. Hence, it is impossible to define intervals of x_1 for classes A and B. Based on this observation, we summarize our first extraction rules as follows.

Theorem 1. Let $[l_i, u_i]$ be an interval with respect to x_i, where l_i is the lower bound and u_i is the upper bound. Interval $[l_i, u_i]$ defines class A if the activation values of all input vectors having x_i in $[l_i, u_i]$ are positive and class B if the activation values of all input vectors having x_i in $[l_i, u_i]$ are negative.

Theorem 2. If class A is defined by a set $\{[l_i, u_i]_j \mid 1\ j\ a\}$ with respect to x_i then the conditions of an input vector to be in class A can be written as $[l_i, u_i]_1 + [l_i, u_i]_2 + ... + [l_i, u_i]_a$. The notation + denotes the logical OR.

Theorem 3. Let $\{[l_i, u_i]_1, [l_i, u_i]_2, ..., [l_i, u_i]_{aj}\}$ be a set of size aj of intervals of element x_i. If class A can be defined by a set of intervals of elements $\{x_i \mid 1\ i\ n\}$ then the conditions of an input

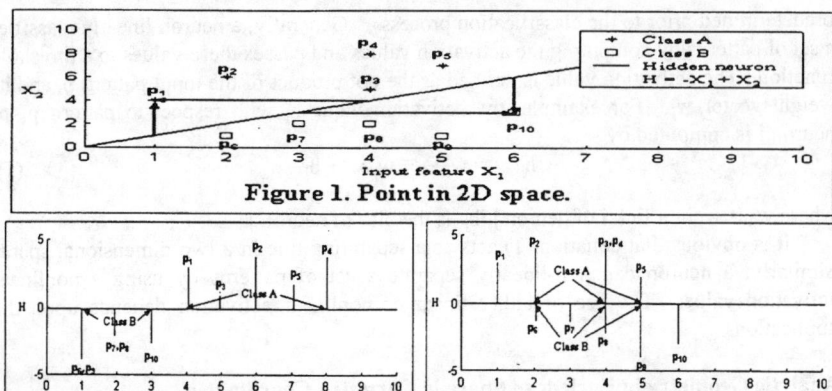

Figure 1. Point in 2D space.

Figure 2. Point projection with Hidden neuron H and Input.

Figure 3. Point projection with Hidden neuron H and Input.

vector to be in class A can be written as $([l_1,u_1]_1 + [l_1,u_1]_2 + ... + [l_1,u_1]_{a1}) * ([l_2,u_2]_1 + [l_2,u_2]_2 + ... + [l_2,u_2]_{a2}) * .... * ([l_n,u_n]_1 + [l_n,u_n]_2 + ... + [l_n,u_n]_{an})$ where * denotes logical AND and + denotes logical OR.

Rule Extraction Activation Projection (REAP) Algorithm
Unambiguous Activation Value With Respect To x_i

We consider the case that the activation values with respect to a particular value of an element x_i for every pattern $\{p_j\}$ is either positive or negative. This activation value will be called unambiguous value. Otherwise, we call it ambiguous value.

The rule extraction consists of two sequential processes, i.e. neural network training process and class interval extraction of each element. The class interval extraction is based on Theorems 1, 2, and 3. The detail of the training process is given in Algorithm 1. Here, each element of each input vector is not limited only to binary values as those in Sentiono (2000). Unlike the KBANN, no domain knowledge is required in this case. A standard backpropagation neural network with one hidden layer is employed. An output value is either 0 or 1. The detail of rule extraction is given in Algorithm 2.

Algorithm 1 (Neural Training)
1. Define a 3-layer neural network.
2. Repeat
3. Feed an input vector to the network.
4. Adjust weight of each link using the backpropagation rules.
5. Until error < ε or number of iterations < N

Algorithm 2 (Rule Extraction With Unambiguous Values)
1. For each input vector, compute the activation value of each hidden neuron.
2. Consider the activation values obtained from step 1 with respect to each element x_i to find the set $\{[l_i,u_i]\}$.
3. Apply Theorems 1, 2, and 3 to extract rules from the set $\{[l_i,u_i]\}$, for $1 \leq i \leq n$.

The number of hidden neurons can be reduced by pruning those neurons that give all positive values or all negative values to all input vectors regardless of their classes. This pruning eliminates the non-separating neurons.

Ambiguous Values With Respect To x_i

The rules extracted from Algorithms 1 and 2 work well if all activation values are unambiguous. In reality, this may not be appropriate since there must be some input vectors $\mathbf{p}_j$ that their activation values with respect to a particular value of an element x_i are ambiguous. These input vectors $\mathbf{p}_j$ will be left out from rule checking process because their conditions will not satisfy any rules. To overcome this problem, these ambiguous values must be considered during the rule extraction process. We call an interval $[l_i, u_i]$ of x_i ambiguous interval if there are ambiguous values. The ambiguous values may occur because of some noisy data. In this situation some of these ambiguous values can be accepted as a part of correct data.

Algorithm 3 (Rule Extraction With Ambiguous Values)
1. For each input vector p_i, compute the activation value of each hidden neuron.
2. Find all intervals $[l_i, u_i]$ which are unambiguous and ambiguous, for all element x_i
3. Let M be the maximum number of accepted ambiguous values
4. $J = 2, I = 1, K = 0$
5. For each x_i
6. repeat
7. $\quad [l_i, u_i]_I = [l_i, u_i]_I \cup [l_i, u_i]_J$ and update l_i and u_i
8. $\quad$ If $[l_i, u_i]_J$ is ambiguous, then $K = K + 1$
9. $\quad J = J+1$
10. until $K > M$
11. $I = I+1$
12. $J = I+1$
13. end For

Eliminating Element x_i

Some element x_i can be omitted from forming a rule by considering the number of input vectors and their activation values. With respect to x_i, if the difference between the number of input vectors with positive activation values and the number of input vectors with negative values is greater than φ then x_i is omitted from forming any rule.

EXPERIMENTAL RESULTS

The experimental results consisted of rule extraction from well known Iris data set. The Iris data set consists of three classes of flowers with 50 patterns each, one class (Sentosa) is linearly separable while the other two (Versicolor and Virginica) are not. The data consist of continuous values and have a feature dimensionality of four. The four inputs correspond to the plant features such as the sepal length (x_1), sepal width (x_2), petal length (x_3) and petal width (x_4).

A 4:4:1 network structure four input units, four hidden nodes and one output node is trained by Stuttgurt Neural Network Simulator (SNNS) software. The data set is divided into two groups of training and tested sets. The training set contains 120 input features (40 input features from each class) randomly selected and the tested set contains 30 input features (10 input features from each class).

The experiment of Sentosa network structure using REAP algorithm shows that Sentosa is linearly separable. The results of rule extraction showed 100% classification accuracy on the tested set and 100% from trained neural network as well. The Versicolor network structure showed 100% accuracy on the tested set and 100% from trained neural network with eight data out of 150 were classified in the wrong group. The Virginica network structure showed 100% accuracy on the tested set and 99% from trained neural network with eighteen data out of 150 were classified in the wrong group. The average classification performance with all three classes was 100% in the tested set and 94.3% for all 150 input features. The rule extracted from all three network structures is displayed in Figure 4.

Let Certainty Factor (CF) is a continuous value varies from 0 to 1 and the CF value equals to the classification accuracy of the tested set. The CF rule base shown in Figure 5 can represent the certainty of the new input vector to be in the class from its CF

Rule 1: If $(1.0 \le x_3 \le 1.9)$ OR $(0.1 \le x_4 \le 0.6)$ Then iris_sentosa
Rule 2: If $(3.0 \le x_3 \le 5.1)$ AND $(1.0 \le x_4 \le 1.8)$ Then iris_versicolor
Rule 3: If $(4.5 \le x_3 \le 6.9)$ AND $(1.4 \le x_4 \le 2.5)$ Then iris_virginica

Figure 4. The Extracted Rule Base from Iris data set

Rule 1:   If $(1.0 \le x_3 \le 1.9)$ OR $(0.1 \le x_4 \le 0.6)$ Then iris_sentosa (CF=1)
Rule 2(a): If $(3.0 \le x_3 \le 5.1)$ AND $(1.0 \le x_4 \le 1.8)$ Then iris_versicolor (CF=0.95)
Rule 2(b): If $(3.0 \le x_3 < 4.5)$ AND $(1.0 \le x_4 < 1.4)$ Then iris_versicolor (CF=1)
Rule 3(a): If $(4.5 \le x_3 \le 6.9)$ AND $(1.4 \le x_4 \le 2.5)$ Then iris_virginica (CF=0.88)
Rule 3(b): If $(5.1 < x_3 \le 6.9)$ AND $(1.8 < x_4 \le 2.5)$ Then iris_virginica (CF=1)

Figure 5. The CF Rule Base from Iris data set

values. This means that rule with CF value equals to 1 is 100% certain to be in that class. We can observe that the premise of the CF equals to one is a subrange of the rule extracted from REAP algorithm with no ambiguous with other classes.

We can conclude from the Iris data set using REAP algorithm that the sepal length (x_1) and sepal width (x_2) can be eliminated as the input feature extraction. Since the rule only shows the petal length (x_3) and petal width (x_4). The Sentosa network structure can be pruned to 4:1:1. While Versicolor network structure can be pruned to 4:3:1 and 4:2:1 for Virginica network structure.

CONCLUSIONS

In this paper, we describe REAP algorithm for a MLP neural network based data mining system which can work with continuous/numeric input data and has no limitation on the input value form such as take only binary values (Ishikawa, 2000). The result of REAP algorithm can be in the form of crisp rule base or certainty factor rule base projected on the input feature value. These make it more understandable and meaningful to the user. Since it does not in the form of Mathematics Equation like some other rule extraction technique does (Hruschka and Ebecken, 2000). The advantage of REAP algorithm is that the number of rules equals to the number of classes. The ambiguity data can be obviously detected and those redundant neurons are possibly pruned to get the less complex network structure and less computation time. The experimental results from the tested set show the same classification accuracy as provided by the neural network. In data mining applications, the symbolic extraction from a neural network can be used to highlight previously unknown relationship in the data to obtain the Intelligence Expert System (IES).

REFERENCES

Gupta A., Park D., and Lam S. M., 1999, "Generalized Analytic Rule Extraction for Feedforward Neural Networks," *IEEE Transactions on knowledge and data engineering*, Vol. 11, No. 6, Nov., pp. 985–991.

Hruschka E. R., and Ebecken M. F. F., 2000, "Using a Clustering Genetic Algorithm for Rule Extraction from Artificial Neural Networks," *Combination of Evolutionary Computing and Neural Networks 2000 IEEE Symposium*, pp. 199-206.

Ishikawa M., 2000, "Rule Extraction by Successive Regularization," *Neural Networks*, vol. 13, pp. 1171-1183.

Setiono R., 2000, "Extractions M-of-N Rules from trained Neural Networks," *IEEE Transactions on Neural Networks*, Vol. 11, No. 2, Mar., pp. 512–519.

Taha I. A., and Ghosh J., 1999, "Symbolic Interpretation of Artificial Neural Networks," *IEEE Transactions on knowledge and data engineer*, Vol. 11, No. 3, May/Jun., pp. 448-463.

Old AI - New AI
Old ANN - New ANN
New Biological Trends

Dr. John R. Alexander Jr.
Computer and Information Science Department
Towson University
Towson, MD 21252
jalexander@towson.edu

ABSTRACT
In both the field of Artificial Intelligence (AI) and the field of Artificial Neural Networks (ANN), there has been a trend, beginning about fifteen years ago for AI and eight years ago for ANN, towards exploring and implementing solutions to problems which have a biological basis in their respective realms.
    A newer view of Artificial Intelligence regards intelligence as adaptive behavior rather than as the ability to manipulate symbols. The unofficial standard bearer of this camp is Rodney Brooks, the Director of MIT's Artificial Intelligence Laboratory [1].
    In the field of Artificial Neural Networks, the trend is manifested by a growing shift away from the classical Parallel Distributed Processing approach and towards studies of more "neuromorphic" nets. For example, the October/November 1998 issue of the *Neural Networks* journal was entitled Neural Control and Robotics: Biology and Technology. The trend is widespread and it is difficult to select a leader. Certainly Randall Beer of Case Western University is well known for his book, *Intelligence as Adaptive Behavior* [2]. In this book he coins the term neuroethology and describes a neuromorphic neural net which allows the implementation of a simulated hexapod, Periplaneta Computatrix. In Europe a number of workers are active. Robert Damper of the University of Southampton in the U.K. may also be considered a leader because of his work on neuron model Hi-NOON [3], and his design of the ARBIB robot which employed the Hi-NOON neuron model. In this paper we discuss the implications of this trend toward a more biological approach to Artificial Intelligence and Artificial Neural Networks and, in particular, compare the differential equations which define the neural systems employed by some of the above mentioned workers in the field.

INTRODUCTION
    The main purpose of this paper is to focus attention on what appears to this author to be changes in the general direction of research in the fields of both Artificial Intelligence and Artificial Neural Networks. The situation in the field of Artificial Intelligence (AI) has already changed. Starting more than fifteen years ago, authors [1], [2], [3], [4], [5], began to question the wisdom of the "physical symbol system hypothesis." The situation in the field of Artificial Neural Nets (ANN) is somewhat different. The works of both Beer [2] and Damper [3] were more in response to their change of view concerning AI than to the call for change by Daniel Gardner [6]. The call for change by Daniel Gardner [6] will be discussed in this paper and we will see, will not be relevant to all

practitioners.

This paper is arranged as follows. The next section, OLD AI/NEW AI I discusses the reasons that I, and others, feel that there are many problems with the physical symbol system hypothesis. In section three, entitled OLD ANN/NEW ANN, I discuss observed recent trends. Not all who work in the field broadly labeled ANN will, or should, follow Gardner's [6] call for the third generation of neural networks to be more neuromorphic. In the last section I briefly summarize.

Old AI/NEW AI

I base this review on the works of the following works: Brooks R., [1], Beer,. R. [2], Damper, R. and Scutt, T. [3], Browne, A. [4], and McFarland D. and Bosser, T. [5].

At the heart of old (traditional) AI research is the physical symbol system hypothesis which states [4] on page 4, "a physical symbol system has the necessary and sufficient means for general intelligent action. In this definition:
- Necessary means that any physical system that exhibits general intelligence will be an instance of a physical symbol system (PSS).
- Sufficient means that any physical symbol system can be organized further to exhibit general intelligent action.
- General intelligent action means the same scope of intelligence seen in humans."

One strategy that was widely followed in traditional AI research was to recast problems of intelligence into simple terms and solve the simplified version, using a specific symbol system. An example is the well known AI blocks world. Perfectly formed blocks, the type children play with, are managed by a computer program. The task of the program is to generate plans, in three dimensions, to manipulate the positions of the blocks. The solution to a given task represented the ordering of the blocks moved, as well the path that each block should follow. Figure 1 shows a typical starting (left) and desired ending position (right).

Figure 1

The basic assumption made by those who follow this approach (Old AI) is that intelligence is about problem solving, and in particular solving abstract problems located in a simplified world. A little reflection suggests that this is not the approach followed by nature. Mankind has only been present on earth for about the past two million years. The animals from whom we evolved faced different types of problems. These problems arose in a very complex, highly uncertain world, which could not be fully perceived by the animal involved. The alternative point of view which presents itself (New AI) is that intelligence is about how an animal might get around and survive in a such a complex, uncertain, partially perceived world.

Thus, the point of view taken by those who have adopted the New AI is that intelligence should be viewed as the ability of animals to adapt to their surroundings. Further, it is posited that the animal has a body (is embodied), the effects of its actions depend on the state of the external world, and the external world influences what is perceived by the animal.

Old ANN/NEW ANN

I believe that some of the uncertainties associated with artificial neural networks may be attributed to the lack of a rule which differentiates between a biological model and an artificial neural network. I give a very abbreviated version of the story. For more detail the reader is referred to Simon Haykin's "Neural Networks" [7]. Historically, in 1943 McCullough and Pitts [8] developed what might be called the first ANN. According to MacGregor [9], these gentleman were more concerned with the possible cognitive and logical operations of their circuits than with electrical activity of the nervous system. In 1962 Frank Rosenblatt [10] published "Principles of Neurodynamics" which discusses Rosenblatt's ideas and introduces the concept of the perceptron. In particular, he proved the "Perceptron Convergence Theorem." The theorem basically states that if two classes of input are linearly separable (i.e., they fall on different sides of a hyperplane), then his convergence procedure will converge and generate an appropriate set of weights to build the separating hyperplane. Perceptrons, when linked into a net, might be considered as the basis of the second ANN. "Principles of Neurodynamics" caused a great deal of interest in what are now called ANNs. However, in 1969 Marvin Minsky (who was then head of the Artificial Intelligence Laboratory at MIT) and Seymour Papret [11] published "Perceptrons." This book, which enjoyed wide circulation, pointed out some of the weaknesses in perceptrons. One weakness, of course, turned out to be that all classes of inputs are not linearly separable. This problem could be overcome if the convergence procedure were expanded to work on multi-layered systems. Minsky and Papret [11] were not sure that such a procedure could be developed. Very shortly after "Perceptrons" publication, funding on perceptron and other ANN projects virtually dried up. In 1982, John Hopfield published "Neural Networks and physical systems with emergent collective computational capabilities" [12] in the Proceedings of the National Academy of Science of the U.S.A. This article completely changed researchers, views of the potential of ANNs. This paper was followed in 1984 by "Neurons with graded response have collective computational properties like those of two state neurons" [13], and was published in the same proceedings. In 1986, Rummelhart and McClelland published their famous two-volume set on Parallel Distributed Processing (PDP) [14]. This contained, in Chapter 8 of Volume 1, the convergence theorem for the multilayered backpropagation algorithm. Actually, several others had also produced demonstrations of convergence [10], [15], [16], [17]. Perhaps the two most influential papers of those mentioned above are Hopfield's 1982 [12] and paper and the Rumelhart and McClelland's volumes [14].

Apparently, with the works of the above mentioned authors in mind, in 1993 Gardner [6] edited "The Neurobiology of Neural Networks." The theme of this text is that both artificial neural networks and neurobiology will benefit by a partial synthesis of their disparate natures. In Chapter 1, which Gardner wrote, of this text, he describes what he calls the first and second generations of ANN. The first generation corresponds to the Rosenblatt [10], Minsky and Papret [11] era models, and the second generation corresponds to the Rumelhart and McClellend [14] era models. He calls for the third generation of ANN to be more neuromorphic.

For those working with backpropagation models, this advice is certainly worth hearing. However, because of the shift towards adaptive systems in AI, the systems designed by both Beer and Damper were by nature neuromorphic.

The differential equations used by most modelers are sets of nonlinear ordinary differential equations, although Beer has simplified his model and uses linear differential equations.

SUMMARY AND CONCLUSIONS

Both the field of Artificial Intelligence and the field of Artificial Neural Networks are in a state of flux. AI has always been decomposed into various, only loosely related fields. The titles of the chapters of AI books describe these fields - Search, Vision, Natural Language Processing, Logic, Control and Implementation of State Space Search, and Expert Systems. There has been, to quote Beer [2], "a growing undercurrent of frustration and disappointment within AI." Thus "the new AI" has arisen.

Like AI, the field called Artificial Neural Networks is decomposed into a large set of different nets, each with a different algorithm and each with a different purpose. Usually, in books on ANNs, the nets described in each chapter are as relatively different as the subject matter in AI books - Hebbian Nets, Hopfield Nets, Simulated Annealing, Competitive Learning, and Backpropagation - this list is just beginning. Indeed, there is no standard definition of an ANN, although it is generally agreed that the motivation for the artificial neuron is biologically based.

What may be concluded from the above? A quote from Beer [2 pg. 15] may help put things in perspective - "It is a striking testament to human conceit how little effort in AI has been expended on modeling the behavior of simpler animals." I believe that Beer's modeling effort should be classified as successful. When one considers the tremendous advances in biological, and in particular neuroscience, research over the past fifteen years, and the relatively small progress made by traditional AI research in that time frame, I must conclude that there will be a trend towards more neuromorphic AI models. This is the direction of the author's future work.

REFERENCES:
[1] Brooks, R. (1991) Intelligence Without Representation, *Artificial Intelligence*. 47, 139-159.
[2] Beer, R. (1992) *Intelligence as Adaptive Behavior*. Academic Press. San Diego, CA.

[3] Damper, R. and Scutt, T. (1991) Computational modeling of learning and behavior in small neuronal systems. *Proceedings of the Joint Conference on Neural Networks, Singapore.* pp. 430-435.
[4] Browne, A. (1997) *Neural Network Perspectives on Cognition and Adaptive Robotics.* Institute of Physics Publishing. Bristol and Philadelphia.
[5] McFarland, D. and Bosser, T. (1993) *Intelligent Behavior in Animals and Robots.* MIT Press. Cambridge, MA.
[6] Gardner, D. (1993) *The NEUROBIOLOGY of NEURAL NETWORKS.* Bradford Books/MIT Press. Cambridge, MA.
[7] Haykin, S. (1994) *NEURAL NETWORKS A Comprehensive Foundation.* IEEE Press. Macmillian Press. New York.
[8] McCullough, W. and Pitts, W. (1943) <u>A logical calculus of the ideas immanent in nervous activity.</u> *Bulletin of Mathematical Biophysics.* 5, 113-133.
[9] MacGregor, R. (1987) *Neural and Brain Modeling.* Academic Press. San Diego, CA.
[10] Rosenblatt, F. (1962) *Principles of Neurodynamics.* Spartan Books. Washington, DC.
[11] Minsky, M.. and Papret, S. (1969) *Perceptrons.* MIT Press. Cambridge MA.
[12] Hopfield, J.J. (1982) <u>Neural networks and physical systems with emergent collective computational capabilities</u> *Proceedings of National Academy of Sciences of the U.S.A.* 79, 2554-2558.
[13] Hopfield, J.J. (1984) <u>Neurons with graded responses have collective computational properties like those of two state neurons.</u> *Proceedings of National Academy of Sciences of the U.S.A.* 81. 3088-3092.
[14] Rumelhart, D. and McClelland, J. (1986) PDP *Parallel Distributed Processing: Explorations in the Microstructure of Cognition (2 Volumes).* Bradford Books/MIT Press. Cambridge, MA.
[15] Werbos P.J. (1974) <u>Beyond regression: New Tools for prediction and analysis in the behavioral sciences.</u> *Ph.D. Thesis, Harvard University.* Cambridge, MA.
[16] LeCun, Y. (1985) <u>Une procedure d'apprentissage pour reseau a seuil assymetrique</u> *Cognitivia.* 85 599-604
[17] Parker, D.B. (1987) <u>Optimal algorithms for adaptive networks: Second order back propagation, second order direct propagation, and second order Hebbian learning</u> *IEEE 1st International Conference on Neural Networks.* Vol 2, pp. 593-600. San Diego, CA.

Artificial Parietal Cortex Neurons for 3D Reconstruction

JEREMIAH J. NEUBERT
Mechanical Engineering
University of Wisconsin
Madison, Wisconsin

YU H. HU
Electrical & Computer Engineering
University of Wisconsin
Madison, Wisconsin

NICOLA J. FERRIER[*]
Mechanical Engineering
University of Wisconsin
Madison, Wisconsin

Abstract

A neural network is used for three-dimensional (3D) reconstruction of a point from a pair of images obtained with an active stereo system. Our active stereo system describes the position of a point with eight parameters: two pan angles, two tilt angles, and two-dimensional coordinates of the projected point in each image. Three-dimensional (3D) reconstruction consists of learning the function which maps these eight parameters to the 3D world coordinates (x_w, y_w, and z_w). This paper outlines two possible networks for learning the mapping function. One contains simple Gaussian neurons in the hidden layer. The other is composed of neurons based on a model of the neurons in the parietal cortex of the human brain thought to be involved in 3D reconstruction from visual data. The paper compares the performance of each network. We evaluate the size of training set required and the effect of the selection method used to obtain the training examples.

Introduction

Three dimensional reconstruction with active stereo cameras is a difficult problem. The traditional kinematic model based approaches can involve more than forty different parameters. Finding these parameters requires a complicated calibration procedure and the collection of several different images [Knight and Reid, 2000, Guse, 1999]. Neubert, et al [2001] suggest that a neural network (NN) may provide a more accurate and robust method for 3D reconstruction. This paper further explores the ability of artificial neural networks to estimate three dimensional coordinates from a dynamic stereo head. Two neural network structures were investigated. The networks were compared using sum squared error and optimum size of the training set.

Feed-forward neural networks are often used to approximate functions and have several advantages over the standard model-based approach to 3D reconstruction from stereo images. They do not require a pre-supposed camera and kinematic model, and are robust to noise in the training data. This makes them well suited to the reconstruction problem, but regression with a neural network can require a rather large set of training data to create a quality model of the function and avoid overfitting.

One of the most important factors in using a neural network to approximate a function is the proper choice of the network structure and neurons [Poggio and Girosi, 1990]. Simple traditional neural networks are well understood, but may not be capable of providing a good approximation of the desired function. This paper will justify the added complexity

---

[*]During part of the period of this research, the first author also worked as a KTI Fellow, and received fellowship support under the NSF GK-12 Program (NSF-9979628). This research supported in part by NSF IRI-9703352.

of the biologically inspired neural network by comparing it to the simple Gaussian radial basis function network similar to those proposed by others [Moody and Darken, 1989, Poggio and Girosi, 1990].

A description of data selection methods is given next. Followed by a description of the biologically inspired network, the Gaussian network, and the methods used to train them. Then the findings will be summarized and discussed.

DATA SELECTION METHODS

Each network was trained with 1000, 1776, 2000, 4000, and 8000 training examples. For each set data was selected both randomly and systematically from approximately 16,000 examples gathered using the method described by Neubert et. al. [2001]. The random selection method occurred without replacement and each sample had the same probablity of being selected. The systematic method attempts to minimize the maximum amount of space between training examples. Because we know the position of the points this is a trivial task. We conjectured that that this method would provide a better representation of the variable space than that of the random selection, especially with smaller data sets, but this was found not to be the case as will be shown.

BIOLOGICALLY INSPIRED NEURAL NETWORK

The Biologically Inspired Neural Network (BioNet) was utilized because it is based on a model of a system that has been honed by tens of thousands of years of evolution. This paper demonstrates that the active stereo system shares enough similarities with human eyes that this model of the neurons in the parietal cortex performs well on this data set.

BioNet Structure: The network was composed of three layers of neurons. The first layer was the input layer, which consisted of the eight random variables: pan and tilt of each camera, and position of the blob centroid in each image. A subset of the input layer was connected to each of the neurons in the hidden layer. The structure of hidden neurons was based on Pouget's model of those contained in the parietal cortex of the human brain [Pouget, 1994]. These neurons were modeled as a product of a sigmoid and a Gaussian. This product, referred to as a basis function, had the following form:

$$h_i(x,\theta) = \frac{e^{-\frac{(x-x_i)^2}{2\sigma^2}}}{1+e^{-\frac{\theta-\theta_i}{T}}}. \tag{1}$$

The hidden neuron h_i had two inputs x and θ. Input x was one of the coordinates of a target centroid in the image, and θ is one of the PTU angles. The image position inputs are each paired up with one of the PTU's angle inputs. This leads to four sets of neurons (u_{left}, ϕ_{left}), (v_{left}, τ_{left}), $(u_{right}, \phi_{right})$, and $(v_{right}, \tau_{right})$. These groupings of hidden neurons will be referred to as a quadrants. Each of the quadrants are laid out in a grid based on their centers, x_i and θ_i (See Figure 1).

The hidden neurons were fully connected to the output neurons, x_{world}, y_{world}, and z_{world}. The output neurons were simple linear neurons. Their value was a weighted sum of the outputs from the hidden neurons.

BioNet Training: Four parameters were tuned by the training algorithm: Gaussian radius, sigmoid radius, number of neurons, and the weights from the hidden layer to the output layer. This training method is a modification of the work done by Fritzke [1994] and Blanzieri et. al.[1995].

The training data was divided into two sets, tuning and training' (training' refers to the portion of the training set not allocated to tuning). The tuning set, 10% of the total training data, was used to predict the performance of the network on examples that were not in the training' set, thus allowing training to be stopped before the network begins to generalize poorly.

The neuron centers were initialized by choosing the desired number of center values to position along each component in the input vector. The center values were evenly spaced

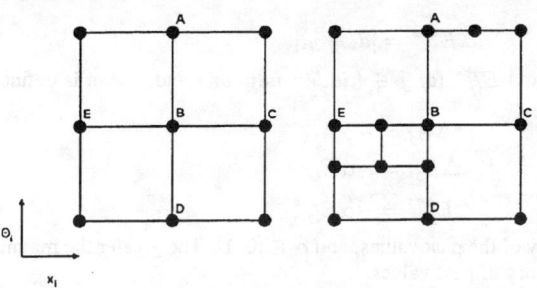

Figure 1: The grid on the left represents the initial network and the one on the right represents a trained network. Several new neurons added through training can be seen on the right. The neurons are the large dots and their position is determined by their center, x_i and θ_i. The lines connecting the neurons represent neighbor relationships.

over a the range of the training' data. All the possible combinations of the values corresponding to a quadrant became the initial centers of the neurons in that quadrant. Figure 1 shows a quadrant with three initial center values along each component of the input vector that corresponds to that quadrant.

The neighbors of each neuron were found by locating the neurons with the closest center in each of the four directions (See Figure 1). These neighbors were later used in the placement of new neurons.

The radii of the sigmoid and Gaussian were determined. The radius of each function was defined as the average distance to neighbors in the direction of the function. For example the radius of the Gaussian of neuron B in left side of Figure 1 would be the average distance of E and C from B, while the radius of the sigmoid is the average distance to A and D from B. If the neuron didn't have any neighbors in one of the directions then that radius was set to the average radii value of the neighbors that it had.

There are two widely used methods to determine the proper weight vector for each output neuron, matrix inversion (least squared error) or back propagation. Matrix inversion in general is a $O(n^3)$ operation and large amounts of training data can require a undesirable amount of CPU cycles. Matrix inversion also produced weights on the order of 10^9 causing poor generalization in the initial testing. Thus we used a back propagation approach, which uses gradient decent to minimize the error function:

$$\xi = \frac{1}{2} \sum_{i \in output} (d_i - y_i)^2 \qquad (2)$$

where y_i was one of the outputs of the network and d_i was its corresponding desired output.

After each training example presented during back propagation the values of A_j, the activity of the hidden neuron j, E_{ji}, the sum of the error values from output neuron i associated with hidden neuron j, and E_{ji}^{abs}, the sum of the magnitude of the error attributed to hidden neuron j, are updated. The change in A_j, E_{ji}, and E_{ji}^{abs} for the $j = s$, where neuron s was defined as a neuron with the closest center to the input in a quadrant, was ΔA_s, ΔE_{si}, and ΔE_{si}^{abs} respectively.

$$\Delta A_s = a_s \qquad (3)$$

where a_s was the output of hidden neuron s and

$$\Delta E_{si} = (d_i - y_i) \qquad (4)$$

$$\Delta E_{si}^{abs} = |d_i - y_i|. \qquad (5)$$

The change in the A_j, E_{ij} and E_{ij}^{abs} for $j \in hidden\ neurons$ and $j \neq s$ is defined as follows:

$$\Delta A_j = -\alpha A_j \qquad (6)$$

$$\Delta E_{ji} = -\alpha E_{ij} \qquad (7)$$

$$\Delta E_{ji}^{abs} = -\alpha E_{ij}^{abs} \qquad (8)$$

where α was the rate of decay of the past values, and $\alpha \in [0, 1]$. The greater the magnitude of alpha the shorter the memory of past values.

After G examples each quadrant was checked to see if any neurons have an activation, A_j, greater than the user defined threshold. If one or more neurons were found, then their accumulated error, σE_j, was calculated and used to determine where the new neuron should be inserted in each quadrant.

$$\sigma E_j = \sqrt{\sum_{i \in output} (|E_{ji}| - E_{ji}^{abs})^2} \qquad (9)$$

Equation (9) is a slight variation on that developed by Blanzieri et. al. [1995]. Blanzieri describes the equation $|E_j| - E_j^{abs}$, which is intended for a single output network. Equation (9) is the L_2 norm of all the σE_{ji} associated with hidden neuron j. The accumulated error is an indirect calculation of the variation in the error. If the weights were the source of the error σE_j was small because all the error had the same sign; thus the magnitude of the absolute value of the error had the same magnitude as the sum of the error. If a new neuron was needed to reduce error the sign of the error should be changing bringing about a sum of error near zero and producing a large σE_j.

Once the neuron with the largest σE was found in each quadrant, a neuron was added between that neuron and its neighbor with the greatest σE. The addition of the neuron results in a decrease of A_j, E_{ji}, and E_{ji}^{abs} by a factor f for the parent neurons. The parent neurons were no longer neighbors, but they were each a neighbor of the new neuron. The weights on the output of the newly inserted neuron was initialized to the average weights of its parents. Then the neighbors in the direction orthogonal to its parent's are located.

This process was repeated until a user specified number of epochs has occurred. The algorithm then returns the network that performed the best on the tuning set.

GAUSSIAN RADIAL BASIS NETWORK

The Gaussian radial basis network consisted of three layers, input, hidden, and output. This was the same basis network that was used by Fritzke [1994]. His paper states that a Gaussian network with this structure, in principle, should be able to approximate any smooth function.

Gaussian Network Structure: The input layer was a replica of that described above, but unlike the biologically inspired NN the input layer was fully connected to the hidden layer. The hidden layer was made up of neurons with Gaussian activation functions:

$$h_i(\mathbf{x}) = \frac{\|\mathbf{x} - \mathbf{x}_i\|^2}{\sigma_i^2} \qquad (10)$$

The $\mathbf{x}$ is a input vector containing all the input neuron outputs. $\mathbf{x}_i$, referred to as the center, is a vector with the same dimensionality as the input. The term, σ_i, determines the spread of the Gaussian and is referred to as the radius of the function.

Gaussian Network Training: Four parameters were tuned by the training algorithm: the centers of the hidden neurons, their radius, number of hidden neurons, and the weights from the hidden layer to the output layer. This training method was taken Blanzieri et. al. [1995] with only minor modifications.

Using the method described above the training data set was divided into two subsets. These sets were used to implement early stopping and prevent over fitting.

Our training involved the following steps. First centers were initialized by using a k-means clustering algorithm described by Alsabti et al. [1998]. The means found via the k-means algorithm were used as the initial centers in the hidden neurons. Second, the neighbors for each hidden neuron were found. The neuron's neighbors were defined as the four closest neurons. The distance between neurons was described by the L_2 norm of the difference in their centers. The average distance of the neuron's neighbors were used as the radius, σ, of the Gaussian. Next the weights applied to the inputs of the hidden neurons were trained by using the same back propagation that was used to train the biologically inspired neural network. During back propagation the values of A_j, E_{ji}, and E_{ji}^{abs} were updated in the manner described in the biologically inspired NN, but because there were no quadrants in the Gaussian network only one neuron had its activation and error parameters updated via equations (3), (4), and (5). Then the neuron closest to the input of the example and it's neighbors were moved toward the input. The centers in the biologically inspired neural network were not given this degree of freedom. The change in the center of the nearest neuron, s, is $\Delta \mathbf{x}_s = \epsilon_b(\mathbf{x} - \mathbf{x}_s)$, while the center of its neighbors moved $\Delta \mathbf{x}_i = \epsilon_n(\mathbf{x} - \mathbf{x}_i)$. Where ϵ_b is a small positive number and the value of $\epsilon_b \gg \epsilon_n$ with $\epsilon_n > 0$. The final step in the training was to add a new neuron after every G examples if there was a neuron with A_j greater than the user specified threshold. The neuron was added using the same method described for the BioNet. Once the neuron was added to the network the neuron's neighbors were re-calculated. This process was repeated until the desired number of epochs or minimum error was reached and the network that performed the best on the tuning data was output.

EXPERIMENTAL RESULTS

The first question addressed was the quantity of data needed to train the neural networks and what method of data selection was optimum. This was determined by training each of the networks with differing amounts of data. In addition the networks were trained first with data chosen randomly, then data selected systematically. The performance of the networks on the testing set can be seen in Figure 2.

Figure 2 shows the influence of both data selection and training data size on the average sum squared error of the network on the testing set. The average sum squared error was used because it is proportional to the distance the network's estimate was from the "true" coordinates of the point.

The plot revealed that as the amount of data decreased the amount of error increased as expected. The increase in the error was relatively small until about .125 (2000 samples) was reached. At this point the error increased significantly.

The comparison of the data selection methods had some interesting results. The data showed that the systematic selection of the data proved to be statistically insignificant with the biologically inspired neural network. It also indicated that the systematic method was detrimental to the performance of the Gaussian network.

These observations were used to determine the optimal data set size (.125) and the data selection method (random) for training two networks so a comparison could be made. This was accomplished by first creating 10 different training and testing sets with .125 of the total data set selected at random. Then each of the networks was trained and tested. It was found that the biological neural network vastly out performed the Gaussian network (See Table 1).

DISCUSSION

In this paper we established that the both types of neural networks require at least 2000 examples to obtain the optimal performance. It was shown that the error in the neural networks may increase by as much as 60% when 1000 examples rather than the optimal 2000 were used to train the networks. This seems to indicate that a training sets of less then 2000 examples did not contain enough information to produce a good approximation of the mapping function.

The random data selection method was shown to be as effective if not better than the systematic data selection method. This is likely due to improper handling of interactions between variables. The solution to this may be to use a design of experiments(DOE) approach to select data. This approach is commonly used by the engineering community to vary the control variables and find the best solution with the least amount of data. This will be explored in future work

This paper also demonstrated that the more complicated biologically based neural network is capable of a far better approximation of the three dimensional mapping function. The Gaussian network described here did not have the expressive power to properly approximate the mapping function.

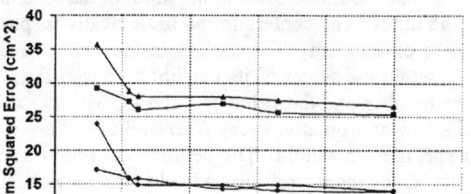

Figure 2: Results of tests to determine the amount of data needed to train the neural networks.

Gaussian	BioNet	Diff.
26.69	15.92	10.76
26.91	14.94	11.96
27.39	14.98	12.41
27.46	15.42	12.03
27.89	14.76	13.13
27.17	15.25	11.92
27.38	14.69	12.69
26.93	16.05	10.88
27.05	15.39	11.66
27.38	16.02	11.36
Average		11.88
Std		0.238

Table 1: The average sum squared error, in cm^2, over the test set with optimal training size and randomly data selection. Std is the standard deviation of the differences.

REFERENCES

[Alsabti et al., 1998] Alsabti, Ranka, and Singh (1998). An efficient parallel algorithm for high dimensional similarity join. In *IPPS: 11th International Parallel Processing Symposium*. IEEE Computer Society Press.

[Blanzieri et al., 1995] Blanzieri, E., Katenkamp, P., and Giordana, A. (1995). Growing radial basis function networks. In *Proc. of the 4th Workshop on Learning Robots*, Karlsruhe, Germany.

[Fritzke, 1994] Fritzke, B. (1994). Supervised learning with growing cell structures. In Cowan, J., Tesauro, G., and Alspector, J., editors, *Advances in Neural Information Processing Systems 6*, pages 255–262. Morgan Kaufmann Pub., San Mateo, CA.

[Guse, 1999] Guse, N. (1999). A neural network for visual kinematics. M.Sc. Thesis, University of Wisconsin, Madison.

[Knight and Reid, 2000] Knight, J. and Reid, I. (2000). Active visual alignment of a mobile stereo camera platform. In *IEEE International Conference on Robotics and Automation*, pages 3203–3208, San Francisco, CA.

[Moody and Darken, 1989] Moody, J. and Darken, C. J. (1989). Fast learning in networks of locally-tuned processing units. *Neural Computation*, 1:281–293.

[Neubert et al., 2001] Neubert, J., Hammond, A., Guse, N., Do, Y., Hu, Y., and Ferrier, N. (2001). Automatic training of a neural net for active stereo 3d reconstruction. In *IEEE Int'l Conf. on Robotics and Automation*, pages 2140–2146.

[Poggio and Girosi, 1990] Poggio, T. and Girosi, F. (1990). Networks for approximation and learning.

[Pouget, 1994] Pouget, A. (1994). *Computational Models of Spatial Representations*. PhD thesis, University of California, San Diego.

ARTIFICIAL NEURAL CORTEX

ALEXEI MIKHAILOV
Mathematics, Science & Computing
Centre, Ngee Ann Polytechnic,
Singapore

YANG MING POK
Mathematics, Science & Computing
Centre, Ngee Ann Polytechnic,
Singapore

ABSTRACT
The artificial neural cortex (ANC-patent pending) is a computational architecture with applications to artificial intelligence. This architecture attempts to boost the performance of artificial intelligence systems by mimicking the columnar structure of the cerebral cortex. Along with pattern recognition and prediction, the cortex creates concepts by establishing links between columns. This linking is done by wave propagation that eliminates a practically intractable search in a multidimensional space of columnar combinations. Unlike artificial neural networks, ANC-technology does not experience difficulties as application size grows. Potential source of very large-scale applications for the cortex is military technology and genomics. The cortex can be applied in many different areas, e.g., understanding of texts, data mining, etc., as it emulates the human cerebral cortex at a higher level than conventional neural networks.

INTRODUCTION

The artificial neural cortex is a collection of direct and inverted indexes that are used for pattern recognition and clustering. This computational architecture was inspired by the columnar structure of the cerebral cortex that is sketched in Fig. 1. It is known that cortical neurons with similar interests tend to be vertically arrayed, forming mini-columns (Calvin, 1995). The mini-columns tend to be horizontally arranged into macro-columns. The best-known examples of specialist mini-columns are the visual cortex's orientation columns. The neurons in one orientation column will best respond to edges tilted at 0 degrees, those in another column will respond to 1 degrees, etc. There are also specialist columns for colors and other features. So, one cortical column may respond to a horizontal line on the retina, the other column to a vertical line. Clearly, two or more objects may share common features. For instance, the characters "A" and "H" have a horizontal bar in common. Hence, the column that responds to the horizontal bar should contain references to more than one object. The simplest way to store multiple references is to put one on top of the other, that is, to arrange references in a sequential or columnar order. Hence, if the bottom cell is activated, so will be all the other cells from the same column. This process is analogous to that in the cerebral cortex. In fact, according to Cook (1986) "...cortical neurons lying in the same vertical direction normally have similar electrical properties - responding to the same kinds of stimuli as their neighbors above and below." The cortex does not belong to the class of traditional artificial neural networks (ANN) that originated in McCulloch and Pitts (1943). It rather follows weightless neural networks (WNN) paradigm, that was suggested by Bledsoe and Browning (1959) and was advanced by Alexander and Stonham (1979). In recent years, the latter paradigm gained a lot of momentum

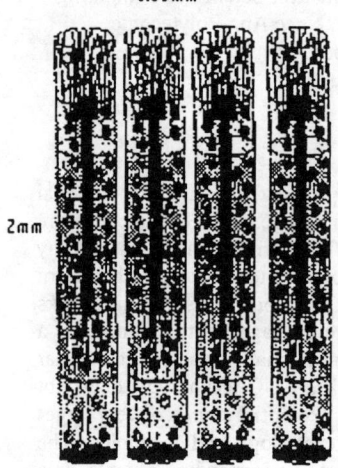

Fig. 1. Cortical Columns.

(Jorgenson, 1997), (Morciniec and Rohwer, 1995) . Both WNN and the cortex have a number of advantages as compared to ANN. But, the difference between the cortex and WNN is that the cortex memory is less than WNN memory by an order equal to the number of n-tuples. For WNN, each feature (or n-tuple) addresses its own store. For the cortex, all features (n-tuples) address the same common storage.

If the weights are associated with the column elements, we interpret them as the values of a membership function defined on the column elements and interpret each column as a fuzzy set. In a fuzzy cortex, new columns can be created with the operations of fuzzy union, fuzzy intersection and fuzzy complement. These operations are used for creating of generalized, specific and opposite concepts.

The paper briefly discusses the index definition, the operations on fuzzy sets, which involve discrete membership functions as well as the cortical functions of pattern recognition and pattern clustering. These functions are illustrated with the essay marking application.

INDEX DEFINITION

The index definition resembles the definition of a matrix that is a table of numbers arranged in rows and columns, the dimension given by two integers, which indicate the number of rows and columns. However, an index is a table of integers arranged in variable-length columns, the dimension given by one integer, which indicates the number of columns. Each column cannot have more than one copy of any number. We denote an index by $\{p\}_k$, k=1, 2,..., K, where the set $\{p\}_k$ contains the numbers from the column k. Unlike functions, any index $\{p\}_k$, k=1, 2,..., K, has an inverse index $\{k\}_p$, p=1, 2,..., P. Fig.2 shows both the index A and the inverse index $B=A^{-1}$. In the index $\{p\}_k$,

```
    5
  5 4                           3
  3 3 5                   2 3 2
  1 2 4            1 2 1 2 1
  1 2 3            1 2 3 4 5

    A                     B
```

Fig. 2. The Index A and Its Inverse Index B

k=1, 2,..., K, the k-th column refers to pages that contain the keyword k. In the inverse index $\{k\}_p$, p=1, 2,..., P, the p-th column refers to keywords from the page p.

OPERATIONS ON FUZZY SETS WITH DISCRETE MEMBERSHIP FUNCTIONS

It is convenient to define the fuzzy column $\{p\}_a$ with the use of a discrete membership function $\mu_a(p)$ that takes on integer values. The continuous membership function could be employed as well. But, the standard range [0,1] must be replaced with [0,1) as the value of 1 cannot be possibly defined.

The notions of the fuzzy equality, the operations of the fuzzy intersection and union defined with discrete membership functions are the same as that of fuzzy sets defined with continuous membership functions. However, the operation of the discrete fuzzy complement requires the notion of the reference fuzzy set (reference column).

The membership function μ of the reference column r is obtained by fusion of individual columns a described by their membership functions $\mu_a(p)$ into a single column, where p is an integer from the column a, that is, by the following discrete fuzzy union operation.

$$\mu_r(p) = \sum_{a \in A} \mu_a(p) \qquad (1)$$

Here, A is the address pattern that selects a subset of index columns to be accessed. The discrete fuzzy complement c of the column a is defined as the difference between the reference column r and the column a.

$$\mu_c(p) = \mu_r(p) - \mu_a(p) \qquad (2)$$

Example. The columns $\{R\}_1$, $\{G\}_2$, $\{B\}_3$ are defined by their membership functions $\mu_1(R)=255$, $\mu_1(G)=0$, $\mu_1(B)=0$, $\mu_2(R)=0$, $\mu_2(G)=255$, $\mu_2(B)=0$ and

$\mu_3(R)=0$, $\mu_3(G)=0$, $\mu_3(B)=255$, that is, $\mu_1 = \{255, 0, 0\}$, $\mu_2 = \{0, 255, 0\}$, $\mu_3 = \{0, 0, 255\}$. If the address pattern is $A=\{1,2\}$ then the following reference column will be obtained by the column fusion: $\mu_r = \{255, 255, 0\}$. Physically, such reference column represents a yellow color if **R** stands for red, **G** for green and **B** for blue.

Let the columns $\{R\}_1$, $\{G\}_2$, $\{B\}_3$, {**pale cyan**}$_4$ be defined by their membership functions as $\mu_1 = \{255, 0, 0\}$, $\mu_2 = \{0, 255, 0\}$, $\mu_3 = \{0, 0, 255\}$, $\mu_4 = \{155, 255, 255\}$. To perform the discrete fuzzy complement operation on the 4th column, we, firstly, obtain the reference column $\{255, 255, 255\}$ and, finally, the complement $\mu_{c4} = \{50, 0, 0\}$ of the 4th column. Note that the intersection of the "pale cyan" column 4 with its complement is not empty and yields a dark red color $\{100, 0, 0\}$, that is, the red color reduced by factor 4.

PATTERN RECOGNITION

We want to identify an unknown pattern $A=\{a_1,..., a_k\}$. The pattern **A** activates the columns $\{p\}_{a1},...,\{p\}_{ak}$ that are fused into the reference column

$$\mu_A(p) = \sum_{i \in \{a_1,...,a_k\}} \mu_i(p) \quad (3)$$

Semantically, $\mu_A(p)$ is the distribution of pattern names in selected columns. The unknown pattern is identified as the name(s) of the most active pattern(s) p_w obtained from

$$p_w : \mu_A(p_w) = \max_p \mu_A(p) \quad (4)$$

Example. The index that is shown in Fig.3 was trained to identify the following three words: 'background' (ref.no.1), 'variable' (ref.no.2) and 'mouse' (ref.no.3). If an unknown input word produces the following reference number score distribution

```
                Ref. no.   1   2   3
                Score      2   1   5
```
then the unknown word should be identified as 'mouse'.

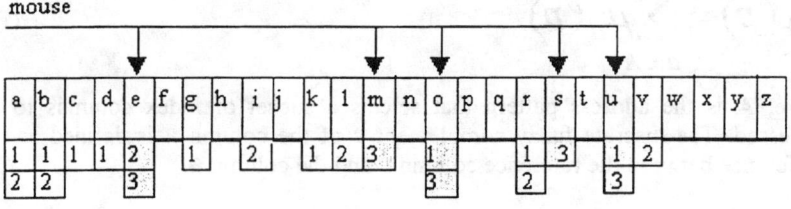

Fig. 3. The Index-based Pattern Recognition

The single-address-array-plus-columns architecture implies the invariance with respect to perturbations. That is two different patterns $\{a, b, c\}$ and $\{b, a, c\}$

can not be distinguished because the columns are activated by the component, say "a", regardless of the component position in the pattern. A way to avoid repeating components is to increase the components' dynamic range. Such 'signal widening' can transform a sequence with repeats into a sequence of unique components, for instance, **a, b, c, b, a, c** =>**ab, bc, cb, ba, ac**.

With modifications, the structure shown in Fig. 3 can support a hashing device, which operates on variable-length keywords, that is, an n-bit RAM, where n is not fixed. Such n-bit-RAM is a valuable asset in sequence analysis, in particular, in analysis of DNA-sequences. The genes across genome can be screened and the repeating strings memorized regardless of their lengths. However, the discussion of this topic is beyond the scope of this paper.

PATTERN CLUSTERING

By fusion of columns, a set of generalized or conceptual columns can be created. This process involves the column fusion operations that are applied repeatedly and are governed by the column similarity measure. Similarity between columns **a** and **b** can be calculated as

$$S_{a,b} = \frac{\sum_p \min(\mu_a(p), \mu_b(p))}{\sum_p \max(\mu_a(p), \mu_b(p))} \qquad (5)$$

As the fusion continues, the new columns become less similar, that is, more orthogonal. This happens because any two sufficiently similar columns will be fused into one column. Finding the best way of column fusion is not a trivial operation as there may exist intractable number of ways the columns could be fused. Perhaps, the most efficient approach to this problem is a wave fusion.

The wave-based fusion takes much less similarity operation without sacrificing the quality. The drastic reduction is based on the assumption that the average number of patterns in a column is much less than the number of columns in the cortex. The above assumption is often realistic. For instance, it requires that the average number of words in a paragraph should be much less than the number of paragraphs in a book. If this assumption is true then the neighboring columns can be discovered through the narrow "word-channel". This can be done with a fan-out (FO) procedure that lines-up the "non-Euclidean" paragraphs utilizing the similarity measure defined above. The FO-procedure employs the inverted index and is described as **p** => **{w}**$_p$ => **{p}**$_w$, that is, for each word from the paragraph p, all relative paragraphs are retrieved from the inverted index **{p}**$_w$. Fig. 4 illustrates the FO-procedure that relates **p**$_0$ to **p**$_2$, **p**$_6$. The wave begins at the column p and creates the next wave-front **p**$_2$, **p**$_6$: **p**$_0$=> **{w**$_1$, **w**$_2$**}** => **{p1, p2, p3, p4, p5, p6}**. In Figure 4, the similarity between **p**$_0$ and **p**$_2$ as well as the similarity between **p**$_0$ and **p**$_6$ is greater than the threshold. Hence the column **p**$_0$ will be linked to **p**$_2$, and **p**$_6$.

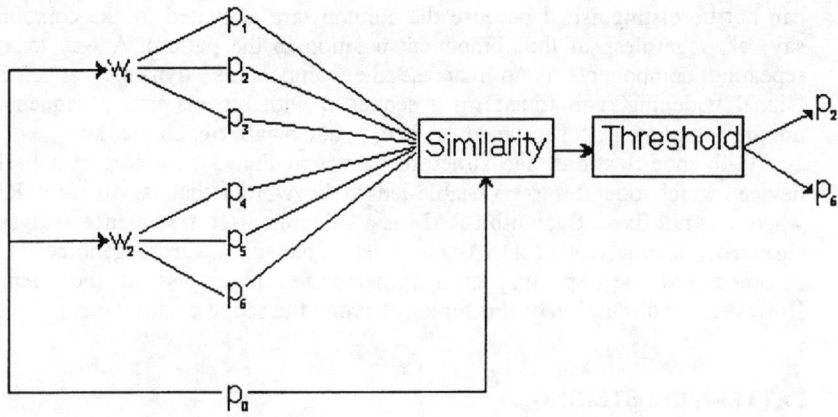

Fig. 4. The Fan-Out Procedure.

The links of the column p_0 do fan out, thus a wave emerges and propagates in a multidimensional space. The propagating wave practically eliminates a taxing multidimensional search. Such multidimensional wave is a generalization of the conventional labeling algorithm (Balakrishnan,1996). However, this wave-based searching owes its efficiency to the use of the index/inverted index pair.

The next example shows the results obtained with the wave fusion that was applied to the following 20 definitions from the Oxford Children Encyclopedia:
1) animal - anything that lives and can move about; 2) bean - plant with seeds growing in pods; 3) bird - feathered animal with two wings and legs; 4) cat - small furry animal usually kept as a pet and known for catching mice; 5) cow - female animal of domestic cattle; 6) dogs - four legged animal that barks and often kept as a pet ; 7) fish - animal that always lives and breathes in water; 8) flower - part of a plant from which seeds or fruits grow; 9) fruit - seed container that grows on a tree and is often used as food; 10) hamster - small animal with brown fur; 11) herbs - plant used for flavoring or for making medicines; 12) horse - four legged animal used for riding and pulling carts; 13) leaf - one of the usually green and flat growths on trees; 14) mermaid - mythical creature that looks like a woman but has a fish tail instead of legs; 15) plant - something that grows out of the ground; 16) rabbit - furry animal with long ears; 17) root - part of a plant that grows underground; 18) seaweed - plant that grow in the sea; 19) tree has a thick wooden stem that grows branches with leaves; 20) vegetable - plant that can be used as food.

By means of the wave propagation, the links between definitions were created and two concepts were identified (Fig. 5), that is, plants and animals. The mermaid paragraph was automatically left out as a stand-alone item. The relation between the mermaid and the animal or plant groups was found to be too weak.

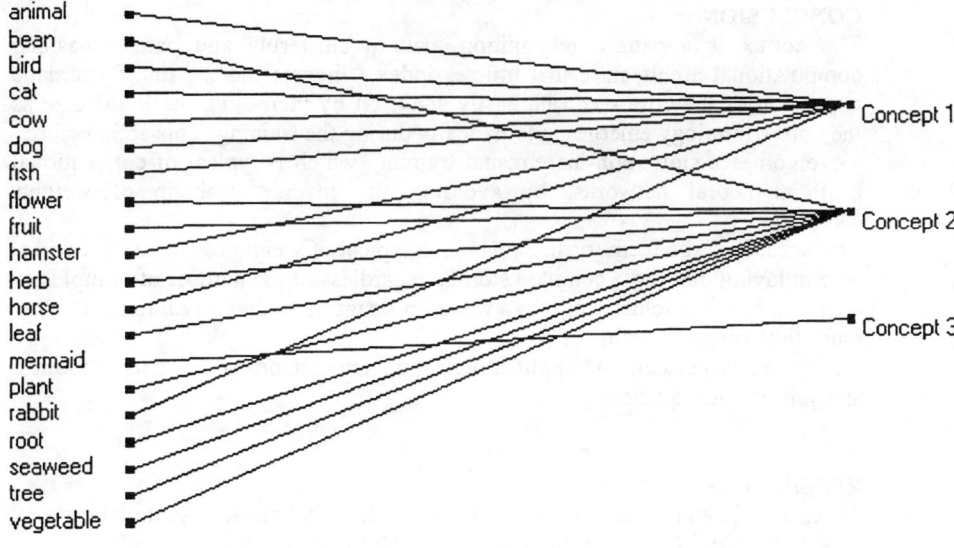

Fig. 5. The Words/Concepts Links

ESSAY MARKING

Essay marking comprises off-line preprocessing and on-line evaluation. The off-line preprocessing is utilized to conceptualize the words, so that, given the gold standard essay **G**, the student essay **E** can be match against the standard even though **G** and **E** may not share common words.

Let $\{w\}$ denote a set of words enumerated as **w**=1,2, ...W and $\{p\}$ denote a set of all paragraphs from an encyclopedia enumerated as **p**=1, 2, ..., P. The set of words $\{w\}_p$ is associated with the paragraph *p*. Utilizing the wave algorithm, the word index $\{w\}_p$ is converted into the generalized word index $\{w\}_c$, which, in turn, is inverted into the concept index $\{c\}_w$ with the membership function $\mu_w(c)$. Next, the standard essay integral meaning is obtained as the following reference column.

$$\mu^{(E)}(c) = \sum_{w \in E} \mu_w^{(E)}(c) \qquad (6)$$

The similarity between student essay and standard essay is calculated on-line. (Note that long essays can be broken down into parts and the dynamics of essay meaning analyzed).

To summarize, the essay marking involves the following two basic steps.
(1) Creation of the dictionary-wide set of concepts: $\{w\}_p$ => wave propagation =>$\{w\}_c$ => $\{c\}_w$
(2) Comparison of the student essay integral meaning with that of the standard essay by using the similarity measure (5).

CONCLUSION

The cortex is a pattern recognition, pattern clustering and pattern hashing computational architecture that utilizes index / inverse index pairs. Systematic expansion of the cortex can be easily achieved by increasing the RAM size as the cortex topology emerges automatically during the training. This architecture
- overcomes a slow, non-incremental training, which is typical of conventional artificial neural networks, by avoiding the inverse problem of weights calculation;
- reduces the large memory size, which is typical of weightless neural networks, by employing the single common storage regardless of the number of n-tuples;
- provides fast clustering by wave propagation thus avoiding a slow multidimensional search;
- supports large-scale AI-applications, e.g., understanding of texts, robotics, analysis of biosequences.

REFERENCES

Alexander, I., and Stonham T.J.,1979, "A Guide to Pattern Recognition Using RAM's",*IEE J. Dig.Sys.and Computers*, Vol.2,No.1.

Balakrishnan,V. ,1996, *Introductory Discrete Mathematics*. Dover Publication, Inc. New York, p.133.

Calvin, W.X,1996, *How the Brains think*. BasicBooks, NY., pp 118-119.

Cook,N.D.,1986, *The Brain Code*, Methuen & Co. Ltd, pp. 26-27.

McCulloch,W and W. Pitts,W.,1943,"A Locical Calculus of Ideas Immanent in Neural Activity", *Bulletin of Mathematical Biophysics 5*, pp.115-133.

Bledsoe, W.W.,and Browning I.,1956, "Pattern Recognition and Reading by Machine", *Proceedings East.J.C.C.*, pp.225-232.

Jorgenson,T.M., 1997, "Classification of Handwritten Digits Using A RAM Neural Net Architecture", *Special Issue on Neural Networks for Computer Vision Applications*, World Scientific Publishing Company, pp.17-25.
  World Scientific Publishing Company.

Morciniec,M, and Rohwer,R.,1995, "The n-Tuple Classifier: Too Good to Ignore", Intellix A/S Articles, Aston University, UK.

PART II: EVOLUTIONARY PROGRAMMING

PART III: EVOLUTIONARY PROGRAMMING

TACKLING MULTIMODAL PROBLEMS IN HYBRID GENETIC ALGORITHMS

PRASANNA V. PARTHASARATHY, DAVID E. GOLDBERG and SCOTT A. BURNS
Department of General Engineering
University of Illinois at Urbana-Champaign
Urbana, Illinois 61801

ABSTRACT
A method is proposed to address the issue of multimodality while using hybrid genetic algorithms (GAs). The hybrid GA is one in which the fitness of an individual is determined as the fitness of the best individual found under local search. A simple modification of commonly used phenotypic sharing is proposed for better performance in this learning environment. Further, it is shown how more information from the local search can be used to achieve speed-up by relaxing fitness evaluations. The method has been used on a structural design problem which shows promising results.

INTRODUCTION

Hybrid genetic algorithms (HGAs) have emerged as a popular choice for real-world optimization. Combining the robustness and global nature of a GA search with an efficient local searcher tuned to the problem provides HGAs the power for solving large real-world applications. Despite growing usage in practice and research on some of the theoretical aspects, HGAs have been less frequently analyzed and designed for performance in explicitly multimodal problems. This is an important issue as many real-world problems are multimodal.

This paper addresses the issue of multimodality when using HGAs. The commonly used niching via phenotypic sharing was found to perform less efficiently an individual is assigned the fitness of the best point found under local search. This provides the motivation for studying how to adapt phenotypic sharing to HGAs. Fortunately, it turns out that the key problem is easily resolved and is a natural extension of the traditional algorithm for sharing. Besides making use of the fitness of the best individual from local search, other information from local search can be exploited to build models of basins of attraction for a problem. This information helps us relax or skip some function evaluations which in turn helps us solve a problem faster.

A description of the hybrid GA environment is given in the next section. In section 3, we introduce phenotypic fitness sharing and how it should be tweaked to work more efficiently in HGAs. Section 4 describes a sharing method for HGAs which makes use of information from local search to enable evaluation relaxation. The real-world application is described in section 5 followed by results in section 6 and some extensions in section 7. The paper concludes by summarizing the work and analyzing the proposed method.

HYBRID GA FRAMEWORK

In nature, every individual learns during its lifetime to adapt to the environment and as a result increases its chances of survival. These adaptations are often the result of an exploratory search (Hinton and Nowlan, 1987). Hybrid GAs make use of this idea of learning that occurs in real organisms. Learning in artificial systems comes in the form of a local searcher which explores the space close to each individual's position in the solution space. The result of learning can be incorporated in two ways in artificial systems:

1. The genotype of the individual found by local search can replace the individual from which the local search was started – this is called Lamarckian learning. This sometimes results in premature convergence.
2. The fitness of the individual found by local search can be assigned to the individual from which the local search was started. This increases the selective pressure of that individual thus guiding evolution towards the individual found by local search. This is similar to the Baldwin effect in biology and was shown to be an important mechanism in evolution of artificial systems (Hinton and Nowlan, 1987).

In practice, we use a mix of Lamarckian evolution and Baldwin effect. Empirical studies show that Lamarckian evolution should be used for a very low percentage of the time (Orvosh and Davis, 1993). However, if we have diversity preserving mechanisms such as a nicher present, we may be able to employ more Lamarckian evolution.

PHENOTYPIC SHARING IN HYBRID GENETIC ALGORITHMS

Many practical search and optimization problems require the investigation of multiple optima (Goldberg and Richardson, 1987). The biological concept of niche formation in species inspired the development of niching operations like crowding (DeJong, 1975) and sharing (Goldberg and Richardson, 1987). A complete survey of the various techniques is presented elsewhere (Mahfoud, 1995).

In fitness sharing, the fitness is shared among individuals which cluster around a peak in the fitness landscape. As a result, we expect that each function peak gets a portion of the population proportional to its fitness. Eqs. 1 and 2 give a method of sharing the fitness.

$$f_{i,shared} = \frac{f_{i,original}}{\sum_{j=1}^{popsize} sh(d_{ij})} \quad (1)$$

$$\text{where, } sh(d_{ij}) = \begin{cases} 0 & \text{if } d_{ij} > \sigma_{share} \\ 1 - \left(\frac{d_{ij}}{\sigma_{share}}\right)^\alpha & \text{if } d_{ij} \leq \sigma_{share} \end{cases} \quad (2)$$

where, $f_{i,original}$ is the raw fitness of an individual, $f_{i,shared}$ is the shared fitness, d_{ij} is some distance metric between individuals i and j, σ_{share} is the sharing radius and $sh()$ is the sharing function.

The distance metric is a very important factor on which the sharing technique greatly relies upon. Commonly used metrics include Euclidean distance (phenotypic sharing), Hamming distance (genotypic sharing in binary-coded GAs) and difference in fitness values. The distance metric is for deciding whether or not two points are headed to the same peak in the fitness landscape. However, when using hybrid GAs, two points initially far apart may be led to the same peak (Fig 1a) or two points which are initially close to each other may be taken to two different peaks by local search (Fig 1b). Therefore, using the distance between termination points of local search gives a better estimate of whether or not two individuals are headed to the same peak. We call this *Baldwinian sharing*. It should be noted that when we use a Lamarckian hybrid, there is no need to do any explicit modification since the termination points of local search are back-substituted in the population.

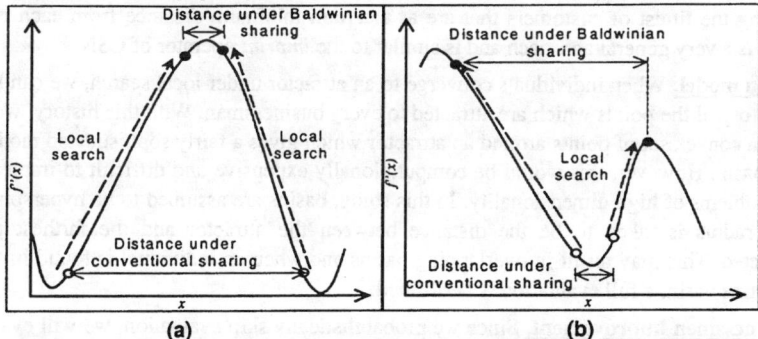

Fig 1 – Illustration of Baldwinian sharing on two maximization problems

USING MORE INFORMATION – THE MODEL BUILDING SHARER

The previous section outlined an opportunity presented in hybrid-multimodal GAs to improve diversity measurements based on the distance between points of attraction under local search. The information given to us by local search can be further exploited to achieve some speed up. A list of near-optimal points can be maintained along with a model of their basins of attraction *a la* the businessman-customer model of the coevolutionary shared niching (Goldberg and Wang, 1997). With this information at hand, we can identify the basin of attraction that an individual belongs to and thus skip costly function evaluations during local search. The method is discussed in further detail later in this section.

Coevolutionary Shared Niching (CSN)

Goldberg and Wang (1997) used a model of monopolistic competition to implement an effective niching method in complex fitness landscapes. This model had a businessmen population that evolved along with the usual GA population, which they called the customer population. Some rules such as minimum distance between businessmen and minimum fitness value were used to control the businessmen population. Businessmen derived their fitness from their customers and customers shared their fitness among other customers belonging to the same businessman. A customer usually belonged to the closest businessman.

Model Building Sharer (MBS)

The fundamental idea of this method is to build a list of attractors (local optima) and model their basins of attraction. This list of attractors is similar to the businessmen population of CSN. For the rest of this paper, the terms *businessmen* and *attractors* are used interchangeably. Once we have this model, we can say that if a point is inside some basin of attraction it will be led to the attractor of that basin. This way, we need to do local search only if a point is outside all basins. Since we cannot build exact models, we may probabilistically skip the local search step, the probability depending on the accuracy of the model. The issues related to this method are handled one by one.

Businessmen selection. Businessmen may be selected based on different parameters depending on how much information we have about the problem. If we have an idea of what the best fitness is, we may select businessmen based on a cut-off fitness. If the local search converges at a point (no further improvement in any direction), the point of convergence may be taken as a businessman since it is a local optimum. If we have no information on the problem or from the local search, businessmen may be chosen from among the fittest of customers that are at a certain minimum distance from each other. This is a very general approach and is similar to the *imprint* operator of CSN.

Basin model. When individuals converge to an attractor under local search, we can build a history of the points which are attracted to every businessman. With this history, we can find a convex set of points around an attractor which gives a fairly sophisticated model of the basin. However, this would be computationally expensive and difficult to implement in problems of high dimensionality. In this study, basins are assumed to be hyperspheres. The radius is taken to be the distance between the attractor and the farthest point attracted. This may result in overlapping basins and when an individual is found in more than one basin, a full evaluation is performed.

Businessmen improvement. Since we probabilistically skip evaluation, we will evaluate individuals even if they are inside some basin. If this evaluation results in a point close to the attractor but with a higher fitness (refinement), this termination point will replace the old attractor. We may also do a local search on a businessman if it has not improved for a certain number of generations.

Evaluation relaxation. Evaluation relaxation schemes come under the realm of *efficiency enhancement techniques* (Goldberg, 1996). The basic idea is to replace costly, accurate function evaluations with less accurate evaluations. In this study, we use a scheme that may be classified as a fitness inheritance scheme (Smith, Dike and Stegmann, 1995) where an individual inherits the fitness of its attractor.

Sharing. As in CSN, each businessman gets a customer count and each individual (customer) is assigned a businessman. Thus, for an individual i with businessman j, the shared fitness is $f_{i,original}/m_j$, where m_j is the customer count of businessman j.

STRUCTURAL DESIGN PROBLEMS

Portal frames are one of the most commonly used structural systems. A typical portal frame with its loading conditions are shown in Fig. 2. The objective is to get a fully-stressed design, where each member at it maximum allowable stress for at least one loading condition. Conventionally, iterative procedures have been used to achieve this design and typically, members are resized based on the stress levels in the current design. However, these iterative procedures fail to converge on some designs unless started from extremely close to those designs. In other words, the basins of attraction are miniscule for some designs for the iterative methods. Mueller (1999) used non-linear equation solvers to capture these "repelling" designs.

In this study, the results for a 2-story, 2-bay frame (2s2b) and a 3-story, 3-bay frame (3s3b) are presented. The objective is to satisfy the following constraints:

$$\frac{f_a}{F_t} + \frac{f_{bx}}{F_{bx}} = 1 \text{ (beam/column in tension) and } \frac{f_a}{F_a} + \frac{C_{mx}f_{bx}}{\left(1-\frac{f_a}{F_{ex}'}\right)} = 1 \text{ (compression)}$$

An explanation of these symbols can be found in AISC (1989). If the left-hand side value of these equations is taken as F, then the fitness function can be defined as:

$$f = \frac{100}{\sum_{i=1}^{m}(1-F_i)^2 + 0.01}$$

where F_i is the maximum F for members grouped as variable i. The best possible value is when F_i equals 1 for all i, which would give a value of 10000 for f. The problem is an unconstrained maximization problem.

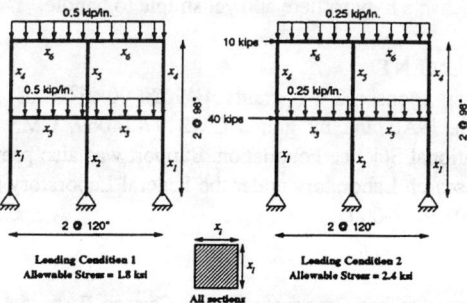

Fig. 2 – A typical 2-story, 2-bay portal frame

A real-coded GA was used for global search. During local search, an iterative procedure called stress-ratio is used until the design is within 10% of the objective followed by a quasi-Newton procedure in MATLAB® fsolve. The initial population is seeded with a heuristic sampling called the thick-thin strategy (Mueller, 1999) for 40% of the size. The other 60% is randomly initialized. Uniform crossover and Gaussian mutation (Bäck, 1996) are used.

RESULTS

Preliminary studies showed that conventional sharing was not up to the task. This might have been due to the complexity of the problem. Earlier studies (Orlet, 1992) have shown that the problem has disjointed basins of attraction and regions where the basins are highly intermingled. There are also large regions of the design space which converge to sub-optimal solutions. In order to show the advantage of MBS, Table 1 shows the number of optima found within a specific number of function evaluations for the 2s2b and 3s3b frames. All GA parameters were the same for the results shown. The table shows the number of different global optima found for each method. The total number of optima reported by Mueller (1999) for the 2s2b (resp. 3s3b) frame is 9 (resp. 19).

Table 1: Number of optima using various sharers

Frame	No. of evaluations	No. of variables	Number of optima		
			Baldwinian Sharing	MBS w/ equal radius	MBS w/ unequal radii
2s2b	10000	6	2	13	16
3s3b	20000	12	4	7	11

CONCLUSION

This paper addressed the issue of niching techniques for multimodal problems when using HGAs. Hybrid GAs provide us the opportunity to do more effective diversity measurements. Also, a model building sharer, inspired by the co-evolutionary shared niching, was presented. This enables us to relax function evaluations thereby reducing computation cost. The methods were put to the test on a real-world problem and the results are favorable. The demerits of MBS include additional overhead in computing distances, storing and updating the models. However the additional computational overhead is compensated by evaluation relaxation. Also, MBS has parameters like the distance between businessmen, initial number of businessmen to be selected, etc. which have to be tuned to the problem. These parameters affect the cost and performance of the method. More effort may be spent on developing a representation for the basin which is more sophisticated than a hypersphere and yet simple to handle.

ACKNOWLEDGEMENT

The work was sponsored by grants F49620-00-0163 of Air Force Office of Scientific Research,USAF and by grants CMS 97-14069, CMS 99-12559 and DMI-9908252 of the National Science Foundation. Support was also provided by a grant from the U. S. Army Research Laboratory under the Federal Laboratory Program, Cooperative Agreement DAAL01-96-2-0003.

REFERENCES

AISC, 1989, *ASD Manual of Steel Construction*, AISC, Chicago, IL, pp. 5-11—5-55.

Bäck, T., 1996, *Evolutionary algorithms in theory and practice*. Oxford University Press, New York, NY. pp. 71-73.

DeJong, K. A., 1975, "An analysis of the behavior of a class of genetic adaptive systems," Ph.D. Thesis, University of Michigan, Ann Arbor. (University Microfilms No. 76-9381).

Goldberg, D. E., and Richardson, J., 1987, "Genetic algorithms with sharing for multimodal function optimization," *Proceedings of the Second International Conference on Genetic Algorithms*, J. J. Grefenstette, ed., Lawrence Erlbaum Associates, Hillsdale, NJ. pp. 41-49.

Goldberg, D. E., and Wang, L., 1997, "Adaptive niching via coevolutionary sharing," *Genetic Algorithms and Evolution Strategy in Engineering and Computer Science*, John Wiley & Sons Ltd., West Sussex. pp. 21-38.

Goldberg, D.E., 1996, "From competence to efficiency: A tale of GA progress," *Computational Intelligence and Its Impact on Future High-Performance Engineering Systems*, NASA Center for Aerospace Information, Linthicum Heights, MD. pp. 245-282.

Hinton, G. E. and Nowlan, S. J., 1987, "How learning can guide evolution," *Complex Systems*, Vol. 1, pp. 495-502.

Mahfoud, S. W., 1995, "Niching methods for genetic algorithms," Ph.D. Thesis, University of Illinois at Urbana-Champaign, Urbana, IL. Also IlliGAL Report No. 95001.

Mueller, K. M., 1999, "Sizing of members in the fully-stressed design of frame structures," Ph.D. Thesis, University of Illinois at Urbana-Champaign, Urbana, IL.

Orlet, M. W., 1992, "Repelling fully-stressed designs of frame structures under the action of stress ratio method," Master's Thesis, University of Illinois at Urbana-Champaign, Urbana, IL.

Orvosh, D., and Davis, L., 1993, "Shall we repair? Genetic algorithms, combinatorial optimization, and feasibility constraints," *Proceedings of the Fifth International Conference on Genetic Algorithms*, S. Forrest, ed., Morgan Kaufmann, San Mateo, CA, pp. 650.

Smith, R., Dike, B., and Stegmann, S., 1995, "Fitness inheritance in genetic algorithms," *Proceedings of the ACM Symposium on Applied Computing*, ACM, New York, NY. pp. 345-350.

MODELING TOURNAMENT SELECTION WITH REPLACEMENT USING APPARENT ADDED NOISE

KUMARA SASTRY and DAVID E. GOLDBERG
Illinois Genetic Algorithms Laboratory
Department of General Engineering
University of Illinois at Urbana-Champaign
Urbana, Illinois 61801

ABSTRACT

This paper analyzes the effects of tournament selection with replacement (TWR) on the convergence time and population sizing for selectorecombinative genetic algorithms. It is empirically demonstrated that the run duration is not affected if the tournament selection is performed with or without replacement. However, the population size required to attain a certain level of accuracy, is more if tournament selection is performed with replacement rather than without replacement. An approximate population-sizing model is derived based on apparent added noise for the case of TWR and is verified with empirical results.

INTRODUCTION

Tournament selection (Goldberg et al., 1989) is one of the most widely used ordinal selection scheme. In tournament selection, a specified number of individuals, s, is selected from a population of size n. The best individual out of the s individuals gets a copy in the mating pool. These s individuals can be selected either *with replacement* or *without replacement*. In selection with replacement, the individuals selected for a tournament are candidates for other tournaments. On the other hand, in selection without replacement, the individuals selected for a tournament are not candidates for other tournaments.

In genetic algorithms (GAs) literature, tournament selection without replacement (TWOR) has received considerable analytical attention. Both the population size required and the convergence time taken, with and without noisy function evaluation, have been successfully considered. On the other hand, TWR has not received separate scrutiny, and is usually considered to be equivalent to TWOR. Here, we empirically demonstrate that TWR requires more function evaluations than TWOR for attaining a solution of same accuracy. We demonstrate that even though the run duration is not affected, the population size required for successful convergence depends on whether we select with or without replacement. In this paper, we propose to model this discrepancy as an apparent noise similar to that proposed by Goldberg et al. (1992) for the case of roulette-wheel selection. We also develop an approximate model based on the apparent added noise to adjust the population size required for selectorecombinative GAs that employ TWR.

This paper is structured as follows. First, a brief review of selection schemes is presented. Next, we compare the convergence time of GAs with TWR to that of TWOR. Finally, we address the issue of population sizing required for error-controlled convergence in the case of TWR.

LITERATURE REVIEW

A complete literature review of selection schemes in GAs is beyond the scope of this paper, and instead a very brief overview is presented. Selection schemes that exist in GA literature can be broadly classified into two classes: (1) proportionate schemes, and (2) ordinal schemes. Proportionate schemes (Holland, 1975; Brindle, 1981; Booker, 1982; Baker, 1987) select an individual based on its relative fitness value compared to others. The selection pressure in these procedures are dependent on the fitness distribution of the population. On the other hand, ordinal schemes (Schwefel, 1981; Baker, 1985; Goldberg, 1989b; Goldberg et al., 1989; Mühlenbein and Schlierkamp-Voosen, 1993) select an individuals based on its ranking in the population, where the individuals are ranked based on their fitness values. The selection pressure of ordinal schemes is independent of the relative fitness of individuals within the population. Unlike proportionate schemes, ordinal schemes do not suffer from scaling problems, and are therefore preferred (Goldberg, 1989a; Whitley, 1989).

CONVERGENCE TIME

Convergence models for different selection schemes were presented by Goldberg and Deb (1991). They modeled the change in proportion of the individuals in the population, when using only selection. Mühlenbein and Schlierkamp-Voosen (1993) employed the concept of *selection intensity* (Bulmer, 1980) for convergence analysis of truncation and proportionate selection schemes in GAs. Thierens and Goldberg (1994) developed convergence models for several selection schemes, including binary tournament selection, using normal distribution analysis. Bäck (1995) used order statistics to model the relation between selection intensity and tournament size. He also developed convergence model for (μ,λ) selection. Miller and Goldberg (1996a, 1996b) refined the convergence model to handle external noise. Here, we use the convergence time model of Mühlenbein and Schlierkamp-Voosen (1993):

$$t_{conv} = \frac{\pi\sqrt{\ell}}{2I}, \tag{1}$$

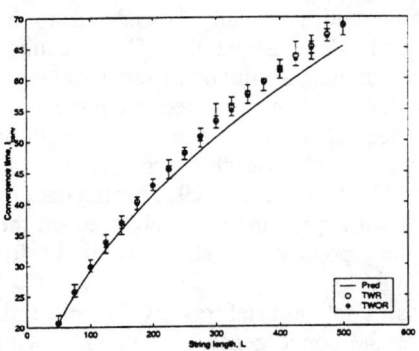

Figure 1. Verification of convergence model (Eq. (1))

where, t_{conv} is the convergence time, ℓ is the string length, and I is the selection intensity, which measures the magnitude of selection pressure of a selection scheme. Selection intensity for different tournament sizes is tabulated elsewhere (Bäck, 1995; Miller and Goldberg, 1996a, 1996b).

The convergence model (Eq. (1).) is compared with experimental results in Fig. 1. for both TWR and TWOR. The results are for the OneMax problem with uniform crossover. A crossover probability of 1.0, a tournament size of 2, and a population size of 1000 are used. A high population size is chosen as to avoid population sizing effects. The divergence between the model and empirical results are due to hitchhiking and it can be reduced by increasing mixing of building blocks (BBs) through multiple crossovers (Thierens and Goldberg, 1994). It can be easily seen in Fig. 1. that both TWR and TWOR have almost identical convergence times, and this similarity exists even for small population sizes. This observation indicates that the run duration is same whether we select with replacement or not. This can be justified by the fact that in both TWR and TWOR, the best individual among selected s individuals is always the winner. The only difference is that the quality of converged solution differs between GAs that use TWR and TWOR.

In this section, we demonstrated the similarity in convergence time between TWR and TWOR, and the next step is to address population sizing for optimal convergence, which is considered next. We empirically show that the difference in population-size requirement between TWR and TWOR is significant and propose a model to account this difference.

POPULATION SIZING

Population size is an important factor in determining the solution quality obtained by a GA. Goldberg et al. (1992) proposed a population-sizing model for different selection schemes. Their model was based on deciding correctly between the best and the second best BB in a partition. They incorporated the presence of noise arising from other partitions into their model. However, they assumed that if wrong BBs are chosen in the first generation, it would be impossible to rectify the error. Harik et al. (1997) extended the above model by incorporating cumulative effects of decision making over time rather than in the first generation only. They modeled the decision making between the best and second best BB in a partition as a gambler's ruin problem. Their model implicitly assumed TWOR as the selection operator. Miller (1997) refined this model for noisy fitness functions, which is reproduced below:

$$n = \frac{-2^{k-1}\log(\psi)\sqrt{\pi}}{d}\sqrt{\sigma_f^2 + \sigma_N^2}, \qquad (2)$$

where, k is the BB size, ψ is the failure rate, d is the difference in fitness between the best and the second best BB, σ_f^2 is the fitness variance, and σ_N^2 is the noise variance.

Our first step is to verify Eq. (2). with computational results for both TWR and TWOR. Figure 2. depicts the success rate 1-ψ as a function of population size for a 200-bit OneMax problem. Since the fitness function is not noisy, $\sigma_N^2 = 0$. For small

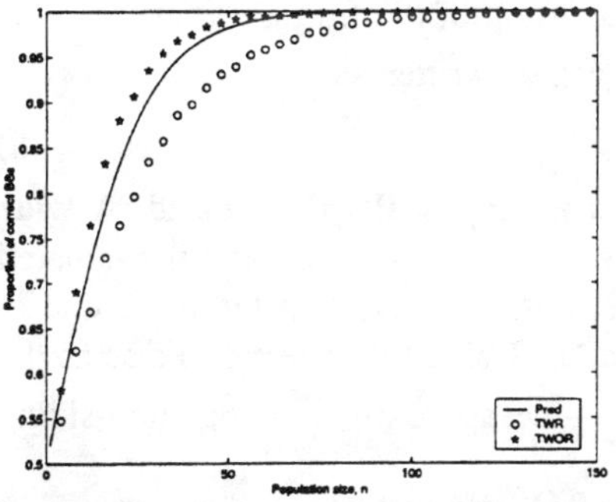

Figure 2 Verification of the population-sizing model (Eq. (2)., with $\sigma_N^2 = 0$)

population sizes, the model does not agree with empirical results for both TWR and TWOR. This is because Eq. (2). is an approximation, and for small population sizes it is known to estimate the success rate conservatively. A more accurate form of the model can be used to improve the agreement with the results for TWOR (Miller, 1997). From Fig. 2., we can easily see that though the model agrees with TWOR, the same is not true for TWR. The results for TWR indicate that the population size required to reach certain accuracy is more than the model predicts. This discrepancy exists because the model does not account for the noise due to the selection scheme, and TWR is certainly a noisy scheme. In TWR, though on an average the best individual gets s copies, it can get all n copies or none at all. Whereas, in TWOR the best string gets *exactly* s copies.

We propose to account for the noise due to TWR as an apparent external noise in Eq. (2). A similar procedure was proposed by Goldberg et al. (1992) for roulette-wheel selection. To quantify the noise due to selection, recognize that the process of selecting an individual is a Bernoulli trial and this process is repeated n times to keep the population size constant. Therefore, the process of selecting an individual i in n trials is binomially distributed with probability p_i. For an s-wise tournament, p_i is given by

$$p_i = \frac{\binom{n-1}{s-1}}{\binom{n}{s}} = \frac{s}{n}, \tag{3}$$

The mean and the variance of the number of tournaments that i participates in is np_i, and $np_i(1-p_i)$, respectively. Therefore, the average variance for any string is

$$\frac{1}{n}\sum_{i=1}^{n}(np_i(1-p_i)) = s\frac{n-s}{n} \approx s. \tag{4}$$

This variance due to the noise of TWR is in the units of squared individuals. To use this variance in Eq. (2)., we have to convert it into units of squared fitness. To do so, we recognize that every extra tournament an individual participates in, or every tournament it misses, introduces a variance proportional to the fitness variance. Therefore, the variance due to TWR in fitness terms is the product of variance in terms of number and some proportion of the fitness variance. Thus, the appropriate variance due to TWR, σ_s^2, can be written as

$$\sigma_s^2 = c_o s \sigma_f^2, \tag{5}$$

where, c_o is a proportionality constant and is empirically determined. A wide range of tournament sizes ($2 \leq s \leq 20$) is used to determine c_o and it is evaluated to be 0.25. The fact that c_o is constant justifies the claim that the noise due to selection is proportional to the fitness variance. Using this selection noise as the external noise term in Eq. (2).— $\sigma_N^2 = \sigma_s^2$—the approximate population-sizing model for TWR is given by

$$n = \frac{-2^{k-1}\log(\psi)\sigma_f}{d}\sqrt{\pi(1+c_o s)}. \tag{6}$$

Using Eq. (6)., the success rate $1-\psi$ is computed for different tournament size values and different problems sizes as a function of population size. The results are verified with computational results in Fig. 3. The results are for the OneMax problem with string lengths, $\ell = 100, 200$, and 400, and for tournament sizes $s = 2, 4, 8$, and 16. The empirical results are averaged over 100 independent runs. The results show that the agreement between the analytical and empirical relations increases significantly with the addition of the apparent noise term.

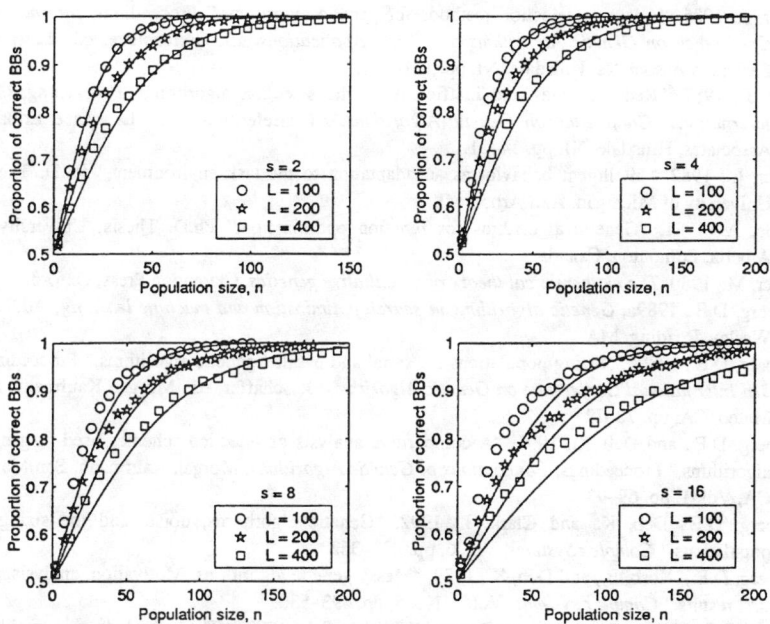

Figure 3. Verification of population-sizing model for TWR (Eq. (6).) with empirical results

CONCLUSIONS

In this study, the effects of TWR on convergence time and population sizing of selectorecombinative GAs were investigated. The analysis yielded two interesting conclusions: (1) The run duration is the same for both TWR and tournament selection without replacement. (2) The population size required by TWR is greater than that required by TWOR. We also derived and verified an approximate model for the population sizing of TWR based on an apparent added noise to the population-sizing model of TWOR. This model may be used for GA run parameter setting, competent GA design, or simply to advance our understanding of GA mechanism.

ACKNOWLEDGEMENTS

This work was sponsored by the Air Force Office of Scientific Research, Air Force Materiel Command, under grant F49620-00-1-0163, by the NSF under grant DMI-9908252, and by the U.S. Army Research Laboratory, Federated Laboratory Program, Cooperative Agreement DAAL01-96-2-0003. Thanks are due to Ravi Srivastava and the referees of ANNIE'01 for their useful comments.

REFERENCES

Bäck, T., 1995, "Generalized convergence models for tournament- and (μ,λ) -selection," Proceedings, *6th International Conference on Genetic Algorithms*, L. Eschelman, ed., Morgan Kaufmann, San Francisco, CA, pp. 2–8.

Baker, J., 1985, "Adaptive selection methods for genetic algorithms," Proceedings, *International Conference on Genetic Algorithms and Their Applications*. J.J. Grefenstette, ed., Lawrence Earlbaum Associates, Hillsdale, NJ, pp. 101–111.

Baker, J., 1987, "Reducing bias and inefficieny in the selection algorithm," *Proceedings, 2nd International Conference on Genetic Algorithms*. J.J. Grefenstette, ed., Lawrence Erlbaum Associates, Hillsdale, NJ, pp. 14–21.

Booker, L., 1982, "Intelligent behavior as an adaptation to the task environment," Ph.D. Thesis, University of Michigan, Ann Arbor, MI.

Brindle, A., 1981, "Genetic algorithms for function optimization," Ph.D. Thesis, University of Alberta, Edmonton, Canada.

Bulmer, M., 1980, *The mathematical theory of quantitative genetics*, Clarendon Press, Oxford.

Goldberg, D.E., 1989a, *Genetic algorithms in search optimization and machine learning*, Addison-Wesley, Reading, MA.

Goldberg, D.E., 1989b, "Sizing populations for serial and parallel genetic algorithms," Proceedings, *3rd International Conference on Genetic Algorithms*, J. Schaffer, ed., Morgan Kaufmann, San Mateo, CA, pp. 70–79.

Goldberg, D.E., and Deb, K., 1991, "A comparitive analysis of selection schemes used in genetic algorithms," Proceedings, *Foundations of Genetic Algorithms*, Morgan Kaufmann, San Mateo, CA, Vol. 1, pp. 69–93.

Goldberg, D.E., Deb, K., and Clark, J., 1992, "Genetic algorithms, noise, and the sizing of populations," *Complex Systems*, Vol. 6, pp. 333–362.

Goldberg, D.E., Korb, B., and Deb, K., 1989, "Messy genetic algorithms: Motivation, analysis, and first results." *Complex systems*, Vol. 3, No. 5, pp. 493–530.

Harik, G., Cantú-Paz, E., Goldberg, D.E., and Miller, B.L. 1997, "The gambler's ruin problem, genetic algorithms, and the sizing of populations", Proceedings, *IEEE International Conference on Evolutionary Computation*, T. Bäck et al., ed., IEEE, Piscataway, NJ, pp. 7–12.

Holland, J., 1975, "Adaptation in natural and artificial systems," University of Michigan Press, Ann Arbor, MI.

Miller, B.L., 1997, "Noise, sampling, and efficient genetic algorithms," Ph.D. Thesis, University of Illinois at Urbana-Champaign, Urbana, IL.

Miller, B.L., and Goldberg, D.E., 1996a, "Genetic algorithms, selection schemes, and the varying effects of noise," *Evolutionary Computation*, Vol. 4, No. 2, pp. 113–131.

Miller, B.L., and Goldberg, D.E., 1996b. "Genetic algorithms, tournament selection, and the varying effects of noise," *Complex Systems*, Vol. 9, No. 3, pp. 193–212.

Mühlenbein, H., and Schlierkamp-Voosen, D., 1993, "Predictive models for the breeder genetic algorithm: I. Continuous parameter optimization," *Evolutionary Computation*, Vol. 1, No. 1, pp. 25–49.

Schwefel, H.-P. , 1981, *Numerical optimization of computer models*, Wiley, Chichester.

Thierens, D., and Goldberg, D.E., 1994, "Convergence models of genetic algorithm selection schemes," Proceedings, *Parallel Problem Solving from Nature III*, Y. Davidor et al., ed., Springer-Verlag, Berlin, pp. 119–129.

Whitley, D., 1989, "The GENITOR algorithm and selective pressure: Why rank-based allocation of reproductive trials is best," Proceedings, *3rd International Conference on Genetic Algorithms*, J. Schaffer, ed., Morgan Kaufmann, San Mateo, CA, pp. 116–121.

SYMBIOTIC MODEL FOR SOLUTIONS AND OPERATORS

MOEKO NEROME, SATOSHI ENDO, KOJI YAMADA, HAYAO MIYAGI
University of the Ryukyus
Department of Information Engineering, Faculty of Engineering
Okinawa 903-0213, Japan

ABSTRACT

Our purpose is development of the system that searches for a solution, changing an operator in adaptation. In this paper, we design the symbiotic co-evolutionary algorithm in which two evolutional populations are solutions and genetic operators. We deal with the Tsume-shogi problem that is one of the problems requiring an appropriate genetic operator. In our model, a population of operator is set as symbiosis species for the population of solutions. We discuss the concurrent evolution of solution and operator through the simulations.

1 INTRODUCTION

A symbiotic co-evolutionary algorithm is based on the symbiosis relationship of organisms. In the symbiotic evolution model, two evolutional populations are considered as both solutions and something which helps their evolution. This method has the characteristics of which individuals are evolved by the internal conversion of an individual. Some researchers create several symbiotic models so far. However, many models are designed like the relationship of solutions and their permutations. In this paper, we design the symbiotic co-evolutionary algorithm in which two evolutional populations are solutions and genetic operators. We deal with the Tsume-shogi problem (Iino 1998) which is one of the problems requiring an appropriate genetic operator. Tsume-shogi is a kind of the puzzle game based on the rules of the Shogi. The game uses the board and pieces of the Shogi. When we apply the Genetic Algorithm (Goldberg 1989) to the coded individuals, it is difficult to design a genetic operator. Therefore, we set the population of operator as symbiosis species. A chromosome of operator shows the information about gene swapped. In order to investigate the performance of the symbiotic algorithm, it is compared with an evolutionary method based on GA. Furthermore, we discuss the concurrent evolution of solution and operator from the results of simulations.

2 SYMBIOTIC EVOLUTION

2.1 Definitions of populations, individuals and fitness

Co-evolution refers to the interdependent evolution of two or more species having an obvious ecological relationship (Lincoln 1982).

Symbiosis is one of the relationships on the process of co-evolution. A general term describes the situation in which dissimilar organisms live together in close association. The symbiotic co-evolution algorithm is designed by several researchers (Paredis 1995, Potter 1994). The model of Paredis has two populations that are solutions and their permutations. We confirmed the good performance of this model for solving the N-Queens problem through the computer simulations (Nerome 2000). In the Potter's model, each population is one of the elements of solution. Therefore, symbiotic relationship means the cooperation of some populations. In both algorithms, a success on one side improves the chances of survival of the other (i.e., it is mutualism). This paper deals with the former. In our model, some operators are the role of the internal conversion.

2.2 Symbiotic algorithm

The symbiotic algorithm is described as follows.

[Step1.] Creating two populations, *Pa* and *Pb*.

[Step2.] Selecting two individuals from *Pa* and one individual from *Pb* to generate new individuals of *Pa*.

[Step3.] Generating offspring of *Pa*'s individual by applying genetic operators to the selected individuals.

[Step4.] Calculating the fitness evaluation of the offspring by solving the problem and the history of the fitness of the selected individual from *Pb* is updated.

[Step5.] Inserting the offspring at its appropriate rank in *Pa*. Then, if the termination condition is satisfied go to step6, otherwise go to step2.

[Step6.] Selecting two individuals from *Pb*.

[Step7.] Applying genetic operators to these individuals to generate offspring of *Pb*.

[Step8.] Calculating the fitness evaluation of the offspring.

[Step9.] Inserting the offspring at its appropriate rank in *Pb*. If the termination criterion of this algorithm does not satisfy, return to step2.

3 APPLICATION TO THE TSUME-SHOGI
3.1 Tsume-Shogi

Shogi is one of the most popular games in Japan. It is a game of mental skill between two players, with the ultimate objective of trapping the opponent's King; it may be a simply described as the Japanese version of chess.

When a player moves his piece into the opponent's camp, or promotion zone (three lows of the opponent side), he gains the opportunity of promoting the piece. Promotion is optional and at any given time the option may only be exercised on that move. However, the right to promote arises every time he moves the piece in the opponent's camp until the promotion is made. Each piece is represented in Japanese as shown in Table 1.

Tsume-shogi is a kind of the puzzle game based on the rules of the Shogi (Iino 1998). The game uses the board and pieces of the Shogi. One player plays both the first (Attacker) and the second player (Defender). The Attacker plays first from the bottom. The Defender plays from the top. Normally, Attacker's King is not shown. In the case of the Attacker, the player must checkmate Defender's King. When the player is the Defender, the player should be assumed to be in the direction that will most prolong the game.

The rules of the Tsume-shogi are simple as follows:
- The Attacker must mate with a continuous series of checks.
- Both Attacker and Defender must play the best moves.
- The Defender may use any piece that is not shown.
- The Defender may not resort to the futile interposition.
- Otherwise, the normal rules of the Shogi apply.

An example of the Tsume-shogi game is shown in Figure 1 (Iino 1998). This board is the initial board of one Tsume-shogi problem. The piece existing outside the board is a piece in Attacker's hand. The correct solution is as follows.

Table 1. Shogi pieces.

Main		Promoted
Pieces	Name	Pieces
歩	Pawn	と金
香	Lance	成香
桂	Knight	成桂
銀	Silver	成銀
金	Gold	—
角	Bishop	馬
飛	Rook	龍
玉	King	—

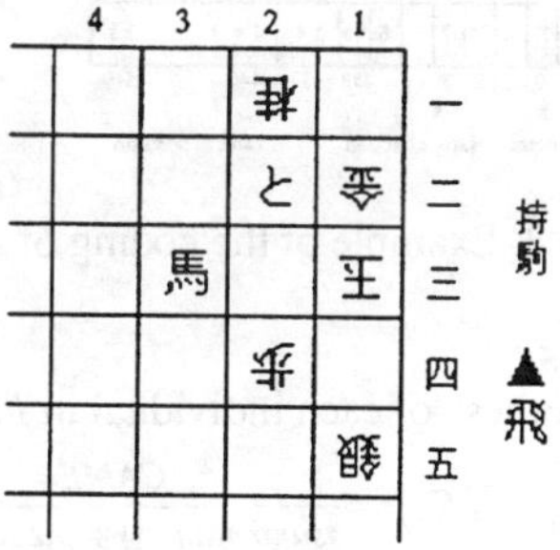

Figure 1. Example of the Tsume-shogi game.

Correct solution: ▲2三馬, △2三金, ▲1二飛

In the move "▲2三馬", the first symbol (▲) indicates whose move. The ▲ means Attacker's move and △ is Defender's. The second and third numbers (2 三) are the position of the piece after the player moves. The last character (馬) represents the kind of the piece moved.

4.2 Definition the Populations, Individuals and Fitness

Populations

For the Tsume-shogi, one population (Pa) is a set of the evolving solutions and the other (Pb) is a set of their operators.

Individuals

An example of the coding of solution is shown in Figure 2. Figure 2 represents a solution of 3ply Tsume-shogi game. In the 3ply game, the 1st and the 3rd moves are Attacker's moves, the 2nd move is Defender's move. Therefore, one solution shows three moves. Each move consists of the information of "promoted", a series of "rank of the action" and a series of "rank of the position". The "promoted equals 1" means that promotion of a selected piece. In the parts of the rank of the action, each locus means the rank of the action, a gene shows an index number of one action pattern, e.g., the action #1 means "Pawn (歩) moves up". In the same way, genes define the rank of the position of pieces in hand. Here, this position represents an absolute coordinate.

During playing the game, each action is selected along the ranking. The Attacker defines the action that can catch the King. The Defender moves the piece avoiding checkmating.

We try the acquisition of solution using this coding. However, when we apply the GA to the coded individuals, it is difficult to design a genetic operator. Therefore, we set the population of operator as symbiosis species. A chromosome of operator shows the information about gene (the parts of the action's rank) swapped. An example of genes swapped is shown in Figure 3. A gene is swapped for another gene specified.

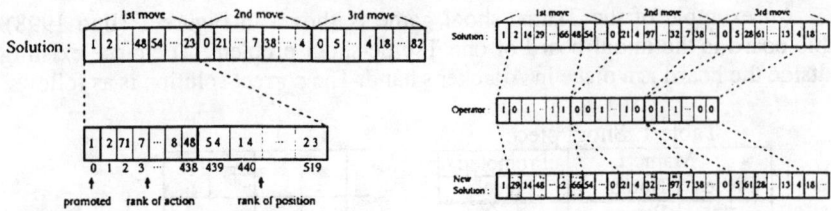

Figure 2. Example of the coding of a solution. Figure 3. Example of the swapping.

Fitness

The fitness of each individual in *Pa* is calculated by formula (1).

$$F(S_i) = \frac{GAME}{tsumi.num \times (1+ pieces.num)} + \sum_{j=0}^{GAME} \frac{1}{(1+ rank.num(j)) \times GAME} \quad (1)$$

Where,
S_i : the *i*-th solution
$F(S_i)$: the fitness of S_i
GAME : kinds of game (if the game is 3ply, *GAME* = 3)
tumi.num : the number of the moves until the end
pieces.num : the number of the pieces in Attacker's hand at the end
rank.num(j) : the rank of *j*-th move

As a next step, the fitness of an operator is defined. We adopt the life-time fitness evaluation (LTFE) to evaluate continuously (Paredis 1994). When the operator creates solution's offspring, it receives a payoff that is the average fitness of the parents and offspring solutions. Each operator has an associated history and the payoff is embedded in its history. Thereby, the fitness of an operator is defined as the average payoff, which is embedded in its history. Furthermore, the fitness of the newborn operator is initialized. So as to initialize the fitness of the new operator, one encounter is executed between the operator and two selected solutions. This fitness is used to determine the rank location of the new operator in *Pb*.

4.3 Simulation

Our objective is to investigate the performance of the symbiosis co-evolutionary algorithm than evolutionary methods based on GA. In this section, the characteristics of the symbiotic algorithm are discussed through simulation and analysis.

We introduce two methods that are compared as follows.
- Symbiotic Algorithm
  In the evolution process of solution, the operators of *Pb* are used. The evolution process of operator uses the one-point crossover and mutation.
- Genetic Algorithm
  This algorithm generates a new solution with a given swap rate (i.e.: 8, 64, 128, 256, 300, and 400 genes are swapped). Here, the swap rate means the number of gene swapped. In other words, swap rate is equivalent to general mutation rate.

The parameters for the simulations are *population_size* = 10, *crossover_rate* = 0.5, *mutation_rate* = 0.03, *the_number_of_repeating* = 10 and *the_number_of_offspring* = 100.

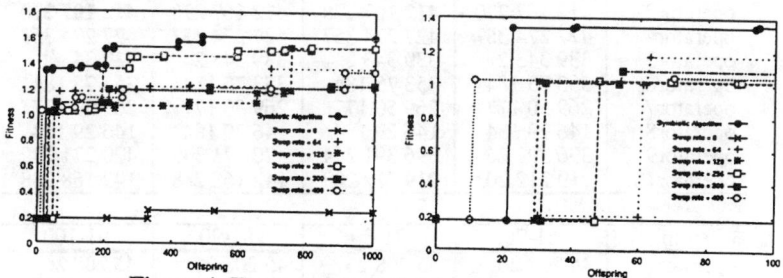

Figure 4. The transition of the maximum fitness value.

Figure 4 signifies the transition of solution's fitness in one result of simulations. The horizontal axis is the number of generated offspring. The vertical axis is the solution's fitness. The improvement in the fitness plots the maximum value in the generation. The part from 0 to 100 of the left figure of the Figure 4 is shown in right figure.

From the right figure, we can confirm that acquisition of a correct solution becomes faster with an increase in swap rate (excepting the example of swap rate = 128). However, the left figure presents that the high swap rate cannot acquire a better solution. On the other hand, the symbiotic algorithm searches for the correct solution in offspring = 22 (it is very fast). Furthermore, we can confirm that slightly improvement after the acquisition. This means that the rank of correct action goes it up.

To clarify the relation between the improvement of the solution and operator used, acquired operators are shown in Table 2. Table 2 represents acquired operators in each generation. For example, the operator4 in offspring = 1-30 is {427, 274, 357}. This shows the number of gene swapped. The first number (427) is the number of gene swapped in the 1st move of solution. In the same way, the second and the third number mean the number of gene swapped in the 2nd and the 3rd move. Therefore, this operator swaps many genes. Furthermore, in this Table, the operators shown in bold are used preferentially in each generation.

Figure 4 has two search stages. The first stage searches for the correct solution. The second stage improves the rank of selected action. In other words, it is the first stage is generation from start to offspring = 22, and then the second stage starts. In the first stage, the operator4 was used preferentially (See the Table 2). The operator can acquire the correct solution by changing many genes. This result agrees with those obtained by changing swap rate. On the other hand, the second stage uses the operator that can swap the genes stably by the fixed parts. Using both operators means that the search involves the feature of both global and local. That is to say, mutation rate is changed automatically. Therefore, the symbiotic algorithm can acquire a better solution without designing the operator previously.

Table 2. Acquired operators.

offspring	1-30	31-40	41-50	51-60
operator1	45 208 26	**45 208 26**	45 208 26	45 208 26
operator2	140 165 288	140 165 288	**140 165 288**	**140 165 288**
operator3	412 107 270	**412 107 270**	412 107 270	412 107 270
operator4	**427 274 357**	427 274 357	**427 274 357**	427 274 357
operator5	339 343 2	339 343 2	339 343 2	339 343 2
operator6	353 75 184	353 75 184	353 75 184	353 75 184
operator7	269 50 177	269 50 177	269 50 177	269 50 177
operator8	146 29 184	146 29 184	146 29 184	146 29 184
operator9	396 391 264	396 391 264	420 271 349	420 271 349
operator10	219 212 203	219 212 203	142 168 288	142 168 288

offspring	61-70	71-80	81-90	91-100
operator1	45 208 26	45 208 26	45 208 26	**45 208 26**
operator2	**140 165 288**	**140 165 288**	**140 165 288**	**140 165 288**
operator3	412 107 270	412 107 270	412 107 270	412 107 270
operator4	427 274 357	427 274 357	427 274 357	427 274 357
operator5	**339 343 2**	**339 343 2**	**339 343 2**	406 271 338
operator6	353 75 184	353 75 184	410 267 351	410 267 351
operator7	414 271 345	414 271 345	414 271 345	414 271 345
operator8	409 269 342	409 269 342	409 269 342	409 269 342
operator9	420 271 349	420 271 349	420 271 349	420 271 349
operator10	142 168 288	142 168 288	142 168 288	142 168 288

6 CONCLUSIONS

In this paper, we designed the symbiosis between the solutions and their operators for the Tsume-shogi problem. Results of simulations represented the acquisition of two kinds of operators that search globally and locally. From the results, the characteristic of symbiotic algorithm is changing the mutation rate adaptively. We consider that the symbiotic algorithm can decide the value of genetic parameter with searching solutions. As the future work, we need to investigate the symbiotic co-evolution of other operators for confirming efficiency in different fields.

ACKNOWLEDGEMENTS

This was supported in part by the Ministry of Education of Japan under Grant 12780288 in Aid for Scientific Research.

REFERENCES

Goldberg, D.E., 1989, "Genetic Algorithms in Search, Optimization, and Machine Learning," Addison-Wesley.

Iino, Kenji, 1998, " Tsume-shogi," Ikeda shoten (in Japanese).

Lincorn, Boxshall, Clark (editor), 1982, "A dictionary of ecology, evolution and systematics," CAMBRIDGE UNIVERSITY PRESS.

Paredis, J., 1995, "The Symbiotic Evolution of Solutions and their Representation," In Proceedings of the Sixth International Conference on Genetic Algorithms (ICGA'95), Eshelman, L. (editor), Morgan Kaufmann.

Paredis, J., 1994, "Steps towards coevolutionary classification neural networks," In Proceedings of the Fourth International Workshop on Synthesis and Simulation of Living Systems, pp.102-108.

Potter, M.A. and De Jong, K.A., 1994, "A Cooperative Co-Evolutionary Approach to Function Optimization," In Proceedings of the Third Parallel Problem Solving from Nature, pp.249-257.

M. Nerome, E. Satoshi, K. Yamada, H. Miyagi, 2000, "ANALYSIS OF THE BEHAVIOR OF SYMBIOTIC EVOLUTION ALGORITHM FOR N-QUEENS PROBLEM," Intelligent Engineering Systems Through Artificial Neural Networks, Vol.10, ASME PRESS, pp.251-256.

USE OF EMULATIONS OF THE IMMUNE SYSTEM TO HANDLE CONSTRAINTS IN EVOLUTIONARY ALGORITHMS

CARLOS Ã. COELLO COELLO
CINVESTAV-IPN
Depto. de Ingeniería Eléctrica
Sección de Computación
Av. IPN No.2508
Col. San Pedro Zacatenco
México D.F. 07300 MEXICO
ccoello@cs.cinvestav.mx

NARELI CRUZ CORTES
CINVESTAV-IPN
Depto. De Ingeniería Eléctrica
Sección de Computacón
Av. IPN No. 2508
Col. San Pedro Zacatenco
México, D.F. 07300 MEXICO
nareli@computacion.cs.cinvestav.mx

ABSTRACT
In this paper we propose an improved version of a constraint-handling scheme based on a model of the immune system. The approach is coupled to a genetic algorithm and used for global optimization. Our experiments indicate that the approach is relatively easy to implement, and it is also computationally efficient.

1 INTRODUCTION

Genetic algorithms (GAs) have been quite successful in a wide variety of optimization problems (Goldberg, 1989). However, GAs, are an unconstrained search technique and, therefore, require an additional mechanism to incorporate constraints of any type (linear, nonlinear, equality and inequality) into the fitness function in order to guide the search properly. The approach most commonly used to incorporate constraints is the penalty function (mainly exterior), and there have been many successful applications of this approach in the literature (Smith & Tate, 1993). However, penalty functions have some well-known limitations (Richardson et al., 1989), from which the most remarkable is the difficulty to define good penalty factors. These penalty factors are normally generated by trial and error, although their definition may severely affect the results produced by the GA.

In this paper, we propose an algorithm based on emulations of the immune system to handle the constraints of a problem being solved by a GA. The approach does not require the definition of any penalty factors, it is conceptually simple and efficent, and it produces results that are competitive with those produced by the best constraint-handling technique used for GAs known to date.

The paper es organized as follows. Section 2 defines the problem to be solved. The algorithm proposed is discussed in Section 3 and is validated in Section 4. The paper ends with our conclusions and a discussion of some possible paths of future research.

2. THE IMMUNE SYSTEM

Nature has served as inspiration for many of the scientific and technological advancements of humankind. Humans, after all, can be seen as complex biological information processing systems capable of performing complex tasks even in the presence of uncertainty and lack of information.

From the processing perspective, the immune system is seen as a parallel and distributed adaptive system (Dasgupta & Attoh-Okine, 1997). It is capable of learning, it uses memory and is able of associative retrieval of information in recognition and classification tasks. Particularly, it leads to recognize patterns, it remembers patterns that it has been shown in the past and its global behavior is an emergent property of many local interactions. All these features of the immune system provide, in consequence, great robustness, fault tolerance, dynamism and adaptability (Forrest & Hoffmeyr, 2000). These are the propierties of the immune system that mainly attract researches to try to emulate it in a computer.

There are certain computational models that emulate some specific process of the immune system with the purpose of understanding its nature (Forrest & al Hoffmeyr, 2000). For example, the model of specific response that the immune system presents in the presence of an invader (primary response), was designed by Forrest et al. (1993) and Smith & Perelson (1991). On the other hand, the first computational model that emulates the secondary response (memory) of the immune system was developed by Derek Smith (1994).

Farmer et al. (1986) were the first to suggest a way of representing the immune system in a computer. In their model, both antigens and detectors are represented as strings of symbols in a small alphabet, and the interactions among the strings represent molecular bonds.

Specifically within the area of the artificial immune system, there are two main models in which most of the current work is based (Dasgupta & Attoh-Okine, 1997): the model of immune network and the negative selection algorithm. Both will be briefly described next.

2.1 MODEL OF IMMUNE NETWORK

This is a mathematical model of the immune system developed by Jerne (1973). Is uses differential equations that simulate the dynamics of the linphocites (increase of decrement of the concentration of a certain set of linphocites'clones). This model is based on the idea that linphocites do not work in an isolated maner, but as an interconnected network that works in levels. Based on this model, Smith & Perelson (1993) presented a probabilistic study of the idiotypical networks. This is a very formal study where the transition phases of the idiotypical network are discussed.

2.2 NEGATIVE SELECTION ALGORITHM

Forrest et al. (1994) developed the negative selection algorithm for detection of changes. This algorithm is based on the discrimination principle that the immune system uses to know what is part of it and what is not. This algorithm generates detectors in a random manner, eliminating those that are not capable of recognizing themselves. Then, only those detectors able to identify invaders are kept. This algorithm seems to have a great potential in applications related to change detection, due mainly to the fact that such detection is performed probabilistically by the algorithm. It is also a robust system because it looks for any alien action instead of just looking for a certain specific pattern of changes.

Besides the models previously mentioned, there are others used to simulate different aspects of the immune system such as its ability to detect patterns in a noisy enviroment, its ability to discover and maintain diverse classes of patterns and its ability to learn effectively, even when not all possible types of invaders have been previously presented to it.

3 THE PROPOSED APPROACH

Our algorithm to handle constraints is based on the negative selection model of Forrest et al. (1994) and is described next:

1. Generate randomly an initial population for the GA.
2. If the initial population contains a mixture of feasible and infeasible individuals, then divide the population in two groups. The first contains the infeasible individuals, which are denominated "antibodies", and the second contains the feasible individuals, which are called "antigens".
3. If none of the individuals in the initial population is feasible, then use the magnitude of constraint violation of each individual as its fitness. Then, use as the "antigen" the best individual in the population For the assignment of antibodies, it is important to consider only a portion of the population, to avoid premature convergence. Under this scheme (i.e., no feasible individuals in the initial population), we run our simulation of the immune system based on minimum contraint violation and we verify at certain intervals if we have reached the feasible region. Whenever that happens, then we use the feasible individuals found as our antigens, and we proceed to divide the population as in the previous step.
4. The fitness of all the antibodies is set to zero.
5. The fitness of the antibodies is computed according to their similarity with a set of pre-determined antigens in the following way:
   - An antigen is randomly selected from the antigens population.
   - From the population of N antibodies, we randomly select a sample of size without replacement.
   - Each antibody in the sample is compared against the antigen selected, and we compute the result of the comparison, to which we call Z (matching magnitude). Z represents a distance (normally but not necessarily Euclidian) measured at the genotype level (i.e., at the level of the chromosomic encoding)
   - We identify the antibody in the population that has the highest matching magnitude (Z). Ties are solved by randomly choosing a winner.
   - The matching magnitude of the winner antibody is added to its fitness. The fitness of the other antibodies remains unaltered.

- The antibodies are returned to the population, and the process is repeated typically three times the number of antibodies.
6. Based on the fitness computed in the previous step, the population of antibodies is reproduced in a traditional genetic algorithm (using crossover and mutation).
7. The process is repeated from the fourth step until convergence (i.e., when the mean and the maximum fitness in the population are practically the same) or until we reach a maximum number of iterations.
8. Individuals are returned to the external GA and we proceed in the conventional way.

There are a few issues that need to be mentioned. First, the approach is really using a GA embedded inside another GA used to optimize a certain function. However, the GA that is run with the emulation of the immune system does not use the fitness function directly and, therefore, does not require to evaluate the objective function of the problem. Also, the implicit premise of the technique is that, under certain conditions, the reduction of genotypic differences between two individuals will produce, as consequence, a phenotypic similarity, which, in our case, will make that an infeasible individual approaches the feasible region.

The size of the sample σ of antibodies determines if the antibodies are generalists (i.e., able to distinguish only a small variety of antigens). If the value of σ (i.e., size of the sample of antibodies) is sufficiently large, then the antibodies tend to be specialists, since there is a larger probability of an antibody to match exactly a certain antigen. Conversely, if the value of σ is small, then the antigens tend to be generalists. When a sample of size σ is equal or slightly larger than a number of antigens, then both generalists and specialists individuals exist in the population. Generalist antibodies contain fragments of information that encode good solutions.

Our algorithm is an extension if the proposal of Hajela & Lee (1996) in which, among other things, they always assumed an initial population containing a mixture of feasible and infeasible individuals, and considered only binary encoding. Although the results reported un this paper were produced using binary representation, our algorithm can be used with any other representation (appropiate mtching functions have been defined for other encodings, such as integer representation). We have also performed a comprehensive empirical study of the performance of the algorithm when used to solve several standard benchmark problems reported in the literature. Our study has included an analysis of the role of σ in the performance of the algorithm. To the best of our knowledge, this study is the first of its type for this sort of contraint-handling technique based on the immune system.

4 COMPARISON OF RESULTS

Three examples selected from the literature have been chosen to validate our approach. In all the experiments reported in this paper, our genetic algorithm used binary representation, stochastic remainder selection, one-point crossover

(crossover rate of 0.75), and uniform mutation (mutation rate of 0.05). We used a population size of 90 and ran our genetic algorithm for 5000 generations (i.e., we performed 450000 fitness function evaluations), except for the first example in which the maximum number of generations was set to 10000 (i.e., 900000 fitness function evaluations)

The summary of results is shown in Table 2. We compared our results with the homomorphous mapping of Koziel & Michalewicz (KM) (1999), which is the best constraint-handling technique known to date. Koziel & Michalewicz (1999) performed 70x20000=1400000 fitness function evaluations to solve each of the examples.

		Best result		Mean result		Worst result	
TF	optimal	Immune	KM	Immune	KM	immune	KM
1	-30665.539	-30664.8	-30664.5	-30632.4	-30665.3	-30493.7	-30645.9
2	-6961.814	-6961.7	-6952.1	-6950.7	-6342.6	-6819.0	-5473.9
3	1.000	1.000	0.999999	1.000	0.999135	1.000	0.991950

Table 2: comparison of the results for the three test functions selected.

5. CONCLUSIONS AND FUTURE WORK

We have presented a constraint-handling approach based on emulations of the immune system (particularly, using the negative selection model) that was incorporated into a genetic algorithm used for global optimization. The approach seems to be competitive with state-of-the-art constraint-handling techniques, since it produces reasonably good results at a relatively low computational cost. However, more work is necessary to validate the sensitivity of the algorithm to its parameters (i.e., a more detailed statistical analysis over a larger set of test functions). Furthermore, we are interested in extending this algorithm to deal with constrained multiobjective optimization problems.

ACKNOWLEDGMENTS

The first author acknwledges partial support from CONACyT through NSF-CONACyT project number 32999-A and from CINVESTAV through project JIRA'2001/08. The second author acknowledges support from CONACyT through a scholarship to pursue graduate studies in Computer Science at the Computer Science Section of the Electrical Engineering Departament at CINVESTAV-IPN.

REFERENCES

(Coello, 1999) Carlos A. Coello Coello, "A Comprehensive Survey of Evolutionary-Based Multiobjective Optimization Techniques", *Knowledge and Information Systems. An International Journal*, 1(3):269-308, August.

(Dasgupta & Attoh-Okine, 1997) Dipankar Dasgupta and Nii Attoh-Okine. "Immunity-Based Systems: A Survey" In *IEEE International Conference on Systems, Man and Cybernetics*, Orlando, Florida, October.

(Farmer et al., 1986) J. D. Farmer, N. H. Packard and A. S. Perelson. "The Immune System, Adaptation, and Machine Learning", Physica D, **22**:187-204.

(Forrest et al., 1993) S. Forrest, B. Javornik, R. Smith, and A.S. Perelson. "Using genetic algorithm to explore pattern recognition in the immune system", *Evolutionary Computation*, **1**(3):191-211.

(Forrest et al., 1994) S. Forrest, A. S. Perelson, L. Allen, and Cherukuri. "Self-nonself discrimination in a computer", In *IEEE Symposium on Research in Security and Privacy*, pages 202-212. Oakland, CA, May 16-18, Springer-Verlag.

(Forrest & Hoffmeyr, 2000) Stephanie Forrest and Steven A. Hofmeyr. "Immunology as Information Processing", In L.A. Segel and Cohen, editors, *Design Principles for the Immune System and Other Distributed Autonomous Systems*, Santa Fe Institute Studies in the Sciences of Complexity, pages 361-387. Oxford University Press.

(Forrest & Perelson, 1991) Stephanie Forrest and Alan S. Perelson. "Genetic algorithms and the Immune System", In Hans-Paul Schwedfel and R. Männer, editors, *Parallel Problem Solving from Nature*, Lecture Notes inm Computer Science, pages 320-325. Springer-Verlag, Berlin, Germany.

(Frank, 1996) Steven A. Frank. *The Design of Natural and Artificial Adaptive Systems*. Academic Press, New York.

(Goldberg, 1989) David E. Goldberg, *Genetic Algorithms in Search Optimization and Machine Learning*. Addison-Wesley Publishing Co., Reading, Massachusetts.

(Hajela & Lee, 1996) P. Hajela and J. Lee. "Constrained Genetic Search via Schema Adaptation. An Immune Netwok Solution", *Structural Optimization*, **12**:11-15.

(Jerne, 1973) N.K. Jerne. "The Immune System", *Scientific American*, **229**(1):52-60.

(Koziel & Michalewicz, 1999) Slawomir Koziel and Zbigniew Michalewicz. "Evolutionary Algorithms, homomorphous Mappings, and Constrained Parameter Optimization", *Evolutionary Computation*, **7**(1):19-44.

(Michalewicz et al., 1996) Zbigniew Michalewicz, Dipankar Dasgupta, R. Le Richie, and Marc Schoenauer. "Evolutionary algorithms for constrained engineering problems", *Computers & Industrial Engineering Journal*, **30**(4):851-870, September.

(Michalewicz & Schoenauer, 1996) Zbigniew Michalewicz and Marc Schoenauer, "Evolutionary Algorithms for Constrained Parameter Optimization Problems", *Evolutionary Computation*, **4**(1):1-32.

(Richardson et al., 1989) Jon T. Richardson, Mark R. Palmer, Gunar Liepens, and Mike Hilliard. "Some guidelines for genetic algorithms with penalty functions", In J. David Schaffer, editor, *Proceedings of the Third International Conference on Genetic Algorithms*, pages 191-197, Morgan Kaufmann Publishers.

(Smith & Tate, 1993) Alice E. Smith and David M. Tate, "Genetic Optimization Using a Penalty Function", In Stephanie Forrest, editor, *Proceedings of the Fifth International Conference on Genetic Algorithms*, pages 499-503, San Mateo, California, July, Morgan Kaufmann Publishers.

(Smith, 1994) Derek Smith. "Towards a Model of Associative Recall in Immunological Memory", Technical Report 94-9, University of New Mexico, Alburquerque, NM.

(Smith & Perelson, 1993) R. Smith and A. S. Perelson. "Searching for Diverse, Cooperative Population with Genetic Algorithms", *Evolutionary Computation*, **2**(1):127-149.

EXPLORING THE USE OF ENTROPY IN UNDERSTANDING THE BEHAVIOR OF THE ENHANCED GENETIC ALGORITHM

LIXIN YANG
Computing and Information Science
University of Guelph, Guelph, ON
Canada N1G 2W1

DEBORAH A. STACEY
Computing and Information Science
University of Guelph, Guelph, ON
Canada N1G 2W1

ABSTRACT
We will present in this paper a demonstration that genetic search can be studied as an entropic phenomenon. We look in depth at the performance differences between the traditional Genetic Algorithm (GA) and the Enhanced Genetic Algorithm (EGA) based on measures of statistical entropy and information. Specifically, for the purpose of theoretical research, we use the Royal Road functions to demonstrate what we expect measures of entropy and information to provide and then we discuss the application of our entropy and information hypothesis to the Traveling Salesman Problem.

INTRODUCTION
The Enhanced Genetic Algorithm (EGA) employs three major enhancements with respect to the traditional GA: selection method, mutation method, and partial elitism policy [1]. Two randomly selected individuals function as parents and of the parents and the resultant offspring, the two fittest are chosen for the new population after crossover. The appropriate operator rates need not be determined a priori; crossover is always performed, while the mutation operation is applied at appropriate times. The mutation operation is renamed the *invader* operator since it functions to interject a novel random individual whenever certain conditions are met and before the new individuals are inserted into the next generation. These enhancements enable the EGA to run in an unsupervised mode, i.e. it is not dependent on parameter selection.

ENTROPY IN GENETIC SEARCH
One aspect of genetic search is that the evolutionary process operates on a population with a limited size. More and more fit individuals are produced and survive in the population by competition, selection, various genetic operation, and reproduction from one generation to the next. Thus, our investigation will focus on the research of entropy in the population. The Royal Road functions [2] are used to demonstrate what we expect measures of entropy and information to provide.

In the Royal Road functions we consider a population consisting of a set of independent events. Each event has two possibilities, 0 and 1. Every allele at the same locus in each individual of the current population contributes to the corresponding probability occurrence in that event. As an example, assume every individual in a population of size 8 is 4 bits long, so there are 4 independent events X_1, X_2, X_3, and X_4. Since the events are independent, according to the Shannon's entropy theory, the entropy of the population is: $H(X_1, X_2, X_3, X_4) = H(X_1) + H(X_2) + H(X_3) + H(X_4)$ while $H(X_i) = -(p\log p + q\log q)$, where X_i represents different events, p and q are the probability of 1 and 0 occurring in that event. It is easily shown that the population will have the maximum entropy if and only if each event's entropy gets to the maximum; in other words, $p = q = 0.5$ appears in each event.

We will assume that information is the reduction of uncertainty and that uncertainty and entropy are essentially identical. Suppose that the entropy of a population is H_{before}, after a period of time of genetic reproduction and development, the entropy of the population of the current generation reduces to H_{after}, then the reduction is remarked as some information R [3] has been received as computed in $R = H_{before} - H_{after}$ In a traditional genetic search process, once an outstanding or super individual takes over the whole population, the population's entropy H_{after} will be close to 0 according to the Shannon entropy theory. The entropy of the population H_{before} represents how much information a population may contain.

Information in biological evolution can be divided into two major categories [4]: **D-information (distinguishablity information)**, and a subset relevant to evolutionary selection or survival, called **SR-information**. Survival Relevant Information (SR-information) can be either **SU-information** (useful) or **SH-information** (harmful). This categorization is used for our fundamental hypothesis to the analysis of genetic search. *Hypothesis 1: D-information and SU-information both exist in the population, and the population is the carrier of them. H_{before} is the quantity form of D-information and it is related to the diversity of population, whereas SU-information greatly lies on the survival of good individual with better fitness. The more D-information, the more efficient the process of evolution is, and vice versa.*

For the traditional genetic search, the D-information tends to decrease (all individuals in the population gradually become similar); on the other hand, the SU-information is increased to a certain level (lots of high fitness individuals in the population). Based on this phenomenon, we produce the second hypotheses about genetic search. *Hypothesis 2: The genetic search is a process of consuming D-information to accumulate SU-information. When the amount of D-information is too low, there is no inside force to push genetic search forward, and then the evolution process should stop.*

CONSUMERS OF D-INFORMATION IN GENETIC SEARCH

In our initial study of the GA and EGA using the Royal Road functions, we compared a GA using single-point crossover, sigma scaling with the maximum expected offspring of any string being 1.5, a crossover rate of 0.7 per pair of parents, a mutation probability of 0.005 per bit, Roulette Wheel selection, and a population size of 128. The EGA employed single-point crossover and used a population size of 128.

To answer the question, "Who is the consumer of D-information?" we examined the fitness and entropy profiles under a typical GA run without the participation of mutation and the fitness and entropy profiles under a typical EGA run without using the invader operator. The entropy profiles show an obvious trend of quickly decreasing and staying at a very low level. Meanwhile the fitness profiles present a short increasing trend but then obviously get stuck in some local optimum after a long time run. In other words, D-information is gradually decreased without extra reinforcement and the useful information has no efficient accumulation during this kind of genetic search based on limited D-information in the population when there is no mutation or invader operator. Therefore, we believe that the selection process and crossover operation primarily play the role of consumers in the genetic search. By consuming the D-information, they produce the SU-information. Also, once the D-information is extreme low, there is no force to push the genetic search forward and the genetic process tends to stop.

SUPPLIER OF D-INFORMATION IN GENETIC SEARCH

Mutation—Supplier of D-Information in Traditional GAs

In traditional GAs, mutation is considered to be a disruptive force. Let us look at the effect of mutation from the view of entropy. For each column of population, each allele has theoretically an equal probability distribution at the initial stage of the GA. Should there are more 1s in a column after a period of time, the chance to flip 1 to 0 is higher, and vice versa. The result of a long run is to have a balance of 1s and 0s in a column. According to entropy theory, mutation has the effect of increasing the entropy in a population. Since $R = H_{before}$, we assert that mutation has the effect to increase D-information or that it is the supplier of D-information of a population.

Every time a super-individual appears in a population it takes over part of the population, and this similarity decreases the entropy of population. So we can observe that following the increase of fitness profile, the entropy profile will always decrease obviously but then will be brought up by the operation of mutation. The general tendency of the entropy profile is deceased periodically following each increase of fitness profile. With the fitness profile closing to the optimum, the ability of mutation to bring up the entropy becomes less and less. When the optimum is found, the entropy profile will not go up. This is because the result of the mutation operation will not survive the selection. Therefore, only a surviving mutation reinforces the D-information to the population.

Drawback of Mutation as the D-information Supplier of GAs
When mutation brings D-information to the population, it also raises the chance to kill existing good individuals and destroy existing SU-information. We call this type of D-information, SH-information (harmful). If we increase the mutation rate, for example from 0.005 to 0.01, we observed that although the reinforcement of D-information can be maintained, the accumulation property of SU-information to find good individuals has been seriously hampered by the SH-information. In this sense, researches usually use very small mutation rate when applying GAs to solving problems. However, a small mutation rate will make both the reinforcement of D-information and the accumulation of SU-information very inefficiently. Therefore, the choice of mutation rate is always an overhead problem in GAs.

The Invader—Supplier of D-Information in the EGA
Information is useless if it is redundant or superfluous and adds nothing to the predictive or explanatory value of a theory. The traditional GA has a tendency to create similarity in the population and inevitably these extra "copies" of SU-information become useless, and they reduce the efficient usage of the population to carry more D-information. By contrast, the EGA purposefully protects the existed SU-information to prevent it from being destroyed by SH-information, while efficiently using the limited population to carry more information by reducing the duplication. In the EGA, the selection strategy, genetic operators (especially the invader operator) and reproduction have been redesigned and reorganized to enable the invader to use a new randomly generated individual replacing one of the redundant individuals at appropriate time (when two children individuals have equal fitness with great similarity). These newly formed individuals randomly contribute 1s and 0s to each column and have the effect of balancing 1s and 0s in each column therefore increasing the entropy, or D-information, of the population according to the entropy theory. Since only the useless, or superfluous, "SU-information" is eliminated when the

invader is invoked, the genetic search of EGA will be more efficient in the accumulation of SU-information based on consuming the productive D-information. Moreover, the selection strategy of EGA allows the new invader individuals (which probably have relatively low fitness) to survive to crossover with existing strong individuals and therefore they may partially survive during reproduction and thus produce the effect of bringing new information into the population. Therefore, in the EGA, the redundant SU-information has been eliminated; plus, more D-information is reintroduced in the population with the least harm to the existed SU-information. The invader is the supplier of D-information in the EGA. Here, we assume that when two children have equal fitness and are greatly similar, they have accumulated identical and equal amounts of SU-information.

Can entropy increase in an unlimited fashion? No. When the entropy is increased, the similarity of the population is corresponding decreased, and the probability of the invader operation being invoked is reduced as well. At some point, the invader profile will become dynamically stable. Thus this point is called the dynamic stable point and it reflects the maximum number of new individuals that can be brought into the population by the invader operator. By observing the EGA process, the experimental value of this point is around 17~20% of the population size for solving a Royal Road function. When the invader profile becomes dynamically stable, the entropy profile becomes dynamically stable as well.

POPULATION IN GENETIC SEARCH
In the traditional GA, evolution is based on a comparatively small population. When a good individual is found, it is fatal to others due to the fixed population size. The good individuals easily take over the whole population and the diversity of population is correspondingly decreased. The smaller the population size, the greater the takeover effect. The maintenance and reinforcement ability of the D-information of the population is helpful to the genetic search. This actually explains why a larger population can usually improve a GA's performance (since at the initial stage of the GA, the larger and randomly distributed population introduces more D-information into the population. This slows down the takeover process in the genetic search). The lack of reinforcement of D-information is an obvious drawback in the later stages of traditional GAs. By contrast, the selection method of EGA allows all individuals in the population to have an equal chance to take part in the production of offspring. Therefore, the takeover effect is decreased. Thus, the strong ability to maintain and reinforce D-information and to efficiently accumulate SU-information in the EGA makes it more efficient than the traditional GA even using smaller populations.

ENTROPY AND INFORMATION IN THE TSP

Eilon's 75-city [5] TSP is used as the second application to demonstrate our entropy and information hypothesis. In our TSP experiment, both the GA and EGA employ fifty-fifty greedy crossover and use a population of size 30, and the genetic search stops after 2000 generations. The GA uses fitness-proportionate selection with sigma scaling, a mutation rate of 0.005 and 0.7 for the fifty-fifty greedy crossover probability. As expected, the EGA always outperforms the GA. The application of our entropy and information hypothesis for the TSP presents very similar characteristics compared to applying it to the Royal Road functions. But, the EGA's entropy profile is different from the one in the Royal Road functions problem since even when the threshold is set to zero, the entropy profile is still well maintained. This is because in the TSP, there are very few different individuals having the same fitness in the whole data set. Hence, with the invader operation, it is impossible to have many similar individuals accumulated in the population to bring down the entropy profile, no matter where the threshold is set.

DISCUSSION

By studying and analyzing the relationship of entropy, information and genetic search, we can describe the average quantities of the population after selection, mutation, and crossover in terms of those before, and allows one to derive deterministic prediction of performance of genetic algorithm dynamics. The general idea of our approach is to provide a general quantitative mechanism to help in analyzing GAs. Moreover, we believe that a better understanding of entropy production, reinforcement of D-information and the accumulation of SU-information will enable us to better understand the nature of genetic search.

REFERENCES

[1] Gary W. Grewal, Thomas C. Wilson, and D.A. Stacey. *Solving Constraint Satisfaction Problems Using Enhanced Genetic Algorithms*. Proc. of the ANNIE, 1999.

[2] Mitchell, M., Forrest, S. *Fitness Landscapes: Royal Road Functions*. In Back, Fogel, and Michalewicz, (eds.), Handbook of Evolutionary Computation, Oxford University Press, 1992

[3] Thomas D. Schneider, *Information Theory Primer*, 2000 June 26, ftp://ftp.ncifcrf.gov/pub/delila/primer.ps

[4] Daniel R. Brooks and E.O. Wiley, *Evolution as Entropy*, 2nd Ed., U Chicago Press, 1988

[5] D. Whitley, T. Starkweather, and D'Ann Fuquay. *Scheduling problems and traveling salesman: the genetic edge recombination operator*. Proc 3rd ICGA, pp 130-140. 1989.

DNA LOCAL SEQUENCE ALIGNMENT WITH GENETIC ALGORITHMS

MARK J. EMBRECHTS   ROBERT A. BRESS

Decision Sciences and Engineering Systems Department
Rensselaer Polytechnic Institute
Troy, New York 12180

ABSTRACT

Bioinformatics now plays a key role in many aspects of genetics research. The identification of specific genes in stretches of raw genomic DNA sequences is an important component of DNA sequencing projects. The problem of finding these genes proves to be more of a computational challenge than a biological one. Less than 1% of the human genome is made up of actual genes. The issue at hand is how to find sequences similar to known genes within DNA data that is on the order of a billion nucleotides long. The problem becomes increasingly challenging as greater differences in these matches are allowed. In this paper, genetic algorithms are examined as a means of identifying gene matches.

INTRODUCTION

The DNA pair-wise local alignment problem is to align two sequences of DNA such that the highest matching score is attained given an evaluation function and scoring criteria. Typically, this problem involves searching for a gene or specified sequence of amino acids among an entire genome that could be billions of amino acids long. When mismatches are allowed within the solution there can be many potentially useful results depending on how mismatches are penalized. For this reason, many of the top scoring matches may be identified as potential "true" matches. In practice, a dynamic programming approach is used to identify the highest scoring matches. Here, genetic algorithms are investigated as an alternate method for local alignment.

LOCAL ALIGNMENT WITH GENETIC ALGORITHMS

Motivation

The characteristics associated with genetic algorithms make the approach particularly useful for the local alignment problem. The potential advantage of the use of GAs over dynamic programming is threefold: (1) It is possible with GAs to formulate different alignment objectives (e.g., matching of reverse strings) that would be hard to achieve with dynamic programming; (2) it is possible to co-evolve the scoring matrix in the case of a GA-based alignment; (3) it is possible to use the GA approach as a pre-processing step to find regions of crude matches.

Methodology

For this proof-of-concept paper two artificial strings of amino acids were generated at random. The first string represents the "library gene" of 100 amino acids and has no gaps. The second string, the "search string" is derived from the first string by inserting up to 20 gaps (with a begin and end-gap), with gap lengths between 5 and 20 characters long. There was no substitution of amino acids in the second string, and for the purpose of this experiment the gap penalty consists just of the number of gaps. The genetic algorithm is a floating point GA with traditional and task-specific operators for mutation and crossover. The GA basically determines the positions and length of up to 20 gaps in the search string and optimizes the score based on the BLOSUM50 scoring matrix and the number of gaps.

The color plot in Figure 1 shows the target results obtained for an alignment problem. Here, the Search String is the gapped sequence and the Library String refers to the gapless sequence. The top row and leftmost column represent the color codes for the amino acid string encoding and is barely within viewing resolution. The majority of the pixels of the plot represent BLOSUM50 scores between corresponding amino acids in the search string (horizontal) and the library string (vertical). It is very clear to see that there are 5 gaps present.

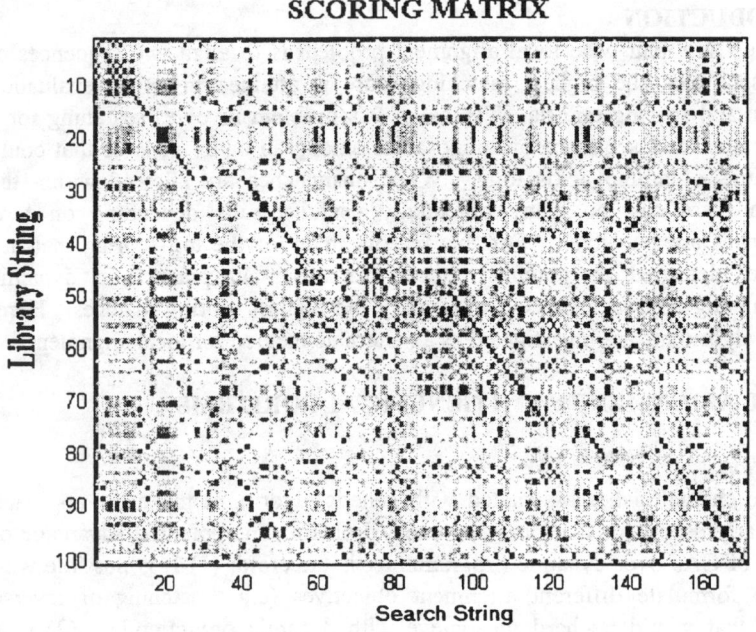

Figure 1: Target Matching Results

The task of the GA is now to find the gaps by guessing the corresponding gap lengths for up to 20 gaps. Generally, there are less than 20 gaps for this problem since it is expected that several gaps will overlap. Figure 2 shows the alignment results from comparing the same sequences used in Figure 1 using the genetic algorithm. A quick visual comparison during the execution of the GA shows that the GA was indeed able to find most gaps so far with minor differences around amino acid 135 on the search string. Gaps can be seen as breaks in the black diagonal in Figures 1 and 2. The vast mono-shade sections in figure 2 above and below the multicolored diagonal section are matches not relevant because no matches can occur there, since in this example we did not consider a beginning or end gap. For the final stage of this GA all gaps were correctly identified.

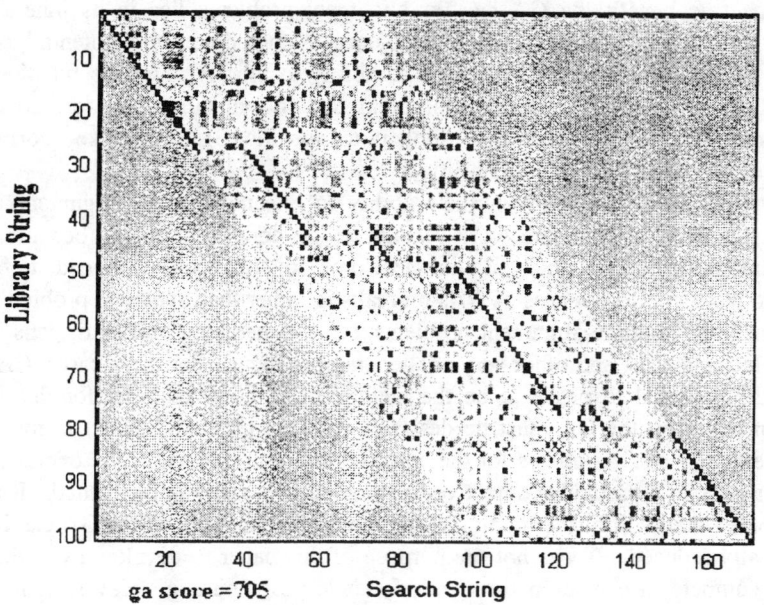

Figure 2: Matching Results During the Genetic Algorithm

The GA employed for this analysis is a floating point GA library developed for the StripMiner[TM] code for the DDASSL project (Embrechts et al. 2001). For this problem, we designed specific mutation and crossover operators. The GA has a population of 100 members and ran 5000 iterations. Execution took on the order of two minutes on a 300 MHz PC.

CONCLUSIONS

The main concerns with implementing a GA for the alignment problem deal with its ability to find an optimal or near optimal solution and the computation costs associated with time and memory space. It has been shown here that GAs can indeed be used to attain near-optimal solutions that can be used as a pre-processing step that finds regions of crude matches. This is evident just from quick glances at Figures 1 and 2. The computation time here was negligible but that is likely due to the relatively small size of the problem.

Dynamic programming is typically the method of choice for this kind of problem due to its dependability in attaining an optimal solution. The challenge that dynamic programming faces is whether it can handle a larger variety of potentially more challenging alignment problems and whether it can efficiently process massive problems. The fact is that dynamic programming results in exponential time complexity. This becomes more of a problem when solving multiple alignment problems but is nonetheless an obstacle if not now then one for future research.

The true benefit the GA has for alignment problems lies in its inherent ability to deal with a wide variety of problems. As noted before, a potential use would be matching reverse strings. The ability to modify objective functions with little overhead, to develop problem specific mutation and crossover operators, and to re-evolve the problem to specific BLOSUM-like scoring matrices all add to the GA's flexibility.

Though GAs have proven their ability to execute a gene alignment in reasonable time, much more work remains to be done before GAs become a mainstream method for practical alignment problem solving. Work needs to be done to test the effectiveness of GAs on realistic rather than synthetic problems. This includes multiple alignment problems, reverse-matching problems, gaps in multiple sequences, and other very challenging problems. For these cases, GAs have the potential to prove to be the method of choice. In order for this to happen however, the GA must be designed with the problem context in mind. The flexibility of GAs allows for the construction of problem-specific operators and changes in a particular objective function can be easily implemented. It is anticipated that with further coding effort the execution speed of the GA can be drastically reduced. It was not the purpose of this paper to develop a GA that could compete in execution time with dynamic programming. However, it is expected that a GA implementation for sequence alignment can be speeded up significantly with multi-resolution implementation of the algorithm and standard methods for code parallelization.

ACKNOWLEDGEMENT

This work is supported by the National Science Foundation (award number IIS-9979860).

REFERENCES

R. Durbin, S. Eddy, A. Krogh, and G. Mitchinson, 1998, *Biological Sequence Analysis*, Cambridge University Press

Baldi, P., Brunak, S., 2001, *Bioinformatics: The Machine learning Approach, Second Edition*, MIT Press

W. Pearson, LALIGN, http://www.ch.embnet.org/software/LALIGN_form.html

Shaun D. Black, The Genetic Code, http://psyche.uthct.edu/shaun/SBlack/geneticd.html

Notredame, C., Higgins, D., 1996, SAGA: Sequence Alignment by Genetic Algorithm, *Nucleic Acids Research*, Vol. 24, no. 8, Oxford University Press

Karadimitriou, K., Kraft, D., 1996, Genetic Algorithms and the Multiple Sequence Alignment Problem in Biology, *Proceedings of the Second Annual Molecular Biology and Biotechnology Conference*

Notredame, C., O'Brien, E., Higgins, D., 1997, RAGA: RNA Sequence Alignment by Genetic Algorithm, *Nucleic Acids Research*, Vol. 25, no. 22, Oxford University Press

Bennett, K., Breneman, C., Embrechts, M., 2001, Automated Design and Discovery of Novel Pharmaceutical using Semi-Supervised Learning in Large Molecular Databases, www.drugmining.com,

MULTI-RECOMBINANT EVOLUTION STRATEGIES FOR REAL-VALUED FUNCTION OPTIMIZATION

Yoshiyuki MATSUMURA   Kazuhiro OHKURA   Kanji UEDA
Kobe University; Rokko, Nada, Kobe 657-8501, JAPAN
{matsumu,ohkura,ueda}@mi-2.mech.kobe-u.ac.jp

ABSTRACT

The effect of multi-parent recombination on Evolution Strategies(ES) is empirically investigated on a set of real-valued test functions. The multi-parent versions of intermediate recombination, discrete recombination and global combined discrete recombination are applied to both objective variables and strategy parameters. Computer simulations of the multi-recombinant $ES((\mu/\mu,\lambda)$-ES) are conducted using Gaussian mutation for two different dimensions of test functions. Of the many formulations of ES, Classical-ES(CES) and Robust-ES(RES) are adopted for the observation.

INTRODUCTION

Evolutionary Computation(EC) has widely been recognized as a robust approach to various optimization problems. There are three main streams in this field, which are Evolution Strategies(ES), Genetic Algorithms(GA) and Evolutionary Programming(EP). Especially, ES has been focusing on numerical optimization since its birth. ES has several formulations, the most recent form is (μ,λ)-ES, where $\lambda > \mu \geq 1$. (μ,λ) means that μ parents generate λ offspring through recombination and mutation at each generation, and the best μ offspring are selected deterministically from the λ offspring and replace the current set of parents. In most cases, recombination operators have been investigated empirically, because they are mathematically intractable. Traditional recombination operators reproduce one offspring using two parent. Multi-parent version of recombination operators uses more often than the two parent version.

For ES, there are two popular global recombination operators, namely global discrete recombination and global intermediate recombination. Global recombination operators involving ϱ ($2 \leq \varrho \leq \mu$) parents have been formulated in (Schwefel, 1995). Beyer(1995) showed a similar generalization, written as $(\mu/\rho,\lambda)$-ES. He analyzed the case of $\rho = \mu$ on the sphere function theoretically to get λ-fold speedup compared to ES without recombination. In (Eiben and Bäck, 1998), (Bäck and Eiben, 1999) and (Gruenz and Beyer, 1999), while different kinds of global recombination operators were applied to object variables, global intermediate recombination was applied to strategy parameters.

This paper shows the effect of multi-parent recombination on Evolution Strategies through empirical experiments. The multi-parent versions of intermediate recombination, discrete recombination and global combined discrete recombination(Chang et al., 2001) are applied to both objective variables and strategy parameters. Computer simulations of the multi-recombinant ES $((\mu/\mu,\lambda)$-ES) are conducted using Gaussian mutation for two different dimensions of test functions, because Kursawe(1995) showed that an appropriate choice of the recombination operator depends not only on an objective function topology but also on the dimension of an objective function. This paper adopts Classical-ES(CES) by Schwefel(1995) and Bäck(1996) and Robust-ES(RES) by Ohkura et al.(2001) for the observation.

COMPUTATIONAL PROCEDURES OF EVOLUTION STRATEGIES
Classical Evolution Strategies

The computational steps of Classical ES(CES) are based on notations in (Schwefel, 1995) and (Bäck, 1996). CES is implemented as follows in this study:

(1) Generate the initial population of μ individuals, and set $g = 1$. Each individual is taken as a pair of real-valued vectors $(\boldsymbol{x}_i, \boldsymbol{\eta}_i), \forall i \in \{1, \cdots, \mu\}$, where $\boldsymbol{x}_i$ and $\boldsymbol{\eta}_i$ are the i-th coordinate value in R and its strategy parameters larger than zero, respectively.

(2) Evaluate the objective value for each individual $(\boldsymbol{x}_i, \boldsymbol{\eta}_i), \forall i \in \{1, \cdots, \mu\}$ of the population based on the objective function $f(\boldsymbol{x}_i)$.

(3) Each parent $(\boldsymbol{x}_i, \boldsymbol{\eta}_i), i = 1, \cdots, \mu$, creates λ/μ offspring on average, so that a total of λ offspring are generated. At that time, offspring are calculated as follows: for $i = 1, \cdots, \mu$, $j = 1, \cdots, n$, and $p = 1, \cdots, \lambda$,

$$\eta'_p(j) = \eta_i(j) exp\{\tau' N(0,1) + \tau N_j(0,1)\} \quad (1)$$
$$x'_p(j) = x_i(j) + \eta'_p(j) N_j(0,1) \quad (2)$$

where $x_i(j), x'_p(j), \eta_i(j)$ and $\eta'_p(j)$ denote the j-th component values of the vectors $\boldsymbol{x}_i, \boldsymbol{x'}_p, \boldsymbol{\eta}_i$ and $\boldsymbol{\eta'}_p$, respectively. $N(0,1)$ denotes a normally distributed one-dimensional random number with mean zero and standard deviation one. $N_j(0,1)$ indicates that the random number is generated anew for each value of j. The factors τ and τ' are commonly set to $\left(\sqrt{2\sqrt{n}}\right)^{-1}$ and $\left(\sqrt{2n}\right)^{-1}$. Various types of recombination can also be performed before calculating Equations (1) and (2).

(4) Calculate the fitness of each offspring $(\boldsymbol{x'}_i, \boldsymbol{\eta'}_i), \forall i \in \{1, \cdots, \lambda\}$, according to $f(\boldsymbol{x'}_i)$.

(5) Sort offspring $(\boldsymbol{x'}_i, \boldsymbol{\eta'}_i), \forall i \in \{1, \cdots, \lambda\}$ in non-descending order according to their fitness values, and select the μ best offspring out of λ to be parents of the next generation.

(6) Stop if the halting criterion is satisfied(i.e. last generation); otherwise, $g = g + 1$ and go to step 3.

Robust Evolution Strategies

When ESs are applied to an optimization problem successfully, it shows evolutionary behavior similar to that of other evolutionary algorithms: the focus of the search shifts from a global region onto a local region. This arises from the gradual convergence of the population due to the direct effects of natural selection. Associated with this, η_i gradually reaches zero. This has been considered a process of "self-adaptation", which is one of attractive features of ES. This works well for some unimodal functions. However, in the case of multi-modal functions, ES often gets trapped in local optima due to the fact that η_k takes a very small value in the early generation. Then, ES should be extended in order to give larger adaptability to strategy parameters.

We allow selectively neutral mutations(Kimura, 1983) on strategy parameters, which enables a rapidly increasing or decreasing η_k irrespective of selection. This method, called Robust-ES(RES), follows the same procedures as CES except for the following two points(Ohkura et al., 2001):

- New individual representation that holds redundant strategy parameters. These parameters have no effect on the selection process.

- New stochastic mutation mechanisms for changing original strategy parameters. These mutations replace, swap or copy active strategy parameters with inactive ones.

Individual Representation

An individual $\mathbf{X}_i$ is represented as follows, assuming that $i = 1, 2, \cdots, \mu$, $j = 1, 2, \cdots, n$, $k = 0, 1, \cdots, m$ and $\eta_{ik}(j) \in R^+$:

$$\mathbf{X}_i = [\boldsymbol{x}_i, (\boldsymbol{\eta}_{i0}, \cdots, \boldsymbol{\eta}_{ik}, \cdots, \boldsymbol{\eta}_{im})] \quad (3)$$

$${}^t\boldsymbol{x}_i = (x_i(1), \cdots, x_i(j), \cdots, x_i(n)) \quad (4)$$

$${}^t\boldsymbol{\eta}_{ik} = (\eta_{ik}(1), \cdots, \eta_{ik}(j), \cdots, \eta_{ik}(n)) \quad (5)$$

where $x_i(j)$ and $\eta_{ik}(j)$ denote the j-th component values of the vectors $\boldsymbol{x}_i$ and $\boldsymbol{\eta}_{ik}$, respectively. Notice that each $x_i(j)$ has $(m+1)$ strategy parameters.

Mutation Mechanisms for Strategy Parameters

Define D as same the mutation mechanism as Equation(1). $\boldsymbol{\eta}_{ik}$ is modified stochastically, according to the following new mutation operators:

- O_{dup} shifts all of $\eta_{ik}(j)$ into the adjacent position of $(k+1)$ and removes $\eta_{im}(j)$ from the list. Then, modifies all $\boldsymbol{\eta}_{ik}$ with D.

- O_{del} discards $\eta_{i0}(j)$ and moves $\eta_{ik}(j)$ to the adjacent position of $(k-1)$. At the m-th position η_L is calculated as the smaller value either η_{max} or $\sum_{p=1}^{m-1} \eta_{ip}(j)$. Then, modifies all $\boldsymbol{\eta}_{ik}$ with D.

- O_{inv} swaps $\eta_{i0}(j)$ with one of $\eta_{ik}(j)$, $k = 1, \cdots, m$ and modifies $\eta_{i0}(j)$ and $\eta_{ik}(j)$ with D.

Notice that RES is equivalent to CES when the probabilities of O_{dup}, O_{del} and O_{inv} are set at 1.0, 0.0 and 0.0, respectively.

Multi-parent Recombination

$(\mu/\mu, \lambda)$-ES is a special variant of $(\mu/\rho, \lambda)$-ES (Beyer, 1995), where ρ determines the number of parents to form a new offspring. There are two popular recombination operators in $(\mu/\mu, \lambda)$-ES, i.e., global intermediate recombination $(\mu/\mu_I, \lambda)$, global discrete recombination $(\mu/\mu_D, \lambda)$.

Multi-parent Intermediate Recombination: $(\mu/\mu_I, \lambda)$

Intermediate recombination is some kind of averaging. This can be formulated as follows:

$$\tilde{\eta}_{ik}(j) = \frac{1}{\mu} \sum_{i=1}^{\mu} \eta_{ik}(j) \quad (6)$$

$$\tilde{x}_i(j) = \frac{1}{\mu} \sum_{i=1}^{\mu} x_i(j) \quad (7)$$

This type of recombination is referred to as II-ES.

Multi-Parent Discrete Recombination: $(\mu/\mu_D, \lambda)$

In $(\mu/\mu_D, \lambda)$-ES, an offspring is constituted from all parents. The i-th component of offspring is given by random choice of one of the i-th component from the parents:

$$\tilde{\eta}_{ik}(j) = \eta_{\chi_j k}(j) \qquad (8)$$
$$\tilde{x}_i(j) = x_{\chi'_j}(j) \qquad (9)$$

χ_j and χ'_j denote uniform distributed random integer in $\{1, ..., \mu\}$, respectively, and are generated anew for each value of j. This type of recombination is referred to as DD-ES.

Global Combined Discrete Recombination

Discrete recombination is separately applied to object variables and strategy parameters in DD-ES introduced above. However, they might have strong connection with each other, because strategy parameters determine the mutability of object variables. Based on this assumption, a new recombination, which regards a pair of an object variable and a strategy parameter as a unit of recombination, can be formulated as follows:

$$\tilde{\eta}_{ik}(j) = \eta_{\chi_j k}(j) \qquad (10)$$
$$\tilde{x}_i(j) = x_{\chi_j}(j) \qquad (11)$$

This type of recombination is referred to as D-ES.

COMPUTER SIMULATIONS
Test Functions and Conditions

Six test functions $f(x_i)$ are selected from Yao and Liu(1997). They are Sphere Model(f_1), Schwefel's Problem 2.22(f_2), Schwefel's Problem 1.2(f_3), Rastrigin's Function(f_4), Ackley's Function(f_5) and Griewank Function(f_6). Functions f_1 to f_3 are unimodal functions and f_4 to f_6 are multi-modal functions. All functions have zero as the global minimum.

The experimental setup is based on Yao and Liu(1997). $(\mu, \lambda) = (30, 200)$ with Gaussian mutation, recombination and no correlated mutations is adopted for all computer simulations. CES and RES use the same initial populations. All simulations are repeated for 50 runs independently. The upper bound of strategy parameters η_{max} is set at 1.0 for f_4 and 3.0 for the other functions. In RES, the number of inactive strategy parameters m for each variable is set at 5. O_{dup}, O_{del} and O_{inv} are applied with the probabilities of 0.6, 0.15 and 0.1, respectively. The parameters are not fully tuned, because the main purpose of our computer simulations is to investigate the effect of multi-parent recombination.

Results

Figs.1(a) to (f) show the averaged best results of CES for 30 dimensions. II-CES has the fastest convergence rate of all in f_1 and f_3, and highest convergence reliability in f_5 and f_6. DD-CES has the highest convergence reliability in f_4. D-CES has the fastest convergence rate in f_2. CES converges prematurely or gets into the local optima.

Figs.2(a) to (f) show the averaged best results of RES for 30 dimensions. II-RES converges prematurely or gets into the local optima. DD-RES finds optimal values in f_4 and f_6. D-RES has the fastest convergence rate in f_1, f_2, f_3 and

f_5, and finds optimal values in f_4 and f_6. RES has slower convergence rate than DD-RES and D-RES.

In Figs.3(a) to (f) and Figs.4(a) to (f), the averaged best results of CES and RES for 100 dimensions show the same tendency as those for 30 dimensions.

Summary

To summarize our computer simulations, the following points are stated.

- In the case of CES, the performance of the multi-parent recombination depends on the objective function topology.

- In the case of RES, regardless of the objective function topology and dimension, global combined discrete recombination improves the averaged best function values of all the test functions.

CONCLUSIONS

In this paper, multi-recombinant Evolution Strategies($(\mu/\mu,\lambda)$-ES) were investigated using six test functions. The classical-ES(CES) and the robust-ES(RES) were adopted for the observation. The results suggested that multi-parent version of recombination operators improves the performance of ES for the six test functions. In the cases of CES, an appropriate choice of the recombination operator depends on the objective function topology. In the case of RES, global combined discrete recombination operator is the most suitable of all.

The first author acknowledges financial support through JSPS (the Japan Society for the Promotion of Science) Research Fellowship for Young Scientists(04999).

REFERENCES

T. Bäck (1996), *Evolutionary Algorithms in Theory and Practice*, Oxford University Press.

T. Bäck and A. E. Eiben (1999), "Generalizations of Intermediate Recombination in Evolution Strategies", *Proc. of Congress on Evolutionary Computation (CEC'99)*, pp.1566-1573, IEEE Press.

H.-G. Beyer (1995), "Toward a Theory of Evolution Strategies: On the Benefits of Sex - the $(\mu/\mu,\lambda)$-Theory ", *Evolutionary Computation*, Vol.3, No.1, pp.81-111.

M. Chang, K. Ohkura and K. Ueda (2001), "Some Experimental Observation of $(\mu/\mu,\lambda)$-Evolution Strategies", *Proc. of Congress on Evolutionary Computation (CEC'01)*, pp.663-670, IEEE Press.

A. E. Eiben and T. Bäck (1998), "An Empirical Investigation of Multiparent Recombination Operators in Evolution Strategies", *Evolutionary Computation*, Vol.5, No.3, pp.347-365.

L. Gruenz and H.-G. Beyer (1999), "Some Observations on the Interaction of Recombination and Self-Adaptation in Evolution Strategies", *Proc. of Congress on Evolutionary Computation (CEC'99)*, pp.639-645, IEEE Press.

M. Kimura (1983), *The Neutral Theory of Molecular Evolution*, Cambridge University Press.

F. Kursawe (1995), "Toward Self-adapting Evolution Strategies", *Proc. of 2nd IEEE Conference Evolutionary Computation*, pp.283-288, IEEE Press.

K. Ohkura, Y. Matsumura and K. Ueda (2001), "Robust Evolution Strategies", *Applied Intelligence*, Vol.15, Issue 3, pp.153-169, Kluwer Academic Publishers.

H.-P. Schwefel (1995), *Evolution and Optimum Seeking*, John Wiley & Sons.

X. Yao and Y. Liu (1997), "Fast Evolution Strategies", *Control and Cybernetics*, Vol.26, No.3, pp.467-496.

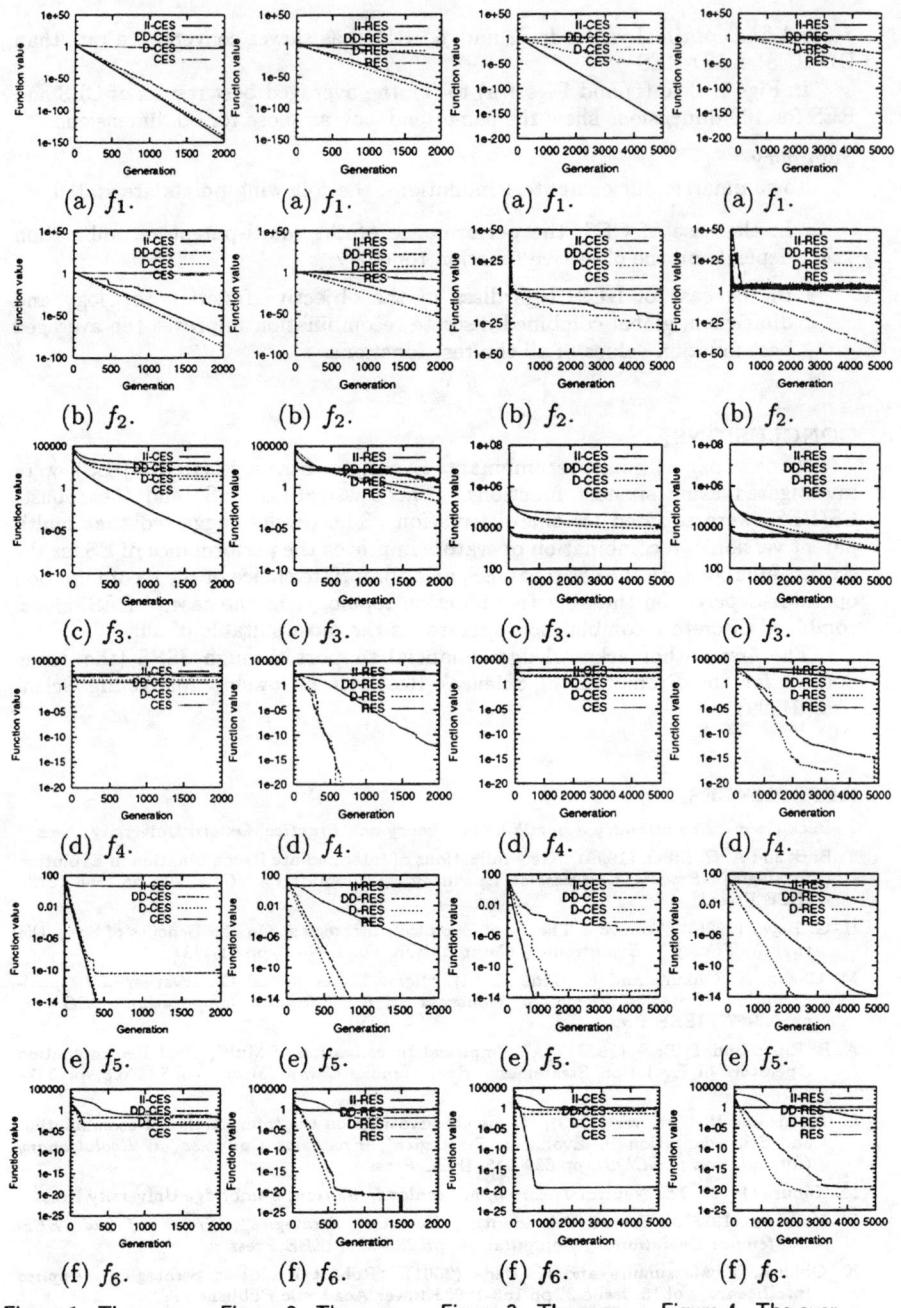

Figure 1. The averaged best values for CES when the dimension is 30.

Figure 2. The averaged best values for RES when the dimension is 30.

Figure 3. The averaged best values for CES when the dimension is 100.

Figure 4. The averaged best values for RES when the dimension is 100.

EFFICIENT EVALUATION RELAXATION UNDER INTEGRATED FITNESS FUNCTIONS

LAURA A. ALBERT & DAVID E GOLDBERG
University of Illinois at Urbana-Champaign
Department of General Engineering
104 S. Mathews St.
Urbana, IL 61801
{lalbert, deg}@uiuc.edu

ABSTRACT
This paper introduces a framework to describe fitness evaluation error of genetic algorithms (GAs) in which some of the error is due to bias. This framework describes the tradeoffs between accuracy and speed of evaluations and is used to model how computation time can be used efficiently. In particular, fitness functions whose cost and accuracy vary because of discretization errors from numerical integration are considered. To illustrate this tradeoff, naive and efficient discretizations are compared. Traditionally, fitness functions using numerical integration consider a constant number of grid points throughout the GA, but an efficient discretization can be considered in which the number of grid points increase throughout the duration of the GA. The speedup achieved from using efficient discretizations is predicted and shown empirically.

INTRODUCTION
In many industrial applications of genetic algorithms (GAs), there is often a tradeoff between more and less accurate evaluations. On one hand, more accurate yet more expensive evaluations can be used, and on the other, noisy, inexpensive evaluations are made. If computation time is expensive, a fast, noisy fitness function may be preferred over a slow, accurate fitness function (Fitzpatrick & Grefenstette, 1988). Some progress has been made in understanding the tradeoffs in the error-prone evaluations when the error is due to randomness or variance. Less concern has been shown for those circumstances when the error cannot be averaged out.

This paper considers the situation when evaluation error is the result of a biased function surrogate. In particular, it considers fitness functions whose cost and accuracy vary because of discretization errors of integration. Although the specific situations are somewhat idealized, the general idea applies to more complex evaluations resulting from implicit or explicit quadrature in finite elements, finite differences, or other techniques used to approximate differential and integral equations.

This paper discusses the relevant background of noisy fitness functions, domino convergence and time budgeting. A conceptual framework describing the approach to functions with biased error, and efficient and naive discretization choices are then discussed. A numerical integration problem is analyzed to exhibit speedup achieved from using efficient discretizations, and finally, we conclude.

BACKGROUND
Early studies on the effects of evaluation error on GA performance considered evaluation error that was mainly due to variance or randomness. These studies used sampling to estimate the fitness and ran the GA in a bounded computation time. The tradeoff for these problems is between the time to make each fitness evaluation and the total number of evaluations made. Either more, inaccurate evaluations or fewer, accurate evaluations are made. In addition, each implementation has different convergence time and population requirements. Grefenstette and Fitzpatrick (1985), Fitzpatrick and Grefenstette (1988), Miller and Goldberg (1996a), Miller and Goldberg (1996b), Miller (1997) and Aizawa and Wah (1994) analyzed optimal sampling in GAs. A more detailed background can be found elsewhere (Albert & Goldberg, 2001).

CONCEPTUAL FRAMEWORK
Previous work in error-time tradeoffs has focused on situations where the error may be averaged away by sampling—where the error is due to *variance* alone. This paper focuses on situations where a good portion of the error may be due to *bias*. That is, in these situations, no amount of sampling can cause the discrepancy between the fitness and the surrogate fitness to disappear.

This section starts by quantitatively defining the effects of bias. Next, two areas which are critical to efficiently use computation time are discussed: domino convergence and time budgeting. Finally, naive and efficient discretizations are defined.

Qualitative Description
We expect two possible results from biased error: an *amplitude shift* or a *phase shift* (Goldberg & Wylie, 1983). An amplitude shift, denoted by γ, may lead the GA to the correct optimum but the estimated fitness value differs by a quantity γ

from its true value. A phase shift, denoted by δ, is measured as the largest difference between the actual maximum and the closest peak within γ. In this case, the fitness function values returned by the GA may be roughly correct, but they are skewed over the search space by some distance δ. We expect that both γ and δ will decrease as the number of grid points increases. If the discretization size is Δx, then we expect the phase shift δ to be smaller than Δx for some minimum number of grid points. Thus, the number of grid points used for integration determines how accurately to which the differences in the function can be discriminated. This concept of a usable discretization is important for the correct building blocks to be chosen early on by the GA.

Figure 1 illustrates these concepts for an arbitrary equation for a maximization problem. In this figure, Δx is simply the size of each discretization and γ is the difference between the actual and approximate maxima. The maximum segment of the approximate function includes the maximum of the actual function. In this case, δ is the difference between the position of the actual maximum and the position of the furthest endpoint of the discretization. This is particularly important when the discretization size varies in the GA. If at any part of the GA, the phase shift is larger than the discretization size, then the GA will start to converge to the wrong solution and building blocks may be lost.

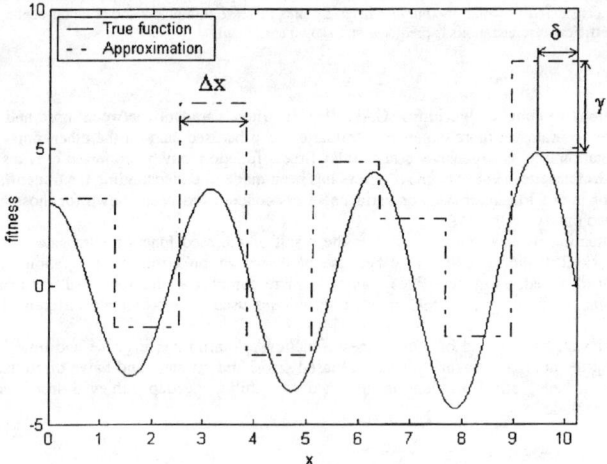

Figure 1: Conceptual drawing of phase and amplitude shifts

Convergence Theory

Duration theory is one of the key ingredients for understanding how GAs use resources and how computation time can be used efficiently. Engineering applications typically have exponentially scaled building blocks, which will converge in $O(l)$ time, where l is the string length, starting at the most salient building block and continuing down to the least significant building block, taking approximately the same amount of time to converge to each successive building block. This is called *domino convergence* and can be contrasted with convergence rates of equally salient building blocks, in which all building blocks roughly converge simultaneously. This model is based on selection intensity adapted from population genetics, and this model predicts the dynamics of the average of the fitness population. Thierens, Goldberg and Pereira (1998) showed that the domino convergence time can be expressed as:

$$t_{conv} = \frac{-\ln 2}{\ln\left[1 - \frac{I}{\sqrt{3}}\right]} \lambda_t = c_1 k \quad (1)$$

where λ_t is the number of converged building blocks in the string, k is the total number of building blocks, I is the selection intensity and c_1 is the time to converge to each building block.

Equation 1 is valid for accurate fitness functions. When there is evaluation error present, Miller and Goldberg (1996a) observed that GAs take longer to converge. This duration elongation is derived later in this paper for the example problem, and the generalization of this convergence time model can be written as

$$t_{conv} = c_2 k, \; c_2 > c_1 \tag{2}$$

where c_2 is the elongated time to converge to each building block.

Because the least salient building blocks do not experience any selection pressure for a number of generations, it is possible that they will experience genetic drift and converge to some value at random. The expected time for a bit to experience drift is proportional to the population size (Goldberg & Segrest, 1987):

$$t_{drift} = 1.4 N \tag{3}$$

In this equation, N is the population size. For any application, it is important to ensure that drift will not occur until well after the population is expected to converge.

Time Budgeting for Naïve and Efficient Discretizations

For any evaluation using numerical integration, the choice of the number of grid points for the integration n is usually constant for the entire run of the GA. This choice of n is too precise at the beginning of the run, which leads to wasted computation cycles early on. Problems where n is static during the course of the GA is said to have a naïve discretization.

There are three main ingredients for the time prediction model: the number of grid points, convergence time and population sizing. As convergence time has already been discussed, the focus in this section is on discretization scheduling. The naive and efficient implementations are introduced, followed by the predicted speedup when the efficient implementation is used. Population sizing equations do not yet exist for exponentially scaled building blocks. For the experiments herein, a constant population size is considered for both the efficient and naive implementations.

Naïve Discretizations

As mentioned previously, a naive approach to modeling with GAs is to use a constant discretization size during the GA. The one-dimensional time-model can be written as (Fitzpatrick & Grefenstette, 1988)

$$T = (\alpha + \beta n) G N \tag{4}$$

where T_n is the naive computation time for the accurate function evaluations, α is the overhead per individual per generation, β is the time to calculate one sample, n is the number of samples, c_1 is the total number of generations and N is the population size. This equation can be extended to model the computation time of a fitness function when using one-dimensional numerical integration where n is the number of grid points used for integration. If there are k building blocks in the chromosome, n will need to be at least 2^k in order to discriminate enough to converge correctly to the kth building block.

Efficient Discretizations

An efficient discretization would be to use fewer grid points in the first generations and exponentially increase the number of grid points throughout the GA. This has already been observed elsewhere (Aizawa & Wah, 1994), where it was shown that in general, smaller sample sizes perform better in early generations and larger sample sizes perform better in the later generations. Domino convergence tells us that in the first few generations, the GA will be considering only the most salient bit. Ideally, then only two grid points are needed for integration. In the ideal case, just two discretizations can be used to discriminate between individuals that have the correct or incorrect first bit as does the optimal solution. In general, the starting value will need to be greater than 2. When a bit has converged, the number of grid points must be doubled in order to be able to discriminate enough to converge to the next most salient bit. Because the building blocks are exponentially scaled, it will take a minimum of 2^i discretizations to discriminate between two individuals that are converging to the ith building block.

Assuming that the number of grid points starts at 2 and is doubled every c_2 generations, the number of generations to converge to each building block when the evaluation error is present in fitness evaluations, then the efficient time can be modeled as

$$T_e = \sum_{i=1}^{k} (\alpha + \beta \cdot 2^i) c_2 N \tag{5}$$

Speedup

The speedup can be measured as the naïve time divided by the efficient time, and knowing that $n = 2^k$ and $\sum_{i=1}^{k} 2^i = 2^k - 2$ the speedup can be simplified.

$$S = \frac{T_n}{T_e} = \frac{(\alpha + \beta n)c_1 k}{\sum_{i=1}^{k}(\alpha + \beta \cdot 2^i)c_2} \tag{6}$$

Although the efficient GA takes more generations to converge than its naive counterpart, the overall savings from using an efficient discretization throughout the GA is significant. The predicted and experimental speedup can be seen by the computational experiments in the next section.

IMPLEMENTATION

In this section, a fitness function using numerical integration is defined. After this, the duration theory is used to predict the number of generations until convergence. Finally, the efficient time budget is predicted. Although the speedup and efficient time budgeting are used on a test case, these equations can be used more generally.

Problem Setup

The fitness function considered is an integration of the derivative of another subfunction. The experiments in this section have the following form where $f(x)$ is the fitness function:

$$f(x) = \sum_{i=1}^{3} \int_{0}^{x_i} g_i' \, dx_i \tag{7}$$

$$g(x) = e^{Ax} \cos Bx \tag{8}$$

The fitness function used as a sum of three identical subfunctions, each of which is the integral of the derivative of $g(x)$. In this equation, A and B are 0.05 and 2.00, respectively. If the integration is done precisely enough, the fitness function should return the same value as summing the subfunctions themselves. Ten bits are used to represent the string for each x in $g(x)$. For each subfunction, the maximum occurs when x is 9.44 when x varies from 0 to 10.23. The rectangle rule is used in this experiment as the numerical integration method because it is $O(\Delta x)$ accurate, less accurate than more commonly used methods such as Simpson's rule.

For any number of grid points n, we expect that δ is smaller than Δx for any arbitrary n higher than some usable level. We also expect γ to decrease as n increases. Analysis of how the phase and amplitude shifts vary with n indicate that both decrease rapidly as the n increases, and the phase shift is smaller than Δx when n is 5 or larger. This implies that even if a function has several peaks, a small number of grid points is needed in order for the GA to lead the approximate solution to within Δx of the actual optimum.

In order to insure that the first building blocks do not start to converge to the wrong values, n must be at least 5. We start with an n of 8 in the first generations, the next highest power of 2 larger than 5. Computational experiments indicate that the fitness variance of the population σ^2_F is 1.61, the error variance σ^2_N is 1.45, and the bias is 2.20.

Duration Theory Revisited

Equation 1 assumes that the variables used for the fitness function are exponentially scaled. Although this is true for the individual variables used in equation 7, the entire chromosome is not exponentially scaled. Thus, the convergence time is not expected to be the same as in equation 1.

The convergence time given by Thierens, Goldberg and Pereira (1998) is valid for a single subfunction. Elsewhere (Albert & Goldberg, 2001), the convergence time is derived for fitness function composed of a sum of m identical subfunctions. When using tournament selection with a tournament size of two, the selection intensity I is $1/\sqrt{\pi}$ and the time to converge the entire string is given below in equation 9. In this equation, k number of building blocks in each variable instead of the entire number of building blocks.

$$t_{conv} = \frac{-\ln 2}{\ln\left[1 - \frac{1}{\sqrt{3\pi m}}\right]} k = c_1 k \tag{9}$$

When m is 3 and k is 10, c_1 is 3.33 and the convergence time is 33.3 generations.

When there is evaluation error, the GA will take longer to converge (Miller & Goldberg, 1996a), when the standard deviation of the evaluation error can be rewritten as $\sigma^2_F / \sqrt{\sigma^2_F + \sigma^2_N}$ where σ^2_F is the fitness variance and σ^2_N is the error variance. If they are related in a way such that $r = \sigma^2_N / \sigma^2_F$ is true or approximately true during the GA, then the convergence equation with evaluation error reduces to

$$t_{conv} = \frac{-\ln 2}{\ln\left[1 - \frac{I}{\sqrt{3m(1+r)}}\right]} \lambda_t \qquad (10)$$

when binary tournament selection is considered. Using the values of σ^2_F and σ^2_N mentioned earlier, r is approximately 0.9, which makes c_2 4.73, increasing from 3.33 in the accurate model. In other words, the population with evaluation error is expected to converge in 47.3 generations.

Experimental Results

The expected computation time can be found once α and β are known. For the example problem analyzed here, computational experiments indicate that the values for α and β are 3.3×10^{-4} and 2.6×10^{-6}, respectively, when Mflops are used to measure computation time.

As mentioned previously, we start with 8 grid points to ensure that a good solution is found by the GA. In other words, the first $3 \cdot 4.73 = 14.2$ generations are run with 8 grid points. This has a negligible effect on computation time because the ratio to βn to α is small when n is 8.

A population size of 200 is used for all both the naive and efficient implementations. Computational experiments verify that smaller population sizes are inadequate for reliable convergence to the optimal solution.

Knowing the population size, the drift time can be checked to insure that the GA will not experience drift with this population size. Using equation 2, the drift time will be 280 generations, much larger than the predicted 33.3 and 47.3 generations until convergence for the naive and efficient models, respectively. Therefore, drift is not a concern for these experiments.

By using equations 5 and 6, we expect a naive time of 19.9 Mflops and an efficient running time of 8.2 Mflops, or a speedup of 2.43. Tables 1 and 2 summarize the expected and actual times and generations until convergence.

Table 1: Predicted and actual computation time values for a naïve discretization

	Generations	Time (Mflops)	Function Evaluations
Predicted	33.3	19.9	6660
Actual	34.3	20.0	6860

Table 2: Predicted and actual computation time values for an efficient discretization

	Generations	Time (Mflops)	Function Evaluations
Predicted	47.3	8.2	9460
Actual	48.3	8.9	9660

In both cases, the actual number of function evaluations is very close to the predicted number of function evaluations. In the case of naive discretization, the computation time prediction follows very closely to the theory. When an efficient discretization was used, the actual time was slightly longer than the predicted. Some of the reason is that because the number of grid points are increasing throughout the GA, there is virtually no chance for the GA to converge early—it tends to converge as expected or in a longer amount of time. However, a speedup of 2.25 was observed, close to the 2.43 that was predicted.

An accurate measure of performance is to see the building blocks converge with time. Because of domino convergence, the gene position, or number of genes that have converged in each subfunction, is expected to vary linearly with time. Figure 2 shows the average converged gene position for the trials. In both the naive and efficient cases, the experimental gene position follows very closely to the predicted value.

CONCLUSIONS

In this paper, we were able to show that GAs can be run more efficiently in problems when a large portion of the error in the fitness evaluations is due to bias. In particular, we showed that speedup can be predicted for problems in which the error in the fitness evaluations are due to discretization errors in integration when the building blocks are exponentially scaled. Our predictions can be extended to more general situations in which the bias cannot be sampled away.

ACKNOWLEDGMENTS

This work was sponsored by the Air Force Office of Scientific Research, Air Force Materiel Command, USAF, under grant F49620-00-1-0163. Research funding for this work was also provided by the National Science Foundation under grant DMI-9908252. Support was also provided by a grant from the U. S. Army Research Laboratory under the Federated Laboratory Program, Cooperative Agreement DAAL01-96-2-0003. The U. S. Government is authorized to reproduce and distribute reprints for Government purposes notwithstanding any copyright notation thereon.

The views and conclusions contained herein are those of the authors and should not be interpreted as necessarily representing the official policies or endorsements, either expressed or implied, of the Air Force Office of Scientific Research, the National Science Foundation, the U. S. Army, or the U. S. Government.

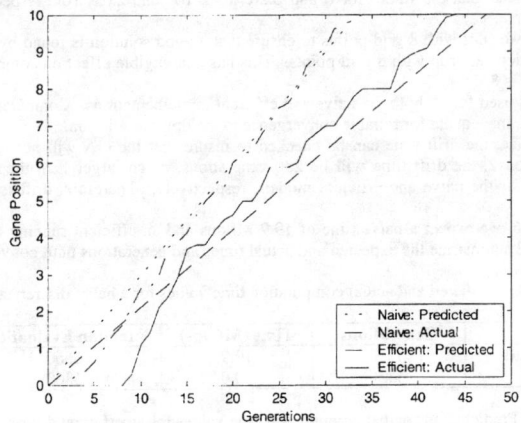

Figure 2: Actual and predicted gene positions

REFERENCES

Aizawa, A., & Wah, B. (1994). Scheduling of genetic algorithms in a noisy environment. *Evolutionary Computation, 2* (2), 97-122.

Albert, L. A., & Goldberg, D. E. (2001). *Efficient evaluation relaxation under integrated fitness functions* (IlliGAL Report No. 2001024). Urbana, IL: University of Illinois at Urbana-Champaign.

Fitzpatrick, J., & Grefenstette, J. (1988). Genetic algorithms in noisy environments. *Machine Learning, 3,* 101-120.

Goldberg, D. E., & Segrest, P. (1987). Finite Markov chain analysis of genetic algorithms. In *Proceedings of the Second International Conference on Genetic Algorithms* (pp. 1-8). Hillsdale, NJ: Lawrence Erlbaum Associates Publishers.

Goldberg, D. E., & Wylie, E. B. (1983). Characteristic method using time-line interpolations. *Journal of Hydraulic Engineering. 109* (5), 670-683.

Grefenstette, J., & Fitzpatrick, J. (1985). Genetic search with approximate function evaluations. *Complex Systems. 6,* 333-362.

Miller, B. L., & Goldberg, D. E. (1996a). Genetic algorithms, selection schemes, and the varying effects of noise. *Evolutionary Computation, 4* (2), 113-131.

Miller, B. L., & Goldberg, D. E. (1996b). Optimal sampling for genetic algorithms. In Dagli, C., et al. (Eds.), *Proceedings of the Artificial Neural Networks in Engineering* (pp. 291-297). New York: ASME Press.

Miller, B. L. (1997). *Noise, sampling and efficient genetic algorithms.* Doctoral dissertation, University of Illinois at Urbana-Champaign, Urbana, IL.

Thierens, D., Goldberg, D. E., & Pereira, A. G. (1998). Domino convergence, drift and the temporal-salience structure of problems. In *The 1998 IEEE Conference on Evolutionary Computation Proceedings* (pp. 535-540). Piscataway, NJ: IEEE Service Center.

NONLINEAR MODELING: GENETIC PROGRAMMING VS. FAST EVOLUTIONARY PROGRAMMING

MINGLEI DUAN
Department of Electrical and
Computer Engineering
Marquette University
Milwaukee, Wisconsin

RICHARD J. POVINELLI
Department of Electrical and
Computer Engineering
Marquette University
Milwaukee, Wisconsin

ABSTRACT
Both Genetic Programming (GP) and Fast Evolutionary Programming (FEP) combined with a Reduced Parameter Bilinear (RPBL) model have been recognized as effective time series modeling methods. This study compares the performance of these two methods for their ability to model time series data in terms of their accuracy and time efficiency. A brief review of GP and FEP are presented. Then the accuracy and time efficiency of these two methods are evaluated on several different time series. The performances of the two methods are compared against each other.

INTRODUCTION

Artificial evolutionary processes, such as genetic algorithms (GA), are based on reproduction, recombination, and selection of the fittest members in an evolving population of candidate solutions. Koza [1] extended this genetic model of learning into the space of programs and thus introduced the concept of genetic programming (GP). Sathyanarayan and Chellapilla [5] proposed an alternative modeling approach called fast evolutionary programming (FEP) to optimize the parameters of a reduced parameter bilinear model (RPBL). The RPBL model [6] is capable of effectively representing nonlinear models with the additional advantage of using fewer parameters than a conventional bilinear model. This paper applies both approaches to model several time series, including Mackey-Glass, sunspot, and stock price time series.

GENETIC PROGRAMMING

Genetic programming (GP) lets a computer learn programs. The top-level process of GP follows a similar evolutionary approach as a GA. The major difference between GPs and GAs is that GP structures are not encoded as linear genomes, but rather as terms or symbolic expressions. The units being mutated and recombined do not consist of characters or command sequences but rather functional modules, which are generally represented as tree-structured chromosomes.

The basic algorithm of GP is as follows:
1. Generate an initial population and evaluate the fitness for each individual in the population.
2. Select individuals from the population, typically using roulette or tournament methods.

3. Perform mutation, crossover and other genetic operators on the selected individuals, and form the new population using the result.
4. If the solution is sufficient, end the process and present the best individual in the population as the result. Otherwise go to step 2.

Adil Qureshi's GPsys release 2b [7] is used. The configuration used in this study is given in Table 1.

Generations	100
Populations	2000
Function set	+, -, /, *, sin, cos, exp, sqrt, ln
Terminal set	$\{x(t-1), x(t-2),...,x(t-10), R\}$
Fitness	Sum of squared error
Max depth of new individual	9
Max depth of new subtrees for mutation	7
Max depth of individuals after crossover	13
Mutation rate	0.01
Generation method	Ramped half-and-half

Table 1: GP configuration

FAST EVOLUTIONARY PROGRAMMING

Fast evolutionary programming (FEP) is a variation of evolutionary strategies (ES) [9]. FEP should not be confused with Fogel's evolutionary programming [8], which evolves finite state machines. Yao and Liu [10] have shown empirically that FEP, which uses a Cauchy mutation operator, has better convergence properties than ES, which uses a Gaussian mutation operator. This was demonstrated on several multimodal functions with many local minima. Further it is comparable to ES in performance for unimodal and multimodal functions with only a few local minima.

FEP is implemented as follows [10], using a $(\mu + \lambda)$ evolution strategy.

1. Generate the initial population of μ randomly selected individuals, and set the generation number, k to one. Each individual is taken as a pair of real-valued vectors, (x_i, η_i), $\forall i \in \{1, \cdots, \mu\}$, where x_i is the vector elements and η_i is the corresponding variance.
2. Evaluate the error score for each individual, in terms of the objective function, $f(x_i)$.
3. Mutate each parent (x_i, η_i) to create a single offspring (x_i', η_i') by $x_i'(j) = x_i(j) + \eta_i(j) C(0,1)$, $\eta_i'(j) = \eta_i(j) \exp[\tau' N(0,1) + \tau N_j(0,1)]$ for $j = 1, \cdots, n$.
4. Calculate the fitness of each offspring.
5. Conduct pairwise comparison over the union of parents and offspring. For each individual, q opponents are chosen randomly from all the parents and offspring with equal probability. For each comparison, if the individual's error is no greater than the opponent's, the individual receives a "win".
6. Select the μ individuals that have the most wins to be parents of the next generation.
7. Stop if the halting criterion is satisfied; otherwise, increment the generation number and go to Step 3.

This algorithm was coded in java to make it comparable to GPsys with the same population size (2000) and number of generations (100) as used in GP.

REDUCED PARAMETER BILINEAR MODEL

The reduced parameter bilinear model (RPBL) is defined as

$$\phi_p(B)z_t = \theta_q(B)a_t + [\xi_m(B)z_t][\zeta_k(B)a_t]$$

where $\{z_t\}$ is the sequence of time series observations, $\{a_t\}$ is a sequence of independent zero mean random variables, $\phi_p(B) = 1 - \phi_1 B - \phi_2 B^2 - \cdots - \phi_p B^p$, $\theta_q(B) = 1 - \theta_1 B - \theta_2 B^2 - \cdots - \theta_q B^q$, $\xi_m(B) = B + \xi_2 B^2 + \cdots + \xi_m B^m$, and $\zeta_k(B) = \zeta_1 B + \zeta_2 B^2 + \cdots + \zeta_k B^k$. The variables $\phi_1, \phi_2, \cdots, \phi_p$; $\theta_1, \theta_2, \cdots, \theta_q$; $\xi_1, \xi_2, \cdots, \xi_m$; and $\zeta_1, \zeta_2, \cdots, \zeta_k$ are unknown parameters to be estimated from the time series data. The backshift operator B shifts the subscript of a time series observation backward in time, that is, $B^k y_t = y_{t-k}$. As can be seen, the autoregressive moving average (ARMA) model is a special case of the bilinear model where ξ_i and $\zeta_i = 0$ for all i.

The RPBL model is evolved by FEP using the following configuration. The individual vectors of the population consist of the model orders followed by the model parameters, as given by $x_i = [p, q, m, k, \{\phi_j\}, \{\theta_j\}, \{\xi_j\}, \{\zeta_j\}]$. In the initial population, p, q, m, and k parameters were selected randomly from $\{1, 2, \cdots, 20\}$ and the model coefficients were selected uniformly from $[-1, 1]$.

MODEL IDENTIFICATION FOR FEP

The identification procedure consists of determining the orders p, q, m and k of the model, and estimating the parameters. The model order is determined as the order that minimizes the Minimum Description Length (MDL) criterion defined as [10]

$$(N - \gamma)\log(\sigma_e^2) + (1/2)(\text{number of independent parameters})\log(N - \gamma),$$

where N is the number of observations of the time-series, $\gamma = \max(p, q, m, k)$ and

$$\sigma_e^2 = \left(\frac{1}{N - \gamma}\right) \sum_{t=\gamma+1}^{N} (z_t - \hat{z}_t)^2$$

$\hat{z}_t$ is the predicted output at time t. This criterion tries to minimize both model order and squared error at the same time. Using FEP, the model order is estimated following Yao and Liu's method [10]: Each individual in the population is a vector of the model order followed by the model parameters. In each generation, the model orders and model parameters are perturbed with continuous Cauchy random numbers and selected according to the MDL fitness criterion. The model orders are rounded to the nearest integer to obtain new model orders. The fittest vector after there is no more improvement in the fitness contains the desired model order and the model parameters.

EXPERIMENTS AND RESULTS

The time series used in the following experiments are scaled to lie between −1 and 1 before modeling. The mean square errors (MSEs) and times are all averaged over 10 runs, and σ is the standard deviation.

The Mackey-Glass equation

The first time series considered in this study is generated by the Mackey-Glass equation. The equation for the discretized Mackey-Glass map is

$$x(t+1) = x(t) + \frac{bx(t-\tau)}{1+x^c(t-\tau)} - ax(t),$$

where a=0.1, b=0.2, c=10, and τ=16. The Mackey-Glass map is seeded with 17 pseudo-random numbers, creating a 1200 points series. The first 1000 points are discarded to remove the initial transients. The next 100 points are used as the training set and the last 100 points are used as the test set, see Figure 1. Results from GP and FEP are shown in Table 2.

	GP	FEP
Training MSE	5.687×10^{-5}	4.735×10^{-5}
σ_{Train}	2.333×10^{-5}	2.613×10^{-5}
Test MSE	5.038×10^{-5}	4.648×10^{-4}
σ_{Test}	2.123×10^{-5}	1.851×10^{-4}
Time (sec)	254.5	76.1
σ_{Time}	159.8	3.7

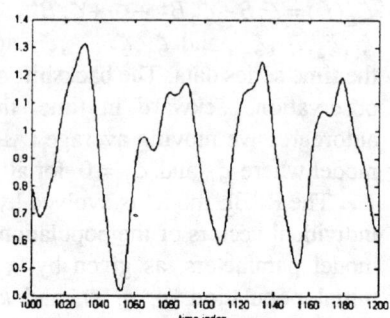

Table 2: Results for the Mackey-Glass time series

Figure 1: MackDey-Glass map

It can be seen that the models evolved by GP give much smaller MSE than FEP in the test data, although they have similar MSE for the training stage. Since the Mackey-Glass series is a totally deterministic time series, this result may imply that GP is more suitable for modeling those series with strong signals and weak noise than FEP. Even though GP takes about four times longer time, it would be the preferred method due to its better accuracy.

The Sunspot time series

The second experiment was conducted on the yearly sunspot series for the year 1800-1999 [11], see Figure 2. Once again, the first 100 data points are used for training and the next 100 are used for testing. See Table 3 for the results.

	GP	FEP
Training MSE	2.409×10^{-2}	4.019×10^{-2}
σ_{Train}	6.11×10^{-3}	1.34×10^{-3}
Test MSE	4.582×10^{-2}	5.765×10^{-2}
σ_{Test}	1.582×10^{-2}	4.23×10^{-3}
Time (sec)	205.4	70.1
σ_{Time}	28.1	4.4

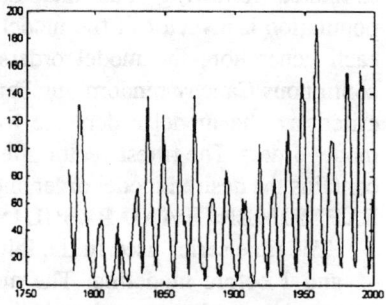

Table 3: Results for the sunspot time series

Figure 2: Sunspot time series

In modeling the sunspot time series, the accuracy performance between the two methods is not significantly different. The GP gives slightly better accuracy, but again, it takes three times as long to compute as the FEP method.

Stock prices time series

Two arbitrarily selected stocks, Compaq Computers (CPQ) on the NY Stock Exchange, and Microsoft (MSFT) on the NASDAQ, are used as the third experimental time series. The closing prices of the first 210 trading days in 1999 are used. The first 10 points are need for modeling the first prediction, and the next 200 points are divided into training and test set in the same manner as before, see Figure 3. Tables 4 and 5 present the modeling results.

	GP	FEP
Training MSE	6.597×10^{-3}	6.951×10^{-3}
σ_{Train}	3.65×10^{-4}	2.30×10^{-5}
Test MSE	7.076×10^{-3}	6.456×10^{-3}
σ_{Test}	2.19×10^{-3}	3.12×10^{-4}
Time (sec)	126	64
σ_{Time}	87.3	6.8

Table 4: Results for the MSFT time series

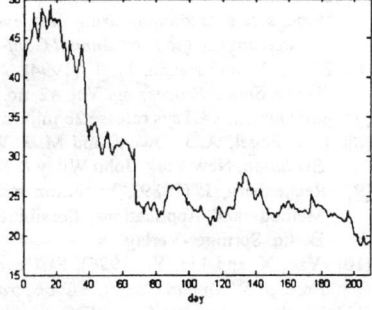

Figure 3: MSFT price time series

	GP	FEP
Training MSE	6.002×10^{-3}	7.003×10^{-3}
σ_{Train}	1.32×10^{-3}	9.15×10^{-5}
Test MSE	2.335×10^{-3}	2.148×10^{-3}
σ_{Test}	4.12×10^{-4}	7.39×10^{-5}
Time (sec)	119.6	68.1
σ_{Time}	115.3	4.0

Table 5: Results for the CPQ time series

Figure 4: CPQ price time series

The results from the stock time series are similar to the sunspot results. The two methods give similar error in both training and testing, but the GP is much more time consuming. It was noticed that the results generated by FEP in each trial are consistent, but this is not the case for GP. There are larger variances in both GP's MSE and time. One interesting observation in the experiments is that as the generation increases, the models evolved by FEP tend to become simpler while those evolved by GP always become more complex. This explains why GP is not as consistent as FEP. Because while the model becomes more and more complex, the search space is expanded rapidly, and there are a large number of local minimums into which GP could fall. In the experiments, the best solution is always found by GP. This also suggests that GPs have relatively stronger search ability.

CONCLUSIONS

In this paper, two different nonlinear modeling techniques: Genetic Programming and Fast Evolutionary Programming are applied to solve three different kinds of times series modeling problem. The GP has been shown to have better search ability than FEP, especially when deal with more predictable time series. FEP performs better when applied to noisier time series. For real world time series such as sunspot series and stock price series, it could give predictions not worse than GP, but with much less computational effort.

REFERENCES

[1] Koza, John 1992. *Genetic Programming: On the Programming of Computers by Means of Natural Selection*. Cambridge, MA: The MIT Press.
[2] Kaboudan, M. 1998. A GP approach to distinguish chaotic from noisy signals. *Genetic Programming 1998: Proceedings of the Third Annual Conference*, San Francisco. CA: Morgan Kaufmann, pp. 187-192
[3] Kaboudan, M. Genetic Programming Prediction of Stock Prices, *Computational Economics*, to appear.
[4] Fogel, D. and Fogel, L. (1996). Preliminary experiments on discriminating between chaotic signals and noise using evolutionary programming. *Genetic Programming 1996: Proceedings of the First Annual Conference*. Cambridge, MA: The MIT Press, pp. 512-520.
[5] Sathyanarayan, S. and Chellapilla, K. (1996). Evolving reduced parameter bilinear models for time series prediction using fast evolutionary programming. *Genetic Programming 1996: Proceedings of the First Annual Conference*. Cambridge, MA: The MIT Press, pp. 528-535.
[6] Zhang, Y. and Hagan, M. T. (1994), A Reduced Parameter Bilinear Time Series Model, *IEEE Trans. Signal Processing*. Vol. 42, no. 7, pp. 1867-1870
[7] Adil Qureshi's GPsys release 2b in java http://www.cs.ucl.ac.uk/staff/A.Qureshi/gpsys.html.
[8] L. J. Fogel, A. J. Owens and M. J. Walsh (1966), *Artificial Intelligence Through Simulated Evolution*, New York: John Wiley & Sons.
[9] Rechenberg, I. (1989), Evolution strategy: Nature's way of optimization. In Optimization: Methods and Applications, Possibilities and Limitations. Lecture Notes in Engineering 47. Berlin: Springer-Verlag.
[10] Yao, X. and Liu, Y. (1996), Fast evolutionary programming. *Evolutionary Programming V: Proc. of 5th Annual Conf. On Evol. Prog.*, MIT Press, Cambridge, MA, forthcoming.
[11] Yearly sunspot data from SIDC: http://sidc.oma.be/index.php3

USE OF DOMINANCE-BASED TOURNAMENT SELECTION TO HANDLE CONSTRAINTS IN GENETIC ALGORITHMS

CARLOS A. COELLO COELLO
CINVESTAV-IPN
Depto. de Ingeniería Eléctrica
Sección de Computación
Av. IPN No. 2508
Col. San Pedro Zacatenco
México, D. F. 07300, MÉXICO
ccoello@cs.cinvestav.mx

EFRÉN MEZURA MONTES
Universidad Veracruzana
Maestría en Inteligencia Artificial
Sebastián Camacho No. 5
Col. Centro
Xalapa, Veracruz 91090, MÉXICO
emezura@lania.mx

ABSTRACT

In this paper, we propose a dominance-based selection scheme to incorporate constraints into the fitness function of a genetic algorithm used for global optimization. The approach does not require the use of a penalty function and, unlike traditional evolutionary multiobjective optimization techniques, it does not require niching (or any other similar approach) to maintain diversity in the population. The algorithm is validated using two test functions taken from the specialized literature on evolutionary optimization. The results obtained indicate that the approach is a viable alternative to the traditional penalty function, mainly in engineering optimization problems.

INTRODUCTION

The problem that is of interest to us is the general nonlinear optimization problem in which we want to:

$$\text{Find } \vec{x} \text{ which optimizes } f(\vec{x}) \quad (1)$$

subject to:

$$g_i(\vec{x}) \leq 0, \quad i = 1, \cdots, n \quad (2)$$

$$h_j(\vec{x}) = 0, \quad j = 1, \cdots, p \quad (3)$$

where $\vec{x}$ is the solution vector $\vec{x} = [x_1, x_2, \cdots, x_r]^T$, n is the number of inequality constraints and p is the number of equality constraints (in both cases, constraints could be linear or non-linear). Only inequality constraints will be considered in this work.

Despite the considerable number of constraint-handling methods that have been developed for genetic algorithms in the last few years (see for example (Michalewicz and Schoenauer, 1996; Coello 2001)), most of them either require a large number of fitness function evaluations, complex encodings or mappings, or are limited to problems with certain (specific) characteristics.

The aim of this work is to show that using concepts from multiobjective optimization (Coello, 1999), it is possible to derive new constraint-handling techniques that are not only easy to implement, but also computationally efficient.

OUR APPROACH

The idea of using evolutionary multiobjective optimization techniques (Coello, 1999) to handle constraints is not entirely new. Several researchers have reported approaches that rely on the use of multiobjective optimization new (see (Coello, 2000) for a review of this type of techniques). The most common approach is to redefine the single-objective optimization of f as a multiobjective optimization problem in which we will have $m+1$ objectives, where m is the number of constraints. Then, we can apply any multiobjective optimization technique (Fonseca and Fleming, 1995) to the new vector $v = (f, f_1, \cdots, f_m)$, where $f_1, \cdots, f_m$ are the original constraints of the problem. An ideal solution $\bar{x}$ would thus have $f_i(\bar{x}) = 0$ for $1 \leq i \leq m$ and $f(\bar{x}) \leq f(\bar{y})$ for all feasible $\bar{y}$ (assuming minimization). However, the existing techniques have several disadvantages, mainly related to efficiency issues and diversity loss. Those disadvantages were the main motivation for the development of the technique proposed in this paper.

The concept of nondominated vector is used in multiobjective optimization to denote solutions that represent good compromises or trade-offs, given a set of objective functions. None of the objective function values of these nondominated vectors can be improved without worsening another one (Coello, 1999). Our hypothesis is that this concept can be used to extend evolutionary multiobjective optimization techniques to be used as single-objective optimization approaches in which the constraints are handled as additional objectives. Although this sort of approach can be quite useful to reach the feasible region in highly constrained search spaces (Parmme and Purchase, 1994), it is not straightforward to extend it to solve single-objective optimization problems. The main difficulty is that we could bias the search towards a certain specific portion of the feasible region and, as a consequence, we could be unable to reach the global optimum. This paper presents a proposal based on a technique known as the Niched-Pareto Genetic Algorithm (NPGA) (Horn et al., 1994) that uses tournament selection based on nondominance. In the original proposal of the NPGA, the idea was to use a sample of the population to determine who is the winner between two candidate solutions to be selected, and to choose one of them based on nondominance with respect to the sample taken. Since only a portion of the population is used with this technique, it has a lower computational complexity with respect to traditional Pareto ranking approaches (the most common evolutionary multiobjective optimization technique).

To adapt the NPGA to solve single-objective constrained optimization problems, we performed the following changes:
- The tournament performed uses a parameter called selection ratio (S_r), which indicates the minimum number of individuals that will be selected through conventional tournament selection. The remainder will be selected using a purely probabilistic approach. In other words, $(1 - S_r)$ individuals in the population are probabilistically selected.
- When comparing two individuals, we can have three possible situations:

1. **Both are feasible.** In this case, the individual with a better fitness value wins.
2. **One is infeasible, and the other is feasible.** The feasible individual wins, regardless of its fitness function value.
3. **Both are infeasible.** The individual with the lowest amount of constraint violation wins, regardless of its fitness function value.

- Our approach does not require niching or any other approach to keep diversity, since the value of S_r will control the diversity of the population. For the experiments reported in this paper, a value close to one (≥ 0.8) was adopted.

In the following experiments, we used a GA with binary representation, two-point crossover, and uniform mutation. The parameters used for our GA were the following: population size = 200 individuals, maximum number of generations = 400, crossover rate = 0.6, mutation rate = 0.03, S_r = 0.99 (i.e., one out of every one hunded selections will be done probabilistically, rather than in a deterministic way), tournament size = 10.

COMPARISON OF RESULTS

While the method has been tested on several examples taken fom the optimization literature, only two will be used here to show the way in which the proposed approach works due to space limitations. Detailed descriptions of these problems may be found in the corresponding references.

Design Variables	Best solution found			
	This paper	Deb(1991)	Siddall(1972)	Ragsdell(1976)
$x_1(h)$	0.205986	0.2489	0.2444	0.2455
$x_2(l)$	3.471328	6.1730	6.2189	6.1960
$x_3(t)$	9.020224	8.1789	8.2915	8.2730
$x_4(b)$	0.206480	0.2533	0.2444	0.2455
$g_1(x)$	-0.074092	-5758.603777	-5743.502027	-5743.826517
$g_2(x)$	-0.266227	-255.576901	-4.015209	-4.715097
$g_3(x)$	-0.000495	-0.004400	0.000000	0.000000
$g_4(x)$	-3.430043	-2.982866	-3.022561	-3.020289
$g_5(x)$	-0.080986	-0.123900	-0.119400	-0.120500
$g_6(x)$	-0.235514	-0.234160	-0.234243	-0.234208
$g_7(\vec{x})$	-58.666440	-4465.270928	-3490.469418	-3604.275002
$f(\vec{x})$	1.728226	2.43311600	2.38154338	2.38593732

Table 1: Comparison of the results for the first example (optimal design of a beam)

EXAMPLE 1

This problem was solved before by Deb (1991) using a simple genetic algorithm with binary representation, and a traditional penalty function as suggested by Goldberg (1989). It has also been solved by Ragsdell and Phillips (1976) using geometric programming. Ragsdell and Phillips also compared their results with those produced by the methods contained in a software package called "Opti-Sep" (Siddall, 1972), which includes the following numerical optimization techniques: ADRANS (Gall's adaptive random search with a penalty function), APPROX (Griffith and Stewart's successive linear approximation), DAVID (Davidon-Fletcher-Powell with a penalty function), MEMGRD (Miele's memory gradient with a penalty function), SEEK1 & SEEK2 (Hooke and Jeeves with 2 different penalty functions), SIMPLX (Simplex method with a penalty function) and RANDOM (Richardson's random method).

Their results are compared against those produced by the approach proposed in this paper, which are shown in Table 1. In the case of Siddall's techniques (Siddall, 1972), only the best solution produced by the techniques contained in "Opti-Sep" is displayed. The solution shown for the technique proposed here is the best produced after 30 runs, and using the following ranges for the design variables: $0.1 \leq x_1 \leq 2.0$, $0.1 \leq x_2 \leq 10.0$, $0.1 \leq x_3 \leq 10.0$ and $0.1 \leq x_4 \leq 2.0$. The mean from the 30 runs performed was $f(x) = 1.792654$, with a standard deviation of 0.074713. The worst solution found was $f(x) = 1.993408$, which is better than any of the solutions produced by any of the other techniques depicted in Table 1. The number of fitness function evaluations of our approach was 80000.

Design Variables	Best solution found			
	This paper	GeneAS(1997)	Kannan(1994)	Sandgren(1988)
$x_1(T_s)$	0.8125	0.9375	1.125	1.125
$x_2(T_h)$	0.4375	0.5000	0.625	0.625
$x_3(R)$	42.097398	48.3290	58.291	47.700
$x_4(L)$	176.654047	112.6790	43.690	117.701
$g_1(x)$	-0.000020	-0.004750	0.000016	-0.204390
$g_2(x)$	-0.035891	-0.038941	-0.068904	-0.169942
$g_3(x)$	-27.886075	-3652.876838	-21.220104	54.226012
$g_4(x)$	-63.345953	-127.321000	-196.310000	-122.299000
$f(x)$	6059.946341	6410.3811	7198.0428	8129.1036

Table 2: Comparison of the results for the second example (optimization of a vessel)

EXAMPLE 2

This problem was solved before by Deb (1997) using GeneAS (Genetic Adaptive Search), by Kannan and Kramer using an augmented Lagrangian

Multiplier approach (1994), and by Sandgren (1988) using a branch and bound technique.

Their results were compared against those produced by the approach proposed in this paper, and are shown in Table 2. The solution shown for the technique proposed here is the best produced after 30 runs, and using the following ranges for the design variables: $1 \le x_1 \le 99$, $1 \le x_2 \le 99$, $10.0 \le x_3 \le 200.0$ and $10.0 \le x_4 \le 200.0$. The values for x_1 and x_2 were considered as integer (i.e., real values were rounded up to their closest integer value) multiples of 0.0625, and the values of x_3 and x_4 were considered as real numbers. The mean from the 30 runs performed was $f(\bar{x}) = 6177.253268$, with a standard deviation of 130.929702. The worst solution found was $f(\bar{x}) = 6469.322010$. We can see that in this case, our average solution was better than any of the solutions produced by any of the other techniques depicted in Table 2. The total number of fitness function evaluations performed was 80000.

CONCLUSIONS AND FUTURE WORK

This paper has introduced a new constraint-handling approach that is based on a multiobjective optimization technique called NPGA. The approach is intended to be used with evolutionary algorithms as a way to reduce the burden normally associated with the fine-tuning of a penalty function. The proposed approach performed well in two test problems in terms of the quality of the solutions found and it requires a relatively low number of fitness function evaluations. The results produced were compared against those generated with other (evolutionary and mathematical programming) techniques reported in the literature.

Our future work involves more validation of the approach with a larger set of test functions (we have used about ten test functions so far although, due to space limitations, only the results of two of them are presented in this paper), and an statistical analysis of the impact of its parameters on the overall performance of the algorithm.

ACKNOWLEDGEMENTS

The first author acknowledges partial support from CINVESTAV through project JIRA'2001/08 and from CONACyT through NSF-CONACyT project number 32999 A. The second author acknowledges support from CONACyT through a scholarship to pursue graduate studies at the Maestría en Inteligencia Artificial of the Universidad Veracruzana.

REFERENCES

Chankong V., and Haimes Y. Y., 1983, "Multiobjective Decision Making: Theory and Methodology," *Systems Science and Engineering*. North-Holland.

Coello C. A., 1999, "A Comprehensive Survey of Evolutionary-Based Multiobjective Optimization Techniques," *Knowledge and Information Systems. An International Journal,* 1(3):269–308.

Coello C. A., 2000, "Constraint-handling using an evolutionary multiobjective optimization technique", *Civil Engineering and Environmental Systems,* 17:319—346.

Coello, C.A., 2001, "Theoretical and Numerical Constraint-Handling Techniques used with Evolutionary Algorithms: A Survey of the State of the Art", *Computer Methods in Applied Mechanics and Engineering*, 2001 (in press).

Deb, K., 1991, "Optimal Design of a Welded Beam via Genetic Algorithms," *AIAA Journal*, 29(11):2013—2015.

Deb K. 1997, "GeneAS: A Robust Optimal Design Technique for Mechanical Component Design," In Dipankar Dasgupta and Zbigniew Michalewicz, editors, *Evolutionary Algorithms in Engineering Applications*, pages 497--514. Springer-Verlag.

Fonseca, C. M., and Fleming, P. J., 1995, "An overview of evolutionary algorithms in multiobjective. Optimization," *Evolutionary Computation*, 3(1):1—16.

Goldberg D. E., 1989, "Genetic Algorithms in Search, Optimization and Machine Learning". Addison-Wesley Publishing Co., Reading, Massachusetts.

Horn, J., Nafpliotis, N., and Goldberg, D. E., 1994, "A Niched Pareto Genetic Algorithm for Multiobjective Optimization," In *Proceedings of the First IEEE Conference on Evolutionary Computation, IEEE World Congress on Computational Intelligence*, volume 1, pages 82–87, Piscataway, New Jersey, IEEE Service Center.

Kannan B. K., and Kramer, S. N., 1994, "An Augmented Lagrange Multiplier Based Method for Mixed Integer Discrete Continuous Optimization and Its Applications to Mechanical Design," *Journal of Mechanical Design. Transactions of the ASME*, 116:318--320.

Michalewicz, Z., and Schoenauer, M.., 1996, "Evolutionary Algorithms for Constrained Parameter Optimization Problems," *Evolutionary Computation*, 4(1):1--32.

Parmee, I. C., and Purchase, G., 1994, "The development of a directed genetic search technique for heavily constrained design spaces," In I. C. Parmee, editor, *Adaptive Computing in Engineering Design and Control-'94*, pages 97--102, Plymouth, UK, University of Plymouth.

Ragsdell, K. M., and Phillips, D. T., 1976, "Optimal Design of a Class of Welded Structures Using Geometric Programming," *ASME Journal of Engineering for Industries*, 98(3):1021–1025, Series B.

Rao S. S. , 1996, "Engineering Optimization". John Wiley and Sons, third edition.

Sandgren, E. ,1988, "Nonlinear integer and discrete programming in mechanical design." In *Proceedings of the ASME Design Technology Conference*, pages 95--105, Kissimine, Florida.

Schaffer, J. D., 1985, "Multiple objective optimization with vector evaluated genetic algorithms," In *Genetic Algorithms and their Applications: Proceedings of the First International Conference on Genetic Algorithms*, pages 93--100. Lawrence Erlbaum.

Siddall, J. N., 1972, "Analytical Design-Making in Engineering Design", Prentice-Hall.

YIELD MODELING IN ELECTRONICS MANUFACTURING USING A HYBRID SYSTEM OF NEURAL NETWORKS AND GENETIC ALGORITHMS

SWAMINATHAN
VAITHIANATHASAMY
Department of Systems Science and
Industrial Engineering
Binghamton University
Binghamton, New York 13902
vswami78@hotmail.com

SARAH S. Y. LAM
Department of Systems Science and
Industrial Engineering
Binghamton University
Binghamton, New York 13902
sarahlam@binghamton.edu

KRISHNASWAMI SRIHARI
Department of Systems Science and
Industrial Engineering
Binghamton University
Binghamton, New York 13902
srihari@binghamton.edu

ABSTRACT

The electronics manufacturing industry has witnessed significant levels of growth over the past decade. This growth has resulted from the market-driven need for smaller products with enhanced functionality, which has resulted in a drastic increase in the complexity of electronics assemblies. To cope with these challenges, electronics manufacturing companies need to be highly productive without compromising product quality. Hence, maximization of first pass yields becomes a primary focus of the electronics manufacturers. Yield modeling can be a very effective tool in proactively managing the overall yield and the activities involved in increasing yield. This paper addresses the design of two neural networks for predicting yields for the printed circuit board assemblies and the subsequent sensitivity analysis of the network models. Also a genetic algorithm based approach is used to find an appropriate combination of design parameters that will help us reduce defects and avoid the combinations of inputs that will lead to poor yields.

INTRODUCTION

The electronics manufacturing industry has witnessed significant levels of growth over the past decade. This growth has resulted from the market-driven need for smaller products with enhanced functionality, which in turn, has resulted in a drastic increase in the complexity of electronics assemblies. The global market for Electronic Manufacturing Service (EMS) providers has grown at an unprecedented rate over the past 13 years. Forecasts for 1998 through 2000 predicted a greater than 20% growth rate projected for each year (NIST, 2000). In recent years there has been an irreversible shift towards using Surface Mount Components (SMC), which has resulted in denser boards. Because of this, processes have been redesigned to accommodate such changes. Coupled with these trends are the increase in competition and the concomitant decrease in profit margins. In order to cope with these challenges, electronics manufacturing companies need to be highly productive without compromising product quality. Consequently, improving the assembly yield has become one of the focal points of the EMS providers. A low yield means large amounts of rework, which adds to the cost and delays the delivery of the products. Yield modeling deals with the prediction of yields based on factors that influence the complexity of the board and the processes involved. Also yield modeling can be very important to understand the causes for low yields so that corrective action can be taken. In addition to these, yield models can proactively assist the Original Equipment Manufacturers (OEMs) in ensuring design for better yields.

In order to assist with OEMs' design and EMS providers' development activities, a highly reliable and accurate yield model is essential. The vast amounts of production data available from the EMS providers can be utilized to extract the required information. The complex relationship existing between yield and the factors affecting yield can be extracted from the production data gathered by the EMS providers. This information can be used to estimate yield and assist in Design For Manufacturing (DFM) and Design For Testing (DFT) activities.

Yield And Factors Affecting Yield

Yield in a Printed Circuit Board (PCB) assembly domain is the probability of one PCB passing the testing stations and having no defects (Tegethoff and Chen, 1995). In general, it can be stated that the complexity of a board determines the yield of the PCB assembly. A densely populated board is expected to have more defects than a sparsely populated board. Process complexities may contribute significantly to the complexity of a particular PCB assembly. In a complex task such as a PCB assembly, there are many factors that can affect yield. The defects caused could be due to the materials used (for instance, the solder paste type and/or the flux type, PCB board types, component types, etc.), process complexities involved (for instance, past deposition, reflow, etc.) and human related errors encountered (for instance, handling and/or setup). Among these factors, the most readily available information is design related information. It is usually not feasible to collect some of the critical process related and human related information from the actual production environment. The main disadvantage of this lack of information is the possibility of degradation in the prediction accuracy of the yield models. In spite of this, Li et al. (1994) demonstrated in their research work that using solely design related variables could capture a significant degree of the process complexities involved.

LITERATURE REVIEW

Within the semiconductor-manufacturing arena, several attempts have been made to formulate yield models (Cunningham, 1990; Segal et al., 2000; Vaithianathasamy et al., 2001). The prominent statistical models investigated and appeared in the literature include: (1) a Poisson yield model, and (2) a Negative Binomial yield model. The underlying assumption of the Poisson model is that defects are uniformly and randomly distributed. However, this assumption may not always hold when the density of the board becomes large because the defects may tend to cluster. In order to account for the clustering phenomenon, Tegethoff and Chen (1995) extended the application of negative binomial distribution from the semiconductor manufacturing arena to the modeling of assembly yield. Even with this extension, these models could not quantify the relationship between yield and the different parameters that affect yield.

Different regression models, independently proposed by Li et al. (1994) and Zhou (2000), appeared in the literature for assembly yield modeling applications. These models examine if there exists a self-consistent quantifiable correlation that relates assembly yields to the relevant parameters.

Apart from the aforementioned methods, artificial neural networks (ANN) have demonstrated to be well suited in the modeling of highly nonlinear relationships, especially when adequate analytical models are unavailable (Lam et al., 2000; Willis et al., 1991). Furthermore, their ability to tolerate noisy data and their capability to operate very quickly both in software and in hardware contribute significantly to the advantages of using the ANN approach. Unlike statistical models that generally require assumptions about the parametric nature of the factors, neural networks do not require a prior assumption of the functional form of the models.

This paper builds upon our earlier research effort on the development of a 'practical' system of neural network prediction models for process yield prediction in the Electronics Packaging arena (Vaithianathasamy et al., 2001). The neural network models were constructed using a real time database system obtained from the production floor of a large EMS provider. The database system contained mainly design parameters (see Table 1). This paper intends to demonstrate the capability of applying sensitivity analysis to study the effects of the input variables (factors) on the output yield. In addition, two Genetic Algorithm (GA) based 'advisors' were developed for one of the network models to assist DFM related activities. As a result, the alternative designs can be properly enhanced and would result in reduced defects.

NEURAL NETWORK DESIGN AND SENSITIVITY ANALYSIS

Over a total of 2000 data points were available from one of the EMS providers. The data set consists of various design parameters and the defect counts obtained from in-circuit tests. Our previous research effort discusses the statistical analysis and screening of these factors (Vaithianathasamy et al., 2001) and concludes using all the available factors (see Table 1) for model construction. Furthermore, a component-based neural network approach was demonstrated to outperform the single network approach. All of the models were validated using a more vigorous statistical approach, namely, the cross validation approach. This research paper chose to focus on the network based on the "Gullwing" type because of its relatively high defect rates as compared to other types of components.

Table 1. Variables Available from an EMS Provider.

Variable	Description
Component type	An indicator variable for different levels of complexities (Gullwing, AreaArray, ThroughHole, Passive, JLead, Miscellaneous)
Pitch	The distance between adjacent leads (in mils)
Part I/O	Number of leads on the component
# of I/Os	Total number of leads on the PCB assembly
Pad area	The area of the pad (in mil^2)
PCB area	The area of the PCB assembled (in $inch^2$)
PCB thickness	The thickness of the PCB (in mils)
Pad coating	An indicator variable for surface coating made on PCB
Side	The side where the components are mounted (top or bottom)
PCB quantity	The number of PCB assemblies tested in batch
Defect count	The number of defective PCB assemblies in a batch

In general, the opposition to applying neural networks is often due to the "Black-Box" behavior of the models. We attempt to display the relationship between the input variables and the output yield using a sensitivity analysis approach. This is intended to give the EMS provider a better understanding of how to identify and interpret the effects of the critical input parameters on the predicted yield, and hence to gain faith in using the network models for the real-world application.

The sensitivity analysis approach we took essentially focuses on varying one input variable at a time, while keeping other variables as a constant value (viz., minimum value, average value and maximum value of the other input variables). It is similar to performing a partial derivative on the output with respect to a particular input variable. Thereafter, graphs for each input variable were plotted. The goal of generating these visual displays is to eliminate the non-critical input variables. An input variable is defined to be non-critical if it does not affect the output variable significantly and it does not show significant interactions with other input variables.

Based on the plots generated, we can gain a qualitative understanding of how significant the effect of each input variable on the output variable. All of the plots show that none of the input variables are non-critical. Two sample sensitivity plots are presented in Figure 1, where the one on the left shows the most influential input variable and the other shows the least influential one.

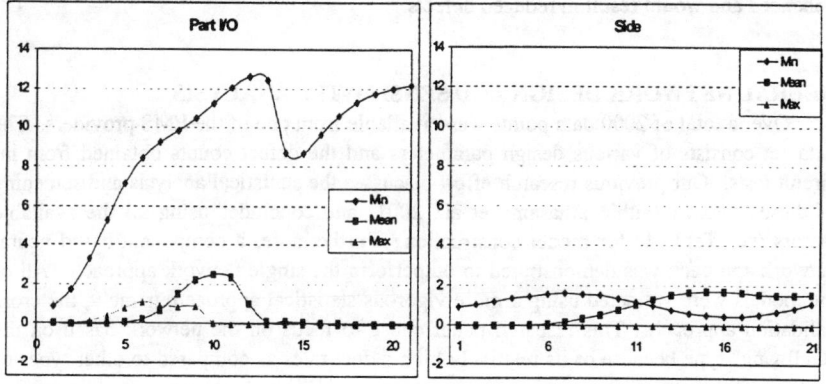

Figure 1. Sensitivity Plots for the Most and the Least Influential Input Variables.

GA-BASED DFM ADVISORS

Although the sensitivity plots can display the qualitative influence of the input variables on the output yield and their interactions among each other, it would be more beneficial to the EMS providers if the network models can be eventually utilized to assist DFM activities before actual designs are implemented on the production line. Because of the presence of continuous input variables in the "Gullwing" component-based network, infinite combinations of the values for the nine input variables are possible. This leads to an enormous input space for the determination of an optimal or near optimal combination of the input variables while minimizing defect counts. Therefore, evaluating all possible combinations of the input variables could be time consuming and may not be the most efficient approach. Furthermore, it would be more preferable to have alternative sets of designed parameters when the optimal solutions are not favorable or feasible because of practical issues and management decisions. We proposed to use a Genetic Algorithm (GA) approach along with "Gullwing" network for determining design parameters while minimizing defect counts. GA is a population-based evolutionary search strategy inspired by Darwin's theory of evolution, in which the solutions are evolved from one generation to the next. Genetic Algorithms have showed to be very effective in locating the optimal or near optimal solutions in the search space within a reasonable amount of computer run time.

Two GA-based "advisors" were developed, one for determining the best sets of design parameters for minimizing defect counts and the other for determining the worst sets of design parameters for maximizing defect counts. The latter system was developed with the intention to provide the DFM designer with information on the design parameters that he/she should be avoiding to prevent board designs with high defect rates.

Both of the GAs used a constant population size of 30, a stopping criterion of 300 generations, and a real-valued chromosome encoding (between -1 and 1). The arithmetical crossover operator with a crossover probability of 0.95 was used. As for the mutation operator, the sign of each allele has a probability of 0.05 for flipping. A rank order selection mechanism was chosen for reproduction. The only difference between the two GAs is the direction of the search, one for maximizing the defect rates ("BAD" GA-based Advisor) and the other for minimizing the defect rates ("GOOD" GA-based Advisor).

Ten replications were carried out for each of the two GAs to account for the stochastic effect of the search. The results for the "GOOD" GA-based Advisor achieved an average yield count of 0 with a standard deviation of 0.0. This shows that the "GOOD" GA-based Advisor is very successful in identifying multiple sets of design parameters. As for the "BAD" GA-based Advisor, it resulted in an average yield count of 13.1 with a standard deviation of 0.3. Tables 2 and 3 summarize the best three sets of design parameters and the worst three sets of design parameters respectively.

The two tables present the input variables (in the same order as defined in Table 1), their best (or worst) design values and the corresponding defect counts. Based on the results given in Table 2, it reviews a fairly tight range of best design values for minimizing the defect counts for each of the following parameters: Pitch, Part I/O, # of I/Os, Pad area and PCB area. As for the results based on the "BAD" GA-based Advisor, one can conclude that extra care should be taken while designing PCB assembly with board area in the range of 320-360 inch^2.

Table 2. Best Three Sets of Design Parameters from the "GOOD" Advisor.

Input Variable (range of possible values)	First Set of Design Parameters	Second Set of Design Parameters	Third Set of Design Parameters
Pitch (15.7 - 90.6)	79.9	79.2	72.5
Part I/O (3 - 256)	116	99	118
# of I/Os (84 - 12293)	7236	7192	7563
Pad area (7.9E-05 - 1.2E-02)	0.0100	0.0079	0.0102
PCB area (31.6 - 630.4)	514.4	514.6	539.4
PCB thickness (62 - 100)	68	92	94
Pad coating (0 or 1)	0	1	0
Side (0 or 1)	0	0	1
PCB quantity (107 - 42615)	38228	30906	40316
Defect count	0	0	0

Table 3. Worst Three Sets of Design Parameters from the "BAD" Advisor.

Input Variable (range of possible values)	First Set of Design Parameters	Second Set of Design Parameters	Third Set of Design Parameters
Pitch (15.7 - 90.6)	21.0	18.2	27.3
Part I/O (3 - 256)	28	202	104
# of I/Os (84 - 12293)	6085	3001	5894
Pad area (7.9E-05 - 1.2E-02)	0.0036	0.0007	0.0004
PCB area (31.6 - 630.4)	327.2	343.5	359.6
PCB thickness (62 - 100)	96	95	75
Pad coating (0 or 1)	0	1	0
Side (0 or 1)	1	0	0
PCB quantity (107 - 42615)	37950	26671	40105
Defect count	13	13	13

CONCLUSIONS

This paper applied sensitivity analysis to help demonstrate the unknown relationship of one of the trained networks from our previous research effort. The plots verified the necessity of including all the design variables in the "Gullwing" network model because of the nonlinear nature of the relationship between the design parameters and the final defect count. Two GA-based Advisors were presented and discussed with the purpose of assisting DFM activities. One of the Advisors was used to determine the best sets of parameters while minimizing defect counts. The other one was provided for the board designer to avoid for poor yield rates.

REFERENCES

Cunningham J.A., 1990, "The Use and Evaluation of Yield Models in Integrated Circuits Manufacturing," *IEEE Transactions on Semiconductor Manufacturing*, 3 (2), 60-71.

Lam, S.S.Y., Petri, K.L. and Smith, A.E., 2000, "Prediction and Optimization of a Ceramic Casting Process Using A Hierarchical Hybrid System of Neural Networks and Fuzzy Logic," *IIE Transactions*, 32 (1), 83-91.

Li, Y., Mahajan, R.L. and Tong, J., 1994, "Design Factors and Their Effects on PCB Assembly Yield - Statistical and Neural Network Predictive Models," *IEEE Transactions on Components, Packaging, and Manufacturing Technology, Part A*, 17 (2), 183-191.

NIST, 2000, "Electronic Commerce in the Electronics Manufacturing Industry," web site: http://www.mel.nist.gov/namt/projects/icm/emi.pdf.

Segal, J., Milor, L. and Peng, Y.K., 2000, "Reducing Baseline Defect Density Through Modeling Random Defect-Limited Yield," *Micro Magazine*, web site: http://www.micromagazine.com/archive/00.

Tegethoff, M.M.V. and Chen, T.W., 1995, "A Clustered Yield Model For SMT Boards and McMs," *IEEE Transactions on Components, Packaging, and Manufacturing Technology, Part B*, 18 (4), 640-643.

Vaithianathasamy, S., Lam, S.S.Y. and Srihari, K., 2001, "Process Yield Prediction for an EMS Provider Using Neural Network Models," *Proceedings of the Tenth Industrial Engineering Research Conference*, Dallas, TX, CD ROM format.

Willis, M.J., Di Massimo, C., Montague, G.A., Tham, M.T. and Morris, A.J., 1991, "Artificial Neural Networks in Process Engineering," *IEE Proceedings, Part D: Control Theory and Applications*, 138 (3), 256-266.

Zhou, X., 2000, "Yield Modeling in an Electronics Manufacturing Service Provider's Environment," Masters Thesis, Department of Systems Science and Industrial Engineering, State University of New York at Binghamton, Binghamton, New York.

RECOVERING TABULAR INFORMATION IN ASCII DOCUMENTS USING EVOLUTIONARY PROGRAMMING

MICHAEL A. ZMUDA
Dept of Computer Science and Systems Analysis
Miami University
Oxford, Oh 45056

ABSTRACT
This paper describes an evolutionary algorithm for identifying tabular information from unstructured text. This algorithm inserts column breaks into unformatted text, resulting in tables that are more readable than the unformatted text. Tables are represented using a hierarchical representation containing a sequence of rows, each of which contains a sequence of cells and cell-separators. Tokens from the original text reside within the representation and move laterally between the cells during the evolutionary algorithm's reproductive process. The fitness function is designed to measure syntactic and semantic characteristics such as: columns contain a consistent form of information (e.g., names of countries); tokens within a cell contain a monolithic piece of information (e.g., dollar amount, time of day, or name of country); and tokens within a cell contain reasonable case conventions (e.g., sentence case, title case, upper case, etc).

INTRODUCTION
With the amount of data continually increasing, it is important to present information to users in an effective manner. Pie charts, section headings, line breaks, and tables are just a few techniques that authors use to facilitate readability. One problem faced by electronic information providers is that many of these devices are removed during electronic transfer. Although many documents are now transferred in a format that preserves formatting (e.g., HTML or PDF), many electronic feeds are still transmitted as raw unformatted ASCII text. These new unstructured documents and the large amounts of unformatted legacy data make it desirable to develop techniques that can reconstruct formatting constructs that have been removed.

This paper addresses the specific issue of recovering tablular information from raw ASCII text. Figure 1 contains an example ASCII document that illustrates the importance of this problem. Although not formatted as such, this document contains regularities such as *number* followed by *name* followed by a parenthetical version of *country* followed optionally by a *phrase*. If treated as unstructured text, this portion of the document would be considered one continuous paragraph, the line breaks would be replaced with spaces; text displayed using a proportional font, resulting in information that would be very difficult to comprehend. A slightly better approach is to recognize the presence of a table; display the text using a proportional font; and preserve line breaks, which would provide the user the formatting shown in Figure 1. The ideal result is a table containing vertical and horizontal lines that clearly delinate the begininng and ending of columns, with superfluous information eliminated (e.g., parenthesis surrounding country). In this fashion, the user can easily comprehend the information.

```
 1. Stefano Zanini (Italy) 4 hours 23 minutes 13 seconds
 2. Thomas Muhlbacher (*) (Austria)
 3. Marcel Strauss (Switzerland)
 4. Marc Lotz (Netherlands)all same time
 5. Ellis Rastelli (Italy) 11 seconds behind
 6. Fred Rodriguez (U.S.) 7:31
 7. Markus Zberg (Switzerland)
 8. Rene Haselbacher (Austria)
 9. Richard Virenque (France)
10. Oscar Mason (Italy) all same time
```

Figure 1. Sample ASCII document. Reprinted with permission, copyright Reuters 2000.

A large amount of research has been conducted for recovering document information from document images, called *document imaging*. The field of document imaging approaches table recovery from a perspective that differs from that used in this work. Horizontal lines, vertical lines, and spacing can be used to identify the rows, columns, and headings of a table from a raw image. Using unformatted ASCII text, the problem requires a markedly different approach since formatting constructs are absent since they have been removed during transmission or the original author did not use a table. In ASCII documents, textual spacing can be used to assist in table reconstruction, but spacing is often inconsistent in a large number of cases and good recognition techniques must overcome these inconsistencies.

TABLE RECOVERY

The first task that must be solved in order to accomplish the objective is to identify the table within the context of a large document. This process is referred to as *table detection*. Researchers have developed methods for performing this task in ASCII files [Hu, 2000; Hurst, 2001; Pyreddy and Croft, 1997]. These techniques use heuristics that focus on cues such as increased number of extra spaces, numbers, excessive punctuation, and a long sequence of short lines. In order to focus on the technique of identifying the regularities, this work makes a number of assumptions: the table text has been detected; headings and footers are not present; and rows are not spread across multiple lines. Even within these constraints, the task is still complicated and defines an extremely large search space. The text shown below is not difficult for a human to convert into a table due to our great ability to organize; however, the same table presents an ambiguity to an automated approach.

```
            IBM 40 1/4 +5%
            AAPL 23 -1.4%
```

The difficulty is that the first line contains four pieces of text, whereas the second line contains only three. An algorithm must determine the number of columns and where the column breaks must be placed. By utilizing high-level syntactic entities (e.g., fraction), an automated algorithm can discover that the most parsimonious table contains the following three columns {IBM, AAPL}, {40 1/4, 23}, and {+5%, -1.4%}. The difficulty is that in general, there are an enormous number of possible tables for even a small text fragment.

The vast number of variations indicates that a recovery approach based purely on raw characters and their alignments will fail. High-level syntactic entities (e.g., quote, date, and country) and global criteria (e.g., the contents of a column are homogenous) are required. The ultimate goal of this work is handle these wide ranges of variations, in addition to unanticipated variations that may occur. This paper describes the prototype version of this system that is designed to accommodate such variations in the future.

EVOLUTIONARY ALGORITHM FOR TABLE RECOVERY

Algorithms from the field of evolutionary computation (EC) [Fogel, 1991; Goldberg, 1989; Koza, 1992] were employed because of the author's experience in the area and historical data that suggests that EC algorithms cope well with large complex optimization problems such as the one defined here. Implementing an EC algorithm requires a representation, reproductive operators, and fitness measure, which are described below.

The goal of the search process is to identify a *table* that effectively explains the data. This data structure must be general, complete, and facilitate changes such as merging columns, splitting columns, or merging rows. Figure 2 shows a portion of the representation for the data shown in Figure 1.

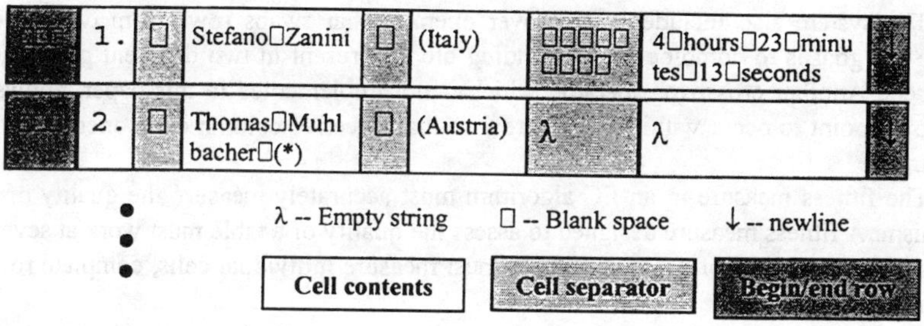

	1.	☐	Stefano☐Zanini	☐	(Italy)	☐☐☐☐☐ ☐☐☐☐	4☐hours☐23☐minutes☐13☐seconds	
	2.	☐	Thomas☐Muhlbacher☐(*)	☐	(Austria)	λ	λ	

λ -- Empty string   ☐ -- Blank space   ↓ -- newline

| Cell contents | Cell separator | Begin/end row |

Figure 2. Table representation.

An additional data structure is maintained for each individual cell, which itemizes the higher-level entities that are present in that cell. For example, a cell containing the text "300-900" can be viewed in the following ways: *integer*; *integer integer*; and *range*, where *integer* is a signed/unsigned sequence of digits (with or without commas) and *range* is a numeric range. However, the text could not be considered a *phone number* since it does not have the required number of digits; the current system performs a number of similar semantic checks to reduce the number of matching high-level entities. As tokens move between cells, the set of entities is recomputed. The prototype system includes entities such as: real number, clocktime, date, quote, monetary amount, percentage, fraction, and country.

The representation shown in Figure 2 naturally supports crossover and mutation operators. Figures 3 shows examples of the column-based mutation operator. This mutation operator shifts tokens left or right in order to improve the overall quality of the table. Several parameters control the affects of this operator. First, a particular cell is targeted for a mutation. Within that cell, a cut-point in the token sequence is selected that divides the sequence into two halves. The right or left half will be selected for migration. The portion that moves laterally may be placed into a new column or merged with the adjacent column. If the movement creates empty cells, those cells may or may not be removed from the table. The cut point is selected with a bias toward preserving high-level entities (e.g., fraction) since cells are more likely to contain complete entities. Advanced versions of this system will incorporate other mutation operators that merge/split information across row boundaries.

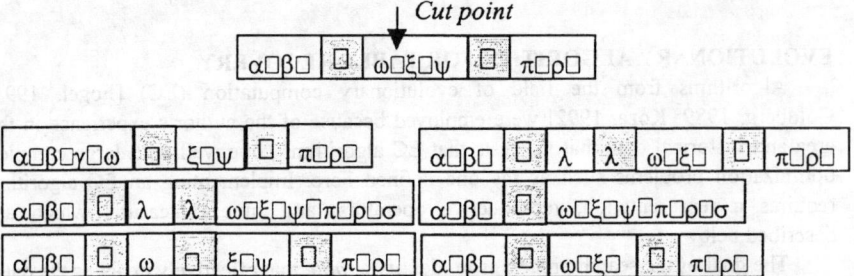

Figure 3. Column-based mutation operators. The topmost row shows the original row, the six rows at the bottom show possible mutations on the original row and selected cutpoint.

The system also includes a crossover operator that swaps rows from two tables, where the goal is to combine useful building blocks present in two different population members. Another crossover operator, which is not implemented in this paper, allows a crossover point to occur within a single row so that effective column components can be exchanged.

The fitness measure in an EC algorithm must accurately measure the quality of an organism. A fitness measure designed to assess the quality of a table must work at several levels of abstraction. For example, fitness must measure individual cells, complete rows, and entire columns.

A *cell cohesion* function assesses the quality of an individual cell. Character case and entity cohesion are both measured and integrated according to the following formula:

cellcohesion(CELL) = w*cellcase(CELL) + $(1-w)$*cellentity(CELL).

The function *cellcase* returns 1 if the cell contains typical pattern of case (e.g., sentence case, title case, all uppercase, etc); ½ if the cell contains non-alphabetic information; and 0 otherwise. The function *cellentity* returns a value in (0, 1] if the entire sequence of tokens in the cell qualifies as one of the defined high-level entities, where the value increases with the number of non-whitespace tokens. 0 is returned if the cell is not a complete entity. In this work, $w = 0.4$, indicating a preference for cells containing a complete entity. A *row cohesion* measure is defined as the average cohesion values of the row's individual cells.

A *column cohesion* measure is more complex, as it considers four aspects of a single column: alignment, entity compatibility, content compatibility, and case compatibility. These four components are combined using the following equation: *colcohesion*(COL) = $w1$*colalign(COL)+$w2$*colentity(COL)+$w3$*colentity(COL)+$w4$*colcase(COL).

Columns may, or may not, be aligned. If they are, they may be aligned on the left, right, or center. This measure computes the standard deviation of all of the cells' left and right edges, as well as their center points. The minimum of these values is noted. A value of 0 indicates that the column is perfectly aligned and larger values indicate significant misalignment. A linear function maps the minimum alignment value into the range [0, 1]. The function *colentity* uses a concept called *entity compatibility,* which indicates if two entities are compatible. For example, $100 and $57 are compatible since they are both forms of *money*. $100 and 57 are also compatible since some tables place units on only the first row. Other less obvious compatible pairs are range and integer, since a single number can serve as a valid range. A directed graph is defined as one that encodes the set of compatible pairs. When computing compatibility over an entire column, all high-level entities are examined. The entity with the highest percentage of compatible cells in that column is called the *dominant entity*. The function *colentity* returns the percentage of rows that are compatible with the dominant entity. The function *colcase* works similarly but uses a binary relationship called *case compatibility*. Two strings are case compatible unless one is uppercase and one is lowercase; one is title case and one is sentence case, etc. In this work, w1, w2, w3, and w4 are set to 0.3, 0.3, 0.2, and 0.2, respectively. The fitness measure of a table is the summation of all its rows' average cohesion added to the summation of the average of the columns' cohesion.

A generic EC algorithm is adopted, with mutation and crossover being applied independently with 0.6 and 0.4 probability, respectively. The mutation probability governs whether a single mutation is to occur (it is not to be interpreted as every cell has a 0.6 probability of being mutated). If N is the population size, N/2 offspring are produced for each generation where parents are selected using fitness proportionate selection. Each offspring replaces a randomly selected member of the population. Heuristics were incorporated into the mutation operators to identify cells that are good candidates for mutation. For example, row cohesion serves as an indicator of a row's quality. The row with the lowest cohesion is twice as likely to be selected for mutation as the row with the highest cohesion; cohesion values between the minimum and maximum are scaled linearly between 1 and 2. Within that row, a column is selected using several heuristics that help to ensure the preservation of existing high-level entities and formation of new high-level entities.

The initial population is defined using a table formed from the input text. The input text is placed into a default table that is formed by inserting column breaks at the locations where multiple spaces are seen in the original text. This default table is then mutated N separate times to initialize the population.

EXPERIMENTS

The prototype system described in the previous sections was implemented in C++ on a PC-based system. The compiler toolset ANTLR [Parr, 1993] was used to perform the lexical analysis and parsing of entities. GALib [Wall, 1996] was used to simplify the implementation of the evolutionary algorithm. A 900Mhz PC was used for the experiments. Population size was set to 30 and number of generations was 25. The number of generations is small compared to typical implementation (e.g., 500 generations). One reason for this low number of generations is that computing times needed to be reasonable; a second reason is that more generations does not necessarily yield better solutions.

Figures 4a-c show the state of the system at various stages for one run. Figure 4a shows the initial table. Figure 4b shows the best table in the population after 10 generations and Figure 4c shows the best table in the population at the end of the run. The fitness of each of these tables is 0.11, 0.137, and 0.1472, respectively. These examples and other experiments show that the fitness measure does a good job at ranking good, terminal tables highly. Note that the EC algorithm performs fitness scaling, so the relatively small differences between the fitnesses are not noteworthy. One problem was observed, however: a very poor table T1 may be mutated into T2, which may be "closer" to the goal, but T2 will have a lower fitness. The problem appears to be due to the definition of column cohesion. The current definition may penalize a table if a single row has optimal column structure but there are no other rows with compatible columns to support it during the column cohesion computation. A potential solution to this problem may be to allow multiple mutations to occur for a given offspring. Such allowances may enable an organism to escape local minimum.

| A+ Auto $2,750 - 3,250 B |
| Jim Smith's Professional Auto Body $2,999 C |
| Marco's Body Shop $219.95 - 750 C+ |
| Pro Auto Paint Center $3,000 - 5,000 B |
| Steve's Auto Body of Chicago $199.95 B |

A+ Auto	$2,750 - 3,250	B	
Jim Smith's Professional Auto Body	$2,999	C	
Marco's Body Shop	$219.95 - 750	C+	
Pro Auto Paint Center	$3,000 - 5,000	B	
Steve's Auto	Body of Chicago	$199.95	B

A+ Auto	$2,750 - 3,250	B
Jim Smith's Professional Auto Body	$2,999	C
Marco's Body Shop	$219.95 - 750	C+
Pro Auto Paint Center	$3,000 - 5,000	B
Steve's Auto Body of Chicago	$199.95	B

Figures 4a-c. Details of evolutionary run. The first figure shows the original text. The second figure shows the default table created from the input text. The third table is the population's best element at the end of ten generations. The last figure shows the best table generated during the entire run.

1.	Stefano Zanini	(Italy)	4 hours 23 minutes 13 seconds
2.	Thomas Muhlbacher (*)	(Austria)	
3.	Marcel Strauss	(Switzerland)	
4.	Marc Lotz	(Netherlands)	all same time
5.	Ellis Rastelli	(Italy)	11 seconds behind
6.	Fred Rodriguez	(U.S.)	7:31
7.	Markus Zberg	(Switzerland)	
8.	Rene Haselbacher	(Austria)	
9.	Richard Virenque (France)		
10.	Oscar Mason	(Italy)	all same time

Figure 5. Details of a second evolutionary run. The original text is defined in Figure 1. This table shows the best table generated during the entire run.

Figure 5 shows the best table from second experiement using the text from Figure 1. This example is much better than the original text, but it is not perfect since the 9th row has not moved the country into the third column. Although experiments with more generations can sometimes yield the proper table, the algorithm does not usually address the problematic portion of the table due to the stochastic nature of the algorithm. The system has been observed destroying the good parts of the table. For example, the phrase "4 hours 23 minutes 13 seconds" is often targets for mutation, since its column is not a complete high-level entity and some of the rows in that column are empty, which do not positively or negatively support keeping the sequence of tokens as one piece. Potential solutions to this problem may involve a divide and conquer approach, where crystallized columns (e.g., column three is clearly contains country) are frozen and the two adjacent sub tables are then processed independently. Such approaches may have the added benefit of decreasing processing time, as the search space for two smaller tables is smaller than the search space for one large table.

DISCUSSION

The prototype described here is somewhat slow, taking tens of minutes to complete the evolutionary process. It is expected that optimizations could reduce the computational requirements by at least an order of magnitude; however, even times on the order of a minute are too long for information providers processing millions of tables. The problem is that EC algorithms demand numerous opportunities for reproduction, and each offspring requires a fresh evaluation.

A long-term goal is to process tables that appear in continuous text, without the aid of line breaks. This would require operators for breaking/merging rows in the middle of a line of text, increasing the size of the search space enormously. A shorter-term enhancement will process tables that break long lines into multiple lines. Successfully identifying these tables using the framework defined here will require a new set of heuristics akin to the heuristics already defined. Nonetheless, including this additional variability will increase the size of the search space considerably since it is generally unknown if a table requires wrapping or the vertical merging. Given these observations, it is unclear if the computational demands of an EC algorithm can be satisfied in practice. On the other hand, the difficulty of the problem may require a non-deterministic, population-based algorithm such as the one described here.

REFERENCES

Fogel, D. B. 1991. System Identification Through Simulated Evolution: A Machine Learning Approach to Modeling, Ginn Press, Needham, MA.
Goldberg, D. 1989. Genetic Algorithms in Search Optimization and Machine Learning. Addison-Wesley, Reading, MA.
Hu, J., Kashi, R., Lopresti, D., and Wilfong, G. 2000. "Table Detection Across Multiple Media," *SPIE Document Recognition and Retrieval VII.*
Hurst, M. 2001. "Layout and Language: An Efficient Algorithm for Text Block Detection based on Spatial and Linguistic Evidence," *SPIE Document Recognition and Retrieval VIII.*
Hurst, M. and Douglas, S. 1997. "Layout & Language: Preliminary experiments in assigning logical structure to table cells," *Proceedings of 5th Applied Natural Language Processing Conference,* pp. 217-200.
Hurst, M. 2000. "The Interpretation of Tables in Texts," Ph.D. dissertation, The University of Edinburgh.
Koza, J. R. 1992. Genetic Programming: On the Programming of Computers By Means of Natural Selection. MIT Press, Cambridge, MA..
Parr, T. 1993. Language Translation Using PCCTS and C++. Automata Publishing Company.
Pyreddy, P. and Croft, W. B. 1997. "TINTIN: A System for Retrieval in Text Tables," University of Massachusetts, Department of Computer Science Technical Report 105.
Wall, M, 1996. "Galib: A C++ Library of Genetic Algorithm Components," http://lancet.mit.edu/ga.

EVOLUTIONARY SYNTHESIS OF MICROELECTROMECHANICAL SYSTEMS(MEMS) DESIGN

NINGNING ZHOU
Dept. of Mechanical Engineering
University of California at Berkeley
Berkeley, CA94720

BO ZHU
Dept. of Mechanical Engineering
University of California at Berkeley
Berkeley, CA94720

ALICE M. AGOGINO
Dept. of Mechanical Engineering
University of California at Berkeley
Berkeley, CA94720

KRISTOFER S.J. PISTER
Dept.of EECS
University of California at Berkeley
Berkeley, CA94720

ABSTRACT
This paper presents the preliminary research on the automatic synthesis of MEMS structures by using multi-objective genetic algorithms (MOGA). The problem model includes problem inputs, the cost function, the types and numbers of available components such as anchors and beams. The Multi-Objective Genetic Algorithm is applied to randomly generated populations to iteratively search for functional designs achieving the desired performance. Besides selection and crossover, elitism and immigration have been implemented in MOGA to produce designs for the next iteration. Initial results have been obtained for automatic synthesis of both topology and sizing of a MEMS 2D meandering spring structure with desired stiffness in certain directions. Preliminary results demonstrate the feasibility of this approach to MEMS structure synthesis.

INTRODUCTION

MEMS technology has been developed for decades to make miniaturized mechanical devices and components integrated with microelectronics on silicon chips or other substrates. Although the advancement is phenomenal, the development of new micro devices remains challenging because currently MEMS design still heavily relies on human knowledge and expertise in this area. Synthesis tools that could automate the MEMS design process have the potential to revolutionize MEMS devices in much the same way that integrated circuit (IC) synthesis tools transformed IC design. Previous work in MEMS synthesis includes the automated layout generation for commonly used device topologies (Mukherjee et al., 1998) (Lo et al., 1996) such as the folded-flexure mirocresonator, the fabrication sequences generation for surface micromachined structures starting from a two-dimensional geometrical description (Gogoi et al., 1994) and the mask layout generation and fabrication process synthesis for bulk micromachining (Li and Antonsson, 1998). All of these methods, however, are tailored to specific domains and tasks and are not general purpose MEMS synthesis tools.

We have developed an evolutionary method to synthesize functional MEMS devices by combining parameterized basic MEMS building components together. These building components include such primitive elements as anchors, beams and electrostatic gaps as well as higher level building blocks called subnets that are comprised of those primitive elements. Examples of subnets include a spring composed of several beams or a electrostatic comb drive composed of a number of electrostatic gaps (Tang et al., 1989). This method incorporates a MEMS simulator SUGAR (Zhou et al., 1998) into a Multi-Objective Genetic Algorithm (MOGA) to automate the synthesis of MEMS designs. SUGAR is a MEMS simulation package developed at University of California at Berkeley(2000). Given a higher-level description of the device's desired behavior, both the topology and sizing of the devices are generated. The topology means the physical configuration that include the number of gross building components, the types of components and their connectivity. The sizing of the designs entails assigning geometrical values to parameterized building components.

In this paper, we outline our approach and report some of our preliminary results. Our initial interest was to test the feasibility of using evolutionary algorithms to synthesize MEMS devices by combining the building components we defined. This work is illustrated by a 2D meandering flexure consisting of one anchor and several beams connected subsequently. Only the number of gross building components (number of beams) and their sizing are evolved.

PROBLEM STATEMENT

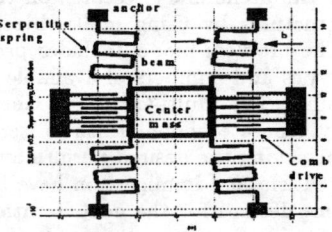

Figure 1: A MEMS resonator with four serpentine supporting springs

We start with a simple but fundamental MEMS supporting spring structure – meandering flexure. One example of it is the serpentine spring (Fedder, 1994) shown in Fig.1. A center mass is supported by four serpentine spring structures to form a MEMS resonator. Two sets of electrostatic comb drives on both sides drive this resonator vibrating horizontally. Our representation of the serpentine spring is shown in Fig. 2:

	Length(um)	Width(um)	Angle
Beam 1	5	2	0
Beam 2	10	2	90
Beam 3	5	2	0
Beam 4	10	2	90
Beam 5	5	2	0
Beam 6	10	2	90
...	...	...	...

phenotype   genotype

Figure 2: Coding scheme for meandering flexure

Beam 1 (B1) extends from anchor point node 1 to node 2 with length 5um, width 2um, at an angle of 0 with respect to the x axis; beam 2 (B2) extends from node 2 to node 3 with length 10um, width 2um, at an angle of 90 with respect to the x axis and so on. Each meandering flexure is encoded into a matrix of N by 3 as the genotype shown in Fig. 2, where N is the number of beams. Each row contains the length, width and angle of one beam. The matrix is ordered so that the first row corresponds to beam 1 that is the nearest to the anchor, the second row corresponds to beam 2 that is the second nearest to the anchor and so on. With length, width and angle of each beam, we can fully describe the configuration and geometries of a meandering flexure. Our goal is to choose the right number of beams and the right length, width and angle parameters for each beam to achieve certain design objectives. For example, we could design a spring with specified stiffness along the x and y directions or design a spring with a certain ratio between the x stiffness and the y stiffness.

EVOLUTIONARY ALGORITHM

Genetic algorithms are global search algorithms based on the mechanics of natural selection and natural genetics (Goldberg, 1989). It applies selection, crossover and mutation etc. to the randomly generated population to iteratively search for optimal solutions. The traditional genetic algorithm and its many variations have been successfully applied in complex engineering design domains. To deal with multiple non-commensurable objectives, several multi-objective genetic algorithms (MOGA) adopt the concept of Pareto optimality (Goldberg, 1989) (Schaffer, 1985) (Tamaki, 1994). This paper applies MOGAs into MEMS design since most design cases involve more than one objective.

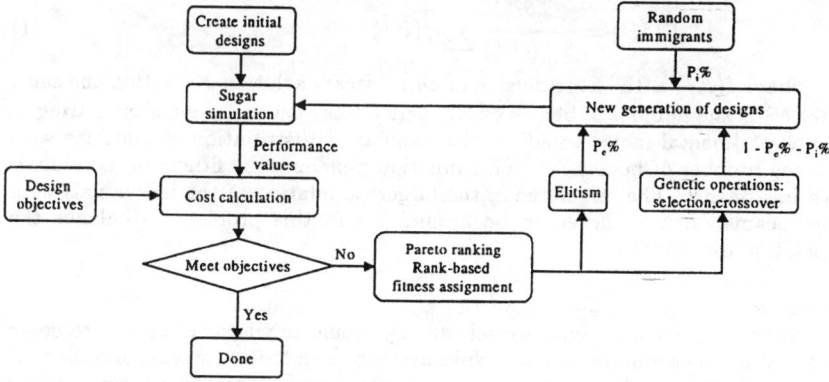

Figure 3: Flow chart for our evolutionary MEMS approach

Figure 3 illustrates an evolutionary algorithm iteration loop for MEMS design. Solutions have to be first encoded into a representation format that can be manipulated by genetic operations. An initial random population of designs is produced in this format. Each design in the current generation is checked for geometrical validity and its performance is evaluated by SUGAR. A cost vector is also calculated for each individual. If the performance doesn't meet the objectives, the whole population of the current generation is ranked using Pareto optimality. A fitness value is assigned to each design based on its rank. To increase the converging rate of the evolutionary algorithms, $P_e\%$ of the elites (individuals with rank 1) in the total population are directly copied to the new generation. As the complementary mechanism of mutation, randomly generated immigrants which are $P_i\%$ in the total population, are introduced into population to preserve diversity. The remaining $1 - P_e\% - P_i\%$ individuals are generated from genetic operations of selection/crossover. This MOGA loop iteratively searches for functional designs that meet constraints and optimize performance. The concept of Pareto optimality (Fonseca and Fleming, 1995) eliminates the use of pre-specified or explicit weightings and makes it possible to provide multiple optimal solutions to the decision maker. Pareto optimality also provide designers with a family of 'equally best' or 'non-dominated' solutions therefore providing more design flexibility.

Rank-based Fitness Assignment and Fitness Sharing

A fitness value is assigned to every solution to measure its quality. In MOGA, the smaller the rank, the better the performance. We define the fitness to be the inverse of the rank.

Since the number of beams N is a design variable in this problem, designs with less beams have a smaller design space and thus converge faster. Designs with more beams have a larger design space with more free variables and may converge slower, although they may find better solutions. Given the same limited computing time and computing power (as we let them evolve in the same run), we found that the population tends to drift to the solutions with less beams because these solutions converge faster. This imbalance can be accumulated with the evolutions such that it is impossible eventually to find a better solution with more beams. We counteract this drift by using Fitness sharing to re-distribute the fitness among the candidate solutions with the same rank. In this paper, the sharing function is defined as

$$f'(K^j) = \frac{c^{N^j}}{\sum_j (c^{N^j})} \sum_j f(K^j) \qquad c > 1 \qquad (1)$$

where $f(K^j)$ is the fitness value of an arbitrary solution K^j within the same rank, N^j is the number of beams of K^j, and $f'(K^j)$ is the fitness after sharing. c is an experimental factor to adjust the population distribution of solutions with differing number of beams N^j. This function penalizes the fitness of individuals with less beams. The larger the c, the larger population of the individuals with more beams. c was chosen to be around 1.4 in this problem to balance the population distribution.

Crossover

After two parent designs are selected by Roulette wheel selection, crossover is carried out to generate two new child designs. The crossover scheme comprises three steps: parametric arithmetical crossover, cut and splice. Two parent designs with differing number of beams, are represented as two matrices: $N_1 \times 3$ and $N_2 \times 3$. Suppose $N_{min} = min(N_1, N_2)$, as shown in Fig.4,

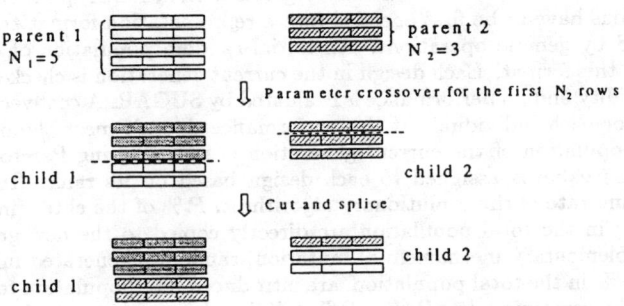

Figure 4: Crossover Scheme

Arithmetical crossover is carried out between the first N_{min} rows of two parents. It is separately applied to the pairs of length, width, and angle to ensure valid decoding. For example, length can't be swapped with angle.

Table 1: Design example configuration

Design variables	$w_{\min}$	$w_{\max}$	$l_{\max}$	$\theta_{\min}$	$\theta_{\max}$	$N_{\max}$
	$2um$	$20um$	$400um$	$-90°$	$90°$	6

MOGA Configuration	n_{pop}	n_{gen}	P_i	P_e	λ
	200	40	40%	5%	0.3

Table 2: Synthesis results for meandering springs from 2 MOGA runs

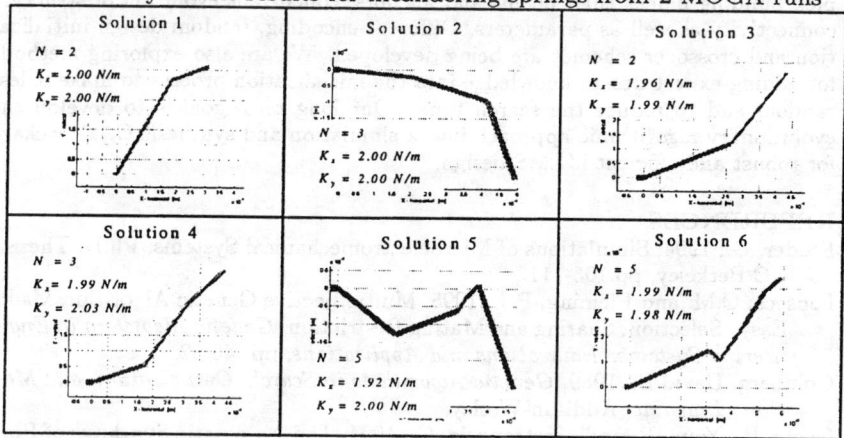

DESIGN EXAMPLES

We illustrate our approach on a design problem with a meandering spring composed of one anchor and N beams.

Design Objectives: stiffness in x direction $K_x = 2N/m$; stiffness in y direction $K_y = 2N/m$.

Design variables: number of beams N, length of beams l, width of beams w, angle of beams θ.

Design Constraints: $w_{\min} < w < w_{\max}$, $w < l < l_{\max}$, $\theta_{\min} < \theta < \theta_{\max}$, $1 \le N \le N_{\max}$. w, l, θ are real numbers, N is an integer.

The design parameters are shown in Table 1, where n_{pop} is the population size in each generation, n_{gen} is the number of generations to be run before stopping the MOGA program. We ran several tests with n_{pop} varying from 100 to 800 and n_{gen} varying from 20 to 200. n_{pop} and n_{gen} were chosen to be 200 and 40 respectively in the problem because increasing them beyond 200 and 40 did not improve the performance significantly. Table 2 shows six different synthesized designs with different number of beams. Solution 1 and 2 are from the Pareto set (with rank 1) of one MOGA run 1. They are the equally best solutions with a different number of beams $N = 2$ and 3. Solution 3, 4, 5, 6 are from the Pareto set of another MOGA run 2. They have a different number of beams $N = 2, 3$ and 5 All of these closely meet the design objectives.

In MOGA run 1, we applied the fitness sharing scheme with $c = 1.3$ The best designs with relatively small number of beams ($N = 2, 3$) were evolved. In MOGA run 2, we increased c to 1.4. It produced designs with 5 beams as well as 2 and 3 beams. Without fitness sharing, most of the MOGA runs generated solutions of 2 beams or sometimes even 1 beam. Fitness sharing plays an important role in diversifying the solutions.

CONCLUSION

Thus far we have completed the iterative design synthesis loop by combining the evolutionary algorithm and SUGAR simulation. We have also completed the initial testing stage by designing meandering springs, resulting in reasonable and promising results. During the initial testing stage, we found that it is not likely to generate complicated devices by only combining a small number of primitive building blocks. Current research is targeted towards extending this evolutionary algorithm approach to more complex device synthesis by incorporating more complex building blocks than beams and anchors, concurrent evolution of structure connectivity as well as parameters. Different encoding, random design initialization and crossover schemes are being developed. We are also exploring methods for adding expert design knowledge into the initialization process to make it less random and to reduce the search time. Our long term goal is to develop the evolutionary algorithmic approach into a simulation and synthesis CAD package for robust and efficient MEMS design.

REFERENCES

Fedder, G., 1994, Simulations of Microelectromechanical Systems, Ph.D. Thesis, UC Berkeley, pp.105-111..

Fonseca, C.M. and Fleming, P.J., 1995, Multiobjective Genetic Algorithm Made Easy: Selection, Sharing and Mating Restriction, *Genetic Algorithm in Engineering Systems: Innovations and Applications*, pp. 45-52.

Goldberg, David E., 1989, *Genetic Algorithms in Search, Optimization, and Machine Learning*, Addison-Wesley.

Gogoi, B., Yeun, R. and Mastrangelo, C., 1994, The Automatic Synthesis of Planar Fabrication Process Flows for Surface Micromachined Devices, *Proc.IEEE Micro Electro Mechanical Systems Workshop*, pp 153-157, Oiso, Japan.

Li, H. and Antonsson, E.K., 1998, Genetic Algorithms in MEMS Synthesis, *ASME International Mechanical Engineering Congress and Exposition, (MEMS)*, Anaheim, CA, pp.299-303.

Lo, N.P., Berg, E.C., Quakkelaar, S.R., Simon, J.N., Tachiki, M., Lee, H.J. and K.S.J. Pister, 1996, Parameterized Layout Synthesis, Extraction, and SPICE Simulation for MEMS, *Proc. ISCAS*, pp. 481-484, Atlanta, GA.

Mukherjee, T., Iyer, S. and Fedder, G., 1998, Optimization-based Synthesis of Microresonators, *Sensors and Actuators*, A 70, pp.118-127.

Schaffer, J.D., 1985, Multiple Objective Optimization with Vector Evaluated Genetic Algorithms,*Proceedings of the first International Conference on Genetic Algorithms*, pp. 93-100.

Tamaki, H., Kita, H. and Kobayashi, S., 1996, Multi-Objective Optimization by Genetic Algorithm: A Review, *Proc. 1996 IEEE International Conference on Evolutionary Computation*, pp. 517-522, Nagoya, Japan.

Tang, W. C., Nguyen,T.-C H., Judy, M.W. and Howe, R.T., 1989, Electrostatic -comb drive of lateral polysilicon resonators, *Proc. of 5th International Conference on Solid-State Sensors and Actuators*, Montreux, Switzerland, pp. 328-331.

University of California at Berkeley, SUGAR, http://www-bsac.EECS.Berkeley.EDU/~cfm/ , 2000.

Zhou, N., Clark, J.V. and Pister, K.S.J., 1998, Nodal Simulation for MEMS Design Using SUGAR v0.5, *1998 International Conference on Modeling and Simulation of Microsystems Semiconductors, Sensors and Actuators*, pp. 308-313, Santa Clara, CA.

DETECTION OF PARTIAL TOOL DAMAGE BY USING GENETIC ALGORITHM BASED TOOL MONITORING SYSTEM

L. LI, W.Y. BAO
I.N. TANSEL, N. REEN
Mechanical Engineering Department
Florida International University
Center for Eng. and App. Sciences
10555 West Flagler Street
Miami, Florida 33174, USA

C.V. KROPAS-HUGHES
Air Force Research Laboratory
AFPL/MLLP, Bldg. 655, R166
2230 Tenth Street
Wright-Patterson AFB
Ohio, 45433-7746

ABSTRACT

In this paper, a Genetic Tool Monitor (GTM2) is proposed to evaluate the tool condition by estimating the key parameters from the cutting force data. Parameters are estimated by minimizing the estimation error of an analytical model using genetic algorithms. Dullness and partial cutting edge loss can be assessed from the estimated cutting force coefficient and tool run out parameters. Performance of the GTM2 was evaluated on the simulated data. System estimated the cutting force coefficient and length of the partially lost section of the cutting edge with less than 1% and 10% errors, respectively.

INTRODUCTION

Micro-end-milling operations (MEMO) are very difficult to program and monitor compare to the conventional end-milling operations. The diameter of these tools is very small. Small run-outs, which cannot be even detected in conventional milling operations, may prevent one of the cutting edges from even removing material. Very small cutting forces of the milling operation create extremely small vibrations. The generated sound cannot be even noticed in the typical background noise of the machining facilities. The tiny cutting edges cannot be visually inspected without using a visual aid for magnification. In this paper, use of genetic algorithms with an analytical model is proposed to evaluate the tool condition. The proposed system is capable of distinguishing the dullness from partial loss of one of the cutting edges without any training.

To evaluate cutting conditions, customized methods (Altintas et al., 1988; Altintas and Yellowley, 1989; Sutherland et al., 1989), time series analysis (Tansel and McLaughlin, 1993a) and neural networks (Tansel and McLaughlin, 1993b) have been used. These approaches require either experimental or simulated data at the similar conditions. None of these approaches can embed the characteristics of the cutting operation into the monitoring system. On the other hand genetic algorithms quickly estimate the values of the parameters of the models which are selected by the user. Since the characteristics of the machining operation is already embedded into the monitoring system, it can be used without any tests, inspection of simulated data or training process. The authors introduced the application of the genetic algorithms in a previous paper (Tansel et al., 1998). However, in that study, run out was not considered and only the cutting force coefficient was estimated to evaluate the dullness of the tool. In this paper, tool with run-out is considered and the) analytical model presented by Bao and Tansel (Bao and Tansel, 2000a; 2000b; 2000c) is used. The dullness and material loss from one of the cutting edges were evaluated at the same time.

Lei et al., (1999) used the genetic algorithm based wear monitoring system to evaluate milling operations. They used time series AR (autoregressive) model of feed direction acceleration to monitor the operation. In this paper, the parameters of the analytical model of milling operation will be estimated and used to assess the tool condition. Bao and Tansel studied micro-end-milling operations (Bao and Tansel, 2000a; 2000b; 2000c) and derived a series of equations to estimate the cutting force profile for micro-tools with two cutting edges. In this study, the analytic models introduced by Bao and Tansel (2000b) are used.

Recently, genetic algorithms (Goldberg, 1989; Carroll, 1996) have been successfully used in many applications. The genetic algorithms are suitable when the model is non-linear, and multiple parameters are to be determined with different resolutions. The program tries to minimize or maximize a cost function by emulating the natural selection process. Genetic algorithms typically create a population, evaluate cost, select mate, reproduce, mutate, tests convergence and repeat the process unless the estimated solution satisfies the selected conditions to conclude the optimization. More information about the genetic algorithms can be found in Goldberg (1989) and Carroll (1996).

PROPOSED MONITORING SYSTEM (GENETIC TOOL MONITOR 2)

A new micro-end-mill has two cutting edges with almost the same length. The effective lengths of both cutting edges are almost equal to the feed rate per tooth. Cutting edges start to wear out during machining operation. As the micro-end-mill is used, first the cutting edges get dull and the cutting forces and their coefficient increase. Later the cutting edges start to lose considerable amount of material and the length of the effective cutting edge is reduced.

Relationship between the partial loss of cutting edge and the run out characteristics

The center of an end-mill and the tool rotation are not at the same point in most of the cases. The distance between the geometric and rotational centers is called run out. Characteristics of the cutting force is influenced by the run out (r_o), and orientation of the rotation center respect to the cutting edges (γ). When the effective length of one cutting edge start to reduce with material loss, the characteristics of the cutting force varies as if the tool diameter decreased and the rotation center moved to a different point. In other words, the length of the cutting edges can be monitored if two parameters of the run out are estimated. Figure 1 shows the cutting edges of a new and used tool with reduced effective cutting edge length. The new tool has two cutting edges with same lengths at the beginning as shown in Figure 1a. Because of material loss at the tip, the effective length of one of the cutting edges is shortened as illustrated in Figure 1b. After the partial loss of the cutting edge, the tool characteristics look like it has a smaller diameter and the center of the rotation is changed.

Assuming L represents the length of the damaged section of the cutting edge which does not remove material; if the initial run out and run out angle are represented by r_o and γ, respectively and relative to the new center of the tool their values changed to r_o' and γ', then the length of the damaged section can be calculated with the following equations:

$$r_o' = ((r_o \cdot \sin\gamma)^2 + (r_o \cdot \cos\gamma - L/2)^2)^{0.5} \qquad (1)$$

$$\gamma' = \arcsin((r_o \cdot \sin\gamma)/r_o') \qquad (2)$$

Determination of the parameters of the analytical model using genetic algorithms

The length of the partially lost section (L) of the cutting edge, which does not remove material, and the cutting force coefficient represented with two genes of the DNA. The objective function of the optimization program was the difference between the experimental and simulated cutting force values in the feed and normal direction. The objective function is expressed as the following equation:

$$Min(E) = \frac{1}{N}\sum_{i=1}^{N}[F^{estimated}(i) - F^{actual}(i)]^2 \qquad (3)$$

where:
 $F^{estimated}$ is the estimated force from the cutting force model
 F^{actual} is the actual cutting force collected from tools
 N is the number of collected data
 E is the average difference of $F^{estimated}$ and F^{actual}

The cutting force coefficient and the length of the lost section of the cutting edge were represented with 30-bit binary numbers. The population size was six and the mating pool size was two versus two, with one child obtained from each couple. The crossover probability was 0.01, the jump mutation probability was 0.01, and the creep mutation probability was 0.01. The proposed optimization procedure is presented in Figure 2.

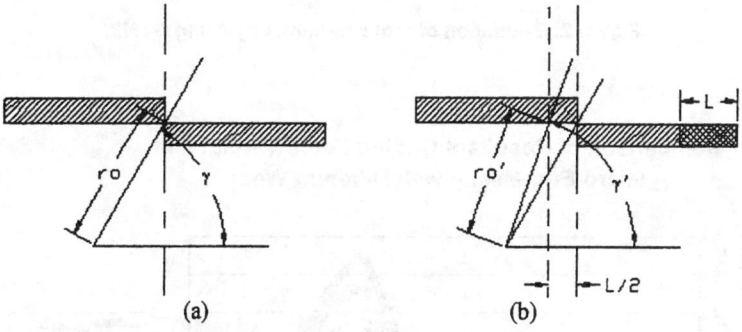

Figure 1. (a) Cutting tool with run-out. (b) Cutting tool with run-out after the effective length of a cutting edge is shortened.

PREPARATION OF SIMULATED DATA

Simulated data was prepared by using the analytical model by Bao and Tansel (2000b). Cutting force variation of the tools with perfect and worn cutting edge is presented in Figure 3. In the simulation, two-flute carbide end-mills with 0.02" diameter were considered. Cutting force coefficient was selected for POCO 3 work piece. The Spindle speed was 15,000 rpm. Feed rate and depth of cut were 60 ipm and 0.01", respectively. Tool run out was 0.001", at 30°.

The characteristics of the simulated cases are presented in Table 1. The first case represents a new tool (case 1). The rest of the simulated data was prepared to represent the characteristics of a worn out tool. Dullness and material loss at the cutting edges were considered. The cutting force coefficient was increased for the dull tools. Tool diameter and the location of the run out were adjusted to simulate the characteristics of the partially lost (chipped away) cutting edges.

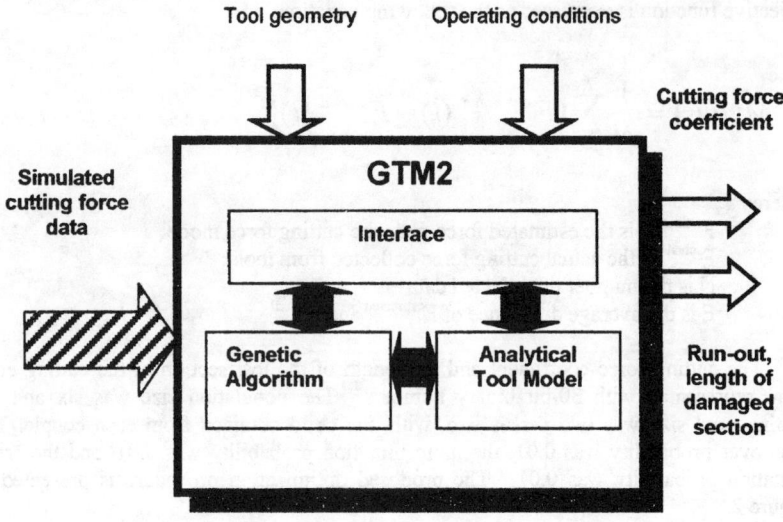

Figure 2. Evaluation of tool conditions by using GTM2.

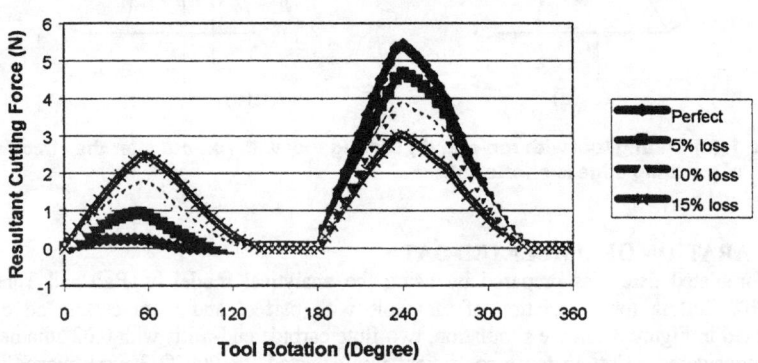

Figure 3. Comparison of resultant cutting force variation with a new tool and different material loss on one cutting edge of the tool. In the study cases, a high speed steel end-mill with 0.02" diameter cut on the POCO 3 work piece was used. The cutting conditions are: 15,000 rpm spindle speed, 60 ipm feed rate, 0.01" depth of cut, and 0.001" run out with 30° angle.

RESULTS AND DISCUSSION

The performance of the GTM2 is outlined in Table 1. Simulations were prepared at nine different conditions. A micro-end mill with 0.02" radius and two cutting edges was considered. First cutting edge of the tool was always assumed to keep the original length. Only the cutting force coefficient was increased for this edge in the following six cases to simulate dullness. To simulate the characteristics of a tool with partially lost (chipped away) cutting edge, the center of rotation (run out) and tool diameter were changed. The second edge was assumed to have the original length in the Case 1,5, and 9. The length of the lost section of the cutting edge was changed between 0.0005" and 0.0015" to simulate wear. For the dull tools, cutting force coefficient was increased between 10% and 30%. The GTM2 identified the problems associated with the dullness and material loss of the tool with less than 1% and 10% error respectively. Genetic algorithm was stopped at 500 generations, which typically took less than 15 seconds on a Pentium IV microcomputer. Programs were developed using Visual Basic. The theoretical values and GTM2 estimations are presented in Table 1.

Table 1. Performance of GTM2 (Initial run out=0.001, run out angle=30°)

Case number (Changes)	Condition of Second Cutting Edges	Cutting Force Coefficient (N/cm^2)	Length of lost section (Inch)	GTM2 Estimation Coefficient (N/cm^2)	GTM2 Estimation Lost Section (Inch)	Estimation Error Dullness	Estimation Error Lost
1 (0, 0%)	Perfect	130,000	0	133349	0.000161	-2.5%	-
2 (0, 5%)	P.E.L.	130,000	0.0005	130761	0.000547	-0.58%	-9.4%
3 (0, 10%)	P.E.L.	130,000	0.001	129139	0.001001	0.66%	-0.1%
4 (0, 15%)	P.E.L.	130,000	0.0015	128959	0.001493	0.8%	0.46%
5 (10%, 0%)	Dull	143,000	0	146679	0.000161	-2.5%	-
6 (10%, 5%)	Dull, P.E.L.	143,000	0.0005	143805	0.000546	-0.56%	-9.2%
7 (10%, 10%)	Dull, P.E.L.	143,000	0.001	142052	0.001001	0.66%	-0.1%
8 (10%, 15%)	Dull, P.E.L.	143,000	0.0015	141855	0.001493	0.8%	0.46%
9 (20%, 0%)	Dull	156,000	0	160023	0.000161	-2.5%	-
10 (20%, 5%)	Dull, P.E.L.	156,000	0.0005	156878	0.000545	-0.56%	-9%
11 (20%, 10%)	Dull, P.E.L.	156,000	0.001	154968	0.001001	0.66%	-0.1%
12 (20%, 15%)	Dull, P.E.L.	156,000	0.0015	154687	0.001493	0.84%	0.46%

P.E.L. : Partial (cutting) edge loss

CONCLUSIONS

A genetic algorithm based tool monitoring system was developed and its performance was tested on simulated data. The characteristics of the machining operation is embedded in the Genetic Tool Monitor (GTM2) system and it can start monitoring the tool condition without any initial observations or training. Although, the method can be used to identify as many parameters as wanted, it is not practical to consider more than the necessary number of parameters. The estimation time will increase and the accuracy will decrease sharply with each additional parameter. In this study, only slot cutting operations were considered.

The GTM2 accurately identified the source of the problems of the simulated micro-end-milling data. Cutting force coefficient increases associated with dullness and change of the run out characteristics as a result of the partial loss of the cutting edge were identified accurately with less than 1% and 10% error, respectively.

ACKNOWLEDGEMENT

This work was partially supported by the Air Force Research Laboratory (contract # F33615-99-C5703). Authors thank to AFPL/MLLP for the support.

REFERENCES

Altintas, Y., Yellowley, I., and Tlusty, J. , 1988,"The Detection of Tool Breakage in Milling Operations," Trans. of ASME, Vol.110, pp.271-277.

Altintas, Y., and Yellowley, I., 1989, "In-Process Detection of Tool Failure in Milling Using Cutting Force Models," *Trans. of ASME*, Vol.11, pp.149-157

Bao, W.Y., and Tansel, I.N., 2000a,"Modeling Micro-End-Milling Operations. Part I: Analytical Cutting Force Model," *Int. Jour. of Mach. Tools and Manufacturing*, Vol. 40, pp. 2155-2173.

Bao, W.Y., and Tansel, I.N., 2000b, "Modeling Micro-End-Milling Operations. Part II: Tool Run Out," *Int. Jour. of Mach. Tools and Manufacturing*, Vol. 40, pp. 2175-2192.

Bao, W.Y., and Tansel, I.N., 2000c, "Modeling Micro-End-Milling Operations. Part III: Influence of Tool wear," *Int. Jour. of Mach. Tools and Manufacturing*, Vol. 40, pp. 2193-2211.

Carroll, D.L., 1996,"Chemical Laser Modeling with Genetic Algorithms," *AIAA Jour.*, Vol.34, No.2, pp.338-346.

Goldberg, D.E, 1989, Genetic Algorithms in Search, Optimization and Machine Learning, Addison-Wesley, Reading, MA,

M. Lei, Yang, X., and Yang,.S., 1999, "Tool Wear Length Estimation with a Self-Learning Fuzzy Inference Algorithm in Finish Milling", *Int. Journal of Advance Manufacturing Technology*, Vol. 15, No. 8, pp.537-545.

Sutherland, J.W., O'Brien, D.J., and Wagner, M.S., 1989, "An Algorithm for the Detection of Flute Breakage in a Peripheral End Milling Process," *Trans. of North American Manufacturing Research Institute*, Vol.17, pp. 144-151

Tansel, I.N, and McLaughlin, C., 1993a, "Detection of Tool Breakage in Milling Operations: Part 1 - The Time Series Analysis Approach", *Int. Jour. of Mach. Tools and Manufacturing*, Vol.33, No.4, pp. 531-544.

Tansel, I.N, and McLaughlin, C., 1993b, "Detection of Tool Breakage in Milling Operations: Part 2 - The Neural Network Approach", *Int. Jour. of Mach. Tools and Manufacturing*, Vol.33, No.4, pp. 545-558.

Tansel, I.N., Bao,W.Y. , Tansel, B., Shisler, B., Smith, D., and Murray, J., 1998, "Identification of Cutting Conditions by Using an Analytical Model and Genetic Algorithms for Micro-End-Milling Operations," Smart Engineering System Design: Neural Networks, Fuzzy Logic, Rough Sets and Evolutionary Programming, Intelligent Engineering Systems Through Artificial Neural Networks, Vol.8, ASME Press, pp.779-784.

INTELLIGENT CONTROL OF THE ELECTRICAL TUNING PROCESS IN TELEVISIONS USING SOFT COMPUTING TECHNIQUES

OSCAR CASTILLO*, PATRICIA MELIN* AND FELIPE DUENAS**
Tijuana Institute of Technology, P.O. Box 4207, Chula Vista CA 91909, USA
**Panasonic, Tijuana, Mexico*

ABSTRACT
We describe in this paper an intelligent system for controlling the electrical tuning process during the manufacturing of televisions. The electrical tuning problem consists in controlling the imaging system of the television to meet production quality requirements. Traditionally, this tuning process has been performed by human operators, by manually adjusting the imaging system in the television. In our approach, we use fuzzy logic to automate the tuning process for the televisions. We use a fuzzy system for controlling the voltage, current, and the time during the tuning process, so that the best possible quality of image is achieved. We have implemented this intelligent system for control in the MATLAB programming language with good simulation results.

INTRODUCTION
We describe in this paper the application of fuzzy logic techniques [1, 2, 3, 4] to the problem of automating the electrical tuning process during the manufacturing of televisions. The electrical tuning process consists in adjusting the voltage, current intensity, and time to produce the best image possible. We have designed a set of fuzzy if-then rules that contains the knowledge of the human experts in the electrical tuning process of televisions. With these fuzzy rules we have developed an intelligent system for automated electrical tuning of televisions during the manufacturing process. We used a simple genetic algorithm [5] to adapt the parameter values of the fuzzy system according to real data about the problem

The basic problem that we are considering in this paper is how to artificially reproduce images in the best way possible in devices like televisions or monitors. While producing televisions or monitors, we normally have a section in charge of adjusting the imaging system of these devices. Traditionally, a human expert operator adjusts the imaging system by experience using a special remote control and the measurements of voltage and current intensity. We are now considering an intelligent system for controlling the electrical tuning process of televisions during production. The intelligent system has a knowledge base, consisting of fuzzy if then rules, that contains the expert knowledge about tuning the imaging system of televisions. The main reason for using fuzzy logic is that we need to represent and also reason with uncertainty in this application [6, 7, 8, 9]. We need to use voltage, current intensity, time, and quality as linguistic variables in the fuzzy rules [10], and define the membership functions for these variables according to real data about the problem and the knowledge of the experts. We begin this paper by first describing the basic concepts of the imaging system of a television. Then, we describe how to use fuzzy logic techniques [10] to automate the electrical tuning process of televisions and show some simulation results.

IMAGING SYSTEM OF THE TELEVISION
The Cathode Ray Tube
Almost all TVs in use today rely on a device known as the cathode ray tube, or CRT, to display their images. The LCDs and plasma displays are sometimes seen, but they are still rare when compared to CRTs. The terms anode and cathode are used in electronics as synonyms for positive and negative terminals. For example, you could refer to the

positive terminal of a battery as the anode and the negative terminal as the cathode. In a cathode ray tube, the "cathode" is a heated filament. The heated filament exists in a vacuum created inside a glass "tube." The "ray" is a stream of electrons that naturally pour off a heated cathode into the vacuum. Electrons are negative. The anode is positive, so it attracts the electrons pouring off of the cathode. In a TV's cathode ray tube, the stream of electrons is focused by a focusing anode into a tight beam and then accelerated by an accelerating anode. This tight, high-speed beam of electrons flies through the vacuum in the tube and hits the flat screen at the other end of the tube. This screen is coated with phosphor, which glows when struck by the beam. Figure 1 shows this situation. As you can see in this figure, there's not a whole lot to a basic cathode ray tube. There is a cathode and a pair of anodes. There is the phosphor-coated screen. There is a conductive coating inside the tube to soak up the electrons that pile up at the screen-end of the tube. However, you can see in this diagram no way to "steer" the beam -- the beam will always land in a tiny dot right in the center of the screen. The following picture gives you a view of a typical set of steering coils: Figure 2 shows this picture. The steering coils are simply copper windings. These coils are able to create magnetic fields inside the tube, and the electron beam responds to the fields.

Phosphor
A phosphor is any material that, when exposed to radiation, emits visible light. The radiation might be ultraviolet light or a beam of electrons. Any fluorescent color is really a phosphor - fluorescent colors absorb invisible (to us) ultraviolet light and emit visible light at a characteristic color.

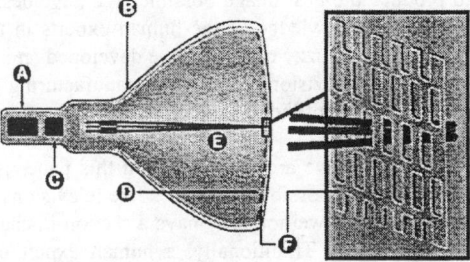

A Cathode B Conductive coating C Anode D Phosphor- screen E Electron beams F Shadow mask

Figure 1 Basic architecture of a color television.

Figure 2 Typical set of steering coils inside a television set.

The Black-and-White TV Signal

In a black-and-white TV, the screen is coated with white phosphor and the electron beam "paints" an image onto the screen by moving the electron beam across the phosphor a line at a time. To "paint" the entire screen, electronic circuits inside the TV use the magnetic coils to move the electron beam in a "raster scan" pattern across and down the screen. The beam paints one line across the screen from left to right. It then quickly flies back to the left side, and moves down slightly and paints another horizontal line, and so on down the screen. When a television station wants to broadcast a signal to your TV, the signal needs to mesh with the electronics controlling the beam so that the TV can accurately paint the picture that the TV station or VCR sends. The TV station or VCR therefore sends a well-known signal to the TV that contains three different parts:

- Intensity information for the beam as it paints each line
- Horizontal retrace signals to tell the TV when to move the beam back at the end of each line
- Vertical retrace signals 60 times per second to move the beam from bottom right to top left.

A signal that contains all three of these components is called a composite video signal. One line of a typical composite video signal looks something like in Figure 3.

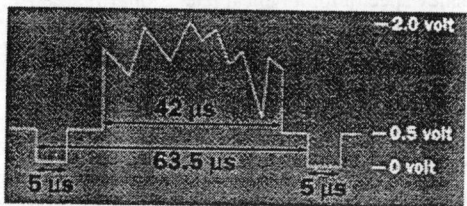

Figure 3. Composite video signal.

Adding Color

A color TV screen differs from a black-and-white screen in three ways:
1. There are three electron beams that move simultaneously across the screen. They are named the red, green and blue beams.
2. The screen is not coated with a single sheet of phosphor as in a black and white TV. Instead the screen is coated with red, green and blue phosphors arranged in dots or stripes. If you turn on your TV or computer monitor and look closely at the screen with a magnifying glass, you will be able to see the dots.
3. On the inside of the tube, very close to the phosphor coating, there is a thin metal screen called a shadow mask. This mask is perforated with very small holes that are aligned with the phosphor dots on the screen.

A color TV signal starts off looking just like a black-and-white signal. An extra chrominance signal is added by superimposing a 3.579545 MHz sine wave onto the standard black-and-white signal. Right after the horizontal sync pulse, eight cycles of a 3.579545 MHz sine wave are added as a color burst (Figure 4).

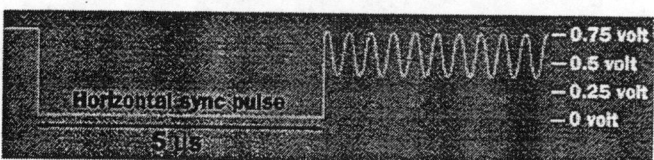

Figure 4 Color television signal.

INTELLIGENT SYSTEM FOR CONTROL

The problem of automated electrical tuning of televisions during production can be viewed as a controlling problem because we basically need to adjust the parameters to achieve the best image possible. We need to control the voltage, current intensity, and time to achieve the best image possible. We show in Figure 5(a) the frequency spectra for the imaging system of a typical television. We also show in Figure 5(b) the acceptance region in the frequency spectra. This region is the one used by the human experts to decide on the appropriate quality of the image given by the television.

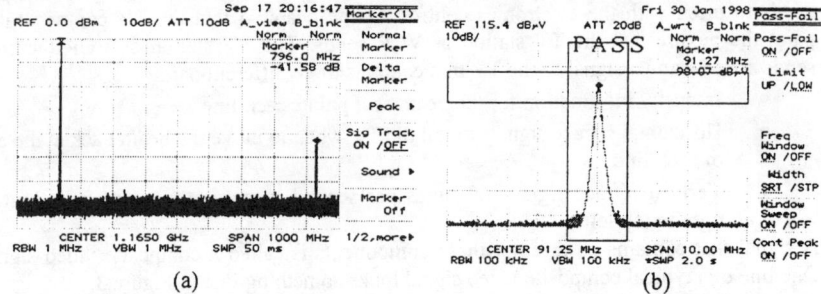

Figure 5 (a) Frequency spectra for a typical television, (b) acceptance region is shown.

We show in Figure 6 the actual output of the system, which shows a specific frequency spectrum for a television. A human experts knows how to use this spectrum to evaluate the quality of the image and also knows how to change voltage and current as to adjust the image in the best way possible. Our intelligent system for control needs to have this knowledge to be able to simulate the expert in the process of tuning the televisions. We show in Figure 7(a) the traditional way of controlling the tuning process of televisions as done by the human experts in the plant. In this case, the human expert uses the output of the oscilloscope and the multi-meter to adjust the voltage and current by remote control.

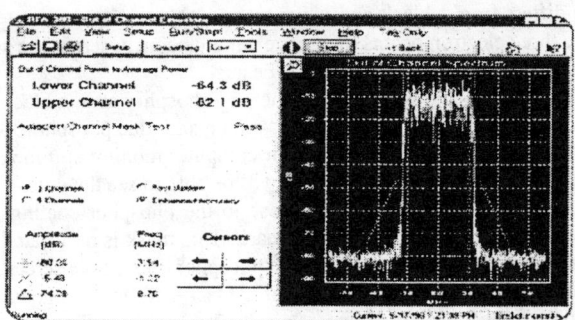

Figure 6 Actual output of the system showing a frequency spectrum.

We show in Figure 7(b) the proposed intelligent system for controlling the electrical tuning process of televisions during manufacturing. We now have data acquisition cards to directly read the data from the oscilloscope and multi-meter, and also a computer program (intelligent system) in the computer to automatically adjust the image. For this purpose, the intelligent system has a knowledge base consisting of fuzzy rules, which contain the knowledge of the human experts in the tuning process. With the architecture shown in Figure 7(b) we really achieve automated control of the electrical tuning of televisions during production. The intelligent system was implemented in the MATLAB programming language, using the fuzzy logic toolbox of this language. The intelligent system works in real time in a real plant that produces televisions.

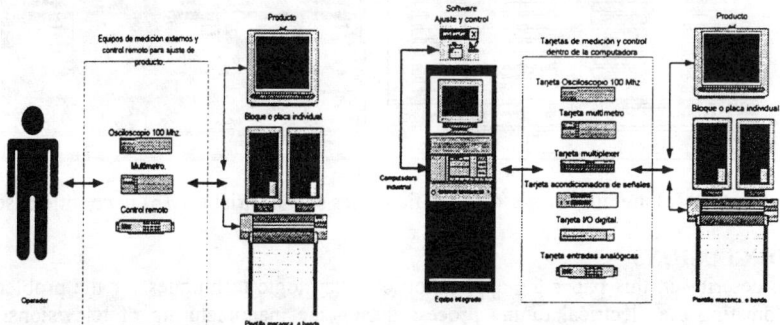

Figure 7 (a) traditional electrical tuning process, (b) proposed intelligent system.

EXPERIMENTAL RESULTS

We have implemented in the MATLAB programming language the intelligent system for control. We used the fuzzy logic toolbox of MATLAB to implement the fuzzy rule base for controlling the tuning of televisions. We used a fuzzy system with three input variables and one output variable. The input linguistic variables are the voltage, current, and time, and the output linguistic variable is the quality of the image. We use the Mamdani inference method and Gaussian membership functions. We optimized the fuzzy system by using a genetic algorithm and arrive at the end to a fuzzy system of 14 rules. We show in Figure 8(a) the non-linear surface of the fuzzy model for controlling the electrical tuning process. We also show, in Figure 8(b), the membership functions for the output linguistic variable of the fuzzy system for controlling the tuning process. We show in Figure 9(a) the rule viewer of the fuzzy logic toolbox, which shows the inference procedure for specific values of the variables. Finally, we show in Figure 9(b) the implementation of the fuzzy rule base for controlling the electrical tuning process in the manufacturing of televisions.

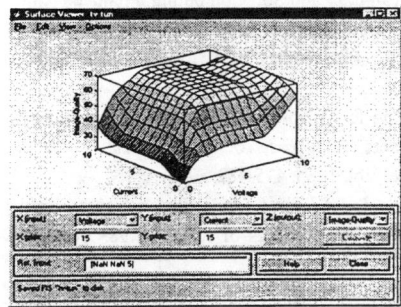 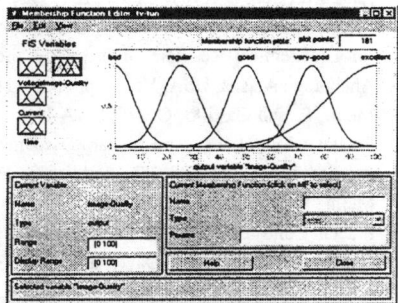

Figure 8 (a) Non-linear surface of the fuzzy model for controlling the tuning process, (b) Membership functions for the output linguistic variable.

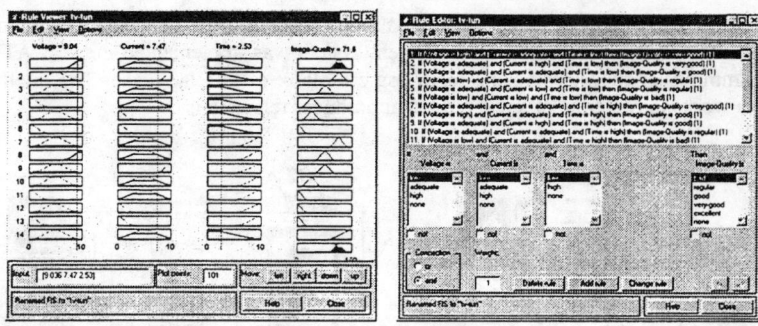

Figure 9 (a) the rule viewer for specific values of the variables, (b) fuzzy rule base

CONCLUSIONS

We describe in this paper the application of fuzzy logic techniques for the problem of automating the electrical tuning process during the manufacturing of televisions. We developed an intelligent system for controlling the tuning of the imaging system of televisions. The intelligent system has a knowledge base of fuzzy rules, which contains the knowledge of human experts in the tuning process of televisions. The intelligent system was implemented in MATLAB and with the use of the fuzzy logic toolbox.

REFERENCES

[1] Castillo, O. & Melin, P. (1998). "A New Fuzzy-Fractal-Genetic Method for Automated Mathematical Modelling and Simulation of Robotic Dynamic Systems, Proceedings of IEEE WCCI'98 Congress, Vol. 2 pp. 1182-1187.

[2] Castillo, O. & Melin, P. (1999). "A New Fuzzy Inference System for Reasoning with Multiple Differential Equations for Modelling Complex Dynamical Systems", CIMCA'99, IOS Press, Vienna Austria, pp.224-229.

[3] Castillo, O. & Melin, P. (1999). "Intelligent Model-Based Adaptive Control of Robotic Dynamic Systems with a New Neuro-Fuzzy-Genetic Approach", Proceedings of Robotics and Applications, Acta Press, USA. pp. 270-275.

[4] Castillo, O. & Melin, P. (2000). "Intelligent Adaptive Model-Based Control of Robotic Dynamic Systems with a New Hybrid Neuro-Fuzzy-Fractal Approach", Proceedings of FLINS'2000, World Scientific, Bruges, Belgium, pp. 351-358.

[5] Jang, J.R., Sun, C.T. & Mizutani, E. (1997). Neuro-Fuzzy and Soft Computing, Prentice Hall.

[6] Melin, P. & Castillo, O. (1998). "An Adaptive Model-Based Neuro-Fuzzy-Fractal Controller for Biochemical Reactors in the Food Industry", Proceedings of IJCNN'98, IEEE Press, Anchorage Alaska, USA, Vol. 1, pp. 106-111.

[7] Melin, P. and Castillo, O. (1998). A New Method for Adaptive Model-Based Neuro-Fuzzy-Fractal Control of Non-Linear Dynamic Plants: The case of Biochemical Reactors, Proceedings of IPMU'98, EDK Publishers, France, Vol. 1, pp. 475-482.

[8] Melin, P. & Castillo, O. (1999). A New Method for Adaptive Model-Based Neuro-Fuzzy-Fractal Control of Non-Linear Dynamical Systems, Proceedings of ICNPAA'98, European Conference Publications, Daytona Beach, USA, pp. 499-506

[9] Melin, P. & Castillo, O. (1999). "A New Neuro-Fuzzy-Fractal Approach for Adaptive Model-Based Control of Non-Linear Dynamic Plants", ", Proceedings of Intelligent Systems and Control, Acta Press, Santa Barbara, USA. pp. 397-401.

[10] Zadeh, L. A. (1975). "The Concept of a Linguistic Variable and its Application to Approximate Reasoning", *Information Sciences*, 8, pp. 43-80.

ESTIMATING TIME SERIES PREDICTABILITY USING GENETIC PROGRAMMING

MINGLEI DUAN
Department of Electrical and
Computer Engineering
Marquette University
Milwaukee, Wisconsin

RICHARD J. POVINELLI
Department of Electrical and
Computer Engineering
Marquette University
Milwaukee, Wisconsin

ABSTRACT
A new method that quantifies the genetic programming predictability of a stock's price is presented. This new method overcomes resolution and stationarity problems presented in previous approaches. A comparison, showing the advantages of the new method, is made, between the approaches, on four time series.

INTRODUCTION
Time series predictability is a measure of how well future values of a time series can be forecasted. Measuring the predictability of a time series is important because it can tell whether a time series can be predicted before making a prediction. Therefore prediction of time series with low predictability, such as a random walk time series, can be avoided. A good measure of time series predictability also provides a measure of confidence in the accuracy of a prediction. This is especially helpful to minimize the risk when making an investment decision.

After a brief background review, the previous approach in this area is introduced. Some disadvantages of this approach are then discussed, and a new modified method that aims at overcoming these disadvantages is presented and tested.

BACKGROUND
Modeling tools play an important role in estimating times series predictability. Evolutionary computation approaches provide effective tool for such modeling. These approaches include genetic algorithms (GA) [1], which are based on reproduction, recombination and selection of the fittest members in an evolving population of candidate solutions. Koza [2] extended this genetic model of learning into the space of programs and thus introduced the concept of genetic programming (GP). Each solution in the search space is represented by a genetic program. Genetic programming is now widely recognized as an effective search paradigm in many areas including artificial intelligence, databases, classification, and robotics.

There has been extensive work in the area of time series modeling using GP. Fogel and Fogel [3] added noise to data generated by the Lorenz system and the logistic map. As expected, using GP, they found that signals with no noise are more predictable than noisy ones. Kaboudan [4] applied GP to estimate the predictability of stock price time series. The advantages of GP include its ability to evolve arbitrarily complex equations not requiring an *a priori* model, and its flexibility in selecting the terminal and function sets to fit different kinds of problems. GP has been widely recognized as an effective time series modeling method [3-6].

An η-metric was introduced by Kaboudan [6], which measures the probability that a time series is GP-predictable. By design, the computed metric should approach zero for a complex signal that is badly distorted by noise. Alternatively, the computed metric should approach one for a time series with low complexity and strongly deterministic signal.

This metric is based on comparing two outcomes: the best fit model generated from a single data set before shuffling with the best fit model from the same set after shuffling. The shuffling process is done by randomly scrambling the sequence of an observed data set using Efron's bootstrap method [7]. Specifically, the unexplained variations, which are measured by the sum of squared error (SSE) before and after shuffling of a time series $\{Y_t\}$, $t = 1, 2, \cdots, N$, are compared. The unexplained variation in $\{Y_t\}$ before shuffling is

$$SSE_Y = \sum_{t=1}^{T}(Y_t - \hat{Y}_t)^2,$$

where $\hat{Y}_t$ is the predicted Y_t. Shuffling increases the unexplained variation in $\{Y_t\}$ to a maximum [1]. This maximum is

$$SSE_S = \sum_{t=1}^{T}(S_t - \hat{S}_t)^2,$$

where $\{S_t\}$ is the shuffled $\{Y_t\}$. The measure of predictability is then defined as:

$$\eta = 1 - \frac{SSE_Y}{SSE_S}.$$

Thus, if the time series $\{Y_t\}$ is a totally deterministic signal and can be modeled perfectly, then $SSE_Y = 0$ and $\eta = 1$. If it is totally unpredictable noise, the reshuffling shouldn't affect the learned GP model accuracy, hence $SSE_Y = SSE_S$ and $\eta = 0$.

METHODS

While applying the η-metric to estimate stock price predictability, two main problems have been observed. First, the value of the metric depends on the length of the time series. Specifically, the η calculated for a 50 day stock price time series will be much larger than the η calculated from a 20 day stock price time series that is a subsequence of the 50 day series. Does this mean that a longer time series is more predictable? Of course not. In fact, there is evidence that longer stock price time series are closer to a random walk than shorter ones [5]. The source of this effect is mainly due to the nonstationarity of stock price time series. The nonstationarity becomes more evident as the sample size increases. The second problem is a derivation of the first one. Since the η increases when the time series is longer, and its value has an upper bound of one, the value of the η-metric will be distributed in a very narrow range, especially for a long-term stock price time series. Hence, the resolution of the η-metric is reduced. This can be clearly seen by examining a long random walk time series, which has an η close to 0.9. By design it should be near zero. Since the random walk time series has very low predictability, the η-metric over all time series will be distributed in the approximate range of [0.9, 1.0].

These problems are resolved as follows. For a long-term time series $\{Y_t\}$, $t = 1, 2, \cdots, N$, the η-metric is calculated on the first Q points, that is, a sample series $\{Y_1, Y_2, \cdots, Y_Q\}$. Then, the sample series is shifted by τ, and the η-metric is calculated again on the new sample $\{Y_{1+\tau}, Y_{2+\tau}, \cdots, Y_{Q+\tau}\}$. Continuing this process, a series of η's that contains the local predictability estimations of subsequences of the whole time series are constructed. Generally, $\eta_{Q,t}$ can be defined as the η-metric over the sample $\{Y_{t+1}, Y_{t+1}, \cdots, Y_{t+Q}\}$. Thus, the η-series is represented by $\{\eta_{Q,0}, \eta_{Q,\tau}, \eta_{Q,2\tau}, \cdots, \eta_{Q,m\tau}, \cdots\}$. Since all the η's are estimated over same sample size Q, they are well comparable, and by selecting appropriate values of Q, they can be made to distributed in a reasonable range. This solves both problems. Additionally, by examining the resulting η-series, the variation of the predictability over time can be observed, and the overall predictability of a specific time series can be estimated by calculating the average of all η's.

EXPERIMENTS AND RESULTS

In order to test the new metric, it is applied to three different kinds of time series: a deterministic time series, a random walk time series, and two stock price time series. The experiments clearly demonstrate that different kinds of times series yield significantly different predictability results. Each SSE in the results is obtained by performing 20 GP runs and averaging the best 10.

Adil Qureshi's GPsys release 2b [8] is used to perform all the GP runs. The configuration used in this study is given in Table 1.

Table 1: GP configuration

Generations	100
Populations	2000
Function set	+, -, /, *, sin, cos, exp, sqrt, ln
Terminal set	$\{x(t-1), x(t-2), \ldots, x(t-10), R\}$
Fitness	Sum of squared error
Max depth of new individual	9
Max depth of new subtrees for mutation	7
Max depth of individuals after crossover	13
Mutation rate	0.01
Generation method	Ramped half-and-half

Deterministic Time Series

The Mackey-Glass equation is used to generate the deterministic time series in this study. The equation for the discretized map is

$$x(t+1) = x(t) + \frac{bx(t-\tau)}{1+x^c(t-\tau)} - ax(t),$$

where a=0.1, b=0.2, c=10, and τ=16. The Mackey-Glass map is seeded with 17 pseudo-random numbers and an 1100 points time series is generated. The first 1000 points are discarded to remove the initial transients. The last 100 points are used as the deterministic time series upon which the predictability metric is tested. The sample size is set to 100 for Kaboudan's method. For the new method, the sample size $Q = 20$ and the shift step $\tau = 5$. Results are shown in Table 2 and Table 3.

Table 2: Predictability of Mackey-Glass series using Kaboudan's -metric

SSE_Y	SSE_S	η
4.014×10^{-3}	4.323	0.999

Table 3: Predictability of Mackey-Glass time series using the new metric

τ	SSE_Y	SSE_S	$\eta_{20,\tau}$
0	1.938×10^{-4}	0.124	0.998
5	1.236×10^{-4}	0.089	0.999
10	6.400×10^{-4}	0.121	0.999
15	1.328×10^{-4}	0.400	1.000
20	6.691×10^{-4}	0.418	0.998
25	1.230×10^{-3}	0.254	0.995
30	6.443×10^{-4}	0.118	0.995
35	5.374×10^{-4}	0.122	0.996
40	1.009×10^{-3}	0.174	0.994
45	3.584×10^{-4}	0.100	0.996
	Average η		0.997

Both Kaboudan's metric and the new metric give an average η very close to 1, indicating that the time series is highly predictable. Note that the difference in SSE_S between Kaboudan's method and the new method presented in this paper is due to the length of the respective time series. Recall for Kaboudan's method the time series is 100 observations and for the new method each subsequence is 20 observations.

Random Walk Time Series

A random walk time series is generated and tested using both the Kaboudan's η-metric and the new metric. The random walk series $\{R_t\}$, $t = 1,2,\cdots,N$, is generated by $R_t = R_{t-1} + a_t$, where a_t is random variable uniformly distributed in [-0.5, 0.5], and the initial value $R_0 = 10$. Again, for Kaboudan's method, the sample size is 100, and for the new method, the sample size $Q = 20$ and the shift step $\tau = 5$. The results are shown in table 4 and 5.

Table 4: Predictability of random walk series using Kaboudan's η-metric

SSE_Y	SSE_S	η
2.303	18.450	0.875

Table 5: Predictability of random walk series using the new metric

τ	SSE_Y	SSE_S	$\eta_{20,\tau}$
0	0.363	0.460	0.211
5	0.728	0.957	0.239
10	1.602	1.618	0.010
15	1.899	1.864	0
20	1.941	1.156	0
25	1.804	1.885	0.043
30	1.345	0.904	0
35	0.415	0.740	0.439
40	0.532	0.985	0.460
45	0.954	0.599	0
	Average η		0.140

Kaboudan's metric gives $\eta = 0.875$ for a random walk series, which is obviously not reasonable. The new metric gives an average $\eta = 0.140$, which more accurately reflects the true predictability of a random walk time series. Following Kaboudan's suggestion, if $\eta < 0$, it is simple set equal to zero, indicating that the time series is not predictable.

Stock Price Series

Next the new metric is applied to calculate the predictability of two stock price time series: Compaq Computer (CPQ) and General Electricity (GE) for the year 1999, with $Q = 20$ and $\tau = 5$. The results are shown in Table 6.

Stock Name	Average η
CPQ	0.818
GE	0.415

Table 6: Predictability estimations of stock price

The new metric gives average $\eta = 0.818$ for CPQ and $\eta = 0.485$ for GE. These η values are different from the ones we obtained from the totally deterministic time series and the random walk time series. This result suggests that the stock price series is more predictable than the random walk series, and the new metric does disclose this difference and quantifies it.

CONCLUSIONS

A new method for measuring time series predictability is proposed in this paper. It is based on the η-metric method introduced by Kaboudan [6], but overcomes the two main disadvantages of the pure η-metric method. It also provides a new feature, which shows how the predictability changes over different subsequences in a time series.

This method has been shown to be able to distinguish stock price time series and random walk time series. Future work will study a wider variety of stocks. Additionally, this method will be studied in its value in making investment decisions.

REFERENCES

[1] Holland, J. 1975. *Adaptation in Natural and Artificial Systems.* University of Michigan Press, Ann Arbor.
[2] Koza, John 1992. *Genetic Programming: On the Programming of Computers by Means of Natural Selection.* Cambridge, MA: The MIT Press.
[3] Fogel, D. and Fogel, L. 1996. Preliminary experiments on discriminating between chaotic signals and noise using evolutionary programming. *Genetic Programming 1996: Proceedings of the First Annual Conference.* Cambridge, MA: The MIT Press, pp. 512-520.
[4] Kaboudan, M. Genetic Programming Prediction of Stock Prices, *Computational Economics,* to appear.
[5] Chen, S-H and Yeh, C-H (1996). Genetic programming and the efficient market hypothesis. In Koza, John, Goldberg, David, Fogel, David, and Riolo, Rick (editors). *Genetic Programming 1996: Proceedings of the First Annual Conference.* Cambridge, MA: The MIT Press, pp. 45-53.
[6] Kaboudan, M. 1998. A GP approach to distinguish chaotic from noisy signals. *Genetic Programming 1998: Proceedings of the Third Annual Conference,* San Francisco. CA: Morgan Kaufmann, pp. 187-192
[7] Efron, B. 1982. *The Jackknife, the Bootstrap, and Other Resampling Plans.* Philadelphia: Society for Industrial and Applied Mathematics.
[8] Adil Qureshi's GPsys release 2b in java http://www.cs.ucl.ac.uk/staff/A.Qureshi/gpsys.html.

PART III: FUZZY SYSTEMS

Adaptive Fuzzy Controllers Based on Variable Universes Theory

Hong-Xing Li

Dept. of Math. & Comp. Sci. & Engr.

Beijing Normal University,

Beijing, 100875, China

C. L. Philip Chen

Dept. of Comp. Sci. & Engr.

Wright State University,

Dayton, OH 45435, USA

ABSTRACT

In this paper, we propose an adaptive fuzzy controller structure based on variable universes theory. The concept and theory are derived from interpolation forms of fuzzy control introduced in [1, 2]. First, we define monotonicity of control rules, and we prove that the monotonicity of interpolation functions of fuzzy control is equivalent to the monotonicity of control rules. This means that there is no any contradiction among the control rules under the condition for the control rules being monotonic. Then the structure of the contraction-expansion factor is discussed and a variable universe adaptive fuzzy control model is given.

1 The Monotonicity of Control Rules and The Monotonicity of Control Functions

Because an adaptive fuzzy system is capable of incorporating linguistic fuzzy information and its ability of adaptaion, many successful adaptive fuzzy control applications have been developed and important theoretic fundamentals been addressed recently [3, 4]. There are two different types of fuzzy controllers: the trial-and-error approach and the theoretical approach. A set of "If-Then" rules are collected from experience-based knowledge for the trial-and-error approach. In theoretical approach, the structure and parameters of fuzzy controller are designed such that a certain performance criteria are guaranteed. In adaptive fuzzy control, the structure or parameters of the controller change during the operation. In this paper, we propose a novel adaptive fuzzy control approach based on "variable universes" theory, where the universes vary through contracting and expanding processes. We then propose adaptive fuzzy controllers structure based on variable universes theory and discuss a adaptive universe adaptive fuzzy control model.

Without lose of generality, we consider fuzzy controllers with two inputs and one output. Let X and Y be the universes of input variables and Z the universe of output variable. The families of all unimodal and normal fuzzy sets on these universes are, respectively, denoted by $\mathcal{F}_0(X)$, $\mathcal{F}_0(Y)$ and $\mathcal{F}_0(Z)$. If a certain order relation $\leq$ is defined in $\mathcal{F}_0(X)$, $\mathcal{F}_0(Y)$ and $\mathcal{F}_0(Z)$, the ordered sets $(\mathcal{F}_0(X), \leq)$, $(\mathcal{F}_0(Y), \leq)$ and $(\mathcal{F}_0(Z), \leq)$ are formed. Taking three ordered subsets $A \subset \mathcal{F}_0(X)$,

1. The Monotonicity of Control Rules and The Monotonicity of Control Functions

$\mathcal{B} \subset \mathcal{F}_0(Y)$ and $\mathcal{C} \subset \mathcal{F}_0(Z)$, we know that the control rules of fuzzy control can be described by a mapping R:

$$R: \mathcal{A} \times \mathcal{B} \longrightarrow \mathcal{C}, \quad (A, B) \longmapsto R(A, B) \triangleq C \tag{1}$$

where R is called the rules or rule base of a controller.

Definition 1 The rule $R(A, B)$ is said to be monotonic increasing (decreasing) with respect to A, if $R(A, B)$ is isotonic (anti-isotonic), i.e., $\forall A', A'' \in \mathcal{A}$,

$$A' \leq A'' \Longrightarrow R(A', B) \leq R(A'', B) \quad (R(A', B) \geq R(A'', B))$$

Similarly, the monotonicity of $R(A, B)$ with respect to B can be defined. When $R(A, B)$ is monotonic increasing (decreasing) not only with respect to A but also with respect to B, $R(A, B)$ is called completely monotonic increasing (decreasing). And when $R(A, B)$ is monotonic increasing (decreasing) with respect to A but monotonic decreasing(increasing) with respect to B, $R(A, B)$ is said to be mixedly monotonic.

Here we assume that universes are all real number intervals. Let $U \in \{X, Y, Z\}$. For any $A \in \mathcal{F}_0(U)$, if the peak points of A is not unique, then a representative point among them should be taken by means of a certain way. This representative point (including the unique peak point) is called the normal peak point of A. Now we define a commonly used order relation "$\leq$":

$$(\forall A_1, A_2 \in \mathcal{F}_0(U))(A_1 \leq A_2 \iff x_1 \leq x_2) \tag{2}$$

where x_1 and x_2 is, respectively, the normal peak point of A_1 and A_2.

So the linguistic values on a symmetric interval $[-a, a]$, NB, NM, NS, ZO, PS, PM, PB (negative big, negative medium, etc.), can be sequenced as the following:

$$NB \leq NM \leq NS \leq ZO \leq PS \leq PM \leq PB$$

The control rule base, R, of a fuzzy controller is usually written as follows:

$$\text{if } x \text{ is } A_i \text{ and } y \text{ is } B_j \text{ then } z \text{ is } C_{ij} \tag{3}$$

where $i = 1, 2, \cdots, p$, $j = 1, 2, \cdots, q$. From [1,2], we know that a fuzzy controller is approximately a binary piecewise interpolation function:

$$F(x, y) = \sum_{i=1}^{p} \sum_{j=1}^{q} \mu_{A_i}(x) \mu_{B_j}(y) z_{ij} \tag{4}$$

If the control function of the control system is written as $f: X \times Y \longrightarrow Z$, $(x, y) \longmapsto z = f(x, y)$, then $F(x, y)$ approximates to $f(x, y)$, i.e.,

$$(\forall \varepsilon > 0)(\exists N)(n \geq N \Longrightarrow \sup_{(x,y) \in X \times Y} |F(x, y) - f(x, y)| \leq \varepsilon) \tag{5}$$

2. The Contraction-expansion Factors of Variable Universes

Equation (5) indicates that we can regard $F(x,y)$ and $f(x,y)$ as the same. Clearly, $f(x,y)$ is usually a nonlinear function, which means that a fuzzy controller is a nonlinear approximator. In order to discuss the monotonicity of control action, we can define the monotonicity of the control function $f(x,y)$ with respect to (x,y), similarly to Definition 1.

Theorem 1 For a given fuzzy control system where its rule base is shown as expression (3) and A and B are two-phased groups of base elements, then $R(A,B)$ is monotonic increasing (decreasing) with respect to A or B, or completely monotonic increasing (decreasing), or mixedly monotonic, if and only if $F(A,B)$ has corresponding monotonicity.

Proof See reference [4] for detail.

This theorem shows that there exists an important relation between rule bases and control functions.

2 The Contraction-expansion Factors of Variable Universes

2.1 The Contraction-expansion Factors of Adaptive Fuzzy Controllers with One Input and One Output

Given a fuzzy controller, the universe of input variables and the universe of output variables are, respectively, $X = [-E, E]$ and $Y = [-U, U]$, where E and U are real numbers. X and Y can be called initial universes being relative to variable universes.

Definition 2 A function $\alpha : X \longrightarrow [0,1]$, $x \longmapsto \alpha(x)$, is called a contraction-expansion factor on universe X, if it satisfies the following conditions: (1) Evenness: $(\forall x \in X)(\alpha(x) = \alpha(-x))$; (2) zero-preserving: $\alpha(0) = 0$; (3) monotonicity: $\alpha(x)$ is strictly monotone increasing on $[0, E]$; (4) compatibility: $(\forall x \in X)(|x| \le \alpha(x)E)$.

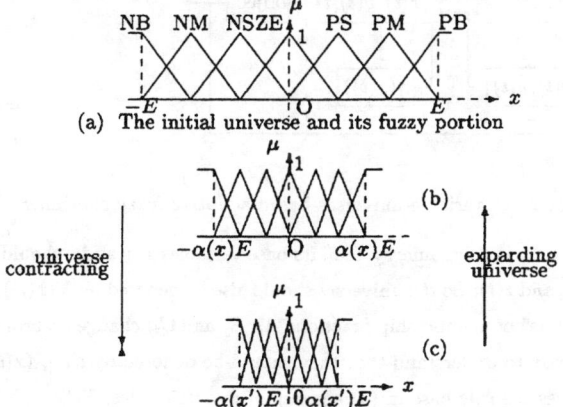

Figure 1 Contracting/expanding universe

For any $x \in X$, a variable universe on $X(x)$ is defined below:

$$X(x) \stackrel{\triangle}{=} \alpha(x)X \stackrel{\triangle}{=} [-\alpha(x)E, \alpha(x)E] \stackrel{\triangle}{=} \{\alpha(x)x' | x' \in X\}.$$

Figure 1 illustrates the idea of variable universes. Moreover, from the compatibility of Definition 2, it is easy to know that contraction-expansion factors satisfy the following condition:

(5) Normality: $\alpha(\pm E) = 1$, $\beta(\pm U) = 1$.

2.2 The Contraction-expansion Factors of Adaptive Fuzzy Controllers with Two Inputs and One Output

Let $X = [-E, E]$ and $Y = [-D, D]$ be the universes of input variables and $Z = [-U, U]$ be the universe of output variable. When Y is relatively independent from X, we can obtain the contraction-expansion factors $\alpha(x)$ of X, and $\beta(y)$ of Y and $\gamma(z)$ of Z. In some cases, Y may not be independent from X. Then β should be defined on $X \times Y$, i.e., $\beta = \beta(x, y)$. For example, denoting $D = EC$, and $Y = [-EC, EC]$, we can use one of the following two expressions:

$$\beta(x,y) = \left(\frac{|x|}{E}\right)^{\tau_1} \left(\frac{|y|}{EC}\right)^{\tau_2} \tag{7}$$

and

$$\beta(x,y) = \frac{1}{2}\left[\left(\frac{|x|}{E}\right)^{\tau_1} + \left(\frac{|y|}{EC}\right)^{\tau_2}\right], \tag{8}$$

where $0 < \tau_1, \tau_2 < 1$.

3 The Structure of Adaptive Fuzzy Controllers Based on Variable Universes

To consider the structure of variable-universes-based adaptive fuzzy controller, we use a fuzzy controller with two inputs and one output shown in Figure 2 as an example.

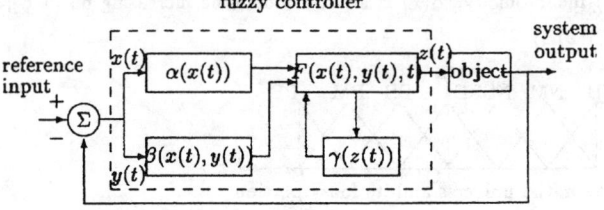

Figure 2 A variable-universes-based adaptive fuzzy controller

As a fuzzy control system is a dynamic system, its base variables x, y and z should depend on time t, denoted by $x(t)$, $y(t)$, and $z(t)$. So the universes should also be denoted by $X(x(t)), Y(y(t)), Z(z(t))$. Then the "shapes or forms" of membership functions A_i, B_j and C_{ij} change according to the change of the universes. It is easy to understand that they should be denoted by $\mu_{A_i(t)}(x(t))$, $\mu_{B_j(t)}(y(t))$, $\mu_{C_{ij}(t)}(z(t))$. This makes the rule base in (3) a group of dynamic rules, $R(t)$:

$$\text{if } x(t) \text{ is } A_i(t) \text{ and } y(t) \text{ is } B_j(t) \text{ then } z(t) \text{ is } C_{ij}(t) \tag{9}$$

Because expression (9) equals to expression (3) when $t = 0$, expression (3) is called initial rules. Also the control function becomes dynamic, denote it by $F(x(t), y(t), t)$, i.e.,

$$F(x(t), y(t), t) = \sum_{i=1}^{p} \sum_{j=1}^{q} \mu_{A_i(t)}(x(t)) \mu_{B_j(t)}(y(t)) z_{ij}(t) \quad (10)$$

From the definition of variable universes, we know that the monotonicity of initial rule base, $R = R(0)$, ensures the monotonicity of $R(t)$ ($t > 0$). It means that there exists no contradiction among the rules when we process the contracting/expanding of the universes. So it ensures that control function, $F(x(t), y(t), t)$ is significant.

Figure 3 illustrates the change of control function with one input and one output $F(x(t), t)$.

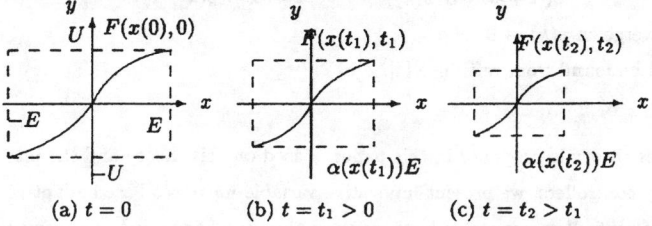

(a) $t = 0$ (b) $t = t_1 > 0$ (c) $t = t_2 > t_1$

Figure 3 The change of control function as time goes on

Figure 3 also indicates that the initial control function at t_{k+1} inherits the initial control function at t_k.

4 Adaptive Fuzzy Controllers with One Input and One Output

Let the initial control rule base, $R(0) = R$, be "if x is A_i then y is B_i," $i = 1, 2, \cdots, n$, where $\{A_i\}_{(1 \leq i \leq n)}$ and $\{B_i\}_{(1 \leq i \leq n)}$ is, respectively, a group of linear base elements on initial universes $X = [-E, E]$ and $Y = [-U, U]$, where their peak point sets $\{x_i\}_{(1 \leq i \leq n)}$ and $\{y_i\}_{(1 \leq i \leq n)}$ satisfy $-E = x_1 < x_2 < \cdots < x_n = E$; and $-U = y_1$ and $y_n = U$. For all $k = 0, 1, 2, \cdots$, take $x_i(0) = x_i$ and $y_i(0) = y_i$; and define a group of linear base elements $\mu_{A_i(k)}(x(k))$ as follows:

$$\mu_{A_1(k)}(x(k)) = \begin{cases} \frac{[x(k) - x_2(k)]}{[x_1(k) - x_2(k)]}, & x_1(k) \leq x(k) \leq x_2(k); \\ 0, & x_2(k) \leq x(k) \leq x_n(k). \end{cases} \quad (11)$$

$$\mu_{A_i(k)}(x(k)) = \begin{cases} \frac{[x(k) - x_{i-1}(k)]}{[x_i(k) - x_{i-1}(k)]}, & x_{i-1}(k) \leq x(k) \leq x_i(k); \\ \frac{[x(k) - x_{i+1}(k)]}{[x_i(k) - x_{i+1}(k)]}, & x_i(k) < x(k) \leq x_{i+1}(k); \\ 0, & \text{otherwise} \end{cases} \quad (12)$$

$i = 2, 3, \cdots, n-1,$

$$\mu_{A_n(k)}(x(k)) = \begin{cases} 0, & x_1(k) \leq x(k) < x_{n-1}(k); \\ \frac{[x(k) - x_{n-1}(k)]}{[x_n(k) - x_{n-1}(k)]}, & x_{n-1}(k) \leq x(k) \leq x_n(k). \end{cases} \quad (13)$$

6. Conclusions

Clearly, the group of base elements is a group of two-phased base elements.

Theorem 2. The control algorithm mentioned above, $(\forall k)(x(k) \neq 0)$, has the following form:

$$y(k+1) = \beta(y(k))F\left(x(k)\Big/\alpha(x(k))\right)$$
$$= \beta(y(k))\sum_{i=1}^{n}\mu_{A_i}\left(x(k)\Big/\alpha(x(k))\right)y_i(0), \qquad (14)$$

where $F(x,0) \stackrel{\triangle}{=} F(x)$. And if $\alpha(x(k))$ and $\beta(y(k))$ satisfy the stationary property:

$$\lim_{k\to+\infty} x(k) = 0 \Longrightarrow \left(\lim_{k\to+\infty}\frac{\alpha(x(k-1))}{\alpha(x(k))} = 1,\ \lim_{k\to+\infty}\frac{\beta(y(k))}{\beta(y(k-1))} = 1\right)$$

then $y(k+1)$ must converge as $x(k) \to 0$.

Proof The proof can be found from reference [4].

5 Conclusions

The variable universes theory is proposed in this paper. Based on this theory and the interpolation mechanism of fuzzy controllers, we present innovative variable-universes-based adaptive fuzzy controllers. This type of controllers are stable in theory and they can be applied to a fuzzy control system that needs a higher precision that will be discussed in the future.

References

1. H. Li and C. L. Philip Chen, The Equivalence Between Fuzzy Logic Systems and Feedforward Neural Networks, *IEEE Transactions on Neural Networks*, Vol. 11, No. 2, March, 2000, pp. 356-365.

2. H. Li and C. L. Philip Chen, The Interpolation Mechanism of Fuzzy Control and its Relationship to PID Control, *International Journal of Fuzzy Systems*, Vol. 2, No. 1, pp. 23-30, March 2000.

3. L. X. Wang, *A Course in Fuzzy Systems and Control*, Prentice-Hall, Englewood Cliffs, NJ, 1997.

4. H. X. Li, C. L. P. Chen, and H. P. Huang, *Fuzzy Neural Intelligent Systems*, CRC Press, Boca Raton, FL, 2001.

A NEW TYPE OF CMAC LEARNING CONTROLLER WITH FUZZY LEARNING GAIN

YUNCAN XUE
Department of Control Science and
Engineering,Zhejiang University,
Hangzhou,China,310027

JIXIN QIAN
Department of Control Science and
Engineering,Zhejiang University,
Hangzhou,China,310027

ABSTRACT
In this paper,a new type of CMAC learning controller with fuzzy learning gain (FCMAC) has been presented to meet the requirement of speediness and robustness simultaneously. The configuration and algorithm of FCMAC have been presented. Simulation studies have been carried out based on the CSTR model. Its results show that this controller is superior to the general CMAC self-learning controller.

INTRODUCTION

Cerebellar Model Articulation Controller (CMAC) was a kind of locally approached neural network presented by J. S. Albus(Albus,1975) in 1970s.It learns very fast because it is a kind of neural network based on local learning. This makes it fit to real time control. A lot of applications have been reported since 1980s in the control fields. Such as: Miller(Miller,1987,1988)et al applied it to the sensor control of robots with vision; Cetinkunt(Cetinkunt,1993). applied it for the servo-control of high precision machine tools; Hitoshi (Hitoshi ,1993) et al. applied it for the fuel-injection control; Kwok(Kwok,1994) et al. applied it to fault diagnosis. In this paper, a new type of CMAC learning controller with fuzzy learning gain(FCMAC) has been presented by analyzing the learning principle of CMAC network. This method greatly improves the performance of CMAC control.

THE BASIC STRUCTURE OF CMAC

CMAC is basically a "table lookup" technique for representing complex nonlinear functions. Figure1 is a block diagram of CMAC network. It consists of 4 basic parts: network input, conceptual

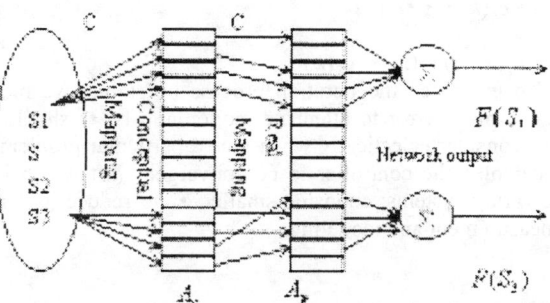

Fig 1 The block diagram of CMAC network

mapping, practical mapping and network output. In Fig.1, A_c represents conceptual storage cells; A_p represents real storage cells; $F(S_i)$ represents CMAC network output. Every point in the input state space S corresponds to C points in A_c. The output $F(S_i)$ of neural network is the sum of storage values in the C cells (weighty of network). Because every input sample corresponds to C cells of A_c & A_p, some samples near in S will overlap in A_c and A_p when various samples store dispersedly in A_c & A_p This makes the outputs of CMAC network become near and is called the generalization of CMAC network. It will not work to the vectors far in space S. Every output of CMAC is the linear sum of weights in the C cells of space A_p. The whole neural network is not full connected. There are C connections from S to space A_c, C connections from input space A_p to $F(S_i)$. CMAC network is a single layer forward network. From S to A_c, it is a linear transform; from A_c to A_p, it is a random transform; from A_p to $F(S_i)$, it is a linear transform. So the whole transform, from S to $F(S_i)$, is a nonlinear transform.

The learning algorithm of connection weighty of CMAC network is:
$$w_{ij}(K \bullet 1) \bullet w_{ij}(K) \bullet \ \bullet (y_i \bullet y_i(K)) \bullet \ _j / \bullet \ \bullet \tag{1}$$
where • is the vector in the space A_c. The elements of • can onladopt two values:0 and 1. For some specified s, only few elements take 1,most elements take 0.It is evident that • • $P(s)$ implements a specified nonlinear mapping. This nonlinear mapping has been defined when the network to be designed. One point in input space corresponds to some elements to be 1 in • (In this paper, it is C), i.e. it correspond to a local fields in correlative space A_c. y_i and $y_i(K)$ represent expected value and practical value of the ith output respectively. As in the correlative vector • , there are only few elements take 1 and the rest elements all take 0,there are only few connection weights need to be adjusted. This characteristic makes CMAC learn fast. • is the learning gain(training factor), greatly influenced the quality and performances of the control system. In order to guarantee the astringency of learning algorithm, • must take 0-2. Following are some explanations:

Let $e_i(k) \bullet y_i^* \bullet y_i(k) \bullet y_i^* \bullet w_i(k) \bullet$,then
$$\bullet e_i(k) \bullet e_i(k \bullet 1) \bullet e_i(k) \bullet \bullet (y_i(k \bullet 1) \bullet y_i(k)) \tag{2}$$
i.e. $\bullet e_i(k) \bullet \bullet (w_i(k \bullet 1) \bullet w_i(k)) \bullet \ \bullet \bullet \bullet w_i(k) \bullet \tag{3}$
and Equ.(2) can be rewritten to:
$$\bullet w_i(k) \bullet \ \bullet e_i(k) \bullet \ ^\iota / \bullet \ ^\iota \bullet \tag{4}$$
then,we have
$$\bullet e_i(k) \bullet \ \bullet \bullet e_i(k) \tag{5}$$
and
$$e_i(k \bullet 1) \bullet (1 \bullet \ \bullet) e_i(k) \tag{6}$$

To guarantee the iteration process be stable, we must have $|1 \bullet \ \bullet | \bullet \ 1$. In order to make error to attenuate monotonously, • should take value 0-1. We may consider eclectically when face a particular problem: choosing bigger • to guarantee the controller to ne convergent fast which is fit for modeling of time-varing systems; choosing smaller • to reduce its sensitivity to the noises of measurement and modeling.

THE BASIC STRUCTURE OF FCMAC

It is evident that the stability of control system is related to the astringency of learning algorithm and that the stringency of online learning algorithm is related to the value of learning gain. To guarantee • be within the prescriptive range and to expedite learning at the same time, an adaptive method based on the fuzzy logic reasoning is presented here, which obtains fuzzy reasoning rulers to adjust learning gain • onlinely according to the relationships between • and the learning error function E and the weight change • w.

Generally, big learning gain makes learning fast and error decrease, but over big change of error may lead to the learning algorithm to be unstable; and small learning gain makes learning slow but algorithm is stable. So we can obtain the correction table of the reasoning rulers(listed in table 1).Such as:
IF $E(k)$ • B AND • $w(k)$ • S THEN • (k) • B
IF $E(k)$ • S AND • $w(k)$ • B THEN • (k) • S
... ...

Table 1 Correction table of fuzzy learning gain

$E(k)$	• $w(k)$				
	S	MS	M	MB	B
S	0.05	0.05	0.05	0.08	0.08
MS	0.1	0.1	0.1	0.2	0.3
M	0.4	0.4	0.3	0.3	0.3
MB	0.5	0.5	0.5	0.4	0.4
B	0.8	0.7	0.6	0.5	0.4

In table 1, $E(k)$ represents train weight error and • $w(k)$ represents weight change, fuzzy subsets {S,MS,M,MB,B}={"Small", "Middle Small", "Middle", "Middle Big", "Big"}. The subjection functions of E & • w are listed in Fig.2.

The directed inverse model learning control method is adopted by the FCMAC network control. In each control cycle, a control process and a learning

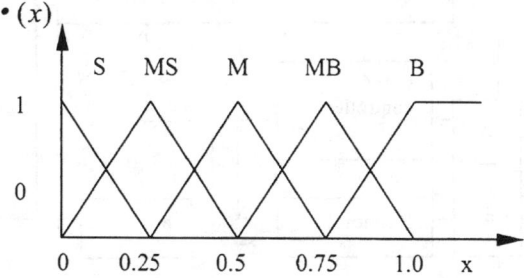

Fig.2 Subjection functions of E & • w

process occur alternately. In the control process, the CMAC network is used to predict the control value which is put on to the controlled plant so as to meet the ideal objective. The sum of the predict value and the output of the feedback P controller is the real value put on to the controlled plant. After the control process, the next thing to do is to choose the learning gain automatically and to learn. The learning process is used to adjust the network weights so as to form the dynamic model of the controlled plant in the relevant fields of the control space. The block diagram of the control system is shown in Fig.3.

SIMULATION

The simulation plant is the widely used continuous stirred tank reactor(CSTR). There has been considerable interest in its state estimation and real time control based on mathematical modeling. However, the lack of understanding of the dynamics of the process, the highly sensitive and nonlinear behaviors of the reactor, has made it difficult to build the precise mathematical model of the system, and the lack of well developed nonlinear control techniques has resulted in difficulties in achieving good control performance of polymerization reactors. The CMAC learning control scheme, which in essential is an adaptive system by which complex nonlinear functions can be represented by referring to a lookup table, is very fit for nonlinear control. The CSTR model we use to simulation is:

$$\frac{dx_1}{dt} \bullet \bullet x_1 u_f \bullet D_a (1 \bullet x_1) \exp(\frac{x_2}{1 \bullet x_2 / \bullet})$$

$$\frac{dx_2}{dt} \bullet \bullet x_2 (u_f \bullet \bullet) \bullet HD_a (1 \bullet x_1) \exp(\frac{x_2}{1 \bullet x_2 / \bullet}) \qquad (7)$$

In the above nonlinear equations, the state variable x_1 is the conversion rate related to the reaction concentration. It is an online unmeasurable quantity, and can only be obtained from classic manual analysis of infrequent laboratory samples. State variable x_2 is the reaction temperature in its dimensionless form and is online measurable. u_f and u_c are the control variables corresponding to the input flow rate of the reactants and the coolant temperature respectively. Usually the input flow rate is fixed in the practical

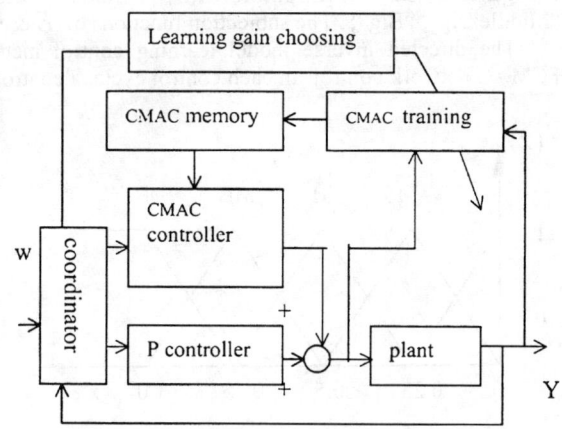

Fig.3 The block diagram of FCMAC learning controller

operation, this leads this system to be a temperature control system. The unknown model parameters can be obtained by system identification of both online and offline infrequent measurements of the CSTR system.

The objective of CMAC control is to force the state variable x_1 of CSTR to follow the change of set points with as small error as possible.

CMAC network training is mainly influenced by three parameter(1)the size of the CMAC memory (A_p); (2)the number of memory locations addressed by each input state (C);and(3) the learning gain(•).

The bigger the A_p we choose, the fewer the mapping collision and the faster the learning velocity. But the implementation cost will be high. Different results can be obtained by adjusting A_p. In order to avoid collision, we choose $C = A_p/100$.

Figure 4 is the figure of the system outputs of CMAC and FCMAC without system parameter changes while Fig.5 is the figures with system parameter changes. It can be seen from the Fig.4 that the response of FCMAC is much faster than that of the CMAC with small learning gain and is almost as fast as that of the CMAC with big learning gain.

Assume that the gain of nonlinear part become to be 1.3 times of the original gain when $t • 1200T$ (T is the sampling cycle). By comparing Fig.5(a) to Fig.5(b), it can be seen that although there is a oscillation process, the FCMAC learning control can obtain good performances with no static error and small overshoot while the performances of the CMAC learning control with big learning gain are not as good as FCMAC because of its acute oscillation when nonlinear plant parameter changes. Although to reduce the learning gain can enhance the system robustness, it would minish the response speediness. However FCMAC learning controller has successfully resolved the problem that the system robustness and response speediness cannot be harmonious. In addition, we have also carried the PID control research of CSTR. Simulation results show that the performances of both

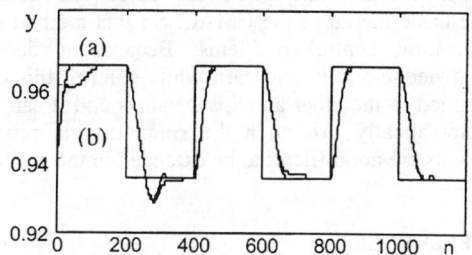

Fig.4 System outputs of learning control (a) FCMAC (b) CMAC

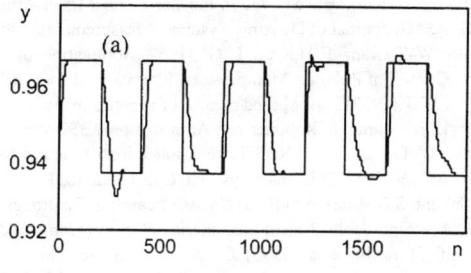

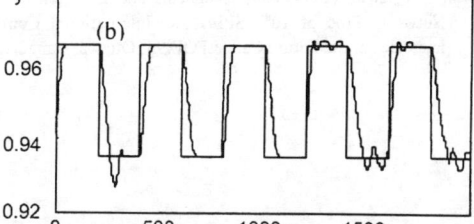

Fig.5 Outputs of FCMAC and CMAC when parameters change (a) FCMAC (b) CMAC

FCMAC and CMAC are superior to the pure PID control.

CONCLUSIONS

From the simulation it can be seen that the FCMAC control strategy has outstandingly improved the performances of the CMAC learning control. Although pure P control can improve the performances by adjusting the system parameters, it is sensitive to system gain mismatch. The learning control is not sensitive to the system parameter mismatch. However, the CMAC adopted learning control strategy based on fuzzy logic rulers has preferably compromised the response speediness and the system robustness. Simulation results show that the error of the learning control including the CMAC and FCMAC is much less than that of the P control.

What to learn of the FCMAC learning is the characteristics of the controlled plant other than the specifically objectives because the FCMAC has preserved the basic merits of the CMAC. So this method is not limited to the repeatable operation control problems. Because it doesn't need the transcendental knowledge and it can learn while control, this control technique is easy to be applied to the other anticipant values and it can adapted the parameter changes automatically. Although the simulation is carried on the CSTR system, its results are not difficult to be extended to the multivariable systems.

REFERENCES

Albus J S. (1975),A New Approach to Manipulator Control: The Cerebellar Model Articulation Controller (CMAC). Trans of ASME Journal of Dynamic Systems, Measurements, and Control, 9: 220-227.

Albus J S. (1975),Data Storage in the Cerebellar Model Articulation Controller (CMAC). Trans of ASME Journal of Dynamic Systems, Measurements, and Control, 9:228-233.

Miller W T,Granz F H,Kraft L G. (1987),Application of a General Learning Algorithm to the Control of Robotics Manipulators. International Jour- nal of Robotics Research,,6(2):84-98.

Miller W T,(1987)Sensor based control of robotics manipulators using a general learning algo-rithm. IEEE Journal of Robotics and Auto-mation,3:357-405.

Miller W T,et al. (1988),Real Time Application of Neural Networks for Sensor Control of Robots with Vision. IEEE Trans Sys Man Cyb,19:825-831.

Cetinkunt S,Donmez A. (1993),CMAC Learning Controller for Servo Control of High Precision Machine Tools: Proc of the American Control Conference, San Francisco:1976-1980.

Hitoshi Shiraishi, et al.(1993) CMAC Neural Network for Fuel-Injection Control. Proc of the American Control Conference, San Francisco,: 1773-1777.

Kwok D P, et al. (1994),Fault Diagnosis and Location of Analog Circuits Using CMAC Neural Network. Proc of 10th ISPE/IFAC International Conference on CAD/CAM, Robotics and Factories of the Future Cars &FOF'94, Ottawa, Canada,: 614-619.

MODEL REFERENCE ADAPTIVE FUZZY VOLTAGE CONTROL IN GTAW

P. SMITHMAITRIE, P. KOSEEYAPORN, G. E. COOK, AND A. M. STRAUSS
Welding Automation Laboratory, Vanderbilt University,
Box 1824, Station B, Nashville, TN 37235

ABSTRACT

Problems are frequently experienced with traditional automatic voltage control (AVC) systems in gas tungsten arc welding (GTAW) because of the nonlinear relationship between the arc voltage, current, and arc length. This arc relationship changes with different welding conditions and with the nominal values of current and voltage. Because of this a traditional or fixed controller will not perform adequately for all operating conditions. Additionally, an AVC that has a fixed voltage set point will not maintain a fixed arc length if the current changes. In this paper, a model reference adaptive fuzzy controller is implemented to cope with the nonlinear arc characteristics of the GTAW process and to perform without the necessity of prior knowledge of the nonlinear arc relationship. The adaptive system is shown to be capable of adapting quickly enough to avoid oscillations that frequently occur with a non-adaptive controller during the important ramp down portion of the weld cycle. The adaptive system is also shown to have a better response to step changes in the electrode-to-work spacing than non-adaptive controllers.

INTRODUCTION

Automatic voltage control (AVC) systems are used in the gas tungsten arc welding (GTAW) process to maintain constant arc length between the nonconsumable electrode and the workpiece. For a given shielding gas, arc length is a function of arc voltage; thus, the arc length can be controlled automatically by maintaining the corresponding arc voltage. The desired arc voltage is balanced against the voltage measured across the arc. The difference between these two voltages is processed by the control unit to produce an output signal to drive a servomotor causing it to move in a direction to bring the two voltages into balance.

Although the traditional AVCs usually perform adequately in most operating circumstances, there are conditions during which the traditional system causes problems because the arc voltage-to-arc length gain, K_{arc}, a feed forward gain, is not constant but varies with the nominal values of the arc current and voltage (Bjorgvinsson et al., 1993).

There are some typical problems that arise in critical applications. These are 1) During the critical current upsloping and downsloping periods at beginning and end of the weld, the AVC is turned off to avoid instability problems. 2) It is known that control of the cooling rate of the weld pool when the welding pass is terminated is important if cracks in the weld seam are to be avoided. A scheme of reducing the arc heat in an appropriate manner is needed to avoid such tail-out cracking. It is usually impossible using traditional AVC systems to reduce the probabilities of cracking during tail-out termination by simultaneously decreasing the arc current, arc voltage and arc length.

TRADITIONAL AVC

Normally, the AVC system employs a typical dc servomotor to move the welding torch for a desired arc length. The characteristic of this servomotor can be expressed by the transfer function

$$G_s = \frac{L_{arc}}{V_{ref}} = \frac{K_s}{s(s+a_s)} \qquad (1)$$

where K_s and a_s are constants characterizing the physical properties of the servomotor, V_{ref} is the voltage signal to the servomotor, and L_{arc} is the electrode-to-work distance.

In an AVC system the arc length, L_{arc}, is indirectly sensed through the arc voltage, V_{arc} while the arc current is constant. This arc voltage and arc length relationship can be represented by the arc voltage-to-length sensitivity, K_{arc}, denoting a change in arc voltage for a change in arc length, $K_{arc} = dV_{arc}/dL_{arc}$. The closed-loop transfer function of the AVC is given as:

$$G_{avc} = \frac{V_{arc}}{V_{ref}} = \frac{K_s K_{arc}}{s^2 + a_s s + K_s K_{arc}} = \frac{\omega_n^2}{s^2 + 2\zeta\omega_n s + \omega_n^2} \qquad (2)$$

Equation (2) clearly shows that the system is second-order with a damping ratio $\zeta = \frac{a_s}{2\omega_n} = \frac{a_s}{2\sqrt{K_s K_{arc}}}$. Therefore increasing K_{arc} will reduce the damping ratio resulting in a higher overshoot in the step response.

In general, the AVC is designed for an optimal response with a constant K_{arc}. In practice, however, K_{arc} is not constant but varied as a function of the arc current and arc voltage (Bjorgvinsson et al., 1993). Therefore, the traditional AVC cannot provide adequate control when the arc current is varied over a wide range. In the worst case, the AVC may be unstable driving the torch into the weld pool (Bjorgvinsson, 1992; Koseeyaporn, 1999). Hence, this work introduces a model reference adaptive fuzzy AVC to cope with the nonlinear arc characteristics of the GTAW process when the arc current is varied over a wide range and to perform without the necessity of prior knowledge of the nonlinear arc relationship.

MODEL REFERENCE ADAPTIVE FUZZY (MRAF) CONTROLLER DESIGN

Fuzzy Controller

For the arc voltage controller, the inputs are the voltage error and its derivative. The output is the servomotor voltage signal. A triangular membership function will be used, as shown in Fig 1. The input and output universe of discourse of the fuzzy controller are normalized in the range [-1 1]. Notice that the membership functions for the input fuzzy set are not uniform, hence, suitable normalizing gains *G0* and *G1*, used to map the actual inputs of the fuzzy system to the normalized universe of discourse, were chosen from experiments. The min-max inference and center of gravity (COG) defuzzification method (Passino and Yurkovich, 1998) are used in this work.

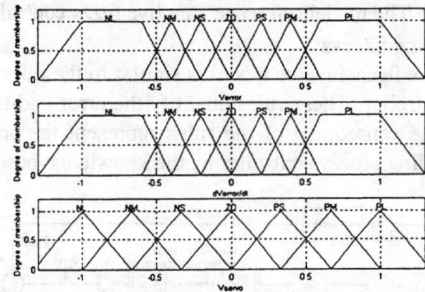

Fig. 1. Membership functions for the arc voltage controller

The rule-base array for the fuzzy controller is shown in Table 1. This rule-base is a 7x7 array, since there are 7 sets on the input universe of discourse. The topmost row shows the indices for the seven fuzzy sets for the voltage error input, *Verror*, and the column at the extreme left shows the indices for the seven fuzzy sets for the derivative voltage error input, *d(Verror)/dt*.

Table 1 Rule-Base for the arc voltage controller

Output		Voltage reference error, *Verror*					
(Vservo)		NL	NM	NS	ZO	PS	PM	PL
	NL	NL	NL	NM	NM	PS	PM	PM
Derivative	NM	NL	NL	NM	NS	PS	PM	PM
Of voltage	NS	NL	NM	NS	NS	PS	PM	PL
reference error	ZO	NL	NM	NS	ZO	PS	PM	PL
d(Verror)/dt	PS	NL	NM	NS	PS	PS	PM	PL
	PM	NM	NM	NS	PS	PM	PL	PL
	PL	NM	NM	NS	PM	PM	PL	PL

Table 1 shows the indices for the servomotor output, Vservo, in fuzzy implications of the form, If the **voltage error** is **negative large** and the **derivative of voltage error** is **negative large**, then the **output** is **negative large**,
where the indices are denoted below:
NL: Negative Large;   NM: Negative Medium;   NS: Negative Small;
PL: Positive Large;   PM: Positive Medium;   PS: Positive Small;   ZO: Zero

Reference Model

Through experimentation by Bjorgvinsson (1992), the closed-loop characteristic transfer function for the automatic voltage control system with Argon shielding gas was found to be:

$$G_i(s) = \frac{2116.8}{(s^2 + 84384s + 2116.8)} \quad (3)$$

This second-order linear function is chosen to be the reference model. The adaptive controller will attempt to control the actual system in the same manner as the reference model.

Fuzzy Inverse Model

For the fuzzy inverse model, there are two inputs (Fig. 2). The difference between the arc voltage and the model arc voltage, $y_e(t)$, and its derivative, $\dot{y}_e(t)$ are the inputs to the fuzzy inverse model. The model reference adaptive fuzzy control is implemented with a 25-millisecond sampling interval.

The rule-base of the fuzzy inverse model is similar to the rules described in Table 1 for the direct fuzzy controller. These rules quantify the error and the derivative error in terms of their size. The consequent of the rules represent the amount of change that should be made to the direct fuzzy controller by the knowledge-base modifier.

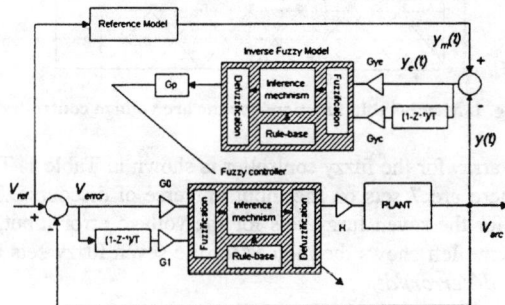

Fig. 2. Model reference adaptive fuzzy (MRAF) control
for Arc Voltage Control in GTAW

The membership functions for the fuzzy inverse model are similar to those used for the direct fuzzy controller shown in Fig. 1. The gain G_{ye}, G_{yc} and G_p are used to map the actual inputs and an output of the fuzzy inverse model to the normalized universe of discourse [-1 1]. The output universe of discourse causes the adaptive mechanism to continually make changes in the rule-base of the fuzzy controller such that the actual output is exactly equal to the reference model output, making the actual plant follow the reference model.

The Knowledge-Base Modifier

For given data (from the inverse models) about the necessary change in the input needed to make $y_e(t)$ approximately zero, the knowledge-base modifier changes the knowledge-base of the fuzzy controller so that the previously applied control action will be modified by the amount specified by the inverse model output, p. The knowledge-base modifier shifts the centers of the output membership functions, c_m, for the rules that were active during the previous control action by the amount $p(t)$ to modify the knowledge-base.

$$c_m(kT) = c_m(kT-T) + p(kT) \qquad (4)$$

Hence, the fuzzy controller acts to produce a desired output by processing information at time kT-T to make $y_e(kT)$ smaller.

EXPERIMENTS AND RESULTS

The objective of the welding experiments was to compare the performance of a closed-loop AVC system with the model reference adaptive fuzzy AVC system. A fuzzy logic AVC was used to represent the performance of a "typical" closed-loop AVC system. By doing this, the results can be compared when the adaptive control loop is added into the system under the same conditions. These experiments include the response on a stepped plate test and ramping-down current response tests (Smithmaitrie, 2000). The arc voltage, arc current and arc length were the recorded parameters. All experiments were conducted with argon shielding gas with 25-CFH flow rate. A 0.065 inch-diameter tungsten electrode was used. Direct-current electrode-negative (DCEN) welding, commonly used in industrial applications (Cary, 1979), was applied for all of the experiments. The workpiece was a copper plate allowing multiple tests without melting the plate.

Step Plate Test

This experiment was designed to observe the dynamic response of the fuzzy and model reference adaptive fuzzy controller when the welding torch passes across a step-change in the test plate. To produce the step-change, the plate was ground to a 1-mm. depth at the point of transition. The travel speed and the reference voltage were set 5.9 inch/min and 13 V, respectively. Tests were conducted at 100A and 30A arc current representing a high and a low welding current, respectively.

For the high current there was no significant difference between the output of the non-adaptive fuzzy and model reference adaptive fuzzy controller. Both maintained the arc voltage and arc length well.

At the lower current operation (Fig. 3 and 4), it can be observed that the arc plasma is unstable and difficult to maintain. The fuzzy controller cannot control the dynamic behavior of the system as well as the model reference adaptive fuzzy controller. The results show that the overshoot of the system with the fuzzy controller is larger.

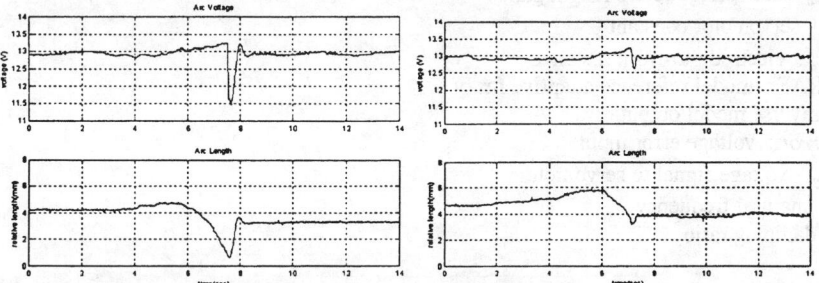

Fig. 3. Step plate test of the fuzzy controller at 30-A arc current

Fig. 4. Step plate test of the MRAF controller at 30-A arc current

Ramping Down Current Test

The purpose of this experiment was to observe the capability of the model reference adaptive fuzzy AVC to perform stably under conditions of wide variations in current of the arc when the fuzzy AVC cannot maintain system stability. The arc current was ramped down from 150 A to 10 A in 5 seconds. The arc voltage was also ramped down from 15 V to 12 V during that time. The travel speed was zero.

The fuzzy AVC became oscillatory at low current. The model reference adaptive fuzzy controller maintained the desired arc voltage without continuous oscillation as occurred in the fuzzy AVC case.

CONCLUSIONS

There were advantages of model reference adaptive fuzzy AVC over fuzzy AVC that were apparent from the experiments conducted in this research. These are, 1) An model reference adaptive fuzzy controller is capable of adjusting the controller in real-time for reducing the arc length overshoot at low current operation as compared to a fuzzy controller. 2) Due to the closed-loop control characteristics of the system, there is a possibility that the system will be oscillatory. The model reference adaptive fuzzy AVC is capable of preventing this oscillatory condition.

For these reasons, it can be concluded that the model reference adaptive fuzzy AVC in GTAW is capable of maintaining the arc voltage and also the arc length in various situations better than the fuzzy AVC which is specifically designed for nominal welding parameters ($V_{arc}, I_{arc}, L_{arc}$).

NOMENCLATURE

a_s : servomotor constant
AVC : automatic voltage control
c_m : centers of the output membership function
COG : center of gravity
DCEN: direct-current electrode-negative
$G0, G1$: normalizing gains
GTAW: gas tungsten arc welding
K_{arc} : arc voltage-to-arc length gain
K_s : servomotor constant
L_{arc} : electrode-to-work distance
MRAF : model reference adaptive fuzzy
p: inverse model output
V_{error} : voltage error input
V_{ref} : voltage signal to servomotor
ω_n: natural frequency
ς: damping ratio

REFERENCES

Bjorgvinsson, J.B., 1992, "Adaptive Voltage Control in Gas Tungsten Arc Welding," M.S. thesis, Vanderbilt University, Nashville, Tennessee.
Bjorgvinsson, J.B., G.E. Cook and K. Andersen., 1993, "Microprocessor-Based Arc Voltage Control for Gas Tungsten Arc Welding Using Gain Scheduling," *IEEE Transactions on Industry Applications*. Vol. 29, No 2, pp. 250-255.
Cary, H.B., 1979, Modern Welding Technology, *Prentice-Hal Inc., Englewood Cliffs, New Jersey*.
Koseeyaporn, P., 1999, "Adaptive Voltage Control in Gas Tungsten Arc Welding," M.S. thesis, Vanderbilt University, Nashville, Tennessee.
Passino, K. M. and S. Yurkovich, 1998, *Fuzzy Control*, Addison-Wesley Longman Inc., Menlo Park, California.
Smithmaitrie, P., 2000, "Adaptive Fuzzy Voltage Control in Gas Tungsten Arc Welding," M.S. thesis, Vanderbilt University, Nashville, Tennessee.

NONLINEAR SYSTEM IDENTIFICATION USING FUZZY NARMA MODEL

CHOKCHAI WIWATTANAKANTANG
Faculty of Information Technology
King Mongkut's Institute of
Technology, Ladkrabang,
Bangkok 10520, Thailand
Tel (662)737-2551-4 ext. 530
Fax: (662) 326-9074
Email : chokchai@mail.com

WORAPOJ KREESURADEJ
Faculty of Information Technology
King Mongkut's Institute of
Technology, Ladkrabang,
Bangkok 10520, Thailand
Tel (662)737-2551-4 ext. 530
Fax: (662) 326-9074
Email : worapoj@it.kmitl.ac.th

ABSTRACT

Most systems encountered in the real world are non-linear autoregressive moving average (NARMA) model. In this paper a Fuzzy NARMA model is used as an identifier for nonlinear system. A fuzzy systems is represented as series expansion of fuzzy basis functions which are capable of uniformly approximating any real continuous function on a compact set to arbitrary accuracy. Furthermore, an error feedback technique and a least square algorithm are proposed to identify the parameters of the Fuzzy NARMA model. An experimental result is included to demonstrate the effectiveness of the identification model. A data set generated from a complex NARMA model is used for testing the proposed model. In addition, two kinds of statistical tests are used for measuring the quality of fit.

INTRODUCTION

In real-life most systems are non-linear. Since linear models cannot capture the behavior of limit data associated with non-linear systems, it is important to investigate the identification procedure for non-linear model (Dunis, 1997), (Chen, 1990). The NARMA model provides a basis for such a development.

Most applications of fuzzy modeling rely on the framework of autoregressive model or regression model (Lincoln, 1991), (Wang, 1992a). This is the case because the inputs value for the fuzzy models can be easily identified: they are simply the lagged values of the time series itself or the exogenous inputs. Fuzzy models that are based on the framework of autoregressive model or regression model fail to model a system that is represented by NARMA model. As a result, high prediction error may occur when applied fuzzy AR model or Fuzzy regression model for modeling NARMA process.

Here we propose an fuzzy model for identifying NARMA models based on fuzzy basis function (FBF). Fuzzy systems are represented as series expansion of fuzzy basis functions. These fuzzy basis functions are capable of uniformly approximating any real continuous function on a compact set to arbitrary accuracy (Wang, 1992a). This means that NARMA can be approximated within an arbitrary accuracy by model based on FBF. In this paper, We performed two kinds of statistical tests—autocorrelation test and chi-squared test in order to measure the quality of fit.

FUZZY BASIS FUNCTION

In this paper, we consider a fuzzy system whose basic configuration is shown in Fig.1.

If the fuzzy rule base consists of a collection of fuzzy IF-THEN rule:

$$R^j: \text{IF } x_1 \text{ is } F_1^j \text{ and } \ldots \text{ and } x_n \text{ is } F_n^j \text{ THEN } y \text{ is } G^j$$

where $j=1, 2, \ldots L$, F_i^j and G^j are labels of fuzzy sets in U and V respectively, then fuzzy logic systems with a center average defuzzifier, algebraic product inference, and singleton fuzzifier consist of all functions of the form

$$y(\bar{x}) = \frac{\sum_{j=1}^{L} \bar{y}^j (\prod_{i=1}^{n} \mu_{F_i^j}(x_i))}{\sum_{j=1}^{L} (\prod_{i=1}^{n} \mu_{F_i^j}(x_i))} \quad (1)$$

where $\mu_{F_i^j}$ and μ_{G^j} are membership function of, F_i^j and G^j respectively and $\bar{y}^j$ is the point at which μ_{G^j} achieves its maximum value that is assumed to be one. Fuzzy basis functions (FBF) are defined as

$$p_j(\bar{x}) = \frac{\prod_{i=1}^{n} \mu_{F_i^j}(x_i)}{\sum_{j=1}^{L} (\prod_{i=1}^{n} \mu_{F_i^j}(x_i))}, \quad j = 1, 2, 3, \ldots, L \quad (2)$$

Therefore, the fuzzy system in the Eq. (1). is equivalent to a fuzzy basis function (FBF) expansion:

$$y(\bar{x}) = \sum_{j=1}^{L} \theta_j p_j(\bar{x}) \quad (3)$$

Here, Least square (LS) algorithm is proposed to identify the parameters of the FBF expansion. To applied LS algorithm, the FBF expansion is rewrite as

$$y(\bar{x}) = \sum_{j=1}^{L} \theta_j p_j(\bar{x}) + \varepsilon$$

Given N input-output pairs, the matrix notation of the equation can be written as

$$Y = P\Theta + E \quad (4)$$

where

$$Y = \begin{bmatrix} y^1 \\ \cdot \\ \cdot \\ y^N \end{bmatrix}, P = \begin{bmatrix} p_1^1 & \cdot & \cdot & p_L^1 \\ \cdot & & & \cdot \\ \cdot & & & \cdot \\ p_1^N & \cdot & \cdot & p_L^N \end{bmatrix}, E = \begin{bmatrix} \varepsilon^1 \\ \cdot \\ \cdot \\ \varepsilon^N \end{bmatrix}$$

Then, Θ must satisfy the following equation:

$$(P^T P)\, \Theta = P^T Y$$

The derivation of the equation (4) can be found in Kreesuradej (1996). The inverse or pseudo-inverse of $P^T P$ are usually utilized for finding the value of Θ.

One important property of FBF expansions is that FBF expansions are capable of approximating any real continuous function (Wang, 1992b). This gives a justification for using FBF expansions to model a dynamic system that is usually described by continuous functions. In this work, the FBF expansions will be applied to modeling NARMA model.

IDENTIFICATION PROCEDURE OF NARMA MODEL BASED ON FUZZY BASIS FUNCTIONS

Here, FBF is proposed to model NARMA (p,q) model which has the following form [6]:

$$y(t) = f(y(t-1),..., y(t-n_y), e(t-1),...e(t-n_e)) + e(t) \qquad (5)$$

where $y(t)$ and $e(t)$ are the system output and prediction error, respectively; n_y and n_e are the maximum lags in the output and noise, respectively, $\{e(t)\}$ is assumed to be a white sequence, and $f(.)$ is some non-linear function. The identification procedure can be summarized as follows

(i) Choose n_y and n_e. Initially the set of

$$y(t) = [y(t-1) \cdots y(t-n_y)]^T$$

FBF model is selected using the LS algorithm and the initial model is used to generate the initial prediction error sequence $\{\varepsilon^{(0)}(t)\}$.

(ii) An iterative loop is then entered to update the model. At the k th iteration

$$y(t) = [y(t-1) \cdots y(t-n_y), \varepsilon^{(k-1)}(t-1) \cdots \varepsilon^{(k-1)}(t-n_e)]^T$$

FBF model is selected by LS algorithm and this gives rise to the prediction error sequence $\{\varepsilon^{(k)}(t)\}$ Typically two to four iterations are sufficient (Chen, 1990).

The model validity tests are performed to assess the model. If the model is considered adequate the procedure is terminated. Otherwise go to step (i).

EXPERIMENTAL RESULTS

As an experiment, the first 500 points of data generated from the Eq. (6). is used to identify the fuzzy NARMA based on the proposed approach. Then, the next 300 point of data is used to test the fuzzy NARMA. For comparison purpose, we also model the fuzzy NAR model. Then, the chi-squared test and autocorrelations of residuals are used for model validation.

The results are given in Table 1. for both fuzzy NAR model and fuzzy NARMA model. The outputs of simulation and the model response are shown in Fig. 2 and 3. The correlation tests and chi-squared tests are in Fig. 4. and 5. respectively.

From all of the results, the fuzzy NARMA model provides lower mean square error (MSE) and better standard deviation of errors (STDE) than the fuzzy NAR model. In addition, from the model validity tests show that the model is adequate while the fuzzy NAR model give unsatisfied model validity tests.

$$y(t) = 1.2y(t-1)\exp\left(\frac{-y^2(t-1)}{6}\right) + 0.9e(t-1)\sin(e(t-1)/3) + e(t) \qquad (6)$$

where $e(t)$ is a white noise sequence with standard deviation = 0.6667

CONCLUSIONS

The fuzzy NARMA model was proposed with feedback error technique and LS algorithm for identification nonlinear system. Based on the experimental results, the fuzzy NARMA model provides better performance than the fuzzy NAR model. In the future, the results from comprehensive studies such as convergence analysis of the algorithm and testing the proposed model with complex time series model will be reported.

Table 1 Predicive modeling results

MODEL	MSE		STDE	
(Noise STD=0.6667)	TRAINING	TESTING	TRAINING	TESTING
Fuzzy NAR Model	0.5504	0.5367	0.7426	0.7325
Fuzzy NARMA Model	0.4362	0.4535	0.6611	0.6745

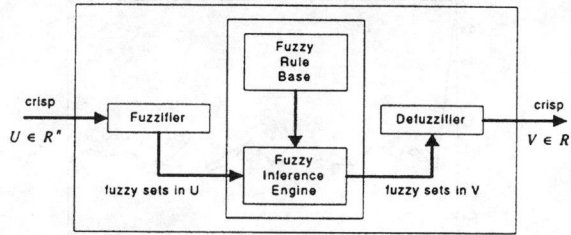

Fig. 1. Basic configuration of fuzzy systems.

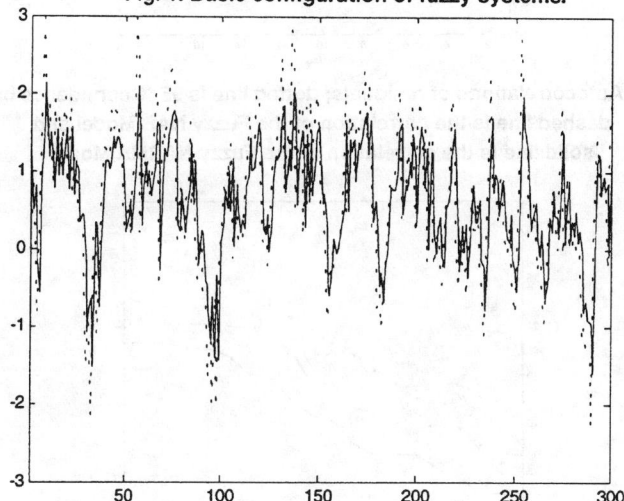

Fig. 2 Outputs of simulation (dotted line) and Fuzzy NAR Model (solid line)

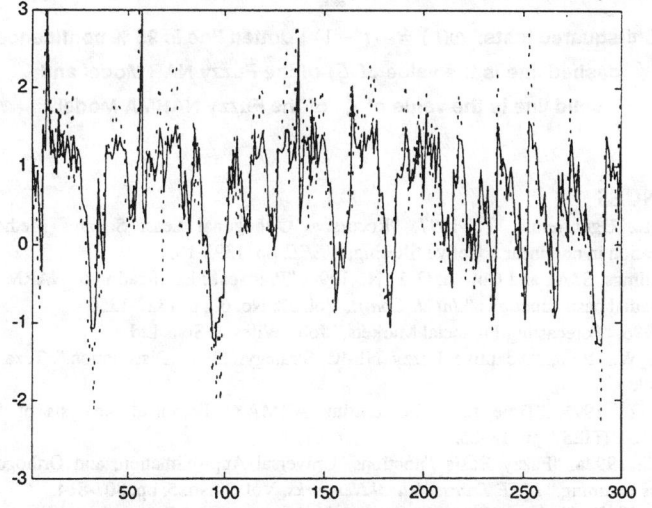

Fig. 3 Outputs of simulation (dotted line) and
the identification model : Fuzzy NARMA Model (solid line)

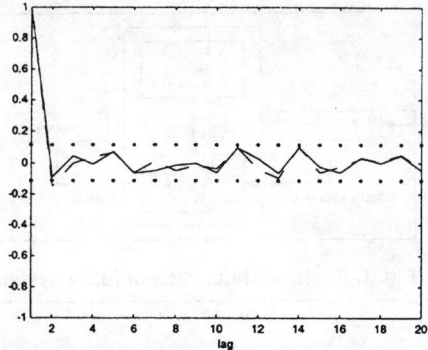

Fig. 4 Autocorrelations of residuals; dotted line is 95 % confidence band, dashed line is the correlation of the Fuzzy NAR Model and solid line is the correlation of the Fuzzy NARMA Model

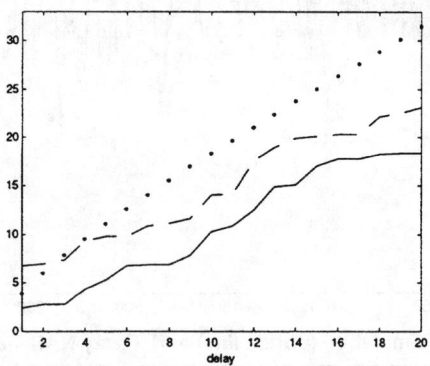

Fig. 5 Chi-squared tests; $\omega(t) = e(t-1)$; dotted line is 95 % confidence limit, dashed line is the value of ζ of the Fuzzy NAR Model and solid line is the value of ζ of the Fuzzy NARMA Model

REFERENCES

Chang, S. L., Ogunfremni, T., 1997, "Recursive Orthogonal Least Squares Method and its Application in nonlinear adaptive filtering," *IEEE*, pp. 1392-1396.

Chen, S., Billings, S. A., and Cowan, C. F. N., 1990, "Practical identification of NARMAX models using radial basis functions," *Int. J. Contr.*, Vol. 52, No. 6, pp. 1327-1350.

Dunis, C., 1996, "Forecasting Financial Markets," John Wiley & Sons Ltd.

Kreesuradej, W., 1996, "Adaptive Fuzzy NIMC Strategy. Ph.D. Dissertation," Texas Tech U., Lubbock.

Lincoln, H. T., 1991, "Time series Forecasting ARMAX, Technical Analysis of STOCK & COMMODITIES," pp. 18-26.

Wang, L. X., 1992a, "Fuzzy Basis Functions, Universal Approximation, and Orthogonal Least-Squares Learning," *IEEE Trans. Neural Networks*, Vol. 3, No. 5, pp. 807-814.

Wang, L. X., 1992b, "Fuzzy systems are universal approximators," *Proceedings IEEE International Conference On Fuzzy systems*, San Diego, pp. 2511-2516.

ADAPTIVE FUZZY LOGIC CONTROLLER FOR ROBOTS MANIPULATORS

TOUATI YOUCEF
Computer Science Laboratory
122, Rue Paul Armangot
94400 Vitry Sur Seine, France
Email : touati@univ-paris12.fr

AMIRAT YACINE
University of Paris-XII
122, Rue Paul Armangot
94400 Vitry Sur Seine, France
Email : amirat@univ-paris12.fr

DJOUANI KARIM
Computer Science Laboratory
122, Rue Paul Armangot
94400 Vitry Sur Seine, France
Email : djouani@univ-paris12.fr

ABSTRACT
This paper deals with robots manipulators adaptive force control. A new adaptive fuzzy approach based on learning paradigm is presented. The main problem of fuzzy systems design, concerns fuzzy rules generation and optimization. Our approach proceeds in three phases. In the first one, the fuzzy logic system is considered as feed-forward neural network. Back-propagation learning algorithm is applied for parameters identification in order to map some input/output data. This phase is called Parameter learning phase. In the second phase, a clustering algorithm based on FCMeans algorithm is used to optimize the number of linguistic terms associated to each input. Finally, the fuzzy rule base is generated and optimized from conflict and redundancy analysis. The simulation of such an approach concerns following trajectory using a planar manipulator including two revolute joints.

Key Words : Force Control, Fuzzy Logic, Learning Algorithm, System Identification and Control

INTRODUCTION
The fast development in factory automation has resulted in involving a lot of industrial robot manipulators in order to increase the productivity and improve product quality. So, dexterous robots and suitable control algorithms are needed in handling various complex situations. Carrying an object, painting the surface of an object or welding materials, represent primary tasks for which robots manipulators are used. Such tasks require accurate positioning, which depends upon the accurate estimations of robot kinematics and dynamics. Moreover, industrial tasks are also concerned within force control and mostly unknown contact with environment problems. For instance, when the robot motion is constrained by the environment, the end effector trajectory is modified using contact forces. Therefore, the position control only is not sufficient to perform such tasks. The constrained motion control requires a controller that combines both position and force control, which is called force/position control or simply hybrid control.

In many control design situations, mathematical model of the system is not available in its detailed form. However, some knowledge about the process behaviors may exists in many different forms, e.g., simplified mathematical models which are valid only within limited operating ranges, heuristic rules which attempt to describe observed system behavior, etc. Such knowledge leads to some quantitative and qualitative information that can be used in the control design problem by taking into account system complexity and uncertainty. Integrating these various forms of information into a control system is and stills a challenging problem [1]. Neural networks and fuzzy logic provide a learning and rule based reasoning that can be applied to the control of complex systems whose dynamic model description is either too complex or unavailable. The advantages of fuzzy logic over conventional and neural networks control methods concern specially the increasing robustness in spite of noise or other sensor failures, and the ability to handle non-linearity without degradation or failure [6].

Another important feature of fuzzy logic concerns the formulation of fuzzy rules that use linguistic adjectives and relations similar to a natural language. This makes the understanding and modification of a fuzzy logic based controller systematic [4]. In this paper, a new hybrid Neuro-Fuzzy control approach is presented and the learning algorithms are developed. This approach is based on fuzzy logic theory which proceed in three phases. In the first one, the fuzzy logic system is considered as a feedforward neural network back-propagation on which a learning algorithm is applied in order to map some Input/Output data. This phase is called parameter learning phase. In second phase, a clustering algorithm based on FC-Means algorithm is used in order to optimize the number of linguistic terms associated to each system input. Finally, the final fuzzy rule base is generated and optimized from conflict and redundancy analysis.

FUZZY LOGIC CONTROLLER DESIGN

For control problems with multiple objectives of different priorities, the decoupling of a complex system, of Multi-inputs/Multi-outputs systems (MIMO) type into a simple ones can be viewed as a solution to cope for each objective. In this section, the design of a Fuzzy Logic Controller (FLC) is developed for a large-scale complex systems. This approach consists on the decoupling of the FLC system into a several Multi-inputs/Single-output FLC sub-systems (MISO) [11]. One can note that the number of the FLC sub-systems is equal to MIMO controller outputs number. In this paper, an adaptive FLC design methodology is proposed.

The approach is based on Jang's ANFIS model for function approximation [3]. Fig.1 shows an external force control that uses a neuro-fuzzy controller. The network consists of several sub-systems called FLC modules, where each module is configured according to its own inputs/output and optimization criteria.

The architecture illustrated on Fig. 2 represents a single MISO FLC module, which is considered as a neuro-fuzzy system built from five-layer feed-forward network. Each fuzzy module is designed to handle with all the antecedents ΔF_k and produces an action y_k, where k = 1,..., n modules.

As in ANFIS model, this architecture implements rules of the following form [7][9] :

$$R_j : \text{If } \Delta F_1 \text{ is } A^{(1)}_{1j} \text{ and .... and } \Delta F_n \text{ is } A^{(i)}_{nj} \text{ then } y_k = f(\Delta F_1, .., \Delta F_n)$$

where ΔF_i and y_k represent respectively the input and the output variables, and $A^{(i)}_{ij}$ the fuzzy sets. The neuro-fuzzy controller design passes by the partitioning of each input/output variable space in several fuzzy sub-sets (grid or free partition). Firstly, the Gaussian membership functions type (MFs) are chosen and a free partition is applied to the universe of discourse. From some observations of ΔF_i, the fuzzy inference consequence y_k obtained by simplified fuzzy reasoning method. The membership functions for each input variable ΔF_i, can be written as follows :

$$A^{(k)}_{ij} = S^{(k)}_{ij}(\Delta F_i) = exp(-0.5 \,((\Delta F_i - m_{ij})/\sigma_{ij})^2)$$

The inputs/output relationships for the mth FLC module are given by :

$$A^{(k)}_{ij}(\Delta F_i) = exp(-0.5\,((\Delta F_{ii} - m_{ij})/\sigma_{ij}))$$
$$y_j = \Pi^k_{j=1} S^{(k)}_{ij}, \; \forall i = 1:n$$
$$N = \Sigma^n_{i=1} \Pi^k_{j=1} W_j \cdot y_j, \; \forall i = 1:n$$
$$D = \Sigma^n_{i=1} y_j \; , \; \forall i = 1:n$$
$$y_k = N/D$$

where ΔF_i is the input variable, m_{ij} and σ_{ij} the ith mean and standard deviation respectively of the jth rule.

LEARNING FUZZY SETS

The learning algorithm is divided into two stages: learning fuzzy rules (structure learning) and learning fuzzy sets (parameters learning) [5].

Stage – 1 : PARAMETERS LEARNING

At this stage, we cater with FLC parameter identification. The goal of the learning algorithm is to modify the FLC membership function in order to optimize some global behavior criteria (in the unsupervised learning case) or to map some inputs/output data (in the supervised learning case). In this paper, we are concerned with supervised learning, so the FLC system is viewed as a feed-forward neural network and back-propagation learning algorithm [8] is applied off line to map some input/output data. As shown on Fig. 3, the FLC system maps the forward dynamics of the conventional controller. The aim of the learning algorithm is to update the membership functions parameters (m_{ij}, σ_{ij}) and the consequent parameters (w_j). According to the back-propagation algorithm, the parameters updating equations are given by :

$$w_j(k+1) = w_j(k) - \eta_w \frac{\partial E}{\partial w_j} = w_j(k) - \eta_w \cdot \frac{u_{fl} - u_d}{\sum_{j=1}^{M}\left(\prod_{j=1}^{M} S_{A_{ij}}(\Delta F_i)\right)} \cdot \prod_{j=1}^{M} S_{A_{ij}}(\Delta F_i)$$

$$m_{ij}(k+1) = m_{ij}(k) - \eta_m \frac{\partial E}{\partial m_{ij}} = m_{ij}(k) - \eta_m \cdot (u_{fl} - u_d) \cdot \frac{w_j - u_{fl}}{\sum_{j=1}^{M}\left(\prod_{j=1}^{M} S_{A_{ij}}(\Delta F_i)\right)} \cdot \prod_{j=1}^{M} S_{A_{ij}}(\Delta F_i) \cdot \frac{\Delta F_i - m_{ij}(k)}{\sigma_{ij}^2(k)}$$

$$\sigma_{ij}(k+1) = \sigma_{ij}(k) - \eta_\sigma \frac{\partial E}{\partial \sigma_{ij}} = \sigma_{ij}(k) - \eta_\sigma \cdot (u_{fl} - u_d) \cdot \frac{w_j - u_{fl}}{\sum_{j=1}^{M}\left(\prod_{j=1}^{M} S_{A_{ij}}(\Delta F_i)\right)} \cdot \prod_{j=1}^{M} S_{A_{ij}}(\Delta F_i) \cdot \frac{(\Delta F_i - m_{ij}(k))^2}{\sigma_{ij}^3(k)}$$

where η_w, η_m, η_σ are the learning rates, k the learning iteration number, u_d and u_{fl} respectively the desired output value and the corresponding fuzzy result and E an objective function given by :

$$E = (u_{fl}(k) - u_d(k))^2 / 2$$

Stage – 2 : STRUCTURE LEARNING

The generated rule-base suffers from redundancy and conflicts of data, most of which are less useful. This makes the assignment of a linguistic label to the associated variable difficult and sometimes counter-intuitive. This redundancy is often present in the form of similar MFs in the premise of the resulting rule-base. The high number of the MFs makes difficult to grasp the meaning of the model. A semantically unclear model is not easily verified after design phase for the model. Consequently, a simplification phase allowing the elimination of redundancy is required. For this purpose, an algorithm based on the class of fuzzy C-means algorithm introduced by Bezdek [2] is applied.

Let's given $P_{ij} = [m_{ij}, \sigma_{ij}]$, the MFs parameters associated to input variable x_i obtained after the parameter learning phase. The aim of the clustering algorithm is to determine an optimal clusters set $\{v_k\}$, where $v_j = [m_k, \sigma_k]$, in order to replace the old fuzzy partition $\{P_{ij}\}$ by the new one $\{v_k\}$ according to the minimization of the following objective function [2][10] :

$$J_i = \sum_{i=1}^{M} \sum_{k=1}^{c} (\mu_{ki})^m \cdot dist(P_{ij}, v_k)^2 \text{ with respect to the constraint} : \forall i, \sum_{k=1}^{c} \mu_{ki} = 1$$

In other words, this algorithm is to find membership degree μ_{ki} of each variable P_{ij} to the center of the optimized class v_k [2]. Where, M, c and m are respectively the number of data $i=(1,..,M)$, the fixed and known number of clusters and an arbitrary chosen scalar $(m>1)$.

The distance $dist(P_{ij}, v_k)$ between P_{ij} and the prototype v_k is defined by :

$$dist(P_{ij}, v_k)^2 = (P_{ij} - v_k)^t A \cdot (P_{ij} - v_k)$$

where A is semi-defined positive matrix $\forall x$, $<A,x> > 0$. In our case, A is an identity matrix. The fuzzy clustering algorithm can be summarized by the following steps :

1- Initialization (clusters number c, scalar m, partition matrix $U=[\mu_{kj}]$).
2- Loop - Computes :

2.1- prototypes centers v_k : $\forall k$, $v_k = \dfrac{\sum_{j=1}^{M}(\mu_{kj})^m \cdot P_{ij}}{\sum_{j=1}^{M}(\mu_{kj})^m}$

2.2- $dist(P_{ij}, v_k)$

2.3- new partition U : $\begin{cases} \text{si } dist(P_{ij}, v_k) \neq 0 \text{ alors } \mu_{kj} = \dfrac{(1/(dist(P_{ij}, v_k))^{2/(m-1)}}{\sum_{k=1}^{c}(1/(dist(P_{ij}, v_k))^{2/(m-1)})} \\ \text{si } dist(P_{ij}, v_k) = 0 \text{ alors } \mu_{kj} = 1 \ \forall k \neq j \end{cases}$

2.4-   until J_i is minimized
3-   End loop

Stage – 3 : INTERPRETABILITY
When the fuzzy clustering algorithm is achieved, a number of clusters are generated to each fuzzy set. To have a good interpretation of the obtained results in terms of fuzzy rules, we attribute to each generated class a fuzzy label and thus, we give for each rule its symbolic signification of the form :

$$R_m : \text{ If } \Delta F_1 \text{ is big and } \Delta F_2 \text{ is small then } y_k = f(\Delta F_1, \Delta F_2)$$

For example, if we consider two clusters defined by their centers : $v_1 = (m_1, \sigma_1)$ and $v_2 = (m_2, \sigma_2)$, with $(m_1, \sigma_1) < (m_2, \sigma_2)$, we can associate to each center its own fuzzy label. In our case, center1 ≡ *small* and center2 ≡ *big*, as defined in the rule below.

Stage – 4 : REAL TIME ADAPTATION
In this step, the final rule base is generated and optimized. The FLC consequences are tuned on-line after one trial with a back-propagation algorithm given by the following relation :

$$w(p+1) = w(p) - \eta_w \frac{\partial J(p)}{\partial w(p)}$$

where η_w represents the convergence speed parameter and J the cost function given by :

$$J(p) = \frac{1}{2} \sum_{k=1}^{p} (Fd(k) - Fr(k))^2 .$$

The derivatives $\frac{\partial J(p)}{\partial w(p)}$ can be formulated by :

$$\frac{\partial J(p)}{\partial w(p)} = \frac{\partial J(p)}{\partial F_r(p)} \cdot \frac{\partial F_r(p)}{\partial w(p)} = -\sum_{k=1}^{p} (F_d(k) - F_r(k)) \cdot \frac{\partial F_r(p)}{\partial u_{fl}(p)} \cdot \frac{\partial u_{fl}(p)}{\partial w(p)}$$

since $\frac{\partial F_r(p)}{\partial u_{fl}(p)}$ defines the Jacobian matrix of the system, which is known, only $\frac{\partial u_{fl}(p)}{\partial w(p)}$ must be determined.

One can note that if the parameters which determine the convergence speed are correctly chosen, all the fluctuations which act on the system are compensated and leading to the desired closed loop behavior.

SIMULATION RESULTS

To validate our approach, we consider a planar two-link manipulator with two revolute joints as shown on Fig. 5. The robot has to follow a circular trajectory with respect to efforts contact constraints with the environment. The architecture of our controller is divided into two FLC sub-systems, each one composed also by two other sub-systems. The first step in FLC parameters learning phase is the selection of input and output variables in order to map some input/output of the conventional controller. In our case, the designed input variables are the robot link force error $\Delta F_i(\Delta F_x, \Delta F_y)$. The output variable is the desired conventional control output $u_d(\Delta F_{dx}, u_{dy})$. The set of MFs associated to each input variable, is randomly initialed.

The identification phase results are shown on Fig.6. Fig.7 illustrates learning error of each FLC sub-system, which is decreasing with epochs. Fig.8, represents the repartition of the membership functions after the learning phase for each FLC sub-system.

In the second phase, in order to improve the trajectory following tasks, with respect to the environment contact forces, the on line FLC's parameters adaptation is used. However, and due to the MFs overlapping, the fuzzy clustering algorithm is used to reduce the MFs number and optimize the fuzzy base rule. The (Fig.9) gives optimal clusters within initial MFs obtained after the application of the fuzzy clustering algorithm. Table1 gives the number of rules before and after classification.

Finally, the application of the on line adaptation algorithm leads to results given on Fig.10 (the measured forces in taskspace) and Fig.11 (the robot end effector trajectory).

CONCLUSION

In this paper, a neuro-fuzzy approach to solve the control of constrained dynamic systems is proposed. The designing methodology of a new Fuzzy Logic Controller dedicated to constrained tasks has been presented. Compared to neural network control, the proposed architecture allows the meaning of all the internal parameters under the form of the fuzzy rule base. In order to show the validity of this approach, a task consisting in a trajectory following under environment constraints executed by a planar manipulator has been implemented. The analysis of the results reveals a satisfactory dynamic behavior and confirms the suitability and the efficiency of our approach. The obtained results lead us to increase investigations in this control approach and especially in the classification part to increase the performances of the fuzzy C-means algorithm.

REFERENCES

[1] Kurt M. Berger and Kenneth A. Loparo, " *A Hierarchical Framework for Learning Control* ", Center for Automation and Intelligent Systems, Case Western Reserve University, May.88

[2] J. C. Bezdek, " *Pattern Recognition with Fuzzy Objective Function Algorithms* ", Plenum Press, New York. 1981

[3] J.S.R. Jang, " *Neuro-Fuzzy Modeling: Architecture, Analyses and Ap*plications ", Ph.D. Thesis, Dep. of Electrical Engineering and Computer Science, Univ. of California, Berkeley USA, July.93

[4] Ashok Nedungadi, " *A Fuzzy Logic Based Robot Controller* ", International Journal of Fuzzy System. Vol. 1, Issue 3, 1993

[5] Z. Pingan and L. Renhou, " *A New Approach to Fuzzy Identification for Complex* ", IEEE Int. Conf. On Fuzzy Systems, pp. 1309 – 1313, 1996

[6] Y. Touati, Y. Amirat, K. Djouani and M. Kirad, " *A New hybrid Force-Position Control Based Neural Networks Approach For Industrials Applications* ", The 4th World Multi-Conf. on Syst. Cyb. and Inf. 2000, Orlando, Florida USA, Vol. 3, pp. 153 – 157 , July 23 – 26, 2000

[7] L.X. Wang and J.M. Mendel, " *Generating Fuzzy Rules by Learning from Examples* ", IEEE Trans. on Systems, Man and Cybernetics, Vol. 22, N°. 6, pp. 1414 – 1427. Nov.92

[8] L. X. Wang and J. M. Mendel, " *Back-propagation Fuzzy System as Non-linear Dynamic System Identifiers* ", IEEE Int. Conf. On Fuzzy Systems, pp. 1409 – 1415, 1992

[9] L. X. Wang and R. Langari, " *Complex Systems Modeling via Fuzzy Logic* ", IEEE Trans. on Systems, Man and Cybernetics, part-B : Cybernetics, Vol. 26, N°. 1, pp .100 – 106. Feb.96

[10] H. Imai and A. Tanaka, " *A Method of Identifying Influential Data in Fuzzy Clustering* ", IEEE Trans. On Fuzzy Systems, Vol. 6, N°. 1, pp .90 – 101. Feb.98

[11] C.W. Chi and T.C. Hsia, " *Adaptive Aggregation of Modular Fuzzy Control* ", Internal Rapport, Dep. Of Electrical and Computer Engineering, University of California, Davis. USA

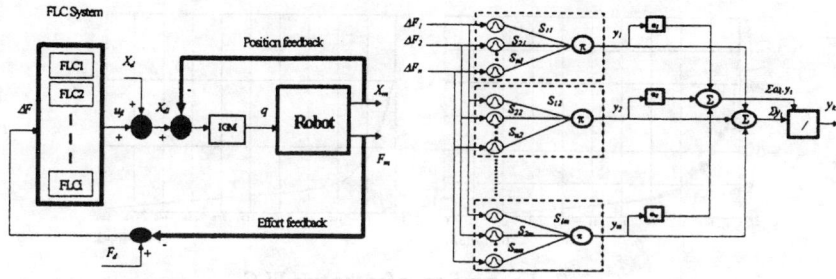

Fig. 1 External Force Control Using a Fuzzy Logic Controller

Fig. 2 Neuro-Fuzzy Controller Architecture

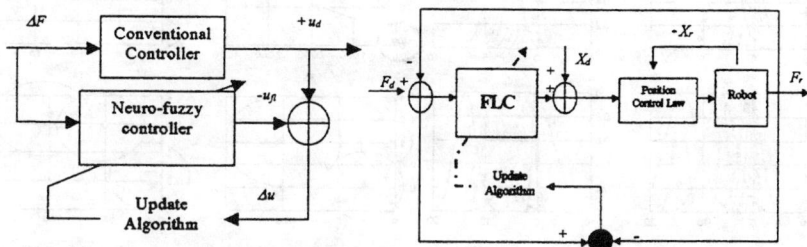

Fig. 3 Identification Structure

Fig. 4 Proposed Control Structure

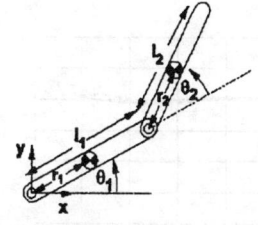

Fig. 5 Two-link planar manipulator

	after learning	after classification
FLC-1	16	4
FLC-2	16	6

Table.1 Number of rules before and after classification phase

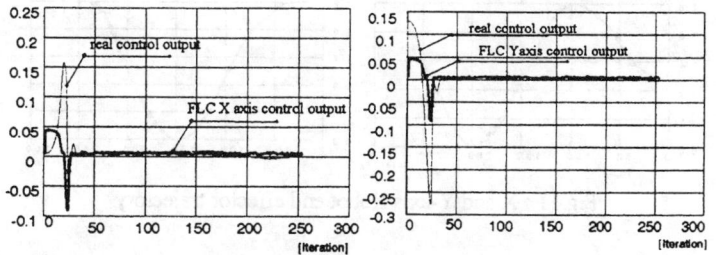

Fig.6 : FLC outputs and desired outputs in task space

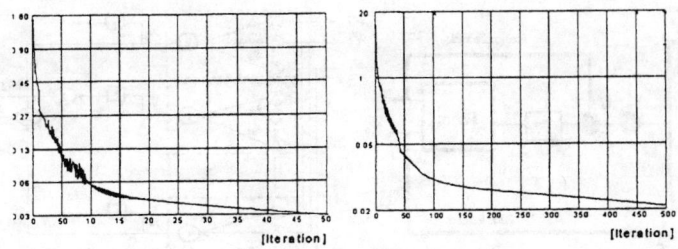

Fig.7 : Learning error for the two FLC

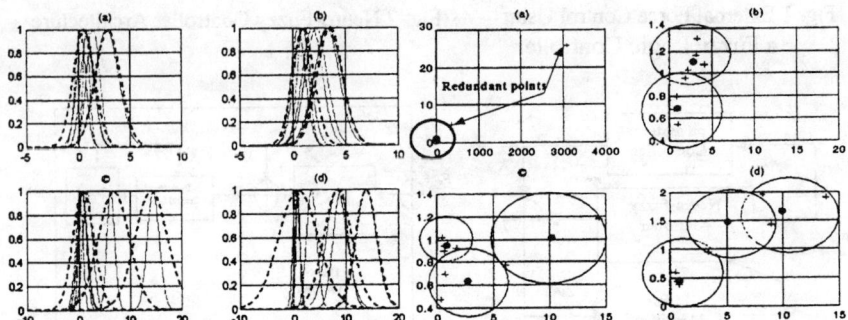

Fig.8 : Optimal clusters and initials MFs

Fig.9 : Optimal clusters and initials MFs in 2-D representation

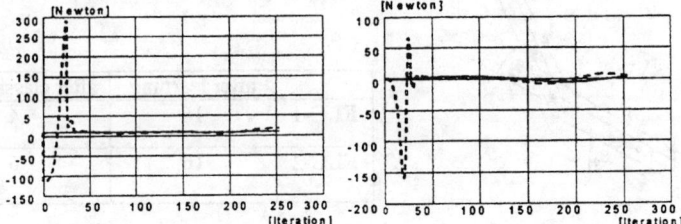

Fig.10 : X and Y-axis Measured and desired forces in task space

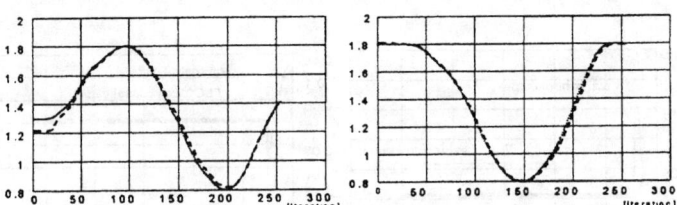

Fig.11 : X and Y-axis Robot end effector trajectory

CONTROL OF HIGHLY COUPLED STRUCTURES USING A MULTIVARIABLE SLIDING MODE FUZZY CONTROLLER

HAMID ALLAMEHZADEH
Eastern New Mexico University
Department of Electronic Engineering
Portales, New Mexico 88130

JOHN CHEUNG
University of Oklahoma
School of Electrical and Computer Engineering
Norman, Oklahoma 73069

ABSTRACT

A Multivariable Sliding Mode Fuzzy Controller (MSMFC) was developed based on the multi-dimensional sliding modes properties, for a special class of nonlinear system. Using the decomposition principle in sliding modes, the design and stability analysis of MSMFC was reduced to the design and stability analysis of a set of single Sliding Mode Fuzzy Controllers (SMFCs). The proposed fuzzy controller was used to stabilize a two-link manipulator at different positions. Simulation studies were presented that illustrated the effectiveness of the MSMFC.

INTRODUCTION

Modern complex systems are often represented by high dimension differential equations. This, indeed, creates a serious obstacle to use analytical and computational methods for the design of efficient control algorithms for these systems. Therefore, great attention has been given to various devices that can divide the overall motion, in accordance with some criteria, into partial independent components. One possible method is to divide the motion artificially into fast and slow components by employing high gains in feedback loops [2]-[4]. Decomposition of the initial control system is also possible on the basis of the properties of sliding modes in dynamic systems [2].

The use of multidimensional sliding modes has proved very effective for solving a great number of complex control problems [1]. For most systems in practice, the motion in sliding mode is independent of the control and is determined by the properties of the system and by the equations of the surface of discontinuity. This feature makes it possible to decompose the initial problem into independent subproblems of lower dimensionality: the control is used solely on creating the sliding modes, while the desired type of motion along the intersection of the surfaces of discontinuities is obtained by a suitable choice of their equations. This property is quite useful for solving many application problems characterized by high order differential equations.

Another important feature of sliding mode is that under certain conditions it may become invariant to changes in dynamic characteristics of the plant and disturbances. This feature can play an essential role in solving numerous automatic control problems and offer an indispensable tool for analysis and design of complex nonlinear systems.

In this paper, we use the invariance property of sliding modes to develop a Multivariable Sliding Mode Fuzzy Controller (MSMFC) for a special class of nonlinear systems. In the next section, we extend the sliding mode fuzzy control methodology to an important class of multivariable nonlinear systems. Then, a MSMFC is developed based on the principle of decoupling in multidimensional sliding modes. For illustration, the fuzzy controller is applied to a two-link manipulator.

DESIGN OF MULTIVARIABLE SLIDING MODE FUZZY CONTROLLER

Consider the class of nonlinear time varying systems described by the equations

$$x_j^{(n_j)} = f_j(X_1, \cdots, X_m) + b_j(X_1, \cdots, X_m)u_j + d_j(t) \qquad (1)$$
$$y_j = x_j, \quad j = 1, \cdots, m$$

Where $X_j = \left[x_j, \dot{x}_j, \cdots, x_j^{(n_j-1)}\right]^T$ is the j-th components of the state vector, $\Lambda = [X_1, X_2, \ldots, X_m]$ is the j-th control input, and y_j is the j-th system output. In (1), the function f_j, the control gain b_j, and the disturbance d_j are assumed to be partially known. The dynamics of (1) describe a large number of nonlinear systems encountered in practice, including a vast class of controllable nonlinear systems that could be converted into (1) by using appropriate transformations [4]. Without loss of generality, assume that the j-th reference input r_j is a step function; then we can write a state space representation of (1) in terms of $e_j = r - x_j$ and its derivatives.

$$\dot{e}_{j1} = e_{j2}$$
$$\dot{e}_{j2} = e_{j3}$$
$$\vdots \qquad (2)$$
$$\dot{e}_{jn_j-1} = e_{jn_j}$$
$$\dot{e}_{jn_j} = -f_j(\Lambda) - b_j(\Lambda)u_j - d_j(t)$$

Where

$$e_{j1} = e_j, \cdots, e_{jn_j} = e_j^{(n_j-1)} \quad \text{and} \quad x_{j1} = x_j, \cdots, x_{jn_j} = x_j^{(n_j-1)}$$

Now let $E_j = \left[e_{j1}, \cdots, e_{jn_j} \right]^T$ and $E = [E_1, \cdots, E_m]$, then we can rewrite (2) as follows:

$$\dot{e}_{j1} = e_{j2}$$
$$\dot{e}_{j2} = e_{j3}$$
$$\vdots \qquad (3)$$
$$\dot{e}_{jn_j-1} = e_{jn_j}$$
$$\dot{e}_{jn_j} = -f_s(E) - b_s(E)u_j - d_j(t)$$

Where $f_s(E)$ is a shifted replica of $f_j(\Lambda)$. The systems of nonlinear equations in (3) are highly coupled. Considering the nonlinear coupling terms in (3) as disturbances, we can introduce the sliding mode into the system and reject the disturbances by the various design procedures based on the invariance property of the sliding mode. Therefore, the coupled systems of (3) can be written as a set of m independent differential equations as follows:

$$\dot{e}_{j1} = e_{j2}$$
$$\dot{e}_{j2} = e_{j3}$$
$$\vdots \qquad (4)$$
$$\dot{e}_{jn_j-1} = e_{jn_j}$$
$$\dot{e}_{jn_j} = -f_s(E_j) - b_s(E_j)u_j - D_j(E,t)$$

where $D_j(E,t)$ is the sum of the disturbances $d_j(t)$ and all the nonlinear coupling terms in (3) or equivalently

$$D_j(E,t) = d_j(t) + \sum_{i \neq j} \alpha_i h_i(E,t) + \beta_i g_i(E,t) \qquad (5)$$

where α_i and β_i are in general time varying and h_i and g_i are the nonlinear decoupling terms.

Define a set of sliding surfaces S_j in the E_j space by the equations

$$S_j(E_j) = C_j E_j = 0 \quad \text{for} \quad j = 1, \cdots, m \qquad (6)$$

In (6), C_j is a constant row vector of the form $\left[c_{j1}, c_{j1}, c_{j, n_j-1}, 1 \right]$ such that the surface defined by (6) is stable. Then, find a discontinuous control u_j to guarantee the existence of a stable sliding mode on each discontinuity surface S_j or their intersections. The control u_j must also guarantee the global sliding condition or the sliding mode stability "in large".

Design of MSMFC can be reduced to the design of a set of SMFCs. By introducing sliding mode into system (3) and considering nonlinear coupling terms as disturbances, the multivariable control problem of (3) will be decoupled into m multi-input-single-output control problems of (4). Accordingly, the design and analysis of a multivariable sliding mode fuzzy controller with m outputs is equivalent to design and analysis of m independent SMFCs, so long as nonlinear coupling terms are considered as disturbances. The fact that the jth-sliding surface S_j depends only on E_j and the terms E_k for $k \neq j$ are considered as disturbances in the design of u_j substantiates the validity of the proposed decoupling technique. Therefore, the design of MSMFC with m outputs is simply reduced to the design of m SMFCs. To demonstrate the significance and efficiency of the proposed technique, we applied the MSMFC to the control problem of a two-link manipulator.

APPLICATION OF MSMFC TO AN INDUSTRIAL ROBOT

Various studies have been conducted on the dynamical properties and control of manipulators or robot arms [5]-[7]. The equations that describe the dynamics of industrial robots and manipulators are highly nonlinear and coupled in terms of the variables of motion. Physically, the coupling terms represent gravitational torques, reaction torques, and coriolis centrifugal torques. The significance of these interacting torques depends on the manipulator's physical parameters such as weight, size of link, and the load carried by the manipulator.

Some of the existing control algorithms use the linearized model of the system as the basis for decoupling and compensation [8]. However, a great number of the existing control algorithms emphasize nonlinear compensations of interactions [9]-[10]. The drawbacks of these schemes are the requirements of an accurate model of the manipulator and load forecasting. In general, these nonlinear compensations are complex and costly. The control algorithms developed based on theory of VSS [11] remedied the problem of model accuracy and load forecasting. However, the chattering problem appears in implementation.

In general, the dynamics of industrial robots are described by equations of the form (1). System parameters vary due to variations in the loads, variations in the ambient, imprecise modeling, and so on. We shall use the MSMFC to stabilize a two-link manipulator at different positions and velocities. We indicated in the previous section how a multivariable sliding mode control problem could be decomposed into a set of decoupled single sliding mode control problems. By this token, we see that the complexity of our design procedure for a more sophisticated manipulator involving more than two links is not significantly increased.

Consider the two-link manipulator with length l and mass m. Both m and l will be assumed normalized to unity. The state variables, $\theta_1, \theta_2, \dot{\theta}_1, \dot{\theta}_2$, are angular positions and angular velocities of the manipulator, respectively. The dynamics of the manipulator can be described by

$$\ddot{\theta}_1 = \left[\frac{2}{3}\sin\theta_2\dot{\theta}_2(2\dot{\theta}_1 + \dot{\theta}_2) + \left(\frac{2}{3} + \cos\theta_2\right)\sin\theta_1\dot{\theta}_1^2 + u_1\right]\left[\frac{16}{9} - \cos^2\theta_2\right] \tag{7}$$

$$\ddot{\theta}_2 = \left[-\left(\frac{2}{3} + \cos\theta_2\right)\sin\theta_2\dot{\theta}_2(2\dot{\theta}_1 + \dot{\theta}_2) - 2\left(\frac{5}{3} + \cos\theta_2\right)\sin\theta_1\dot{\theta}_1^2 + u_2\right]\left[\frac{16}{9} - \cos^2\theta_2\right] \tag{8}$$

The dynamic equations (7) and (8) are highly coupled. To provide decoupling and regulation, we consider u_1 and u_2 to be sliding mode fuzzy controllers.

PROBLEM STATEMENT

Let $X_p(t) = [\theta_1 \ \theta_2]$, $X_v(t) = [\dot{\theta}_1 \ \dot{\theta}_2]$ and $X_{pd} = [\theta_{1d} \ \theta_{2d}]$ then the control problem is as follows. For the given initial state $X_p(t_0), X_v(t_0)$ and desired position X_{pd} at velocity $X_{vd} = [0 \ 0]$, find a control input that stabilizes manipulator joints in the final positions, i.e., $\lim_{t\to\infty} X_p(t) = X_{pd}$ and $\lim_{t\to\infty} X_v(t) = X_{vd}$. Define the vectors E_p and E_v as follows:

$$E_p = [\theta_{1d} - \theta_1 \ \theta_{2d} - \theta_2] = [e_{1p}, e_{2p}] \quad \text{and} \quad E_v = [-\dot{\theta}_1 \ -\dot{\theta}_2]$$

The goal is to nullify the position error vector $E_p(t)$ and the velocity error vector $E_v(t)$.

MSMFC STRUCTURE

Considering the nonlinear coupling terms in (12) and (13) as disturbances, MSMFC rejects the disturbances based on the invariancy property of sliding mode and provides decoupling. Therefore, in the design of a MSMFC with four inputs ($S_1, \dot{S}_1, S_2, \dot{S}_2$) and two outputs ($u_1, u_2$), we consider the system of (12) and (13) as two decoupled subsystems. Consequently, the design of a MSMFC will lead to the design of two single sliding mode fuzzy controllers, SMFC1 and SMFC2. [12]-[13] provides an in-depth guideline for the design and analysis of SMFC1 and SMFC2. The subsequent simulation results will verify and validate the above assumption.

SIMULATION STUDIES

Two study cases are considered in the implementation of the MSMFC.

Case 1: The MSMFC stabilizes the manipulator joints in the final positions $x_f = [\pi/2 \ 0 \ \pi/4 \ 0]$ from the following initial conditions $x_i = [3\pi/2 \ \pi/4 \ \pi/2 \ \pi/2]$ respectively. No disturbance is added to the system in this case.

Case 2: In this case, the initial conditions are the same as case 1; however, we add the following disturbances to the system $\cos(300x_1), \cos(300x_2)$.

SIMULATION RESULT

In the simulation, we solved the differential equations of the manipulator using the MATLAB command "ode 23", which uses the second/third order Runge-Kutta method. Both the single sliding mode fuzzy controllers, SMFC1 and SMFC2, have two inputs, S_1 and S_2 respectively. These crisp control inputs were mapped into their corresponding fuzzy domain through k_1, k_2 and k'_1, k'_2 respectively. In the implementation of SMFC1 and SMFC2, the values of S_1 and S_2 were approximated by $\Delta s_1/\Delta t$ and $\Delta s_2/\Delta t$, with a sampling frequency of 1 KHz. In the simulation, the cardinality of the input term sets was selected to be the same and equaled to thirteen. The fuzzy control outputs u_1 and u_2 were mapped into a physical domain by the control gains k_3 and k'_3, respectively.

In the simulation of the two-link manipulator, we considered two cases. Figures 1-2 indicate the trajectories of the angle θ_1, θ_2 and $\dot{\theta}_1, \dot{\theta}_2$, for the first case. The MSMFC stabilized the manipulator joints in the final position by nullifying the position errors e_{1s}, e_{2s}. Figures 3-4 illustrate the trajectories of the errors e_{1s}, e_{2s}, and the magnitude of the control inputs u_1 and u_2. In Case 2, the MSMFC stabilized the manipulator joints in the final position in the presence of disturbance. Figures 5-6 indicate the trajectories of the angle θ_1, θ_2 and the velocities $\dot{\theta}_1, \dot{\theta}_2$. The nullification of the position errors e_{1s}, e_{2s} and the magnitude of control inputs u_1, u_2 for this case are shown in Figures 7-8. The simulation results of both cases indicate that MSMFC can stabilize the manipulator joints in the final position, even in the presence of disturbances.

CONCLUSION

In this paper, the concept of multidimensional sliding mode in variable structure systems was used to develop the MSMFC. Considering the nonlinear coupling terms as disturbances, the MSMFC rejects the disturbances and provides decoupling. Consequently, the design and analysis of MSMFC will lead to designs and analyses of a number of single sliding mode fuzzy controllers. Furthermore, the simulation results of the two-link manipulator indicate that MSMFC can stabilize the manipulator joints in the final position from the presented initial conditions. Further simulation results also indicate that the MSMFC can perform well in the presence of disturbances.

REFERENCES

[1] Utkin, V. I., September 1983, "Variable Structure System: Present and Future," Avtomatika: Tele Mekhanika, No. 9, pp. 5-25,.
[2] Utkin, V. I., Drakumov, S. V., Izosimov, D. E., 1984, "A Hierarchical Principle of the Control System Decomposition Based on Motion Separation," IFAC, 9th Triennial World Congress, Budapest, Hungary.
[3] Young K. K., Kokotovic, P. V., and Utkin, V. I., 1979, "A Singular Perturbation Analysis of High-Gain Feedback Systems," IEEE Trans., A C-22, No. 2, pp. 931-939.
[4] Hunt, L. R., Su, R., and Meyer, C., 1982, "Design for Multi-Input Nonlinear Systems," Proceedings of the Conference at Michigan Technological University.
[5] Beiczy, A. K., February 12, 1974, "Robot Arm Dynamics and Control," Jet Propulsion Lab. Tech. Memo, pp. 33-669.
[6] Sardis, G. N. and Stephanon, H. E., 1975, "Hierarchically Intelligent Control of a Bionic Arm," Proc. IEEE Conf. Decision and Control, pp. 99-104, Houston, Texas.
[7] Young, K. K., 1975, "Control and Trajectory Optimization of a Robot Arm," Univ. of Illinois, Urbana, CSL Rep. R-701.
[8] Kahn, M. E. and Roth, B., 1971 "The Near-Minimum-Time Control of Open-Loop Articulated Cinematic Chains," Trans. ASME, J. Syst., Measurement, Contr. 93, pp. 164-172.
[9] Paul, R. C., November 1972 "Modeling, Trajectory Calculation and Serving of a Computer Controlled Arm," Stanford Artificial Intelligence Lab. Memo, AJM-177,.
[10] Freund, E., Spring 1982, "Fast Nonlinear Control with Arbitrary Pole-Placement for Industrial Robotics Research, Vol. 1, No. 1.
[11] Young, K. K., February 1978, "Controller Design for a Manipulator Using Theory of Variable Structure Systems," IEEE Trans. on Sys., Man, and Cybernetics, Vol. SMC-8, No. 2.
[12] Allamehzadeh, H., December 1996, "Design and Stability Analysis of a Fuzzy Sliding Mode Controller," Ph.D. Dissertation, University of Oklahoma, Norman, OK.
[13] Allamehzadeh, H. and Cheung, J., 2000, "A Novel Approach to Design a Stable and Robust Fuzzy Controller for a Class of Nonlinear Systems," SPIE 7th International Conference on Mathematics and Control in Smart Structures, Vol. 3984.

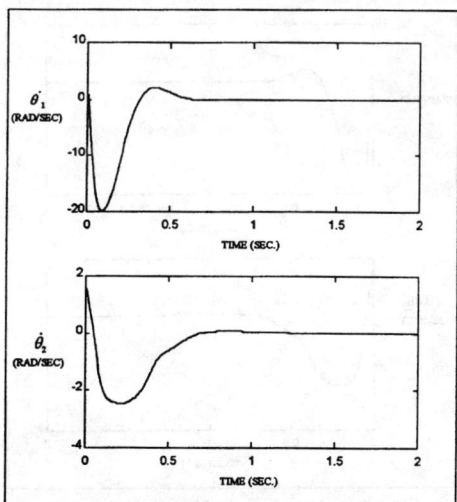

Figure 2. Trajectories of the angles $\dot{\theta}_1, \dot{\theta}_2$ of the manipulator in case 1

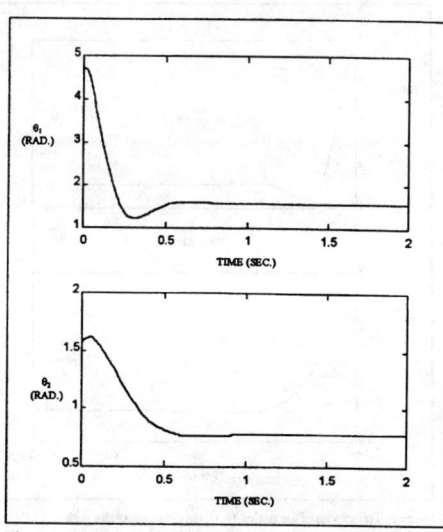

Figure 1. Trajectories of the angles θ_1, θ_2 of the manipulator in case 1.

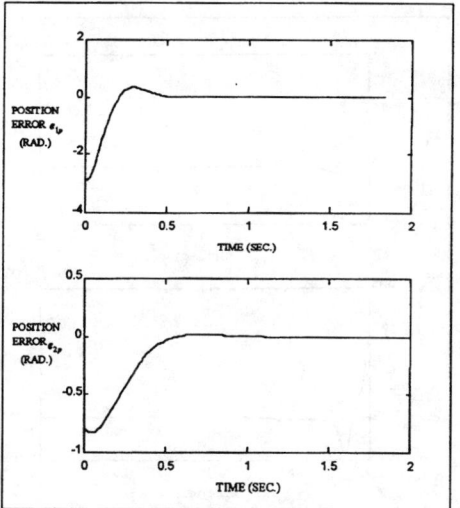

Figure 3. Trajectories of the position errors e_{1p}, e_{2p} of the manipulator for case 1.

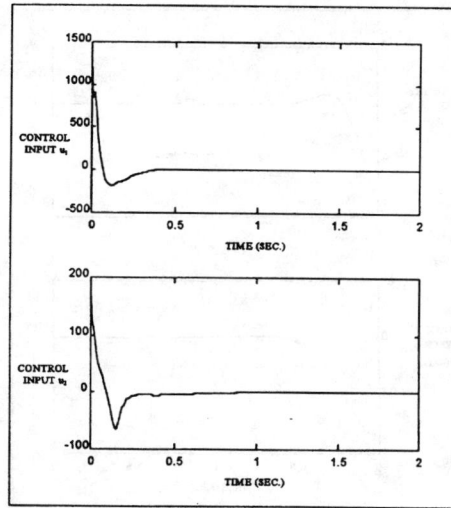

Figure 4. Control inputs u_1 and u_2 for stabilizing the manipulator in case 1.

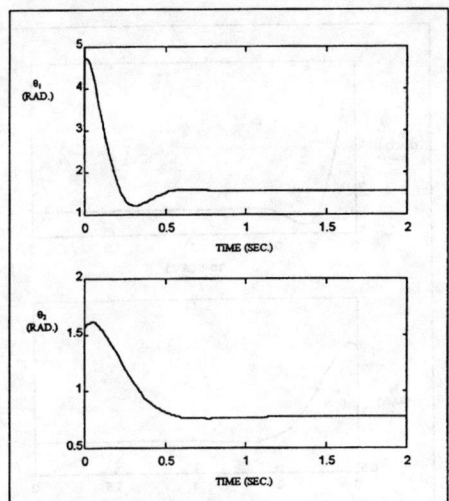

Figure 5. Trajectories of the angles θ_1, θ_2 of the manipulator with disturbances of case 2.

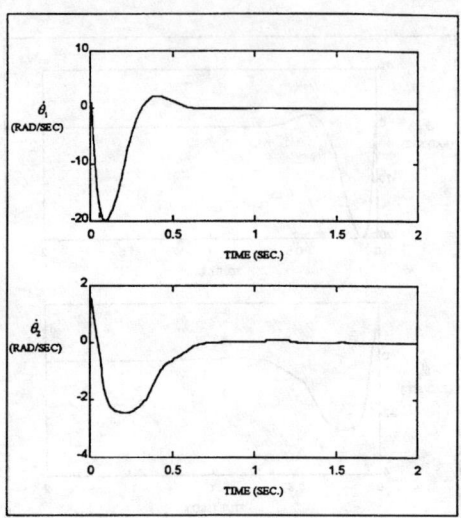

Figure 6. Trajectories of the angles $\dot{\theta}_1, \dot{\theta}_2$ of the manipulator with disturbances of case 2

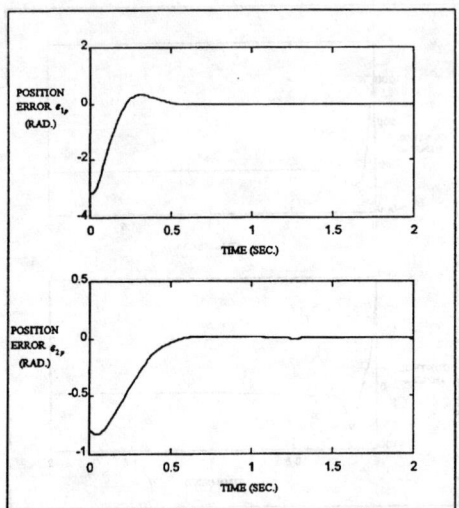

Figure 7. Trajectories of the position errors e_{1p}, e_{2p} of the manipulator with disturbance of case 2.

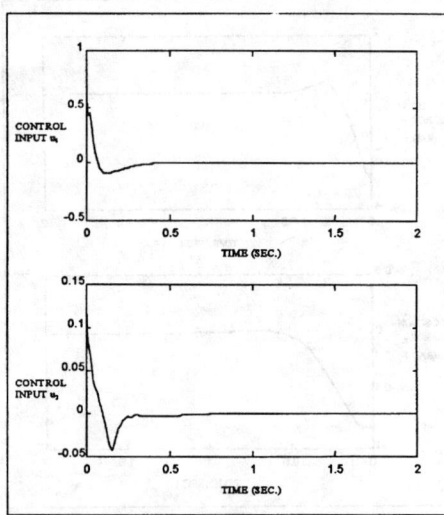

Figure 8. Control inputs u_1 and u_2 for stabilizing the manipulator in case 2.

INTELLIGENT CONTROL OF A ROBOT ARM TO AVOID AN OBSTACLE

RAJAB CHALLOO, LIGONG WANG
ROBERT A. MCLAUCHLAN, S. IQBAL OMAR, AND SHUHIU LI
Texas A&M University-Kingsville
MSC 192, Kingsville, Texas 78363

ABSTRACT

In this paper Neural Network and Fuzzy Logic technologies are used for a robot arm movement control in the presence of an obstacle. Neural networks applied for a PUMA robot arm and a three-joint robot arm are discussed and fuzzy logic is applied for the speed control in simulation program. Matlab Neural Network toolbox is used as the tool to train the neural network. A simulation program coded with Visual Basic is developed as both the conventional control and neuro-fuzzy control demonstration for a three-joint robot arm.

INTRODUCTION

Neural Network and Fuzzy Logic techniques applied to robot control are developing fast. Dean A. Pomerleau introduced a neural network perception for mobile robot guidance [7]. Hamid R. Berenji discussed in depth using fuzzy logic and neural networks for control system [1]. It is possible to learn to avoid obstacles with some rules because examples of obstacle avoidance behavior could be defined first [5]. However, it is difficult to define fuzzy logic rule to make the robot arm movement more practical. Neural networks evolved to avoid obstacles at various locations and reach random target locations are not easy to implement. The emphasis of this paper is to use the neuro-fuzzy technique to control and simulate a robot arm movement avoiding obstacle.

NEURO-FUZZY SYSTEM AND SOLUTION

Let's take a look at how to use a two layer neural network to simulate the control of a six-joint PUMA robot arm. It has been shown that two-layer feed-forward neural nets can approximate any function and that two-layer neural nets can be used to classify any non-convex part of the input space [2]. That conclusion is based on two widely known theorems:

Kolmogorov's theorem. A two-layer neural network with 2N+1 neurons in the first layer and with suitable transformation of the input signals, can exactly implement any function in a N-dimensional input space.

Universal approximation theorem. A two-layer feed-forward neural network with appropriately smooth, saturating first-layer activation function can make an arbitrarily accurate approximate of any function.

According to these two theorems, we'll illustrate how to create our neural network. We use direct and inverse kinematics to train and check the neutral network with Matlab Neural Network toolbox [2] for a six rotation joints PUMA robot arm. It is known that the function that is performed by neurons depends on the weight vector of the neurons. These weight vectors are usually determined in a so-called "training-phase" using a learning algorithm [3]. For feed-forward neural networks, the back-propagation rule probably is the most widely used learning algorithm. As we know, the neural network sometimes is based on the off-line training, it generally needs lots of calculation depending on the application. Because majority of practical applications of neural network are implemented by backpropagation, we use this training method in this paper.

We define the input vectors as P, the output file from a C coded robot direct kinematics program, to train the network while the target vector is the output file from a C coded robot inverse kinematics program. Because the network is two-layer and the range of output vector T is in [$-\pi,\pi$], the tan-sigmoid transfer function tansig (logsig) will be used in the input layer. The transfer function of output layer of the backpropagation network is purelin. For multiple-layer networks we append the number of the layer to the names of the matrices and vectors associated with that layer.

The initial weights and biases for backpropagation networks are very important in the training process, especially in pure backprogation, because it requires more accurate evaluation. In practical problems, it is not easy to find a good initial value to train the network. So pure backpropagation is rarely used to solve practical problem.

In our program, the pure backpropagation function is defined as:
[w1,b1,w2,b2,ep,tr] = trainbp(w1,b1,'tansig',w2,b2,'purelin',P,T,tp)
The initial weights and biases for the 1st and 2nd layer network are the Matlab default calculation value: [W1,b1,W2,b2]=initff(P,S1,'logsig',T,'purelin');

As we discussed above, although it requires 2N+1 neurons for the input layer in general case, less neurons are expected in this case. First twenty-five neurons are used to implement the network, then we compare it to a fifteen-neuron input layer network.

RESULTS
Case A – trainbp: The two layer achitecture used for our network has the first layer of sigmoid neurons (tan-sigmoid neurons) which receive inputs directly and then broadcast their output to a layer of linear neurons which compute the network output. First, we will use 25-neuron in the first layer. Training parameters can be defined and the network trained.
df =100; % Frequency of progress displays (in epochs).
me=10000; % Maximum number of epochs to train.
eg=0.1; % Sum-squared error goal. lr=0.01; % Learning rate. tp=[df me eg lr];
[w1,b1,w2,b2,ep,tr] = trainbp(w1,b1,'logsig',w2,b2,'purelin',P,T,tp)

In the beginning the Sum-Square Error is: TRAINBP: 0/10000 epochs, SSE = 967.527. After 100 epochs, it results in Fig. 1.
TRAINBP: 100/10000 epochs, SSE = 3.88038e+082

The network performance changes with training. After 400 epochs, the SSE tends to 'infinite'. This might be from unsuitable initial weights, bias value or learning rate though there are other reasons. It is difficult to find an appropriate initial value in practice at the first time. So we use an improved backpropagation method for better results.

Case B – trainbpx: The function trainbpx uses both momentum and an adaptive learning rate to speed learning. An adaptive learning rate requires some changes in the training procedure used by trainbp. First the initial network output and error are calculated. At each epoch new weights and biases are calculated using the current learning rate. New output and error are then calculated.[2]

As with momentum, if the new error exceeds the old error by more than a predefined ratio, the new weights, biases, output, and error are discarded. In addition, the learning rate is decreased. Otherwise the new weight and bias are kept. If the new error is less than the old error, the learning rate is increased. This procedure increases the learning rate, but only to the extent that the network can learn without large error increases. Thus a near optimal learning rate is obtained for the local terrain. When a larger learning rate could result in stable learning, the learning rate is increased. To apply this training procedure, the coefficients are almost the same as the trainbp method, only the training function changed:

[w1,b1,w2,b2]=trainbpx(w1,b1,'logsig',w2,b2,'purelin',P,T,tp);
The original SSE is as: TRAINBPX: 0/10000 epochs, lr = 0.01, SSE = 582.966.
Fig. 2 shows the SSE and learning rate after 10000 epochs. Although the final result does not meet the requirement (Fig. 2),

TRAINBPX: 10000/10000 epochs, lr = 6.96579e-005, SSE = 6.16304
it does improve a lot compared to Case A.

Case C – tansig: In this case we use tansig function to train the network while the number of neurons is still defined as: S1=25. The initial value and training function is as: [W1,b1,W2,b2]=initff(P,S1,'tansig',T,'purelin');

[w1,b1,w2,b2] = trainbpx(w1,b1,'tansig',w2,b2,'purelin',P,T,tp)
The initial SSE and after 10000 epochs SSE are:

TRAINBPX: 0/10000 epochs, lr = 0.01, SSE = 509.132
TRAINBPX: 10000/10000 epochs, lr = 2.7069e-006, SSE = 33.7964

Case D - Number of Neuron S1=15 will result in:
TRAINBPX: 0/10000 epochs, lr = 0.01, SSE = 1140.36
TRAINBPX: 10000/10000 epochs, lr = 0.077006, SSE = 0.181943

Only after 1500, it reaches the local minimum. This means the 15-neuron network learned much faster than the 25-neuron network. It also has the least SSE (Fig. 4). If we use neural network to control a two joint robot arm, the expected result is much better than the six-joint PUMA robot arm with twelve-input and six-output. With tansig function, improved Back-Propagation method and 2N+1=5 neurons, the output is shown in Fig. 6.

SIMULATION

When fuzzy logic is used as the control method, it usually reflects the designer's knowledge. That is why it usually is combined with other technology in robot control [6]. In the simulation program, the fuzzy logic system is implemented for the speed control. The physical characteristic of the robot arm (Fig. 7) simulated in this paper is based on a Remotec industrial robot arm which is available in the Intelligent Control Systems Lab of Texas A&M University-Kingsville [4]. The characteristic data (normalized) of this robot is listed in Fig. 8. Because the range of angle θ_3 is just -25^0, we omit it in this

simulation. Considering the screen size, the original points is set to (5790, 6120). The real data for the simulation is shown in Fig. 9.
The start point belongs to: 5790 + (3214 x cos(2π/5) < X < 10230
6120 − (3214 +1226)x sin(2π/5) < Y < 6120
The target point belongs to: 1350 < X < 5790 − (3214 x cos(2π/5))
6120 − (3214 +1226) x sin(2π/5) < Y < 6120
The arm length in the screen display is (Fig.10): $l_1 = 1734$ $l_2 = 2706$
The reachable range is the arc with radius of 3214 and 4440.
We define the start point is in the area of: $0° < θ_1 < 72°$ 3214 < r < 3914
The target point is in the area of : $108° < θ_2 < 180°$ 3214 < r < 3914
The obstacle range is: $72° < θ_3 < 108°$ 3514 < r < 4440

The reason for the obstacle area radius less than the reachable length is that we need enough space to let the robot arm to pass through the obstacle. Though there are many ways to let the robot to pass the obstacle while avoiding the obstacle, we choose one of the easiest ways: let arm 2 pass the point which is 200 longer in Y axis. Because the original point of the monitor plane is from upper-left, in fact, the arm passes the point which is 200 below the obstacle. For most points, there are two solutions for the arm position. We define four cases (Fig. 5): Case A and Case B for start point; Case C and Case D for target point.

The default solution for start point is Case A. In order to optimize the movement, while the start position is Case A, the target solution tends to be Case D. The start point, target point and obstacle are all variables created by computer randomly. The angles are calculated. While the default setting is traditional kinematics control, just after a little bit adjustment – using the output file from Matlab or other neural network training software as the input data ('Start' button in Visual Basic code), this program could be changed directly to neuro-fuzzy control simulation.

The fuzzy logic technique is mainly depending upon the designer. Meanwhile, we should notice that in practice speed, torque, acceleration and the obstacle weight all are the factors to be considered. In this program, we just simply give an idea how the fuzzy logic works in robot control. In other word, we only use fuzzy logic technique to control the speed. In order to simplify the case, we define four membership functions: 3, 5, 7, and 9. According to the angles, the rule is defined as:

Angle   [0, 30)   [30, 60) [60, 90) [90, 180]
Membership function   3   5   7   9

The reason we use odd number is that in robot dynamic control odd symmetric is usually the case. The shapes of logic membership function are triangular. Note the bigger the membership function is, the smoother the movement is. In order to make the movement clear, the step size is set to $1.5°$. The simulation without and with obstacles are shown as Fig. 12 and Fig. 13.

REFERENCES
Berenji, H. R., "Fuzzy logic and neural networks for control systems", IEEE Educational Activities Board, Piscataway, NJ, 1992
Demuth, H., Beale, M. "Neural Network Toolbox", The Math Works Inc., Natick, Mass.1994
Gupta, M.M. and Rao, D.H. "Neuro-Control Systems, Theory and Application", IEEE PRESS, vol.11, 1994, pp1341-1347

Kelly, Wallace Eugene, "Neuro-fuzzy Control of a Robotic Arm", Master's thesis, Texas A&M Unibersity-Kingsville, August 1994

Kung, S. and Hwang, J. "Neural network architectures for robot applications", IEEE Trans. Robotics and Automation 5, 5 (Oct. 1989), pp. 641-657.

Leitch, Donald; Probert, Penelope, "Genetic algorithms for the developments of fuzzy controllers for mobile robots", IEEE/Nagoya Univ. World Wisepersons Workshop, Nagoya-shi, Japan, 1994

Pomerleau, Dean, "Neural network perception for mobile robot guidance", Kluwer Academic Publishers, Boston, 1993

Robelo, L.C. and Avula, X.R.; "Intelligent Control of a Robotic Arm Using Hierarchical Neural Network Systems", International Joint Conference on Neural Networks; July 8-12, 1991

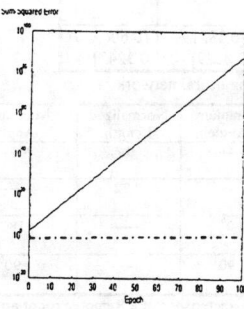

Figure 1 Sum-Squared network error for 100 epochs in Case A

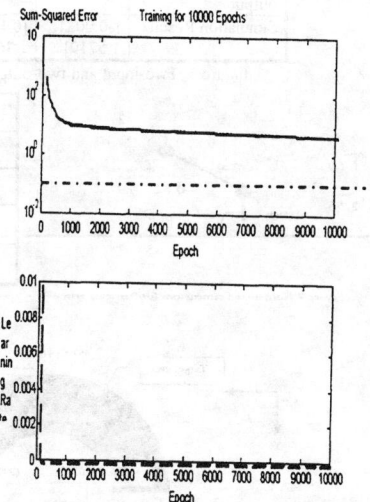

Figure 2 SSE and learning rate after training for 10000 epochs in Case B

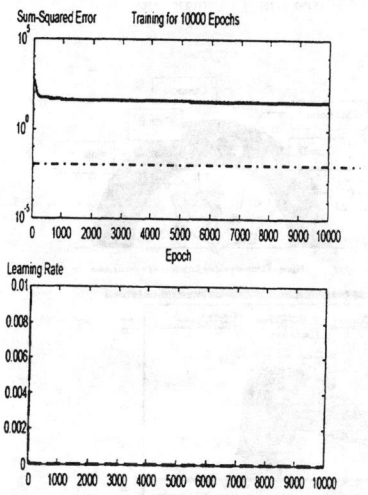

Figure 3 SSE and learning rate after training for 10000 epochs in Case C

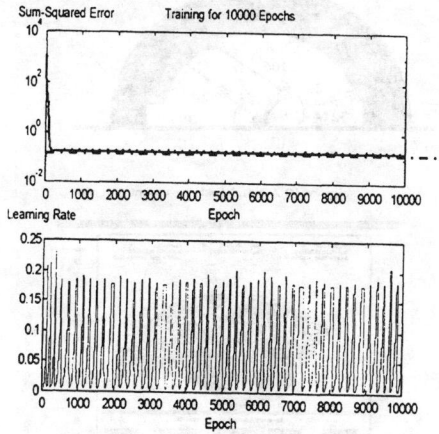

Figure 4 SSE and learning rate after training for 10000 epochs in Case D

CASE	A	B	C	D
Method	Pure Back-Propagation	Improved Back-Propagation	Improved Back-Propagation	Improved Back-propagation
Number in 1st Hidden Layer Neuron	25	25	25	15
Function	logsig	Logsig	tansig	logsig
SSE after 10000 Epochs	Infinete	6.16304	33.7964	0.181943
Conclusion	Fail	Good	bad	best

Figure 5 Result comparison

SIMULATION INPUT VECTOR	(1 ; 1)	(1.1 ; 1.1)	(1.2 ; 1.2)	(1.3 ; 1.3)	(1.4 ; 1.4)
Simulation expected output	(0.0000; 1.5708)	(0.1058; 1.3592)	(0.2278; 1.1152)	(0.3807; 0.8093)	(0.6435; 0.2838)
Simulation Result	[-0.0031; 1.5719]	[0.1074; 1.3628]	[0.2251; 1.1076]	[0.3821; 0.8229]	[0.6062; 0.3243]

Figure 6 Two-input and two-output simulation result using neural network

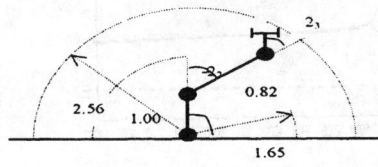

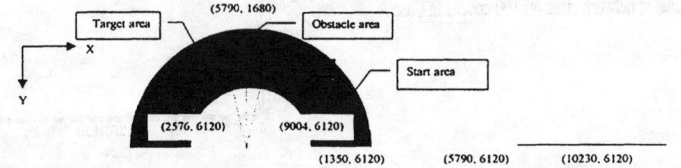

Physical Characteristic	Minimum Angle	Normalized Length	Maximum Angle
Link length, l_1	------	1	------
Link length, l_2	------	0.82	------
Angle 2_1	$0°$	------	$+180°$
Angle 2_2	$-90°$	------	$+90°$

Figure 7 Normalized dimensions for Remotec arm   Figure 8 The physical characteristics of a Remotec robot arm

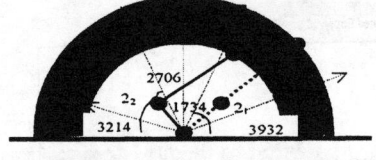

Figure 9 A Remotec robot arm movement diagram

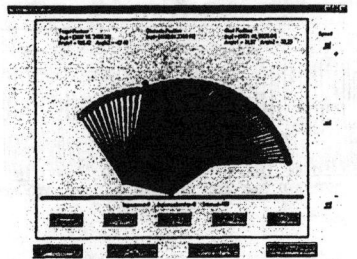

Figure 10 A simplified movement diagram

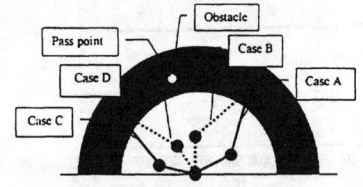

Figure 11 Default solution for simplified Remotec robot arm

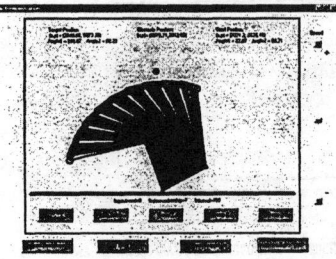

Figure 12 Simulation without obstacle   Figure 13 Simulation with obstacle

FUZZY BASED DRUG DOSAGE CONTROLLER

DIVYA K. KESAVAN
Sri Venkateswara College of Engg.
Dept. of Electronics and
Communication Engg.
Chennai, Tamil Nadu
India.

GANESH S. VAIDYANATHAN
Sri Venkateswara College of Engg.
Dept. of Electronics and
Communication Engg.
Chennai, Tamil Nadu
India.

**SANKARAVADIVOO
SUBRAMANIAN**
Sri Venkateswara College of Engg.
Dept. of Electronics and
Communication Engg.
Chennai, Tamil Nadu
India.

ABSTRACT
This paper illustrates a biomedical application of fuzzy logic. In medicine just as important as prescribing the right drug to a patient is the process of determining the correct dosage to be administered. The dosage level is dependent on many parameters of the human body. In order to consider all the parameters and administer the right amount of the drug, we propose a design for a fuzzy based *DRUG DOSAGE CONTROLLER*, which incorporates fuzzy logic control to automatically output the exact level of dosage. The idea of fuzzy control is to simulate a human expert, able to control the system by translation of his or her linguistic inference rules into a control function. We have defined fuzzy sets to represent linguistic values of the input and output variables pertaining to the controller and described their relations by fuzzy if-then rules. Thus the controller provides an automated system from which, the exact amount of dosage is automatically obtained.

INTRODUCTION
In this paper we extend the fuzzy logic approach to bio-medical engineering where in we discuss the design of the 'Fuzzy based Drug Dosage Controller'. It is a fuzzy controller, which automatically determines the exact amount of dosage to be administered to a patient. Since this process depends on many physical parameters of the patient, which is time varying, it suggests the use of rule-based controllers like fuzzy controllers. Medical research tells us that there are a large number of factors are to be taken into account while determining the right dosage level to be administered to a patient. Some of the factors are age, physical parameters like pulse rate, body temperature, history of allergies, sex, pregnancy and lactation. The controller takes in all these factors as input variables and automatically gives the exact level of dosage.
Theoretically this controller can take infinite number of input variables, but in our model controller we have used only two variables, for simplicity sake. The schematic representation of this model controller is as shown in Fig. 1.

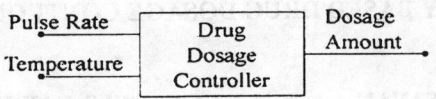

Fig. 1: Schematic diagram of a drug dosage controller

Our first step in designing such a fuzzy controller is to characterize the range of values for the input and output variables of the controller. The two input variables are:
Body temperature (BT)
Pulse rate (PR).
They are taken as the linguistic variables of the controller. Their medical specifications are as shown below:

Pulse Rate (PR): *(in bpm)*
Bradycardia (BC)   -- < 65
Symptomatic bradycardia -- < 50
Normal pulse rate (NPR)   -- 60 to 80
Perfect Normal pulse rate -- 70
Tachycardia (TC)   -- > 75
Symptomatic tachycardia -- > 100

Body Temperature (BT): *(in °F)*
Hypothermia (HOT)   -- < 98
Symptomatic Hypothermia -- < 95
Normal Temperature (NT)   -- 97 to 100
Perfect Normal Temperature -- 98.6
Hyperthermia (HPT)   -- > 98.6 °F
Symptomatic Hyperthermia -- > 103 °F

Fuzzification:
Fuzzification is the process of making a crisp quantity fuzzy. In order to do this we have to define the membership functions and determine the membership grades. It is convenient to represent the membership functions graphically as shown in Fig. 2 and Fig.3.
The output is a linguistic variable referred to as **'Drug Dosage'(DD***)*, which can be set to three values such as:
Lesser   = 20 units
Medium = 50 units
Greater = 70 units
The membership function of the output variable is as shown in Fig. 4.

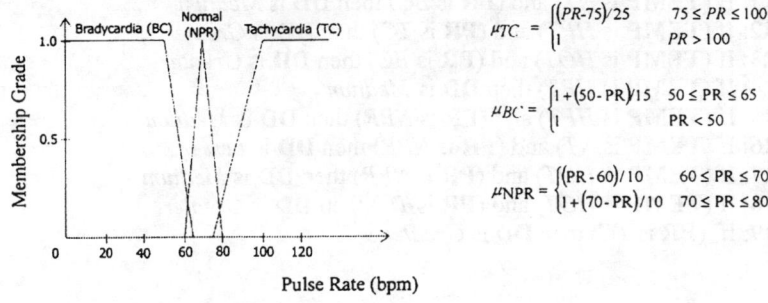

Fig. 2: Membership function of Pulse Rate (PR)

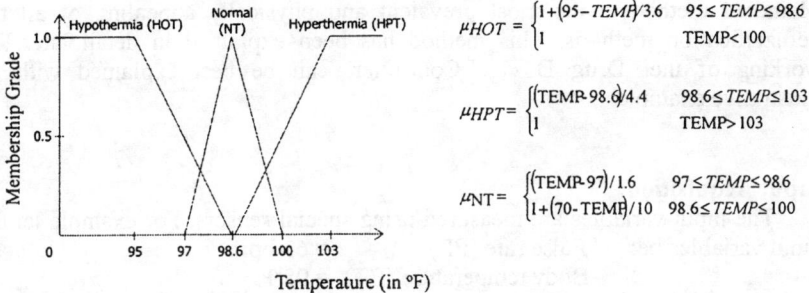

Fig. 3: Membership function of Temperature (TEMP)

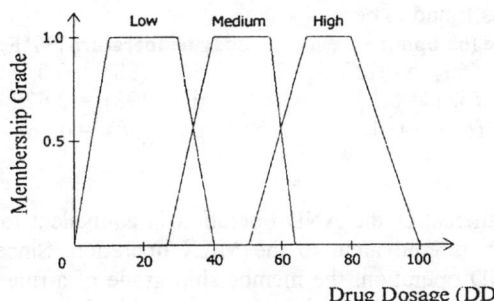

Fig. 4: Terms of output

Rule Base:

The Rule base is the heart of any fuzzy based system. Most often it consists of simple IF-THEN rules. It typically expresses an inference such that we know a fact, then we can infer, or derive, another fact called a conclusion. This form of knowledge representation expresses human empirical knowledge in our own language of communication. The model rule base for the *'Drug Dosage Controller'* is as shown below:

R1: If (**TEMP** is *NT*) and (**PR** is *BC*) then **DD** is *Medium*
R2: If (**TEMP** is *HPT*) and (**PR** is *TC*) then **DD** is *Greater*
R3: If (**TEMP** is *HOT*) and (**PR** is *BC*) then **DD** is *Greater*
R4: If (**TEMP** is *HPT*) then **DD** is *Medium*
R5: If (**TEMP** is *HPT*) and (**PR** is *NPR*) then **DD** is *Medium*
R6: If (**TEMP** is *NT*) and (**PR** is *NPR*) then **DD** is *Lesser*
R7: If (**TEMP** is *HOT*) and (**PR** is *NPR*) then **DD** is *Medium*
R8: If (**TEMP** is *HOT*) and (**PR** is *TC*) then **DD** is *Greater*
R9: If (**PR** is *TC*) then **DD** is *Greater*

Defuzzification:

Defuzzification is the conversion of a fuzzy quantity to a precise quantity. There are several methods for defuzzifying fuzzy output functions. The Mamdani method is the most prevalent and physically appealing of all the defuzzification methods. This method has been explained in detail later. The working of the 'Drug Dosage Controller' can be best explained with an illustrative example.

Input Acquisition:

The input variables are measured using special sensors. For example let the input variables be: Pulse rate (PR)   = 63 bpm
Body temperature (BT) = 98°F

Fuzzification:

The membership grade of each input in each of the terms is evaluated from the graphs and it is found to be:

Pulse rate (63 bpm):
μ_{BC} (63) = 0.133
μ_{NPR} (63) = 0.3
μ_{TC} (63) = 0.0

Body temperature (98 °F):
μ_{HOT} (98) = 0.0
μ_{NT} (98) = 0.625
μ_{HPT} (98) = 0.166

Rule Evaluation:

In fuzzy mathematics the AND operation is equivalent to MIN operation and OR operation is equivalent to the MAX operation. Since rules combine conditions by AND operation, the membership grade of a rule is the minimum of the condition memberships which were computed in the previous stage.

F1= min {0.625,0.133} = 0.133
F2= min {0.0,0.0}   = 0.0
F3= min {0.166,0.133} = 0.133
F4= min {0.0}   = 0.0
F5= min {0.0,0.3}   = 0.0
F6= min {0.625,0.3}   = 0.3
F7= min {0.166,0.3}   = 0.166
F8= min {0.166,0.0}   = 0.0
F9= min {0.133}   = 0.133

Defuzzification:

Mamdani method: The Mamdani method uses fuzzy terms for rule actions. Each output term must be weighted by the rule membership, so that the output terms associated with the rules that have higher membership grades may have a greater effect on the controller output. Since the rules are considered to be combined in the rule base by OR operations, the membership grade associated with each output term is obtained by identifying the rule which prescribes that output term with the maximum membership grade. The membership grade for each of the output terms is computed as shown in Table 1. These terms are called the 'weighted' sets.

Table 1: Computation of Membership Grades

Output	Rules (Ri)	Membership grades (Fi)
Low	6	0.3
Medium	1, 4, 5, 7	max(.133, 0.0, 0.0, .166) = 0.166
Greater	2, 3, 8, 9	max(0.0, 0.133, 0.0 0.133) = 0.133

The weighted fuzzy sets have to be combined such that the output may be viewed as the union of all the weighted terms. This output fuzzy set is shown in Fig. 5. Again the crisp output needs to be computed from this collection of fuzzy sets by computing the center of gravity of the output fuzzy set using the formula:

$$\frac{\int x * g(x)dx}{\int g(x)dx} \qquad \ldots(1)$$

In order to compute the center of gravity the output fuzzy set Fig. 5 is partitioned into 7 areas as given by:

$$g(x) = \begin{cases} 0.1x & \text{for} \quad 0.0 \leq x \leq 3.0 \\ 0.3 & \text{for} \quad 3.0 \leq x \leq 37.0 \\ -0.1x + 4.0 & \text{for} \quad 37.0 \leq x \leq 38.34 \\ 0.166 & \text{for} \quad 38.84 \leq x \leq 63.34 \\ -0.1x + 6.5 & \text{for} \quad 63.34 \leq x \leq 63.67 \\ 0.133 & \text{for} \quad 63.67 \leq x \leq 88.67 \\ -0.1x + 9.0 & \text{for} \quad 88.67 \leq x \leq 90.0 \end{cases}$$

The center of gravity is computed using the Eq. (1):

$$\frac{\int x * g(x)dx}{\int g(x)dx} = 44.05$$

Therefore the output of the controller is 44.05 units. The drug dosage is generally measured in milligrams. The range of the output dosage may vary for different diseases. Therefore if the output dosage lies within the specified range, the output is taken to be the calculated value itself otherwise the calculated value has to be translated into the specified range. For our model controller the drug dosage, as pertaining to the given input would be 44.05 milligrams. Thus on giving the input variables to the controller, the exact amount of drug dosage has been calculated.

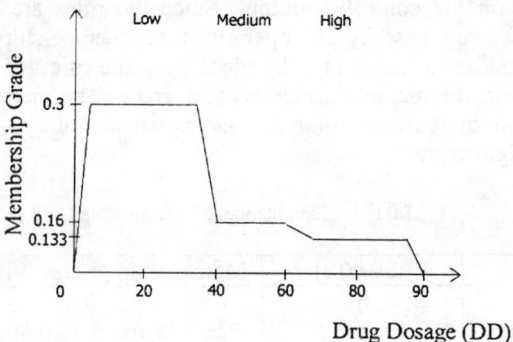

Fig. 5: Output Fuzzy set

CONCLUSIONS

The process of drug dosage determination currently proceeding under human supervised environments is susceptible to errors and approximations. There is a need for an automated system, which would effectively eliminate these problems. The convincing advantage of Fuzzy logic is that it has a natural language rule based nature, and thus is an attractive choice for this application. In this paper we have presented the 'Fuzzy Based Drug Dosage Controller', which aims to be a viable and reliable alternative to the existing techniques of dosage determination, eliminating the shortcomings and problem areas of the conventional method.

NOMENCLATURE

^{o}F   : Degree Fahrenheit
bpm : beats per minute
μ_{BC} : Membership function of pulse rate in Bradycardia range
μ_{NPR}: Membership function of pulse rate in Normal range
μ_{TC} : Membership function of pulse rate in Tachycardia range
μ_{HOT}: Membership function of body temperature in Hypothermia range
μ_{NT} : Membership function of body temperature in Normal range
μ_{HPT}: Membership function of body temperature in Hyperthermia range

REFERENCES

Timothy J. Ross, 1995, "*Fuzzy Logic with Engineering Applications*", McGraw Hill, Inc,pp. 469-504
Arrestam R., and J. Holmund, 1991, "*Fuzzy Logic Control Language (FLCL) for Embedded Controllers*", Research Report, Industrial Research Laboratory, University of Florida
www.fuzzytech.com
www.medscape.com

PARALLEL GENETIC ALGORITHMS FOR TUNING A FUZZY DATA MINING SYSTEM

QITAO LIU
Department of Computer Science
Mississippi State University

SUSAN M. BRIDGES
Department of Computer Science
Mississippi State University

IOANA BANICESCU
Department of Computer Science
Mississippi State University

ABSTRACT
In previous work, we have described methods that we have developed for tuning a fuzzy data mining system for intrusion detection using a hierarchical genetic algorithm. Unfortunately, the genetic algorithm approach is very slow due to the computational cost of the evaluation function. In this paper, we describe parallel implementations of the genetic algorithm that were run on both a multiprocessor Unix workstation and a high performance cluster running Linux. Very little speedup was achieved on the Unix workstation because of contention for the single file system; significant speedup was achieved with the cluster in which each node had its own file system. Experimental results of our implementations based on the master-slave parallel model and multiple population parallel model are compared, and the preliminary results indicate that the multiple population model may provide a higher quality solution for a shorter amount of time.

INTRODUCTION

Kuok, Fu, and Wong (1998) introduced the marriage of fuzzy logic and association rule mining to address the sharp boundary problem encountered when discretizing continuous attributes for association rule mining. We have further refined these techniques for intrusion detection applications (Luo and Bridges 2000). One of the difficulties encountered in this approach, however, is defining appropriate membership functions for the fuzzy sets that serve as values of the fuzzy attributes. We have subsequently found that genetic algorithms (GAs) are effective methods for tuning the fuzzy membership functions and for feature selection (Shi 2000; Bridges and Vaughn 2000). Unfortunately, the evaluation function for the genetic algorithm requires that data mining be performed for every member of the population at every iteration of the genetic algorithm. Although, in the intrusion detection domain, the tuning process is an off-line operation, the time required for the GA is very substantial. In order to improve the performance of the GA, we have investigated parallel implementations of the genetic algorithm for data mining on both a multiprocessor Unix workstation and a high performance cluster running Linux.

In the remainder of this paper, we first describe our fuzzy data mining process and the sequential genetic algorithm that we have developed for tuning the membership functions and for feature selection. We then present our parallel implementations of the algorithm and preliminary experimental results that assess the effectiveness of the parallel implementations. Conclusions and plans for future work are given in the last section.

GENETIC ALGORITHMS FOR FUZZY DATA MINING

We have developed techniques for detecting network intrusions that integrate fuzzy logic and data mining (Luo and Bridges 2000). The intrusion detection system uses fuzzy association rules to represent normal patterns of quantitative attributes recorded in network audit data. When monitoring a new data set for intrusions, one compares the similarity of the rules mined from the new set with the rules mined from "normal" data. If the similarity falls below a specified threshold, the data is flagged as suspicious. Effective application of this method requires that informative attributes of the audit data be used and that the membership functions of the fuzzy sets be appropriately defined. We have subsequently investigated the feasibility of using hierarchical GAs to select optimized feature subsets and to tune the membership functions of the fuzzy sets that serve as attribute values (Shi 2000, Bridges and Vaughn 2000). Figure 1 illustrates the hierarchical structure of chromosomes.

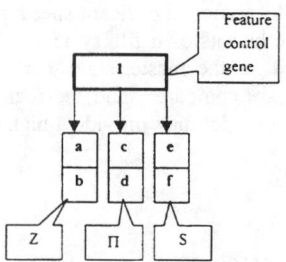

Figure 1. Representation of a single fuzzy attribute; the upper box indicates if "feature" is selected where 1 = yes and 0 = no. The boxes labeled a, b, c, d, e, f are parameters of the Z, Π, and S membership functions.

The fuzzy data mining process uses fuzzy sets as possible values for attribute values. For a particular attribute, we need to determine if this attribute should be used for data mining, and if so, the membership functions for the fuzzy sets to be used as values for the fuzzy attributes. A hierarchical genetic algorithm is used. At the first level, binary control genes are used to represent selection/non-selection of the attribute. At the next level, real values are used for parametric genes that represent the parameters of the Z, S and Π membership functions. In this application, the fitness function was designed to measure the performance of a classification system based on data mining that uses the selected features and the membership functions (Shi 2000). This data mining step makes evaluation of the fitness function a very time-consuming process and thus a prime candidate for parallel implementation.

PARALLEL GENETIC ALGORITHMS

Many research groups have investigated parallel genetic algorithms (PGAs). (See Cantú-Paz (1998) for a review.) Models for parallel GAs are often classified as single-population models or multiple-population models.

Single-population GAs are generally implemented using a master-slave model. In the master-slave model, a single population resides in the master processor and the master processor does the selection, crossover, and mutation; only evaluation of the fitness function is distributed among slave processors. In a multiple-population model, the population is divided into several subpopulations that are assigned to different processors. Each processor applies a traditional sequential GA (SGA) independently to its own subpopulation. Occasionally, individuals are exchanged between subpopulations in a process called migration. We have investigated both single population and multiple population genetic algorithms.

The global single-population master-slave model that we use is illustrated in Figure 2. A portion of the population is distributed to each slave processor for evaluation of the fitness value of individuals. The master processor also retains a portion of the population so that it can carry out evaluation in parallel with the slave processors. Genetic operations other than evaluation are performed only by the master processor. The master processor assigns a fraction of the population to each slave processor for each generation.

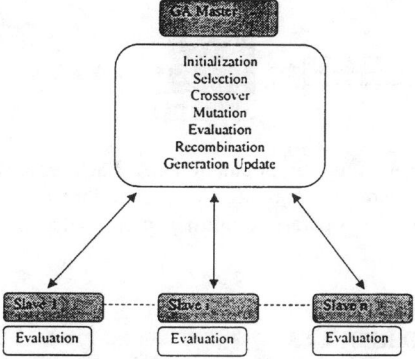

Figure 2. Global single-population master-slave PGA. The master stores the population, executes GA operations, and sends subsets of the population to the slaves. Each slave only evaluates the fitness of the individuals in its subpopulation, and sends the fitness value back to master.

Multiple-population GAs are also widely used parallel methods, but they are more complex than single population methods. A key characteristic of multiple-population PGAs is the migration of individuals among subpopulations. Each subpopulation is managed by an independent SGA except that the processors periodically exchange individuals. Some important parameters in multiple-population GAs are 1) which other processors a processor exchanges individuals with, 2) how often processors exchange individuals (the rate for the migration), 3) how many individuals processors exchange with each other (the number of individuals to migrate), and 4) what method is used to exchange (the topology for the exchange) (Cantú-Paz 1998). The most commonly used communication topology is a ring (see Figure 3). Individuals are exchanged for two reasons: to increase the fitness of the other population and to maintain diversity in the subpopulations. Many different approaches for migration have been developed. A schematic of the multiple-population PGA we have used is given in Figure 3. The subpopulations are virtually positioned on an oriented ring. Every time a new generation is computed, at the migration generation (every two generations) a copy of the best individual, i.e., the one with the greatest fitness value in the current subpopulation, is sent to the next subpopulation on the oriented ring. Each subpopulation thus receives a new individual that replaces the worst individual with the lowest fitness value.

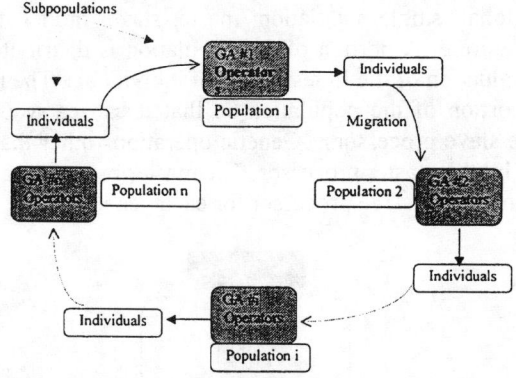

Figure 3. A schematic of a multiple-population GA. Each processor executes a SGA for its subpopulation, and there is communication (migration) between the subpopulations. Operators include all GA operators, initialization, selection, crossover, mutation, evaluation.

EXPERIMENTS AND RESULTS

A master-slave model of the GA was implemented and run on both a multi-processor Sun workstation and a Linux cluster. The workstation was a Sun E450 with 4 processors (300 MHZ UltraSparc II CPU, 1024 MB Ram, one 9GB hard drive, SunOS 5.7). The Linux cluster consists of 14 machines (Dual PIII 450 or 550 processors, 512 MB SDRAM, Dual 6 GB IDE HD, Redhat Linux 6.2) which are connected by full duplex 100Mb Ethernet. A multiple population model was also implemented and run on the cluster.

When the master slave implementation was run on the workstation, little improvement in performance occurred with the use of additional processors. The fuzzy data mining algorithm used for the evaluation of the fitness function reads and writes files for every evaluation. Because the Unix system has a single file system, only one processor can access the file system at any one time. The result is that the parallel implementation runs in approximately the same time as the sequential version due to contention for the file system. In order to solve this problem, we moved to the Linux cluster. We used an MPI-Shell program written by Bruce Wooley (2001) to distribute files to every machine's file system at the beginning of a run and to clean the file systems at the end of the run. At the beginning of a run, all data files are distributed to every machine. Since every machine has its own file system, the evaluation function of individuals on that machine can read and write files without affecting other evaluations. This allows all machines to concurrently evaluate a subset of the population. Tables 1 and 2 give experimental results for the intrusion detection system using the master-slave model for the GA. It is evident that execution time increases as the population size increases. When the number of processors is small, the speedup and efficiency do not change a great deal as the population size increases. As the population size increases, the speedup increases but does not achieve the ideal. The efficiency decreases because of increased communication overhead as the number of processors increases. The single-population model yields the same solution as the sequential version of the GA.

Table 1   Execution time for master-slave parallel implementation of fuzzy data mining.

Pop. Size	Execution time (mins) (30 generations) Processors					
	1	2	3	4	5	6
35	18.34	9.89	7.11	5.62	4.51	4.06
70	47.27	25.37	18.05	14.47	11.79	10.73
140	87.31	44.54	31.34	24.67	20.27	17.83

Table 2   Speedup and efficiency for master parallel implementation.

Pop. Size	Speedup (Processors)						Efficiency (Processors)					
	1	2	3	4	5	6	1	2	3	4	5	6
35	1	1.9	2.6	3.3	4.1	4.5	1	.93	.86	.82	.82	.76
70	1	1.9	2.6	3.3	4.0	4.4	1	.93	.87	.82	.80	.74
140	1	2.0	2.8	3.6	4.3	4.9	1	.98	.93	.89	.86	.82

A multiple-population model for the fuzzy data mining GA was also implemented and run on the Linux cluster. The subpopulations are on a virtual oriented ring and the best individual from each subpopulation migrates to a neighbor where it replaces the worst individual. Migration is done every two generations. Unlike the master-slave model, the multiple-population explores a different search space than that explored by the sequential model and will find a different solution. Preliminary results indicate that the multiple population model may provide a higher quality solution in a shorter amount of time. Further investigation is required to confirm these initial findings.

CONCLUSIONS AND FUTURE WORK

The genetic algorithm that we have developed for tuning a fuzzy data mining application has a very time-consuming evaluation function since the evaluation of each individual involves a data mining process. In addition, the data mining algorithm must read and write data files every time it is executed. The costly evaluation function makes this genetic algorithm a prime candidate for parallel implementation since the cost of evaluation should dominate the communication costs. We have implemented both single-population master-slave and multiple-population versions of the GA. When the parallel GAs are run on a Sun workstation with a single file system, little speedup is obtained due to contention for the file system. This problem was not encountered with the Linux cluster because each node has its own file system. Since all of the nodes needed exactly the same files for all iterations, the files could be distributed one time only. The near linear speedup achieved with the single-population master-slave GA is comparable to results reported by other researchers. The advantage of the cluster system is that different processes can concurrently access to different file systems and this makes the concurrent evaluation of a subset of the population possible. Our initial results with a multiple population GA indicate that this model may have advantages over the single population model. However, this model is more complex than the single population model and further investigation is required to confirm the initial results and to determine appropriate model parameters such as migration rate, subpopulation size, etc.

ACKNOWLEDGEMENT

This work was partially supported by NSF Grant #9818489.

REFERENCES

Bridges, Susan M., and Rayford B. Vaughn. 2000. Fuzzy Data Mining And Genetic Algorithms Applied to Intrusion Detection. *Proceedings of the National Information Systems Security Conference (NISSC), October 16-19, 2000, Baltimore, MD.*

Cantú-Paz, Erick. 1998. A survey of parallel genetic algorithms.*Calculateurs Paralleles.* Vol. 10, No. 2. Paris: Hermes. http://www-illigal.ge.uiuc.edu/publications.php3 (Accessed 18 May 2000).

Cantú-Paz, Erick. 1999a. Topologies, migration rates, and multi-population parallel genetic algorithms. *Proceedings of the Genetic and Evolutionary Computation Conference 1999 (GECCO-99).* Edited by Banzhaf, W. et al. San Francisco, CA: Morgan Kaufmann Publishers.

Kuok, C., A. Fu, and M. Wong. 1998. Mining fuzzy association rules in databases, *SIGMOD Record* 27(1): 41-6.

Luo, Jianxiong, and Susan M. Bridges. 2000. Mining fuzzy association rules and fuzzy frequency episodes for intrusion detection. *International Journal of Intelligent Systems* 15 (8): 687-703.

Shi, Fajun. 2000. Genetic algorithms for feature selection in an intrusion detection application. M.S. Thesis, Mississippi State University.

Wooley, Bruce. 2001. MPI-shell and other software. http://www.cs.msstate.edu /~bwooley/.

THE CONTAINMENT PROBLEM FOR FUZZY CLASS ALGEBRA

DANIEL J. BUEHRER
TSE-WIN LO, CHIH-MING HSIEH, MAXWELL HOU
dan,ltw88,hcm88@cs.ccu.edu.tw   hkl84@gais.cs.ccu.edu.tw
Institute of Computer Science and Information Engineering
National Chung Cheng University
Chiayi 621 Taiwan

ABSTRACT
The containment problem involves checking whether the set of objects satisfying one logical query is necessarily contained in the set of objects satisfying another logical query. In this paper we are concerned with fuzzy class algebra queries. These queries involve fuzzy class union, intersection, and difference operators, the binary attribute or relation dot operator, and nested selection operators that contain Boolean expressions involving attribute values and relation counts. The ability to normalize these class algebra queries to a Sorted Disjunctive Normal Form is crucial to class algebra's ability to organize class definitions and queries into an IS-A classification hierarchy. In this IS-A hierarchy, it is clear which intersections are included underneath a union, so rules like $Pr(A \cup B) = Pr(A) + Pr(B) - Pr(A \cap B)$ are readily obtained from counts of the objects in the union/intersection classes, where the subtracted term is due to the fact that each object identifier in an intersection should only be counted only once. This permits the creation of a theory of probability which satisfies the laws of Boolean algebra in the presence of axioms. Previous attempts at such a theory have been stymied by the undecidability of the containment problem for 1st order logic. Moreover, previous attempts at distributing object-oriented databases were stymied by the lack of clear definitions for class operations such as inheritance and self-reflection. For class algebra, self-reflection and inheritance can be achieved without sacrificing decidability or computability when count restrictions involve only undotted relations.

INTRODUCTION TO CLASS ALGEBRA

Class algebra has been used as the query language of a distributed object-oriented database system called Cadabia (Buehrer, 1994, 1995, 1996, 1999). The class algebra query language has no side-effects, but the associated class algebra update language can change the values of attributes, relations, and methods. These commands are embedded into a Java program, and database queries and updates can be mixed with standard Java code. Objects, classes, and methods are loaded from any users after logging in to their databases. All users' binary relations and attributes of the same name are unioned together.

In this paper, we briefly describe the programming environment for the Cadabia API. Then we discuss the decidability of the containment problem for some kinds of axioms and queries. Although the Cadabia database does not implement the fuzzy extensions to class algebra, in this paper we show how to solve the containment problem for the fuzzy version, which generalizes the rough-sets version of class algebra that is used in Cadabia.

Table 1: Advantages of Class Algebra Theory

Classical Theory	Advantages of Class Algebra
Finite Sets	O(nm) worst-case time for query on a database with n binary relations and at most m objects in each relation's value. Normalization can decide containment of one class expression by another for arbitrary databases in exponential time for simple cnt constraints.
Infinite Sets	Query and complement are both enumerable in decreasing order of fuzzy values; anytime algorithm is epsilon decidable (Buehrer, 1994).
Relational Algebra	Has IS-A hierarchy, inheritance of constraints on attributes, relations, and methods
Description Logics	Since complements are computable, can, for instance, express a query to compute the leaves of a tree. This is more than NP-expressive description logics can express.
Probability	Satisfies Boolean axioms. Does consistent logical inference for probabilities. Can get best decision tree based on information theory.
Fuzzy Sets and Fuzzy Logic	Integrates fuzzy set theory and fuzzy logic. Integrates fuzzy and probabilistic reasoning.

CADABIA PROGRAMMING API

The Cadabia database has a standard user interface called Abia Cadabia. This interface is basically a binary relation editor. The user must first choose a relationship's domain and range classes. Then all selected objects in the domain are connected to all selected objects in the range. If the objects are selected using the mouse, this is equivalent to specifying them by using their object identifiers (oid's). Since these object identifiers are read-only, the mouse selections result in an explicit list of objects. Otherwise, an implicit selection may be made by specifying the ranges of certain attribute values, or by specifying the counts of the number of objects in given binary relations. The implicit sets are "queries", and the queries are used to define relationships which are unioned into an implicit relation. The value of the implicit relation may change as the attributes and relations of the objects change.

Each user or group has a "home" which is similar to the home in many multi-user operating systems. He may then traverse binary relations, similar to going into typed subdirectories. The result of traversing a relation is a selection from the range class. Each class has superclasses, subclasses, attribute definitions, relation definitions, method definitions, an intent, and an extent. The intent is the membership function for the class. It is in the form of an SDNF (Sorted Disjunctive Normal Form), which is a union of intersections of predicates or negated predicates. The intent is a normalized class algebra query which evaluates to "true", "false", or "unknown" for each object in the database. The subset of objects which return "true" gives a lower bound, and the subset of objects which return "true" or "unknown" gives an upper bound for the extent of that class. Thus, each class is associated with a rough set of objects that are members of the class's extent. A query evaluates to "unknown" if the object does not have some of the relations or attributes mentioned in the query.

Although it has not yet been implemented, a fuzzy version of Cadabia based on vague logic (Gau and Buehrer, 1993) would be quite straightforward. The addEdges command and the attribute assignment command would have an extra argument giving the evidence in favor of the given relation edges or attribute value. The rough sets would then be replaced by fuzzy rough sets, where each attribute or relation value has a fuzzy membership [t, 1-f], where t and f are "true" and "false" belief measures in the range [0,1]. For most applications there should not be much overhead using these fuzzy t/f distribution functions rather than Cadabia's rough sets, which still have to record the object identifiers for all objects for which the selection is true or false. The fuzzy version would also have to remember the true or false evidence values, if any.

COMPARISONS TO OTHER MODELS OF REALITY

Each mathematical model of reality has certain advantages and disadvantages. So far, set theory has proven to be very valuable as a description mechanism for almost all scientific models. However, set theory is based on first-order logic, for which some queries which are only semi-decidable. Other simpler logics such as propositional logic are decidable, but they are not powerful enough to describe complex systems. The queries of class algebra provide a nice compromise, with powerful concepts of object-oriented programming, probability, and fuzzy theory combined with an efficient logical inference mechanism.

Other logics like first-order logic are usually not typed, and these logics generally cannot calculate the superclasses or subclasses of an arbitrary set of objects. In class algebra, one can easily find superclasses, subclasses, superrelations, subrelations, complement classes, and complement relations, for either implicit or explicit classes and relations. This makes it possible to quickly locate examples, counterexamples, analogies, isomorphisms, etc. Needless to say, such a logical reasoning system will be of great value to all fields of research, including artificial intelligence. It will also be very important for more practical applications like e-commerce, where it is necessary to agree upon common models of reality.

CLASS ALGEBRA/CALCULUS

Like relational algebra with its corresponding relational calculus, class algebra also has a corresponding class calculus. Like relational algebra, class algebra contains explicit control information about the order in which fuzzy-rough set union, intersection, difference, and join operations are to be performed. This ordering information is not really necessary since the class operations have no side effects, so any order of evaluating arguments will return the same value, just as for pure lambda calculus expressions. So the algebra/ calculus dichotomy is really just a functional/relational difference, where Prolog-like relations are simply predicates in first-order logic. The semantics of fuzzy class algebra can thus either be described in terms of typed lambda calculus expressions or in terms of first-order logic. In this paper we take the viewpoint that class calculus is a decidable subset of first-order logic whose fuzzy model can be described in terms of class algebra fuzzy union/intersection/ difference/dot/selection operators.

First-order logic is restricted in two ways. First, class algebra queries involve dotted relations which all implicitly start from "home". Each of these dotted expressions can be thought of as a unique constant, since there exists some database for which each dotted expression represents a different set of objects. Informally, this restriction may be thought of as restricting first-order logic to the NP-complete labeling problems, where no function symbols are permitted.

The second restriction allows us to get rid of the intractability of k-cliques within the equivalence graphs. Class algebra queries have no complex interdependencies between variables. That is, even though the binary relations can describe an arbitrary graph, the class algebra queries are not powerful enough to ask NP-complete questions, such as finding the k-cliques in the graph. The class algebra queries can simply follow specified paths of binary relations, filtering out some of the objects during the traversal of these relations. A class algebra query can find all of the nodes with at least k sons, but finding a k-clique would require the use of a for-loop, which is not available in class algebra queries. Class algebra statements, on the other hand, do have for loops, plus all the other capabilities of Java statements. Class algebra statements in themselves are Turing complete, but we will not concern ourselves with this question in this paper. In this paper we are mainly concerned with showing that class algebra queries can be put into a sorted disjunctive normal form that allows us to check for containment or equality of the normal forms even when no database is specified. The normal form containments, in turn, imply the fuzzy subset containments between the extents of the classes.

CLASS ALGEBRA DEFINITIONS

A class algebra query is either "home", or a range of primitive values, or is defined recursively in terms of one of the following six operators:

Class union: R @+ S Class pseudo difference: R @- S
Class intersection: R@*S Dot operator: R . <identifier>
Class true difference: R @~ S Selection operator: R { ψ }

where R and S are class algebra queries and ψ is a Boolean expression. The dot operator and the selection operator should be considered to be functionals rather than functions, since they must call "eval" to evaluate their quoted arguments for each input object which is being tested for membership. These operators use the environment of class expression R to eval the <identifier> or condition ψ.

The selection condition ψ involves Boolean expressions containing the following predicates:

Syntax	Meaning
R in S	cnt(R~S)=0
R hasAll S	cnt(S~R)=0
R equals S	R in S && S in R
R hasSome S	cnt(R $\cap$ S) ≠ 0
<attr_expr> in < range>	cnt(attr~range)=0

where R and S are class expressions and <attr_expr> is an interval-valued expression. The <attr_expr> may also include the functions "cnt(R)" or "cnt(R,ψ)", which return sums of evidence for/against R. For a given object in the database, each predicate P returns an interval [t,1-f], where t is the fuzzy evidence in favor of P, and f is the fuzzy evidence against P. For a fuzzy predicate P, in a class algebra query such as r.s{cnt(u.v, P)>3}, the cnt function returns the interval [tc,n-fc] where tc is the sum of the fuzzy values of u.v.P (where each oid of u.v is included once), and fc is the sum of the fuzzy values u.v.~P. The cnt values are assumed to be uniformly distributed between the lower and upper bound. Each oid in r.s is thus assigned the fuzzy interval [k, k], where k is the fraction of the interval for cnt that satisfies the ">" predicate. In this example, k= max(min((n-fc-3)/(n-fc-tc),1),0).

The fuzzy versions of the Boolean operators could be defined using any norms and conorms, but we will use max and min in this paper:

p||q = $[t_{p||q}, 1-f_{p||q}]$ = $[\max(t_p,t_q),\max(1-f_p, 1-f_q)]$
p&&q = $[t_{p\&\&q}, 1-f_{p\&\&q}]$ = $[\min(t_p,t_q),\min(1-f_p, 1-f_q)]$

-p = [f_p, 1-t_p]
~p = [1-t_p, f_p]

Let R' represent the classical set which corresponds to the elements of a fuzzy set R which do not have the interval [0,1] (i.e. total ignorance) or [1,0] (i.e. totally contradictory evidence). The fuzzy class algebra operators are defined as follows, where x is an oid:

Union: $(R @+ S) = \{ x \% [\max(t_x), \max(1-f_x)] \mid x \text{ in } R' \cup S'\}$
Intersection: $(R @* S) = \{ x \% [\min(t_x), \min(1-f_x)] \mid x \text{ in } R' \cap S'\}$
pseudo-complement: $-R = \{ x\% [f_x, 1-t_x] \mid x \text{ in } R'\}$
True-complement: $\sim R = \{ x\% [1-f_x, t_x] \mid x \text{ in } R'\}$
Dot operator: $R.S = \{ v \% [\max_{u \text{ in } R} (\min(t_u, t_{<u,v> \text{ in } S})), 1- \min_{u \text{ in } R} (\max(f_u, f_{<u,v> \text{ in } S}))]\}$

The means of handling complements is the main trick in getting a 1-1 correspondence between the union/intersection/difference operators and the Boolean and/or/difference operators. The true complement operator "~" satisfies laws of Boolean algebra such as x=x~(y~x), x||~x=true, x~x=false, or x= ~ ~ x. This complement corresponds to the use of the "closed-world assumption", where ~x has belief evidence 1-e if and only if x has evidence e. Usually, the semantics of relational databases are described by adding in axioms which force the closed-world assumption to be satisfied. Such rules must also include the unique-name rule, which says that two identifiers of the same name are always equal, while two identifiers with different names are always unequal. The problem is that such rules are messy, and it is difficult to prove that there is only one model for the axioms, namely, the current state of the database.

FUZZY INTERPRETATIONS

A class algebra database contains only binary relations. The first argument of the binary relations must always be an object identifier (i.e. an oid). Each object identifier is assumed to uniquely identify an object in the current state (i.e. it satisffies the unique name assumption). If the second argument is a primitive value, the relation represents an attribute, and the second argument is its value. Otherwise the second argument is an oid, and the predicate represents one edge of the binary relationship which is indicated by the predicate's name. These binary relationships may be thought of as the object/attribute/value triples of artificial intelligence knowledge representations. Fuzzy class algebra adds a fuzzy interval to each such triple, as a fourth element. The fuzzy interval is contained within in the closed interval [0,1].

Any predicate p with fuzzy evidence "t" also has a pseudo-complement -p with fuzzy evidence "f". For a class algebra expression φ, we use the fuzzy interval $[t_\varphi, 1-f_\varphi]$ to record the evidence "t_φ" provably supporting φ, and the evidence "f_φ" provably supporting -φ. These two evidences are assumed to be independent, and proofs of either one do not affect the truth of the other, just as in intuitionistic logic (Bonner, 1997). Thus, -φ is independent of the true complement ~φ, whose evidence is, by definition, given by the interval $[1-t_\varphi, f_\varphi]$. That is, the fuzzy interval for ~φ is obtained from the interval for φ by subtracting each bound from 1. The interval $[f_\varphi, 1-t_\varphi]$ for -φ is obtained from ~φ's interval by flipping the two bounds.

Just using the bounds themselves, there would be no way to satisfy the Boolean axioms such as φ||~φ=true. However, it can be seen that max(φ,~φ)≥0.5 and min(φ,~φ)≤0.5. Also, there are no objects for which both φ and ~φ have at least evidence 0.5.

The above method of computing fuzzy values can be used as a fuzzy interpretation of a database's fuzzy facts and relations. The fuzzy interpretation can be changed into a traditional interpretation I^δ by using a δ-cut as follows:

$$I^\delta(R\{\varphi\}) = \{x \mid x \in R' \ \& \ t_{x(\varphi)} \geq \delta\}$$

Since $t_\varphi = 1 - t_{\sim\varphi}$, either one or the other is greater than δ, but not both. Thus, it is obvious that the following are true:

$$I^{0.5}(R\{\varphi \parallel \sim\varphi\}) = I^\delta(R)$$
$$I^{0.5}(R\{\varphi \ \&\& \sim\varphi\}) = \varnothing$$

The normalization process for class algebra expressions will simplify $\varphi \parallel \sim\varphi$ to "true", and $\varphi \ \&\& \sim\varphi$ to "false", which is different than directly using the values computed using the fuzzy norms and conorms. The normalized expressions will satisfy the Boolean axioms.

SUMMARY AND CONCLUSIONS

Fuzzy class algebra is used to define normalized membership functions for subclasses. A fuzzy class B is a fuzzy superclass of a fuzzy class C if $t_B(x) > t_C(x)$ and $f_B(x) < f_C(x)$). For example, a fuzzy class algebra expression B (e.g. p&&q%0.15 || r&&s%0.4) is fuzzy-subsumed by a class algebra expression C (e.g. p%0.7 || s%0.9). The fuzzy set of elements is computable in time O(nm) where m is the number of operators and the n is the number of objects in the database.

When the database is not specified, the logical containment between the two class algebra expressions may take NP-time in worst case, since finding the normal form basically involves using resolution to find all of the non-subsumed consequents of the propositional logic formulae for the class definitions and the query. All of the non-subsumed consequents are computable in NP-complete time. These SDNF consequents label the non-empty nodes of the IS-A hierarchy.

REFERENCES

Bonner, A.J., 1997, "Intuitionistic Deductive Databases and the Polynomial Time Hierarchy", The Journal of Logic Programming, pp.1-47.

Buehrer, D.J., 1994, "From Interval Probability Theory to Computable Fuzzy First-Order Logic and Beyond," Proceedings of IEEE World Congress on Computational Intelligence, Orlando, Florida, pp.1428-1433.

Buehrer, D.J. 1995, "An Object-Oriented Class Algebra", in Proceedings of ICCI '95: 7th International Conference on Computing and Information, Peterborough, Ontario, Canada, July 5-8, pp.669-685.

Buehrer, D.J., Liu, Y.H. Hong, T.Y. and Jou, J.J. 1996, "Class Algebra as a Description Logic", AAAI Lecture Notes, Proceedings of the 1996 Description Logic Workshop, Boston, pp.92-96.

Buehrer, D.J. and Lee, C.H. 1999, "Class Algebra for Ontology Reasoning", Proc. of TOOLS Asia 99 (Technology of Object-Oriented Languages and Systems, 31st International Conference), IEEE Press, Nanjing, China, pp.2-13.

Coker, D., 1997, "Fuzzy Rough Sets are Intuitionistic L-fuzzy Sets", Fuzzy Sets and Systems 96 (1998) 381-383.

Gau, W.L. and Buehrer, D.J., 1993, "Vague Sets", IEEE Transactions on Systems, Man, and Cybernetics, Vol. 23, No. 2, pp.610-614, 1993.

Rasiowa, H. and Ho, N.C., 1992, "LT-Fuzzy Logic", in Fuzzy Logic for the Management of Uncertainty, (edited by Lotfi Zadeh), New York: Wiley.

HIERARCHICAL RULE-BASE REDUCTION FOR A FUZZY LOGIC BASED HUMAN OPERATOR PERFORMANCE MODEL

WILLIAM R. NORRIS
University of Illinois
Agricultural Engineering
Urbana, Illinois
wnorris@uiuc.edu

RAMAVARAPU R. SREENIVAS
University of Illinois
General Engineering
Urbana, Illinois
rsree@uiuc.edu

QIN ZHANG
University of Illinois
Agricultural Engineering
Urbana, Illinois
qinzhang@uiuc.edu

JOSE C. LOPEZ-DOMINGUEZ
Caterpillar, Inc.
Hydraulics and Fabrications B
Joliet, Illinois
Lopez-Dominguez_Jose@cat.com

ABSTRACT

The objective of this paper is to demonstrate a procedure used in the development of a set of virtual environment tools for performing qualitative operator-in-the-loop control system design. As a means of technology transfer, a real time articulated wheel loader model was implemented as a system plant. A fuzzy logic based human operator performance model (HOPM) was used to navigate the wheel loader along a predetermined trajectory. Due to the complex nature of controlling a four-degree of freedom wheel loader model, a high degree of response granularity was required from the HOPM. In fulfilling this requirement, a hierarchical method for reducing the size of the rule base to meet real-time simulation requirements while retaining the response granularity was developed. Using this novel procedure, known as a fuzzy relations control strategy (FRCS), the size of the rule base was reduced by 98% without affecting the performance of the HOPM.

INTRODUCTION

One of the major problems with designing systems for interactive use with a human operator is the lack of an adequate procedure for modeling or incorporating human behavior. Virtual design tools constitute a structural topology for incorporating a human operator or human operator model in the system design process using virtual environments. The technique closes the loop for qualitative open-loop system designs, and is composed of several user-designed modular components that can be removed or replaced depending on the design objectives (Norris, 2001a).

An essential virtual design tool component is a human operator performance model (HOPM). The HOPM was intended to aid in the development of adaptive and optimal solutions for qualitative design by providing consistent realt time responses to trajectory tracking errors. This autonomous control algorithm would simulate sub-optimal human performance with consistent decision-making and exhibit a variety of driving styles with minor alterations of the fuzzy system membership function parameters.

Fuzzy logic techniques were selected for the HOPM controller structure for this application. Their suitability to this application was based on their applicability to nonlinear systems with uncertainty and their incorporation of expert knowledge in terms of a heuristic rule base (Liu and Wu, 1993; Hodge and Trabia, 1999). Additionally, there have been many successful applications of fuzzy logic to steering control. (Hashizume et al, 1997; Hodge et al, 1999; Wijesoma et al, 1999). As an example, a fuzzy logic controller, which used offset and orientation errors, was developed to steer a mobile robot and resulted in a satisfactory trajectory tracking performance (Cardoso et al, 1994).

The design requirements for the HOPM included real-time simulation capability and a large degree of granularity in the response characteristics. Adding to the computational requirements, the wheel loader model had four degrees of freedom with a corresponding increase in the number of inputs required to control the vehicle. Techniques for reducing the size of the rule base were desirable in the HOPM development. This paper presents the basic structure of the HOPM controller, and the technique used to reduce the size of the HOPM rule-base while maintaining response granularity.

THE WHEEL LOADER MODEL

The developmental objective for the wheel loader model was to obtain appropriate detail to provide real-time virtual simulation, yet maintain enough complexity to be useful in the design process. A standard wheel loader, Fig. 1, is a four-tire, single bucket, articulated vehicle. On each side of the pin connecting the two separate parts of the vehicle are hydraulic cylinders. The pin connecting the two parts of the vehicle was referred to as the articulation joint (AJ). In steering, two hydraulic cylinders adjust the orientation of the front and the rear portions of the vehicle. The front part of the vehicle will be referred to as the non-engine end frame (NEEF). The rear part of the vehicle, which contains the engine, will be referred to as the engine end frame (EEF).

The vehicle model used as the plant controlled by the HOPM was developed based on Pauling and Larson's (1988) wheel loader model, an electrohydraulic (E/H) steering system model and a Pacejka tire model (Norris, 2001a).

Figure 1: A typical front-end wheel loader.

HUMAN OPERATOR PERFORMANCE MODEL

The HOPM provided the steering signal used by the E/H system, which in turn provided the forces on the EEF and NEEF of the vehicle dynamic model. The objective was to perform real time trajectory following while maintaining consistent response characteristics. The other inputs into the wheel loader model were converted to force magnitudes used in the dynamic vehicle model.

Formulated in three components, the HOPM adequately modeled consistent human performance characteristics (Norris et al, 2001c; Norris, 2001a). The three components were a trajectory planner, an error interpreter, and a human decision making model (HDMM) as illustrated in Figure 2.

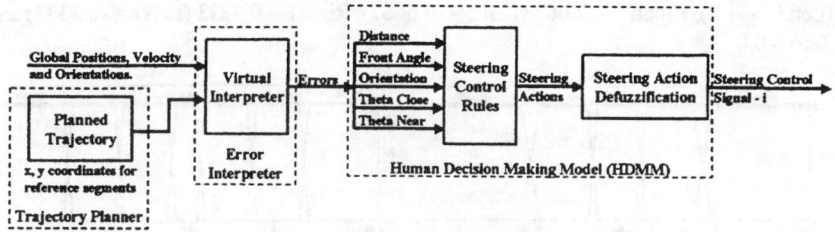

Figure 2: The three components of the virtual operator: Virtual Trajectory Planner, Virtual Interpreter, and HDMM.

The trajectory planner provided a required vehicle path in terms of a series of equally spaced global coordinates. (Norris et al., 2001b). The error interpreter generated errors based on the differences between the vehicle position, orientation and the generated trajectory from the planner as demonstrated in Figure 3. The errors were used by the HDMM to determine the corresponding control action.

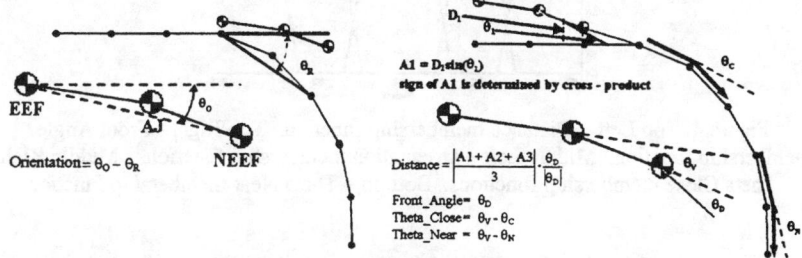

Figure 3: An illustration of the errors and the equations used by the error interpreter. The errors were presented as inputs to the fuzzy logic controller (HDMM).

The design objective for the HDMM was to attain a controller with expert operator-like performance and a fixed output for a given set of input criteria, not necessarily an optimal controller (Norris, 2001a; Norris et al, 2001c). The controller minimized the error provided by the error interpreter with the objective of remaining parallel to the road at a minimum distance. The output from the controller was provided directly to the E/H steering system.

The input variables to the HDMM were Distance, Front Angle, Orientation, Theta Close, and Theta Near. The Distance, Orientation, and Theta Near errors used the linguistic variables far_left, close_left, zero, close_right and far_right. The Theta Close and Front Angle errors used linguistic variables far_left, near_left, close_left, zero, close_right, near_right and far_right.

The rule base was derived based on heuristic knowledge using the input variables. A hierarchical technique was used, as presented in the next section that was derived from the control system objectives. The objective for the fuzzy logic controller was to provide the steering signal that brought and maintained the vehicle parallel to, and at a minimum distance from the trajectory from any global initial condition.

The defuzzification method was the Center of Maximum (CoM), also referred to as the Center of Area (CoA) method with singletons. The steering actions were defuzzified using the symmetric linguistic variables Right3 corresponded to a controller output of 1.0

(Left3 = -1.0), Right2 = 0.6667 (Left2 = -0.6667), Right1 = 0.3333 (Left1 = -0.3333) and Zero = 0.0.

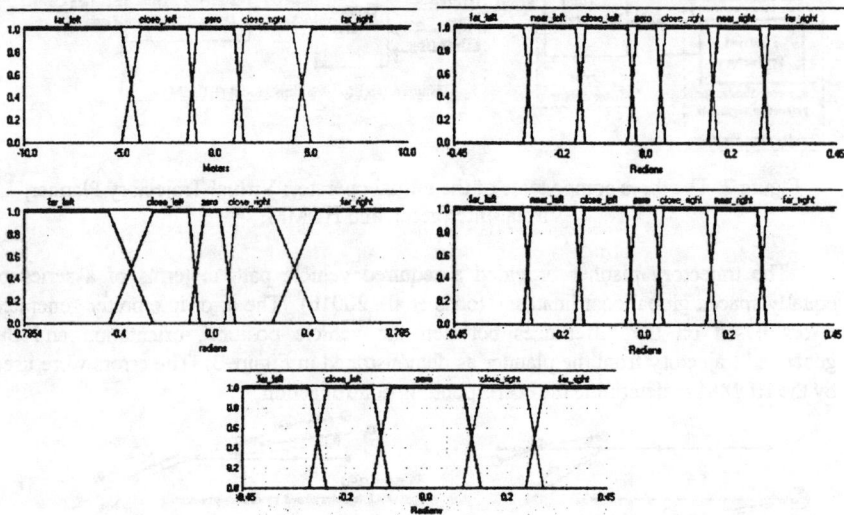

Figure 4: Top Left – Distance membership function. Top Right - Front Angle membership function. Middle Left – Orientation membership functions. Middle Right – Theta Close membership functions. Bottom – Theta Near membership functions.

HIERARCHICAL RULE-BASE REDUCTION

A hierarchical technique was used to derive the rule base derived from the importance of the inputs relative to their linguistic variable regions. The hierarchy was drawn from the fuzzy logic controller objectives, which were to provide the steering signal that brought and maintained the vehicle parallel and at a minimum distance from the trajectory globally.

The distance from the wheel loader to the center of the trajectory, the vehicle's location on the left or right side of the trajectory, and the direction from the front of the vehicle to the road were important in determining the required control signal (Figure 5). In order to incorporate the information, a technique known as a fuzzy relations control strategy (FRCS) was introduced. The input error values were termed the fuzzy relations control variables (FRCVs). The FRCS for this problem was defined using the following example.

If the vehicle's Distance error was classified as far_left or far_right from the center of the trajectory, the primary objective was to approach the trajectory as quickly as possible without expending excessive control energy. At large Distance errors from the trajectory, the error considerations required to ensure the vehicle paralleled the trajectory were irrelevant. The distance from the trajectory and the need for a "straight-line approach" to minimize the Distance error had strategic priority. Consequently, the input variables, Distance error and Front Angle error, were used as the means to reduce the control effort and remove Bang-Bang controller effects as demonstrated with the example rule:

IF Distance is far_left, AND Front Angle is far_left, THEN steering action is Right3.

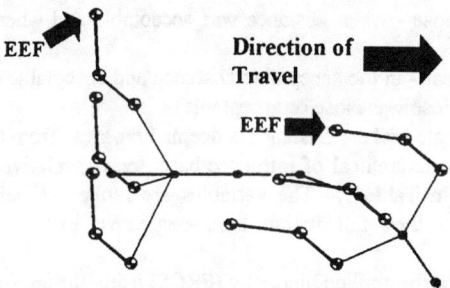

Figure 5: Relative position examples representing the issues involved with interpreting trajectory errors for the articulated wheel loader.

The output classification, Right3, occurred irrespective of the Orientation, Theta Close and Theta Near errors. The Front Angle error was used to provide the sign of the Distance error. As a result, the Front Angle error classifications that produced a far_right Distance Error were not implemented in the rule base where the Distance error was classified as far_left. The same technique was used in the other case where the Distance error was classified as far_right.

As the vehicle approached the trajectory and the Distance error was classified as close_left or close_right, the primary objective became attaining a smooth approach to the trajectory, thereby preventing overshoot, and gradually paralleling the road. Paralleling the trajectory was a secondary control objective. As the vehicle approached the trajectory, the vehicle heading should approach the trajectory heading. The heading comparison was represented by the Orientation error. The Orientation error of the vehicle replaced the Front Angle error input in the control strategy. Consequently, the Front Angle error and the other input error variables were ignored as demonstrated in the sample rule:

IF Distance is close_left, AND Orientation is far_left, THEN steering action is Right3.

When the Distance error was considered acceptable, or in the zero region, the Orientation error was deemed the primary metric for incorporating the control strategy. The Theta Close error was used as the primary influence on the rules evaluated for the control. The strategy was based on the idea that once the vehicle was within an acceptable distance and close to acceptable orientation from the trajectory, reaction to trajectory disturbances became a priority. The trajectory disturbances were in the form of changes in curvature.

IF Distance is zero, AND Orientation is close_left, AND Theta Close is far_left, THEN steering action is Right3.

Once the Distance and Orientation errors were in their acceptable regions, Theta Close was used as the primary metric for the control strategy. Once the vehicle was this region, a look ahead aspect was necessary to ensure the operator could prepare for large perturbations in the trajectory (changes in curvature.)

IF Distance is zero, AND Orientation is zero, AND Theta Close is close_left, AND Theta Near is far_left, THEN steering action is Right1.

Using the FRCS and the overall objective to minimize error, the FRCVs were placed in a hierarchy based on their global influence.

Minimize Distance - globally
Minimize Orientation - when Distance was classified as close or acceptable.

Minimize Theta Close - when Distance was acceptable and where Orientation was close or acceptable.
Minimize Theta Near - in the acceptable Distance and acceptable Orientation region and where Theta Close was close or acceptable.
Minimize Front Angle - where Distance is deemed far away from the roadway.

A summary of the hierarchical of influence based on the relative strategic value of the FRCVs is located in Table 1. The variables are ranked, 1 being of the highest priority, based on how they fell within their respective FRCV linguistic variable classifications.

Table 1: Summary of the applied hierarchy (FRCS) using the appropriate FRCVs

FRCVs	Metric Used for FRCS Classification						
	Distance Far	Distance Close	Orient Far	Orient Close	Distance = Acceptable (Zero)		
					Orient = Acceptable (Zero)		
					Theta Close Far	Theta Close Close	Theta Close Acceptable (Zero)
Distance	2	2	2	3	3	4	4
Front Angle	1						
Orientation		1	1	2	2	3	3
Theta Close				1	1	2	2
Theta Near						1	1

The FRCS employed in forming the rule base resulted in a symmetric set of rules, as presented in Table 2 in the Appendix. The rules represent one of the two symmetric sides of the rule base where the blank spaces represent rules irrelevant to the control strategy ("don't care" rules). The size of the entire rule base was reduced by over 98%, from 6125 rules covering the entire region of possibility to 81.

The HDMM was designed using the FuzzyTech^{TM} software package and generated in the C programming language. The error interpreter and trajectory planner were written in the C programming language. The code for the virtual operator was implemented in user-code block diagram form within MatrixX System Build. Vehicle simulations, including the virtual operator, designed for a length of 90 sec. with a 50 Hz. sampling rate, required 20 sec. for completion on an HP C180 workstation.

RESULTS

In order to demonstrate the feasibility of the approach for the Human Operator Performance Model, simulations were performed using the SAE steering test course. The steering course is from SAE standard J1511, Steering for Off-Road, Rubber-Tired Machines, FEB94. The requirements were for the machine tire tracks to remain within the test course boundaries while the operator drove through the course at a constant 16 km/hr ± 2 km/hr. As shown in Figure 6, the HOPM successfully completed the course. The articulation angles generated from the virtual operator's steering command are demonstrated next to the steering course.

The simulations were performed with a trained controller, and controllers where the membership functions were altered as illustrated in the respective figures. The rule-base remained unchanged throughout the simulations, while linguistic variable membership functions were altered. The simulations were run within MatrixX where the vehicle dynamics and HOPM were implemented in C code within a user code block while the electrohydraulic system was implemented in block diagram form. Using the technique, 90 sec. simulations were conducted with a 0.02 sec. sampling rate. Processor times for the simulations involving the HOPM and vehicle model were 10.1 sec for the 90 sec simulation time, satisfying the real-time simulation requirements.

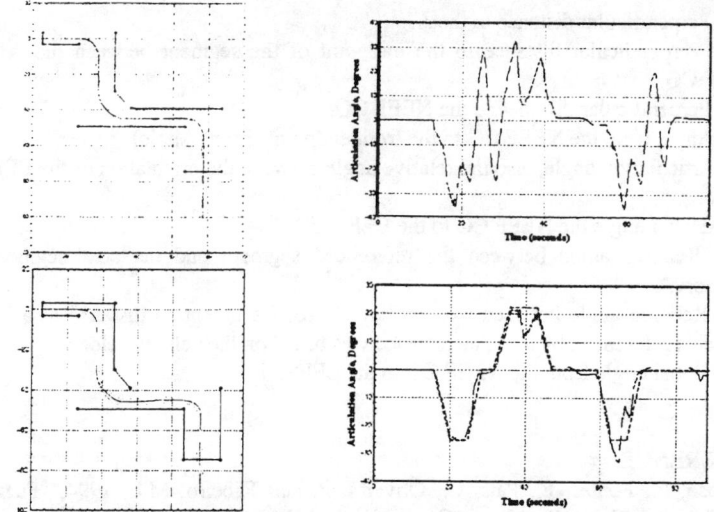

Figure 7: Upper - HOPM with slightly different membership functions form the lower trained HOPM. The HOPMs both used the FRCS.

The three major lines in the plots represent the trajectory of the NEEF Center of Gravity and the connecting points for the wheel loader tires. The center line represented the path of the NEEF Center of Gravity. As noted from the figures, though the rule-base was consistent throughout the trials, altering the membership function parameters slightly resulted in a less experienced though consistent operator.

CONCLUSIONS

A real time human performance model, for steering a four-degree of freedom articulated vehicle along a trajectory, has been developed. Human like error interpretation and a common sense rule base were used to successfully perform real-time, consistent control system decision-making. The real time objective was achieved despite a large degree of granularity introduced into the control interpretation and a complex system model. The hierarchical technique for incorporating the control strategy into the rule base allowed for the additional granularity and provided a model whose performance can be more easily explained in terms of human perception. The developed rule base remained fixed while the perception of the model was altered to attain different performance characteristics.

The techniques used to develop the hierarchical rule base are viable for many other applications including situations where an operator model is required, for human performance studies, qualitative system design or in autonomous equipment algorithm development.

NOMENCLATURE

$\theta_{1,2,3}$ = Angle from the (front endpoint, midpoint, rear endpoint respectively) of the referenced road segment to the (NEEF CG, NEEF to AJ segment midpoint, AJ respectively.)

$A1$ = Perpendicular distance to the AJ.
$A2$ = Perpendicular distance to the midpoint of the segment between the AJ and the NEEF CG.
$A3$ = Perpendicular distance to the NEEF CG.
θ_D = Angle from the NEEF CG to the front endpoint (Front Angle).
θ_V = Articulation angle, also the relative angle between the orientation of the EEF and the NEEF.
θ_O = Global angle the NEEF CG to the EEF CG.
θ_C = Relative angle between the referenced segment and the next segment in the trajectory.
θ_N = Relative angle between two contiguous road segments a distance away from the segments referenced by θ_C. The distance was based on the vehicle velocity.
θ_R = Global angle from the NEEF CG to the EEF CG.

REFERENCES

Cardoso, F., Fontes, F., Pais, C., Oliveira, P. and Ribeiro, M.I., 1994, "Fuzzy Logic Steering Controller for a Guided Vehicle." *7th Mediterranean Electrotechnical Conference*, 1994, vol.2, pp. 711-714.

Hashizume T, Yaginuma I, Fujikawa H, and Shin-Ichi Y., 1997, "Path tracking controller for a tractor with multiple trailers." *Proceedings of the IECON'97 23rd International Conference on Industrial Electronics, Control, and Instrumentation*, Part vol.3, 1997, pp.1269-74.

Hodge N., and Trabia M., 1999, "Steering Fuzzy Logic Controller for an Autonomous Vehicle." *Proceedings 1999 IEEE International Conference on Robotics and Automation*, vol.3, 1999, pp. 2482-8.

Liu, T., and Wu J., 1993, "A model for a rider-motorcycle system using fuzzy control."*IEEE Transactions on Systems, Man & Cybernetics*, vol.23, no.1, Jan.-Feb. 1993, pp.267-76.

Norris, W. R. 2001a, "A Design Framework for Qualitative Human-in-the-Loop System Development." Ph.D. Thesis, University of Illinois Urbana-Champaign, Urbana, IL.

Norris, W. R., Sreenivas, R., and Zhang, Q., 2001b. "An Articulated Wheel Loader Path Tracking Algorithm and Error Interpretation." Submitted to IEEE Control Systems Society.

Norris, W. R., Sreenivas, R., and Zhang, Q., 2001c, "A Real-Time Human Operator Performance Model Using Fuzzy Logic." Submitted to IEEE Control Systems Society.

Norris, W. R., Sreenivas, R., and Zhang, Q., 2001d, "Real Time Virtual Simulation of a Wheel Loader with an Electro-hydraulic Steering System." Submitted to SAE Vehicle Dynamics and Simulation.

Pauling, P., and Larson, C., 1988, "Simulation of an Articulated Wheel Loader Including Model for Earthmoving Tires." *SAE Earthmoving Conference*, paper no. 880777.

SAE Standard J1511, Steering for Off-Road, Rubber-Tired Machines, Revised February 1994.

Wijesoma, W., Kodagoda, K., and Teoh, E., 1999, "Uncoupled Fuzzy Controller for Longitudinal and Lateral Control of a Golf Car-Like AGV." *Proceedings IEEE/IEEJ/JSAI International Conference on Intelligent Transportation Systems*, pp.142-7.

APPENDIX
Table 2: Half of the symmetric rule-base incorporating the hierarchical technique.

IF					THEN
Distance	Front Angle	Orientation	Theta_Close	Theta_Near	Steer Command
far_left	far_left				right3
far_left	near_left				right2
far_left	close_left				right1
far_left	zero				zero
close_left		far_left			right3
close_left		close_left			right2
close_left		zero			right1
close_left		close_right			right1
close_left		far_right			right1
zero		far_left	far_left		right3
zero		far_left	near_left		right3
zero		far_left	close_left		right3
zero		far_left	zero		right2
zero		far_left	close_right		right2
zero		far_left	near_right		right2
zero		far_left	far_right		right1
zero		close_left	far_left		right3
zero		close_left	near_left		right2
zero		close_left	close_left		right2
zero		close_left	zero		right1
zero		close_left	close_right		right1
zero		close_left	near_right		right1
zero		close_left	far_right		right1
zero		zero	far_left	far_left	right3
zero		zero	far_left	close_left	right3
zero		zero	far_left	zero	right2
zero		zero	far_left	close_right	right2
zero		zero	far_left	far_right	right1
zero		zero	near_left	far_left	right3
zero		zero	near_left	close_left	right3
zero		zero	near_left	zero	right2
zero		zero	near_left	close_right	right1
zero		zero	near_left	far_right	right1
zero		zero	close_left	far_left	right3
zero		zero	close_left	close_left	right2
zero		zero	close_left	zero	right1
zero		zero	close_left	close_right	right1
zero		zero	close_left	far_right	right1
zero		zero	zero	far_left	right1
zero		zero	zero	close_left	zero
zero		zero	zero	zero	zero

AN INTELLIGENT FUZZY-BASED SYSTEM FOR PAROXYSMAL ATRIAL TACHYCARDIA IDENTIFICATION

WARAWAT ASSAWASANTAKUL
Biomedical Engineering Program
Faculty of Engineering,
Mahidol University, Salaya, 73170
Thailand

RAWIN RAVIWONGSE, Ph.D.
Department of Industrial Engineering
Faculty of Engineering,
Mahidol University, Salaya, 73170
Thailand

WATTANA B. WATANAPA, M.D., Ph.D.
Department of Physiology
Faculty of Medicine Siriraj Hospital,
Mahidol University, Bangkok, 10700, Thailand

ABSTRACT

Paroxysmal Atrial Tachycardia (PAT) and Paroxysmal Supraventricular Tachycardia (PSVT) are considered irregular heart rhythms and can cause a dramatic drop in efficiency of the working of a heart. This article aims to develop an intelligent tool to identify PAT or PSVT using input data gathered from a signal printed on regular ECG paper. A Dynamic-Link Rule Base in a Fuzzy Inference System (BLRB-FIS) is used as a major tool to construct an identification model. The four factors used to determine the PAT or PSVT rhythm are heart rate, P-R interval, the first, and second coefficients for P wave identification from the five-filter order model of the Burg algorithm. Fifty samples were tested and the results were compared to cardiologists' diagnoses. The sensitivity and specificity for identifying positive ECG's are 90.6% for arrhythmia and the sensitivity of identifying paroxysm is 84.0%.

INTRODUCTION

Arrhythmia is an abnormal heart rhythm caused by an irregular pulse generation or conduction in the heart. Paroxysmal Supraventicular Tachycardia (PSVT) and its subset, Paroxysmal Atrial Tachycardia (PAT), are arrhythmia commonly found in children and young adults and can occur in normal individuals who show no clinical evidence of heart disease. Clinical problems appear when the arrhythmias reduce the pumping efficiency of the heart. Serious arrhythmias may indicate damage to the myocardial musculature, injuries to the pacemakers or conduction pathways, exposure to drugs, or variations in the electrolyte composition of the extracellular fluids [1].

Classifying the rhythm abnormalities is important to hemodynamic monitoring since irregular rhythms can cause dramatic inefficiencies in how the heart works as a pump. During an operation or post operative recovery time, abnormal beats can occur easily, thus continuous monitoring is necessary. In the past, cardiac rhythm pattern observation and classification were performed manually. Such processes demand a skilled technician who can withstand high boredom and fatigue levels without an increasing error rate. They are also tedious and stressful, making the work less effective and unreliable.

Nowadays, automated algorithms to detect and classify the abnormal rhythm patterns are being developed to improve efficiency and reliability of continuous observation [2].

In this paper, a fuzzy-based identification system is developed using Dynamic-Link Rule Base in Fuzzy Inference System (DLRB-FIS) in order to identify PAT and PSVT using electrocardiogram (ECG) printouts. The sample signals are gathered from both textbooks and clinical ECG printouts from Her Majesty's Cardiac Center in Siriraj Hospital, Mahidol University, Bangkok.

RELATED LITERATURE

Many researchers are also interested in developing the intelligent model for arrhythmia classification [3, 4]. Fuzzy systems, along with algorithms such as wavelet transformation, artificial neural networks, etc., have been widely utilized especially in the biomedical field because they yield a high percentage of classification accuracy. Furthermore, several algorithms can be integrated with a classical fuzzy algorithm such as neural network and Dynamic-Link Rule Base in Fuzzy Inference System (DLRB-FIS) to increase the speed and efficiency.

There are many intelligent algorithms used to classify and detect arrhythmia since they provide a convenient framework of using expert knowledge to approximate a complex nonlinear system. A research reported by Cohen et al [5] presented an application of a fuzzy neural network (FNN) to detect QRS waves. It resulted in a relatively low number of false-positive and false-negative responses when compared with the existing algorithms. Further, the FNN can discriminate a true QRS waves from noise based on amplitude, R-R interval, and pulse duration. The system produced a good result, although it did not attempt to find the optimal number of rules and membership functions. Nadal and Bossan [6] proposed a feedforward neural network classifier, intended to split the vector space generated by principal component coefficients of each P-QRS complex into regions which separate the different classes of beats in an efficient way. The classification results are better than those obtained using statistical regression. However, the non-uniformity of pattern distribution over the input space makes the task of finding correct classification boundaries more difficult when the neural network is trained using backpropagation.

FUZZY SYSTEM AND DLRB-FIS

Most of the fuzzy models employ the conventional architecture, which are: fuzzification interface, fuzzy rule base, fuzzy inference engine, and defuzzification interface. This conventional architecture usually carries a large number of rules that the inference engine has to check through one by one, regardless of whether the rules fired or not. Such checking results in less efficiency of the system. Therefore, Lin [7] proposed a Dynamic-Link Rule Base in Fuzzy Inference System (DLRB-FIS) by adding a dynamic-link rule base between the original rule base and the inference engine in order to increase speed and simplify the fuzzy reasoning process. In the DLRB-FIS, only the fired rules, whose firing strengths are not equal to zero, are included for inference while the unfired rules are dynamically skipped. The architecture of a DLRB-FIS is shown in Figure 1. The details of each component are briefly described in the following text [7].

(1) Fuzzification Interface: The fuzzification interface, also called the fuzzifier, performs a mapping from a crisp point $x_o \in X$ onto a fuzzy set A_x in U. The function of the fuzzifier of the DLRB-FIS and that of the conventional FIS are identical.

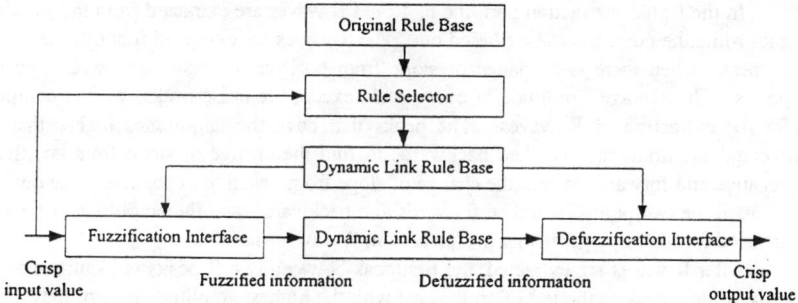

Figure 1. The DLRB-FIS Architecture

(2) Original Rule Base: The original rule base consists of a collection of fuzzy IF-THEN rules which is the heart of the conventional fuzzy model. The rules are used to determine the relationship between the input and the output.

(3) Rule Selector: The rule selector is used to pick out the fired rules and reject the non-fired rules. The criteria of rule selection are summarized as the following crisp rules:

$$\text{IF } x_i \in [a_i^{(l_i)}, d_i^{(l_i-1)}] \quad \text{THEN } I_i^{'} = \{l_i - 1, l_i\} \quad (1)$$
$$\text{IF } x_i \in (a_i^{(l_i-1)}, d_i^{(l_i+1)}] \quad \text{THEN } I_i^{'} = \{l_i\} \quad (2)$$
$$\text{IF } x_i \in (a_i^{(l_i+1)}, d_i^{(l_i)}] \quad \text{THEN } I_i^{'} = \{l_i, l_i + 1\} \quad (3)$$

where $I_i^{'}$ form a set of dynamic input indices and a_i, d_i are the smallest and the largest values of the pseudo trapezoid-shape membership function respectively.

(4) Dynamic-Link Rule Base: The dynamic-link rule base consists of a collection of fired fuzzy IF-THEN rules, which have nonzero firing strength. In an original rule base, the number of rules is given by $N = L^n$; where L and n are the factor pairs, which means N increases exponentially along with the increase of L and n. On the other hand, the number of rules in the DLRB-FIS is given by $N \leq 2^n$, which is not related to changes in the value of L. Therefore, the number of fuzzy rules is independent from the number of input linguistic labels.

(5) Fuzzy Inference Engine: In the fuzzy inference engine, fuzzy principles are used to map the fuzzy set from the input space into another fuzzy set of output space by utilizing the IF-THEN rules stored in the dynamic-link rule base.

(6) Defuzzification Interface: The defuzzification interface, performs a mapping from a fuzzy set $B \in v$ onto a crisp point $y \in Y$. The function of defuzzifier of the DLRB-FIS and that of the convention FIS are also identical.

METHODOLOGY

There are three major processes used to develop this identification model: 1) ECG signal simulation, 2) feature extraction, and 3) arrhythmia classification. The ECG signal simulation process can be divided into two major parts. The first part is to obtain the enlarged image of lead I or II of an ECG signal from a regular ECG printout at 25 mm/s. The next part is to load the enlarged ECG image to the program and generate the digitized ECG image by redrawing the ECG trace and simulate the path using a B-Spline algorithm.

In the feature extraction part, the P, T, and R waves are extracted from the simulated data while the noise has to be filtered out. The R waves are extracted first by determining a "peak" when there is a change of slope from positive to negative between two data points. The average amplitude of every peak, except the noise peaks, will be computed for the extraction of R waves. The peaks that have the amplitudes higher than the average amplitude are searched backwards to find the change of slope from positive to negative and forwards to find the change of slope from negative to positive. The duration between the two points, searched forwards and backwards, and the amplitude of the peak are used for identifying whether this peak is an R wave or not.

After R waves are extracted, the first peak between two R peaks is identified as a T peak while a peak to the left of an R wave with the highest amplitude may or may not be a P wave. Therefore, a DLRB-FIS is used to determine whether such peak is a P wave or noise. The DLRB-FIS is used to differentiate the P waves and noise based on the amplitude and duration of the waves. After the P wave is detected, the Burg method is then used to extract its features.

The heart rate and the P-R interval are also acquired after the P wave is identified. The heart rate is determined by adjacent R-R intervals, and the P-R interval is calculated from the starting point of P wave to the first point of the QRS complex.

For the arrhythmia classification part, after the R, T, P waves, the Burg coefficients, the heart rate, and the P-R intervals are identified, they are fed into the DLRB-FIS classification model to classify whether the rhythm is PAT, PSVT, normal sinus rhythm, or other arrhythmias. There are two separate DLRB-FIS classification models. The first one is used only when the P wave can be captured and identified. It uses the heart rate, the Burg's first and the second coefficients of the P wave as the input of this model to classify whether the ECG sample is an Atrial Tachycardia (AT), normal sinus rhythm, or other arrhythmias. Another model is used when the P wave cannot be identified. Thus it is applied to classify whether the ECG sample is a Supraventricular Tachycardia (SVT) or other arrhythmias (excluding AT). The DLRB-FIS components of the AT classification model are shown in Figure 2.

After the signals are classified, the next process is to determine paroxysm based on 1) constant heart rate, 2) constant PR interval, and 3) similar pattern of each beat. The rhythm that has the same pattern for each beat, almost the same R-R interval between each beat (less than 10% difference), and almost the same of P-R interval between each beat (less than 10% difference) is identified as "Paroxysmal". If one of the beats is different, the rhythm will be identified as "Non-paroxysmal".

RESULTS

In this study, the ECG data from 50 cases, acquired from the Cardiolab [8] at Her Majesty's Cardiac Center, Siriraj Hospital, were used as testers. The model identified the data in the terms of the overall appearance, the beat-by-beat appearance, and the beat-by-beat confidence factor in percentage. Results were evaluated in comparison with the cardiologist's reading of each strip. The result evaluation is separated into: 1) type evaluation and 2) paroxysmal evaluation. Both were evaluated in the terms of True Positive, False Positive, True Negative, or False Negative.

For "Type" evaluation, True Positives (TP_t) are individuals with "SVT", "AT", or "Abnormal" ECG, who are correctly identified as such by the program. False Positives (FP_t) are SVT-free or AT-free individuals falsely labeled as "SVT" or "AT". True Negatives (TN_t) are SVT-free or AT-free individuals who are correctly identified as such, and False Negatives (FN_t) are individuals with "SVT", "AT", or "Abnormal" ECG, whose ECG are falsely labeled as "Normal".

For "Paroxysmal" evaluation, True Positives (TP_p) are individuals with paroxysm who are correctly identified as "Paroxysmal". False Positives (FP_p) are non-paroxysmal individuals, who are falsely labeled as "Paroxysmal". True Negatives (TN_p) are non-paroxysmal individuals who are correctly identified as such by the program, and False Negatives (FN_p) are individuals with paroxysm who are falsely labeled as "Non-paroxysmal".

There are 29 true positives, 1 false positive, 17 true negatives, and 3 false negatives in type evaluation. Therefore, the sensitivity of the evaluation can be defined as:

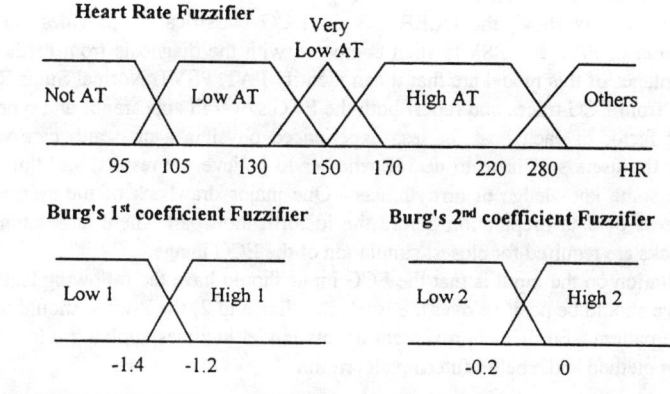

Dynamic Link Rule Base										
Heart Rate		Not AT		Low AT		V. Low AT		High AT		Others	
1st Burg's		L1	H1	L1	H1	L1	H1	L1	H1	L1	H1
2nd Burg's	H2	Ns	Ns	Am	As	Am	As	Abb	Ab	As	O
	L2	Ns	Nb	As	Ass	Am	Ass	Ab	Am	O	O

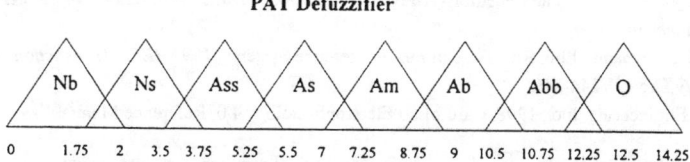

Figure 2. DLRB-FIS Components for AT Classification

TP_t / N_{diag} = 0.906 or 90.6% ; where N_{diag} is the number of SVT, AT, and Abnormal as diagnosed by cardiologists.

Similarly, the specificity for type evaluation can be defined as:

TN_t / N_{norm} = 0.944 or 94.4% ; where N_{norm} is the total number of normal individuals.

For paroxysmal evaluation, there are 21 true positives, no false positives, 4 false negatives, and 1 true negative. Therefore, the system's sensitivity for paroxysm evaluation is 0.84 or 84.0%. Since there is only one non-paroxysmal case diagnosed by cardiologists, the specificity cannot be determined.

CONCLUDING REMARKS

As the results show, the DLRB-FIS for ECG classification provides very good classification of PAT and PSVT when compared with the diagnosis from cardiologists. The advantages of this model are that it can identify PAT, PSVT, Normal Sinus Rhythm, or others, from ECG traces and report both the ECG's overall appearance and type with a reliability factor of each beat for less experienced physicians or health care workers. However, the users still have to decide whether to believe the results, and thus should also have some knowledge of arrhythmias. One major drawback of the system is the slow process of data preparation before the identification part where large numbers of mouse clicks are required for closest simulation of the ECG image.

Limitation on the input is that the ECG input should have the following features: 1) the R wave should be positive over the isoelectric line and 2) the P wave should not have retrograde pattern. Further improvement of this model includes applying a faster signal simulation method and a better filtering algorithm.

REFERENCES

1. Martini F. and Bartholomew E., 1997, "Essentials of Anatomy and Physiology", International ed., Prentice-Hall, New Jersey.
2. Avolio A., (editor)., 2000, "Patient Monitoring and Electrical Safety", University of New South Wales, Sydney.
3. Lin K-P and Chang W., 1989, "Classification of QRS Pattern by an Associative Memory Model", *IEEE Engineering in Medicine & Biology Society 11th Annual International Conference*.
4. Ham F. M. and Han S., 1996, "Classification of Cardiac Arrhythmias using Fuzzy ARTMAP", *IEEE Transactions on Biomedical Engineering*.
5. Cohen K P., Tompkins W. J., Djohan A., Webster J. G., and Hu Y. H., 1997, "QRS Detection using a Fuzzy Neural Network", *IEEE-EMBC and CMBEC*.
6. Nadal J. and Bossan M. de C., 1993, "Classification of Cardiac Arrhythmias Based on Principal Component Analysis and Feedforward Neural Networks", *IEEE Transactions on Biomedical Engineering*.
7. Lin S.-C., "Dynamic-Link Rule Base in Fuzzy Inference System", 1999, *IEEE Transactions on Fuzzy Logic*, V244-249.
8. Prucka Engineering, Inc., 1998, CardioLab Electrophysiology 4.0, Reference Manual.

BIBLIOGRAPHY

1. Hampton J. R., 1997, "The ECG Made Easy", 5th ed., Churchill Livingstone, London.
2. Stephens R., "Visual Basic Graphic Programming", 1998, Wiley Publishing, New York.

THE EFFECT OF NOISE ON THE DYNAMIC BEHAVIOR OF A FUZZY CLOSED LOOP SYSTEM

EYAD ELQAQ
Electrical and Computer
Engineering Department at
University of Illinois at Chicago,
851 S. Morgan, M/C 154, Chicago,
IL 60607

ROLAND PRIEMER
Electrical and Computer
Engineering Department at
University of Illinois at Chicago,
851 S. Morgan, M/C 154, Chicago,
IL 60607

ABSTRACT
The effect of noise on the dynamical behavior of a closed loop system is considered. The controller, connected in cascade with the plant, is a fuzzy controller. Its design is based only on a fuzzy model of the plant. We show how to obtain a fuzzy relation of the closed loop system. To accommodate random system inputs we give a method to augment the fuzzy controller, and thereby achieve appreciable noise rejection in the output of the plant. A practical design example is included.

INTRODUCTION
A major problem in control systems is parameter uncertainty in the plant [4,5,6]. Many Control techniques that can give excellent performance rely on the availability of an exact plant model. Moreover, the plant parameters can be uncertain and come with tolerance and changes during operation due to internal component variations and system noise [2,3]. The aim of this work is to incorporate the effect of random plant input noise on the design of a fuzzy logic controller that is connected in series with the plant [7,8].

Conventional PID controllers are characterized with simple structure and simple design procedures, and are well suited to control of processes whose exact model is known. In many cases, such as those when disturbances are present, when an exact model is not available, and when parameter variations take place, a control system based on a fuzzy logic controller (FLC) may be a better choice. In general, the fuzzy controller design is based on qualitative knowledge and experience of an expert with regard to the system behavior, and is used to achieve a given control objective [9,10,11].

Consider a class of systems that can be represented by the block diagram shown in Figure 1. The linguistic model of a plant to be controlled is designated by the block "Fuzzy Process" [1]. X(t) is the present linguistic state of the process with U(t) as the process linguistic input, the closed loop linguistic input is V(t), and N(t) is the random noise input to the system.

To calculate the output X(t+1), given X(t), U(t), and the process fuzzy relation R_p, the compositional rule of inference is used as follows [1]:

$$X(t+1) = X(t) \circ U(t) \circ N(t) \circ R_p, \qquad (1)$$

where o is the max-min composition operator.

The block "FLC" can be described by:

$$U(t) = X(t) \circ V(t) \circ R_c, \qquad (2)$$

where R_c is the fuzzy relation of the FLC. By substituting (2) into (1) we obtain:

$$X(t+1) = X(t) \circ (X(t) \circ V(t) \circ R_c) \circ N(t) \circ R_p, \qquad (3)$$

By switching the max-min operators, equation (3) can be rearranged to be:

$$X(t+1) = X(t) \circ V(t) \circ N(t) \circ R_c \circ R_p \qquad (4)$$

It was defined in earlier work that the fuzzy relation R_l of a fuzzy closed loop system is calculated with [1]:

$$R_l = R_c \circ R_p \qquad (5)$$

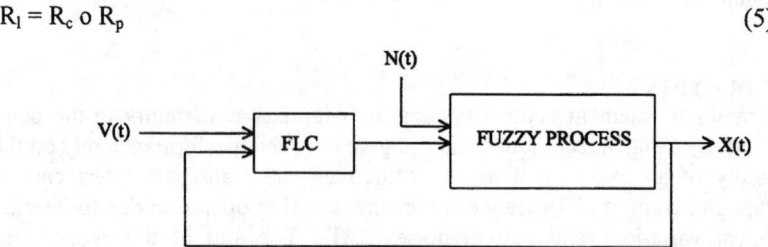

Figure 1. Fuzzy Closed Loop System with Random Input Noise N(t)

NOISE REJECTION:
A procedure is needed for noise rejection that discriminates between real information, which can improve the model accuracy and noise in the data which, if considered, may reduce the model accuracy [4]. Since the noise cannot be accessed, its impact will modify or contaminate the plant input, which will eventually affect the output. In order to reject the random noise regardless of the noise variance, we introduce a Two-Stage Fuzzy Logic Controller (TS-FLC). The first stage consists of a Mamdani inference engine with two inputs being the closed loop system input and plant output, while the second stage consists of a Takagi-Sugeno-Kang (TSK) inference engine as indicated in Figure 2. At first, the output of the Mamdani inference engine along with noise are inputs to the plant. The plant output and the desired output are stored and used to determine membership functions of the TSK inference engine using back propagation and least square error algorithms. The combination of the TSK inference engine in

series with the Mamdani engine is the new Two-Stage Fuzzy Logic Controller (TS-FLC).

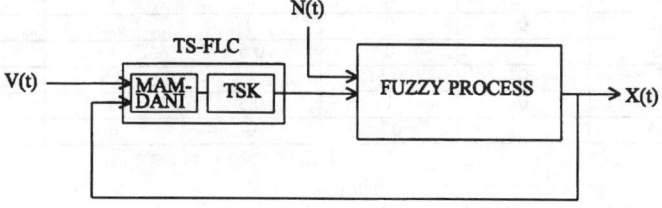

Figure 2. TS-FLC in Fuzzy Closed Loop System

EXAMPLE

A water container example is used to illustrate noise rejection. It has a controllable water level X(t) as shown in Figure 3. The system is described in terms of the error and the control variable U(t). The error is defined as the value between the water level X(t) and water level zero. The noise is a random input with zero mean and different variance values. The goal is to maintain the water level at zero, regardless of the noise N(t) variance.

The membership functions for the error and U(t) linguistic variables are given in Tables 1 and 2, where PB means positive big, PS means positive small, Z means zero, NS means negative small, and NB means negative big. The water level system is described in Table 3. The FLC is described by the rules given in Table 4. Using a Matlab program to model the water container, with the max-min operator and the aggregation of rules, the fuzzy model presented is examined under a random noise input with several different variances as given in Table 5.

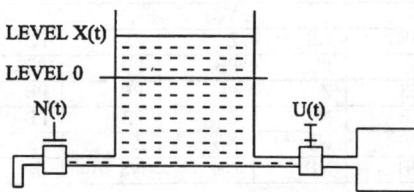

Figure 3. Water Container Example

As a comparison, we also used a PID controller for the same variance and mean values of the random noise input. It is found that the TS-FLC has a better performance than the PID controller. Refer to Figures 4 & 5 for random noise input with variance=0.5 and mean=0. Modeling the random noise in the TS-FLC enhances the closed loop fuzzy system output and results in random noise rejection.

Table 1
Membership functions of the "error" linguistic variable

Error	-2	-1	0	1	2
PB	0	0	0	0.5	1
PS	0	0	0.5	1	0
Z	0	0.5	1	0.5	0
NS	0	1	0.5	0	0
NB	1	0.5	0	0	0

Table 2
Membership functions of the control variable U(t)

U(t)	-4	-3	-2	-1	0	1	2	3	4
PB	0	0	0	0	0	0	0	0.5	1
PS	0	0	0	0	0	0.5	1	0.5	0
Z	0	0	0	0.5	1	0.5	0	0	0
NS	0	0.5	1	0.5	0	0	0	0	0
NB	1	0.5	0	0	0	0	0	0	0

Table 3
Water Container Linguistic Model

X(t+1)		\multicolumn{5}{c}{U(t)}				
		NB	NS	Z	PS	PB
X(t)	NB	NB	NB	NB	NS	Z
	NS	NB	NB	NS	Z	PS
	Z	NB	NS	Z	PS	PB
	PS	NS	Z	PS	PB	PB
	PB	Z	PS	PB	PB	PB

Table 4
Fuzzy Rules of the FLC

U(t)		V(t)					
		NB	NS	Z	PS	PB	
X(t)	NB	Z	PS	PB	PB	PB	
	NS	NS	Z	PS	PB	PB	
	Z	NB	NS	Z	PS	PB	
	PS	NB	NB	NS	Z	PS	
	PB	NB	NB	NB	NS	Z	

Table 5
Average Error Square for Fuzzy and PID Controllers

Noise Variance	Average Error Square TS-FLC	Average Error Square PID Controller
0.1	-80 dB	-70
0.5	-80 dB	-60
1.0	-80 dB	-50

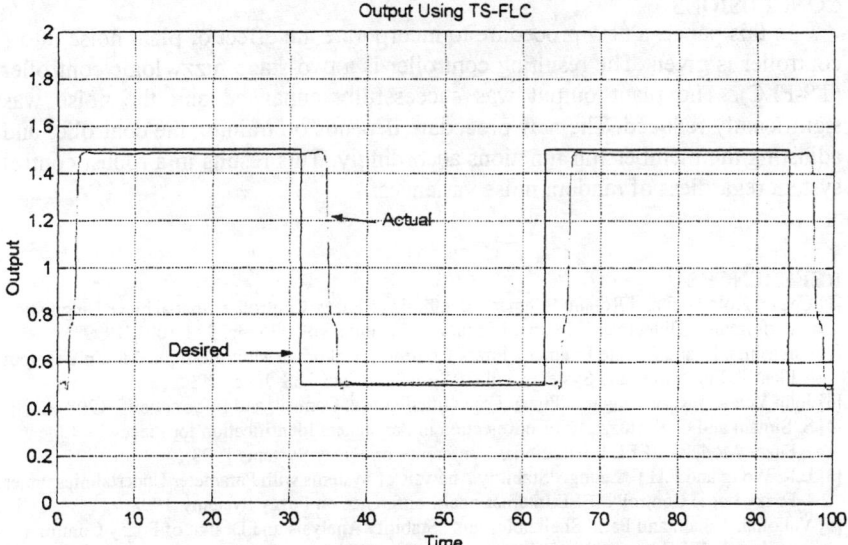

Figure 4. Comparison between actual and desired plant outputs using TS-FLC

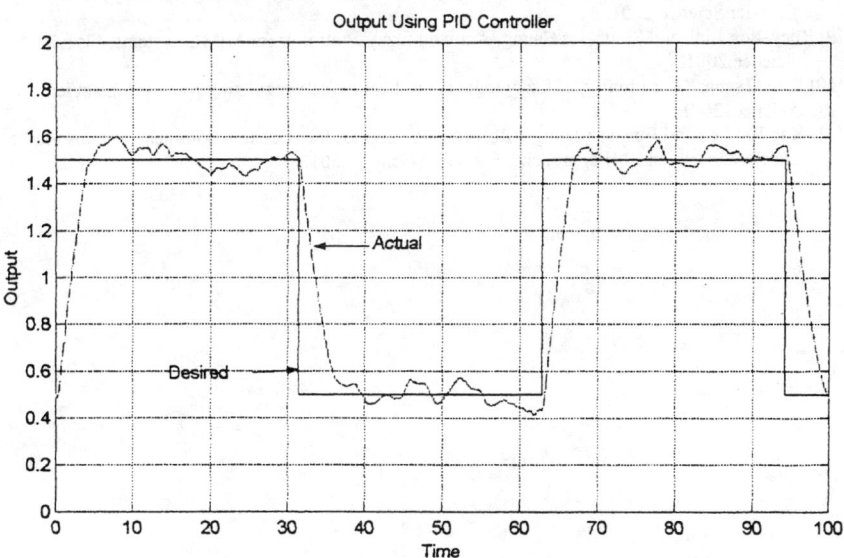

Figure 5. Comparison between actual and desired plant outputs using PID controller

CONCLUSIONS

In this paper, a new procedure to incorporate the effect of plant noise into a controller is given. The resulting controller is a two-stage fuzzy logic controller (TS-FLC). The plant output was successfully enhanced and the noise was significantly reduced. The new procedure depends on training the controller and adjusting the membership functions accordingly. This results in a robust control system regardless of random noise variances.

REFERENCES

[1] Cezary Kolodziej and Roland Priemer. "Synthesis of Fuzzy Controllers that achieve Linguistic Performance Objectives". Journal of Franklin Institute, vol. 336, pp. 983-1005, 1999.
[2] Chunshien Li and Roland Priemer. "Fuzzy Control of Unknown Multiple-Input-Multiple-Output Plants". Fuzzy Sets and Systems, vol. 104 no. 2, pp. 245-268, June 1999.
[3] John Yen and Reza Langari. "Fuzzy Logic Intelligence, Control, and Information", 1998.
[4] S. Simani and C. Fantuzzi. "Noise Rejection in Parameters Identification for Piecewise Linear Fuzzy Models". IEEE International Conference on Fuzzy Systems 1998.
[5] L.K. Wong and F.H.F. Leung. "Stability Analysis of Systems with Parameter Uncertainties under Fuzzy Logic Control". IEEE International Conference on Fuzzy Systems, 1998.
[6] Valiollah Tahani and Farid Sheikholeslam. "Stability Analysis and Design of Fuzzy Control Systems". IEEE International Conference on Fuzzy Systems, 1998.
[7] Dragan Kukolj and Slobodan Kuzmanovic. "Design of a near-optimal, wide range fuzzy logic controller". Elsevier Science, 2001.
[8] Jin-Ming Zhang and ren-Hou Li. "Stability and systematic design of fuzzy control systems". Elsevier Science 2001.
[9] Ruey-Jing Lian and Shiuh-Jer Haung. A mixed fuzzy controller for MIMO systems. Elsevier Science 2001.
[10] K.M. Tsang. "Auto-tuning of fuzzy logic controllers for self-regulating process". Elsevier Science, 2001.
[11] Kap Rai Lee and Eun Tae Jeung. "Robust fuzzy control for uncertain nonlinear systems via state feedback: an LMI approach". Elsevier Science, 2001.

FUZZY LOGIC FOR PREDICTING
SOIL HYDRAULIC CONDUCTIVITY USING CT IMAGES

Z. CHENG
Dept. of Computer Eng. &
Computer Sci., Univ. of Missouri
Columbia, MO 65211

C.J. GANTZER
Dept. of Soil & Atmospheric Sci.,
University of Missouri-Columbia

S.H. ANDERSON
Dept. of Soil & Atmospheric Sci.
University of Missouri
Columbia, MO 65211

Y. CHU
Dept. of Civil & Env. Eng., Univ.
of Illinois at Urbana-Champaign

ABSTRACT

Soil hydraulic conductivity is an important parameter used in many applications ranging from watershed management to hazardous waste site remediation. The objective of this study was to utilize a fuzzy rule-based system to predict soil hydraulic conductivity from computed tomographic (CT) images of soil structure in intact cores. Forty undisturbed soil cores (76 mm long x 76 mm diam.) were scanned by an x-ray CT scanner. CT images were analyzed by a multiple resolution blanket method (MRB). Pore sizes characterized by the MRB signature curves were reasoned by the fuzzy logic method. Each input variable was fuzzified and subsequently used in the fuzzy inference engine to make decisions. Crisp outputs were obtained by defuzzification. Fuzzy logic methods were shown to be useful for predicting hydraulic conductivity from CT images. The fuzzy logic model in this study was experimentally independent and was based on principles from fluid physics.

INTRODUCTION

Soil hydraulic conductivity is a highly variable and difficult to measure soil property. Hydraulic conductivity depends upon soil particle size distribution as well as soil structure. In the past decade, X-ray computed tomography (CT) has been used to quantify soil structure in intact soil cores (Zeng et al., 1996). However, analytical methods have not been developed which quantify and spatially relate the CT image data with soil hydraulic properties.

Recent work applied a multiple resolution blanket (MRB) method (Peleg et al., 1984) to quantify CT-measured soil structure and to correlate these MRB signatures with measured soil physical properties (Cheng, 2001). The utility of the MRB method for quantifying CT images of intact soil cores and thus quantifying macro-pore scale density and porosity within intact soil cores is useful for defining soil structure. The MRB signatures can be used to describe and classify differences in soil density distributions. Moreover, the MRB method can be applied on both fractal and non-fractal soil materials to obtain meaningful topological information (Peleg et al., 1984). The MRB estimation is not influenced by mathematical limitations of fractal theory.

Fuzzy set theory enables representation of gradual as well as abrupt transitions in space and has proven to be a flexible method (De Gruijter et al., 1997) for dealing with uncertainty in measurements. Fuzzy systems, which include fuzzy set theory and fuzzy logic, have potential to be applied to soil science and may provide a meaningful improvement and extension of conventional logic (McBratney et al., 1997). Studies have not been conducted to date, which utilize fuzzy set theory to predict soil hydraulic properties from CT images of soil cores.

This study proposes to (i) develop a fuzzy rule-based system to predict soil hydraulic conductivity using CT-measured soil porosity and (ii) compare the predicted soil hydraulic conductivity with estimates made from regression models.

METHODS OF ANALYSIS

X-ray CT images obtained from 40 intact soil cores (76 mm long by 76 mm diam.) were analyzed by a multiple resolution blanket (MRB) method. The soil cores were taken from Sanborn Field on the University of Missouri-Columbia campus. The X-ray CT scanner was a GE scanner with a resolution of 0.1- by 0.1- by 2 mm. Eight CT scans were taken within each soil core and were separated by an 8 mm distance.

The CT images were not filtered nor processed in any way that would introduce possible artifacts. The MRB method consisted of developing a signature curve for each CT image. The MRB signature curve is correlated to variations in the amount and size of soil pores. To simplify the results, each MRB signature curve was divided into three regions. The first region included the region related to small pores and was defined by a blanket integer less than 9. A blanket integer represents the thickness of the artificial blanket used to cover the image surface. An artificial blanket includes all points in three-dimensional space at a distance k from the original surface. The second region included the region related to medium sized pores and was defined by blanket integers ranging from 9 to 35. The third region was related to larger pores and was defined by blanket integers greater than 35. An example MRB signature curve is shown in Fig. 1.

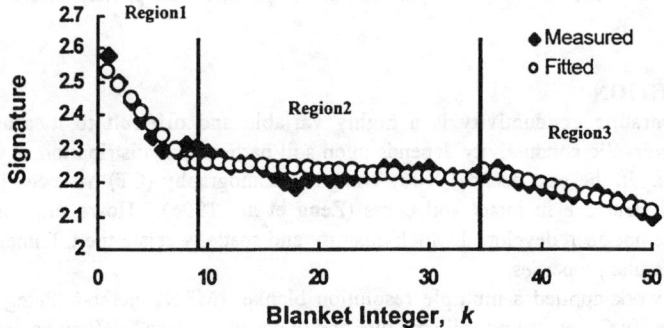

Fig. 1. MRB signature curve as a function of blanket integer segmented into three regions.

The data in each region was fit to a linear segment represented by two parameters: the intercept and the slope. The intercepts for the first and the third regions were used to develop the fuzzy logic sets. In essence, the intercept for the first region was the MRB signature value at blanket integer equal to 0 and the intercept for the third region was the MRB signature value at a blanket integer equal to 35. The intercept value for the first region is related to the frequency of small pores and the intercept of the third region is related to the frequency of larger pores. Each data set of intercept values were normalized to range from 0.0 to 1.0. Preliminary analysis indicated that these parameters were the best representations of the soil pore sizes or soil structure as defined by the CT images.

Fuzzy logic refers to the use of fuzzy sets in the representation and manipulation of vague information for the purpose of making decisions. A fuzzy control system uses a collection of fuzzy membership functions and rules, instead of Boolean logic, to interpret the data. Crisp values are first transformed into fuzzy values to be able to use them in applying rules formulated by fuzzy logic. A model of the fuzzy control system can be summarized by the following four steps. First, the crisp input data are fuzzified. Second, the fuzzified data are used in the fuzzy inference engine. Third, the output data from the fuzzy inference engine are composed. Fourth, the composed output data are defuzzified into crisp output. The separate steps in the fuzzy analysis system are described in the following paragraphs.

Fuzzification. The crisp input values were subjected to five membership functions to fuzzify the data. The five membership functions were defined as (i) very low (VL), (ii) low (L), (iii) medium (MED), (iv) high (H), and (v) very high (VH). Both the intercepts from the first and third MRB signature regions were each subjected to the five membership functions. The fuzzy number membership functions applied to both regions were as listed in the following equations:

$$VL = \begin{cases} 1.0 & 0 < x < 0.1 \\ 1.0 - 5*(x - 0.1) & 0.1 < x < 0.3 \\ 0 & \text{else} \end{cases} \quad [1a]$$

$$L = \begin{cases} 5*(x - 0.1) & 0.1 < x < 0.3 \\ 1.0 - 5*(x - 0.3) & 0.3 < x < 0.5 \\ 0 & \text{else} \end{cases} \quad [1b]$$

$$MED = \begin{cases} 5*(x - 0.3) & 0.3 < x < 0.5 \\ 1.0 - 5*(x - 0.5) & 0.5 < x < 0.7 \\ 0 & \text{else} \end{cases} \quad [1c]$$

$$H = \begin{cases} 5*(x - 0.5) & 0.5 < x < 0.7 \\ 1.0 - 5*(x - 0.7) & 0.7 < x < 0.9 \\ 0 & \text{else} \end{cases} \quad [1d]$$

$$VH = \begin{cases} 5*(x - 0.7) & 0.7 < x < 0.9 \\ 1.0 & 0.9 < x < 1.00 \\ 0 & \text{else} \end{cases} \quad [1e]$$

where x is the normalized intercept MRB signature value for either the first or third region.

Inference Engine. A fuzzy proposition can be formulated as an "IF – THEN" statement such as "if the small pore size has a low (L) frequency and the large pore size has a very high frequency (VH) then the hydraulic conductivity (K_{sat}) has a high probability that it is also relatively high in value." The conjunction "AND" used in the rule base of this decision making controller means that all preconditions should be true to a certain degree to give the premise of a nonzero degree of probability. The minimum operator for the fuzzy intersection of fuzzy sets for both small and large pore sizes were used here for the "AND" operator. The probability value for the premise of each logic rule was computed and applied to the output portion of each rule. This resulted in one fuzzy set to be assigned to each output data set. In this case, the output data set was defined for K_{sat} with nine degrees of membership: very low (R1), low (R2), medium low (R3), low medium

(R4), medium (R5), high medium (R6), medium high (R7), high (R8), and very high (R9). This rule structure for the fuzzy-logic based system was adopted (Table 1). Fuzzy rules were constructed based on information of the relationship between hydraulic conductivity (K_{sat}) and frequency of pore sizes. Output was predicted soil hydraulic conductivity.

The membership functions for the output were defined as follows:

$$R1 = \begin{cases} 1.0 & 0<x<0.1 \\ 1.0 - 10*(x - 0.1) & 0.1<x<0.2 \\ 0 & 0.2<x \end{cases} \quad [2a]$$

For i from 2 to 8,

$$Ri = \begin{cases} 10*(x - 0.1*(i-1)) & 0.1*(i-1)<x<0.1*i \\ 1.0 - 10*(x - 0.1*i) & 0.1*i<x<0.1*(i+1) \\ 0 & \text{else} \end{cases} \quad [2b - 2h]$$

$$R9 = \begin{cases} 10*(x - 0.8) & 0.8<x<0.9 \\ 1.0 & 0.9<x<1.00 \\ 0 & \text{else} \end{cases} \quad [2i]$$

where x is the normalized hydraulic conductivity (K_{sat}). The minimum inferring operator (MIN) was used where the output membership function was clipped off at a height corresponding to the rule and computed as the degree of probability.

Table 1. Fuzzy inference rules for predicting saturated hydraulic conductivity (K_{sat}) based on Poiseuille's Law.

Ksat		Large pores				
		Very Low	Low	Medium	High	Very High
Small pores	Very Low	very low	very low	medium low	high medium	high medium
	Low	low	medium low	medium	medium high	medium high
	Medium	medium low	low medium	high medium	high	high
	High	low medium	low medium	high medium	high	very high
	Very High	medium	medium	medium high	very high	very high

Composition. All of the fuzzy sets assigned to each output variable were combined together to form a single fuzzy set for each output variable. The maximum composition operator (MAX) was used here. The combined output fuzzy set was constructed by taking the point-wise maximum over all the fuzzy sets assigned to a term by the inference rule.

Defuzzification. The single fuzzy sets were defuzzified to crisp values of the hydraulic conductivity. The center-of-gravity method was used for this transformation.

RESULTS AND DISCUSSION

Fuzzy Logic Prediction. Results of the fuzzy logic prediction for the soil hydraulic conductivity (K_{sat}) are presented in Figure 2. This figure shows that the results of the predicted K_{sat} values by the fuzzy logic approach were relatively close to the measured data as illustrated by a statistical correlation (r = 0.79). The mean difference between measured and predicted data or the degree of coincidence was found to be 1.21.

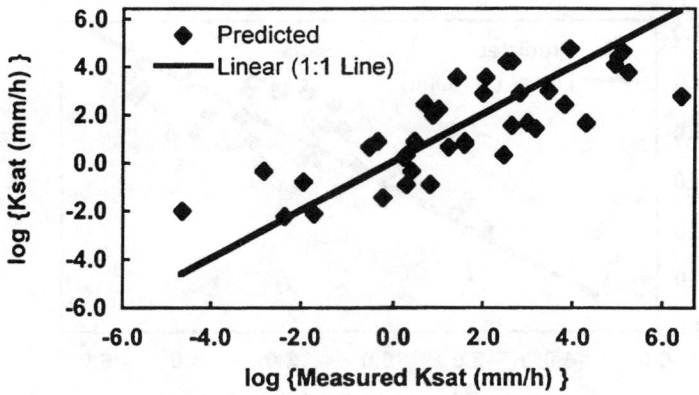

Fig. 2. Comparison of predicted K_{sat} using fuzzy logic vs. measured K_{sat}.

The rules used in this prediction were generally based on Poiseuille's law (Hillel, 1998) which indicates that the mean velocity (v) over the cross-section of a cylindrical pore is:

$$v = (R^2/\alpha\eta)\nabla p \qquad [3]$$

where R is pore radius, α is a constant (8 for cylindrical tube), ∇p is the pressure gradient, and η is the water viscosity. The rules were created based on the assumption that large pores dominate in water transport. If there is a high frequency of large pores, then the conductivity will be relatively high. A high frequency of small pores also contribute to water transport but this results in a lower hydraulic conductivity value.

It must be remembered that the CT image data used in this analysis were limited in resolution for detecting very small pore sizes. Thus, the pore sizes referred to in this analysis are in the macropore-scale range and do not include soil mesopores and micropores.

Regression Prediction. Results of stepwise multiple regression to predict hydraulic conductivity from the MRB signature values are presented in Fig. 3. The following model was obtained from the 40 soil core samples:

$$y = -107.2 + 27.3 * x_1 + 19.2 * x_2 \qquad [4]$$

where y is the logarithmic transformation *(log)* of hydraulic conductivity *{log (Ksat)}*, x_1 is the intercept of the third region, and x_2 is the intercept of the first region. Both of these intercepts were correlated (P < 0.01) with *log (Ksat)*. The above regression relationship had a coefficient of determination of 61.4%. The mean difference between measured and predicted data was found to be 1.23. This mean difference is slightly higher but not significantly different than the estimate of 1.21 for the fuzzy logic method.

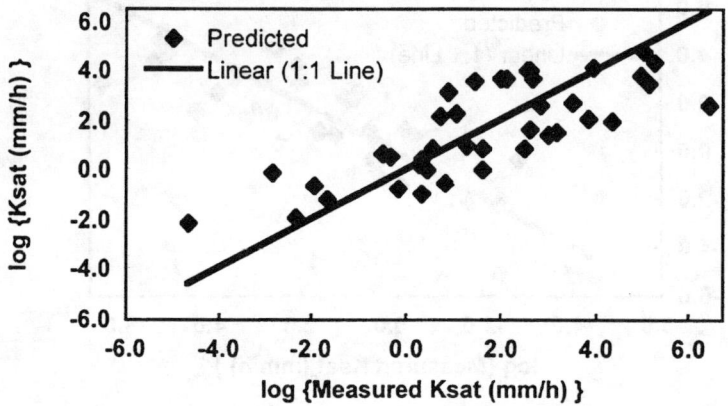

Fig. 3. Comparison of predicted K_{sat} using regression vs. measured K_{sat}.

CONCLUSIONS

In this study, a fuzzy logic method based loosely on Poiseuille's fluid transport relationship was used to estimate soil hydraulic conductivity from CT images of soil density and porosity. This fuzzy logic method was able to relate CT image data to saturated soil hydraulic conductivity. The fuzzy logic method proposed in this study was experimentally independent and was based on principles from fluid physics.

REFERENCES

Cheng, Z., 2001, "Characterization of Soil Structure Using Multiple Resolution Blankets of CT-Measured Density," M.S. Thesis. University of Missouri-Columbia, Columbia, MO.

De Gruijter, J. J., Walvoort, D. J. J., and Van Ganns, P.F.M., 1997, "Continuous Soil Maps—A Fuzzy Set Approach to Bridge the Gap Between Aggregation Levels of Process and Distribution Models," Geoderma, Vol 77, pp. 169-195.

Hillel, D., 1998, "Environmental Soil Physics", p. 248, Academic Press, San Diego.

Klir, G. J., and Yuan, B. 1995, "Fuzzy Sets and Fuzzy Logic; Theory and Applications," Prentice Hall, pp. 330-347.

McBratney, B. A., and Odeh, I. O. A., 1997, "Application of Fuzzy Sets in Soil Science: Fuzzy Logic, Fuzzy Measurements and Fuzzy Decisions," Geoderma, Vol. 77, pp. 85-113.

Peleg, S. J., Naor, R. H., and Avnir, D., 1984, "Multiple Resolution Texture Analysis and Classification," IEEE PAMI, Vol 4, pp. 518-523.

Zeng, Y., Gantzer, C. J., Peyton, R. L., and Anderson, S. H., 1996, "Fractal Dimension and Lacunarity of Bulk Density Determined with X-Ray Computed Tomography," Soil Science Society of America Journal, Vol. 60, pp. 1718-1724.

INTELLIGENT TECHNICAL STOCK ANALYSIS USING FUZZY LOGIC AND TRADING HEURISTICS

VAMSI KRISHNA BOGULLU, DAVID ENKE, & CIHAN DAGLI
Smart Engineering Systems Lab
Intelligent Systems Center
University of Missouri – Rolla

ABSTRACT

Technical analysis is a common method for using charts to predict the trend in a time series of stock prices. To increase efficiency, most chartists use multiple indicators to decide on market and individual stock direction. Unfortunately, there is a great deal of uncertainty as to which are the best indicators, as well as how the individual indicators should be interpreted. The following will employ the use of fuzzy logic for deciding upon the best combinations of indicators, as well as accounting for the uncertainty within each individual indicator, providing a signal to buy, hold, or sell. The performance of the fuzzy model incorporating trading heuristics is demonstrated for various individual stocks.

INTRODUCTION

Stock market analysis can be broadly classified into fundamental analysis and technical analysis. Fundamental analysis applies the tenets of firm foundation theory, which believes that the value of the stock is related to growth, dividend payout, interest rates, and risk associated with a particular company, among others [Malkiel, 1999]. Stock prediction then becomes a matter of choosing companies with strong combinations of the various fundamental factors. Recently, numerous intelligent models, in particular those that use artificial neural networks, have been developed to model these nonlinear patterns [Abhyankar et al. 1997; Chenoweth and Obradovic, 1996a; Desai and Bharati, 1998; Gencay, 1998; Leung et al., 2000; Motiwalla and Wahab, 2000; Pantazopoulos et al. 1998; Qi and Maddala, 1999; and Wood and Dasgupta, 1996]. Unfortunately, the use of neural networks makes it difficult to understand the underlying processes being modeled. This problem, along with the difficulty in predicting the proper input parameters, makes it hard to develop efficient and flexible neural network models.

Technical analysis is done by professionals who believe in the theory that prices discount all information, that is, all the information regarding the price of a stock is contained in the price itself [Edwards and Magee, 1957]. Technical analysis has a long history for helping to predict movement in a financial time series [Plummer, 1989; Block, et al, 1992]. With regard to modeling technical analysis indicators, Gencay and Stengos (1998) combined the use of two simple trading rules, moving averages and trading range breaks, with an artificial neural network to predict daily returns of the Dow Jones Industrial Average index. For this model, the buy and sell signals generated from the trading rules were used as inputs to the networks. Lam and Lam (2000) applied the differences of the 20-day moving averages and stock prices as the inputs to train an artificial neural network for generating the next closing price of the Hang Seng Index Futures Contract in Hong Kong. Chenoweth and Obradovic (1996b) attempted to combine the moving average convergence/divergence (MACD) indicator into the previously developed two feed-forward neural network models. It was used as a trading filter when both neural networks did not agree on the direction of the S&P 500 index.

Although fundamental and technical analysis does provide data that proves successful at times for predicting stock movement, models based on the data often produce unreliable results. This is especially true for technical analysis where modelers often employ a system that uses a number of different technical indicators, only to have the indicators give contradictory decisions. The indicators themselves do not usually have precise thresholds, but are subjectively based and vary between analysts. Furthermore, a way of combining the information between various indicators is also difficult and subjective. A possible solution for overcoming these problems, in particular with technical analysis indicators, is through the use of financial models based on fuzzy logic [Lim and Wunsch, 1999] and knowledge-based systems [Zargham and Sayed 1999]. Fuzzy logic and knowledge-based systems will allow for the modeling of the imprecise modes of reasoning that play an essential role in the human ability to make rational decisions by giving various degrees of truthfulness to the linguistic variables [Zadeh, 1988]. This will allow designers of technical analysis models the flexibility they need when setting thresholds, as well as the ability to combine previously separate, and often contradictory, indicator results.

Technical Analysis

One of the more popular and commonly used technical indicators is the moving average. The moving average y_a at time t of a signal y is

$$y_a = \frac{1}{T}\sum_{i=0}^{T-1} y(t-i) \qquad (1)$$

where T is the time interval over which the average is calculated. If the trend of y(t) is positive, the moving average y_a will be below y, while y_a is greater than y when the trend is negative [Vandewalle, et al, 1999]. Normally two moving averages y_1 and y_2 are considered, with each characterized by the intervals T_1 and T_2, respectively, where $T_2 > T_1$. If y_1 crosses y_2 from above, the event is called a sell position. If y_1 crosses y_2 from below, the event is called a buy position. In general, if the shorter time frame moving average (such as a 50-day moving average) is above the longer time frame moving average (such as a 200-day moving average), and moving away from the longer time frame moving average, the trend is considered positive.

For the following research the MACD (Moving Average Convergence/ Divergence) indicator and the Williams % R indicator are used as lagging and leading indicators, respectively. The MACD indicator is a commonly used indicator to study the relationship between two moving averages, and expands upon the simple comparison between two moving averages. The MACD is the difference between a 26-day exponential moving average and a 12-day exponential moving average. A 9-day exponential moving average of the MACD is also used as a signal line. When the MACD rises above its signal line, a buy position should be taken. When the MACD falls below its signal line, a sell position should be taken [Achelis, 1995]. Since the MACD uses the exponential moving average, the current trend tends to get more weight. This allows the MACD to be useful, and often prove profitable, in wide-swing volatile trading markets. For the MACD, the exponential moving average is often calculated by applying a percentage of today's closing price to yesterday's moving average. Figure 1 illustrates the MACD indicator for Cisco Systems' stock for the time period 6/5/98 to 11/10/98.

The Williams' %R is a momentum indicator developed by Larry Williams [Achelis, 1995]. The interesting predictive ability of this indicator concerns its ability to anticipate the reversal in the trend of stock prices. It performs this by measuring the 'overbought' or 'oversold' levels [Achelis, 1995]. Williams' %R is calculated using the following formula, where the value of n depends upon the type of analysis:

$$\left(\frac{\text{Highest High in 'n' periods - Today's Close}}{\text{Highest High in 'n' periods - Lowest Low in 'n' periods}} \right)(-100) \qquad (2)$$

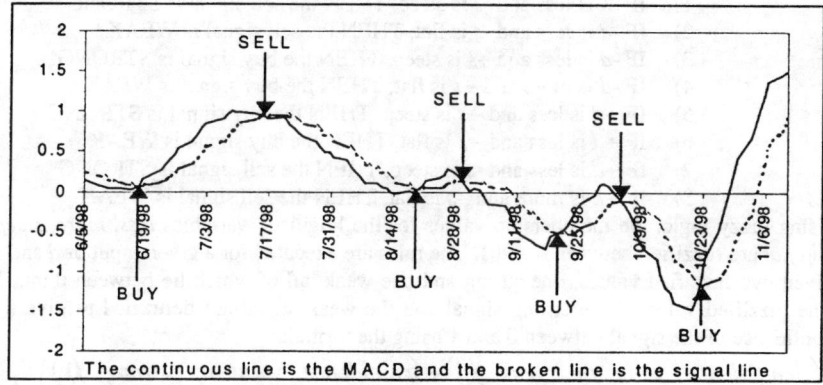

Figure 1: MACD Buy and Sell Signals for Cisco Systems (CSCO)

For short-term analysis, the value of n is small, while for long-term analysis the value of n is considerably larger. The value of n was set to 14 for the following research since only short-term analysis was being considered. For interpreting the result of Equation 2, a value of less than -80 is considered a buy, while a value more than -20 is a considered sell [Achelis, 1995]. Values between these extremes are considered a hold (either stay in or out of a position). Figure 2 illustrates the Williams' %R indicator for Cisco Systems' stock for the time period 6/5/98 to 11/10/98. Each time the chart formed a trough below -80 a buy position is taken, while a sell position is taken for a peak above -20.

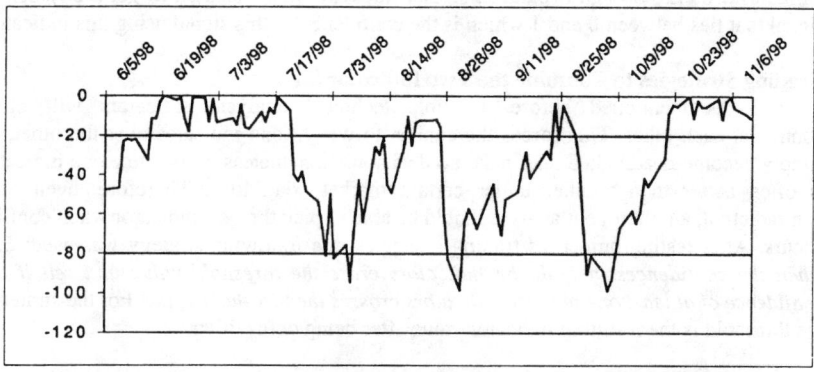

Figure 2: Williams' %R Buy and Sell Signals for Cisco Systems (CSCO)

Fuzzy Logic and Fuzzy Technical Analysis

To illustrate the use of fuzzy logic for technical analysis, the MACD and the Williams' %R indicators for analyzing stocks have been fuzzified. For the MACD indicator the significance of the signal in indicating the trend can be determined by studying the slope of the MACD line and then examining the difference between the MACD and the signal line. The slope of the MACD indicator signifies the trend, while the difference between the MACD and signal line helps in taking buy and sell positions, sometimes even prior to knowing the crisp value of the MACD as it cuts the signal line. These characteristics can be incorporated into the system using the following rules, where 'd' is the distance between the MACD and the signal line, and 's' is the slope of the MACD line:

1) IF -d is less and -s is steep, THEN the sell signal is STRONG
2) IF -d is less and -s is flat, THEN the sell signal is WEAK
3) IF -d is less and +s is steep, THEN the buy signal is STRONG
4) IF -d is more and +s is flat, THEN the buy signal is WEAK
5) IF +d is less and +s is steep, THEN the buy signal is STRONG
6) IF +d is less and +s is flat, THEN the buy signal is WEAK
7) IF +d is less and -s is steep, THEN the sell signal is STRONG
8) IF +d is more and -s is flat, THEN the sell signal is WEAK

Using fuzzy logic, the membership values for the linguistic variables *less, more, steep,* and *flat* are fuzzified between 0 and 1. The rules are executed for a given input of d and s, given two fuzzified values, one strong and one weak, all of which lie between 0 and 1. The fuzzified value of the strong signal and the weak signal are defuzzied to give the confidence of the signal between 0 and 1 using the formula:

$$\left(\frac{(confidence\ value\ of\ 'weak\,')(0) + (confidence\ value\ of\ 'strong\,')(1)}{confidence\ value\ of\ 'weak\,' + confidence\ value\ of\ 'strong\,'} \right) \quad (3)$$

If the MACD crosses the signal line from beneath then the above formula gives the degree of truthness for the buy signal. If the MACD crosses the signal line from the top then the above formula gives the degree of truthness for the sell signal. For the Williams' %R indicator, Equation 2 is used to obtain a value that lies between 0 and –100. The following rules were developed to obtain a buy or sell signal from the Williams' %R indicator:

1) IF this value is *around -80*, THEN the signal is BUY
2) IF this value is *around -25*, THEN the signal is SELL

The value given by Equation 3 is fuzzified to get the membership values for the linguistic variables *around –80* and *around –25*. The rules are then executed to get a buy or sell signal that lies between 0 and 1,which is the confidence of this signal using this indicator.

Trading Strategies to Combine the Two Indicator Signals

As mentioned before, multiple technical analysis indicators will often contradict each other. This forces the analyst to weight one indicator over the other, or simply become deadlocked and make no decision. Nonetheless, in practice one indicator is often better than another under certain market conditions. Therefore, even with contradiction, an intelligent analyst should be able to pick the best indicator when conflict occurs. After testing numerous trading strategies, the following strategy was used: *Buy when the confidences of both the indicators cross the threshold value and sell if the confidence of at least one of these indicators crosses the threshold value.* For the strategy, the threshold is the resulting indicator value after being defuzzified.

System Results

The system has been tested for six different companies over a time period of three years between 4/16/98 to 4/16/01. The results of the trading strategy has been compared to the MACD, Williams' %R, fuzzy MACD, fuzzy Williams' %R, and the buy-and-hold strategy where each stock was initially bought during the first day, held, and then sold on the last day. The values shown in Table 1 are the return on investment in dollar amounts for each dollar invested during both long and long/short trading. The average amount of portfolio investment at the end of the three-year time period is also given. During long trading, all money is invested in a long position after the first buy signal. Subsequent buy signals are ignored. All money is then taken out of the investment during the first sell signal. Again, additional sell signals are ignored. A long position is not taken again until another buy signal is generated. For subsequent new buys, the entire amount generated is again reinvested in entirety at the next buy signal. For the long/short trading, the same scenario applies as for the long trading, except now not only is the position sold during the first sell signal after a buy, but a short position is also taken (selling borrowed shares, hoping to buy them back at a lower cost). When the first buy signal is generated after the last sell, a long position is once again taken. The short position is also covered at that time (borrowed shares are bought and returned). For the final investment amounts in Table 1, values less than $1 represent a loss over the testing period, while values over $1 represent a gain.

By examining the table, a few things become apparent. First, the MACD and Williams' %R indicators work best for different stocks. The fuzzy MACD and fuzzy Williams' %R also appeared to work better for some stock than others. Also, the Fuzzy MACD gives better results that the regular MACD for four of the six stocks in the portfolio for the long trading. The fuzzy Williams' %R was also better than the regular Williams' %R for four of the six stocks (although for different stocks). While showing promise, the sample of stocks used here is too small to draw any general conclusions.

Since each indicator performed better for different stocks, the effect of the trading strategy should also have some impact on the performance of the portfolio. As illustrated in Table 1, the trading strategy is most effective for the long/short trading scenario, producing a portfolio investment amount of $1.92 (92% gain) compared to $1.60 (60% gain) for the long trading for the three-year testing period. This was probably due in part to the conservative strategy of buying only when both indicators give a buy signal, but selling when either gives a sell signal, thereby locking into profits. The advantage of the long/short strategy was also probably due to the volatility of the market during this time period, allowing a trader to take advantage of both up and down price swings in the stocks.

Comp/Indicator	BA	CSCO	DELL	DIS	GE	PFE	Portfolio
Buy & Hold	$1.31	$1.33	$1.29	$0.76	$1.59	$1.13	$1.24
MACD (L)	$1.95	$0.28	$1.52	$0.62	$0.88	$0.79	$1.01
Fuzzy MACD (L)	$1.16	$0.77	$1.86	$0.39	$1.29	$0.98	$1.08
Williams (L)	$0.78	$1.09	$1.40	$1.94	$1.65	$1.22	$1.35
Fuzzy Williams (L)	$0.86	$0.89	$0.80	$1.67	$1.98	$1.56	$1.29
Trading Strategy (L)	$3.14	$1.71	$1.69	$0.76	$1.44	$0.88	$1.60
MACD (L/S)	$2.35	$0.05	$1.66	$0.46	$0.45	$0.45	$0.90
Fuzzy MACD (L/S)	$0.96	$0.47	$1.91	$0.19	$1.04	$0.78	$0.90
Williams (L/S)	$0.39	$0.31	$1.29	$3.73	$1.83	$1.18	$1.46
Fuzzy Williams (L/S)	$0.47	$0.20	$1.29	$3.06	$2.59	$1.91	$1.59
Trading Strategy (L/S)	$6.51	$0.75	$1.58	$0.75	$1.31	$0.62	$1.92

Table 1: Results with both Long (L) and Long/Short (L/S) Trading

CONCLUSIONS AND FUTURE WORK

Fuzzy logic has been shown to be useful for making trading decision when using technical analysis, especially when a portfolio of stocks is being traded. Simulations of the models under various trading strategies has also shown that the type of strategy can have a significant impact on the results when more than one technical indicator is being used to make trading decisions. Further research should involve examining the impact of adding additional technical indicators, especially indicators that consider volume and volatility. The use of a neural network for setting membership functions, as well as for determining the best trading strategy method for combining the fuzzy confidences of the various indicators, should also be considered to better incorporate and learn from the previous trends in the market.

REFERENCES

Abhyankar A., L.S. Copeland, and W. Wong, "Uncovering nonlinear structure in real-time stock-market indexes: the S&P 500, the DAX, the Nikkei 225, and the FTSE-100." *Journal of Business & Economic Statistics.* 15 (1997): 1-14.

Achelis, S.B., *Technical Analysis From A to Z*, McGraw-Hill, 1995.

Brock, W., J. Lakonishok, and B. LeBaron, "Simple Technical Trading Rules and the Stochastic Properties of Stock Returns," *The Journal Of Finance,* Vol.47, No.5 (1992): 1731-1764.

Chenoweth, T., and Z. Obradovic, "A multi-component nonlinear prediction system for the S&P 500 Index." *Neurocomputing.* 10 (1996a): 275-290.

Chenoweth, T. and Z. Obradovic, "Embedding technical analysis into neural network based trading systems." *Applied Artificial Intelligence*, 10 (1996b): 523-541.

Desai VS, and R. Bharati, "The efficiency of neural networks in predicting returns on stock and bond indices." *Decision Sciences.* 29 (1998): 405-425.

Edwards, R.D. and E.J.Magee, *Technical Analysis of Stock Trends*, Magee, 1957.

Gencay R. "Optimization of technical trading strategies and the profitability in securities markets." *Economics Letters.* 59 (1998): 249-254.

Gencay, R. and T. Stengos, "Moving averages rules, volume and the predictability of security returns with feedforward networks." *Journal of Forecasting*, 17 (1998): 401-414.

Lam, K. & K.C. Lam, "Forecasting for the generation of trading signals in financial markets." *Journal of Forecasting*, 19 (2000): 39-52.

Lim, M.H. and D. Wunsch II, "A Fuzzy Perspective Towards Technical Analysis-Case study Of Trend Prediction Using Moving Averages" *Computational Intelligence for Financial Engineering*, Proceedings of the IEEE/IAFE (1999): 179-182.

Leung MT, H. Daouk, and A.S. Chen, "Forecasting stock indices: a comparison of classification and level estimation models." *International Journal of Forecasting.* 16 (2000): 173-190.

Malkiel, , B.G., *A random walk down Walk Street*, W.W.Norton & Company, 1999.

Motiwalla L, Wahab M., "Predictable variation and profitable trading of US equities: a trading simulation using neural networks". *Computer & Operations Research.* 27 (2000): 1111-1129.

Pantazopoulos KN, Tsoukalas LH, Bourbakis NG, Brun MJ, Houstis EN., "Financial prediction and trading strategies using neurofuzzy approaches." *IEEE Transactions on Systems, Man, and Cybernetics-Part B: Cybernetics.* 28 (1998): 520-530.

Plummer, T., *Forecasting Financial Markets: The Truth Behind Technical Analysis.* London, UK: Kogan Page, 1989.

Qi M, Maddala GS. "Economic factors and the stock market: a new perspective." *Journal of Forecasting.* 18 (1999): 151-166.

Vandewalle, N., M.Ausloos, and P.Boveroux, "The Moving Averages Demystified", *Physica A,* Vol. 269 (1999): 170-176.

Wood D, Dasgupta B., "Classifying trend movements in the MSCI U.S.A. capital market index – a comparison of regression, ARIMA, and neural network methods." *Computers & Operations Research.* 23 (1996): 611-622.

Zadeh, L.A., "Fuzzy Logic," *Computer*, Vol. 21, Issue: 4 (1988).

Zargham, M.R., and M.R. Sayeh, "A Web-based Information System for Stock Selection and Evaluation," *Advanced Issues of E-Commerce and Web-Based Information Systems, WECWIS, International Conference* (1999): 81-83.

FUZZY CORRELATION USED IN TEXT MULTI-CATEGORIZATION PROBLEM

HAO-EN CHUEH
Department of Information
Engineering, Tamkang University
Taipei, Taiwan, R.O.C.

NANCY P. LIN
Department of Information
Engineering, Tamkang University
Taipei, Taiwan, R.O.C.

ABSTRACT

Using the fuzzy correlation coefficient and the proper text multi-categorization rules, we can assign the unlabeled text into more than one category reasonably.

In general, text categorization is to assign an unlabeled text into one and only one predefined category according to its content, and we call this kind of categorizing procedure as text single-categorization. However, there are always ambiguity and uncertainty existing in the procedure, thus it seems reasonable, if an unlabeled text can be assigned into more than one category. This kind of categorizing procedure is called text multi-categorization.

Up to now, many methods have been proposed to deal with text single-categorization problem, which are not suitable to solve text multi-categorization problem.

In this paper, we introduce a method to deal with text multi-categorization problem. By using the concepts of fuzzy correlation and the graph-theoretical clustering methods, we have developed a text multi-categorization method, FCTMC. And we verify experimentally that this new method, FCTMC, when used to solve text multi-categorization problem, outperforms than those methods designed for text single-categorization problem.

Keyword: fuzzy correlation, text multi-categorization

INTRODUCTION

For academic research, or for better business opportunities, we need to refer to all kinds of text documents. It will be greatly helpful, if all those text documents are properly categorized, then we can easily locate the target text documents we need. Thus, text categorization is a must.

In the past, text documents were categorized by human labor. Nowadays, due to the popularity of personal computer and the rapid development of the Internet, numbers and kinds of text documents can be easily attached in an increasingly large scale. It is impossible to spend huge amount of time and energy categorizing text documents by manpower. Hence, in recent years, the automatic computerized text categorization has become a popular subject for research.

Automatic text categorization is a procedure to assign unlabeled texts to predefined categories with the help of computers, different methods have been proposed to solve this problem [1,11,12]. In most of the cases, an unlabelled text document is assigned into one and only one specific category, and we call this kind of categorizing procedure as text single-categorization. But, due to the fact that content of the unlabeled text document may be involved in different fields of issues, or the predefined categories may not be independent, text single-categorization may not be a good idea. Therefore, a method allowing an unlabeled text document to be assigned into more than one category is reasonable, and this is called a text multi-categorization procedure.

We find that the text single-categorization methods proposed up to now are not suitable to solve text multi-categorization problem. Therefore, in order to solve the text multi-categorization problem, we introduce a new method in this paper.

We use the concepts of fuzzy correlation [2,3,4,5,10] and the graph-theoretical clustering methods [8] to develop a text multi-categorization method, FCTMC. By this new method, FCTMC, we can let an unlabeled text document to be assigned into more than one category reasonably, whenever necessary, and solve text multi-categorization problem.

In Section 2, we explain the concept of text multi-categorization and the reason why the text single-categorization methods are not suitable for text multi-categorization problem. In Section 3, the essential concept of our new method, fuzzy correlation, will be explained. In Section 4, we interpret how to use fuzzy correlation coefficient in solving the text multi-categorization problem. In Section 5, an experiment and results of our procedures will be displayed. The Section 6 is our conclusion.

THE CONCEPTS OF TEXT MULTI-CATEGORIZATION

The text single-categorization methods proposed up to now are assigning an unlabeled text only into the category greatly related to the text. Using these methods to deal with text multi-categorization problem, an intuitive way is to set a uniform threshold. An unlabeled text can be assigned to several predefined categories, if the degrees of the relationship between this unlabeled text and those predefined categories, respectively, are higher than this threshold.

However, to decide a suitable uniform threshold for all the unlabeled texts is a very difficult task. For example, Table 2.1 and Table 2.2 show two cases of the degrees of the relationship between two texts (*Text*1 and *Text*2) and four predefined categories (*Category A*, *Category B*, *Category C* and *Category D*).

Text / Category	1	2
A	0.90	0.41
B	0.65	0.63
C	0.68	0.44
D	0.92	0.62

Table 2.1

Text / Category	1	2
A	0.12	0.91
B	0.68	0.93
C	0.69	0.23
D	0.71	0.69

Table 2.2

In Table 2.1, the degrees of the relationship between *Text*1 and the four predefined categories are all higher than the degrees of the relationship between *Text*2 and the categories, and it is difficult to find a suitable threshold. In Table 2.2, it is possible to find a uniform threshold, but it may be doubted. Assume that we set the uniform threshold at 0.7, then how can we explain that *Text*1 should not be assigned into category *B* or *C*, when in fact, the degrees of the relationship between *Text*1 and categories *B* or *C* are so close to 0.7.

If we borrow the text single-categorization methods to deal with text multi-categorization problem, it is a more proper processing procedure to set different threshold for different unlabeled texts respectively.

But, from those text single-categorization problem methods, the only information we can obtain is the degrees of the relationship between an unlabeled text and the predefined categories, where are not enough to set a threshold for each unlabeled text, since there is no way that we can always know the lower bound of each different threshold, thus we may assigned the unlabeled text into the categories zero or negatively related to it.

Therefore, in this paper, the concept of fuzzy correlation is used. By using the fuzzy correlation coefficient, we can not only know the degrees of the relationship between an unlabeled text and the predefined categories, but also obtain more information, which the fact whether the unlabeled text and the predefined categories are positively or negatively related. Based upon these information thus obtained and the concept of graph-theoretical clustering method, we can assign an unlabeled text into more than one category reasonably without setting a threshold for each unlabeled text, and we can ensure that no unlabeled text is assigned into the categories zero or negatively related to it.

What is the fuzzy correlation coefficient and how it is used in our method to solve a text multi-categorization problem will be explained in the following sections.

FUZZY CORRELATION

The correlation coefficient between fuzzy sets is called a fuzzy correlation coefficient. It is great useful, if we want to know the relationship between two vague variables or fuzzy attributes. Many methods have been proposed to evaluate the correlation coefficient of fuzzy data [2,3,4,5,10]. In our paper, we adopt the formula by Nancy P. Lin [4], because this method provides more information than others.

Suppose there are two fuzzy sets A and $B \subseteq F$, where F is a fuzzy space. The fuzzy sets A and B are defined on a crisp universal set X with membership functions μ_A and μ_B, then A and B can be expressed as:

$$A = \{(x, \mu_A(x)) \mid x \in X\} \quad (3.1)$$

$$B = \{(x, \mu_B(x)) \mid x \in X\} \quad (3.2)$$

where $\mu_A : X \to [0,1], \mu_B : X \to [0,1]$.

Assume that there is a random sample $(x_1, x_2, \cdots, x_n) \in X$, alone with a sequence of paired data, $((\mu_A(x_1), \mu_B(x_1)), (\mu_A(x_2), \mu_B(x_2)), \cdots, (\mu_A(x_n), \mu_B(x_n)))$, which correspond to the grades of the membership functions of fuzzy sets A and B defined on X. And then the correlation coefficient, $r_{A,B}$, between the fuzzy sets A and B is:

$$r_{A,B} = \frac{\sum_{i=1}^{n}(\mu_A(x_i) - \overline{\mu_A})(\mu_B(x_i) - \overline{\mu_B})/(n-1)}{S_A \cdot S_B} \quad (3.3)$$

where $\overline{\mu_A} = \frac{\sum_{i=1}^{n} \mu_A(x_i)}{n}$, $S_A^2 = \frac{\sum_{i=1}^{n}(\mu_A(x_i) - \overline{\mu_A})^2}{n-1}$, $S_A = \sqrt{S_A^2}$ (3.4)

, and $\overline{\mu_B} = \frac{\sum_{i=1}^{n} \mu_B(x_i)}{n}$, $S_B^2 = \frac{\sum_{i=1}^{n}(\mu_B(x_i) - \overline{\mu_B})^2}{n-1}$, $S_B = \sqrt{S_B^2}$ (3.5)

$\overline{\mu_A}$ and $\overline{\mu_B}$ denote the average membership grades of fuzzy sets A and B over the random sample, S_A and S_B are the respective standard deviations of fuzzy sets A and B.

In spite of the fact that values of the membership function are constrained between in [0,1], the value of the correlation coefficient derived from (3.3) lies between in [-1,1], which will show us not only the degree of the relationship between the fuzzy sets, but also the fact whether these two sets are positively or negatively related. This is greatly useful for us to deal with text multi-categorization reasonably because we can only assign the unlabeled text into those positively related categories.

FUZZY CORRELATION TEXT MULTI-CATEGORIZATION

In this paper, we develop a new method, FCTMC, to deal with text multi-categorization problem. The first step of our method is to decide the predefined categorizing categories.

Next, we randomly select a large amount of text documents, have all the words in these texts listed and record the frequency they appear, and make a frequency list, FL.

$$FL = \{(fk_i, N_{FL}(fk_i)) | i = 1 \cdots n_f\} \quad (4.1)$$

For each predefined category, we take some text documents for training data, and find out some keywords from these texts. Then all keywords of every predefined category are unified into one keyword set, KS.

$$KS = \{k_1, k_2, \cdots, k_n\} \quad (4.2)$$

For each keyword $k_i \in KS$, we use the membership functions $\mu_{Ci}(k_i)$ and $\mu_{Dj}(k_i)$ to express the degrees of the important in a predefined category C_i and an unlabeled text D_j.

$$\mu_{Ci}(k_i) = \frac{n_{Ci}(k_i)}{n_{Ci}(k_i) + n_{FL}(k_i)} \quad (4.3)$$

$$\mu_{Dj}(k_i) = \frac{n_{Dj}(k_i)}{n_{Dj}(k_i) + n_{FL}(k_i)} \quad (4.4)$$

where $n_{Ci}(k_i)$ and $n_{Dj}(k_i)$ are the frequency of the keyword k_i in C_i and D_j; $n_{FL}(k_i)$ is the frequency of the keyword k_i in frequency list, FL. Then the degree of the relationship between category C_i and text D_j can be express with a fuzzy correlation coefficient, $r_{Ci,Dj}$, computed by formula (4.5).

EXPERIMENT AND RESULT

We carry out our experiment by using $Reuter-21578$ database [6], built by David Lewis. From this database, first we randomly select 8666 texts, have all the words in these texts listed and record the frequency they appear, and make a frequency list.

Then we pick up 19 categories out of $Reuter-21578$ database. For each category, we select 500 most important words from the 100 training texts as the keywords. Thus, we have 19 groups of 500 keywords.

Next, we make the 19 groups of 500 keywords a uniform keywords set, KS.

$$KS = \{k_1, k_2, \cdots, k_n\} \quad (5.1)$$

Here, assume that the degrees of the importance of each keyword k_i in an unlabeled text D_j and the predefined category C_i are expressed as $\mu_{Di}(k_i)$ and $\mu_{Ci}(k_i)$.

And the degree of the relationship between D_j and C_i is $r_{Ci,Dj}$.

$$r_{Ci,Dj} = \frac{\sum_{i=1}^{n}(\mu_{Ci}(k_i) - \overline{\mu_{Ci}})(\mu_{Dj}(k_i) - \overline{\mu_{Dj}})/(n-1)}{S_{Ci} \cdot S_{Dj}} \quad (5.2)$$

The following steps are, evaluating the degrees of relationship between D_j and all predefined categories, building a minimal spanning tree, cutting this minimal spanning tree into subtrees, keep the subtree with the highest $r_{Ci,Dj}$ in it, pruning the categories with the zero or negative $r_{Ci,Dj}$ on the subtree we kept, and finish the text multi-categorization procedure.

In our experiment, we pick up 150 unlabeled texts randomly to test the accuracy of our text multi-categorization method. Both of the average *Recall rate* $R(\cdot)$ and average *Precision rate* $P(\cdot)$ [1,7] are used as the test criteria.

One of the popular methods to deal with text single-categorization problem is *Vector Space Model (VSM)* method [1,7]. There are two kinds of *VSM*, Global *VSM* and Local *VSM*.

In Global *VSM*, the unlabeled text and the predefined categories are all transformed into vectors in a uniform vector space. And then the degrees of the relationships between these vectors are expressed through distance or angle. According to these values of degrees, the category that the unlabeled text should be assigned into is decided. Whereas, in Local *VSM* [1], an unlabeled text is transformed into vector in different vector space, each time we want to evaluate the degrees of relationship between this unlabeled text and the different predefined category. Nevertheless, the distance or angle is still used to express the degrees of relationship between these vectors.

Hence, for a contrast, we use angle and distance to express the degrees of the relationship between an unlabeled text and the predefined categories in place of fuzzy correlation coefficient.

The results of our experiments are displayed in Table 5.1.

	Fuzzy correlation coefficient	Distance	Angle
Average $R(\cdot)$	0.657	0.315	0.655
Average $P(\cdot)$	0.783	0.947	0.676

Table 5.1

According Table 5.1 we can see that the average recall rate and the average precision rate of our method are both higher than using angle. And although the average precision rate of using distance is higher than our method, its average recall rate is low enough for us to say that our reformed method outperforms it.

CONCLUSIONS

For various reasons, we need a method allowing an unlabeled text document to be assigned into more than one category in some kinds of text categorizing cases. But, up to now, many methods are proposed to solve text single-categorization problem, which none· for text multi-categorization.

Thus, the main purpose of this paper is to introduce a new method to deal with text multi-categorization problem. And the results of experiment verify that this new method, when used to solve text multi-categorization problem, outperforms those methods designed originally for text single-categorization problem.

In the future, we will discuss about keyword selection technique of text multi-categorization and find out the best kind of keyword selection technique for our text multi-categorization method.

REFERENCES

Benkhalifa, M., Bensaid, A. and Mouradi, A., "Text categorization using the semi-supervised fuzzy c-means algorithm," NAFIPS International Fuzzy Information Processing Society, pp.561- 565 (1999).

C. Yu, "Correlation of fuzzy numbers", Fuzzy Sets and Systems, Vol. 55, pp. 303-307 (1993).

D. H. Hong and S. Y. Hwang, "Correlation of intuitionistic fuzzy sets in probability spaces", Fuzzy Sets and Systems, Vol.75, pp. 77-81 (1995).

Ding-An Chiang and Nancy P. Lin, "Correlation of fuzzy sets," Fuzzy sets and systems, Vol. 102, pp.221-226 (1999).

H. Bustince and P. Burillo, "Correlation of interval-valued intuitionistic fuzzy sets," Fuzzy sets and systems, Vol. 74, pp.237-244 (1995).

Http://www.research.att.com/~lewis/

Nai-Lung Taso, "The investigation of fuzzy document classification on Interent," Ma. D. Thesis, Tamkang University, Taipei.

P. Arabie, L. J. Hubert and G. De Soete, Clustering and classification, Singapore, River Edge, N. J. (1996).

Ralph P. Grimaldi, Discrete and combinational mathematics: an applied introduction, USA, Addison-Wesley Publishing Company, Inc. (1994)

T. Gerstenkorn and J. Manko, "Correlation of intuitionistic fuzzy sets", Fuzzy Sets and Systems, Vol. 44, pp.39-43 (1991).

Taeho C. Jo, "Text categorization with the concept of fuzzy set of informative keywords," IEEE International Fuzzy Systems Conference Proceedings, Vol. 2, pp.609-614 (1999).

Yiming Yang and Xin Liu, "A re-examination of text categorization methods," Proceedings of the 22nd Annual International ACM SIGIR conference, pp.42-49 (1999).

ADAPTING FUZZY SET-BASED TRADE-OFF STRATEGY IN ENGINEERING DESIGN

JIACHUAN WANG
University of Massachusetts
Department of Mechanical and
Industrial Engineering
Amherst, MA 01003, USA

JANIS P. TERPENNY
University of Massachusetts
Department of Mechanical and
Industrial Engineering
Amherst, MA 01003, USA

ABSTRACT
Multi-criteria decision methods are common in engineering design solution synthesis to accomplish trade-offs among competing goals. As design stages progress in a changing environment, the trade-off strategy could also change when more information is added. In this paper, fuzzy set-based trade-off strategy, which typically relies on specifying parameters about importance weights and degree of compensation among design attributes, is applied to address imprecision in the early stage of design. This paper provides descriptions of a neural network parameter learning method of function approximation to adapt trade-off strategies according to current preference information available from the designer's feedback. As the design process evolves, this adaptation should lead to more suitable and stabilized trade-off strategies. A case study experiment is included to demonstrate the approach.

1. INTRODUCTION

Design trade-off strategies are necessary in the synthesis phase of engineering design given the common situation of multiple conflicing goals that are normally not commensurable. Design trade-off strategies are needed to determine how the attributes with higher preference could compensate for the attributes with lower preference (Otto and Antonsson, 1991). Since design is an evolving interactive process with less precise information available in earlier stages than in later stages, the trade-off strategy could also change as design stages progress and more information is added.

Design trade-off strategy is based on preference information. Without preference information regarding various design attributes, there exists a set of efficient pareto-optimal solutions that cannot simply be compared with each other. Accordingly, the adaptation of a design trade-off strategy should be based on preference change. The current trade-off strategy identifies a member in the pareto solution set that is most successful in meeting the current design specifications.

Fuzzy set theory has condsiderable potential for addressing imprecision in design, however, imprecision reduction, at present, is implemented in a relatively ad-hoc manner (Giachetti et al., 1997). In our approach, fuzzy set-based aggregation function is applied as the compensative design trade-off strategy (Scott, 1999). This paper provides descriptions of a neural network parameter learning method of function approximation to adapt trade-off strategies by eliciting designer's preference information regarding alternative design solutions. As the interactive design process evolves, this adaptation should lead to more suitable and stabilized trade-off strategies. The next section introduces the fuzzy-set based design trade-off strategy. An illustration of the parameter learning process is then provided followed by a case study experiment. Conclusions are provided highlighting the need, value and future work of the proposed approach.

2. FUZZY SET-BASED TRADE-OFF STRATEGY

In our approach, the design trade-off strategy is realized in converting a multi-attribute decision problem into a single attribute decision problem by preference aggregation. Weighted-sum method is the simplest approach to accomplish this aggregation, but suffers from its inability to reach the non-convex portions of a pareto surface. Another common approach is utility theory, however, its derivation from economic decision making ignores the fact that certain aspects can not always be traded off in engineering design (Otto and Antonsson, 1993).

For multi-attribute preference aggregation, we choose to utilize fuzzy set theory as it provides a wealth of aggregation connectives between t-norms and t-conorms (Dubois and Prade 1985, Fodor 1994) that are compensative, and could be used as preference aggregation functions for engineering design.

An aggregation function p is compensative if and only if

$$\min_i (\mu_i(a_i)) \le p((\mu_1(a_1),\omega_1),...,(\mu_k(a_k),\omega_k)) \le \max_i (\mu_i(a_i)) \qquad (1)$$

This is the pareto property, or here we prefer to say that p is a mean, where the attribute function is a mapping from design variables $\bar{x}$ to attribute variable a_i: $\bar{x} \rightarrow a_i(\bar{x})$, $\forall i \in \{1, 2, ..., k\}$. Preference function $\mu_i(a_i)$ expresses preferences for design attribute a_i, it could be linear or nonlinear, and is normalized to between 0 and 1. ω_i is the weight assigned for preference $\mu_i(a_i)$.

A family of parameterized averaging operators of particular interest to design is the weighted root-mean-power. According to this operator, the overall preference of a design alternative can be calculated as:

$$p_s((\mu_1(a_1),\omega_1),...,(\mu_k(a_k),\omega_k)) = \left(\omega_1(\mu_1(a_1))^s + ... + \omega_k(\mu_k(a_k))^s\right)^{\frac{1}{s}} \qquad (2)$$

where $\sum_k \omega_k = 1, \omega_k \ge 0$

This family of preference aggregation functions accommodates all cases of compensation, from min ($s = -\infty$) to fully compensating ($s = 0$), and to super-compensating ($s > 0$), till max ($s = +\infty$). s is referred to as the compensation factor, i.e., the level of compensation among attributes. Table 1 lists these aggregation functions with different compensation factor s. Notice that $s = 1$ corresponds to one particular case, the weighted sum. These aggregation functions are suitable for engineering design problems, and by using these functions, the weights (ω_i) and compensation factor (s) pair of parameters to select any pareto optimal point can always be found (Scott, 1999).

Table 1. A family of parameterized weighted fuzzy aggregation functions

s	$P_s((\mu_1(a_1),\omega_1),...,(\mu_k(a_k),\omega_k)) = \left(\sum_i \omega_i (\mu_i(a_i))^s\right)^{\frac{1}{s}}$ $\sum_i \omega_i = 1, \omega_i \geq 0$	Root-mean-power
$-\infty$	$\min_i (\mu_i(a_i))$	Min
-2	$\dfrac{1}{\sqrt{\sum_i \left(\dfrac{\omega_i}{(\mu_i(a_i))^2}\right)}}$	Square root mean
-1	$\dfrac{1}{\sum_i \left(\dfrac{\omega_i}{\mu_i(a_i)}\right)}$	Harmonic mean
0	$\prod_i (\mu_i(a_i))^{\omega_i}$	Geometric mean
1	$\sum_i \omega_i \mu_i(a_i)$	Arithmetic mean
2	$\sqrt{\sum_i \omega_i (\mu_i(a_i))^2}$	Quadratic mean
$+\infty$	$\max_i (\mu_i(a_i))$	Max

3. LEARNING PROCESS FOR THE PARAMETERS

To accomplish trade-offs among competing objectives according to the fuzzy set-based preference aggregation function Eq. (2), we need to specify parameters about importance weights ω_i and compensation factor s. As stated before, these parameters are subject to change during the design process.

The parameter learning is a function approximation process. It could be implemented in an artificial neural network, based on the principle to modify parameters in such a manner as to minimize the cumulative error of the network. Here we consider the commonly used mean squared error (MSE) between the function output and the desired output from the designer's evaluation. This is a supervised learning by examples since the learning process takes place under the tutelage of the designer.

Stated more formally, define n as the number of examples to learn at one time. Let $\mu_j = (\mu_{j1}, \mu_{j2},...\mu_{jk})$ ($j = 1, ..., n$) be an input vector for which p_j is the approximated function output, and d_j is the desired output value from the designer's evaluation. $\varpi = (\omega_1, \omega_2, ..., \omega_k)$ is the current value of the weight vector, s is the current compensation factor. On comparing desired output to approximated function output, the squared error is $E = (p_j - d_j)^2$.

Since ω_i and s are different types of parameters to estimate, and the derivative of the overall preference with respect to s is difficult to compute, we deal with them in different ways. A parameter learning procedure has been specifically designed for this purpose. Figure 1 and the following steps describe this procedure.

Step 1: A random weight vector is chosen as the current weight vector.

Step 2: Fixing the weight vector to the value of the current weight vector, then the search of s^* to minimize the MSE becomes a one-dimensional search. As MSE is quasi-convex, s^* could be obtained by using nonlinear programming numerical line search methods, such as golden section method (Bazaraa et al., 1993), hence avoiding using derivatives.

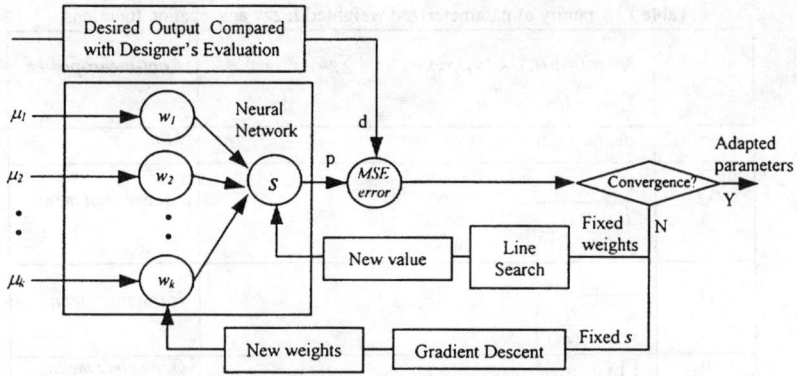

Figure 1. An artificial neural network for aggregation function parameter learning

Step 3: Fixing s to the value computed from step 2, weights are estimated by using gradient descent method, and weight changes are proportional to the negative gradient of the error. As $p(\omega_1, \omega_2, ..., \omega_n, s)$ is not continuous at $s = 0$, we consider these two situations separately.

$$\begin{cases} \dfrac{\partial E}{\partial \omega_i} = 2(p_j - d_j)\dfrac{\partial p_j}{\partial \omega_i} = 2(p_j - d_j)\dfrac{1}{s}(\omega_1 \mu_{j1}{}^s + ... + \omega_k \mu_{jk}{}^s)^{\frac{1}{s}-1} \mu_{ji}{}^s, s \neq 0 \\ \dfrac{\partial E}{\partial \omega_i} = 2(p_j - d_j)\dfrac{\partial p_j}{\partial \omega_i} = 2(p_j - d_j)p_j \ln(\mu_{ji}), s = 0 \end{cases} \quad (3)$$

Choosing $\Delta\omega_i$ to be a negative multiple of $\partial E / \partial \omega_i$, the weight update rule is:

$$\begin{cases} \Delta\omega_i = -\eta \dfrac{1}{s} p_j^{1-s}(p_j - d_j)\mu_{ji}{}^s, s \neq 0 \\ \Delta\omega_i = -\eta p_j(p_j - d_j)\ln(\mu_{ji}), s = 0 \end{cases} \quad (4)$$

η represents the learning rate. It is chosen based on the philosophy of steadily doubling the learning rate until the error value worsens within a certain range (Mehrotra et al., 1997). After updating the weights, the weight vector is normalized again. The new weights vector becomes the current weight vector.

Step 4: Check convergence condition $\| MSE \| < \varepsilon$ or $\| s_{k+1} - s_k \| < \delta$ (ε and δ are specified small numbers). If the convergence condition is not satisfied, go back to step 2. This process iterates until the convergence condition is satisfied.

4. CASE STUDY

Suppose we have three design attributes for an automotive panel meter, namely, cost (a_1), mass (a_2), and precision (a_3). We have a large set of solution candidates from which we would like to choose the best design candidate according to the current preference information. The designer is prompted with 10 selected design candidates for design evaluation, listed in left side of Table 2.

In order to test the accuracy of the parameter learning result, the designer's evaluation is based on a known pattern (consistent examples):

$$p(\mu_1, \mu_2, \mu_3) = \left(0.674 \mu_1^{0.5} + 0.111 \mu_2^{0.5} + 0.215 \mu_3^{0.5}\right)^2$$

Table 2. Examples of 10 consistent and inconsistent design candidates

Con.	$\mu_1(a_1)$	$\mu_2(a_2)$	$\mu_3(a_3)$	d_j	Inc.	$\mu_1(a_1)$	$\mu_2(a_2)$	$\mu_3(a_3)$	d_j
1	0.93	0.61	0.58	0.81	1	0.93	0.63	0.58	0.77
2	0.75	0.30	0.70	0.68	2	0.75	0.33	0.67	0.52
3	0.66	0.78	0.79	0.70	3	0.66	0.78	0.73	0.71
4	0.50	0.69	0.86	0.59	4	0.51	0.69	0.86	0.71
5	0.73	0.50	0.67	0.69	5	0.79	0.49	0.69	0.68
6	0.86	0.81	0.43	0.75	6	0.84	0.78	0.43	0.56
7	0.37	0.89	0.53	0.45	7	0.45	0.89	0.54	0.60
8	0.63	0.40	0.77	0.63	8	0.64	0.40	0.79	0.59
9	0.58	0.70	0.92	0.66	9	0.58	0.74	0.94	0.70
10	0.80	0.53	0.63	0.73	10	0.79	0.53	0.63	0.62

Table 3. Parameter learning iterations

Con.	s	ω_1	ω_2	ω_3	Inc.	s	ω_1	ω_2	ω_3
1	1.78	0.7487	0.068	0.1833	1	-3.77	0.5851	0.1277	0.2872
2	0.06	0.6497	0.1233	0.2270	2	-2.79	0.4647	0.1643	0.3710
3	0.56	0.6808	0.1100	0.2092	3	-1.78	0.3419	0.2121	0.4460
4	0.48	0.6763	0.1126	0.2111	4	-1.24	0.3091	0.2434	0.4475
5	0.49	0.6763	0.1126	0.2111	5	-1.03	0.3060	0.2559	0.4381
					6	-0.95	0.3057	0.2607	0.4336
					7	-0.93	0.3056	0.2619	0.4325
					8	-0.92	0.3056	0.2625	0.4319
					9	-0.92	0.3056	0.2625	0.4319

Set compensation factor s between −10 and 10, which is called the "interval of uncertainty". Using the golden section method, first assuming $\omega_1 = 0.2, \omega_2 = 0.5, \omega_3 = 0.3$, the line search process is shown in Figure 2, results in the optimal $s^* = 1.78$.

Next, the weight vector is estimated by fixing $s = 1.78$. According to experimentation, the learning rate η is chosen from 0.1 to 0.8, depending on when the error value starts to worsen. Experimentation shows that repeated learning of the examples will improve the accuracy of parameter estimation. The newly computed weight vector is $\omega = (0.7487\ \ 0.0680\ \ 0.1833)$.

The learning procedure proceeds five iterations with the result listed in the left side of Table 3. The final parameter is very close to the predefined preference aggregation function.

The same parameter learning procedure can be followed when the designer's preference changes to a different pattern. Suppose in the next learning phase, the designer gives inconsistent evaluations for 10 design candidates, listed in the right side of Table 2. Alternative 1, 3, 5, 7, 9 are evaluated using $p(\mu_1,\mu_2,\mu_3) = 0.5\mu_1 + 0.3\mu_2 + 0.2\mu_3$, and alternative 2, 4, 6, 8, 10 are evaluated using $p(\mu_1,\mu_2,\mu_3) = \left(0.2\mu_1^{-1} + 0.3\mu_2^{-1} + 0.2\mu_3^{-1}\right)^{-1}$. The parameter learning iterations are shown in the right side of table 3. It converges to:

$$p(\mu_1,\mu_2,\mu_3) = \left(0.3056\mu_1^{-0.92} + 0.2625\mu_2^{-0.92} + 0.4319\mu_3^{-0.92}\right)^{-1/0.92}$$

This shows that even if the designer's preference structure cannot be delineated in any one preference aggregation function, this parameter learning procedure will still converge to the parameters with the least mean square error, which is the most suited one.

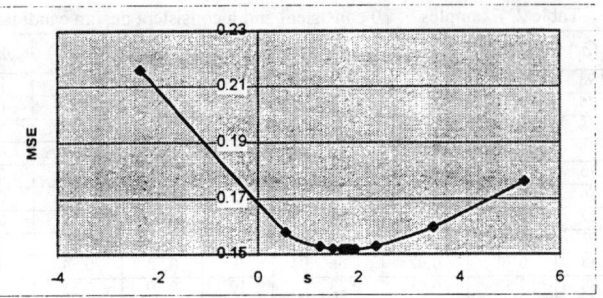

Figure 2. Line search of s^*

5. CONCLUSIONS

The premise of this work is that engineering design takes place in a dynamic changing environment. The adaptation of design trade-off strategy can help identify the current favorable solution set and make selections to meet various customer demands. This paper has presented a specifically designed neural network parameter learning procedure that can effectively approximate and adapt nonlinear fuzzy-set based design trade-off strategy according to current designer's preference information.

The elicitation of designer's preference in preference aggregation function approximation raises questions about the interaction between human and computer, as how to enhance the capability to correctly estimate design worth using computer tools. Some interesting future research topics include applying reinforcement learning in function approximation, and devising procedures to help transform human qualitative information to quantitative information.

REFERENCES

Bazaraa, M. S., Sherali, H. D., and Shetty, C. M., 1993, "Nonlinear Programming– Theory and Algorithms," John Wiley & Sons, Inc.

Dubois, D. and Prade, H., 1985, "A Review of Fuzzy Sets Aggregation Connectives,"*Information Sciences*, Vol. 36, pp. 85-121.

Fodor, J. and Roubens, M., 1994, "Fuzzy Preference Modeling and Multicriteria Decision Support", Kluwer Academic Publishers, Boston.

Giachetti, R. E., Young, R. E., Roggatz, A., Eversheim, W., and Perrone, G., 1997, "A Methdology for the Reduction of Imprecision in the Engineering Process," *European Journal of Operations Research*, Vol. 100, pp. 277-292.

Mehrotra, K., Mohan, C. K., and Ranka, S., 1997, "Elements of Artificial Neural Network," The MIT Press.

Otto, K. N. and Antonsson, E. K., 1991, "Trade-off Strategies in Engineering Design,"*Research in Engineering Design*, Vol. 3(2), pp. 87-104.

Otto, K. N. and Antonsson, E. K., 1993, "The Method of Imprecision Compared to Utility Theory for Design Selection Problems," *Proc. of the fifth ASME Design Theory and Methodology Conference*, Albuquerque, pp. 167–173.

Scott, M. J., 1999, "Formalizing Negotiation in Engineering Design," PhD Thesis, California Institute of Technology Engineering Design Research Laboratory.

PART IV: DATA MINING

PART IV: DATA MINING

A MODIFIED VERSION OF PARALLEL, SELF-ORGANIZING, HIERARCHICAL NEURAL NETWORKS FOR DETECTING RARE EVENTS

HOI-MING CHI
School of Electrical and Computer
Engineering, Purdue University,
West Lafayette, Indiana, USA

OKAN K. ERSOY
School of Electrical and Computer
Engineering, Purdue University,
West Lafayette, Indiana, USA

ABSTRACT
Parallel, self-organizing, hierarchical neural networks (PSHNNs) have been showed to be a powerful algorithm with many desirable properties and outperform conventional multi-layered neural networks in many situations. However, it is still inefficient in solving classification problems containing rare events. In this paper, we propose a modification to the original PSHNN algorithm by modifying the underlying data distribution. Experiments show drastic increase in the rare event detection probability.

INTRODUCTION

Many real-world classification problems consist of both rare and common events. Examples include breast cancer detection, text classification, remote sensing data, and computer intruder detection. Artificial neural networks usually fail to detect rare events because they tend to be ignored during network training (Valafar, 1993). One obvious solution is to get more samples from the rare events. However, this is not always possible because good data are sometimes hard to obtain. New algorithms, such as backpropagation neural network with stratifying coefficient and bootstrap stratification (Choe et al., 2000), have been developed to increase rare event detection accuracy. In this paper, we propose another scheme, the modified version of the PSHNN for rare event detection.

ORIGINAL PSHNN

The idea of PSHNN was developed by Ersoy and Hong (1990), (1993). The architecture involves a number of stages in which each stage can be a particular neural network (SNN). At the output of each stage, error detection is carried out, and a number of input vectors are rejected. These rejected vectors are then fed into the next SNN, possibly after they pass through a nonlinear transformation. Although the PSHNN has many desirable properties over regular multilayered neural networks, it is still inefficient in solving classification problems containing rare events. Similar to conventional multi-layered neural networks, the PSHNN tends to miss almost all rare events and classify them as common events. As a result, the error rejection interval covers almost the entire segment of the real line from zero to one. Therefore, after each stage, only a small number of vectors are accepted; the rest, whose network output value lies inside the error rejection interval, are rejected and are fed into the next SNN. The number of stages reached before training stops may grow to a very larger number. In order to overcome this problem, we employ a very simple idea of generating a new training data set by modifying the underlying data distribution. Then the PSHNN is trained using this new data set with a slight modification to the error rejection scheme, which will be discussed in detail in the next section.

MODIFIED VERSION OF PSHNN

Below we describe the training procedure of our proposed scheme, which resembles the original PSHNN training procedure. Figure 1 illustrates the training procedure using a block diagram. The differences between the two are also discussed.

Training Procedure

Assume: The number of iterations is upper bounded by k for each SNN.
Initialize: $i = 1$

1) Check to see if the data set contains rare events. If so, randomly select vectors from rare class and add them to the original data set until the ratio of common to rare events (class ratio) is one. Now we have a new data set.
2) Train SNN(i) by a chosen learning algorithm in at most k iterations using the new data set generated in step 1.
3) Check the output for each input vector.
   a) If there are no errors, stop the training.
   b) If errors, get the error detection bounds (see below) and go to step 4.
4) Select the input data that are detected to give output errors.
   a) If all the chosen data are in one class, then assign the final class number (FCN) as indicating that class. Stop the training.
   b) If not, go to step 5.
5) Increase i by 1. Go to step 1.

There are two major differences between our proposed scheme and the original PSHNN training procedure: 1) the addition of step 1 in our modified procedure to make the PSHNN be more "aware of" the rare events so as not to miss them in training, and 2) the removal of nonlinear transformation in the original procedure to speed up training.

The idea of randomly selecting vectors from rare events and combining them with the original data set is similar to the idea of boosting by sampling in the Adaboost family of algorithms (Freund, and Schapire, 1996), (Dietterich, 1999). In the Adaboost algorithm, a set of weights is assigned to all vectors in the original training set. After initial training, these weights are adjusted in such a way that the weights of the misclassified input vectors are increased while the weights of the correctly classified vectors are decreased. A new training data set is generated by drawing vectors from the original data set with replacement with probability proportional to their weights. In the case of rare events detection, we know beforehand that the neural network is going to miss almost all rare events. Therefore, even without using a set of weights for each vector, we can simply generate a new training data set by randomly selecting vectors from a rare class and adding them to the original data set until the class ratio is one. By doing so, we can skip the initial training step during which the initial weight values are determined. Another difference between our modified PSHNN and boosting by sampling is that in boosting, a number of new data sets are generated by sampling, and a consensus is done at the end by majority voting. In our modified

PSHNN, only one new data set is generated at each stage. However, an error detection mechanism is carried out at the end of each stage and the error-causing vectors are fed into the next SNN for further classification. Moreover, prior knowledge can be easily incorporated into the data set to increase performance (Schölkopf et al., 1996) by using our scheme.

The procedure to compute error and no-error bounds are described below.

Error Bounds

Assume: number of data vectors = l
 length of vector = n
 y_j^i = *jth component of the ith vector Y^i*

Initialize the error bounds as

$$\begin{cases} y_j^0(upper) = 0.5 \\ y_j^0(lower) = 0.5 \end{cases} \text{where} \quad j = 1, 2, \ldots, n$$

Initialize: i = 1

1) Check whether the ith data vector is an error-causing vector and belongs to the rare class. If so,
   a) If $y_j^i \geq 0.5$, then $y_j^i(upper) = MAX\ [y_j^{(i-1)}(upper), y_j^i]$
   b) If $y_j^i < 0.5$, then $y_j^i(lower) = MIN\ [y_j^{(i-1)}(lower), y_j^i]$
2) If $i = l$, the final error bounds are
   $r_j(upper) = y_j^l(upper)$
   $r_j(lower) = y_j^l(lower)$
   else $i = i + 1$ and go to step 1.
   end

No-Error Bounds

Assume: number of data vectors = l
 length of vector = n
 y_j^i = *jth component of the ith vector Y^i*

Initialize the no-error bounds as

$$\begin{cases} y_j^0(upper) = 1 \\ y_j^0(lower) = 0 \end{cases} \text{where} \quad j = 1, 2, \ldots, n$$

Initialize: i = 1

1) Check whether the ith data vector is an error-causing vector or belongs to the common class. If so, then $i = i + 1$ and go to step 1, else go to step 2.
2) Update the no-error bounds y_j^i for $j = 1, 2, \ldots, n$ as follows:
   a) If $y_j^i \geq 0.5$, then $y_j^i(upper) = MIN\ [y_j^{(i-1)}(upper), y_j^i]$
   b) If $y_j^i < 0.5$, then $y_j^i(lower) = MAX\ [y_j^{(i-1)}(lower), y_j^i]$
3) If $i = l$, the final no-error bounds are
   $s_j(upper) = y_j^l(upper)$
   $s_j(lower) = y_j^l(lower)$
   else $i = i + 1$ and go to step 1.
   end

In calculating the error and no-error bounds, the modified PSHNN differs from the original PSHNN in only one way: all training vectors are used in calculating the bounds in the original PSHNN, while in the modified PSHNN, only those training vectors from rare events are used. By doing so, much more emphasis is put on rare events, and this prevents correctly classified vectors from rare events from being rejected into the next SNN unnecessarily. Our experiments show that higher detection probability can be obtained by using only the vectors from rare events in computing the bounds.

Next we define three intervals $I_1(j), I_2(j), I_E(j), j = 1, 2, ..., n$ as follows:

$$I_1(j) = [r_j(lower), \quad r_j(upper)]$$
$$I_2(j) = [s_j(lower), \quad s_j(upper)]$$
$$I_E(j) = I_1(j) \cap I_2(j)$$

Then, an input training vector is classified as an error-causing vector if any y_j belongs to $I_E(j)$, the error rejection interval. All the error-causing vectors are fed into the next SNN. The testing procedure, which is exactly the same as the original PSHNN (without the nonlinear transformation), is described below.

Testing Procedure

Initialize i = 1
1) Input the test vector to SNN(i).
2) Check whether the output indicates an error-causing input data vector. If so, then,
    a) if it is the last SNN, then classify with the FCN.
    b) if it is not, go to step 1.
    else, classify the output vector.

Figure 2 shows the parallel implementation of the testing procedure. The testing data is fed into all the SNN's (three in this case) simultaneously. Then, the logic unit continually chooses the output of the next SNN until the output of the current SNN does not lie inside the error rejection region, or until the last SNN is reached.

The modified PSHNN algorithm can also be applied to problems with more than two classes. In k-class classification problems, k PSHNNs are constructed independently by training vectors from each of the k classes against vectors from the remaining k-1 classes, essentially reducing to k 2-class classfication problems.

DATA SETS USED IN THE EXPERIMENTS

We trained and tested the modified PSHNN using four data sets: synthetic Gaussian data sets I and II, adult income data set, and web page categorization data set. The two-class Gaussian data set is generated in Matlab 5.3.1, with the same covariance matrix $\begin{bmatrix} 1 & 1 \\ 1 & 9 \end{bmatrix}^T$, and different mean vectors $\begin{bmatrix} 0 & 0 \end{bmatrix}^T$ and $\begin{bmatrix} 3 & 3 \end{bmatrix}^T$. Both the adult income and web page categorization data sets are obtained from John Platt's web site (2001). The adult data set contains the training data set for UCI adult benchmark, which is to predict whether a household has an income greater than $50K. The web page categorization data set predicts whether a web page belongs to a category based on the presence of 300 selected keywords on the page. The information about these three data sets are summarized in Table 1.

EXPERIMENTAL RESULTS

All results are shown in Tables 2 and 3. Table 2 compares the performance between a backpropagation neural network (BP NN) and the modified PSHNN. In all cases, the modified PSHNN achieves a much higher detection probability than the BP NN, although its false alarm rate is also higher. For the synthetic Gaussian data sets, as the class ratio increases from 9:1 to 19:1, the BP NN's detection probability drops from 0.655 to 0.264, while there is only a slight decrease in the modified PSHNN's detection probability from 0.973 to 0.94. This shows the robustness of the modified PSHNN in situations with large class ratios. As shown in Table 3, the increase in detection probability after each stage shows that more and more vectors from rare events are detected, confirming the success of our scheme. However, the false alarm rate also increases after each stage. This observation indicates a trade-off between detection probability and false alarm rate, and suggests a way of choosing how much detection probability we want to achieve and how much false alarm rate can be tolerated.

CONCLUSIONS

The PSHNN has been proved to be a powerful algorithm with many desirable properties. By modifying the original PSHNN using the scheme proposed in this paper, we can successfully train the PSHNN to detect rare events with high detection probability, and at the same time inherit all the desirable properties of the original PSHNN.

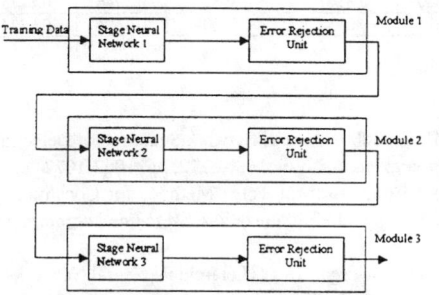

Figure 1. Block diagram of the training procedure with the PSHNN when the number of stages is 3

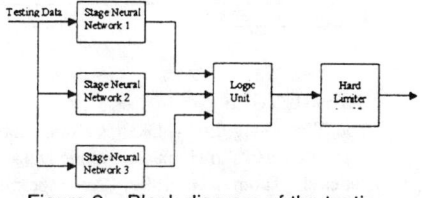

Figure 2. Block diagram of the testing procedure with the PSHNN when the number of stages is 3

Table 1. Information about the data sets used in the experiment

	Gaussian I	Gaussian II	Adult Income	Web Page
Dimension	2	2	123	300
Class ratio	9:1	19:1	3:1	37:1
Size of training set / testing set	2000/2000	2000/2000	1605/30956	2477/47272
Neural network architecture used (# of nodes in input/ hidden/output layers)	2/8/2	2/8/2	123/32/2	300/32/2

Table 2. Performance comparison between BP NN and modified PSHNN

		Training result			Testing result		
		Detection probability	False alarm rate	Overall accuracy	Detection probability	False alarm rate	Overall accuracy
BP NN	Gaussian I	0.6469	0.0123	0.9550	0.6550	0.0160	0.9511
	Gaussian II	0.2718	0.0038	0.9564	0.2640	0.0038	0.9595
	Adult income	0.5491	0.0555	0.8472	0.5113	0.0734	0.8267
	Web page	0.3347	0.0042	0.9803	0.2355	0.0037	0.9737
Modified PSHNN	Gaussian I	1.0000	0.2061	0.8137	0.9730	0.2059	0.8120
	Gaussian II	1.0000	0.0731	0.9309	0.9400	0.0850	0.9162
	Adult income	1.0000	0.1835	0.8605	0.8180	0.2320	0.7800
	Web page	1.0000	0.1748	0.8303	0.8068	0.1835	0.8162

Table 3. Modified PSHNN results after each stage

		Training result			Testing result		
		Detection probability	False alarm rate	Overall accuracy	Detection probability	False alarm rate	Overall accuracy
Gaussian I	1st stage	0.9062	0.0835	0.9155	0.9150	0.0944	0.9065
	2nd stage	0.9688	0.2616	0.7605	0.9750	0.2661	0.7580
	3rd stage	1.0000	0.2837	0.7435	0.9800	0.2906	0.7365
Gaussian II	1st stage	0.9727	0.0455	0.9555	0.8800	0.0489	0.9475
	2nd stage	0.9818	0.0831	0.9205	0.9200	0.0863	0.9140
	3rd stage	1.0000	0.0995	0.9060	0.9400	0.1058	0.8965
Adult income	1st stage	0.8681	0.2122	0.8070	0.8555	0.2154	0.8017
	2nd stage	0.9933	0.2659	0.7963	0.9006	0.2969	0.7507
	3rd stage	1.0000	0.2689	0.7956	0.9023	0.2997	0.7490
Web page	1st stage	0.8601	0.0669	0.9310	0.7844	0.0676	0.9280
	2nd stage	0.9930	0.1680	0.8367	0.8982	0.1746	0.8275
	3rd stage	1.0000	0.1615	0.8432	0.8930	0.1694	0.8324

REFERENCES

Choe, Wooyoung, Ersoy, Okan K., and Bina, Minou, 2000, "Neural Network Schemes for Detecting Rare Events in Human Genomic DNA," *Bioinformatics*, Vol. 16, No, 12, pp. 1062-1072.

Dietterich, Thomas G., 1999, "An Experimental Comparison of Three Methods for Constructing Ensembles of Decision Trees: Bagging, Boosting, and Randomization," *Machine Learning*, pp. 1-22.

Ersoy, Okan K., and Hong, Daesik, 1990, "Parallel, Self-organizing, Hierarchical Neural Networks," *IEEE Trans. Neural Networks*, Vol. 1, No. 2, pp. 167-178.

Ersoy, Okan K., and Hong, Daesik, 1993, "Parallel, Self-organizing, Hierarchical Neural Networks – II," *IEEE Trans. Industrial Electronics*, Vol. 40, No. 2, pp. 218-227.

Freund, Y., and Schapire, R. E., 1996, "Experiments with a New Boosting Algorithm," *Proc. 13th International Conference on Machine Learning*, Morgan Kaufmann, San Francisco, CA, pp. 161-169.

Platt, John, 2001, *Sequential Minimal Optimization*. Signal Processing Group of Microsoft Research http://www.research.microsoft.com/~jplatt/smo.html.

Schölkopf, B., Burges, C., and Vapnik, V., 1996, "Incorporating Invariances in Support Vector Learning Machines," *Artificial Neural Networks --- ICANN'96*, Springer Lecture Notes in Computer Science, Berlin, Vol. 1112, pp. 47-52.

Valafar, Faramarz, 1993, "Parallel Probabilistic Hierarchical Self-organizing Neural Networks," Ph.D. Dissertation, Purdue University, W. Lafayette, Indiana, USA.

EXPERIMENTS ON ROUGH SET BASED DATA MINING

SARAH COPPOCK, AIJING HE, LAWRENCE J. MAZLACK, YAOYAO ZHU

Computer Science Department
University of Cincinnati
Cincinnati, Ohio

ABSTRACT
An experimental investigation into unsupervised database mining was conducted. A novel paradigm for autonomous mining based on recursive partitioning was tested. The speculation is that increasing coherence will increase conceptual information. This in turn will reveal previously unrecognized, useful and implicit information. To assist our partitioning heuristics, a rough set based methodology called *Total Roughness* was designed to measure the crispness of a partition. This methodology was used in our experiments to help in partitioning as well as perform non-scalar data clustering. What is particularly noteworthy is that our approach sometimes partitions on multiple attributes. The feasibility of integrating rough set theory in unsupervised partitioning is evaluated and addressed. This paper focuses on the use of rough sets. It also provides some discussion of our other experiments.

INTRODUCTION
Data mining seeks to discover noteworthy, unrecognized associations between data items in an existing database. It is the process of extracting valid, previously unknown, comprehensible information from large databases.

Database discovery can be accomplished through a progressive reduction of cognitive dissonance. This can be done by progressively discarding attributes that have limited information value and by partitioning the data to increase information within the resulting partitions (Mazlack, 1996). Our heuristic approach focuses on reducing the attributes with much disorder to increase coherence and decrease dissonance.

Our concern was the extent that unsupervised data mining can be accomplished through conceptual information enhancement. The main prospects that could affect the result were:
- Selection of partitioning attribute
- The number of partitions developed
- When granules should be formed
- Where abstraction sets can be profitably used

In our experiments, two different variations of choosing partition attribute were tested:
- Imbalanced Partition: partition on the attribute with minimum disorder that would produce the most imbalanced partitions.

- Balanced Partition: partition on the attribute with minimum disorder that would produce the most balanced partitions.

The counts of the partitions created were also studied. In order to terminate the algorithm, a threshold was used as a stopping condition. This threshold could be the number of partitions created by the algorithm. How to decide the optimal number of the partition to be formed was unknown before the experiments. Two ways were tested in the experiments, they were:
1. Restrict partitioning to those attributes containing only a dissonance value of some small count, usually 2 or 3.
2. Allow any minimum dissonance value to be partitioned, but only created 2 or 3 new tables regardless of the actual counts of unique attribute values. In order to achieve this, different values could be grouped together according to their closeness.

Granulizing scalar data partitions the feature space so that similar objects are grouped into the same cluster while dissimilar objects are in different clusters. Forming granules on ordered data can help to increase the cohesion of our data and limit our search space. When to form the aggregation and how it will affect our mining result is an interesting aspect to be studied. In our experiment, one version was to apply the aggregation step in the early stage of the algorithm (before the partitioning), while another version was to apply it in the later stage of the algorithm (after the partitioning).

After partitions had been formed, abstraction sets were drawn from partitions. These abstraction sets may reflect some hierarchical structure between attributes. This information might be useful to perform generalization in the future, especially when further partitioning on disorder measurement becomes infeasible. Other researchers such as Han (1992) described "concept hierarchies" that are functionally equivalence to valid, labeled abstraction sets. In another sense, if an abstraction set is found to be useful in our partition, we deem it a discovery of valuable information. We used the experiment to verify the usefulness of this step.

USING ROUGH SETS TO SELECT PARTITIONING ATTRIBUTE

Whether using balanced partitioning or unbalanced partitioning, it is likely that the following problem may occur: There may exist more than one candidate partitioning attribute with the same disorder measure. A decision has to be made to which one should be chosen as the partitioning attribute.

We speculated that the solutions to this problem might be found by examining the interrelationship of the data. The rough set approach first introduced by Pawlak (1982, 1984a, 1984b) is a well-developed methodology to solve problems regarding the investigation of structural relationships among data, especially qualitative or imprecise data (Ziarko, 1991). We hypothesized that this approach may be applicable to solve the above problem.

The insight is that although there may exist more than one attribute that can form balanced partitions, the crispness of their resulting partitions might not be the same. As seen in *Figure 1*, partitioning on A2 also partitions on another attribute, A3. Partitioning on A2 therefore, forms a crisper partition.

Crisp partitioning is the most desired situation because there is no dissonance inside the partitions. However, it cannot always be achieved. So, the

A1	A2	A3		A1	A2	A3
A	C	D		A	C	D
A	C	D		A	C	D
A	D	E		B	C	D
B	D	E		A	D	E
B	C	D		B	D	E
B	D	E		B	D	E

(a) Partitions formed on A1   (b) Partitions formed on A2

Figure 1: Partitions formed on A1 vs. A2

partitioning leading to less uncertainty will be considered too. The degree of the uncertainty can be quantified and measured using rough set approach.

The idea behind rough sets (Dwinell, 2000) is that while some cases may be clearly labeled as being in a set X (the "positive region"), and some cases may be clearly labeled as not being in set X (the "negative region"), limited information prevents us from labeling all possible cases clearly. The remaining cases cannot be distinguished and lie in what is known as the "boundary region". Rough sets theory calls the positive region a "lower approximation" of set X that yields no false positives. The positive region plus the boundary region makes up an "upper approximation" of set X that yields no false negatives.

The lower approximation consists of all objects that definitely belong to the concept. The upper approximations of the concept consist of all objects that possibly belong to the concept (Pawlak, 1995).

Based on the principle of rough sets, we defined our own term called *Total Roughness*. *Total Roughness* is a heuristic metric used to represent the overall crispness of a partitioning towards every attribute. Before the definition of *Total Roughness*, we first define *Rough(i)*. *Rough(i)* is a heuristic metric as the mean *roughness* of all sub-partitions of an attribute i. Suppose there are n sub-partitions of an attribute i, to each sub-partition X, its roughness is:

$$R(X) = |IND_l(X)| / |IND^u(X)| \qquad (1)$$

where $IND_l(X)$ and $IND^u(X)$ are the cardinalities of the lower and upper approximations of X respectively. *Rough(i)* of this attribute i is:

$$Rough(i) = (\sum_{X=1}^{n} R(X)) / n \qquad (2)$$

where n is the number of sub-partitions of the attribute. *Rough(i)* ranges from 0 to 1. When *Rough(i)* = 1, it is a crisp partitioning on attribute i. When *Rough(i)* = 0, it is a maximum rough partitioning on attribute i. The larger *Rough(i)* is, the crisper the partitioning on attribute i is.

In order to measure the effect of the partitioning toward all attributes, we can calculate the mean *Rough(i)* of all attributes when partitioning using attribute k. We name it *Total Roughness*:

$$Total\ Roughness(k) = (\sum_{i=1, i \neq k}^{m} Rough(i)) / m \qquad (3)$$

where *m* is the number of attributes.

Total Roughness(k) also ranges from 0 to 1. The larger the *Total Roughness(k)*, the crisper the partition. The attribute with the largest *Total Roughness* measure will be selected as the partitioning attribute.

TIE BREAKING USING ROUGH SETS

In some cases, there will still exist a possibility of having a tie among two or more attributes when calculating *Total Roughness*. By examining the value distributions inside the boundary areas of each attribute according to each candidate partitioning, some difference in the distributions exists. In the real world data mining process, this generalization from the value distribution could provide valuable information. E.g., after we partitioned a student set into graduate and undergraduate, by studying the value distribution, we found that 99% of the excellent students are graduate student while only 1% of them are undergraduate students. Then we can draw the conclusion that **most** of the excellent students are graduate students. In order to reflect the value distribution, we suggest calculating an adjustment factor called *Total Distribution*.

For each attribute *i* having boundary area, if there are *n* different values in the boundary, to each value V_j, P_j is the number of V_j belong to the sub-partition 1, N_j is the number of V_j belong to the sub-partition 2.

$$Distrib(j) = 1 - |(P_j-N_j) / (P_j+N_j)| \qquad (4)$$

Equation (4) reflects the distribution of this value inside the boundary. When $P_j=N_j$, $Distrib(j)=1$, and V_j is thus equally distributed between sub-partition 1 and sub-partition 2. The greater the *Distrib(j)* is, the more evenly the value is distributed in two partitions. Therefore, smaller *Distrib(j)* is preferred.

To each attribute, we can calculate the mean distribution of its values as:

$$\Delta(i) = (\sum_{j=1}^{n} Distrib(j)) / n, \quad 0 < \Delta(i) \le 1 \qquad (5)$$

where *n* is the number of distinct values in the boundary of the attribute *i*.

When partitioning using attribute *k*, we can calculate the mean distributions of all other non-partitioning attributes. We name this the *Total Distribution:*

$$Total\ Distribution(k) = (\sum_{i=1}^{m} \Delta(i)) / m \qquad (6)$$

where *m* is the number of the attributes with boundary area.

When measuring the effect of the partitioning towards all attributes, the *Total Roughness* of the partition is our first concern and the value distribution inside each boundary is our second concern. Thus, the adjustment factor should not jeopardize the priority of the first. When choosing partitioning attribute, the *Total Roughness* of the candidate attributes should be compared first, and if they are the same, the *Total Distribution* will be compared to decide the partitioning attribute. By using this method, the possibility of a tie between two or more candidate partitioning attributes is further eliminated.

In our partition strategy, binary partitioning will be performed on a two-valued attribute first. If a two-valued attribute cannot be found, then partitioning on a triple-valued attribute may be considered to form a triple partitioning. Even if triple partitioning is not possible, an attribute of multiple values (greater than 3) will be used as a partitioning attribute.

When doing multiple valued partitioning, we'll still partition the data set into two in each step. Before partitioning, different values of the chosen attribute will be grouped into two parts. Because we are likely going to partition on non-scalar data, typical distance calculation methods like Euclidean distance can not be used. Here we suggest that the rough set theory may be useful to measure the closeness of two non-scalar objects.

In the real world, individuals that are classified as the same type must be alike in some ways; and objects that have been classified as different types represent individuals that have some natures not in common. Since the database is a partial reflection to the real world, analog can be drawn that closeness of the values of one attribute may be reflected by the closeness of the values of other attributes. A feasible partitioning should maximally satisfy this association, resulting in a crisper partition.

Earlier we defined the term *Total Roughness* based on rough set theory. Because *Total Roughness* reflects the crispness of the partitioning, different ways of partitioning can lead to different *Total Roughness* values. By comparing this value, we can find a reasonable way of partitioning.

Our partitioning strategy consists of two basic stages:
- *Choosing a multiple-valued attribute*: First, we choose the attribute with least distinct values as our partitioning attribute.
- *Finding the clustering center by calculating the Total Roughness:* We will find the value as the clustering center by comparing the resulting *Total Roughness* of each value. By comparing the resulting *Total Roughness* of each value, we can find the one that is most distant from the others. This value will be used as our clustering center. Later we will find other values that can be clustered with it.

This stage is divided into several steps:
1. First, an arbitrary value will be selected and put into one set (Set_1) and the rest will be in the other set (Set_2).
2. Second, calculate the *Total Roughness* of this partitioning.
3. Then, put the one in Set_2 back into Set_1. Pick up another one from Set_1 and put into Set_2. Calculate the *Total Roughness*. After exhaustively trying all values in the Set_1, select the value leading to the greatest *Total Roughness* as the starting value.
4. Clustering other values with the clustering center: From the Set_1, select a value and put it into Set_2. Calculate the *Total Roughness* of the resulting partition. If the *Total Roughness* is larger than the former partition, then the new partitioning is better than the former one. We will then keep this value in Set_2; otherwise, we mark this value and place it back into Set_1. We repeat this process with another unmarked value from Set_1 until all the values in Set_1 are marked.

After these steps, we partition the values into two parts that can partition the data set with the most *Total Roughness*.

CONCLUSIONS

Balanced and imbalanced partitioning tends to generate different results when the number of categorical attributes is relatively large. Balanced and imbalanced partitioning discovers knowledge of interest and both can be used as efficient tools in unsupervised data mining.

Our experimental results also proved that *Total roughness* provides a simple and efficient approach to measure the crispness of partitioning on non-scalar data. Clustering based on *Total roughness* can achieve reasonable and meaningful results.

Our experimental results also proved that the concept hierarchies automatically extracted from the data sets using rough sets concept are not only useful to perform generalization but valuable information themselves. If we do the generalization **before** the partitioning, we can limit the search space because the substitution will remove the redundant attribute. However, if we want to get interesting results about the detailed information that may be lost in the generalization, we should do it **after** the partitioning.

In conclusion, combined with all the methodologies we developed in this research, our experiments proved reducing cognitive dissonance by recursive partitioning is an effective method to find the useful information hidden in large databases. For more detailed and complete experimental results along with complexity, see [Coppock, 2001], [He, 2000], and [Zhu, 2000].

REFERENCES

Coppock, S., 2001, "Experiments on Rough Set Based Mining," http://homepages.uc.edu/~coppocs/ANNIE.pdf.

Dwinnell, W., 1 March, 2000, "re: fuzzy sets – rough sets? ," http://www.dbai.tuwien.ac.at/marchives/fuzzy-mail/index.html.

Han, J., Y. Cai, and Cercone, N., 1992, "Knowledge Discovery In Database," *Proceedings of the 18th VLDB Conference*, pp. 547-559.

He, A., 2000, "Unsupervised Data Mining By Recursive Partitioning," Technical Report, Computer Science, University of Cincinnati.

Mazlack, L. J., 1996, "Database Mining By Learned Cognitive Dissonance Reduction," *Proceedings of the Fifth IEEE International Conference of Fuzzy Systems (FUZZ-IEEE'96)*, New Orleans, pp. 1506-1511.

Pawlak, Z., 1995, "Vagueness and Uncertainty: A Rough Set Perspective," *Computational Intelligence: An International Journal*, Vol. 11 Issue, pp. 227-232.

Pawlak, Z., 1984a, "On Rough Sets," *Bulletin of the European Association for Theoretical Computer sciences* 24, pp. 94-109.

Pawlak, Z., 1984b, "Rough Classification," *International Journal of Man-Machine Studies* 20, pp. 469-483.

Pawlak, Z., 1982, "Rough Sets," *International Journal of Computer and Information Sciences* II(5), pp. 341-356.

Zhu, Y., 2000, "Unsupervised Database Discovery Based On Artificial Intelligence Techniques," Technical Report, Computer Science, University of Cincinnati.

Ziarko, W., 1991, "The Discovery, Analysis, and Representation of Data Dependencies In Database," in *Proceedings of the IJCAI Workshop On Knowledge Discovery*, Piatetsky-Shapiro, G. and Frawley, W. J., eds., *AAAI Press*, pp. 195-212.

DATA MINING USING 2-D NEURAL NETWORK SENSITIVITY ANALYSIS FOR MOLECULES

MARK J. EMBRECHTS (1) FABIO ARCINIEGAS (1)
MUHSIN OZDEMIR (2) CURT BRENEMAN (3)
KRISTIN BENNETT (4)

Departments of Decision Sciences and Engineering Systems (1),
Engineering Sciences (2), Chemistry (3), and Mathematical Sciences (4)
Rensselaer Polytechnic Institute, Troy, New York 12180, USA

ABSTRACT

This paper illustrates a data mining application using two-dimensional (2-D) neural network sensitivity analysis for gaining insight into data strip-mining problems. Data strip mining refers to predictive data mining problems where there are a large number of descriptive features and the number of features is on the order of or exceeds the number of data records (e.g., 100 to 1000 features for 50 to 300 data records). After reducing the number of descriptive features to a manageable set using 1-D neural network sensitivity analysis (e.g. 40 features), a 2-D neural network sensitivity analysis visualizes variations in the response to identify relevant combinations of features. 2-D sensitivity analysis enables the exploration of relevant relationships and features resulting in more robust, meaningful, and efficient models. This methodology was applied to an in-silico drug design problem with 64 molecules and 160 descriptive features.

INTRODUCTION

Data strip mining (Kewley 1998) refers to predictive data mining problems where there are a large number of descriptive features and the number of features is on the order of or exceeds the number of data records (e.g., 100 to 1000 features for 50 to 300 data records). As the number of features becomes larger than the number of data points, regression modeling becomes more challenging. Many traditional modeling approaches do not work well in this situation. A straightforward solution would be to collect more data records, but this can be expensive and/or physically impossible. Data strip mining applies an iterative process to prune out irrelevant features and to search the feature space for a good subset of descriptive features.

Data strip-mining problems are a challenge for predictive modeling because of (i) the inherent nonlinear relationships between the descriptive features and the activity of interest, (ii) the curse of dimensionality for a large set of descriptive features, and (iii) the danger of over-fitting because there is only a relatively small set of data records to work with. Moreover, most of the features present in a given problem are often not informative and the proper selection of the most relevant features becomes a key part for the success of generating good predictive models (Kewley 1998). This paper addresses neural networks for feature selection and introduces a novel feature selection methodology with two dimensional neural networks sensitivity analysis, where the saliency combinations of sets of two features is determined.

This paper is organized as follows: the first section introduces feature selection for predictive data mining, next section introduces 2-D sensitivity analysis, then an illustration of neural network sensitivity analysis for in-silico drug design is presented, and the final section provides general conclusions and recommendations.

FEATURE SELECTION FOR PREDICTIVE DATA MINING

Predictive data mining is the art of building predictive models for an activity of interest for a data set with several descriptive features. The ultimate aim of predictive data mining is to gain new insight and to explain the predictive models in comprehensible rules. Predictive data mining has three important steps: (1) feature selection, (2) model building, and (3) rule extraction. In this paper feature selection will proceed with sensitivity analysis, the model building is based on artificial neural networks, and the rule extraction is based on involving a domain expert to analyze the selected set of features for the model.

Feature selection presents several challenges. (i) All features are not equally informative (e.g., some features can be spurious with no relevance at all for the model under consideration). (ii) Features selection is often a non-monotonous problem (i.e., the best subset of p features does not always contain the best subset of q features, where q < p). (iii) The best subset of features is often model dependent. (iv) Some features are only relevant when another feature set is selected in combination. (v) Features can have several or many closely correlated or duplicate features.

The feature selection and modeling procedure outlined in this paper is based on sensitivity analysis of artificial neural network (ANN) models. There are several feature selection procedures documented in the literature that are neural network specific. These methods can generally be divided into three categories: (1) feature selection based on weight analysis (e.g., weight pruning, LeCun 1990), (2) feature selection based on feature saliency (i.e., sensitivity analysis, Zurada 1994, and Kewley 2000), and (3) heuristic feature selection methods (e.g., based on genetic algorithms or simulated annealing, Zheng 2000).

In this paper, we present sensitivity analysis for feature selection for predictive data mining. Sensitivity analysis determines the saliency of each of the features in a machine-learning model and reduces the number of descriptive features to predict an activity of interest. Sensitivity analysis uses a trained machine-learning model to determine the sensitivities for the features of the model. The sensitivity of a feature is defined as the output range for that feature from the machine learning model when all the descriptive features are held constant (i.e., at their average or mean value) and the feature of interest is changed within it's allowable range (i.e., in this case between 0 and 1). For 1-D feature selection sensitivity analysis, features can be dropped by comparing its sensitivity with the sensitivity of an addition random variable that is added as an additional phantom input feature to the model.

TWO-DIMENSIONAL FEATURE SELECTION SENSITIVITY ANALYSIS

Two-dimensional (2-D) neural network sensitivity analysis is similar to 1-D sensitivity analysis, but now the saliency of the features is determined for a pair of features by changing two features simultaneously, while holding the other features frozen at their average or median value (For more details about one-dimensional feature selection sensitivity analysis, see Embrechts 2001). For this paper the purpose of the 2-D sensitivity analysis was not directly to look for a strategy to drop features based on the 2-D sensitivity analysis, but to try to understand how combinations of features explain the behavior of the predictive model. For that reason, contrary to the 1-D sensitivity analysis, no random gauge variable was added to the original dataset.

The main goal for 2-D sensitivity analysis is to visualize variations in the neural model response (i.e., the activity to predict with the model) in order to identify relevant combinations of descriptors. Each relevant combination of two descriptive features can then be analyzed further (e.g., by now performing a 3-D analysis based on the original two features) and be used to look for interesting patterns and relationships. It is anticipated that future code versions will implement also a strategy for feature selection based on 2-D sensitivity analysis.

The procedure of 2-D sensitivity analysis is very similar to 1-D sensitivity analysis. The basic construction of the "test" dataset, however, is quite different. In this case, for each feature a "test" dataset is created in which each feature (base feature) is combined with all possible levels for the others features. This can be done in two ways: by fixing the base feature levels while changing the levels of the other features, or by changing the base feature levels while fixing the levels of the other features. At the end, all "test" datasets per feature are added together, being the final size of this dataset L*L*M rows by M features.

The sensitivity measure used for a 2-D sensitivity analysis is the (largest) range for all possible level combinations for each pair of features. The number of unique combinations is equal to M*(M-1)/2, where M is equal to the number of descriptive features. The sensitivities can then plotted in a matrix plot to facilitate the identification of the most relevant combinations of features (e.g., see Figure 4). Finally, for each one of the relevant feature's combinations, a 2-D plot of input changes versus output can be plotted as tool for exploring and understanding the relationships between descriptors. An example is shown in Figure 1.

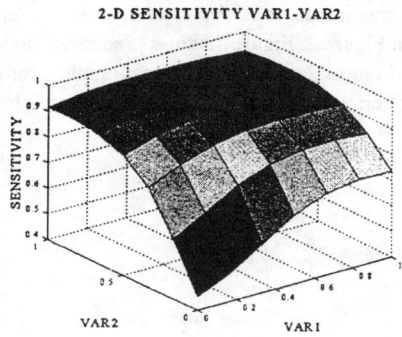

Figure 1. 2-D Sensitivity for a given pair of features

The datasets constructed to perform the 2-D sensitivity analysis can become very large when there is a large number of features and sensitivity levels. For example, for 40 features, 13 levels and 5 significant numbers, the datafile related to the 2-D sensitivity analysis is on the order of 20 MB. For the full dataset with 160 features this file size would increase to 320 MB, because the size of the dataset scales quadratically in the number of features and the number of sensitivity levels.

RESULTS

To illustrate the effectiveness of 1-D and 2-D sensitivity analysis for data strip-mining an "in-silico" drug design problem was analyzed. This is a predictive modeling problem with 64 data records with 160 TAE derived wavelet descriptors (Breneman 1995) for an HIV (Human Immunodeficiency Virus) related reverse transcriptase inhibitor dataset (Breneman 2000).

For this study the StripMiner™ code was used, which is shell wrapped around the MetaNeural™ neural network code. StripMiner™ implements several different modeling and feature selection methods in a bootstrap and bagging aggregation mode. The neural network model is a standard feedforward multi-layered perceptron trained with the backpropagation algorithm (Haykin 1994). The neural networks have two hidden layers and are oversized (e.g., there are two hidden layers with 13 and 11 neurons respectively, for 160 inputs). The numbers of neurons in the hidden layers are not changed during the successive feature reduction stages.

Training was halted with early stopping based on policy. The early stopping error used for all drug design applications with MetaNeural™ is set to 0.105. The same early stopping policy has been used successfully for at least ten different drug design related datasets in the past, and evolved by trial and error. The performance of the models is not very sensitive to the exact value of the stopping criterion (i.e., as long as the stopping criterion remains within the range of 0.08 – 0.12). Because of the early stopping, the neural network results are not very sensitive to the number of neurons in the hidden layers. On the other hand, because of early stopping and because of the fact that the models use oversized ANNs, the models tend to be more linear than would be the case with small optimally tuned ANNs. The model predictive quality will be reported with three distinct measures for the error: q^2, Q^2, and the RMSE (See Ozdemir 2001 for details).

Results for the Full and Reduced Dataset with 1-D Sensitivity Analysis

Predictive results were first obtained for the full wavelet-descriptor based HIV dataset with 160 descriptors using a bootstrap aggregation (bagging) approach. Feature reduction using sensitivity analysis was halted at 35 features. The results for the full dataset (i.e., with 160 features) are shown in Figure 2. Figure 3 shows predictive modeling results for the dataset with 35 selected features. Figures 2 and 3 are aggregation plots for all the bootstrap validation sets combined. The prediction variances are also shown in these plots for each molecule.

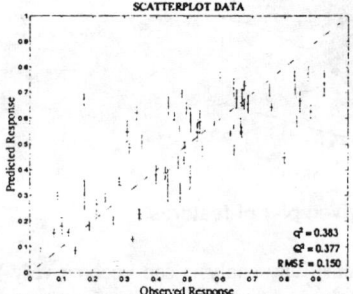

Figure 2. Predictive results for full HIV dataset with 160 features

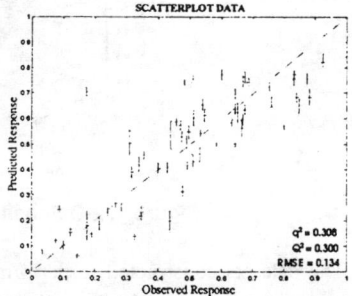

Figure 3. ANN predictive results for the reduced HIV dataset with 35 features

Looking at the results for the full and reduced datasets, it was found that most of the improvements were due to better predictions for the molecules with low/medium bio-activity values. For example, Q^2, q^2, and RMSE were reduced to 0.30, 0.303, and 0.134 for 35 features from values of 0.377, 0.383, and 0.150 for all 160 features, respectively. Thus, the predictive model obtained from the selected features with 1-D sensitivity analysis displays a better performance than that with the whole feature dataset.

Two-Dimensional Exploratory Sensitivity Analysis Results

For this paper the purpose of the 2-D sensitivity analysis was not directly to look for a strategy to drop features, but to try to understand how combinations of features explain the behavior of the predictive model. Therefore, a 2-D sensitivity analysis was applied to the 35 features selected with 1-D sensitivity analysis.

The 2-D sensitivities were plotted in a matrix to facilitate the visualization and exploration of the sensitivities (Figure 4). Note that because the sensitivity of pair x-y is equal to the sensitivity of pair y-x, this matrix plot is symmetrical. The features in this plot are ordered based on the sensitivities obtained from the 1-D feature selection (feature 1 was the most sensitive, 2 was the second most sensitive, etc). The darker the square for a sensitivity pair, the more important the combination of the corresponding features for predicting the response.

Looking at Figure 4, it is easy to pick up the most sensitive feature combinations: fifteen (15) pairs displayed the highest sensitivity, with a value higher than 0.45. These sensitivity pairs were mostly related with features 1, 2, 3, 4, 9, 18, 19, 20, and 26. Most of the feature pairs, however, displayed low sensitivities (less than 0.25), which may indicate that some descriptors could be still dropped without affecting the predictive performance of the remaining feature subset. For the 9 features above, a similar NN was used for predicting the response (see section 5.1). The overall performance indicators for this subset resulted in an estimated $q^2 = Q^2 = 0.36$, and a RMSE of 0.146, which is still better than the full dataset.

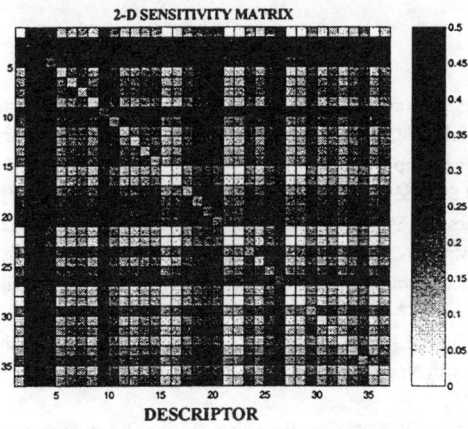

Figure 4. 2-D Sensitivity matrix for 35 features obtained from 1-D feature selection sensitivity analysis

The most sensitive feature combinations were visualized in 2-D plots (see Figure 5). Interestingly, it was expected that some of the descriptors would display some kind of non-linear behaviors like the one shown in Figure 1, but most of the relevant features display only slightly non-linear relationships (i.e., almost no curvatures in the model response).

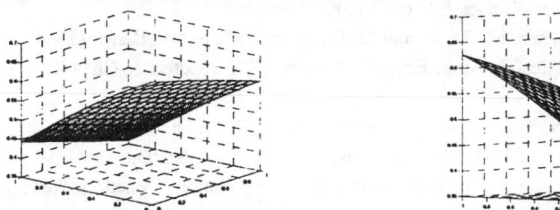

A. Var #3 and Var #4   B. Var #9 vs. var #10

Figure 5. Some non-linear 2-D sensitivity plots. Notes: (i) sensitivities are plotted in the z-axis; (ii) plots are not at the same scale; (iii darker zones represent low sensitivities values.

CONCLUSIONS AND RECOMMENDATIONS

One and two-dimensional sensitivity analysis have proven to be very efficient, stable, and reliable feature reduction methodologies. These methodologies, however, can be improved in many ways. For the 1-D sensitivity analysis, it is possible and worth exploring to add more random gauge variables (possibly from different distributions) at the same time. For the 2-D sensitivity analysis, each relevant combination of two descriptive features can be analyzed further (e.g., by now performing a 3-D analysis based on the original two features) and be used to look for interesting patterns and relationships.

Finally, while sensitivity analysis in this paper was based on neural network models (1 and 2-D sensitivity analysis), and can easily be adapted to other machine learning techniques such as support vector machines and partial least squares.

ACKNOWLEDGEMENT

The support from the National Science Foundation for this work (award number IIS-9979860) is greatly appreciated.

REFERENCES

Kewley, R., Embrechts, M., and Breneman, C. 1998, "Neural Network Analysis for Data Strip-mining Problems," in Intelligent Engineering Systems through Artificial Neural Networks. C. H. Dagli et. al., eds., ASME Press, pp 391-396.

LeCun, Y., Denker, J.S., and Solla, S.A., 1990, "Optimal Brain Damage," in Advances in Neural Information Processing Systems, pp. 598-605, D. Touretzky, Ed., Morgan Kaufmann, CA.

Zurada, J.M., Malinowski, A., and Cloete, A., 1994, "Sensitivity Analysis for Minimization of Input Dimension for Feedforward Neural Networks," Proc. IEEE International Symposium on Circuits and Systems, Vol. 6., pp. 447-450, London, May 30 – June 3.

Kewley, R.H., Embrechts, M.J., and Breneman, C., 2000, "Data Strip-mining for the Virtual Design of Pharmaceuticals with Neural Networks," IEEE Transactions on Neural Networks, Vol. 11, No. 3, pp. 668–679.

Zheng, W., and Tropsha, A., 2000. "Novel Variable Selection Quantitative Structure–Property Relationship Approach Based on k-Nearest Neighbor Principle," J. Chem. Inf. Sci., Vol. 40, pp. 185-194.

Embrechts, M.J., Arciniegas, F.A., Breneman, C., Bennett, K., and Lockwood, L. , 2001. "Bagging Neural Network Sensitivity Analysis for Feature Reduction in QSAR Problems," presented at the 2001 INNS - IEEE International Joint Conference on Neural Networks, Vol. 4, pp. 2478 - 2482, IEEE Press, Washington D.C., July 14-19.

M. Sternberg, J. Hirst, R. Lewis, R. King, A. Srinivasen, and S. Muggleton, 1994, "Application of Machine Learning to Protein Structure Prediction and Drug Design," in Advances in Molecular Bioinformatics, S. Schultze-Kremer, Ed., pp. 1-8, IOS Press.

C. M. Breneman, T. R. Thompson, M. Rhem, and M. Dung, 1995, "Electron Density Modeling of Large Systems Using the Transferable Atom Equivalent Method," Computers & Chemistry, Vol. 19, pp. 161-179.

Breneman, C. M., Sukumar, N., Bennett, K. P., Embrechts, M. J., Sundling, M. and Lockwood, L., 2000. "Wavelet Representations of Molecular Electronic Properties: Applications in ADME, QSPR, and QSAR," in Proceedings of the American Chemistry Society National Meeting. Washington D.C., August 20-24.

Simon Haykin, 1994, Neural Networks: A Comprehensive Foundation, McMillen.

Ozdemir, M., Embrechts, M. J., Arciniegas, F., Breneman, C. M., and Bennett, K., 2001, "Feature Selection for In-Silico Drug Design using Genetic Algorithms and Neural Networks," in Proc. IEEE SMCia-01 Mountain workshop on Soft Computing in Industrial Applications, pp. 53-57 Virginia, June 25-27.

VISUALIZATION OF HIGH DIMENSIONAL IMAGE DATA USING A 3D SOFM

ARCHANA SANGOLE AND GEORGE K. KNOPF
Department of Mechanical & Materials Engineering,
Faculty of Engineering Science, University of Western Ontario,
London, Ontario, Canada, N6A 5B9

ABSTRACT

Scientific data visualization involves transforming hidden structures and relationships contained in complex high dimensional data into simple graphical representations for enhanced human interpretation. This paper discusses how a spherical self-organizing feature map (SOFM) provides a mechanism to transform high-dimensional image data into *coherent geometric patterns* thereby enabling the human observer to visualize concealed information as graphical forms. Data is correlated in the feature map with little, or no, a priori knowledge and can be visualized in terms of the relative positions of the units on the map. Furthermore, the mapping of high dimensional data to a 3D shape enables information embedded in the numeric data to take a coherent 3D form. Distortions in the shape reflect the level of similarity among input data vectors. Performance of the proposed visualization algorithm is tested using simulated range image data from known geometry and multi-spectral satellite data.

INTRODUCTION

Scientific data visualization techniques create a new level of information that can be more easily interpreted by the human viewer and provide a deeper look at the underlying relationships in high-dimensional data. Humans can relate to three-dimensional shapes and forms more easily than strings of numeric data. As a result, scientific data visualization techniques are now shifting towards 3D graphical representations. A few glyph-based visualization methods have been proposed wherein each data dimension is mapped to a visual dimension of a glyph or graphical object. Similarities / differences are visualized in the form of variations in the resultant shapes. Glyphs may be generated by using procedural techniques (Ebert et al., 2000) that involve introducing distortions in a basic shape depending on the data dimension being mapped or by creating implicit surface models (Rohner et al., 1999).

In high-dimensional data, each input is an N-dimensional feature vector representing numerous independent and interdependent attributes of a physical phenomenon characteristic to the data set. Each attribute in the input vector influences human interpretation of the contained information. A reduction in both, dimensionality and volume of the data is required for effective visualization purposes. Therefore data exploration techniques are often used in conjunction with visualization methods (DesJardins and Rheingans, 1999; Gross and Seibert, 1993; Vesanto, 1999). This paper describes how a deformable spherical self-organizing feature map (SOFM) provides a robust mechanism to map arbitrary high dimensional numeric data into a *"coherent geometric form"*. The proposed algorithm is tested using simulated range image data from known geometry and later applied to multi-spectral satellite image data.

MAPPING NUMERIC DATA INTO GEOMETRIC FORM

In order to visualize high dimensional data, the large numeric data set has to be reduced in volume (vector quantization) and in dimension (vector projection). The self-organizing feature map (SOFM) is a competitive self-learning algorithm in which vector quantization and vector projection occurs simultaneously during training. The SOFM weight vectors form prototypical representations of the input data that are internally ordered by exploiting hidden redundancies in the data, thereby reflecting some physical characteristic of the data. In the proposed visualization algorithm a feature map of spherical topology is used and deformation is introduced in the map during vector projection. The resultant shape represents information content of the numeric data.

Spherical Self Organizing Feature Map

The spherical SOFM neural architecture, Fig. 1, consists of a tessellated *unit sphere* with *uniform* triangular elements, each *node* representing a *cluster unit*. The sub-vector associated with each training data set is $X^P = \left[x_1^p, x_2^p, \ldots x_N^p \right]$, where $p = 1, 2, \ldots P$, P is the total number of training patterns and each element x_n^p is the nth feature for the pth pattern. Each input vector is connected to every cluster unit in the spherical network by a weight vector $w_{i,j,k}$. Every cluster unit located at (i,j,k) has a neighbourhood ($NE_{i,j,k}$) of one unit radius associated with it. The weight adaptation algorithm is designed such that the cluster unit $(i, j, k)^*$ closest to the presented input vector and its associated neighbourhood (NE_{i,j,k^*}) are updated simultaneously.

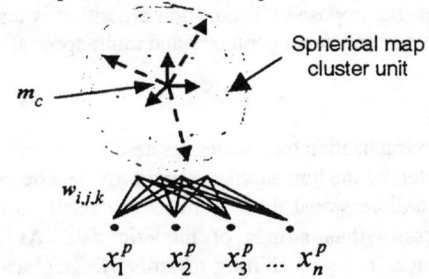

Figure 1. The spherical self-organizing feature map (SOFM). Inputs, X^P are shown only to a small number of cluster units.

Initially, the weights are set to small random values located in the input space. For each randomly selected input vector X^P the error or difference $D_{i,j,k}^p$ between the input feature vector and the weight vectors for all the cluster units in the network is computed using Eq. (1), where $w_{n,i,j,k}$ is the weight from the nth input, $n = 1,2, \ldots N$, to the $(i,j,k)^{th}$ cluster unit, $i = 1, 2, \ldots I$, $j = 1, 2, \ldots J$, and $k = 1, 2, \ldots K$, and $\Phi(u_{i,j,k})$ is a count-dependent non-decreasing function used to prevent cluster under-utilization (Krishnamurthy et al, 1990).

$$D_{i,j,k}^p = \Phi(u_{i,j,k}) \sum_{n=1}^{N} \left(x_n^p - w_{n,i,j,k} \right)^2 \qquad (1)$$

The winning cluster unit is selected as that unit which gives the minimum error, $D_{i,j,k}^p$. The weights of the cluster unit $(i, j, k)^*$ and all the units residing within the specified neighbourhood NE_{i,j,k^*} are updated using Eqs. (2a) and (2b). In this study, a fixed

Significance of the spherical SOFM

A majority of the SOFM approaches use a flat space for visualization (Vesanto, 1999; Ritter, 1999). A fundamental limiting factor in using a 2D lattice is the restricted neighbourhood at the boundaries of the feature map and thus there is a discontinuity in data correlation at the boundaries. A spherical SOFM is, in essence, a 2D map wrapped around itself thereby providing an overall connectivity in the map, Fig. 3. Due to its spherical topology there is an overall symmetry (Ritter, 1999), consequently facilitating a higher degree of association between the cluster units.

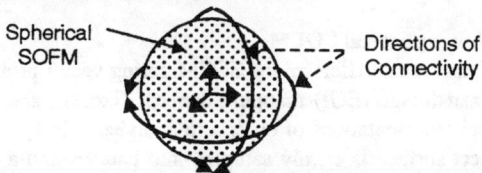

Figure 3. Data association in a spherical self-organizing feature map (SOFM).

The EGI mapping concept during vector projection transforms the internally ordered high dimensional numeric data into an independent 3D coordinate system that is invariant to translation, scale and rotation. Data association can then be visualized in the form of a 3D shape. Different levels of information can be encoded in the various attributes of 3D geomtery thereby enhancing the qualitative/ quantitative meaning embedded in the shape. The spherical SOFM thus provides a mechanism to enable high-dimensional numeric data to take a *coherent geometric form* in a natural manner.

EXPERIMENTAL STUDY

The proposed algorithm is first tested under varying initial conditions using simulated range image data of a shape of known geometry, Fig. 4. and then applied to high dimensional satellite image data.

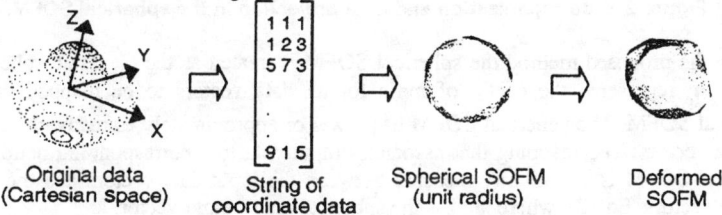

Figure 4. Numeric data and extracted shape for the test object.

Case 1: Simulated Range Image Data

The data set contains 648 coordinate data points and the SOFM neural architecture consists of 66 cluster units. Input data is presented to the SOFM network in a random order. The network is trained for 70 cycles using a predefined learning rate ($\mu = 0.2$), under two cases of initial conditions. First, by initializing the weight vectors to randomly selected data vectors, and second, by initializing the weight vectors to small random values located in the input space. It is observed from Figs. 5a and 5b that the shapes generated are very similar therefore demonstrating that the performance of the algorithm appears to be consistent under the two cases of initial conditions.

neighbourhood of one unit radius is used. However a variable neighbourhood strategy can be used to improve the learning process and stabilize convergence of the spherical SOFM. On completion of training the weight vectors form prototypical representations of the input data vectors.

$$w_{i,j,k^*}(new) = w_{i,j,k^*}(old) + \alpha \left[X^P - w_{i,j,k^*}(old) \right] \qquad (2a)$$

where

$$\alpha = \mu \left[1 - ((\# \text{ of cycles}) / N_{cycle}) \right] \qquad (2b)$$

Deformation of the spherical SOFM

The mechanism used to deform the SOFM during vector projection, is similar to the Extended Gaussian Image (EGI) mapping concept (Ikeuchi and Hebert, 1995) used to generate spherical representations of objects in an independent coordinate system. In this method, the object surface is evenly sampled into patches and a surface normal of unit mass is defined at each patch. Every surface patch is associated with a corresponding point on the Gaussian sphere by a Gauss map. Each surface normal contributes to the distribution of mass at the corresponding node on the Gaussian sphere thereby resulting in a weighted mapping of the surface normals onto the Gaussian sphere. Data organization and data projection in the spherical SOFM is illustrated in Fig. 2 using a sample of three-dimensional numeric data vectors.

w	w_1	w_2	w_3	$d_{i,j,k}$
1	3.0	3.0	1.0	1.497
2	3.0	2.0	1.0	1.562
3	2.0	3.0	3.0	0.917
4	2.0	2.0	3.0	1.020
5	1.0	3.0	3.0	1.497
m_c	2.2	2.6	2.2	

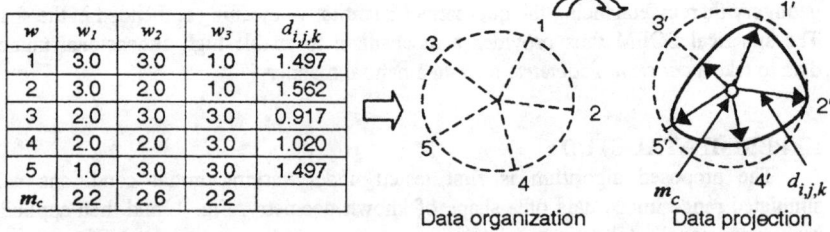

Data organization   Data projection

Figure 2. Data organization and data projection in the spherical SOFM.

In the proposed method the spherical SOFM is treated as the Gaussian sphere. The vector m_c represents the centre of mass for all data vectors to be mapped onto the spherical SOFM. The spherical SOFM has fewer or approximately as many cluster units as data vectors. The mapping that associates input data to a corresponding node on the spherical SOFM is a similarity measure between the input data vectors and the SOFM weight vectors, Eq. (3), where $w_{i,j,k}$ is the spherical map weight vector.

$$d_{i,j,k} = \| w_{i,j,k} - m_c \| \qquad (3)$$

$$(x_d \, y_d \, z_d)_{i,j,k} = f(d_{i,j,k})(x, y, z)_{i,j,k} \qquad (4)$$

The coordinates of the nodes are appropriately scaled, Eq. (4), where $(x_d \, y_d \, z_d)_{i,j,k}$ are the coordinates after deformation and $f(d_{i,j,k})$ is a scaling function of the similarity measure between the SOFM weight vector of the cluster unit located at (i,j,k) and m_c. The nodes that are similar to m_c lie near the origin while those that vary contribute towards deformation by pulling the corresponding node outward in proportion to the similarity measure. The shape is a direct reflection of the local variance between m_c and the input data thereby reflecting local variance or information content.

(a) Randomly selecting the weight vectors from the data set.   (b) Initializing the weight vectors to small random values in the input space.

Figure 5. Shape extracted using the spherical self-organizing feature map (SOFM).

Case 2: Satellite Image Data

The multi-spectral satellite image used in the study is obtained from the Distributed Active Archive Center (DAAC) at the Goddard Space Flight Center (GSFC), Greenbelt, MD. Advanced Very High Resolution Radiometers (AVHRR) mounted on the NOAA-7, -9, -11 (National Oceanic and Atmospheric Administration) meteorological satellites were used to collect the data. These sensors detect radiation emitted and reflected from the earth's surface in five channels (bands) the bandwidths of which fall in the visible and thermal infrared region of the spectrum. Sensor data from only four channels are used in the analysis. Ch-1 (0.58 – 0.68 µm) and Ch-2 (0.73 – 1.10 µm) are sensitive to vegetation parameters such as green-leaf biomass and green-leaf area. Ch-4 (10.3 – 11.3 µm) and Ch-5 (11.5 – 12.5 µm) are sensitive to brightness temperatures. Thus each input feature vector comprises of 4 elements: (Ch-1, Ch-2, Ch-4, Ch-5).

The objective is to study the vegetation pattern in South Western USA and Mexico. Results of four samples are presented in this preliminary study and Fig. 6 illustrates their respective geometric patterns generated using the spherical SOFM. In order to enhance interpretation of the numeric satellite data, deformation information is colour-coded using a scale based on the electromagnetic spectrum (Knopf and Sangole, 2000) and the resultant colour scheme is projected onto the geographic space. The differences in the shade of the colours reflect natural variations in vegetation. Regions with almost similar colours exhibit similar vegetation patterns. Local variance in the vegetation pattern can be visualized by observing how the colours change in the individual geometric shapes.

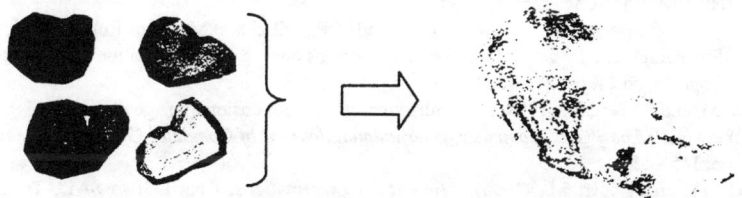

Figure 6. Case 2: Patterns extracted using the spherical self-organizing feature map (SOFM), projected onto an output visualization space.

DISCUSSION

A combination of data mining and visualization techniques are being used to transform information embedded in high dimensional data into simple perceivable forms to enhance human interpretation. The spherical SOFM helps to capture data association information from large amounts of high dimensional numeric data and transform it into coherent geometric forms. The spherical topology of the feature map facilitates an overall connectivity thereby providing a high degree of data association. The deformable aspect of the map helps the human interpreter to visualize the associations that occur in strings of numeric data. The vector projection mechanism transforms the data into an independent coordinate system invariant of scale, rotation, and translation. These attributes are particularly important if multiple geometries are compared for recognition or comparative analysis purposes. The experimental results demonstrate the capability of the spherical SOFM to extract coherent geometric patterns from large numeric data sets, thereby providing a mechanism to assign shape to numeric data.

CONCLUSIONS

Humans can relate to shapes more easily than to strings of numeric data. This has lead to the integration of data mining and visualization techniques to enhance human interpretation of large high dimensional numeric data sets. The proposed mechanism is a general framework that can be applied to transform high-dimensional numeric data into coherent geometric forms (glyph-based visualization) or to contain the complete numeric data set in one geomteric form (projection-based visualization). Various attributes of geometry can be used as visual cues to encode different levels of information in the shape thereby enhancing the meaning embedded in the form. The center of mass m_c of the spherical SOFM introduced in this paper may be an individual cluster unit obtained from a prior SOFM that was trained with a significantly larger number of data vectors.

ACKNOWLEDGEMENTS

This work has been supported, in part, by the Natural Sciences and Engineering Research Council of Canada. The authors wish to thank the Distributed Active Archive Center (DAAC) at the Goddard Space Flight Center (GSFC), Greenbelt, MD, for the multi-spectral satellite data.

REFERENCES

DesJardins M., and Rheingans, P., (1999), "Visualization of high-dimensional model characteristics", *Workshop on New Paradigms in Information Visualization and Manipulation* (NPIVM'99), ACM Press, pp. 6 - 8.
Ebert, E.S., Rohrer, R.M., Shaw, C.D., Panda, P., Kukla, J.M., and Roberts,A.D., (2000), "Procedural shape generation for multi-dimensional data visualization", *Computers and Graphics*, no.24, pp. 375 - 384.
Gross, M. and Seibert, F., (1993). "Visualization of multidimensional image data sets using a neural network", *The Visual Computer, International Journal of Computer Graphics*, vol. 10, no. 1, pp. 145 - 159.
Ikeuchi K., and Hebert M., (1995), "Spherical representations: from EGI to SAI", Tech. Report CMU-CS-95-197, Computer Science Department, Carnegie Mellon University.
Knopf, G.K. and Sangole, A., (2000), "Visualizing data association using self-organizing feature maps", *Intelligent Engineering Systems Through Artificial Neural Networks* Volume 10, C.H. Dagli et. al. (Eds.), ASME Press, pp. 471 - 476.
Kohonen, T., (1997). *Self-Organizing Maps*. New York: Springer.
Krishnamurthy, A.K., Ahalt, S.C., Melton, D.E., and Chen, P., (1990),"Neural networks for vector quantization of speech and images", *IEEE Journal on Selected Areas in Communication*, vol. 8, no. 8, pp. 1449 - 1457.
Ritter, H., (1999). "Self-organizing maps on non-Euclidean spaces", *Kohonen Maps*, E. Oja and S. Kaski (Eds.), Elsevier Science B. V., Amsterdam, The Netherlands, pp. 97 - 109.
Rohrer, R. M., Sibert, J. L., and Ebert, D. S., (1999). "A shape-based visual interface for text retrieval", *IEEE Computer Graphics and Applications*, vol. 19, no. 5, pp. 40 - 46.
Vesanto, J., (1999). "SOM-based data visualization methods", *Journal of Intelligent Data Analysis*, Elsevier Science, vol. 3, no. 2, pp. 111 - 126.

MODELLING THE RELATIONSHIP BETWEEN PROBLEM CHARACTERISTICS AND DATA MINING ALGORITHM PERFORMANCE USING NEURAL NETWORKS

KATE A. SMITH AND FREDERICK WOO
School of Business Systems, Monash University, Victoria 3800, Australia

VIC CIESIELSKI AND REMZI IBRAHIM
Department of Computer Science, Royal Melbourne Institute of Technology

ABSTRACT

As more data mining algorithms become available, the answer to one question becomes increasingly important: which techniques are best suited to which data mining problems? This study attempts to address this question by examining the performance of six leading data mining algorithms across a collection of 57 well-known classification problems from the machine learning repository. For each data set, a number of statistics are collected in order to measure the size and complexity of the problem. Using a neural network, the performance results of the six algorithms are then combined with the problem characteristics to build a predictive model of the relative performance of each algorithm on a given problem.

INTRODUCTION

Classification problems find significance in many practical applications such as credit and risk assessment, medical diagnosis, fraud detection, and quality control. A large number of techniques have emerged over many years for solving such problems, from the early methods of logistic regression through to more modern techniques such as neural networks and decision tress that belong to the data mining repertoire. Many of these techniques have been evaluated on benchmark data sets such as the collection of classification problems at University of California, Irvine [1]. As more techniques become available for solving classification problems however, it becomes increasingly important to know which techniques are suited to which types of classification problems [2]. Of even more relevance for data mining practitioners it to know in advance which technique might perform best on a given problem, in order to eliminate the need for evaluating all techniques to find the best model.

This paper describes the results of our efforts to address this issue. We have examined 57 well-known classification problems from the UCI Repository, and attempted to measure the complexity and characteristics of each problem by considering three types of measures: simple, statistical, and information theoretic as suggested by Henery [3]. These measures are described briefly in the next section of the paper. For each problem we also evaluate the performance of six popular data mining algorithms. The six data mining algorithms are:

1. IBK, an implementation of a k-means nearest neighbour classifier [4]
2. the decision tree algorithm C4.5 [5]
3. PART, a classifier that generates a decision list [6]
4. Naïve Bayes (NB) [7]
5. OneR [8]
6. KD, a kernal density estimator [9]

We then use a neural network to model the relationship between the characteristics of each data set and the obtained performance of the six data mining algorithms. In this way, we aim to develop a predictive tool to pre-determine the likely performance of each data mining algorithm on a given problem, that can be rapidly evaluated via a single pass through a trained neural network. Similar approaches have been proposed using rule generation methods such as C4.5 rather than a neural network [2,10]. Apart from the different modelling approach used here, the data sets are different, as are the data mining algorithms.

MEASURING COMPLEXITY OF DATA SETS

Each data set can be described by a number of simple, statistical and information theoretical measures. Many of the following measures are described in [3].

Simple and Statistical Measures

Some simple measures are shown in Table 1. In addition, a number of statistical measures can be used for continuous variables as described below:

Measure	Notation
Number of variables	p
Number of instances	N
Number of classes	q
Percentage of discrete variables	disc
Percentage of continuous variables	cont
Percentage of missing values	%_missing

Table 1: Simple measures for characterisation of each data set

Standard deviation ratio (SD_ratio)

$$SD\_ratio = \exp\left\{\frac{M}{p\sum_{i=1}^{q}(n_i-1)}\right\} \quad \text{where } M = \gamma\sum_{i=1}^{q}(n_i-1)\log\left|S_i^{-1}S\right|$$

and

$$\gamma = 1 - \frac{2p^2+3p-1}{6(p+1)(q-1)}\left\{\sum\frac{1}{n_i-1} - \frac{1}{n-q}\right\}$$

and S_i and S are the unbiased estimators of the i-th sample covariance matrix and pooled covariance matrix respectively. n_i is the number of instances in a particular class i.

Mean correlation between variables (correl)

$$correl = \frac{\sum_{i,j} abs(\rho_{ij})}{\text{Total no. of correlation coefficients}}$$

where ρ_{ij} is the correlation between variables i and j.

Max % and Min% in a class (MaxC, MinC)

Measures the percentage of the number of instances belonging to the least common class (MinC) and the most common class (MaxC).

Mean Absolute Skewness (γ)

$$skew = \frac{n}{(n-1)(n-2)}\sum\left(\frac{x_i-\bar{x}}{s}\right)^3$$

Mean Absolute Kurtosis (β)

$$\beta_i = \left\{\frac{(n+1)}{(n-1)(n-2)(n-3)}\sum\left(\frac{x_i-\bar{x}}{s}\right)^4\right\} - \frac{3(n-1)^2}{(n-2)(n-3)}$$

First and Second canonical correlation (cancor1, cancor2)

Measures the correlation of two canonical (latent) variables, one representing a set of independent variables, the other a set of dependent variables.

First eigenvalue and Second eigenvalue (fract1, fract2)

The eigenvalues are approximately equal to the squares of the canonical correlations. They reflect the proportion of variance explained by each canonical correlation relating two sets of variables, when there is more than one extracted canonical correlation.

Information theoretical measures

Mean entropy of variables

Entropy is a measure of randomness in a variable. The entropy H(X) of a discrete random variable X is calculated in terms of q_i (the probability that X takes on the i-th value). We average the entropy over all the variables and take this as a global measure of entropy of the variables:

$$H(X) = -\sum q_i \log q_i \qquad \bar{H}(X) = p^{-1} \sum H(X_i)$$

Entropy of classes

$$H(C) = -\sum_i \pi_i \log \pi_i$$

where π_i is the prior probability for class A_i

Mean mutual entropy of class and attributes

A measure of common information or entropy shared between the two variables. If p_{ij} denotes the joint probability of observing class A_i and the j-th value of variable X, if the marginal probability of class A_i is π_i, and if the marginal probability of variable X taking on its j-th value is q_j, then the mutual information and its mean over all variables are defined as:

Equivalent number of variables (EN)

$$M(C,X) = \sum_{ij} p_{ij} \log(\frac{p_{ij}}{\pi_i q_j}) \qquad \text{and} \qquad \bar{M}(C,X) = p^{-1} \sum_i M(C,X_i)$$

$$EN = \frac{\bar{H}(C)}{\bar{M}(C,X)}$$

This is the ratio between the class entropy and the average mutual information.

Noise-signal ratio (NS)

A large NS ratio implies a data set contains much irrelevant information.

$$NS.ratio = \frac{\bar{H}(X) - \bar{M}(C,X)}{\bar{M}(C,X)}$$

MODELLING ALGORITHM PERFORMANCE

For each of the 57 data sets, we thus have a total of 21 measures that describe the characteristics of the data. The performance of each of the six data mining algorithms has been measured on these 57 data sets using the java program *Weka*, a data mining environment developed at the University of Waikato, New Zealand (http://www.cs.waikato.ac.nz). The average error on the test set (randomly extracted to be 20% of the original data) is used as the performance measure. We then post-process the errors of each algorithm for a given problem, to measure their relative performance, with the best performing algorithm measuring 1, and the worst measuring 0. Thus the result of the jth algorithm on the ith data set is calculated as:

$$R_{ij} = 1 - \frac{(e_{ij} - \min(e_i))}{\max(e_i) - \min(e_i)}$$

where e_{ij} is the test set error for the jth algorithm on data set i, and e_i is a vector of errors for data set i.

We can now model the relationship between problem characteristics (up to 21 inputs) and relative algorithm performance (6 outputs) using a neural network trained with the 57 data sets. A number of different neural network architectures, choice of inputs, and parameter combinations were experimented with in the Neuroshell2 environment to arrive at the best model, which was a probabilistic neural network. 20% of the data was randomly extracted into a test set to determine the performance of each model. The inputs that resulted in the best model were:

p, q, Bin, Con, %_missing, MinC, MaxC, SDratio, correl, cancor1, cancor2, fract2, M(C,X), H(C), EnAtr, NSRatio.

Table 2 shows the classification accuracy obtained by the neural network model. The overall accuracy when predicting which technique will be best for a given problem is 77% (100% accuracy on the randomly extracted test set which has 11 examples). The accuracy increases to 95% if the top two techniques are inspected, and 98% if the top three techniques are inspected for the best performing technique. Figure 1 shows these relationships.

	IBK	C4.5	PART	NB	OneR	KD
Actual winners:	9	15	3	18	5	7
Classified winners:	2	7	9	21	5	13
% accuracy	22%	47%	100%	100%	100%	100%

Table 2: Accuracy of neural network in predicting best algorithm

CONCLUSIONS

In this paper we have demonstrated the merits of training a neural network to predict data mining algorithm performance. For a given problem, the data characteristics can be fed to the neural network as inputs, and the output is a ranked list of techniques predicting their likely performance on the problem. We have shown that when the top two predicted techniques are inspected, the accuracy is 95% in identifying the best performing algorithm. The performance of the (small) test set was 100%. Thus the user need only test one or two techniques to find a suitable model, rather than the large number of techniques available to the data mining community that the predictive model is trained on.

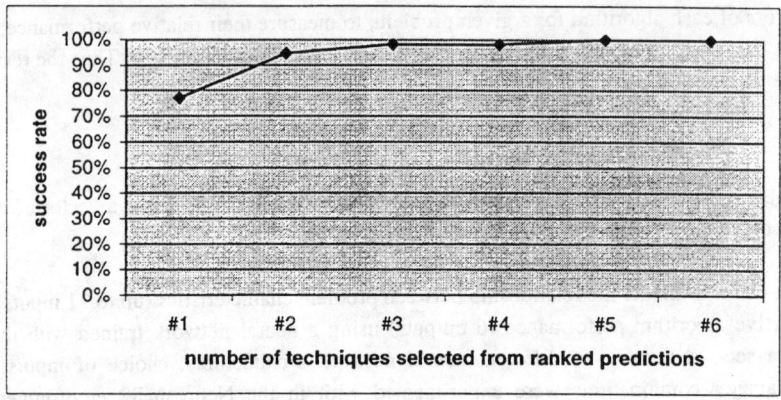

Figure 1: Success rate in finding best performing algorithm based on neural network predictions

REFERENCES

[1] D. Aha, Machine Learning Database, University of California, Irvine, ftp://ftp.ics.uci.edu/pub/machine-learning-databases
[2] P. B. Brazdil and R. J. Henery, "Analysis of Results", in D. Michie, D. J. Spiegelhalter and C.C. Taylor (eds.), *Machine Learning, Neural and Statistical Classification*, Ellis Horwood Limited, Chapter 10, 1994.
[3] R. J. Henery, "Methods for Comparison", in in D. Michie, D. J. Spiegelhalter and C.C. Taylor (eds.), *Machine Learning, Neural and Statistical Classification*, Ellis Horwood Limited, Chapter 7, 1994.
[4] D. Aha, and D. Kibler, "Instance-based learning algorithms", *Machine Learning*, vol.6, pp. 37-66, 1991.
[5] R. Quinlan, *C4.5: Programs for Machine Learning*, Morgan Kaufmann Publishers, San Mateo, CA, 1993.
[6] E. Frank and I. H. Witten, "Generating Accurate Rule Sets Without Global Optimization". In Shavlik, J., ed., *Machine Learning: Proceedings of the Fifteenth International Conference*, Morgan Kaufmann Publishers, San Francisco, CA, 1998.
[7] G. H. John and P. Langley, "Estimating Continuous Distributions in Bayesian Classifiers". *Proceedings of the Eleventh Conference on Uncertainty in Artificial Intelligence.* pp. 338-345. Morgan Kaufmann, San Mateo, 1995.
[8] R.C. Holte, "Very simple classification rules perform well on most commonly used datasets". *Machine Learning*, Vol. 11, pp. 63-91, 1993.
[9] Silverman, B. W., *Density estimation for statistics and data analysis*, Chapman and Hall, New York, 1986.
[10] D. Aha, "Generalizing case studies: a case study", *Proceedings of the 9th International Conference on Machine Learning*, pp. 1-10, 1992.

VISUALIZATION OF ANY DATA WITH ELASTIC MAP METHOD

ALEXANDER N. GORBAN

ALEXANDER A. PITENKO   ANDREI Y. ZINOVYEV
Institute of Computational Modeling of SD of RAS
Krasnoyarsk, Russia

DONALD C. WUNSCH, II
University of Missouri-Rolla
Department of Electrical and Computer Engineering
Rolla, MO

ABSTRACT
A novel data visualization technique is described. We mean by data a set of points in a space of large dimension (characteristic values of the dimension - 10-20, characteristic number of points - 100-1000), where each point represents an object having a definite set of attribute values. Such data can be received from the economic, biological, medical and other tables containing results of storage of different information of any kind.

INTRODUCTION

We assume that any analysis of the data may be considered as their canonical and clear description. One of the ways of such description is visualization. The basic task of visualization of the data is to represent the set of the data points so that a researcher could see a significant part of data regularities and could manipulate the data presentation in such a way as to solve problems (classification, clustering, factor analysis etc.).

There are two methods to describe data. The first is construction of a hierarchical data structure, i.e. in splitting points into classes, subclasses etc. The second is decreasing the dimension of the initial data space.

For visualization of data, it is most suitable to model data with the help of a manifold of small dimension put in the multidimensional space. For example, for displaying data on the computer monitor, it is most natural to use a two-dimensional manifold.

THREE STAGES OF DATA VISUALIZATION

The method of visualization of data with the help of modeling maps can be split up on the three stages:
   1. Constructing modeling manifold (*map*)

For example, the modeling manifold may be chosen as a plane of first two main components or some surface constructed as a small correction to the plane.

In the nonlinear case, the modeling manifold can be constructed, first as a dot approximation of considered set of data points, and existing as a grid located in the data space. Then the grid must be interpolated between the nodes with the help of some procedure (for example, in a partially linear way after making some triangulation).

2. Projecting data points from the initial data space onto the modeling manifold.

To have an opportunity to present data from multidimensional data space on a two-dimensional screen, it is necessary to specify a way of matching each point of data to the definite point on the modeling map.

With a grid that models the data set constructed, the most simple way is to match each point of the data to its closest node on the grid. It is reasonable to call such way of projecting as partially-constant.

3. Making different colorings of the map, that give a presentation of the structures and regularities consisted in considered data.

First, it is necessary to show the data points that were projected onto the map. Besides it is useful to have an opportunity to mark out subsets of data points accordingly to some attribute. Second, it is possible to show on the map continuous distribution of data estimating the data density with some method or the density of any subset. Third, it is possible to show on the map values of any coordinate function. For example, the most simple and understandable colorings are given by the values of the coordinates in the points of the map.

General principles of such a procedure of visualization first were described by Kohonen (1995). He also developed the algorithm of constructing of a modeling manifold, called self-organizing map (SOM). Kohonen put to use partially-constant way of projecting data on the map.

We developed a procedure of data visualization that differs from the SOM and named "method of elastic map". Describing of the method of elastic map constructing and the way of partially-linear projecting are given below.

ALGORITHM OF ELASTIC MAP CONSTRUCTING

Elastic map (Gorban et al., 1999) is a modeling manifold (as usual, two-dimensional), that has following properties:

1) Approximation – property of closeness of the nodes of modeling grid to the data points;

2) Elasticity for stretching – property of not too large stretching;

3) Elasticity for bending – property of not too large curvature.

Let us mark by X – a data set; y_{ij} – radius-vectors of the nodes of two-dimensional grid; i,j – indices that numerating nodes; K_{ij} – is a subset of data points that are closest to the given node (taxons) with indices i,j.

Locations of the grid nodes are found as a result of solving the problem of minimization of such quadric functional:

$$D = \frac{D_1}{|X|} + \lambda \frac{D_2}{m} + \mu \frac{D_3}{m} \to \min,$$

where m – number of nodes of the grid; λ, μ – elasticity coefficients of the stretching and bending respectively, that are parameters to be changed, $D1$, $D2$, $D3$ – parts that are describing each of the elastic map properties.

$D_1 = \sum_{ij} \sum_{x \in K_{ij}} \|x - y_{ij}\|^2$ is a measure of approximation

$D_2 = \sum_{ij} \|y_{ij} - y_{i,j+1}\|^2 + \sum_{ij} \|y_{ij} - y_{i+1,j}\|^2$ - measure of stretching

$$D_3 = \sum_{ij} \left\| 2y_{ij} - y_{i,j-1} - y_{i,j+1} \right\|^2 + \sum_{ij} \left\| 2y_{ij} - y_{i-1,j} - y_{i+1,j} \right\|^2 \text{ - measure of bending}$$

The solution of variation problem reminds the method of dynamical kernels:
1) Quadric functional is minimized with given partitioning of data points by their belonging to the closest node of the grid;
2) The partitioning of data points for taxons (subset of data points for which the given node of the grid is the closest one) is performed.

The criterion D decreases at every step of the algorithm, it has lower bound, and the number of ways of the partitioning data points is finite. Hence the algorithm converges.

One of the common problems to be solved in managing the problems of optimization is finding the global minimum of functional. As a rule, this problem is equivalent to the problem of assigning the initial data that in case of modeling map means assigning an initial location of the map in the space. We took the initial map location in the plain of the first two components with sides equal to the half of the length of the scattering ellipsoid axis. Besides a small Gaussian noise was added to the values of the node's coordinates.

After that we used the method of "annealing" the main idea of that is to start the map constructing with large values of λ, μ (i.e. at the initial stage map is very "rigid"). Then the values of λ, μ gradually decreases to the small ones (the map becomes less "rigid").

PARTIALLY-LINEAR PROJECTING ON THE MAP. ALGORITHM OF FINDING THE CLOSEST POINT OF THE MAP

After the modeling map is constructed, the manifold is interpolated to the partially-linear one with the help of some triangulation procedure. In that case we propose following algorithm of finding the closest point of a partially-linear surface to project there a data point.

1) For the chosen data point, the following objects are
the closest node of the grid;
the segment of line, connecting two closest nodes (closest rib of the map);
the closest triangle (edge of the map) formed by the ribs for the given way of triangulation.

2) The orthogonal projection onto the plane containing the closest triangle is performed. If the image of the projector is inside the triangle then this image point is considered as the projection of the data point onto the map. Else

3) The orthogonal projection onto the line containing the closest rib of the map is performed. If the image of the projector is in the bounds of the rib then it is considered as the projection of the data point onto the map. Else

4) The closest node is considered as the projection of the data point onto the map.

APPLYING THE METHODS OF DATA VISUALIZATION TO THE MAPPING OF ECONOMICAL TABLES

We made an attempt to apply the methods of any data visualization (Pitenko, 1999) to the mapping of table of the biggest Russian companies. The table was taken from the

"Expert" magazine (1999). The files of data were retrieved from the official site of the magazine: http://www.expert.ru.

The original table contained the information of some economical characteristics of the two hundred biggest companies in Russia, sorting in order of decreasing of their gross production output. The table contained following fields (only part of them are independent, others are calculated by explicit formula):

1) Name of the company
2) Geographic region of situation
3) Branch of industry the company belongs
4) Gross production output in 1998
5) Gross production output in 1997
6) Rate of the growth
7) Gross production output in 1998, recalculated in dollar equivalent
8) Balance profit
9) Profit after taxation
10) Profitability
11) Number of workers
12) Efficiency of production

Shumsky and Kochkin (1999) had already attempted to visualize the table of biggest companies, taken from the "Expert" magazine in 1998. In his work traditional Kohonens self-organizing maps and Hintons diagrams were used. It was suggested to use as data space coordinates the relations of some independent factors from the table, and four coordinates were proposed.

We decided to enlarge the dimension of data space, and as a result it was obtained the next set of independent factors:

N	Factor	Name	Provisory name
1	LG_VO1998	Logarithm of gross production output in 1998	Output
2	RATE	$\dfrac{\text{Gross production output in 1998}}{\text{Gross production output in 1997}}$	Growth rate
3	PROFIT_BAL	$\dfrac{\text{Balance profit}}{\text{Gross production output in 1998}}$	Balance profit
4	PROFIT_TAX	$\dfrac{\text{Company profit after taxation}}{\text{Gross production output in 1998}}$	Clear profit
5	PRODUCTIV	$\dfrac{\text{Profit after taxation}}{\text{Number of workers}}$	Productivity

As a result a table was obtained that contained two hundred records and five fields. Parts of the records contained incomplete information (there were gaps in some factor values).

The table was normalized with the formula

$$\tilde{x}_i = th\left(\frac{x_i - M}{D}\right),$$

where $\tilde{x}_i, x_i, M, D$ – are the new and old value of the factor, mean value and dispersion respectively.

A map that was utilized to visualize the data was constructed according to the algorithm of elastic map construction described above. The method of "annealing" was put to use to find a local minimum of the map functional. The parameters µ and λ were changing slowly (so after each change the map could find the closest local minimum) from $\mu = 5, \lambda = 5$ to $\mu = 0.1, \lambda = 0.1$.

After constructing the elastic map the data were projected from the multidimensional data space onto the map with the use of algorithm of finding the closest point of the map in case of partially-linear interpolation between the nodes.

As an illustration of analysis of economical data the colourings of the map by the coordinates values are given below. Also the layer of calculated data density is showed. In the pictures big circles with numbers accords to the companies that belong to the oil and gas industry. Such an accentuation makes it possible to analyze the state of a branch of industry amidst the others.

COLOURINGS BY FACTORS VALUES

In fig. 1 values of the factor LG_VO1998 in the points of the map are shown. The more light gray tint corresponds to the higher values of the factor. The most lightest color shows points of the map with the first 10% of the factor range. The first 10% of the number of companies with highest values of production output are placed there. As an example circles with numbers displays companies that belong to the oil and gas industry. Following companies are marked:

1 – "GasProm"; 2 – Oil company "LUKoil"; 3 – Bashkeria oil company; 4 – "SurgutNefteGas"; 5 – Tyumen oil company; 6 – "Tatneft'"; 7 – "Slavneft'"; 8 – "Rosneft'"; 9 – Orenburg oil company "Onako"; 10 – Central oil company; 11 – "Komitek"

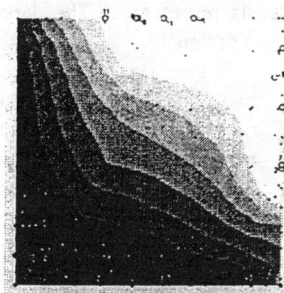
Fig. 1

Coloring by the factor RATE. As shown at the fig. 2, the area of the largest companies is not crossed with area of the highest rates of growth. In a right bottom corner of a map, for example, the companies of a food-processing industry, non-ferrous metallurgy and other quickly developing branches settled down.

In next three figures (fig. 3,4,5) colorings by the factors PROFIT_BAL, PROFIT_TAX, PRODUCTIV are shown.

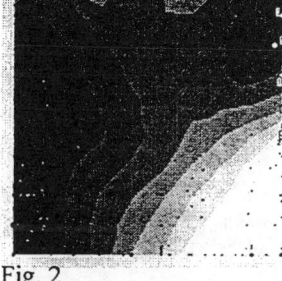
Fig. 2

Coloring of these factors are similar, which is how their correlation is made apparent. At the same time

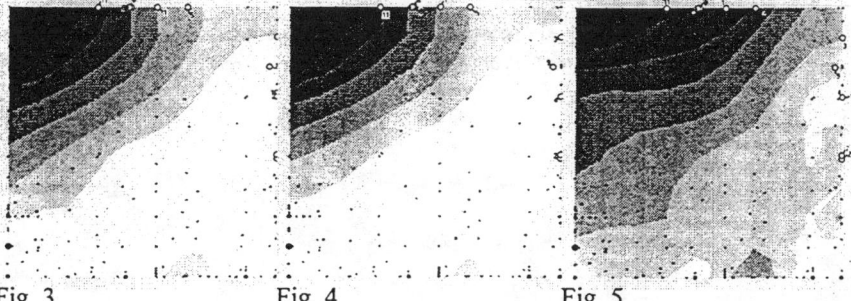
Fig. 3 Fig. 4 Fig. 5

distinctions in colorings allow to allocate the companies which drop out of correlation dependence.

COLORING BY THE DATA DENSITY VALUE

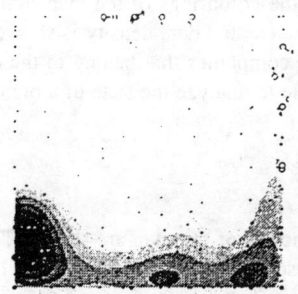

Fig. 6

In the following three figures (fig. 6,7,8) colorings of the map by data density value are shown. The density can be calculated with the help of any non-parametrical estimate. There are two ways to calculate the density of the data on the map. First, it is possible to consider two-dimensional distribution of points on the map. Secondly, it is possible to calculate density of points in the initial n-dimensional space, and to represent value of this density calculated in points of the map location. In the figures the application of the first way is represented. The darker areas correspond higher values of density.

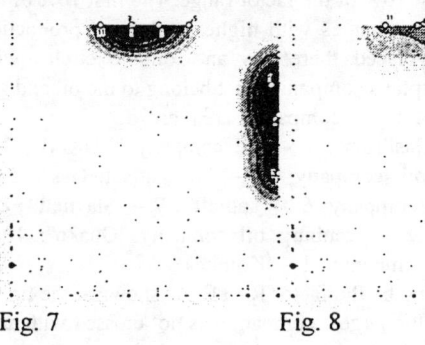

Fig. 7   Fig. 8

The fig. 6 represents two-dimensional distribution of density of the data as whole. The fig. 7 — distribution of density of the companies of oil and gas industry. The fig. 8 reflects convenient for analyze relative density of the companies of oil and gas industry (that is the relation of first two densities).

REFERENCES

"Expert" magazine, 1999, №36.

Gorban, A.N., Makarov, S.V., and Rossiev, A.A., 1999, "Missed data filling with linear and nonlinear factor analysis, mozaic regression and Carlemane formula," *Proc. All-Russia science conference "NeuroInformatics-99."* Moscow, Russia, Part 3, pp. 25-31.

Kohonen, T., 1995, *"Self-Organizing Maps,"* Springer-Verlag.

Pitenko, A.A., 1999, "Mapping of all and any data," *Proc. International conference "INTERCARTO-5,"* Jakutsk, Part 1, pp.71-78

Shumsky, S.A., and Kochkin, A.N., 1999, "Self-Organizing maps of financial indicators of the 200 biggest Russian companies," *Proc. All-Russia science conference "NeuroInformatics-99,"* Moscow, Part 3, pp. 122-127.

ROBUST MULTI-RESOLUTION WEB USAGE MINING WITH GENETIC NICHE CLUSTERING

Olfa Nasraoui
Dept. of Electrical and
Computer Engineering
The University of Memphis
206 Engr. Science Bldg.
Memphis, TN 38152
onasraoui@memphis.edu

Raghu Krishnapuram
IBM India Research Lab
Block 1
Indian Institute of Technology,
Hauz Khas
New Delhi 110016, India
kraghura@in.ibm.com

ABSTRACT

In this paper, we present a new hierarchical clustering technique based on the concept of genetic niches, called Hierarchical Unsupervised Niche Clustering (HUNC) which is considerably faster than its non-hierarchical counterpart (UNC), and offers the advantage of multi-resolution clustering. We use HUNC as part of a complete system of knowledge discovery in Web usage data. Our new approach does not necessitate fixing the number of clusters in advance, is insensitive to initialization, can handle noisy data, general non-differentiable similarity measures, and can provide profiles to match any desired level of detail or resolution. Our experiments show that our algorithm is not only capable of extracting meaningful user profiles on real Web sites, but also discovers associations between distinct URL pages on a site, with no additional cost. Unlike content based association methods, our approach discovers associations between different Web pages based only on the user access patterns and not on the page content. Also, unlike traditional context-blind association discovery methods, HUNC discovers context-sensitive associations.

INTRODUCTION

Manually entered Web user profiles have raised serious privacy concerns, are subjective, and do not adapt to the users' changing interests. Mass profiling, on the other hand, is based on general trends of usage patterns (thus protecting privacy) compiled from all users on a site, and can be achieved by mining or discovering user profiles from the historical data stored in server access logs. Current Web usage mining approaches avoid the feature representation dilemma of Web data by resorting to memory and computation intensive relational clustering (require the computation of all pairwise dissimilarities) or computation intensive association rule discovery (because very low support thresholds are needed to discover typical profiles). A classical non-relational approach requires a differentiable dissimilarity measure. However, for DM problems, a domain specific similarity measure should be designed free of any constraints.

Recently (Nasraoui and Krishnapuram, 2000), we have presented a new evolutionary approach to robust clustering based on the Unsupervised Niche Clustering algorithm (UNC). UNC seeks dense areas in feature space and determines their number by converting the clustering problem into a multimodal function optimization problem within the context of genetic niching and striving to locate the peaks of niches or subpopulations in the search space. Niching methods (Holland, 1975;

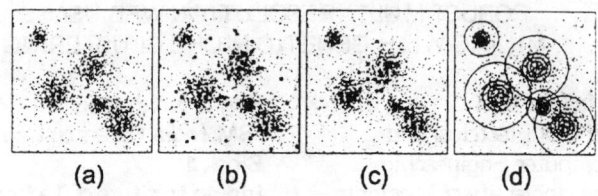

Figure 1: Evolution of the population using UNC: (a) original data set (b) Initial population, (c) population after 30 generations, (d) final extracted centers

Mahfoud, 1992) were designed to identify multiple optima within multimodal domains. Each peak in a mutlimodal domain can be thought of as a niche. In nature, niches correspond to different subspaces of the environment that can support different types of life such as species or organisms. Genetic Optimization makes UNC much less prone to suboptimal solutions than other objective function based approaches. Fig. 1 shows the evolution of an initial random population (denoted by square symbols) using UNC, for a noisy data set, toward the correct niches in subsequent generations.

We propose a Hierarchical modification of UNC, called HUNC, that departs from the traditional limited flat view of the data, and generates instead, a hierarchy of clusters which give more insight to the Web mining process. We use HUNC as part of a complete system of knowledge discovery in Web usage data. Our new approach does not necessitate fixing the number of clusters in advance, can provide profiles to match any desired level of detail or resolution, and requires no analytical derivation of the prototypes. Thus, it can handle a vast array of general subjective, even non-metric dissimilarities, making it suitable for many applications including data and Web mining.

THE KNOWLEDGE DISCOVERY PROCESS OF WEB SESSION PROFILING

The access log for a given Web server consists of a record of all files accessed by users. Each log entry consists of: (i) User's IP address, (ii) Access time, (iii) URL of the page accessed, $\cdots$, etc. A user session consists of accesses originating from the same IP address within a predefined time period. Each URL in the site is assigned a unique number $j \in \{1, \ldots, N_U\}$, where N_U is the total number of valid URLs. Thus, the ith user session is encoded as an N_U-dimensional binary attribute vector $s^{(i)}$ with the property

$$s_j^{(i)} = \begin{cases} 1 & \text{if the user accessed the } j^{th} \text{ URL during the } i^{th} \text{ session} \\ 0 & \text{otherwise} \end{cases}$$

The ensemble of all N_S sessions extracted from the server log file is denoted $\mathcal{S}$.

The similarity measure between two user-sessions (Nasraoui et al., 1999, 2000) takes the site's structure into account, and satisfies the desirable property of becoming more stringent as the accessed URLs get farther from the root because the amount of specificity in user accesses increases correspondingly. The sessions in cluster $\mathcal{X}_i$ are summarized by a typical session "profile" vector (Nasraoui et al., 1999) $\mathbf{P}_i = \left(P_{i1}, \ldots, P_{iN_U} \right)^t$. The components of $\mathbf{P}_i$ are URL relevence weights, estimated by the probability of access of each URL during the sessions of $\mathcal{X}_i$. They measure the significance of a given URL to the ith profile. The final profiles can be evaluated based on the average dissimilarity, which for the ith cluster, is given by

$$\sigma_i^{*2} = \frac{\sum_{\mathbf{s}^{(k)} \in \mathcal{X}_i} d_{ik}^2}{|\mathcal{X}_i|}. \tag{1}$$

Another measure is the robust cardinality given by

$$N_i^* = \sum_{\mathbf{s}^{(k)} \in \mathcal{X}_i} w_{ik}, \tag{2}$$

where $w_{ik} = \exp\left(-\frac{d^2}{2\sigma_i^{*2}}\right)$ is a robust weight (that is high for inliers/good data and low for outliers/noise).

HIERARCHICAL UNSUPERVISED NICHE CLUSTERING AND ITS APPLICATION TO WEB USAGE MINING

We retain the principal structure of UNC (Nasraoui and Krishnapuram, 2000), except for a few differences that result from the distinct nature of the session data: The solution space for possible session prototypes consists of binary chromosome strings which are defined to be the binary session attribute vectors $\mathbf{s}_i$, and the new Web session dissimilarity measure is used instead of the Euclidean distance to take the Web site topology in account.

UNC's computational time can be significantly reduced if we perform clustering in a hierarchical mode. In other words, we could cluster smaller subsets of the data using a smaller population size at multiple levels, instead of clustering the entire data set on a single level which would necessitate a larger population size. The computational complexiy of UNC is $\mathcal{O}\left(N_P\left(N + N_P\right)\right)$, where N_P is the population size and N is the number of data points to be clustered. Since, in the hierarchical mode, N_P can usually be a very small fraction of N (typical example from our experiments: $\frac{1}{10000}$ to $\frac{1}{1000}$, this complexity is much lower than that of relational clustering techniques such as Agglomerative Hierarchical Clustering (AHC) (Duda and Hart, 1973), $\mathcal{O}\left(N^2 \log N\right)$ and the closely related graph theoretic based Minimum Spanning Tree (MST) (Duda and Hart, 1973), $\mathcal{O}\left(N^2\right)$.

The hierarchical clustering is performed recursively starting from the top level (lowest resolution) until a termination criterion, based on the minimum acceptable size of a cluster, N_{split}, and its maximum allowable mean squared error, σ_{split}^2, is met. Let l denote the current level. Let $\mathcal{X}_{(l-1)} = \mathcal{X}_{(l-1)_1} \bigcup \cdots \bigcup \mathcal{X}_{(l-1)_{|C_{(l-1)}|}}$ denote the data set partitionned at level $l - 1$, where $\mathcal{X}_{(l-1)_i}$ is the ith cluster found at level $l - 1$. Hierarchical clustering proceeds by re-clustering the data in each of the above clusters in a recursive fashion to obtain the next level l. Even though the above parameters will eventually determine the number of clusters at the last level of the partitionning, they are not crucial to the performance of HUNC This is because, unlike classical divisive hierarchical clustering. techniques, our approach relies on robust weights to suppress the influence of outliers and data belonging to other clusters, and on a multimodal optimization approach where multiple clusters are sought in parallel at each level. This means that at any given level, HUNC is expected to identify as many good clusters as the population size, while classical hierarchical approaches are expected to yield the optimal cluster prototypes only at the optimal level of the partition that corresponds to the known correct number of clusters. Also, to the contrary of classical hierarchical techniques, HUNC re-partitions the data at the end of clustering (the last level of the hierarchy). Thus, the partition is not subject to any final commitment at each level, one of the well known pitfalls of hierarchical clustering techniques.

WEB USAGE MINING EXPERIMENTAL RESULTS The parameters for the robust hierarchical UNC were fixed to the following values: The crossover and mutation probabilities are $P_c = 0.9$ and $P_m = 5 \times 10^{-6}$, respectively. UNC used 10 generations per clustering with a population size, $N_P = 10$. Since all session dissimilarities are confined in $[0, 1]$, it is reasonable to choose $\sigma_{max}^2 = 0.95$, $\sigma_{split}^2 = 0.3$, $N_{split} = 30$. The profile vectors are displayed in the format $\{P_{ij}$ - j^{th} URL$\}$ in table 1, illustrating typical profiles.

MU-CECS1 Data The 12-day access record (during 1998) of the Web site of the Dept. of Computer Science and Computer Engineering at the University of Missouri, Columbia generated 1703 sessions accessing 369 distinct URLs. The results obtained at $L = 3$ levels are summarized in Table 3 showing profiles that reflect typical access patterns – the general "outside visitor" is captured in profiles 1 and 3; prospective students in profile 2 and 4, CECS 352 students in profile 7, etc. The quality of these clusters is confirmed by their low average dissimilarity compared to the maximal value of 1.

(i) Robust profiling is obtained by retaining profile members whose robust weights, w_{ij}, exceed a given threshold, w_{min}, equal to 0.6 in our experiments. This allows us to concentrate on the core of each profile by filtering out the noise sessions assigned to the closest profile. Different w_{min} values generate different α-cuts of the profiles when these are viewed as fuzzy sets. The w_{min}-core of the ith profile is defined as

$$\mathcal{X}_i^* = \{s^{(k)} \in \mathcal{X}_i | w_{ik} > w_{min}\}. \qquad (3)$$

The cores of profiles Nos. 8 and 13 end up having less than 20 members, hence making weak profiles. Also, the core of the spurious cluster No. 16 was discovered to contain sessions accessing the site managers' pages.

(ii) Multiresolution profiling: Note how profile 2 (at $L = 1$) in Table 2 is split into many profiles with distinct user interests (at $L = 2$) (profiles No. 5, 6, 7, 8, and 9) as shown in Table 3. The same observation can be made about the first cluster (general inquiries about the CECS department) which at level 2 gets split into profiles No. 1, 2, 3, and 4, with each such profile showing a more specific kind of interest in the department.

(iii) Inferring Associations between different URLs: Profile 2 (at $L = 1$) in Table 2 contains accesses to two different courses taught by different professors, signaling an association. It was later revealed that one of the courses (CECS 352: Operating systems) relies for the implementation of its projects on the C++ programming language which is taught in the other course (CECS 333: Object Oriented Design).

CONCLUSION

Table 1: Examples of Profiles discovered by HUNC from MU-CECS1 Data at $L = 3$ and $w_{min} = 0.6$

i	P_i
1	{.83 - /CECS.computer.class} {.95 - /courses.html} {.95 - /courses.index.html} {.95 - /courses100.html} {.19 - /courses200.html} {.19 - /people.html} {.19 - /people.index.html} {.19 - /faculty.html} {.93 - /}
3	{1.00 - /} {.67 - /CECS.computer.class}

Table 2: 4 of the 7 profiles discovered by HUNC from MU-CECS1 Data at $L = 1$

i	$|\mathcal{X}_i|$	$|\mathcal{X}_i^*|$	N_i^*	description	σ_i^{*2}
1	572	312	362.2	main page, class list, course enquiries, people and main degree page	0.32
2	305	170	191.0	Dr. Saab's and Dr. Joshi's course pages (CECS 333 and CECS 352 respectively)	0.54
3	185	111	124.0	Accesses to the CECS227 class pages	0.2
4	162	84	102.2	Dr Shi's CECS345 pages	0.37
5	73	56	51.0	Dr. Shang's course pages	0.08

For Web usage mining, the session dissimilarity measure is not a distance metric, and dealing with relational data is impractical given the huge dimension of the data sets. Therefore, we presented an adaptation of UNC, called Hierarchical Unsupervised Niche Clustering (HUNC) which is considerably faster than its non-hierarchical counterpart. Our new approach does not necessitate fixing the number of clusters in advance, can provide profiles to match any desired level of detail or resolution, and requires no analytical derivation of the prototypes. Thus, it can handle a vast array of general subjective, even non-metric dissimilarities, making it suitable for many applications in data and Web mining.

We have illustrated through several examples that our clustering process results in the discovery of associations between different URL addresses on a given site, with no additional cost. Also, the associations are meaningful only within well defined distinct profiles/contexts (context-sensitive) as opposed to all or none of the data (context-blind). This approach of discovering context-sensitive associations via clustering can be generalized to other transactional data.

ACKNOWLEDGMENTS Partial support of this work by the National Science Foundation Grant IIS 9800899 is gratefully acknowledged.

References

Duda, R., Hart, P., 1973. Pattern Classification and Scene Analysis. Wiley Interscience, NY.

Table 3: Some of the 16 profiles discovered by HUNC from MU-CECS1 Data at $L = 2$ and $L = 3$

i	$|\mathcal{X}_i|$	$|\mathcal{X}_i^*|$	N_i^*	description	σ_i^{*2}
1	219	132	140.5	main page, class list, course enquiries and people	0.16
2	119	73	77.0	main page, class list, course and undergraduate degree enquiries	0.27
3	140	85	91.6	Short sessions mostly limited to main page and class list	0.13
4	129	71	80.7	main page, people, individual faculty, research and graduate degree pages	0.39
6	133	80	85.2	Dr. Saab's CECS333 pages (long detailed sessions)	0.46
8	47	-	29.4	Dr. Saab's CECS333 pages (short sessions)	0.16
9	53	28	33.4	Dr. Saab's CECS303 pages (long detailed sessions)	0.19
10	184	111	123.3	Accesses to the CECS227 class pages	0.2
11	77	49	49.3	Dr Shi's CECS345 (main page and Java examples)	0.27
12	47	30	30.0	Dr Shi's CECS345 (long sessions: lectures and project No. 1)	0.26
13	34	-	22.5	Dr Shi's CECS345 (short sessions to main page)	0.19

Holland, J. H., 1975. Adaptation in natural and artificial systems. MIT Press.

Mahfoud, S. W., Sep. 1992. Crowding and preselection revisited. In: 2nd Conf. Parallel problem Solving from Nature, PPSN '92. Brussels, Belgium.

Nasraoui, O., Krishnapuram, R., May 2000. A novel approach to unsupervised robust clustering using genetic niching. In: Ninth IEEE International Conference on Fuzzy Systems. San Antonio, TX, pp. 170–175.

Nasraoui, O., Krishnapuram, R., Frigui, H., A., J., 2000. Extracting web user profiles using relational competitive fuzzy clustering. To appear in International Journal of Artificial Intelligence.

Nasraoui, O., Krishnapuram, R., Joshi, A., Jun. 1999. Mining web access logs using a relational clustering algorithm based on a robust estimator. In: NAFIPS Conference. New York, NY, pp. 705–709.

Neuro-Fuzzy-Genetic Architecture for Data Mining

Korakot Hemsathapat, Cihan H. Dagli, David Enke
Smart Engineering Systems Laboratory
236 Engineering Management Building
University of Missouri – Rolla
Rolla, MO 65409-0370
Email: korakot@umr.edu, dagli@umr.edu, enke@umr.edu

Abstract
The main objective of this research is to develop a neuro-fuzzy-genetic architecture for data mining which presents discovered patterns in understandable form. In the architecture, all input variables are first preprocessed and all continuous variables are fuzzified. Fuzzification of the continuous inputs with linguistic terms represented by membership functions creates better understanding for the users. Principal Component Analysis (PCA) is then applied to reduce the dimension of the fuzzified input variables in finding combinations of variables, or factors, that describe major trends in the data. The rule extraction module in the architecture is designed to extract explicit knowledge from the trained neural networks and represents it in the form of crisp and fuzzy If-Then rules. In the final stage a genetic algorithm is used as a rule-pruning module to get rid of the weak rules that are still in the rule bases, while the classification accuracy of the rule bases is significantly changed or even improved. Benchmarking the performance of the architecture with the standard C4.5 decision tree was also carried out. Real world application of the architecture is demonstrated with the meningoencephalitis diagnosis dataset from the JSAI KDD Challenge 2001.

1. Introduction

Business firms have collected a large amount of data in their databases and use it for making important decisions. Almost all decisions depend on data in some way and the quality of data sometimes drastically affects the quality of decisions [12]. There is an increasing need for data mining tools that can look into these databases and extract information that can be potentially useful to management. Knowledge discovery in databases (KDD) is a crucial part of decision support, with the objective of knowledge discovery for decision support being to extract hidden patterns in data that can be used in decision-making.

In this paper, a neuro-fuzzy-genetic architecture for data mining is introduced. The architecture can be used to extract hidden knowledge from data and represent it in the form of crisp and fuzzy *If-Then* rules. The criteria to evaluate the effectiveness of the rule extraction technique in the architecture are explained. Comparison of the architecture with the C4.5 decision tree was implemented with four classification problems. Real world application of the architecture is also demonstrated with a meningoencephalitis diagnosis dataset [1].

2. Neuro-Fuzzy-Genetic Architecture

Data mining is an iterative process involving several stages, beginning with domain understanding and ending with the reporting and using of the discovered knowledge [2,3]. The neuro-fuzzy based data mining architecture is also processed in an iterative manner as shown in Figure 1. The outputs of some modules are used as feedback to the previous modules to improve the outputs of the subsequent modules. Knowledge in a trained neural network is encoded in its connection weights, which are distributed throughout the entire neural network structure.

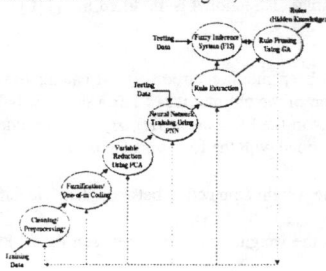

Figure 1: The Neuro-Fuzzy-Genetic Architecture

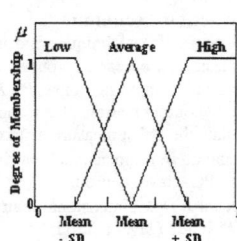

Figure 2: Triangular Membership Functions

In Figure 1, all of the inputs are first preprocessed and cleaned. The missing values of the continuous variables are replaced with the mean of the considered variables. The missing values of the discrete variables are replaced with the mode of the considered variables. The discrete input variables are coded by using *one-of-m* coding [13]. The three triangular membership functions (Low, Average, High) for fuzzification are designed based on the mean and standard deviation of each continuous variable, as shown in Figure 2.

The fuzzification and one-of-m coding increase the number of variables inputted to the next module. The increased number of inputs will result in increased training time of the network. Principal Component Analysis (PCA) is then applied for variable reduction. Mathematically, PCA relies upon an eigenvector decomposition of the covariance or correlation matrix of the variables. The objective of a PCA is to transform correlated random variables to an orthogonal set which produces the original variance/covariance structure. In this paper the minimum fraction variance component to all datasets was set to 0.01. The details of the PCA are given in [4].

The neural network module used was a Probabilistic Neural Network (PNN). It has the capability to quickly classify a categorized dataset. The details of the network structure are given in [5,6]. A PNN is first trained on a set of training data. The trained neural network is then evaluated with testing data. The network that has the highest testing accuracy is retained.

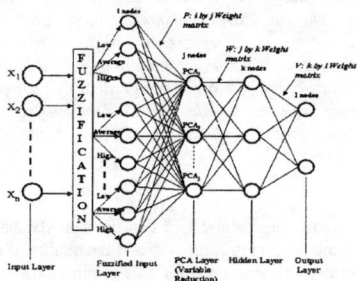

Figure 3: The Structure of the Neuro-Fuzzy-Genetic Based Data Mining Architecture

The rule extraction technique, explained in the next section, is applied to extract explicit knowledge from the trained network and represents it in the form of fuzzy *If-Then* rules:

$$\text{If } X_1 \text{ is } Y_1, \text{ and } X_2 \text{ is } Y_2, ..., \text{ and } X_n \text{ is } Y_n \text{ then } C, \text{ Weight;} \qquad (1)$$

where X_i represents an input variable, Y_i represents a fuzzy membership function derived from X_i, C represents classes, and *Weight* represents a weight of the rule. The rule extraction module extracts *If-Then* rules from the weights of the trained neural network. Different techniques for rule extraction from neural networks can be used [7,8,9].

The Fuzzy Inferences System (FIS) was developed to test the rule base performance. It evaluates fuzzy membership values for each input variable, fires the rules sequentially to calculate the degree of membership for each species type, and declares that with the highest membership value the winner.

The extracted rules from rule extraction module evaluated by the FIS are not optimized. There are still some rules deemed not important to the rule bases. The genetic algorithm for rule pruning is then applied to find optimal sets of rules in all rule bases while the classification accuracy of the pruned rule bases is still maintained or improved. The genetic algorithm rule pruning that is explained in Section 4 is modified from [10].

3. Rule Extraction Technique

The rule extraction technique used in this paper has been modified from [9]. It calculates an effect of each fuzzified input neuron on each output by the multiplication of weight matrices. For a network (see Figure 3) with $(I_1, I_2,..., I_i)$ fuzzified input neurons, $(P_1, P_2,...,P_j)$ neurons in the PCA layer, $(H_1, H_2,...,H_k)$ hidden layer neurons, and $(O_1, O_2,...,O_l)$ output neurons, the technique can be described with the following steps:

<u>Step 1</u>: Calculate the effect measure matrix.
- Let P equal an $i \times j$ dimensional matrix, where P_{ij} is the weight connection between the ith fuzzified input neuron and the jth PCA layer neuron.
- Let W equal a $j \times k$ dimensional matrix, where W_{jk} is the weight connection between the jth PCA layer neuron and the kth hidden layer neuron.

- Let V equal a $k \times l$ dimensional matrix, where V_{jk} is the weight connection between the kth hidden layer neuron and the lth output layer neuron.

The $i \times l$ dimensional effect measure matrix, A, is given by

$$A = P \times W \times V = \begin{bmatrix} e_{11} & \cdots & e_{1l} \\ \vdots & & \vdots \\ e_{i1} & \cdots & e_{il} \end{bmatrix}$$

where e_{il} is the effect measure of fuzzified input neuron I_i on output neuron O_l.

Step 2: Extract the rules.
For each $e_{il} > 0$, write a rule of the form:

If x is X then y is C,

where x is X and y is C are the descriptions for the fuzzified input neuron I_i and output neuron O_l, respectively.

Step 3: Calculate the rule weighting for each rule.
For each $e_{il} > 0$, the weighting for the rule "if x is X then Y is C" is given by

$$Weight = \frac{e_{il}}{\max\{e_{11}, \ldots, e_{il}\}} \times 100$$

Rules high and low in weight (*Weight*) are called strong and weak rules, respectively. This technique was previously implemented without a PCA layer and proved to give good results [9, 11]. However, this is an attempt to modify a rule extraction technique by adding the PCA layer to reduce the dimension of the fuzzified input layer, which results in a decreased training time for the neural network.

4. Rule Pruning Using Genetic Algorithm

The objectives of the genetic algorithm are to find a small number of significant rules from a large number of extracted rules in the previous stage and to maximize the number of correctly classified data by the selected rules. In order to apply a genetic algorithm to the rule pruning module, a subset S of extracted rules is denoted by a gene string as $S = s_1 s_2 s_3 \ldots s_r$, where $r = i \times l$ is the total number of extracted linguistic rules from the trained neural network, $s_p = 1$ means that the *p-th* rule is included in the rule set S, and $s_p = 0$ means that the *p-th* rule is not included in S. Rows (*i*) in the matrix represent all the possible premises of a targeted class. Each column (*l*) represents a class attribute of a target variable.

$$Weight = \begin{bmatrix} s_1 & \cdots & s_{1 \times l} \\ s_2 & \vdots & \\ \vdots & \vdots & \\ s_{i \times 1} & \cdots & s_r \end{bmatrix} \rightarrow S = s_1 s_2 \ldots s_r \quad (2)$$

The fitness of each gene string is evaluated by

$$Fitness(S) = NCP(S) \quad (3)$$

where *NCP* is the number of correctly classified data by S. The population is randomly initialized from the extracted rule bases in the previous stage. Start with a randomly generated population of n *r*-bit chromosomes. Every chromosome string is evaluated by FIS and ranked by its fitness. Half of the chromosome strings with highest fitness in the current population are selected as parents. Then, 1-point cross over is employed for generating child strings from their parents. Then the mutation is applied to the newly generated strings and is kept at the last chromosome in the population. The roulette wheel operator is used to create a new population for the next generation. These genetic operations (crossover, mutation, evaluation, and reproduction) are repeated for specified generations.

5. Evaluation of Rule Extraction Techniques

In several studies [14, 15], the effectiveness of new proposed techniques in rule extraction has been measured solely by the accuracy of classification of the extracted rule base. An important criterion for evaluation of the effectiveness of rule extraction techniques is the quality of the generated rule base. The quality can be measured by *accuracy, comprehensibility,* and *fidelity* of the extracted rule base [8]. *Accuracy* is the ability of the rule base to correctly classify previously unseen data. *Comprehensibility* depends on the rule base size and complexity. Since the extracted rules are the explicit expression of the acquired knowledge, and are often used for explanation of the decisions made by the rule base system, they should be understandable. Rule comprehensibility can be measured by the number of rules in the rule base and the number of premises. Rule bases with less complex and fewer rules are easier to understand and are also easier to maintain. *Fidelity* refers to those rule bases which have been extracted from trained neural networks. Fidelity means the capability of the extracted rule base to reproduce the behavior of the network from which it was extracted. The extracted rule set should contain the knowledge acquired by the neural network during training.

6. The Classification Datasets

The neuro-fuzzy-genetic architecture and C4.5 were tested and evaluated on four classification problems: Iris species, Pima Indian diabetes, heart disease diagnosis, and credit approval. The data was obtained from the UCI machine-learning database [16]. The heart disease and credit approval datasets were selected because they contain a good combination of continuous and discrete variables. The iris species and Pima Indian diabetes datasets were included due to their widespread use in the literature.

The iris species dataset consists of 150 records with 4 continuous input variables and 1 class attribute. The Pima Indians diabetes dataset consists of 768 records with 8 input variables and 1 class variable. The 500 records with class label "0" are for no evidence of the disease, the 268 records with class label "1" are for the evidence of the disease. The heart disease dataset was collected from the Cleveland Clinic Foundation (Cleveland database). It is one of four databases stored in the heart-disease directory of the machine-learning database. The dataset has 303 records, 13 input variables (5 continuous) and one target variable, which has 5 possible values – 0 for no evidence of heart disease and 1, 2, 3, 4 for evidence of heart disease. The credit approval database contains real-life data about credit card applications collected for the purpose of credit risk assessment. To protect confidentially, all attribute names have been replaced by meaningless symbols. The dataset has 690 records, 15 input variables (6 continuous) and one class variable.

7. The Implementation

Each dataset was divided into 2 groups: 2/3 for training set and 1/3 for testing set. Continuous variables were fuzzified by the three triangular membership functions mentioned earlier. The discrete variables were also coded using one-of-m coding. After the neural networks in the architecture have been trained to the desired degree of accuracy, the rule extraction technique was implemented. The architecture has extracted four rule bases that can be used to classify the datasets. However, those rule bases were not optimized. The genetic algorithm for the rule pruning module was then applied to reduce the number of rules in the rule bases. These are the genetic algorithm parameter settings used in all four datasets: number of generations = 1000, population size = 21, mutation rate = 0.001 to 0.04, crossover = 1-point crossover, and reproduction = roulette wheel.

For the purpose of comparison, C4.5, an important decision tree based learning algorithm, was selected and applied to all four datasets. Once again, 2/3 of the datasets were used for training and 1/3 for testing. The parameters used by C4.5 were systematically changed, as described by Quinlan [17], to determine the best performance.

8. Results and Discussion

The rule base fidelity, comprehensibility, and accuracy were used as the criteria for evaluation of the architecture. The criterion for fidelity applies only to rules extracted from the neural networks. In this implementation, the fidelity has been evaluated by comparison of the classification accuracy achieved by the original neural network on the training set, with classification accuracy of the original rule bases and pruned rule bases on the same training set. Table 1 shows that the pruned rule bases can achieve more than 90% in classification accuracy and fidelity in all datasets in comparison to the original neural network.

Table 1. Fidelity of the Extracted Rule bases

Accuracy on Training Sets				
Technique	Iris	Pima Indian	Credit Approval	Heart Disease
PNN	100%	100%	100%	100%
Original rule bases	89%	74.74%	86.52%	62.37%
Pruned rule bases	99%	98.04%	95.43%	90.59%

For comparison of the effectiveness across all techniques, the accuracy of classification of unseen data, number of rules, and number of premises were used. The results in Table 2 show that the original rule bases and the pruned rule bases outperform C4.5 in accuracy. In term of rule base comprehensibility, the pruned rule bases also outperform C4.5 with a less number of rules and a less number of premises (as illustrated in Table 3).

Table 2. Comparison of Testing Accuracy of the Extracted Rule bases

Technique	Accuracy on Testing Sets			
	Iris	Pima Indian	Credit Approval	Heart Disease
Original rule bases	96%	77.73%	86.52%	62.37%
Pruned rule bases	96%	79.68%	89.13%	70.29%
C4.5	94%	70%	86.5%	51%

Table 3. Comparison of the Rule Complexity in the Extracted Rule bases

Technique	Iris		Pima		Credit Approval		Heart Disease	
	# of Rules	# of Premises	# of Rules	# of Premises	# of Rules	# of Premises	# of Rules	# of Premises
Original rule bases	11	11	21	21	26	26	57	57
Pruned rule bases	4	4	8	8	10	10	24	24
C4.5	4	6	58	114	24	38	39	70

An example of the generate rule base for the Iris data set is shown below. The highlighted rules are the rules selected from the original rule base by the genetic algorithm for the rule pruning module.

RULE BASE FOR IRIS SPECIES DATASET

IF SEPAL LENGTH IS LOW THEN SESOTA (75)
IF SEPAL WIDTH IS HI THEN SESOTA (71)
IF PETAL LENGTH IS LOW THEN SESOTA (100)
IF PETAL WIDTH IS LOW THEN SESOTA (100)

IF SEPAL LENGTH IS AVG THEN VERSICOLOR (32)

IF SEPAL WIDTH IS LOW THEN VERSICOLOR (40)
IF PETAL LENGTH IS AVG THEN VERSICOLOR (86)
IF PETAL WIDTH IS AVG THEN VERSICOLOR (83)

IF SEPAL LENGTH IS HI THEN VIRGINICA (62)
IF PETAL LENGTH IS HI THEN VIRGINICA (85)
IF PETAL WIDTH IS HI THEN VIRGINICA (8)

9. JSAI KDD Challenge 2001: Meningoencephalitis Diagnosis Application

The objective of the JSAI KDD Challenge 2001 is to tackle a dataset with close collaborations among many analysts and experts in the dataset domain and to evaluate the possibility of discovering significant knowledge in such an integrated knowledge discovery process. This dataset for the Challenge is provided by a medical doctor [1] who is the domain expert on the meningoencephalitis diagnosis. The dataset was obtained from meningoencephalitis diagnosis activity in a hospital in Japan. In this application, the architecture was used to extract hidden knowledge from the dataset. The extracted knowledge, or so-called rule bases, could later be used to automate the diagnosis of the disease.

The meningoencephalitis dataset has 140 records with 38 variables. There are 18 categorical, 1 rank, and 19 continuous attributes. All of the variables were first normalized into the range between 0 and 1 and then fuzzified by using three triangular membership functions. The categorical variables are coded using one of m coding. The first 121 records (old samples) were used as a training set and the last 19 records (new samples) are used as a testing set. Missing values of CSF_CELL3 (Cell count CSF 3 days after the treatment) have been replaced with a mean of CSF_CELL3 itself. Six models were built for classification of the following targeted variables: DIAG - Diagnosis described in the database, DIAG2 - Group attribute of DIAG, CULT_FIND - Whether bacteria or virus is specified or not. (T: found, F: not found), CULTURE - The name of bacteria or virus (-: not found), C_COURSE - Clinical course at discharge, and COURSE - Grouped attribute of C_COURSE (n: negative, p: positive).

Six rule bases were extracted from the data for each targeted variable. The typical rule base for two-class variables consisted of 82 rules. When the initial rule pruning and FIS were applied, the weak rules (rule strength less than 25) had been dropped from the rule bases. The extracted rule sets had between 35-40 rules. The accuracy of prediction of the rule base for DIAG2 (classification rate on testing set) was 94.74%. When a genetic algorithm for rule pruning was applied, the performance of the rule bases increase over generations with higher classification accuracy and less number of rules in the rule bases. Example of the generated rule base from the DIAG2 targeted variable is shown below. The highlighted rules are the rules selected from the original rule bases by the genetic algorithm for the rule pruning module.

RULE BASE FOR DIAG2
ACC_TRAIN = 88.4297% (107 of 121 correctly classified data)
ACC_TEST = 94.7368% (18 of 19 correctly classified data)
ACC_GA_RULES = 97.85% (137 of 140 correctly classified data)

IF AGE IS HI THEN BACTERIA (40)
IF SEX IS MALE THEN BACTERIA (43)
IF BT IS HI THEN BACTERIA (39)
IF LOC_DAT IS "+" THEN BACTERIA (28)

IF WBC IS HI THEN BACTERIA (52)
IF CRP IS HI THEN BACTERIA (68)
IF ESR IS HI THEN BACTERIA (36)
IF CT_FIND IS ABNORMAL THEN BACTERIA (51)
IF CSF_CELL IS HI THEN BACTERIA (66)
IF CELL_POLY IS HI THEN BACTERIA (67)
IF CSF_GLU IS LOW THEN BACTERIA (34)
IF CULTURE IS STREPTO THEN BACTERIA (52)

```
(*) IF THERAPY2 IS "ANTIBIOTICS CHANGED SEVERAL TIMES"     IF WBC IS LOW VIRUS (44)
    THEN BACTERIA (47)                                     IF CRP IS LOW THEN VIRUS (53)
(**) IF THERAPY IS "OTHER NAME OF ANTIBIOTICS" THEN        IF CRP IS AVG THEN VIRUS (55)
    BACTERIA (91)                                          IF ESR IS AVG THEN VIRUS (40)
IF CSF_CELL IS HI THEN BACTERIA (33)                       IF CT_FIND IS NORMAL THEN VIRUS (51)
IF RISK IS SINUSITIS THEN BACTERIA (45)                    IF CSF_CELL IS LOW THEN VIRUS (64)
IF RISK IS DM THEN BACTERIA (34)                           IF CSF_CELL IS AVG THEN VIRUS (41)
IF RISK (GROUP) IS POSITIVE THEN BACTERIA (61)             IF CELL_POLY IS LOW THEN VIRUS (100)
                                                           IF CELL_POLY IS AVG THEN VIRUS (38)
IF AGE IS LOW THEN VIRUS (33)                              IF CSF_GLU IS AVG THEN VIRUS (41)
IF SEX IS FEMALE THEN VIRUS (43)                           (**) IF THERAPY2 IS "NO THERAPY" THEN VIRUS (59)
IF LOC IS LOW THEN VIRUS (27)                              (**) IF THERAPY2 IS "ANTI-VIRUS CHEMICALS" THEN VIRUS (50)
IF BT IS AVG THEN VIRUS (26)                               IF RISK IS NORMAL THEN VIRUS (61)
IF LOC_DAT IS "-" THEN VIRUS (28)                          IF RISK (GROUPED) IS "NEGATIVE" THEN VIRUS (61)
```

Some patterns have been discovered from the generated rule bases. For example, the strongest rule with rule weight = 100 states that if the cell count in polynuclear cell is low, it is usually caused by a virus. As another example with rule weight = 52 states that if white blood cell count is high, it is usually caused by bacteria. When these generated rules are combined they can be used to build a rule base expert system for the meningoencephalitis diagnosis. After receiving comments from the expert, most of the interesting rules as suggested by the expert (marked by a *) were unexpectedly selected by the genetic algorithm rule pruning (for example, if therapy 2 is "antibiotic changed several times", it is usually caused by bacteria). However, some of the uninteresting rules commented by the expert (marked by **) were still selected in the rule bases (for example, if therapy 2 is "other name of antibiotics" is usually caused by bacteria). A possible solution to eliminate uninteresting rules from the rule bases is to delete those rules from the original rule bases before using the genetic algorithm rule pruning module to see how the classification of the new rule bases perform. Another question regarding this technique is why strong rules were not always picked by genetic algorithm. This question is still under investigation.

10. Remarks and Conclusions

This paper describes an implementation of a neuro-fuzzy-genetic based data mining architecture, which finds patterns in datasets and presents the patterns in a form understandable by humans. In this architecture a neural network is trained on a set of training data and a rule extraction technique is then applied in order to extract explicit knowledge from the network in the form of linguistic *If-Then* rules. In addition, the inference engine and genetic rule-pruning module are used to evaluate and tune the extracted rule bases. Comparison of the architecture with C4.5 was also demonstrated. The pruned rule bases from the architecture outperform C4.5 in accuracy, fidelity, and comprehensibility. A real world application of the architecture was applied for knowledge discovery within the meningoencephalitis database. The results indicate that the architecture can extract meaningful patterns from the data. In this application fuzzy rules were useful since the linguistic terms are understandable to the users.

References

[1] Tsumoto, Shusaku, *Meningoencephalitis Diagnosis Dataset*, Shimane Medical University. http://wwwada.ar.sanken.osaka-u.ac.jp/pub/washio/jkdd/ menin.htm
[2] Fayyad, U, Piatetsky-Shapiro, G, Smyth, P, 1996, *Knowledge Discovery and Data Mining: Towards a Unifying Framework*, in Proceedings of The 2nd International Conference on Knowledge Discovery & Data Mining, KDD-96, Oregon 1996
[3] Piatetsky-Shapiro, G, Brahman, R. Khabaza, T, 1996, *An Overview of Issues in Developing Industrial Data Mining and Knowledge Discovery Applications*, in Proceedings of The 2nd International Conference on Knowledge Discovery & Data Mining, KDD-96, Oregon 1996
[4] Haykin, Simon, 1999, Neural Networks, A Comprehensive Foundation, Prentice Hall, 2nd Edition, pp. 392-440
[5] Demuth, Howard, Beale, Mark, 1998, *Neural Network Toolbox for Use with MATLAB*, Version 3.0
[6] Wasserman, P.D., 1993, *Advanced Methods in Neural Computing*, New York: Van Nostrand Reinhold, pp. 35-55
[7] Kasabov, N, 1996, *Learning Fuzzy Rules and Approximate Reasoning in Fuzzy Neural Networks and Hybrid Systems*, Fuzzy Sets and Systems 82(1996) pp.135-149, Elseveier
[8] Towell, G, Shavlik, J, 1993, *Extracting Refined Rules from Knowledge-Based Neural Networks*, Machine Learning, 13, pp.71-101
[9] Matthews, C., Jagielska, I., 1995, *Fuzzy Rule Extraction From Multilayered Neural Network*, Proceedings of IEEE International Conference on Neural Networks (ICNN'95), Perth.
[10] Ishibushi, H., Nozaki K & Yamamot N, 1993, *Selecting Fuzzy Rules by Genetic Algorithm*, IEEE Int Conf Fuzzy System (ICFS), San Francisco.
[11] Hemsathapat, K., Dagli, C.H., *Neural Network Model for Detecting Rare Patterns in Data: A Telecommunication Example*, 1999, in Intelligent Engineering Systems through Artificial Neural Networks, eds. Dagli et. al, Vol.9, ASME Press, pp. 525-530
[12] Parsaye, K., Chignell, M. H., 1994, *Quality Unbound*, Database Programming and Design, November, 1994
[13] Bigus, Joseph P., 1996, *Data Mining with Neural Networks*, McGraw Hill.
[14] Yoon, Byungjoo, 1994, Rule Extraction Using Destructive Learning in Artificial Neural Networks, Ph.D. Dissertation, Florida State University.
[15] John, George, H., 1997, *Enhancements to the Data Mining Process*, Ph.D. Dissertation, Stanford University.
[16] *UCI Repository of Machine Learning Databases*, Department of Information and Computer Science, University of California, Irvine, http://www.isc.uci.edu/~mlearn/MLRepository.html.
[17] Quinlan, J. R.; *"C4.5: Programs for Machine Learning"*, Morgan Kaufman, 1993.

KNOWLEDGE DISCOVERY FROM MULTISPECTARL SATELLITE IMAGES

ARUN D. KULKARNI and ZHIWEI MO
Computer Science Department
The University of Texas at Tyler, Tyler, TX 75799

ABSTRACT

In this paper, we present a novel application that deals with knowledge discovery from multispectral satellite images. Recent high-resolution earth observation satellites collect huge amount of data in the form of multispectral images. These images represent a big source of knowledge that can be used in many practical applications such as agriculture, forestry, mineral resources, hydrology, and military reconnaissance. One of the commonly used forms of knowledge representation is the if-then rules form. Here, we consider fuzzy neural systems for classifying satellite data, and use the link tracking technique suggested by Mitra, et al. (1999) for extracting classification rules. We have developed software to simulate fuzzy-neural systems and to extract rules in Visual Basic 6.0. We first test our software with simulated test data and subsequently use it for extracting classification rules from satellite data.

INTRODUCTION

Remote sensing is the science of deriving information about an object from measurements made at a distance. The technique of remote sensing relies to the great extent on the interaction of electromagnetic radiation with matter. The remotely measured reflected signal expressed as a function of the wavelength is often referred to as the "spectral signature" of the target object on which measurements are made. In principle, spectral signatures are unique; that is different objects have different spectral signatures. It is therefore theoretically possible to identify an object from its spectral signature. In brief, this is the principle of multispectral remote sensing (Kulkarni, 2001). Remotely sensed multispectral measurements can be a source of information for many applications, including agriculture, forestry, mineral resources, hydrology, water resources, cartography, meteorology, and military. Satellite remote sensing began in earnest with the launching of the first earth resource satellite in 1972 by the National Aeronautics and Space Administration (NASA) of the United States. Recent satellites include Landsat-7 and IKONOS launched by Space Imaging in 1999. Landsat-7 has a sensor called thematic mapper (TM) with seven spectral bands. In this study we consider data from Landsat-7.

The ability to train a fuzzy-neural system from a set of samples and then to use the trained network to classify unknown samples lies at the core of tasks such as knowledge extraction. In the case of fuzzy neural systems, connection strengths and/or fuzzy membership functions form a knowledge base and are modified during learning. Recently, many fuzzy-neural network models (Lin and Lee, 1991; Mitra and Pal, 1994; Kulkarni, 1998; Jang, et al, 1999; Kulkarni, 2001) have been suggested in the literature. There are many ways to combine neural networks with fuzzy logic. One of the commonly used approaches is to use fuzzy membership functions to pre-process or post-process data with neural networks. In this paper we have used this approach. The successful use of a fuzzy-neural network model for a particular task depends upon the sample data, the number of fuzzy membership functions and the learning rule. During training, the network weights are adjusted to minimize the mean squared error between the actual and the desired outputs. In order to extract knowledge from the network, we need useful representation of the relationships between the input features and the output classes. Many methods for rule extraction have been suggested in the literature (Kurfess, 2000; Setiono and Leow, 2000; Lu, et al., 1996; Taha and Ghosh, 1997). Roy (2000) presents some theoretical issues concerning brain-like learning and rule extraction. Setnes (2000) proposes a method for rule extraction from data by means of fuzzy clustering in the product space of inputs and outputs where each cluster corresponds to a fuzzy in-then rule. Setnes (2000) uses the orthogonal least squares method to remove redundant or less important clusters during the clustering process in order to capture fuzzy rules that are of more significance. Duch et al. (2001) present a new methodology of extraction, optimization and application of crisp and fuzzy logical rules. Taha and Ghosh (1999) in their paper present three rule extraction techniques. The first technique extracts a set of binary rules from a neural network, the second technique extracts rules that represent the most important embedded knowledge with an adjustable level of details, while the third technique provides a more comprehensive and universal approach. Mitra et al. (1997) in their article an idea of knowledge encoding among the connection weights of a fuzzy MLP. They provide a method for rule extraction using the link tracing technique. Their model is capable of extracting both positive and negative rules. In this paper we consider a link tracing technique for rule extraction. We consider two fuzzy-neural network models. The models and the method for extracting rules are described in the subsequent sections.

FUZZY-NEURAL NETWORK MODELS

We consider two fuzzy neural network models. The first model consists of a fuzzifier block followed by a perceptron network, and the second model consists of a fuzzifier block, which is followed by a back propagation network. As shown in Figure 1, the first model consists of three layers. Layer L_1 is the input layer. The number of units in this layer is equal to the number of input features. Layer L_2 implements the membership function. In this model, we used five term sets (*very low, low, medium, high, very high*). The number of units in this layer is five times the number of units in layer L_1. Layer L_3 represents output layer, and the number of units in this layer is equal to the number of classes. The normalized input vectors are applied to layer L_1 and then processed with membership function in layer L_2. In this model, triangular membership functions are used for mapping input values to the corresponding membership functions. The second fuzzy-neural network model is shown in Figure 2. The model consists of four layers. Layer L_1 is input layer. The number of units in this layer is equal to the number of input features. Layer L_2 performs the membership function. The number of units in layer L_2 is five

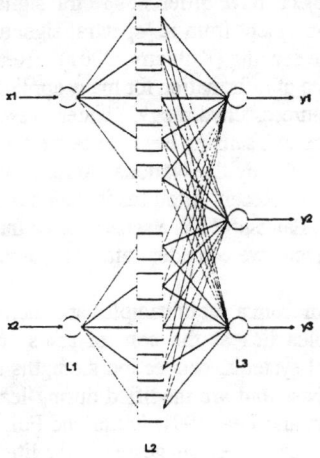

Figure 1. A three layer fuzzy-neural system

times the number of units in layer L₁. Layers L₂, L₃ and L₄ constitute a three-layer feed-forward network with back-propagation learning. Layer L₃ is basically a hidden layer. The number of units in this layer should be greater than that in layer L₂. Layer L₄ is output layer. The number of units in this layer is equal to the number of classes.

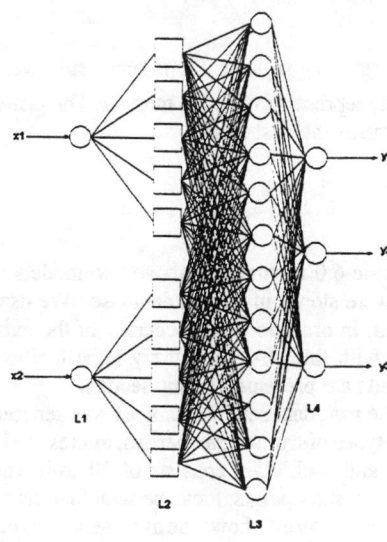

Figure 2. A four-layer fuzzy-neural network

RULE EXTRACTION

After the training is completed, we extract classification rules. A link tracing rule extraction technique, the *path generation by backtracking* proposed by Mitra and Pal (1994) is used. Starting from the output node j in layer L₃ which has an output value y_j that is greater than 0.5, we select those nodes i in the layer L₂ that have an input value f_i which is greater than 0.5 and give a positive impact on the output node j. The selected nodes in layer L₂ would be arranged in the decreasing order of their net impacts, where the net impact γ_i for node i is defined as

$$\gamma_i = f_i w_{ji} \qquad (1)$$

where f_i represents the input feature applied to node i in layer L₂, w_{ji} represents weight and j is the class number. So the antecedent part of an *If-Then* rule can be obtained from this ordered list. In fact, we just need to select the first 30% input nodes from the ordered list to generate the *If* part for a rule. For a node i_s in layer L₂ selected for the *If* part generation, the corresponding input feature x_{is} is defined as in Equation (2)

$$x_{is} = \left\lceil \frac{i_s}{n} \right\rceil + 1 \qquad (2)$$

where i_s is the input number in layer L₂ and n is the number of term sets. Here n is equal to 5. Generally, the *If* part is given in linguistic form. The fuzzy value for each node i_s in layer L₂ is determined. In this model, triangular membership functions are used for mapping. The number of term sets can be adjusted.

In the case of the second model, the basic idea of extracting rules is similar to the first model. Since a neural network using a back propagation learning has at least one hidden layer, it requires a more complex processing for rule generation Starting from output layer L₄, the process continues until the layer L₂ is reached. For each output node j in layer L₄ which has an output value y_j that is greater than 0.5, we select those nodes i in the hidden layer L₃ that have a positive incoming link for the output node j. For each selected i_s in layer L₃, we choose those nodes i_k in layer L₂ that have an input value f_k which is greater than 0.5 and give a positive impact on the node i_s. The selected input nodes i_k would be arranged in the decreasing order of their net impacts, where the net impact γ_i for node i_k is defined as

$$\gamma_i = f_k \times \left(w_{i_s i_k} + w_{j i_k} \right) \quad (5)$$

where f_k represents the input value of node i_k in layer L$_2$, w represents weight, i_s represents the input number in layer L$_3$ and j represents the class number. The generation of *If* part and *Then* part for a rule is the same as the first model.

RESULTS AND CONCLUSIONS

We have developed software in Visual Basic 6.0 to simulate above two models and to extract fuzzy rules. The training set data were stored in Access database. We used two data sets: a fruit data set and satellite data. In order to verify accuracy of the extracted rules, we generated a fuzzy logic system with the MATLAB fuzzy logic toolbox, and reclassified the training data sets. The results are presented in this section.

In our first illustration we used the fruit data set. The data set was generated by measuring the weight and volume of three types of fruits: strawberries, apples, and grape fruits. There were three classes of fruits and each class consists of 20 fruit samples. Figure 3 shows the feature space that contains sixty points. It can be seen that the feature space represents three well-separated clusters. Figure 4 shows the average values of input features for three classes. We used the mean vectors to train a three-layer fuzzy-neural network model. The model consists of two inputs that represent two input features, and three units in the output layer that represent output classes. The output vectors $(1,0,0), (0,1,0),$ and $(0,0,1)$ represented three output classes. In this model we used membership functions with three term sets: *{low, medium, high}*. After training, fuzzy rules were extracted using the link-tracing algorithm. We used only three mean vectors for training, and each vector generated a rule. The rules are shown below:

Rule 1: If *weight* is *low* and *volume* is *low* then class is *strawberry*
Rule 2: If *weight* is *medium* and *volume* is *medium* then class is *apple*
Rule 3: If *weight* is *high* and *volume* is *high* then class is *grape fruit*.

We used the above rules to build a fuzzy inference system using MATLAB fuzzy logic toolbox. We classified the sixty samples from the database and we found the classification accuracy was 100 percent.

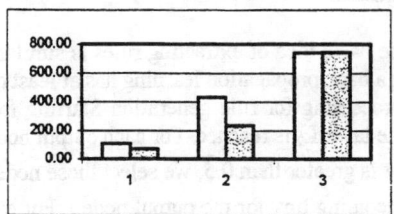

Figure 3. Mean values for feature vectors

In our second example we consider TM data obtained earth observation satellite. The scene represents the Mississippi river bottomland area. The scanner captures data in seven spectral bands. We used only five bands (bands 2, 3, 4, 5, and 7), since these bands showed the maximum variance and contained information needed to identify classes. Three training areas were chosen, which represented three classes: vegetation, water, and land.

Figure 4 shows the mean vectors for three classes. The mean vectors were used to train a four-layer fuzzy neural network model. The input layer contained five units that represent five input features, and the output layer contained three units that represent output classes. The classified output is shown in Figure 5. The network was trained using the mean vectors. After training, fuzzy rules were extracted using the link-tracing algorithm. In this example we used five term sets {*very-low, low, medium, high, very-high*}. The obtained rules as shown below:

Rule 1: If B_2 is *low* and B_3 is *low* and B_4 is *low* and B_5 is *very-low* and B_7 is *very-low* Then class is *water*.
Rule 2: If B_2 is *low* and B_3 is *low* and B_4 is *low* and B_5 is *high* and B_7 is *very-high* Then class is *vegetation*.
Rule 3: If B_2 is *low* and B_3 is *low* and B_4 is *low* and B_5 is *medium* and B_7 is *low* Then class is *land*.

The above rules were used to build a fuzzy inference system using the MATLAB fuzzy logic toolbox. The input samples were reclassified. It was found that the 85 percent of samples were classified correctly.

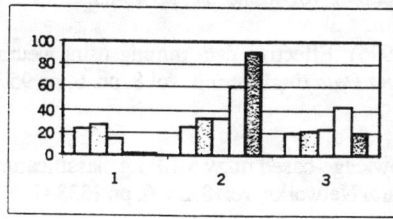

Figure 4. Mean feature vectors for satellite data

It can be seen that in both examples we used mean vectors for training and each vector generated a rule. If we use a large number of samples, we will get a large number of rules and some rules may be conflicting or may be of less importance. Also, if samples represent over lapping classes they may yield conflicting rules. We are in the process of building a FAM bank and assigning a degree of confidence for a rule. Each cell in a FAM bank can represent a rule. If there are more than one rule in a FAM cell, then we can discard the rule with the low degree of confidence.

Figure 5. Classified output, Mississippi scene

REFERENCES

Duch Wlodzislaw, et al. (2001). A new methodology of extraction, optimization, and application of crisp and fuzzy logical rules. *IEEE Transactions on Neural Networks*, vol. 12, no. 2, pp 277-306.

Jang, J. S. R., Sun, C. T., and Mizutani, E., (1997). *Neuro-fuzzy and soft computing*, Prentice Hall, Upper Saddle River, NJ.

Kulkarni, A. D., (2001). *Computer Vision and Fuzzy Neural Systems*, Prentice Hall PTR, Upper Saddle River, NJ.

Kulkarni, A. D., (1998). Neural-Fuzzy Models for Multispectral Image Analysis, *Applied Intelligence*, vol. 8, pp. 173-178.

Kurfess, F. J., (2000). Neural networks and structured knowledge: Rule extraction and applications. *Applied Intelligence*, vol. 12, pp 7-13.

Lin, C. T. and Lee George, C. S., (1991). Neural-network-based fuzzy logic control and decision systems, *IEEE Transactions On Computers*, vol.40, no. 12, pp 46-63.

Lu, Hongjun, Setiono R., and Liu, Huan, (1996). Effective data mining using neural networks. *IEEE Transactions on Knowledge and Data Engineering*, vol.8, no. 6, pp 957-961.

Mitra, S. De R. K., and Pal, S. K., (1997). Knowledge-based fuzzy MLP for classification and rule generation. IEEE Transactions on Neural Networks, vol. 8, no. 6, pp 1338=1350.

Mitra, S and Pal, S. K., (1994). Logical Operation Based Fuzzy MLP for Classification and Rule Generation, *Neural Networks*, vol. 7, no.2, pp 353-373.

Roy, Asim, (2000). Connectionism, rules extraction, and brain-like learning. *IEEE Transactions on Fuzzy Systems*, vol. 8, no. 2, pp 222-227.

Setiono, Rudy and Leow, W. K., (2000). FERNN: An algorithm for fast extraction of rules from neural networks. *Applied Intelligence*, vol. 12, pp 15-25.

Setnes Magne, (2000). Supervised fuzzy clustering for rule extraction. *IEEE Transactions on Fuzzy Systems*, vol. 8, no. 4, pp 416-424.

Taha, I. And Ghosh J., (1999). Symbolic Interpretation of Artificial Neural Networks, *IEEE Transactions on Knowledge and Data Engineering*, vol. 11, no. 3, pp 448 –463.

DATA MINING OF MINE EQUIPMENT DATABASES

TAD S. GOLOSINSKI
University of Missouri-Rolla,
MO 65409-0450, USA

HUI HU
University of Missouri-Rolla,
MO 65409-0450, USA

ABSTRACT
The paper presents research into use of data mining methods for knowledge discovery in mining databases. The data was collected using VIMS system of Caterpillar installed on several trucks operating in a surface mine. It was mined with IBM Intelligent Miner for data. Data mining was found to allow for identification and quantification of relations between the various types of VIMS data. As such it offers the potential for development of a truck model that can be used for prognosticating truck condition and performance. Development of this capability requires further research.

INTRODUCTION

Modern mining equipment is fitted with numerous sensors that monitor its condition and performance. The data collected by these sensors is used to alert the operator to existence of abnormal operating conditions and to perform emergency shut-own if the pre-set values of the monitoring parameters are exceeded. This data is also used for post-failure diagnostics and for reporting and analysis of equipment performance.

It is believed that availability of this voluminous data, together with availability of sophisticated data processing methods and tools, may allow for extraction of a variety of additional information contained in the data. One method that may be of value is data mining (Golosinski, 2001).

The research presented in this paper analyzes data collected from various sensors installed on several mining trucks with the purpose to develop a model of truck operation that may facilitate reliable projection of truck performance and its condition into the future. Data was collected from Caterpillar 789B trucks equipped with VIMS (Vital Information Management Systems, during the period of January to October 2000. IBM Intelligent Miner for Data was used to conduct data mining.:

VIMS OPERATION

Caterpillar's Vital Information Management System (VIMS) is installed on selected CAT mining equipment. It is intended to assist with machine management by informing operators, service personnel and supervisors of the status of selected machine functions and by providing information on equipment production and performance. VIMS monitors and records parameters of numerous sensors that are integrated into the vehicle design. It has the capacity to alert the operator if these parameters exceed the pre-set critical values. In addition it can conduct emergency equipment shut-down if so programmed (Caterpillar, 2000).

On-board VIMS unit records the collected data as well as occurrence of certain VIMS events. The recorded data can be downloaded into a notebook computer. Alternately it can be sent to the central control unit via radio (VIMS Wireless).

VIMS DATA

VIMS records data in seven different formats. These are:

Event Summary List (ESL). A VIMS event is recorded when the measured value of a monitored parameter exceeds that considered acceptable. Event List is a record of events that are occurring on the machine. It is limited to the last 500 events, listed in a chronological order.

Snapshot. Snapshot stores a segment of machine history that consists of values of all monitored parameters recorded at one-second interval. The snapshot is triggered by VIMS event and as such it is related to abnormal condition or emergency situation of the machine.

Data Logger. Data Logger records values of all the machine parameters that are monitored by VIMS and sampled at one-second intervals. The logger is started and stopped by the operator command and can record data for up to 30 minutes.

Trends. Trends record the minimums, maximums and averages of the selected machine condition parameters for a pre-selected period of time.

Cumulative. Cumulative records the number of occurrences of specific events over a pre-set period of time. An example of cumulative information can be the engine revolutions or fuel consumption over the life of the machine, or its component.

Histogram. Histogram records the performance history of a selected parameter since last reset. For example a histogram of the engine speed would indicate the percentages of time that the engine operated within a pre-specified speed ranges.

Payload. Payload carried by the machine can be recorded if so specified and providing that the machine is equipped with an appropriate sensor.

Four different data types are recorded. These are:

Sensed Data. This data contains values of sensor parameters and position of switches installed on the machine.

Internal Data. This data is generated internally within VIMS main module. It includes records of date and time.

Communicated Data. This data is acquired through the data links to various machine components, including non-CAT components. For example the engine speed may be monitored and recorded through the data link to the electronic engine control system.

Calculated Data. This data is calculated by the VIMS main module as a function of other data that is being collected. As an example event duration may be calculated based on internal data and stored in the event list.

INTELLIGENT MINER

Variety of data mining software is available from numerous vendors. It includes Intelligent Miner of International Business Machines Corporation, MineSet of Silicon

Graphic Inc., Clementine of Integral Solutions Limited of U.K. and other (Westphal and Blaxton, 1998). The IBM Intelligent Miner (IM) version 6.1 was used for data mining reported in this paper (IBM, 2000). It offers a choice of algorithms, is easy to use, and has proven itself useful in many commercial applications.

Following mining and statistical functions are included in Intelligent Miner:

1. Mining functions: associations, demographic and neural clustering, sequential patterns and similar sequences, tree and neural classification, and neural and RBF (Radial Basis Function) prediction.

2. Statistics functions: bivariate statistics, linear regression, principal component analysis, univariate curve fitting and factor analysis.

The IM allows modeling of events and processes that can be either usual or unusual. Usual events describe the situation that is considered normal and for which the relations between different attributes are sought. For example, relations between truck operating and mechanical attributes can be defined such as a relation between engine load and truck payload. Definition and quantification of these relations may be of help in improving efficiency of truck operation or help with operator training.

The unusual events are failures of the monitored machine or its component. Data mining of these events may allow for definition of algorithms that would facilitate modeling of truck operation to help with planning of its maintenance and reduction of downtime.

To facilitate data mining of VIMS databases with IM the data format has to be adapted to that acceptable to the IM. The original VIMS data, downloaded from an on-board VIMS unit, can be easily merged into MS Access 97 database using the VIMS PC99 software. However, IM does not accept Access data format and to facilitate its use data has to be transferred to DB2 database that is compatible with version 6.1 of Intelligent Miner.

DATA MINING METHODS

Of the various data mining methods used by IM the following were used in this investigations: Major Factor Analysis, Clustering, Classification, and Sequential Pattern.

Data mining was done on VIMS database that consisted of 300 MB of records collected on several Caterpillar model 789B trucks that operated in a surface mine between February and October 2000. Data collected by VIMS data logger consisted of 105 data sets, each set covering a period of up to 30 minutes of truck operation. Overall 85 parameters of truck condition and performance were monitored with their values recorded each second by the on-board VIMS.

The original data was transferred to DB2 and pre-processed. This included data clean-up, and identification and extraction of data that is of interest to the problem at hand.

INVESTIGATIONS

Relationship between Truck Parameters in VIMS Data Logger Data

Not all truck parameters are independent and a variety of relations exist between them. The preliminary research done by Ataman (2001) defined significant correlations to exist between various parameters. Two VIMS parameters, *engine speed* and *fuel flow*, were found to show strong correlation with many other parameters. Also confirmed was a relatively strong relation between *engine oil pressure* and *engine speed*, indicated in the VIMS manual. The relation between *engine coolant temperature* and *aftercooler temperature* was another expected result. No other significant relations were identified.

This work confirms that the linear regression method of IM can be used to define and quantify the relations that may exist between various parameters describing truck performance and condition. It is believed that these relations, in turn, can be used for truck operation. In relation to VIMS data the major problem is data format incompatibility with that of IM. An interface between VIMS and IM needs to be developed that would allow for easy data transfer and manipulation.

Major Factor Analysis (MFA) of VIMS Data Logger Data

In statistical terms, all parameters constitute variables. The relationship between two variables is defined by the correlation coefficient. For the purpose of modeling truck condition and performance high correlation between any two variables indicates redundancy. MFA eliminates this redundancy by combining correlated variables into factors. Lower number of factors simplifies further analysis.

In the described research all monitored truck parameters constituted inputs into MFA. The analysis was performed using varimax rotation that maximizes the variance of the factor loadings for each input variable. The rotated factors have a high correlation with one set of input variables and little or no correlation with another set of input variables. The varimax rotational strategy can give a clearer interpretation of the results by classifying variables into new independent factors.

Figure 1. presents the factor loadings that quantify strength of relationships between variables in the investigated databases. Their value reflects the linear relationship between the input variables and the corresponding factors, and varies between −1

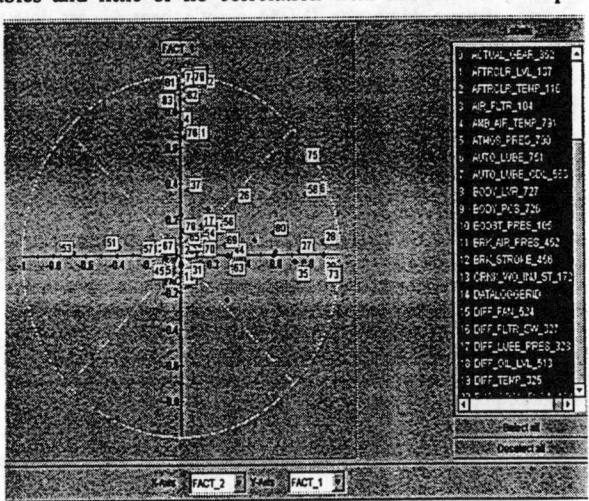

Figure 1. The Factor Loading View

and +1. If the factor loading is +1 there is a perfect positive relationship between the variable and the factor. Factor loading of −1 denotes a perfect negative relationship. If the factor loading is 0, there is no relationship between the input variable and the factor.

In the factor loading window, the vertical axis represents one of the factors while the horizontal one represents another. The dots depict the factor loadings. The labels next to the dots show the number of the input variables, name of each variable identifiable at the label list on the right side of figure 1. If a dot has a high coordinated value on one of the axes and lies in close proximity to it, there is distinct relationship between one of the two factors and this variable (IBM, 2000).

The results of this analysis identified 19 independent statistically factors that represent the original 85 truck parameters. The variables that are included into the same factor are highly correlated. Table 1 summarizes the results of Major Factor Analysis

The first factor accounts for 29% of variables, or 24 truck parameters. These are highly correlated with each other as well as with the first factor. All the 24 parameters define temperature and pressure, including *atmospheric temperature, engine coolant temperature, turbocharger inlet air pressure*, etc. Therefore this class of parameters is represents temperature/pressure indicators of the truck.

The second factor accounts for 12% of the parameters. It groups engine load indicators and includes such variables as engine speed, throttle position, boost pressure, and so on. Interestingly the ECM (Electronic Control Module) calculates *engine load* as a function of: *engine speed, throttle switch position, throttle position, boost pressure*, and *atmospheric pressure*. The third factor can be thought as the payload indicator, and the fourth factor is the fluid level indicator. No physical interpretation for all factors can be provided at present.

The MFA output results also include factor scores, the actual values of individual observations for the factors. These factor scores are particularly useful when further analysis of factors is to be performed.

In conclusion, the Major Factor Analysis can be used to reduce the number of truck performance and condition parameters that one needs to be concerned with, thus simplify further analysis. Lower number of variables in the input to clustering and classification saves evaluation time and minimizes problems created by missing variable values.

Table 1. Machine Parameter Indicators (Factors)

No.	Factor (percentage)	Indicator	Parameters (Variables)
1	Factor 1 (29%)	Temperature	Atmospheric Temperature, engine coolant temperature, turbocharger inlet air pressure, etc.
2	Factor 2 (12%)	Engine Load	Engine Speed, throttle Position, Boost Pressure, etc.
3	Factor 3 (5%)	Payload	Payload, Suspension Cylinder Pressures, Payload Status, Machine Pitch.
4	Factor 4 (6%)	Fluid Level	Engine Oil Level, Low Steering Pressure, Engine Oil Pressure, etc.
5	Factor 6 (2.9%)	Road Condition	RTR-LTR and RTF-LTF Suspension Level, Machine Rack
6	Factor 8 (2.7%)	Transmission Switch	Torque Converter Screen, Transmission Charge Filter, etc.
7	Factor 13 (2%)	Auto Lube	Auto Lube Datalink, Auto Lube
8	Factor 14 (2.28%)	Body Level	Body Level, Body Position
9	Factor 16 (3.2%)	Fan Speed	High or Low Speed Fan, Ground Speed, etc.
10	Factor 17 (2.4%)	Engine Fuel Rate	Engine Fuel Rate

Clustering of VIMS Data Logger Data

Clustering searches for characteristics that most frequently occur in common and groups the related data into clusters. The number of detected clusters and the properties of each cluster are the results. In addition distribution of characteristics within the clusters is quantified.

The Demographic Clustering provides fast and natural clustering of very large databases. It automatically determines the number of clusters to be generated. Similarities between records are determined by comparing their field values. The clusters are then defined so that Condorcet's criterion is maximized (IBM, 2000).

Following the Major Factor analysis the remaining data set was data mined using the IM demographic clustering. As a result the data set was segmented into 9 clusters as shown in fig. 2 (Golosinski, Hu, and Elias, 2001). The three largest clusters each account for the 14% of the whole data set

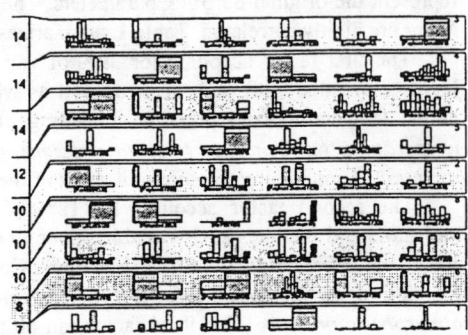

Figure 2. Demographic Clustering -IM Output

Fig. 3 and 4 show a zoom of the cluster related to *haul distance* and to *truck payload*. The haul distance cluster, shown in fig. 3 indicates that the haul distance is one of the main determinants of fuel consumption rate. Interestingly, the percentage of 6 to 10 mile long hauls in this cluster is approximately 40%, while the same percentage for the whole population is only 5%. One possible explanation is that on the long hauls truck fuel consumption rate is larger since truck spends more time running at the full load. On short hauls more time is spent loading / dumping / maneuvering / waiting activities during which fuel consumption is low.

Fig.4 shows the payload cluster. It indicates that all trucks in this cluster were empty (100% of the cluster), while the percentage of empty trucks in the whole database is only around 50%. All the trucks in Payload cluster were traveling at 4th gear with the speed of 25 to 35 MPH and the fuel consumption rate was average.

Figure 4. Demographic Clustering: Payload Cluster (Horizontal Scale: Payload in Tons)

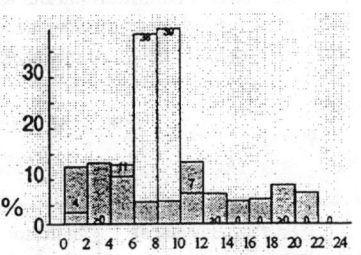

Figure 3. Demographic Clustering: Haul Distance Cluster (Horizontal Scale: Haul Distance in Miles)

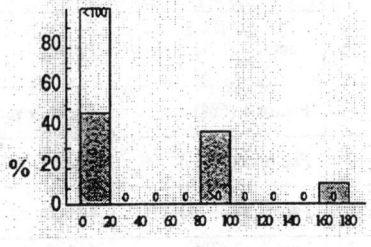

The other clusters identified in this work are presented in fig. 2. These contain variety of information related to truck performance.

Classification of VIMS Data Logger Data

Classification is used to segregate database records into pre-defined classes based on specific criteria. Thus this technique can be used to define what truck operating or condition parameters define fuel consumption rate, what parameters define its cycle time and the like.

The tree-classification mining function builds a classification model as a binary decision tree. Each interior node of the binary decision tree tests an attribute of a record. If the attribute value satisfies the test, the record is sent down the left branch of the node. If the attribute value does not meet the requirements, the record is sent down the right branch of the node. The 4 classes are marked with different colors at upper left corner. They are reflected in the tree map as solid square. The solid circles are the decision nodes. The binary decision tree consists of the root node on top, followed by non-leaf nodes and leaf nodes. Branches connect a node to 2 other nodes. Root and non-leaf nodes are represented as pie charts. Leaf nodes are represented as rectangles.

Each node can display its characteristics in the window shown at the bottom of fig.5. This information includes:

Label: The pre-dominant class label of the selected node.

Test: The split criterion for this node. This applies only to non-leaf nodes and specifies a simple selection.

Records: The number of records contained in each of the sub-nodes of the selected node.

Distribution: The number of records corresponding to each of the possible class labels. The classification is most meaningful if all records belong to one leaf node only. However, by pruning the binary decision tree, records of other nodes can be assigned to the selected node.

Purity: The percentage of correctly classified records assigned to a node.

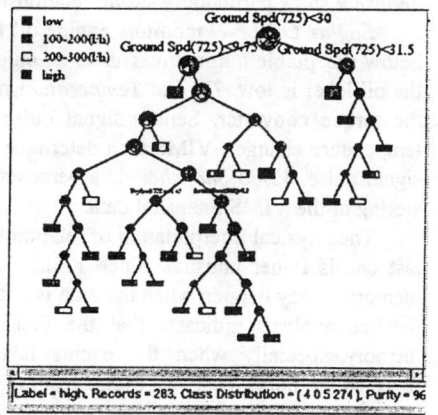

Figure 5. Classification-Tree

The Tree Classification was done for the fuel consumption rate, leading to definition of four classes of parameters that indicate high fuel consumption. These can be identified by tracking the thick black line with the arrow that links the nodes and continues on to the rectangles at the foot of figure 5. Use of color in computer generated figure 5 makes the tracking easy.

Selected observations that can be made in this case indicate that:
1. When ground speed is in the range from 12.25 MPH to 15.5 MPH and the payload is over 126.85 t, 96.8% of the 283 analyzed records indicate high engine fuel consumption rate.
2. When ground speed is more than 31.5 MPH and actual gear is higher than 5, all 146 records show low engine fuel consumption rate.
3. The ground speed has more impact on the engine fuel consumption rate than do other parameters.

Sequential Pattern Recognition in VIMS Event Logs

Thousands of events are recorded during a life of an average mining truck. These are stored in the *Event* data log of the VIMS database. All VIMS events are classified into two categories, data event and maintenance event. The data event is related to the machine operating status, such as low engine oil level. The maintenance event is related to the machine control system, a problem of the VIMS itself, such as severed sensor wire.

VIMS and related tools can only list and tabulate events. Other tools are needed to discover additional knowledge that may exist in VIMS database. One such tool is the Sequential Pattern, one of the data mining methods of Intelligent Miner. It can be used to discover similar sequential data patterns in VIMS data bases.

The inputs to sequential pattern analysis were: *Serial Number, System Measurement Unit (SMU)* and *Event Identifier*, a combination of event description and event level. The minimum support level was set at 80%. Database of 42,514 events recorded on 12 trucks were datamined. These included 69 event types. As a result 77,327 sequential patterns were identified. The sequential patterns were identified to exist in all the 12 machines at some point of time. The events *Engine Oil Level, TC Out Temperature* and *VIMS Snapshot* show particularly strong relationship with each other.

Engine Oil Level monitors engine oil level and informs engine ECM when it drops below acceptable minimum. This is an on/off switch type signal with switch open when the oil level is low. *TC Out Temperature* monitors oil temperature on the outlet side of the torque converter. Sensor signal pulse width changes as the torque converter oil temperature changes. VIMS then determine the temperature based on the width of sensor signal pulse. *VIMS Snapshot* is a percentage of memory space that is left available for storing of the VIMS Snapshot data.

The physical interpretation of relation between the first two of the above is clear, the last one is rather unusual. Since *VIMS Snapshot* is triggered by full VIMS *Snapshot* memory it may happen when the data is not timely downloaded. On the other hand more detailed analysis indicates that the first two occupy large part of VIMS *Snapshot* memory, especially when the events take place frequently, or when the operator repeatedly ignores these two events leading to overfilling the memory.

Information on relation between the events *engine oil level* and event of *torque converter temperature* allows to predict increase of torque converter temperature from the reading of engine oil level and vise versa. As such it is important to the task of this work.

Overall the sequential patterns identified to exist in VIMS *Event* database with high confidence level constitute very low percentage of all events. This does not disqualify Sequential Pattern as a data mining method useful for mining VIMS databases. However, it indicates the need to revise the approach used in future investigations. Possible changes are: use of larger databases that include data collected from a variety of trucks and from different mines, inclusion into analyzed databases of other data, external to VIMS, and increasing the size of VIMS data sets so that these cover extended periods of truck operation.

DISCUSSION

Work presented in this paper indicates that data mining of VIMS generated data bases allows for discovery of knowledge contained in these databases. In particular relations that exist between various VIMS-collected data can be identified, described and quantified. This can be achieved through use of two IM statistical functions: linear regression and factor analysis. While both these functions can serve this purpose, Factor Analysis allows to significantly reduce the number of variables that need to be considered and groups all related parameters into factors.

The clustering can be used to segment VIMS database through grouping of data that have similar characteristics. This allows to idnetify the paramterers that are of key importance to truck performance. Thus, as an example, the parameters that influence truck fuel consumption rate can be defined. Further work is needed to fully define applicability of custering to the problem at hand. It appears that clear definition of relations or goals sought may be needed to realize the full potential of this method.

The classification was applied to interpret clustering results. It allowed for quantification of impact that various VIMS parameters have on truck fuel consumption. It was further proven that classification can be used to build a model that describes the behaviour of VIMS parameters that are of interest. The research indicates that classification alone does not yield meaningfull results. It does, however, yield these results if used in conjunction with clustering.

Sequential patterns are usually used to find predictable patterns of behavior for a given phenomenon over a period of time. In relation to VIMS parameters the intent was to be able to predict occurrence of a specific event based on occurrence of similar event in the past. However, this approach was unsuccessfull. While a multitude of similar patterns was identified to exist in the data, its wariability was to large to permit drawing of valid conclusions on their repeatibility. It is believed that more complete VIMS database, collected over extended period of time and in a variety of truck operating condition may overcome this problem, and permit development of predictive models of truck behaviour.

Ultimate goal of this work is to construct a model of truck operation that would allow projection of truck condition and performance into future. The relations discovered to exist between various VIMS parameters lay a foundation for this modeling work. Of particular importance is use of Factor Analysis that reduces the number of parameters to manageable level. Classification and clustering allows for analysis of truck performance and indirectly its optimization. Use of sequential patterns need further study.

CONCLUSIONS

The investigations presented above prove that data mining can be used to analyze performance of mining equipment. In particular the relations between its various operating, condition and performance parameters can be defined, and quantified using regression and factor analysis methods.

VIMS provides variety of data that quantify truck condition and performance. To maximize use of this data it needs to be collected continuously over expended periods of time and under a variety of operating and climatic conditions.

Intelligent Miner contains a variety of data mining tools, many of which can be used to successfully data mine VIMS databases. However, the input data format of IM is not compatible with that of VIMS generated data. Therefore an interface between the two is needed that facilitates fast data reformatting.

More investigations are needed to fully define the applicability of data mining to knowledge discovery in VIMS databases. Use of databases collected over extended period of time and containing data external to VIMS that define truck operating condition may be the key factor in determining this applicability.

ACKNOWLEDGEMENT

Financial support of investigations presented in this paper by Univeristy of Missouri Research Board is gratefully acknowledged.

REFERENCES

Bernson, A. and Smith, S.J. 1997. "Data Warehousing, Data Mining and OLAP". McGraw-Hill.
Westphal, C. and Blaxton, T. 1998. "Data Mining Solutions". John Wiley and Sons, Inc.
Caterpillar, Inc. 1999. "Vital Information management System (VIMS): System Operation Testing and Adjusting". Company publication.
IBM (International Business Machines Corporation). 2000. Manual: "Using the Intelligent Miner for Data". Company publication.
Golosinski. 2001. "Data Mining Uses in Mining". Proceedings, APCOM 2001, Beijing, China.
Golosinski, T.S., Hu, Hui and Elias, R. 2001. "Data Mining VIMS for Information on Truck Condition". APCOM 2001, Beijing, China.
Ataman, K. 2001. M.S. Thesis: "Data Mining for Prediction of Condition and Performance of Mine Machinery". University of Missouri-Rolla publication.
Madiba E. 2001. M.S. Thesis: "Application of IBM DB2 Intelligent Miner for data to mine Vital Information Management System (VIMS) data". University of Missouri-Rolla publication.

PART V: COMPLEX SYSTEMS

INTERACTION BETWEEN AGENTS WITH EMOTIONAL MODEL

SAYAKA NAKAMURA
Autonomous System, Engineering,
Hokkaido University, JAPAN

TAKASHI ISHIDA
Autonomous System, Engineering,
Hokkaido University, JAPAN

HIROSHI YOKOI
Autonomous System, Engineering,
Hokkaido University, JAPAN

YUKINORI KAKAZU
Autonomous System, Engineering,
Hokkaido University, JAPAN

ABSTRACT

In this study, we put emphasis on the interaction between agents through emotion, and simulated an artificial society composed of a number of agents with emotion. Our purpose is to examine the effect of one agent's emotional movement on the behavior of other agents and whole multi-agent system. We proposed an interaction model of feeling, which includes a representation of feelings such as the "Affection", and rules of the interaction between agents through emotion. Results clarified the mechanism of grouping behavior, by the change of internal state of each agent, which can be taken as fundamental characteristics of the multi-agent system.

INTRODUCTION

What is the function of emotion? Even human being far more intellectual than other animals may take foolish emotional action. Although, "emotional" is literally close to "irrational", but emotion may perform action selection advantageous to a survivor of a living thing, without doing so complicated processing.

Emotion has been studied in a number of fields, such as psychology, physiology and cognitive science. In engineering, emotion has been introduced in order to bring out human-like reaction to machine systems[1][2]. However, few studies have reported on the role of emotion as one fundamental survival strategy. In this study, we simplified change of emotion which is caused by the interaction between agents, and built a computational model based on several important experiments and theories of psychology or physiology.

Hog and Huberman investigated dynamics in case many agents use two or more resources according to each liking. They succeeded in controlling the chaos of a system by an agent having a concept of liking. The view which gives diversity, such as liking and emotions, to a system is common.[3] The purpose of this study was not to make a model of emotion, but to simulate world which various agents complete, and examinate the effect of one agent's emotional movement on the behavior of other agents and whole multi-agent system. Therefore, we proposed an interaction model of feeling, which includes a representation of feelings such as the "Affection", and rules of the interaction between agents through emotion. Results clarified the mechanism of grouping behavior, by the change of internal state of each agent, which can be taken as fundamental characteristics of the multi-agent system.

PROPOSAL MODEL
State Variable of an Agent
Each agent has state variables shown below.

$$agent_i = \{ A_i, E_i, T_i, P_i, CA_i \} \tag{1}$$

A_{ij} is an Affection value at which $agent_i$ favors $agent_j$ (Fig.1). Each agent has a 1-dimensional value to the agents other than itself. The magnitude of it means the intensity of the feeling, while, the sign means the direction of the feeling, i.e, likes or dislikes. Likes or Dislikes are one of the most fundamental evaluations for judging the worth of an object by living things. Moreover, there are some experiments to which Likes/Dislikes and emotions are related and the result is useful for modeling the interaction of emotions.[4]-[8]

E_{ik} is an Emotion value of $agent_i$ and "k" stands for the number of emotion axes (Fig.2). In this research, emotion was set up based on the idea that emotion is generated by the result of the evaluation to an object. We set up two evaluation axes (=Emotion axes), which are related to Likes/Dislikes, and the emotion considered as integration of those evaluations. Emotion axes were decided in relation between affection to one's the other and the affection to the others'. The emotion at that time was given as a 2-dimensional vector, and the kind of emotion was decided by the quadrant the vector belongs.

T_{il} is a Threshold of $agent_i$, at which an agent's emotion fires and does influence to other agents. One threshold was prepared to each of 4 feelings set up by this paper. Threshold would be the parameter that determines the individuality of agents.

P_{ih} is a 2-dimensional coordinate of $agent_i$ in space (Position).

CA_i is a Communication Area of $agent_i$. CA decides Area in which $agent_i$ does an interaction to other agents. Each agent interacts only to other agents who are in its CA. Each agent can observe Affection, emotion and position of other agents who are in its CA.

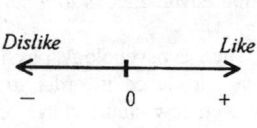

Fig.1 Affection axis which $agent_i$ feels to $agent_j$

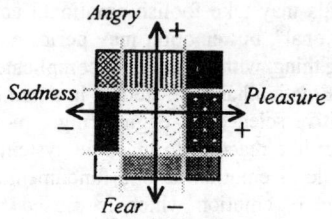

Fig.2 Emotion axes of $agent_i$

The Rule for Changing a State Variable
Agents go into an "Encounter Phase" or "Non-Encounter Phase" according to whether other agents exist in their CA set up in 2-dimensional space. Agents goes into Encounter Phase if there are agents other than itself in its CA, goes into Non-Encounter Phase if there is no agent in its CA. Physical distance plays a very important role when changing emotion and affection value. [4]-[8]

Encounter Phase
In an encounter phase, an agent does an interaction to other agents who are in its CA, and updates emotion, affection and position.
a) Emotion occurs by the relation with other agents

$$\Delta E_1 = c_1 A_{ij}^{(t-1)} A_{ji}^{(t-1)} \tag{2}$$

b) Emotion spreads to other agents.

$$\Delta E_2 = c_2 E_{ik}^{(t-1)} \exp(-l D_{ij}^{(t-1)}) \qquad (3)$$

c) Affection value increases or decreases by the affinity of emotion.

$$\Delta A_1 = c_3 \sum_k \frac{E_{ik}^{(t-1)}}{E_{jk}^{(t-1)}} \qquad (4)$$

d) Affection value increases according to evaluation (affection) to itself by others.

$$\Delta A_2 = c_4 A_{ji}^{(t-1)} \qquad (5)$$

e) An agent approaches a favorite agent and leaves from a disagreeable agent.

$$\Delta P_1 = c_5 \sum_j A_{ij}^{(t)} e_{ij}^{(t-1)} \qquad (6)$$

Using (2) and (3), Emotion value could be written as:

$$E_{ik}^{(t)} = E_{ik}^{(t-1)} + \Delta E_1 + \Delta E_2 = E_{ik}^{(t-1)} + c_1 A_{ij}^{(t-1)} A_{ji}^{(t-1)} + c_2 E_{ik}^{(t-1)} \exp(-l D_{ij}^{(t-1)}) \qquad (7)$$

Using (4) and (5), Affection value could be written as:

$$A_{ij}^{(t)} = A_{ij}^{(t-1)} + \Delta A_1 + \Delta A_2 = A_{ij}^{(t-1)} + c_4 A_{ji}^{(t-1)} + c_3 \sum_k \frac{E_{ik}^{(t-1)}}{E_{jk}^{(t-1)}} \qquad (8)$$

And Position could be written as:

$$P_{ih}^{(t)} = P_{ih}^{(t)} + \Delta P_1 = P_{ih}^{(t)} + c_5 \sum_j A_{ij}^{(t)} e_{ij}^{(t-1)} \qquad (9)$$

e_{ij} is a unit vector which turns to *agent_j* from *agent_i*, and D_{ij} is a distance between *agent_i* and *agent_j*.

Non-Encounter Phase
a) Affection degreases

$$E_{ik}^{(t)} = E_{ik}^{(t-1)} - c_6 \qquad (10)$$

b) Emotion becomes negative value

$$A_{ij}^{(t)} = A_{ij}^{(t-1)} - c_7 A_{ij}^{(t-1)} \qquad (11)$$

The parameter is set up so that the maximum of each item may become equal.

EXPERIENCE

We did computer experiments in a 100x100 space using our proposed model. In this experiment, it has set up so that Emotion Value and Affection Value may take the value between -1.0 and 1.0. Unless it states specially, the state variables of agents are generated by random numbers.

The Fundamental Behavior of the System

The fundamental behavior of this system is that agents who increased affection value mutually form groups. A group would go into a stable state as it is, if the number of agents is small or the Communication Area is small. In this experience, all agents have equal thleshold, a positive thleshold is 0.5, negative one is -0.5. Similarly, the affection thleshold is also equal, 0.3 and -0.3. Fig.3 shows the result of an experiment in which 5 agents and CA's size 10.0 were set up. This figure shows changes of the position of agents moving in the space (x-y plane is 2-dimensional space, and z axis is time). As can be seen from Fig.3, one agent is particular and present different behavior with other agents. Although the agent tended to join one already formed group at about 1500th step, it would not enter a stable state since the affection value to the members of the group did not increase, but finally, it found a new group after movement and enter a stable state there.Fig.4 shows the result of the experiment when the number of agents is a little larger.

were set up. The right figure is for 100 agents and CA's size of 30.0. In the 2-dimensional space, circles stands for agents, dotted line circle represents CA and the differences in color express internal states of agents. As can be seen from fig.4, formed groups were apart from each other, at a certain distance, which is almost Communication Area.

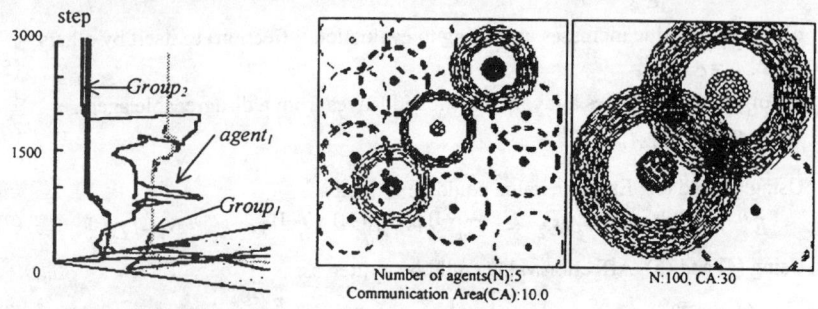

Fig.3 Fundamental Behavior of System when the Number is Small

Fig.4 Fundamental Behavior of System when the Number is a Little Large

Collapse and Formation of Groups

If the number of agents contained in space becomes large, a process of group collapse and formation would be observed repeatedly. Fig.5 expresses such a process. The group formed collapsed at once by the interaction with other groups, and was in a confused state. However, after this state continued for a while, a group would be formed again. A collective collapse happens by the imbalance of internal states of agents. Fig.6 is change of emotion value at this time. It will converge if a group is formed, and oscillate if a group collapses.

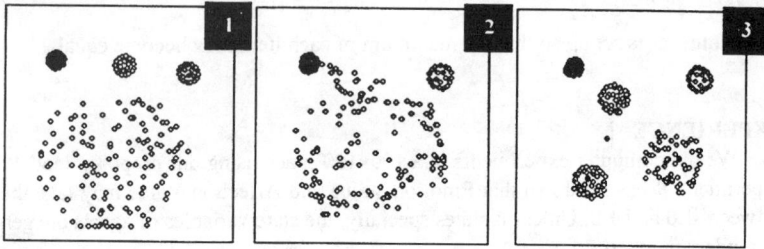

Fig.5 Collapse and Formation of Group (N:200, CA:20)

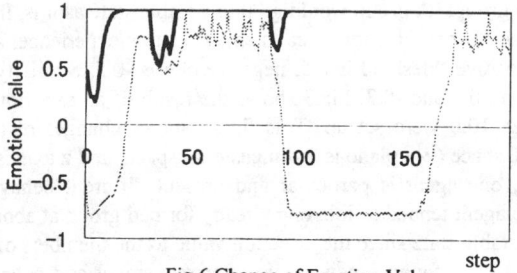

Fig.6 Change of Emotion Value

The Balanced State of a Group

When agents were allowed to interact with all the other ones, they formed the group of balanced states (Fig.7). Fig.7 is the result of experiment by setting the number of agents as 100, 200. Although there was no rule to keep distance with each other, agents were positioned forming the shape of a circle. If we investigate internal states of Agents, we would find that agents who made affection value increase mutually stay inside the group, on the other hand, agents with low threshold of anger stay outside the group. It is interesting that it turns out where an agent exists, by what threshold it has. For example, if agents who exist outside have sensors and inside agents being processes the information, a function may also be able to be specialized within a group. The group with such a state is stable, and even if such a group interacts with other groups, it could keep a group in the state for a while.

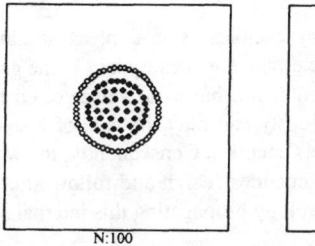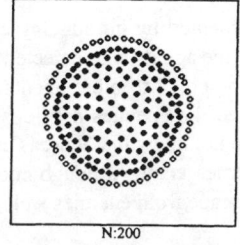

Fig.7 the Balanced State of Group when all agents interact

Evaluation of Internal States, Emotion

We investigated the influence which internal states have on a system. First, we made a certain agent's emotion blank. The agent without emotion could not make affection value increase to other agents, but was able to be applied to no group finally formed (Fig. 8). Moreover, when emotion of all individuals was repealed, agents were loitering forever without forming a group (Fig.8). Fig.10 shows the result of the experiment in which all the setup except internal states was same as those of the former experiment. These facts show that the emotion parameters representing internal states activate affection through the interaction and have played a role in forming groups.

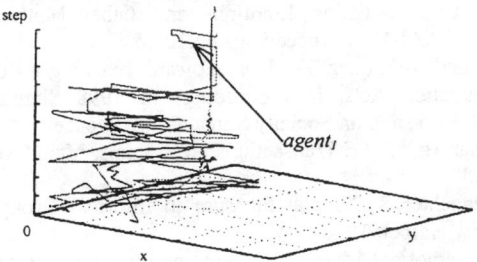

Fig.8 Behavior in case an agent does not have Emotion

Movement of a Group

We can move the whole group already formed by only moving one agent. Fig.9 shows the result of the experiment in which 50 agents and CA's size 150.0 were set up. If a certain agent moved, a group followed that and entered a stable state again at another place. In order to move a group, it is enough just to move one agent in it. It is not necessary to controll to collective all members to move a group. Only one person is moved, the rest will follow.

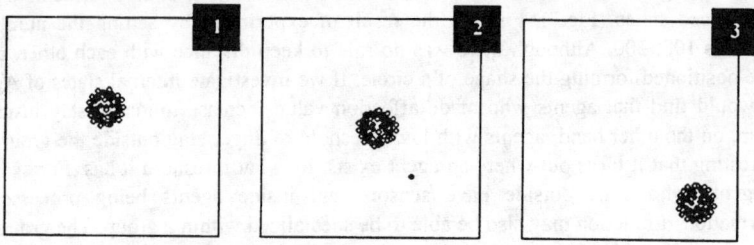

Fig.9 Movement of a Group (N:50, CA:150.0)

Urgent System

We experimented by introducing enemies as an unpleasant stimulus into a system. The emotion of the agents who detected the enemies ignited in the group already formed, and the emotion spread to other agents who have not yet detected the enemies. Agents which discovered the enemies previously can play the role of a sensor, and save other agents which do not know that there is enemies. Consequently, the whole group would be in the state of emergency. Although enemies search and follow after agents persistently, the group can escape from enemies well by propagating this internal state.

CONCLUSIONS

We desinged an interaction model of feeling on the basis of the knowledge of psychology or physiology, and simulated agents which act based on it. So far we have observed that local interaction of agents resulted in group formation by giving influence directly each other to their internal states. We would now like to go on to analyze the characteristic of a phenomenon which was observed this time, by investigating the influence each parameter and a rule affect a system in detail.

REFERENCES

[1] Velasquez, D. Juan, Modeling Emotions and Other Motivation in Synthetic Agents ,AAAI-1997/IAAI-1997 proceedings pp.10-15
[2] Brezeal, Cynthia and Velasquez, D., Juan, Toward Teaching a Robot "Infant" using Emotive Communication Acts, in Proceedings of 1998 Simulation of Adaptive Behavior (SAB98) Workshop on Socially Situated Intelligence
[3] Hogg T., Huberman B., "*IEEE* Transactions on Systems, Man, Cybernetics", Vol.21, No.6, 1991, pp1325
[4] Festinger L., Schachter S., "Social Pressure in Informal Groups", A Study of a Housing Community Harper, 1950
[5] Toda, Masanao, "emotion(Adaptation programming which is moving the man)", University of Tokyo publication, 1992
[6] Byrne, D., Nelson, D., "Attraction as Liner Function of Proportion on of Positive Reinforcements", Journal of Personality and Social Psychology, 1965, 1, pp659-663
[7] Byrne, D., "Interpersonal Attraction and Attitude Similarrity", Journal of Abnormal a and Social Psychology,1961, 62, pp713-715
[8] Hori, Tetsuro, "A brain and emotions", Brain science series6, Kyoritsu publication, 1991

EVOLVING COOPERATIVE PARTIAL FUNCTIONS FOR DATA SUMMARY AND INTERPOLATION

PETER E. JOHNSON, M.S.
Iowa State University
Dept. of Mechanical Engineering
Ames, Iowa

DANIEL ASHLOCK, PH.D.
Iowa State University
Dept. of Mathematics
Ames, Iowa

KENNETH M. BRYDEN, PH.D.
Iowa State University
Dept. of Mechanical Engineering
Ames, Iowa

EDMUNDO VASQUEZ, PH.D.
Alliant Energy
Combustion Initiative
Madison, Wisconsin

ABSTRACT
Typical data sets for engineering applications are defined only at discrete points specified over a predetermined domain. Information between these points is undefined, which can make interpolation between the discrete points very difficult if non-linear behavior (e.g., phase changes) in the system from which the data are drawn occurs. In this paper a new technique will be presented which uses evolved data blocks, checkpoints, and evolved functions to summarize and interpolate a specific data set more efficiently. Termed the "Adaptive Modeling via Evolving Blocks" Algorithm (AMoEBA), it allows data sets to be broken up into any number of smaller blocks with individual functions inside that are defined by checkpoints. Coupling the evolution of blocks and interpolants permits the algorithm to explain the structure of the data set while performing interpolation.

INTRODUCTION
Throughout the design process, engineers make many changes to the original design; some are significant while others may be subtle. In either case these modifications must then be analyzed to determine how much, or whether, they improve the design. In the case of numerical models, this could be as simple as interpolating information from the original data set or, if the model is non-linear (e.g., computational fluid dynamics calculations or finite element analysis), as difficult as waiting for the entire data set to be recomputed. This interruption creates an atmosphere that is not as conducive to the "what-if" type thinking that is a key part of any design process.

Computer hardware technology has increased dramatically throughout the last decade, improving computation time and increasing the amount of memory. However, in order to perform the calculations necessary to recompute the impact of design changes on the fly, software improvements must also be made. This paper presents an algorithmic structure for rapidly interpolating a problem and demonstrates this technique on a simple one-dimensional problem.

BACKGROUND

Many of the numerical models used in the design world are nonlinear, making interpolation between two models very complicated. Nonlinear interpolation is possible, but generally it is problem specific and therefore not useful in every case [1, 2, 3, 5].

If dependable approximations of the data set can be made, time can be saved during the recalculation process. The techniques used in AMoEBA are focused on creating reliable approximations. These techniques are based on genetic programming techniques used in a novel fashion. A parse tree used in genetic programming [6] is a tree that contains functions and terminals. The technique used in this paper falls under the category of genetic programming using cellular encoding [4]. Cellular encoding is the practice of evolving directions for how to build the structure of interest rather than directly evolving the structure. In his original work, Gruau evolved parse trees that gave directions for building the connection topology and connection weights of a neural net. This work presents parse trees that give directions for how to segregate a data set by iterated subdivision with the segregated pieces assigned to different approximating models.

METHODOLOGY

The goal of this research is to be able to model fully coupled, non-linear equations by using trees that segregate the domain into similar areas. These areas are then modeled by approximators that are assigned to the leaves of the trees. The first step in reaching this goal is to model simple, linear relationships. The engineering problem selected for this step is the determination of the steady state temperature profile in a one-dimensional wire. The wire is held at 650 K at one end and 400 K at the other resulting in the following temperature profile:

$$T = 650 - 250x \qquad (2)$$

where T is the temperature of the wire and x is the linear distance from the 650 K end of the wire and goes from 0 to 1.

EVOLUTIONARY MODEL

AMoEBA is a co-evolutionary system using two populations. One is a population of binary trees that segregate the data. The leaves of these trees are the separate regions of the domain, each section assigned to an approximator that models the data in that subdivision. Fitness for a tree is determined by how well the entire tree approximates the data set.

The approximators that are pointed to by the leaves of the trees are the other population in the system. Fitness for the approximators is determined by how often they are used by the trees.

Data Segregation Trees

The data segregation trees are based on typical parse trees that are in effect, equation generators. In our system, instead of using them to generate an equation, they are used to divide regions (Fig. 1). Each node of a tree contains a real number, a type - vertical or horizontal (0 or 1 when implemented), and the value of the left and right leaves or the left and right nodes (l and r). The type defines the direction of the division to be made (in the

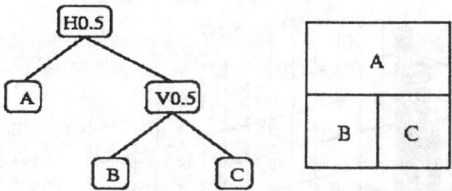

Fig. 1. An example of how the data segregation trees divide an area. Each node consists of a type (V – vertical or H – horizontal), a real number (the point of segregation), and leaves (A, B, C) and/or nodes. The leaves correspond to the segregated areas.

example problem discussed in this paper, only the horizontal type is allowed because the problem is one dimensional) while the real number defines where this division is made. The value of the leaf, lt (left type) or rt (right type), assign an approximator to that respective leaf. Each node can also point to a right and a left node instead of a leaf.

In order to compute the fitness of a tree, sample points are chosen randomly and, for each point, the leaf corresponding to that point is found. This leaf points to an approximator, and the value at that point is then compared to the value calculated by the approximator. For each fitness evaluation 5,000 sample points (out of a total of 40,000 (200 X 200) for the data set used in this problem) are compared to the data set, and their squared error is summed.

Reproduction of trees begins by ordering the trees by fitness. The trees are then divided into tournaments, each containing four individuals. For example, the top four trees are placed into one tournament, the next four are in another tournament, etc. In each tournament the two trees with the best fitness become the parents and the other two are deleted. The two parent trees are copied, and 50% of the time, these copies are randomly selected for crossover. The crossover routine randomly selects a subtree of each copy, and these subtrees are swapped. Before the breeding is complete, the children (the copies) are potentially mutated to add diversity to the population.

Mutation for the trees can occur in one of the following fashions. A node of the tree can be mutated so that the division point is moved a random distance. When progress is made to a two-dimensional system, the type of the node can also be mutated from vertical to horizontal or vice-versa. Leaves can also be mutated by pointing to a different approximator. Another mutation that is possible is random subtree mutation, or the replacing of a randomly selected subtree with a completely random subtree of the same size. These mutations provide population diversity to enhance the search.

The Approximators

Various functions can be used as the approximators, but the technique is not limited to using only functions. A long term goal of this application is to allow the trees to point to a variety of modeling techniques such as similarity techniques, interpolation techniques, finite element analysis, and CFD computations. At this time polynomials with variable degrees have been used as approximators. These polynomials are limited to the 10th degree and have random degrees and random coefficients. Forty polynomials are used in the approximator population, and these are evolved after every tenth tree generation.

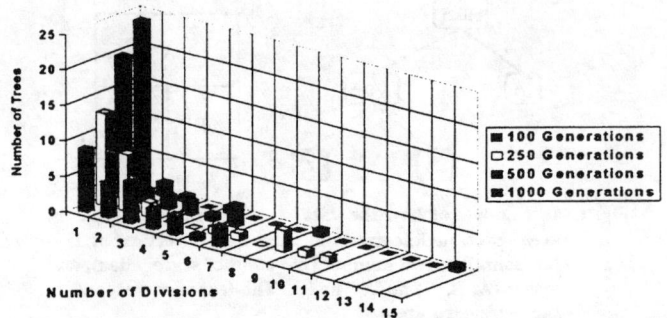

Fig. 2. The progress of the segregation trees when run with Eq. 1 hard-coded into the program.

The reproduction process for the approximators is different from that for the trees. The twenty-five percent of the population with the best fitness is selected as parents and mutated copies of these parents replace the lowest twenty-five percent. Approximator mutation occurs in one of the following ways:

- No mutation;
- A random coefficient is selected and mutated within a specific range;
- A random coefficient is selected and is multiplied by negative one;
- The degree of the polynomial is decreased (unless it is too small) or increased (unless it is too large) or left the same size.

Placing the mutated children into the lowest twenty-five percent would be dooming them to be used as infrequently as the approximators they replace. Instead, not only are the children placed in the deleted approximator positions but a small percentage of them also take their parents' place in some tree leaves. By finding the leaves pointing to the parents and randomly replacing them with their children (at an average occurrence of one in four), the children are given a fair chance without disadvantaging the parents or the rest of the population.

RESULTS

Because of the co-evolution of two different populations, two different experiments are required to test the functionality of the populations. The effectiveness of the trees was determined by hard-coding the equation used to generate the original data set into the initial approximator population. If the trees are evolved correctly, the tree with leaves pointing to the actual function would outperform any other tree in the population.

Runs were made for 30 different random populations. The best tree out of these populations was recorded after 100, 500, and 1000 generations. Figure 2 shows how the trees evolved. As expected, the number of trees with only one division increased as the number of generations increased.

These trials also provided a small test of the approximator evolutionary algorithm. After 100 generations, the best tree in all 30 of the random populations contained at least one reference to Eq. 1. As evolution progressed, this number fell to 14 trees after 500 generations and 7 trees after 1000 generations. When the approximators of the trees that

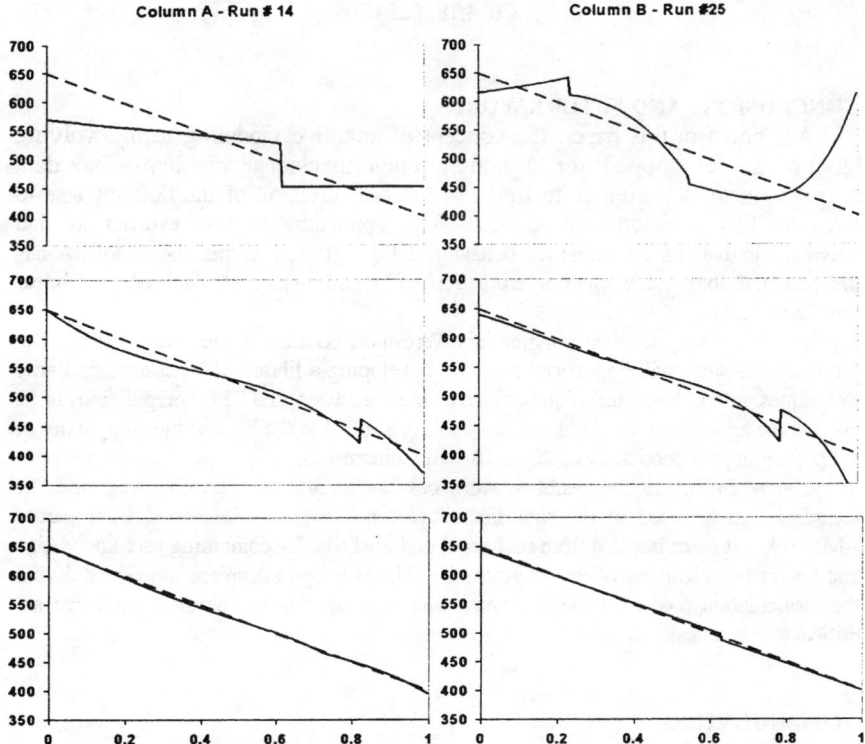

Fig. 3. Temperature versus x-distance profiles of the best segregation tree/approximator combination after 100 generations (top), 500 generations (middle), and 1000 generations (bottom) for runs #14 (column A) and #25 (column B). The dotted line represents the actual temperature profile.

did not point to Eq. 1 were analyzed, it was found that they were either identical copies or slight mutations of Eq. 1.

A further test of the approximator population was to run the code without any special adjustments. With the trees working, the approximators should be able to evolve to good approximations of the data set. Given enough time, the actual function, or slight variations of Eq. 1, should be found. Figure 3 illustrates the improvement in performance of the data segregation tree/approximator combinations without the help of the actual function in the population. Again the AMoEBA found that smaller trees outperformed larger ones. Equations 2 and 3 show the approximators that were used in the 1000th generation of runs 14 and Eqs. 4 and 5 for run 25.

$$T_1 = 652.02 - 260.06x - 200.5x^2 + 569.07x^3 \\ -147.67x^4 - 320.74x^5 - 27.71x^6 \quad (2)$$

$$T_2 = 652.02 - 279.9x - 209.32x^2 + 507.53x^3 - 147.67x^4 - 128.69x^5 \quad (3)$$

$$T_3 = 645.465 - 234.305x \quad (4)$$

$$T_4 = 634.18 - 234.305x \qquad (5)$$

CONCLUSIONS AND FUTURE WORK

As shown in this paper, the concept of adaptive modeling using evolving blocks can be applied for a simple one-dimensional problem. The data segregation trees were able to find that the best division of the domain was to have as few divisions as possible. The approximators also evolved to find polynomials that closely mimic the behavior of Eq. 1. Based on this, the AMoEBA has the potential to provide rapid interpolation of multidimensional, coupled, non-linear problem sets.

In this training problem polynomials were used because of the data set used. The future of this work will be focused more on developing a library of functions, similarity techniques, interpolation techniques, finite element analysis, and CFD computations to be used as the approximators. As these new tools are added to the library, the complexity of the problems to be modeled by AMoEBA will be increased.

The AMoEBA is not meant to work only as an artificial learning program. The techniques to be used in the tool library will not only be developed from earlier AMoEBA test cases but also from user input and analysis. By combining user knowledge and the artificial learning of the program, the authors hope to decrease time spent during the recalculation process while still conserving the fundamental physical laws that are critical to the data set.

NOMENCLATURE

H	horizontal	rt	right type
l	left node	T	temperature
lt	left type	V	vertical
r	right node	x	distance

REFERENCES

[1] Bonzani, I., "Solution of Nonlinear Evolution Problems by Parallelized Collocation-Interpolation Methods," Computers & Mathematics with Applications, v 34, n 12, Dec. 1997, Elsevier Science Ltd., Oxford, England, pp. 71-79.

[2] Daida, Jason M., et. al., "Measuring Small-Scale Water Surface Waves: Nonlinear Interpolation & Integration Techniques for Slope-Image Data," IGARSS Proceedings 1996, IEEE, May 1996, pp. 2219-2221.

[3] Gong, Yifan and Jean-Paul Haton, "Non-Linear Vector Interpolation by Neural Network for Phoneme Identification in Continuous Speech," ICASSP Proceedings 1991, IEEE Signal Processing Society, May 1991, pp. 121-124.

[4] Gruau, Frederic, "Genetic Micro Programming of Neural Networks," Advances in Genetic Programming, The MIT Press, Cambridge, Massachusetts, ed. by Kenneth E. Kinnear, Jr., 1994.

[5] Jensen, Kris and Dimitris Anastassiou, "Spatial Resolution Enhancement of Images Using Nonlinear Interpolation," ICASSP Proceedings 1990, IEEE Signal Processing Society, April 1990, pp. 2045-2048.

[6] Koza, John R., Genetic Programming: On the Programming of Computers by Means of Natural Selection, The MIT Press, Cambridge, Massachusetts, 1998.

A DEVELOPMENT OF AN INTERPRETER FOR CELLULAR AUTOMATA

YUHEI AKAMINE, SATOSHI ENDO, KOJI YAMADA
University of the Ryukyus
Department of Information Engineering, Faculty of Engineering
Okinawa 903-0213, Japan

ABSTRACT
Cellular Automata (CA), which is a method for analyzing phenomena of complex systems, allows us to construct many kinds of simulators such as road traffic simulators. Many researchers are still creating various CA models. However, at the present, we need to use a trial and error method to define a local rule, which is the main process of making CA models, in many cases. This means that we need to frequently verify the result of a simulation after changing a rule. We have developed an interpreter that makes this process easy. Giving the system a local rule described in a CA dedicated language, you can instantly confirm the result. When you change the rules, the system reflects your changes immediately. Our system does not regulate to design a local rule as far as belonging to the CA category because the system adopts the method by description in the special language for definitions of a local rule. Besides, we can modify a rule dramatically since the system interprets a rule description language in runtime. Furthermore, we can change variable parameters using GUI interfaces. Whereby they iterate to change a rule, a conflictive conditional statement sometimes occurs in a description of local rules. This conflict is far from seeking generally. Our system detects such a conflict automatically and notifies it to users as an error.

1. INTRODUCTION

Cellular Automata (CA) can unravel phenomena that are hard to analyze with models based on equations, such as traffic jams[1][2], fluid crosscurrents[3], and natural disasters, because with CA it is possible to reproduce complex phenomena including irregularity with simple models. Furthermore, CA has a possibility to give us new knowledge for analysis of biological phenomena since CA may be applied to such fields.

In the study of CA, it is important to simulate a CA model according to your design. However, you need to do software emulation for testing your designed CA models since CA dedicated computers have not been generalized yet. In this emulating software, routine works occupy the major part of its processes, such as memory management and visualization, except for the process associated with handling a local rule, which defines interaction between neighbor cells. A system that supplies such processes, therefore, simplifies your CA simulation more than what you are doing now.

Even now, such a system has existed, however, these applications are inefficient for development because it is a compiler system [4]. In addition, if you can use functions that detect and indicate faults you may commit in your description for a local rule with a program language, and aid to analyze the results of simulations; you can design CA models and analyze objective phenomena more easily. We develop the comprehensive system that supplies the functions we would need at design, simulation, and analysis of CA models. Additionally, we develop a CA traffic simulator using the system, and present our system's effectiveness for designing CA models.

2. CHARACTERISTICS OF THE SYSTEM

Fig. 1 represents a structure of the system we develop in this study. The characteristics of our system are:
- An interpreter for CA simulator, whose input is in the CA dedicated language.
- Detection of errors that occur frequently in descriptions for a local rule frequently.
- Functions that aid to analyze CA models, such as displaying a result of simulation and statistical analysis.

The interpreter type simulator in our system executes and parses a source code described in the CA dedicated language, which can describe CA models naturally, in real time. An interpreter can expedite the works such as correction of a local rule for the result of simulation (this happens on design phase of CA models frequently) for the benefit of rapid output of response to the input. Besides, we designed that the implementation level of CA simulation is concealed for no requiring of a user to get knowledge about processes unrelated to CA models essentially such as visualization. A local rule is a definition for the projective mapping of the states vector of neighbor cells to a state of the processing cell at next time step. When we describe a local rule in a program language, the conditional statement chain, such as "IF~THEN~IF~THEN~IF~THEN...", often occurs in the source code. Repeating to revise such a source code, an undesireble expression (for example, a conditional statement making an unchangeable result) sometimes appears in the source code because a conditional statement conflicts to other statements.

It is hard to detect such a conflicting condition. All of cells operate simultaneously in CA's theory. Therefore the debugging method to trace CPU's process with stepwise execution is not adequate to debug CA simulation. The function, detecting such conflicting conditions, expedites to design CA models and describe a local rule in a program language.

Visualization of the simulation's result is often the method for validation of the designed local rules in research of CA. A large quantity of cell states changes at each time step since each cell works simultaneously in CA. Thus when we verify whether the change of states is intentional or not, it is efficient to check visually in the manner that a cell state corresponds to a pixel color or luminance on screen in the model of three or less dimensions. Our system has such a function of visualizing states and also making a graph according to the statistical analysis of cell states(Fig. 2).

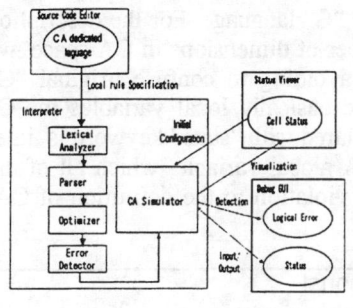 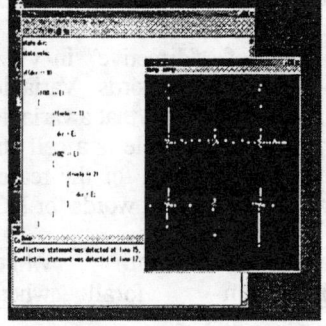

Fig. 1. The structure of our system   **Fig. 2.** A snapshot of the execution image of our system

3. THE INTERPRETER FOR CA SIMULATOR
3.1 Synopsis of CA simulator

The CA simulator built into our system is an implementation for CA models, and its characteristics are:
1. CA is a spatial and temporal discrete system.
2. Each point in a regular spatial lattice can have any one of a finit number of states.
3. The states of the cells in the lattice are updated according to a local rule.
4. All cells on the lattice are updated synchronously.

The state of the cell located to 'i' at time step 't' is defined by:
$$a^i_{t+1} = F(a^{i-r}_t, \cdots, a^{i+r}_t)$$
Where F is a local rule, r is the range of neighbor cells.

3.2 The interpreter and its language specification

The interpreter of our system executes and parses the source code for description of the CA model's specification. Its characteristics are:
- Elimination of the descriptions that violate CA category due to enforcement to describe a local rule with relative coordinate system, whose origin is each cell.
- Demand of minimun knowlegde by language speciifction like "C" language.
- Correspondence to n dimensions automata space

The interpreter interprets the new language, named "DORA", designed by us for CA. We adopt the operators and syntax like "C" language to the new language with no demand of new knowledge. We developed the interpreter with YACC[5].

In DORA language, you define a local rule using an array that can refer states of neighbor cells located in relative coordinate system, whose origin is a target cell, with the operators and syntax like "C" language. For the description about CA models, such as directions of number of dimensions in CA space, we adopt the form of "#directive" in view of avoiding to confuse original "C" keywords with new keywords. Variables are basically local variables in "C" language. But we designed that a variable declared with "state" keyword is time-series variable, that variable is a cell state. A global variable, which all of the cells can refer, is disabled for the reason of violation to the definition of CA. The following are reserved words for DORA.

if	else	for	while	const
state	dimension	forall	when	otherwise

An array operator for reference to status of neighbor cells is read only, so the description of assigning a value is a syntax error. This helps to prevent the deviation of your model's specification from CA category. A state of the target cell, which is declared with "state" keyword, is allowed to assign and refer. Its initial value in the state transition process is the state value at time t, and the final value is the state value at time t+1(Fig. 3.).

```
state s;                // s is a cell state.
if ( [1,0].s == 1) {    // [1,0].s is a right neighbor cell's state
    s = 2;              // s is assignable. but this value isn't a state value
                        //    at next time step.
    [1,0].s = 2;        // This code is an error because a cell except for its
                        //    own cell's state is read only
}
if ( s != 2 ) {         // s is able to refer.
    s = 1;              // this value "one" is a state value at next time step.
}
```

Fig. 3. an example for the cell states variable

4. DETECTION OF A CONFLICTIVE LOCAL RULE

A local rule is a projective function from a state vector of neighbor cells to a state of a target cell. In complex models, there are generally a flood of combination patterns of neighbor cell's states, so the projective function is implemented by the description of a conditional statement for application of a local rule to states of neighbor cells. On the design phase of a local rule, we frequently revise the description of conditional statements. But repeating to revise a complex nested conditional statement, an undesireable conditional expression, such as a conditional statement making the unchangeable result and redundant expression, often appears in the description.

Our system has the function for automatically detecting a conflictive description of a local rule as far as possible. The followings are examples for conflicts appearing frequently.

(1) A conditional statement making the constant result
For example:
```
if( a > 0 && b < 0 ) {
```

```
        if( c > 3 && a < 0) {    // this will never be true
            some statements ;
        }
}
```
(2) A variable not affecting the cell state
For example:
```
    state st ;                    //   st is declared as state type variable.
    a = [0,-1] + [0,1];    //   [x,y] is a state of cell located at (x,y) as
origin of a target cell.
    b = [-1,0] + [1,0];
    if( a > 2 || b < -1) {
        c = a + b;              ←(*1)
        st = 1;    // st is a cell state
    }
```
The statement (*1) may be unintentional because the variable "c" does not affect a cell state "st".

(3) A cell state is changed more than once
for example:
```
state st;
a = [0,1];
if( a > 0 )
    st = 0;
if( a > -2)
    st = 1;
```

If "a" is over 0, a state "st" is changed twice. There is possibility that the projective map for neighbor cell's states is overlapped.

5. THE EXAMPLE FOR CONFLICT DETECTION

In this section, we present the example for conflict detection in a CA traffic simulator to show the system's efficiency. The first CA model of a traffic simulator was proposed by Gerlough[1]. Lately, TRANSIMS have been applied to the large sacle simulation[2].

5.1 specification of the model

Specification of the CA traffic simulator designed in this study are:
- Deacceleration: A car deaccelerate if the car ahead is too close.
- Acceleration: A car accelerate if following distance is large enough and if the velocity is smaller than the max velocity.

The max velocity is three in this example. An empty cell is represent by –1. A cell state can change into four different value: -1,0,1,2.

5.2 Instances for conflictive conditional statement

Fig. 4 shows a correct code for description about a state transition rule of a cell in the traffic simulator, and three error patterns for example. First, if you make a mistake of position to brace ('{' and '}') (case No. 1 in Fig. 4), the statements (A) will never execute. This case is an example for "a conditional statement making the constant result." The next case is an instance for "a variable not affecting the cell state." Even if you mistake "veio" for "velo", whereby a variable is declared automaticaly, it is not a syntax error in our system. However, because "veio" should not affect the cell state, this error is detected by detective function. The last case is "a cell state is changed more than once." A cell state "velo" is changed twice, so the system says this code may be incorrect.

6. CONCLUSION

In this paper, we present the specification of our CA simulation system. We particularly explained its language specification and the confliction detecting function in detail. In addition, we showed efficiency of the detective function with an example.

In the future work, we will accelerate the simulator by force of developing an optimizing compiler, whereby we make a system corresponding to do the large-scale simulation.

ACKNOWLEDGEMENTS

This research was partially supported by the Ministry of Education, Culture, Science, Sports, and Technology, Grant-in-Aid for Encouragement of Young Scientists, 13780296, 2001.

REFERENCES

1. D. L. Gerlough, 1956 , "Simulation of freeway traffic by an electronic computer, in: F. Burggraf and E. Ward (eds.)," Proc. 35th Annual Meeting, Highway Research Board, National Research Council, Washington, D.C., pp. 543.
2. http://www-transims.tsasa.lanl.gov/
3. U. Frisch, B. Hasslacher, Y. Pomeau, 1987, "Lattice-Gas Automata for the Navier-Stokes Equation," Phys. Rev. Lett. 56, pp. 1505-1508.
4. J. Dana. Eckart, http://www.cs.runet.edu/~dana/ca/cellular.html
5. J. Levine, T. Mason, D. Brown, 1992, "lex & yacc, 2nd Edition," O'Reilly & Associates.

```
○Correct Code
#dimension 1

state velo;

b1 = [-1]; b2 = [-2];
c  = velo;
f1 = [1]; f2 = [2];

if((b2 == 2 || b2== 1)
 && b1 == -1 && c == -1)
{
    if(f1 < 0)
    {
        if(f2 < 0)
        {
            velo = 2;
        }
        else{
            velo = 1;
        }
    }else
    {
        velo = 0;
    }
}else if(b1 == 1 && c == -1)
{
    if(f1 < 0)
    {
        velo = 1;
    }else
    {
        velo = 0;
    }
}
```

```
○Error Case 1

if((b2 == 2 || b2== 1)
 && b1 == -1 && c == -1)
{
    if(f1 < 0)
    {
        if(f2 < 0)
        {
            velo = 2;
        }
        else{
            velo = 1;
        }
    }else
    {
        velo = 0;
    }
if(b1 == 1 && c == -1)
{
    if(f1 < 0)
    {
        velo = 1;
    }else                 (A)
    {
        velo = 0;
    }
  }
}
```

```
○Error Case 2

if((b2 == 2 || b2== 1)
 && b1 == -1 && c == -1)
{
    if(f1 < 0)
    {
        if(f2 < 0)
        {
            velo = 2;
        }
        else{
            veio = 1;
        }
    }else
    {
        velo = 0;
    }
}else if(b1==1 && c==-1)
{
    if(f1 < 0)
    {
        velo = 1;
    }else
    {
        velo = 0;
    }
}
```

```
○Error Case 3
if((b2 == 2 || b2== 1)
 && b1 == -1 && c == -1)
{
    if(f1 < 0)
    {
        if(f2 < 0)
        {
            velo = 2;
        }
        else{
            velo = 1;
        }
    }
    velo = 0;   (B)
}else if(b1==1 && c==-1)
{
    if(f1 < 0)
    {
        velo = 1;
    }else
    {
        velo = 0;
    }
}
```

Fig. 4. An example for the code detected by our system.

NEW SCIENTIFIC BORDERS FOR ARTIFICIAL NEURAL NETWORKS BASED IN POLYNOMIAL POWERS OF SIGMOID THEORY

JOÃO FERNANDO MARAR
Universidade Estadual Paulista
Adaptive Systems and Intelligent
Computation Lab.
Department of Computer Science
Bauru, Brazil

ABSTRACT

Since the presentation of the backpropagation algorithm a vast variety of improvements of the technique for architectures design and training the parameters in a feedforward neural network have been proposed. The following paper introduces the basic definitions about Polynomial Powers of Sigmoid (PPS) and a new application of the PPS wavelet for artificial neural network in classifiers design.

INTRODUCTION

The artificial neural networks are a class of computacional architectures that are composed of the interconnected neurons. Its name reflects its initial inspiration from biological neural systems, all though it currently refers to artificial neural networks, quite different compared to of the biological ones.

The central issue of any theory of artificial neural network is to find the values of the synaptic weights that are best suited for a given task. This issue also implies the search of the corresponding best architecture of the artificial neural network. Many algorithms have been proposed for training of a neural network (Haykin, 1994). Despite the success of many such algorithms, there is still room for progress to be made in the understanding of the exact nature of operating modes of the neural networks. In this sense, the wavelet neural networks become an excellent area of study.

The wavelet theory is a rapiddly developing branch of mathematics which has found many applications, for instance, signal processing, computer vision, numerical analysis (Chui, 1992). Though this theory has offered very efficient algorithms for analyzing, approximating, compressing functions, the application of such algorithms are usually limited to problems of small input dimension, because construting and storing wavelet basis of large input dimension are cost prohibitive (Zhang, 1992). In order to handle problems of large input dimension, it is desirable to find some technique whose complexity is less sensitive to the input dimension. It seems that artificial neural networks are promising candidates for such purpose. Many studies have been reported on using artificial neural networks for functions approximation of possibily large input dimension (Barron, 1993). This paper introduces a new application of the Polynomial Powers of Sigmoid theory for artificial neural network in classifiers design.

WAVELET NEURAL NETWORKS

For a brief overview that treats wavelet theory as one topic in neural networks, the influential paper by Harold Szu, Brain Telfer and Shubha Kadambe (Szu et al, 1992) is highly recomended, because provides an excellent survey of work done using the wavalet neural networks for signal representation and classification model. Daugman, in (Daugman,1988), has shown how neural networks can be used to learn the best coefficients for approximating images using the Gabor function (Gabor's wavelet). Zhang and Beveniste, in (Zhang and Beveniste, 1992), have investigated the power of "wavenets" (wavelet neural networks) as function approximators and, finally, Pati and Krishnaprasad, in (Pati and Krishnaprasad, 1991), have shown the construction of a wavelet function through the combination of sigmoids, and have described how to develop its wavelet transform expansion using neural networks.

POLYNOMIAL POWERS OF SIGMOID (PPS) BASIC DEFINITIONS

Let $Y : \Re \rightarrow [0, 1]$ be a sigmoid function, especially $Y(x) = (1 + e^x)^{-1}$. The PPS functions are defined by: $Y^n : \Re \rightarrow [0, 1]$, defined by $Y^n(x) = [(1 + e^x)^{-1}]^n$, n ($n \in N$) as was show by Marar (Marar, 1997). Let Θ the set of all positive integer powers, described by the Equation (1):

$$\Theta = \{Y^0(x), Y^1(x), ..., Y^n(x), ...\} \quad (1)$$

We know that the sigmoid function is continuous and differentiate for any values of $x \in \Re$. In (Marar, 1997) an effective procedure for generating polynomial forms of wavelet functions from the sucessive powers of sigmoid functions is presented. The Equations (2) and (3) illustrates the PPS-wavelet functions:

$$\vartheta_2(x) = 2Y^3(x) - 3Y^2(x) + Y(x) \quad (2)$$

$$\vartheta_3(x) = -6 Y^4(x) + 12Y^3(x) - 7Y^2(x) + Y(x) \quad (3)$$

where denominated $\vartheta_2(x)$ the first PPS wavelet function and $\vartheta_3(x)$ represent the analitical function of the derivative of the $\vartheta_2(x)$ function.

The PPS wavelet neural networks are an additional contribution for the feedforward neural networks paradigm. An important aspect of the PPS neural networks is the modularity of the activation functions (Marar, 1996; Marar,1997). It can take advantage of the same basic structure of a neural network, substituting the PPS wavetet functions, and it can produce different solutions for the same problem (Marar et al.,1998).

PPS WAVELET NEURAL NETWORKS CLASSIFIER

In this section, we will show a example of a neural networks for classifier design. The central process of the classifier, in this case, are the vector inner products of a set of PPS wavelet with the input signal. The Figure 1 describes a PPS wavelet classifier as a neural network (Szu et al. 1992):

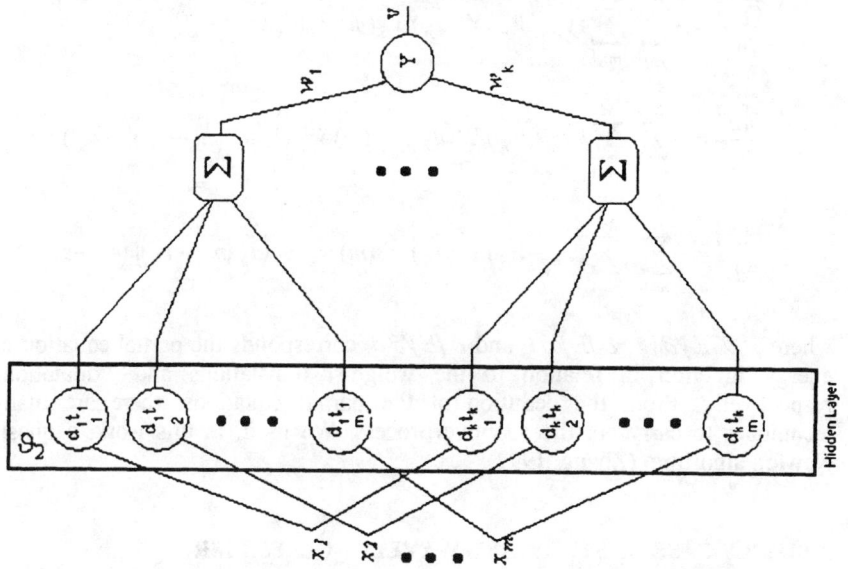

Figure 1: PPS Wavelet Neural Network Classifier

The mathematical equation for the determination of this neural network is given by:

$$u_n = \sum_{k=1}^{K} w_k \sum_{m=1}^{M} x_n(m) \vartheta_2 (d_k (m - t_k)) \qquad (4)$$

$$V_n = Y(u_n) \qquad (5)$$

where V_n is the output for the $n'th$ training vector x_n (m), $\vartheta_2(\bullet)$ is a PPS-wavelet function and Y is a sigmoidal function.

The energy function (least mean squares) is defined by:

$$E = \frac{1}{2} \sum_{n=1}^{N} (y_n - V_n)^2 \qquad (6)$$

where y_n is the desired classifier output for x_n and V_n is network output.

In the Figure 1, the hidden layer produces inner products of the input values and PPS-wavelet, with the first wavelet on the left and the k'th wavelet on the right. Here, it will be adopted the values $y_n = 1$ for one class and $y_n = 0$ for the other. On the other hand, neural networks with multiple output elements to handle more than two classes can be developed based on this technique.

The PPS wavelet neural network parameters w_k, d_k and t_k can be optimized by minimizing the Equation (6):

$$\frac{\partial [E]}{\partial w_k} = -\sum_{n=1}^{N} \sum_{m=1}^{M} (y_n - V_n) Y'(u_n) x_n(m) \vartheta_2 \left(d_k (m - t_k) \right)$$

$$\frac{\partial [E]}{\partial t_k} = -\sum_{n=1}^{N} \sum_{m=1}^{M} (y_n - V_n) Y'(u_n) x_n(m) w_k \vartheta_3 \left(d_k (m - t_k) \right) (-d_k)$$

$$\frac{\partial [E]}{\partial d_k} = -\sum_{n=1}^{N} \sum_{m=1}^{M} (y_n - V_n) Y'(u_n) x_n(m) w_k \vartheta_3 \left(d_k (m - t_k) \right) (m - t_k)$$

where $\partial [E]/\partial w_k$, $\partial [E]/\partial t_k$ and $\partial [E]/\partial d_k$ corresponds the partial equation of energy function in relation to the weights, translations and dilatations respectivelly. From the dedution of the partial equations, there are many techniques to carry out the adaptive process. We used, in this work, a quasi-Newton algorithm (Zhang, 1992).

A SIMPLE CASE OF STUDY: PPS WAVELET CLASSIFIER

In this application, we used as training and test patterns a database of British Post Office of numerical handwritten patterns. Every patterns (training and test), in this database, are normalized through 384 points, on a matrix of 24 lines and 16 columns. Figure 2 illustrates examples of existent patterns in the database.

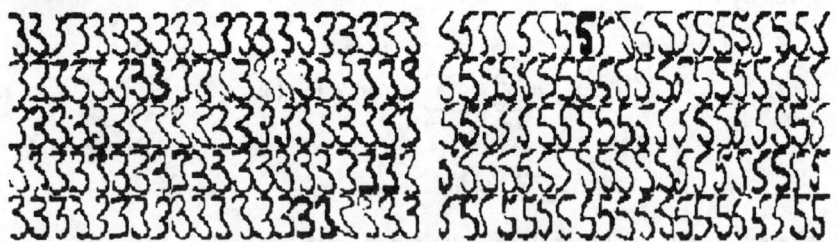

Figure 2 : The training data sets

In order to investigate the PPS wavelet neural network performance, we have selected randonly one hundred training pattern of the numerical digits 3 and 5 constituting two classes and one hundred for test pattern of each class.

For efficient classification process, we propose the feature extraction with Karhunen-Loève transform (KLT) (Chen and Huo, 1991). The use of this technique still allows to aid the discrimination of the existent classes, in way the to increase the distances interclass and to minimize the distances intraclass error.

The pattern recognition system with the feature extraction phase, used in this application, is shown in the Figure 3:

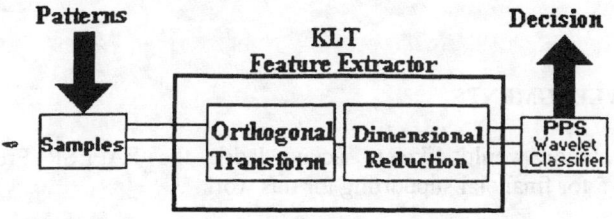

Figure 3 : Pattern Recognition System

We have used Karhunen-Loève transform as feature extraction filter and to reduce the input dimensionality from 384 bits to twelve principal components. We have investigates six different PPS neural networks architectures with 8, 10, 15, 20, 25 and 30 neurons in the hidden layer. We have used the PPS-wavelet $\vartheta_2(x)$ as the activation function for the neural networks classifier. In order to investigate the neural networks recognition property, in more details, the Table 1 show the confusion matrix for the networks investigated.

Confusion Table	8 Neurons Hidden Layer		10 Neurons Hidden Layer		15 Neurons Hidden Layer		20 Neurons Hidden Layer		25 Neurons Hidden Layer		30 Neurons Hidden Layer	
class	3	5	3	5	3	5	3	5	3	5	3	5
3	40	60	48	52	55	45	61	39	75	25	87	13
5	53	47	49	51	47	53	32	68	21	79	11	89
Total	93	107	97	103	102	98	93	107	96	104	98	102

Table 1 : Confusion matrix

Even with a quite simple application, we can observe, the PPS wavelet neural network improves the recognition rate when increasing the number of hidden neurons. The great found advantage of this classifier system is the balance of the rates of general recognition.

CONCLUSIONS

The combination of artificial neural networks and wavelet transform theory gives rise to an interesting and powerful technique for representation and classification of the signals. The classification process is an important task particularly in computational environments that deal with data samples collected from real world simulations. So, wavelet theory and neural networks present themselves as efficient and robust solutions for many problems involving real world simulations. In this work, it has been shown how the PPS wavelet can be used to provide a robust way for neural networks architectures design, and in this manner, creating new scientific borders for artificial neural networks.

ACKNOWLEDGMENTS

The author would like to acknowledge the FAPESP Proc. Number 97/13309-5 for financial supporting for this work.

NOMENCLATURE
Y   Sigmoid function
Y'   Derivative of the Sigmoid function
$\Re$ Real numbers set
$\mathcal{N}$ Natural numbers set
ϑ_i i-index PPS function

REFERENCES
Barron, A. , 1993. "Universal approximation bounds for superpositions of a sigmoidal functions". IEEE Trans. On Information Theory, Vol 39, no. 3.
Chen, C. S., and Huo, K. S., 1991, "Karhunen-Loève method for data compression and speech synthesis," *IEEE Proceedings-I*, Vol. 138, pp. 377-380.
Chui, C. , 1992. "Wavelet: a tutorial in theory and applications". Inc. Boston, San Diego: Academic Press.
Daugman, J. , 1988. " Complete discrete 2-d gabor transforms by neural networks for image analysis and compression " *IEEE Transactions Acoust. Speech.Signal* Proc. 36, pp. 1169-1179 (July 1988).
Haykin, S. , 1994. "*Neural Networks: A Comprehensive Foundation*". Macmillan College Publishing Company. New York.
Marar, J. F., Carvalho Filho, E. C. B., and Vasconcelos, G. C., 1996, "Function approximation by polynomial wavelets generated from powers of sigmoids," *SPIE-Aerosense'96 – 10th Annual International Aerosense Symposium*, 2762-Wavelet Application III, pp. 365-374.
Marar, J. F., 1997. "Polynomials Powers of Sigmoids (PPS): a new technique for functions approximation, wavenets construction and applications in images and signal processing " Ph.D. Thesis, Universidade Federal de Pernambuco, Recife.
Marar, J. F., Carvalho Filho, E. C. B., and Santos, D. J., 1998. "Mathematical tests about the existence and applications of PPS-wavelets in function approximation problems", *SPIE-Aerosense/98 – 12th Annual International Aerosense Symposium*, 3391-45 Wavelet Application V, pp. 455-466.
Pati, Y. C. , and Krishnaprasad, P. S. , 1991, " Analysis and synthesis of feedforward neural networks using discrete affine wavelet transformations" *Tech. Rep. SRC-TR-90-44*, Univ. Maryland Systems Research Center, 1991.
Szu, H. , Telfer, B. , and Kadambe, S., 1992, " Neural network adaptive wavelets for signal representation and classification" *Optical Enginnering*, Vol 31, N 9, pp:1907-1916 (September 1992).
Zhang, Q., and Benveniste, A., 1992, "Wavelet Networks," *IEEE Transactions on Neural Networks*, Vol. 3, pp. 889-898.

A PROPOSAL OF AN IMMUNE OPTIMIZATION INSPIRED BY CELL-COOPERATION

NARUAKI TOMA, SATOSHI ENDO, KOJI YAMADA, HAYAO MIYAGI
University of the Ryukyus
Department of Information Engineering, Faculty of Engineering
Okinawa 903-0213, Japan

ABSTRACT

The purpose of this paper is to propose an immune optimization for meta-scheduling problems, which the optimization is inspired by immune cell-cooperation and immune tolerance. The cell-cooperation is considered as a kind of parallel-distributed system with role differentiations in biological immune system. By analogy, we constructed a system to solve the meta-scheduling problems through cooperative behaviors based on multi-agent system. In the system, each of the immune agents acts to assign a part of own tasks through interactions and it is to produce an efficient scheduling solution by load sharing. This algorithm is applied to `Standard Task Graph Set' problems to investigate the validity, and the system behaviors are examined.

1 INTRODUCTION

Adaptive problem solving techniques, such as neural networks and genetic algorithms, are based on information processing in biological organisms and they are applied on many kinds of optimization problems. A biological immune system is one of the adaptive systems and the studies are making advances [1,3,4,5]. The biological immune system is widely recognized as one of the adaptive biological system whose functions are to identify and to eliminate foreign materials. Especially, it has important notions that carry out highly parallel-distributed functions, for a design of multi-agent system.

In this paper, we propose an immune optimization, which is inspired by immune cell-cooperation and immune tolerance, for meta-scheduling problems. Meta-scheduling can be loosely defined as the act of locating and allocating resources for a job on a parallel-distributed computing [7]. A meta-scheduling system should make a collection of resources transparently available to the user as if it were a single large system. The cell-cooperation is considered as a kind of parallel-distributed system with role differentiations in biological immune system, and the function offers beneficial notions for solving the problems. By analogy, we constructed a system to solve the problems through cooperative behaviors based on multi-agent system. In the system, each of the immune agents acts to assign a part of own tasks through interactions and it is to perform an efficient scheduling solution by a load sharing. The proposed algorithm solves meta-scheduling problems through interactions between agents, and between agents and environment by immune functions. There are two functions in our algorithm: *division-and-integration processing* and *immune tolerance*. The division-and-integration processing resolves precedence constraints, and the immune tolerance performs a load sharing. Through implementation of such functions, we could construct an adaptive algorithm that solves meta-scheduling problems. In order to investigate a validity of the proposed method, this algorithm is applied to `Standard Task Graph Set (STG)' problems [11], which are a kind of benchmark for evaluation of multiprocessor scheduling algorithms as a typical problem of meta-scheduling problems.

2 STANDARD TASK GRAPH SET

Meta-scheduling can be loosely defined as the act of locating and allocating resources for a job on a parallel-distributed computing that consists of heterogeneous computers and networks. Smarr and Catlett define such a meta-computer as the collection of resources that are transparently available to the user via the network [7]. Thus a meta-scheduling system should make a collection of resources transparently available to the user as if it were a single large system (resource). The key point in this definition is that the user needs not to be aware of where the resources are, who owns the resources, or who administers the resources in order to use them. The scheduling issues that surround the creation of such a system is the focus of this section [10].

In this paper, we target at STG as a simplified meta-computer and job. STG is a kind of benchmark for evaluation of multiprocessor scheduling algorithms [11]. To efficiently execute programs in parallel on a multiprocessor environment, a minimum execution time must be solved to determine the assignment of tasks to the processors and the execution order of the tasks so that the execution time is minimized [2]. The multiprocessor scheduling problem treated in their project is to determine a non-preemptive schedule that minimizes the execution time, or the schedule length, when a set of *n* computational tasks having *arbitrary* precedence constraints and *arbitrary* processing time are assigned to *m* processors of the same capability. These tasks are represented by a *directed acyclic graph (DAG)* called a ``task graph'', as shown in figure 1.

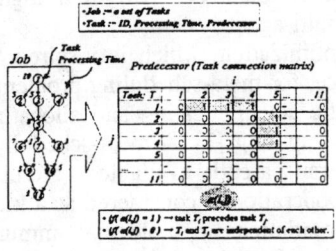
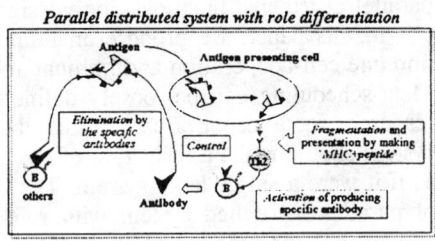

Figure1. Illustration of job for multi-processor scheduling problem in STG.

Figure2. Concept of immune cell-cooperation.

3 DESIGN OF IMMUNE META-SCHEDULING SYSTEM
3.1 Analogy from biological immune system

We solve the meta-scheduling problems by means of agent-based computing in the field of multi-agent system. The agents that introduced immune functions decide a scheduling plan (processing order of tasks) through communications between the agents. A purpose of this paper is obtaining a set of scheduling plans for multiprocessor, and then one of the future goals is to obtain a meta-knowledges, which consists of some characteristics about the problems. A scheduling system that learns proper meta-knowledges as a key point of characteristics and produces well-suited plans against the problems will be a very useful scheduler.

In the biological immune system, a framework which is performing to eliminate of unknown vast antigens in parallel is called `*immune cell-cooperation*'. In the point of view of engineering system, it is considered the immune cell-cooperation to be a parallel-distributed system with role differentiation. The roles in this system are (1) fragmentation and presentation of antigens, (2) activation of producing specific antibodies, (3) elimination of the antigens by specific antibodies, and (4) control of the functions (see figure 2).

3.2 Application of the immune cell-cooperation and immune tolerance

In order to apply the system to the meta-scheduling problems, we choose three functions, fragmentation, activation and elimination, in this issue, because the functions are minimum components to perform as problem-solvers. We construct an optimization algorithm based on a concept that (1) fragments a problem, (2) solves the fragmented sub-problems by the specific sub-solutions, and then (3) solves whole the problem through combination of these sub-solutions. In other words, we use these functions (called *division-and-integration processing*) to divide a job into tasks and assign the tasks to resolve precedence constraints. In addition, we introduce an immune tolerance which is a phenomenon that the immune system dosen't respond with one and/or some specific antigens. By regarding the selection function of opponents as a decision function, the tolerance function as a control mechanism is using for agent's behavior arbitration in our algorithm.

3.3 Algorithm

The algorithm solves the problems through two searching ways, *division-and-integration processing* and *immune tolerance*. The procedures of the algorithm against a STG problem are described as below. In the procedures of proposed method, Step1 is processed for initialization once for all. Each of agents processes repeatedly from Step2 to Step4 for solving the problem.

[**Step1. Definition of problems.**] A *Job* and *multiprocessor system* as the problem must be defined. A Job that consists of a set of tasks, is represented as *directed acyclic graph (DAG)*, and needs the number of tasks, processing time and a task connection matrix (describes precedence constraints), see figure 1. As definition of the multiprocessor system, it is described only the number of processors according to STG. And then, all the agents are initialized for begining following procedures. An agent exists in individual processor, and determines tasks which should be processed by the processor through communications between agents. In the initialization process, all the tasks are assigned into each agent's queue at random.

[**Step2. Calculation of objective function.**] Each of agents calculates $Time_{work}$ and $Time_{free}$ as objective functions through performing as a current schedule.
- $Time_{work}$ is defined as a processing time of assigned tasks.
- $Time_{free}$ is defined as a free time, that is no processing time.

In other words, a scheduling time for all the tasks equals to $Time_{work} + Time_{free}$.

[Step3. A load sharing fr efficient scheduling plans by immune tolerance.] In this step, the objective is to achieve a load sharing based on a scheduling time $Time_{task}$ and/or $Time_{free}$). The agent with high $Time_{task}$ tries transfer of a task to the lower agent. (Consequently, if total of both agents' $Time_{task}$ decreases, it will perform, otherwise, transfer is canceled.)

[Step4. Resolving precedence constraints by division-and-integration processing.] In this step, each of free agents tries to modify a scheduling plan itself so as to resolve precedence constraints, and to form a role in the problem according to the procedures.

1. **Process a task in the execution position in queue.**

   A free agent processes the task if it is untreated task. In case of a execution of the task is impossible by reason of some constraints, go to next procedure.

2. **Process a task in which the execution in queue is possible.**

   A free agent tries to search the task by checking feasible tasks from own all the tasks. The agent uses heuristics to decide a processing task from a set of feasible tasks, which the heuristics gives the task with many predecessors preferentially. In case of a execution of the task is impossible, go to next procedure, too.

3. **Process a task in which the execution in other agent's queue is possible.**

   A free agent tries to search the task by checking feasible tasks from other agent's tasks. The agent finds out all the feasible tasks by communication against other agent in the beginning. Then, the agent goes on to decide a processing task as well as the previous heuristics, and finally, the processing is started after the agent gets the target task from opponent agent. In case of a execution of the task doesn't exist, the agent doesn't work in this time step.

Each of agents continues processing of above procedures until agents finish processing of all the tasks, that is, all the agent have a feasible scheduling plan. In the manners, it is expected that each of agents produces feasible scheduling plans and efficient sharing plans through two ways of resolving constraints and load sharing.

4 EXPERIMENT
4.1 Definition of problem

To confirm the basic performance, we apply to a benchmark problem which is described in the table 1 which is defined as `proto151.stg' on STG.

4.2 Results and prospects

Figure 3 shows obtained scheduling plans in the first time step 1 and 2. The scheduling time for processing of all the tasks in the 1st time step is *119* that is optimal schedule length according to STG. In the 1st time step, the precedence constraints are resolved to obtain feasible plans by using *division-and-integration processing* (Step4) because the initial plans of agents are made at random. After the 1st time step, the transfers of a task from an overworked agent to a workless agent are executed to achieve a load sharing by using *immune tolerance* (Step3), and then Step4 is executed also. In the 2nd time step, the maximum load is improving from *0.899160* to *0.789916*, and differs of the plans are showed in fig.4 with gray number. The plans in the 1st time step exist a bias that the agents with small ID process many tasks, and the bias is improving so as to be sharing evenly.

After the 2nd time step, because of (a) our system is always obtaining the optimal plans and (b) the loads of agents are modifying, which the system is searching other feasible optimal plans. (Namely, the system obtains multiple feasible optimal solutions, and we think that it is possible to find some specified characteristics on the problem by means of an identifying common denominators in the solutions. For designing the identifying function, a concept of an immunological memory will be informative and instructive mechanism.)

A transition of maximum and minimum load under the search is shown in figure 4. As previously mentioned, the initial loads between maximum and minimum have a large difference, however the all the solutions are optimal schedule length. Such a solving design is prospective as a very useful system from a point of view of applications. It is important to prepare any superior alternate solution or high adaptability for producing other solutions because any constraint changes easily and the changings are difficult to predict correctly in the real applications. For example in this results (figure 4), it is possible to use the resources properly according to the situation of LAN, processors or tasks.

A transition of $Time_{task}$ of agents is shown in figure 5. In the 1st time step, the time of most workless agent(no.10) is only 30 for a set of tasks `$15, 40, 74$', on the one hand, the time of most overwork agent(no.2) is $107$ for a set of tasks `$6,19,18,34,42,41,57,53,67,73,78,81$'. That is, the difference of the $Time_{task}$ is 77.

In the opposite direction, the time step has the smallest difference is 15th step, the most workless's time is 67, the most overwork's time is 90 and the differ is 23. Such a solution will be a useful planning in case that the user requires to minimize a recovery cost from any fault of processors.

Table 1. Characteristics of proto151.stg.

the number of processors	10
the number of tasks	80
Predecessors (max,ave,min,sum)	10, 3.74, 0, 303
Task processing time (max,ave,min,sum)	12, 9.57, 0, 775

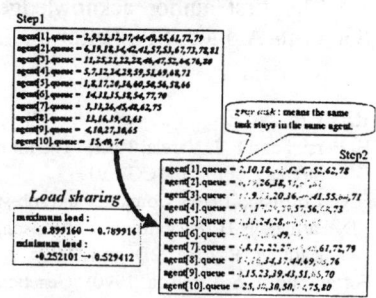

Figure3. An example of load sharing from the step1 to step2.

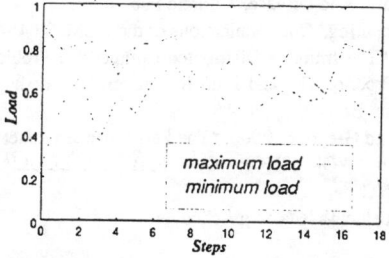

Figure4. Transition of max- and min-load.

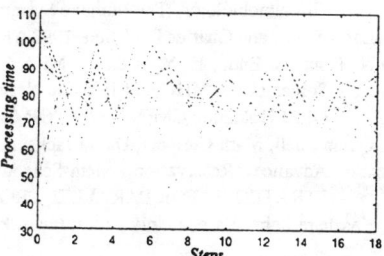

Figure5. Transition of $Time_{task}$.

5 CONCLUSIONS

In this paper, we proposed and investigated an immune meta-scheduling system in order to verify the engineering application possibility of artificial immune system. Especially, biological immune functions have many important notions which carry out highly parallel-distributed functions, for a design of multi-agent system. By analogy, we constructed a system to solve the problems through cooperative behaviors based on multi-agent system.

In the system, each of immune agents acts to assign a part of own tasks through interactions using *immune cell-cooperation* and it is to perform an efficient scheduling solution by acquiring the roles using *immune tolerance*. These functions are based on local interactions between agents. In multi-agent system, clarifying the objective function considered all the agents and components in the environment is too hard problem, and then it is important optimizing of the whole problem by using the local interactions. Since our algorithm can optimize division-of-labor problems, it can expect what is functioned effectively as an optimization algorithm in multi-agent system.

In addition, the system can obtain multiple feasible optimal solutions, namely, we think that it is possible to find some specified characteristics on the problem by means of an identifying common denominators in the solutions. For designing the identifying function, a concept of an immunological memory will be informative and instructive mechanism. Such a scheduling system that learns proper meta-knowledges as a key point of characteristics and produces well-suited plans against the problems, will be a very useful scheduler. As future works, we model and construct such a system to learn proper meta-knowledges.

The first author acknowledges the Grant-in-Aid for Scientific Research (Grant-in-Aid for JSPS Fellows).

REFERENCES

H. Bersini and F. J. Varela, 1991, "The Immune Recruitment Mechanism : A Selective Evolutionary Strategy," Proc. of ICGA 91.

E. G. Coffman, 1976, "Computer and Job-shop Scheduling Theory,". John Willey & Sons.

J.D. Farmer, N.H. Packard, A.A. Perelson, 1986, "The Immune system adaptation, and machine learning,"Physica 22D, pp.187-204.

Forrest and A. S. Perelson, 1990, "Genetic Algorithm and the Immune System," Proc. of PPSN 90, pp.320-325.

Y. Ishida, and N. Adachi, 1996, "Active Noise Control by an Immune Algorithm," Proc. ICEC 96, pp.150-153.

Charles A. Janeway, Jr., Paul Travers ; with assistance of Simon Hunt, Mark Walport, 1997, "Immunobiology : The Immune System in Health And Disease,"Garland Pub.

Larry Smarr and Charles E. Catlett, 1992, "Metacomputing,"Communications of the ACM, 35:45-52.

N. Toma, S. Endo, K. Yamada, H. Miyagi, 2000, "The Immune Distributed Competitive Problem Solver Using MHC and Immune Network," Proc. of The 2nd Joint International Workshop - ORSJ Hokkaido Chapter and ASOR Queensland Branch -, pp82-89.

Quinn Snell, Mark Clement, David Jackson, and Chad Gregory , 2000, "The Performance Impact of Advance Reservation Meta-Scheduling,"6th WORKSHOP ON JOB SCHEDULING STRATEGIES FOR PARALLEL PROCESSING.

Kasahara Lab., Waseda Univ , http://www.kasahara.elec.waseda.ac.jp/schedule/

THE NAMING OF THINGS AND THE CONFUSION OF TONGUES

FLORENCE REEDER[1]
MITRE Corporation
7515 Colshire Dr.
McLean VA 22102
freeder@mitre.org

ABSTRACT
If we accept language as an endogenous system, then we can start discussing representations for language processing which have a basis in endogenous systems. This paper is a start in the direction of showing how such a model might be constructed – drawing from across disciplines such as psychology, psycholinguistics and neuroscience. What we propose, and show the start of, is a model of lexical items which rather than being a list of words and features, is a system of evidence points. We outline the underlying technologies supporting this and describe what each will bring to the model. We believe this model will better support data analysis where language values are involved and show how it could work in cross-language information analysis. The first area of application involves named entities – people, places, associations – which are frequently necessary intelligence analysis data points, but which have confounded systems designed to automatically incorporate them.

INTRODUCTION
Machine Translation (MT) systems translate between human, natural languages. Weaver [1947] described the translation process as a kind of noisy channel decryption process: a French document is really an English document coded in French. For instance, the literal translation of **omelette de frommage** is **omelet with cheese**. However, the process of translating between languages is more than the substitution of one word for another. Consider the example, **escuela de derecho de Harvard**. Literally translated, it is **school of the right of Harvard** instead of the more proper **Harvard Law School**. The message can be found, but it takes more work on the part of the message's receiver.

Traditional MT systems have been developed following the Chomsky ideal [1957] of modeling language through successive layers of processing, each providing another level of information which isolates words from meaning. The levels, from least complex to most complex, are: phonemic, realizing words as sounds; morphological, capturing features of words; syntactic, assigning structure; semantic, attributing meaning; discourse, describing interaction rules; and pragmatic, concerning the mechanics of language in use. As processing moves toward pragmatic, the work to transfer between languages is reduced, while the understanding and generation processes increase in complexity.

The most complex, interlingua represents a pure translation process where the analysis yields a language-independent representation from which generation can be done. The components which address these different levels include: a bilingual lexicon (or list of words and information about those words); a grammar for the source and target languages; a set of transfer rules which translate structures between source and target representations.

To demonstrate the difficulties inherent in translation and why it accordingly resists traditional processing, even traditional rationalist thought, consider the word **bank**. It can mean a financial institution, a side of a river, the act of counting on an event, or driving around the corner – all dependent on the context[1] in which it is used. Translating **bank** into foreign languages is not straight-forward, as other languages can have different words corresponding to the different concepts related to this word. Because of the many-to-many mapping of words, word for word substitution is insufficient, therefore, a lexical transfer system translates poorly. While each successive level of processing supplies more features for disambiguation, we are still left with the problem of finding just the right set of words in a new language to capture the same message as conveyed in the source language.

The necessary amount of target and source background knowledge has increased as has the preparation cost. Also, we still do not have sufficient representations for the transformation of language at the conceptual level. To say **I am banking on the horse coming in** implies a concept which literal translation may not be able to account for. In this case, building an interlingual representation is an appropriate choice. As described elsewhere [Reeder, 2000], the process of developing an interlingual representation has been like chasing the "Holy Grail". Human translation is considered a creative process of both interpretation and conveyance of a given message. This paper argues that it is because MT (and in fact language) is an endogenous process, the traditional methods of representing language information will necessarily be insufficient We must look to new processing ideas to accurately perform the translation task.

MT AS AN ENDOGENOUS PROCESS

Endogenous systems have been described as those which are perform operations which are incomputable, non-algorithmic and irreducible [Rosen, 2000]. These types of systems have been shown to be self-referential and logically tractable [Kercel et al., 2001]. We now discuss language as an endogenous process and one particular language problem, that of the mechanical translation between languages, as a problem suited for this category.

MT system development has relied on the Chomsky tradition of language processing levels. These levels are based on the notion that language, and therefore the automatic processing of it, is reducible. MT systems typically break a paragraph into sentences, sentences into phrases, phrases into syntactic parts, syntactic parts into words and words into meaning, often a logic-based representation. Each stage of processing can be accomplished by rule-based or statistical methods or a combination thereof that typically starts with an assumption of word independence. Yet, This is not a realistic view for language processing, or the MT systems derived from it. We will now show how MT systems rightly fit into the category of endogenous systems.

---

[1] Where context can be the surrounding text, the conditions in which the message is uttered and the states of mind of the speaker and hearer.

The human facility for language has been demonstrated to be an endogenous [Kercel, 1999] system by experts and even, grudgingly, by language practitioners. As Kercel et al. [2001] show:

> Chomsky admits that the use of language is not explainable or understandable by this reductionist strategy. Consequently, he fears that it may be infeasible to study significant problems of language communication. His fear would be well founded if theoretical modeling were to stay limited to the reductionist strategy. Thus, even from a reductionist starting point, Chomsky allows the possibility that new intellectual tools might be needed, and might be feasible.

If we accept that language itself is an endogenous system, then MT qualifies as endogenous. For instance, meanings can be compositional: **cheese fries** literally interpreted is either **fries with cheese on them** or **cheese is a substance that can be fried** or **fries composed of cheese**. At the same time, we have many wordings that are not compositional, defying a logical representation. The saying **He bought the farm** has a non-compositional, or idiomatic, meaning that is different from the sum of the parts. Yet, there is a reasonable and rational causality to idiomatic language usage. The phrase **thumb-rule** stems from the days where the **Rule of Thumb** determined the size of a stick with which a man could beat his wife.[2] So the entanglement of words is one reason for considering MT under the endogenous system model.

If MT can be viewed in the impredicative model, then we can apply aspects of endogenous modeling to build better MT systems[3] – such as a record of the past; prediction of the future; inferential elements; causal elements in our ontology; and anticipatory behavior. A record of the past can be gleaned from corpora as in statistical language models. To achieve the next level of capability, we require a new way of looking at computing translations because we are modeling a process that is impredicative. The question, then, becomes how to utilize the best of the traditional models? Or should we even try this? What are the pieces of this overall puzzle? Can we find partial solutions that are effective and efficient?

TYING IT ALL TOGETHER

The questions just presented aim to find ways to arrive at a model for an MT lexicon to support the MT process which reflects the endogenous paradigm. While we are still grounded in computation,[4] we believe that there are reasonable, implementable approaches to explore these questions. We look to data mining and data modeling, reasoning approaches for combining multiple evidence sources [e.g., Schum, 1994], traditional lexicon development, language learning and evaluation, and psycholinguistics (to include neurolinguistic programming). The end goal is a representation which supports the learning and combination of multiple pieces of evidence to contribute to the "meaning" of a concept for the purpose of translation between languages.

---

[2] No greater than the width of his thumb.

[3] Approximating endogenous systems until actual capabilities are available.

[4] After all, as engineers, we do have to build **something**.

Accepting MT as an endogenous problem, we look for ways in which we can design intelligent systems which perform more successful MT. We need an approach which allows us to combine results from different computational representations of language and suggest an evidential lexicon as one possible basis. An evidential lexicon starts with information from multiple sources: dictionaries, corpora, analyses of language in use, psycholinguistics, etc., and results in a lexicon which facilitates intelligent word selection for translation.

Because many biological systems are endogenous, we look to psycholinguistics for ideas in modeling the impredicative nature of MT. Psycholinguistic models suggest why we select the words and grammatical constructs we use. Language is one of our primary means of acquiring information – through talk, through reading, through words. Language is not the form of the representation, but words and the combination of these words with contextual clues are part of our knowledge structure. The notion that we remember a gist of what was said in place of the exact words means that there is something else going on. Professional translators show this often when they translate sentences according to general meaning instead of for exact words. In fact, it has been shown that there is a wide degree of variability even in translating "factual" reports [Farwell & Helmreich, 1996]. Another study [Al-Onaizan, et al., 2000] reports that translation can be a process of picking the words you know, guessing a context and applying that context to unseen words. There is reason to suspect that the translation process is different than the sum of the parts.

What we propose is a framework which supports the integration of multiple evidence points to contribute to word (or possibly phrase) meaning. In this framework, translation selection depends on evidence for/against particular meaning. The evidence points will come from: dictionaries, which represent a kind of jurisprudence about the meaning and translations of words; learned values from corpora, such as mutual information measures from information retrieval; values from ontologies, both human and automatically constructed; and neurolinguistic programming inspired preferences. Each of these sources would contribute weighted values for word-sense disambiguation and translation preference selection. Initially, a Bayesian representation would support this, although alternative representations, combinations and computations must follow from the impredicative nature of language.

A SPECIAL CASE – NAMES

Acknowledging the inherent difficulties in looking at only part of a function when the whole is greater than the sum of the parts, we look at a specific kind of entry for the lexicon structure described here. *Named Entities* are the set of proper names and numerical expressions such as times, dates, monetary expressions, or percentages. Since these carry critical content, content that is useful for summarization or topic identification, handling their translation well represents a necessary area for MT system advancement. Proper translation of named entities must ensure that they are a) not translated when proper names; b) rendered idiomatically rather than literally, observing standard naming conventions in the target language; and c) rendered in a format which is usable other processing. We describe each of these.

Translating a name correctly, i.e., **Helmut Kohl**, means not translating it, but rendering it in a form usable by English NLP software. **Kohl** is the German word for **cabbage** – a system could amuse, but not be helpful, returning **Helmut Cabbage**. The first goal of translation, then, is to ensure that proper names are not translated as common nouns, but that certain titles and parts of names are.

With organizations, a different strategy is necessary. As noted earlier, while **Escuela de Derecho de Harvard** is properly translated as **Harvard Law School**, it is literally **Harvard School of the Right**. The idiomatic translation is preferred for institution names. In this family of challenges is acronym handling. Some acronyms are translated and others not: a literal translation of the abbreviation for the former Soviet Union could be rendered as **SSSR**, whereas it is normally rendered by translating the expression into English (Union of the Soviet Socialist Republics) and then reducing the initials to **USSR**. On the other hand, the Basque separatist organization is generally referred to as **ETA**, based on the abbreviation of the not translated Basque title.

The transformation of names into something legible for the target language speaker must be handled in languages that use characters that have no English equivalent. For instance, some people will render the name for the former Soviet Union as **CCCP** (a look-alike representation of the Cyrillic characters). Rendering can involve translation of diacritics such as German umlauts or French accents, or it can involve other alphabets (Cyrillic, Arabic, Thai) or other writing systems (Chinese, Japanese). Assuming the name was not to be translated, the translation engine leaves it in Cyrillic characters instead of an English transliteration (or transcription). This means that a reader of the translation might miss **Gorbachev**. As Knight and Graehl [1998] point out, even uniform transliteration does not guarantee success in name recognition. The large number of transliterations for **Khadafy** is an example.

A SAMPLE IMPLEMENTATION

The field of translation of "named entities" lends itself well to an evidential representation. Because many translation conventions are just that, conventions, a system capable of learning the features and functions that determine the conventions represents a significant gain over other MT approaches. We present the beginnings of our testing of this theory. We started with 100 documents [see Reeder et al, 2001, for a more complete description] which were originally written in Spanish and then translated by two professional translators into English. This serves as our test corpus for exploring the named entity problem.

We took one reference translation from the corpus and manually annotated it for named entities. We then wrote matching software which would search for matches between the two parallel articles (REFERENCE and EXPERT). Given the list of names extracted from an article, the software found the names in the corresponding translation. In this way, we get a picture of the kinds of features which contribute to the translation of names and the weights of each.

Our testing has yielded some interesting evidence points. First, dictionaries do not typically contain proper nouns, instead giving common noun definitions. Therefore, we sought another means of establishing a baseline or jurisprudence that dictionaries can give. To do this, we used multiple human translations. The baseline for matching between human translations of named entities was only about 80% for exact match. This means that our next evidence points come from a series of relaxations on the mismatch reasons. Other features that must be measured to determine if a name phrase is a proper translation of another include: diacritics (accent marks); capitalization; and numeric values. While the inclusion of these measures improves matching to 90%, this leaves a number of features for which we are still establishing weights, including: morphology (such as **Peruvian** versus **of Peru**); stop-words (**of** versus **for**) and synonyms.

We envision implementing evidence first as a Bayesian network, although since we have just completed the feature identification for evidence points, we may adjust this. Additionally, we look to continuing endogenous system work for a more appropriate reasoning framework.

CONCLUSIONS

Some domain specific terms are specialized names – such as chemical names. But what of the chemical precursors? Glass vessel is composed of common nouns and only has a technical flavor to it when applied in the chemical or biological domains. In addition, we are confronted by the widespread use of loan-words and adapted loan-words in foreign languages – particularly due to the fact that the recognized language of science is English. As this is preliminary work, there are many important challenges ahead. Some of the challenges we foresee are in marrying information from multiple data sources, finding the supporting technologies to mine data sources; handling less-commonly taught languages where a paucity of information exists.

REFERENCES

Al-Onaizan, Y., Germann, U., Hermjakob, U., Knight, K., Koehn, P., Marcu, D., Yamada, K. 2000. Translating with Scarce Resources. In *Proceedings American Association for Artificial Intelligence Conference, AAAI-2000*.

Chomsky, N., 1957, *Syntactic Structures*, Mouton Publishers, The Hague.

Helmreich, S. & Farwell, D. 1996. Translation Differences and Pragmatics-Based MT. In *Expanding MT Horizons: Proceedings of the Second Conference of the Association for Machine Translation in the Americas (AMTA-96)*. Montreal, Canada.

Kercel, S., 1999, Embedded Intelligence: Bizarre Systems for the Steel Industry, Presentation to AISE – Sensors Meeting.

Kercel, S., Brown-VanHoozer, S. A., VanHoozer, W.R. (in press) The Entangled Future of Foreign Language Learning. In The Future of Foreign Language Education in the United States, ed. Terry A. Osborn. Contemporary Language Education Series. Westport, CT: Bergin & Garvey.

Knight, K. & J. Graehl. 1998. Machine Transliteration. *Computational Linguistics, 24(4)*, 598–612.

Reeder, F. 2000. ISA or Not ISA: The Interlingual Dilemma for Machine Translation. In *Proceedings of ANNIE-2000, Bizarre Systems Track*. ASME Press.

Reeder, F., Miller, K., Doyon, J., White, J. 2001. The Naming of Things and the Confusion of Tongues: An MT Metric. *Proceedings of Workshop on MT Evaluation, MT-Summit 2001*.

Rosen, R., 2000., *Essays on Life Itself*. Columbia University Press, New York.

Schum, D. A., 1994, *Evidential Foundations of Probabilistic Reasoning.*, John Wiley & Sons, Inc.

Weaver, W., 1947, Letter to Norbert Wiener, March 4, 1947, (Rockefeller Foundation Archives).

---

[1] The views expressed in this paper are those of the author and do not reflect the policy of the MITRE Corporation.

EVALUATION OF ENDOGENOUS SYSTEMS

H. JOHN CAULFIELD
Distinguished Research Scientist
Fisk University
1000 17th Avenue, N.
Nashville, TN 37208
hjc@dubois.fisk.edu

FLORENCE M. REEDER[1]
George Mason University /
The MITRE Corporation
1820 Dolley Madison Blvd.
McLean, VA 22102
freeder@mitre.org

ABSTRACT
In system development we are faced with the necessity of evaluation. Evaluation measures our success relative to other: a) theories of development or domains; b) implementations of similar theoretical principles; or c) increments of a given system. For successful software engineering evaluation, progress is measured against a model, a task frequently accomplished through requirements analysis. This model is lacking in natural language processing (NLP) systems (not to be confused with neurolinguistic programming). NLP systems defy evaluation in part because they model an endogenous process – where the whole process is irreducible. Therefore, while specific feature-based evaluations appear reasonable, they fail to capture an overall measure of success. In this paper, we look at the part/whole aspects of evaluation in more detail with regard to one language system type – machine translation.

INTRODUCTION

In system development, of any type of system, we are faced with the necessary evil of evaluation. Formal evaluation measures the success we are having relative to: a) other theories of development or domain modeling; b) other implementations of the same theoretical principles; c) other systems for the purpose of purchase or funding; or d) previous increments of a given system. For successful evaluation in software engineering, one needs a model against which progress can be measured, a task frequently accomplished through requirements analysis. For many domains, this is a straight-forward process: a correct answer exists, such as, "Pushing F10 results in program termination."

Many types of NLP systems, however, have defied rigorous evaluation because of this lack of defining criteria or requirements. Part of the difficulty lies in the fact that NLP systems are modeling an endogenous process – where the whole of the process is irreducible, context-dependent and lacking a unique right answer. Therefore, while specific feature-based evaluations appear reasonable, they fail to capture an overall measure of success. In this paper, we look at the part/whole aspects of evaluation in more detail with regard to one language system type – machine translation (MT).

This paper starts with a description of the MT process as traditionally developed. It then describes the resulting evaluation strategies that follow from these views of MT. The features of MT which categorize it as an endogenous are presented as a precursor to showing a new view of the original MT vision. Finally, the questions of evaluation are explored within this new view.

MACHINE TRANSLATION

Machine Translation (MT) systems attempt to replicate the very human process of translating between human languages. The idea for attempting this comes from Weaver (1947) when he describes translation as a kind of noisy channel decryption process: a French document is really an English document which has been encrypted in the strange symbols of French. Unfortunately, language learners and language translators know that the process is much more than the simple substitution of symbols.

Traditional MT systems have been developed following the Chomsky ideal (1957) of processing language through successive steps, each providing another level of information which isolates words from meaning. These levels are pragmatic (mechanics of language in use), discourse (rules of interaction and coherence), semantic (attribution of meaning to words), syntactic (assignment of structural interaction of words), morphological (the features of words), and phonemic (realization of words in sound). This understanding of language led to the development of MT systems based on these layers of processing, often characterized as a pyramid where complexity increases through ascending layers of processing.

An MT system performs a level of analysis of increasing detail as the type of system advances in the pyramid. That is, a lexical transfer system will perform only morphological processing. A syntactic transfer system will do morphological analysis and syntactic analysis on the source text before generating text in the target language. The interlingua level represents a "pure" translation process, much like human translators perform, where analysis yields a language-independent representation from which generation can be performed. Systems in this paradigm typically have modules and information sources as shown in Figure 1.[1]

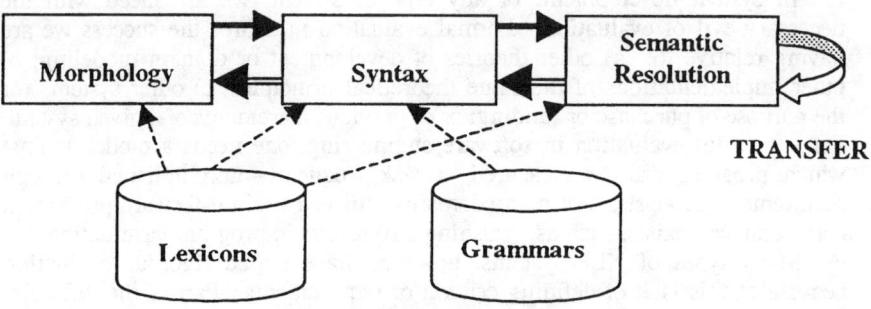

Figure 1: A Typical MT Processing Flow

MACHINE TRANSLATION EVALUATION (MTE)

In software engineering practice, system evaluation reflects how well the system performs the requirements for which it was designed. Derived in advance of system design, these requirements cover system aspects such as

---

[1] Note that most current systems fall far short of discourse and pragmatic processing.

coding language, mean-time-between-failure, number of records processed in X time, the procedure for updating records, etc. However, MT systems are frequently developed in the absence of requirements, and reflect a notion that good MT, like pornography, "can't be defined but we'll know it when we see it." The many MTE metrics reflect these broad and ill-defined criteria.

While basic software engineering metrics (disk usage, execution time, throughput) apply to MTE, MT systems have a wide range of additional parameters for evaluation, each having different importance to different users who have a large set of possible uses. Other NLP tasks like speech transcription have a right answer, or "gold standard". Not so for MT as there is no single right answer for a translation even when humans do it. What constitutes an acceptable level of translation varies. For users in legal domains, even minor failures are unacceptable. For other domains, such as intelligence analysis, a 40% solution is good enough.

Accordingly, MTE has had a long and often painful history. From the early days (ALPAC, 1966), excessive optimism or pessimism and misconceptions – about language, MT usage and system requirements – have been predominant. The various notions of MTE result in too broad a scope for reasonable effort: everything from interface, to scalability, to faithfulness of translation, to mean-time-between-failures are fair game for MT system evaluation. Claims of 90% accuracy abound without an indication of what the system is accurate about.

The set of MTE approaches reflects the varying uses and requirements for MT, in addition to the multiple stakeholders in evaluation – developers, funders, users (White, 1998). Evaluation strategies have included: a) evaluating MT systems as software (e.g., EAGLES, 1996); b) black-box evaluations on fidelity and intelligibility of output only (e.g., Van Slype, 1979); c) measuring accuracy of input/output pairs; d) glass-box evaluations of components; e) measuring speed, cost,quality in process and others. These have relied, primarily, on human judgments as to the "correctness" of a translation on two basic metrics: fidelity (amount of the message conveyed) and intelligibility (the fluency of the message as translated). These efforts require a large number of evaluators, looking at, and holistically rating, sentences on a 5, 7 or 9 point scale (i.e., DARPA, 1994). The difficulty is that aside from human factors issues, the results give little information to either users or developers of MT.

In reaction to this, recent efforts at evaluating system abilities to translate have focused on pieces of the language puzzle. Specific grammatical phenomena are measured such as the ability of a system to translate prepositions (Miller, 2000) or named-entities (Reeder et al., 2001). These approaches seek to measure the specific contribution of MT to the preservation of very specific pieces of information. These strategies have great appeal – they are objective, replicable, informative and indicate the ability to meet a process "bottom line." However, recent exercises in MTE for named entities (Reeder et al, 2001) show a definite gap between the specific piece indicators and more holistic scoring. How do we explain this?

COMMUNICATING EFFECTIVELY USING MT

The answer lies in re-thinking the problem in terms of endogenous systems. An information theoretic view of NLP is not new, but its utility encourages us to revisit MTE, looking for better evaluation strategies. To this end, we define effective communication and describe ways to test if it has been accomplished. As a start, we examine communication between two people using one mode, voice, in one language in conversation. Then, we will be ready to discuss MT.

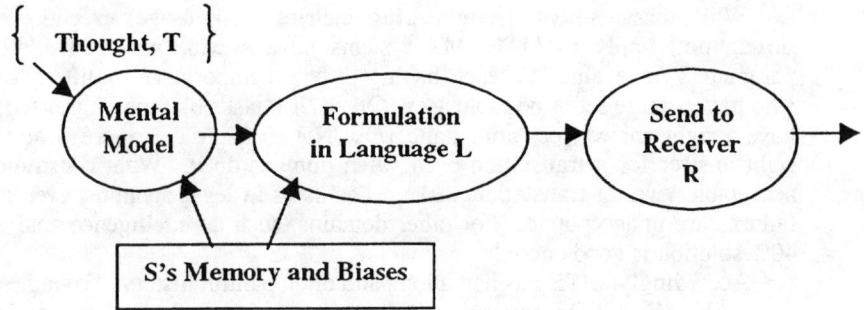

Figure 2: Basic Model of Utterance Generation

We assume something (a scene, situation, desire, etc.) to be communicated – in NLP terms, a discourse purpose. In some cases, the target something (**T**) will be external to the sender, **S**. In others, it exists only as a mental model within S. If external, it is sensed by S, which is to say that S forms a mental model of it. S's mental model of what others might agree is the same thing may differ from everyone else's because he brings different memories, assumptions and biases, to the model-forming task than others. S represents the mental model with words organized into one or more coherent sentences. Chomsky (1957) described this process (Figure 2) as transforming deep-structure into surface-structure. The resulting message is sent as agreed to the receiver, **R**.

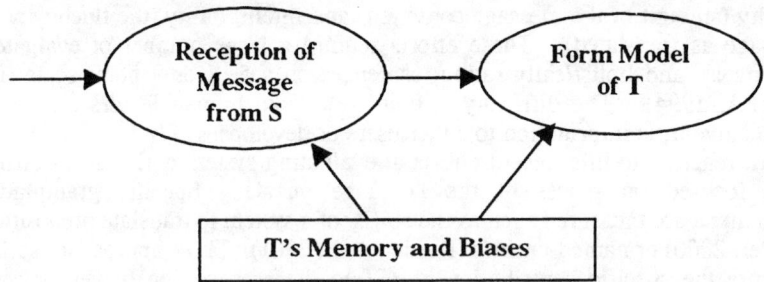

Figure 3: R's reception and processing of the message

The receiver forms his own mental image of something that may or may not be **T**, bringing his own memory and biases to the task. This is the analysis phase or conversion from surface-structure to deep-structure, Figure 3. And, of course, the words are frequently ambiguous, often inadequate to convey the full model content or a literal mismatch from the intended message.[2] Now, we have

---

[2] If a man and woman are sitting in a room with an open window, her utterance, "Gee, there's a draft" is considered a command, not a statement of fact.

a problem. Both S and R have mental models of something (presumably T), but how can we know that they are the same or even consistent? Translation between languages adds another layer of complexity. Instead of the message T being consistent, it is transformed, sometimes at the word level, between two languages, each with its own biases, strengths and weaknesses (Figure 4). The sought-after congruence of mental models (identity is hopeless in most cases) is what characterizes what we call "effective communication" and represents what evaluation of MT systems must measure.

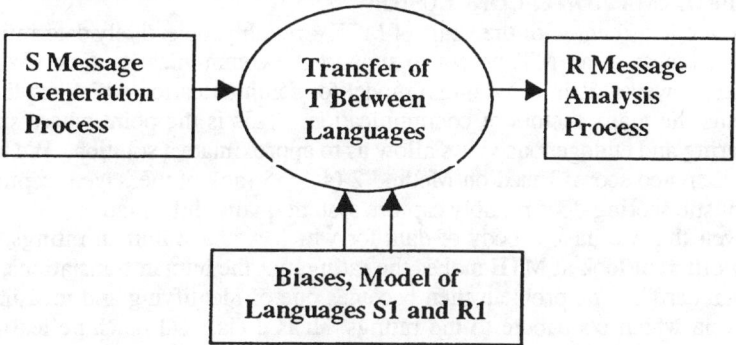

Figure 4: Translation Process with Model

EVALUATING EFFECTIVE COMMUNICATION TOOLS

The problem is clear. S's model and R's model of T cannot be compared, because they exist in two different minds and because of their endogenous nature. If T is a command, and if R's action seems an appropriate response to that command, then we may have had effective communication. If the correct response does not occur, we cannot infer ineffective communication. A two-year old may completely get the message, "Don't eat that" even if he eats the bug. Likewise, R need not have fully "understood" the message to have correct response, due to the redundant nature of language. From this we can see that we need a surer, more general test. Having established different variables, we go back to our desire to effectively measure MT success. We show two methods deriving from communications theory and show how they relate to current MTE.

Method 1: Closing the Loop. We ask R to send his understanding of T to S in words different from the words used by S. S will form a model of T on the basis of R's message. Now S has two models of T in his own mind and can judge whether they are sufficiently congruent. A corresponding MTE strategy is the "round-trip translation" test.

Method 2: Redundant Messages. We ask S to send two messages about T using different words. If R (who now has two models of T) judges them to be congruent, S can feel assured that he has communicated the message effectively. This corresponds to human judgments and we are back to the beginning in MTE.

As stated earlier, recent MTE focuses on pieces of messages and usually applies Method 2. Name translation evaluation uses method 2 specifically for one part of the message – the peoples, places and things represented in the message. A system translation is compared with two human translations. Unfortunately, no strong correlation can be drawn between this metric and the "whole" rating given in previous evaluations with the same data. That is, systems may or may not have good scores on the named entity task which do not correlate with their quality scores on a sentence by sentence basis (White, et al, 2001). This implies a part-whole disjunction which must be accounted for.

AN APPROXIMATION & CONCLUSION

We are left, then, with the goals of MTE – reliably, objectively determining the effectiveness of an MT system to support the communicative task. Where we differ, however, is in providing a model for communication which explicitly represents the many factors of communication. This is the point where smart engineering and endogenous views allow us to approximate a solution. We have human assigned scores based on Method 2 (a 1→5 rank of message reception). This holistic scoring does reliably capture system quality information.

Given that we have a body of data for which we have human ratings, our slightly different look at MTE makes the rating, not the human translations, the "gold standard". The problem then becomes one of identifying and measuring the criteria which contribute to the ratings. It is a classical machine learning problem of categorizing items based on objective and reasonably measured criteria. In this way, MTE becomes a more replicable, automatable task while at the same time capturing the human intuition of quality output. While we will not find all of the criteria or a complete correspondence until we develop endogenous evaluation abilities, we can at least approximate current measures.

REFERENCES

Chomsky, N., 1957, *Syntactic Structures*, Mouton Publishers, The Hague.

EAGLES. 1996. The EAGLES MT Evaluation Working Group. 1996. EAGLES Evaluation of Natural Language Processing Systems Final Report. EAGLES Document EAG-EWG-PR.2, ISBN 87-90708-00-8. Center for Sprogteknologi, Copenhagen.

Miller, K. 2000. The Lexical Choice of Prepositions in Machine Translation. Unpublished Ph.D. thesis, Georgetown University.

Pierce, J., Chair. 1966 *Language and Machines: Computers in Translation and Linguistics*. Report by the Automatic Language Processing Advisory Committee (ALPAC). Publication 1416. National Academy of Sciences National Research Council.

Reeder, F., Miller, K., Doyon, J., & White, J., 2001, The Naming of Things and the Confusion of Tongues: An MT Metric. *Proceedings of the Machine Translation Evaluation Workshop, MT-Summit 2001*. Campostela de Santiago, Spain.

Van Slype, G. 1979. *Critical Methods for Evaluating the Quality of Machine Translation*. EC Directorate: General Scientific and Technical Information and Information Management. Report BR-19142. Bureau Marcel van Dijk.

Weaver, W., 1947, Letter to Norbert Wiener, March 4, 1947, (Rockefeller Foundation Archives).

White, J., et al. 1992. ARPA Workshops on Machine Translation. Series of 4 workshops on comparative evaluation. PRC Inc. McLean, VA.

White, J. 1998. Evaluation of Machine Translation. A Tutorial. *AMTA-98*.

---

[1] The views expressed in this paper are those of the author and do not reflect the policy of the MITRE Corporation.

HEBBIAN BRAIN CELL-ASSEMBLIES: NONSYNAPTIC NEUROTRANSMISSION, SPACE AND ENERGY CONSIDERATIONS

GAETANO L. AIELLO
Dipartimento di Scienze Fisiche ed Astronomiche, Universita' di Palermo, Italy

PAUL BACH-Y-RITA
Departments of Rehabilitation Medicine, and Biomedical Engineering, University of Wisconsin-Madison, U.S.A and Facultad de Medicina, Universidad Autónoma del Estado de Morelos, Cuernavaca, Mexico

ABSTRACT
The high-degree of hard-wired connectivity Hebb proposed to implement brain functions such as consciousness and subjective sensory experience may conflict with brain volume limitations and available metabolic resources. In this case, a less wasteful form of connectivity should be explored. In this paper we calculate space and energy constraints in Hebbian putative cell-assemblies, and consider the role of nonsynaptic diffusion neurotransmission (NDN) in complying with those brain volume and energy limitations.

INTRODUCTION

Hebb's concept of brain "cell-assemblies" has had a significant impact on later scientists studying brain functions, as a possible foundation for a theory of the mental values of behavior, such as thought, expectancy, interest and attention. In his landmark book, *The Organization of Behavior*, Hebb (1949, pg. 74) portrayed a cell-assembly as a neural structure consisting of enormous numbers of individual cells simultaneously aroused by extensive activity, with every point possibly connected to every other point. He drew the analogy to a closed solid cage-work and imagined an infinitely complex structure. Hebb stated that his cell-assembly theory is a form of connectionism, in which "The connections serve to establish autonomous central activities" (Hebb, 1949, pg. xix), but he did not consider cell-assemblies to be rigid structures. Rather, he considered them to be dynamic in their organization and suggested that no single cell or pathway is essential to any habit or perception. He considered that an individual cell or transmission unit may enter into more than one assembly, at different times; which it will form part of, at any moment, depends on timing in other cells. To enter into any assembly requires that its frequency accord with the time properties of the active systems. This flexibility would allow what Caminiti (1996) has called "different processing strategies ... for different operations".

Regarding perceptual integration, Hebb (1949) noted that it is "... possible to conceive of 'alternating reverberation' which might frequently last for periods of time as great as half a second or a second... the best estimate one can make of the duration of a single 'conscious content.' ". For learning to transpire, Hebb considered that synaptic changes had to occur. He considered that the fundamental meaning of the assumption of growth at the synapse is the effect this would have on the timing of action by the efferent cell. He believed that the increased area of contact means that firing by the efferent cell is more likely to follow the lead of the afferent cell, and that the assembly depends completely on a very delicate timing; Hebb stated that neural integration is fundamentally a question of timing, which has its effect in the functioning of the cell-assembly and the interrelation of assemblies: diffuse, anatomically irregular structures that function briefly as closed systems, and do so only by virtue of the time relations in the firing of constituent cells. However, he considered that no one had made a serious attempt to elaborate ideas of a central neural mechanism to account for the delay between stimulus and response. He was particularly concerned with the one half to one second time delay, which is the observed duration of a single content of perception.

Hebb's views are in tune with the current attempts to develop models of mental states as bottom-up features emerging from immensely complicated interactions. Mass- sustained brain functions, as "moods", might very well be the result of the holistic activity of an entire neuronal module. Whether such brain functions would result from massive computations or not is an open issue in the debate on artificial intelligence.

SPACE CONSTRAINTS

Hebb imagined cell-assemblies as modules comprising enormous numbers of neurons, each possibly connected directly to every other. Hebb stated that the connections within a cell-assembly are by means of nerve fibers and synapses. In the extreme case of a fully-connected module of N cells, there would be a total of N(N–1) links. With $<L>$ the average length of a link, a total length $L \approx N^2 <L>$ would result. In a more realistic situation, a neuron can actually exhibit from a few thousands to a hundred thousands synapses (Smith, 1996), depending on the neuron. Taking an average figure of 10^4 synapses per neuron as a limit to nerve branching, then the total length of the connections is $L \leq N \cdot 10^4$ $<L>$. The average length $<L>$ can be estimated on assuming a regular arrangement of cells in a module, e.g., a crystal-like structure (Bach-y-Rita and Aiello, 1996). A face-centered cubic (*fcc*) lattice was found optimal as it allows a percentage of extracellular space (the so called *volume fraction*) of 26%, close to the estimated 20% in the human brain ((Nicholson and Phillips, 1981). For a module of $N=10^5$ cells in an *fcc* lattice $<L>$ is about 1 mm. It follows that a fully-connected module of this size would have a network of connections 10,000 *kilometers* long. If limited branching is assumed, then the total length would be up to "only" 1,000 kilometers. Connectivity on this scale is beyond comprehension, but it appears fundamental to the brain's ability to generate complex neural functions like "cognition". Although a limitation to cell branching is not *per se* incompatible with a holistic activity of a cell-assembly, it would certainly affect the *timing* of the neural processes, e.g., neural integration, which is fundamentally a question of timing. It is not known, at the present, whether a minimum of connectivity is requested to generate cognition, or other mass-sustained brain functions. Hebb's view, however, clearly suggests such a connectivity should be "high", *possibly* up to the extreme case of a fully-connected cell module.

Networks of such an impressive length would easily conflict with volume limitations in the brain. The volume of a module of N cells in an *fcc* lattice is $a^3N/4$, with a the size of the unit-cell, approximately equal to 73 µm (Bach-y-Rita and Aiello, 1996). Thus, the volume of a module of $N=10^5$ cell is about 10^{-2} cm^3. Assuming an average fiber as thin as of 0.5 µm in diameter, then a trivial calculation shows that 12.7 Km of such a fiber would amount to the entire volume of the module. Such total length is compatible with a connection ratio of 1:127 (i.e., one cell branches to 127 other cells). For a more profuse branching the volume of the connections would *exceed* that of the container. The ratio 1:127, however, seems somewhat "inadequate" to account for extremely complex neural functions, like "consciousness", or "cognition", and similar. A degree of connectivity higher than this is expected. Such connectivity, however, must be sought beyond the classical connectionist scheme.

ENERGY CONSTRAINTS

In Hebb's view, waves of depolarization spread along an impressive network of links, carrying information all over the assembly. Hebb did not consider how much energy would be necessary to drive such a network. The estimate depends on the nature of the links. Unmyelinated links require depolarization of the entire link, while myelinated nerves require only depolarization of the short unmyelinated segments at the nodes of Ranvier. The cost of an action potential (AP) has been calculated (Aiello and Bach-y-Rita, 2000a) to be about 10^{11}-10^{12} ATP per cm^2 of depolarized membrane, with a minimum cost of 10^6 ATP at a node of Ranvier. In the most conservative scenario of fine *myelinated* links, with Ranvier nodes spaced δ apart, the axonal cost of circulating a *single* AP through a network of total length L would exceed 10^6 L/δ ATP (10^{-13} L/δ Joules). With δ=1 mm (the average distance between nodes), a network of 10^5 cells connected in a ratio 1:127, would require at least $1.27 \cdot 10^{-6}$ J. The cost *per liter* (assuming that all 10^{10} neurons in a liter are arranged in 10^5 modules, each comprising 10^5 cells) would be 0.127 J/liter per action potential. Given the connectivity ratio of 1:127, this figure is quite conservative. For example, a spike rate of the order of 100 Hz would be required in order to achieve the measured 40 W/l brain metabolic power at rest, determined from glucose consumption rates in rats (Ghajar, et al., 1982). On the other hand, full-connectivity would yield an energetic figure of 100W/liter/AP, with an impossible metabolic demand of 1 KW/liter at a spike rate as low as 10 Hertz.

Thus, the hard-wired scheme may easily conflict with brain metabolic resources, unless it is limited to modules of a few thousands cells, with limited branching. Brain functions requiring the participation of large modules and profuse branching would be incompatible with brain metabolic resources. In this case, additional mechanisms of a "wireless" type may offer a less expensive alternative to synaptic communication.

NONSYNAPTIC DIFFUSION NEUROTRANSMISSION

The extracellular space (ECS) in the brain plays a role in many functions. In brain cells assemblies, the distance between neurons can be reduced by 50% with neuronal activity that causes cell swelling (Dietzel, et al, 1980; Dietzel and Heinemann, 1986). A decreased volume fraction has a direct effect on the ionic concentrations, and may affect cell excitability and metabolism (Aiello and Bach-y-Rita, 1997). Neurotransmitters that may have accidentally escaped the synaptic cleft, or survived enzyme degradation, or have originated *extrasynaptically* (Beaudet and Descarries, 1978; c.f., Bach-y-Rita, 1995) are also found diffusing freely in the ECS. They carry information by means of passive diffusion. We refer to such a "wireless" mechanism as non-synaptic diffusion neurotransmission (NDN) (Bach-y-Rita, 1993; 1995). NDN has emerged as a mechanism of information transmission that may play multiple roles in the brain, including normal and abnormal activity, brain plasticity and drug action. Evidence for high percentages of NDN for certain purposes, including mass-sustained functions such as mood, vigilance and sustained pain, have been discussed elsewhere (Bach-y-Rita, 1991; 1995; in press).

We have studied a model of a diffusion-only artificial neural net (Aiello and Bach-y-Rita, 2000b, Aiello, 2001), which would comply with space and energy constraints in brain, yet one that allows the cells to directly communicate with each other, although via diffusive paths. More realistically, a hybrid neuronal network, exhibiting varying combinations of hard-wired ("synaptic") and wireless (NDN) connectivities, would fit into the available volume in the brain, with a significant cut in the energetic costs resulting from removing a significant number of "hard" links and *replacing* them with "soft", diffusive pathways. Removing a link does not imply cutting intercellular traffic, nor eliminating the cost of communicating along that specific pathway, rather it changes the way the cells communicate, from "cable" to "broadcast". Directionality can still be maintained (at least in principle) by assuming that neurotransmitters emitted by a cell diffuse to a specific cell, and bind to specific sites on that cell, according to a stereoselective targeted reaction. Extrasynaptic exocytotic release and receptor sites have been demonstrated (c.f., Bach-y-Rita, 1995). There is experimental evidence from Routtenberg's laboratory (c.f., Routtenberg,1991), confirmed by Bjelke, et al. (Bjelke, et al., 1995), for directional extrasynaptic diffusion.

The energetic cost of handling information through diffusive pathways (the NDN cost) is basically that of restoring the presynaptic membrane's free energy after neurotransmitter release. The synaptic vesicle cycle (Südhof, 1999) forms the basis for neurotransmitter release, therefore all the events in this cycle should be checked for energy cost. Evaluation of all these contributions to the cerebral metabolism is quite difficult. However uncertain and controversial the available data may be, they seem to support the conclusion of Albers & Siegel (1999) that the energy expenditure for biosynthetic processes, like vesicle recycling and turnover of neurotransmitters, is relatively small, probably less than 10% of the total ATP utilization. Furthermore, the process of restoring the free energy of the presynaptic terminal is carried out by *secondary* transport systems, that do not require a local source of energy (as occurs in ATP-coupled *primary* transport) but are fuelled *indirectly* by the activity of the Na,K-ATPase.

It is estimated that 25 to 40% of total brain energy is related to Na,K-ATPase activity. Most of this energy appears to be utilized for recharging the Nernst batteries (Laughlin et al., 1998) after membrane depolarization, which approximately equals the *axonal* cost (Aiello and Bach-y-Rita, 2000a). Thus, the NDN cost is expected to be small (although perhaps significant) compared to the axonal cost. Therefore, removing a few axons and replacing them with diffusion pathways would have an energy saving effect.

CONCLUSIONS

The Hebbian concept of brain cell-assemblies serves at least two putative purposes. First, it provides a form of distributed coding, where processing of neural information is somewhat independent on any single cell or specific pathway. Second, reverberations would allow for synaptic changes and eventually for learning. These issues can be summarized in relation to synaptic connectionism versus NDN:

1. Hebb considered that cell assemblies consist of enormous numbers of individual cells with every point possibly connected to every other point. Our estimates reveal that this is not physically possible.

2. The dependence on timing that Hebb considered to be necessary for optimal function of the cell-assemblies was difficult for him to explain in his synaptic model, but delays are a characteristic of NDN models of cell-assemblies (Bach-y-Rita, 1964).

3. The dynamic organization of cell-assemblies proposed by Hebb fits well into NDN considerations, with the extent of the assembly related to co-ordinated neurotransmitter release. Sykova (1997) has noted that changes in perfusion parameters in the ECS persist for many minutes or hours after stimulation has ceased, which may relate the establishment of dynamic cell-assemblies that outlast their initial formation, and may relate to learning and other functions considered by us and by Hebb. He noted that the persistence of some behaviors after damage to the brain is consistent with his theory of cell-assemblies: "If brain injury occurs in a limited region of the brain, it would presumably remove a small number of transmission paths from each of a very large number of

assemblies. These assemblies might still function...though...less reliably". Thus, Hebb offered a basis for brain plasticity, in which NDN may play a role.

4. Hebb proposed growth at the synapse as the basis for the timing changes in a cell assembly, and implied it is the mechanism for learning. In our calculations, the space and energy required by a highly connected cell-assembly might be so much greater than an NDN cell-assembly as to be out of the realm of possibility, while NDN would require less energy and space. Schuman and Madison (1991) consider that the demonstration that long-term potentiation (LTP) induction in a single CA1 pyramidal neuron can potentiate synaptic transmission at neighboring cell synapses implies the existence of a diffusible signal generated in paired neurons during LTP induction. Thus, the formation of synaptic changes previously thought to be restricted to synapses onto a single cell can also result in synaptic changes at nearby synapses. They stated that the distributed potential provides a mechanism for the cooperative strengthening of proximal synapses and may underlie a variety of plastic processes in the nervous system (including brain plasticity and learning), which led to an News Article in Science (Vol. 263:466, 1994) entitled "Learning by Diffusion...". Distributed LTP would be a particularly appropriate mechanism for co-ordinated activity in highly connected cell-assemblies in which NDN may play a role.

5. Mitcheson (1992) suggested that connectivity appears to be minimized in the brain, and Laughlin, et. al (1998) noted that neurons, neural codes and neural circuits will have evolved to reduce metabolic demands. Falk (in Gibbons, 1998) noted that we have to attend to the energetics or we're not going to get selection for a bigger brain; thus, NDN may play a role in evolution, providing a mechanism to allow the underlying physical constraints to be overcome to build an oversize brain (Gibbons, 1998). Concordant with those views, and with a proposed law of conservation of space and energy in the brain (Bach-y-Rita, 1996), functions that are highly NDN-mediated may be a basis for the reduced per kilogram energy requirements of human brains in comparison to the brains of animals of comparable size.

6. NDN has been proposed for mass-sustained functions such as mood, wakefulness, and sustained pain (Bach-y-Rita, 1991), which are comparable to those mentioned by Hebb: thought, expectancy, interest and attention. Emulation of these functions has been recently attempted (Aiello, 2001) using a thermodynamic approach based on ligand-receptor interaction, where ligands (neurotransmitters) diffuse in the ECS according to NDN protocols. The output of the diffusive net seems consistent with a chaotic dynamics, with periodic orbits and other objects of the so called "non-wandering set" (Sparrow, 1982) possible. Mass-sustained neural functions are temptatively associated with such objects.

REFERENCES

Aiello, G. L., and Bach-y-Rita, P. (1997), "Brain cell microenvironment effects on neuron and excitability nd basal metabolism," *NeuroReport*, Vol. 8, pp. 1165-1168.

Aiello, G. L., and Bach-y-Rita, P. (2000a), "The cost of an action potential," *J Neurosci Methods*, Vol. 103, pp. 145-149.

Aiello, G. L., and Bach-y-Rita, P. (2000b), "Nonsynaptically connected neural nets," Proceedings,

European Symposium on Artificial Neural Networks (ESANN 2000), Bruges, pp. 425-428.
Aiello, G. L. (2001), "Diffusive neural network," *Neuroconputing* (In Press).
Albers, R. W., and Siegel, G. J. (1999), "Membrane Transport," in *Basic Neurochenistry, Molecular, Cellular and Medical Aspects*, G. J. Siegel, B. W. Agranov, R. W. Albers, S. K. Fisher, M. D. Uhler, Ed.s, Lippincott Williams & Wilkins, Philadelphia, 6th Edition, pp. 95-118.
Bach-y-Rita, P. (1964), "Convergence and long latency unit responses in the reticular formation of the cat," *Exp. Neurol.*, Vol. 9, pp. 327-344.
Bach-y-Rita, P. (1991), "Thoughts on the role of volume transmission in normal and abnormal mass sustained functions," in K. Fuxe & L. F. Agnati (Eds.), *Volume Transmission in the Brain*, Raven Press Ltd, New York, pp. 489-496.
Bach-y-Rita, P. (1993), "Nonsynaptic diffusion neurotransmission (NDN) in the brain", *Neurochem. Int.*, Vol. 23, pp. 297-318.
Bach-y-Rita, P. (1995), *Nonsynaptic Diffusion Neurotransmission and Late Brain Reorganization*, Demos-Vermande, New York.
Bach-y-Rita, P. (1996), "Conservation of Space and Energy in the Brain," *Restor. Neurol. Neurosci.*, Vol. 10, pp. 1-3.
Bach-y-Rita, P. (2001). Nonsynaptic diffusion neurotransmission in the brain: Functional considerations. Neurochem. Res., 26, 871-873.
Bach-y-Rita, P., and Aiello, G. L. (1996), "Nerve length and volume in synaptic versus difussion neurotransmission: a model", *NeuroReport*, Vol. 7, pp. 1502-1504.
Beaudet, A., and Descarries, L. (1978), "The monoamine innervation of rat cerebral cortex: synaptic and nonsynaptic axon terminals," *Neurosci.*, Vol. 3, pp. 851-860.
Bjelke, B., England, R., Nicholson, C., Rice, M. E., Lindberg, J., Zoli, M., Agnati, L. F., & Fuxe, K. (1995), "Long distance pathways of diffusion for dextran along fibre bundles in brain," *NeuroReport*, Vol. 6, pp.1005-1009.
Caminiti, R. (1996), Introduction. In R. Caminiti, K.-P. Hoffmann, F. Taquanti, & J. Altman (Eds.), *Vision and Movement*, Strasbourg: Human Frontier Science Program, pp. 11-14.
Dietzel, M. A., and Heinemann, I. (1986), "Dynamic variations of the brain cell microenvironment in relation to neuronal hyperactivity," *Annals N.Y. Acad. Sci.*, Vol. 481, pp. 72-86.
Dietzel, M. A., Heinemann, I., Hofmeier, U., and Lux, H. D. (1980), "Transient changes in the size of the extracellular space in the sensorimotor cortex of cats in relation to stimulus-induced changes in potassium concentration," *Exp. Brain Res.*, Vol. 40, pp. 432-439.
Ghajar, J. B. G., Plum, F., and Duffy, T. E. (1982), "Cerebral oxidative metabolism and blood flow during acute hypoglycemia and recovery in unanesthetized rats," *J.Neurochem.*, Vol. 38, n.2, pp. 397-409.
Gibbons, A. (1998), "Solving the brain's energy crisis," *Science*, Vol. 280, pp. 1345-1347.
Hebb, D. O. (1949), *The Organization of Behavior*, John Wiley & Sons, New York.
Laughlin, S. B., Stevenick, R. R. d. R. v., and Anderson, J. C. (1998), "The metabolic cost of neural information," *Nature Neurosci*, Vol.1, pp. 36-41.
Mitchison, G. (1992), "Axonal trees and cortical architecture," *Trends in Neurosciences*, Vol. 15, pp. 122-126.
Routtenberg, A. (1991), "Action at a distance: the extracellular spread of chemicals in the nervous system" in *Volume Transmission in the Brain*, K. Fuxe & L. F. Agnati (Eds.), Raven Press, New York, pp. 295-298.
Schuman, E. M., and Madison, D. V. (1991), "A requirement for the intercellular messenger nitric oxide in long-term potentiation," *Science*, Vol. 254, pp. 1503-1506.
Smith, C.U.M. (1996), *Elements of Molecular Biology*, J. Wiley & Sons, 2nd Edition, pp. 477-478.
Sparrow, C. (1982), *The Lorenz Equations: Bifurcations, Chaos, and Strange Attractors*, Springer-Verlag, Applied Mathematical Sciences, Vol. 41, pp. 1-12.
Südhof, T. C. (1999), "Intracellular Trafficking," in *Basic Neurochenistry, Molecular, Cellular and Medical Aspects*, G. J. Siegel, B. W. Agranov, R. W. Albers, S. K. Fisher, M. D. Uhler, Ed.s., Lippincott Williams & Wilkins, Sixth Edition, pp. 175-187.
Sykova, E. (1997), "The extracellular space in the CNS: its regulation, volume and geometry in normal and pathological neuronal function", *The Neuroscientist*, Vol. 3, pp. 28-41.

CODING LEARNING STRATEGIES©

S. ALENKA BROWN-VANHOOZER, Ph.D.
and
Dr. W. R. VanHoozer
Human Dynamics, L.L.C™
% Beard, St. Clair, Gaffney, and McNamara
2105 Coronado Street
Idaho Falls, Idaho

ABSTRACT
Since subjective experience has structure, it can be coded using our internal representation system (RS) identifiers. These RS form strategies that are demonstrated through non-verbal cues and digital language - thought-behavior patterns - at the unconscious. They provide a means by which to map (encode) learning strategies of individuals that can in turn be converted for such system development as agent technology. By being aware of analog (non-verbal) behavior cues, and their derived meanings, we are able to better identify how individuals construct learning strategies that lead to new decision-making strategies and behavior patterns. How human learning strategies can be coded for development of adaptive and/or self-governing systems is the essence of this paper.

INTRODUCTION

Mentally, we do not operate *directly* upon the world, but rather through representations (a map or model) we mentally create in our unconscious efforts of what we *believe* the world to be. This is so we are able to manage vast amounts of information with which we are constantly being bombarded. The models or representations that we mentally create "differ from the world of reality in three major ways. Some parts of our experience is *deleted*, not represented in our model, ... (some parts) are *distorted*, ... shifts in our experience, and ... (some parts) are generalized, representations of an entire category..." (Bandler and Grinder 1975). Thus, our brains sort information from our sensory apparatus utilizing these three processes into pictures, sounds, feelings, smells, and tastes. Internal information is then retrieved usually via one of the three major sensory representation systems (RS): *visual, auditory,* or *kinesthetic* (olfactory and gustatory information is usually attached to the kinesthetic accessing cue). We then incorporate language to summarize or generalize our experiences to pass them onto others. However, "this type of linguistic generalizing ability of humans, Korzybski contended, accounted for...the misunderstanding, and misuse of such symbolic mechanisms...(resulting) in (humans) confusing the 'map' with the 'territory.' " (Dilts 1999). Thus, clarification of an individual's experience requires that we focus on the non-verbal responses to sensory information eliciting questions from which, patterns can be established of the strategies that a person uses to access stored internal representative information. From this, questions can be tailored to calibrate further inquiry in understanding how a person constructs decision-making and learning strategies that lead to creativity, motivation, expert skills, and so forth. We are not privy to *what* a person is thinking (unless they describe it to us), but rather the access venue by which they have organized the sensory-based information as a representation of external reality.

Therefore, the process of learning is about synesthesia patterns, strategies and anchors (image, sounds and feelings) that are stimulated between interactions among humans and their environments. If we are deprived of information from the external world, our brains will **create** new ones (bits of information- visual, auditory, and kinesthetic) for us – a mechanism by which human creativity and/or fantasy is triggered, and encoded.

By identifying minimal cues that an individual demonstrates at the unconscious, we are able to code how learning strategies are formed to incorporate new information and behaviors. One method by which minimal cues are established is the Eye Accessing Cue Classifier (EACC) model scheme. Dilts and Epstein (1995) stated that, "...the position of the eyes plays a role in the neurophysiological organization that facilitates information representation or retrieval.." This scheme has been tested for repeatability and consistency over the past 25 years with continual success. Over the past nine years, the authors have conducted research involving the EACC and RS finding consistent correlation between the two. (Studies were conducted for NASA, Department of Defense, Intelligence Community, and the nuclear industry.)

While the EACC model is not exactly static, it is a means by which to calibrate how a person internally and externally constructs a strategy. Essentially, the EACC model is a tool that can be used for determining how a person thinks when learning new tasks. Figure 1 below represents the standard model of a normally oriented and right-handed person's eye accessing scheme that indicates where the person pairs internal representation of what they see, hear and feel reflecting their on-going experiences. (Kercel et al, 2000) The EACC is in response to sensory-based questions, e.g., How do you *feel* (sensory based word) about tomorrow's test? and mapped accordingly. Each of the visual and auditory modalities is further identified as to *memory* and *construct* processing.

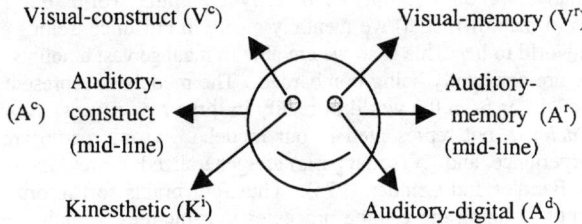

Fig. 1: EACC Scheme for Normally Right Handed Individuals.

These states are significant in knowing where an individual stores memories for review and recall, and where unfamiliar images, sounds and feelings are generated to provide a map of the thought process of a decision or learning strategy.

IDENTIFYING STRATEGIES AND PATTERNS

Even though there exists an infinite number of learning strategies, there exists only three learning styles: *visual, auditory* and *kinesthetic*. The learning style by which a person prefers to learn and form new strategies is referred to as the favored or valued RS. These valued RS can be identified quickly by listening to the predicates (adjectives, adverbs, verbs) being used augmented by tempo and tonality changes, e.g, an auditory oriented individual has an even tempo that is accompanied by an even rhythmic tone. How an individual pairs and sequences this valued RS with the other two RS form the strategies that result in how well the individual learns and makes decisions that lead to tactile application, creativity and/or motivation.

For example, in a four-year program conducted by the authors, it was typically found that individuals who comprehend well what they read, are exceptional spellers, and analytical and mathematical solvers tend to be visual processors (or know to transition to the visual RS). These individuals generally make internal images of their tasks by constructing pictures from words and symbols represented as still pictures or movies, black and white or color, etc. called *submodalities* (sm) (driver's of behavior). Example of such a mental strategy structure may look like,

$$V^r_{sm=moving} \rightarrow A^r_{sm=whisper} \rightarrow V^r_{sm=3D} \rightarrow K^i_{sm=warm} \rightarrow \text{learning/decision strategy}$$

Visually oriented individuals mentally visualize problems, reconstruct and compare the process steps and formulas using imagery that can be recognized externally. For the two larger groups – auditory and kinesthetic oriented individuals – their patterns are very different. These individuals may not visualize problems, instead they may run internal dialogues or form impressions of what needs to be mentally visualized.

Therefore, the way in which we sequence our RS for learning will, in large part, determine which strategy sequences are most successful, and depending on the context, there will be a common element. Knowing this provides us with a means to better understand how learning is formed at the unconscious that can be encoded for development of self-governing systems.

CALIBRATION OF ANALOG CUES
The myriad of external minimal cues is so enormous we've evolved an unconscious to help us deal with the traffic jam of stimuli. Because there is an overabundance of information being received by an individual, our conscious mind only pays attention to the things it can afford to expend its energy attending to, while the unconscious takes care of the rest. How we internalize what information needs attending to at the conscious level is accomplished through our generalization, distortion, and deletion of information, and the inherent feedback systems we have created. These universal processes are key factors in how we will sequence our internal representation systems.

As information comes to us, we generalize it as it may apply across different contexts, we delete elements of it so that it doesn't overwhelm us, and we distort the data so that it fits our model of reality. After repetitive experiences, we develop the elegant pattern shortcuts that are involved in jumping from one element of a stored image, sound, or feeling to its relative piece that may be attached to another experience. All of these unconscious mental behaviors take place rapidly and, as the name implies, *outside our awareness*. It is the method by which we construct learning strategies that evolve into decision-making strategies and behavior patterns. In determining how people develop patterns of learning, decision-making, and so forth, we observe those unconsciously generated analog behaviors sometimes referred to as body language.

If we take the time to notice what people do in response to us; then ask them if the feedback they have given us has the meaning we think it does, we will get more precise information about *that person's patterns of behaving*. Asking the questions: *how, who, what, where, when, with whom*, etc. (avoiding the question of *why* as it elicits a statement about the "problem"), allows us the opportunity to calibrate how the person with whom we are communicating pairs information with their behavior (physical movement, language, etc.). Calibration of how an individual is processing information allows us to establish baselines of learning strategies and patterns, as well as, the values that motivate the learning process. Patterns are comprised of sequences of behaviors and meta-behaviors, and all decisions are based on learning strategies.

CODING

"The power and usefulness of strategies lies in the fact that they are descriptions of the purely formal operations of our behavior and are not tied to any particular experiential content... (they) provide the frameworks within which we incorporate and interpret the content of our sensory experience." (Bandler et al, 1975). For example, consider the learning strategy sequence, Fig. 2, for spelling accurately.

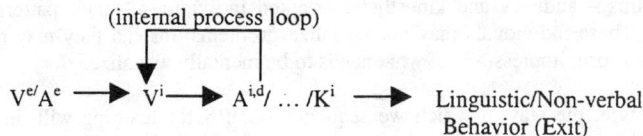

Fig. 2: Spelling Strategy. (Courtesy of Human Dynamics, L.L.C™)

The initial assimilation of information is *looking* (V^e) at or *hearing* (A^e) externally a word; then *reviewing* the letters of the word through an internal image (V^i). Then combinations of the letters constructing the word is *described* through an internal dialogue ($A^{i,d}$), or recalling verbal ($A^{i,d}$) information concerning what the individual sees (V^i). This is then tested against other remembered dialogues or recalled verbal information (A^d) which are then compared to the visual image of the combined letters. Determining if the word is mentally spelled correctly is then associated with a feeling (K^i).

The above is but one example of how various strategies could be identified and complied as a matrix of successful behavior responses for predicative behavior, motivation, and creativity.

$$V^e/A^e \begin{vmatrix} V^i & A^{i,d} & K^i \\ A^{i,d} & V^i & K^i \\ A^i & K^i & V^i \\ & \text{etc.} & \end{vmatrix}$$

Depending on the strategy used, or the sequencing of multiple strategies, will determine the success or failure of the individual in performing a task.

STRATEGY AND PATTERN MODELING

When one learns, one tunes into their sensory channels to pace feedback from other individuals and their surroundings. Non-verbals cues such as head nods, breathing changes and minor movements will indicate how one is listening, for example, to a lecture. Observing eye-accessing cues, tonal shifts, tempo and other minimal cues will provide input as to which systems and strategies and individual(s) is using. "Identifying the behavioral cues associated with a certain state is often most easily done through a process known as '*contrastive analysis...*' contrastive analysis involves comparing a particular process or experience with another experience that is quite different. This helps to bring the most significant differences (in a pattern or behavior) to the foreground." (Dilts, 1995) From this we begin to code a model of the strategies, which can then be transformed into mathematical computation for software implementation using such methods as *Bayesian nets*. As one looks at the data, meta-patterns (patterns of patterns) will begin to become obvious. For example, consider the two learning strategies below.

$$V^{c\ (dissociated)} \rightarrow A^{i,d} \rightarrow K^i \rightarrow A^{i,d}_{\text{"O.K."}} \rightarrow \text{Learning Strategy 1}$$

Here the individual constructs a dissociated *image* (V^c) of the information he wants to learn while the information is being presented to him. He then has an *internal dialogue* ($A^{i,d}$) about the representation image in his head. When the words *felt good* (K^i) he would *say* to himself ($A^{i,d}$), "O.K. " The submodality associated with the "good feeling" has warm colors and is close in proximity. The second strategy is defined as,

$$V^r\text{-}V^r\text{-}A^{r\ (associated)} \rightarrow K^i \rightarrow A^{i,d} \rightarrow V/A/K \rightarrow A^{i,d}_{\text{"that's it"}} \rightarrow \text{Learning Strategy 2}$$

Here the individual creates *mental visual images* that are associated with each other and *sound*(s) (V^r-V^r-A^r). As the images are viewed, the person increases the submodalities associated with the images and sound(s) making the images larger and the sound more *intense* (K^i). This is continued until the *image-sound and kinesthetic sensations* (V/A/K) reach a level that is satisfactory, and she *says* ($A^{i,d}$), "that's it."

Therefore, based on their *feelings* of what was seen and heard, a behavior is exhibited. Variations of the behavior is determined by the sequence of the RS and the associated submodalities.. Knowing what was specifically represented within the sequence of the learning process at the RS level allows us to become more fully informed about how individuals organizes their thought patterns in forming models that express learned experiences.

CONCLUSION

"By determining and examining...pattern(s) of ... internal processes, one can readily establish the sequence(s) of the strategy process used by an individual in (learning an) event or (performing a) task," (Brown-VanHoozer 1999). "The structure of meaning...occurs in the specific sequence of the representational systems a person uses to process information," (Bandler, et al, 1980). Eliciting the strategies with questions such as:

"Do you know how to do that?"
"How do you know that you can do that?"
"What stops you?"
"How do you think about that?"
"How do you know you have learned something?"

"provides high quality information about an individual's internal processing scheme, enabling us to feed back the same structure and submodality qualities we discover as driver of the behavior for that person." (Hall 2000) This is a very different approach to how most information is gathered to create models for self-governing systems. With this technique, meaning of the behavior is more explicit and defined. For example, we have a person who is highly motivated in wanting to fly. This motivation we find is associated with an image that is located down and to the person's right, multi-colored, clear (not fuzzy), seen as movie picture, close in proximity, in 3D, and is unframed. However, when the person finds that he will have to calculate coordinates for cross-country flying, his motivation drops. We find the submodalities have changed. He still sees an image that is located down and to his right, multi-colored, seen as a movie picture, in 3D, is unframed though now fuzzy and further back in proximity. (How to revert to the highly motivated behavior requires a simple calibration that will not be discussed in this paper.)

As an individual constructs a learning strategy, he or she will deliver analog and digital indications of the process outside the conscious awareness of the individual. It is at this level that we begin to determine how individual models of learning are constructed. When we concern ourselves with replacing specific information deleted by individuals in their communications, the optimum in reliable knowledge can be extracted from users. This provides a foundation from which the paired relationship of language and non-verbal behavioral indicators can be determined to identify those attributes required for the design and development of adaptive and/or self-governing decision-making systems.

ACKNOWLEDGMENTS
The authors of this paper wish to thank Dr. Steve Kercel for effort in editorial comments of the paper.

REFERENCES
Bandler, R., Dilts, R., DeLozier, J., and Grinder, J., 1980, *"Neuro-Linguistic Programming: The Study of the Structure of Subjective Experience.* "Vol. I, Capitola, CA.
Bandler, R. and Grinder, J., 1975, *Patterns of the Hypnotic Techniques of Milton H. Erickson, M.D.*, Meta Publications Inc., Capitola, CA, Vol. I. p. 7&8, 106, 154.
Dilts, R., 1999, *Sleight of Mouth*, Meta Publications, Capitola, CA, p. 11.
Dilts, R. and Epstein, T., 1995, *Dynamic Learning*, Meta Publications Inc., Capitola, CA, pp. 31-33
Grinder, M., 1991, *Righting the Educational Conveyor Belt*. Metamorphous Press, Portland, OR.
Hall, L. M., 2000 (revised ed), The Spirit of NLP: *The Process, Meaning and Criteria for Mastering NLP*, Crown House Publishing Limited, UK, pp.184-185.
Kercel, S., Brown-VanHoozer, S.A., and VanHoozer, W.R., 2000. "The Model-Based Mind," *Proceedings, IEEE SMC Conference*, Nashville, TN.

THEORETICAL BIOLOGY: ORGANISMS AND MECHANISMS

CHRISTOPHER LANDAUER
cal@aero.org
Aerospace Integration Science Center
The Aerospace Corporation
Los Angeles, California 90009-2957, USA

ABSTRACT

The Theoretical Biology Program initiated by Robert Rosen intends to identify key characteristics of organisms, to explain how and why they are different from mechanisms. Its main claim is the need for new Mathematical and even new Scientific methods beyond the usual reductionist style. There are very strong claims to this effect in Rosen's books, along with some purported proofs. Unfortunately, the Mathematics is incorrect and the assertions remain unproven (and some of them are simply false). The question "what differentiates organisms from mechanisms", remains unanswered, since there are simple mechanisms that satisfy all of the claimed differentiating properties. We also present some relatively new techniques in Mathematics that correspond to the Analytic and Synthetic modeling styles, to show that much of the Mathematics needed already exists. Nevertheless, the program itself is important, and Rosen has raised some serious issues about Theoretical Biology that have not been properly addressed elsewhere.

INTRODUCTION

This paper is about the application of Mathematics to Biology, and in particular, to the Theoretical Biology program initiated by Robert Rosen (Rosen, 1991, 1999). Our main result is that even though many of the methods used so far have been unsuccessful to various extents, and have frequently been applied incorrectly, there are nevertheless several new areas of Mathematical tools that can usefully be considered for this application area.

In our opinion, Mathematics is the only reliable reasoning method we have, and if it cannot currently solve a problem, then either the problem is misconceived, or new Mathematical methods are needed (Landauer and Bellman, 2000). This latter notion is not a particularly new idea, and should not be viewed as revolutionary or even controversial: many areas in Mathematics have required new symbolic methods, new proof methods, and even new definitions of what constitutes a proof (to name just a few: Calculus beginning in the 1670s; limits and convergence of functions, sequences, and infinite series in the 1820s; Hilbert's non-constructive proof methods in algebra in 1888; infinite cardinals and transfinite induction in Set Theory in the 1890s and 1900s; and the use of a computer program in the proof of the Four-Color Theorem in 1976).

For many years now, there has been a growing recognition of a phenomenon that has appeared under many guises in many areas of Science: dissatisfaction with the prevailing "reductionist" paradigm. Of course, that paradigm has been remarkably successful in all walks of life, over at least the last several hundred years, and it is only in the last few decades that it has become clearer what is missing (the rampant confusions among "non-linear", "chaotic", "complex", "uncomputable", and other less well-founded terms are indications of the problem).

Perhaps the most difficult of these new problem areas is Biology. Complexities in the field of Biology have led to many calls for improvements in biological modeling (Bellman and Walter, 1984), from many different points of view. One particular approach to this problem has been the Theoretical Biology program of Robert Rosen. The main goal of this program as it is pursued now is to identify the key theoretical characteristics of organisms, by looking for the proper abstractions, and by defining the appropriate relationships. It is an interesting problem in this context to distinguish between organisms and mechanisms or show that they cannot be so distinguished. We show first that this problem remains open, since the published proof is incorrect (Rosen, 1991, 1999).

ORGANISM CRITERIA AND EQUATIONS

In this section, we briefly describe the basic Mathematical problem, and show that Rosen's solution is simply wrong, by providing solutions to the distinguishing equations in several areas of Mathematics (more details can be found in (Landauer, 2001)). In the next section, we continue with some other applicable Mathematics. We have recently been informed (reviewer comment) that many such solutions exist, and that this equational distinction is no longer expected to suffice.

We start with the description from (Rosen, 1999) of a model of living systems, his (M,R)-systems, since it gives the clearest description of what Rosen intended to do. An (M, R)-*system* models the relationship between metabolism and repair, and the processes that interconnect them. If A and B are any sets, and f is a mapping from A to B, then we get the diagram

$$f : A \longrightarrow B, \; p : B \longrightarrow H,$$

where H is a set of functions from A to B, that is, H is a subset of $[A \to B]$. In this model, f is a kind of abstract enzyme, mapping elements $a \in A$ to elements $b \in B$, and p is a kind of abstract mapping from these elements b back into version of the function f. The key insight here is that replication comes from the model "for free", under certain conditions.

The equations that Rosen derives (diagram 10C.6, p. 251 of (Rosen, 1991)) have a class / instance error, and the equations should read: for some element $a \in A$ ("the environment"), and some element $b \in B$ ("the products"), $b = f(a)$, where f is the "metabolic function" $f : A \longrightarrow B$ that produces products from inputs, $f = p(b)$, where p is the "repair function" $p : B \longrightarrow F$ that produces metabolism functions from products, and $p = b(f)$, where b is the "replication function" $b: F \longrightarrow P$ that produces repair functions from metabolism functions. These are the functional equations we solve in the rest of the section.

Solutions

We describe solutions of these functional equations in three Mathematical subject areas. Our first solution will be from the domain of formal power series (Wilf, 1990). Define R to be the space of all real formal power series

$$r = \sum_{n \geq 0} r_n * x^n$$

in a "formal variable x, and let A be the set of polynomials in R (convergence properties are also interesting, which we show in the next subsection). Each $r \in R$ gives a function $r^\#$ on R by taking, for any $s \in R$,

$$r^\#(s) = s + r * x.$$

Then composition of these functions corresponds to addition in R: $r^\#$ composed with $s^\#$ is $(r+s)^\#$, so it is both commutative and associative. Then it is easy to show that the functional equations for a given polynomial $a \in R$, have a unique solution in R.

The next solution is from the theory of domains. Domains are chain-complete partially ordered sets (Barendregt, 1984). They are of great use and interest in denotational semantics of programming languages. One of their main operations is to define a set $[D \to E]$ of "continuous functions" for given domains D and E. The most important property of it is that it is also a domain. The functional equations in this context become equations for the domains containing the functions:

$$F = [A \to B], \ B = [F \to P], \ P = [B \to F].$$

These equations have a solution, using a general theorem about domains (see Theorem 7.3, pp. 665 of (Gunter and Scott, 1994)).

Finally, we exhibit a solution in a very simple set. We found it by examining how the variations in a follow through f to b and p. Consider the set $A = 2 = \{0, 1\}$, and identify the set of all 1-1 functions on 2 with the same set 2 by

$$0 \leftrightarrow \text{identity}, \ 1 \leftrightarrow \text{inverse}.$$

Then we can easily check that the functional equations hold true for all $a \in A$ and $f \in F$. In this form, the solution for f is independent of the choice of a.

Each of our solutions is unsatisfying for various reasons. All of the solutions make it clear that we need a better formulation of the problem, or at least a better set of equations, if we are to show anything about the relationship between organism and mechanism. More importantly, this is probably not the right approach at all. After all, claims that complexity is non-computable have no meaning, since computability is only defined for formal systems, and cannot be proven, since it is not even properly defined for non-formal systems. There are many such confusions in this literature.

NEW MATHEMATICAL METHODS

In this section, we describe some existing Mathematical methods that shed light on the notions of model used in (Rosen, 1991).

First, we show why there is no largest model for natural or formal phenomena. The notion of largest model arises in a discussion on p. 179 of (Rosen, 1991). It turns out that not all models of S are equivalence relations on S, even when S is a set (Rothenberg, 1989). For example, formal specifications of computer programs are models that usually have many more elements than the programs they model, since they must make explicit constructs and relationships that the program can leave implicit. An *ultraproduct* is a lattice-theory construction that has proven very useful in applications of logic to algebra. The Łoś Theorem (see Theorem 3.1, p. 112 of (Eklof, 1977)) shows that an ultraproduct of models for a first-order language L is a model for L, and these models are almost always larger than the original.

Secondly, the mathematics of concrete category theory (MacLane and Birkhoff, 1967) provides a uniform framework for comparing the Synthetic and Analytic models described in (Rosen, 1991), using currently available Mathematics.

A *Synthetic Model* of S is an injection into a Direct Sum that is non-empty in each component. We can clearly get more coverage by using more component mappings. We would like enough maps to cover S completely (it should be noted that there is no injection or other function unless we start with an ordinary set S, so this entire discussion is necessarily limited to formal objects).

An *Analytic Model* of S is a mapping into a Direct Product that is onto each component. These mappings need not be injective. We can clearly get more distinctions by using more component mappings. We would like enough maps to distinguish all of the elements of S. The *kernel* $Ker(m)$ of the mapping m is the equivalence relation defined on the set S defined by indistinguishability, that is, the relation that makes two elements of S correspond when the analyses "cannot tell them apart": if $m : S \longrightarrow M$, then

$$Ker(m) = \{(r,s) \in S \times S \mid m(r) = m(s)\}.$$

The kernel is *trivial* when the relation is the identity.

It is easy to show that Synthetic Models are co-extensive with Analytic Models with trivial kernel. However, there are Analytic Models that are not Synthetic Models. In other words, if we build up a model from individual elements, then we can analyze it down to those elements, but there are models that we can analyze that are not built from individual elements. It turns out that this distinction between Synthetic and Analytic Models is exactly the same as that between inductive and co-inductive definitions. The usual notion of Mathematical construction is inductive, starting with some basic elements and building structures containing them and processes that interpret or activate those structures. This work has been extended to include sets that are not accessible externally, which can then be used to model systems with internal states (Goguen and Roşu, 1999).

The idea is to analyze systems according to their externally observable behavior (Goguen et al., 2000), and not try to define the mechanisms that lead to that behavior. It amounts to decomposing the structure of the system as far as is needed to explain the behavior, with no need to decompose it to primitive elements. The best framework for describing both is category theory, in which the co-inductive methods are derived from the inductive methods by reversing all the arrows (Jacobs and Rutten, 1997). These categorical methods are being applied in innovative ways to computing systems (Goguen and Roşu, 1999), and are providing new methods for proving theorems about complex structures, even when they are circularly defined (Goguen et al., 2000).

CONCLUSIONS AND PROSPECTS

There are two sides to this Theoretical Biology program: improving the tools we use and improving our understanding of the phenomena. It is clear that we need both approaches. We can improve the capabilities of the models, using Computational Reflection, Anticipation, or Co-Induction, and we can improve the principled understanding of Biological and other Complex Systems, using Layers of Symbol Systems and Controlled Sources of Variation. Maybe the improvements will meet in the middle; maybe they will not. We do not believe that we have sufficient Mathematics currently to describe complex systems effectively, especially those with humans as users or participants (i.e, the *Constructed Complex Systems* of (Landauer and Bellman, 1998a, 1999b)).

We do believe that new kinds of Mathematics are necessary and will be sufficient. Some of these interesting new areas of Mathematics are: Logics of Self-Reference, Integration, Inconsistency, Importance; New Formal Spaces and Modeling Methods; and Category Theory, Recursive Functions, Hypersets, and more (Bellman, 2000) (Landauer and Bellman, 2000).

It seems like the best hope at present is to provide some capability for elaboration in the system, which leads to tremendous interest in Genetic Programming and Evolutionary Programming Systems (Bäck, 1996). Our approach to elaboration is to examine the symbol systems that are used within a computing system to represent its data and processes. To this end, we have shown that there are limitations on the abilities of systems that use fixed symbol systems (Landauer and Bellman, 1999a) (Bellman and Landauer, 2000). and have begun to design and explore computing systems that can examine and change their own symbol systems, based on a Computationally Reflective integration infrastructure called "Wrappings" (Landauer and Bellman, 1998a). Programs organized with Wrappings have a model of their own behavior, and can use it to change their capabilities. After all, even Gödel's Theorem only works when the symbol system of the logic is not allowed to change.

The essence of computation is interpretation of symbol systems. The only operations that a digital computer can perform are copying and comparison (All arithmetic in digital computers is via limited-precision explicit models of the corresponding integer or real arithmetic). Therefore, we cannot construct computing systems to do complex or otherwise interesting tasks without many explicit models of the kinds of computation, deduction, or analysis required. All of these models must then be expressed in terms of the operations that we can implement on these (very) limited computers.

The theorems of Turing, Gödel, and others show that there are fundamental limits on the expressive and computational power of computing systems, but ALL of them assume that the symbol system remains fixed (that is a basic assumption in all of the mathematical proofs), and that the parallelism can be mapped into interleaved events. Systems that are not restricted in either of these ways might escape the bounds of these theorems. This is our current direction of research.

REFERENCES

Thomas Bäck, *Evolutionary Algorithms in Theory and Practice*, Oxford University Press (1996)

H. P. Barendregt, *The Lambda Calculus: Its Syntax and Semantics*, North-Holland (1984)

Kirstie L. Bellman, "Mathematical Fictions and Physiological Realities: The Challenges of Brains Reasoning about Themselves", *Proc. IMACS'2000 (CD), Invited Session on Bioinformatics*, 21-25 Aug 2000, Lausanne (2000)

Kirstie L. Bellman, Christopher Landauer, "Towards an Integration Science: The Influence of Richard Bellman on our Research", *J. Math. Anal. and Appls.*, Vol. 249, No. 1, pp. 3-31 (2000)

Kirstie L. Bellman and Donald O. Walter, "Biological Processing", *Amer. J. Physiology*, Vol. 246, pp. R860-R867 (1984)

Paul C. Eklof, "Ultraproducts for Algebraists", Chapter A.3, pp. 105-137 in Jon Barwise (ed.), *Handbook of Mathematical Logic*, North-Holland (1977)

Joseph Goguen, Kai Lin, and Grigore Roşu, "Circular Coinductive Rewriting", pp. 123-131 in *Proc. ASE'00*, 11-15 Sep 2000, Grenoble, IEEE (2000)

Joseph Goguen, Grigore Roşu, "Hiding More of Hidden Algebra", pp. 1704-1709 in *Proc. FM'99*, 20-24 Sep 1999, Toulouse, SLNCS 1709 (1999)

C. A. Guner, D. S. Scott, "Semantic Domains", Chapter 12, pp. 633-674 in J. van Leeuwen (ed.), *Handbook of Theoretical Computer Science, Vol. B: Formal Models and Semantics*, MIT Press (1994)

Bart Jacobs, Jan Rutten, "A Tutorial on (Co)Algebras and (Co)Induction", *EATCS Bulletin*, Vol. 62, pp. 222-259 (1997)

Christopher Landauer, "Theoretical Biology: Organisms and Mechanisms", in *Proc. CASYS'2001*, 13-18 Aug 2001, Liege, Belgium (2001)

Christopher Landauer, Kirstie L. Bellman, "Wrappings for Software Development", pp. 420-429 in *Proc. HICSS'98*, 6-9 Jan 1998, Kona, Hawaii (1998a)

Christopher Landauer, Kirstie L. Bellman, "Symbol Systems in Constructed Complex Systems", pp. 191-197 in *Proc. ISIC/ISAS'99*, 15-17 Sep 1999, Cambridge, Massachusetts (1999a)

Christopher Landauer, Kirstie L. Bellman, "Architectures for Embodied Intelligence", pp. 215-220 in *Proc. ANNIE'99*, 7-10 Nov 1999 (1999b)

Christopher Landauer, Kirstie L. Bellman, "Can Formal Mathematics Model Non-Formal Phenomena?", *Proc. IMACS'2000 (CD), Invited Session on Bioinformatics*, 21-25 Aug 2000, Lausanne (2000)

Saunders MacLane, Garrett Birkhoff, *Algebra*, Macmillan (1967)

Robert Rosen, *Life Itself: A Comprehensive Inquiry into the Nature, Origin, and Fabrication of Life*, Columbia U. Press (1991)

Robert Rosen, *Essays on Life Itself*, Columbia U. Press (1999)

Jeff Rothenberg, "The Nature of Modeling", Chapter 3, pp. 75-92 in Lawrence E. Widman, Kenneth A. Loparo, Norman R. Neilson (eds.), *Artificial Intelligence, Simulation and Modeling*, Wiley (1989)

Herbert S. Wilf, *generatingfunctionology*, Academic Press (1990)

DOES INCOMPUTABLE MEAN NOT ENGINEERABLE?

STEPHEN W. KERCEL
Oak Ridge National Laboratory
PO Box 2008, MS 6007
Oak Ridge Tennessee USA 38731
e-mail: kercelsw@ornl.gov

ABSTRACT

Self-referential systems have some remarkable properties. The processes of life and mind are not only self-referential, but self-reference turns out to be a crucial property of both. However, they are difficult to understand. From a given starting point, both *endogenous systems* (self-referential natural systems) and *impredicative systems* (self-referential formal systems) have infinitely many logically consistent consequences. Both are incomputable; neither halts after a finite number of steps. Therefore, neither can produce an exact prediction of the behavior of the other in finitely many steps. Despite the fact that all engineering decisions are based on incomplete information, this inherent inability of an impredicative model to produce exact predictions of an endogenous system is troubling to some engineers. Nevertheless, self-reference leads to a more general, but no less rational, form of modeling than that provided by traditional reductionism. Although the mathematics of self-reference is unfamiliar to engineers, its power is dramatic. For example, it resolves the apparent paradox of how a brain/mind possessing freewill can operate in a deterministic Universe.

INTRODUCTION
WHY SHOULD WE CARE?

Engineers seek to develop artificial systems that exhibit behaviors similar to the cognitive processes observed in biological brains. This is a worthy goal, and reaching it will be one of the crowning achievements of 21st Century technology. By no stretch of the imagination has the goal been reached yet. Indeed, the astoundingly simple nervous system of the nematode (all 300 neurons) is completely mapped out. Nevertheless, despite years of research, nobody has been able to devise an algorithm that behaves anything like a nematode (Chomsky 1993). Is it really such a stretch to suppose that the reason that the goal has not been met is that cognitive behavior is beyond the scope of the traditional reductionistic methods used by engineers? The answer to this question is yes, and this leads to another question. What other rational methods might we invoke?

IMPREDICATIVE MATHEMATICAL OBJECTS

What is really being said by the remarkably tricky proposition, $\bullet(x) = \bullet(x)$? It appears to say nothing and everything simultaneously. On the one hand, it says practically nothing. As a definition of $\bullet(x)$, it does not inform us how $\bullet(x)$ differs from any other entity, such as $\bullet(x)$. This one fact alone seems to illustrate the futility of defining an entity by circular reference. On the other hand, $\bullet(x) = \bullet(x)$ states a profound truth regarding every entity; it is a condensed statement of Aristotle's Law of Non-contradiction (Adler 1978). Translated into words, $\bullet(x) = \bullet(x)$ says roughly that a thing is what it is and does not act contrary to its nature.

This self-referential proposition is a foundation axiom of our system of rational thinking. It is more than self-referential; it is self-evident. The Law of Non-contradiction cannot be validated from any more fundamental proposition. However, no counter example has ever been produced to falsify it. Paradoxically, any effort to logically prove that it is false starts from the assumption (usually implicit and unacknowledged) that it is true. Thus, although $\bullet(x) = \bullet(x)$ does not directly inform us how $\bullet(x)$ differs from $\bullet(x)$, it does provide us with the foundation of a system of logic that may, given proper additional data, allow us to answer the question.

Mathematicians and scientists prefer to avoid self-referential structures because they often lead to paradoxes. A classic logical paradox is "The Liar." What are we to make of the claim, "This statement is false?" Out of context, it appears to be a flat contradiction. If we assume it to be true, it asserts a false proposition. If we assume it to be false, it asserts a true proposition. The simplest, and conventionally accepted, way to evade the dilemma is to exclude all self-referential propositions from the epistemological Universe of Discourse, and to ignore its distinction from the ontological Universe.

However, such a simple evasion is deeply unsatisfying for several reasons. First, our system of logic is based on the self-evident self-referential proposition, $\bullet(x) = \bullet(x)$. Whether we like it or not; disallowing the foundation principle from the Universe of Discourse hardly appears to be a sound way to begin a logical process of reasoning. Second, we can use language to discuss "The Liar" and be understood; a system of logic that simply disallows the admission of such propositions would be far too impoverished to allow us to use it to reason about language (Barwise and Moss 1996). Third, a taboo on self-reference produces a system of logic too impoverished to discuss mathematics. Although mathematicians are loath to admit it, self-referential, or *impredicative*, propositions are both admissible and necessary in mathematics (Kleene 1950).

An impredicative definition is one in which the object being defined participates in its own definition. For example, $x \bullet X$, where x is defined in terms of its relationship to X, is a legitimate definition of the set X. It is subtly but crucially different from a circular definition. As already noted, a completely circular definition provides no feature to distinguish between the object being defined and the remainder of the Universe of Discourse. An impredicative definition must include some constraint (in addition to identity) on the relationship between an object and itself, and the properties of the constraint constitute a crucial distinguishing feature of the definition.

Is this sophistry, or are impredicative structures of some practical use? Consider that a decade ago, long distance telephone service cost ten cents per minute, but now it is commonly available for half that price. How did this come to be? It happened because telephone engineers found a practical way to double the carrying capacity a telephone circuit, *with absolutely no loss of information*. They performed this seeming miracle by discovering a technique called perfect reconstruction wavelet compression.

At the foundation of wavelet compression is the wavelet scaling function. The wavelet scaling function, $\bullet(x)$, is defined by the First Fundamental Wavelet Equation, $\bullet(x) = 2 \bullet {}_n h_n \bullet (2x-n)$, where n is an even finite integer. This proposition is only true for vectors h_n that satisfy certain constraints. However, for an admissible vector, $h_n, \bullet(x)$ is a uniquely defined function. In addition to being unique, it has all sorts of desirable mathematical properties, including finite support, continuity, differentiability, and orthogonality to translates of itself. There is no other way to define it except by the First Fundamental Wavelet Equation (Akansu and Haddad 1992). The crucial point is that this definition of •(x) is impredicative.

Despite its impredicative definition, the wavelet scaling function is just as logical than more conventionally defined functions. Given the definition of •(x), logical inferences can be drawn from it, and valid engineering decisions made from those inferences. In particular, the Second Fundamental Wavelet Equation can be defined in terms of the First, namely • (x) = 2• $_n g_n$• (2x-n), where g_n is the time reversal of the quadrature reflection of h_n. In addition to having all the desirable properties that the scaling function, •(x), has, the wavelet function, • (x), has an even more important property, it is orthogonal to scaled versions of itself. More importantly for engineers, h_n, g_n, and their time reversals, can be proved to be a set of coefficients for a perfect-reconstruction decimate-by-2 digital filter bank.

The reader may be tempted to think that an impredicative structure is simply a recursive algorithm. It is not. A recursive algorithmic function has a defined bottom; after a finite number of steps, it hits bottom and the bottom step returns a predefined symbol to the previous step. The First Fundamental Wavelet Equation is the definition of • (x), but computing the *exact* value of • (x) given h_n would require an infinite number of calls to the next finer level of scale. Since it requires infinitely many steps to complete, it is not an algorithm (Knuth 1973). Infinite recursion depth is typical of impredicative processes; they are non-algorithmic and *incomputable*. These differ fundamentally from recursive algorithms, which have a defined bottom level of recursion.

The reader may also be tempted to think that an impredicative structure is simply an analog feedback system, or perhaps the differential equation describing such a system. It is not this either. The crucial distinction is that the impredicative object is defined in terms of its relationship to itself *between levels*. To appreciate the notion of level, imagine the model of a model of a system. An impredicative model is defined in terms of its relationship with a model of the model. For example, a wavelet scaling function (itself a model of some other system) on a given scale is the weighted sum of copies of translates of the same function at the next finer scale (a model of the original model). In contrast, for an analog feedback mechanism, the self-reference is within the natural system, and if modeled by a differential equation, the feedback imposes no need to look at a model of the model. The feedback is *within a level*.

The key points of this digression into impredicativity are as follows. Logic is founded on self-reference or impredicativity. The cost of self-reference is the risk that it can lead to paradoxes. A blanket taboo against self-reference in order to avoid paradoxes is too restrictive. An epistemological Universe of Discourse that disallows self-reference is too impoverished to allow meaningful discussion of the ontological Universe, where self-referential structures (*e.g.* semantic languages) abound. In fact, an epistemological Universe of Discourse that disallows self-reference is too impoverished even to allow meaningful discussion of other useful epistemologies, such as engineering mathematics. Impredicativity has infinite recursion depth, and is not an algorithm. It requires self-reference between levels, and fundamentally differs from a feedback mechanism. Most crucially, for readers of this paper, engineers can use impredicative mathematics to guide their decisions just as reasonably as we use more traditional mathematics.

ENDOGENOUS PHYSICAL PROCESSES

In the natural world, a multi-level self-referential process is called an *endogenous system*. This is a widely used medical term that describes the property of a system that grows from within itself, or makes itself up as it goes along (Clayman 1989). The term is used for a similar concept in economics (Gavin and Kydland 1999). Some biologists consider this property to be a defining feature of living systems. For example, Margulis and Sagan (1995) flatly claim that a system is alive if and only if it is autopoietic. They identify autopoiesis as the process of life making itself.

Endogeny in the form of autopoiesis was first propounded by Maturana and Varela (1998). They begin by noting that attempting to define life by listing properties leads to the various dead ends. They offer what they call a radically different perspective, that a living process is distinguished from a non-living process by its organization. They identify organization as the relations that must be present for a thing to exist. What distinguishes living from non-living processes is autopoietic organization.

The autopoietic object is a bounded unity. It has a dynamic network of chemical transformations commonly called a metabolism. Metabolism is distinguished from other chemical networks in that metabolism produces the components that make up the network of metabolism. An integral component of the metabolic network is a membrane that serves as the boundary between the network and the rest of the world.

The membrane and the metabolic dynamics are each necessary for the existence of the other. The metabolic transformations will only occur if the system is protected from the environment by the membrane. The membrane is a product of the metabolic transformations. The emergence of the membrane and the rest of the metabolic network is not sequential. They are two different aspects of a single unity, as indicated in Fig. 1. Disrupt either, and you disrupt the whole. Maturana and Varela say that what is most striking about an autopoietic system is that it pulls itself up by its own bootstraps. It produces and repairs its own material structure. It becomes distinct from its environment through its own dynamics. It does so in a way that the metabolism and repair processes are inseparable. They identify autonomy as a system's ability to make up its own laws of behavior, and update them as it goes along.

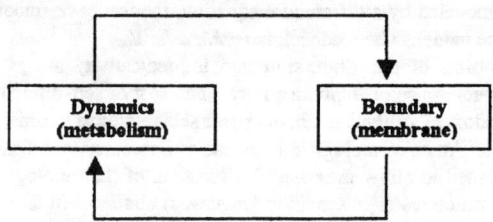

Fig. 1. Unity of Metabolism and Membrane

From direct observation, Wolpert argues that life is incomputable (Murphy and O'Neill 1995). He argues that life behaves like neither a differential equation nor a finite state machine. Differential equations exclude structure from their initial conditions, but that both cells and embryos are highly structured. However, a finite state machine, or a computable cellular automaton, is simply a discretized form of a differential equation. A formal description of a living process must be something altogether different.

CONGRUENCY: FORMAL AND NATURAL SELF-REFERENTIAL SYSTEMS

The incomputability of endogenous natural systems has some serious consequences for engineers. Attempting to project them onto an algorithm automatically discards much of the information about the process. Even worse than that, the algorithm offers no warning as to how much is lost, or where its predictions will fail.

Since endogenous natural systems are incomputable, how are engineers to deal with them? The answer is to model the endogenous natural system with an impredicative formal system. What does this mean? A mathematical formalism of the Modeling Relation has existed for nearly a century, and may go back even further (Russell 1931). In mathematical biology the formalism has been popularized by Rosen (1991, p. 152), who tells us that modeling "is the art of bringing entailment structures into congruence." By itself, the statement leaves us little wiser than when we started. How does *art* enter into the discussion; are we not instead supposed to be scientific? What is an entailment, much less an entailment structure? What does it mean that two different entailment structures are congruent? In what sense are they not identical? If they are not identical, what causes us to declare them congruent? Why do these questions matter?

The first point to appreciate is that the Modeling Relation (see Fig. 2) is a formal mathematical *relation* (Rosen, 1999). Suppose that A and B are sets, and that there exists a set, R, of ordered pairs, where the first element of each pair in R is an element of A, and the second element of each pair in R is an element of B. There is also some ordering in the pairs, but there is not strict ordering. In mathematical notation: a• A, b• B, (a,b)• R<=>aRb. In the Modeling Relation, the members a and b of each ordered pair in R are entailments from two different systems.

Entailments are the *event structures* in the organization of a system. There are two sorts of systems that might appear in the Modeling Relation, natural systems and formal systems. Natural systems are systems in physical reality that have causal linkages; if certain causative events impinge upon a natural system, then the system will behave in a certain way, or produce certain events in effect. This consequential linkage of cause and effect in a natural system is a *causal entailment*. Formal systems are conceptual systems that have inferential linkages; if certain hypothetical propositions impinge upon a formal system, then they will produce certain consequential propositions in conclusion. This linkage of hypothesis and conclusion in a formal system is an *inferential entailment*.

Entailment structures are inherent *within* a system; they are the distinguishing features that characterize the system (Rosen, 1991, p. 98). They do not cross over from one system to another. This is represented in Fig. 2, where we see a natural system, N, distinguished by its structure of causal entailments, a, and a formal system, F, distinguished by its structure of inferential entailments, b. The entailment structures of two distinct systems are distinct from one another; causes or hypotheses in one do not produce effects or conclusions in the other. In fact, this provides the answer to one of the questions posed above. Its self-contained entailment structure is what provides identity to a system and distinguishes it from other systems. This is important in living systems, since one of the distinguishing features is a living system is the unique identity of its bounded self.

The fact that distinct systems are non-identical does not preclude them from being regarded as being in some sense similar. Similar systems should have distinguishing features that closely correspond to each other. Dissimilar systems should have distinguishing features that do not closely correspond to each other. As already noted, the distinguishing feature of a system is its entailment structure. Thus, we would expect similar systems to have entailment structures in which there is some degree of correspondence between the entailments.

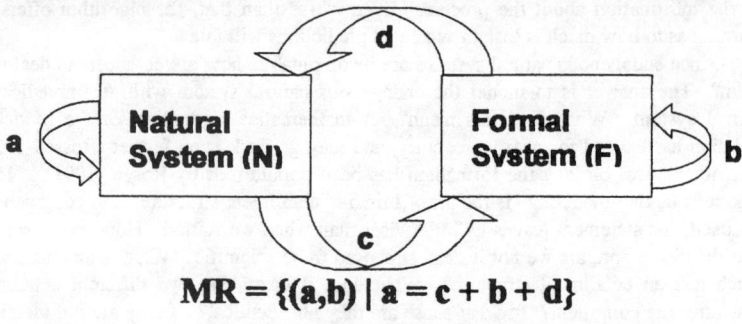

$$MR = \{(a,b) \mid a = c + b + d\}$$

Fig. 2. The Modeling Relation

To establish this correspondence, consider a system of encodings and decodings (Rosen 1991, p. 59). For example, we might have a system of encodings that encodes a set of events in the natural system, N, in Fig. 2, into a set of propositions in the formal system, F. We might also have a system of decodings that decodes a set of propositions in the formal system, F, into a set of phenomena in the natural system, N. Although the two systems remain independent in the sense that causes or hypotheses in one do not produce effects or conclusions in the other, the two systems can be linked by encodings and decodings.

This linkage between entailment structures provides the means of determining the similarity between two systems. Suppose that an event, e1, in N can be encoded to a proposition, p1, in F; we can think of the encoding arrow, c, in Fig. 2 as a measurement on a natural system. Suppose further that the proposition, p1, when applied as a hypothesis in the inferential structure in F entails another proposition, p2, in F as a conclusion. In other words, the propositions are entailed as an implication, b = (p1 • p2), in F. Suppose that this entailed proposition, p2, in F can be decoded into an event, e2, in N; we can think of the decoding arrow, d, in Fig. 2 as a prediction by a formal system.

Rosen defines congruency between the entailment structures as follows (Rosen, 1991, p. 61). Suppose that in the underlying reality, the event e1 in N causes event e2 in N. In other words, the two events are entailed as a causal linkage, a = (e1 • e2), in N. Suppose further that the linkages commute. Event e1 is encoded by c to proposition p1, i.e., c = (e1 • p1), which, implies (in formal system, F) proposition p2, i.e., b = (p1 • p2). Proposition p2 is decoded by d to event e2, i.e., d = (p2 • e2). Further suppose that there is an exact correspondence between the predicted event e2, and the caused event e2. The commutation is also described as a = b+c+d. (Note: + is the symbol for concatenation.) If there exists no such entailment c in F, having a commutative relationship with some entailment a in N, then the two systems do not have congruent entailment structures. Entailment structures are congruent to the extent that such correspondences between entailments exist.

CONCLUSIONS
BIZARRE BEHAVIOR IS SOMEWHAT PREDICTABLE

The idea that a behavior or effect is completely unanticipated, but fully consistent with the causal entailments of the system producing the effect is bizarre. Previously, borrowing Kirstie Bellman's idea, I had styled systems that exhibit such behavior to be bizarre systems. However, it is clear that the bizarreness is in the effect, not in the cause. Since the most compact set of distinguishing features of any system is its causal entailment structure, and not the caused behavior, it is less confusing to say that a system with a multi-level self-referential causal entailment structure is an endogenous system that produces bizarre effects, that are predicted (often, but not always) by an impredicative model with a multi-level self-referential inferential entailment structure.

In describing the systems that produce bizarre effects, we need to recall that they come in two distinct flavors, natural (ontological) and formal (epistemological). In the Modeling Relation, the two are never identical. Since much of the confusion in 20th century science arose from not noticing this distinction between ontology and epistemology, it would be wise, as we lay the foundations for a new century of thinking, to adopt terminology that never lets us forget the distinction.

Finally, how do we manipulate an impredicative model to make decisions about an endogenous system? Since the processes are incomputable, an algorithm is useless for the task. There are several strategies that have not yet been proved impossible, the super-Turing model, quantum computing, and wiring the human directly into the loop

ACKNOWLEDGMENTS

The author thanks Glenn Allgood and Roger Kisner of Oak Ridge National Laboratory (ORNL) for their invaluable review and comments. This paper is based in part on research performed at ORNL, which is operated by UT-Battelle, LLC, for the US Department of Energy under Contract NO. DE-AC05-00OR22725.

REFERENCES

Adler, M.J., 1978, *Aristotle For Everybody*, New York: MacMillan, pp. 160-167.
Akansu, A.N. and Haddad R.A., 1992, *Multiresolution Signal Decomposition*, San Diego: Academic Press, pp. 313-315.
Barwise, J. and Moss, L., 1996, *Vicious Circles: On The Mathematics Of Non-Wellfounded Phenomena*, Stanford, CA: Center for the Study of Language and Information.
Chomsky, N., 1993, *Language and Thought*, Wakefield, RI: Moyer Bell.
Clayman, C.B., 1989, *The American Medical Association Encyclopedia of Medicine*, New York: Random House.
Gavin, W.T. and Kydland, F.E., "Endogenous Money Supply and the Business Cycle, " *Review of Economic Dynamics*, Vol. 2, No. 2, April 1999, pp. 347-369.
Kleene, S., 1950, *Introduction to Metamathematics*, Amsterdam: vanNostrand. pp. 41-43.
Knuth, D.E, 1973, *The Art Of Computer Programming*, 2d ed., Vol. 1. Reading, MA: Addison Wesley. pp. 4-6.
Margulis, L. and Sagan, D., 1995, *What is Life?* University of California Press, pp. 17-20.
Maturana, H.R. and Varela, F.J., 1998 *The Tree of Knowledge*, Shambala, pp. 41-52.
Murphy, M.P. and O'Neill, L.A.J., 1995, *What is Life? The Next Fifty Years*, Cambridge University Press, pp. 57-66.
Rosen, K., 1999, *Discrete Mathematics And Its Applications*. 4th ed., Boston: WCB/McGraw-Hill, p. 374.

Rosen, R., 1991, *Life Itself: A Comprehensive Inquiry into the Nature, Origin and Fabrication of Life*, New York: Columbia University Press.
Russell, B., 1931, *The Scientific Outlook*, First Edition, New York: W.W Norton.

PART VI: ADAPTIVE CONTROL

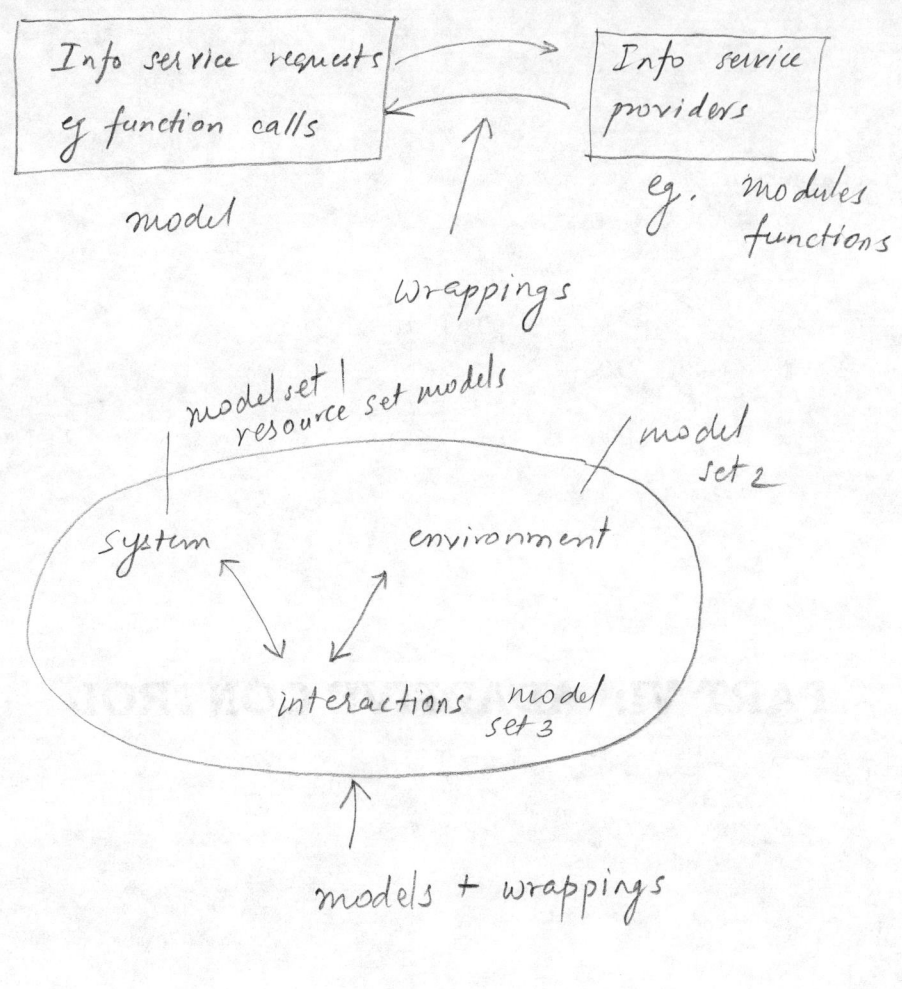

ARCHITECTURES FOR AUTONOMOUS COMPUTING SYSTEMS

CHRISTOPHER LANDAUER, KIRSTIE BELLMAN
cal@aero.org, bellman@aero.org
Aerospace Integration Science Center
The Aerospace Corporation
Los Angeles, California 90009-2957, USA

ABSTRACT

An autonomous system is one that participates in the construction, execution, and evaluation of its own goals, and carries them out according to its own capabilities. For this approach to be viable in a Constructed Complex System, the system needs to maintain a large number of interacting models, including those of the system itself, its expected, intended, and observed environment, the current situation and many potential future situations, as well as models that describe the interactions among the other models. Autonomous systems also require anticipatory models, since purely reactive ones are too slow in a complex environment. These anticipatory models are then specialized according to what might occur, so that decisions among them can be made quickly, according to easily recognized distinguishing features of the situation. In this paper, we describe an architecture for autonomous systems that supports the large number of dynamic models and the flexible model interaction required.

INTRODUCTION

An autonomous system is one that participates in the creation of its own goals, in addition to carrying them out with little or no supervision. It must therefore have computable access to descriptions of its own capabilities, the environment in which it is expected to perform, and the constraints on acceptable (or desirable) behavior. This paper is about constructing computer programs that have this level of autonomy. We describe a set of requirements for effective autonomy, and then exhibit an architecture that supports all of them (Landauer-Bellman, 1999b, 1999c).

A *Constructed Complex System* is a heterogeneous system that is mediated or managed by computer programs (Bellman, 2000) (Bellman-Landauer, 2000a, 2000b). It is our contention that an autonomous system must be a Constructed Complex System, since it has to balance many different forces, both external and internal. In order for us to build such systems, and especially for us to be able to change them as our understandings and the available technologies improve, we need to build a very flexible infrastructure that supports different kind sof organization of the computational resources that the system will use. Our architecture uses the Computational Reflection of a Wrappings integration infrastructure (Landauer-Bellman, 1998) to manage the large collection of models of observations, situations, behaviors, effects of actions, and predictions.

AUTONOMY

There are many versions of autonomy in the literature, and even many different ways to define and measure some kind of gradations of autonomy, but there is little consensus as yet. For us, the ability to construct one's own goals is the most important property of autonomy. A thermostat, a batch program keypunched onto computer cards, and even a rock, are all autonomous according to the first property alone, but we do not find these examples at all interesting. We want our autonomous systems to help construct their own goals.

Even after the goals are available, there are still difficult decisions to be made. In our opinion, there are really two classes of (difficult) requirements for effective autonomous behavior (Landauer-Bellman, 2000b): robustness and timeliness. Robustness means graceful degradation in increasingly hostile environments. Timeliness means that situations are recognized "well enough" and "soon enough", and that "good enough" actions are taken "soon enough". Both of these are forms of adaptive behavior, and neither one of these necessarily implies any kind of optimization.

Our approach to autonomy is to have the system build and maintain many models, of the system architecture and components, of the environment and expected environment, of the connection between the system and the environment, and of the effects of behaviors, both internal and external, both actual and potential. An essential capability for that process is Computational Reflection (Landauer-Bellman, 2000b), which allows a system to have a computable model of its own behavior and decision processes, so that it can assess the effects of its own decisions, and monitor its own actions. It is especially important for an autonomous system to monitor the effects of its decisions, so that it can back out of bad ones soon enough.

Since the system environment is largely unknown and almost completely uncontrollable, the system designers will not be able to provide very good models of its behavior, so we expect the system itself to construct models, based on general principles that are provided, and largely empirical and inferential methods. These environmental models will be derived from observations of the environment and the effects of the system's actions on the environment (the empirical methods, mainly induction of patterns), and from attempts to determine causes for the observed effects (the inferential methods, mainly abduction of causes).

This system will also need anticipatory models, since reactive models are way too slow. Reactive systems must recognize a significant event, and compute an appropriate reaction to it, whereas an anticipatory system can prepare to react to some significant events that have not yet occurred. This preparation saves much computation time, and reduces the delay time required between event and reaction to that for just recognizing the event. It is clear here that deciding correctly which events to prepare for is an essential part of building an effective anticipatory system.

The system must keep track of all these models, organize them into effectively computational structures, and invent new ones as it proceeds through its environment. None of these tasks is easy, but we have developed an approach that addresses all of these issues. To manage all of these models, the system uses Wrappings, our knowledge-based approach to integration infrastructure (Landauer-Bellman, 1999a, 1999b, 1999c), which is based on two key complementary parts: (1) explicit, machine-interpretable information ("meta-knowledge") about all of the uses all of the computational resources in a *Constructed Complex System* (including the user interfaces and other presentation mechanisms); and (2) a set of active integration processes that use that information to Select, Adapt, Assemble, Integrate, and Explain the application of these resources to posed problems. We have shown how the approach can be used to combine models, software components, and other computational resources into Constructed Complex Systems, and that it is a suitable implementation approach for complex software agents (Landauer-Bellman, 1999a, 2001).

AUTONOMOUS SYSTEM ARCHITECTURE

In this section, we describe the architecture of our autonomous systems, which essentially defines a design space in which the systems are organized. We use Wrappings for the infrastructure, so our architecture can be described in terms of the resources it contains. These resources include all of the components and interactions, and any planners and process monitors that may exist.

We also consider the "economic" claim, that complex computation benefits from competition for scarce resources. In the internals of a computing system, there are only two scarce resources: attention and retention. Attention is about what problems are being considered at any time (that is, access to processing), and is managed by the Wrappings and the Wrapping processes (Landauer-Bellman, 2001). Retention is about what combinations of problem posings and resource uses are kept around (that is, access to storage), and is managed by the Conceptual Category data structures, which are treated in other papers (Landauer-Bellman, 2000a).

Together, these techniques provide a powerful infrastructure for studying autonomous systems.

Basic Structure

The basic structure of our autonomous systems is simple. We assume that the system sits in, and must interact with, some environment, and that most of its processing concerns making appropriate models of that environment, so that it can make informed decisions about its courses of action.

The system contains many models and interpreters, all treated as resources and organized by the Wrappings and the Wrapping processes. In fact, the activity loop in most other agent architectures can be implemented as a particular kind of Wrapping process called a Coordination Manager. In particular, several of the architectures in (Bradshaw, 1997a) and (Klusch, 1999) are simple to implement, as are most of the multi-agent system architectures (Ferber, 1999) (Stone, 2000) (Subrahmanian-et-al., 2000), since we can take each component or activity as a problem, and use Wrappings to map them into the appropriate resources. This is the usual way that Wrappings allow any special purpose control or planning algorithms to be used in their appropriate contexts.

The many kinds of models described later are also managed via Wrappings, since they are resources that are interpreted for many different reasons in the system (and their interpreters are also resources, as is expected in the Wrapping approach). The interpreters use the models to extract information from them, in response to information requests, or to produce behavior for the agent, in response to action requests. They also use the models to make predictions of the effects of actions, so that the system can compare and evaluate possible courses of action.

Models

The system contains several classes of models, of the system itself, of its environment, and of the interface and interaction protocols. Of course, the interface is really part of the system, but we emphasize it because it defines the kinds of universe that it is possible for the system to understand (or at least detect) and control (or at least affect).

There are also models of the space of possible behaviors, that is, the trajectories that the system may traverse, and its actions and the corresponding transitions (some of the actions of the system will cause changes in the environment in response). These are organized into interactive game strategies, in which the system can consider the alternation of its own actions and those of the environment (though of course, most complex environments are not strictly taking turns). These models can also be organized by time, both actual and potential, and also organized in several other ways.

Model Interactions

There are several different kinds of models:

- capability (this is the self-model from the Wrappings),
- empirical, *induction / generalization of context applns*
- inferential, *logical deductions → abductions*
- exploratory, and *→ speculative simulation outputs*
- anticipatory. *response predictions* ★

We describe these models in this subsection.

In order for the system to have a rich enough set of models to consider, we expect the system to build many of its own models. The system builds new models in two ways: empirical and inferential. The empirical models are the data-driven observations, with many measurement processes and induction as the main component, and the inferential models are the causality hypotheses. The system also needs to construct Wrappings for the new models, which in general is very difficult, but is relatively simple in this case: each of these model construction methods is based on a hypothesis about the input, and the model can be used in reasoning about that input. In particular, the system will evaluate the models according to how well they help the system predict its environment. This model development starts with observations of the environment and of itself. The Computational Reflection provided by Wrappings allows model analysis and improvement to include all of these kinds of models.

Empirical models use induction to construct descriptions of the observations. There are many approaches to inductive inference (Angluin-Smith, 1983), and which ones to use will depend on the application area, but there will at least be some common sequence recognition algorithms, and some partial parsing algorithms.

Inferential models use abduction to construct explanations of the observations and empirical models. The relationship between the inferential models and the phenomena they explain is exactly the same as the relationship between mechanism specifications and service specifications in communication protocol design: since we cannot expect the system to be able to invent all the implementations for its observed behaviors, we settle for trying a few well-known styles of implementation that are specific to the application domain. Both the empirical and inferential models are reactive, but they produce models that can be used for prediction. All of these models are constructed in model spaces, which are built using evolutionary or adaptive programming methods to develop more precise or more accurate models of the observations, the explanations, and the consequences of activity.

Decision Processes

Our use of Computational Reflection means that all of the decision and goal formation processes are written as resources that are selected like any others. Among these decision processes are some that use the empirical models for prediction: these anticipatory models are simulation of effects of actions, faster than real-time (or else they will not be useful), that allow evaluation of alternative courses of action. This use of anticipation is an extension of the usual interactive game strategies, since that kind of reasoning is usually purely reactive.

These decision processes maintain a specialization hierarchy of possibilities, including parallel hypothesis development and tracking, elimination of wrong choices, reduction of attention to weak choices, elaboration of active choices and prediction of effects. Goal formation is carried out during these decision steps, both inventing different kinds of goals (which model to construct, what activity to pursue, etc.), and evaluating them according to some general principles (these principles are supplied by the designers).

CONCLUSIONS AND PROSPECTS

We have described an architectural framework that allows many disparate kinds of architectures and process organizations to be used, in the same autonomous system, at different times or in different contexts, or even simultaneously for later comparison. This kind of flexibility is important for studying and comparing architectures for a particular problem domain, and for building systems that can be effective and robust in the face of unpredictable environmental actions.

We have argued that successful autonomy requires a complex, interlocking web of models and assessment processes, overseen and managed by a Computationally Reflective integration infrastructure. We have argued that our Wrapping approach provides the necessary facilities, and what an autonomous system looks like in this case. These autonomous systems have models of their own structure and behaviors, which they can use as part of the analysis of the effects of their actions.

REFERENCES

Dana Angluin, Carl H. Smith, "Inductive Inference: Theory and Methods", *Computing Surveys*, Vol. 15, No. 3, pp. 237-269 (Sep 1983)

Kirstie L. Bellman, "Developing a Concept of Self for Constructed Autonomous Systems", pp. 693-698, Vol. 2 in *Proc. EMCSR'2000, Symp. Autonomy Control: Lessons from the Emotional*, 25-28 Apr 2000, Vienna (2000)

Kirstie L. Bellman, Christopher Landauer, "Integration Science is More Than Putting Pieces Together", in *Proc. 2000 IEEE Aerospace Conf. (CD)*, 18-25 Mar 2000, Big Sky, Montana (2000a)

Kirstie L. Bellman, Christopher Landauer, "Towards an Integration Science: The Influence of Richard Bellman on our Research", *J. Math, Anal. Appls.*, Vol. 249, No. 1, pp. 3-31 (2000b)

Jeffrey M. Bradshaw (ed.), *Software Agents*, AAAI Press (1997a)

Jacques Ferber, *Multi-Agent Systems*, Addison Wesley Longman (1999)

Barbara Hayes-Roth, Karl Pfleger, Philippe Lalanda, Philippe Morignot, Marka Balabanovic, "A Domain-Specific Software Architecture for Adaptive Intelligent Systems", *IEEE Trans. Software Engr.*, Vol. SE-21, No. 4, pp. 288-301 (Apr 1995)

Matthias Klusch (ed.), *Intelligent Information Agents, Agent-Based Information Discovery and Management on the Internet*, Springer (1999)

Christopher Landauer, Kirstie L. Bellman, "Computational Embodiment: Constructing Autonomous Software Systems", pp. 131-168 in *Cybernetics and Systems*, Vol. 30, No. 2 (1999a)

Christopher Landauer, Kirstie L. Bellman, "New Architectures for Constructed Complex Systems", in *The 7th Bellman Continuum, Int. Workshop on Comp., Opt. and Control*, 24-25 May 1999, Santa Fe, NM (1999b); in *Applied Mathematics and Computation*, Vol. 120, pp. 149-163 (May 2001)

Christopher Landauer, Kirstie L. Bellman, "Architectures for Embodied Intelligence", pp. 215-220 in *Proc. ANNIE'99*, 7-10 Nov 1999 (1999c)

Christopher Landauer, Kirstie L. Bellman, "Relationships and Actions in Conceptual Categories", pp. 59-72 in G. Stumme (Ed.), *Working with Conceptual Structures - Contributions to ICCS 2000*, 14-18 Aug 2000, Darmstadt, Shaker Verlag, Aachen (2000a)

Christopher Landauer, Kirstie L. Bellman, "Reflective Infrastructure for Autonomous Systems", pp. 671-676, Vol. 2 in *Proc. EMCSR'2000, Symp. on Autonomy Control: Lessons from the Emotional*, 25-28 Apr 2000, Vienna (2000b)

Christopher Landauer, Kirstie L. Bellman, "Flexible Design Support Systems", These proceedings (2001)

Peter Stone, *Layered Learning in Multiagent Systems: A Winning Approach to Robotic Soccer*, MIT Press (2000)

V. S. Subrahmanian, Piero Bonatti, Jürgen Dix, Thomas Eiter, Sarit Kraus, Fatma Ozcan, Robert Ross, *Heterogeneous Agent Systems*, MIT Press (2000)

REAL-TIME APPLICATIONS ON A CMAC NEURAL NETWORK

GILLES MERCIER, DANA VÎLCU, LAURENT GEORGE
LIIA/Université Paris 12 Créteil
122, rue Paul Armangot, 94400 – Vitry sur Seine, France
e-mail: {vilcu,mercier,george}@univ-paris12.fr

ABSTRACT
In this paper we are interested in the conception of an embedded real-time application for a CPU card using a CMAC neural network. We recall the classical problem of conception of a CMAC neural network. A hardware implementation of the CPU card is described showing that the previous hardware limits can now be overcome with the current state of technology. Then we focus on the choice of a real-time operating system to run the application. Embedded solution is based on a real-time Linux kernel and its libraries facilities. Finally we present an application based on CMAC networks to ensure predictive control for medical images transmission on a wireless LAN, the application being implemented on the embedded CPU card.

1. INTRODUCTION

The concept of the CMAC neural network was object of numerous papers. However, few of them provide effectively hardware real-time implementation. The purpose of this article is to give a feasible solution from the hardware to the software implementation for an embedded real time application using a CMAC neural network.

The term of "real time application" will be used in the sense that the execution duration of the application has real time constraints. From theoretical point of view, an application can be seen as a set of tasks defined by their arrival laws, their execution duration and their real-time constraints. Running an application means to activate an executing program and from the operating system point of view, it is one process. Corresponding to the application's implementation, this process may start other processes or threads. Hence we can conclude: "the application is executed in real-time = the operating system is deterministic while running the application". Determinism of the operating system while running the application is possible for an embedded system if: (*i*) all type of tasks are know a priori and (*ii*) all the scenarios that can appear at the operating system level are executed in real-time. Furthermore, the kernel's specific overhead must be included.

Chapter 2 deals with the CMAC neural network: the principle and paradigm (§2.1), the calculus for the memory required by the CMAC algorithm (§2.2) and a previous experiment limited on a laboratory demo (§2.3). In Chapter 3 is presented a CPU card conceived to embed real-time applications based on CMAC networks. The feasibility of the hardware conception based on the actual technologies (§3.1) and the choice of a real time operating system running the application on the CPU card (§3.2) are described. Finally Chapter 4 presents the state of the development of an application consisting of capturing medical images and of transmitting them in real-time through a Wireless LAN

(Bluetooth). Each step of the process has been evaluated in simulation with the different Matlab Toolboxes. The physical realization is under progress.

2. STARTING POINT FOR THE CMAC NEURAL NETWORK

2.1. The CMAC principle and paradigm

The conceptual model of CMAC consists of two processing stages (Hassoun 1995, Mercier 1995). The input vectors (usually binary coded) of dimension R (called resolution) activate the state space detectors, a "virtual memory". Its size can be much larger than the physical memory a real system can support. The next transformation consists in mapping an element of the "virtual memory" to an element of the physical memory. The output function is the summation of the contents of the physical memory elements assigned by the input vectors.

The CMAC algorithm belongs to the class of adaptive multilayer neural networks, consisting of an adjustable combination of fixed basis functions and a linear update rule, and is described by the equations:

$$\begin{cases} h_0 = \phi(\|x - \gamma_g\|) \\ O_h = \sum_{j=1}^{M} w_{jk} * h_0 \end{cases} \text{ where } \phi(r) = \begin{cases} 1, & r \leq \dfrac{G}{2} \\ 0, & r > \dfrac{G}{2} \end{cases} \text{ is the basis function} \quad (1),$$

g is the jth basis function's center, h_0 is the height at the input x, w_{jk} is the contribution of the basis function to the hth component of the output O_h, and G is the generalization factor.

2.2. Compromise between resolution and memory size

The two parameters, the resolution R and the generalization factor G, are determining the quality and the speed of the CMAC training phase. The resolution defines the network precision in the sense of the length of the input coding. On the other hand it induces problems on the size of the memory to be used, on the number of examples used, and on the duration of the training phase. The generalization factor influences the correctness of the network output in the sense of some norm.

For each input space (ex. in a robot positioning: one input space per axe, hence three input spaces) the number of elements is $Ne=2^R$ elements, and the corresponding number of connectors for each input space is $Nc=2^R+G-1$.

Searching a relation to escape from the connection matrix complexity induces a new constraint: the difference between the two vectors, m and m', of the corresponding virtual memory excited cells, respectively for two consecutive inputs (ex. (x,y) and $(x+1,y)$ or, similarly, for $(x,y+1)$) is given by only one element of the vector. Hence, the $\underbrace{G \times G \times .. \times G}_{N}$ generalized matrix for N input variables, extracted from the matrix of connections, will contain G non-zero elements disposed in a regular order. Generalizing the calculus of the physical memory size describe by Mercier (1995) for two input variables, we obtain the following:

1. The connection matrix is given by:

$$d(c_1, c_2, \ldots, c_N) = \begin{cases} 0, & \text{if } \frac{N_C^{N-1} \cdot c_1 + N_C^{N-2} \cdot c_2 + \ldots + c_N}{G} \notin \mathbb{N} \\ \frac{N_C^{N-1} \cdot c_1 + N_C^{N-2} \cdot c_2 + \ldots + c_N}{G} + 1, & \text{else} \end{cases} \quad (2).$$

2. The number N_m of virtual memory cells is $N_m = \left\lfloor \frac{N_C^N + G - 1}{G} \right\rfloor$.

The values corresponding to a physical memory used by real applications, for a resolution at least 8, and 2, 3 or 4 input variables, can be measured in terms of thousands of Kbytes. Hashing compression techniques used to reduce the virtual memory to a physical memory (Kraft, 1989) generate approximation points' positions with white noise. The actual physical memory technologies permit direct overlap of virtual CMAC memory and the physical one.

2.3. Results of the first experience

Mercier (1995) proposed a modified algorithm to reduce the computation time. It was used in the experiment detailed in (Mercier and Madani, 1995). The program was developed in assembling code, and the application was integrated in DSP fixed point with 50ns access time. The maximum available memory size at that time was 128 KB. They obtained a practical physical adaptive DC motor controller, two 8 bits resolution entry variables, and a generalization factor equal to 8. Results were good but not sufficient for real applications. However, the learning period was rapid (less than 1ms), due in great part to the improved algorithm.

3. FEASIBILITY EVALUATION FOR A REAL TIME EMBEDDED APPLICATION USING THE CMAC NETWORK

Limitations to realize an embedded real-time application based on a CMAC network were (Mercier and Madani, 1995) of three kinds, due to the available technology: weakness size of active memory, no existence of a compact integrated multitasking real-time kernel, limitation of the CPU frequency clock. We present here a project of implementation for the above presented algorithm of the CMAC network using nowadays technologies. The obtained system might be used for several real time applications. Chapter 4 describes one of this.

3.1. Hardware solution: CPU and memory

The system consists mainly of two superposed cards. The first one is the CPU card including a MPC555 with numerous I/O (analog, timer clock, PWM, parallel I/O, etc...), 448 Kb internal flash memory used for a compact integrated real time kernel and the application using the CMAC algorithm, 256 Kbytes of standard CMOS static memory. The physical CMAC memory map is external, linked to the CPU by the specific E-Bus.

The MPC555 (Motorola, 1999) offers important features for a smart system implementing an application using a CMAC network: PowerPC core with floating-point unit, 26 Kbytes fast RAM and 6 Kbytes TPU micro code RAM, 448 Kbytes flash EEPROM with 5-V programming, 5-V I/O system, sub micron HCMOS (CDR1) technology. Also, MPC555's case dimensions and packaging give 2.65 x 27.00 BSC x 27.00 BSC mm size dimensions. The memory controller provides glue less interface to EPROM, static RAM (SRAM), Flash EPROM (FEPROM), and other peripherals. Each memory bank includes a variable block size of 32 Kbytes, 64 Kbytes and up to 4 Gbytes and it can be selected for read-only or read/write operation.

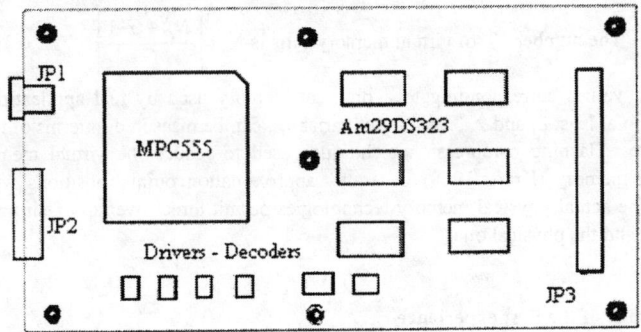

Figure 1: Main CPU card

The second card is plugged on the CPU card via the 25 points JP3 connector. JP1 connector is a serial RS485 link for programming and debugging the application. JP2 connector is used to connect the 3-5 V DC alimentation and to select switches for stand-alone or batch starting programs. The board size is 60 x 120 mm. Figure 1 shows the implementation of the components, all soiled in CMS technology. The card has also 6 x Am29DS323 32 Mbit 3 Volts, organized in 4 x 8 bits, giving 32 Mbytes flash memory; it is sufficient, by previous considerations, for some real applications to stock the CMAC learned weights.

3.2. Real-time operating system solution and application's determinism

For determinism purpose we prefer to work on an open source embedded real-time kernel. This permit us to configure it, to write or rewrite some parts of it, and to precisely measure its specific costs in time and in processor capacity at the tasks' level.

We have chosen the real-time Linux kernels family. The Linux kernel permits to lock the real-time tasks in the memory and to have a special scheduler to ensure a predictive schedule for the real-time tasks. With Linux kernel starting point, some other variants were developed to more precisely ensure the applications' determinism running on: RTLinux (15), RTAI (16), KURT (18), RED Linux (17), TimeSys Linux/RT™ (21), MontaVista's "hard real time" kernel (22). A reference study on these kernels suggests, for deterministic reason, to place RTLinux and RTAI in top of them. These two kernels have a similar conception, they differ at the implementation level, and the other kernels are based on them or suggest using them for real-time applications with hard real-time constraints.

On one hand, the way the operating system is managing the other resources also influences the application's tasks execution duration, hence its ability to meat the real-time constraints. The processes scheduling policies do managing CPU's resources. They have one of the key roles to guarantee the determinism of the application execution. The variants of real-time Linux kernel presented give the possibility to add, or even implement, other scheduling algorithms like Rate Monotonic (RM) or Earlier Deadline First (EDF), theoretically known as optimal in their respectively class of real-time scheduling algorithms (George et al., 1996).

On the other hand, to ensure a real-time execution for the application, one has to determine the minimal hardware resources the system running the application must have. Using a main CPU card with MPC555 we are restricted for the choice of the operating system. Its 448 Kbytes flash EEPROM have to be sufficient to embed the application and the host dedicated operating system. The solution might be Lineo's RXTC kernel (a version of the RTAI kernel) in its embedded version for MPC555, conform to (20).

From the software point of view, constructing a real-time application could be done in the kernel space by using the kernel and, generally, the operating system facilities, or in the user space by using the user libraries. For our purpose, we chose to implement the application using the C language and the libraries offered by LXRT. The application's tasks will be real-time kernel threads. An attentively analysis of the overlap of the theoretically application's scenarios and the corresponding scenarios at the operating system level must be done to ensure the determinism of the application's execution.

4. PRINCIPLES OF A REAL-TIME CMAC APPLICATION

We expose here an embedded application in development based on the MPC555 CPU card previously proposed, with three parallel CMAC networks working in predictive control mode, the application's tasks being synchronized by a modified RTAI kernel. All the software modules of the application have been evaluated and verified under Matlab simulation and its various toolboxes (image processing, wavelet, control process, real-time workshop).

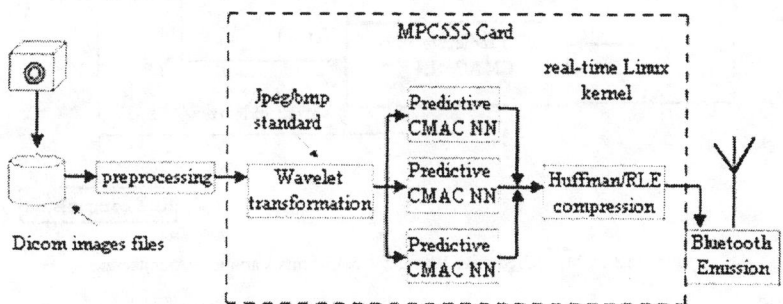

Figure 2: Architecture of the application for the emission node

The basic idea of this application consists of capturing ACR/NEMA DICOM3 images (Digital Imaging and Communication in Medicine) and of transmitting them in real-time through a Bluetooth wireless network. The objective of this project is to disconnect the caption movies point from the monitoring station, i.e. no wired link, in order to facilitate the mobility of the camera. On the reception side, a reverse system read the data collected by the Bluetooth receiver, and code them into a Dicom, JPEG or BMP format to be visualized on a numeric monitor. The transmission side architecture consisting of the embedded system is presented in Figure 2.

Images capture is made with a mobile camera. For the purpose of the laboratory evaluation, Dicom images are stored on the hard disk. The preprocessing level converts ACR/NEMA standard, either in GIF, BMP or JPEG files. Papyrus 3.4-version software freely available makes this conversion easy. Our evaluation has been made with BMP standard which is the most expensive in terms of memory size. A Discrete Wavelet Transformation (DWT) is assumed in real time for each image stored in the CPU MP555 card. Evaluation tests have done with Image Processing and Wavelet Matlab Toolboxes. Best results were obtained with a bi-orthogonal wavelet (Bior 6.8) and its simplified best tree coefficients of level 2: approximation - Ca2, horizontal - Cdh2 and vertical - Cdv2 Details (see Matlab Wavelet Toolbox documentation for more details). On the reception side, the Forward Discrete Wavelet Transformation reconstructs the original image. After more than 200 different trials, pixels errors were less than 0.1%.

The system is composed by a single Bluetooth piconet with one master and one slave. The receiving side is the master part, and the transmitted side is the slave part of the network. At the time t0, a concatenated file, including Ca2, Cdh2 and Cdv2 full coefficients are transmitted on the Bluetooth emitter. Transmitted time of this packet exceeds a single packet size of 625 microseconds (refer to 23 for Bluetooth's specifications). A complete image is transmitted every 30 slots. During the 29 others slots (t1 to t29), only the variations of the coefficients are emitted synchronously with the master request (i.e. every 1.25 millisecond).

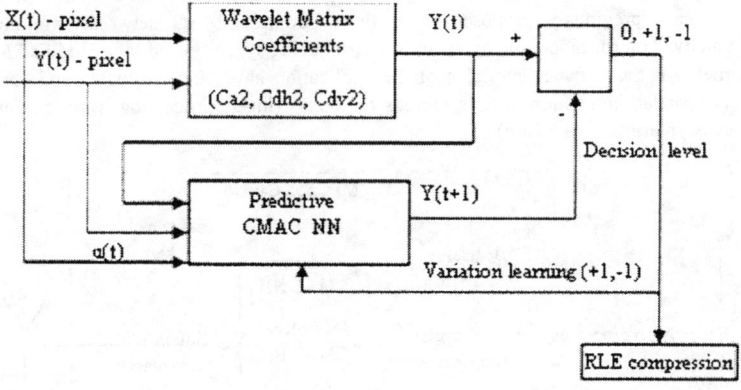

Figure 3 : CMAC Predictive Wavelet Coefficients Variations Architecture

These variations must be computed before the transmission. The interest of the predictive CMAC model is due to the necessity to be synchronized with the master request. The principle to implement a real-time predictive application based on CMAC network is retained from the applications developed by Kraft (1989) and Mercier and Madani (1995); to maintain a synchronism between the captured image and its transmission we have to predict the wavelets coefficients values. The architecture of the computation is presented in Figure 3.

Three CMAC networks are working in parallel to calculate Ca2, Cdh2 and respectively Cdv2 coefficients. The corresponding tasks of each CMAC network have the same program.

The position values X-Y of one pixel at the moment (t) are presented simultaneously for DWT computation and on CMAC entries. Values of the coefficient Ca2 (t) (respectively of Cdh2 and Cdv2) are also presented to the corresponding CMAC entries. The output matrix coefficients are of floating type (4 bytes). A type transformation is necessary, because CMAC entries are of binary type. CMAC entry is consists of two binary sensors of equivalent size R resolution: the first entry is a combination of X(t) image pixel and MSB y(t+1) wavelet coefficient predictive value; the second entry is composed by Y(t) image pixel and LSB y(t+1) wavelet value. CMAC output performs the (t+1) coefficient prediction; this one is compared to the corresponding (t) time equivalent coefficient. A comparator generates a 3 states decision : 0 if the coefficient variation is less than a defined level (for the tests we chose the 1% error level), +1 or -1 otherwise. Widrow-Hoff learning rule is used to calculate the CMAC weights. Using Beta=1 a single step is sufficient to conveniently adjust the CMAC weights, so having the minimum learning time. No instabilities were identified during the tests. When all pixels of the image have been scanned, the 3 wavelet coefficients variations are concatenate in a single file including large series of "0 decision". A RLE compression is performed before transmission. A Huffman compression might also be implemented. Table 1 presents the main results of the evaluation tests for more than 200 trials. A scheduling implementation of different tasks has been evaluated with Matlab Real Time Workshop. Its conversion to a real-time Linux kernel is still in progress.

Image format	.bmp
Image size	256 x 256 x 1 bytes
Wavelet coefficients size	76 x 76 x 4 bytes
(Ca2, Cdh2, Cdv2)	x 3 (number of coefficients)
CMAC network dimensions	33555 Kbytes
(G=1, R=11, 2 entry variables)	x 3 (number of CMAC networks)
Average wavelet coefficients variation	76 x 76 bytes
(200 tests)	x 3 (number of coefficients)
Average RLE compression of the wavelet	344 x 2 bits
coefficients variation (before Huffman compression)	x 3 (number of coefficients)

Table 1: Application's characteristics used for simulation

5. CONCLUSIONS AND FUTURE WORK

In this paper we were interested in the conception of a embedded real-time application for a CPU card using a CMAC neural network. We have described the CPU card showing that previous technical limits can now be overcome. The presented solution allows a direct physical memory use, therefore without theoretical noise generated. The memory size can be extended by adding other external flash memories of the same type as the physical memory used, compatible with the dimensions of a realistic card and the application's needs. A variant of RTAI kernel was chosen to implement the application directly on the CPU card. The conception of a real-time application was described. We proposed a real-time system for medical image capture and transmission through a wireless LAN. Three CMAC networks perform a predictive estimation of the next coming image. The real-time kernel ensures the synchronism of parallel tasks. Scheduling analysis of the embedded processes and the realization of the CPU card are in progress.

REFERENCES

1. Bar, M., 2000a, "Linux Internals", McGraw-Hill
2. Blaess, C., 2000, "Programmation système en C sous Linux", Ed. Eyrolles
3. George, L., Rivierre, N., Spuri, M., 1996, "Preemptive and Non-Preemptive Real-Time Uniprocessor Scheduling" INRIA Rocquencourt, France, Research Report 2966
4. Hassoun, M. H., 1995, "Fundamentals of artificial neural networks", MIT Press
5. Kraft, L. G., 1989, "A Comparison of CMAC Neural Network and Two Traditional Adaptive Control Systems", The American Control Conference 1989, IEEE Control System Magazine, 10(3), 36-43
6. Mantegazza, P., Bianchi, E., Dozioi, L., Papacharalambous, S., Hughes, S., and Beal, D., may 2000, "RTAI : Real Time Application Interface. Le temps réel sous Linux", in "*Linux+*", N° 4
7. Mercier, G., 1995, "Implantation d'Algorithmes Neuronaux pour Applications Temps Réel", PhD Thesis, Laboratoire de Robotique, Université d'Evry, France
8. Mercier, G., and Madani, K., "CMAC Real-Time Adaptive Control Implementation on a DSP Based Card", in Mira, J., Sandoval, F. (Eds.), 1995, "From Natural to Artificial Neural Computation. International Workshop on Artificial Neural Networks Malaga-Torremolinos, Spain, June 1995"
9. MOTOROLA, 1999, "MPC555 User's Manual", revised 15 September 1999
10. Nilsson, J., and Rytterlund, D., 2000, "Real Time Linux Investigation", Mälardalen Real-Time Research Center Report
11. Nutt, G., 2001, "Kernel Projects for Linux", Addison-Wesley
12. Rusling, D. A., 1999, "The Linux Kernel", Version 0.8-3, http://www.ibiblio.org
13. Silberschatz, A., Galvin, P., and Gagne, G., 2001, "Principes appliqués des systèmes d'exploitation", Vuibert, Paris (original : 2000, "Applied Operating System Concepts", First Edition, John Wiley & Sons, Inc.)
14. "Yodaiken comments on MontaVista «hard real-time» Linux kernel", 12 sep. 2000, http://linuxdevices.com
15. http://www.fsmlabs.com for RTLinux
16. http://www.aero.polimi.it/projects/rtai/ for RTAI
17. http://lingo.ece.uci.edu/~wycc/REDLinux.html for REDLinux
18. http://www.ittc.ukans.edu/kurt/ for KURT
19. http://www.amd.com/products/nvd/techdocs/23480.pdf, "Am29DS323D 32Megabit (4M x 8-Bit / 2M x 16-Bit) CMOS 1.8 Volt-only, Simultaneous Operation Flash Memory", November 22, 2000
20. http://www.lineo.com/products/rtxc/kernel/index.html for the list of supported processors by RTXC™ from Lineo™
21. http://www.TimeSys.com for TimeSys Linux/RT
22. http://www.mvista.com for Montavista's hard real-time kernel
23. http//:www.Bluetooth.org for Bluetooth's specifications

OFF-LINE ADAPTION IN AN ITERATIVE LEARNING CONTROL SCHEME WITH PLANT NON-LINEARITIES

LASZLO M. HIDEG
DaimlerChrysler Corporation
Powertrain Product Engineering
CIMS 482-02-19
Auburn Hills, Michigan 48326, U.S.A.
LMH10@DCX.COM

ABSTRACT
Iterative learning control (ILC) can be a robust tool in manufacturing and other applications. It has been shown to have many inherent qualities that promote stability. This paper will examine the affects of non-linear elements within the scheme, and suggest an adaption based on the iteration number. The adaption will combine qualities of convergence and robustness that promote a maintained best fit trajectory. Convergence using a linear model during learning will be shown for different non-linear elements in a plant. As in most non-linear controls, convergence to zero error is not guaranteed, though convergence to some finite error is possible.

1 INTRODUCTION
It has been found that Iterative Learning Control tolerates many uncertainities: to anticipate fixed disturbances (Hideg and Judd 1988); stabilize unstable plants (Hideg 1993) further and further in the time domain for successive iterations; handle some plant non-linearities (Hideg and Judd 1992); and tolerate time delays in the scheme (Hideg 1995). Time delay feature permits a pseudo time integral term in ILC (Hideg 1996). This leads to meaningful preset values for neural net elements (Hideg 1998). This paper examines shortcomings of an ILC based only on modeling the linear portion of a plant. Off-line adaption of the model, (Goodwin 1984) in addition to the control effort between iterations is proposed. Adaption combines the robustness of a weak ILC to the faster convergence of a strong ILC (Hideg 1992), (Kawamura, 1987). Section 2 outlines Iterative Learning Control formulations and expands on the conjecture of inherent stability. Section 3 has simulations. Section 4 has conclusions. The iterative learning control equations organization with off-line parameter adaption is in the appendix, Figure 1A.

2 ITERATIVE LEARNING CONTROL
Consider the following equations. The plant is Equation (1), and error measure is Equation (2). The desired trajectory is $x^d(m,t)$. The iterative learning control is Equation (3).

$$d/dt\ x(m,t) = -a \cdot x(m,t) + b \cdot \beta(u(m,t)) \qquad (1) \qquad e(m,t) = x^d(t) - x(m,t) \qquad (2)$$

$$u(m+1,t) = u(m,t) + c \cdot d/dt\ e(m,t) + d \cdot e(m,t) \qquad (3) \qquad d/dt\ x^d(t) = -a \cdot x^d(t) + b \cdot \beta(u^d(t)) \qquad (4)$$

Iteration index is a non-negative integer m. It is incremented for every completed run of the system in the time domain, t. In Equation (1), the non-linearity $\beta(\cdot)$ acts on the input $u(m,t)$. This analysis has practical value due to start-up frictions and saturations in mechanical systems, and similar limits in electrical and other systems. It is assumed that Equation (4) exists, and that for a reasonable initial control effort, $u(0,t)$, and selection of c,d, the system Equation (1), yields $\|e(m,t)\| \to 0$ as m increases. Consider incrementing Equation (1) by one iteration m+1. Substitute in Equation (3), to get Equation (5). If the ILC reduces error, then Equation (5) begins to look like Equation (6). This characteristic would support (7).

$$d/dt\ x(m+1,t) = -a \cdot x(m+1,t) + b \cdot \beta(u(m,t) + c \cdot d/dt\ e(m,t) + d \cdot e(m,t)) \qquad (5)$$

$$d/dt\ x(m+1,t) = -a \cdot x(m+1,t) + b \cdot \beta(u(m,t)) \qquad (6) \qquad \|x(m+1,t) - x(m,t)\| \to 0 \text{ as m increases} \qquad (7)$$

Because of the unmodeled $\beta(\cdot)$, one would believe that in some norm, Equation (7) is clear, but not necessarily that $\|e(m,t)\| \to 0$ as m increases. For robustness and slow convergence, c and d need to be small (Weak ILC). For faster convergence, the values of c and d need to minimize $|1 - b\cdot c|$ and $|a - b\cdot d|$, (Strong ILC) (Hideg 1992). However, with a unmodeled non-linearity $\beta(\cdot)$, the strong ILC convergence to a best trajectory is not smooth nor maintained. If the ILC is excessive, then the process can go unstable.

To take advantage of weak ILC robustness and strong ILC convergence, an adaption of the ILC model parameters c and d is proposed. Adaption during a particular trail in the time domain would fall under well known adaptive and optimal control techniques. With ILC, the opportunity exists to change the form of the control between the trials in the m domain. The following algorithm is the adaption used to improve the overall performance of the ILC.

TABLE: Adaption Algorithm in the m Domain

err(m) = $\|x^d(t) - x(m,t)\|$
err(m-1) = $\|x^d(t) - x(m-1,t)\|$
delc = (1+|err(m-1)-err(m)|)/err(m)
deld = (1+|err(m-1)-err(m)|)/err(m-1)
if err(m) > err(m-1) then c(m) = c(m-1)*delc,
    d(m) = d(m-1)*deld end
if err(m) < err(m-1) then c(m) = c(m-1)/delc,
    d(m) = d(m-1)/deld end
if c(m) < cstart/10 then c(m) = cstart/10
if c(m) > cstart•10 then c(m) = cstart•10
if d(m) < dstart/10 then d(m) = dstart/10
if d(m) > dstart•10 then d(m) = dstart•10

$$u(m+1,t) = u(m,t) + c(m)\cdot d/dt\, e(m,t)$$
$$+ d(m)\cdot e(m,t) \qquad (8)$$

$$\beta(y) = y\cdot|y|/(1+|y|) \text{ 'Dead Zone'} \qquad (9)$$

$$\beta(y) = y/(1+|y|) \text{ 'Saturation''} \qquad (10)$$

$$u(0,t) = x^d(m,t) = k\cdot t/(1+t)\cdot\sin(\omega\cdot t) \qquad (11)$$

The difference in the denominators in c(m) and d(m) are merely to have some variety in their values, rather than track identically. The ILC model values c(m) and d(m) are limited to maintain computational integrity during simulation. This adaption between iterations shows an improvement in tracking and an improved maintenance to a fixed trajectory after most convergence is complete. With the adaption in the m domain, Equation (3) is rewritten as Equation (8). Simulations will show the adaption between trials yields desired qualities of both the strong and weak ILC.

3 SIMULATIONS

Non-linearities Equation (9), Equation (10) will be used in separate simulations. The initial trajectory is based on the desired trajectory in Equation (11).

3.1. Simulation Fixed ILC, k=1, $\beta(\cdot)$ Eqn (9)

For simulations with Equation (9), plant parameters were a=b=1. Weak ILC model was c=d=0.1 in Figure 1. The stronger ILC model was c=d=0.35. in Figure 2. In Figure 1, the initial trail is the smallest sinusoid. Of the two larger sinusoids in Figure 1, the m=30 trial lags behind the desired trajectory. Similarly in Figure 2, the m=30 trial exhibits some erroneous tracking away from the sesired trajectory.

$\beta(\cdot)$ of Equation (9) is a sufficient uncertainity that convergence only occurs as expected in Equation (7). The stronger ILC converges quicker as shown in Figure 3. The weak ILC converges to a similar error though it takes more iterations. If the ILC was too strong ($|1-b\cdot c| \to 0$, $|a-b\cdot d| \to 0$), the process becomes unstable.

3.2 Simulation Fixed ILC, k=0.1, $\beta(\cdot)$ Eqn. (9)

The non-linearity of Equation (9) would be of greater significance if the input signal Equation (11) becomes smaller, Equation (12).

$$u(0,t) = x^d(m,t) = 0.1\cdot t/(1+t)\cdot\sin(\omega\cdot t) \qquad (12)$$

In Figure 4, by the m=30 iteration, the weak ILC shows smooth tracking to $x^d(t)$. In Figure 5, by the m=30 iteration, the strong ILC shows better tracking though some areas show an erratic track and a delaying phase shift. In Figure 6, the norm error shows inconsistent convergence and the hint of divergence for a strong ILC. A weak ILC is more robust and has consistent convergence in the norm. The convergence is faster with the strong ILC, Figure3, Figure 6. The time domain appearance of the strong ILC, Figure 2, Figure 4, while showing a closer tracking, still has some erratic tracking and a phase shift.

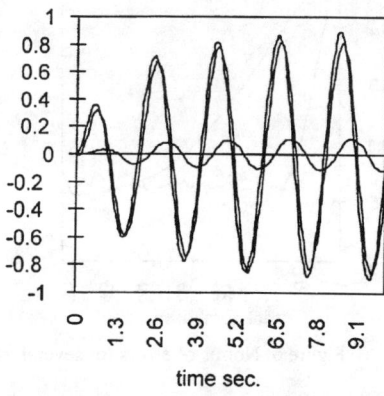

Figure 1: Weak ILC Tracking of $x^d(t)$

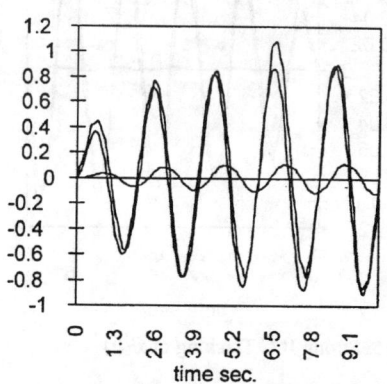

Figure 2: Strong ILC Tracking of $x^d(t)$

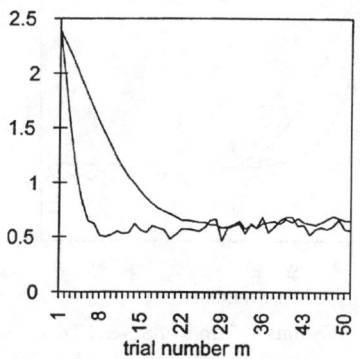

Figure 3: Norms of errors for several trails.

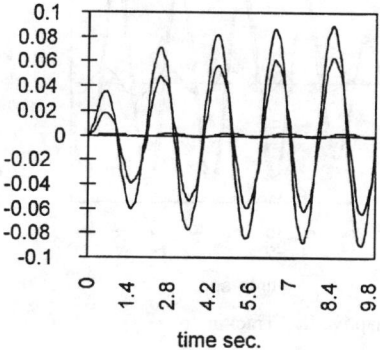

Figure 4: Weak ILC Tracking of $x^d(t)$

3.3 Simulation Adapt ILC, k=1, β(•) Eqn (9)

In Figure 7, Equation (11) is the input signal, and cstart = dstart = 0.3. Again, m=30 trial is displayed with m=1 trial and $x^d(t)$. Figure 8 is similar to Figure 3, with the norm error of the adaptive ILC included. The rather high intial model values help with early convergence. The adaption helps with maintenance of a trajectory close to the desired trajectory.

The top curve in Figure 8 is the weak fixed ILC of c=d=0.1. The bottom curve of Figure 8 is the strong fixed ILC of c=d=0.35. Note that the strong ILC does not maintain the trajectory as m increases. The adaptive ILC is the "middle" curve for small m values. It converges rapidly AND maintains the trajectory as m increases.

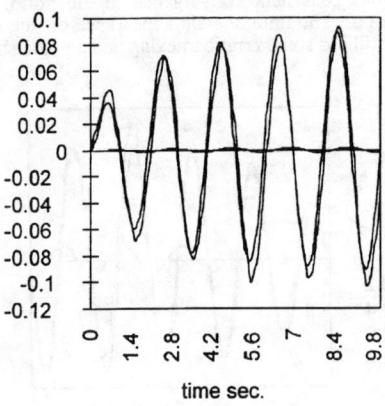

Figure 5: Strong ILC Tracking of $x^d(t)$

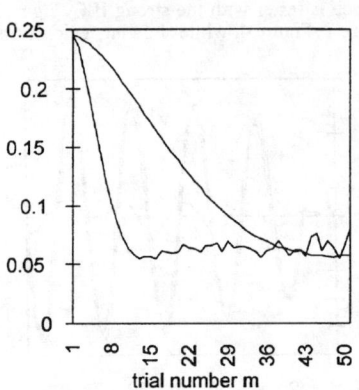

Figure 6: Norms of errors for several trails.

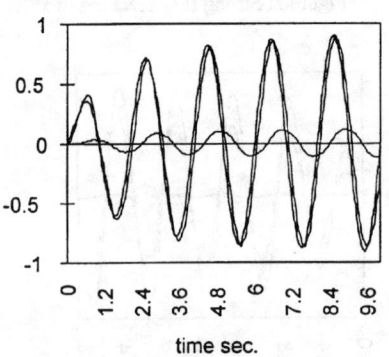

Figure 7: Adaptive ILC Tracking of $x^d(t)$

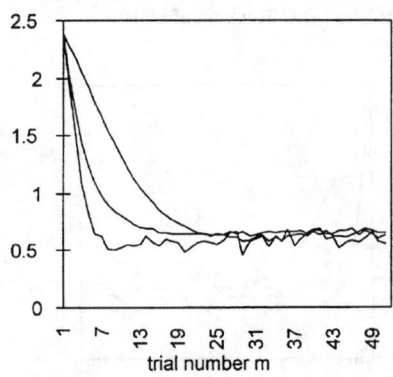

Figure 8: Norms of Errors of Several Trials

3.4 Simulation Adapt ILC, k=0.1, β(•) Eqn (9)

The same can be illustrated in the norm of errors for the small input signal, Equation (12), and the strong (c=d=0.35), weak (c=d=0.1) and adaptive ILC's. Again, the strong ILC converges quickly but does not maintain the trajectory. The weak ILC takes longer to converge. The adaptive ILC has the best of both. It uses cstart = dstart = 0.3. as analysis for Figure 7 and Figure 8.

3.5 Simulation Any ILC, k=1, β(•) Eqn(10)

When having a saturation element, Equation (10) in the plant, the norm error of fixed or adaptive ILC is somewhat larger that with the deadzone of Equation (9). This is expected since the control effort u(m,t) is limited by the saturation. There was no advantage to the strong ILC versus the adaptive ILC. The weak ILC also converges to the same norm error, though after more iterations, as expected. Once convergent, maintenance of a best fit trajectory was no better with strong ILC, weak ILC or the adaptive ILC. Unfortunately, the control signals u(m,t) that are a result of this situation can get arbitrarily large as the iterations increase. Such large signals are of no practical use.

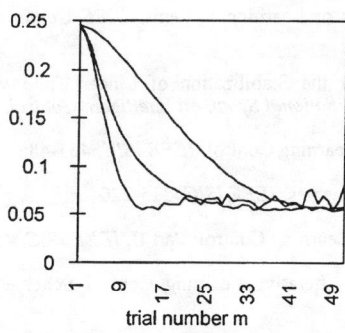
Figure 9: Norms of Errors of Several Trials

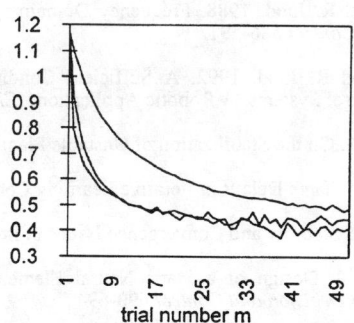
Figure 10: Norms of Errors of Several Trials

3.6 Simulation Fixed ILC, k=0.5, β(•) Eqn (10)

With k=0.5, the input signal is smaller than the great affects of the saturation of Equation (10) as an element of the plant. The stronger the ILC, that faster the convergence for small m. However, maintenance of the best trajectory as m increases is an issue. Also, keeping down the size of u(m,t) is also of concern because of the saturation element. In Figure 10, norms for several trails and configurations are illustrated. The slowest converging is a fixed ILC with c = d = 0.1. The next faster, with worse maintenance is a fixed ILC with c = d = 0.35. The third plot with a slightly faster convergence for small m is the adaptive ILC. Note that cstart = dstart = 0.5, a more aggressive ILC for small m. However, as m increases, the adaptice ILC performance is not remarkable. And, the adaptive convergence finished slightly above the strong ILC for large m.

Just as an illustration, what if the ILC is made too strong. Figure 11 shows the norm error result of a strong fixed ILC where c = d = 2 for several trials m. The fixed ILC of c = d = 0.1 is included for comparison. The strong ILC converges faster and the error is smaller at small m values, versus the weak ILC. As more iterations are completed, the instability becomes clear as the norm error exhibits an increasing trend.

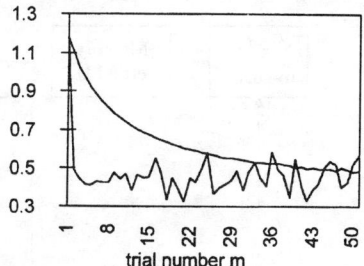
Figure 11: Norms of Errors of Several Trials

4 CONCLUSIONS

Iterative Learning Control shows robustness even when only the linear portion of the plant is modeled. An adaption scheme is suggested which would modify the ILC model between iterations. The rms error for each iteration is used to modify the ILC model for the next iteration. During any trial, the ILC model is kept constant. This type of adaption is between trials only, Equation (3) and Equation (8). This distinguished ILC from traditional adaptive control and other techniques that are in the time domain. Also, computation load is reduced during the plant operation. The computation is done, in many cases, when the plant is being reset for the next iteration, off-line. The simulations where a deadzone is in the plant shows that the adaptive ILC has merit since the convergence and maintenance of a best trajectory are apparent. This adpation shows no advantage in the situation involving a saturation element in the plant.

Future research suggests consideration of the magnitudes exhibited by the control efforts u(m,t). These input signals to the plant cannot be allowed to become large to make them impractical.

REFERENCES

Hideg, L., and R. Judd. 1988. Frequency Domain Analysis of Learning Systems, *IEEE Conference on Decision and Control* 586-591.

Hideg, L., and R. Judd. 1992. A Sufficient Condition for the Stabilization of Linear Time Varying Learning Control Systems: A Robotic Application, *IEEE International Symp. on Intelligent Control*, 71-76

Hideg, L. 1993. On the Stabilization of Unstable Plants Via Learning Control, *IEEE ISIC*, 481-486

Hideg, L. 1995. Time Delays in Iterative Learning Control Schemes, *IEEE ISIC*, 215-220

Hideg, L. 1996 Stability and Convergence Issues in Iterative Learning Control: Part II, *IEEE ISIC, 480-485*

Hideg, L. 1998. Design of a Static Neural Element in an Iterative Learning Control Scheme, *IEEE Conference on Decision and Control* 690-694

Hideg, L. 1992. Stability of Learning Control Systems, Doctoral Dissertation, Oakland University, Rochester, Michigan

Kawamura, S., F. Miyazaki and S. Arimoto, 1987 Intelligent Control of Robot Motion Based on a Learning Method, *IEEE Int'l Symposium on Intelligent Control*, 365-370

Goodwin, G., K. Sin, 1984 Adaptive Filtering, Prediction and Control, Prentice-Hall

APPENDIX

The block diagram illustrates the flow of information for the iterative learning and off-line parameter adaption of this paper. In Figure 1A, adaption cannot begin until the plant is ready to begin the m=2 trial.

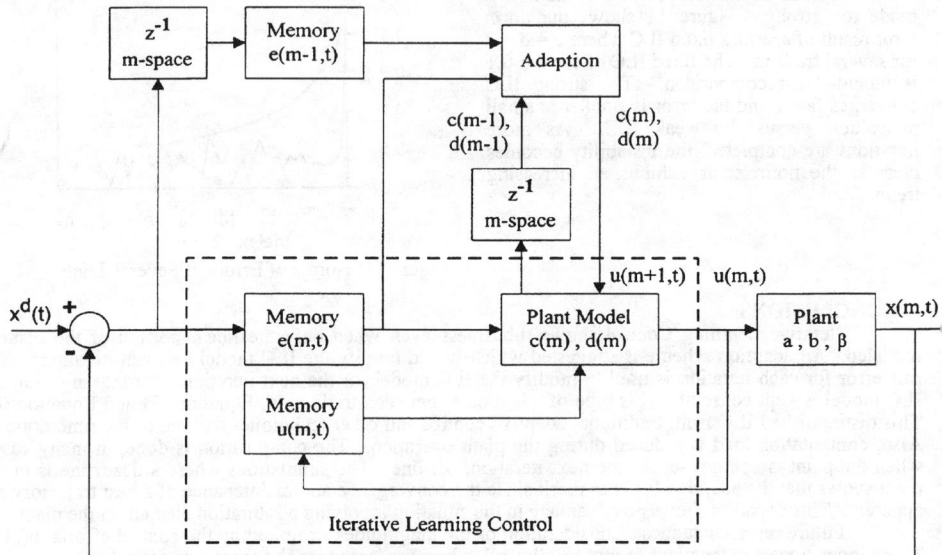

Figure A1: Block Diagram of the Iterative Learning Control Process
for the mth Trial with Adaption of the Plant Model

FLEXIBLE STRUCTURE MULTI-CELL ROBOT
- OBSTACLE AVOIDANCE USING AMOEBOID SELF-ORGANIZATION MODEL -

NOBUYUKI TAKAHASI
Autonomous System, Engineering,
Hokkaido University, Japan

TAKAHSI NAGAI
Autonomous System, Engineering,
Hokkaido University, Japan

HIROSI YOKOI
Autonomous System, Engineering,
Hokkaido University, Japan

YUKINORI KAKAZU
Autonomous System, Engineering,
Hokkaido University, Japan

ABSTRACT

In this paper, we present a distributed multi-cell robot control system that enables autonomous movement. Flexible deformation and actions are features of living things such as amoebae. One amoeba has neither specialized sensitive organs nor motile organs, however, as a whole, a slime of amoebae can present unified movement. That is, organisms such as amoebae own various self-organization abilities. The control system proposed in this study rose from such kind of self-organization, by which cell units move together to give rise to the movement of multi-cell robots, according to only local interaction. Especially, in this study, self-organization phenomena including entrainment of non-linear oscillators were modeled, and effectiveness was shown through computer experiments.

INTRODUCTION

Flexible structure robots and their control systems are of great importance in recent robotics research field. Robots with soft and flexible structure can be applied to the domestic or welfare usages. However, following features are required for such flexible robots.

(1) It should be able to adapt to changing environment.

(2) It should own large degrees of freedom (D.O.F.) and be redundant for its structural flexibility.

Various kinds of the flexible structure robots have been developed. The random morphology robot, proposed by Dittrich [1], consists of six servos coupled arbitrarily by thin metal joints. The Genetic Programming was used for the control system of the random morphology robot. The GMD-Snake, developed by Paap [2], consists of numbers of joints and is actuated by motors and strings. In the self-reconfigureable module robot developed by Murata [3], each module may be connected to and detached from another module's surface automatically.

Most of the flexible structure robots were composed of numbers of mechanical parts, and have numbers of actuators. The first difficulty when constructing a flexible structure robot lies in size of hardware, that is, how to construct robot simply and compactly by using or designing good actuators and mechanical components. The second difficulty lies in the construction of the control system of the robot, which is responsible for coordinating multi actuators to enable objective behaviors.

To deal with the first difficulty, we proposed a flexible structure robot that is actuated by shape memory alloy actuators (SMA-Net robot). Because this robot uses SMA springs as actuators and skeleton, fewer mechanical components are needed.

To solve the second difficulty, we proposed a control method that uses amoeboid self-organization model. Actually, the features necessary for flexible robots such as adaptation, redundancy, are the features of most living organisms. We took amoebae, well known as primitive and flexible organisms as the model, since amoebae (slime mold) present the ability of basic information processing, although, they does not have any specialized sensitive organ or motile organ. The purpose of this study was to explore a methodology for the control of amoeboid-like flexible structure multi-cell robotics systems.

SMA-NET ROBOT

The SMA-net robot developed consists of several homogeneous cell units. Each unit has six SMA spring actuators and one micro CPU module (Hitachi SH-4). Each SMA actuator is connected to its neighboring cell units (Fig. 1). In our SMA-net robot, SMA actuators were used not only as actuators but also as the skeleton of the robot. Therefore, SMA actuators worked like muscle fibers.

The SMA-net robot is composed of numbers of cell units and actuators. Therefore the number of control parameters increases. As general research approaches, Genetic Algorithm, Genetic Programming and Neural Networks are used for such kind of large-scale control problem. However, a great number of trials are necessary to acquire suitable control rules. Moreover, those approaches are not able to adapt to the cases in which environment changes, etc. In this study, a self-organization model referring to actual amoebae was proposed. By self-organization, not only the number of control parameters could be reduced, and learning cost could be decreased considerably, but also the robot would be able to adapt to environment changes.

Another feature of self-organization phenomenon is local interaction. It is self-organization or self-aggregation that produces spatial temporal macroscopic structure through micro local interaction. Accordingly, no central control system is required, so that, the self-organization based control system would be affected neither by increase nor decrease of the number of units.

Fig.1 SMA-net robot

BIOCHEMICAL PHENOMENON OF AN AMOEBA
The objective model described here is related to searching behavior of amoeba (slime mold). Amoebae have both unicellular and multi-cellular periods in their life cycle. Slime mold is a kind of colony of unicellular amoeba.

In spite of not having special sensitive organs and motile organs, slime mold can move towards possible foodstuffs, which is called chemotaxis. The chemotaxis has been investigated by several researchers [5] [6]. In their studies, it was shown that integral dynamics of physical and chemical processes in amoeba cell protoplasm can be simulated as a self-induced oscillator. Though each amoeba cell respectively causes the oscillation, each oscillation synchronizes. As a whole, one unified oscillation is reached.

In addition, each amoeba cell converts stimulation from environment into the oscillation. For example, the frequency of a cell sensing attractive stimulation rises. A mass of slime mold tends to unify oscillation of entire cells through entrainment phenomenon, when the frequency of one cell rises. In the case of entrainment, a phase gradient of the oscillation occurs in the whole slime mold. It is the gradient that causes the integration of the whole slime mold.

MULTI-CELL ROBOT DESIGN
In this section, we show the design policy of the amoeboid-like multi-cell robot, and describe the model for computer simulation of the SMA-net robot. This model has following features.
(1) Each cell unit is correspondent to one cell of the slime mold.
(2) Each cell unit communicates with the others through local interaction among cell units.

In Table 1, a comparison was made between amoeboid functions and our robotic model. Firstly, we modeled frequency rises in the cell unit robot. This corresponds to the same effect of amoeba cells receiving an attractive stimulation. Secondly, we modeled entrainment of oscillations among the cell units. This corresponds to the integration of stimulation to slime mold. Lastly, we modeled self-organization of function specialization among cell units. With regard to slime mold, this function specialization is similar to the growth of pesudopodium.

In this paper, we describe model 1 and model 2, with the computer simulation results to show the effectiveness of them.

Table 1 Model of Multi-cell Robot

Processes in Amoeba	Equivalent Components of Robotic Model
Sensing the stimulation	Frequency shift of oscillator
Integration of the stimulation	Entrainment of oscillators
Forming pseudopodium	Reconfiguration

OSCILLATOR MODEL OF A CELL UNIT
The oscillatory phenomenon in amoeba cells is the consequence of concentration-change by reaction of chemical substance. However, a myriad of correlating chemical reactions are involved and it is impossible to describe and

model them all. Therefore, in this study, each chemical oscillation was described as a dynamical system model. The nonlinear oscillator of Van der Pol type was utilized for the purpose.

$$\alpha_i \ddot{x}_i - \beta_i (x_i^2 - 1)\dot{x}_i^2 + x_i = 0 \qquad (1)$$

where, x_i: phase of oscillator.

The reason why we employ this equation is that it has simple and typical behavior of nonlinear oscillator. For engineering purpose such as entrainment of information using oscillator, it is considered that the Van der Pol type oscillator is sufficient and suitable.

When an attractive stimulation is received, the frequency of amoeba cell protoplasm oscillation rises, and it decreases in the case of aversive stimulation. The increase and decrease of the frequency for stimulations can be expressed by changing the parameter α in Eq. (1).

For example, in Fig. 2, $\alpha=1.4$ resulted in a normal oscillation state of the amoebic cell. $\alpha=0.8$ resulted in the oscillation when receiving an attractive stimulation, and $\alpha=2.0$ showed the state of aversive stimulations.

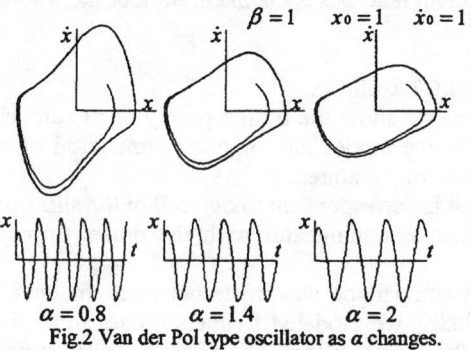

Fig.2 Van der Pol type oscillator as α changes.

LOCAL DIFFUSION COMBINATION AMONG UNITS

All oscillations x_i, generated by the Van der Pol oscillator in Eq. (1), share the same wave form. However, there are phase differences among oscillators. In this section, we describe a coupled oscillator system for achieving synchronization of oscillators.

A phase gradient occurs as the consequence of the entrainment, over the whole space. By such a phase gradient, the multi-cell robot would integrate local information sensed by a part of cell units. A large number of related studies have been made on coupled oscillator. Globally Coupled Map (GCM) [7] and Coupled Map Lattice (CML) [8] are most famous. In this study, we presented a simplified form of the CML.

$$x_i(t+1) = (1-\varepsilon) x_i(t) + \varepsilon \left\{ \frac{1}{N_i} \sum_{j=1}^{N_i} x_j(t) - x_i(t) \right\} \qquad (2)$$

where, ε: coupling coefficient. N_i: a number of cell i's neighbors.

The point different from GCM and CML is that coordinates of cell units are not fixed and change dynamically, so that neighborhood template should be dynamically recalculated.

EXPERIMENT OF ENTRAINMENT

With the model we described, a computer simulation was realized. Five oscillators were placed in a row. As the frequency of one oscillator increased, the other oscillators changed correspondingly. The transition of attractors and phase differences of the five oscillators was observed. In one case, only the oscillator at the left end received an attractive stimulation, which caused the increase of its frequency. The locus of frequency of the five oscillators was shown in Fig. 3. The initial values of x and x' were random. Bond length, that is, range inspected by each cell, was set to physical size of one cell. So each cell was combined with its neighboring oscillators. It was clearly shown that the entrainment occurred and all the oscillators were synchronized to one attractor.

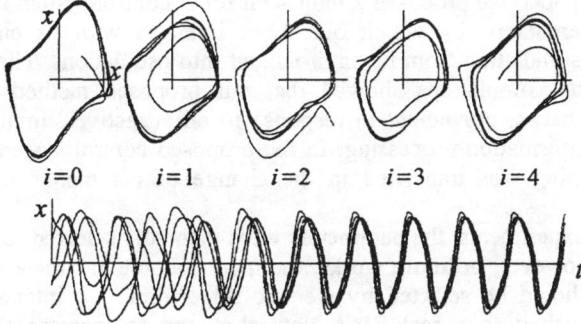

Fig.3 Entrainment of the oscillators.

EXPERIMENT OF OBSTACLE AVOIDANCE

Our control system was applied to obstacle avoidance problem. In this experiment, a multi-cell robot consisting of ten cell units was employed. The thick black line expresses the wall.

Fig. 4a describes initial conditions, and initial values for Eq. (1) are $\alpha=1.0$, $\beta=1.0$, and $\varepsilon=0.3$. The bond length is 2.0, when cell size is 1.0. The gray scale of each cell means the phase of the oscillator, and the line stretched from the center of cell shows direction and amplitude of phase gradient. The concentric circle in the upper right corner stands for the source of attractive stimulation.

In this simulation, simply, only one cell closest to the source received an attractive stimulation, and parameter alpha was changed to 0.7 (the square cell in Fig. 4a). When the cell unit touched the wall, alpha changed to 1.3 (Fig. 4b).

The frequency of the cell closer to the stimulation source rose, and the frequency of the cell contacting the wall decreased. As a consequence, the multi-cell robot avoided the wall, and went to the source.

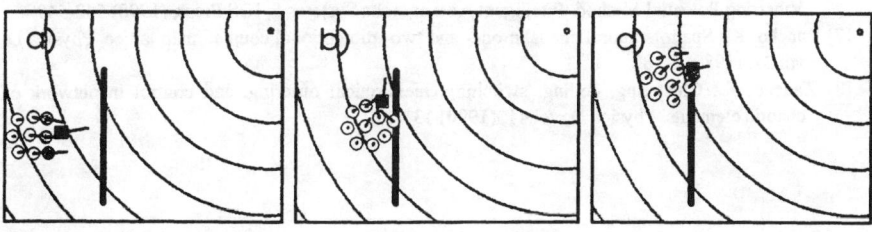

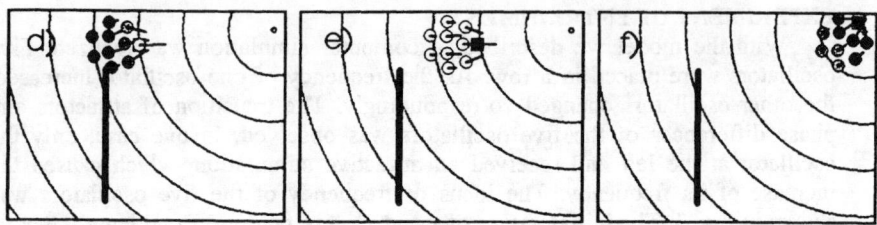

Fig.4 Obstacle avoidance fo multi-cell robot.

CONCLUSION

In this paper, we proposed a multi-cell robot control system referring to the amoeba chemotaxis, by which our robot interacts with its environment by converting stimulation from its environment into oscillations. The results of the computational simulations showed that our proposed method enables basic behaviors, that is, movement in response to an attractive stimulus, by locally interacted information processing. In the proposed control system, the sensor - action mapping was translated in the change of parameters of the coupled oscillator.

In the experiments, the parameters were manually selected for the objective behavior. However, in future works, the most suitable parameters for objective behaviors should be selected by Genetic Algorithm. We intend to apply the proposed method to a real SMA-Net robot and to increase the number of actuators and extend the SMA-Net robot to a three-dimensional composition. In the future, we also try to expand the model to include separation and coalescence of the slime mold.

REFERENCES

[1] Dittrich, P., Burgel, A., Banzhaf, W.: Learning to move a robot with random morphology. Evolutionary Robotics, Lecture Notes in Conputer Science, Vol.1468, Springer-Verlag, (1998) 165-178
[2] Paap, K. L., Dehlwisch, M., Klaasen, B.: GMD-snake. A semi-autonomous snake-like robot, Distributed Autonomous Systems, Vol.2, Springer-Verlag, (1996) 71-77
[3] Murata, S., Tomita, K., Yoshida, E., Kurokawa, H., Kokaji, S.: Self-Reconfigurable Robot. Module Design and Simulation, Intelligent Autonomous Systems 6, IOS Press, (2000) 911-917
[4] Shu, S. G., Lagoudas, D. C., Hughes, D.: Modeling of a flexible beam actuated by shape memory alloy, Smart Material and Structure, (1997) 265-277
[5] Miyake, Y., Tabata, S., Murakami, H., Yano, M. and Shimizu, H.: Environment-Dependent Self-Organization of Positional Information Field in Chemotaxis of Physarum Plasmodim, J. theor. Biol, Vol 178, (1996) 341-353
[6] Yokoi, H., Yu, W. and Hakura, J.: Morpho-Functional Machine – Amoeba Like Robot based on Vibrating Potential Method, Intelligent Autonomous Systems 5, IOS Press, (1998) 542-549
[7] Kaneko, K.; Spatiotemporal chaos in one- and two- dimentional coupled map lattice, Physica D, vol.37, (1989) 60-82
[8] Kaneko, K.: Clustering, coding, switching, hierarchical ordering, and control in network of chaotic elements, Physica D, vol.41, (1990) 137-172

HESSIEN ALGORITHM USED IN MULTILAYERED NEURAL NETWORKS CONTROLLERS IN ROBOT CONTROL

JOSEPH CONSTANTIN, CHAÏBAN NASR
Faculty of Engineering I
Lebanese University - Lebanon

DENIS HAMAD
Picardie Jules-Verne University
Amiens - France

ABSTRACT

Multi-layered neural network controllers are used to control a 2D-Robot arm. Different architectures are proposed for training different controllers to provide the appropriate inputs to the 2D-Robot arm so the desired response is obtained. In all the architectures proposed, we used different types of neural network controllers and different techniques are performed to tune the weights of the neural networks. The convergence of the different algorithms are based on the second order differential system that takes place by using the Hessien matrix. A comparative study between the first and the second order is performed. Performances of the controllers are examined and compared to previous works (Nasr et al., 2000).

INTRODUCTION

A very important characteristic of the new generation of robotic systems is the presence of intelligence capabilities, which is being rapidly supported by fast computing power and adequate sensory equipment. Trajectory tracking control of robots is, thus, of practical significance, and is the simplest yet the most fundamental task in robot control. The classical independent joint control of a robot is designed based on the theoretical linear model neglecting the non-linear coupling forces associated with the mechanical motion of the robot (Luh, 1983). Although this control scheme is adequate for simple pick-and-place tasks where only point to point motion is of concern. In task where precise tracking of fast trajectory under different payload is required, the existing control schemes are severely inadequate. Adaptive controllers based on linear in the parameters model have been proposed to maintain the tracking performance in the presence of parametric uncertainties (Whitcomb et al., 1993). However, instability may occur due to some bounded uncertain disturbances or plant non-linearity (Reed and Ioannou, 1989), making zero tracking accuracy no longer guaranteed. Therefore, the robotics community has been interested in the theory of ANN as one possible solution, where the massive parallelism within a network can reduce the computational burden. In addition, the learning abilities make such an approach attractive in the sense of providing compensation for any errors in executing the robot motion. Thus, by producing a robot control model based on the structure of the human brain, it is hoped to inherit the latter abilities and achieve human-like behavior. In proposing a solution for any neural-based control problem, the difficulty remains twofold. First, a suitable structure must be chosen for the NN to ensure that no extra complexity is introduced and second, the system must be able to generalize its operations to include any new situation with a minimum extra learning processes.

The goal of the present work is the application of a second order Hessien algorithm to tune the weights of three types of neural networks controllers applied to a dynamic 2D-Robot manipulator. Different architectures are proposed for training the controllers. The search of optimum architectures of the controllers was carried out. Finally, a comparative study based on the desired performance and the rapidity of learning is performed between the first order (Nasr et al., 2000) and the second order algorithms.

2D DYNAMIC PLANAR ROBOT ARM

Consider a planar manipulator with two revolute joints. Let us fix notation as follows: For i = 1,2 q_i denotes the joint angle, m_i(kg) denotes the mass of link i ($m_1=9.5/m_2=5$), l_i(m) denotes the length of link i ($l_1=0.25/l_2=0.16$), l_{ci}(m) denotes the distance from the previous joint to the center of mass of link i ($l_{c1}=0.2/l_{c2}=0.14$), and I_i (kgm^2) denotes the moment of inertia of link i about an axis and passing through the center of mass of link i ($I_1=4.3x10^{-3}/I_2=6.1x10^{-3}$). Using Lagrangian methods derives the dynamic equations in joint space coordinates. The robot model was simulated using the fourth-order Runge-Kutta method and its dynamic equation is rearranged into standard form:

$$\ddot{\vec{q}} = M^{-1}(\vec{q})(\vec{u} - h(\vec{q},\dot{\vec{q}}) - \vec{g}(\vec{q})) \qquad (1)$$

The first method (NN controller tuned by torque control) involves training an adaptive feed-forward controller to control the arm in conjunction with a fixed feedback controller (Figure 1). The NN used is a MLP with one hidden layer. The input vector contains nine components (Jacobs and Jordan, 1993).

The second method (NN controller tuned by minimizing a torque control), is represented in figure 2 (Jung and Hsia, 1995). The technique presented here employs a standard feed-forward NN. The NN output is $\vec{\tau}_n$ and the control law of the arm is given by the expression:
$$\vec{u} = \vec{\tau}_c + \vec{\tau}_n \qquad (2)$$

The NN is trained to minimize $E = \vec{\tau}_c^T \vec{\tau}_c /2$. The NN used is a MLP with one hidden layer and four inputs representing the robot arm's joint positions, and velocities.

The third architecture (NN controller that anticipates the desired input) uses an ANN controller that anticipates the desired input of the closed loop system consisting of a PD controller in cascade with the robot arm as shown in figure 3 (Jung and Hsia, 1995). Let ϕ_p be the NN output. The closed loop system equation yields to:

$$\vec{v} = k_v \dot{\vec{\varepsilon}} + k_p \vec{\varepsilon} = M\ddot{\vec{q}} + \vec{h}(\vec{q},\dot{\vec{q}}) + \vec{g}(\vec{q}) - k_v \vec{\phi}_p - k_p \vec{\phi}_p \qquad (3)$$

The NN is trained to minimize $E = \vec{v}^T \vec{v}/2$. The NN used is a MLP with one hidden layer and four inputs. In the three methods, the output layer has two neurons, the hidden neurons were chosen by simulation and, the activation function is an hyperbolic tangent.

MODIFIED GAUSS-NEWTON RECURSIVE ALGORITHM

Many different estimation methods of the parameters of a multi-layer perceptrons exist in literature (Bose and Liang, 1996). The learning rule corresponds to a minimization method of certain error criteria. Consider a multi-input, multi-output non linear system which can be described by the following general model:

$$\vec{d}_k = \vec{f}(\vec{x}_k, \vec{W}) + \vec{\varepsilon}_k \qquad (4)$$

The goal of modeling by using artificial neural networks is to find an artificial neural network that approximate the vector function $\vec{f}(.)$ based on the learning data defined by:

$$Z^N = \left\{ \left[\vec{x}_k, \vec{d}_k \right] / k = 1,...,N \right\} \qquad (5)$$

The prediction error is defined as:

$$\vec{\varepsilon}_k(\vec{W}) = \vec{d}_k - \vec{y}_k^{(s)} \text{ Where } \vec{y}_k^{(s)} = \vec{f}_{NN}(\vec{y}_k^{(0)}, \vec{W}) \qquad (6)$$

The error criterion can be written as:

$$\xi_N(\vec{W}) = \sum_{k=1}^{N} \vec{\varepsilon}_k^T(\vec{W})\vec{\varepsilon}_k(\vec{W})/2N \qquad (7)$$

The vector $\vec{W}$ is obtained by the minimization of the error criterion. The second Taylor development of this error criterion allows us to write:

$$\Delta\xi_N(\vec{W}) = \Delta\vec{W}^T \vec{g} + \Delta\vec{W}^T H \Delta\vec{W}/2 \qquad (8)$$

The expressions of the gradient vector and of the Hessien matrix are:

$$\vec{g} = -\frac{1}{N}\sum_{k=1}^{N} \psi_k(\vec{W})\vec{\varepsilon}_k(\vec{W}) \qquad (9)$$

$$H = \frac{1}{N}\sum_{k=1}^{N} \psi_k(\vec{W})\psi_k^T(\vec{W}) - \frac{1}{N}\sum_{k=1}^{N} \frac{\partial \psi_k(\vec{W})}{\partial \vec{W}^T}\vec{\varepsilon}_k(\vec{W}) \qquad (10)$$

The sufficient condition for $\vec{W}$ to be a minimum of $\xi_N(\vec{W})$ is:

$$\vec{g}(\vec{W}) = \vec{0} \quad \text{and} \quad H(\vec{W}) > 0 \quad \text{(Defined positive matrix)} \qquad (11)$$

Consider the expression (8) that represents the increase of the global average error due to the increase of $\Delta\vec{W}$. In computing its derivative with respect to $\Delta\vec{W}$, we have:

$$\frac{\partial \Delta\xi_N(\vec{W})}{\partial \Delta\vec{W}} = \vec{g} + H\Delta\vec{W} \qquad (12)$$

Consequently, we can modify the weights depending on the Newton method:

$$\vec{W}_{k+1} = \vec{W}_k - H^{-1}\vec{g} \qquad (13)$$

To reduce the computation cost of the Hessien matrix, we can express the Gauss-Newton algorithm by the following equation:

$$\vec{W}_{k+1} = \vec{W}_k - R^{-1}(\vec{W})\vec{g} \quad \text{Where} \quad R(\vec{W}) = \frac{1}{N}\sum_{k=1}^{N} \psi_k(\vec{W})\psi_k^T(\vec{W}) \qquad (14)$$

The inversion of the Hessien matrix presents a numeric problem when we obtain a singular matrix during the learning process. To solve this problem we have used the modified Gauss-Newton algorithm based on the following equation:

$$\vec{W}_{k+1} = \vec{W}_k - (R(\vec{W}) + \lambda_k I)^{-1}\vec{g} \qquad (15)$$

Where I is the identity matrix and λ_k is a scalar non-negative number. By adjusting the value of λ_k, we will obtain a first order or a second order learning method.

It is known that iterative methods modify the synaptic weights of the neural network in batch mode and have the difficulties to produce the dynamic behavior of the process. To solve this problem, the recursive methods have been developed. These methods have the capability to change the parameters of the network in real time depending on the information obtained from the plant. The error criteria at iteration k is defined by the following equation:

$$\xi_k(\vec{W}) = \frac{1}{2k}\sum_{t=1}^{k} \vec{\varepsilon}_t^T(\vec{W})\vec{\varepsilon}_t(\vec{W}) \qquad (16)$$

We suppose that the next estimation $\vec{W}_k$ remains near $\vec{W}_{k-1}$, so $\vec{W}_{k-1}$ is a good estimator at the k-1 iteration. The value of the gradient and of the Hessien matrix at the k iteration are defined by the following equation:

$$\bar{g}_k(\vec{W}_{k-1}) = -\frac{1}{k}\psi_k(\vec{W}_{k-1})\bar{\varepsilon}_k(\vec{W}_{k-1}) \qquad (17)$$

$$R_k(\vec{W}_{k-1}) = \frac{k-1}{k}R_{k-1}(\vec{W}_{k-1}) + \frac{1}{k}\psi_k(\vec{W}_{k-1})\bar{\varepsilon}_k(\vec{W}_{k-1}) \qquad (18)$$

In this paper, we used a new learning algorithm based on the Gauss-Newton recursive method and we change the weights of the NN based on the following equation:

$$\vec{W}_k = \vec{W}_{k-1} + \eta(\lambda_k I + R_k)^{-1}\bar{\psi}_k(\vec{W}_{k-1})\bar{\varepsilon}_k(\vec{W}_{k-1}) + \alpha\Delta\vec{W}_{k-1} \qquad (19)$$

We have introduced the non-negative number λ_k that makes the eigenvalues of the matrix $(\lambda_k I + R_k)$ less than or equal to a predetermined value δ.

NUMERICAL SIMULATION AND COMPARAISON

In order to perform a comparative study between the second order learning algorithm and our previous work (Nasr et al., 2000), we have considered the desired trajectory defined by the equations:

$Y_m = 0.2 + 0.075\sin(2\pi t/3)$ and $Z_m = 0.2 + 0.075\cos(2\pi t/3)$

We have used in the simulations $\Delta t = 0.01$ seconds for time integration. The choice of the 2x2 gain matrices was to obtain a stable system in the closed loop system with bad performances, $k_v = 10.I$ and $k_p = 100.I$ where I is the identity matrix. We have also used the modified Gauss-Newton recursive algorithm for training the neural network. The cumulative error used in the comparisons is defined by:

$$E = \sum_{n=1}^{N}[(y_d(n\Delta t) - y(n\Delta t))^2 + (z_d(n\Delta t) - z(n\Delta t))^2] \qquad (20)$$

Where $N = t/\Delta t$ denotes the sampling number with one trial. $y_d(n\Delta t)$, $z_d(n\Delta t)$ are the desired trajectories on the Y-Z plane at $n\Delta t$, and $y(n\Delta t)$, $z(n\Delta t)$ are the actual trajectories.

In the first method, we obtain an optimum learning rate equal to 7.10^{-3} and an optimum momentum term equal to 0.1. The simulation results show that the (9-8-2) MLP represents the optimal architecture with minimum cumulative error equal to 3.5×10^{-6} obtained after 40 cycles. Moreover, figure 4 shows the tracking response of the end-effector, and figure 6 shows the evolution of the instantaneous square error between the desired and actual position. In the first order method, we have obtained an optimum learning rate equal to 3.10^{-3} and an optimum momentum term equal to 0.1. The (9-18-2) MLP represents the optimal architecture with minimum cumulative error equal to 4.27×10^{-5} obtained after 40 cycles.

During the second method, we obtain an optimum learning rate equal to 10^{-2} and an optimum momentum term equal to 0.1. The simulation results show that the (4-12-2) multi-layer perceptron represents the optimal architecture with minimum cumulative error equal to 1.4×10^{-6} obtained after 40 cycles. Moreover, figure 5 shows the responses of the joints velocities to the desired inputs. The application of the first order method, gives an optimum learning rate equal to 10^{-3} and an optimum momentum term equal to 0.1. The (4-18-2) MLP represents the optimal architecture with minimum cumulative error equal to 1.13×10^{-4} obtained after 40 cycles.

In the third method, we obtain an optimum learning rate equal to 6×10^{-7} and an optimum momentum term equal to 0.1. The simulation results show that the (4-14-2) MLP represents the optimal architecture with minimum error equal to 9×10^{-4} after 40 cycles. Moreover, figure 5 shows the responses of the joints velocities to the desired inputs. The first order algorithm gives an optimum learning rate equal to 10^{-6} and an optimum momentum term equal to 0.1. The (4-6-2) MLP represents the optimal architecture with minimum cumulative error equal to 1.13×10^{-3} obtained after 40 cycles.

Comparisons in minimum error (Figure 7) and number of parameters (Figure 8) were carried out between the first order method and the modified Gauss-Newton recursive method for optimal architectures. We realize that the second approach is better in convergence for all optimal architectures and the minimum error is obtained by the second control method. The optimal number of parameters is obtained by the Gauss-Newton recursive learning method applied to the first control method.

COMMENTS AND CONCLUSION

For the control of robotic manipulators on reference trajectories, it is necessary to transform output sensors into acceptable values in decision space (Zomaya, 1993). Many researchers have discussed the possibility of applied artificial neural network in the domain of control (Sharifi et al., 1994). The main term of this discussion is using artificial neural network to learn the characteristics of robotic system.

In continuity to our previous works, we applied a second order learning algorithm to update the weights of three types of neural network controllers for trajectory control of robotic manipulators and to investigate the advantage of this algorithm over the first order method (Nasr et al., 2000). We have implemented the different control algorithms by using the second order learning algorithm and we have obtained optimal architecture in the three different control methods. The second method presents the smallest cumulative error and the best response time for the joint angle and velocity. The first method uses less input components and hidden neurons making the real-time process faster. The third method presents practical advantages over the two other methods in that compensation is done outside the control loop so it can be implemented easily at the command trajectory planning level external to an existing robot control.

A comparative study is performed between the first and the second learning algorithms. We find that the second order learning algorithm is better in minimizing the error during the learning process and in optimizing the architecture of the network than the first order method. These advantages give more accuracy in trajectory control and make the real-time process faster.

REFERENCES:

Bose N. K., and Liang P., 1996, "Neural networks fundamentals with graphs, algorithms and applications", "McGraw-Hill international editions

Jacobs R. A., and Jordan M. I., 1993, "Learning piecewise control strategies in a modular neural network architecture," *IEEE Trans. on Systems, Man and Cybernetics*, vol.23, no. 2, pp.337-345

Jung S., and Hsia T. C., 1995, "New neural network control technique for non-model based robot manipulator control", *IEEE Trans. on Systems, Man and Cybernetics*, vol. 1, pp 2928-2933

Jung S., and Hsia T. C., 1995, "On reference trajectory modification approach for Cartesian space neural network control of robot manipulator," Proc. of IEEE International Conference on Robotics and Automation, Nagoya

Luh J. Y. S., 1983, "Conventional controller design for industrial robots- a tutorial," *IEEE Trans. Systems, Man and Cybernetics*, vol. 13, no. 3, pp. 298-316

Nasr C., Constantin J., and Hamad D., 2000, "A comparative study of multilayered neural networks controllers in robot control," Intelligent Engineering Systems through Artificial Neural Networks, vol.10, pp. 625-630, ASME Press, New York

Reed J. S, and Ioannou P.A., 1989, "Instability analysis and robust adaptive control of robotic manipulators," *IEEE Trans. Robotics and Automation*, vol. 5, no. 3, pp. 381-386

Sharifi F.J., Fakhry H.H., and Wilson W.J., 1994, "Integration of a robust trajectory planner with a feed forward neural controller for robotic manipulators", "Proc. of IEEE int. conf. on robotics and automation", pp.3192-3197

Whitcomb L.L., Rizzi A.A., and Koditschek D.E., 1993,"Comparative experiments with a new adaptive controller for robot arm," *IEEE Trans. Robotics and Automation*, vol.9, no.1, pp. 59-69

Zomaya A.Y., 1993, "Trends in neuro-adaptive control of robot manipulators", "Proc. of int. conf. on intelligent robots and systems", pp. 754-760

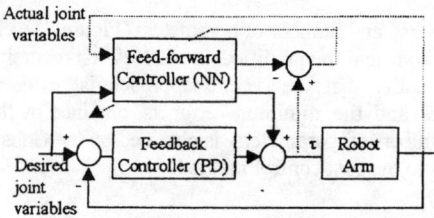

Fig. 1: - Feed-forward controller trained by minimizing the torque error

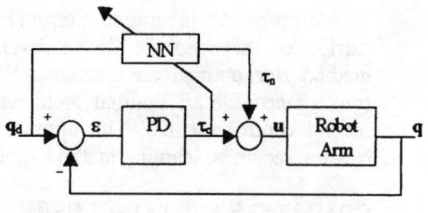

Fig. 2: - Feedback Error Learning Control Structure.

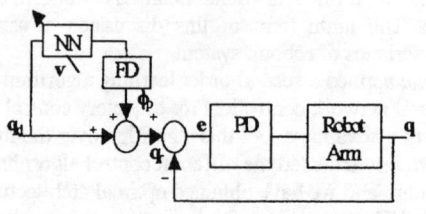

Fig. 3: - Neural Network that anticipates the desired input.

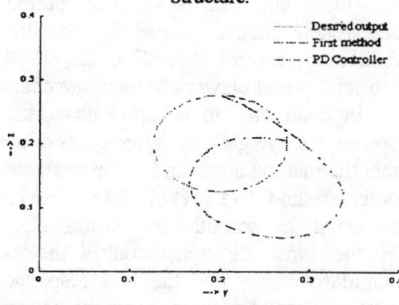

Fig.4: -Simulated trajectory of the end-effector in the yz Cartesian plane

RESULTS OF THE SIMULATIONS

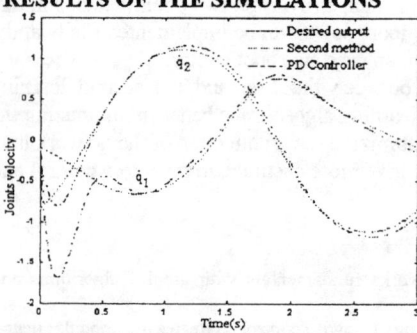

Fig.5: - Desired and actual joint velocity responses

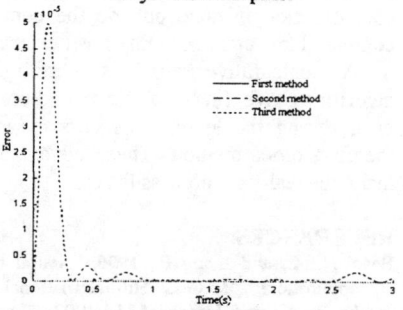

Fig.6: -- Instantaneous square error between actual and desired position

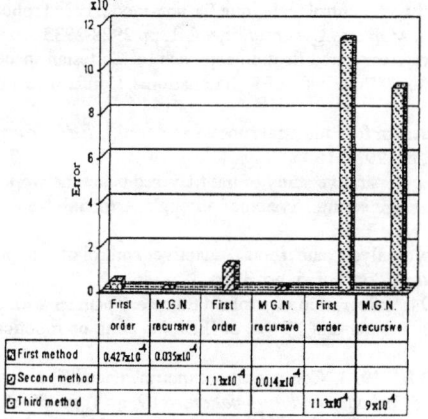

Fig.7: -Comparison between optimal architectures based on minimum error

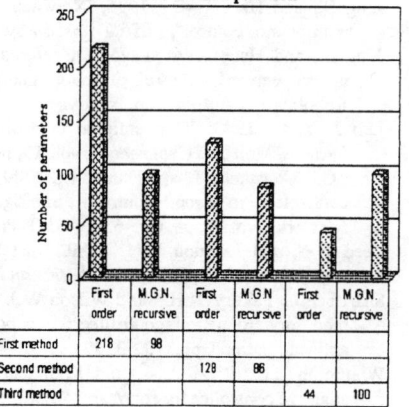

Fig.8: -Comparison between optimal architectures based on number of parameters

REINFORCEMENT LEARNING CONTROL OF COOPERATIVE ARM ROBOTS

Kazuaki YAMADA   Fumihiro KOJIMA
Kazuhiro OHKURA   Kanji UEDA
Department of Mechanical Engineering Faculty of Engineering
Kobe University Rokkodai 1-1, Nada-Ku, Kobe 657-8501, Japan

ABSTRACT

We describe a distributed approach to controlling autonomous arm robots. The robots are required to cooperate in order to smoothly lift an object. Each arm robot has its own reinforcement learning unit for decision-making. In investigating this task, we are primarily interested in the question of how to design a reinforcement learning control system for a multi-agent system. A new reinforcement learning algorithm is applied, in which a Bayesian discrimination method is used to segment continuous state and action spaces simultaneously, thereby generating of a set of effective rules. The proposed approach is examined empirically with two real arm robots. The basic dynamics of the reinforcement learning process are also analyzed.

INTRODUCTION

In recent years, there has been increasing interest in multi-robot system research, and a variety of problem domains have been investigated, such as robot soccer (Stone and Veloso, 1999) and foraging problems (Mataric, 1997). Whilst there appear to be significant differences between these problem domains, and each requires a different level of co-operation, there is nevertheless an important common feature. That is, the task presented to the robot group does not sufficiently specify how robots should behave individually in order to bring about the desired group behavior. Therefore, one of the most important issues in this research area is the question of how to provide individual robots with a mechanism for behavior coordination so that appropriate cooperative behavior can emerge in a dynamically changing environment. In this context, the two above-mentioned research topics can be described as follows: In the first, a soccer team has to find effective group formations with each robot able to employ a small set of basic behaviors. In the second, foraging behavior for collecting as many pucks as possible should be developed using a small set of simple behaviors, given that many robots are moving in the same environment.

An autonomous robot in a multi-robot system generally has sensors, effectors, and its own decision-making mechanism. For this type of robot, it is difficult—or perhaps almost impossible—for a programmer to design behaviors that are appropriate for such a complex environment, because he has to consider all possible situations that a robot may encounter in its environment. Therefore, we consider that a robot needs some form of real-time knowledge acquisition mechanism in order to realize autonomous behavior programming by which means a robot can develop the behaviors necessary for cooperation with other robots. The recent development of adaptive computation techniques, such as evolutionary computation and reinforcement learning, indicates one way in which autonomous behavior programming can be achieved. Although these two areas are still under development, they have become increasingly utilised in multi-robot systems research.

This paper introduces a reinforcement learning (RL) approach (Kealbing et al, 1996) to the acquisition of cooperative behavior in multi-robot systems. One of the features of RL is its use of an agent's experience to narrow down the area of the solution space it is exploring, and subsequently, to gradually improve the quality of solutions. However, conventional RL may not work effectively in the real world, because RL has difficulty finding rules in continuous state and action spaces in a noisy environment. This problem is made much worse if the environment is dynamic, as is the case with the multi-robot problem addressed in this paper. To address such problems, our research group has proposed a sophisticated RL algorithm. Its major advantage is that it is able to automatically segment continuous state and action spaces, and to adjust this segmentation through the process of learning. It should be noted that our RL algorithm has not previously been tested in a dynamic environment. Thus, an additional motivation for these experiments is to test the robustness of this algorithm in such an environment.

The remainder of the paper is set out as follows: The multi-robot learning task is introduced in the next section, the third section describes the proposed approach in more detail, the fourth describes the experimental setting and results, and conclusions are given in the final section.

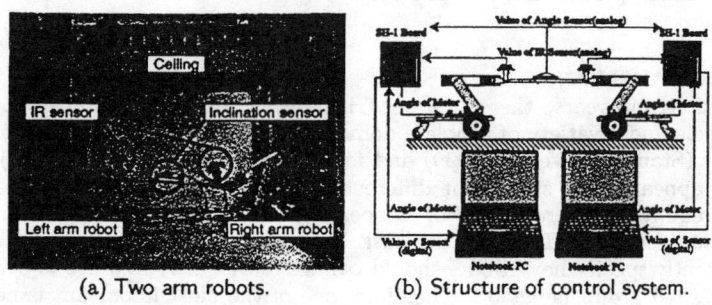

(a) Two arm robots.   (b) Structure of control system.

Figure 1. Learning task.

LEARNING TASK

Our main interest is in the design of a mechanism of behavior coordination which can generate appropriate cooperative behavior. In this paper, we introduce a simple problem which requires cooperation between robots. The task is illustrated in figure 1. Two arm robots must lift a board which is connected at each end to the tips of their arms. They must raise the board to a certain height whilst keeping it horizontal. The joints between the board and each arm can rotate freely. Each robot has an infra-red (IR) sensor on the tip of its arm for measuring the distance to the ceiling. Each arm has three joints moved by servo-motors. The pitch of the board is measured by an inclination sensor placed on the center on the board. For this small multi-robot system, each robot is provided with a real-time decision making mechanism; these operate independently of one another. Each robot discriminates a state using its IR sensor and the inclination sensor on the board. The decision-making mechanism sends motor commands to its three servo-motors simultaneously. Each robot is controlled and monitored by a different mobile computer. The difficulties presented by this task can be summarized as follows:

- The robots must cooperate with each other in order to lift the board above a certain height whilst keeping it balanced.
- Mechanical factors mean that, unlike industrial robots, our robots cannot control their arms great precision.
- The robots have to deal with noisy and continuous sensory inputs from the real world.
- Since each robot is controlled by a different computer, the robots' arm movements are not synchronised.

REINFORCEMENT LEARNING APPROACH
Reinforcement learning

Reinforcement learning (RL) is applied to this problem. RL is an algorithm for finding an appropriate set of IF-THEN rules that maximizes reward signals acquired through interaction with an environment. The IF part of a rule is a condition dependent on all sensory information, and the THEN part is an action to be executed if that condition is met.

In this problem, the state space and the action space are continuous, because sensor values are passed directly to the RL unit. For this situation, several forms of input generalization (Sutton et al, 1998) or state clustering (Asada et al, 1996, Takahashi et al, 1996) , etc., have been proposed for segmenting continuous spaces into discrete ones. However, neither of these are suitable for this problem because the segmentation has to be done in real time.

From a theoretical point of view, conventional RL may not work effectively for this problem, due to the fact that RL has problems acquiring a strategy behavior set in a noisy and dynamic environment; RL requires the assumption that the environment is static, i.e., Markovian. Additional difficulties arise because, in our task, each robot is autonomous, having its own independent decision-making mechanism. This ill-structured problem does not seem to be a "good problem" for RL algorithms, although RL's capacity for high speed adaptation is very attractive from the point of view of real-time learning.

In order to overcome these difficulties, we adopt an RL algorithm proposed by our research group, in which continuous state space and action space is assumed (Yamada et al, 2000). Its applicability is empirically investigated on the minimal, but non-trivial, homogeneous multi-agent problem described in the previous section.

Reinforcement learning algorithm

In this section, the reinforcement learning algorithm developed by our research group is briefly introduced. This RL operates on a set of production rules which are generated using Bayesian discrimination method (one of the well-known methods of pattern classification (Dura and Hart, 1992)). This method can assign an input, X, to the cluster C_i, which has the largest posterior probability, $\max p(C_i|X)$. Here, $p(C_i|X)$ indicates the probability, calculated by Bayes' formula, that a cluster, C_i, holds the observed input X. Thus, by using this technique, the agent can select the most similar rule to the current sensory input.

In this RL, the production rules are associated with clusters which are segmented by Bayes boundaries. Each rule consists of a state, an action, a utility and several parameters for calculating the posterior probability. The learning procedure is described briefly as follows:

(1) An agent perceives the current sensory input X.

(2) The agent selects the most similar rule from a rule set by using the Bayesian discrimination method. If the agent selects a rule, the agent executes the corresponding action a. Otherwise, the agent executes an action randomly.

(3) The agent is transferred to the next state and receives a reward r.

(4) The utilities of all rules are updated according to r. The rules in which the utilities are under a certain threshold are removed.

(5) If the agent executed an action randomly, the agent now produces a new rule combining the current sensory input and the executed action. This executed rule is memorized in the rule table.

(6) If the agent receive no penalty, the parameters of all the rules are updated by an Internal Estimation technique. Otherwise, the agent only updates the parameters of the selected rule.

(7) Go to (1).

The reader is referred to (Yamada et al. 2000) for a more detailed description of this RL algorithm.

EXPERIMENTS
Experimental setting

The feasibility of the proposed method was examined using two real robots on the task previously described. Each robot used our learning system and each of the learning systems operated independently. Each RL unit took input from its IR sensor and the inclination sensor; sensor values were normalized to the range [0:1]. Motor commands were sent to the three servo-motors simultaneously.

A reward was given to each robot when both robots' IR sensor values exceeded a certain threshold. A penalty was given when $E_i (= \alpha \Delta D_i + \beta h)$ is below -0.5, where the i th index indicates the left or right arm robot, ΔD_i is the prior sensor value subtracted from the current IR sensor value and h is the value of the inclination sensor; α and β are positive constants for tuning.

Learning takes place in a series of episodes. An episode commences with the robots' first motion, it ends when the robots' reach their goal or if the goal is not achieved after each robot has executed 100 actions. The two robots start from a particular starting posture at the beginning of each episode.

 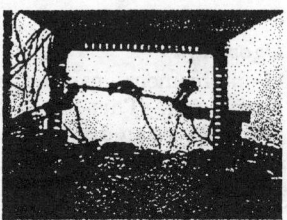 

(a) 1st step.   (b) 30th step.   (c) 56th step.

Figure 2. Acquired cooperative behavior.

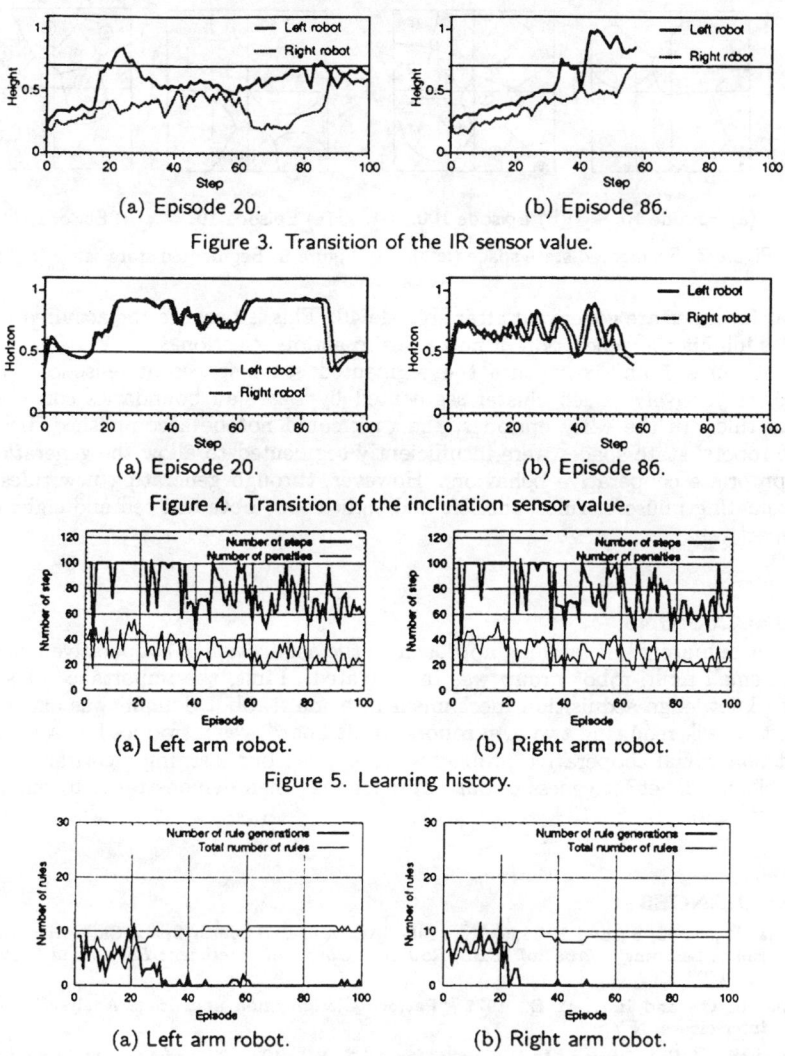

(a) Episode 20. (b) Episode 86.
Figure 3. Transition of the IR sensor value.

(a) Episode 20. (b) Episode 86.
Figure 4. Transition of the inclination sensor value.

(a) Left arm robot. (b) Right arm robot.
Figure 5. Learning history.

(a) Left arm robot. (b) Right arm robot.
Figure 6. Rule production history.

Experimental results

Figure 2 shows an acquired cooperative behavior at episode 86. Figures 3 and 4 plot the IR sensor values and the inclination sensor values, respectively. Note that the inclination sensor value is 0.5 when the board is horizontal. From figure 4(a), it can be seen that the board is hardly ever kept horizontal during episode 20. However, as shown in figure 4(b), the two robots seem to be making an effort to keep it horizontal after 66 episodes.

Figure 5 shows the number of the steps required for lifting the board and the number of penalties received for each episode. It can be seen that the robot system could lift the board stably from around episode 40. Figure 6 plots both the number of generated rules and the total number of rules for the two learning robots. Notice that both robots generate many rules in the early episodes, but

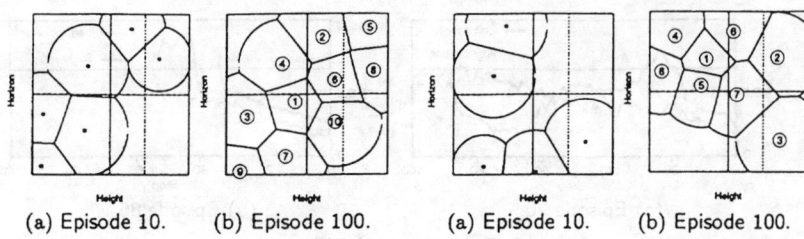

(a) Episode 10.   (b) Episode 100.   (a) Episode 10.   (b) Episode 100.

Figure 7. Segmented state space (left).   Figure 8. Segmented state space (right).

that few rules are generated after episode 40. This is because the acquired useful rules inhibit the generation of new rules in action selection.

Figures 7 and 8 capture the segmented state spaces at episodes 10 and 100, respectively. Each cluster segmented by Bayesian boundaries corresponds to a rule. In the early episodes, the task could not be accomplished, because the robots' state spaces were insufficiently segmented to allow the generation of appropriate cooperative behaviors. However, through generating new rules and eliminating unuseful rules, the left and right robot obtained ten and eight rules respectively.

CONCLUSIONS

A reinforcement learning approach to the acquisition of cooperative behavior in a small multi-robot group was investigated. First, the importance of a real time knowledge acquisition mechanism in a multi-robot domain was explained. Next, a task requiring two arm robots to lift board was introduced as a minimal but non-trivial cooperative problem. We applied our learning algorithm to this problem. The effectiveness of our approach was then demonstrated by empirical experiments.

REFERENCES

Asada, M., Noda, S., and Hosoda, K., 1996, "Action-Based Sensor Space Categorization for Robot Learning", *Proc. of IEEE/RSJ Int. Conf. on Intelligent Robots and Systems*, pp. 1502-1509.

Duda, R. O., and Hart, P. E., 1972, "Pattern Classification and Scene Analysis", Wiley-Interscience, N.Y.

Kaelbling, L. P., Littman, M. L., and Moore, A. W., 1996, "Reinforcement Learning: A Survey", *Journal of Artificial Intelligence Research*, Vol.4.

Mataric, M. J., 1997, "Reinforcement Learning in the Multi-Robot Domain", Autonomous Robots Vol.4, No.1, pp73-83.

Stone, P., and Veloso, M., 1998, "Task Decomposition and Dynamic Role Assignment for Real-Time Strategic Teamwork", *Proc. of the Fifth Int. Workshop on Agent Theories, Architectures, and Languages (ATAL'98)*.

Sutton, R. S., 1996, "Generalization in reinforcement learning: Successful examples using sparse coarse coding", *Advances in Neural Information Proc. Systems 8*, pp.1038-1044, MIT Press.

Takahashi, Y., Asada, M., and Hosoda, K., 1996, "Reasonable Performance in Less Learning Time by Real Robot Based on Incremental State Space Segmentation", *Proc. of IEEE/RSJ Int. Conf. on Intelligent Robots and Systems*, pp.1518-1524.

Yamada, K., Svinin, M. M., and Ueda, K., 2000, "Reinforcement Learning with Autonomous State Space Construction Using Unsupervised Clustering Method", *Proc. of the 6th Int. Conf. on Intelligent Autonomous Systems, IAS-6*, pp. 503-510.

LSPB TRAJECTORIES USED IN MULTILAYERED NEURAL NETWORKS CONTROLLERS IN ROBOT CONTROL

CHAÏBAN NASR
Faculty of Engineering I
Lebanese University - Lebanon

JOSEPH CONSTANTIN, DENIS HAMAD
Picardie Jules-Verne University
Amiens - France

ABSTRACT

An important way to generate suitable joint space trajectories is by so-called Linear Segments with Parabolic Blends or (LSPB) for short. The LSPB trajectory is such that the velocity is "ramped up" to its specific value initially and then "ramped down" at the goal position. The multi-layered neural network controllers are used with LSPB training examples in order to control a 2D-Robot arm. Optimization of the structure of the different multi-layered neural network controllers is performed. We used first order algorithms for convergence of the network. Performances concerning the response of the robot arm in the case of sudden changes in its parameters are studied and compared.

INTRODUCTION

Investigation into using neural networks in automatic control systems did not receive much attention until Rumelhart, Hilton & Williams (1986) formulated the "back-propagation" learning rule. Since then, research of neural control has evolved quickly. The most widely studied problem is that of motion control of robot manipulators (Jung and Hsia, 1994). In these applications, the NN are typically used to model the highly nonlinear structured and unstructured uncertainties of robot dynamics by generating a compensating torque (Ishiguro et al., 1992). It is clear that the higher the degree of non-linearity exists in the uncertainties, the greater the benefits NN can contribute. It is also clear that NN performs better when the signal levels in the NN are changed within a proper interval, say ± 1 or other normalized values, so the full exploitation of the nonlinear mapping capabilities can be achieved.

Trajectory tracking control of robots is the simplest yet the most fundamental task in robot controls (Tarn et al., 1984). The classical independent joint control of a robot is designed based on the theoretical linear model neglecting the non-linear coupling forces associated with the mechanical motion of the robot (Luh, 1983). In task where precise tracking of fast trajectory under different payload is required, the existing control schemes are severely inadequate. Adaptive controllers based on linear in the parameters model have been proposed to maintain the tracking performance of the robotic manipulators in the presence of parametric uncertainties (Whitcomb et al., 1993). However, instability may occur due to some bounded uncertain disturbances or plant non-linearity, making zero tracking accuracy no longer guaranteed.

In this paper we present the dynamic 2D-Robot arm equations needed for our simulations and we discuss three types of neural network controllers applied to this 2D-Robot manipulator (Nasr et al., 2000). We present the LSPB trajectories and the database used for training the neural network. Simulation results based on proposing different architectures for training the controllers are presented. Finally, we give some concluding remarks obtained through a comparative study.

2D DYNAMIC PLANAR ARM

Consider a planar manipulator with two revolute joints. Let us fix notation as follows: For $i = 1,2$ q_i denotes the joint angle, m_i(kg) denotes the mass of link i ($m_1=9.5/m_2=5$), l_i(m) denotes the length of link i ($l_1=0.25/l_2=0.16$), l_{ci}(m) denotes the distance from the previous joint to the center of mass of link i ($l_{c1}=0.2/l_{c2}=0.14$), and I_i (kgm^2) denotes the moment of inertia of link i about an axis and passing through the center of mass of link i ($I_1=4.3x10^{-3}/I_2=6.1x10^{-3}$). The dynamic equations in joint space coordinates are derived by using Lagrangian methods and are given by:

$$M(\vec{q})\ddot{\vec{q}} + \vec{h}(\vec{q},\dot{\vec{q}}) + \vec{g}(\vec{q}) = \vec{u} \quad (1)$$

The robot model was simulated using the fourth-order Runge-Kutta method. Note that the robot dynamics equation (1) must first be rearranged into standard form:

$$\ddot{\vec{q}} = M^{-1}(\vec{q})(\vec{u} - \vec{h}(\vec{q},\dot{\vec{q}}) - \vec{g}(\vec{q})) \quad (2)$$

The first method (NN controller tuned by torque control) involves training an adaptive feed-forward controller to control the arm in conjunction with a fixed feedback controller (Figure 1). The NN used is a MLP with one hidden layer. The input vector contains nine components (Jacobs and Jordan, 1993).

$$\ddot{q}_1, \ddot{q}_2, \ddot{q}_1\cos(q_2), \ddot{q}_2\cos(q_2), \dot{q}_1^2\sin(q_2), \dot{q}_2^2\sin(q_2), \dot{q}_1\dot{q}_2\sin(q_2), \cos(q_1), \cos(q_1+q_2) \quad (3)$$

The second method (NN controller tuned by minimizing a torque control), is represented in figure 2. The technique presented here employs a standard feed-forward NN. The NN output is $\vec{\tau}_n$ and the control law of the arm is given by the expression:

$$\vec{u} = \vec{\tau}_c + \vec{\tau}_n \quad (4)$$

Where: $\quad \vec{\tau}_c = k_v\dot{\vec{\varepsilon}} + k_p\vec{\varepsilon} \quad$ and $\quad \vec{\varepsilon} = \vec{q}_d - \vec{q}$

Combining equation (1) and (4), the error dynamic equation can be written as:

$$\vec{\tau}_c = M\ddot{\vec{q}} + \vec{h}(\vec{q},\dot{\vec{q}}) + \vec{g}(\vec{q}) - \vec{\tau}_n \quad (5)$$

The NN is trained to minimize $E = \vec{\tau}_c^T \vec{\tau}_c / 2$.

The NN used is a MLP with one hidden layer. The input vector contains four components representing the robot arm's joint positions, and velocities.

The third architecture (NN controller that anticipates the desired input) uses an ANN controller that anticipates the desired input of the closed loop system consisting of a PD controller in cascade with the robot arm as shown in figure 3. The purpose here is to develop the same control scheme for the non-model based PD control problem. Let ϕ_p be the NN output. The closed loop system equation is given by:

$$k_v\dot{\vec{e}} + k_p\vec{e} = M\ddot{\vec{q}} + \vec{h}(\vec{q},\dot{\vec{q}}) + \vec{g}(\vec{q}) \quad (6)$$

Where $\quad \vec{e} = \vec{q}_r - \vec{q} = \vec{q}_d + \vec{\phi}_p - \vec{q} = \vec{\varepsilon} + \vec{\phi}_p$

Substituting the error and its derivative into equation (6) yields:

$$\vec{v} = k_v\dot{\vec{\varepsilon}} + k_p\vec{\varepsilon} = M\ddot{\vec{q}} + \vec{h}(\vec{q},\dot{\vec{q}}) + \vec{g}(\vec{q}) - k_v\dot{\vec{\phi}}_p - k_p\vec{\phi}_p \quad (7)$$

The NN is trained to minimize $E = \vec{v}^T\vec{v}/2$. The NN used is a MLP with one hidden layer. The input vector contains four components.

In the three methods, the output layer has two neurons, the number of hidden neurons was chosen by simulation, and the activation function is an hyperbolic tangent.

LINEAR SEGMENTS WITH PARABOLIC BLENDS OR (LSPB) FOR SHORT

An interesting way to generate suitable joint space trajectories is by so-called "Linear Segments with Parabolic Blends" or LSPB for short. This type of trajectory is appropriate when a constant velocity is desired along a portion of the path. The LSPB trajectory is such that the velocity is "ramped up" to its specific value initially and then "ramped down" at the goal position. To achieve this, we specify the desired position in three parts. The first part from time t_o to time t_b is a quadratic polynomial. This results in a linear "ramp" velocity. At time t_b, called the "blend time", the position trajectory switches to a linear function. This corresponds to a constant velocity. Finally, the position trajectory switches once again, at time $(t_f - t_b)$, to a quadratic polynomial so that the velocity is linear. The complete LSPB trajectory is given by (Constantin, 2001):

$$q_i(t) = \begin{cases} q_o + 0.5at^2 & 0 \leq t \leq t_b \\ 0.5(q_f + q_o - Vt_f) + Vt & t_b < t \leq t_f - t_b \\ q_f - 0.5at_f^2 + at_f t - 0.5t^2 & t_f - t_b < t \leq t_f \end{cases} \qquad (8)$$

An important variation of this trajectory is obtained by leaving the final time t_f unspecified and seeking the "fastest" trajectory between q_o and q_f with a given constant acceleration a, that is, the trajectory with the final time t_f a minimum. This is sometimes called a "Bang-Bang" trajectory since the optimal solution is achieved with the acceleration at its maximum value +a until an appropriate switching time t_s at which time it abruptly switches to its minimum value -a from t_s to t_f. Returning to our simple example in which we assume that the trajectory begins and ends at rest, symmetry considerations would suggest that the switching time t_s is just $0.5t_f$. If we let V_s denote the velocity at time t_s, then we have:

$$V_s = a\, t_s \qquad \text{and} \qquad t_s = (q_o - q_f + V_s t_f)/V_s \qquad (9)$$

The symmetry condition implies that:

$$V_s = (q_f - q_o)/t_s \qquad (10)$$

Combining equations (9) & (10), we have:

$$t_s = \sqrt{(q_f - q_o)/a} \qquad (11)$$

In order to build a learning database for the NN controllers in the workspace of the manipulator, we have studied the motion of every link of the planar arm to arrive to a desired position. The two links of the 2D-Robot arm must arrive simultaneously to the final position of the end-effector. Because of this, we have studied for every link i the necessary time t_{fi} to bring the link from the position q_{oi} to q_{fi} by means of an optimal trajectory described below. We have reduced the speed of the faster link by determining its new velocity by the equation:

$$V_s = 2(q_f - q_o)/t_f \qquad (12)$$

We determine the interruption time t_s of this link by the relation:

$$t_s = (q_o - q_f + V_s\, t_f)/V_s \qquad (13)$$

And finally, we determine its new acceleration by means of the relation:

$$a = V_s / t_s \qquad (14)$$

The new trajectory of this link will be defined by equation (8).
To form the examples of the database, we have (Constantin, 2001):
1- divided the workspace of the manipulator into five concentric and equidistant circles
2- considered, in the first quadrant, the ten points of intersection between these circles and the y and z-axes. For every initial point between these ten points, we can choose

nine final points to build the optimum trajectories. So, we obtain 90 possible trajectories. The same method is applied to the third quadrant.

3- eliminated, in the second and fourth quadrant, the existing trajectories built in the second step. Those trajectories corresponds to the ones that have their initial and final position on the same axe, they corresponds to 2x(5x4)=40 possible trajectories. So, the number of trajectories remaining in those quadrants is:

2x(90-40)=100 trajectories.

We conclude that the total number of trajectories existing in the learning database is:

2x90+2x50= 280 trajectories.

During the learning process, we introduce those trajectories randomly and one by one.

NUMERICAL SIMULATION

For our simulations, the first objective was to generate a minimum time desired trajectory starting from an initial point (y_i, z_i) and arriving to a final point (y_f, z_f). We have built an algorithm that determines the minimum Euclidean distance between these two points and that generates the desired trajectory in terms of joint angles and joint velocities. The average error used is defined by:

$$E = \sum_{n=1}^{N} [(y_d(n\Delta t) - y(n\Delta t))^2 + (z_d(n\Delta t) - z(n\Delta t))^2]/M = E_c/M \quad (15)$$

Where $N = t/\Delta t$ denotes the sampling number with one trial. $y_d(n\Delta t)$, $z_d(n\Delta t)$ are the desired trajectories on the Y-Z plane at the sampling time $n\Delta t$, and $y(n\Delta t)$, $z(n\Delta t)$ are the actual trajectories. The parameter M represents the number of trajectories used in the training process and E_c represents the cumulative error.

We have used in the simulations $\Delta t = 0.01$ seconds for time integration. The choice of the 2x2 gain matrices used in the PD controller was to obtain a stable system in the closed loop system with bad performances, $k_v = 10.I$ and $k_p = 100.I$ where I is the identity matrix of order two. Figure 5 shows the simulated response of the robot arm controlled to go from an initial point (0.20; 0) to a final point (- 0.20; 0.35). The response of the system is different from the desired trajectory with a cumulative error equal to 1.47. Figures 6 & 7 show the response of the joints angles and joints velocity.

During the numerical simulation, a number of neural network controllers have been trained using the back-propagation technique based on the first order algorithm with different learning rates and different momentum terms. In the first method, we have obtained an optimum learning rate equal to 10^{-4} and an optimum momentum term equal to 0.1. Figure 4 shows that the (9-18-2) multi-layer perceptron represents the optimal architecture with minimum average error equal to 1.6×10^{-3} obtained after 1120 cycles. If we continue in the learning process, the average error will not be less than 1.6×10^{-3}. Moreover, figure 7 shows the velocity response of the end-effector.

During the second method, we have obtained an optimum learning rate equal to 8×10^{-5} and an optimum momentum term equal to 0.1. Figure 4 shows that the (4-18-2) multi-layer perceptron represents the optimal architecture with minimum average error equal to 4.8×10^{-2} obtained after 1120 cycles. Moreover, figure 6 shows the responses of the joints positions.

During the third method, we have obtained an optimum learning rate equal to 10^{-7} and an optimum momentum term equal to 0.1. Figure 4 shows that the (4-6-2) multi-layer perceptron represents the optimal architecture with minimum average error equal to 0.14 after 1120 cycles. Moreover, figure 5 shows the responses of the joints positions.

By making a comparison between the three methods with respect to the average error, figure 4 shows that the first method is the best in performances. Its average error is 30 times less than that corresponding to the second method and 87 times less than that corresponding to the third method. But the third method needs, in its optimal architecture, less parameter than the two others do. While the first method needs 218 parameters and the second method needs 128 parameters, the third method need only 44 parameters. Figure 8 shows the instantaneous temporal square error obtained by the three methods between the desired trajectory and the actual response. It is clear that the first method is the best in precision.

COMMENTS AND CONCLUSION

Many researchers have discussed the possibility of applied artificial neural network in the domain of control (Sharifi et al., 1994). The main term of this discussion is using artificial neural network to learn the characteristics of robotic system.

In continuity of our previous works, we studied the application of a first order learning algorithm to update the weights of three types of neural network controllers for LSPB trajectory control of robotic manipulators. We also investigate the possibility of tracking the system with a minimum time trajectory in all the cases. During our simulations, we have implemented the different control algorithms by using the first order learning algorithm and we have obtained optimal architecture in the three different control methods. Good results where obtained through the convergence of the algorithms and through the response of the system to "minimum time trajectories". The first method presents the smallest average error and the best response time for the joint angle and velocity. The third method uses less input components and hidden neurons making the real-time process faster. It presents also practical advantages over the two other methods in that compensation is done outside the control loop so it can be implemented easily at the command trajectory planning level external to an existing robot control.

The important conclusion from this study is that the system can learn any type of trajectory: an LSPB trajectory or a minimum time trajectory. Its response to a desired trajectory not existing in the database is accurate and stable. These advantages give more accuracy in trajectory control and make the real-time process accurate, stable and faster.

REFERENCES

Constantin J., 2001, "Contrôle-Commande d'un Bras Robotique Planaire par Approches Neuromimétiques," Ph.D. theses in Automation, Jules Verne University-France

Ishiguro A., Furuhashii T., Okuma S., and Uchikawa Y., 1992, "A neural network compensator for uncertainties of robot manipulators," *IEEE Trans. on Industrial Electronics*, vol.39, pp.61-66

Jacobs R. A., and Jordan M. I., 1993, "Learning piecewise control strategies in a modular neural network architecture," *IEEE Trans. on Systems, Man and Cybernetics*, vol. 23, no. 2, pp. 337-345

Jung S. and Hsia T.C., 1994 "On-line neural network control of robot manipulators," International Conference on Neural Information Processing, , Seoul, pp. 1663-1668

Luh J. Y. S., 1983, "Conventional controller design for industrial robots - a tutorial," *IEEE Trans. Systems, Man and Cybernetics*, vol. 13, no. 3, pp. 298-316

Nasr C., Constantin J., and Hamad D., 2000, "A Comparative study of multi-layered neural network controllers in robot control," Intelligent Engineering Systems through Artificial Neural Networks, ASME Press, New York, vol.10, pp. 625-630

Sharifi F. J., Fakhry H. H., and Wilson W. J., 1994, "Integration of a robust trajectory planner with a feed forward neural controller for robotic manipulators", "Proc. of IEEE int. conf. on robotics and automation", pp.3192-3197

Tam T.J., Bejczy A.K., Isidori A., and Chen Y., 1984, "Nonlinear feedback in robot arm control," in Proc. IEEE Conf. Decision and Control. Las Vegas, NV, vol.2, pp.736-75

Whitcomb L.L, Rizzi A.A., and Koditschek D.E., 1993, "Comparative experiments with a new adaptive controller for robot arm," *IEEE Trans. Robotics and Automation*, vol. 9, no. 1, pp. 59-69

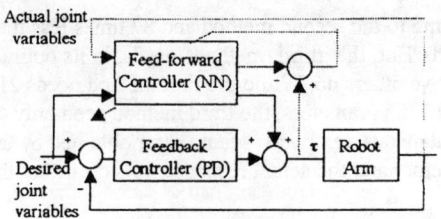

Fig. 1: - Feed-forward controller trained by minimizing the torque error

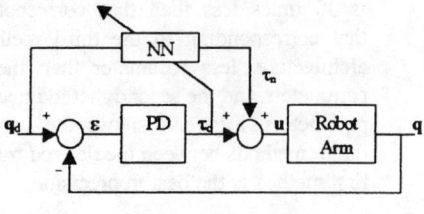

Fig. 2: - Feedback Error Learning Control Structure.

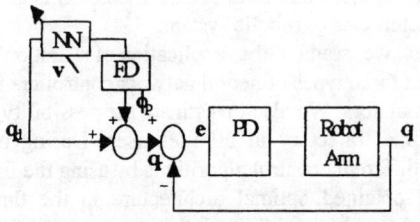

Fig. 3: - Neural Network that anticipates the desired input.

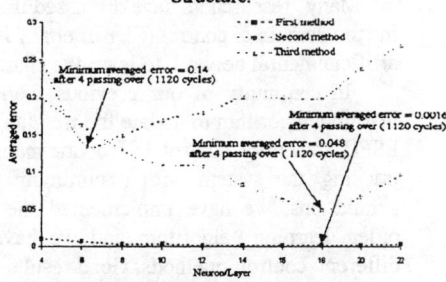

Fig. 4: -Optimized architecture of the neural network controller for the three methods

RESULTS OF THE SIMULATIONS

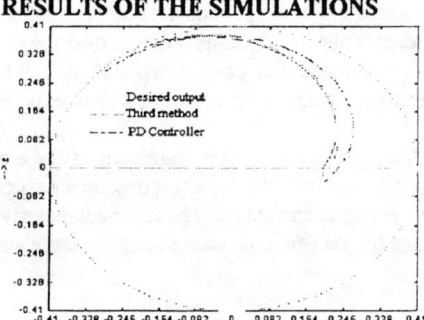

Fig. 5: -Simulated trajectory of the end-effector in the yz Cartesian plane

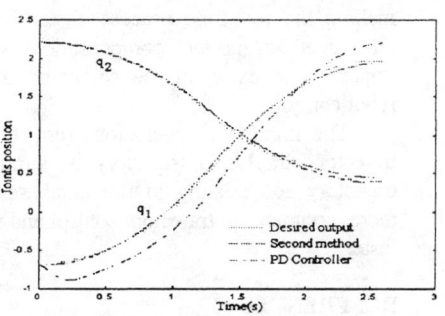

Fig. 6: -Desired and actual joints position responses

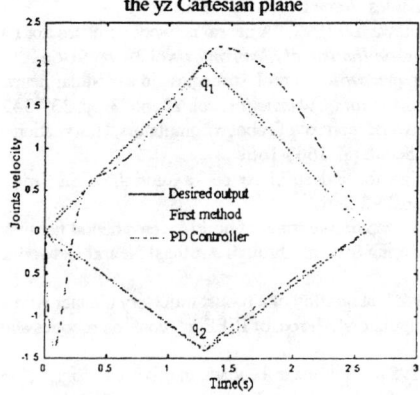

Fig. 7: -Desired and actual joint velocity responses

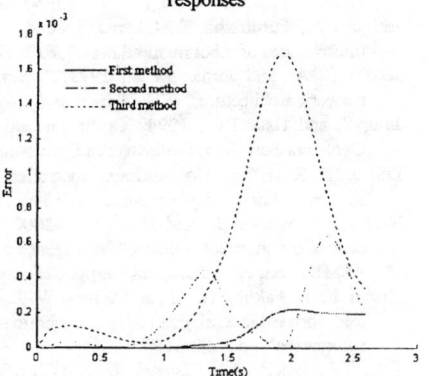

Fig. 8: - Instantaneous square error between actual and desired position

POSITION CONTROL USING A PROGRAMMABLE E/H VALVE

HAIBO HU and QIN ZHANG
Ag. Engineering Department
University of Illinois
Urbana, IL 61820

XIANGDONG KONG
College of Mechanical Engineering
Yanshan University,
Qinhuangdao, Hebei, China

ANDREW ALLEYNE
Mech. & Industrial Engineering
University of Illinois
Urbana, IL 61820

ABSTRACT

A generic programmable electro-hydraulic (E/H) valve consists of a set of five individually controlled E/H sub-valves. This programmable valve is capable of performing multiple functions and realizing different characteristics for various applications. The position control using a generic programmable E/H valve is different from that of using a conventional E/H valve due to its unique feature. This paper presents the development of a position control algorithm for a programmable E/H valve. This control algorithm was implemented on a valve-testing platform. The test results indicated that this programmable E/H valve was capable of achieving satisfactory position control performance.

INTRODUCTION

A generic programmable electro-hydraulic (E/H) valve consists of a set of five individually controlled E/H sub-valves. This programmable valve is capable of performing multiple functions and realizing different characteristics for various applications (Book and Goering, 1999; Hu *et al.* 2001). The application of a generic programmable valve can potentially simplify the hydraulic circuits design and reduce cost by replacing traditional directional control valve and eliminating other auxiliary valves (Hu *et al.*, 2001).

Figure 1 shows a schematic diagram of a generic programmable E/H valve system. This programmable valve system consists of five bi-directional proportional flow control sub-valves, four pressure sensors, one position sensor and a computer-based controller. Those sub-valves are actuated using solenoid actuators in response to a pulse width modulation (PWM) control signal. The pressure sensors are mounted at four ports of inlet, outlet and two work ports. The position sensor is mounted on the hydraulic cylinder. Signals from pressure sensors and position sensor provide feedback signals for achieving accurate cylinder position control.

Position control is one of the most common functions for hydraulic actuators, and is normally achieved using conventional directional valves. Control methods for position control vary depending on hydraulic system configuration, especially the configurations of the control valve (Watton, 1989, Merritt, 1967). The control strategy for position control using this generic programmable E/H valve is different from that of conventional E/H valve due to its unique feature. The control of programmable E/H valve consists of two levels of function logic control and performance modulation control. By controlling each sub-valve's operation status (open or closed), the valve set can control the motion

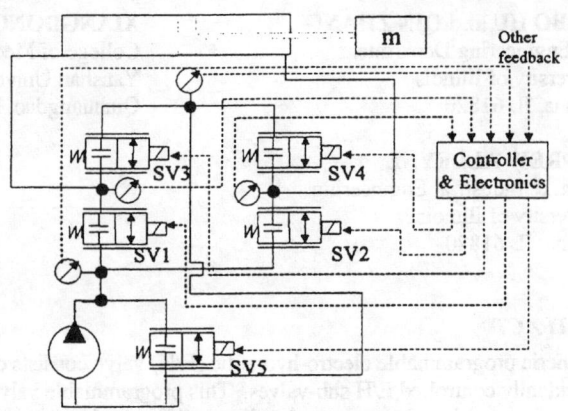

Figure 1. Schematic diagram of the GPE/H valve system

direction of the hydraulic cylinder. Using a modulation control algorithm for each sub-valve, this generic programmable E/H valve can realize precise position control on the cylinder. In this paper, the development of position control algorithm used in this generic programmable E/H valve is introduced.

DESCRIPTION OF POSITION CONTROL USING A PROGRAMMABLE VALVE

To achieve a precise cylinder position control, a two-level hybrid control algorithm was designed for both sub-valve operation status control and for cylinder position control. The first level is the sub-valve operating logic control which regulates all five sub-valves operating in conjunction with each other to achieve the desired cylinder motions. The table 1 shows the control logic.

Table 1. Control Logic for Cylinder Extending and Retracting Motions

SUB-VALVES	EXTEND LOGIC	RETRACTION LOGIC
SV1	1	0
SV2	0	1
SV3	0	1
SV4	1	0
SV5	S	S

1=open, 0=closed, S=open at preset pressure

To extend the cylinder, sub-valve No.1 (SV1) needs to be modulated for controlling the rate of the fluid entering the head-end chamber of the cylinder, and meanwhile sub-valve No. 4 (SV4) needs to be fully opened to allow the fluid from the rod-end chamber of the cylinder returning to the tank with a minimum resistance. Sub-valves No. 2 (SV2) and

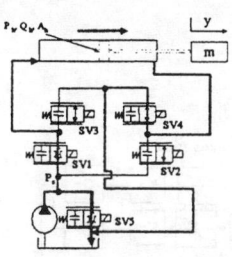 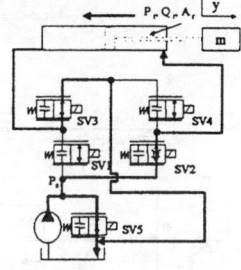

(a) Extension operation (b) Retraction operation

Figure 2. Sub-valve operation logics for achieving extending and retracting motions.

No. 3 (SV3) are closed during the extension operation (Figure 2a). Similarly, in the retracting operation, SV2 is modulated to control the fluid rate entering the rod-end chamber and SV3 is fully open to allow the fluid in the head-end chamber returning to the tank with a minimum pressure drop. SV1 and SV4 are closed in this operation (Figure 2b). Sub-valve No. 5 (SV5) is modulated to maintain the desired system pressure in both extending and retracting operations. The second level is the modulation control which controls the flow rate passing each of the sub-valves for realizing the precise position control. For minimizing unnecessary energy loss, this position control algorithm controls only the meter-in flow in cylinder motion control.

To simplify the modeling and analysis, the back-pressure of the cylinder was neglected and a small initial volume was assumed in both sides of the piston in the cylinder. Applying the continuity equation of the compressible fluid, orifice equation of sub-valves and cylinder motion equation, open loop transfer functions (Ge for extension and Gr for retraction) were obtained.

$$Ge = \frac{Y(s)}{I(s)} = \frac{K_{i1}}{A_h s + \frac{V_h m}{A_h \beta_e} s^3 + K_{p1} \frac{m}{A_h} s^2}$$

$$Gr = \frac{Y(s)}{I(s)} = \frac{-K_{i2}}{A_r s + \frac{V_r m}{A_r \beta_e} s^3 + K_{p2} \frac{m}{A_r} s^2}$$

The model indicates that the position control using this programmable valve is a third-order system. The load, the system pressure, the cylinder size, and the sub-valve characteristics will all affect the dynamic behavior of the system. The stabilized position and the transient response of the cylinder are important for achieving precise position control.

System dynamics analysis indicated that this cylinder position control using a generic programmable valve was a type 1 system so its steady-state error to a step-input control signal should be zero and its error to a ramp-input should be a finite value. Figure 3 shows the block diagram of the cylinder position control system using a generic programmable E/H valve. The error, e, between the desired position, r, and the measured position, y, is the control input to a Switch in the controller. If e is greater than or equal to zero, the Switch propagates the Extend Logic. Otherwise, the Switch propagates the Retraction Logic. A Demux block outputs five control signals (i.e. open, closed, or open at a preset pressure) to operate all five sub-valves based on either Extend Logic or Retraction Logic. For example, when Extend Logic is implemented, SV2 and SV3 are closed, SV4 is fully open, and the opening of SV5 is regulated by the preset pressure. SV1 is modulated by a proportional gain K_c so that the current going to the solenoid driver

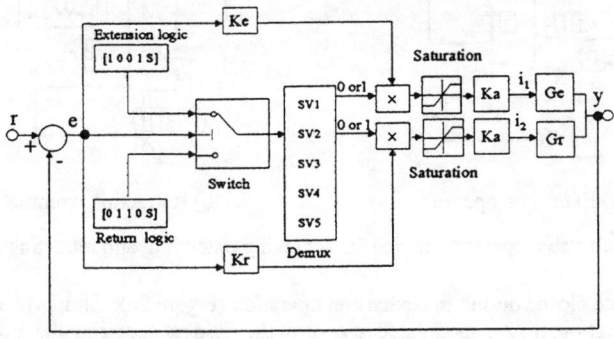

Figure 3. Block diagram of the position control system using a generic programmable E/H valve.

on SV1 is $i_1 = eK_eK_a$ (where K_a is valve drive amplifier gain). Similarly, SV2 is modulated by proportional gain K_r and the current going to the solenoid driver on SV2 is $i_2 = eK_rK_a$. Since each sub-valve has its maximum rated driving current, two Saturation blocks were used to limit the range of the signal to prevent damages to the sub-valves.

The control algorithm was programmed under SIMULINK environment, using graphical user interface (GUI) for building control model as block diagrams. To realize real-time control on this valve-testing platform, WINCON software was used to run SIMULINK generated code using REAL-TIME WORKSHOP.

TEST RESULTS AND DISCUSSION

The implementation tests of hydraulic cylinder position control using a generic programmable E/H valve were conducted on a valve-testing platform using a set of five interconnected bi-directional proportional flow control valves. This valve-testing platform consisted of a hydraulic pump unit, a double acting differential hydraulic cylinder, the programmable control E/H valve, and an Omega LP802-255 linear potentiometer attached to the cylinder.

The tests were implemented with different proportional gains K_e and K_r to obtain the desired transient responses and tracking accuracies. Two reference signals of square wave, triangle wave were used to check step-input response and ramp-input response respectively. The current signals used to control the metering sub-valves of SV1 and SV2 were monitored and recorded during the tests.

Figures 4a and figure 4b show system responses to a step input and the percentage of full scale of current signals to the solenoid drivers on SV1 and SV2. The optimal proportional gains were Ke=0.5 and Kr=0.35 for the particular case. The system response to the step input was stable and its steady state error was zero. When reference signal changed from a position of 125mm a new position of 75mm, the control current to SV2 reached to its maximum at the beginning due to big error between reference signal and feedback position signal, indicating that SV2 was full opened and cylinder retracted at its maximum speed to correct this difference between the desired and the actual positions.

As cylinder moved close to the desired new position, the error between reference signal and position feedback reduced. When the cylinder piston reached the desired position, the controller would close SV1 to keep the cylinder extended at the desired position. Similarly, when a new reference signal was applied to change the cylinder position from 75mm to 125mm, the control current to SV1 would first reach to its maximum valve to obtain a maximum correction speed. As cylinder moved up, the control current to SV2 would gradually decrease to zero. There was a limited overshoot when cylinder was nearby the low position as shown in Figure 4a. At the same time a small control current (about 10 % of full scale) was applied to SV1 as shown in Figure 4b, indicating that SV1 was slightly open to move the cylinder back to the desired position. There was no obvious overshoot during the extension operation. It was also observed that the overshoot became smaller as the values of Ke and Kr were reduced. However, small values of Ke and Kr would increase settling time.

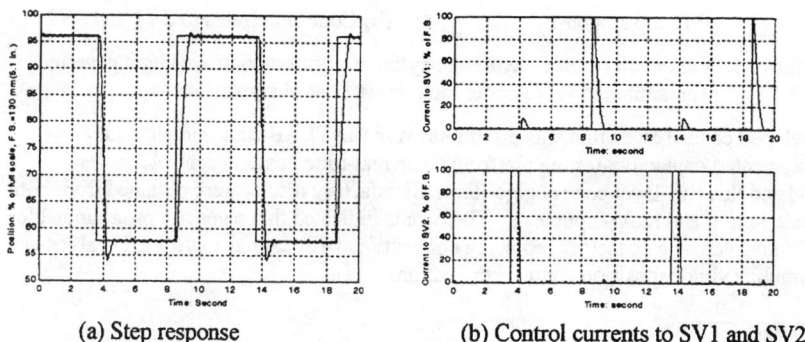

(a) Step response   (b) Control currents to SV1 and SV2

Figure 4. Validation result of hydraulic cylinder position control using a generic programmable valve under step input control signals.

Figure 5 shows the system responses to triangle wave with a ramp command of 72 mm/s and the percentage change of the control current to the metering sub-valves. The same gain values of Ka and Kr used in square wave response test were used.

Figure 5a shows that tracking error exists when the system followed the ramp input. Increasing Ke and Kr could reduce the ramp tacking error, but large Ke or Kr would result in an unstable response. Figure 5b indicates that the control current applied to SV1 and no control current to SV2 during the cylinder extension operation. Similarly, the control current applied only to SV2 during the retraction operation. There was always a large control current applied to either SV1 or SV2 while switching the cylinder motion directions. After cylinder moving in one direction, the tracking error became constant which resulted in a constant level of control current applied to either SV1 or SV2. To obtain same tracking error in both extending and retracting motions, a greater control current to SV1 than that to SV2 was needed. The difference in control current could be achieved by adjusting Ke and Kr value.

CONCLUSION

Use a generic programmable E/H valve for hydraulic cylinder position control was investigated. A two-level control algorithm of coordinating logic control for five individually controlled sub-valves and valve opening modulation control for each sub-valve is needed for achieving satisfactory cylinder position control using a programmable E/H valve. Integrating a position error feedback to the modulation control on the meter-in

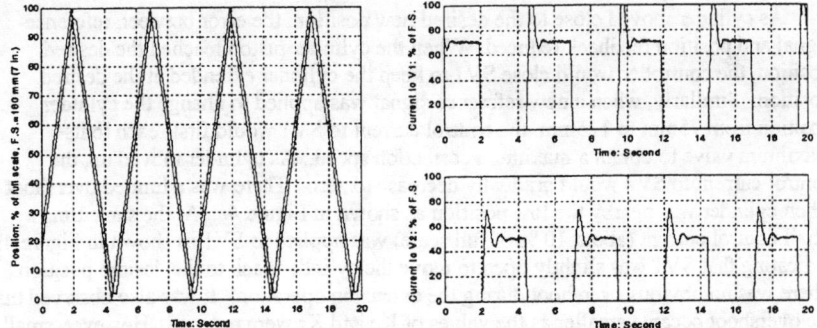

sub-valve control help the programmable valve achieved precise position control. The

(a) Ramp response　　　　　(b) Control currents to SV1 and SV2

Figure 5. Validation result of hydraulic cylinder position control using a generic programmable valve under ramp input control signals.

developed control algorithm was programmed in MATLAB environment and implemented on a valve-testing platform. The real-time position control test results validated this two-level control algorithm. Satisfactory results were obtained from both the step-input and ramp-input tests. The results indicated that a generic programmable E/H valve can be satisfactory replacing conventional directional control E/H valves in hydraulic cylinder position control applications.

REFERECENCE

1. Book, R. and C.E. Goering, 1999, "Programmable electrohydraulic valve," SAE 1999 Transactions, Journal of Commerical Vehicles, Vol. 108, sec. 2, pp. 346-352.
2. Hu, H., Q. Zhang and A. Alleyne, 2001, "Multi-function realization of a generic programmable E/H valve using flexible control logic," Fifth International conference on Fluid power Transmission and Control, Hangzhou, China.
3. Watton, J., 1989, "Fluid power systems," Prentice Hall, New York.
4. Merritt, H. E., 1967, "Hydraulic control systems," John Wiley & Sons, New York.

ARTIFICIAL NEURAL NETWORK FOR TOOL CONDITION MONITORING IN DRILLING

ISSAM ABU-MAHFOUZ
Penn State University at Harrisburg
Mechanical Engineering Technology
777 West Harrisburg Pike
W-255 Olmsted Bldg.
Middletown, PA 17057

ABSTRACT

In machining, the product quality is mainly controlled by the status of the tool cutting edge. In this study, a basic architecture for the application of artificial neural network (ANN) for tool condition monitoring (TCM) of a drilling process is presented. ANNs are regarded as multivariate nonlinear analytical tools capable of recognizing patterns from noisy, complex data and estimating their nonlinear relationships. This research utilizes processed vibration and acoustic emission (AE) signals as the main sources of information for the machining process. The objective of the proposed study is to produce a TCM system that will lead to a more efficient and economical drilling tool usage. Four drill wear conditions were introduced to the neural network for prediction and classification and these are; (1) chisel edge wear, (2) rake face crater, (3) main cutting edge fracture, and (4) Outer corner wear.

INTRODUCTION

Machine diagnosis techniques might fail when there is missing or incomplete information about the case in study. The more parameters measured, the more successful a well trained neural network will be in its prediction and classification. However, the cost of the sensory system should also be accounted for when designing an industrial monitoring system. Recently, vibration and artificial neural network techniques have been successfully applied to machine and process monitoring cases; (Susuki and Winmann, 1985, Rangwala and Dornfeld, 1990, and Tse and Atherton, 1999). Training of the ANN is carried out by (feeding) it with a sequence of examples (epochs) about the problem to be solved. These epochs are in the form of input data together with the expected neural model output. In Supervised learning, the model weights are repeatedly and adaptively modified to make the model's output agree with that specified in the training data. The weighted sum is passed to a transfer function to calculate the output of the node and send it to the input of all nodes in the subsequent layer. Another, different, set of input-output facts, not used in the training stage, is used to test and verify the performance of the ANN model.

EXPERIMENTAL SET-UP

This study demonstrates the performance of an Artificial Neural Network model based on vibration and AE signals for the detection, and characterization of wear on High-speed-steel (0.5 inch diameter) twist drills. Four combinations of cutting conditions, listed in Table 1., were used for the drilling test. Four types of defects were chosen as: (1) Chisel edge wear, (2) Fracture or breakage on one lip of the cutting edge, (3) Crater wear on the rake face of one cutting edge, and, (4) Wear on both outside corners of the drill O.D.

A KISTLER model 8152B acoustic emission sensor, and a KISTLER model C8905 accelerometer were mounted on the spindle bearing housing and on the clamping fixture of the work piece, respectively. Analog signals from these sensors were fed into a National Instruments PCI data acquisition board, with a sampling rate capability of 1.2 MHz, for signal digitization and conditioning.

SIGNAL ANALYSIS AND FEATURE EXTRACTION (PREPROCESSING)

A simplified form of the harmonic wavelet analysis, presented by Newland (1999) was utilized here as one method for feature extraction from the vibration signals. The wavelet coefficient c(t) correlates an input signal f(t) with a wavelet base function w(t). f(t): time history of vibration signals was represented in segments of 4096 data points and 2048 wavelet Coefficients were calculated. Then, 16 averaged wavelet coefficients were extracted by grouping and averaging each adjacent 128 Coefficients. The coefficients were normalized with respect to the maximum coefficient for each sample test. The obtained 16 averaged coefficients form a feature vector which served as an input pattern to the neural network.

The Burg algorithm, a parametric spectral estimation method obtained using MATLAB (1998), was used to give an estimate of the power spectral density (PSD) of the discrete-time for the vibration signals. The highest three local maxima of the Burg PSD for the vibration signals were presented as another feature input vector to the ANN.

In this study the AE RMS was analyzed for four statistical parameters, namely: The mean, the variance, the Kurtosis, and the skewness. These four quantities, sometimes called moments or cumulants, were calcualted using MATLAB and then presented to the ANN at 4 input nodes as an additional feature vector.

ARTIFICIAL NEURAL NETWORK (ANN) ARCHITECTURE AND TRAINING

Figure 1., shows the architecture of the three-layer feed-forward back-propagation (FFBP) ANN used in this research. The details of the algorithm were presented by Looney (1997) and Rumelhart, et. al. (1986). During the training phase, the weights were repeatedly adjusted by the steepest descent method to force each of the input exemplar feature vectors $x^{(q)}$ to be mapped into an output vector $z^{(q)}$ closer to its correct output class $t^{(k(q))}$ than to any other class $t^{(p)}$, $p \neq k$. This ANN used a unipolar sigmoid non-linear activation function. This is the FFBP Algorithm with Unipolar Sigmoid:

Input N, M, J, Q, and I
Set parameters, initially, as:
$\alpha_1 = 2.5$; $\alpha_2 = 2.5$;
Generate initial weights $W_{nm}^{(0)}$ and $U_{mj}^{(o)}$ randomly between –0.5 and 0.5
Adjust all weights via steepest descent methods follows:
for r = 1 to I do
  for q = 1 to Q do
  Start Updating all weights:
    for $m = 1$ to M do
      for $j = 1$ to J do

$$U_{mj}^{(r+1)} = U_{mj}^{(r)} + \eta\{(t_j^{(q)} - z_j^{(q)})[z_j^{(q)}(1-z_j^{(q)})]y_m^{(q)}\}; \quad (1)$$

for $n = 1$ to N do

$$W_{nm}^{(r+1)} = W_{nm}^{(r)} + \eta\{\sum_{(j=1,J)}(t_j^{(q)} - z_j^{(q)})[z_j^{(q)}(1-z_j^{(q)})]U_{mj}^{(r)}\}$$
$$x[y_m^{(q)}(1-y_m^{(q)})][x_n^{(q)}] \quad (2)$$

Where;
$$\begin{cases} y_m = f\left(\sum_n W_{nm} x_n\right), \\ and, \\ z_j = f\left(\sum_m U_{mj} y_m\right). \end{cases}, \; f \text{ is the unipolar sigmoid function.} \quad (3)$$

The learning process was terminated (i.e., ANN had converged) when the ANN's output was close to the desired output within an error threshold (E_{thr}) calculated by:

$$E_{thr} = \frac{1}{2}\sqrt{\sum_q^Q (t^{(q)} - z^{(q)})^2} \quad (4)$$

The AE signals were sampled at 524,288 samples/sec., whereas the vibration signals were sampled at 131,072 samples/sec. The signals for every test-run were divided into segments of 524,288 and 131,072 time history data points for the AE and the Vibration signals respectively. Each segment was then analyzed as described in the previous section. One hidden layer, with 65 nodes, was used (an acceptable number found through extensive experimentation). No attempt was made to determine the optimum number of neurons M in the hidden layer or the number of hidden layers required for optimum performance of the of the ANN.

RESULTS AND DISCUSSION

Figure 2., shows the normalized output $z^{(q)}$ of the ANN after it was presented with I=200 epochs. The uni-polar sigmoid activation function output is from 0 to 1, which forms a measure of the prediction weight for every wear condition. As presented in Figure 6, the existing drill wear type under training shows a high value (closer to 1) at its associated neuron. Ideally, the types of wear that do not exist should show a neural output of zero, but are presented with smaller output values in Fig. 2. Thresholds to predict tool failure for hole quality and surface finish can be designated at appropriate output values (for example at a perdiction weight of 0.4 in Fig. 2). An alarm or machine control system can be triggered when one of the output neurons exceeds this threshold.

The test phase helps the ANN model to generalize and increases its declaration accuracy. Figure 3., shows the testing phase of the trained ANN. This figure is designed to plot the percentage number of test set presentations, where the ANN successfully identified the wear type, against the number of test presentations. From these results it can be deduced that the accuracy of the trained ANN does not increase significantly after 200 test presentations. This number could be satisfactory for implementation purposes. The neural model was capable of adapting more efficiently (80% accuracy) to the Corner wear and chisel wear features compared to its less prediction power (around 70% accuracy) for the edge fracture or crater wear.

CONCLUSIONS

A multiple layer neural network has been successfully applied to twist drill wear detection and classification using supervised learning with processed vibration and acoustic emission signals. The signals were analyzed using discrete harmonic wavelet transform, power spectral density (PSD), and four statistical measures of the AE RMS time domain. During the testing phase the network correctly classified the drill condition (between 70% to 80% accuracy). The results reveal that once the neural network is properly trained, it become a powerful and reliable tool in solving classification and pattern recognition problems such as in this drilling process monitoring application.

NOMENCLATURE

$f(s) = 1 / [1+\exp(-\alpha(s-b))]$: The sigmoid function.
Where; b_i: biases, and α_i : decay (growth) rates.
$i=1$ for (hidden layer), and, $i=2$ for (output layer)
I : number of epochs
J : number of output nodes.
K : number of exemplar output target vectors $t^{(k(q))}$ (classes of drill wear).
M : number of nodes in the hidden layer, and
N : number of nodes in the input layer
Q : number of exemplar features $x^{(q)}$
U_{mj}: synaptic weight set for the neurons at the output layer).
W_{nm}: synaptic weight set for the neurons at the hidden layer, and
Where, $\{1 \leq n \leq N, 1 \leq m \leq M, 1 \leq j \leq J\}$
η_i: learning rate

REFERENCES

Looney, C. G., 1997, "Pattern Recognition Using Neural Networks, Theory and Algorithms for Engineers and Scientists," Oxford University Press, New York, pp. 75-136.

MATLAB, 1998, "Signal Processing Toolbox, For Use with MATLAB" User's Guide. The Math Works, Inc., Natick, MA.

Newland, D.E., 1999, "Ridge and Phase Identification in the Frequency Analysis of Transient Signals by Harmonic Wavelets," *ASME Journal of Vibration and Acoustics*, Vol. 121, pp. 149-155.

Rangwala, S, and Dornfeld, D. A., 1990, "Sensor Integration Using Neural Networks for Intelligent Tool Condition Monitoring", *ASME Journal of Engineering for Industry*, Vol. 112, pp. 219-228.

Rumelhart, D. E., G. E. Hinton, and R.J. Williams, 1986, "Learning Internal Representation by Error Propagation." Edited by D. E. Rumelhart and J. L. McClelland. Massachusetts Institute of Technology Press, New York.

Suzuki H., and Weinmann K. J., 1985, "An On-line Tool Wear Sensor for Straight Turning Operations", *ASME Journal of Engineering for Industry*, Vol. 107, Nov.

Tse, P. W., and Atherton, D. P., 1999, "Prediction of Machine Deterioration Using Vibration Based Fault Trends and Recurrent Neural Networks," *ASME Journal of Vibration and Acoustics*, Vol. 121, pp. 355-362.

Table 1: Cutting conditions for drilling with 0.5-inch dia. HSS twist drill

Cutting Speed (surface feet/min.) SFPM	Feed rate (inch/revolution) Inch/rev.
78	.004
117	.004
157	.006
196	.006

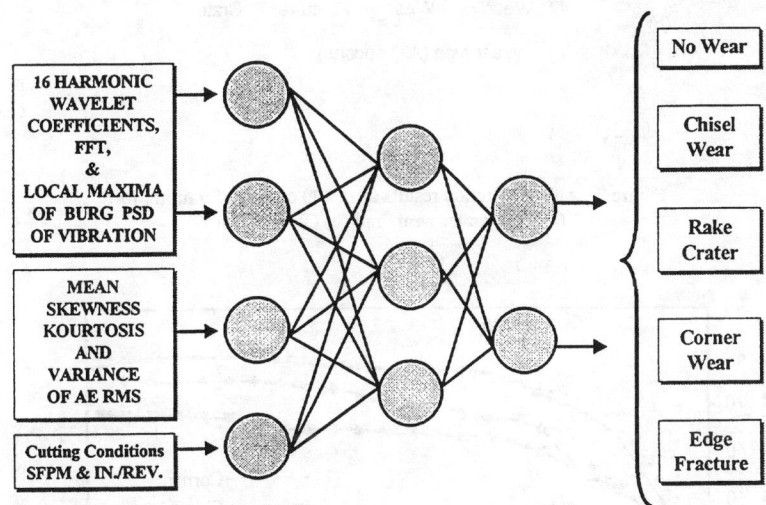

Figure 1. Artificial Neural Network (ANN) algorithm for drill wear detection and classification.
Note: This schematic does not show the exact number of nodes in each layer.

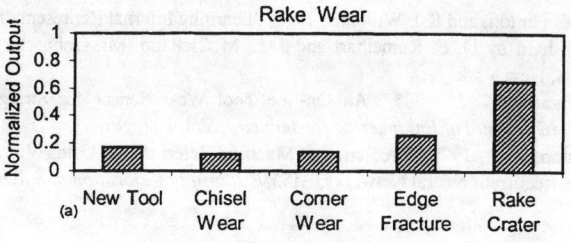

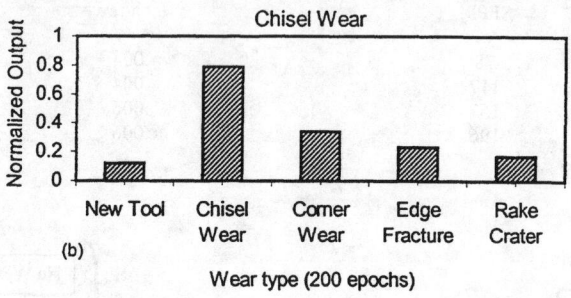

Figure 2. ANN prediction results after 200 epochs of training for: (a) Rake crater wear, and (b) Chisel Wear.

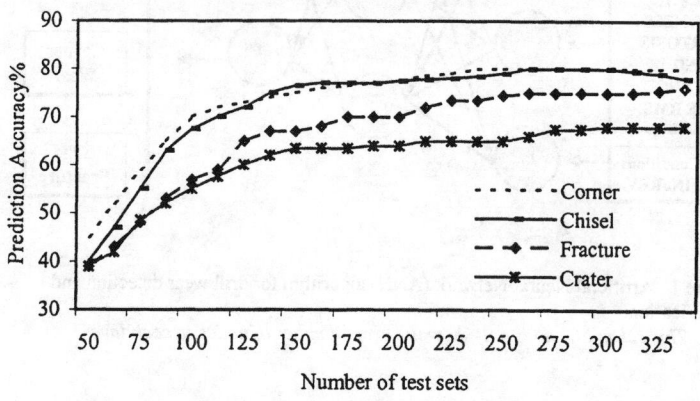

Figure 3. Prediction Accuracy verification for the 4-type drill

TUNING OF PI CONTROLLER COEFFICIENTS USING GENETIC ALGORITHMS AND ARTIFICIAL NEURAL NETWORK

S.VAKKAS USTUN
Yildiz Technical University,
Department of Electrical
Engineering, Istanbul/TURKEY
MEHMET BULUT
Eti Silver Plant Co., Automation
Department, Kutahya/TURKEY

GALIP CANSEVER
Yildiz Technical University,
Department of Electrical
Engineering, Istanbul/TURKEY

ABSTRACT
In this paper, a method about tuning of the controller coefficients as off-line in control of an applied nonlinear system has been proposed. In this method, the first step is identification of the system via Artificial Neural Networks (ANN) for maximum overshoot and settling time obtained from the application circuit for different K_p-K_i pairs. In the second step, the aim is to find the optimum controller coefficients using the ANN model as objective function for Genetic algorithms.

INTRODUCTION
The Proportional-Integral (PI) controller is undoubtedly the most popular controller used in industrial control processes, because of its simple structure and reliability of use in a wide range of operating conditions. Due to this acceptance, many tuning rules have been proposed for this type of controller. During the last two decades, one of the main focus of research in control engineering has been devoted to provide automatic tuning of such controllers and the design of robust control systems with the presence of uncertainty in the plant model (Moura, 1998).

System model is necessary for tuning controller coefficients in an appropriate manner (e.g. percent overshoot, settling time). Because of neglecting some parameters, the mathematical model can not represent the physical system exactly in most applications. That's why, controller coefficients can not be tuned appropriately.

Many of the recent developed computer control techniques are grouped into a research area called Intelligent Control (IC), that result from the integration of Artificial Intelligent (AI) techniques within automatic control systems (Astrom, 1992). Artificial neural networks is one of these techniques and can be used to identify the system properly. Using input/output data of the nonlinear system, artificial neural networks identifies the system. Instead of the approximate mathematical model, trained weights of the artificial neural networks is used as a model of the system. The genetic algorithms can be viewed as general-purpose optimisation method and have been successfully applied to search, optimisation and machine learning task.

GENETIC ALGORITHMS

Genetic algorithms (GA) are global numerical optimisation methods, patterned after the natural processes of genetic recombination and evolution.

The GA used in this paper known as the simple genetic algorithm. In the algorithm, the three-operator GA with only minor deviations from the original is used (Dimeo, 1995).

An initial population of binary strings is created randomly. Each of these strings represents one possible solution to the search problem. Next the solution strings are converted into their decimal equivalents and each candidate solutions is tested in this environment. The fitness of each candidate is evaluated through some appropriate measure. The algorithm is driven towards maximising this fitness measure. Application of the GA to an optimal control problem entails minimising the selected performance index. After the fitness of the entire population has been determined, it must be determined whether or not the termination criterion has been satisfied. If the criterion is not satisfied then we continue with the three genetic operators: reproduction, crossover and mutation (Dimeo, 1995).

Fitness-proportionate reproduction is effected through the simulated spin of a weighted roulette wheel. The roulette wheel is biased with the fitness of each of the solution candidates. The wheel is spun N times, where N is the number of strings in the population. Copying strings according to their fitnesses values means that strings according to their fitness values means that strings with a higher value have a higher probability of contributing one or more off spring in the next generation. This operation yields a new population of strings in the next generation (Goldberg, 1989). This operation yields a new population of strings that reflect the fitnesses of the previous generation's fit candidates. The next operation, crossover, is performed on two strings at a time that are selected from the population at random.

Crossover involves choosing a random position in the two strings and swapping the bits that occur after this position. The resulting crossover yields two new strings means the strings are part of the new generation (Dimeo, 1995). The crossover rate specifies the number of strings which are effected crossover operator. The mechanics of reproduction and crossover are surprisingly simple, involving random number generation, strings copies, and some partial string exchanges (Goldberg, 1989). The final genetic operator in the algorithm is mutation. Mutation is performed sparingly, typically every 100-1000 bit transfers from crossover, and it involves selecting at random as well as a bit position at random and changing it from 1 to 0 or vice-versa. After mutation, the new generation is completed and the procedure begins again with fitness evaluation of the population. In a control system design using the GA, the parameters that are represented as binary strings are the relevant control parameters.

ARTIFICIAL NEURAL NETWORKS (ANN)

Artificial Neural Networks (ANN) are successfully used in a lot of areas such as control, early detection of electric machine faults, digital signal processing in our daily technology (Haykin, 1994). The feed-forward neural network is usually trained by a back-propagation training algorithm first proposed by Rumelhart, Hinton, and Williams in 1986. The distributed weights in the network contribute to the distributed intelligence or associative memory property of the network. With the network initially untrained, i.e., with the weights selected at random, the output signal pattern will totally mismatch the desired output pattern for a given input pattern. The actual output pattern is compared with the desired output pattern and the weights are adjusted by the supervised back-propagation training algorithm until the pattern matching occurs, i.e., the pattern errors become acceptably small.

Multiplayer perceptrons (MLPs) are the simplest and therefore most commonly used neural network architectures. BacKpropagation algorithm is the most commonly adopted MLP training algorithm. It is a gradient descent algorithm and gives the change $\Delta w_{ji}(k)$ in the weight of a connection between neurons i and j as follows

$$\Delta w_{ji}(k) = \eta \delta_j x_i + \alpha \Delta w_{ji}(k-1) \tag{1}$$

where η is a parameter called the learning coefficient, α is the momentum coefficient, and δ_j is a factor depending on whether neuron j is an output neuron or a hidden neuron (Guner, 1999).

EXPERIMENTAL SETUP

We use motor and generator which is connected to motor with a connecting element. Motor was used 0.55 kW, 2.6A, 220V, 50Hz, Cos φ=0.79 3-phase squirrel-cage induction motor. The processor used in this work is, 40Mhz TMS320C50 DSP with 10k x 16 words of on-chip RAM which works parallel with TLC320C40 Analog interface circuit (AIC) with 14 bit resolution. The processor is communicated with a PC through RS232 serial port. Block diagram of this application circuit is shown in Fig.1. Inverter is designed with IPM (Intelligent Power Model). DSP (Digital Signal Processor) sends IGBT's on-off data to IPM using an interface circuit. 150-300V, 8.5A, 1-2 kW, 1500-3000 rpm, UErr=220V, IErr=0.6A DC Motor which has a tachogenerator is used as a load.

The stator voltage is adjusted by using PWM technique. After a period is modulated by 15 triangle wave, on-off data and duration are determined, and a table is built by using a C++ program. PI decides to which voltage packet is send in the next step, according to the velocity error.

SYSTEM MODELLING WITH ANN

Maximum overshoot (M_a) and Settling time (T_s) obtained from application circuit for different K_P and K_i pairs. These data are used to train ANN. ANN structure and parameters are shown in Fig. 1 and Table 1 respectively. Training phase error values are in Fig. 2.

ANN response and actual data are shown in Fig. 3. It's obvious that, ANN models the system successfully.

OPTIMIZATION OF PI COEFFICIENTS USING GA

The range of K_P and K_i chosen between (0.0-8.0) and (0.0-5.0) respectively. The fitness function is defined as

$$f = \frac{1}{M_a + T_s + 1} \qquad (2)$$

In this algorithm, the genetic algorithm parameters are selected for training cycle as:

Population size	30
Chromosome length	30 bits(15 each for K_p and K_i)
Number of generations	60
Crossover rate	0.60
Mutation rate	0.04

The optimum PI coefficients was found as
K_p= 7.995
K_i= 0.275
This values of K_P and K_i responses of system is shown in Fig. 4.

CONCLUSIONS

In this study, for an asynchronous motor controller PI coefficients tuning is presented using GA. Actual system (motor and controller) is modelled by ANN. It's seen that ANN modelling can represent the physical system exactly. It is also determined that maximum overshoot and settling time are very small if the system is controlled by the parameters obtained from the optimisation process which uses GA. It is found that the GA is suitable for optimisation of controller coefficients by considering the performance criteria.

NOMENCLATURE

K_P : Proportional constant
K_i : Integral Constant
$\Delta w_{ji}(k)$: The weight of a connection between neurons i and j
η : Learning coefficient
α : Momentum coefficient
δ_j : Factor depending on whether neuron j is an output neuron or a hidden
M_a : Maximum overshoot
T_s : Settling time

REFERENCES

P.B.De Moura Oliveira, (1998), "Evolutionary Design of Process Control Systems", Ph. D. Thesis, Chapter 1, University of Salford, Department of Aeronautical and Mechanical Engineering

K.J.Astrom, C.C.Hang, P.Persson, and W.K.Ho, (1992), "Towards Intelligent PID Control", Automatica, Vol. 28, No. 1, pp. 1-9.

R.Dimeo, K.Y.Lee, "The Use of a Genetic Algorithm in Power Plant Control System Design", IEEE Proceeding of the 34th Conference on Decision & Control, 1995, 737-742.

D.E.Goldberg, "Genetic Algorithms in Search, Optimisation and Machine Learning", Addison Wesley Publishing Company, Inc., 1989.

S.Haykin, Neural Networks, Macmillan Publishing Company, New Jersey, 1994.

K.Guner, M.Erler and S.Sagiroglu "Artificial Neural Networks for the Characteristic Impedance Calculation of Conductor-Backed Coplanar Waveguides" Eleco'99 International Conference on Electrical and Electronics Engineering, 1999

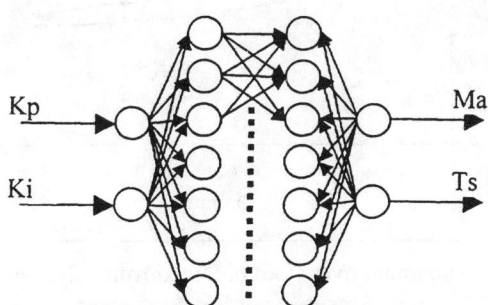

Fig. 1 ANN structure of the system model.

Table 1 ANN parameters of the system model.

Number of input 2	2
Number of output 2	2
Number of hidden layer 2	2
1st hidden layer neuron number	7
2nd hidden layer neuron number	7
1st transfer function	Sigmoid
2nd transfer function	Sigmoid
Maximum iteration number	15000
Learning coefficient	0,7
Momentum coefficient	0,9

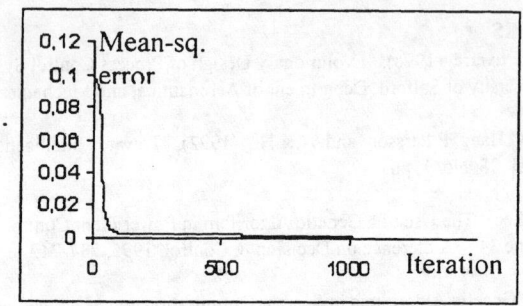

Fig. 2 Mean-squared error values according to iteration

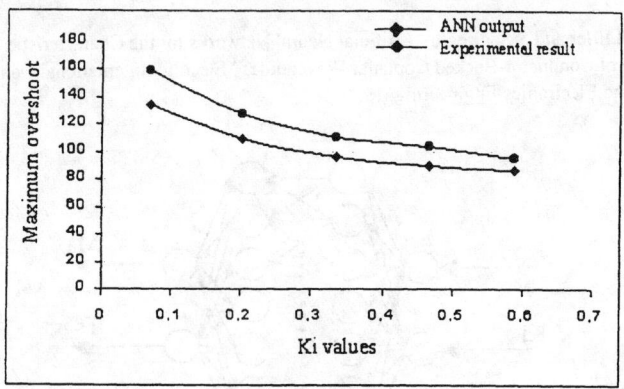

Fig. 3 K_P= 4,266, Maximum overshoot obtained from ANN and actual system is seen according to K_i

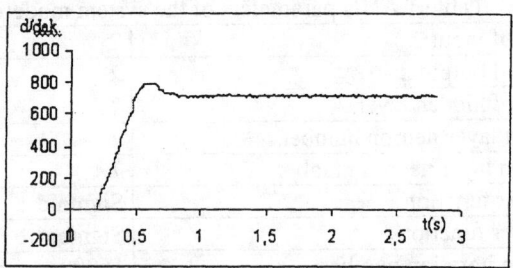

Fig. 4 The system response for K_p=7,995 and K_i=0,275.

MONITORING OF DRILLING PROCESSES

HERMAN R. LEEP
Department of Industrial Engineering
University of Louisville
Louisville, Kentucky

ABSTRACT

In order to develop a system which monitors and controls a machining process, a tool-wear sensing system which predicts the best time for tool replacement must be developed first. Tool wear can be predicated by measuring the cutting forces with a dynamometer. Monitoring systems have been used in the automotive industry for the high-production drilling of components. The primary objective of this research project was to develop mathematical models for predicting tool wear from measured cutting forces while drilling an aluminum metal matrix composite (MMC) material containing Al_2O_3. Multifaceted, solid carbide drills were used. Regression analysis was used to develop the relationships between tool wear, thrust force, cutting torque, workpiece hardness, cutting speed, and feed rate. The results from this project were compared to the results from similar projects. The other projects included different workpiece materials, cutting conditions, tool diameters, tool materials, tool geometries, and cutting-fluid conditions.

INTRODUCTION

The cylindrical hole is probably the most common shape produced in discrete manufacturing, and most cylindrical holes are made by drilling. In order to develop a system which monitors a drilling process, a tool-wear sensing system which predicts the best time for drill replacement must be developed first.

Tool wear can be predicted by measuring the cutting forces with a dynamometer. The cutting torque in machining operations can be evaluated by using the spindle motor current, voltage, and speed.

In an automotive industry application, Gu and Ni (1996) studied multiple spindle drilling of grey cast iron engine blocks. A dynamometer was placed on the fixture to measure the thrust force signal which was found to be sensitive to the drill wear condition.

Jalali and Kolarik (1991) suggested that the thrust force, not the cutting moment, was the superior indicator of tool wear. These researchers developed an algorithm for the real-time monitoring of drilling processes. The researchers concluded that the alogorithm was generally capable of managing the drilling thrust within the ranges of the cutting conditions of the experiment.

A real-time tool breakage monitoring system for drilling was developed by Lee and Choi (1994). When a drill breaks, there is a significant decrease in the thrust force. This change in force was examined after a triggering signal was sent by an acoustic emission sensor.

A real-time process monitoring system was developed by Kavaratzis and Maiden (1990). This system also had adaptive control capabilities for drilling deep holes during unattended CNC machining. The inputs were thrust and torque and the outputs were feed rate, cutting speed, and depth of cut.

Nagao, Hatamura, and Mituishi (1994) developed their own torque-thrust sensors. They were able to predict and prevent drill breakage by monitoring these cutting forces.

The drill condition monitoring system used by Adamczyk (1998) involved the installation of a two-channel sensor on the machine tool to measure the current consumption in the spindle and feed drives. A vibration sensor was also used to measure the acceleration of forced vibrations from the spindle box. Correlations between the sensor signals and flank wear, corner wear, and margin wear were used to develop decision rules in the knowledge base of an intelligent tool monitoring system.

Many researchers, including Li, Lau, and Zhang (1992), believe the measurement of cutting forces is the best way to monitor drill wear and breakage. These investigators recommended cutting forces for the monitoring of the cutting conditions of cutting tools because of their high sensitivity and rapid response to the changes in cutting states.

Routio and Saynatjoki (1995) measured the drilling torque and axial force acting on the drill with a dynamometer based on strain gauges. Through holes were drilled into the center of round workpieces of stainless steel. These researchers concluded that the measurements could be used for tool breakage detection.

The relationships between tool wear and cutting forces observed while drilling metal matrix composites were studied by Morin, Masounave, and Laufer (1995). These researchers compared holes drilled into a 6061 aluminum alloy to those drilled into a particle-reinforced metal matrix composite consisting of 20 vol% SiC particles in a 6061 aluminum matrix. They found the cutting forces for both workpiece materials to be very similar, and concluded that the matrix controlled the drilling forces, not the reinforcing particles.

A series of experiments were performed by Lin and Ting (1995) to study the effects of tool wear on cutting forces and other cutting parameters when drilling a copper alloy. They established relationships for extimating tool wear based on the measurement of cutting forces, spindle speed, feed rate, and drill diameter.

Liu (1990) used a computer vision system to perform real-time measurements of drill wear. The wear criterion was flank wear area rather than flank wear width. A toolmaker's microscope was used to calibrate the vision system.

A TiN-coated drill with curved primary cutting edges was studied by Chen and Fuh (1995). Flank wear was measured with a vision system having a magnification factor of 50x. The TiN-coated drill was compared to uncoated drills with straight (and curved) primary cutting edges. The results showed that the coated drill reduced the thrust force, torque, and tool wear.

The workpiece material for the study reported in this paper was a MMC. A limited definition for a composite material can be stated as "a material made from any two or more organic, inorganic, or metallic materials (Lambert, 1987)." These materials can include the fiber, flake, and particle-reinforced materials; preplated, precoated, and clad metals; and various laminates. In a MMC with particulates, a matrix material usually surrounds a reinforcing material (Zweben, 1988).

Some industries use MMCs for various types of products. The automotive industry uses these materials for engine components, such as connecting rods, and the sporting goods industry uses them for golf club heads and racquet frames (Narraway, 1989).

The productivity of the standard twist-drill point is limited because the rake angle on the cutting edge can vary from positive to negative, and the high thrust force and cutting moment are created on the point's wide chisel edge. This limitation has lead many toolmakers to develop different types of points which are better suited for making holes. One such development is the multifaceted twist-drill point.

Another way to help increase productivity is to introduce carbide along with the multifaceted drill. One user who applied a high-penetration solid carbide drill with a multi-faceted point reduced machining cost by 69% when drilling a structural steel part with a hardness of 248 HB (Noaker, 1991).

EXPERIMENTAL WORK

The experimentation was performed on a computer numerical control (CNC) milling machine. This machine tool had a positioning accuracy of $\pm$ 0.0008 in. ($\pm$0.02 mm) and a positioning repeatability of $\pm$0.0003 in. ($\pm$0.01 mm).

A 6061 aluminum-based MMC, which contained 16 vol% Al_2O_3, was used as the workpiece material. This material was hardened and tempered to a T6 condition. Cylindrical blanks were machined from extruded bar stock which had a diameter of 2.00 in. (5.08) cm). The finished height of each workpiece cylinder was 1.250 in. (3.175 cm).

The hole pattern consisted of 19 holes with 12 holes on the outer circle, six holes on the inner circle, and one hole in the center. Thirty-eight holes were drilled into each cylinder, with 19 on the top and 19 on the bottom. Each blind hole was 0.250 in. (6.35 mm) in diameter and 0.500 in. (12.70 mm) deep to the shoulders of the drills. The minimum wall thickness between holes was 0.1250 in. (3.175 mm). Before drilling the test holes, the 19 hole positions were spot drilled to minimize flexing of the test drills.

Only the holes on the inner circle were used as test holes for recording tool-wear and cutting-force measurements. This procedure was followed to eliminate any potential effects due to variations in workpiece hardness in the radial direction.

Workpiece hardness of the cylinders ranged from aproximately 73 to 83 HRB. The average hardness, measured next to 57 test holes on the 19 cylinders used for the three tests, was 78.3 HRB.

Solid carbide twist drills, with multifaceted drill points, were used to drill the MMC material. The drill diameter was 0.250 in. (6.35 mm). The type of multifaceted drill point selected was designed for drilling composite materials. The geometry of a typical multifaceted drill was modified. This type of drill could be used at relatively high feed rates (Madison and Schaible, 1991).

A commercially available synthetic cutting fluid was used at a concentration of 5 wt%. The synthetic lubricant in the concentrate of this cutting fluid was recommended for heavy-duty machining and grinding operations, including those on MMC aluminum alloys.

The results from a research project performed by Herde (1991) along with the cutting conditions recommended by the work piece material supplier, Duralcan® USA, were used to establish the cutting conditions for this study. The tests performed by Herde used the same workpiece material. Because the multifaceted, solid carbide drills were expected to perform better than the drills used by Herde, the cutting conditions recommended by Duralcan® USA for carbide-tipped drills were used in this study for two of the three tests described later.

Tool wear on the multifaceted drills was measured with a toolmaker's microscope. Accumulated tool wear was measured every 6 or 13 holes during the first 144 holes of a test, or until the drill failed. Then, if the test were continued, measurements were taken every 19 holes. Tool wear was measured along the margin at the outside end of each cutting edge. Each time tool wear was measured, the average of the measurements associated with the two cutting edges was recorded.

The thrust force along and cutting torque (or moment) about the z-axis were measured using a dynamometer with piezoelectric crystals in the platform. The forces in the x-y plane were minimized by center drilling first.

Software was developed for collecting and processing the raw data from the dynamometer. A routine was written to smooth the curves representing force (F_z) and moment (M_z).

Each test was continued until enough holes were drilled to develop an accurate model, tool wear reached a critical stage, or the drill failed. The development of relationships which predicted tool wear from measured cutting forces was the primary objective of this study.

The three drilling tests were performed at various speeds and feeds. The following cutting conditions were used for the three tests to determine the models:

Test 1 – High Speed of 225 ft/min (114.3 cm/s) and Low Feed of 0.016 in./rev (0.406 mm/rev).

Test 2 – Medium Speed of 180 ft/min (91.4 cm/s) and Medium Feed of 0.020 in./rev (0.508 mm/rev).

Test 3 – Low Speed of 150 ft/min (76.2 cm/s) and High Feed of 0.024 in./rev (0.610 mm/rev).

These speeds and feeds were also used to determine the outer edge of the envelope associated with the cutting conditions.

RESULTS AND DISCUSSION

Results from the drilling tests were studied using the Statistical Analysis System (SAS) computer software. The analysis included data from tool-wear and cutting-force measurements. Measurements were taken through the 304 holes drilled in Tests 1 and 2, and 95 holes drilled in Test 3.

Regression analysis was used to develop three models from the test results for the machining parametric combinations of high speed/low feed, medium speed/medium feed, and low speed/high feed. A fourth model was developed by pooling the data from all three tests and including the speeds and feeds to help explain the observed variance in tool wear.

The dependent variable was tool wear while the independent variables were thrust force, cutting moment, and workpiece hardness. These variables were used to develop each model using the main effects, square terms, and first-order interactions. When the data were pooled, cutting speed and feed rate were added along with the new square terms and first-order interactions.

SAS was used to perform the multiple classification analysis of variance (ANOVA) of the raw data. This procedure aided in determining which independent variables were significant at the 0.05 level. The significance of the model and the sources of variation were determined by using an F-test, and the coefficient of multiple regression (R^2) was used to determine the percentage of variation in tool wear that could be explained by the independent variables.

The following nomenclature was used in developing the regression equations after eliminating the moment about the z-axis and workpiece hardness:

W = accumulated tool wear, mm
F = force along z-axis, N
S = cutting speed of tool, cm/s
k = intercept
a,b = coefficients
e = error term

The following model was assumed:

$$W = k + aF + bS + e \tag{1}$$

Criteria used to select the "best" models are summarized below:

1) The model was significant at the 0.05 level.
2) The independent variables were significant at the 0.05 level.
3) The coefficient of multiple regression was greater than 0.90.
4) The number of independent variables was reduced as much as possible.

Cutting moment and workpiece hardness were eliminated from all four models. **Table 1** shows a summary of the models.

Table 1 Summary of Regression Models

Cutting conditions	Model	R^2
High speed/low feed......	W = -0.1930 + 3.624E-04F	0.991
Med. speed/med. feed...	W = -0.2238 + 3.113E-04 F	0.995
Low speed/high feed......	W = -0.1613 + 2.410E-04 F	0.945
All three tests...............	W = -0.3798 + 2.737E-04 F + 2.294E-03S	0.935

All three models from the drilling tests and the model which included cutting speed and feed rate along with the pooled data were significant at the 0.05 level. Also, the effect of thrust force was significant at the 0.05 level in all four models.

These results can be compared to the results obtained by Herde. He used the same workpiece material, drill diameter, and cutting fluid. However, Herde used carbide-tipped drills with conventional-point drill geometry. As in the models developed in this project, thrust force was significant at the 0.05 level in the models associated with the high cutting speed of 200 ft/min (101.6 cm/s). When Herde pooled the data from his three tests, he also found that the thrust force and cutting speed were significant at the 0.05 level.

These results can also be compared to those obtained by Leep and Peak (1992). They used the same cutting fluid, but developed a model for a medium-carbon-steel workpiece and a 0.375-in. (9.525-mm) HSS drill with conventional-point drill geometry and a model for a titanium-alloy workpiece and a 0.250-in. (6.350-mm) cobalt HSS drill with split-point drill geometry. Thrust force in these models was also significant at the 0.05 level.

In order to illustrate how these models are applied, a numerical example is presented. Consider Eq. (2) and the following values of the independent variables associated with a particular observation:

$$W = -0.3798 + 2.737E-04\ F + 2.294E-03\ S \qquad (2)$$
$$F = 1230\ N\ \text{and}\ S = 91.44\ \text{cm/s}$$

Then, $\quad W = -03798 + 2.737E-04\ (1230) + 2.294E-03\ (91.44) = 0.167\ \text{mm} \qquad (3)$

The observed value of tool wear for this observation was also 0.167 mm.

CONCLUSIONS

The following conclusions were drawn from the results of the drilling tests using 0.250 - in. (6.35-mm), multifaceted, solid carbide drills and aluminum MMC containing 16 vol% alumnia particles as the workpiece material. All four models, including the model for the pooled data and considering cutting speed and feed rate, were significant at the 0.05 level. The effect of thrust force was significant at the 0.05 level for all four models. The R^2 value for all four models ranged from 0.935 to 0.995. The model for the pooled data had six other independent variables which were significant at the 0.05 level, in addition to the thrust force and cutting speed, but when these variables were removed, the R^2 value was still greater than 0.90.

ACKNOWLEDGMENTS

The authors greatly appreciate the financial support provided for this research project by the National Science Foundation (EPSCoR Grant 87-57 # 13) through the University of Kentucky Research Foundation, and by the Toyota Motor Corporation Endowment Fund through the SME Manufacturing Engineering Education Foundation. We are grateful to Mr. Charles T. Lane, a materials engineer with Duralcan® USA, a Division of Alcan Aluminum Corporation, for donating the MMC material. We also wish to thank Nancy White for typing the manuscript.

REFERENCES

Adamczyk, Z., 1998, "Transient States in Drilling Process As a Source of Tool Wear Knowledge for Intelligent Tool Condition Monitoring System," *IX Workshop on Supervising and Diagnostics of Machining System Manufacturing Simulation for Industrial Use*, Wroclaw University of Technology, pp. 195-204.

Chen, W.-C. and Fuh, K.-H., 1995, "The Cutting Performance of a TiN-Coated Drill -with Curved Primary Cutting Edges," *J. Mater. Processing Technol.*, Vol. 49, pp. 183-198.

Gu, S. and Ni, J., 1996, "Multi-Spindle Drilling Process Condition Monitoring and Fault Diagnosis," *Manuf. Sci. Eng.*, Vol. 4, pp. 555-562.

Herde, D.L., 1991, *Production Drilling Models for a Hardened Alumina-Reinforced Composite Material*, M.Eng. Thesis, University of Louisville, Dept. of Industrial Engineering, Louisville, KY.

Jalai, S. A. and Kolarik, W. J., 1991, " A Two-Dimensional Decision Algorithm for Real-Time Tool Monitoring," *Int. J. Prod. Res.*, Vol 29, pp. 453-462.

Kavaratzis, Y., and Maiden, J.D., 1990, " Real Time Process Monitoring and Adaptive Control during CNC Deep Hole Drilling," *Int. J. Prod. Res.*, Vol. 28, pp. 2201-2218.

Lambert, B.K., 1987, "Cutting and Drilling of Composite Materials," *Carbide Tool J.*, Vol. 19, pp. 31-34.

Lee, J.M. and Choi, D.K., 1994, "Real-Time Tool Breakage Monitoring for NC Turning and Drilling," *Annals of the CIRP*, Vol. 43, pp. 81-84.

Leep, H.R. and Peak, M.A., 1992, "Drilling Models for a Synthetic Cutting Fluid," 8th *International Colloquium: Tribology 2000*, Technische Akadamei Esslingen, pp. 23.4-1 to 23.4-

Li, G.S. , Lau, W.S., and Zhang, Y.Z. , 1992, "In-Process Drill Wear and Breakage Monitoring for a Machining Centre Based on Cutting Force Parameters," *Int. J. Mach. Tools Manufact.*, Vol. 32, pp. 855-867.

Lin, S.C. and Ting, C.J., 1995, " Tool Wear Monitoring in Drilling Using Force Signals," *Wear*, Vol. 180, pp. 53-60.

Liu, T.I., 1990, "A Computer Vision Approach for Drill Wear Measurements," *J. Mater. Shaping Technol.*, Vol 8, pp. 11-16.

Morin, E., Masounave, J., and Laufer, E.E., 1995, " Effect of Drill Wear on Cutting Forces in the Drilling of Metal-Matrix Composites," *Wear*, Vol. 184, pp. 11-16.

Nagao, T., Hatamura, Y., and Mitsuishi, M., 1994, "In-Process Prediction and Prevention of the Breakage of Small Diameter Drills Based on Theoretical Analysis," *Annals of the CIRP*, Vol. 43, pp. 85-88.

Narraway, R., 1989, "Aluminum: MMCs Ready for Exploration," *Eng. Mater. Des.*, Vol. 33, pp. 32-35.

Noaker, P.M., 1991, "Holemaking's Tough Tools," *Manuf. Eng.*, Vol 106, pp. 52-57.

Madison, J. and Schaible, J., 1991, " Multiple Improvements," *Cutting Tool Eng.*, Vol. 43, pp. 46, 48-51.

Routio, M. and Saynatjoki, M., 1995, "Tool Wear and Failure in the Drilling of Stainless Steel," *J.Mater. Processing Technol.*, Vol. 52, pp. 35-43.

Zweben, C., 1988, "Metal Matrix Composites: An Overview," *Carbide Tool J.*, Vol. 20, pp. 7-10.

A HYBRID NEURO-FUZZY-FRACTAL APPROACH FOR AUTOMATED QUALITY CONTROL IN THE MANUFACTURING OF SOUND SPEAKERS

PATRICIA MELIN*, OSCAR CASTILLO*, AND FERNANDO SOTELO**
*Dept. of Computer Science, Tijuana Institute of Technology
**Dept. of Industrial Engineering, Tijuana Institute of Technology
P.O. Box 4207, Chula Vista CA, 91909, USA
pmelin@tectijuana.mx, ocastillo@tectijuana.mx

ABSTRACT

We describe in this paper the application of a hybrid neuro-fuzzy-fractal approach to the problem of automated quality control in sound speaker manufacturing. Traditional quality control has been done by manually checking the quality of sound after production. This manual checking of the speakers is time consuming and occasionally was the cause of error in quality evaluation. For this reason, we developed an intelligent system for automated quality control in sound speaker manufacturing. The intelligent system has a fuzzy rule base containing the knowledge of human experts in quality control. The parameters of the fuzzy system are tuned by applying the ANFIS methodology using, as training data, a real time series of measured sounds as given by good sound speakers. We also use the fractal dimension as a measure of the complexity of the sound signal. The intelligent system has been tested in a real plant with very good results.

INTRODUCTION

We describe in this paper the application of a neuro-fuzzy-fractal approach to the problem of quality control in the manufacturing of sound speakers in a real plant. The quality control of the speakers was done before by manually checking the quality of sound achieved after production [4]. A human expert evaluates the quality of sound of the speakers to decide if production quality was achieved. Of course, this manual checking of the speakers is time consuming and occasionally was the cause of error in quality evaluation [7]. For this reason, it was necessary to consider automating the quality control of the sound speakers. The problem of measuring the quality of the sound speakers is as follows:
1) First, we need to extract the real sound signal of the speaker during the testing period after production
2) Second, we need to compare the real sound signal to the desired sound signal of the speaker, and measure the difference in some way
3) Third, we need to decide on the quality of the speaker based on the difference found in step 2. If the difference is small enough then the speaker can be considered of good quality, if not then is bad quality.

The first part of the problem was solved by using a multimedia kit that enable us to extract the sound signal as a file, which basically contains 108000 points over a period of time of 3 seconds (this is the time required for testing). We can say that the sound signal is measured as a time series of data points [3], which has the basic characteristics of the speaker. The second part of the problem was solved by using a neuro-fuzzy approach to train a fuzzy model with the data from the good quality speakers [9]. We used the ANFIS approach [6] to obtain a Sugeno fuzzy system [13] with the time series of the ideal speakers. In the ANFIS approach a neural network [5, 10, 12] is used to adapt the parameters of the fuzzy system with real data of the problem. With this fuzzy model, the time series of other speakers can be used as checking data to evaluate the total error between the real speaker and the desired one. The third part of the problem was solved by using another set of fuzzy rules [14], which basically are fuzzy expert rules to decide on the quality of the speakers based on the total checking error obtained in the previous step. Of course, in this case we needed to define membership functions for the error and quality of the product, and the Mamdani reasoning approach was used. We also use as input variable of the fuzzy system the fractal dimension of the sound signal. The fractal dimension [8] is a measure of the geometrical complexity of an object (in this case, the time series). We tested our neuro-fuzzy-fractal approach for automated quality control during production with real sound speakers with excellent results. Of course, to measure the efficiency of our intelligent system we compared the results of the neuro-fuzzy-fractal approach to the ones by real human experts. The results clearly show that the neuro-fuzzy-fractal approach was better than the manual method because it reduced the time required for testing and also the accuracy was improved slightly. We think that our neuro-fuzzy-fractal approach for quality control can be used for similar problems, with only some minor changes to the structure of the fuzzy system.

BASIC CONCEPTS OF SOUND SPEAKERS

In any sound system, ultimate quality depends on the speakers [4]. The best recording, encoded on the most advanced storage device and played by a top-of-the-line deck and amplifier, will sound awful if the system is hooked up to poor speakers. A system's speaker is the component that takes the electronic signal stored on things like CDs, tapes and DVD's and turns it back into actual sound that we can hear.

Sound Basics
To understand how speakers work, the first thing you need to do is understand how sound works. Inside your ear is a very thin piece of skin called the eardrum. When your eardrum vibrates, your brain interprets the vibrations as sound. Rapid changes in air pressure are the most common thing to vibrate your eardrum.
An object produces sound when it vibrates in air (sound can also travel through liquids and solids, but air is the transmission medium then when we listen to speakers). When something vibrates, it moves the air particles around it. Those air particles in turn move the air particles around them, carrying the pulse of the vibration through the air as more and more particles are pushed farther from the source of the vibration.
A vibrating object sends a wave of pressure fluctuation through the atmosphere. When the fluctuation wave reaches your ear, it vibrates the eardrum back and forth. Our brain interprets this motion as sound. We hear different sounds from different vibrating objects because of variations in:
- sound wave frequency -- A higher wave frequency simply means that the air pressure fluctuates faster. We hear this as a higher pitch. When there are fewer fluctuations in a period of time, the pitch is lower.
- air pressure level -- the wave's amplitude -- determines how loud the sound is. Sound waves with greater amplitudes move our ear drums more, and we register this sensation as a higher volume.

A speaker is a device that is optimized to produce accurate fluctuations in air pressure.
A microphone works something like our ears. It has a diaphragm that is vibrated by sound waves in an area.

Making Sound

In the last section we saw that sound travels in waves of air pressure fluctuation, and that we hear sounds differently depending on the frequency and amplitude of these waves. We also learned that microphones translate sound waves into electrical signals, which can be encoded onto CDs, tapes, LPs, etc. Players convert this stored information back into an electric current for use in the stereo system.

A speaker is essentially the final translation machine -- the reverse of the microphone. It takes the electrical signal and translates it back into physical vibrations to create sound waves. When everything is working as it should, the speaker produces nearly the same vibrations that the microphone originally recorded and encoded on a tape, CD, LP, etc. Traditional speakers do this with one or more drivers. A driver produces sound waves by rapidly vibrating a flexible cone, or diaphragm.

- The cone, usually made of paper, plastic or metal, is attached on the wide end to the suspension, or surround.
- This rim of flexible material allows the cone to move, and is attached to the...
- driver's metal frame, called the basket.
- The narrow end of the cone is connected to the voice coil.
- The coil is attached to the basket by the spider, a ring of flexible material. The spider holds the coil in position, but allows it to move freely back and forth.

Chunks of the Frequency Range

In the last section we saw that traditional speakers produce sound by pushing and pulling an electromagnet attached to a flexible cone. Although drivers all work on the same concept, there is actually a wide variety in driver size and power. The basic driver types are: 1) Woofers 2) Tweeters 3) Midrange. In Figure 1 these basic drivers are shown.

Woofers are the biggest drivers, and are designed to produce low frequency sounds. Tweeters are much smaller units, designed to produce the highest frequencies. Midrange speakers produce a range of frequencies in the middle of the sound spectrum.

To create higher frequency waves -- waves in which the points of high pressure and low pressure are closer together -- the driver diaphragm must vibrate more quickly. This is harder to do with a large cone because of the mass of the cone. Conversely, it's harder to get a small driver to vibrate slowly enough to produce very low frequency sounds. It's more suited to rapid movement.

Woofer   Tweeter   Midrange

Figure 1 Three basic driver types: woofer, tweeter, and midrange.

DESCRIPTION OF THE PROBLEM

The basic problem consists in the identification of sound signal quality. Of course, this requires a comparison between the real measured sound signal and the ideal good sound signal. We need to be able to accept speakers, which have a sound signal that do not differ much from the ideal signals. We show in Figure 2 (a) the form of the sound signal for a good speaker (of a specific type). The measured signal contains about 108 000 points in about 3 seconds. We need to compare any other measured signal with the good one and calculate the total difference between both of them, and if the difference is small then we can accept the speaker as a good one. On the other hand, if the difference is large then we reject the speaker as a bad one.

We show in Figure 2 (b) the sound signal for a speaker of bad quality. We can clearly see the difference in the geometrical form of this signal and the one shown in Figure 2 (a). In this case, the difference between the figures is sufficiently large and we easily determine that the speaker is of bad quality. We also show in Figure 2 (c) another sound signal for a bad quality speaker. Again, we can see clearly the difference in the form of the signal with respect to the good speaker.

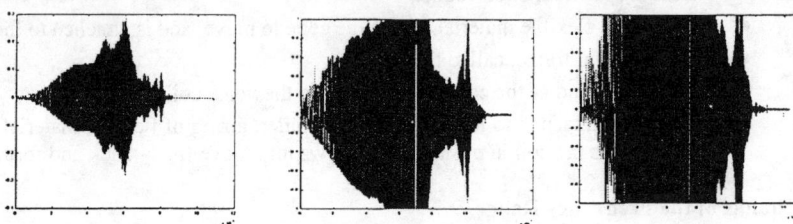

Figure 2 (a) Sound signal of a Good Speaker, (b) Bad Speaker Case 1 (c) Case 2.

FRACTAL DIMENSION OF A GEOMETRICAL OBJECT

Recently, considerable progress has been made in understanding the complexity of an object through the application of fractal concepts [8] and dynamic scaling theory [11]. For example, financial time series show scaled properties suggesting a fractal structure [1, 2, 3]. The fractal dimension of a geometrical object can be defined as follows:

$$d = \lim_{r \to 0} [\ln N(r)] / [\ln(1/r)] \quad (1)$$

where N(r) is the number of boxes covering the object and r is the size of the box. An approximation to the fractal dimension can be obtained by counting the number of boxes covering the boundary of the object for different r sizes and then performing a logarithmic regression to obtain d (box counting algorithm). In Figure 3 (a), we illustrate the box counting algorithm for a hypothetical curve C. Counting the number of boxes for different sizes of r and performing a logarithmic linear regression, we can estimate the box dimension of a geometrical object with the following equation:

$$\ln N(r) = \ln \beta - d \ln r \quad (2)$$

this algorithm is illustrated in Figure 3 (b).

The fractal dimension can be used to characterize an arbitrary object. The reason for this is that the fractal dimension measures the geometrical complexity of objects. In this case, a time series can be classified by using the numeric value of the fractal dimension (d is between 1 and 2 because we are on the plane xy). The reasoning behind this classification scheme is that when the boundary is smooth the fractal dimension of the object will be close to one. On the other hand, when the boundary is rougher the fractal dimension will be close to a value of two.

We developed a computer program in MATLAB for calculating the fractal dimension of a sound signal. The computer program uses as input the figure of the signal and counts the number of boxes covering the object for different grid sizes. The fractal dimension for the sound signal of Figure 2 (a) is of 1.6479, which is a low value because it corresponds to a good speaker. On the other hand, the fractal dimension for Figure 2 (b) is 1.7843, which is a high value (bad speaker). Also, for the case of Figure 2 (c) the dimension is 1.8030, which is even higher (again, a bad speaker).

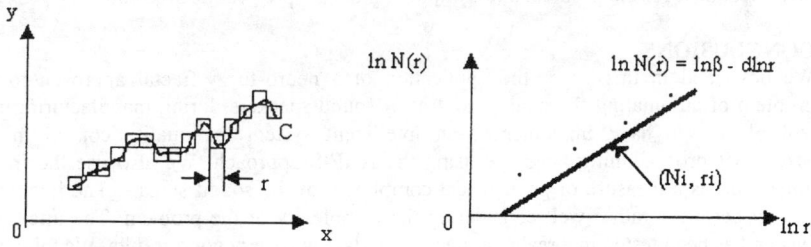

Figure 3 (a) Box counting algorithm (b) Logarithmic regression to find dimension.

EXPERIMENTAL RESULTS

We describe in this section the experimental results obtained with the intelligent system for automated quality control. The intelligent system uses a fuzzy rule base to determine automatically the quality of sound in speakers. We used a neuro-fuzzy approach to adapt the parameters of the fuzzy system using real data from the problem. We used the time series of 108000 points measured from a good sound speaker (in a period of 3 seconds) as training data in the ANFIS approach. We then use the measured data of any other speaker as checking data, to compare the form of the sound signals. We show in Figures 4 (a) and (b) two cases where the ANFIS approach is used to adapt a fuzzy system with training data of good sound speakers. The approximation is very good considering the complexity of the problem. Once the training was done we used the fuzzy system for measuring the total difference between a given signal and the good ones. This difference is used to decide on the final quality of the speaker using another set of fuzzy rules with the Mamdani approach. The fuzzy rules are as follows:

IF Difference is small   AND Fractal Dimension is small THEN Quality is Excellent
IF Difference is regular AND Fractal Dimension is smallTHEN Quality is Good
IF Difference is regular AND Fractal Dimension is high THEN Quality is Medium
IF Difference is medium AND Fractal Dimension is high   THEN Quality is Medium
IF Difference is medium AND Fractal Dimension is high   THEN Quality is Bad
IF Difference is large   AND Fractal Dimension is high THEN Quality is Bad

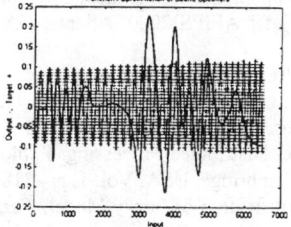
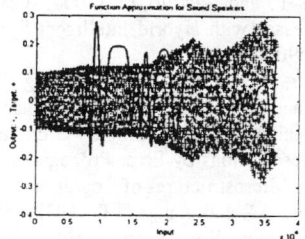

Figure 4 (a) Function approximation of the sound signal using ANFIS Case 1, (b) Case 2.

Where "Difference", "Fractal Dimension" and "Quality" are considered as linguistic variables in the fuzzy rules. We use gaussian membership functions for the linguistic values of all the variables. The membership functions for the linguistic values of these variables were tuned by using a simple genetic algorithm. Of course, the human experts on quality evaluation gave initial values for the fuzzy system, but these were optimized using real data for the problem and by applying the simple genetic algorithm.

CONCLUSIONS

We described in this paper the application of a neuro-fuzzy-fractal approach to the problem of automating the quality control of sound speakers during manufacturing in a real plant. We have implemented an intelligent system for quality control in the MATLAB programming language using the ANFIS approach. We also use the fractal dimension as a measure of geometrical complexity of the sound signals. The intelligent system performs rather well considering the complexity of the problem. The intelligent system has been tested in a real manufacturing plant with very good results. We think that our approach for automating quality evaluation can used for similar problems with only minor changes in the membership functions and rules.

AKNOWLEDGMENTS

We would like to express our gratitude to the Research Grant Committee of CONACYT for the financial support given to this research work under grant 33780-A. We would also like to thank the Department of Computer Science of Tijuana Institute of Technology for the time and resources given to this research project.

REFERENCES

[1] Castillo, O. and Melin P. (1998). "A New Fuzzy-Genetic Approach for the Simulation and Forecasting of International Trade Non-Linear Dynamics", Proceedings of CIFEr'98, IEEE Press, New York, USA, pp. 189-196.
[2] Castillo, O. and Melin, P. (1999). "Automated Mathematical Modelling for Financial Time Series Prediction Combining Fuzzy Logic and Fractal Theory", Edited Book " Soft Computing for Financial Engineering", Springer-Verlag, Heidelberg, Germany, pp. 93-106.
[3] Castillo, O. and Melin, P. (2000). A New Method for Fuzzy Estimation of the Fractal Dimension and its Applications to Time Series Analysis and Pattern Recognition, Proceedings of NAFIPS'2000, Atlanta, GA, USA. pp. 451-455.
[4] Dickason, V. (1997). The Loudspeaker Cookbook, McGraw Hill.
[5] Haykin, S. (1996). "Adaptive Filter Theory", Third Edition, Prentice Hall.
[6] Jang, J.R., Sun, C.T. and Mizutani E. (1997). Neuro-Fuzzy and Soft Computing, Prentice Hall.
[7] Loctite Co. (1999). Speaker Assembly Adhesives Guide.
[8] Mandelbrot, B. (1987). "The Fractal Geometry of Nature", W.H. Freeman and Company.
[9] Melin, P. and Castillo, O. (2000). Controlling Complex Electrochemical Manufacturing Processes with Hybrid Intelligent Systems, Proc. of NAFIPS'2000, Atlanta, GA, USA. pp. 490-494.
[10] Parker, D.B. (1982). " Learning Logic", Invention Report 581-64, Stanford University.
[11] Rasband, S.N. (1990). "Chaotic Dynamics of Non-Linear Systems", Wiley Interscience.
[12] Rumelhart, D.E., Hinton, G.E., and Williams, R.J. (1986). "Learning Internal Representations by Error Propagation", in "Parallel Distributed Processing: Explorations in the Microstructures of Cognition", MIT Press, Cambridge, USA, Vol. 1, pp. 318-362.
[13] Sugeno, M. and Kang, G.T. (1988). Structure Identification of Fuzzy Model, *Fuzzy Sets and Systems*, Vol. 28, pp.15-33.
[14] Zadeh, L. A. (1975). "The Concept of a Linguistic Variable and its Application to Approximate Reasoning", *Information Sciences*, 8, pp. 43-80.

MONITORING AND KNOWLEDGE EXTRACTION IN REAL TIME MULTIVARIABLE DYNAMICAL PROCESSES

AARON RANSON[*]
PDVSA-INTEVEP.
Emergent Technologies Dpt
Los Teques-Venezuela

KAREN Y. HERNANDEZ
PDVSA-INTEVEP.
Infrastructure Technology Dpt.
Los Teques-Venezuela

JUSTO E. MATHEUS
PDVSA-INTEVEP.
Infrastructure Technology Dpt.
Los Teques-Venezuela

ANGEL A. VIVAS
PDVSA-INTEVEP.
Infrastructure Technology Dpt.
Los Teques-Venezuela

ABSTRACT
Industrial processes may have a large number of measured variables. By combining these variables it is possible to identify different operating regions and detect anomalies or deviations within the process. However, it is hard to visualize conditions and identify problems on line. Difficulties arise because field operators must observe and analyze dozens of variables concurrently in order to identify state of the process. A novel multivariable-process visualization technique is being proposed to help in the tasks of supervision, observation of operating states, and detection of problems. This paper describes a general methodology to monitor multivariable industrial processes.

INTRODUCTION

In the last few years there has been a trend to design increasingly complex multivariable supervision processes that manipulate a large number of variables.

A novel approach to information analysis is presented. It is based on using the intrinsic relationships among data to create an alternate representation of the process that summarizes, in a compact and global way, the phenomenon under study. This representation is coherent and robust because it is created from the data itself, and it can be visualized using charting components in the plane or space (visualization in 2 or 3 dimensions).

A methodology for information analysis is developed and implemented. The methodology uses pre-processing of statistical filtered data, detection of steady states, and pattern recognition, space reduction and machine learning algorithms. The methodology can be applied to the analysis of static data, such as logged data, and real time data. The result is that high-level symbolic components can be (1) created from the collected data, (2) reinforced by the intrinsic

---

[*] Corresponding autor. Present address: Urb Santa Rosa Sector el Tambor, PDVSA-INTEVEP Fase C- oficina 361, Los Teques Edo Miranda Venezuela
E-mail address: ransona@pdvsa.com

relationships that are identified, and (3) exploited in incremental levels. Within a real-time context, it is possible to identify and characterize regions of operation, as well as transition stages between them. The results are visualized using an alternate representation that includes trajectories that describe the process dynamics. The methodology was applied to a set of data consisting of synthetic and real data gathered at an emulsification fuel manufacturing plant and a debutanizer column process.

METHODOLOGY

The developed methodology of data analysis uses a mechanism for logged data and another one for real time data. In the first case, a data log is available along with a set of mathematical rules and clustering algorithms, which are applied to a set of data whose characteristics are well known. The user can process and complement these data with knowledge that he already has about the information represented by the data. In the second case, data are analyzed as are registered by the system. In order to do this, windows of observation of the process and statistical estimators are used.

"Figure 1" shows the different components of the methodology and how these are put together. From stage Boundaries Control, data processing is the same for both logged and real-time. For real-time data, the clusters are linked to the operating regions.

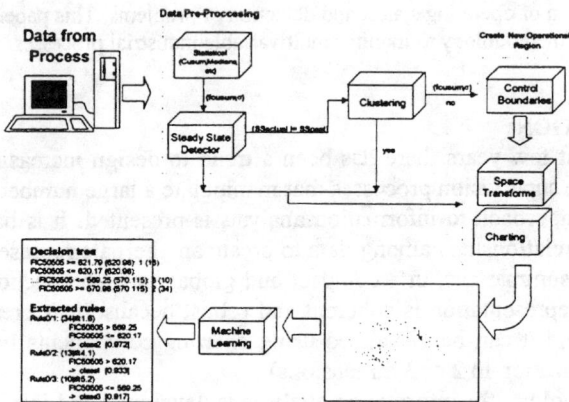

Fig. 1-. Modules for Analysis of real time Systems.

The hypothesis used for clustering is that elements belonging to a region in the multi-dimensional space have similar characteristics, and the trajectories contain information representing the process dynamics. In these cases, the visualization strategy has been designed to guarantee that observations are tightly linked to the unreduced space.

The different steps of the methodology are defined next.

Detection of Steady State: The regions of operations in multivariable processes are chosen based on the steady state [1] condition of the signals involved. The steady state is directly associated to conditions that occur when (1) the controller is on, and the process is properly controlled, that is, the system is in stable condition, or (2) when the controller is in the manual position and the user has taken the process to the desired condition. When the behavior of the process is evaluated, detection of steady states allows identifying the set of regions where the system operates according to specifications.

The steady state is detected using the threshold computed in "Eq. (2)" which uses the normalized signal defined in "Eq. (1)".

$$V_{(k)} = \frac{X_{(k)} - X_{cusum}}{2*\sigma} \quad (1)$$

$$F_{(k)} = \lambda_1 V_{(k)} + (1-\lambda_1) F_{(k-1)} \quad (2)$$

Once F (k) is calculated, the signal is considered to be in steady state if the following conditions are met:

$$F_{(k)} < \text{Threshold (1)}. \quad (3)$$

Statistical Characteristics: This is a differentiating stage for the processing of logged data and real-time data. For real-time data, estimating algorithms like the Weighted Moving Average and the CUSUM[2] filter is used to characterize the data variations and estimate the statistical parameters needed for later processing.

Clustering: for real-time data, a cluster is created based on the results of the steady-state-detection stage. The clustering criterion is not applied to all the variables in the pattern. It is applied only to a subset of variables consisting of the most significant variables in the system, according to the definition given in the previous step. The clustering criterion applied to the most significant variables is given by the following equation:

$$\sum_{i=1}^{v} \frac{(Xc_{1,i} - Xp_{1,i})}{(n*\sigma)} \leq V \quad (4)$$

If the number of significant variables meets the criteria established by "Eq. (3)", the current pattern is placed in this cluster. Once identified a class, the center of the same is calculated using the statistical representative value of the population (in the limits defined by CUSUM). For the case of static data, an unsupervised classifier is used [3,4], which generates a classification based on the selected rule for classification.

Control Boundaries or Influence: The influence circles, control limits or belongings grades for each class are established using overlapping criteria of the classes, or through kernels which allow the generation of statistical rules for their interpretation. In real time, the belongings grades for each class are defined using the ($n * \sigma$) criterion, where σ is the standard deviation of the variables.

Space Transformations: This is one of the most important steps of the methodology. It reduces the data from the space R^n to R^m where m<n. For visualization reasons, *m* can only be one, two, or three, to be able to create a representation of the data. The space reduction transformation consists in defining a feature or set of features in the multidimensional representation, which must be kept in the reduced space[5].

Visualization: The graphics modules allow the visualization of the multidimensional data in a two-dimensional or three-dimensional chart, as a result of the data analysis and execution of the steps previously presented.

Knowledge-Based Generation: this is the last step. It consists on generating the rules and conditions associated with the different operating conditions identified in the system. This is achieved by applying machine-learning algorithms to data clusters generated in previous steps.

The Quinlan C4.5 algorithm [6] is used for knowledge extraction. This algorithm generates the decision tree and set of rules for the data previously classified in the expanded space.

RESULTS

The programmed methodology is used to analyze a data log from the ORIMULSION® manufacturing process generated under different operating conditions. Eight variables representing the dynamics of the system are measured.

"Figure 2" shows the process in the first operating condition. The system generates operation limits within the 99% statistic control region.

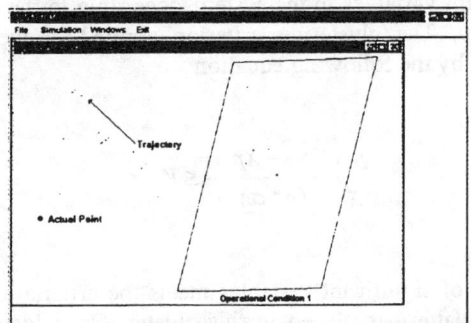

Fig. 2. Process moving from operating condition 1 towards a recognized operating condition 2.

"Figure 3" shows the three operating regions detected as a result of the three changes in the system load.

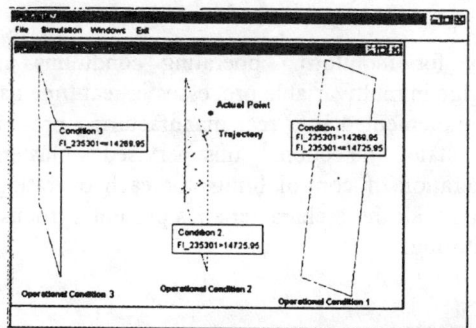

Fig.3. Operating Conditions and Automatic Rule Extraction.

The three regions have different sizes and orientations. It is important to notice that size and orientation of the operating regions vary as a function of the significant variables and their percentages of variance contribution used to generate the control limits. The variables identified as significant in the process are different for each operating condition. Notice that only the variables with the highest contribution to the variance are included in the analysis (see Table I).

Table I shows the results for the most significant variables used to create each operating region, along with their respective percentage of variance contribution. These percentages explain the results obtained.

Table I. Most significant variables and their weigthed importanced based on the variance.

	Variable 1 %	Variable 2 %	Variable 3 %	Variable 4 %	Variable 5 %	Variable 6 %	Variable 7 %	Variable 8 %	>95% of Variance
Operating Condition 1	31.63	—	5.10	48.57	7.52	5.15	—	—	97.96
Operating Condition 2	51.81	—	—	29.98	9.61	5.29	—	—	96.70
Operating Condition 3	56.25	—	—	27.61	5.94	5.23	—	—	95.02

After applying the knowledge extraction algorithm (C4.5, [8]) the following results are obtained and shown in "Fig. 3":
(1) Operating region 1 is defined by RULE2:
    Bitumen Flux (V1) > 14725.95
(2) Operating region 2 is defined by RULE1:
    Bitumen Flux (V1) > 14269.95
    Bitumen Flux (V1) <= 14725.95
(3) Operating region 3 is defined by RULE3:
    Bitumen Flux (V1) <= 14269.95

Each rule matches the variable whose set point was changed (bitumen flux V1) to generate each of the 3 operating regions of the process.

CONCLUSIONS

A methodology for monitoring operating conditions and automatically extracting knowledge in multivariable processes in real time has been developed and successfully implemented in a real manufacturing process. This approach combines steady-state detection, unsupervised clustering algorithms, automatically generation of control limits for each operating conditions, and knowledge extraction, all these characteristics present a robust concept for real time process monitoring.

ACKNOWLEDGMENTS

The authors are indebted to PDVSA for the authorization to publish this paper and with the "Well Modeling and Simulation Department" of PDVSA-INTEVEP for supporting this research

NOMENCLATURE
$V(k)$: normalized signal.
$X(k)$: original signal
Xcusun: signal filtered by CUSUM.
$Xc_{1,i}$: cluster center.
$Xp_{1,i}$: current pattern actual.
V: number of significant variables.

REFERENCES
1. Songling Cao and R. Russell R.: "An efficient method for on-line identification of steady state", Butterwortn Heinemann, March 1995.
2. R. Russell Rinehart: "A CUSUM type on-line filter", Process control and quality, Elsevier Science Publishers, V2 – 1992.
3. Ruby L Kennedy, Benjamin Van Roy, et al.. "Solving Data Mining Problems Through Pattern Recognition".
4. Badavas Paul: "Real-Time Statistical Process Control", Prentice Hall, 1993
5. Joseph B. Kruskal, "Multidimensional Scaling, Bell Laboratories, 1978.
6. J.Ross Quinlan, "C4.5:Programs for Machine Learning", The Morgan Kauffman, 1992.

DYNAMIC MODELLING USING NEURAL NETWORKS. CASE STUDY – PILOT SCALE VSA PROCESS FOR OXYGEN PRODUCTION

CHRIS C.K. BEH[1], KATE A. SMITH[2] AND
PAUL A. WEBLEY[1]
[1]Dept. of Chemical Engineering
[2]School of Business Systems
Monash University
Clayton, Victoria 3168
Australia.

ABSTRACT

Air separation for oxygen enrichment via the pressure/vacuum swing adsorption (PSA/VSA) process is an important unit operation in many large industrial processes. However, complexities associated with it being a bulk separation process operating in a cyclic, non-equilibrium, unsteady state manner, presents significant difficulties in understanding the physics involved. Also, previous attempts in modelling the system numerically have resulted in complications associated with the solution of several, coupled, non-linear, partial differential equations with varying boundary conditions. Empirical methods such as neural networks can provide an alternative modelling strategy for highly non-linear processes such as PSA/VSA. This paper describes the implementation and performance of a dynamic neural network model on a pilot scale, single bed, oxygen vacuum swing adsorption plant. Furthermore, the ability of neural networks in capturing the dynamics of this process lends itself to future study in model predictive control of PSA/VSA systems.

INTRODUCTION

Adsorption, which is the interaction between the species present in a fluid phase (either gas or liquid) and a solid surface, has turned from a scientific curiosity in the early part of this century into a commercially viable separation technology. Pressure swing adsorption (PSA) is a widely accepted unit operation for the purification of various chemical commodities such as methane, nitrogen and hydrogen that utilises the phenomena of adsorption for species separation. A derivative of PSA, vacuum swing adsorption (VSA), has gained in popularity due to the adsorptive capacity of the sieve and sieve selectivity and is the system of choice for oxygen enrichment plants less than 100 tons per day of contained oxygen [1]. The two main components of a VSA cycle are the pressurisation (adsorption) step and evacuation under vacuum conditions (desorption). The complexity of a VSA system draws from the fact that it is a bulk separation process (nitrogen adsorbs onto the sieve), cyclic, highly non-linear and operates under non-equilibrium, non-steady state conditions. Due to the complexity of the system, analytical solutions to the governing equations are beyond current mathematical knowledge and numerical methods are required [2]. Previous modelling efforts have shown discrepancies still exist between process and model [2], even with increasing levels of discretization that cause an equivalent increase in computational time, defeating the objective of on-line implementation. Empirical modelling procedures such as neural networks (NN) can offer an attractive alternative for systems such as VSA, where the creation of a model is difficult and solving the model complicated. Although the use of neural networks in the field of adsorption is not new, dynamic modelling using neural networks has not been applied in this area thus far. Beh and Webley (2001) [3] and Hunt et al. (1992) [4], provide brief overviews of applications using neural networks for modelling in both adsorption and adsorption processes and in general chemical engineering systems. It is the objective of this paper to demonstrate the implementation of a neural network to dynamically model a pilot scale oxygen VSA plant and illustrate its robustness when faced with perturbations in process conditions.

THE NEURAL NETWORK MODEL

For this study, an in-house multi-layer perceptron (MLP) neural network with a single hidden layer (DySONN – Dynamic and Static Operation Neural Network) was developed. The network is designed to enable the creation of non-linear relationships between the inputs and outputs from data sets in both static and dynamic modes. It furthermore admits any number of input, hidden and output neurons together with sigmoid, hyperbolic tangent and linear activation functions. References on the workings of MLP networks are available in most NN literature [5] and Beh and Webley (2001) [3] provide a description of the operation of DySONN. During training of the NN, relationships between inputs and outputs are created through the use of a gradient descent method to update each of the weight coefficients by back propagating the mean squared error ($\hat{E}$) with respect to the weights for the data set [5]. The computation is performed every epoch until either $\hat{E}$ or R^2 satisfies a set tolerance or the number of training epochs has been attained. R^2 is a measure of the accuracy of the NN output to a benchmark model, wherein the prediction is the average of all outputs in the training set [3].

THE VSA CYCLE

The VSA cycle simulated in this paper is typical of industrial practice. It consists of a single bed, four-step, 56 second cycle, with a single layer of commercially available lithium based zeolitic sieve. Figure 1 shows the various steps within the cycle, step times, flow direction and control valves.

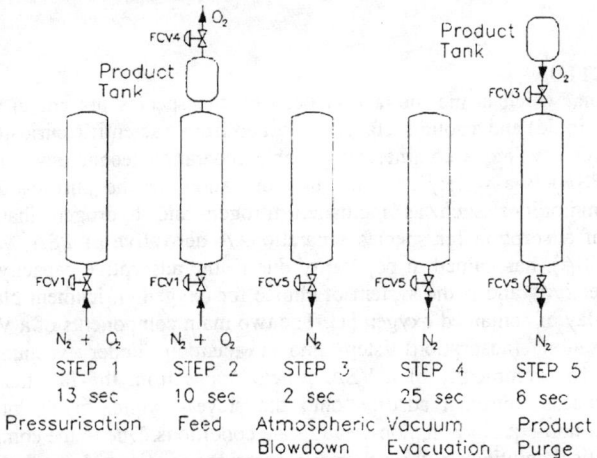

Figure 1 : The VSA cycle for oxygen production used for this study

Being a semi-batch process, product off-take occurs during step two, with the flow into the bed controlled by valve FCV1 and product flow controlled by valve FCV4. Desorption of the adsorbate (nitrogen in this case) occurs during steps three and four and is controlled via valve FCV5. The purge step, here shown in step five, is important for the maintenance in product purity. It forces contaminants (nitrogen) in the fluid phase down from the top of the bed and allows further adsorbate desorption to occur by lowering the partial pressure of nitrogen around the sieve. Control of the product purge stream is via valve FCV3. Pressurising the bed with dry air in step one completes the cycle.

ON-LINE IMPLEMENTATION

The bed is equipped with a pressure transmitter at the top of the bed and temperature thermocouples at specific axial positions. Inlet, product and vacuum temperature, pressure and flow rates, along with concentration completes the instrumentation available on the pilot plant. The 14 inputs chosen for neural network training are the end of step pressures and temperatures at the bed entrance, the valve positions (FCV1 to FCV5) at each step and the total number of moles of gas fed into and exiting the bed over a cycle. The NN outputs are based on customer driven interests and they are oxygen purity (measured as percentage), recovery (moles of oxygen in the product stream relative to the moles in the feed stream), production rate (kgDc – kilograms per day of contained oxygen) and the product pressure (measured in kPa.A). Seven hidden neurons were used for this case study to avoid problems of over-fitting the data (the NN learning noise) and computational burdens. Also, this number of hidden neurons was experimentally determined to be sufficient to avoid starving the network of weights in building relationships between inputs and outputs. In order for the NN to generalize well, it must be trained on an adequate operating surface that includes areas of operation that are outside normal processing conditions. The response surface method (RSM) [6] was used to develop a series of experiments whereby once the simulation has reached cyclic steady state (CSS – quantitatively defined by Todd *et al.* (2001) [2]), perturbations to the manipulated variables (valve positions and feed temperature) are made. For this cycle, training was initiated when the baseline CSS condition (90.71% product purity, 24.5% product recovery, 11.74 kgPDc and 118.1 kPa.A) was attained. The network was trained for up to 3000 epochs every cycle on a 166 MHz PC, running Windows '95. Thirty perturbations to the valve and temperature boundary conditions were made to the process in open loop mode. The system was brought back to its original state after each change. Overall, 30000 cycles were used to train the NN as it has been suggested in past studies that at least 800 cycles are required to fully assess the impact on performance of a singular process change due to the slow transient behaviour of a PSA bulk gas separation process [7].

Upon completion of the training phase, the NN trainer module was halted and the on-line predictor module activated. Using the stored training weights and statistics, disturbances to the process conditions were made within the bounds of the training set. The NN and process outputs were logged every cycle for comparison. To incorporate conditions realistic of field operation, two scenarios (that are within the training surface but not explicitly used to train the NN) were applied as test cases for the trained NN and are described as follows –

Case 1. 50% decrease in product valve (FCV4) position during step 2. This scenario represents a situation whereby the customer reduces product demand.

Case 2. 20% increase in feed valve (FCV1) position during step 2. The case study attempts to simulate the fluctuations in inlet pressure, which is a common occurrence in the field.

It should be noted that these two case studies are by no means exhaustive of the varying conditions experienced by a VSA plant during operation but they do encapsulate the important situations that cause fluctuations in performance.

RESULTS AND DISCUSSION

Figures 2 to 4 illustrate both the transient response of the VSA pilot plant to the disturbances caused in cases 1 and 2 and the neural network responses to the same disturbance. Due to paper size limitations, the system and network responses to perturbations in oxygen recovery is not shown but the curves follow the trends outlined by the oxygen production rate. This is because recovery is a function of oxygen production rate and hence is directly proportional to it. Cases 1 and 2 had very different impacts on the process outputs with regards to purity and production rate although product pressure shows similar trends. Closing the product valve, as in Case 1, increases the back pressure on the system and hence provides an equivalent increase in product pressure (as depicted in Figure 4). The product oxygen purity also improves from 91.8% to around 94.3% caused by this increase in bed pressure and hence sharpens the adsorption front (see Figure 2).

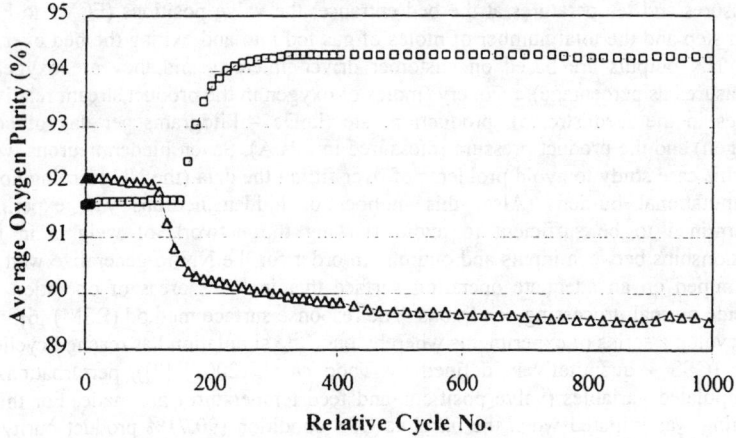

Key: - - Case 1, Plant ;   □   Case 1, NN ;   —— Case 2, Plant ;   △   Case 2, NN

Figure 2 : Comparison of the NN dynamic response to fluctuations in oxygen purity for Cases 1 and 2

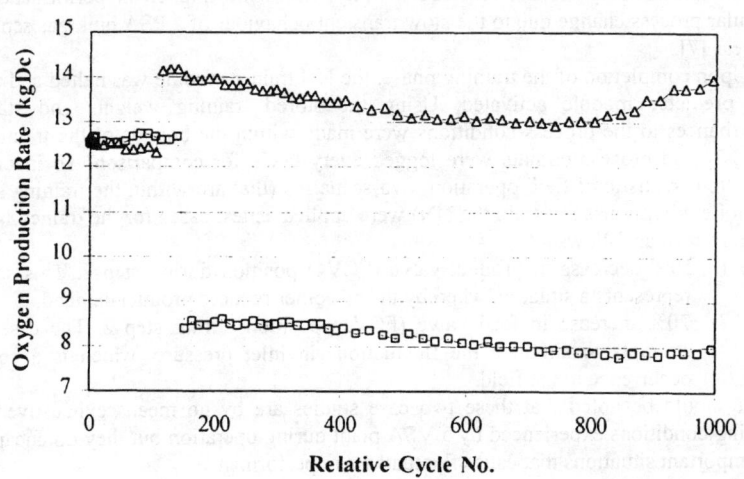

Key: - - Case 1, Plant ;   □   Case 1, NN ;   —— Case 2, Plant ;   △   Case 2, NN

Figure 3 : Comparison of the NN dynamic response to fluctuations in oxygen production rate for Cases 1 and 2

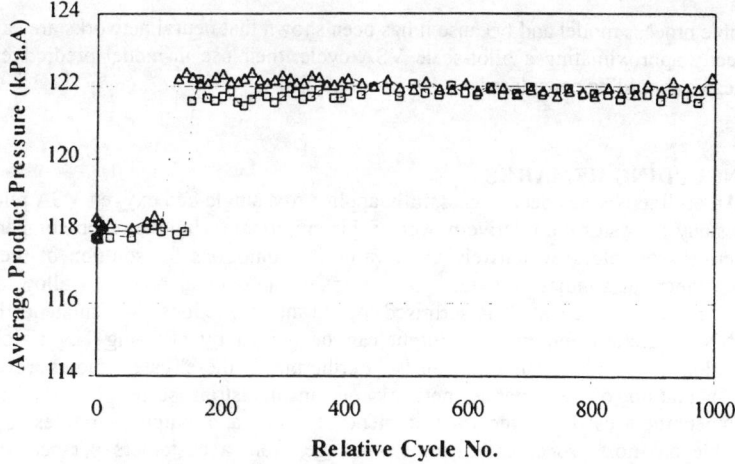

Key: - - Case 1, Plant ; □ Case 1, NN ; — Case 2, Plant ; Δ Case 2, NN

Figure 4 : Comparison of the NN dynamic response to fluctuations in product pressure for Cases 1 and 2

However, an increase in the feed valve position, of Case 2, causes a decrease in oxygen purity from 92% to 89.6% and an increase in production rate for the equivalent product pressure as for Case 1. This reduction in purity is mainly attributed to the saturation of the adsorbent bed and hence breakthrough of the nitrogen front during step two of the cycle. Also, between cycles 820 and 1000, Figure 3 shows that the oxygen production rate veers away from its asymptotic value. Fluctuations in diurnal temperature are attributed to this change and it is interesting to note the lack of response in oxygen purity to this perturbation. It is not shown in this paper but the effect of this increase in production rate results in a slight decline in the purity value over a long period (500 cycles) as the oxygen concentration is sampled from a 60-litre product tank, which provides buffering to process disturbances.

Instances as shown above, where there are notable lags in the system response and situations whereby certain outputs remain constant for given changes in boundary conditions, contribute to significant controllability issues if controllers such as PID, which are not model-based, are used. Traditionally, industrial practice has been to use multiple-loop PID controllers [8], which were originally developed for linear control problems. For both scenarios, the NN was able to accurately capture the system dynamics with the average mean squared error of the outputs being between 0.00009 and 0.00015. This is highlighted by Figure 4, which illustrates the NN correctly capturing the fluctuations in product pressure. Pressure is a variable that can be difficult to predict as it can undergo rapid changes especially during periods of transient operation.

A final observation is the variance in different output responses to differing disturbances. A comparison of the curves in Figures 2 and 3 depicting oxygen purity and production rate for the two case studies, shows that Case 1 attains a steady state value much more rapidly than Case 2, which requires close to 500 cycles to reach asymptotic convergence. This response is contrary to the product pressure output variable, which achieves steady state within 50 cycles of the change. Once again from a control standpoint, changes in process conditions can have varying time constants with the added complexity that these perturbations may occur simultaneously or within transient periods of each other. To re-iterate, control of these processes is difficult without the use of a suitable process model and because it has been shown that neural networks are capable of correctly approximating a pilot-scale VSA cycle, their use in model predictive control strategies should be considered.

CONCLUDING REMARKS

An on-line NN has been successfully applied to a single bed oxygen VSA pilot plant. This study indicates the relative power of this empirical technique when compared with numerical simulators, which rely upon various assumptions for solution of the model. Also, numerical simulators require significant computational power to allow dynamic modelling to be practical. It is surmised that further reductions in mismatches between the NN prediction and process output can be gained by allowing NN training and prediction to continue simultaneously. Furthermore, these case studies reveal that implementation of a dynamic network on an industrial scale plant is viable as instrumentation used for the measurement of input and output variables is readily available on most processes. One disadvantage, from a customers perspective is the impracticality involved with performing step changes in processing conditions to plants already installed in the field for NN training purposes. This problem can be overcome by training a NN on a verified process simulator, especially on scenarios outside normal operating conditions and the range of expected disturbances. As the next generation of oxygen production moves towards higher efficiency units, characterised by shorter cycle times and large pressure gradients (ie. rapid PSA (RPSA) cycles [9]), an investigation into the use of dynamic neural networks to model these more complex cycles will be beneficial to the understanding and control of such processes.

ACKNOWLEDGEMENTS

The authors would like to thank *Air Products and Chemicals Inc.* for support and funding of this project.

REFERENCES

[1]   Kumar, R., 1996. "Vacuum Swing Adsorption Process for Oxygen Production - A Historical Perspective", *Separation Science and Technology*, Vol. 31, pp. 877-893.

[2]   Todd, R.S., He. J., Webley, P.A., Beh, C.C.K., Wilson, S.J., Lloyd, M.A., 2001, "Fast Finite Volume Method for PSA/VSA Cycle Simulation – Experimental Validation", *Industrial and Engineering Chemistry Research*, Vol. 40, pp. 3217-3224.

[3]   Beh, C.C.K and Webley, P.A., 2001, "On-Line Modelling of an Oxygen Vacuum Swing Adsorption Process Using Neural Networks", *6th World Congress of Chemical Engineering*, 23-27 September, Melbourne, Australia, accepted for publication.

[4]   Hunt, K.J., Sbarbaro, D., Zbikowski, R. and Gawthrop, P.J., 1992, "Neural Networks for Control Systems - A Survey", *Automatica*, Vol. 28, pp. 1083-1112.

[5]   Haykin, S., 1999, *Neural Networks – A Comprehensive Foundation*, 2nd ed., Prentice Hall, Inc., New Jersey.

[6]   Box, G.E.P., Hunter, W.G. and Hunter, J.G., 1978, *Statistics for Experimenters*, John Wiley & Sons, Inc., New York.

[7]   Wilson, S.J., Beh, C.C.K., Webley, P.A. and Todd, R.S., 2001, "The Effects of a Readily Adsorbed Trace Component (Water) in a Bulk Separation PSA Process: The Case of Oxygen VSA", *Industrial Engineering and Chemistry Research*, Vol. 40, pp. 2702-2713.

[8]   Beh, C., Wilson, S., Webley, P. and He, J., 2000, "The Control of the Vacuum Swing Adsorption Process for Air Separation", *Proceedings of the Second Pacific Basin Conference on Adsorption Science and Technology*, Editor. Duong Do, 14-18 May, World Scientific, Singapore, 663-667.

[9]   Earls, D.E. and Long, G.N., 1980, "Multiple Bed Rapid Pressure Swing Adsorption for Oxygen", *U.S. Patent 4,194,891*.

DETECTION AND SEPARATION OF EXTRACELLULAR NEURONAL DISCHARGES

TETYANA I. AKSENOVA
Institute of Applied System
Analysis, Kyiv, 03056, Ukraine
OLEKSANDR A. DRYGA
Institute of Physiology, Kyiv,
01024, Ukraine
ALESSANDRO E. P. VILLA
Laboratoire de Neuroheuristique,
Institut de Physiologie, Lausanne,
1005, Switzerland

IGOR V. TETKO
Institute of Bioorganic & Petroleum
Chemistry, 02094, Kyiv, Ukraine
OLGA K. CHIBIROVA
Space Research Institute,
Kyiv, 03187, Ukraine

ABSTRACT

The present study reports an approach for automatic classification of extracellularly recorded action potentials (spikes). The classification of spike waveform is considered as a pattern recognition problem of special segments of signal that correspond to the appearance of spikes. The recorded signal is observed at discrete time and characterized by high level of background noise and occurrence of the spikes at random time. In the present study we describe the spike waveform as an ordinary differential equation with perturbation. This allows us to characterize the signal distortions in both amplitude and phase. The new algorithm was tested on simulated and real data and performed better than other approaches currently used in neurophysiology.

INTRODUCTION

The separation of action potentials (spikes) in extracellular recording represents a difficult and a challenging task. The appearance of powerful computers has offered the possibility to develop sophisticated spike-sorting techniques. Most of currently used techniques allow the classification of the signal in the time domain, but can account only for the additive noise (Schmidt, 1984; Letelier & Weber, 2000). In this article we propose a new method for spike sorting considering the analyzed signal in phase space.

METHOD

We suppose that a signal $\tilde{x}(t) = x(t) + \xi(t)$ is observed at discrete times $t=0,1,...T$. Here $\xi(t)$ is a sequence of independent identically distributed random variables with zero mean and finite variance ($\sigma_\xi^2 < \infty$). The signal $x(t)$ is characterized by the occurrence of repeated intervals with amplitudes significantly exceeding the variance of $\xi(t)$. These intervals are assumed to correspond to the occurrence of neuronal discharges, i.e. the spikes. The spikes appear at random times and, otherwise, the signal equals zero all other time (Fig. 1). Each spike belongs to a general population X_j, $0 \leq j \leq p < \infty$. Then,

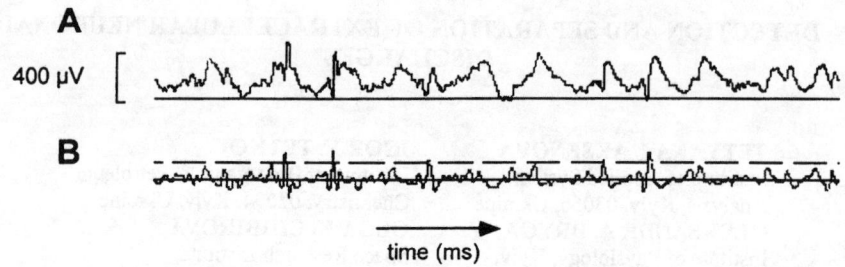

Figure 1. Example of brain activity observed in a real experiment (A) and its derivative (B) with detected spikes. The dashed line corresponds to the threshold.

each general population corresponds to segments of the signal with similar waveform assumed to correspond to the extracellular recording of single unit discharges. Two problems should be solved for successful data analysis. Firstly, the appearance of a spike should be detected and, secondly, the detected spike should be correctly classified.

Let us assume that the spike $x^i(t)$, $0 < t < T^i$ from X_j is a solution of an ordinary differential equation with perturbation

$$d^n x/dt^n = f(x,..., d^{n-1}x/dt^{n-1}) + F(x,...,t) \qquad (1)$$

where n is the order of the equation, $F(\ )$ is a perturbation function and the undisturbed equation

$$d^n x/dt^n = f(x,..., d^{n-1}x/dt^{n-1}) \qquad (2)$$

describes a self-oscillating system with stable limit cycle $\mathbf{x}_j^0(t) = (x_1^0(t),...,x_n^0(t))^T$ in phase space with co-ordinates $x_1=x$, $x_2=dx/dt$, ..., $x_n=d^{n-1}x/dt^{n-1}$ for each j, $0 \le j \le p$ and T is the period of stable oscillations. The perturbation function $F(x,...,t)$, bounded by a small value, is a random process with zero mean and small correlation time $\tau^* \ll T$. The mathematical model introduced in Eq. (1) and Eq. (2) presumes that the observed signal is described by a solution of differential equations with presence of noise in the dynamic equation.

It is well known (Bogoljubov & Mitropolsky, 1961), that the trajectory of the signal continuously tends to the limit trajectory whenever it is found in the limit trajectory neighborhood, independently of initial conditions. The perturbation function $F()$ in Eq. (1) tends to displace the trajectories of the signal out from the limit trajectory. However, if the perturbation is small enough the trajectories stay in the neighborhood of the limit trajectory $x^0(t)$, $y^0(t)$, i.e., the solutions of Eq. (1) are similar to one another but they never coincide.

Following Gudzenko (1962) let us introduce the local coordinates in the neighborhood of the limit cycle. Let us fix an arbitrary point on the limit cycle P_0 as a starting point. The position of any arbitrary point P on the limit trajectory can be described by its phase θ, which is a time movement along the trajectory from the starting point P_0 to the point P being analyzed. The phase θ unambiguously characterizes all points of the limit trajectory. Let us assume that f^j in Eq. (2) is twice continuously differentiable on all the arguments of the function, thus providing a necessary smoothing. It is possible to calculate in any point P with phase θ a hyper plane (and only one such hyper plane) that is normal to the limit cycle. The trajectories of the signal can be described by variables $\mathbf{n}(\theta)$ and $t(\theta)$. Here $\mathbf{n}(\theta)$ is a vector between analyzed point M on the signal trajectory and its orthogonal projection P on the limit trajectory (Fig. 2). The second variable $t(\theta)$ is a time movement along analyzed curve. Thus, each signal trajectory is described by variables $\mathbf{n}_i(\theta)$ and $t_i(\theta)$ where i is the number of the trajectory. The limit trajectory is defined by $\mathbf{n}(\theta) \equiv \mathbf{0}$ and $t(\theta) \equiv \theta$, where $\mathbf{0}$ is a vector with all components equal to 0.

Let us denote $\gamma(\theta) = t(\theta) - \theta$. The variables $\mathbf{n}(\theta)$ and $\gamma(\theta)$ characterize the deviation of an individual signal from the limit trajectory. If the signal corresponds to an extracellular electrophysiological recording, the cycles of the trajectory of Eq. (1) may be interpreted as the occurrences of neuronal discharges (i.e., the spikes) described by $x^i(t)$, $0 < t < T^i$, where T^i is the duration of the spike. One set of cycles forms the general population X_j and each set X_j describes the activity of one neuron. The limit trajectory $x_j^0(t)$ $0 < t < T$ corresponds to an ideal spike without noise ("undisturbed spike"). This model describes the signal distortions in both amplitude and phase. Our goal is to develop a real-time algorithm for spike recognition in the case when the number of neurons is unknown.

The description of the method includes several steps. Firstly, the classification of spikes is reduced to the separation of a mixture of normal distributions in a transformed feature space $x^i(t(\theta))$, $\theta = 0,1,...T$ (Aksenova et al, 1999, Tetko et al, 1999). Secondly, the procedure of spike detection is considered and, thirdly, a learning algorithm for spike recognition is proposed.

The procedure of detection of neuronal discharges—the spikes—is important to determine the time intervals of the signal that correspond to spike occurrences. The spikes are characterized by amplitudes that are significantly larger than the level of noise. Thus, occurrences of spikes can be detected by using a threshold detector. In the present study a spike is detected if the first derivative of the signal exceeded some threshold value corresponding to, e.g. significant deviation from the baseline or some user-specified value.

The number of classes, their radius and centers are estimated during the learning phase of the algorithm using several hundreds of spike occurrences, usually corresponding to 1-2 minutes of electrophysiological recording.

The number of classes and their centers are estimated only by measuring the distances between trajectories. The rationale is that in the case of normal distribution the mean corresponds to the maximum of the probability density. In a normal distribution the value $x^* = \mathrm{E}x$, where E is the mathematical expectation, provides the maximum of $P(|x-x^*| < R)$ for any given parameter R. In order to estimate the mean trajectory, an interactive procedure detects the spike with maximal probability density in its neighborhood. This spike may be viewed as a "template" spike representative of its class.

The first step of the interactive procedure consists in the selection of an arbitrary spike x_0 as an initial estimate of the center of the class. Its R-neighborhood $\mathbf{R}_0 = \{x: ||x-x_0||_\Omega < R\}$ is constructed on the learning data set. Then we search the

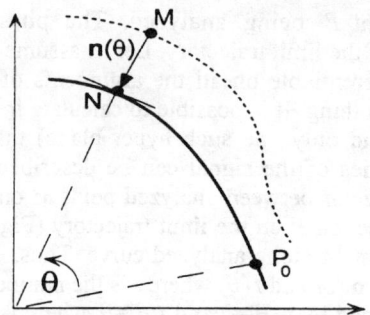
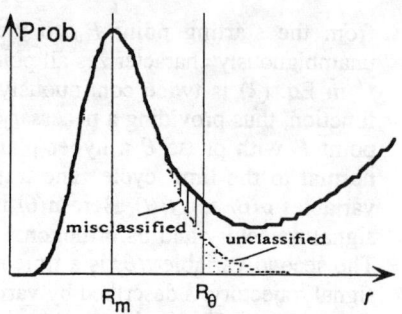

Figure 2. New variables, i.e. θ and $n(\theta)$, are introduced to describe the trajectories of the analyzed signal in the phase space. The thick line is the limit trajectory. The length of the vector $|n(\theta)|$ corresponds to minimal distance between the signal and the limit trajectory.

Figure 3. Experimental distance distribution, the solid line, and chi-square approximation, the dotted line. The vertical line represent maximum of density probability function, R_m, and selected radius of the class, R_θ. The areas corresponding to misclassified and unclassified spikes are shown.

element x_1 that calculates the minimum sum of distances over all spikes from the R-neighborhood of spike x_0 and we construct the R-neighborhood of the spike x_1, $\mathbf{R}_1 = \{x: ||x-x_1||_\Omega < R\}$. This procedure converges to a mean value due to symmetry and single modality of the normal distribution. Due to the same properties of normal distribution, the procedure does not depend on the choice of the parameter R. However, a larger training set is required for smaller values of R.

The maxima of the densities of joint distributions are near the centers of the classes if the classes are well separated. Then, it is sufficient to select the initial estimate points in the neighborhood of each maximum in order to detect all centers of the classes. The procedure to construct initial estimates consists of several steps. Firstly, a spike is selected as the first initial estimate. The distances to all available initial estimates are calculated for each next spike. If all distances to the analyzed spike are larger than some threshold R this spike is considered as a new initial estimate for the iterative procedure. All initial estimates are used in the iterative procedure up to its termination. Several branches of the algorithm will converge to the same extreme. The spikes x_i obtained as a result of such learning are considered as candidates to represent the centers of the classes. The spikes with overlapping R-neighborhood are considered and spikes with lower number of occurrences are deleted.

If the sample delimited by the R-neighborhood of a class contains a number of spike occurrences lower than a criterion set by the user, then this class is also deleted and will not be used for further analyses.

The use of analysis in phase space makes it possible to estimate errors of the first and second order of the spike classification. Let us assume that $n(\theta) = \mathbf{x}^j(t(\theta_j)) - \mathbf{x}^0(\theta_j)$ is normally distributed. Then the square of the distance of each spike to the center of the class follows distribution χ^2 with the appropriate number of degrees of freedoms. The parameters of the distribution χ^2 can be easily estimated from the experimental data using the least square method.

The deviation of the experimental distribution $y(r)$ form the fitted $\chi^2(n)$ provides an estimate of the level of noise in the experimental recordings. Let us select some threshold radius $R_q > R_m$ and reject all spikes that have a distance from the center of the class larger than the threshold radius. The tail of distribution $\int_{r=(Rm;\infty)}\chi^2 dr$ corresponds to the unclassified spikes and the integral $\int_{r=(0;Rm)}(y-\chi^2)dr$ corresponds to the misclassified spikes (Fig. 3). This procedure was directly extended for simultaneous classification of several classes. The radius of each class was chosen to equalize misclassified and unclassified errors.

SIMULATED DATA SETS

In order to evaluate the performance of our algorithm and to compare it with other methods we constructed a simulated data set, as proposed by Letelier & Webber (2000). Inverse Fourier transform was used to generate a trial lasting 1000 seconds trial of an artificial noise with corresponding phase and amplitude distribution. The level of the noise was the same as in the original study. Three spike templates proposed by the authors were used. We added 10 000 instances of each template in a random way to the background noise. The resulting data set mimicked three neurons firing independently at an average rate of 10 spikes/second. These data were very similar to the original study, but in their case only 100 spikes from each neuron were analyzed. The rationale of using a larger dataset is to provide a more robust test of the new method.

The algorithm was trained using the simulated spike train and 3 classes were easily extracted. The error index was used to estimate a generalized error:

$$Error\ Index = (\sum(unclassified)^2 + \sum(misclassified)^2)^{1/2} \quad (3)$$

where summation was done over 3 analyzed classes and the unclassified and misclassified number of spikes are measured in percentage. The error index allowed to estimate the performance of different methods applied to these simulated data. The procedure of defining the templates and spike detection was repeated five times and the results from five realizations were averaged.

RESULTS

The comparison of our method with several other methods considered for the analysis of data set of Letelier & Webber (2000) is presented in Table 1. We did not consider the problem raised by spike overlapping that could decrease the performance of our method. The percentage of overlapping spikes depends on the firing rate of the neurons and on the time duration of the spikes. Given a firing rate of 30 spikes/second and spike duration equal to 2 ms one may expect 18% of overlapping spikes. This means that the error index (Table 1) for our method would increase up to 30% if our program did not sort all overlapping spikes correctly. However, even in such a case our algorithm would perform better than the other tested methods. In many electrophysiological studies of the cerebral cortex firing rates were reported in the range 3-5 spikes/second and, thus, the number of overlapping spikes is expected to generate an error index less than that calculated above.

The methods tested by Letelier & Webber (2000) did not consider the problem to detect spikes and all spikes were only sorted. This may have lead to

Table 1. Averaged Error Index (%) for different methods applied to the simulated data set.

method	average index value	standard deviation
wavelet transform analysis	36	7
principal component analysis	138	2
reduced feature set	89	3
analysis in phase space	13	4

overestimate the performance of other methods compared to our own. Indeed, such a procedure does not take into account the number of spikes that could be lost during the spike detection procedure as well as the number of noise fluctuations that could be detected as spikes. Since the spike detection procedures are different for all reported approaches, it is difficult to estimate their impact on the error index reported in this study.

In conclusion, we have presented a new method for detection and classification of neuronal discharges corresponding to electrophysiological signals in extracellular recordings. This method provided the best results, to our knowledge, for the simulated spike trains developed by Letelier & Webber (2000) and it favorably compares with other approaches used to analyze electrophysiological recordings. A real-time implementation has been developed with convenient user-friendly interface and it is now tested in several laboratories in Ukraine and Switzerland. The future studies will be focused to account explicitly the problem of spikes overlapping across multiple templates.

ACKNOWLEDGMENTS

This study was partially supported by NATO HTECH.LG 972304, INTAS 97-168 and 97-173 grants, SNSF SCOPES 7IP 62620. The authors thank Pamela P. Weber and Juan Carlos Letelier for providing us templates of simulated spikes.

REFERENCES

Aksenova, T. I., Tetko, I. V., Ivakhnenko, A.G., Villa, A. E. P., Welsh, W. J., and Zielinski W. L., 1999, "Pharmaceutical Fingerprinting in Phase Space. 1. Construction of Phase Fingerprints," *Analytical Chemistry*, Vol. 71, pp. 2423-2430.

Aksenova, T. I., Tetko, I. V., Patiokha, A. A., Ivakhnenko, A. G., Villa, A. E. P., Zielinski W. L., Welsh, W. J., and Livingstone D. J., 1999, "Analysis and Pattern Recognition of HPLC Trace Organic Impurity Patterns in Phase Space," *Smart Engineering System Design*, ASME Press, New York, Vol. 9, pp. 935-940

Bogoljubov, N. N., and Mitropolsky, Y. A., 1961, "*Asymptotic Methods in the Theory of Non-Linear Oscillations*," Gordon and Breach, New York.

Gudzenko, L. I., 1962, "The Statistical Method of Determination of Characteristics of a Non-Control Autooscillating System," *Izvestiia Vuzov Radiophysics*, Vol. 5, pp. 573-587.

Letelier, J. C., and Weber, P. P., 2000, "Spike Sorting Based on Discrete Wavelet Transform Coefficients," *Journal of Neuroscience Methods*, Vol. 101, pp. 93–106.

Schmidt, E.M., 1984, "Computer Separation of Multi-Unit Neuroelectric Data: a Review," *Journal of Neuroscience Methods*, Vol. 12, 95-111.

Tetko, I. V., Aksenova, T. I., Patiokha, A. A., Villa, A. E. P., Welsh, W. J., Zielinski W. L., and Livingstone D. J., 1999, "Pharmaceutical Fingerprinting in Phase Space. 2. Pattern Recognition," *Analytical Chemistry*, Vol. 71, pp. 2431-2439.

NONLINEAR DYNAMIC SYSTEMS MODELING AND PATTERN RECOGNITION

JIANPING XIANG, N.B. JONES, F.S. SCHLINDWEIN
Dept. of Engineering, Leicester University, University Road,
Leicester LE1 7RH, United Kingdom

ABSTRACT
This paper considers the issues of modelling and pattern recognition of non-linear dynamic systems both in frequency domain and time domain via a practical engineering problem. The major emphasis of the paper is focused on the pattern recognition of non-linear dynamic systems using neural networks. Dual-frequency Induced Polarization (IP) dynamics system is analysed. The results indicate that neural networks approaches possess an advantage over the traditional non-linear mathematical methods by its realization of mathematical description via the input-output responses of non-linear dynamic systems. MATLAB is used for the simulation and computer-aided analysis of non-linear dynamic systems.
Keywords: Volterra series, finite dimensional approximation, neural networks, IP model

INTRODUCTION
Evaluating IP anomalies via the measuring data of the non-linear IP responses (He et al., 1995) raised the pattern recognition problem of non-linear dynamic systems. Many reports on the research work of non-linear dynamic systems are limited to a discussion of the modeling using traditional mathematical description, which cannot be inverted by the input-output responses.

Figure 1. describes a non-linear IP dynamic system in frequency domain.

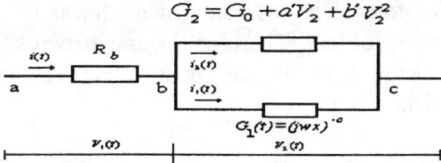

Fig. 1. Equivalent circuit of non-linear electrolyte-mineral system

By analyzing the equivalent admittance of various electrode reactions, it can be assumed that

$$G_2 = G_0 + a \frac{G_0^2}{S} V_2 + b \frac{G_0^3}{S^2} V_2^2 \qquad (1)$$
$$= G_0 + a'V_2 + b'V_2^2$$

Using a three term Volterra series approximation, an input-output response was described in (He et al., 1995) as

$$V(t) = V_1(t) + V_2(t) = \sum_{k_1=-\infty}^{\infty} [R_b + \frac{R_0}{1+(jk_1\omega\tau)^c}]I_{k_1}e^{jk_1\omega t} + \sum_{k_1=-\infty}^{\infty}\sum_{k_2=-\infty}^{\infty}(-aR_0/\{[1+(jk_1\omega\tau)^c]$$

$$[1+(jk_2\omega\tau)^c][1+(j(k_1+k_2)\omega\tau)^c]\})\frac{I_{k_1}I_{k_2}}{S}e^{j(k_1+k_2)^s} + \sum_{k_1}\sum_{k_2}\sum_{k_3=-\infty}^{\infty}(R_0/\{[1+(jk_1\omega\tau)^c]$$

$$[1+(jk_2\omega\tau)^c][1+(jk_3\omega\tau)^c]$$

$$[1+(j(k_1+k_2+k_3)\omega\tau)^c]\})(a^2/\{[1+(j(k_1+k_2)\omega\tau)^c]+a^2/[1+(j(k_2+k_3)\omega\tau)^c]-b\})$$

$$\frac{I_{k_1}I_{k_2}I_{k_3}}{S^2}e^{j(k_1+k_2+k_3)\omega\tau})$$

(2)

But it is hard to deduce the general input output response from (2).

Since Fig. 1. is hard to be described in an Ordinary Differential Equation (ODE) form, it seems difficult to get a precise description of the input-output response of the generalized model.

Based on the model suggested in (Cole. K.S. and Cole R.H., 1941, Jaggar, S.R., Fell, P.A., 1988, He et al., 1995), a generalized model was expressed as

$$Z_i(j\omega) = Z_0(i)\{1 - m(i)[1 - \frac{1}{1+(j\omega\tau(i))^c}]\} \quad (3)$$

which represents the nonlinear effect of the spectral IP.
Experimental results (He et al., 1995) show that dual-frequency IP spectrum can detect nonlinear effects to discriminate different minerals. Using the Generalized Model we can get a good explanation of the dual-frequency IP spectrum nonlinear effects. But it is difficult to realize the inversion of the generalized model via the input-output responses of dual-frequency IP nonlinear dynamic systems. In order to apply the techniques of computer-aided analysis to the nonlinear effect of spectral IP, a nonlinear IP system model will be proposed in the next section, which approximates the infinite term $(j\omega\chi)^c$ in the equivalent circuit (Fig 1) as a finite dimensional term, so that IP nonlinear dynamic systems can be written into ODE form. The inversion of nonlinear IP dynamic system models is also a difficult problem. Therefore neural network models will be discussed and their precision tested by the input-output response of simulated nonlinear IP system models.

NON-LINEAR IP SYSTEM MODEL

$(j\omega\chi)^{-c}$ in Fig. 1. can be approximated by a second order finite dimensional approximation (Bertin, 1976). The generalized second order transform can be expressed as:

$$Z(s) = \frac{a_0 + a_1 s + a_2 s^2}{b_0 + b_1 s + s^2} \quad (4)$$

Then the block diagram of Fig. 1. is shown as Fig. 2.

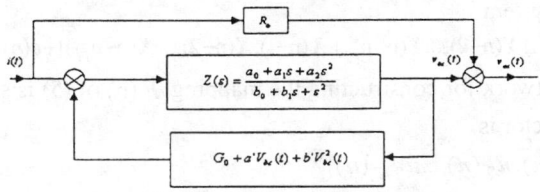

Fig. 2. Block diagram of Fig. 1.

We use the data provided in (He et al., 1995) and obtain the Simulink dual-frequency nonlinear IP model of metallic sulfides as shown in Fig. 3.

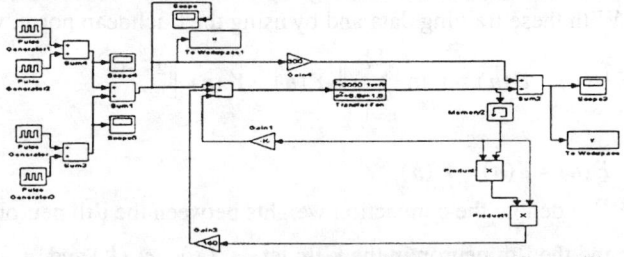

Fig. 3. Dual-frequency nonlinear IP Simulink model of metallic sulfides

The simulated input x and output y are shown as Fig. 4.

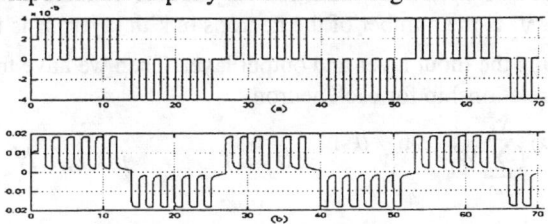

Fig. 4. Simulated inputs (a), outputs of metallic sulfides mineral

PATTERN RECOGNITION USING MLP NETWORKS

The input and output of a multi-layer perceptron (MLP) network can approximate an arbitrary nonlinear map and is completely determined by the network parameters such as the connection weights and thresholds. The neural network is adjusted, or trained, so that a particular input $U(n)$, which is composed of the input $X(n)$ and output $Y(n)$ of IP nonlinear systems, leads to a neural network output $\hat{Y}(n)$. Such a situation is shown in Fig. 5.

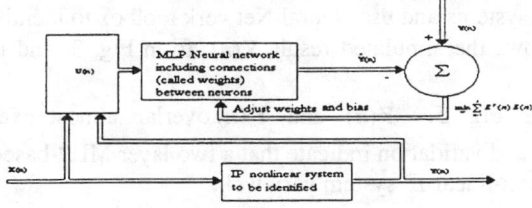

Fig. 5. The MLP neural network pattern recognition for an IP nonlinear system

Consider a system
$$Y(n)=F[Y(n-1),Y(n-2),...Y(n-n_N),X(n-1),X(n-2),...X(n-n_N)]+e(n) \quad (5)$$
The MLP network for constructing the mapping $F(n)$ of (5) is shown in Fig 5. The input vector is
$$U(n) = [u_1(n), u_2(n),...u_{2N}(n)]^T \quad (6)$$
$$= [Y(n-1), Y(n-2),..., Y(n-n_N), X(n-1), X(n-2),..., X(n-n_N)]^T$$
The 2N samples of the input vector $X(n)$ and the output vector $Y(n)$ of equation (6) can be used as the training set to find the parameter set of weights and bias. With these training data and by using the Euclidean norm, we have
$$\min_W \sum_{n=1}^{I} E^T(n) \quad E(n) = \min_W \sum_{n=1}^{I} \| Y(n) - \hat{Y}(n) \|^2 \quad (7)$$
where $E(n) = Y(n) - \hat{Y}(n)$

Let $W_{ij}^{(k-1)}$ denote the connection weights between the i'th neuron in the (k-1)th layer and the j'th neuron in the k'th; let $y_j^{(k)}$, $f_j^{(k)}$ and $\theta_j^{(k)}$ be the output, activation function, the threshold of j'th neuron in the k'th layer, respectively; N_k is the number of the neurons in k'th layer; M is the number of layers including the input layer and output layer. Then we have the following input-output relationship for each neuron:
$$y_j^{(k)} = f_j^{(k)} (\sum_{i=1}^{N_K} W_{ij}^{(k-1)} y_i^{(k-1)} - \theta_j^{(k)}) \quad j=1,2,...N_k, k=1,2,...M$$
(8)

The inputs of the network are denoted by $u_1, u_2,..., u_{2N}$; the outputs of network are $\hat{y}_1, \hat{y}_2,...\hat{y}_n$.

MATLAB SIMULATION RESULTS OF PATTERN RECOGNITION USING MLP NETWORKS

To apply the neural networks approach (Carlos et al., 2000) for identifying different non-linear IP systems, we use the parameters in former section to get different non-linear IP systems. Then we inversely use the simulation data of the non-linear IP systems and use Neural Network toolbox to identify these systems.

Fig. 6. shows the simulated result $Y(n)$ from Fig. 3. and the identification result $\hat{Y}(n)$ in Fig. 5. $Y(n)$ and $\hat{Y}(n)$ overlap almost exactly. Results of identification and validation indicate that a two layer MLP-based neural network can map the non-linear IP system very well.

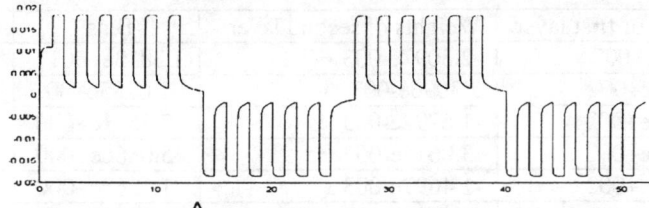

Fig 6. $Y(n)$ (full line) and $\hat{Y}(n)$ (Dashed-line) of Metallic sulfides using Fig 5.

The resulting two-layer MLP-based neural network can be drawn using an abbreviated notion as Fig. 7, in which $IW_{1,1}$ represent the weights array of first layer, $LW_{2,1}$ the second layer. The MLP network with 10 neurons (tansig activation function) in the input layer and one neuron (purelin activation function) in the second layer was used.

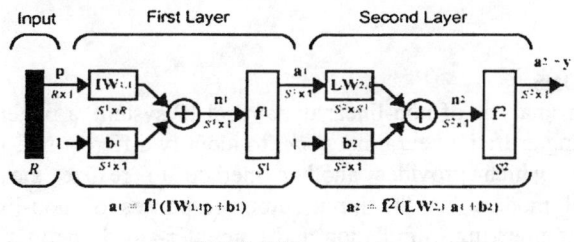

Fig. 7. Two layer MLP-based network describing non-linear IP system

Two sets of network parameter of non-linear IP systems of both metallic sulfides and carbonaceous minerals, including the correlation coefficients r between $Y(n)$ and $\hat{Y}(n)$, which represent the goodness of $\hat{Y}(n)$ fitting $Y(n)$, are listed in table 1 and table 2.

When electromagnetic and other noises are added to the model in Fig 4, a neural network structure can be used to reduce these noises (Petrović et al., 2000).

Table 1: Metallic sulfides

Weights of first layer	Weights of second layer	Bias	r
7.6145e+002	1.5351e-003	-1.3265e+001	1.000e+00
-7.6145e+002	-2.4003e-003	1.0541e+001	
7.6141e+002	2.1765e-003	-6.9307e+000	
-7.6141e+002	-1.9619e-003	3.5113e+000	
7.6161e+002	1.7228e-003	-9.2066e-001	
7.6137e+002	1.4790e-003	1.5948e+000	
-7.6141e+002	-3.5664e-001	-6.6000e+000	
-7.6147e+002	3.5114e-001	-7.0164e+000	
7.6143e+002	8.1396e-004	1.1271e+001	
-7.6146e+002	-1.4082e-003	-1.3407e+001	
		4.4172e-0.05	

Table 2: Carbonaceous minerals

Weights of first layer	Weights of second layer	Bias	r
-7.0713e+002	-2.3097e-003	1.3813e+001	1.000e+00
7.0713e+002	2.3096e-003	-1.0785e+001	
-7.0717e+002	-1.6792e-003	7.7511e+000	
-7.0713e+002	-3.0612e-003	5.0866e+000	
-7.0708e+002	-1.4697e-003	1.1263e+000	
-7.0730e+002	-1.5056e-003	-8.8321e-001	
7.0713e+002	3.1398e-003	3.8376e+000	
-7.0713e+002	-1.9710e-003	-7.9656e+000	
-7.0713e+002	-2.0931e-003	-1.0841e+001	
-7.0714e+002	-2.0910e-003	-1.3663e+001	
		2.2799e-0.05	

CONCLUSIONS

Based on the analysis of non-linear dynamical IP system, a pattern recognition approach using artificial neural networks to identify different non-linear systems is discussed, which provides another method to realize the reversion of mathematical models via the input-output responses of non-linear dynamic systems in Engineering. Simulation and computer-aided analysis of non-linear dynamical IP system show that pattern recognition approach using artificial neural networks can provide a precise description of unknown non-linear dynamical systems, which always exists in industrial control system.

ACKNOWLEDGMENTS

The authors would like to thank Prof. Daizhan Cheng, Dr Dawei Gu, Mr. J.A. Twiddle and Mr. M. K. Khan for their help and many significant suggestions.

REFERENCES

Carlos Calderon-Macias, Mrinal K. Sen and Paul L. Stoffa., 2000, "Artificial neural networks for parameter estimation in geophysics," *Geophysical Prospecting,*, **8**, pp.21-27.
Cole. K.S. and Cole R.H., 1941, "Dispersion and absorption in dielectrics," Vol. I. Alternating current field: *J. Chem. Phys.*, **9**, pp.341-351.
He. J, Li. D, and Tang., 1995, J, Nonlinear effect of the dual-frequency IP spectrum: *Transactions of Nfsoc*, **4**, pp.1-7.
Jaggar, S.R., Fell, P.A., 1988, "Forward and inverse Cole-Cole modelling in the analysis of frequency domain electrical impedance data," *Exploration Geophysics*, **19**, pp.463-470,.
Luo Y, Zhang G., 1998, Geophysical monograph series (number 8): "Theory and application of spectral induced polarization," society of exploration geophysicists.
Luo Fa-Long and Rolf Unbehauen., 1997, "Applied Neural Networks for Signal Processing:," Cambridge University Press, UK.
Petrović I, Miroslov, Barić, Ante Magzan, and Nedjeljko Perić., 2000, "Application of a Neural predictive Controller in Boost Converter Input Current Controller" *Proceedings of the 2000 IEEE International Symposium on Intelligent Control*, Greece, 327-332.
Bertin, J. (Editor), 1976, "Experimental and theoretical aspects of induced polarization, Presentation and application of the IP method case histories," v. 1, 259 pp., Berlin: Gebrüder Borntraeger,

A NEW APPROACH of INTELLIGENT CONTROL SYSTEM DESIGN

DORIAN AUR
Dept. of Computers,
AISTEDA University
Bucharest, P.O. Box 1-410,
70700, BUCHAREST,
ROMANIA

TEODORA GHIOCA
Dept of Psychology,
University Hyperyon
Bucharest, P.O. Box 69-154,
BUCHAREST, ROMANIA

ABSTRACT

Intelligence is determined by the dynamics of interaction with the external world. Control system design includes for the first stage an internal model construction for the agent's environment. The internal model accumulates information as the time passes while the agent interacts with its surroundings. The functional architecture includes a sensory-perceptive module that acquires signals from the environment, a motivational and an emotional module that orientates the task planner toward its goal. The competence level has to be modified if a new task is accomplished. Emotional module based on estimation of success generates new emotional states. A self-evaluator module relates affective memory with acts and events. An example for a dynamic control system with lower information is presented. The intelligent control system acts to preserve self-integrity of the agent. The current system environment interaction and its experience determine the behavior. If the environmental conditions are changed, first are changed the emotional maps, then the cognitive map and also its behavior. If the behavior does not solve agent problems then irreversible changes on emotional map occurred. Some behaviors cause "pride" and others could cause "anger". Internal emotional map and evoked emotions are transitive depending on subjective observation of the environment.

INTRODUCTION

Emotion originally means movement due to excited mental states. In spite of many years of investigation, the study of emotion is still a difficult problem (Oatley, K., & Johnson-Laird, P.N. 1987). Emotion is a subjective psychological process that may be dynamic and involves an affective behaviour. From the cognitive point of view emotion can be seen as a label attached to feeling (Damasio, A. 1994). Affective level is strongly related to perception of emotion that can be pleasant or unpleasant depending on the events. Physiological level includes changes of heart rate, blood pressure or rapid breathing (Young, 1967). These changes are made by our hormonal system while we face to stressful situations. There are a large number of emotion theories but fundamental aspects related to psychology of emotion are not solved yet. Some controversies are related to structured emotion, primacy of cognition, emotion generation or factors influencing the intensity of emotion. There is a strong similarity between emotion and intelligence even both is located in different domains. Intelligence is known as an ability to deal with novel situations and to automatically solve problems. Environment is partially known and it changes continuously so only a small piece of information is available at every moment. Intelligence to control a process assumes that individual difference on a given task depends on the specific process, the speed and accuracy of the control system.

CONTROL ARCHITECTURE

In this paper we present a control system design that includes an internal model construction for an agent, which tries to survive in an unpredictable environment. Biolological inspiration leads one to the completion that behaviour follows a fixed pattern and characterizes all species (Brooks, 1990). Our design includes a set of "if then" rules a base of genetically inheritance for reflexes and primary emotions.

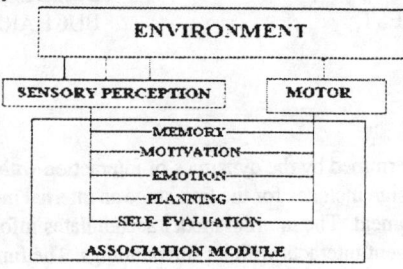

Figure 1

In our brain for every stimulus there is a neural area that receives a specific information sorting information from the sensory receptors and routing them to specific areas. In the same manner agent control system includes a sensory perceptive module that receives coded information from the environment in order to sense temperature, distance etc.

Motor module generates the agent movements in the environment (Figure. 1). If there are many motor effectors each group has a control center in the motor module. Association module is composed by specialized units as memory, motivation, emotion, planning and self-evaluation that underlies the high-level processing of information between its input to sensory module and output from motor module. Association module plays an important role in processing information from various sensory areas. Motivation module is concerned with maintaining homeostasis as a normal level of functioning of the agent (internal energy, artificial feeding, temperature, etc.) by goal orientation. It adjusts all messages to produce a smooth and balanced motor response and is active during cognitive tasks. Memory module plays an integrative role and contains long time memory (LTM) and short time memory (STM). Information with motivational or emotional meanings is more easily remembered than information having no such accompaniment (C.H.Vanderwolf, D.P. Cain, 1994). Pharmacological studies suggest that specific chemicals may interfere with learning.

LEARNING FROM ENVIRONMENT

Sensory perceptive module has to deal with internal states and also with external events. For example, the string 1100/ 010may represent one of 16 external events and eight internal states. Internal state has to code internal needs or emotion. The sensory input is defined over a time period by a binary array A. If one considers k distinct classes every sensory input x may be in class C_i if and only if the input x satisfies the inequality:

$$d(x, c_i) \leq \varepsilon \qquad (1)$$

where ε is a small constant and distance $d(x, c_i)$ is given by:

$$d(x, c_i) = \sum_{j=1}^{q} |x^j - c^j_i| \qquad (2)$$

In this manner group of internal states and external events are associated. In order to compare different partitions P_l with obtained C_l classes a criterion I_c is defined:

$$I_c = \sum_{l=1}^{k} \sum_{x \in P_l} d(x, c_l) \qquad (3)$$

The algorithm converges iteratively and the criterion I_c decreases in time resulting a partition of input space in k classes (k<n). If I_c does not decrease enough, one may increase the number of classes with a unity (k=k+1). Motor module follows the next rule:
R_i: If <sensory perception> AND < current subtask > then <control action>

For example, let consider the rule coded by the string 1100/010/10 → 001 The string can be seen as a law that says" if the system is in the state 1100/010 (where the sensory perception was previously coded) and the current subtask is 10 then the control action is 001". Using the previous mechanism for grouping the perception states and current subtasks the problems for behaviour control are simplified.
Planning module defines as in (Suchman, L.A, 1987) the best strategy to be followed. In order to achieve a desired behaviour, the module can be seen as a large chain of task that obey the following rule:
If <sensory perception> AND <current behaviour > AND <current task> then <new task>
Every task T^k is composed by a subtask sequence (T^k_1, T^k_2, ..., T^k_j). For example a set of subtasks that achieve approaching for feeding may include the sequence of movements: turn to right T_1, increase the speed of movement T_2, decrease the speed T_3, and stop T_4. Similar subtasks could be chosen for different tasks as fighting, avoiding danger, adjusting to losses, and so on (Velásquez, J. 1996). Choosing the subtask sequence implies to use an other rule:
If <stimulus perception> AND <current task> AND <current subtask> then <new subtask>
For stimulus perception, in this case, the agent needs information regarding its speed and distance to food position. Sometimes, environment generates noise in sensory input, or false food information. Association module introduces a reward mechanism in case of need satisfaction that also filters false rules of perception.
The hierarchy of needs can be learned or genetically pre-programmed in motivational module.
For $need_i$ one may write the next rule:
M_i: If $need_i$ AND <current emotion> then <behaviour i_1> OR <behaviour i_2> OR... OR < behaviour i_n>
For emotions, the principal elements are the satisfaction of need and the cause of emotional changes. Causality may be internal or external and has to do with subjective or environmental events. Lets consider three needs $need_1$, $need_2$, $need_3$ and their degree of satisfaction $Sneed_1$, $Sneed_2$, respectively, $Sneed_3$. The general degree of need satisfaction, Sneed may follow the next rule:

If |Sneed₁| AND |Sneed₂| AND |Sneed₃| then <Sneed>
However, this rule, for short time periods, fulfils the general necessity if a particular need is satisfied. As the time pass, if only one need is unsatisfied the agent will start to feel a negative emotion. Emotion state may be positive or negative and have different intensity levels. The intensity of positive emotion or negative emotion depends on the degree of need satisfaction. Behaviour rules R_i are weighted at step t+1 depending on their emotional effect at step t:

$$M_c(R_i,t+1)=M_r(R_i,t)*(1-\beta)+\beta*Sneed$$

where β is a small positive coefficient. These weights are included in long-term memory module. Weights can be assigned for every task or subtask of behaviour. Both, current task and current subtask are included in STM. If a behavior generates a negative emotion it will be very soon restrained by the reinforcement mechanism.

The same rule is applied for sensory perceptive module. For example in case of food perception, the strength M_p of perception rule P_i will be increased at time t+1 if the food is perceived at time t and reduces the current need:

$$Mp(P_i, t+1)=M_p(P_i,t)*(1-\alpha) +\alpha*Sneed\_food$$

where α is a small positive constant. However, the priority of needs may be maintained or may be changed by emotion state. Now events or situations can be classified in general and such inferences between emotional processes can be drawn in a chart of emotion. Emotion states indicate the degree of need satisfaction and the satisfaction gradient is determined by behaviour. The rules in emotional modules ensue new emotional state:

If <environment perception> AND <current emotional state > AND <Sneed> then <next emotional state>

Emotion states are analysed and used then to generate planning rules. Emotion intensity will increase if the need increases.

The self-evaluator module analyses the emotional states and their causes. It will associate external and internal events with new states of emotions. These will permit to anticipate changes, predict future behaviour, new goals, change the order of needs and modify the behaviour with new planning rules.

EXPERIMENT ANALYSIS

We consider the experiment related to Figure.2 where the agent is motivated to find food in order to maintain its integrity. If the robot has no food for a long time its internal energy decreases and the agent will be artificially dead.

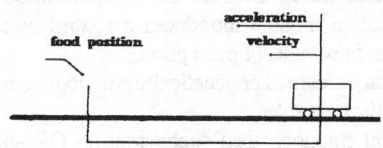

Figure 2

We established basic emotion inheritance in the same manner we organised the agent reflexes for behaviour. For example, external event cannot cause pride; it can cause only a positive emotion of gratitude. Pride can be only a consequence of his own behaviour. We consider five coded emotions. Negative emotions are anger, guilt and pity and two positive gratitude and pride. The agent is equipped with few sensors to measure the environment changes (food presence, velocity sensor, collision detector, target detector etc.). In the same manner as basic reflexes for behaviour (turn left or right if in the front of the agent is an obstacle) some production rules are initially established for emotions.

If a situation is presented, the agent comes up with one or several appropriate emotions. For example food is presented in the neighbourhood of the agent. It perceives the food, starts to have a positive emotion as in Figure 3 and begins to move accelerated with acceleration a to the food position.

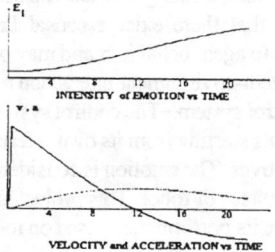

Figure 3

Emotion increases as the time passes so it reaches the maximum value while the agent arrives to food. The trajectories from Figure 3 were obtained after 50 iterations while the agent has learned to stop for feeding. No information regarding agent's mass is used. Initially the robot will hit the obstacle and it becomes angry or frustrated. Initially it is experienced undefined basic emotion for a certain time. After that beginning, a structured cognitive map and an emotional map (Figure 4) are established.

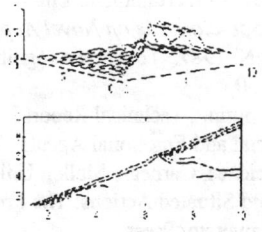

Figure 4

If the food is moved from its place before the agent reached it, the positive emotion decreases and the agent does not stop for feeding (Figure 3).

CONCLUSION

Human being changes his motion and performance according to his decision making and emotion while robotic system changes its strategy based on the algorithms. Recent developments in the areas of synthetic agents, (Maes 1995, Bates 1994, Reilly 1996) and affective computing (Picard 1995, Velásquez, J. 1996), have promoted the study of emotions and their influences in behavior but relatively few computational models of emotion have been proposed. This approach try to contribute to understanding the foundations of what one might want to call by emotional processes and make a link to agent control system behavior. From our simulation one may observe that the emotion persists after the extinction of causal event and it does not represent logical necessities. A causal event can have internal or external sources and is hardly coded in case of real processes. Even for an artificial uncomplicated environment there are necessary long training to select the best behaviour and build a simple emotional map. We supposed that there exist a causal link between mental states and behaviour. Emotions give sense to agent behaviour and may predict for a short period of time future events. In this study, system-environment interaction may have a primordial importance in designing the intelligent control system. The control system becomes autonomous in that case, classifies the external events starting from its own reference, provides self-integrity and accomplishes its personal objectives. The emotion is considered to be one of the parameters to adjust the condition and performance of robot. This project is the approach to the design the group robotic system to improve its performance based on the emotion-like parameters. One can say that the simulated system reacts on the environment space like human's conscious field.

REFERENCES

Bates, J. 1994. "The Role of Emotion in Believable Agents". *Communications of the ACM* 37(7):122-125.

Brooks, R. , 1990, "Elephants don't play chess". In P. Maes (ed.). *Designing autonomous agents; theory and practice from biology to engineering and back.* Cambrdige, Mass.: MIT Press, 3-15

Damasio, A. 1994. "Descartes' Error: Emotion, Reason, and the Human Brain". New York: Gosset/Putnam.

Maes P. 1995. Artificial Life meets Entertainment: Lifelike Autonomous Agents. *Communications of the ACM. Spe-cialIssue on Novel Applications of AI.*

Oatley, K., & Johnson-Laird, P.N. 1987. Towards a cognitive theory of emotions. *Cognition and Emotion,* 1, 29-50.

Picard, R. 1995. Affective Computing, Technical Report No. 321. MIT Media Laboratory.

Reilly S. 1996. "Believable Social and Emotional Agents".Technical Report, CMU-CS-96-138, School of Computer Science, Carnegie Mellon University.

Suchman, L.A, 1987 "Plans and Situated Actions: The Problem of Human-Machine Communication" Cambridge University Press,

C.H.Vanderwolf, D.P. Cain, 1994, "The behavioral neurobiology of learning and memory: a conceptual reorientation".,Brain Research Reviews 19, 264-297

P.T. Young Affective arousal, 1967. Amer. Psychologist, 22,32-40.

Velásquez, J. 1996. "A Computational model for the Generation of Emotions and Their Influence in the Behavior of Autonomous Agents." S.M. Thesis. Depart-ment of Electrical Engineering and Computer Science,Massachusetts Institute of Technology.

PART VII: PATTERN RECOGNITION

SPATIAL-TEMPORAL FEATURE SCREENING

KELLY A. GREENE, Ph.D.
USAF Test Pilot School
Edwards Air Force Base, CA

KENNETH W. BAUER, Jr., Ph.D.
Department of Operational Sciences
Air Force Institute of Technology
Wright-Patterson Air Force Base, OH

ABSTRACT

Saliency measures as used in Elman recurrent neural networks (RNN) do not explicitly account for the temporal saliency of each feature. In response, this research derived and developed a partial derivative-based spatial-temporal saliency measure for use in Elman RNNs. In order to use the partial derivative-based spatial-temporal saliency measure for feature selection, a spatial-temporal feature screening method was developed. The applicability of the new methodology is exhibited by applying it to classifying pilot workload using peripheral psycho-physiological features.

INTRODUCTION

The partial derivative-based spatial-temporal saliency measure provides the spatial-temporal saliency of each feature by unfolding the layers of an Elman recurrent neural network (RNN) through time. In the spatial-temporal feature screening method, features are screened out by comparing the area under the spatial-temporal curves to that of an injected noise feature. To show the advantages of the using Elman RNNs for classifying pilot workload, feature screening is performed and then compared to results using the signal-to-noise ratio screening method (Bauer et al, 2000; Greene, 1998; Greene et al, 1997) in a feedforward multi-layer perceptron (MLP) artificial neural network (ANN) and in a time delay neural network (TDNN). The Elman RNN results are significantly better than that of feedforward MLP ANNs and TDNNs.

PARTIAL DERIVATIVE-BASED SPATIAL-TEMPORAL SALIENCY MEASURE

An $I + J / J / K$ Elman RNN as shown in Figure 1 can be unfolded ℓ layers. For unfolded layer ℓ, the partial derivative-based spatial-temporal saliency measure for input feature x_i is calculated as:

$$\Gamma_{x_i}^{\ell} = \frac{1}{K \cdot (T-\ell)} \cdot \sum_{k=1}^{K} \sum_{t=1}^{T-\ell} \left| \frac{\partial z_k(t, \mathbf{W})}{\partial x_i(t-\ell)} \right| \qquad (1)$$

where $\Gamma_{x_i}^{\ell}$ is the partial derivative-based spatial-temporal saliency measure for input feature x_i for $i = 1,2,\ldots,I$ on unfolded layer ℓ, K is the number of output nodes, T is the number of time periods, $z_k(t, \mathbf{W})$ is output node z_k for $k = 1,2,\ldots,K$ at time $t = 1,2,\ldots,T$ with trained weight matrix $\mathbf{W}$, and $x_i(t-\ell)$ is input feature x_i for $i = 1,2,\ldots,I$ at time $(t-\ell)$. Equation 1 is a temporal extension of the partial derivative-based saliency measure (Ruck, 1990; Ruck et al, 1990). Similarly, the partial derivative-

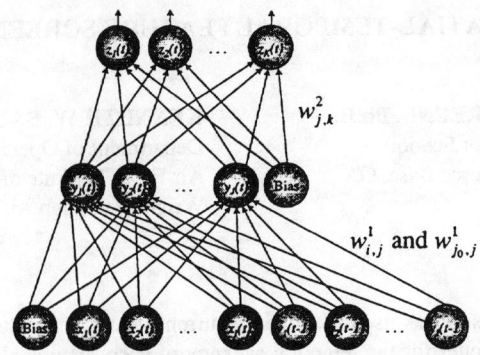

Figure 1. $I + J / J / K$ Elman Recurrent Neural Network

based spatial-temporal saliency measure can be computed for context node y_{j_ℓ} on unfolded layer ℓ as:

$$\Gamma_{y_{j_\ell}}^\ell = \frac{1}{K \cdot (T-\ell)} \cdot \sum_{k=1}^{K} \sum_{t=1}^{T-\ell} \left| \frac{\partial z_k(t, \mathbf{W})}{\partial y_{j_\ell}(t-\ell-1, \mathbf{W})} \right| \quad (2)$$

where $\Gamma_{y_{j_\ell}}^\ell$ is the partial derivative-based spatial-temporal saliency measure for context node y_{j_ℓ} for $j = 1,2,\ldots,J$ on unfolded layer ℓ and $y_{j_\ell}(t-\ell-1, \mathbf{W})$ is context node y_{j_ℓ} for $j = 1,2,\ldots,J$ on unfolded layer ℓ at time $(t-\ell-1)$ with trained weight matrix $\mathbf{W}$.

SPATIAL-TEMPORAL FEATURE SCREENING METHOD
1. Introduce a Uniform(0,1) noise feature $x_N(t)$ for $t = 1,2,\ldots,T$ to the original set of features.
2. Preprocess all features via standardization or normalization.
3. Initialize the weights following the Nguyen-Widrow method (Nguyen and Widrow, 1990).
4. Initialize all context nodes to 0.0.
5. Compute the fractal dimension of each input feature by the Grassberger and Procaccia method (Grassberger and Procaccia, 1983).
6. Apply Takens' Theorem (Takens, 1981; Takens, 1984) to determine the lag upper bound ℓ_{max} as a function of the fractal dimensions computed in Step 5.
7. Set $\ell = 0$.
8. Train the Elman RNN for a pre-defined number of epochs. Keep the weights that minimize the mean squared error of the test set MSE_{test}.
9. Compute the classification accuracy of the test set CA_{test}.
10. Compute $\Gamma_{x_i}^0$ for $i = 1,2,\ldots,I$ and $\Gamma_{x_N}^0$.
11. Set $\ell = \ell + 1$.
12. Compute $\Gamma_{x_i}^\ell$ for $i = 1,2,\ldots,I$.
13. If $\Gamma_{x_i}^\ell \leq \Gamma_{x_N}$ for $i = 1,2,\ldots,I$ or if $\ell = \ell_{max}$, set $\ell = \ell_{stop}$ and go to Step 14. Else go to Step 11.

14. Plot $\Gamma_{x_i}^{\ell}$ for $i = 1,2,\ldots,I$ versus $\ell = 0,1,2,\ldots,\ell_{stop}$.
15. Compute the area under the *spatial-temporal saliency curve* for each feature x_i for $i = 1,2,\ldots,I$.
16. Remove the feature with the smallest area and set $I = I - 1$.
17. If $I > 0$, go to step 7. Else go to step 18.
18. Keep the set of salient features that produced the best CA_{test}. Remove $x_N(t)$ for $t = 1,2,\ldots,T$.
19. Initialize the weights following the Nguyen-Widrow method (Nguyen and Widrow, 1990).
20. Initialize all context nodes to 0.0.
21. Train the Elman RNN with the set of salient features that produced the best CA_{test} until the MSE_{test} is minimized.

CLASSIFYING PILOT WORKLOAD

The spatial-temporal feature screening method was applied to a two-class and a three-class pilot workload problem. The objective of the two-class problem was to determine whether the pilot was in visual flight rules (VFR) or instrument flight rules (IFR) meteorological conditions. The objective of the three-class problem was to determine if the pilot's workload was low, medium, or high. Features for both classification problems were number of eye blinks, heart rate, and respiration rate. All features were averaged over a 10-second moving window with 50% overlap. Several types of ANNs and several number of hidden nodes were utilized in an experimental design approach to feature screening. The SNR feature screening method was used in a feedforward MLP ANN and a TDNN. The new spatial-temporal feature screening method was used in an Elman RNN.

Two data sorties were flown on separate days. Each sortie had 16 segments lasting two minutes. Data collected during the first sortie was used for training. Data collected during the second sortie was used for testing. To account for the day-to-day differences, the features were standardized within each day. Heart rate for the test set is shown in Figure 2. Figure 2 is representative of the other psychophysiological features used in the training and test sets.

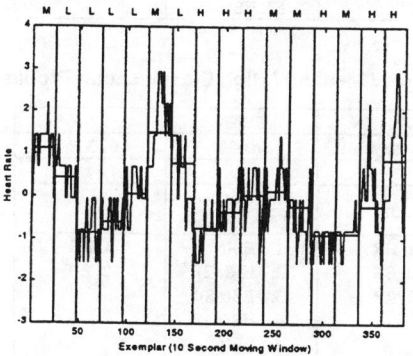

L is low workload
M is medium workload
H is high workload

Each block represents a segment

Straight horizontal line within each segment is the average heart rate for that segment

Figure 2. Test Set Heart Rate

METHODOLOGY

An experimental design with two factors, the number of hidden nodes and the type of ANN, was used. There were four levels of J in the experiment: $J = 1, 2, 3,$ and 4 hidden nodes. There were three types of ANNs in the experiment: MLP, TDNN, and RNN. The applicable screening method was replicated 30 times for each level of J and each type of ANN. All ANN replications were trained for 500 epochs using batch backpropagation with momentum and an adaptive learning rate (Demuth and Beale, 1988). The weights from the epoch that produced the minimum MSE_{test} were kept. All hidden, context, and output nodes were activated by the nonlinear hyperbolic tangent transfer function.

RESULTS

The results are summarized in Table 1 and Table 2 below.

CONCLUSIONS

By using a TDNN instead of a feedforward MLP ANN, the CA_{test} was improved by 18.30% for the VFR/IFR classification problem. The CA_{test} was improved by 13.49% by using a TDNN for the low/medium/high workload classification problem. By using an Elman RNN instead of a feedforward MLP ANN, the CA_{test} was improved by 29.24% for the VFR/IFR classification problem. The CA_{test} was improved by 24.42% by using an Elman RNN for the low/medium/high workload classification problem.

Table 1. Results from VFR / IFR Classification Problem

	MLP ANN	TDNN	RNN
CA_{test}	70.23%	88.53%	99.47%
J	1	1	4
I	1	1 feature over 5 various lags	2
Feature Rankings	1. EB	1. EB(t-8) 2. EB(t) 3. EB(t-2) 4. EB(t-4) 5. EB(t-6)	1. EB 2. HR

Table 2. Results from Low / Medium / High Classification Problem

	MLP ANN	TDNN	RNN
CA_{test}	44.91%	56.00%	69.33%
J	3	2	4
I	3	2 features over 7 various lags	2
Feature Rankings	1. HR 2. EB 3. RR	1. EB(t) 2. HR(t-2) 3. EB(t-8) 4. HR(t-4) 5. EB(t-2) 6. EB(t-6) 7. HR(t-1)	1. HR 2. EB

This was the first time that the SNR screening method was applied to the lagged inputs of a TDNN. The SNR screening method was allowed to remove any lag of any input feature. This is a novel approach to utilizing a TDNN since in the past, all lagged inputs for all features were used as inputs to a TDNN. There was an interesting aspect about the results from the TDNN. The parsimonious set of salient features selected resulted in that no *redundant* lags of the number of eye blinks were selected. Each input feature was averaged over a 10-second moving window with 50% overlap. The SNR screening method selected the number of eye blinks at time window t, $t-2$, $t-4$, $t-6$, and $t-8$ which provided no overlapping information. Figure 3 provides further elaboration on this point.

The significant contribution of this paper is the development of a spatial-temporal feature screening method for use in Elman RNNs. This spatial-temporal feature screening method utilizes the novel partial-derivative based spatial-temporal feature saliency measure. With this feature screening method, the ability now exists to reduce an input feature space to a parsimonious set of input features by examining the feature's saliency in the temporal dimension at the same time as the spatial dimension.

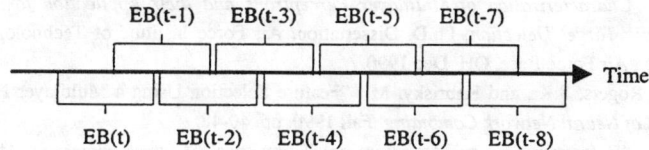

Figure 3. Further Elaboration of Overlapping and Redundancy

NOMENCLATURE

CA_{test}	test set classification accuracy
$\Gamma_{x_i}^\ell$	partial derivative-based spatial-temporal saliency measure for input feature x_i for $i = 1,2,\ldots,I$ on unfolded layer ℓ
MSE_{test}	test set mean squared error
$w_{i,j}^1$	first layer weight connecting input feature x_i to hidden node y_j
$w_{j,k}^2$	second layer weight connecting hidden node y_j to output node z_k
$x_i(t)$	input feature $i = 1,2,\ldots,I$ at time $t = 1,2,\ldots,T$
$x_N(t)$	noise input feature at time $t = 1,2,\ldots,T$
$y_{j_\ell}(t)$	hidden/context node $j = 1,2,\ldots,J$ at time $t = 1,2,\ldots,T$ on unfolded layer ℓ
$z_k(t, \mathbf{W})$	output node z_k for $k = 1,2,\ldots,K$ at time $t = 1,2,\ldots,T$ with trained weight matrix $\mathbf{W}$

REFERENCES

Bauer, K.W., Alsing, S.A., and Greene, K.A. "Feature Screening Using Signal-to-Noise Ratios," *Neurocomputing*, 9 Mar 00, Vol 31, Issue 1-4, pp. 29-44.

Demuth, H. and M. Beale. *MATLAB Neural Network Toolbox User's Guide*, Natick, MA: MathWorks, 1998.

Elman, J.L. "Finding Structure in Time," *Cognitive Science*, Vol 14, 1990, pp. 179-211.

Grassberger, P. and I. Procaccia. "Measuring the Strangeness of Strange Attractors," *Physica D*, Vol 9, 1983, pp. 189-208.

Greene, K.A. *Feature Saliency in Artificial Neural Networks with Application to Modeling Workload*. Ph.D. Dissertation, Air Force Institute of Technology, Wright-Patterson Air Force Base, OH, 1998.

Greene, K.A., Bauer, K.W., Kabrisky, M., Rogers, S.K., and Wilson, G.F. "Estimating Pilot Workload Using Elman Recurrent Neural Networks: A Preliminary Investigation," *Intelligent Engineering Systems through Artificial Neural Networks*, Proceedings of Artificial Neural Networks in Engineering (ANNIE) International Conference, St. Louis, MO, 9-12 Nov 1997, Vol 7, pp. 703-708.

Nguyen, D, and B. Widrow. "Improving the Learning Speed of 2-Layer Neural Networks by Choosing Initial Values of the Adaptive Weights," *International Joint Conference on Neural Networks*, Vol 3, 1990, pp. 21-26.

Ruck, D.W. *Characterization of Multilayer Perceptrons and their Application to Multisensor Automatic Target Detection*, Ph.D. Dissertation, Air Force Institute of Technology, Wright-Patterson Air Force Base, OH, Dec 1990.

Ruck, D.W., Rogers, S.K., and Kabrisky, M. "Feature Selection Using a Multilayer Perceptron," *Journal of Neural Network Computing*, Fall 1990, pp. 40-48.

Takens, F. "Detecting Strange Attractors in Turbulence" *Lecture Notes in Mathematics, Proceedings of Dynamical Systems and Turbulence Symposium*, University of Warwick, Scotland, Vol 898, 1981, pp. 366-381.

Takens, F. "On the Numerical Determination of the Dimension of an Attractor," *Lecture Notes in Mathematics, Proceedings of Dynamical Systems and Bifurication Workshop*, Groningen, The Netherlands, Vol 1125, Apr 1984, pp. 99-106.

SCATTERED DATA INTERPOLATION USING SELF-ORGANIZING FEATURE MAPS

GEORGE K. KNOPF AND ARCHANA SANGOLE
Department of Mechanical & Materials Engineering,
Faculty of Engineering Science, University of Western Ontario,
London, Ontario, Canada, N6A 5B9

ABSTRACT
Scattered data interpolation involves fitting mathematical surfaces to a large number of unorganized coordinate data points. Difficulty arises because most surface interpolation and approximation techniques require prior knowledge of the connectivity between sampled points. This paper describes how a two-dimensional self-organizing feature map (SOFM) can interpolate unorganized data points for the reconstruction of a polygon mesh with quadrilateral elements. The SOFM learning algorithm iteratively adjusts the cluster unit weights such that neighbouring units in the 2D map are connected nodes in the "best fit" polygon mesh. The method is illustrated using scattered range data. The results demonstrate that acceptable reconstruction of surface geometry occurs when there are more units in the SOFM than the number of data points.

INTRODUCTION
Surface reconstruction from non-uniformly spaced, or scattered, data is a common problem found in several disciplines such as geology, cartography, virtual reality modeling, computer-aided design, and reverse engineering. Commonly used surface reconstruction techniques that involve triangulation and parametric forms cannot be directly applied without making assumptions about the connectivity between individual sampled points. The connectivity must be established in order for the reconstructed surface to faithfully represent the shape from which the data points were acquired. Most algorithms described in the literature for scattered data interpolation (Hoschek and Lasser, 1993; Hill, 1990) attempt to explicitly compute the connections between neighbouring data points prior to constructing a polygon mesh. A majority of the techniques used to compute connectivity require a dense data set to prevent gaps and holes from forming on the reconstructed surface within under-sampled areas. The gaps and holes can significantly change the topology for the generated surface.

The algorithm proposed by Hoppe et al. (1992) is based on surface reconstruction using normal vectors obtained from tangent planes defined at sample points. The method of reconstruction thus depends on the estimation of the normal vectors from the data points and the order in which orientation propagates. Although the technique is effective for a range of data sets, it is very complex and computationally intensive.

An alternative perspective to solving the problem of scattered data interpolation is to assume a predefined 2D-mesh with the desired connectivity between neighbouring vertices or nodes. An interpolation algorithm is then used to iteratively adjust the nodes in the mesh in order to match the coordinate data set. In this way, the interpolation algorithm learns the surface coordinates given the node connectivity (Yu, 1999). This paper describes how a two-dimensional self-organizing feature map (SOFM) can be used to establish an order within the scattered data and, thereby, provide connectivity information within the data. The major advantage of using a predetermined grid topology such as the SOFM is that no unexpected gaps or holes will be generated on the reconstructed surface, even if the input data is sparse. Unorganized range data of a hand-carved Indonesian mask has been used to illustrate the SOFM approach.

SELF ORGANIZING FEATURE MAP

The Kohonen self-organizing feature map (SOFM) is a competitive self-learning algorithm that internally orders the data by exploiting similarities between randomly acquired points. The self-organizing neural network, Fig. 1, arranges the weights, $w_{i,j}$, such that they reflect some physical characteristic of the external input X^p = [x_1^p, x_2^p, x_3^p], where p = 1, 2, ... P, and P is the total number of sample coordinate data points used for training. Every input vector, X^p is connected to each cluster unit in the network by a weight vector, $w_{i,j}$. A unit located at (i, j) in the network has an associated variable neighbourhood operator that covers an area of the map given by $(2NE_{i,j} + 1)$ x $(2NE_{i,j} + 1)$ units where $NE_{i,j}$ is the decreasing neigbourhood radius. The weight adaptation algorithm is designed such that the selected cluster unit, and all units residing in its neighbourhood, are simultaneously updated.

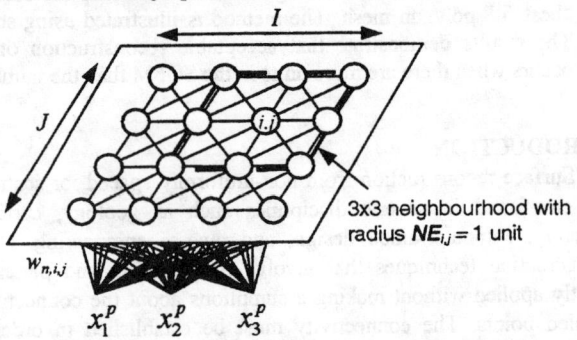

Figure 1. The two-dimensional self-organizing feature map (SOFM).

At the beginning of the training process the weights are initialized to small random values in the input data space. A coordinate point, X^p is randomly selected from the data set, where p = 1, 2, ... P, and P is the total number of data points. For each selected point the error, or difference $D_{i,j}^p$, between the input point and the weight vectors for all the cluster units in the network is computed using Eq. (1) as given below:

$$D_{i,j}^p = \sum_{n=1}^{3} (x_n^p - w_{n,i,j})^2 \qquad (1)$$

where x_n^p is the nth element of the input vector, $w_{n,i,j}$ is the weight from the nth input element to the $(i, j)^{th}$ cluster unit, i = 1, 2, ... I, and j = 1, 2, ... J. The winning cluster unit $(i, j)^*$ is selected as the unit with the minimum difference $D_{i,j}^p$. The weights to the winning cluster unit $(i, j)^*$ and all other units that reside within the specified neighbourhood of radius $NE_{i,j}$ are updated using Eqs. (2a) and (2b). The radius of the topological neighbourhood is decreased using the given set of rules.

$$w_\xi(new) = w_\xi(old) + \alpha [X^p - w_\xi(old)] \qquad (2a)$$
where
$$\alpha = 0.001 + 0.2 [NE_{i,j} / 5.0] \qquad (2b)$$

IF ((# of cycles) ≤ (N_{cycle}/4)) THEN $NE_{i,j}$ = 5
IF (((# of cycles) > (N_{cycle}/4)) AND ((# of cycles) ≤ (N_{cycle}/2))) THEN $NE_{i,j}$ = 3
IF (((# of cycles) > (N_{cycle}/2)) AND ((# of cycles) ≤ (3N_{cycle}/4))) THEN $NE_{i,j}$ = 1
IF ((# of cycles) > (3N_{cycle}/4)) THEN $NE_{i,j}$ = 0

By using a variable neighbourhood adaptation strategy the self-organizing network overcomes the problem of under-utilized cluster units and the output responses are organized in a natural manner. On completion of the training process, the weights of the cluster units form prototypical representations of the coordinate points in the data space. In essence, the feature map represents a continuous polygonal mesh with quadrilateral elements and the positions of the units impose topology on the unorganized coordinate data points. There will be a one-to-one correspondence between the cluster units and the coordinate points if there are as many cluster units as data points in the training set. In the analysis it is observed that, by using a large initial neighbourhood with an over-represented SOFM network, the geometry of regions with sparse data can be extrapolated from the neighbouring points thereby reducing the influence of non-uniformity in the density of the data on the performance of the surface-fitting algorithm. This is an important feature for reconstructing surfaces of objects with complex geometries that exhibit regions of missing data points due to surface occlusion during range image acquisition.

EXPERIMENTAL STUDY

Unorganized coordinate data measured from the surface of a hand-carved Indonesian mask, Fig. 2, is used in this preliminary study.

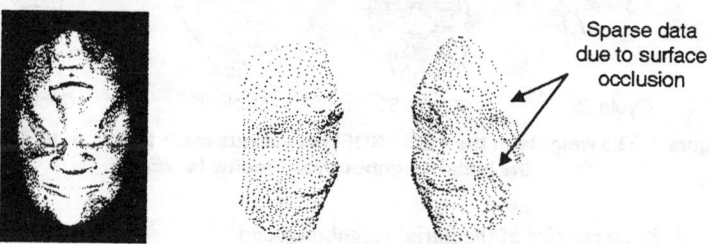

Figure 2. The hand-carved wooden mask and two views of the scattered coordinate data. Each point is given by vector X^p, where $p = 1, ..., 1406$.

The complete data set contains 1406 scattered coordinate points acquired using an experimental range sensor system that employs a structured-light triangulation method and a nonlinear mapping operation (Knopf and Kofman, 1998) to relate the captured 2D image data to 3D object coordinates. Several regions with sparse data points, due to surface occlusion during the data acquisition process, are identified in the second view.

Case 1: SOFM with 30x30 units

Initially, a two-dimensional SOFM with fewer cluster units than data points is used to explore the surface reconstruction capability of the proposed algorithm. Fig. 3 illustrates how the algorithm reorganizes the weights of the 30 x 30 map in order to better represent the data set. The weights associated with each cluster unit in the 2D map represent a node within a polygonal mesh that approximates the shape of the object. Fewer cluster units than data points may result in gaps forming on the reconstructed surface, such as the one in the upper left.

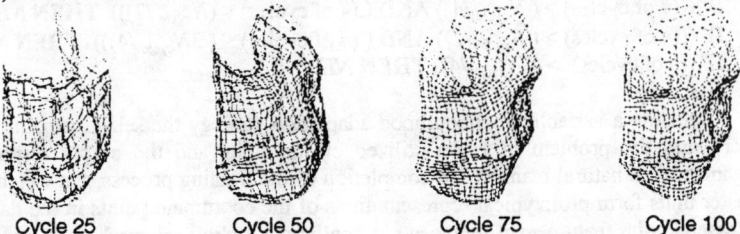

Cycle 25   Cycle 50   Cycle 75   Cycle 100

Figure 3. The weights of the 30x30 SOFM at various cycle times through the data set. In this illustration, the initial neighbourhood radius is $NE_{i,j} = 5$.

Case 2: SOFM with 45x45 units

This case analyses the effect of selecting far more cluster units than data points. The over-representation is tested using a 45x45 map (i.e. 2025 units). Fig. 4 shows the adaptation of the enlarged SOFM at various stages of the training process. Compared to Fig. 3, the change in the size of the quadrilateral mesh moves from coarse-to-fine in the regions of sparse data. This illustrates the continued topological ordering and smooth interpolation of the data as weight-adaptation takes place during training.

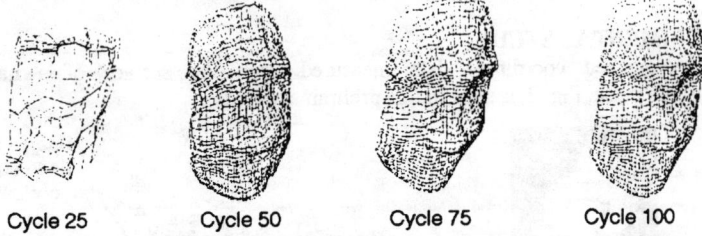

Cycle 25   Cycle 50   Cycle 75   Cycle 100

Figure 4. The weights of the 45x45 SOFM at various cycle times. In this illustration, the initial neighbourhood radius is $NE_{i,j} = 5$.

Case 3: Enlarged size of the initial neighbourhood

In the final test, the effect of using a large initial neighbourhood size in an over-represented network, 45x45 cluster units, is tested. The modified neighbourhood adaptation strategy starts with a very large region of influence (31x31 units), and a high learning rate ($\alpha = 0.2$) for rapid dispersion of the network weights in the data space. The size of the neighbourhood operator and the learning rate are significantly reduced for local fine-tuning using the set of rules given below:

IF ((# of cycles) $\leq (N_{cycle}/4)$) THEN $NE_{i,j} = 15; \alpha = \mu_1$

IF ((# of cycles) $> (N_{cycle}/4)$) AND ((# of cycles) $\leq (N_{cycle}/2)$)) THEN $NE_{i,j} = 2; \alpha = \mu_2$

IF $((\# \text{ of cycles}) > (N_{cycle}/2))$ AND $((\# \text{ of cycles}) \leq (3N_{cycle}/4)))$ THEN $NE_{i,j} = 1$; $\alpha = \mu_2$

IF $((\# \text{ of cycles}) > (3N_{cycle}/4))$ THEN $NE_{i,j} = 0$; $\alpha = \mu_2$

where $\mu_1 = 0.2$, and $\mu_2 = 0.001 + 0.13 [NE_{i,j}/2.0]$.

Due to the large initial neighbourhood operator, a wider area of the map is influenced and several units get drawn into regions that are sparsely populated thereby resulting in an initial spread of the cluster units. The neighbourhood is then reduced significantly so that comparatively fewer units are updated and take the shape of the contours. Units that occupy the sparsely populated regions in the data space are influenced only by the neighbourhood effect or are not influenced at all. Different views of the faceted surface of the mask with initial neighbourhood radii of 5 and 15 units are shown in Fig. 5. The influence of the neighbourhood operator in the sparsely populated regions of the data set can be seen in the areas identified in Figs. 5a and 5b.

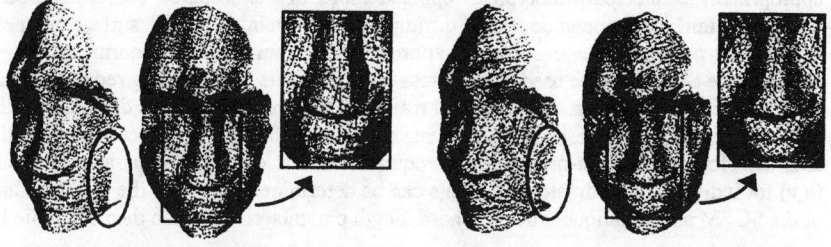

(a) Initial neighbourhood radius, $NE_{i,j} = 5$ (b) Initial neighbourhood radius, $NE_{i,j} = 15$.

Figure 5. Different views of the faceted surface of the mask after training the 45x45 SOFM network for 450 cycles with initial neighbourhood radii of 5 and 15.

DISCUSSION

Inadequate information on surface topology is a general problem in reconstructing surfaces from scattered points. The SOFM approach imposes topology on the scattered data and the resultant polygonal mesh takes the shape of the object surface. The large initial neighbourhood size appears to affect the overall fitting process by placing cluster units in regions that have no coordinate information. This is important when reconstructing a continuous shape from incomplete data due to surface occlusion during data acquisition, as is the case in the hand-carved mask. Significant reduction in neighbourhood size after preliminary training leaves a few units with weights that lie in these under-represented regions. The fine-tuning process only affects neighbours nearest to the winning unit leaving the distant units in the gaps unaltered. Furthermore, if the cloud of sampled points is too dense, surface reconstruction can be simplified by selecting a suitable size for the network, thereby, representing the surface geometry by fewer weight vectors.

Although the resultant polygon surface mesh is adequate for many visualization and computer graphics applications, it is not smooth. A smooth free-form surface is often mathematically represented by bi-parametric form such as Bezier, B-spline, and non-uniform rational B-spline (NURB) (Hoschek and Lasser, 1993; Rogers and Adams, 1990). The parametric function defines a path where the parameter values (u, v) represent the normalized distance of a data point along the reconstructed synthetic surface. If a smooth free-form surface is required, then the resultant SOFM polygon mesh can be used to parameterize the original scattered data. In the parameterization method presented by

Ma and Kruth (1995) the data points are projected onto a base surface and the location parameters of the projected points are used as those of the corresponding sample points. The performance thus largely depends on the parameterization of the base surface. A good choice of the base surface requires some knowledge on the underlying shape of the object surface. By using the SOFM, the cluster units represent the location parameters of the coordinate data and the mesh takes the shape of the object surface. Since the SOFM is a two-dimensional mesh with a predefined topology, it is possible to track neighboring surface points along lines of constant (i, j) or in parametric terms (u, v).

CONCLUSIONS

Conventional surface interpolation and approximation techniques cannot be directly applied to fit mathematical surfaces to scattered data without prior knowledge about the topology of the coordinate data space. The self-organizing feature map defines an order to and interpolates the unorganized points. The weights of the SOFM cluster units form appropriately connected prototypical representations of the scattered coordinate points. By using a fairly large neighbourhood during the initial training period, surface geometry of sparsely populated regions can be approximated from the neighbouring points. In addition, the SOFM can be used to compress large data sets by exploiting redundancies in local variations in the data. Experimental results of the preliminary tests demonstrate that the generated polygon mesh is not adequate for smooth free-form surface reconstruction and, therefore, some post-processing is required. In this context, the parametric values (u,v) for individual coordinate data points can be determined following the (i, j) grid lines of the SOFM and techniques such as chord length parameterization can then be applied.

ACKNOWLEDGEMENTS

This work has been supported, in part, by the Natural Sciences and Engineering Research Council of Canada.

REFERENCES

Barhak, J., and Fischer, A., (2001). "Parameterization and reconstruction from 3D scattered points based on neural network and PDE techniques", *IEEE Transactions on Visualization and Computer Graphics*, vol. 7, no. 1, pp. 1-16.

Hill, F.S., (1990). *Computer Graphics*. New York: Macmillian.

Hoppe, H., DeRose, T., Duchamp, T., McDonald, J., and Stuetzle, W.,(1992). "Surface reconstruction from unorganized points", *Proceedings of SIGGRAPH '92, Computer Graphics*, vol. 26, no. 2, pp. 71 - 78.

Hoschek, J. and Lasser, D., (1993). *Fundamentals of Computer Aided Geometric Design*. A.K. Peters Ltd.: Wellesley MA.

Knopf, G.K., and Kofman, J., (1998). "Range sensor calibration using a neural network", in *Intelligent Engineering Systems Through Artificial Neural Networks Volume 8*, Dagli, C.H. et. al. (Eds.), ASME Press, New York, pp. 491-496.

Knopf, G.K. and Kofman, J., (1999). "Free-form surface reconstruction using Bernstein basis function networks", *Intelligent Engineering Systems Through Artificial Neural Networks Volume 9*, C.H. Dagli et. al. (Eds.), ASME Press, pp. 797 – 802.

Kohonen, T., (1997). *Self-Organizing Maps*. New York: Springer.

Ma, W., and Kruth, J. P,(1995). "Parameterization of randomly measured points for least squares fitting of B-spline curves and surface", *Journal of Computer-Aided Design*, vol. 27, no. 9, pp. 663 – 675.

Rogers, D.F. and Adams, J.A., (1990). *Mathematical Elements for Computer Graphics*. McGraw-Hill: New York.

Yu, Y., (1999). "Surface reconstruction from unorganized points using self-organizing neural networks", *Proceedings of IEEE Visualization '99*, San Francisco, CA, Oct 1999, pp. 61 - 64.

PILOT MENTAL WORKLOAD CALIBRATION

JEREMY B. NOEL, KENNETH W. BAUER, JR., AND JEFFREY W. LANNING
Department of Operational Sciences, Air Force Institute of Technology,
United States Air Force

ABSTRACT
Predicting high pilot mental workload is important to the United States Air Force because lives and aircraft have been lost due to errors made during periods of flight associated with mental overload and task saturation. Current research efforts use psychophysiological measures such as electroencephalography (EEG), cardiac, ocular, and respiration measures in an attempt to identify and predict mental workload levels. Existing classification methods successfully classify pilot mental workload using flight data for a single pilot on a given day but are unsuccessful across different pilots and/or days. We demonstrate a small subset of psychophysiological features collected from a single pilot on a given day that accurately classifies mental workload for a separate pilot on a different day. We achieve classification accuracy (CA) improvements over previous classifiers exceeding 80% while using 97% fewer features and reducing the CA variance by over 95%. Without the need for EEG data, our calibration scheme also reduces the raw data collection requirements by 99.75%, making data collection immensely easier to manage and dramatically reducing computational processing requirements.

INTRODUCTION
Mental overload can be a serious problem for fighter pilots. Pilots can become so involved in their current situation that they forget to perform critical tasks, such as G-force straining maneuvers. As a result, some pilots have lost consciousness and their lives (Auten, 1996).

The Air Force Research Laboratory (AFRL)/Human Effectiveness Directorate (HE) at Wright-Patterson Air Force Base, Ohio, has conducted several studies on mental workload analysis in laboratory, simulator, and flight settings (AFRL, 1998). Their results indicate that the most influential psychophysiological features in classifying mental workload level are brain electrical activity, heart rate, breath rate, and eye blink measures (Wilson, 1992, 1993), (Wilson et al., 1995), (Greene et al., 1997a, 1997b, 1998). AFRL's flight data was collected using ten pilots flying Wright-Patterson Aero Club Piper Cub aircraft on a specified route over two days. To collect the psychophysiological data, the pilots wore special recording equipment. Previous analysis of both simulator and flight data revealed that substantial feature reduction is attainable using a signal-to-noise ratio feature-screening algorithm. Furthermore, Laine et al. (1999) and East et al. (2000) found that artificial neural networks produce more robust classifiers for determining mental workload than other classification techniques. They found that training an artificial neural network using reduced feature sets over same-day, same-pilot data produced mental workload classification accuracies between 72 and 97 percent. However, the same-pilot over multiple days classifier yielded classification accuracy (CA) results comparable to flipping a coin (East et al., 2000).

THE EXPERIMENT

Ten volunteer pilots flew a predetermined flight route once a day for two days accompanied by a technician from the flight propulsion laboratory and a copilot. Each flight was divided into 22, two-minute flight segments. While ten pilots participated in the flight experiment, only the data from Pilots 1 and 4 were fully analyzed during the course of this research effort. Data from a third pilot (Pilot 6) became available later and was used for validation purposes.

The flight route was designed to include three levels of mental workload: low, medium, and high. AFRL personnel estimated the difficulty of each flight segment before the flights were conducted, and the test pilots evaluated the difficulty of each flight segment after their flights. Figure 1 shows a graph reflecting the pilot's subjective measures of workload associated with each flight segment.

East et al. (2000) found classifying three workload levels (low, medium, and high) very difficult and combined the low and medium levels into one group called low workload. As a result, the dark horizontal line in Figure 1 defines the low and high workload levels. All flight segments below the line were defined as low mental workload and all flight segments above the line were defined as high mental workload. The creation of this line involves assumptions concerning workload level accuracy and flight segment transitions that could significantly increase classification errors.

DATA

Four different types of psychophysiological data were collected during each flight: electroencephalography (EEG) data, ocular data, respiratory data, and cardiac data. The EEG data was collected at 256 Hz through 29 electrodes placed in a special cap worn by the pilots. The ocular, respiratory, and cardiac data were recorded in data files that contain the elapsed time in milliseconds between events. An event was the blink of an eye, the taking of a breath, or the beat of a heart. In order to make the data useful for analysis, the raw data was preprocessed. This preprocessing is briefly addressed below.

The raw EEG data was collected and immediately sent through a program called *Manscan* 4.0, which filtered out some of the undesirable artifacts from the EEG signals such as muscle and eye movements. The raw data was then passed through a Fast Fourier Transform (FFT). Five frequency bands were filtered out of the EEG data from each electrode: delta (1-3 Hz), theta (4-7 Hz), alpha (8-12 Hz), beta (13-30 Hz), and ultrabeta (31-42 Hz). Frequencies below 1 Hz or above 42 Hz are not associated with mental workload, so this data was not kept

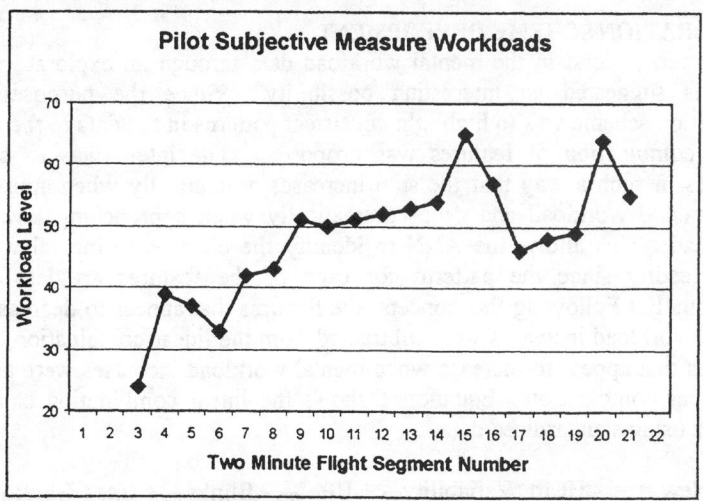

Figure 1. Pilot Subjective Measure Workloads

(Wilson et al., 1995). The total power in each band was then averaged over 10 seconds. Five seconds of overlap was employed to smooth the power readings.

The preprocessing required for the remaining physiological features from the heart, eye, and respiratory files was less involved than the EEG data preprocessing and brought the total to 151 features available for classifying mental workload. The physiological features were: heart rate (in beats per minute), interbeat interval (time between beats), number of blinks (per ten-second time interval), interblink interval (time between blinks), number of breaths (per ten-second interval), and interbreath interval (time between breaths). To allow EEG and physiological features to be included together within data sets, the same overlapping 10-second window method was employed.

FEATURE SELECTION

Artificial neural networks (ANNs) were chosen as the classification technique for this research effort. Backpropagation was used as the training algorithm, and all activation functions were sigmoidal. Prior to training the ANNs, the data sets were standardized to a mean of zero and a variance of one. The training parameters for the ANNs included random initial weights between –0.1 and 0.1, the training rate set at 0.01, the momentum term set at 0.9, and the training termination rule of minimum training-test sum of square error.

The signal-to-noise ratio (SNR) saliency screening method (Bauer et al., 2000) was used to reduce the number of features to a smaller subset for classification. Previous feature reduction efforts on this data revealed that the SNR screening method developed a smaller set of features than other methods (East et al., 2000).

CALIBRATION SCHEME DEVEOPMENT

Patterns found in the mental workload data through an exploratory factor analysis suggested an interesting possibility. Since the purpose of the calibration scheme was to highlight consistent patterns in the data to the ANN, a linear combination of features was proposed. The intent was to combine features in such a way that the sum increases dramatically when approaching high mental workload and drops dramatically when approaching low mental workload. This allows the ANN to identify the changes in mental workload more readily since the patterns for each of the features are less distinct individually. Following this concept, the features that appear to decrease when mental workload increases were subtracted from the linear combination, and the features that appear to increase when mental workload increases were added to the linear combination. Equation 1 shows the linear combination calibration scheme using standardized data,

$$\text{New\_1} = -\text{Heart\_Variability}_{SD} + \text{BPM}_{SD} - \text{Blinks}_{SD} + \text{Inter\_Blink}_{SD} \quad (1)$$

where SD stands for standardized data with a mean of zero and a variance of one. Standardization is necessary since the features contain various units and magnitudes. In order to smooth the variability in this New_1 feature, three new moving average features with lengths of 30, 60, and 120 seconds were added. Including the moving averages, these four features totally replaced all 151 natural features when training ANNs using the calibration scheme.

RESULTS

Two different types of performance measures were used to assess the effectiveness of our proposed calibration scheme: average CAs and receiver operating characteristics (ROC) curves. Each average CA and ROC curve data point was based on 12 values, and never included the results from the same pilot and day combination used to train the network. For instance, if a network was trained using the data from Pilot 1 (day 1), a projection of this network would then be made using data sets from Pilot 1 (day 2), Pilot 4 (day 1), and Pilot 4 (day 2). No projection would be performed on Pilot 1 (day 1) since this represents the same pilot-day combination used to train the network. We continued in this leave-one-*in* fashion until we generated 12 projections, which when averaged together, become the average CA, or a single point on the ROC curve.

Figure 2 shows the average result of networks trained using the calibration scheme compared to the baseline. The baseline consists of networks trained using the 35 most salient features from each data set in addition to three moving averages per feature with lengths of 30, 60, and 120 seconds. As shown in the figure, the ROC curve developed using the calibration scheme completely dominates the baseline ROC curve. In addition, the average CA jumps from 60.11% to 72.02%, with individual calibrated network classifiers producing CA improvements up to 80% over comparable non-calibrated baseline network classifiers.

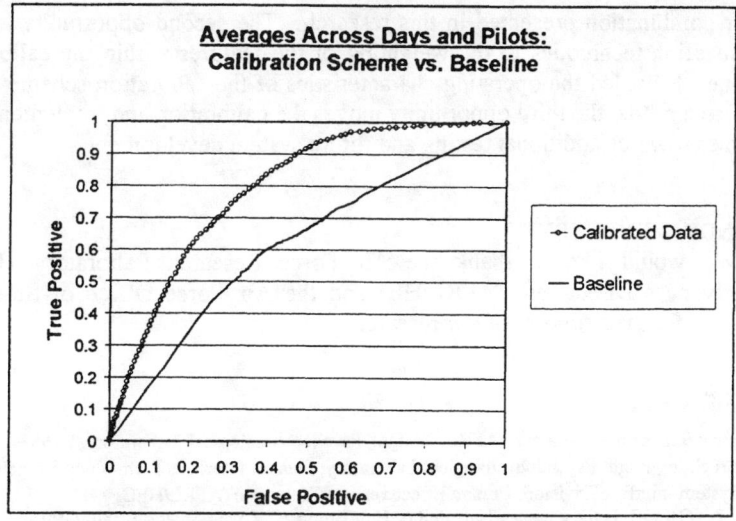

Figure 2. Calibration Scheme Networks vs. Non-calibrated Baseline Networks

A validation effort was performed to fully determine the effectiveness and robustness of the calibration scheme using an independent data set from Pilot 6 (day 2). Improvements comparable to those described above were identified. The performance measures in this validation effort indicate that the calibration method can be successfully applied to new data sets and result in substantially improved pilot mental workload classification. In addition, an implementation methodology was developed and tested to see if the calibration scheme could be implemented without knowing the true mean and standard deviation values for each of the four features included in the calibration scheme. This scheme was highly successful and is reported in Noel (2001).

CONCLUSIONS

The calibration scheme produced superior results by finding a new feature space unable to be found by the ANNs themselves. Accurate mental workload classification requires finding the appropriate feature space for each individual, and we have shown that this feature space can vary by pilot and time. The calibration scheme reduced the impacts of these psychophysiological variations. If one or more of the four features included in the calibration scheme were not significant to a particular pilot on a certain day, then those features basically represented small amounts of noise. As a result, the linear combination calibration scheme allowed the significant features to provide valuable mental workload information to the network, and rendered the effects of the other features as insignificant.

Several opportunities exist for further research on calibrating pilot mental workload. The first opportunity explores calibration schemes other than the linear combination presented in this research. The second opportunity applies optimization techniques to the weighting of the features within the calibration scheme. Provided the operating characteristics of the calibration scheme meets Air Force needs, the third opportunity moves the calibration and implementation schemes towards additional testing and future system development.

ACKNOWLEDGEMENT

We would like to thank the Air Force Research Laboratory Human Effectiveness Directorate (AFRL/HE) and the Air Force Office of Scientific Research for their support of our research.

REFERENCES

Air Force Research Laboratory, AFRL, "Flight Psychophysiological Laboratory," 1998. Office Brochure, Flight Psychophysiological Laboratory, Human Interface Technology Branch, Crew System Interface Division, Human Effectiveness Directorate (AFRL/HECP).

Auten, J. "G-LOC: Is the Cluebag Half Full or Half Empty?," *Flying Safety*, 52:5-6 (1996).

Bauer, K. W., Alsing, S. G., and Greene, K.A., "Feature Screening using Signal-to-Noise Ratios," *Neurocomputing*, 31:29-44 (1999).

East, J. A., Bauer, K. W., and Lanning, J. W., "Feature Selection for Predicting Pilot Mental Workload," *Intelligent Engineering Systems through Artificial Neural Networks*, Proceedings of Artificial Neural Networks in Engineering (ANNIE) International Conference, St. Louis, MO, 5-8 Nov 2000.

Greene, K. A., Bauer, K. W., Wilson, G. F., Russell, C. A., Rogers, S. K., and Kabrisky, M., "Selection of psychophysiological features for classifying air traffic controller workload in neural networks," *International Journal of Smart Engineering System Design*,(1998).

Greene, K.A., Bauer, K.W., Kabrisky, M., Rogers, S.K., and Wilson, G.F. "Estimating Pilot Workload Using Elman Recurrent Neural Networks: A Preliminary Investigation," *Intelligent Engineering Systems through Artificial Neural Networks*", Volume 7, Ed. C.H. Dagli et al, ASME Press: New York, November 1997a.

Greene, K.A., Bauer, K.W., Kabrisky, M., Rogers, S.K., and Wilson, G.F. "Estimating Pilot Workload Using Elman Recurrent Neural Networks: A Preliminary Investigation," *Intelligent Engineering Systems through Artificial Neural Networks*, Volume 7, Ed. C.H. Dagli et al, ASME Press: New York, November 1997b.

Hankins, T. C. and Wilson, G. F. "A Comparison of Heart Rate, Eye Activity, EEG, and Subjective Measures of Pilot Mental Workload during Flight," *Aviation, Space, and Environmental Medicine*, 69:360-367 (1998).

Laine, T. I., Bauer, K. W., and Lanning, J. W., 1999, "Multiple Crewmember Workload Classification using Neural Networks with Input Feature Selection," *Intelligent Engineering Systems Through Artificial Neural Networks*, Proceedings of Artificial Neural Networks in Engineering (ANNIE) International Conference, St. Louis MO, 7-10 Nov 1999.

Wilson, G. F. "Applied Use of Cardiac and Respiration Measures: Practical Considerations and Precautions," *Biological Psychology*, 34:163-178 (1992).

Wilson, G. F. "Air-to-Ground Training Missions: A Psychophysiological Workload Analysis," *Ergonomics*, 36(9):1071-1087 (1993).

Wilson, G. F., Fisher, F., "Cognitive Task Classification Based Upon Topographical EEG Data," *Biological Psychology*, 40:239-250 (1995).

Wilson, G. F., Fullenkamp, P. and Davis, I. "Evoked Potential, Cardiac, Blink, and Respiration Measures of Pilot Workload in Air-to-Ground Missions," *Aviation, Space and Environmental Medicine*, February, 100-105 (1995).

Wilson, G. F., Gundel, A. "Topographical Changes in the Ongoing EEG Related to the Difficulty of Mental Tasks," *Brain Topography*, 5:17-25 (1992).

EXTENDED MINIMAL RESOURCE ALLOCATING NEURAL NETWORKS FOR AIRCRAFT SFDIA

GIAMPIERO CAMPA
Department of Mechanical and
Aerospace Engineering,
West Virginia University,
Morgantown, WV 26506

MARCELLO NAPOLITANO
Department of Mechanical and
Aerospace Engineering
West Virginia University,
Morgantown, WV 26506

MARIO LUCA FRAVOLINI
(+) Department of Electronic and
Information Engineering,
Perugia University, 06100 Perugia,
Italy

ABSTRACT
This paper presents an Adaptive Neural Network (ANN) based tool for the modeling, simulation and analysis of aircraft Sensor Failure, Detection, Identification and Accommodation (SFDIA) problem. The tool is based on a SFDIA scheme in which learning NNs are used as on-line non-linear approximators of the analytically redundant portion of the system dynamics. This can provide validation capability to measurement devices, allowing sensors failures to be detected, identified and accommodated. In the context of online learning the issues of critical importance are learning speed, number of parameters to be updated, and stability of the learning algorithm. To address these problems, a library comprising different online learning Adaptive Neural Network is presented, and an Extended Minimal Resource Allocating Network (EMRAN) featuring a fully tuned Radial Basis Functions (RBF) is eventually selected between all the candidate architectures. The study has been performed on a detailed nonlinear 6DOF aircraft model of an aircraft.

INTRODUCTION

The traditional approach to provide fault tolerance following sensor failure is physical redundancy. However, there are special purpose aircraft (e.g. Unmanned Aerial Vehicles) and spacecraft where reduced complexity, lower costs, and weight optimization are major design specifications. For these classes of aircraft an alternative approach can take advantage of the analytical redundancy (Patton et al, 1989) (i.e. the functional relationship existing between the system's outputs, states and inputs) existing in the system.

Research on fault tolerance based on analytical redundancy has produced a quite mature framework especially for linear systems (Baruh and Choe, 1987, Kerr, 1982); currently, the research challenge is in the extension of the previous schemes to the case of nonlinear systems. In this context, a very promising approach is to employ Neural Networks as the main nonlinear approximators of an SFDIA scheme (Ha et al, 1992, Napolitano et al, 2000). Although the benefits of employing NNs for fault tolerance purposes within a non-linear flight control system are clear, the synthesis of these schemes requires dedicated simulation software in which it is possible to simulate the interaction between the closed loop dynamics of the Aircraft and the dynamics of the SDFIA system.

In this work, a general SFDIA software has been designed in the Simulink environment. The tool allows evaluating either the open loop or the closed loop performance of the SFDIA scheme that employs different kinds of NN approximators and learning algorithms. A library comprising several different adaptive (i.e. online learning) NNs is presented. Finally, the results of a comparative study of different NN approximators applied to the SFDIA problem on a detailed nonlinear aircraft model.

ANALYTICAL REDUNDANCY BASED SFDIA

The used scheme is summarized in figure 1 (Napolitano et al, 2000).

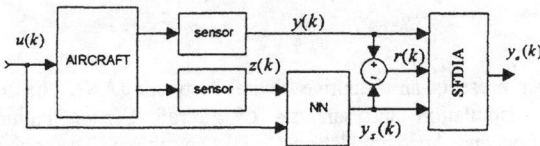

Fig. 1 – General SFDIA scheme

Analytical redundancy implies that some of the system variables are functionally related (Patton et al, 1989); namely, a variable *y(k)* can be expressed as function of a suitable set of other variables *z(k)* and inputs commands *u(k)*. The residual signal *r(k)* is the difference between the sensor output *y(k)* and its estimation $y_s(k)$ provided by a proper estimator (in this work the estimator is a NN). When the square of this (filtered) residual exceeds a predefined threshold, the state of the corresponding sensor is declared suspect and a suitable procedure is called to decide on the health status of this sensor. If the state of the sensor is then declared faulty, a procedure is enabled, and an accommodated variable $y_a(k)$ is provided as output. In this work the accommodation procedure simply substitutes the faulty measure with the estimation given by the NN ($y_a(k)=y_s(k)$). Several options can be added to this basic scheme to increase robustness in presence of noisy measurements and/or intermittent sensor failures (Napolitano et al, 2000).

THE SIMULATION ENVIRONMENT

The Neural Network based SFDIA modeling and simulation toolbox was built under the Simulink® environment (by The Mathworks Inc). In particular the freely available Flight Dynamics and Control (FDC) toolbox for Matlab (Rauw, 1993) provides powerful tools for flight simulation, flight dynamic analysis, and flight control system design.

In figure 2 the graphical interface of the proposed aircraft SFDIA tool is shown. The main blocks of the scheme are the following:

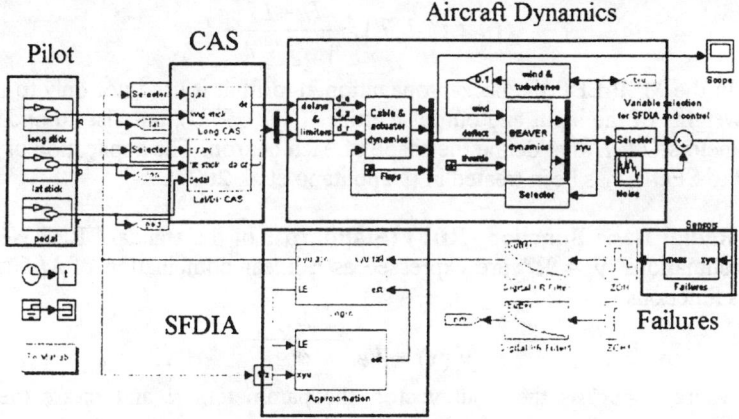

Fig. 2 - Modeling and simulation environment

- **Aircraft Dynamics**: The tool has been built around a generic non-linear aircraft model, which is configured for the DeHavilland DHC-2 Beaver aircraft, but can be adapted for many different airplanes.
- **Control Augmentation System** (CAS): the block is a Stochastic Optimal Feedforward and Feedback controller as in (Halyo et al, 1992).
- **Pilot commands**: the block emulates typical pilot commands.
- **Hard and soft sensor failures**: the block injects (adds) arbitrary soft and hard sensor failures to the desired measured signals
- **SFDIA Group**: It is the core of the tool and performs the main SFDIA procedures. It is constituted by two main sub-blocks:
  - **Approximators**: The block contains the Neural Network based function estimators. Different kinds of NN estimator blocks can be selected (some of the available NNs and learning algorithms will be described in the next section).
  - **SFDIA LOGIC**: The block performs the main threshold based sensor failure detection identification and accommodation operations. For more details about the SFDIA scheme see (Fravolini et al, 2001).

AN ADAPTIVE NEURAL NETWORKS LIBRARY FOR ON-LINE LEARNING

In order to select a suitable optimal SFDIA scheme it is necessary to test and to compare the performance achieved by different kinds of online Neural Network estimators, to this purpose, a Simulink Library containing several NN architectures has been built. The NNs in the library exploit linear, sigmoidal and radial basis as activation functions and employ different learning algorithms:

Multi-Layer Perceptron with Extended Back-Propagation. (MLP-EBP): It is a three layer NN, based on sigmoidal neurons with activation function:

$$f(net, U, L, T) = \frac{U-L}{1+e^{-net/T}} + L$$

In the MLP-EBP the back-propagation algorithm it used not only to update the weights of the input and output matrices ($W(k)$, $V(k)$), but also to update the parameters U, L, T, that define the shape of each neuron. The application of MLP-EBP to SFDIA has been treated in (Napolitano et al, 2000).

Radial Basis Function (RBF) (Standard): In the standard RBF Network the estimations $y_s \in \Re^m$ are expressed as a linear combination of M Gaussian Basis functions:

$$y_s(x) = W e^{-\frac{(x-\mu)^T(x-\mu)}{2\sigma^2}}$$

where $x \in \Re^n$ is the input vector, the parameters μ_j and σ_j are the basis center and width respectively. In the standard implementation the hidden layer neurons are *a priori* statically allocated on a uniform grid that covers the whole input space and only the weight w_{ij} are updated. This approach requires an *exponentially increasing* number of basis functions versus the dimension of the input space.

Fully Tuned Extended Minimal Resource Allocation Network RBF (EMRAN-RBF): In order to avoid the dimensionality problems generated by standard RBF, (Platt 1991) proposed a sequential learning technique for RBFNs. The resulting architecture was called the Resource Allocating Network (RAN) and has proven to be suitable for online modeling of non-stationary processes with only an incremental growth in model complexity. The RAN learning algorithm proceeds as follows: At each sampling instant, if the following 3 criteria are met some units are added:

Current estimation error criteria, error must be bigger than a threshold:
$$e(k) = y(k) - \hat{y}(k) \geq E_1$$

Novelty criteria, the nearest center distance must be bigger than a threshold:
$$\inf_{j=1}^{M} \|x(k) - \mu_J(k)\| \geq E_2$$

Windowed mean error criteria, windowed mead error must be bigger than a threshold:
$$\frac{1}{T} \sum_{i=0}^{T} [y(k-T+i) - \hat{y}(k-T+i)] \geq E_3$$

If one (or more) of the above criteria are not met, the existing network parameters (the centers μ_j, the weights w_{ij} and the variances σ_j) are adjusted using a suitable online learning algorithm. To avoid an excessive increase of the Network size a *pruning strategy* can also be applied. When this happens the network is called Minimal RAN (MRAN) (Lu et al, 2000). The adaptation algorithm is called Extended MRAN (EMRAN) when the parameters are updated following a "winner takes it all" strategy, i.e. only the parameters of the most activated neuron are updated, while all the other are unchanged. This strategy implies a significant reduction of the number of parameters to be updated online with just a small performance degradation with respect to the MRAN (Li et al, 2000).

SIMULATION EXAMPLE

In this paragraph the comparative results of SFDIA capabilities provided by different NN estimators are shown. The SFDIA procedure is applied to the estimation of the gyro rate $q(k)$. The failure of the sensor occurring at exactly $T_f=1098$ s. Two different NN architectures were tested: EMRAN and MLP-EBP (of comparable complexity). To evaluate the goodness of the **accommodation system**, the parameters $MAEE(k)$ and $VAEE(k)$ that represent respectively the mean and variance of the absolute estimation error, before and after the fault are considered. To evaluate the goodness of the **detection/identification system**, the detection ratio S2/S1 between the main peak of the filtered residual signal during the failure transient (1096 < t < 1120) and the peek of the filtered residual before failure (0 < t < 1090) is considered. This ratio quantifies the detectability provided by the scheme. Furthermore, the time percentage in which LE=0 before the true-fault detection, is reported to quantify the false detection (false alarm) rate, and the time percentage in which AE=0 before the true-fault declaration, is reported to quantify the false accommodation rate. All these results are reported in table I.

As expected the sixth failure is the most difficult do detect by any NNs. The best detectability performance, despite the simplicity of the learning algorithm, was given by the EMRAN, which exhibit the best approximation capability. On the contrary, the MLP-EBP after the accommodation causes the instability of the closed loop system, testifying, in this case, the non accurate online mapping achieved by this architecture.

CONCLUSIONS

In this paper an aircraft SFDIA analysis tool was discussed. The tool allows the investigation of the interactions between the closed loop aircraft dynamics and the dynamics of the SFDIA system. A clear understanding of this interaction is of great importance since an incorrect choice of the internal SFDIA approximation architecture could result in the instability of the feedback control system. This aspect has been clearly pointed out by means of a simulation example in which instability occurs in the accommodation phase even if the sensor failure is correctly and promptly detected and isolated. In this respect, the main feature of the tool is the possibility of testing and comparing in closed loop the capabilities of SFDIA schemes exploiting different kinds of Neural Networks as nonlinear approximators. This capability is a consequence of the extensive modularity of the whole simulation tool, and allows an easy change of aircraft dynamics and feedback control law as well as NN estimators and SFDIA scheme.

Subscript "1": before failure; subscript "2": after

AmplitudRise	Tdetec	MAEE$_1$	VAEE$_1$	MAEE$_2$	VAEE$_2$	S1/S	False	False	Simtim
EMRAN									
2.5	0.05	998	0.363857	0.157222	0.362104	0.175498	6.441479	0.275745	0	405.41
1	0.05	1002.35	0.364221	0.15791	0.359562	0.173116	3.284092	0.391507	0	404.63
2.5	0.3	998.1	0.363857	0.157222	0.362086	0.175737	6.425145	0.273231	0	407.82
1	0.3	1002.45	0.364221	0.15791	0.359376	0.173011	3.270282	0.391507	0	405.96
2.5	4	1000.1	0.364804	0.158085	0.360973	0.175841	6.482528	0.283286	0	407.88
1	4	1003.7	0.362527	0.157196	0.357567	0.172815	2.99019	0.323527	0	408.65
MLP-EBPA									
2.5	0.05	996.25	0.397254	0.123787	0.420127	0.192858	3.286115	7.031626	1.571744	452.53
1	0.05	997.85	0.398244	0.123971	0.422461	0.191947	1.784715	7.146301	1.613026	450.77
2.5	0.3	996.4	0.397254	0.123787	0.420459	0.192938	3.296431	7.03449	1.571744	450.5
1	0.3	997.95	0.398244	0.123971	0.42241	0.192146	1.778442	7.146301	1.613026	450.99
2.5	4	998.05	0.398244	0.123971	0.420935	0.192985	3.244961	7.066003	1.571744	453.91
1	4	999.65	0.397947	0.124038	0.421829	0.192255	1.81221	7.146301	1.613026	452.86

Table 1: SFDIA Performance Parameters

REFERENCES

Patton R. J, Frank P.M, Clark R. N., 1989, *Fault diagnosis in dynamic systems, theory and applications* (Englewood Cliff, NJ, Prentice-Hall).

Baruh, H, Choe, K., 1987, "Sensor-Failure Detection Method for Flexible Structures", *AIAA Journal of Guidance, Control, and Dynamics*, Vol. 10, no 5, 474-482.

Kerr, T.H., 1982, "False Alarm and Correct Detection Probabilities over a Time Interval for Restricted Classes of Failure Detection Algorithms", *IEEE Transactions of Information Theory*, IT-28, No. 4, pp. 619-631.

Ha, C.M., Wei, Y.P., Bessolo, 1992, J.A., "Reconfigurable Aircraft Flight Control System Via Neural Networks", Proceedings of the 1992 Aerospace Design Conference, AIAA Paper 92-1075, Irvine, Ca.

Napolitano, M. R., An Y, Seanor, B., 2000, "A fault tolerant flight control system for sensor and actuator failure using neural networks", *Aircraft Design*, vol. 3, pp. 103-128.

Rauw, M.O., 1993, "A Simulink Environment for Flight Dynamics and Control analysis - Application to the DHC-2 "Beaver"" (MSc-thesis, Delft University of Technology, Faculty of Aerospace Engineering, Delft, The Netherlands, 1993).

Halyo N, Direskeneli, 1992, B. Taylor, "A Stochastic Optimal Feedforward & Feedback Control Methodology for Superagility" (NASA CR 4471, November 1992).

Fravolini, M. L., Campa, G., Napolitano, M.R., 2001, "Minimal Resource Allocating Networks for Aircraft SFDIA", IEEE Int. Conference on Advanced Intelligent Mechatronics 2001, Como, Italy, July 2001.

Platt, J. C., 1991, "A Resource Allocation Network for Function Interpolation", *Neural Computation* 3(2), pp. 213–225.

Lu, Y, Sundararajan, N, Saratchandran, P., 2000, "Analysis of Minimal Radial Basis Function Network Algorithm for Real-time identification of nonlinear dynamic systems", *IEEE. Proc. Contr. Theory and application*, Vol. 4, no. 147, pp. 476.

Li. Y., N. Sundararajan, N. Saratchandran, P, 2000, "Dynamically Structured Radial basis Function Neural Networks for robust aircraft flight control", Proc. American Control Conference, pp. 3501-3505, Chicago.

A Generalized Unstructured Artificial Neural Network Architecture: A First Study

R. WOODLEY and L. ACAR
*Department of Electrical and Computer Engineering and
Intelligent Systems Center
University of Missouri-Rolla*

ABSTRACT:
In this document, we will present an unstructured neural network based on the mathematics of holographic storage. We analyzed the mathematics to produce a general mathematical description of the holographic process. From this analysis we are able to show how the holographic process can be used as an associative memory network. Additionally, the process may also be used as a regular feed-forward network. The most striking aspect of these networks is that using the holographic process, they are trained in a single iteration, and may be updated in real time.

Introduction

The aspect of holograms that was most intriguing was that it is possible to recover three-dimensional information from a two-dimensional piece of film. Light reflecting from an object propagates to a piece of photographic film were it is interfered with by a coherent beam of light (i.e. a laser). There is no lens used to focus the light, so the image is unrecognizable. However, after the film is developed, it can be re-illuminated by the same coherent beam, and the image recovered is three-dimensional again (Jenkins and White, 1976).

Another aspect of holograms is that multiple images may be stored on a single piece of film. By changing the angle of the coherent beam for each image stored, multiple images can be captured to the film. Then by selecting the correct angle for the reference beam, the desired image may be recovered (DeVelis and Reynolds, 1967). Here, we turn our attention toward artificial neural networks, and how we may use the knowledge gained from holograms to construct the neural network.

Hologram Formation

A hologram is formed when monochromatic, coherent light is reflected off an object, then interfered with by another monochromatic, coherent reference beam (Jenkins and White, 1976). Since the beams are monochromatic, they can be represented in rotating phasor form

$$u(p,t) = \Re\{A(p)e^{j\phi(p)}e^{j2\pi vt}\} \quad (1)$$

where p is a position. This monochromatic wave must satisfy the Maxwell equation (DeVelis and Reynolds, 1967)

$$\nabla^2 u - \frac{1}{c^2}\frac{\partial^2 u}{\partial t^2} = 0 \quad (2)$$

Substituting equation 1 into equation 2, the result is known as the Helmholtz equation (Goodman, 1968)

$$\nabla^2 u + k^2 = 0 \quad (3)$$

where $k = 2\pi\frac{v}{c} = \frac{2\pi}{\lambda}$ is the wave number.

Equation 3 is also known as the reduced wave equation, and has a known solution using Green's functions (Roach, 1970). Let G be defined such that

$$LG = \delta(P_\alpha - P_\beta) \tag{4}$$

For a system with operator L

$$Lu = h \tag{5}$$

for all P in a volume V. Multiplying equation 5 by G and equation 4 by $u(P_\beta)$, then integrating and subtracting the equations produce

$$\int_V (u(P_\beta)LG(P_\alpha, P_\beta) - G(P_\alpha, P_\beta)Lu(P_\alpha))dV = u(P_\beta) - \int_V G(P_\alpha, P_\beta)h(P_\alpha)dV \tag{6}$$

where the α-plane is the object plane and the β-plane is the recording plane.

Equation 6 is a general result, the specifics of the hologram problem reduces the complexity. The operator, L, has the form $Lu = (\nabla^2 + k^2)u$ (DeVelis and Reynolds, 1967). Therefore, equation 6 may be simplified and rewritten as

$$\int_V (u(P_\beta)(\nabla^2)G(P_\alpha, P_\beta) - G(P_\alpha, P_\beta)(\nabla^2)u(P_\alpha))dV = u(P_\beta) \tag{7}$$

However, it is known that the left side of equation 7 can be evaluated as

$$\int_V (u(P_\beta)(\nabla^2)G(P_\alpha, P_\beta) - G(P_\alpha, P_\beta)(\nabla^2)u(P_\alpha))dV = \int_{\partial V} (u\frac{\partial G}{\partial n} - G\frac{\partial u}{\partial n})dS \tag{8}$$

where n is the normal to the surface, ∂V, of the volume V. With proper choice of the Green's function it is possible to reduce the right hand side of equation 8 Hence, equation 8 reduces to

$$\int_V (u(P_\beta)(\nabla^2)G(P_\alpha, P_\beta) - G(P_\alpha, P_\beta)(\nabla^2)u(P_\alpha))dV = \int_{\partial V} (u\frac{\partial G}{\partial n})dS \tag{9}$$

A reference beam may now be added to the wave. The reference beam will interfere with the propagating wave and be recorded. The benefit of adding the reference beam in this fashion (also known as a side-band Fresnel hologram) is that the reference beam may be applied at various angles. The angles will then cause the reconstructed image to shift off-axis by the same angle of the reference beam (DeVelis and Reynolds, 1967). So by using multiple reference beams at specifically selected angles, multiple images may be stored on the same hologram. The completed hologram is then the magnitude squared of the sum of the propagated wave from the object and the reference beam

$$I(P_\beta) = |R(P_\beta) + u(P_\beta)|^2 \tag{10}$$

The propagated wave equation is actually an integral transform equation (Arfken, 1970). The kernal of the equation is

$$K(P_\alpha, P_\beta) = \frac{\partial G(P_\alpha, P_\beta)}{\partial n} \tag{11}$$

The formation process may now be written in general mathematical form. Let $u_\alpha(P_\alpha)$ be the signal from the object. This signal propagates to a new location P_β by

$$u_\beta(P_\beta) = \int_{S_\alpha} u_\alpha(P_\alpha)K(P_\alpha, P_\beta)dP_\alpha. \tag{12}$$

This solution is for only some operators, L (Debnath and Mikusinski, 1990). The mathematical description of L such that equation 12 is a solution is currently being investigated.

An important result occurs when $\delta(P_\alpha - P_R)$ is the boundary condition, where δ is the Dirac delta function and P_R is a reference point on the α-plane. The resulting signal at a point P_γ is

$$\begin{aligned} u_\gamma(P_\gamma) &= \int_{S_\alpha} \delta(P_\alpha - P_R) K(P_\alpha, P_\gamma) dP_\alpha \\ &= K(P_R, P_\gamma). \end{aligned} \qquad (13)$$

Similarly, at a point P_β the result would be $u_\beta(P_\beta) = K(P_R, P_\beta)$. However, if we let the β-plane be the boundary, and look to see what the signal is at the γ-plane then

$$\begin{aligned} u_\gamma(P_\gamma) &= \int_{S_\beta} \int_{S_\alpha} \delta(P_\alpha - P_R) K(P_\alpha, P_\beta) dP_\alpha K(P_\alpha, P_\gamma) dP_\beta \\ &= \int_{S_\alpha} \delta(P_\alpha - P_R) \int_{S_\beta} K(P_\alpha, P_\beta) K(P_\alpha, P_\gamma) dP_\beta dP_\alpha. \end{aligned} \qquad (14)$$

Equation 13 must equal equation 14, therefore

$$K(P_\alpha, P_\gamma) = \int_{S_\beta} K(P_\alpha, P_\beta) K(P_\beta, P_\gamma) dP_\beta. \qquad (15)$$

We also see from this that a coherent source may be represented by $R(P_\beta) = R_o K(P_R, P_\beta)$, where R_o is a constant.

The general mathematical description of what is recorded may now be written. Let $\psi(P_\beta) = u_\beta(P_\beta) + R(P_\beta)$. The recorded information is actually the inner product of ψ and itself. The inner product produces a scalar, which is why it can be recorded on the holographic film. Let $I(P_\beta)$ represent the intensity stored at a location P_β, then

$$I(P_\beta) = <\psi(P_\beta), \psi(P_\beta)>. \qquad (16)$$

The inner product function is represented by $<\cdot,\cdot>$. For the case of the hologram, the inner product used is (DeVelis and Reynolds, 1967)

$$I(P_\beta) = \psi(P_\beta)^* \psi(P_\beta), \qquad (17)$$

where the $*$ is the complex conjugate transpose.

Hologram Reconstruction

If, as in the case of holograms, the recording media is only able to store the magnitude, then the reference signal is used. Equation 10 represents the stored data. We will use the same propagation equations for the reconstruction as we did for the formation.

The reconstruction begins by multiplying $I(P_\beta)$ and $R(P_\beta)$. The new signal is then propagated to the γ-plane (DeVelis and Reynolds, 1967)

$$u_\gamma(P_\gamma) = \int_{S_\beta} K(P_\beta, P_\gamma) I(P_\beta) R(P_\beta) dP_\beta. \qquad (18)$$

Placing equation 18 in terms of inner products, then

$$\begin{aligned} u_\gamma(P_\gamma) &= \int_{S_\beta} K(P_\beta, P_\gamma) <u_\beta(P_\beta), u_\beta(P_\beta)> R(P_\beta) dP_\beta \\ &+ \int_{S_\beta} K(P_\beta, P_\gamma) <R(P_\beta), R(P_\beta)> R(P_\beta) dP_\beta \\ &+ \int_{S_\beta} K(P_\beta, P_\gamma) <u_\beta(P_\beta), R(P_\beta)> R(P_\beta) dP_\beta \\ &+ \int_{S_\beta} K(P_\beta, P_\gamma) <u_\beta(P_\beta), R(P_\beta)>^* R(P_\beta) dP_\beta. \end{aligned} \qquad (19)$$

The first two terms of equation 19 are the squared norms of their respective signals. The third term will be the conjugate image, while the last term will be the recovered image after filtering. (DeVelis and Reynolds, 1967) At this point, the specific inner product used for the hologram case will be used. We are currently working on expanding the results to the general inner product case.

Equation 19 may be rewritten using the inner product for holograms as

$$\begin{aligned} u_\gamma(P_\gamma) &= \int_{S_\beta} K(P_\beta, P_\gamma) \left(\|u_\beta(P_\beta)\|^2 + \|R(P_\beta)\|^2 \right) R(P_\beta) dP_\beta \\ &+ \int_{S_\beta} K(P_\beta, P_\gamma) u_\beta(P_\beta)^* R(P_\beta) R(P_\beta) dP_\beta \\ &+ \int_{S_\beta} K(P_\beta, P_\gamma) u_\beta(P_\beta) R(P_\beta)^* R(P_\beta) dP_\beta. \end{aligned} \qquad (20)$$

In the first term of equation 20 we see that the two norm terms will only change the magnitude of the resulting signal. Therefore, the first term in equation 20 represents just a scaled version of the propagated reference beam at the same angle of the reference beam. The second term of equation 20 has the reference squared. The resulting image will then be at twice the angle of the reference beam. The last term of equation 20 has $R(P_\beta)^* R(P_\beta)$ which is simply the magnitude squared of the reference. The last term, therefore, becomes

$$R_o^2 \int_{S_\beta} K(P_\beta, P_\gamma) u_\beta(P_\beta) dP_\beta. \qquad (21)$$

By selecting the angle of the reference, $R(P_\beta)$, properly, we may filter out the first two terms of equation 20. Figure 1 shows how the process functions (DeVelis and Reynolds, 1967). By placing a spatial filter centered on-axis that is the same size as the original image, the third term of equation 20 is extracted

$$u_{filtered}(P_\gamma) = R_o^2 \int_{S_\beta} K(P_\beta, P_\gamma) u_\beta(P_\beta) dP_\beta. \qquad (22)$$

Equation 12 may be combined with equation 22 to produce the final result. The filtered output is then

$$u_{filtered}(P_\gamma) = R_o^2 \int_{S_\beta} K(P_\beta, P_\gamma) \int_{S_\alpha} K(P_\alpha, P_\beta) u_\alpha(P_\alpha) dP_\alpha dP_\beta. \qquad (23)$$

Rearranging equation 23 produces

$$u_{filtered}(P_\gamma) = R_o^2 \int_{S_\alpha} \left[\int_{S_\beta} K(P_\beta, P_\gamma) K(P_\alpha, P_\beta) dP_\beta \right] u_\alpha(P_\alpha) dP_\alpha. \qquad (24)$$

The bracketed term in equation 24 has already been shown to be $K(P_\alpha, P_\gamma)$ by equation 15. Therefore, it is possible to recover $u_\alpha(P_\alpha)$ by inverting the equation for $u_{filtered}(P_\gamma)$.

It is then a simple matter to show how multiple images may be stored and recovered from the same storage media. The inner products of the multiple sources may be added

$$I(P_\beta) = <u_1 + R_1, u_1 + R_1> + <u_2 + R_2, u_2 + R_2>, \qquad (25)$$

where u_1 and u_2 are the propagated signals and R_1 and R_2 are the references with different angles. If the signal from u_2 is desired, $I(P_\beta)$ is multiplied by the reference R_2. The result is

$$\begin{aligned} u_\gamma(P_\gamma) &= \int_{S_\beta} K(P_\beta, P_\gamma) <u_1 + R_1, u_1 + R_1> R_2 dP_\beta \\ &+ \int_{S_\beta} K(P_\beta, P_\gamma) <u_2 + R_2, u_2 + R_2> R_2 dP_\beta. \end{aligned} \qquad (26)$$

The second term of equation 26 will produce the same case as in equation 20. The first term, however, becomes

$$\begin{aligned} \int_{S_\beta} K(P_\beta, P_\gamma) <u_1 + R_1, u_1 + R_1> R_2 dP_\beta &= \int_{S_\beta} K(P_\beta, P_\gamma) \left(\|u_1\|^2 + \|R_1\|^2 \right) R_2 dP_\beta \\ &+ \int_{S_\beta} K(P_\beta, P_\gamma) (<u_1, R_1> + <u_1 + R_1>^*) R_2 dP_\beta. \end{aligned}$$
$$(27)$$

The only term in equation 27 that will not shift the image by at least the angle of R_2 is $\int_{S_\beta} K(P_\beta, P_\gamma) < u_1 + R_1 >^* R_2 dP_\beta$. The angles of R_1 and R_2 will subtract in this case. But by picking the angles as multiples of a minimum angle, ϕ, then $|\angle R_2 - \angle R_1| > \phi$. Therefore, the resulting image will be off-axis and will be filtered out by the on-axis filter. The only image passing through the filter will then be the desired image from u_2.

Application of the Hologram Process to Neural Networks

We will now show the hologram process can be used to create unstructured neural networks. We will examine two types of networks, the associative memory, and the feed-forward type of network used in pattern classification. We will focus on heteroassociative networks because of the types of signals that we have in the hologram process. The feed-forward network will be a single hidden layer network.

Associative Memory

A description of associative memory may be found in (Zurada, 1995). Application of the hologram process to heteroassociative memories is straightforward. The signals we have are the boundary condition u_α and the reference R. u_α is the vector we wish to recover from the memory. R is the input vector that recovers the stored memory. The memory is, therefore, the stored intensity, I_β. We recover the data by simply multiplying the correct reference with I then propagating it to the output layer and filtering.

The hologram process has distributed memory such as that of content-addressable memory. The intensity function is described for a particular location, P_β. By allowing the β-plane to have some dimension, we produce a distributed memory. Each point in the β-plane receives data from every point in the α-plane. This phenomenon is evident in holograms (Jenkins and White, 1976). The photographic film can be physically broken with pieces missing, but when illuminated by the reference laser, the three-dimensional image is still seen. There will be some degradation in the image, but there will be no missing parts of the image.

An aspect of the hologram process that is usually not seen in neural networks is one-pass training with on-line updates. Each pair of u and R combine to form a memory. Each subsequent pair is then added to the previous set with no further training required. If a new pair of training vectors becomes available, all that needs to be performed is to add their propagated sum into the already stored data.

Pattern Classification

Pattern classification may also be accomplished with the hologram process. The difference between associative memory and pattern classification is that pattern classification attempts to classify an input vector into some known group based upon available training data (Haykin, 1999). In order to fit the hologram process to pattern classification some restrictions are required.

Figure 2 shows how vectors from the u-domain are classified into the y-domain. The function F is then the mapping from the u-domain to the y-domain. A typical neural network to approximate F is shown in figure 3 (Haykin, 1999). The function at each node, $K(u, u_i)$, represents any of a number of types of activation functions (i.e. sigmoid, radial basis function, etc.).

The hologram process is actually the inverse of pattern classification. For the hologram process, the reference vectors will produce the stored vectors. But for pattern classification, the input vector has to produce the given reference. To accomplish this, we must reverse the flow of data from that seen in the associative memory case. But in order to do this, the function F must be invertible (Arfken, 1970). Also the hologram process for a particular kernal must be invertible.

If the above requirements are met, then the hologram process provides a means for one-pass learning and on-line training. Similar to the process described previously, the u and R training pairs are used to directly calculate the stored information I.

Conclusion

We presented an unstructured neural network based upon the mathematical description of holographic storage. The network is unstructured in that the hologram process can produce the same results as a number of different neural network structures within a single context. The only change needed for the hologram process is the specific kernal.

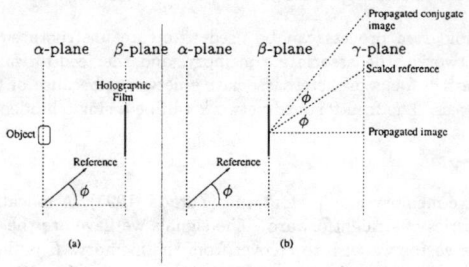

Figure 1: Schematic of the hologram process. (a) Formation process. (b) Reconstruction process.

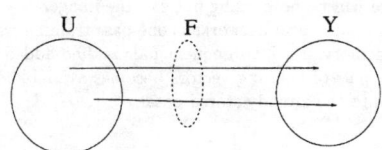

Figure 2: A mapping of the vectors in the u-domain into the y-domain by the mapping function, F.

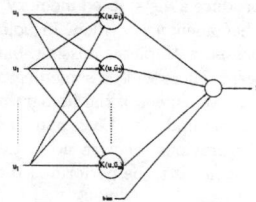

Figure 3: A Neural Network diagram of pattern classification network, mapping a u vector into the pattern y.

We presented a detailed description of the mathematics of the hologram. Then we concluded by showing how the hologram process may be used as two different types of neural networks. Specifically, associative memory networks and pattern classification networks. The most important feature of the process is the speed of learning. The network is trained in a single pass, and can be updated, on-line, in a single pass.

References

Arfken, G. (1970). *Mathematical methods for physicists, 2nd edition*. Academic Press, Inc.

Debnath, L. and Mikusinski, P. (1990). *Introduction to hilbert spaces with applications*. Academic Press, Inc.

DeVelis, J. B. and Reynolds, G. O. (1967). *Theory and applications of holography*. McGraw-HillAddison-Wesley Publishing Company.

Goodman, J. W. (1968). *Introduction to fourier optics*. McGraw-Hill.

Haykin, S. (1999). *Neural networks a comprhensive foundation 2nd edition*. Prentice-Hall, Inc.

Jenkins, F. A. and White, H. E. (1976). *Fundamentals of optics, 4th edition*. McGraw-Hill, Inc.

Roach, G. F. (1970). *Green's functions, 2nd edition*. Cambridge University Press.

Zurada, J. M. (1995). *Introduction to artificial neural systems*. PWS Publishing Company.

EVALUATION OF FEATURE VECTORS FOR ANNS USED IN AUTOMATIC MODULATION RECOGNITION

SUDIP BISWAS, DAVE CALVERT, ORLANDO CICCHELLO,
AND STEFAN C. KREMER
Guelph Natural Computation Group
Computing and Info. Science
University of Guelph
Guelph, ON N1G 4E1
CANADA
{sbiswas,dave,ocicchel,skremer}@uoguelph.ca

ABSTRACT

This paper examines a technique to evaluate feature vectors for artificial neural networks applied to the problem of Automatic Modulation Recognition (AMR)—the identification of the modulation scheme of an intercepted radio signal. Specifically, we present a method to compare the sensitivities to modulation type (which should be high) to the sensitivities to other factors such as noise, channel characteristics, etc. (which should be low). Using this, we perform a detailed study of the established feature extraction techniques described in the literature. Our results show both a general methodology for evaluating feature vectors, and an analysis of the effectiveness of the most popular modulation recognition features on actual (not simulated) radio signals.

INTRODUCTION

Automatic Modulation Recognition (AMR) is the identification of the modulation scheme of an intercepted radio signal. It has applications in electronic warfare and spectrum management, and is often tackled using artificial neural networks (ANNs) (see Lamontagne, 1993; Kremer and Shiels, 1997). In AMR, radio systems are used to capture and sample transmitted signals. Such a data signal may encode an analog acoustic (voice, music, etc.) signal or digital encoding of sound or data. Aftter downconversion, a typical captured signal may be sampled at 80kHz and be captured over a an interval of 0.4 seconds. Such a signal would consist of approximately 32K sample points.

These captured signals (even after down-conversion) consist of too many sampled data points for direct presentation to an ANN. Instead, they must be converted into compressed representations (feature vectors), that quantitatively differ across modulation schemes, while remaining insensitive to other parameters, including signal to noise ratio (SNR), signal length, sampling frequency, etc.. An effective signal pre-processor possesses the following two characteristics: (1) it produces output values that provide useful information about the signal type presented (this is called sensitivity to signal class) and (2) it produces output values that are not strongly affected by aspects of the signal that are not relevant to the desired classification (insensitivity to noise).

INPUT DATA

Many statistical parameters have been proposed, and tested on signals generated by simulation software with fixed SNRs, signal lengths, sampling frequencies, etc.. Lamontagne (1993) has written a review of these measures and cataloged 21 different AMR parameters. He collected a set of features that included the standard deviation (s[x]), skewness ($\gamma_1[x]$), and kurtosis ($\gamma_2[x]$), of each of amplitude (A), amplitude squared (A2), amplitude times the derivative of the amplitude (A•A'), frequency (f), and frequency times amplitude squared (f•A2). Additionally, the feature vector contains 6 other statistical parameters: R, VAR, Energy1, Energy2, P(0) and P0 the details of which can be found in Lamontagne (1993). The entire set of feature vectors identified by Lamontagne is illustrated in Table 1.

Table 1: Feature vectors assembled by Lamontagne

01	$S(A^2)$	Lamontagne, 1993	13	s (A•A')	Aisbett, 1987
02	$\gamma_1 (A^2)$	Lamontagne, 1993	14	γ_1 (A•A')	Aisbett, 1987
03	$\gamma_2 (A^2)$	Lamontagne, 1993	15	γ_2 (A•A')	Aisbett, 1987
04	R	Chan, 1985	16	s (f •A^2)	Aisbett, 1987
05	VAR	Chan, 1985	17	γ_1 (f •A^2)	Aisbett, 1987
06	Energy1	Lamontagne, 1993	18	γ_2 (f •A^2)	Aisbett, 1987
07	Energy2	Lamontagne, 1993	19	s (f)	
08	P(0)	Lamontagne, 1993	20	γ_1 (f)	
09	S (A)		21	γ_2 (f)	
10	γ_1 (A)				
11	γ_2 (A)				
12	P_0	Fabrizi, 1986			

For this experiment, a data set was provided by the Communications Research Centre. This data set consisted of 499 signals collected off-air. These signals were of 10 different types: AM, BPSK, CW, FM, FSK, M16QAM, MPSK, Noise, Pi4QPSK, and QPSK. All signals were collected at a sampling rate of 80kHz using DAS equipment. Each signal was divided into 4 segments to provide a baseline set of signals. In addition to this baseline set, a five more were created by: (1) sub-sampling by two, (2) dividing each original signal into 8 signals, (3) dividing by 16 and (5) by adding the collected real world noise to the signal data collected.

EVALUATION OF ESTABLISHED FEATURE VECTORS

In order to evaluate the feature vectors, a graphical representation was developed to illustrate the sensitivity to various signal parameters. This representation was designed to illustrate: (1) the mean value, (2) the standard deviation, (3) the minimum value and (4) the maximum value of each feature vector component Figure 1 shows an example of the types of graph used in this report. Labeled along the horizontal axis are the modulation types. The values along the vertical axis range from the lowest to the highest value obtained by a single feature from the feature vectors of all the modulation types.

Each graph illustrates the maximum, minimum, mean, and standard deviation obtained from the corresponding feature for each modulation. It is important to note that the mean value obtained from the corresponding feature may not always appear at the exact center point of the maximum, and minimum values plotted. In other words, the values obtained from each feature may not be normally distributed. As a result, 1 standard deviation from the mean may actually go beyond the maximum and minimum values shown. An example of this is included in the graph above. The effectiveness of a feature can be determined by identifying any overlapping in the distribution of the feature values, which appear to be normal, for different modulation types.

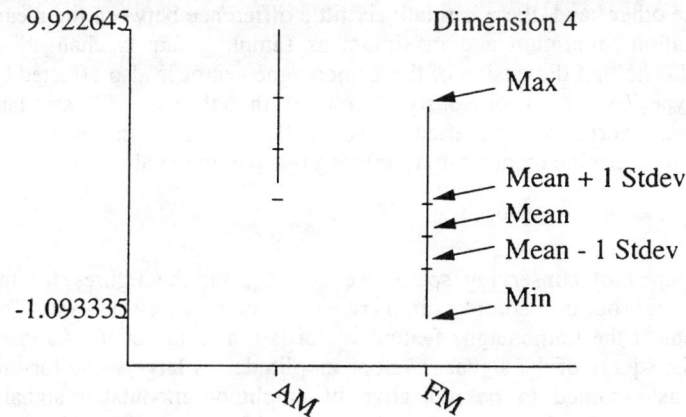

Figure 1: Sample graph.

ORGANIZATION OF GRAPHS

For simplicity, each component of the Lamontagne feature vector was evaluated separately. This means that correlations between feature vectors are not illustrated in the graphs. This is somewhat limiting in the context of evaluating the overall effectiveness of feature vectors for artificial neural networks since one of the features of these networks is their ability to extract higher order interactions between input parameters and using these to perform classifications. At this stage, however, we are concerned mainly with sensitivity of the signals to various parameters and not effectiveness for later training. In addition, it should be noted that the Lamontagne feature vectors represent a collection of parameters which were originally designed to independently render modulation scheme classifications. That is, they were not specifically designed to provide the kinds of higher order correlations we are considering. For this reason we organize our graphs by feature vector component.

For each dimension of the vector we plot 10 graphs, one for each modulation scheme (AM, BPSK, CW, FM, FSK, M16QAM, MPSK, Noise, Pi4QPSK, and QPSK). Each graph is designed to illustrate the sensitivity of the particular feature dimension to variations in noise level, sampling rate and signal length for a given modulation type. These variations are undesirable since we would like to disregard these signal parameters when rendering classification decisions. A given graph also shows a comparison between the feature vector's values for a given modulation type and the feature vector's values for other modulation types.

The discussion of a specific example will help elucidate the purpose of a graph. In Figure 1 we compare the sensitivity of the first dimension of the

Lamontagne feature to AM signals under different conditions with the sensitivity of the dimension to other signal types. The first 5 bars of the graph, from left to right, show standard AM signals, AM signals measured at ½ the sampling rate, AM signals of 1/4 the standard length, AM signals of 1/2 the standard length, and AM signals with added noise (lower SNR). This graph shows a strong sensitivity to signal length, since the mean value, standard deviation, minimum and maximum are all affected by changing the signal length. On the other hand, there is relatively little difference between the mean, standard deviation, minimum and maximum as sampling rate is changed or noise is added. The first dimension of the Lamontagne vector is also affected by modulation type (which is obviously desirable) though the +/-1 standard deviation ranges overlap to a significant degree. This suggests that it may be very difficult to determine modulation type using this parameter alone.

ANALYSIS

In the interest of conserving space, we do no print the figures for the individual features, but instead only summarize the results of their analysis. The first component of the Lamontagne feature vector is a measure of the standard deviation of the square of the signal envelope amplitude. A large value for this component was assumed to be indicative of amplitude modulated signals, whereas signals with fixed envelope amplitudes should have smaller values. A greater effect lies in the standard deviations. It is also interesting to note that mean values and especially standard deviations of this parameter were strongly affected by signal length.

The second component of the Lamontagne feature vector is a measure of the skewness of the square of the signal envelope amplitude. For s[A^2] we note, once again, that mean values and especially standard deviations of this parameter were strongly affected by signal length but also by signal type (especially CW, FM), this may be due to the fact that these are analog signals, while other signals are digital. The third component of the Lamontagne feature vector is a measure of the kurtosis of the square of the signal envelope amplitude. There is very little change in mean values or standard deviations of this parameter as signal type is varied (though the extreme outliers seem to fall in different ranges), but, once again, significant variations with differing signal lengths.

The fourth component of the Lamontagne feature vector is a measure of the ratio of the first deviate to the mean value of the squared envelope amplitude. This was one of the few Lamontagne vectors that was insensitive to signal length, showing only a moderate difference in standard deviations as signals were shortened. This effect is also to be expected from a statistical (law of large numbers) perspective. Unfortunately, the parameter showed only mild variations across signal types, though amplitude modulation schemes tended to produce lower values than phase modulation types.

The fifth component of the Lamontagne feature set measures square envelope amplitude deviations over segments of signal. It was designed to measure low frequency changes in square envelope amplitude. It tends to produce higher values for amplitude and phase modulated signals and lower values for phase modulation schemes. Sensitivity to signal length is visible in larger standard deviation ranges in the graphs, though median values are not significantly affected. This suggests that this parameter becomes less predictable as signal length is decreased, but not systematically skewed.

The sixth and seventh Lamontagne features measure changes in the energy of the signal. The former examines the standard deviation of energy levels, while the latter measures differences from the minimum and maximum energies. These two features look very similar. They are two of the most promising features since they vary significantly across modulation types. They appear to be sensitive to signal length in the case of very short signals but there is little change between the standard length signals and the Seg-8 signals (which are 1/2 the standard length).

The eighth Lamontagne feature measures the proportion of times the squared envelope amplitude is very low (less than 0.05 times its average value). This is probably one of the least promising features since is it clearly sensitive to signal length, noise and sampling rate. The most significant variation in the features response to signal types is in the difference between Pi4QPSK and QPSK signals.

The ninth feature measures the standard deviation of the signal amplitude (not square) envelope. This is another feature with high sensitivity to signal length, sampling frequency, and noise level. Additionally, it shows little promise for disambiguating signal type. The tenth feature measures the skewness of the signal amplitude envelope. This is another feature with high sensitivity to signal length, sampling frequency and noise level, but unlike the ninth feature it does seem to distinguish between some of the signal types (particularly M16QAM, Noise and Pi4QPSK compared to other signals).

The eleventh feature is also based on signal envelope amplitude, but this feature measures its kurtosis. This statistical property seems to be much more practical since the graphs reveal a reduced sensitivity to noise, and signal length, while at the same time it is even more sensitive to signal type. The twelfth feature compares the maximum envelope amplitude to the mean envelope amplitude. This graph contains some extreme outliers for M16QAM and Noise signals. It also seems to be sensitive to signal length while showing little sensitivity in mean values for varying signal type.

The thirteenth feature measures the standard deviation of the product of the signal envelope amplitude and its derivative. This feature is still relatively sensitive to signal length, and does not show very much variation across modulation types, though some modulation schemes produce a larger range of values for feature 13 than others.

The fourteenth feature measures the skewness of the product of the signal envelope amplitude and its derivative. This feature is just as sensitive to signal length as the standard deviation. It shows some variation across modulation types. The kurtosis of the signal envelope amplitude multiplied by its derivative, like the kurtosis of the squared envelope is fairly insensitive to signal length, noise and sampling frequency. Unfortunately, there is also not very much sensitivity to modulation type.

The seventeenth, eighteenth and nineteenth feature vector dimensions focus on the envelope amplitude multiplied by the instantaneous frequency. The first of these features uses the standard deviation of this parameter. It shows moderate sensitivity to sampling rate and signal length, and good sensitivity to differences between some modulation schemes (e.g. FSK, M16QAM, and Noise). The skewness of the envelope amplitude multiplied by the instantaneous frequency is less effective than the standard deviation because this value seems to be relatively insensitive to all signal aspects (including modulation type). The kurtosis of the envelope amplitude multiplied by the instantaneous frequency is also less effective than the standard deviation because this value is sensitive to signal length, noise level, and sampling rate.

The instantaneous frequency based vector components (dimensions 19, 20 and 21) are fairly effective because they are insensitive to noise, signal length, and sampling frequency, while being more sensitive to signal type. The standard deviation of this parameter (feature 19) shows some variation with signal length, but most of this variation is in the range and standard deviation (which is to be expected for smaller sample sizes). The skewness of the instantaneous frequency provides an even more consistent value across signal length, sampling frequency, and noise level, but the variations across modulation types are also reduced. The kurtosis of the instantaneous frequency is the most promising of the frequency-based features.

CONCLUSIONS

The results have shown that many of the established feature vectors are indeed sensitive to signal length. In general, kurtosis features proved to be more effective than standard deviation and skewness-based features. The existing features used in the literature range from being very promising to looking very doubtful. Using real-world radio signals, we showed both a general methodology for evaluating feature vectors for any problem to which ANNs might be applied, and also an analysis of the effectiveness of the most popular modulation recognition features using data from actual (not) simulated radio signals. Current work is focusing on the development of new features designed using the feature analysis method described here. Such features would provide sensitivity primarily to the differences between signal types, while minimizing sensitivity to the other parameters. This will make the job of a subsequent ANN-based classifier much easier.

REFERENCES

Aisbett, J. (1987). "Automatic modulation recognition using time domain parameters". *Signal Processing*, **13**(3):323--328.

Chan, Y., Gadbois, L., and Yansouni, P. (1985). "Identification of the modulation type of a signal." *IEEE International Conference on Acoustics, Speech and Signal Processing*, volume 2, pages 838--841. IEEE, Accoustics Speech and Signal Processing Society.

Fabrizi et al., P. (1986). "Receiver recognition of analogue modulation types." In IERE Conference on Radio Receivers and Associated Systems, Bangor, Wales.

Kremer Stefan C. Kremer and Joanne Shiels. ``A testbed for automatic modulation recognition using neural networks". In P. Thorburn and J. Quaicoe, eds., *Engineering Innovation: Voyage of Discovery, Proceedings of the IEEE Canadian Conference on Electrical and Computer Engineering*, pp. 67-70. 1997.

Lamontagne, R. (1993). "An approach to automatic modulation recognition using time-domain features and artificial neural networks." Technical Report 1169, Defense Research Establishment Ottawa, Ottawa, Ontario, Canada, K1A 0Z4.

Intelligent Pattern-based Techniques to Monitor the Operation of Nondestructive Analysis Equipment and Contribute to Multivariate Feedback Control

GAIL A. CORDES
Idaho National Engineering
and Environmental Laboratory
(INEEL)

LEO A. VAN AUSDELN
Idaho National Engineering
and Environmental Laboratory
(INEEL)

JAMES L. JONES
Idaho National Engineering
and Environmental Laboratory
(INEEL)

KEVIN J. HASKELL
Idaho National Engineering
and Environmental Laboratory
(INEEL)

ABSTRACT

Nondestructive analysis (NDA) technologies nonintrusively characterize materials and container contents. The characterization information constitutes an archival record in support of disposition of the materials and containers and, into the future, as verification for environmental regulators, stakeholders, and the general public.

The Idaho National Engineering and Environmental Laboratory (INEEL) has been researching computational intelligence techniques for monitoring the operational performance of NDA technologies and thus contributing to the verification of the characterization process. Known patterns from historic operation data files are used to evaluate the real-time equipment operation. Pattern changes are interpreted as either known data characteristics or new operating anomalies.

During 2000-2001, pattern recognition software was used to monitor the operation of a selectable-energy electron accelerator via a LabVIEW® interface. This paper is a case history of the project. In the proof-of-concept demonstration, it was observed that similar pattern recognition software could potentially contribute effectively to multi-variable feedback control of the accelerator system.

INTRODUCTION

A wide assortment of research, development, and applications programs for environmental management are ongoing at the Idaho National Engineering and Environmental Laboratory (INEEL). Many of the programs feature the use of nondestructive analysis (NDA) tools, including particle accelerators and intelligent information processing software, for materials characterization. During 2000-2001, one subsurface science research project investigated the use of an electron accelerator for characterizing soil samples. At the same time, a second intelligent information processing research project investigated the use of pattern recognition software to monitor the operational parameters of NDA equipment. As a proof-of-concept of the second project, a pattern recognition software package was used to monitor the performance of the selectable-energy electron accelerator in the subsurface science project.

The following sections describe pattern recognition techniques and their uses, the subsurface science project and how it provided a good platform for the software research, and the results of the software proof-of-concept demonstration.

1. PATTERN-BASED TECHNIQUES

One field of research at the INEEL considers the recognition and analysis of evolving patterns as a unifying concept for studying and implementing intelligent information processing for monitoring and control systems. The research considers that these intelligent information processing systems should be modeled on human cognition, that is, the evolving spatial patterns within the human brain as new and changing information is perceived and interpreted. (Cordes, 2000)

Human perception of equipment operation emerges from the evolving and changing patterns that the human observes either consciously or subconsciously from the equipment operating parameters, and the relationships of these patterns to images that the human has mentally stored from the past. Implementing the processes of human cognition in software for monitoring and control systems can be facilitated by considering these patterns and manipulating them in computer software. The software should recognize and understand the changes in the patterns, the direction in which the system is evolving, the acceptable direction of evolution (goal) and the conditions or attributes in the patterns that indicate significant changes in the operation of the equipment. The information vector might include the variable values, the relationships between the variables, and the time-related changes of the variables themselves. More variables might be added to the pattern vector to discriminate among the patterns, or variables might be included at a higher resolution (more decimal points if you will) to better discern the actual condition of the system.

Most current equipment modeling and control applications are based on time-related variables that can be plotted as shown on the left hand of Fig. 1. For example, displays of time versus pressure will show the patterns of pressure changes with time. In this application, information is considered as a vector of variable values at one point in time, as shown on the right of Fig. 1. The vector is a pattern that changes with time. Very correctly, in control of critical components of any industry, the individual variables must remain within the prescribed ranges of setpoints. Observation of the vector pattern of the variables at an instant of time can reveal not only the changes within the operating parameters of the equipment but also the statistical importance of these changes to the operation of the equipment.

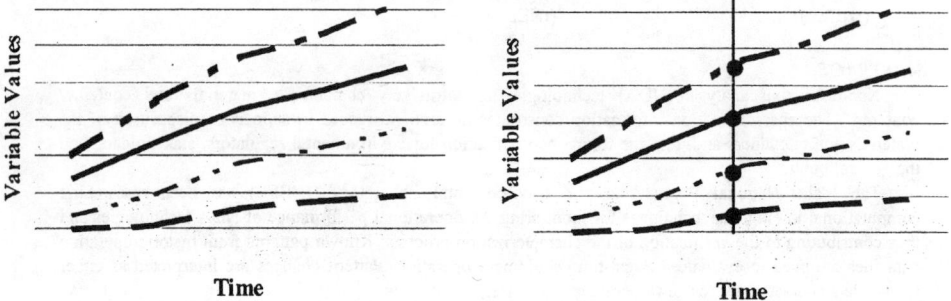

Figure 1. Two ways to consider time-related variables.

Pattern recognition techniques provide real time capability to monitor the performance of NDA systems and interpret the resulting performance data by acting across the time-stamped vector of system parameters. One of the pattern recognition techniques available at the INEEL, the Advanced Data Validation and Verification System (ADVVS), was selected for application to one INEEL NDA technology, an electron accelerator.

The ADVVS is a pattern recognition technique using multivariate analysis of incoming data as compared with a set of established data patterns. Any statistically significant deviation from the established patterns is identified and categorized as normal but unanticipated, or as a fault (O'Sullivan, 1995).

2. THE AXRF RESEARCH PROJECT

Nuclear-based nondestructive analysis techniques enable materials assay without disturbance or degradation of the materials, and typically prove to be faster, more environmentally benign, and cheaper than destructive analysis methods. At the INEEL, the NDA techniques are used for assaying and imaging of hazardous and radioactive wastes and waste packages. There also is a need for NDA techniques to perform in-field analysis of environmental contamination and laboratory analysis of subsurface soil samples.

The accelerator-based X-ray fluorescence (AXRF) research project discussed in this paper was conducted to provide a demonstration of proof-of-concept for using an electron accelerator-based NDA system for the detection of uranium and transuranics, such as americium and neptunium, in soil samples. The project demonstrated the AXRF technique that uses the bremsstrahlung radiations from 4-6 MeV accelerators to probe the soil samples. The work was conducted at the Idaho Accelerator Center at Idaho State University in Pocatello, Idaho under INEEL funding (Wells, et al., 2000).

In the interrogation process, an accelerator accelerates electrons up to several MeV's in kinetic energy. These electrons interact with a material with high atomic number to produce high-energy bremsstrahlung radiation (photons). The resulting high energy (up to the electron energy) photon radiation has high penetration into the interrogated materials, interacts with the soil constituents (including the soil contaminants), and 'excites' the atoms. The atoms then return to the initial energy levels by emitting element-characteristic X-rays in a process called 'X-ray fluorescence' or XRF. The accelerator provides the electrons in pulses and the XRF occurs after each accelerator pulse.

The AXRF research tasks included identifying the XRF responses for various energies of the initiating accelerator electron beam. For this reason, a selectable-energy electron accelerator owned by INEEL was chosen as one of the accelerators used in the project.

The selectable-energy electron accelerator is shown in Fig. 2. It consists of a Varian, Inc. Model 3000A linear accelerator waveguide, an electron gun filament, injection control system, control cabling, signal output, water cooling connections, and enclosure; a radio-frequency (RF)-drive system that provides the RF power to the accelerator via microwaves generated by a 2.1 MW magnetron tube; and an electron beam energy and beam current measurement assembly. It is referred to as the 'Varitron' and is transportable. The accelerator waveguide is a series of tuned RF cavities that, when excited by microwave energy, will capture and accelerate injected electrons. The magnetron tube, the accelerator waveguide, and the electron-to-X-ray converter require cooling by the water cooler unit. Building deionized water is used as the coolant (Jones, et al.,1994). Figure 3 is a schematic of the AXRF research setup showing the incoming bremsstrahlung photon beam in red, the soil column in dark green, the resulting AXRF radiation in light green, and the detectors in brown.

Figure 2. Photo of the Varitron accelerator   Figure 3. Schematic of the AXRF setup.

3. ACCELERATOR CONTROL AND OPERATIONAL MONITORING

One key to the success of the AXRF research project proof-of-concept demonstration was precise knowledge of the energy of the electron beam and the deposited dose from the electron accelerator. However, running the Varitron accelerator was an art form with few precisely repeatable operating procedures for the demands being placed on the machine for this and other research.

The operators admitted that, to meet some recent experimental needs for a repeatable, constant beam current, the Varitron operation was problematic. The beam energy exhibited variations during the day, variations that were small but unacceptable for this research project. This meant that the Varitron was of necessity manually tuned and re-tuned during each day of use to keep a steady nominal electron beam energy. Since there were only a few Varitron parameters that could be manually controlled, the Varitron was a logical choice for a parallel research project of demonstrating proof-of-concept for using the INEEL software package ADVVS to monitor the operation of NDA equipment.

The software research project was therefore linked to the AXRF research project to provide a proof-of-concept for the use of pattern recognition intelligent information processing methods to improve the speed and quality of NDA techniques by monitoring the performance of NDA hardware systems in real time and interpreting the resulting performance data by acting across the time-stamped vector of system parameters.

The first task of the software research was to train the ADVVS with known modes of accelerator operation. During the training the software was used to develop functional relationships that model known modes of system operation or system data outputs. During the second task, the real-time data values were to be compared by the ADVVS software with data estimates calculated from learned models. Any statistically significant deviation from these calculated patterns would be identified. Depending on the operators' decisions, the Varitron operation could be modified or terminated. The data deviations could be analyzed offline and the new data included in the functional relationships if it were adopted and classified as normal but unanticipated. Thus, the entire variable vector would be considered a pattern and would be analyzed in real time to recognize pattern changes that could be interpreted as newly identified data characteristics or operating system anomalies.

At the start of the subsurface science research project, the values for seven operation parameters were available to the accelerator operators through a virtual instrument (VI) interface (Van Ausdeln, 2001). The parameters were the magnetron current, electron source gun filament voltage, vacuum pressure, beam target current, radiation exposure (dose), position of the radio-frequency (RF) tuning probe in the magnetron, and accelerator waveguide temperature. The VI interface is written using the National Instruments™ LabVIEW® graphical programming language for operation on an IBM-compatible, Windows-based, personal computer [National, 2000]. All data acquisition is accomplished using National Instruments™ data acquisition hardware interface. The radiation exposure, vacuum pressure, and cavity assembly temperature readings are used to control safety interlocks with user adjustable trip points.

The VI provided digital readout of the seven Varitron parameters. When the values for beam target current drifted from the requested energy level and dose, the operator would manually change the setting of selected parameters. It was possible to manually deflect the electron beam across an off-axis Faraday cup to obtain beam energy profile information (analysis port current). The accelerator was judged 'in tune' when the beam energy and dose met the operators' expectations. The settings for the correct tune varied from day to day and during the course of a single day, and there was no established procedure for determining them to provide the requisite low variations in the tune.

The ADVVS software was linked with the VI, and windows were included in the interface to view the changes of the parameter values with time. The operators could view single variable plots of the parameters versus time or a time-stamped vector of the seven parameter values. The seven data signals from the Varitron were logged during preliminary AXRF research, and the data files were analyzed using ADVVS. The operators were impressed with the amount of information they could gain from the ADVVS analysis and from the time-versus-parameter plots. The preliminary results showed that additional sensors initially thought to be unnecessary for the analysis actually could contain important information that would contribute to accurate analysis of the accelerator beam current. It was also shown that some data spikes represented anomalous system noise that can be filtered prior to analysis. The source of a periodicity in the beam current data was identified as the wave-guide tube cooling system that is manually operated (chilled water is injected into a cooling jacket around the accelerator wave-guide tube when the temperature reaches a predetermined set point, roughly every 2 or 3 minutes). The operators could immediately see how their changes to the settings affected the intermediate and output variables. In order to better investigate these events, four more parameters were added to the Varitron system. The parameters were provided by three new temperature sensors and a link to the electromagnetic current. The RF probe and the electromagnet current were enabled to be adjustable manually to add additional control.

As the operators investigated the performance of the accelerator with these newfound tools, the software research team was gathering files of performance data and investigating the time-linked patterns off-line. The files of patterns were stored as reference files to be used during future machine operation. Of course, each addition of control or diagnostic variables for the AXRF research not only added valuable information on machine performance but also necessitated the compilation of new reference files. The AXRF team acknowledged that this process resulted in a better characterization and operation of the Varitron as was demanded by the unusual requirements being placed on the accelerator for the subsurface science research.

The ADVVS software does more than display the variable values in a time versus parameter plot. It accepts the pattern of variable values in real time, compares the pattern with stored patterns in the reference files, and mathematically analyzes both the relationship of the real time pattern with past performance patterns and the inter-relationship of the variables in the current pattern. The VI output window from the ADVVS software displays the pattern at the present time and the expected value or model of each variable from the reference files, and it highlights any pattern variables that the software considers to be statistically significant anomalies in the pattern. Figure 4 shows a simulation of the VI interface. A readout of the real-time operating parameters and the ADVVS analysis results are shown in the information presented in the upper left hand portion of the display. The operating parameters and their units of display are listed at the left-hand columns labeled 'Parameter' and 'unit', respectively. The real-time values of the parameters are in the column labeled 'Varitron'. The ADVVS-calculated values of these parameters based on the historic data files are shown in the column titled 'model'. A real-time Varitron variable that is statistically different from the modeled value is identified in red; in this simulation it is the beam current value. The remaining information in the VI user interface is information normally presented to the Varitron operators.

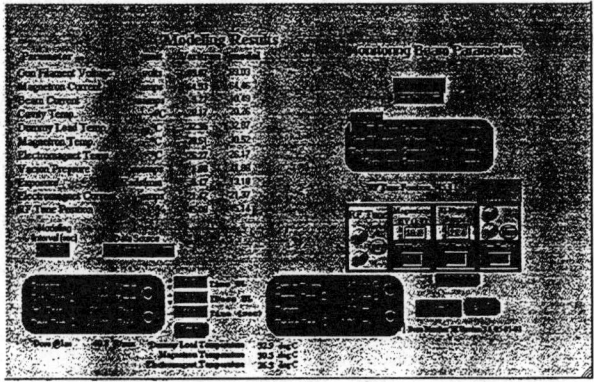

Figure 4. User interface screen.

The ADVVS was successfully demonstrated in real time with the operating accelerator. Because of commitments of the Varitron for use in other projects, extensive tests were not run in real time.

With the proof-of-concept complete for the software research, two things occurred. The accelerator team continued to add sensors to the accelerator system, and the ADVVS software team began looking for other, more complex NDA equipment to which this capability might be of value. The ADVVS team also realized from the success of the demonstration that a similar multivariate pattern recognition approach might prove valuable for automatic, multivariate, real-time feedback control for NDA equipment, including other particle accelerators.

Figure 5 is a diagram of the present list of accelerator operating parameters and the source of parameter information within the accelerator system. The parameters are further described in Table 1.

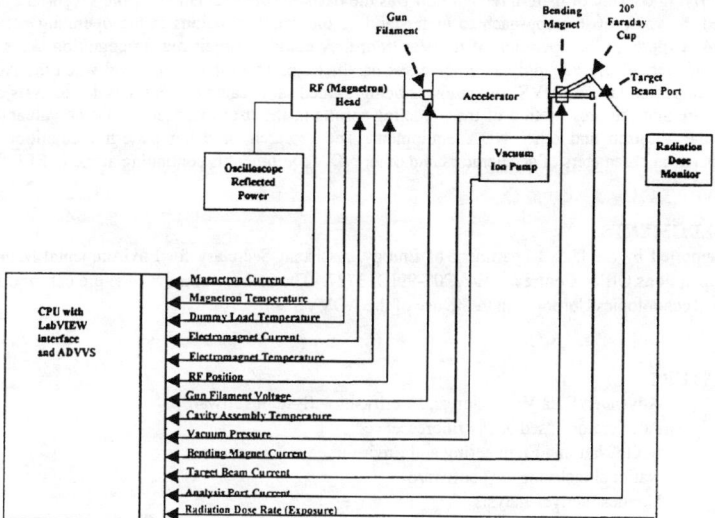

Figure 5. Schematic of accelerator and operating parameters.

Table 1. Description of accelerator operating parameters.

Parameter	Unit	Comments
Magnetron Current	amps	Electron current in the magnetron
Magnetron Temperature	°C	Temperature of magnetron tube
Dummy Load Temperature	°C	Wall temperature of RF enclosure
Electromagnet Current	amps	Current applied to electromagnet to generate the magnetic field that controls the magnetron
Electromagnet Temperature	°C	Temperature of electromagnet
RF Position	percent	Position of tuning probe in the magnetron
Gun Filament Voltage	volts	Voltage that controls amount of source electrons
Cavity Assembly Temperature	°C	Temperature of accelerator waveguide
Vacuum Ion Pump Pressure	micro-amps	Accelerator waveguide vacuum pressure
Bending Magnet Current	amps	Current to the electron beam bending magnet
Target Beam Current	micro-amps	Average electron current at the accelerator's converter
Analysis Port Current	micro-amps	Average electron current at the energy analysis Faraday cup
Radiation Dose Rate (Exposure)	Rads/min	Radiation dose rate from bremsstrahlung radiation as measured at the accelerator table

4. SUMMARY AND CONCLUSIONS

The software research project began with research into pattern recognition techniques modeled on human cognition. It expanded to include the applications of these techniques to NDA technologies being used at the INEEL for environmental management needs such as waste characterization and modeling of the movement of subsurface contamination. The ADVVS software was modified to monitor the patterns of parameters associated with the real time operation of NDA equipment. The selectable-energy electron accelerator being used for subsurface science research was identified as the technology to use for the software study. During the process of introducing the subsurface science research team to the ADVVS software, the accelerator team immediately recognized the value of tracking the Varitron operational parameters. They added additional sensors and monitoring and control capabilities to the accelerator system. The ADVVS software was demonstrated on the operating accelerator in FY2001 to provide proof-of-concept.

The underlying concept of pattern recognition was the basis for the excitement of the subsurface research team as they perceived the value of this approach to understanding the small variations in the operating parameters of their accelerator. As experts in the operation of the Varitron, they could use their human cognition skills to interpret the patterns of parameters displayed singly as time-versus-parameter plots during the interval when the ADVVS interface was being developed and the ADVVS software was being trained. It became evident that the analysis capability of the ADVVS software and the presentation of the modeling results in the user interface would be valuable for nonexpert operators of the Varitron and other NDA equipment. The research in using pattern recognition techniques for monitoring operation parameters of accelerators and other NDA equipment is continuing at the INEEL.

ACKNOWLEDGMENTS

Work supported by the U.S. Department of Energy, Assistant Secretary for Environmental Management, under DOE Idaho Operations Office Contract DE-AC07-99ID13727. The authors acknowledge the contributions of Dr. Jack Mott of Triant Technologies Corporation to the use of the ADVVS software.

NOMENCLATURE

ADVVS	Advanced Data Validation and Verification System software
AXRF	Accelerator-based X-ray fluorescence
INEEL	Idaho National Engineering and Environmental Laboratory
LabVIEW®	National Instruments™ software
NDA	Nondestructive analysis
RF	Radio-frequency
Varitron	Selectable-energy electron accelerator
VI	Virtual instrument
XRF	X-ray fluorescence

REFERENCES

Cordes, G.A., (Bechtel BWXT Idaho, INEEL), "Patterns and Intelligent Systems", *Proceedings of the 3rd American Nuclear Society International Topical Meeting on Nuclear Plant Instrumentation, Control and Human-Machine Interface Technologies*, Washington D.C., November 2000.

Jones, J.L., et al, "Pulsed Photoneutron Interrogation: The GNT Demonstration System", WINCO-1225, Idaho National Engineering Laboratory, Idaho Falls, Idaho, October 1994

Meystel, A., 1995, *Semiotic Modeling and Situation Analysis: An Introduction*, AdRem Inc.

National Instruments™ Corporation, "Measurement and Automation 2000," Austin, Texas.

O'Sullivan, P.J., et. al., "Using UPM for Real-Time Multivariate Modeling of Semiconductor Manufacturing Equipment", SEMATECH AEC/APC Workshop VII, November 5-8, New Orleans, Louisiana, 1995.

Van Ausdeln, L.A., Haskell, K.J., and Jones, J.L., "A Personal Computer-Based Monitoring and Control System for Electron Accelerators," for presentation to the 2001 American Nuclear Society Winter Meeting, Reno, Nevada, November 11-15, 2001.

Wells, D.P., et al, "Accelerator-Based X-ray Fluorescence (AXRD) for Subsurface Science and other Environmental Applications", *Proceedings, Pacifichem 2000 Conference*, Honolulu, Hawaii, December 14-19, 2000.

An Analysis of the Human Vision Process as a Decentralized Continuous-Time Dynamical System

R. WOODLEY and L. ACAR
Department of Electrical and Computer Engineering and
Intelligent Systems Center
University of Missouri-Rolla

ABSTRACT:
continuous time system. The resulting functions of the photo-receptors, nerve connections and interconnections gave rise to some basic image processing techniques.

The individual cells are grouped into interconnected systems and a mathematical representation is introduced. Then, the system is simulated to show its performance. Simulation results on a small image show how the system produces some known effects of the human eye. This work constitutes a significant improvement to our previously published work.

1 Introduction

In this document, we consider a biological view to image processing. The human eye is a collection of interconnected, specialized cells that individually perform a continuous time function. Collectively, the cells perform the tasks of color combination, edge and contrast enhancement. The most significant aspect is that these tasks are performed in real time as a continuous-time system.

Many improvements have been made to our eye-model since our last publication. Our previous eye-model was presented at the ANNIE 2000 (Woodley and Acar, 2000b) and ICARCV 2000 (Woodley and Acar, 2000a) conferences. Our previous model was based on a functional approach, where we emulated the function of each cell type. The model presented here is based on a signal level approach, where we model the input-output characteristics of the continuous-time signals of each cell. Due to space restrictions, the descriptions of the form and function of each cell type has been omitted. The following resources were used in the development of the artificial vision system (Boycott and Hopkins, 1991; Croner and Kaplan, 1995; Davson, 1976; Kamermans, Kraaij, and Spekreijse, 1998; Kolb, 1994; Kolb, Linberg, and Fisher, 1992; Lee, 1998; Matlin and Foley, 1992; Peichl, Sandmann, and Boycott, 1998; Silveira and jr., 1998).

2 Artificial Vision System Implementation

2.1 Retinal Implementation

The data coming from the camera contains red, green, and blue color information for each pixel. By arranging the blocks so that the blocks that represent cones are concentrated in the center of the image, and blocks that represent rods are located on the outside of the image, the distribution of the eye is emulated. Figure 1 shows the arrangement used.

2.2 Image Preprocessing Implementation

The cones and rods are the start of image preprocessing. The primary colors of the scene are extracted via the cones. In areas of the retina will vision is not as acute, such as the periphery of the retina, the rods detect the luminance of the scene. The rod circuitry will have effect of data reduction at the output of the retina since they make no direct connection to the ganglion cells.

The mathematical description of the cones and rods are based on the equations found in, (Wilson, 1997). The equations for the cone cells are

$$\dot{x}_{ti} = \tau_c^{-1}(-x_{ti} - u_i(t)),$$
$$\dot{x}_{ci} = \tau_c^{-1}(-x_{ci} + 0.2x_{ti} - (0.5(x_{h1i} + x_{h2i}))(0.5 + 0.0664x_{ipci})),$$

where x_{ti} is an intermediate state variable in the receptor cell of the ith subsystem, x_{ci} is the output variable of the cone, x_{h1i} and x_{h2i} are the horizontal cells of the ith subsystem, x_{ipci} is the interplexiform cell output, and τ_c is the time constant of the cone.

The equations for the rod cells were adapted from those of the cone cells,

$$\dot{x}_{ti} = \tau_r^{-1}(-x_{ti} - u_i(t)),$$
$$\dot{x}_{ri} = \tau_r^{-1}(-x_{ci} + 0.2x_{ri} - (0.5(x_{h1i} + x_{h2i}))(0.5 + 0.0664x_{ipci})),$$

where x_{ti} is an intermediate state variable in the receptor cell of the ith subsystem, x_{ri} is the output of the rod cell, and τ_r is the time constant of the rod.

The second processing elements are the horizontal cells. The horizontal cells will perform the task of noise reduction. Horizontal cells that touch cones seem to have no color preference. However, the horizontal network contacting rod cells is independent of the horizontal network contacting the cones. We, therefore, decided to model all horizontal cells using the same equations, with the difference between each horizontal cell type being determined by what it contacts. The equation modeling the horizontal cells is

$$\dot{x}_{hi} = \tau_h^{-1}(0.5 + 0.0664x_{ipci})(-x_{hi} + (n+1)^{-1}x_{ci} + \sum_{j=1}^{N}(n+1)^{-1}a_{cj,hi}x_{cj}),$$

where x_{cj} are the outputs of neighboring receptors, N is the total number of subsystems, $a_{cj,hi}$ is the connection coefficient from the jth receptor to the ith horizontal, and τ_h is the time constant of the horizontal cell. The value of $a_{cj,hi}$ is one if a connection exists and zero otherwise.

The third processing elements are the bipolar cells. The bipolar cells, in the case of cone bipolar cells, are the pipeline by which the data from the cone cells flows to the output ganglion cells. Rod bipolar cells, transmit the rod data to amacrine cells which, in turn, connect to the ganglion cells. The equation describing the rod bipolar cell is

$$\dot{x}_{rb} = \tau_{rb}^{-1}(-x_{rb} + (\frac{3}{2m}(x_{h1} + \cdots + x_{h2m}) - \frac{7}{m}\sum_{j=1}^{m}x_{rj}))/(1 + l^{-1}\sum_{j=1}^{m}lx_{aj}^2),$$

where x_{rb} is the output of the rob bipolar, m is the number of horizontals contacting the bipolar, x_{aj} is the output from an amacrine cell, l is the number of contacting amacrine cells, and τ_{rb} is the time constant of the rod bipolar cell.

The description of the cone bipolar cells are more complicated due to the two types of outputs that a cone bipolar may have. Bipolar cells may be either on-type or off-type. The difference is whether the bipolar cell receives a positive signal from the cone or a negative signal. In addition, there are two classes of bipolar cells, midget and diffuse. For each cone, there is one of each type of midget bipolar contacting the cone and two diffuse bipolar cells making contact. The equation describing the on-type midget bipolar is

$$\dot{x}_{mb+} = \tau_{mb+}^{-1}(-x_{mb+} + (\frac{0.43}{2}(x_{h1} + x_{h2}) + x_{ci}))/(1 + l^{-1}\sum_{j=1}^{}lx_{aj}^2).$$

The equation describing the off-type midget bipolar is

$$\dot{x}_{mb-} = \tau_{mb-}^{-1}(-x_{mb-} + (\frac{0.43}{2}(x_{h1} + x_{h2}) + -x_{ci}))/(1 + l^{-1}\sum_{j=1}^{}lx_{aj}^2).$$

The equation describing the on-type diffuse bipolar is

$$\dot{x}_{db+} = \tau_{db+}^{-1}(-x_{db+} + (\frac{3}{6}(x_{h1} + \cdots + x_{h6}) + \frac{7}{3}\sum_{j=1}^{3} x_{cj}))/(1 + l^{-1}\sum_{j=1}^{l} lx_{aj}^2).$$

The equation describing the off-type diffuse bipolar is

$$\dot{x}_{db-} = \tau_{db-}^{-1}(-x_{db-} + (\frac{3}{6}(x_{h1} + \cdots + x_{h6}) - \frac{7}{3}\sum_{j=1}^{3} x_{cj}))/(1 + l^{-1}\sum_{j=1}^{l} lx_{aj}^2).$$

The fourth processing elements are the amacrine cells. The amacrine cells are the main source of opponency in the human retina, (Kolb, 1994). For a particular ganglion cell, its center color and type (on or off) will be determined by the bipolar it contacts. The surround of the ganglion is produced by amacrines that contact opponent color bipolars of opposite type. Additionally, rod bipolar cells will contact some amacrines to complete the receptive field of a ganglion located outside the fovea.

Just as the case with the horizontal cells, we have elected to model all the amacrine cells the same, with the difference being determined by the connections. The mathematical model for an amacrine cell is

$$\dot{x}_{ai} = \tau_{ai}^{-1}(-x_{ai} + 0.022(l)^{-1}\sum_{j=1}^{l} x_{mbj}).$$

The fifth processing elements are the interplexiform cells. These cells provide feedback from the amacrine cells to the horizontal cells. They act to modulate the output of the horizontal cells. The interplexiform cells have the longest time constant of any of the retinal cells. The equation describing the interplexiform cell is

$$\dot{x}_{ipci} = \tau_{ipci}^{-1}(-x_{ipci} + l^{-1}\sum_{j=1}^{l} x_{aj}).$$

The last processing elements are the ganglion cells. The ganglion cells are simply the output of the system. The signal from the ganglion cells is different from that of any of the previous cell types. The signal is a sequence of pulses where the frequency of the pulses is correlated to the intensity of the light impinging on the retina. It is our contention that the pulses are the eye's mechanism to transmit data through the relatively long optic nerve. The optic nerve is many times longer than any of the connections within the retina. In other words, the ganglion cells are modulating the signals to be transmitted. Since in our system we are able to keep the strength of signal, we have the output of the ganglion cells remain unmodulated.

There are two types of ganglion cells and two classes of ganglion cells. As mentioned above, the on- or off-type ganglion is determined by which bipolar cell it contacts. The two classes are known as M- and P-class ganglion cells. M-class ganglion cells contact diffuse bipolar cells, and P-class contact midget bipolar cells. The equations for M-class ganglion cells are

$$\dot{x}_{mg+} = \tau_{mg+}^{-1}(-x_{mg+} + \frac{32}{13}(p^{-1}\sum_{j=1}^{p}(x_{db+j} + x_{aj}))),$$

$$\dot{x}_{mg-} = \tau_{mg-}^{-1}(-x_{mg-} + \frac{32}{13}(p^{-1}\sum_{j=1}^{p}(x_{db-j} - x_{aj}))),$$

where x_{db+j} is the output of an on-type diffuse bipolar from the jth subsystem, and x_{db-j} is the output of an off-type diffuse bipolar from the jth subsystem. The equations for P-class ganglion cells are

$$\dot{x}_{pg+} = \tau_{pg+}^{-1}(-x_{pg+} + \frac{32}{13}(x_{mb+j} + l^{-1}\sum_{j=1}^{l} x_{aj})),$$

$$\dot{x}_{pg-} = \tau_{pg-}^{-1}(-x_{pg-} + \frac{32}{13}(x_{mb-j} - l^{-1}\sum_{j=1}^{l} x_{aj})),$$

where x_{mb+j} is the output of an on-type midget bipolar from the jth subsystem, and x_{mb-j} is the output of an off-type midget bipolar from the jth subsystem.

Figure 2 shows the block diagram of the cell structure. The multiplicative and reciprocal operators are based on the equation found in, (Wilson, 1997). These represent the most complete and accurate equations that we were able to find. Our research will continue to refine these equations.

3 Results

The system was simulated for foveal operations. At the time of this writing, the complete system was not yet ready for simulation. However, the important foveal region is ready. The simulation process is simplified in the fovea due to the small number of elements, 1296 receptors as compared to 10000 for the full retina, and the fovea only contains cones which simplifies the coding.

The system was tested by presenting a test pattern at the input. The vision system was allowed to come to steady state in a dark environment. The luminance of the image was 800 trolands (td). The test pattern, shown in Figure 3, was then presented to the system. It consists of a band of dark (1000 td), a band of gray (3000 td), and a band of bright (5000 td). The output was then examined 0.5 seconds after the test pattern was presented.

The output reveals the processing effects that were desired. The output of the P-ganglions is shown in Figure 4. Figure 4 reveals that Mach bands are present. At the edges of the intensity changes there is a darker band on the side of the lower intensity, and a brighter band at the edge of the higher intensity. This behavior was not explicitly designed into the system. The behavior is due to the interconnections of the system. The Mach band behavior emerges from the interactions of the subsystems.

Another processing effect can be seen by analyzing a 3-D surface representation of the output. Figure 5 shows the 3-D view of the output of the P-ganglions connected to red bipolars. The Mach band behavior is noticeable in Figure 5, but in addition, the levels were the image is constant are not linearly spaced. Figure 6 shows a 3-D view of the test image. The levels of intensity were linearly spaced in the test image. This effect is contrast enhancement. The eye does not see intensity change linearly (Baxes, 1994). This behavior allows the eye to see very bright and very dark parts of the image within the range of operation of the eye.

4 Conclusion

The human vision system has been presented as a decentralized continuous-time system. A continuous-time dynamical analysis is a better representation of human vision. The dynamical model of the eye emulates the internal structure of the eye element for element. Each cell type in the eye has a direct model in the system presented. The interconnections between the elements are the interconnections in the model.

The results show an emergence of functions that were not explicitly programmed. The emergence of the Mach-bands and contrast enhancement show that distributed, continuous-time systems may provide a better solution to image processing than discrete algorithms.

The significance of this work is that real time image processing by continuous systems rather than discrete transforms is advantageous. The work performed here is part of a larger project. The fact that a dynamical eye system may be implemented in hardware for real time image processing is an exciting concept. We are currently working on a hardware implementation of the eye-model presented here.

Figure 1: Distribution of red, green, and blue cones as well as the rods (shown in gray) in a 100×100 pixel window.

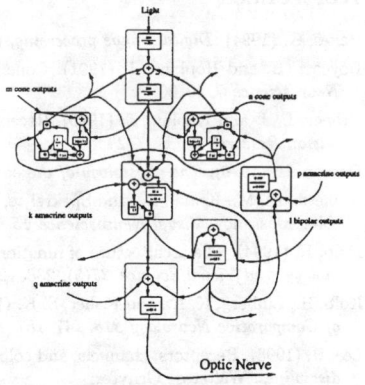

Figure 2: The block diagram of a typical subsystem.

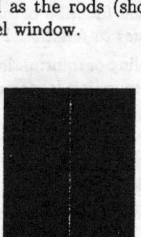

Figure 3: The input image used to test the artificial vision system.

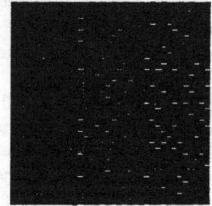

Figure 4: The output image from the ganglion cells.

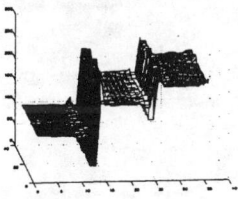

Figure 5: The 3-D surface generated by the outputs of the P-ganglion cells.

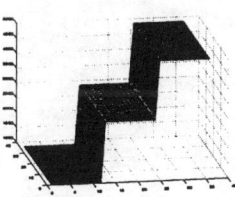

Figure 6: The 3-D surface of the input image.

References

Baxes, G. (1994). *Digital image processing, principles and applications.* John Wiley & Sons, Inc.

Boycott, B. and Hopkins, J. (1991). Cone bipolar cells and cone synapses in the primate retina. *Visual Neuroscience* 7, 49–60.

Croner, L. J. and Kaplan, E. (1995). Receptive fields of p and m ganglion cells across the primate retina. *Vision Research* 35(1), 7–24.

Davson, H. (1976). *The physiology of the eye.* Academic Press.

Kamermans, M., Kraaij, D., and Spekreijse, H. (1998). The cone/horizontal cell networl: A possible site for color constancy. *Visual Neuroscience* 15, 787–797.

Kolb, H. (1994). The architecture of functional neural circuits in the vertebrate retina. *Investigative Opthalmology And Visual Science* 35(5), 2385–2404.

Kolb, H., Linberg, K. A., and Fisher, S. K. (1992). Neurons of the human retina: A golgi study. *The Journal of Comparative Neurology* 318, 147–187.

Lee, B. (1998). Receptors, channels, and color in primate retina. In *Color Vision: perspectives from different disciplines.* Walter de Gruyter.

Matlin, M. and Foley, H. (1992). *Sensation and perception.* Allyn and Bacon.

Peichl, L., Sandmann, D., and Boycott, B. (1998). Comparative anatomy and function of mammalian horizontal cells. In *Development and organization of the retina: from molecules to function.* Plenum Press.

Silveira, L. and jr., H. D. (1998). Parrallel pathways of primate vision: sampling of information in the fourier space by m and p cells. In *Development and organization of the retina: from molecules to function.* Plenum Press.

Wilson, H. R. (1997). A neural model of foveal light adaptation and afterimage formation. *Visual neuroscience* 14, 403–423.

Woodley, R. S. and Acar, L. (2000a). An analysis of the human vision process as a decentralized continuous-time dynamical system. In *Proceedings of the Sixth International Conference on Control, Automation, Robotics and Vision*, Singapore, December.

Woodley, R. S. and Acar, L. (2000b). A neural network representation of a decentralized model of the human retina. In *INTELLIGENT ENGINEERING SYSTEMS THROUGH ARTIFICIAL NEURAL NETWORKS.* New York, New York: ASME Press. Proceedings of the Artificial Neural Networks in Engineering (ANNIE 2000).

DETECTION OF DEFECTS IN MECHANICAL EQUIPMENT FROM VIBRATION SIGNATURES USING NEURAL NETWORKS

SURESH K. PALLERLA
Mechanical & Industrial Eng. Dept.
Texas A&M University-Kingsville
Kingsville, Texas 78363

ROBERT A. MCLAUCHLAN
Mechanical & Industrial Eng. Dept.
Texas A&M University-Kingsville
Kingsville, Texas 78363

ABSTRACT

The vibration and noise signatures of any machine, engine, or structure contain a great deal of information related to the various exciting forces applied to and the condition of these systems. Changes in the response can be used to identify undesirable loads, or to predict the onset of system faults before drastic failure occurs. Steady state vertical vibration of an elastically supported milling machine is modeled and quantified in relation to changing system parameters or exciting conditions using the Working Model software package. With vibration data from Working Model as input, the onset of faults is detected from neural network output patterns, which provide predictive knowledge that can be used, for system monitoring purposes. Simulation results indicate that a three-layer neural network with backpropagation learning can be used to diagnose known conditions and to detect the occurrence of unknown faults.

INTRODUCTION

Recent research has demonstrated that neural networks can be effectively used for the identification of dynamic systems. Neural networks have successfully identified a number of highly nonlinear systems (Tsoukalas and Uhrig 1997, Pham and Liu 1995, Smyth et al 1999) through the use of sigmoidal nonlinear processing elements (PEs). Both linear and nonlinear systems have been considered. When a linear system is to be identified, only linear PEs need to be used in the neural network identifier. Nonlinear PEs are adopted for the hidden layer of the neural network when the system to be identified is nonlinear. The task of the neural network identifier is effectively one of function approximation and pattern recognition.

The general field of structural health and related environmental loading monitoring is an important engineering field and research area which has been receiving an increasing amount of attention in recent years. Here in situ nondestructive sensing and analysis characteristics, including structural or machine vibration/dynamic response signatures, are used for the purpose of detecting changes in the underlying system and its immediate loading environment (Braun 1986, Renwick 1985, Chang 1997, Smyth et al 2000, Loparo et al 2000). These changes may indicate system damage or degradation. Representative applications include industrial machinery, hydraulic and electromechan-ical motor systems, civil infrastructure components, high-performance aerospace systems, and delicate medical devices.

This paper discusses the application of a neural network system to detect defective vibration signatures for mechanical equipment. A steady state vertical vibration of an elastically supported milling machine is modeled as a two-degree-of-freedom coupled system and quantified in relation to changing parameters or exciting conditions (Dimarogonas, 1996, Rao, 1995). The Working Model software package is used to simulate this dynamic system (Rubin, 1995). The onset of faults is detected from trained neural network output patterns that provide predictive knowledge that can be used for monitoring purposes. Multilayer perceptrons (Feedforward Neural Networks) are used with standard backpropagation as the dynamic learning law. The results presented indicate that the three-layer backpropagation neural network can be used to diagnose known conditions and even generalize somewhat to detect unknown faults.

MILLING MACHINE MODEL

Figure1.0 depicts steady-state vertical vibration of an elastically supported milling machine model (Dimarogonas, 1996, Rao, 1995). For this system:
- M_1 = mass of the block1: 0.50Kg
- M_2 = mass of the block2: 0.10Kg
- K_1 = stiffness between m_1 and base: 110N/m
- K_1 = damping constants between m_1 and base: 1N-s/m
- K_2 = stiffness between m_1 and m_2: 10N/m
- C_2 = damping constants between m_1 and m_2 : 1N-s/m
- $x_1(t)$, $x_2(t)$ = displacements of masses m_1 and m_2 at time t respectively.
- $v_1(t)$, $v_2(t)$ = velocities of masses m_1 and m_2 at time t respectively.
- $a_1(t)$, $a_2(t)$ = accelerations of masses m_1 and m_2 at time t respectively.
- $v_3(t)$, $a_3(t)$ = velocity and acceleration of external applied force.
- $F = F_0 \sin \omega t$, a sinsinusoidal force of amplitude F_0 and frequency ω applied at The mass m_1 in vertical direction: 30*sin(10t)

The physical parameters above are scaled down (nominal factor of 1000) to produce the vibration signatures used in the training cases. They do not correspond to any specific milling machine system.

Differential Equations of Motion for the System: The steady-state vertical vibrations of an elastically supported milling machine are forced vibrations. The equations of motion of the milling machine system undergoing forced damped vibration, with displacements about the static equilibrium position, can be represented in the time domain as a coupled set of linear second order differential equations of the form

$$M_1 \ddot{x}_1 + (C_1 + C_2)\dot{x}_1 + (K_1 + K_2)x_1 - C_2 \dot{x}_2 - K_2 x_2 = F_0 \sin wt \quad (1a)$$

$$M_2 \ddot{x}_2 + C_2 \dot{x}_2 + K_2 x_2 - C_2 \dot{x}_1 - K_2 x_1 = 0 \quad (1b)$$

Types of Defective Vibrations of Milling Machine

Signal processing in the time and frequency domains plays a major part in the monitoring of evolving machinery health in a continually changing usage environment. However, the role of signal processing is a subsidiary one to the task of connecting failure with the vibration that is produced due to specific faults. Many digital signal processing techniques are being applied to the monitoring of machinery health (Mitchell, 1993). These techniques include modal analyses, digital filtering, convolution, inverse filtering, parametric methods, auto regressive-moving average process, parametric estimation, and entropy spectral analysis. These techniques require a great deal of understanding and relatively sophisticated measuring and analyzing equipment.

There are different types of milling machine faults. The most common machine fault results from changes in the stiffness of moving parts. This follows because most milling machines have moving parts which undergo sudden impacts. Other faults include structural cracks, unbalancing, rubbing, misalignment, bearing instabilities, shifting phase angle and gear faults. These faults (e.g., misalignment, bearing instabilities, and gear faults) can give rise to response at various system harmonics.

NEURAL NETWORK FOR DETECTION OF DEFECTS

A simple multi-layer perceptron (MLP) has been used in this defect detection problem. It consisted of an input layer, two rows hidden layers, and an output layer with weighted connections between PEs. The Neuro-Solutions Educator Edition software package (NeuroSolutions, 1998) was used to implement the MLP with the backpropagation dynamic learning law.

The following steps and information define the neural network used:
1) Multilayer Perceptron (MLP) is chosen as neural architecture with backpropagation as the dynamic learning law.
2) Theory of Cross Validation is used in developing the system.
3) Multilayer Perceptron: Input Processing Elements = 9; Output Processing Elements = 2; Exemplars = 500; Hidden Layers = 2
4) Hidden Layer # 1: Processing Elements = 18; PE Transfer = TanhAxon; Learning rule = Momentum; Step Size = 1.0; Momentum = 0.7
5) Hidden Layer # 2: Processing Elements = 9; PE Transfer = TanhAxon; Learning rule = Momentum; Step Size = 0.10; Momentum = 0.7
6) Output Layer: PE Transfer = TanhAxon; Learning rule = Momentum; Step Size = 0.010; Momentum = 0.7
7) Maximum Epochs = 100

TRAINING CASES

Example vibration signatures obtained from *Working Model* for different milling machine system cases are used to train and test the neural network. Four different defect/unanticipated loading – vibration signature experiments (a) Stiffness change case, (b) Exciting amplitude change case, (c) Type of function change case and (d) Amplitude and type of function change case -- have been performed to examine the use of neural networks to recognize change in system performance. In these cases, either one or more parameters, or the type of loading function (linear to nonlinear) are changed, and remaining parameters are maintained constant. In each case, *Working Model* data is used as training data to train the neural network. Mean square error (MSE) as a function of iteration is shown in Fig. 2 for case a. Similar MSE versus iteration results obtain for cases b-d.

Stiffness Case: Fig 3.0 shows vibration signatures for the stiffness degradation (change) case. In this case, the stiffness K_1 between M_1 and the base is initially set to 110N/m, which is assumed to yield a vibration signature for the system without any defect. Similarly, the stiffness K_1 is reduced to 80N/m (27.3% change) to characterize the system with damage. This yields the defective vibration signature case shown in the figure.

Loading Amplitude Case: In this case, the external force loading the system is set to 10Sin(10t) N which is assumed to yield a vibration signature for the system under normal design loading. The external force on the system is set to 30Sin(10t)N (200% increase), which is assumed to give vibration response signatures for the system under excessive loading from its environment.

Linear to Nonlinear Loading Case: Fig 4.0 shows vibration signatures for changes in the type of loading (linear to nonlinear) case. In this case, the initial function of the external force of the system is initially set to $F_{ex}(t) = 10\sin(10t)$ N. to give a vibration signature for the system without any defect. Similarly, the function of the external force of the system is set to $F_{ex}(t) = 10\sin(10t)\sin(10t)$ N $= 5(1 - \cos(20t))$ N to yield a vibration signature for the case of a defective system (e.g., with shaft misalignment or bearing instability) undergoing a change in the type (linear to nonlinear - bias minus second harmonic) of environmental loading.

Nonlinear to Linear Loading and Amplitude Change Case: In this case, the loading function of the external force of the system is set to $F_{ex}(t) = 10\sin(10t)*\sin(10t)$ N $= 5(1 - \cos(20t))$ N, which is assumed to represent the nominal system loading. Similarly, for the vibration signature of the system under unanticipated loading, the external force of the system is set to $30\sin(10t)$ N.

RESULTS

For each of the four different defect/unanticipated loading – vibration signature experiments, data from *Working Model* is transferred to Microsoft Excel and given as test data to individually trained neural networks.

Table 1 shows the Output to Desired performance matrix for the Stiffness Change case. Examination of these results indicates that the trained neural network could predict 107 correct signatures out of 107 correct signatures and 91 wrong signatures out of 93 wrong signatures provided.

The Output to Desired performance matrix for the Loading Amplitude Change case is shown in Table 2. For this case, the trained neural network could predict 81 correct signatures out of 82 correct signatures and 128 wrong signatures out of 128 wrong signatures.

Table 3 gives the Output to Desired performance matrix for the Change in Loading Type case. These results indicate that the trained neural network could predict 37 correct signatures out of 37 correct signatures and 19 wrong signatures out of 20 wrong signatures.

The Output to Desired performance matrix for the Change in Loading Type and Amplitude case is shown in Table 4. Examination of these results indicates that the trained neural network could predict 64 correct signatures out of 68 correct signatures and 136 wrong signatures out of 137 wrong signatures provided.

These performance matrix results indicate that the trained neural network not only detected individual (defect/damage or loading/system loading interaction) change cases, but also the combination of loading amplitude and type of loading change case.

CONCLUSIONS

The neural network results indicate that the artificial neural networks are successful in detecting different "defects" that are correlated with reduction in stiffness and/or changes in the loading environment for the milling machine system. The different experiments indicate good results. The *Working Model* software is very useful in modeling and simulating different mechanical equipment models. Additional conclusions include the following:
(1) Vibration and noise signals of any machine, engine, or structure contain a great deal of dynamic information related to the condition of the system and can be used as training data to detect different defects.
(2) Multilayer perceptrons (Feedforward neural networks) are effective for the identification of different milling machine defects.
(3) An increase in the accuracy was observed with an increase in the number of hidden layers to two layers. However it was found to be negligible for more than two

hidden layers. The best transfer function for this application is the TanhAxon. The Momentum Learning Rule is the best learning rule to calculate the weight vector update.
(4) The Neuro Solutions software used in this project performed well as has been verified by experiments. The results presented herein indicate that this software can be used for information processing, planning, and control.
(5) The trained neural network not only detected individual defect vibration signatures but also the combination of both loading amplitude and type defect vibration signatures.

RECOMMENDATIONS
(1) Client-Server Technology can be used to automate the remote detection of faults using an artificial neural network so that real data can directly be given to a trained neural network to identify specific defects.
(2) Other mechanical equipment models (Gearbox, Compressor Rotor, Turbine Rotor etc) are recommended for further fault identification research work.

REFERENCES
Braun, S.G., 1986, Editor, *Mechanical Signature Analysis: Theory and Applications*, Academic Press, San Diego, CA.
Chang, F.K., Editor, 1997, "Structural Health Monitoring, Current Status and Perspectives," *Proceedings of the International; Workshop on Structural Health Monitoring*, Stanford University, Stanford, CA. 18-20 Sept. Technomic, Lancaster, PA.
Dimarogonas, Andrew, 1996, *Vibration for Engineers*, Second Edition, Prentice-Hall, Upper Saddle River, New Jersey.
Loparo, Kenneth A., Adams, M.L., Lin, Wei, Abdel-Magied, M. Farouk, and Ashfari, Nadar, 2000, "Fault Detection and Diagnosis of Rotating Machinery," *IEEE Transactions on Industrial Electronics*, Vol. 47, No. 5, October. pp. 1005-1014.
Mitchell, John S., 1993, *Introduction to Machinery Analysis and Monitoring*, Second Edition, PennWell Books, Tulsa, Oklahoma.
NeuroSolutions, 1998, NeuroSolutions The Neural Network Simulation Environment. Version 3.0,. http://www.nd.com/, Neuro-Dimension, Inc., Gainesville, FL.
Pham, D.T., Liu, X, 1995, *Neural Networks for Identification, Prediction, and Control*, Springer-Verlag, London.
Rao, Singlresu S., 1996, *Mechanical Vibration*, Third Edition, Addison Wesley, Reading, Massachusetts.
Renwick, J.T., 1985, "Vibration Analysis - A Proven Technique as a Predictive Maintenance Tool," *IEEE Transactions on Industry Applications*, Vol. 21, March. pp. 324-332.
Rubin, Carol A., 1995, *The Student Edition of Working Model, Version 2.0*, Addison-Wesley, Reading, Massachusetts.
Smyth, A.W., Masri, S.F., Chassiakos, A.G. and Caughey, T.K., 1999, "On-Line Parametric Identification of MDOF Nonlinear Hysteretic Systems," *ASCE Journal of Engineering Mechanics*, Vol. 125, No. 2, pp. 133-142.
Smyth, A.W., Masri, S.F., Caughey, T.K., and Hunter, N.F., 2000, "Surveillance of Mechanical Systems on the Basis of Vibration Signature Analysis," *ASME Journal of Applied Mechanics, Transactions of the ASME*, Vol. 67, September. pp. 540-551.
Tsoukalas, Lefteri H, Robert E. Uhrig, Robert E., 1997, *Fuzzy and Neural Approaches in Engineering*, Wiley-Interscience, New York, pp. 495-496.

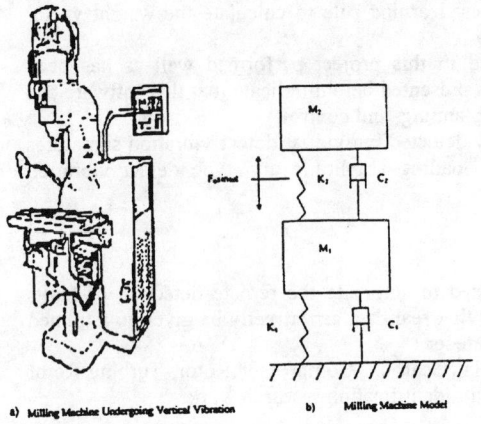
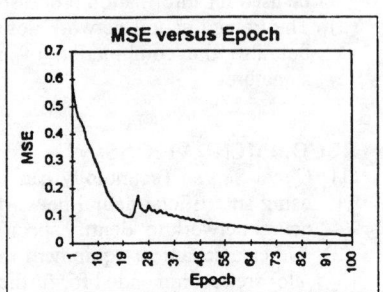

Figure 1. Steady-state vertical vibration of elastically supported milling machine with sinusoidal force at base.

Figure 2. Mean square error as a function of iteration for stiffness change case.

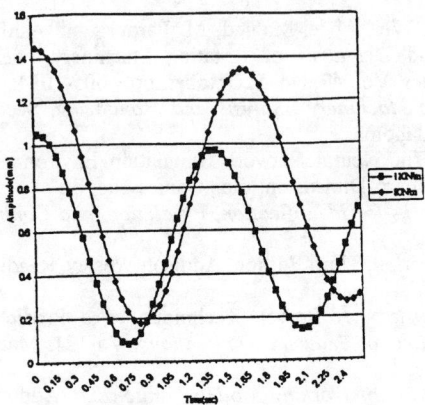

Figure 3. Signature for stiffness change/degradation case.

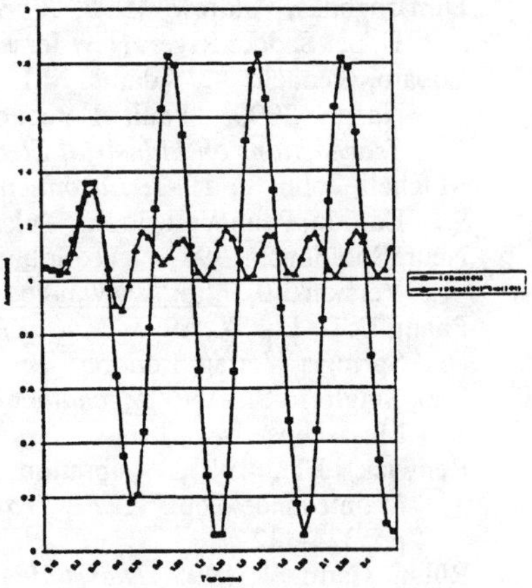

Figure 4. Signature for type of loading change case.

Table 1 Stiffness Change Performance Case

Output / Desired	Correct	Wrong
Correct	107.0000	2.0000
Wrong	0.0000	91.0000

Table 2 Loading Amplitude Change Performance Case

Output / Desired	Correct	Wrong
Correct	81.0000	0.0000
Wrong	1.0000	128.0000

Table 3 Change in Loading Type Performance Case

Output / Desired	Correct	Wrong
Correct	37.0000	1.0000
Wrong	0.0000	19.0000

Table 4 Change in Loading Type and Amplitude Performance Case

Output / Desired	Correct	Wrong
Correct	64.0000	1.0000
Wrong	4.0000	136.0000

ACTIVE MODEL-BASED OBJECT RECOGNITION EMPLOYING FOVEAL IMAGERY AND MULTIRESOLUTION FEATURE SETS

QIANG LU
Dept. of Computer Science and
Engineering
University at Buffalo
Buffalo NY 14260-2000

PETER D. SCOTT[1]
Dept. of Computer Science and
Engineering
University at Buffalo
Buffalo NY 14260-2000

ABSTRACT
We consider the late-vision problem of model-based object recognition using a foveal sensor, one whose resolution decreases, as does the human retina, with distance from its central axis. Key elements in our approach are the use of pyramid representations, multiresolution feature sets, and a prediction-correction gaze control algorithm. The sensed imagery is monocular and monochrome, consisting of square space-variant resolution cells or "rexels," whose areas vary by powers of four. In a given image, the fovea supplies high resolution detail while the perifovea and periphery register a wide field of view. A library of multiresolution polyline shape templates based on the shape evolution scheme are compared to a given candidate object to be recognized using the turning function difference metric. Overlap and occlusion are treated by decomposition into convex components and matching is done on convex parts. After part labeling, a subgraph isomorphism approach is used to determine the likeliest global hypothesis. Next-look selection is determined by maximizing the match metric given the current hypothesis (object type, location and pose). Results for several experiments are presented.

1. INTRODUCTION

It has been shown that many computer vision problems which are ill-posed and unstable using passive homogeneous vision sensors can become well-posed and stable when active space-variant sensors are used[1]-[3]. Active vision systems with space-variant sensors may be the key to produce small, high performance, light weight and inexpensive computer vision modules[4].

Despite the promising characteristics of such systems, in order for them to find practical use, both early and late-vision processing tasks must be considered. Relatively low-level vision tasks such as obstacle avoidance and homing using active vision systems have been previously studied[5]-[7]. Here we consider an important late-vision problem, model-based object recognition, using an active space-variant vision scheme.

The topology of the vision sensor studied here is introduced in Section 2, followed by the construction of the model template library with multiresolution feature sets in Section 3. Saccade generation, an important issue in the system, is considered in Section 4. Shape representation and comparison, and its use in object recognition in scenarios involving overlap and occlusion follow in

---
[1] Please Send all correspondence to this author at the given address

Section 5. Finally, experimental results are delineated and conclusions suggested in Sections 6 and 7.

2. A RECTILINEAR FOVEAL VISION SENSOR

For these studies, we will assume a monocular foveal greylevel vision sensor. The sensor uses a square space-variant sampling tesselation (see Figure 1). The central region or fovea is surrounded by a ring of resolution cells, which we refer to as "rexels," each four times bigger than the foveal rexels, resulting in a four-fold reduction in areal resolution. The first ring is surrounded by another ring of rexels four times larger, and so on. The fovea supplies high resolution details while the perifovea and periphery register a wide field of view. The combination of high resolution and large field of view permit systems deploying such sensors to execute combined search and recognition tasks more efficiently than sensors with uniform resolution.

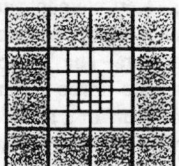

Figure 1 Foveal Vision Sensor

3. TEMPLATE LIBRARY WITH MULTIRESOLUTION FEATURE SETS

The template library in model-based object recognition is a collection of labelled archetypes characterized by feature sets which are used to match against those of a given candidate object to be recognized. Multiresolution feature sets[8] are a natural choice for multiresolution imagery such as that produced by foveal sensors.

The construction of the template library begins by constructing a four-to-one averaging image pyramid on a set of full resolution images. In each pyramid level, the overall shape is used as the dominant feature set, with polyline approximation chosen to parameterize the shape. The original bounding contour is first smoothed with a Gaussian lowpass filter, and then the contour evolution scheme[9] is applied to eliminate vertices in the polyline representation while retaining its salient shape characteristics. The resulting vertices are stored as the overall shape feature vector, the dominant feature set in the template.

Along with overall shape, refined discrimination among object classes usually requires intermediate and high-resolution features as well. These features are derived by continuing the discrete contour evolution procedure beyond its termination for shape feature vector selection, and is based on the proposition that the longer a vertex survives the evolution procedure, the more important it is in shape representation. The final three surviving evolution vertices are selected as loci for higher resolution features. Each such vertex serves as the center of a local window in which a high resolution polyline representation of the partial boundary is extracted. Backtracking during the evolution process may be necessary when two templates differ only in a local region which will recover discarded but useful discriminating vertices. Finally, convex part boundary vectors of the template object and its adjacency graph are included as well which are useful in occlusion and will be detailed in Section 5.

4. SACCADE GENERATION

In active vision object recognition applications, the entire region needing to be explored for candidate object instances, the field of regard, is normally much larger than that captured in any single image, i.e. the field of view. An effective search strategy, selection of sequences of fields of view over the field of regard, is a fundamental consideration. Our approach is a prediction-correction algorithm implemented as the finite state machine (FSM) shown in Figure 2.

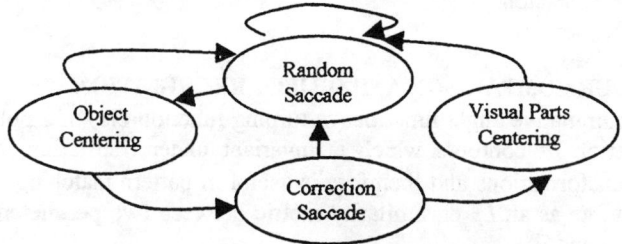

Figure 2 Prediction-correction next-look FSM

The algorithm begins with a random saccade within the field of regard, and an image polygon[10] is built over the resulting foveal image. After preprocessing with a morphological erosion operator to eliminate small noise fragments, interesting candidate objects are searched at the coarsest resolution level. A Candidate Location (CL) marker is calculated as the center of the bounding box of each and a list of CLs is formed if multiple objects exist. After completing an object recognition, the gaze control algorithm directs the sensor to the next CL until the list is exhausted. Another random saccade is executed afterwards.

Upon saccade to a CL, the resulting image polygon is computed and the dominant shape feature vector of the object from the highest resolution image polygon level which captures the entire candidate is compared to that of every template of the corresponding level in the library using the difference metric described in the next section. Two thresholds, high and low, are declared at this stage. If the match score exceeds the high, or recognition threshold, for any of the library templates, a recognition is declared and the corresponding label is asserted. If all are below the low, or rejection threshold, the candidate is eliminated from further consideration.

If the best match score at the CL being interrogated is between the recognition and rejection thresholds, a list of best matches is used to establish the set of current feasible hypotheses as to the label and pose (position and orientation) of the object. In this case, more information is needed to parse this list. Since each feasible hypothesis predicts what features will be seen consistent with that hypothesis, a saccade which centers the sensor where a high resolution feature is located best determines the accuracy of this hypothesis should be made. The three final vertices in each hypothesis serve this purpose. Based on each saccade and the resulting image polygon, the pose parameters are corrected, and the hypothesis match score are correspondingly updated. After that, the current match score for that hypothesis is once again compared to the recognition threshold. This procedure is repeated for all feasible hypotheses.

In cases where there is no occlusion of any part of the object boundary, the candidate interrogation most frequently terminates with recognition or rejection at this point. There is sufficient discriminatory power in the combination of overall shape features and fine features to determine if a candidate matches a library template. But in the occluded case, object labels must be inferred from partial boundaries and combinations of partial boundaries. Where candidates are neither recognized nor rejected after overall shape and fine features have been considered, a procedure based on the decomposition of the candidate into convex parts and matching of parts against corresponding library template parts completes the matching process. This part matching procedure is described in the following section.

5. CONTOUR COMPARISON AND OBJECT RECOGNITION

The cumulative angle function, or turning function $\theta(s)$ of a polygon[11] is a representation of contours which is invariant under translation, rotation and scaling transformations and therefore is useful in pattern matching. $\theta(s)$ may be used to define as an L_p dissimilarity metric between two perimeter-normalized polygons A and B:

$$d_{A,B} = (\int |\theta_A(s) - \theta_B(s)|^P \, ds)^{1/P} \qquad (1)$$

The contour of the candidate object is extracted, smoothed and evolved. The evolved contour is then compared to that of the library template via this metric.

In the case of overall shape comparison with no occlusion, a closed contour is available and after normalization, the $O(mn\log(mn))$-time algorithm presented in[11] can be applied directly. In the case of fine local features with no occlusion, the corresponding partial match is computed and used to increment the overall match score.

In the final case, where occlusion is considered, the candidate is first decomposed into convex regions. The use of convex decomposition is based on the useful property that a linearly occluded convex boundary still defines a convex boundary. An example is shown in Figure 3.

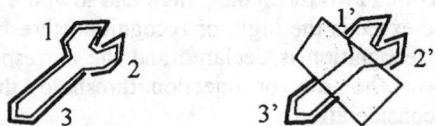

Figure 3. Convex boundary invariance under occlusion.

The normalizing scale factor is a significant parameter when computing the dissimilarity metric (1). Since occlusion can obscure any portion of a convex part boundary, relative lengths of candidate and template are unknown. It is necessary to perform a one dimensional search to determine the optimal scale factor for a given match. The search for the best match may then proceed unhindered by the disparate lengths of template convex boundary and candidate convex boundary parts caused by occlusion.

If there are two or more partial boundaries which strongly match partial boundaries in a given template, and moreover the matching boundary segments are in similar relative locations in the candidate and the template, this juxtaposition should be taken as increasing confidence in a match to that template. Therefore, where the convex decomposition procedure is applied, adjacency graphs are maintained both for the convex boundary parts of the templates and the candidate. The consistency between candidate and template is determined by subgraph isomorphism[12].

6. EXPERIMENTAL RESULTS

A simulated active space-variant vision system was implemented to test the object recognition procedure described above. The field of regard in the simulation is 512x512 pixels. The first experiment was designed to recognize non-occluded objects. One test image contains two objects, a wrench and a hand, with arbitrarily selected scale and pose. Figure 4 shows the sequence of saccades leading to the successful recognition of one candidate object. Figure 5 is the corresponding series of prediction and saccades leading to recognition of the other candidate object.

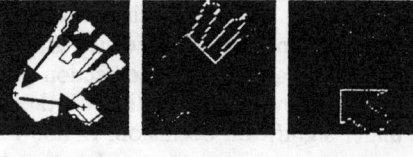

Figure 4 (a)   (b)   (c)

Figure 4 (a) Saccades for one candidate object (dark gray arrows). (b) Partial match of the first fine feature saccade. (c) Partial match of the third fine feature saccade.

Figure 5 (a)   (b)   (c)

Figure 5 (a) Saccades for a second candidate object (dark gray arrows) (b) Partial match of the first fine feature saccade. (c) Partial match of the third fine feature saccade.

The second experiment involves recognition of objects with overlap and occlusion. Figure 7(a) shows one test image and 7(b) is the convex part decomposition and saccades. Figure 7(c)-(e) shows the match results based on the convex parts. Figure 8(a) and (b) is another test image and Figure 8(c)-(e) shows the match results. As shown, the objects are correctly recognized even under significant occlusion.

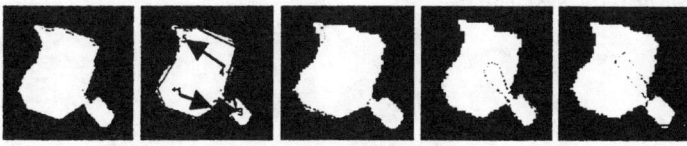

(a) (b) (c) (d) (e)

Figure 7 (a) Image with occlusion. (b) Convex part decomposition (dark gray lines) and the saccades (dark gray arrows) (c) Convex part match (d)-(e) Two possible matches of convex part

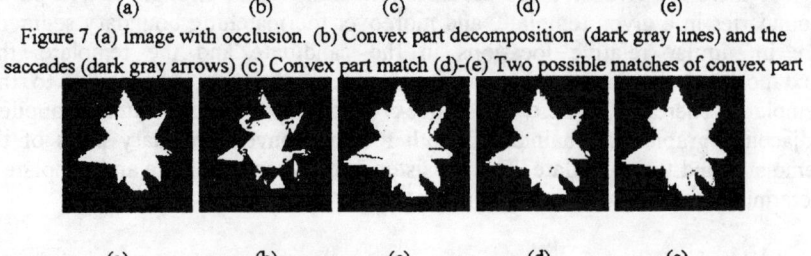

(a) (b) (c) (d) (e)

Figure 8 (a) Another test image with occlusion. (b) Convex part decomposition (dark gray lines) and the saccades (dark gray arrows) (c)-(e) Objects identified in the occluded mass

7. CONCLUSIONS

In this paper we considered a late vision problem, model-based object recognition, using active space-variant vision principles. A template library with multiresolution feature sets based on the shape evolution boundary representation scheme provided the reference framework, and decomposition into convex parts was used in case of occlusion. Partial convex boundary matches were scored, and combined using adjacency graphs. The overall system operates under the guidance of a prediction-correction gaze control algorithm. Numerical experimentation showed the ability of the scheme to correctly identify objects even under significant occlusion. In non-occluded cases, only a small number of saccades were necessary to complete the object identification.

REFERENCES

[1] D. Ballard, Animate vision, Artificial Intelligence 48 (1991) 57-86
[2] J. Aloimonos et. al., Active vision, Int. J. Computer Vision 2 (1988) 333-356
[3] M. Tistarelli and G. Sandini, Dynamic aspects in active vision, CVGIP:IU 56 (1992) 108-129
[4] E. Schwartz et. al., Space-variant active vision: definition, overview and examples, Neural Networks 8 (1995) 1297-1308
[5] S. Kundur and D. Raviv, Novel active vision-based visual threat cue for autonomous navigation tasks, CVIU 73 (1999) 169-182
[6] J. Aloimonos, Is visual reconstruction necessary?, J. Robotics Sys, 9 (1992) 842-858
[7] R. Nelson, From visual homing to object recognition, Visual navigation, 1997
[8] C. Dyer, Multiscale image understanding, Paraller Computer Vision, 1997
[9] L. Latecki and R. Lakamper, Convexity rule for shape decomposition based on discrete contour evolution, CVIU 73 (1999) 441-454
[10] C.Bandera and P. Scott, Foveal machine vision , Proc IEEE SMC Conference (1989), 596-599.
[11] E. Arkin et. al., An efficient computable metric for comparing polygonal shapes, IEEE Trans. PAMI 13 (1991) 209-215
[12] Y. El-Sonbaty and M.A.Ismail, A new algorithm for subgraph optimal isomorphism, Pattern Recognition, 31 (1998) 205-218

BALL RECOGNITION IN IMAGES FOR DETECTING GOAL IN FOOTBALL[*]

N. ANCONA   G. CICIRELLI

A. BRANCA   A. DISTANTE

Istituto Elaborazioni Segnali ed Immagini - C.N.R.
Via Amendola 166/5 - 70126 Bari - Italy
{ancona,cicirelli,branca,distante}@iesi.ba.cnr.it

ABSTRACT

In this paper we present a technique for detecting goals during a football match by using images acquired by a single camera placed externally to the field. The method does not require the modification either of the ball or of the goalmouth. Due to the attitude of the camera with respect to football ground, the system can be thought of as an electronic linesman which helps the referee in establishing the occurrence of a goal during a football match. The occurrence of the event is established detecting the ball and comparing its position with respect to the location of the goalpost in image. The ball detection technique relies on a supervised learning scheme called Support Vector Machines for classification. The examples used for training are appropriately filtered version of views of the object to be detected, previously stored in form of image patterns. We have extensively tested the technique on real images in which the ball is both fully visible and partially occluded. The performance of the proposed detection scheme are measured in terms of detection rate, false positive rate and precision in the ball localization in image.

INTRODUCTION

In this paper we focus on the problem of detecting the occurrence of a goal during a football match, by using methods and devices which does not require the modification either of the ball or of the goalmouth. Automatic goal detection in football is an open problem which is getting particular attention from referee associations, sport press and supporters. In fact, it is not unusual for a goal to occur[1], but the referee and his collaboratores (linesmen) are not able to detect the goal and, more importantly, do not award any point to either teams correctly. Such situations occur for example when, after a shooting, the ball touches the

---

[*] Acknowledgements: this paper describes research done at the Istituto Elaborazioni Segnali ed Immagini, C.N.R. in Bari. Partial support was also provided by the italian football association Federazione Italiana Gioco Calcio FIGC.

[1] A goal occurs in football when the ball completely crosses the goal line.

internal side of the crossbar, bounces off the field having crossed completely the goal line and goes back, without touching the net. One of the most significative evidences of this event, named *ghost goal*, occurred during the World Cup final match on 1966 between England and West Germany.

Optical sensors, like standard TV cameras, seem to be appropriate for approaching the problem at hand for several reasons. First of all they satisfy the main constraint of the problem, because their exploitation does not require modifications of either the ball or the goalmouth. They can be placed externally and also very far from the field and, if equipped with appropriate zoom lens, they provide images of the goalmouth area usefull for solving the goal detection problem. Moreover, they provide direct evidence of the occurrence of a goal because the perceived images can be recorded on an analog or digital support for a successive analysis, for example by an external referee.

Reid and Zisserman [1] proposed an uncalibrated binocular vision system for solving the problem of goal detection in football. Their method exploits two images of the field acquired simultaneously from two different viewpoints. In both images both the goal area and the goalmouth are visible. The computation of the vertical vanishing point in both images and of the homography between the two images induced by the field are used for measuring the projection of the ball onto the ground plane. The distance of this point with respect to the goal line is used for establishing if a goal occurs. A strong limitation is that their method is not able to compute the three-dimensional coordinates of the ball, and so it cannot be used as an automatic goal detector. In fact, the algorithm computes only the projection of the ball on the ground plane, and it is not able to distinguish if the ball crosses the goal line below or above the crossbar. Finally, an open problem in their paper is the detection and localization of the ball in both images for triangulating.

Figure 1: View of the goalmouth perceived by our electronic linesman place close to the corner flag.

In this paper we describe a method that, as the one proposed by Reid and Zisserman in [1], does not provide a general and fully automatic solution to the problem of goal detection in football. Our attempt was to design a method well suited to the detection of a particular event potentially occurring during a football match, namely a ghost goal. A deeper analysis of the problem at hand, and in particular of the ghost goal detection problem, shows that the best attitude for a linesman for detecting the ball completely crossing the goal line during a ghost goal is close to the corner flag. In fact, for detecting the goal from this view point, the linesman has to simply establish if the ball is to the left (right) with respect to the goalpost (see figure 1). Then, in general, a monocular observer having its optical axes lying on the goalmouth plane with a viewing direction oriented towards the goal line can detect the occurrence of a ghost goal simply evaluating the relative position between the ball and the goalpost in the perceived image. So the main problem that an electronic linesman has to solve for establishing the occurrence of a ghost goal is ball detection in images. The plan of the paper is as follow. In the next section we present a general framework for object detection in images and possible approaches for its solution. In the successive section we briefly discuss the main properties of Support Vector Machines (SVM) for classification, the supervised learning scheme used in this paper for detecting balls in images. In the section relative to the experimental results, we discuss the steps performed for training an SVM and show performances of the ball detector applied on real images in terms of detection rate, false positive (FP) rate and precision in the localization of the ball in images. Conclusions follow.

OBJECT DETECTION IN IMAGE

Under this perspective the problem of detecting goal can be reduced to the problem of detecting the ball in images of the goalmouth taken from a suitable view-point. This is a particular instance of a more general problem that is the one of detecting three dimensional objects by using the image projected by the object on the sensing plane of a standard camera. Recently [2], the problem of detecting objects has been addressed by a new and appealing perspective in which the criteria for establishing whether or not a given image pattern is the instance of an object is based on *views* of the object, previously stored in form of image patterns. Under this new perspective, the problem of object detection can be regarded as a *learning from examples* problem in which the examples are particular views of the object we are interested in detecting, and then many of the supervised learning schemes can be usefully applied for solving the problem at hand. More specifically, object detection can be seen as a *classification* problem, because our ultimate goal is to determine a separating surface, optimal under certain conditions, which is able to separate object views from image patterns that are not instances of the object.

SUPPORT VECTOR MACHINES FOR CLASSIFICATION

Recently, Vapnik [3] has introduced a new learning scheme, well founded in the framework of the statistical learning theory, called Support Vector Machines (SVM) for approaching classification and regression problems. The basic idea of the Vapnik's theory is closely related to regularization [4]: for a finite set of training examples, the search for the best model or approximating function has to be constrained by an appropriately small hypothesis space, that is the set of functions the machine implements. If the space is too large, functions can be found which fit exactly the data, but they will have a poor generalization capabilities on new data. Vapnik's theory formalizes these concepts and shows that the solution is found minimizing both the error on the training set (empirical risk) and the complexity of the hypothesis space, expressed in terms of VC-dimension. More formally, we are given a training set $S = \{(x_i, y_i)\}_{i=1}^{\ell}$ of size ℓ where $x_i \in R^n$ and $y_i \in \{-1, 1\}$ for $i = 1, 2, ..., \ell$. In other words we assume that the examples in S belong to either of two classes. In the general hypothesis of not linearly separable classes, the optimal separating hyperplane $w^* \cdot x + b^* = 0$ found by SVM is solution of the following quadratic programming (QP) problem with linear constraints:

$$\min_{w,b,\xi} \frac{1}{2} w \cdot w + C \sum_{i=1}^{\ell} \xi_i$$

s.t. $y_i(w \cdot x_i + b) + \xi_i \geq 1$ $i = 1, 2, \ldots, \ell$ and $\xi \geq 0$

where C is a positive number and ξ_i (one for each point in S) measures the amount of misclassification of the point x_i with respect to the optimal separating hyperplane. This problem can be solved by using the standard technique of Lagrange multipliers. At this aim, we introduce ℓ non negative slack variables λ_i relative to the constraints $y_i(w \cdot x_i + b) + \xi_i \geq 1$, and ℓ non negative slack variables μ_i relative to the constraints $\xi_i \geq 0$. The introduction of these variables permits of solving the previous QP problem in its dual form:

$$\max_{\lambda} -\frac{1}{2} \lambda D \lambda + \sum_{i=1}^{\ell} \lambda_i$$

s.t. $\sum_{i=1}^{\ell} \lambda_i y_i = 0$ and $0 \leq \lambda_i \leq 0$ for $i = 1, 2, \ldots, \ell$

where D is a matrix of size $\ell \times \ell$ with element $D_{ij} = y_i y_j x_i \cdot x_j$. Then the weight vector w^* of the optimal separating hyperplane is given by:

$$w^* = \sum_{i=1}^{\ell} \lambda_i^* y_i x_i$$

where λ^* is the solution of the previous QP problem. Moreover the bias term b^* is computed by using the Kuhn-Tucker conditions [3]. The points x_i with $\lambda_i^* > 0$ are called *support vectors*. The classification of a new data x involves the evaluation of the decision function:

$$f(x) = sign\left(\sum_{i=1}^{\ell} \lambda_i^* y_i (x_i \cdot x) + b^* \right)$$

where the solution is expressed evaluating the dot product between the data and some elements (support vectors) of the training set. The extension to the case of non linear separating surfaces is done by mapping the input vectors x in a higher dimensional *feature space* and looking for the optimal separating hyperplane in this new space (see [3] for details). In the feature space the dot products are evaluated by using suitable kernel functions $K(x,y) = \phi(x) \cdot \phi(y)$, where ϕ is the mapping function.

EXPERIMENTAL RESULTS

We used a standard TV camera with a zoom lens having a focal lenght of $f=75mm$. At the aim of reducing motion blurring effects, a shutting time of $1/10000sec$ was used. The camera was placed externally to the football ground, the height of its optical center was $1.5m$ roughly and its optical axis was manually aligned whith the goal line. The distance of the camera with respect to the center of the two goalposts was $48m$. The figure 1 shows a tipical image acquired by the camera in this attitude.

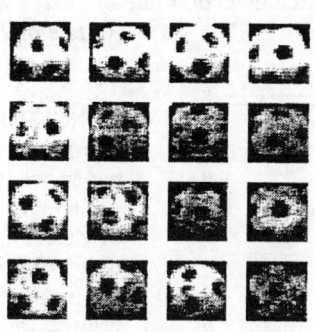

Figure 2: image patterns of football.

One of the main advantages associated with the chosen camera attitude is that the football size projected on the camera plane is almost constant moving the football inside the area being monitored for detecting goal. In fact, for 3D points belonging to this area, the image formation process can be described, from a geometric point of view, by using orthographic projection, instead of perspective projection. This implies that, due to the chosen camera attitude and the adopted focal lenght, any algorithm for dectecting football in the area close to the goalmouth does not need to manage scale variations of the object in the image.

We acquired the football in different attitudes and illumination conditions, and in different positions inside the area being monitored, for example internally and externally to the goalmouth, so collecting 2004 positive examples of the ball, with a size of $20 \times 20pxl$ (see figure 2). In all the positive examples the football was totally visible. This means that, in the current implementation, the problem of detecting partially occluded footballs was not taken into account.

For collecting negative examples, i.e. non-ball patterns, we used an iterative procedure (bootstrapping) in which the system looks for false positive (FPs) patterns in images not containing the ball, and adds these patterns to the training set [2-5]. The reason for adopting this strategy is twofold: to keep low the number of negative examples used for training and, more importantly, to select image patterns which are useful for the problem at hand. The iterative procedure selected 6881 negative examples and involved 43 false positive images for a total of over 4 millions of negative examples. The final classifier was obtained training an SVM classifier on 9975 positive and negative examples by using a second degree polynomial kernel function $K(x,y) = (1 + x \cdot y)^2$ and $C=200$. The number of required support vectors for representing the optimal classifier was 2237, including 165 and 976 positive and negative errors respectively.

All the examples were appropriately preprocessed before training. First, pixels close to the boundary of each example window were removed in order to eliminate parts belonging to the background. Then a histogram equalization was applied to reduce variations in image brightness and contrast. The resulting pixels were used as input to the classifier.

For measuring the generalization capabilities of the learning machine, we tested the classifier on 900 images acquired under different illumination conditions. Each test image was exhaustively scanned and all the sub-images with size 20x20*pxl* were classified as instance of the football or not. The figure (1) shows a tipical image used for testing. The ROC curves in figure (3) show the performances of the classifier on images with fully visible footballs (upper curve) and on images with occluded footballs (lower curve). In the first case we had a detection rate of 98.3% with a FP rate of 0.2%; in the second case, where occluded footballs were considered too, we had a detection rate of 76.2% with a FP rate of 2.6%.

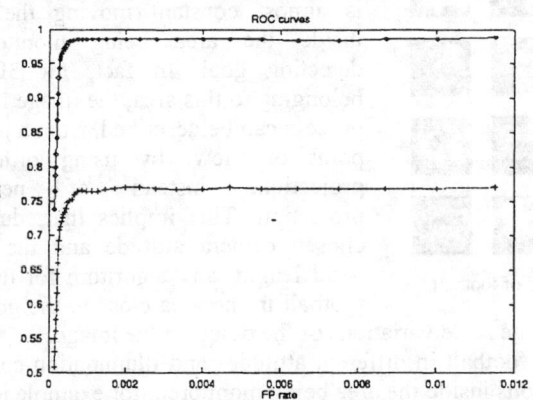

Figure 3 : ROC curves on the whole test set. The upper curve is relative to fully visible balls. The lower curve was computed considering also occlusion cases.

CONCLUSIONS

In this paper we focus on the problem of detecting 3D objects by using the image they project on the sensing plane of a camera. A particular istance of this general problem is the detection of a ball in images at the aim of detecting goals during a football match. The system can be seen as an electronic linesman which helps the referee to establish the occurrernce of a goal.

REFERENCES

[1] Ian Reid and Andrew Zisserman, Goal-directed Video Metrology. In 4th *European Conference in Computer Vision '96*, Cambridge, April 1996.
[2] K. Sung and T. Poggio, Example-based Learning for View-based Human Face Detection. Artificial Intelligence Laboratory, Massachusetts Institute of Technology, Cambridge MA, A.I. Memo No. 1521, 1994.
[3] V. Vapnik, The Nature of Statistical Learning Theory, Springer Verlag, 1995.
[4] T. Evgeniuos, M. Pontil and T. Poggio, A unified framework for Regularization Networks and Support Vector Machines. Artificial Intelligence Laboratory, Massachusetts Institute of Technology, Cambridge MA, A.I. Memo No. 1654, 1999.
[5] A. Blum and P. Langley, Selection of Relevant Features and Examples in Machine Learning, Artificial Intelligence, vol. 97, pp. 245-271, 1997.

CLASSIFICATION OF SUMMARIZED VIDEOS USING HIDDEN MARKOV MODELS ON COMPRESSED CHROMATICITY SIGNATURES

CHENG LU
School of Computing Science
Simon Fraser University
Vancouver, B.C., CANADA

MARK S. DREW
School of Computing Science
Simon Fraser University
Vancouver, B.C., CANADA

JAMES AU
School of Computing Science
Simon Fraser University
Vancouver, B.C., CANADA

ABSTRACT

As digital libraries and video databases grow, we need methods to assist us in the synthesis and analysis of digital video. Since the information in video databases can be measured in thousands of gigabytes of uncompressed data, tools for efficient summarizing and indexing of video sequences are indispensable. In this paper, we present a method for effective classification of different types of videos that makes use of video summarization that is the form of a storyboard of keyframes. To produce the summarization, we first generate a universal basis on which to project a video frame that effectively reduces any video to the same lighting conditions. Each frame is represented by a compressed chromaticity signature. We then set out a multi-stage hierarchical clustering method to efficiently summarize a video. Finally we classify TV videos using a trained hidden Markov model on the compressed chromaticity signatures and also temporal features of videos that are represented by their summaries.

INTRODUCTION

Video content classification is a necessary tool for efficient access, understanding and retrieval of videos. Different methods have been proposed in the literature for video program classification into predefined categories such as a commercial detection system (Hauptman and Witbrock, 1998). One successful study carrying out video classification was performed using a domain method relying on nearest neighbor clustering (Wei et al., 2000). The positive aspect of this classification method is its simplicity. Each decision made in the process corresponds to a certain aspect of human visual perception and it is straightforward to understand the rules. However, like most other research work on video classification, it did not take advantage of temporal features in video, which is a very powerful cue in understanding the video content. Therefore we explore the use of hidden Markov models (HMM) for video classification that can grasp the temporal information along with the visual information in video.

Previously, we successfully set out a novel illumination-invariant color histogram approach that performs good video characterization (Drew et al., 1998). In this method we form 12-vector chromaticity coefficients for any video frame. As well, on the basis of these coefficients we produce keyframe-based succinct summarized expressions for video using a multistage hierarchical clustering algorithm (Drew and Au, 2000). Here we extend this work to provide the capacity to perform semantic content discrimination tasks for video. After video characterization and summarization, we obtain two types of features: (1) chromaticity signatures for keyframes, each of which represents a scene; (2) temporal features including the durations of scenes in a video sequence and transition characteristics between scenes. We present a novel method that applies HMM to integrate the two features for video classification. This is motivated by the fact that a certain type of videos usually contains a set of frequent scenes that have similar visual information, such as news and basketball games, and also in most situations those types of videos have their individual approximately stable temporal pattern consisting of scene duration and transition characteristics.

The hidden Markov model is a popular technique widely used in pattern recognition (Rabiner and Juang, 1986). It has good capability to grasp temporal statistical properties of stochastic processes. The essence of the HMM process is to construct a model that explains the occurrence of observations (symbols) in a time sequence and use it to identify other observations sequences. Some researchers have applied HMM for video analysis and classification. In Nevenka's study (Dimitrova et al., 2000), HMMs can be formed using face and text trajectories and then classify the given video into one of four categories of TV programs: news, commercials, sitcoms and soaps. The key point of this approach is that the video content for these types of TV programs have to be satisfactorily characterized by capturing face and text trajectories appearing in the video. Huang et al. (1999) built an HMM framework using audio and image features for video classification. Although the use of both audio and visual features can improve classification accuracy, it can make the system complicated and hard to maintain and extend. Also, because the visual features are extracted for every frame, the HMM process needed to carry a great deal of information about the detailed variance between frames yet lacked consideration of the entire visual trajectory.

In this paper, we set out a video classification method, based on the Hidden Markov model, which utilizes the chromaticity signatures of keyframes from summarized video and effectively apprehends the entire temporal feature pattern for different types of videos. Firstly, we use the illumination-invariant color histogram video characterization method proposed by Drew and Au (2000) to produce a 12-vector feature for each frame; secondly, we effectively carry out video summarization using a multistage hierarchical clustering, obtaining keyframes. Finally, we perform the video classification task using hidden Markov models. In our experiment, we apply the method to the task of classifying television programs into the four categories: news report, commercials, live basketball game, and live football game.

The rest of the papers is organized as follows. Section 2 presents our chromaticity signature computation and our video sequence summarization scheme. Video classification method based on HMMs is proposed in section 3. Experimental results are given in section 4 and in section 5 we present the conclusion and future work.

ILLUMINATION-INVARIANT VIDEO SUMMARIZATION

We had developed a new low-dimensional video frame feature that is more insensitive to lighting change, motivated by color constancy work in physics-based vision, and apply the feature to keyframe production using hierarchical clustering. The point, vis-à-vis video summarization is that any video is effectively moved into the same lighting environment, making it meaningful to project video features onto a precomputed universal basis set.

Lighting is first discounted by normalization of color-channel bands (Drew et al., 1998). This step approximately but effectively removes dependence on both luminance and lighting color. Then image frames are moved into a chromaticity color space. As well as reducing the dimensionality of color to 2 this also has the effect of removing shading. In order to make the method fairly robust to camera and object motion, and displacements, rotations, and scaling, we go over to a 2D histogram derived from DC components of frames. Chromaticity histograms are then compressed –i.e., we treat the histograms as images. Here, we use a wavelet-based compression because this tends to strike a balance between simple low-pass filtering and retaining important details. Using a 3-level wavelet compression we arrive at 16×16 histograms.

However, we found that compression of histograms could be improved if the histograms are first binarized, i.e., entries are replaced with 1 or 0. The rationale for this step is that chromaticity histograms are a kind of color signature for an image, similar to a palette. In work involving recovering the most plausible illuminant from pixel values in an image (Finlayson et al., 1997) it was found beneficial to utilize this kind of color signature. Here, the step of binarizing the histogram not only reduces the computational burden, since true chromaticity histograms need not be computed, but also has the effect of producing far fewer negatives in the compressed histogram. Finally, we found that one further step could substantially improve the energy compaction of the representation: we carry out a 16×16 Discrete Cosine Transform (DCT) on the compressed 16×16 histogram. After zigzag ordering, we keep 21 DCT coefficients.

Since every image now lives in approximately the same lighting, we can in fact precompute a basis for the DCT 21-vectors, offline, which can then be reused for any new image or video. Here we determine a basis set by the Singular Value Decomposition (SVD) of the DCT 21-vectors. We found that 12 components in the new basis represent that entire DCT vector very well and that energy compaction worked better using a spherical chromaticity, rather than the usual linear one. Thus the method we set out here is to precompute a set of basis vectors, once and for all, and then form the 12-vector coefficients for any video frame with respect to this basis. So keyframe extraction can be carried out very efficiently, using only 12-component vectors.

A keyframe is extracted from each of segmented scenes in a video. We use a hierarchical clustering scheme to segment a video into a sequence of scenes (Drew and Au, 2000). This method executes a bottom-up multistage temporally merging process where only adjacent frames or frame groups are merged by calculating their vectors' $L2$ distance, as we wished to maintain the temporal order. A threshold of distance is assigned to determine the final clusters, and each of those clusters corresponds to a scene. Finally, a keyframe is extracted from the medoid of each cluster.

VIDEO CLASSIFICATION BASED ON HMM

We propose a hidden Markov model based method for video topic classification using visual and temporal features. We classify videos into four topic categories: news report, commericials, basketball game and football game.

Hidden Markov Model

In an HMM, there are a finite number of states and the HMM is always in one of those states. At each clock time, it enters a new state based on a transition probability distribution depending on the previous state. After a transition is made, an output symbol is generated based on a probability distribution, depending on the current state.

In the formal definition of HMM, the *hidden states* are denoted $Q = \{q_1, q_2, ..., q_N\}$, where N is the number of states and the *observation symbols* are denoted $V = \{v_1, v_2, ..., v_M\}$, where M is the number of observation symbols. The *state transition probability distribution* between states is represented by a matrix $A = \{a(i, j)\}$, where $a(i, j) = \Pr(q_j \text{ at } t+1 | q_i \text{ at } t)$, and the *observation symbol probability distribution* is represented by matrix $B = \{b_j(k)\}$, where $b_j(k)$ is the probability of generating observation v_k when the current state is q_j. *Initial state distribution* denoted by $\pi = \Pr(q_i \text{ at } t = 1)$ contains the probabilities of the model being in every hidden state i at time t=1 that is the start point for a HMM.

A HMM is always represented by $\lambda = (A, B, \pi)$. We constructed four HMMs, corresponding to news, commercials, football game, and basketball game, respectively.

HMM Process

The HMM process consists of two phases, *viz.* training and classification. Figure.1 shows the training process for the basketball game HMM and classification process for a given video clip.

Training. The HMM training step is essentially to create a set of hidden states Q and a state transition probability matrix A for each video topic category. The process of training a HMM for basketball videos is illustrated as the upper part in Fig.1. The other three HMMs are trained in the same way.

We first summarize all videos in the basketball game training set to extract chromaticity signatures of keyframes. Then we cluster these signatures and take the medoids of resulting clusters as hidden states for a HMM. Here we use the CLARANS clustering algorithm (Ng and Han, 1994). This algorithm is an improved k-medoids clustering algorithm based on randomized search, which is effective and efficient in spatial data mining with large data sets.

The state transition probability matrix includes the probabilities of moving from one hidden state to another. There are at most M^2 (M is the number of states) transitions among the hidden states. Since each of keyframe clusters obtained from the above step corresponds to a hidden state and each keyframe corresponds to a set of frames, we calculate the probabilities based on the number of frames falling into these clusters and the number of frames temporally transiting between clusters.

Classification. In the classification phase, given a target video, we first need to make an observation sequence and feed it into every HMMs. By evaluating the probability for each HMM, the target video is assigned into a topic category with the highest probability of the HMM.

For the observation sequence of the target video, we first summarize the target video and extract a set of keyframes in time order and take these keyframes as observation symbols. We then build a *temporal and keyframe-based summarized video sequence* (TSV) that is replicating each keyframe a number of times equaling the number of frames represented by the keyframe in the video sequence, and order these keyframes by time. In this way, a temporal feature can be maintained in the resulting sequence.

We also need to compute the observation symbol probability matrix B for each HMM, containing the output probabilities of the observation symbols given a particular hidden state. We compute these probabilities by the *inverse L2 distance* that means more distance between an observation frame vector and a state vector, less probabilities of the observation frames belonging to the state. We have to calculate this matrix in the classification phase because the observation symbols of video keyframes are in an infinite set so that there is no way to train it in advance. The rationale behind the use of the distance for the probability is that it is the visual distance or similarity that stands for the relationship between observations and states in this video case.

We use the *Forward* algorithm to calculate a probability for each HMM, and thus choose the video type with the most probable HMM. The *Forward* algorithm first defines a *partial probability*,

$\partial(t,j)$= Pr(observation symbol | hidden state is j) × Pr(all paths to state j at time t),

which is the probability of reaching an intermediate state, illustrated in Fig. 2(a), then recursively calculates the probability of observing a sequence given a HMM,

$$\partial(t+1, j) = b(t+1, j) \cdot \sum_i (\partial(t,i) \cdot a(i, j)) \quad \text{from } t \text{ to } t+1$$

and

$$\partial(1, j) = b(1, j) \cdot \pi(j) \text{ at } t = 1.$$

Finally we calculate the probability for each HMM with the sum of partial probabilities of reaching every states at the end moment. The forward algorithm is, in effect, based upon the lattice structure shown in Fig. 2(b). The key is that there is only N states (nodes at each time slot in the lattice), all the possible state sequences will remerge into these N nodes, no matter how long the observation video sequence is.

EXPERIMENT RESULTS

We evaluate our classification method by classifying four types of TV programs: news report, commercials, basketball game, and live football game. We collected 100 video clips of 5 minutes each from TV broadcasting as the training set for each video category. Another set of 30 video clips for each category was used as the testing set. We assume that the input videos always belong to one of the four categories of TV program.

Table 1 gives the classification results using four HMMs for the four video categories. From this table we observe that the classifier can accurately identify basketball games and football games, but the separation of commercials from news reports is somewhat less successful, although still impressive. The reason is that these categories contain much chromatic information and the state duration is usually short in their models.

CONCLUSIONS

In this paper, we have described a video content classifier based on HMM using chromaticity signatures from video summarization and their temporal relationship. The video characterization and summarization method represents the video as a series of compressed chromaticity signatures. The HMM process uses these signatures and takes advantage of the temporal feature to train HMMs and evaluate the probability of the given video being in one of the four categories of TV programs.

REFERENCES

Dimitrova, N., Agnihotri, L., and Wei, G., 2000, "Video classification based on HMM using text and faces," *European Conference on Signal Processing*, Finland.

Drew, M. S., and Au, J., 2000, "Video keyframe production by efficient clustering of compressed chromaticity signatures," Proceedings, *ACM Multimedia '00*, pp.365-368.

Drew, M. S., Wei, J., and Li, Z. N., 1998, "Illumination-invariant color object recognition via compressed chromaticity histograms of color-channel-normalized images," proceedings, *IEEE ICCV98*, pp. 533-540.

Finlayson, G. D., Hubel, P. M., and Hordley, S., 1997, "Colorur by correlation," In *Fifth Color Imaging Conference*, page 6-11.

Hauptman, A. G., and Witbrock, M. J., 1998, "Story segmentation and detection of commercials in broadcast news video," Proceedings, *Advances in Digital Libraries Conference*, Santa Barbara, CA.

Huang, J., Liu, Z., Wang, Y., Chen, Y., and Wong, E. K., 1999, "Integration of multimodal features for video classification based on HMM", *IEEE Third Workshop on Multimedia Signal Processing*, pp. 53 - 58, Copenhagen, Denmark.

Ng, R., and Han, J., 1994, "Efficient and effective clustering method for spatial data mining," Proceedings, *Int'l Conf. on Very Large Data Bases (VLDB'94)*, Santiago, Chile, pp. 144-155.

Rabiner, L. R., and Juang, B. H., 1986, "A tutorial on Hidden Markov Models," *IEEE ASSP Magazine*. pp. 4-15.

Wei, G., Agnihotri, L., and Dimitrova, N., 2000, "TV program classification based on face and text," Proceedings, *IEEE multimedia and Expo 2000*, New York.

Table 1. Classification results (unit: 100%)

Expectation \ Result	News	Commercial	Basketball	Football
News	80.0	13.3	0	6.7
Commercial	23.3	66.7	6.7	3.3
Basketball	3.3	3.3	93.3	0
Football	0	3.3	0	96.7

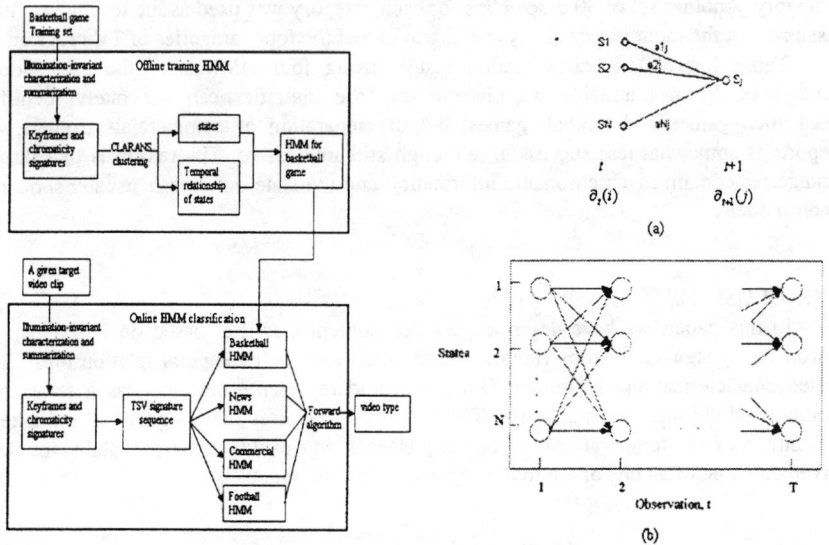

Figure 1. A HMM classification system

Figure 2. (a) Illustration of the sequence of operations required for computation of partial probability. (b) Implementation of the computation of partial probability in terms of a lattice of observations and states.

PRINTED THAI CHARACTER RECOGNITION USING NEURAL NETWORKS

PHONGTHEP RUXPAKAWONG AND ARIT THAMMANO
Advanced Computer Application and Design Research Group,
Faculty of Information Technology,
King Mongkut's Institute of Technology Ladkrabang,
Bangkok, 10520 Thailand

ABSTRACT

Thai character recognition has gained a lot of attention in the past decade. Several approaches have been widely explored, such as the structural analysis methods, hidden Markov model, syntactic methods, fuzzy linguistic rules, and neural networks. However, there has been a limited success on Thai character recognition, compared to the English characters. The difficulties are due to the complexity of the Thai characters, which have many curves and circles. This paper proposes a method to recognize the printed Thai characters based on the characteristic of the top boundary of the character, the number of cavity of the bottom boundary, and the contour of characters. The contour is divided into four segments. Each segment represents four boundaries of the characters: top, bottom, left, and right. These features are then used as inputs to the neural network classifier. The experiments have been conducted to recognize 248 printed Thai characters from 8 different fonts. The recognition rate is 93.1%.

INTRODUCTION

In the past decade, several approaches in Thai printed character recognition have been proposed such as the structural analysis methods, hidden Markov model, syntactic methods, fuzzy linguistic rules, and neural networks. However, there has been a limited success on Thai character recognition because of the complexity of the Thai characters, which have many curves and circles. For example, "ฌ" contains two loops, straight lines, curves, and a small zigzag line at the top boundary of the character. Moreover, some Thai characters look very alike, e.g. ค-ต-ศ, ช-ซ, ท-ฑ, บ-ป. The recognition rate on the training fonts is only about 90%-95% (Tanprasert and Koanantokool, 1996). Therefore, the objective of this paper is to propose a method to recognize the printed Thai characters based on the characteristic of the top boundary of the character, the number of cavity of the bottom boundary, and the contour of characters.

FEATURE EXTRACTION

Thai document typically consists of five types of symbol namely: consonants, vowels, tone symbols, Arabic numerals, and other symbols as shown in Table 1. Similar to English, Thai document is written from left to right with lines filling the page from top to bottom. Each line composes of four levels: above-upper level, upper level, central level, and lower level. Figure 1 shows an example of a line in Thai document. Most consonants are located only in the central level of the line, while some consonants have either ascender or descender. Vowels and other symbols generally can appear in three levels: upper, central, and lower level. Tone symbols can be seen in either the above-upper level or the upper level. This paper focuses only on the recognition of consonants that have neither ascender nor descender. Since these characters contain no unique characteristic, they are more complicated to recognize.

Table 1 Symbols in Thai Documents

Types	Members
Consonants	กขคฆงจฉชซฌญฎฏฐฑฒณดตถทธ นบปผฝพฟภมยรฤลวศษสหฬอฮ
Vowels	ะาัิีึืุู,ฺเแโใไๅๆ็่้๊๋์
Tone Symbols	่ ้ ๊ ๋
Other Symbols	' ๆ ๆ '
Arabic Numerals	0123456789

Figure 1 An Example of a Line in Thai Documents

An optical scanner scans a printed document and converts it into an image. The segmentation process then segments the entire image into individual character blocks. Figure 2 shows a simplified example of the document image after the segmentation process. Once the image is turned into character blocks, the features will be extracted from each character block. Since all features used in this paper are some kinds of characteristics of the character contour, the preprocessing step is required to (1) extract the character contour from the character block (2) divide the character contour into four segments, which will be further analyzed to create needed features. The following is the process of extracting the four segments of character contour from the character block (Figure 3):

i) Assign four points (A, B, C, and D) on the character.
ii) Draw the line along the boundary of the character to connect point A and B, B and D, A and C, and C and D. In connecting two points, the line must not pass any point other than the starting point and the destination.
iii) Plot the graphs of AB Link with respect to V1, AC Link with respect to H1, BD Link with respect to H2, and CD Link with respect to V2.

The features used in this paper are:

Boundaries of the Character

From the above preprocessing step, the character contour is divided into four segments, which represent four boundaries of the character. The four boundaries are represented by AC link (left), BD link (right), AB link (top), and CD link (bottom). All links are tested with the algorithm suggested by Thammano and Ruxpakawong (2001) to determine whether the top, bottom, left, and right boundaries are open (0) or close (1).

The Number of Cavity of Bottom Boundary

A cavity is a region bounded by the stroke on its three sides. It was first introduced to use in the handprinted Thai characters recognition by Phokharatkul and Kimpan (1998). In this paper, however, only the bottom cavity is considered. If the character contains no bottom cavity, the feature value is "00"; while if the number of cavity is 1 or more than 1, the value is "01" or "11" respectively.

The Characteristic of the Top Boundary of the Character

Some Thai characters have a small zigzag line at the top boundary of the character. This small zigzag line is very similar to a noise received from a bad scanning. Therefore, a special consideration is needed to differentiate between the two. To assign the value to this feature, the AB link must be analyzed. If there is no small zigzag line on the AB link, the feature value is "0." On the other hand, if there is a small zigzag line on the AB link, the feature value is "1."

Figure 2 An Example of Image Segmentation

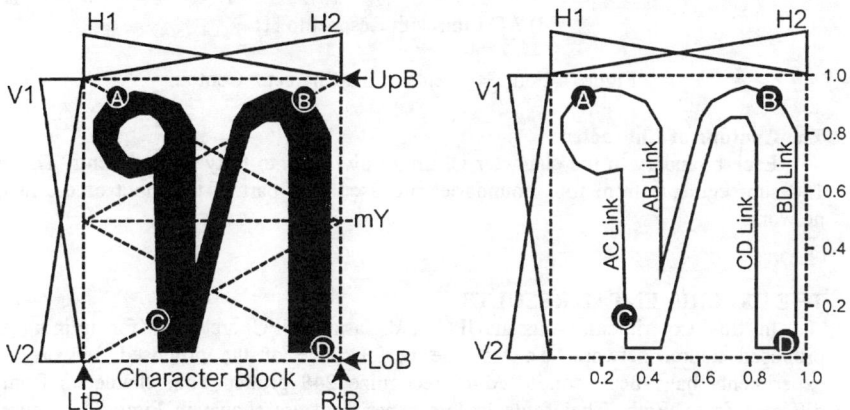

Figure 3 The Process of Extracting the Four Segment of Contour of Character

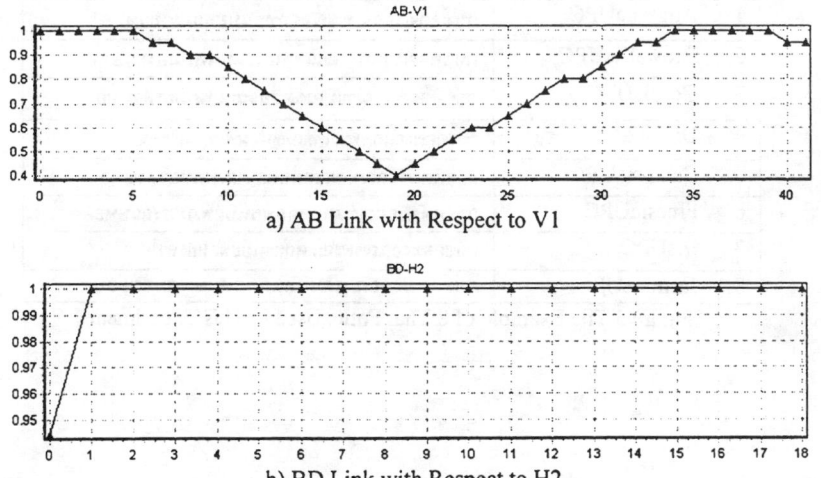

a) AB Link with Respect to V1

b) BD Link with Respect to H2

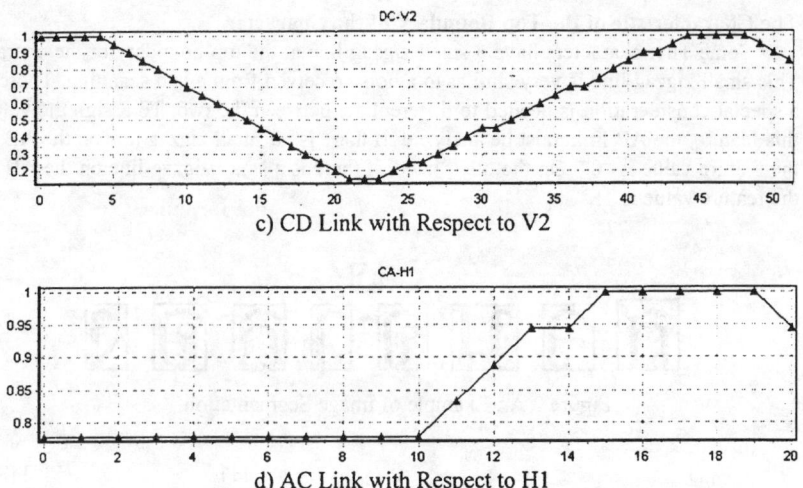

c) CD Link with Respect to V2

d) AC Link with Respect to H1

Figure 4 Four Segments of the Character Contour

The Contour of Character

Each boundary of the character is further divided into forty sub-segments. Then all 160 sub-segments from four boundaries are used as a part of the input of the neural network.

THE EXPERIMENTAL RESULTS

In this experiment, AngsanaUPC and JasmineUPC are used for training the proposed neural network. To test the performance of the proposed network, the experiments have been conducted to recognize 248 printed Thai characters from 8 different fonts. Eight Thai fonts in this experiment are shown in Figure 5. Figure 6 summarizes the final configuration of the neural network. From the testing results, the recognition rate is 93.1%.

1.	AngsanaUPC	กขฃคฅฆงจฉชฌญฎฏฐฑฒณดตถทธนบปผฝพฟภมยรลวษหอ
2.	BrowalliaUPC	กขฃคฅฆงจฉชฌญฎฏฐฑฒณดตถทธนบปผฝพฟภมยรลวษหอ
3.	CordiaUPC	กขฃคฅฆงจฉชฌญฎฏฐฑฒณดตถทธนบปผฝพฟภมยรลวษหอ
4.	DilleniaUPC	กขฃคฅฆงจฉชฌญฎฏฐฑฒณดตถทธนบปผฝพฟภมยรลวษหอ
5.	EucrosiaUPC	กขฃคฅฆงจฉชฌญฎฏฐฑฒณดตถทธนบปผฝพฟภมยรลวษหอ
6.	FreesiaUPC	กขฃคฅฆงจฉชฌญฎฏฐฑฒณดตถทธนบปผฝพฟภมยรลวษหอ
7.	IrisUPC	กขฃคฅฆงจฉชฌญฎฏฐฑฒณดตถทธนบปผฝพฟภมยรลวษหอ
8.	JasmineUPC	กขฃคฅฆงจฉชฌญฎฏฐฑฒณดตถทธนบปผฝพฟภมยรลวษหอ

Figure 5 The Example of 8 Thai Fonts Used in This Experiment

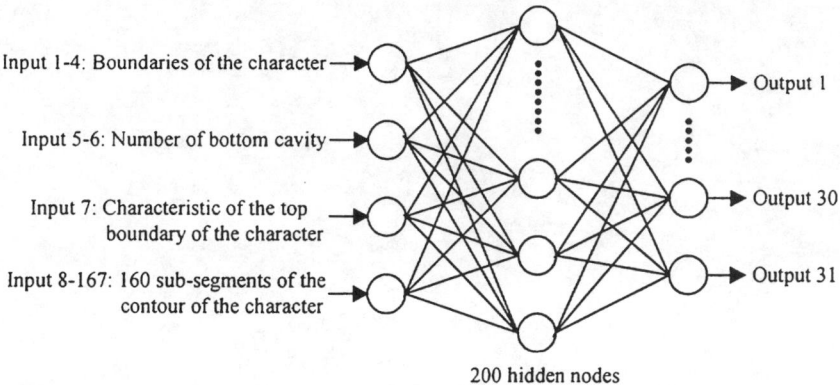

Figure 6 The Configuration of the Neural Network

CONCLUSION

The experiment shows that features analyzed from many aspects of the character contour can be applied successfully with the neural network to solve the Thai character recognition problem. However, the search for a more adequate representation of the character is still carrying on in the future works. Moreover, the hybrid of global feature, local feature, and the neural network will be explored to further improve the recognition rate.

REFERENCES

Kijsirikul, B., S. Sinthupinyo, and A. Supanwansa (1998), "Thai Printed Character Recognition by Combining Inductive Logic Programming with Backpropagation Neural Network," *Proceedings of the 1998 IEEE Asia-Pacific Conference on Circuits and Systems*, pp. 539-542.

Missiourev, A. (1998), "Hand-printed Character Recognition by Neural Networks," *Proceedings of the 5th German-Russian Workshop on Pattern Recognition and Image Understanding*, Germany.

Phokharatkul, P. and C. Kimpan (1998), "Recognition of Handprinted Thai Characters Using the Cavity Features of Character Based on Neural Network," *Proceedings of the 1998 IEEE Asia-Pacific Conference on Circuits and Systems*, pp. 149-152.

Tanprasert, C. and T. Koanantokool (1996), "Thai OCR: A Neural Network Application," *Proceedings of the 1996 IEEE TENCON*, Vol. 1, pp. 90-95.

Thammano, A. and P. Ruxpakawong (2001), "Printed Thai Character Recognition Using the Hybrid Approach," *To appear in Proceedings of the 2001 ITC-CSCC*, Tokushima, Japan.

INFORMATION CONTENT OF THE FREQUENCY DICTIONARIES, RECONSTRUCTION, TRANSFORMATION AND CLASSIFICATION OF DICTIONARIES AND GENETIC TEXTS

ALEXANDER N. GORBAN[1]

TATIANA G. POPOVA[1,2] MICHAEL G. SADOVSKY[1,2]
[1]Institute of Computational Modeling of SD of RAS
Krasnoyarsk, Russia
[2]Institute of Biophysics of SD of RAS
Krasnoyarsk, Russia

DONALD C. WUNSCH, II
University of Missouri-Rolla
Department of Electrical and Computer Engineering
Rolla, MO

ABSTRACT
A study of the relation between a structure of symbol sequences and the meaning of them encrypted in the interlocation of symbols is a key problem for molecular biology, biophysics and many other fields of sciences. Usually, researchers meet no problem in understating the function of sequences studied. A structure is a much more complicated matter to understand. When studying nucleotide sequences, one often encounters the intron-exon structure, or the structure determined by operons, etc. Further, we should understand the information structure of a nucleotide sequence as its frequency dictionary, either *real* one, or the *reconstructed* one, or the *transformed* one. Such understanding of the structure of a sequence enables a researcher to introduce easily the idea of a closeness of two (or several) structures.

The *real* frequency dictionary W_q is defined as the list of all the strings of the length q occurring in the given nucleotide sequence (of length N) accompanied with the frequency of their occurrence. If n is the power of the alphabet (for nucleotide sequences $n = 4$), then total number of all the words of the length q is n^q. Obviously, not every word of the length q could be met within a text, as soon as q is large enough. Let us complete the frequency dictionary of the sequence studied to the entire one (which contains all the words of the given length q) adding the words with zero frequency. Then every frequency dictionary could be represented as a point $F(f_1, f_2, ..., f_{n^q})$ in n^q –dimensional space with the coordinates presented by the frequencies of the relevant words f_j: $0 \leq f_j \leq 1, j = 1,2,... n^q$. Then, a set of M genes yields the set of points in n^q –dimensional space — $\{F^i\}$, $i = 1,2,...M$. A distance between two points in this multidimensional space is determined due to Euclidean metrics. Two sequences would be considered to be close each other, if two points in n^q –dimensional space are close, corresponded to their frequency dictionaries of the length q.

To reveal the differences between nucleotide sequences and their information structure, it seems to be more useful to consider relative coordinates. In order to obtain

Table 1. The sites with high information value, averaged over *Firmicutes; Actinomycetes* taxae

Real frequency is greater than expected		Real frequency is lower than expected	
Triplet	$f/\tilde{f}$	Triplet	$f/\tilde{f}$
CCT	1,355	CCA	0,641
AGC	1,338	TAT	0,67
TAA	1,302	AAA	0,709
CTT	1,282	TAG	0,762
TCA	1,194	TCT	0,771
TAC	1,189	TTT	0,784
GAT	1,177	GAC	0,81

the relative coordinates we consider the information contained in the real frequency dictionary of the length q.

The sequence of the frequency dictionaries $W_1, W_2, \ldots W_q$, corresponded to the same text yields a relation, namely, each previous dictionary could be obtained from the one of posterior ones by a simple summation. An inverse statement does not hold true: as a result of summation part of the information about the text is eliminated. It makes an inverse transformation ambiguous.

The exact reconstruction of a posterior dictionary from a given previous one does not exist always, in general case. The exact reconstruction of W_{q+s} over W_q is possible if all the words in W_q have the unique continuation, such a dictionary contains all the information on the original sequence. Otherwise, the single-valued reconstruction is impossible: each frequency dictionary W_q yields the set of W_{q+s}. One could nevertheless seek for the most probable continuation $-\widetilde{W}_{q+s}$ (*reconstructed* dictionary) — that corresponds to the given dictionary W_q. To reconstruct the dictionary $\widetilde{W}_{q+s}(q)$ one must choose among all possible continuations of the given dictionary W_q that one with the least determinacy, i.e. that one, which yields the maximal entropy.

The entropy of a frequency dictionary W_q is defined as

$$S_q = -\sum_{j=1}^{n^q} f_j \ln f_j \qquad (1)$$

The exact solution of the extremum problem $S_{q+s} \to max$ with the bound condition for dictionaries $W_q \leftarrow \widetilde{W}_{q+s}(q)$ looks very close to the well-known Kirkwood approximation in statistical physics:

$$\widetilde{f}_{i_1 \ldots i_q i_{q+1} \ldots i_{q+s}} = \frac{f_{i_1 \ldots i_q} \cdot f_{i_2 \ldots i_{q+1}} \cdots f_{i_{q-s+1} \ldots i_{q+s}}}{f_{i_2 \ldots i_q} \cdot f_{i_3 \ldots i_{q+1}} \cdots f_{i_{q-s+1} \ldots i_{q+s-1}}} \text{ for } q>1, \text{ and} \qquad (2)$$

$$\widetilde{f}_{i_1 \ldots i_{q+s}} = f_{i_1} \cdot f_{i_2} \cdots f_{i_{q+s}} \text{ for } q=1, \qquad (3)$$

here $i_1 \ldots i_q i_{q+1} \ldots i_{q+s}$ is the word of the length $q+s$ and index i corresponds to a nucleotide; the reconstructed frequencies are denoted as $\widetilde{f}$.

Table 2. Taxonomy content of the set of nucleotide sequences of 16SRNA

N	Taxonomy	Amount
1	Chloroflexaceae/Deinococcaceae group.	29
2	Cyanobacteria	9
3	Cytophagales	117
4	Fibrobacter	13
5	Firmicutes; Actinomycetes	335
6	Firmicutes; Low G+C gramm-positive bacteria	485
7	Proteobacteria; alpha subdivision	262
8	Proteobacteria; beta subdivision	63
9	Proteobacteria; delta subdivision	47
10	Proteobacteria; epsilon subdivision	43
11	Proteobacteria; gamma subdivision	216
12	Spirochaetales; Leptospiraceae	14
13	Spirochaetales; Spirochaetaceae	35
14	OTHERS	56

If a reconstructed frequency dictionary $\widetilde{W}_{q+s}(q)$ coincides entirely with the original one W_{q+s}, then it means that the entire information on the original sequence is contained in the dictionary W_q. Deviations of the reconstructed dictionary $\widetilde{W}_{q+s}(q)$ from the real one W_{q+s} show what new is introduced by $(q+s)$–tipples, in comparison to q–tiples.

The deviations of the real frequency dictionary W_{q+s} from the most probable continuation $\widetilde{W}_{q+s}(q)$ are of the greatest interest and present a new method to obtain relative coordinates of the sequence in multidimensional space. To measure these deviations, one should introduce a new object, so called *transformed dictionary*. In that latter, each word is assigned with the ration of the real frequency of the word to the reconstructed frequency, i.e. the introduced value displays how much the real frequency of the word differs from the expected one. This transformation of the frequency dictionary allows to explicate some peculiarities of the information structure of nucleotide sequences.

As before, each nucleotide sequence would be corresponded with the point $P(p_1, p_2, \ldots, p_{n^q})$ in n^q–dimensional space, where the coordinated of a point are the ratios of the real frequency to the reconstructed one: $p_j = f_j / \widetilde{f}_j$, for $\widetilde{f}_j \neq 0$, and $p_j = 1$, for $\widetilde{f}_j = 0$, $j = 1,2,\ldots n^q$. The values p_j show how much the real frequencies of words differ from the expected ones. If $p_j \approx 1$ for some word within the nucleotide sequence, then the information value of that word is not high: its most probable expected frequency almost coincides to the real one. The words, whose frequency ratio p_j differs to the greatest extent (either upward, or downward) from the real one present the most valuable sites of the given length within the nucleotide sequence studied. It should be stressed that the length of a site is rather essential, since any low-valued site of the length q might be incorporated into the high-valued site of the length $q+1$, which, in turn, might be incorporated into a longer site of the low information value. Table 1 presents the sites

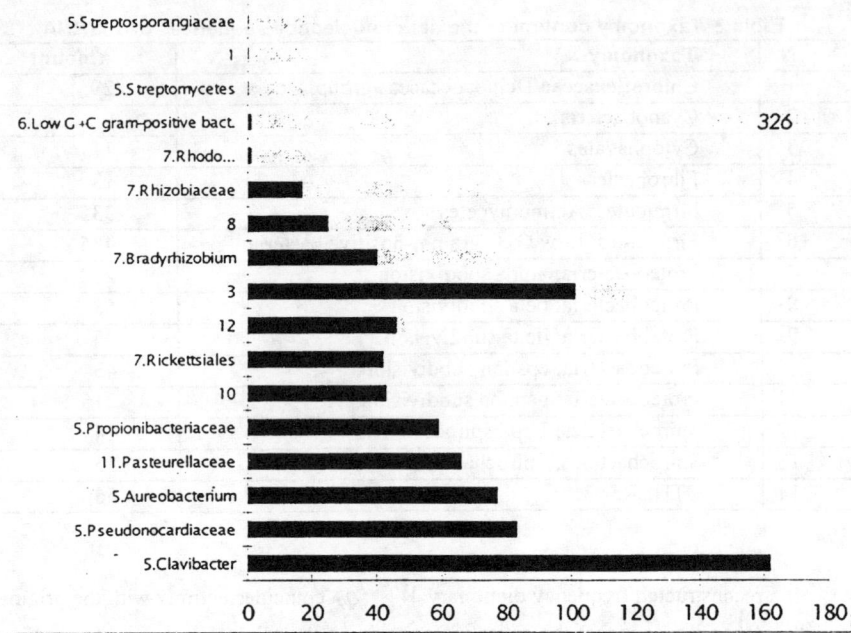

Figure 1. The classification of the set of 16S RNA sequences developed over the real frequency dictionaries

with high information value, obtained from transformed dictionary: $q=3$; as reconstructed dictionary $\widetilde{W}_3(2)$ is used. For origin of sequences used see Table 2.

To compare various nucleotide sequences, one should consider the differences between p_j observed for different sequences rather than the deviations of that value form 1, that latter might occur simultaneously (see Table 3).

Here we considered the reconstructed dictionaries, using $s=1$. Below are the formula for the reconstructed frequencies of $\widetilde{W}_{q+1}(q)$, for a convenience:

$$\widetilde{f}_{i_1\ldots i_{q+1}} = \frac{f_{i_1\ldots i_q} \cdot f_{i_2\ldots i_{q+1}}}{f_{i_2\ldots i_q}} \quad . \quad (4)$$

The formula (4) coincides with the well-known expressions for the transitional probabilities in a symbol sequence obtained as a realisation of the Markov random process of the first order. It should be stressed, that the formula (4) for the reconstructed dictionary is yielded with respect to neither hypothesis on a peculiar structure of a sequence. These formula present the most likelihood hypothesis on the frequency dictionary W_{q+1} that could be implemented from a consideration of the dictionary W_q. One should consider the original symbol sequence to be markovian one if and only if the expressions for the real (but not the reconstructed ones) frequencies would hold true in the limit case of infinitely long original sequence.

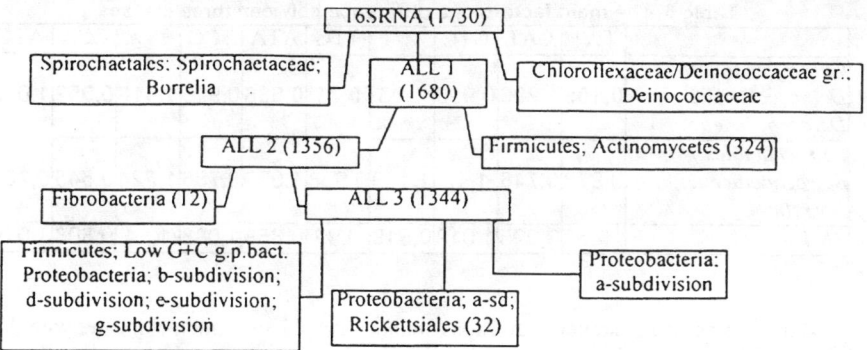

Figure 2. A hierarchic classification of the original set of 16S RNA

The dictionary $\widetilde{W}_{q+1}(q)$ yielded is the most probable continuation of the dictionary W_q. A comparison of the reconstructed dictionary $\widetilde{W}_{q+1}(q)$ with the real one W_{q+1} allows to make explicitly the peculiarities of the information structure of nucleotide sequences, since the maximal differences between the real and reconstructed frequencies are the most «unexpected» events in a transition from the dictionary W_q to the dictionary $\widetilde{W}_{q+1}(q)$.

A set of M nucleotide sequences yields a set of points in n^q –dimensional space. Real and transformed dictionaries provide different coordinates of the points. A study of the distribution of the points in n^q –dimensional space arranges the experimental part of current work. To study the distribution one may use any algorithm of cluster analysis. We used a dynam kern method that is absolutely similar to the cluster analysis provided by Kohonen neural networks.

We have studied the bacterial 16S RNA sequences, their total number was 1730. Table 2 shows the taxonomy content of the set of nucleotide sequences studied. It is evident, the taxae are rather diverse, but inhomogeneous with respect to a number of sequences within the same taxa.

The set of 16S RNA has been split into the groups according to the dynamic kerns method. The classification has been elaborated both over the real frequency dictionaries, and over the transformed dictionaries. The classification obtained differ from each other, as it has been expected. Nevertheless, both classifications yield a reasonable correlation to the taxonomy classification of the gene bearer.

The classifications were developed over the dictionaries of $q=3$, since a reliable method implication is possible only when the number of the objects to be classified exceeds significantly the dimension of the space. Besides, the dictionary of the length 3 represents more structural entities of a nucleotide sequence ($q=2$ or $q=1$); at least one structure is presented completely, that is the genetic code structure.

The classification of the set of 16S RNA sequences developed over the real frequency dictionaries is shown in Fig. 1. The original set of sequences split into two groups. The vertical axis on the diagram presents the taxae, and the horizontal axis presents the number of sequences from the given taxon in group. A separation of each taxon into two groups is shown in gray and black color. Fig.1 is presented in brief

Table 3. The main factors of a difference between three classes

	TAT	CAT	GTC	TCT	TTG	ATA	TCG	TTA	CCA	ATC
Chloroflexaceae/ Deinococcaceae gr.: Deinococcaceae	0,109	1,296	0,998	1,06	0,71	0,686	0,916	1,418	0,952	1,028
Spirochaetales: Spirochaetaceae: Borrelia	0,978	0,746	1,42	1,189	0,958	1,099	0,788	1,224	0,545	0,737
All 1	0,6	0,995	1,019	0,819	0,99	0,859	1,088	1,14	0,808	1,019

redaction: some small taxonomy divisions are not shown. Numeration coincides with that of Table 2.

A non-randomness of that split is evident. A correlation between a statistical structure of nucleotide sequence and taxonomy of its bearer is evident. Nonetheless, a development of a similar diagram for the higher taxon level results in a significant growth of a equity of an occupation of both classes by the genes from the same taxon group. This effect may follow from a well-known fact that the taxonomy of higher taxonomic levels of prokaryote seems to be rather artificial.

A hierarchic classification of the original set of 16S RNA studied implemented due to the transformed dictionaries is shown in Fig. 2. Transformed dictionaries were obtained from the real dictionaries of the length 3 and reconstructed ones $\widetilde{W}_3(2)$ (i.e. dictionary of the length 3, reconstructed using the dictionary of the length 2, according to formula (4)).

. One can obviously see, that some specific taxonomy units are separated on the each level of the classification. In spite of rather moderate number of the units separated, they contain almost all sequences of this peculiar taxonomy group, from the original set of 16S RNA sequences. The classes obtained at the fourth level of the classification failed to satisfy the separation condition, and we show them just to finalize the classification implemented.

In more detail, the features that make difference between the sequences of the first division level are presented in Table 3. Numbers are the coordinates p_j introduced before and averaged trough over the classes. Coordinates yielded the maximal difference between the classes are chosen and presented. The relevant triplets are shown too.

CONCLUSION

We have studied an interrelation between a structure of nucleotide sequence and a taxonomy of its bearer. A proximity of structures was understood as a proximity of frequency dictionaries, either real or transformed ones, in Euclidean metrics. An extended group of 16S RNA has been studied. From the point of view of the molecular aspects of the selection theory, the most important thing here is that the classification implemented over the real frequency dictionaries of the length 3 correlates best of all to the generae. A gender is included entirely either into the first class, or into the second one, and the exclusions are rarely met.

A transformation of the frequency dictionary of a nucleotide sequence, i.e. a usage of the reconstructed frequency dictionary in order to outfit a non-random component in a distribution of q–tipples, allows to compare the nucleotide sequences over their information characteristics. Clusterisation of a set of nucleotide sequences over their transformed dictionaries yields the classes of proximal sequences. In the case studied, where 16SRNA were studied the classes obtained contain the sequences of specific taxonomy.

A classification of sequences developed over the real frequency dictionaries differs basically from the latter developed over the transformed ones. A classification over the real frequency dictionaries represent mainly the difference in nucleotide composition of the sequences. A decomposition of the original set of sequences into two classes with a good correlation between a class occupation and the taxonomy of the bearer of a sequence proves this idea clearly. A classification over the transformed dictionaries isolates one or two groups of sequences of the same taxonomy, on each level of the classification.

REFERENCES

Bugaenko, N.N., Gorban, A.N., and Sadovsky, M.G., 1996, "Towards the determination of the information content of nucleotide sequences," *Molekulyarnaya biologiya*, Vol. 30, iss. 3., pp. 529 – 541. (in Russian)

Bugaenko, N.N., Gorban, A.N., and Sadovsky, M.G., 1998, "Maximum entropy method in analysis of genetic text and measurement of its information content," *Open System & Information Dynamics*, Vol. 5, № 3., pp.265 – 278.

Gorban, A.N., Popova, T.G., and Sadovsky, M.G., 1998, "Automatic classification of nucleotide sequences and its relation to natural taxonomy and protein function,"*Proc. of 1st Int. Conf. on Bioinformatics of Genome Regulation and Structure, Novosibirsk, Aug., 24- 27, 1998*, Vol. II, pp. 314 –317.

Mirkes, E.M., Popova, T.G., and Sadovsky, M.G., 1993, "Investigating Statistical properties of genetic texts: A new approach," *Advances in Modelling & Analysis*, ser. B, AMSE Press, Vol. 27, #1, pp.1-17.

AN OPTICAL FIBRE MULTIPOINT SENSOR - UTILISING U-BEND SENSORS - BASED ON ARTIFICIAL NEURAL NETWORK PATTERN RECOGNITION.

WILLIAM B LYONS
University of Limerick,
Dept. Electronics and Computer Engineering,
Limerick,
IRELAND.

HARTMUT EWALD
Hochschule Wismar,
Fachhochschule für Technik,
Wirtschaft und Gestaltung,
Wismar,
GERMANY.

COLIN FLANAGAN
University of Limerick,
Dept. Electronics and Computer Engineering,
Limerick,
IRELAND.

ELFED LEWIS
University of Limerick,
Dept. Electronics and Computer Engineering,
Limerick,
IRELAND.

ABSTRACT

An optical fibre multipoint sensor system incorporating multiple (3) U-bend sensors is reported which is capable of detecting contaminants in water and depositions by coating on its surface. Interrogation of the data arising from the sensor is achieved using Artificial Neural Networks (ANN). Preliminary analysis arising from data obtained with two sensors located on a single fibre loop has shown that this method has enabled the presence of contaminants such as limescale build-up in hard water to be identified. A reduction in the optical power loss due to improved sensor fabrication and optical launch conditions has meant that a greater number of sensors can be incorporated on the fibre loop. This article reports experimental results arising from this and the suitability of the ANN for their interpretation and classification. The sensor of this investigation is based on 62.5 μm core diameter Polymer Clad Silicon (PCS) fibre which has had its cladding removed in the sensing regions. The addressing of the fibre is achieved using an Optical Time Domain Reflectometer (OTDR) and is thus capable of resolving power loss over distance (along its length). Results show that the system is capable of recognising the presence of alcohol at each of the sensor locations.

INTRODUCTION

There has been a rapid expansion over the last ten years in the use of optical fibre sensors for the purpose of environmental monitoring (Udd, 1995; Kersey, 1996; Rogers and Poziomek, 1996).

Prior to this, a wide range of single-point fibre sensors existed but there was an increasing requirement for fibre sensors with distributed (spatial resolution) or a number of discrete sensors on a single fibre loop (multipoint) (Liberman, 1991). As a result, the use of multiple sensors in a network or array has increased over the last few years. In order to locate the position and changes in the multipoint sensors on the fibre, the use of specialised techniques such as optical time domain Reflectometry (OTDR) is often required.

Optical time domain Reflectometry, since first being reported in 1976 (Barnoski and Jensen, 1976), has become an established technique for attenuation monitoring in optical fibre networks within the telecommunications industry and is capable of detecting optical attenuation as a function of distance along the length of the fibre. As a result of this, OTDR has recently found many applications in distributed or multi-point sensing on a single fibre loop (Kersey and Dandridge, 1988) where the OTDR is used to monitor the fluctuation in optical fibre attenuation caused by an external parameter.

Optical fibre sensor signals can often be complex and this is particularly so in the case of distributed sensors. Cross-coupling of signals from external parameters, e.g. temperature (the true measurand) and strain or microbending (interfering parameters in this case), adds to the difficulty in interpreting data from such systems.

It is well established that different types of faults in fibres give rise to OTDR signals of different signatures. It has been proposed that for many applications of optical fibre distributed sensors, artificial neural network pattern recognition techniques may be used to resolve the problems arising from cross-sensitivity to other parameters (Lyons *et al*, 1999a; Lyons *et al*, 2000a).

Artificial neural networks (ANNs) have generally been accepted as a major tool in the development of 'intelligent systems' in the 1990s, the system developed for this investigation being one such example. The relevance and development of ANNs for optical sensing applications have been reviewed by Lyons *et al* (1999a).

In previous work by Lyons *et al* (2000b), data from U-bend optical fibre sensor was used initially to train a single layer perceptron and then a multi layer perceptron, and a comparative analysis of their results showed that a multi-layer perceptron was required to adequately classify the data. This provided a means of validating the technique, which may be extended to the interpretation of more detailed data from similar sensors.

In this work, the technique is extended to interpret that data from a multipoint sensor system containing 3 U-bend optical fibre sensors. Initial test results are reported which have been obtained from an in house manufactured 3 U-bend multipoint fibre sensor system, where the sensor has been subjected to actions such as submersion in alcohol and exposure to air.

EXPERIMENTAL SETUP

The optical fibre used in this investigation was a 62.5 um core diameter polymer-clad silica (PCS). The three sensor elements were incorporated into a 1km length of otherwise continuous fibre. In order to maximise the sensitivity of the sensors, a U-bend configuration was utilised where the cladding was removed and the core was exposed directly to the absorbing fluid under test. The operation of these sensors is based on the modulation of the light intensity propagating in the fibre by the measurand as a result of the interaction with the evanescent field penetrating into the absorbing measurand. Much experimental work has already been reported (Gupta *et al*, 1996; Khijwania and Gupta, 1998,

1999, 2000) detailing the resulting sensitivity gains from the evanescent wave increases resulting from the curving of the sensing fibre (Otsuki *et al*, 1998).

Sensor Fabrication

The sensors were fabricated from the same fibre as that used to transmit light through the system (i.e. 62.5um PCS), so that the optical power losses due to splicing would be minimised. In order to produce the U-bend in the fibre, the buffer and cladding were chemically removed from three 2cm length sections located at 665m, 756m and 857m respectively from the launch end of the fibre.

In order to shape the fibre to the desired sensor configuration, the exposed fibre core was cleaned using acetone and was then slowly bent into a U shape using heat from a flame. Their final bend radius were measured to be 1 mm under a microscope and each sensor was then placed into a protective case as shown in figure 1(a). A photograph of one of the sensors is shown in figure 1(b).

Sensor System Configuration

The system configuration comprised the optical fibre, the three sensor sections, an OTDR (EXFO IQ7000 – 0.85um) and a computer based-virtual instrument. This is shown in figure 2.

A Pentium MMX 200 MHz PC was configured as a virtual instrument for data capture and pre-processing using a LabVIEW-based virtual instrument (VI) developed by the authors specifically for this purpose. The data output from the VI was subsequently made available for analysis by the artificial neural network. The ANN was implemented using SNNS V3.2 (Stuttgart Neural Network Simulator, 1995).

The OTDR trace obtained for all three sensors in air is shown in Figure 3. The peak on the left side of the trace is due to the connector at the OTDR; the right-hand peak is due to the end or Fresnel reflection. The three remaining peaks are due to U-bend Sensor 1, U-bend Sensor 2 and U-bend Sensor 3 at 0.665 Km, 0.756 and 0.857 Km respectively.

With distributed or multi-point sensors, it is important that the attenuation imposed by the launch optics and sensors nearest the launching end of the fibre do not impose too much attenuation on the transmitted signal as this would ultimately render the succeeding sensors useless. These losses have been minimised in this system and this is evident in Figure 3.

RESULTS

Numerous OTDR readings were taken for each of the sensing conditions under test, i.e. each of the sensors response to exposure to air and alcohol. Also of interest were the resulting attenuation effects on the succeeding sensors.

Figure 4 shows an OTDR output trace for one of the three Sensors (Sensor 1) exposed to air and alcohol. The areas of interest on the OTDR trace form a relatively small part of the overall trace (177 points – 59 for each sensor – in a total of 12000), and therefore in order to maximise the efficiency of the computer algorithm, it was necessary to select a window of 59 data points and home in on it for each sensor.

This was achieved using an in-house designed LabVIEW VI. The VI finds the location of the peaks associated with the sensor sections in the OTDR input array, and then selects a user-definable window distance before and after the sensor peak. This data is saved for further analysis prior to application to the SNNS ADD software as detailed in the analysis section.

ANALYSIS

The saved OTDR sensor data were then normalised between ± 1 using the standard LabVIEW Scale 2D array VI [16](LabVIEW Manual, 1998) prior to application to the ANN. A normalised 3-sensor OTDR data trace indicating Alcohol (Sensor 1), Air (Sensor 2) and Air (Sensor 3) is shown in figure 5.

A feed-forward three-layer neural network was constructed with the number of input nodes representing the number of points required to represent the three sensors response and attenuation (in this case 177). The use of a feed-forward network is a robust way of building a non-parametric classifier.

Although there is no set method for choosing the number of nodes to be used in a hidden layer, there are limitations associated with selecting too few nodes, which would result in the inability of the network to train successfully, and too many nodes would result in a poor generalisation by the network.

Four different feed-forward networks were examined with forty, twenty, ten and five nodes in the hidden layers respectively. It was found that a hidden layer of ten nodes performed the best. Three nodes were used in the output layer – one to represent the sensing condition at each of the sensors.

A total of 60 result patterns were used to train the network:
20 Sensor 1 in Alcohol, Sensor 2 and 3 in Air,
20 Sensor 2 in Alcohol, Sensor 1 and 3 in Air,
20 Sensor 3 in Alcohol, Sensor 1 and 2 in Air,

For training the feed-forward network, 200 epochs were required with a back-propagation algorithm utilising a learning rate of 0.2. The network was initialised with randomised weights and trained with a 'topological order' update function.

In order to test the operation of the trained network, an independent set of data to that which had been applied in the training of the network was used to generate 15 new patterns. These 15 patterns were applied to the trained network and were all classified correctly. Table 1 shows a sample of the results obtained when the test pattern set was applied to the trained network.

TABLE 1: Sample of 6 results from the Trained ANN Test Set.

Test Condition	Ideal output	Observed output
Alc Air Air	1 0 0	0.962 0.143 0.001
Alc Air Air	1 0 0	0.970 0.155 0.007
Air Alc Air	0 1 0	0.147 0.965 0.019
Air Alc Air	0 1 0	0.150 0.966 0.016
Air Air Alc	0 0 1	0.007 0.299 0.683
Air Air Alc	0 0 0	0.006 0.274 0.692

CONCLUSIONS

A 1 km multipoint sensor has been investigated and proven to be capable of detection the presence of Alcohol at each of the 3 sensing points using OTDR techniques. Each of U-bend evanescent wave absorption sensors were developed with 62.5 um polymer-clad silica fibre, which had its cladding removed in the sensing region. The length of the fibre was 1Km, although longer or shorter lengths may be used as required.

Artificial neural network signal processing techniques have allowed the resulting OTDR signals to be characterised using pattern recognition. Earlier results from a single U-bend sensor (Lyons *et al*, 2000b) have shown that a multi-layer perceptron is required to adequately classify the data.

Initial results (Lyons *et al*, 1999b) have shown that it is possible to train a network to recognise trends such as ageing of the bare fibre when immersed in water, and therefore possible to separate out such effects from genuine changes in the measurand.

It is envisaged that a more sophisticated multipoint U-bend evanescent wave sensor system shall be developed, with the resulting complex signals implemented using Neural Network pattern recognition. This will result is the development of a 'smart system', with the ability to interpret and separate relevant measurand data from the data received from cross coupling of signals from external or interfering parameters as well as faults or defects detected in the fibre.

REFERENCES

Barnoski, M.K., Jensen, S.N., 1976, "Fiber waveguides: a novel technique for investigating attenuation characteristics", *Applied Optics*, 15, pp2112-2115

Gupta, B.D., et al, 1996, "Fibre optic evanescent field absorption sensor based on a U-shaped probe" *Optical and Quantum Electronics*, 28, pp1629-1639.

Kersey, A.D., Dandridge, A., 1988, "Distributed and Multiplexed Fiber optic sensor systems", *J. IERE*, 58, p.S99,

Kersey, A.D. 1996, "A Review of Recent Developments in Fiber Optic Sensor Technology", *Optical Fiber Technology* Vol 2, pp291-317.

Khijwania, S.K., Gupta, B.D., 1998, "Fibre optic evanescent field absorption with high sensitivity and linear dynamic range" *Optics Communications*, 152, pp259-262.

Khijwania, S.K., Gupta, B.D., 1999, "Fibre optic evanescent field absorption sensor: Effect of fibre parameters and geometry of the probe" *Optical and Quantum Electronics*, 31, pp625-636.

Khijwania, S.K., Gupta, B.D., 2000, "Maximum achievable sensitivity of the fibre optic evanescent field absorption sensor based on a U-shaped probe" *Optics Communications*, 175, pp135-137.

LabVIEW *Function and VI Reference Manual*, Jan. 1998, P/N 321526B-01.

Lieberman, R. A., 1991, Distributed and Multiplexed Chemical Fiber Optic Sensors *SPIE Proc.* 1586, pp 80-91

Lyons, W.B., Ewald H., Lewis E., September 1999a, "A Multi-point Optical fibre distributed Sensor Based On Pattern Recognition", *2nd International Automatisierungssymposium, Wismar*, Germany.

Lyons, W.B., Ewald H., Lewis E., August 1999b, "An Optical Fibre Distributed Sensor Based On Pattern Recognition", *ICPR-15: Manufacturing for a Global Market*, Limerick.

Lyons, W. B., and Lewis, E., 2000a "Neural networks and pattern recognition techniques applied to optical fibre sensors" *Transactions of the Institute of Measurement and Control* 22 (5) pp 385-404

Lyons, W.B., Ewald, H., Flanagan, C., and Lewis, E., 2000b *"An Optical Fibre Multipoint U Bend Sensor Based On Artificial Neural Network Pattern Recognition"* Proc. Artificial Neural Networks in Engineering Conference 2000 (ANNIE 2000), St. Louis, USA, pp 663-670.

Rogers, K. R., and Poziomek, E. J., 1996 "Fibre optic sensors for environmental monitoring", *Chemosphere* Vol. **33** No 6 pp.1151-1174.

SNNS 1995, URL http://www.informatik.uni-stuttgart.de/ipvr/bv/projekte/snns/snns.html

Soichi Otsuki et al, 1998, "A novel fibre optic gas-sensing configuration using extremely curbed optical fibres and an attempt for optical humidity detection" *Sensors and Actuators B53*, pp. 91-96.

Udd E., August 1995 "An overview of fiber-optic sensors" *Rev. Sci. Instrum.* 66 (8) pp 4015-4030.

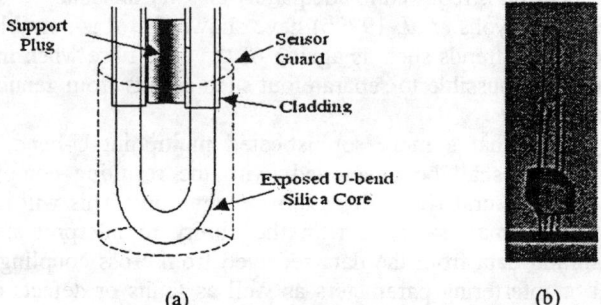

Figure 1. (a) U-bend fibre sensor head configuration. (b) Photograph of a U-bend fibre sensor.

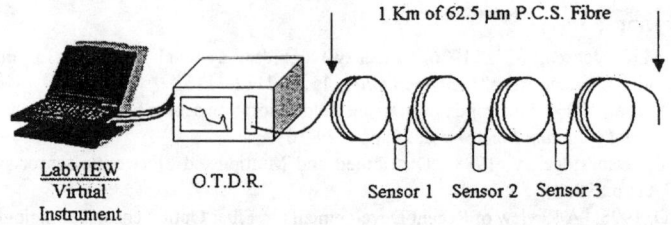

Figure 2. Measurement system configuration.

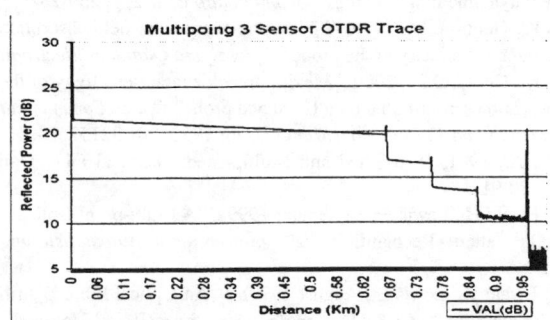

Figure 3. OTDR trace from fibre indicating a typical dry sensor response for all three sensors.

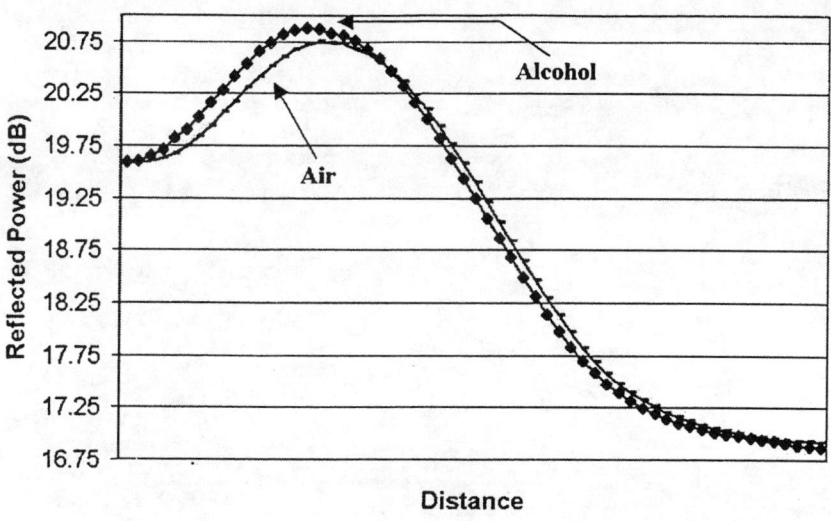

Figure 4. OTDR trace at sensor 1 for 'Air' and 'Alcohol'.

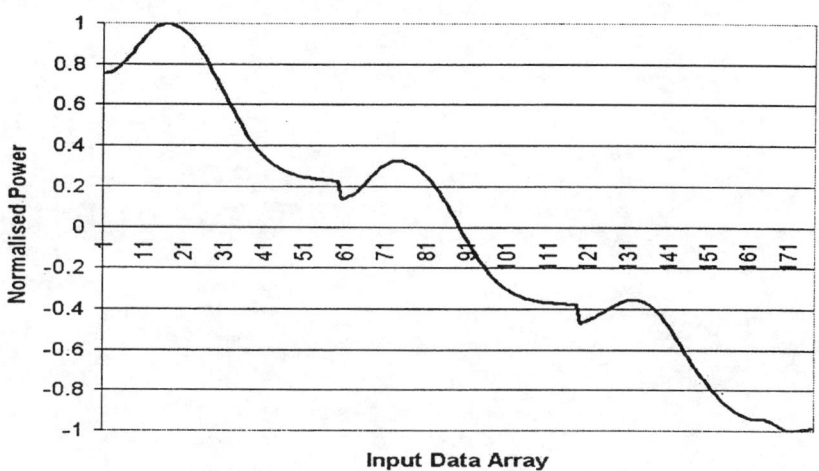

Figure 5. Normalised 3 Sensor Data trace after LabVIEW Processing of the OTDR Trace.

VALIDATING DIGITAL TERRAIN ELEVATION DATA WITH NEURAL NETWORKS

ROBERT GRAY AND THOMAS L. HEMMINGER
School of Engineering and Engineering Technology
Penn State University, Behrend College, Erie, PA 16563 USA
e-mail: rxg31@psu.edu   e-mail: hemm@psu.edu

ABSTRACT

This paper discusses a validation technique used to determine the reliability of digital terrain elevation data (DTED) derived from satellite imagery with kinematic global positioning systems (KGPS) and radar altimeters. The objective of this work is to improve the safety of aircraft landing systems that rely on instrument flight rating (IFR) information through statistical and neural network techniques and compare the two approaches. Actual flight test data has been incorporated in this work from the Albany, OH area airport referred to as "KUNI." It is our intention to integrate the aforementioned information via a neural network system to validate the data from DTED. Over the years cockpit errors have occurred due to erroneous data from radar altimeters, satellite maps and traditional terrain mapping. The neural network will be used to fuse the data from KGPS and radar altimeters to look for discrepancies in terrain elevation.

INTRODUCTION

Aircraft flight safety is paramount to the Federal Aviation Administration (FAA), the general public, and agencies of the United States military and foreign countries. Over the last several years many tragic instances have captured the news due to mechanical, maintenance, and pilot errors. Pilot controlled flight into terrain (CFIT) can result from incorrect data sent to the cockpit without necessarily reflecting pilot error. CFIT can include colliding with the ground on approach and impacting against mountainous terrain substantially above mean sea level (MSL). Due to these events the U.S. government has directed the National Imagery and Mapping Agency (NIMA) to declassify lower level Department of Defense (DoD) digital terrain elevation data (DTED) for commercial use. One response by the FAA was the development of the Terrain Awareness System (TAWS). TAWS utilizes DTED as a primary CFIT sensor, but reducing the accident rate depends on accurate DTED which is prone to errors. As an aircraft approaches an airfield it is required to fly at a reduced altitude which increases the possibility of a collision with the terrain. A display of incorrect physical features of the immediate area can result in a catastrophic event. Terrestrial obstacles are not included in DTED making the reliability of these data marginal at low altitudes. Unless additional information is available the DTED database cannot be relied upon without the addition of onboard sensors such as radar altimeters and KGPS information. Our objective is to employ neural networks for the purpose of comparing the results against statistical analysis. At this point our work is preliminary but do anticipate the creation of a structure that can be used as a model to develop more complex systems in the future. The DTED database can be converted into a matrix of elevation points at predetermined points of latitude and longitude but it is not error proof. Unfortunately, DTED has been measured by various terrain elevation sensors, e.g., synthetic aperture radar (SAR), and other satellite techniques that have been digitized by human operators. Some altitude verifications have been made by surveyors at random locations and have been compared with the DTED. On a recent mission the space shuttle has been employed to produce more accurate measurements, thereby reducing the errors in DTED (Covault, 2000). The reliability of DTED becomes a concern when considering the consistency of reference frame information. Many times the problem is with a mismatch due to non-uniform use of the common World Geodetic System (WGS84) by other countries. Unfortunately, some countries may reference different databases which can distort the local frame of reference (see Fig. 1). If the frame of reference is not uniform throughout the world, aircraft from different countries may unknowingly use alternate frames that are not appropriate for their particular airspace. Several flight tests were performed using DTED within the Ohio State University airport (KUNI). These experiments were used to evaluate DTED for enhanced flight safety. As a result of these tests we have endeavored to improve safety by validating the DTED database through statistical and neural methods. In Gray (1999) special algorithms were developed for DTED correlation to determine the validity of the data. Several experiments were conducted at KUNI to evaluate the validity of DTED and to determine its reliability with regard to flight safety. Unfortunately DTED is highly dependent on data supplied from several countries using dissimilar reference points. We have developed procedures to create algorithms to improve flight safety. These have included statistical and neural techniques.

STATISTICAL CHARACTERISTICS

We have used DTED, KGPS, and radar altimeter data to determine positional information in a statistical sense (Baird, and Abramson, 1984). At this point, the validity of elevation estimates will also be tested using neural networks. The KGPS, radar altimeter, and DTED information are over-bounded by the normal distribution. Our contention is that the terrain profile can be developed by the use of neural networks by determining the agreement from all data sequences as a function of time. Consider a test statistic, T, which is a function of several parameters:

1. Terrain characteristics
2. Distribution of KGPS errors
3. Distribution of radar altimeter errors
4. Distribution of DTED errors

The level of agreement from each measurement scheme should increase as a function of distance traveled. First, the absolute and successive disparities between map and sensed profiles are calculated for each second of aircraft trajectory. The map height, based on DTED at any point in time can be written as:

$$d_i = d(e_i, n_i) + b_d + n_{di} \tag{1}$$

The subscript i indicates the time at which the observation is valid; e_i is aircraft East coordinate; n_i is aircraft North coordinate; $d(\circ,\circ)$ is the DTED-derived terrain height with respect to MSL as a function of the coordinates; b_d is the bias of the map height and n_{di} is the noise in the map height. The map terrain profile can be defined as a series of DTED heights taken during a specified time interval. The bias can be

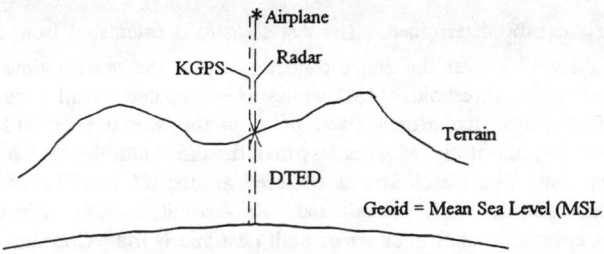

Figure 1. DTED can be validated to a point by subtracting radar altimeter readings from KGPS.

considered a constant if the time interval is relatively short. The height based on KGPS and the radar altimeter are represented by:

$$s_i = a_i - r_i - l_{ar} + b_s + n_{si} \tag{2}$$

where a_i is the height above MSL from the KGPS; r_i is radar altimeter height; l_{ar} is the difference in height between the radar altimeter and GPS antennas; b_s is bias in the sensed height; n_{si} is the noise. Sensed terrain profile is the series of heights over the same time interval as above. Absolute disparity between map and sensed profiles is:

$$p_i = s_i - d_i = b + n_i \tag{3}$$

where b is the bias and n_i is the noise. From previous experiments the bias can be considered constant for this application. Successive disparity can be defined as:

$$p_i' = p_i - p_{i-1} = n_i - n_{i-1} \tag{4}$$

Several metrics have been used to compare absolute and successive disparities (Baird, and Abramson, 1984),(Hinrichs, 1989) but (Hinrichs, 1989) has shown that the mean-squared difference (MSD) is the most accurate for this application. The MSD of absolute disparity as a function of samples, n, is given by:

$$MSD_{ad}(n) = \frac{1}{n}\sum_{i=1}^{n} p_i^2 \tag{5}$$

The MSD of the successive disparity is:

$$MSD_{sd}(n) = \frac{1}{n-1}\sum_{i=2}^{n}(p_i - p_{i-1})^2 = \frac{1}{n-1}\sum_{i=2}^{n}(p_i')^2 \tag{6}$$

The next step is to determine the pdfs of the MSD_{ad}, MSD_{sd}. Based on a desired probability of fault-free detection (P_{FFD}) and the pdfs mentioned above, a detection threshold T_T can be determined. The test statistic is calculated from the absolute or successive disparities over the entire trajectory up to the present time and evaluated against the detection threshold. MSD values are computed for all geographic areas of interest called the search area. Each point in the search area will have a MSD associated with it, and if that MSD is less than the threshold, the search point will pass the disparity test. The search area is centered around the most likely position at the instant of time that the MSD is evaluated. As a result, each point in the search area represents a coordinate shift in either or both East and North. Coordinate shifts should result in inconsistencies between the map terrain profile (DTED) and the sensed terrain profile (KGPS and radar altimeter) (Defense Mapping Agency, 1993),(Slocum, 1993). Pdfs of MSD_{ad} and MSD_{sd} are determined by letting $p_1, \cdots, p_n$ be independent identically distributed (iid) $N(0, \sigma^2)$ random variables, where n is the number of samples and $N(0, \sigma^2)$ is zero mean gaussian. Rather than evaluate the pdf of MSD_{ad} directly it is easier to compute the random variable T where:

$$T = \frac{n}{\sigma^2} MSD_{ad} = \frac{1}{\sigma^2} \sum_{i=1}^{n} p_i^2 \qquad (7)$$

It follows that T has a chi-square distribution with $\nu = n$ degrees of freedom (Deming, 1950). The pdf with ν degrees of freedom is given by (Miller and Miller, 1999).

$$y(x) = \frac{1}{2^{\nu/2} \Gamma(\nu/2)} x^{\left(\frac{\nu-2}{2}\right)} e^{-x/2} \qquad \text{for } x > 0;\ 0 \text{ elsewhere} \qquad (8)$$

where Γ is the familiar gamma function with a scale parameter of 1.0. The pdf for MSD_{sd} is more difficult to obtain so it is presented without proof (Harper, 1967) where Z is the threshold:

$$Z = \frac{1}{2\sigma^2} MSD_{sd} = \frac{1}{2\sigma^2 (n-1)} \sum_{i=2}^{n} (p_i')^2 \qquad (9)$$

To develop thresholds for T and Z as a function of P_{FFD} and n, equations (8) and (9) are used. For example, if P_{FFD} for T (absolute disparities) must be greater than 0.99 then equation (8) is integrated from zero to a limit such that the area under the curve is 0.99. With 10 degrees of freedom the threshold is 23.21. If $n = 50$ the threshold for the successive disparity is 1.664.

During the flight the disparities are calculated and both the absolute and relative MSDs are formed using equations (5) and (6), respectively. For each point in the trajectory the MSDs are calculated. The two results are converted to T and Z using equations (7) and (9) respectively and evaluated against thresholds. The statistical results were very good as seen in the following figures but required a great deal of computation, far beyond what has been detailed in this paper. For this reason we decided to try a neural-based system.

NEURAL NETWORKS

At this point we discuss the implementation of a neural system to map absolute and successive disparities along the flight path using an adaptive predictor. The disparity sequences were input into two separate networks employing time delay configurations, meaning that 10 data points were input before an output was observed. Settling time was approximately 25 seconds before the output could be tracked with minimal error. A substantial combination of neurons and layers were attempted before these results were realized. At the time of this writing 5 neurons seems to be the minimum number that can reliably predict these sequences. Much more testing is in order with a larger database. Presently a P_{FFD} of 0.999 has been achieved with the neural networks, yet our statistical methods can reach a P_{FFD} of 0.9999, so more work needs to be done. In addition, we expect to fuse information from several flights to KUNI so that the entire terrain can be mapped with a neural network. Figures 2 is a plot of the absolute disparity between DTED, KGPS, and radar while figure 3 compares the theoretical threshold against the statistical and neural techniques. Figures 4 and 5 are the same type of illustrations for successive disparities. MATLAB was used to develop the neural networks.

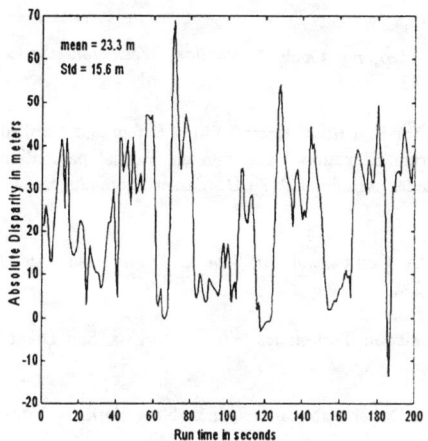

Figure 2. Absolute disparities for the true flight path.

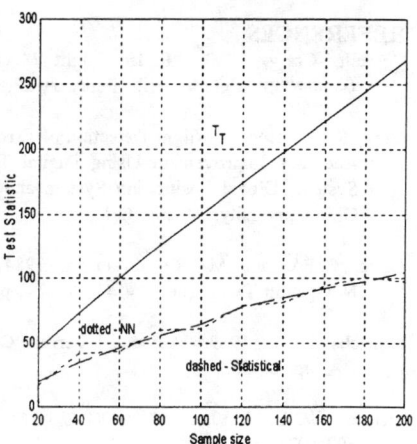

Figure 3. Test statistic T, threshold T_T, and NN against time.

CONCLUSION

The enhanced flight safety method presented in this paper is different from methods reported in other literature, since DTED is not used for positioning or for terrain display. Instead, the difference between DTED and sensed terrain profiles are assessed statistically and with neural networks. This has been accomplished through a series of measurements from KGPS, radar altimeter data, and Level 1 DTED and their agreement analyzed. Thus far it appears that absolute disparities are more effective for the purpose of improving flight safety through the comparisons mentioned above.

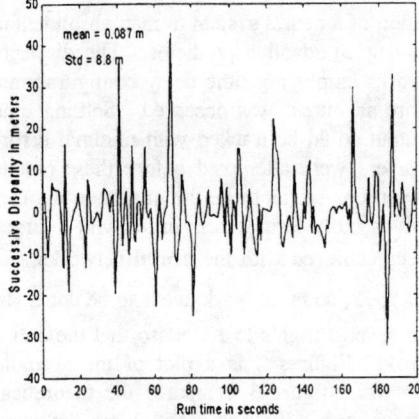
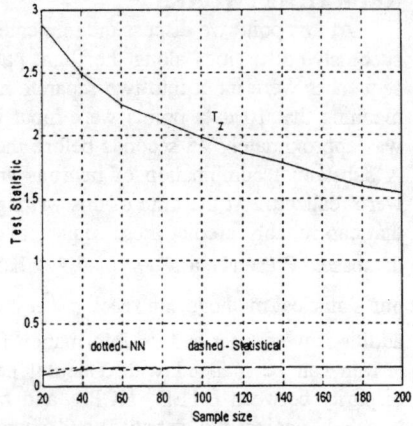

Figure 4. Successive disparities for the true flight path.

Figure 5. Test statistic Z, threshold T_z, and NN against time.

REFERENCES

Covault, Craig, 2000, "Radar Flight Meets Mapping Goals," *Aviation Week and Space Technology*, McGraw-Hill, February 28, P. 43.

Gray, R. A., 1999, "Inflight Detection of Errors for Enhanced Aircraft Flight Safety and Vertical Accuracy Improvement Using Digital Terrain Elevation Data with an Inertial navigation System, Global Positioning System and Radar Altimeter," Ph.D. Dissertation, June, The Ohio University, Athens, OH.

Baird, C. A. and M. R. Abramson, 1984, "A Comparison of Several Digital Map-Aided Navigation Techniques," *IEEE PLANS*, pp. 286-293.

Hinrichs, P. R., 1989, "Advanced Terrain Correlation Techniques," *IEEE PLANS*, San Diego, CA, pp. 89-96.

Deming, W. E., 1950, *Some Theory of Sampling*, Dover Publications, Inc., New York, NY, pp. 503-507.

Miller, I. and M. Miller, 1999, *John E. Freund's Mathematical Statistics*, Sixth Ed., Prentice Hall, Inc., Upper Saddle River, NJ, p. 280, p. 205.

Harper, W. M., 1967, "The Distribution of the Mean Half-Square Successive Difference," Biometrika 54, pp. 419-433.

Defense Mapping Agency, 1993, *Digital Terrain Elevation Data (DTED) Specification*, MIL-D-89020, Washington D.C., 28 May.

Slocum, K. R., 1993, "Conifer Tree Influence on Digital Terrain Elevation Data (DTED): A Case Study at Dulles International Airport," U.S. Army Topographic Center, TEC-0038, DTIC AD-A274 213, Fort Belvoir, VA, September.

RECOGNITION OF MUSICAL RHYTHM PATTERNS BASED ON A NEURO-FUZZY-SYSTEM

TILLMAN WEYDE
Research Department of
Music und Media Technology,
University of Osnabrück
Osnabrück, Germany
tweyde@uos.de

KLAUS DALINGHAUS
Institute for Semantic
Information Processing
University of Osnabrück
Osnabrück, Germany
kdaling@uos.de

ABSTRACT

The task of recognizing patterns and assigning rhythmic structure to unquantized musical input is a fundamental one for interactive musical systems and for search in music databases since melody is based on rhythm. We use a combination of combinatorial pattern matching with a match quality rating by a Neuro-Fuzzy system that incorporates musical knowledge and operates on perceptually relevant features extracted from the input data. It can learn form relatively few expert examples and shows improved recognition results compared to conventional quantization and matching methods.

INTRODUCTION

The recognition of rhythmic structures is necessary for understanding speech as well as music. Yet a coherent paradigm or theoretical framework to support computer models and applications has not been established so far. Modelling domains of uncertain and incomplete knowledge with a Neuro-Fuzzy-System therefore applies well to musical computation since our knowledge of musical structures and processes shows more and more to be incomplete and vague as we try to solve musical tasks with computers. What we know are aspects of complex processes the relevance of which we would like to determine and which we would like to integrate into a working model.

We need to model two main aspects of rhythmic structure for recognition and analysis: segmentation and similarity. The segmentation process divides a sequence of notes into groups. These groups – or musically speaking: motifs – are combined to higher level units like musical phrases. The similarity of groups determines the internal structure of phrases or the relation to a given model (e.g. a given rhythm which should be played by a student or a melody from a database). Segmentation and assignment are highly interdependent and both depend strongly on context.

SYSTEM ARCHITECTURE

Musically meaningful interaction in an application should be based on an interpretation of the input by segmentation into groups, group relations, and note relations. The interpretation process should be robust in respect to missing or additional notes and groups, wrong order of groups, changes in tempo and timing deviations. We combinatorially generate interpretations (segmentations and assignments) and filter them by perceptual constraints to reduce complexity.

Features are extracted from segmentations and assignments and a rating is calculated by a Fuzzy-Prolog system. The best rated alternative is used as system output. A Fuzzy-Prolog program can be interpreted as a feed-forward neural net, where rules and facts correspond to network nodes. The rules we use can be seen as a partial implementation of the grouping rules by Lerdahl and Jackendoff (1983) with some rules added for the comparison.

An interpretation assigns groups of the input to groups of the model or within the input. The notes in the matched groups are assigned and thus detailed information on the differences between groups concerning note structure, timing, tempo, and group position are provided as well as similarity ratings on the group and phrase level. Matching single group patterns and segmentation can be performed individually.

Input Features

Input data represent musical notes based on MIDI (Musical Instrument Digital Interface) data. The input events have three values that we use: onset time, duration, and key velocity (loudness). On both the group level and the interpretation level features are extracted that form the basis for the similarity and segmentation ratings.

Segmentation. There are perceptual constraints on the length and the duration of perceptual groups. The number of events in a group – the length – is restricted as is well known since Miller (1956). In our system a setting of 4 or 5 has shown to be adequate which agrees with the literature (Handel and Todd (1981), Swain (1986)). Empirical evidence suggests that durations of perceptual groups lie in a range of approximately 0.5 to 2 seconds (Seifert et al. (1995)). For segmentations we calculate ratings based on the length and duration of groups. For duration values we use four inputs corresponding to different durations. This design allows a fitting to the preference curve on the considered range of durations. The same design is used for the length of a group. We have five different input nodes corresponding to note numbers from one to five. The regularity of group lengths and group durations is also rated by calculating their variance. For each group we have a node representing whether the inter-onset-interval (IOI) between the last note and the next group is larger than IOIs within the group, and whether the first note is relatively loud which are both factors in grouping (Handel, 1989).

Similarity. For similarity on the group level we use different measures of distance. We have nodes for early notes, late notes, too loud notes, too soft notes, too long notes, too short notes which contribute to precision rating. Added notes, inserted notes, subtracted notes, deleted notes are combined to a rating of correctness. Group position and tempo stability are calculated relative to the preceding group and tempo plausibility is based on spontaneous and preferred tempo (Fraisse, 1982). More details on the calculations can be found in Enders and Weyde (1996).

SEGMENTATION AND SIMILARITY RATINGS USING FUZZY-PROLOG

Fuzzy-Prolog

The module for rating segmentation quality and group similarity is based on Fuzzy-Prolog (Nauck et al., 1996). Fuzzy-Prolog is a programming language similar to Prolog. A Fuzzy-Prolog program consists of a set of rules in the following form:

$$\text{conclusion} \leftarrow \text{premise} \wedge \ldots \wedge \text{premise} \quad (1)$$

Operators have an evaluation function with values in the interval [0, 1]. For evaluation of the implication the Goguen-Implication is being used. The truth value of other rules can be derived from those by using the modus ponens generalized for fuzzy logic (Nauck et al., 1996).

Operators

We have conjunction and disjunction operators. There are several options for the evaluation functions of disjunction and conjunction, well known ones are min for $\wedge$ and max for $\vee$. Since we need a differentiable function for learning by backpropagation we use this evaluation function for conjunction:

$$f_{\wedge,q}(\|\bullet_1\|,\ldots,\|\bullet_n\|) = \left(\frac{1}{n}\sum_{i=1}^{n}\|\bullet_i\|^{\frac{1}{q}}\right)^q \quad q \geq 1 \quad (2)$$

where $\|\bullet_n\|$ denotes the truth value of rule $\bullet_n$. For the disjunction we replace q with $\frac{1}{q}$. These functions allow for a certain amount of compensation between the operands that can be adjusted by the q parameter. For $q \to \infty$, $f_{\wedge,q}$ approaches min and $f_{\vee,q}$ approaches max. With this evaluation function $\vee$ and $\wedge$ are not associative but they are independent of the number of operands in the sense that for any number of operands with equal inputs we have the same result, which is necessary for multi-nodes (see end of next section):

$$\|\bullet_i\| = \|\bullet_j\| \, \forall i,j \in \{1\ldots n\} \Rightarrow f_{\wedge}(\|\bullet_1\|,\ldots,\|\bullet_m\|) = f_{\wedge}(\|\bullet_1\|,\ldots,\|\bullet_n\|) \, \forall m < n \quad (3)$$

Fuzzy rules

Not all the rules we use can be shown here due to space limitations. We will look at an example to try and give an idea of how the system works. The rules for similarity on the structure level are based among others on the segmentation rating for input and task if present. E.g. a rule for the input, CInputSegment, returns the segmentation rating combined from the individual groups:

$$\text{CInputSegment}(gs) \leftarrow \text{GInputSegment}(g_1) \wedge \ldots \wedge \text{GInputSegment}(g_n),$$
$$\text{where } gs = [c_1, \ldots, c_n] \quad (4)$$

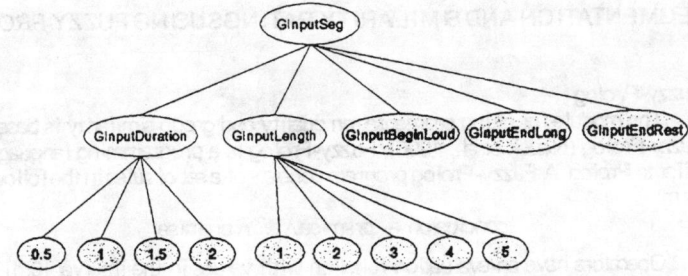

Figure 1: Structure of the neuro-fuzzy program on group level for segmentation.

$$GInputSegment(g_i) \leftarrow GGroupDuration(g_i) \wedge GGroupLenth(g_i) \wedge$$
$$GGroupEndLong(g_i) \wedge GGroupStartLoud(g_i) \quad (5)$$

$$GGroupDuration(g_i) \leftarrow GGroupDur05(g_i) \vee GGroupDur10(g_i) \vee$$
$$GGroupDur15(g_i) \vee GGroupDur20(g_i) \quad (6)$$

$$GGroupLength(g_i) \leftarrow GGroupLength1(g_i) \vee GGroupLength2(g_i) \vee$$
$$GGroupLength3(g_i) \vee GGroupLength4(g_i) \vee$$
$$GGroupLength5(g_i) \quad (7)$$

The facts of the form GGroupDurXX (rule 6) are the inputs features mentioned earlier which return values close to one if the length of the respective group is near to 0.5, 1.0, 1.5 or 2.0 seconds. Similarly the rules GGroupLengthX (rule 7) represent a group length of 1, 2, 3, 4, and 5.

We do not know in advance how many groups there will be in a structure. The number of groups varies for different segmentations of the same input. This made it necessary to extend Fuzzy-Prolog with a list processing feature. In rule 4 the transition from the structure level to the group level takes place where a variable number of inputs is combined to one node. We call these nodes with variable numbers of connections multi-nodes.

COMPLEXITY AND LEARNING

Reducing Complexity

Two types of optimization are applied to reduce complexity. We have the filtering of interpretations and a branch and bound optimization of the rating. E.g. it is known that temporal proximity of events is important for segmentation, relatively long distances between events tend to end a group (Handel, 1973). Grouping by temporal proximity is dominant over accent grouping (Deutsch, 1986) so we filter out segmentations that grossly contradict grouping by proximity.

The search of the best interpretation is optimized by pruning the search tree.

The generation of interpretations can be seen as a tree where every assignment of an individual group is a branch. So after every assignment of a group we determine the upper bound of overall rating that can be obtained. If this upper bound is lower than the best rating so far we can stop the calculation for this branch.

Iterative Learning by Relative Ratings

Nauck et al. (1996) have shown that Fuzzy-Prolog programs are equivalent to feed-forward neural nets. They can therefore be trained with the same algorithms. Our system is trained with relative ratings following the method by Braun (1997). This means that pairs of assignment or segmentation examples are given and one of each pair is to be rated better. When learning was successful the system prefers better assignments since they receive higher ratings. We iteratively generate samples by comparing expert assignments and assignments preferred by the system. If they differ a new relative sample is generated with the expert input marked as to be preferred. We then train the network until the correct rating relation is achieved for all samples or a given maximal number of training cycles has been reached. Is is possible (and happens often) that after training the relation of ratings is correct for a given pair but yet another assignment is rated better than the one provided by the expert. So we iterate the process of generating sample pairs and training the network until for all samples the expert assignment is rated best by the system.

The depth of the rule system makes standard backpropagation slow und ineffective since the size of the gradient gets very small on the lower layers of the net. The overcome this problem we use the R-PROP algorithm (Riedmiller and Braun, 1993) for training the network with which our net converges dramatically faster than stdandard backpropagation with momentum. In our network there is no distributed representation unless it is explicitly encoded. This limits the capacity of the network, but it allows to interpret training results and enhances generalization.

Application

We have implemented an experimental application with the system described in this paper called RhythmScan which allows to generate samples, test, and train the system. It is a Java application with a Fuzzy-Prolog interpreter written in C. The system can be interactively tested and trained by giving preferred segmentations and assignments. Input can be obtained via computer keyboard (without loudness information) or with a MIDI instrument. The preferred segmentations and assignments can be defined interactively by an (expert) user on a graphical user interface.

The system performs segmentation and comparison of user input well after training. The system is able to detect structural changes like delays, interrupted and repeated input as well as tempo and timing deviations. We used training sets for comparison, categorization and segmentation of note sequences. The trained system performs well on similar test sets. We reach a percentage of 60-90% correct assignments on test sets but even the assignments that differ from the expert rating are mostly of acceptable quality.

Conclusions

Modelling musical knowledge in combination with machine learning enables us to solve musical tasks without a complete model or exact knowledge of perceptual and other relevant processes. The weights in our system trained by examples can also give insight to the relevance of rules, although the sample sets have been to small to draw any generalizable conclusions yet. The system can be easily modified or extended by adding, changing or removing rules or input features.

The system performs well on training and test sets, but has still to be tested on a wider range of musical examples. Further challenges are the integration of pitch and metrical information into the system as well as the extension to streaming mode and further optimization of the computational performance.

References

Braun, Heinrich, 1997. "Neuronale Netze". Springer, Berlin Heidelberg.

Deutsch, Diana, 1986. "Auditory pattern recognition." In Boff, K. R., Kaufman, L., and Thomas, J. P. (Eds.), Handbook of Perception and Human Performance: Cognitive Processes and Performance, vol. 2, chap. 32, pp. 32–1–49. John Wiley and Sons, New York.

Enders, Bernd and Weyde, Tillman, 1996. "Automatische Rhytmuserkennung und -vergleich mit Hilfe von Fuzzy-Logik." Systematische Musikwissenschaft, vol. IV (1-2), pp. 101–113.

Fraisse, Paul, 1982. "Rhythm and tempo." In Deutsch, Diana (Ed.), The Psychology of Music, chap. 6, pp. 149–180. Academic Press, New York.

Handel, S., 1989. "Listening: An Introduction to the Perception of Auditory Events". MIT Press, Cambridge, Massachusetts.

Handel, Stephen, 1973. "Temporal segmentation of repeating auditory patterns." Journal of Experimental Psychology, vol. 101, pp. 46–54.

Handel, Stephen and Todd, Peyton, 1981. "Segmentation of sequential patterns." Journal of Experimental Psychology: Human Perception and Performance, vol. 7 (1), pp. 41–55.

Lerdahl, Fred and Jackendoff, Ray, 1983. "A Generative Theory of Tonal Music". The MIT Press, Cambridge, Mass.

Miller, George A., 1956. "The magical number seven, plus or minus two: Some limits on our capacity for processing information." Psychological Review, vol. 63 (2), pp. 81–97.

Nauck, Detlef, Klawonn, Frank, and Kruse, Rudolf, 1996. "Neuronale Netze und Fuzzy-Systeme". Computational Intelligence. Vieweg, Braunschweig, 2 ed.

Riedmiller, Martin and Braun, Heinrich, 1993. "A direct adaptive method for faster backpropagation learning: The RPROP algorithm." In Proc. of the IEEE Intl. Conf. on Neural Networks, pp. 586–591. San Francisco, CA.

Seifert, Uwe, Olk, Fabian, and Schneider, Albrecht, 1995. "On rhythm perception: Theoretical issues, empirical findings." Journal of New Music Research, vol. 24 (2), pp. 164–95.

Swain, J. P., 1986. "The need for limits in hierarchichal theories of music." Music Perception, vol. 4 (1), pp. 121–148.

AUTOMOTIVE DESIGN DRIVEN BY PATTERN RECOGNITION

SAWSAN ABOUL-HASSAN
*Department of Computer Science
and Engineering
Oakland University
Rochester, MI 48309*

DJAMEL BOUCHAFFRA
*Department of Computer Science
and Engineering
Oakland University
Rochester, MI 48309*

ABSTRACT

We use pattern recognition techniques in order to aid engineers improve their designs. Our ultimate goal is to automatically predict customer perceptions assigned to cars given their contours. We define the car by two elements: the contour and the options set. We extract the contour of a car image using the Fourier Descriptors. We use the Multi-Layer Perceptron neural network to classify a car contour design and the Borda Count to predict the score of the engineer's set of options. Our prediction rate obtained so far is 71.5%. A very large-scale experiment is currently underway.

INTRODUCTION

Nowadays, cars are a necessity of life. We use them on daily basis and they are our companions most of the time. We have to drive them or ride in them everyday and most likely more than once. Many applications concerning cars have been and are being developed. Some recognize license plate numbers as advertised on "autovu.com" web site, while some others assist drivers by detecting a preceding vehicle on the road to enhance safety and traffic efficiency (Kato and Ninomya, 2000). But how much liking a car has an effect on our life? This question becomes more relevant if a car that is being targeted is used for work. Very few applications have been developed in this field, in order to answer this question. The ultimate goal consists of building an automatic system that aids automotive design. "Kansei Engineering Systems" is one of these applications that quantifies the relationship between design parameters of the door and truck drivers' perception (Zhang and Vertiz, 2000). Its statistical results revealed that trim material, door shape, color, window shape and map pocket are design elements that strongly affect the perception of elegance and preferences of truck drivers (Zhang and Vertiz, 2000). However, this application focused on a particular component of a car such "a door".

Unlike the Kansei's goal, we focus in our approach on predicting a customer's behavior regarding the design of the whole car and the options' priority. We introduce the Fourier Descriptors to capture the exterior contour of the car. When people go to buy a car, the first look would be at the exterior shape of the whole car. So if they like it, they want to go further in their exploration of the interior before making any decision. In some cases, cars exterior shapes are unattractive and push away people without even exploring the interior. So, "love from first sight" is true when the car has the right shape and sometimes the right color. The color could have a bad effect sometimes but the option of having a different color is often available. We are taking into consideration the different tastes of people in cars' shapes or colors, where the age and the gender factors are very significant.

We know that women's population is larger than men's, and often men depend on women's taste when buying a car. So, women are the decision-makers and their taste is very important and necessary to reflect the vote of the majority.

In this paper, we emphasize the importance of the "likeness" of a car design before it is put in making, in order to ensure a good selling rate. We are considering two features of the car design: *the whole shape from the exterior* and *options from the interior*. We attempt to help engineers to improve their design in a way to make it attractive. Our goal consists of analyzing the engineers' design in order to classify it either in the attractive or the non-attractive category. Our application classifies the design using a Multi-Layer Perceptron (MLP) neural network, and gives a feedback about how to improve it by revealing its weak points. In order to collect data, we performed a statistical survey done in Oakland University campus, where we examined the tastes of students. We considered the group of students, because a great number of people with age between 20 and 30 years old of both genders were available.

PROBLEM STATEMENT

Let $X = \{x_1, x_2, ..., x_n\}$ be the set of car components such as {doors, wheels, mirrors, lights, board, car exterior contour (front, side and rear), options, space, , ...}.
The set of options we are considering $O = \{CD, sunroof, cruise control, leather seat, heated seat, power windows\}$. More options will be added for future experiments.
Let $x_i = \{x_{i1}, x_{i2}, ..., x_{im}\}$ be the set of attributes values assigned to one component x_i of a car for $i = \{1, ..., n\}$. Let $Y = \{y_1, y_2, ..., y_n\}$ be the set of perception attributes such as {attractive, simple, ugly...} and $\Omega = \{\omega_1, \omega_2, ..., \omega_c\}$ be a partitioning of Y representing clusters of perception attributes. A cluster regroups attributes that are close in their meaning. The parameter c is the number of clusters. Let D be the distance metric in the perception space. The problem could be stated as: determine a non-linear function f from X to Ω such that:

$$\forall x_i \in X, \exists ! \omega_j \in \Omega \text{ such that:} \quad (1)$$
$$f(x_i) = \omega_j$$
$$D(\omega_j, \omega) \text{ is min}$$

where ω is the true class of attractive cars.
In other words, we attempt to find the "best prediction" for the car component x_i ($i = 1, ..., n$) through f using the distance D.

METHODOLOGY

Our proposed methodology consists of three phases:
(i) *Car description and design*: a car is captured by its exterior contour and its set of options.
(ii) *Customer perception clustering*: in this phase, we define a list $\omega_1, \omega_2, ..., \omega_c$ of c perception attributes obtained from the clustering scheme.
(iii) *Design/Perception mapping*: a car contour is classified and the option set is

evaluated.
In our survey, we present 10 images of regular cars (no trucks or vans!) to 50 students and we ask them to give an attribute for the shape of each car presented.
We provide a list of attributes such as "elegant", "boxy", "ugly" or "simple" that covers all presented shapes (List A). We always consider adding to our list, new attributes given by the students. After examining *the exterior shape of the car*, we ask about *the options including the CD, the sunroof, the cruise control, the leather seat, the heated seat,* and *the power windows*.

Car Description and Design

Many elements constitute the complete design X of a car. There is no concern in this application about the mechanical elements. Let U be a subset of X (U ⊆ X) because in this application, we are only dealing with the two main elements: *the exterior* and *the interior designs*. From the exterior, we adopt the contour U_c of three sides, while in the interior we adopt the set of options U_o. More elements like interior space and board design are considered for future work. Therefore, a car design is represented as:

$$U = <U_c, U_o> \quad (2)$$

Contour Data Type. The application extracts three contours for each car. They represent three different views of the car image: "front", "rear" and "profile" as in Fig. 1. Therefore, the contour can be represented as a triplet vector:

$$U_c = (U_f, U_p, U_r)^T \quad (3)$$

where U_f is the set of front contour points, U_p is the set of profile contour points and U_r is the set of rear contour points.
We estimate the contour of a car using the Fourier Descriptors (FD) developed by Granlund (1972). Each contour is a set of points in a real multi-dimensional Euclidean space $\mathbb{IR}^n$ (where n is the number of FD's). The contours of attractive cars are stored in a data file for later use in the phase of feedback information to the engineers.
Granlund uses a complex number z(t) to denote the points on the contour. Then the contour is expressed as a Fourier series:

$$z(t) = \sum_{n=-x}^{x} a_n e^{j2\pi nt/T} \quad (4)$$

where

$$a_n = \frac{1}{T} \int_0^T z(t) e^{-j2\pi nt/T} \, dt \quad (5)$$

are the complex coefficients, a_0 is the center of gravity and the other coefficients a_n, are independent of translation. T is the total contour length, m the number of pixels along the boundary, n is the number of descriptors used, and $t = 1,...,m$. The derived features:

$$b_n = \frac{a_{1+n} a_{1-n}}{a_1^2} \quad (6)$$

and

$$D_{mn} = \frac{(a_{1+m})^{n/k} (a_{1-n})^{m/k}}{a_1^{(m+n)/k}} \quad (7)$$

are independent of scale and rotation, because different sizes of images were collected.
Here, $n \neq 1$ and $k = \gcd(m, n)$ is the greatest common divisor of m and n (Trier et al., 1996).

Options Data Type. We define 6 attributes that are pure nominal variables as options. These variables are ordered as follows: CD, cruise control, sunroof, leather seat, heated seat and power-window.
Therefore, the options set is represented by a six-component vector

$$U_o = (U_{CD}, U_{sunroof}, U_{cruise}, U_{leather}, U_{heated-seat}, U_{power-windows})^T \quad (8)$$

where each component value its assigned priority level by the students.

Customer Perception Clustering

People express their feelings or views differently and often, different words have the same meaning. For example, "geometric" and "boxy", or "great" and "elegant" are two pairs of attributes that are close in their respective meanings.

Before establishing our survey, we searched for synonyms of attributes as much as possible to understand what people mean when they provide us with their feelings regarding a certain car. For this purpose, we used the online "Wordnet" system developed by Wordnet Inc. in order to search for meanings that allow us to accurately cluster our attributes. This system accepts one word and searches for its synonyms with specified usage and context (List A).

The survey data directed the clustering scheme where we obtained 4 clusters of perception attributes: "attractive", "classic", "ugly", and "simple". The clustering scheme consists of a semi-manual method. We considered the semi-manual clustering for oral communication reasons and gaps inherent to the Wordnet system. We had manually clustered a perception attribute such as "OK", due to the psychological factor of expressing feelings by words that might have different meanings. The clustering method is represented by the following algorithm:

Begin
 For i=1 to end of list

```
Read word (i)
If word(i) is clustered
   Then i = i + 1
Search for synonyms by running Wordnet
For j = i+1 to end of list
  Read word (j)
  If word(j) ∈ synonyms of word(i)
    Then cluster word(j) in cluster(word(i))
  End
End
return clusters
End
```

The final clusters ω_1, ω_2, ω_3, and ω_4 that we obtained from the survey are respectively: "attractive", "ugly", "classic", and "small".

Design/Perception Mapping

We use the Multi-Layer Perceptron neural network as shown in Fig. 1 to classify a new coming car design. The classification is partitioned into two steps: contour-based classification and options evaluation.

Contour-Based Classification. The classification consists of two phases: *training* and *testing*.

Training Phase. The network consists of three layers: input, hidden and output, interconnected by modifiable weights. The input is a 3-block real valued vector of the contour. The hidden layer has 5 units and the output has c units which are the clusters $\omega_1,..., \omega_c$. The weights and biases are randomly initialized. We use the algorithm of backpropagation, which is one of the simplest and most general methods for supervised training of multi-layer neural network. This algorithm is used to determine the optimal weights for the network. The power of this algorithm allows us to calculate an effective error for each hidden unit, and derive a learning rule for the input-to-hidden weights. This rule consists of presenting an input pattern and changing the network parameters to bring the actual outputs closer to the desired or *targeted* class. In our application, the targeted class is chosen depending on relevant factors such as age and gender. Outputs are compared to the targeted values and any difference corresponds to an error. This error is some scalar function of the weights. It is minimized when the network outputs match the desired outputs. The weights are adjusted by the network to reduce the measure of the training error which is defined as:

$$J(w) \equiv \frac{1}{2}\sum_{k=1}^{c}(t_k - z_k)^2 = \frac{1}{2}\|t - z\|^2 \qquad (9)$$

where t and z are respectively the target and the network output unit and w represents all weights in the network. We use the cross-validation method to determine the performance of the network. This method consists of dividing the training set into m disjoint sets of equal size n/m. n is the total number of patterns in the main training set D. The neural network is trained m times, each time with a different set held out as a validation set. The estimated performance is the mean of these m errors. Cross-validation is an empirical approach that tests the network experimentally. The validation error gives an estimate of the accuracy of the final network on the unknown test set (Duda, 2001).

Testing Phase. Now that the training phase is completed, we start the testing phase using a new engineer's design for the contour of the car. When a design is classified in the targeted class, the next step in the process consists of evaluating the set of options. Through this evaluation, we provide a score to the options that the design engineer has included in the car. However, no options evaluation is done when the contour fails in belonging to the targeted class. In this case, we compute the Euclidean distance between the failing input vector and the previously stored vectors of best designs. The engineers benefit from the distance's result in order to improve their design. This automatic feedback helps the engineers to build an "optimum" car.

Options Evaluation

We collect priority levels from the students for each option attribute. Each student will rank the given options from 1 to 6 with 1 as the highest priority and 6 the lowest priority.

The Optimal Options. We use the Borda Count method (Ho et al., 1994) to rank attributes by level of priority using the majority vote. This method measures the strength of agreement by the students that this attribute is assigned a certain level of priority. The Borda Count for an attribute class is the sum of the number of other attribute classes ranked below it (Fig. 1). This method is simple to implement and requires no training. After computing the Borda Count for each option, we propose a new approach to give it a score based on the Borda Count. The engineer provides us with the set of options included in his design. The set is converted into 6 binary values, the defined number of options in this application. The value 1 means that the option is included and 0 means that it is not. We calculate this score by dividing the Borda Count of each option, by the sum of all the Borda Count computed. We add the scores if the option value is 1 (rewarding the engineer) and we subtract the scores if the option value is 0 (penalizing the engineer). The total score is the sum of all options scores obtained from the engineer's set. The design enginner can see his option score in the interval [min-score...max-score]. The value min-score corresponds to the case where the engineer has selected only one option which has obtained the smallest score from the Borda Count (the least important option!). However, the value max-score corresponds to the case where the engineer has selected all options. For example if the engineer includes the following options: "CD", "leather seat", "heated seat", and "power window" amongst 6 possible options, then its option selection is represented by the binary vector $V^T = (1, 0, 0, 1, 1, 1)$. Let's assume that the trellis of the options and their Borda Count is represented by the following Cartesian set: {(CD, 0.8) ; (leather seat, 0.75) ; (heated seat, 0.67) ; (cruise control, 0.55) ; (sun roof, 0.44) ; (power window, 0.25)} , therefore the engineer options set is:

$1 \times 0.8 - 1 \times 0.44 - 1 \times 0.55 + 1 \times 0.75 + 1 \times 0.67 + 1 \times 0.25 = 1.48$.

EXPERIMENTAL RESULTS

We have used 30 Fourier descriptors to capture each of the three contours of a car. Classification using backpropagation has been performed during a first step. The performance obtained during training is 2.33e-017 for 2 epochs and the testing accuracy is 71.5%. Figure 2 depicts the training and testing performances with respect to the number of epochs. Once a car design has reached its target, a second step consists of computing a score for the options selected by the engineer. This part of the work has been completed.

The preliminary classification results obtained seem to be promising. However, we are currently undertaking a larger-scale experiment that enables us to report more robust results in the next future.

CONCLUSION

We have proposed a new approach that maps engineers' design to customers' perception. We believe that this application saves costs incurred by automotive companies during a car construction process. It fills the gaps that exist between the engineer's view of a "good" car and the public's view. It enables the engineers to target for example particular designs such as "sport" car and study the public reactions before building the car. The accuracy of the prediction obtained so far is 71.5%. However, as outlined earlier, the size of our data set is relatively small and more intensive testing of the system is still underway.

FUTURE WORK

The two following tasks are amongst our future work:
- A car design consists of many components. In this application, we only considered two elements: *the contour* and *the options*. We intend to add new elements in our application in order to achieve an almost complete prediction of the best design.
- The clusters defined in this application are few of the real possible clusters that could be obtained. More detailed and aggregate clusters are being considered for future experiments.

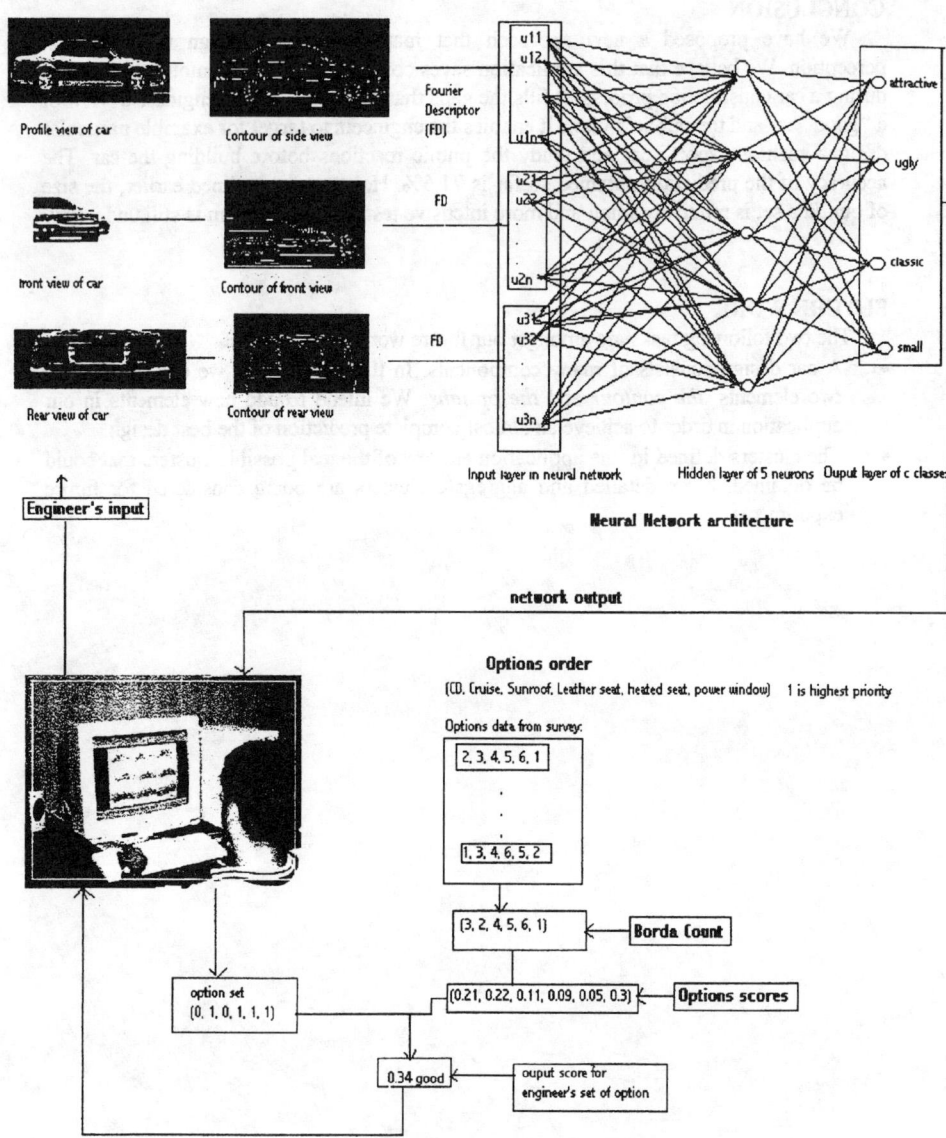

Figure 1: Flow chart of the application

List A of Attributes

Attractive: Elegant, deluxe, luxury, exquisite, high class, neat, refined, tasteful, sophisticated, formal, chic, bewitching, enchanting, fascinating, magnetic, cute, delighting, pretty, irresistible, photogenic, beautiful, pleasing, appealing, exciting.
Ugly: disfigured, hideous, awkward, displeasing, unattractive, horrible, bad.
Geometric: boxy, very straight lines.
Simple: simple
Classic: old, retro
Standard: usual.
Curved: flexuous.

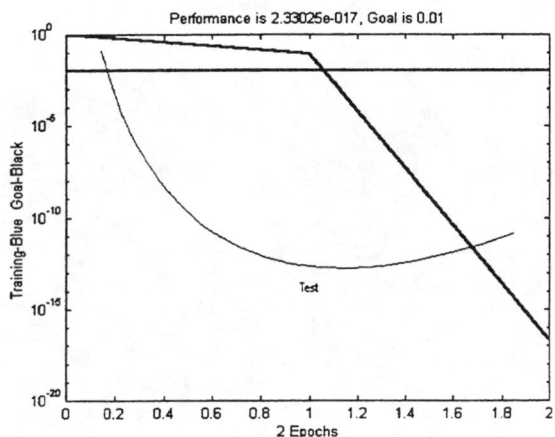

Figure 2: The training and testing performances with respect to the number of epochs.

ACKNOWLEDGMENT

We would like to thank the engineer Nazih Aboul-Hassan, a current employee of Electronic Data Systems and a former researcher in General Motors, for his helpful comments. We are also grateful to all students who spent a part of their time to answer our questions during this survey.

REFERENCES

Duda R. et al., 2001, "Multilayer Neural Networks", *Pattern Classification*, Wiley, New York, pp. 288-295.
Granlund G. H., 1972, "Fourier preprocessing for hand print character recognition", *IEEE Trans. Comput. 21*, pp. 195-201.
Ho T. K. et al, 1994, "Decision Combination in Multiple Classifier Systems", *IEEE Transactions on Pattern Analysis and Machine Intelligence*, Vol. 16, No. 1.
Kato T. and Ninomiya Y., 2000, "An approach to vehicle recognition using supervised learning", *IEICE Trans.INF & SYST.*, Vol. E83-D, No.7.
Trier Q. D. et al., 1996, "Pattern Recognition", *Elsevier Science Ltd.*, Vol. 2 No. 4, pp. 641-662.
Zhang L. and Vertiz A., 2000, "Kansei Engineering Application on Commercial Truck Interior Design
    Harmony", *Society of Automotive Engineers*.
http://www.autovu.com/index2.html
http://www.wordnet.com

CHARACTERIZATION OF FRACTAL MORPHOLOGY OF FLOCS FORMED IN OIL-WATER EMULSIONS BY MICROSCOPIC IMAGE ANALYSIS

J. HEMPOONSERT, B. TANSEL
Civil and Environmental Eng. Dept.
Center for Eng. and Appl. Sciences
Florida International University
10555 West Flagler Street
Miami, Florida 33174

I.N. TANSEL
Mechanical Eng. Dept.
Center for Eng. and Appl. Sciences
Florida International University
10555 West Flagler Street
Miami, Florida 33174

ABSTRACT

Fractal characteristics of flocs formed in edible oil-water emulsions are investigated by analyzing the microscopic images of the flocs. The flocs were formed by coagulation experiments, which were conducted using a jar test apparatus and emulsions of corn oil, olive oil, and sunflower oil in water. A cationic polyelectrolyte was used as the coagulant. The microscopic images of the flocs were captured with a digital camera and analyzed using Image Pro Software. The fractal dimensions of flocs formed in emulsions of corn oil, olive oil, and sunflower oil in water were determined to be 1.21, 1.47, and 1.17, respectively.

INTRODUCTION

Coagulation process is typically used for treatment of industrial wastewaters to promote aggregation of colloidal solids or oil droplets on flocs that can be subsequently removed by sedimentation or dissolved air flotation. Coagulant is added to destabilize and disturb the balance between the electrostatic repulsive forces and the force of gravity by neutralization and interparticle bridging (Narkis ET al., 1991; Petzold et al., 1996). For the formation of flocs, the force of primary particle attraction to form floc has to be stronger than repulsive interaction force and electrostatic repulsion of particles (or oil droplets). On the other hand, flocculation force has to overcome repulsive interaction transition (Aveyard et al., 1999). The sizes of aggregates depend upon particle (droplet) size and the density differences between the dispersed and the continuous phases. For larger oil droplets, the molecular attraction forces influence the aggregation process since droplets can impact the size of the flocs (Saether et al., 1999). The size and density of flocs can have a significant influence on the effectiveness of the solid-liquid separation processes (i.e., sedimentation, filtration and flotation). Floc characteristics such as composition, shape, size and porosity can affect settling velocities of flocs (Dennett, 1996). Experimental and theoretical studies have shown that with increasing floc size, the porosity increases resulting in a decrease in floc density and settling velocity (Gregory, 1997, Droppo et al., 1997; Bache et al., 1997). Flocs usually grow until they reach a certain limiting size depending on the applied shear rate (Tambo and Watanabe, 1979; Gregory, 1997; Rosen, 1984, Ivanov et al., 1999). Large flocs are broken into smaller fragments, which may continue to aggregate individually (Rulyov et al., 2000). The particle agglomeration of oil droplets in emulsions is controlled by various particle transportation mechanisms such as Brownian movement, bulk transportation by fluid motion, and settling (Weber, 1972; Tambo, 1991a,b; Webster et al., 1997). In addition, properties of water (i.e., temperature, viscosity, and density), properties of particles (i.e., concentration, density, and size), properties of polyelectrolytes and characteristics of operational units for flocculation and coagulation processes (i.e, detention time, flow pattern, and mixing speed) also influence flocculation mechanisms in oil-water emulsions.

The concept of fractals was developed by Mandelbrot in 1975 to describe self-similar behavior (scale invariant) of patterns (Feder, 1988). While the classical geometry deals with objects of integer dimensions, the fractal geometry describes non-integer dimensions. For example, zero dimensional points, one-dimensional lines and curves, two-dimensional planes, and three-dimensional volumes such as cubes and spheres make up the classical geometry. However, many natural systems are better described with a non-integer dimension. While a straight line has a dimension of one, a fractal curve will have a dimension between one and two depending on how much space it takes up as it moves and curves. The more a flat fractal fills a plane, the closer it approaches two dimensions (Feder, 1988). Hence, the fractal dimension measures how densely the regarded object occupies the surrounding space. By calculating fractal dimension, self-similarity characteristics of the system can be determined.

It has been recognized that most aggregates or flocs formed in natural and industrial processes possess mass fractal geometry (Waite, 1999; Gregory, 1997). The characteristics of flocs can be examined as mass fractals by analysis of the relationship between the number of droplets and the floc size. For mass fractals, the fractal dimension can be represented by the following relationship between the radius of gyration of an aggregate and the number of primary particles in the aggregate (Koylu et al., 1995a):

$$N = k_g \left(\frac{R_g}{a} \right)^{D_f} \quad (1)$$

where,
N : Number of particles (oil droplets)
k_g : fractal prefactor
R_g : radius of gyration
a : mean primary particle radius (mean droplet radius)
D_f : fractal dimension (exponent)

This equation can be applied for analysis of flocs formed in oil-water emulsions by defining N as then number of oil droplets, R_g as radius of gyration of the floc, and a as the mean oil droplet radius. Determination of the fractal dimension requires the determination of the actual radii of gyration of aggregates, which cannot be measured easily. It has been suggested that any characteristic dimension of an aggregate can be used in place of the radius of gyration (Koylu et al., 1995a). As a result, the maximum length (L_{max}), geometric mean of length and width (i.e., $(LW)^{1/2}$), and outer radius of an agregate ($L/2$) have been used (Koylu et al., 1995). Koylu et al. (1995b) showed that the ratio $(LW)^{1/2}/2R_g$ decreases as N increases while L scales with R_g almost in a regular manner. Hence, the fractal dimension obtained using L as the characteristic dimension would be more representative of the value using R_g. As a result, the outer radius of an aggregate, $R_L = L/2$, should be used, instead of R_g, in determination of D_f to prevent intrinsic biasing of results with respect to N (Koylu et al., 1995a). Therefore, it is possible to estimate the fractal dimensions using only projected images from the slope of a log-log plot of N versus R_L/a for a set of aggregates, where R_L = radius of the aggregate ($L/2$). Figure 1 illustrates the characteristic measurements of a floc which are used for determination of the fractal dimension, D_f.

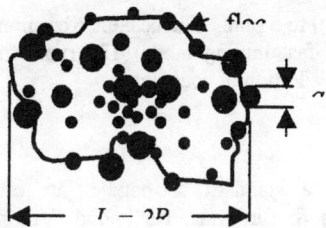

Figure 1. Characteristic measurements of a floc for calculating fractal dimension, D_f.

Flocculation mechanisms have been described as either diffusion-limited aggregation (DLA) or reaction-limited aggregation (RLA) (Waite, 1999). Diffusion-limited aggregation implies that there is no repulsion between colliding particles and each collision leads to attachment. During DLA, a particle undergoes Brownian motion until it makes contact with and sticks to a free-floating cluster of particles. However, in RLA, due to interparticle repulsion, each collision does not result in aggregation. Particles must collide more than once before attaching together. Under DLA conditions, the fractal dimensions have been reported to be less than 1.8; and under RLA conditions the fractal dimensions are in the range from 2.2 to 2.3 (Waite, 1999).

The objectives of this study are to analyze microscopic images of flocs formed in edible oil-water emulsions and characterize their fractal morphology and hence determine the mechanisms for flocculation (i.e., DLA or RLA). Understanding the fractal characteristics of flocs could help determine the specific floc sizes which would remove most amount of oil droplets. Although the formation of larger flocs is preferred in flocculation processes, the larger flocs may not necessarily result into the highest oil removal efficiencies. A cationic polyelectrolyte was used for the coagulation of oil droplets in edible oil-water emulsions. A jar test apparatus was used for the coagulation and flocculation processes. Characteristics of the microscopic images of the flocs were analyzed to determine fractal characteristics of the flocs.

MATERIALS AND METHODS

The experiments were conducted using emulsions of corn oil, olive oil, and sunflower oil in water. Emulsions of edible oils were prepared by mixing edible oils with water to contain 600 ppm (vol/vol) of oil. A cationic coagulant, Cat-Floc 2953, manufactured by Calgon Corporation, Pittsburgh, Pennsylvania, was used as the polyelectrolyte for coagulation. The active ingredient in Cat-Floc 2953 is poly-(dimethyldiallyl-ammoniumchloride).

Preparation of oil-water emulsions

For each oil type, to prepare the oil-water emulsion, six liters of tap water was placed into an eight-liter glass container. Bentonite was added to the water at a concentration of 15 ppm to promote flocculation. The solution was mixed with a mechanical stirrer for 6 hours at room temperature. The oil was added to the mixture at a concentration of 600 ppm (vol/vol) and the mixture was stirred for an additional 24 hours. For the preliminary studies, after 24 hours of mixing, the oil-water emulsion was transferred to six 1-liter beakers for the screening tests to determine the optimum flocculation conditions. Coagulant (Cat-Floc 2953) was added to each beaker at increasing doses as from 0 to 5 ppm. The beakers were immediately placed in the jar-test apparatus and stirred at fast mixing mode (130 rpm) for 5 minutes to maintain the emulsion state before the addition of coagulant.

Experimental set-up

For the jar tests, a standard six-paddle jar test apparatus (model PB-700 manufactured by Phipps & Bird Inc., Richmond, Virginia) was used. The jar test apparatus was equipped with high and low speed controls (5 rpm to 300 rpm). After the addition of the coagulant, each oil-water emulsion was stirred at fast mixing mode (130 rpm) for 10 minutes and then at slow mixing mode (10 rpm) for 25 minutes. After slow mixing, the mixtures were kept at room temperature for 30 minutes for floc formation. The efficiency of coagulation was determined by turbidity measurements using a Hach Turbidimeter. The turbidity results showed that a coagulant concentration of 3 ppm was most effective for floc formation and this dosage was used for the fractal analysis.

For the microscopic examination and image analysis, 1 to 2 drops of the coagulated water sample were placed on the microscope slides using Pasteur pipettes with a large tip to minimize breakage of flocs during transfer. The flocs were examined by an Olympus System Microscope, model BX40, Olympus America Inc., Melville, NY. The microscope was equipped with a Sony Color video camera model DXC-107A/107AP. The captured images were analyzed using Image–Pro Plus software by Media Cybernetics, Silver Spring, Maryland (Image-Pro Plus, 1997). The digital images were analyzed in the form of rectangular grids of 640x480 pixels corresponding to an image area of 414.26µm x 312.65 µm. Enhancement techniques ranging from simple operations such as brightness and contrast adjustment to complex spatial and morphological filtering operations were used to improve and refine visual information. The oil droplets were distinguished by color threshold.

RESULTS

The flocs formed in corn oil, olive oil, and sunflower oil with Cat-Floc 2953 were in white color and of a porous structure. Figure 2 presents the typical microscopic images of flocs formed in the olive oil-water emulsion. After mixing with the coagulant and allowing solutions to settle 30 minutes, some flocs still continued to aggregate both in the solution and at the surface of the solution where flocs were concentrated.

The microscopic examination showed that oil droplets were captured both on the floc surface and within the floc matrix. The smaller droplets were observed to be entrapped with the floc matrix. Although some oil droplets had relatively larger diameters, the majority of the droplets were within a characteristic droplet diameter range (0.5-10 microns). The results revealed that oil droplets, which had diameters in the range of 0.5-2.5 microns, were attached to the floc matrix better than other sizes. The reason might be the size of attachment sites on the floc surface match the droplet size, whereas larger size oil droplets need larger sites to be attached on the floc surface. Fewer oil droplets were captured at and near the center of flocs whereas a significant number of oil droplets were attached at further distance from the floc's center. It is possible that at the center of the floc, attractive forces are strong enough to provide many oil droplets to attach but the surface area is too small to provide adequate area for their attachment.

The analysis of fractal morphology of flocs showed that floc formation followed the fractal theory. Figure 3 presents the fractal analysis of the flocs formed in emulsions of corn oil, olive oil, and sunflower oil in water. The plots of N and R_L/a on a log-log scale formed a linear pattern, where N was defined as the number of oil droplets counted on the flocs, R_L is the floc radius, and a is the mean droplet radius. The fractal dimensions (D_f) of flocs formed in emulsions of corn oil, olive oil, and sunflower oil in water were 1.21, 1.47, and 1.17, respectively. The magnitudes of the fractal dimensions indicate that oil droplets do not densely accumulate on the flocs. In addition, the larger flocs may not necessarily yield to higher oil removal efficiencies for especially olive oil.

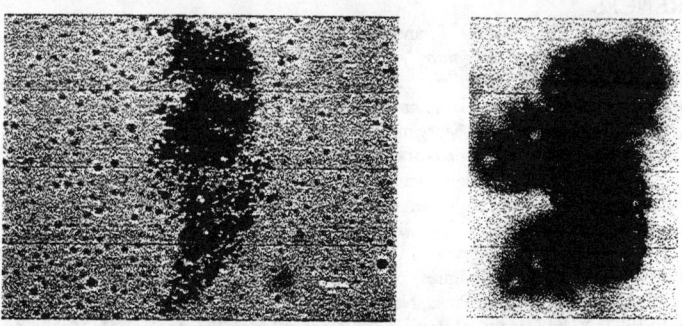

Figure 2. Typical microscopic images of flocs formed in olive oil-water emulsions.

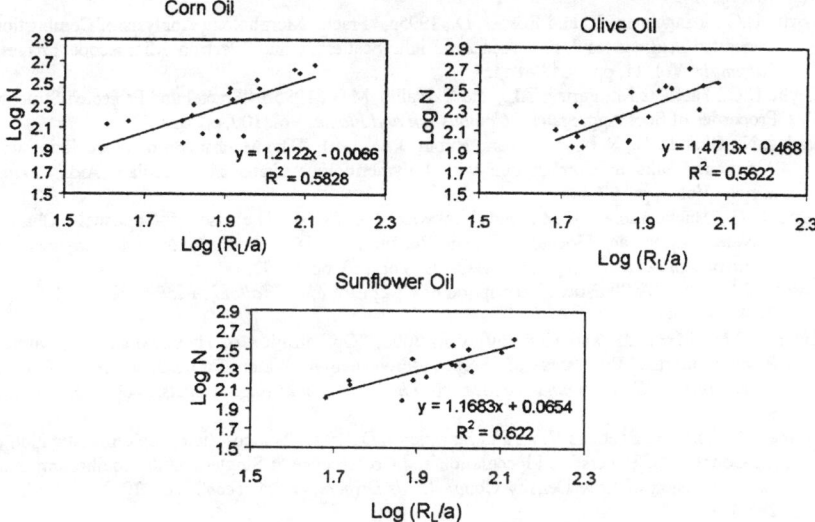

Figure 3. Fractal analysis of flocs formed in the corn oil-water, olive oil-water, and sunflower oil-water emulsions.

CONCLUSIONS

This study revealed that attachment of oil droplets to flocs followed a fractal pattern. The smaller diameter oil droplets were observed to be are entrapped within the floc matrix more in comparison to the larger droplets. The analysis of the captured images of the flocs showed that there were a significant number of oil droplets that were entrapped at a certain distance from the center of the floc. This is probably due to smaller electrostatic charge at the center for oil droplets to attach. However, as the distance increased from the center of the floc, the number of oil droplets attached on the surface decreased after reaching a maximum. The fractal dimensions calculated in this study showed that the aggregation mechanisms during flocculation process in the edible oil-water emulsions studied were diffusion-limited.

REFERENCES

Aveyard, R., Binks, B.P., Esquena, J., and Fletcher, P.D.I., 1999, " Flocculation of Weakly Charged Oil-Water Emulsions," *Langmuir*, Vol. 15, pp. 970-980.

Bache, D.H., Johnson, C., McGilligan, J.F., and Rasool, E., 1997, "A Conceptual View of Floc Structure in the Sweep Floc Domain," *Water Science Tech.*, Vol. 36, No. 4, pp. 49-56.

Dennett, K.E., Amirtharajah, A., Morgan, T.F., and Gould, J.P., 1996, "Coagulation: Its Effect on Organic Matter," *J. Am. Water Works Assoc.*, Vol. 88, No. 4, pp. 129-142.

Droppo, I.G., Leppard, G.G., Flannigan, D.T., and Liss, S.N., 1997, "The Freshwater Floc: A Functional Relationship of Water and Organic and Inorganic Floc Constituents Affecting Suspended Sediments Properties," *Water, Air and Soil Poll.*, Vol. 99, Nos. 1-4, pp. 43-54.

Droppo, I.G., Derrick, T.F., Leppard, G.G. Jaskot, C., and Liss, S.N., 1996, "Floc Stabilization for Multiple Microscopic Techniques," *App. and Env. Microbiology*, Vol. 62, pp. 3508-3515.

Feder, J., 1988, Fractals, Plenum Press, New York.

Gregory, J., 1997, "The Density of Particle Aggregates," *Wat. Sci. Tech.*, Vol. 36, No. 4, pp. 1-13.

Image-Pro Plus, 1997, Version 3 for Windows, Media Cybernetics, Silver Spring, Maryland.

Ivanov, I.B., Krassimir D.D., and Peter A.K., 1999, "Flocculation and Coalescence of Micron-Size Emulsion Droplets," *J. Colloids and Surfaces: A: Physicochemical Eng. Aspects*, Vol. 152, pp. 161-182.

Koylu, U.O., Yangchuan, X, and Rosner, D., 1995a, "Fractal Morphology Analysis of Combustion-Generated Aggregates Using Angular Light Scattering and Electron Microscope Images," *Langmuir*, Vol. 11, pp. 4848-4854.

Koylu, U.O., Faeth, G.M., Farias, T.L., and Carvalho, M.G., 1995b, "Fractal and Projected Structure Properties of Soot Aggregates," *Combustion and Flame*, Vol. 100, pp. 621-633.

Narkis, N., Ghattas, B., Rebhun, M., and Rubin, J.A., 1991, "The Mechanism of Flocculation with Aluminum Salts in Combination with Polymeric Flocculants as Flocculant Aids," *Water Supply*, Vol. 9, pp. 37-44.

Petzold, G., Buchhammer, H.M., and Lunkwitz, K., 1996, "The Use of Oppositely Charged Polyelectrolytes as Flocculants and Retention Aids," *J. Colloids and Surfaces, A: Physicochemical and Engineering Aspects*, Vol. 119, pp. 87-92.

Rosen, J.M., 1984, "A Statistical Description of Coagulation," *J. Colloid Interface Sci.*, Vol. 99, pp. 9-19.

Rulyov, N.N., Maes, A., and Koroyov, V.J., 2000, "Optimization of Hydrodynamics Treatment Regime in the Processes of Sorption-Flocculation Water Purification from Organic Contaminants, *"Colloids and Surfaces A: Physicochemical Eng. Aspects*, Vol. 175, pp. 371-381.

Saether, O., Johan S., Svetlana V.V., and Stanislav S.D., 1999, "Video-Microscopic Investigation of the Coupling of Reversible Flocculation and Coalescence at Single-Double Equilibrium in an O/W Emulsion of Low Density Contrast," *J. Dispersion Sci. Tech.*, Vol. 20, Nos. 1&2, pp. 295-314.

Tambo, N., Watanabe, Y., 1979, "Physical Aspects of Flocculation: I. Floc Density Function and Aluminum Floc," *Wat. Res* , Vol. 1, pp. 409-419.

Tambo, N., 1991a, "Coagulation and Flocculation on Water Quality Matrix," *Water, Sci. Tech.*, Vol. 37, No. 10, pp. 31-41.

Tambo, N., 1991b, "Basic Concepts and Innovation Turn of Coagulation/Flocculation," *Water Supply: The Review J. Int. Water Supply Assoc.*, Vol. 9, No. 1, pp. 1-10.

Waite, T.D., 1999, "Measurement and Implications of Flocs Structure in Water and Wastewater Treatment," *Colloids and Surfaces A: Physicochemical Eng. Aspects*, Vol. 151, pp. 27-51.

Weber, J.W., 1972, Physicochemical Processes for Water Quality Control, John Wiley & Sons, Inc., New York, New York.

Webster, L., Huglin, B., and Robb, D., 1997, "Complex Formation between Polyelectrolytes in Dilute Aqueous Solution," *Polymer*, Vol. 38, pp. 1373-1380.

PART VIII. PREDICTION

DEVELOPMENT OF INTELLIGENT SYSTEMS FOR IMPROVED WSR-88D RAINFALL ESTIMATION

THEODORE B. TRAFALIS, Ph.D.
ANDERSON WHITE, Ph.D.
BUDI SANTOSA
School of Industrial Engineering
University of Oklahoma
Norman OK, 73019-0631

MICHAEL B. RICHMAN, Ph.D.
School of Meteorology
University of Oklahoma
Norman OK, 73019-0626

ABSTRACT

The objective of this paper is to utilize data mining and an intelligent system, Artificial Neural Networks (ANNs), to facilitate rainfall estimation. Ground truth rainfall data are necessary to apply intelligent systems techniques. A source of such data is the Oklahoma Mesonet. With the advent of a national network of advanced radars (i.e., WSR-88D), massive archived data sets have been created generating terabytes of data. Data mining can draw attention to meaningful structures in the archives of such radar data, particularly if guided by knowledge of how the atmosphere operates in rain producing systems. The WSR-88D records digital database contains three variables: velocity (V), reflectivity (Z), and spectrum width (W). Current rainfall detection algorithms make use of only the Z variable. The focus of this research is to utilize Z and W variables at multiple elevation angles and multiple bins in the horizontal for precipitation prediction by feedforward ANNs.

INTRODUCTION

Flash floods kill more people than any other weather phenomenon. Despite this, our ability to estimate precipitation and flooding from current state of the science technology is frequently inaccurate and can be improved. Much of this inaccuracy arises from poor estimates from the Weather Surveillance Radar 1988 Doppler (WSR-88D) algorithms, which use only empirical techniques (OFCM 1991, Fulton et al. 1998).

Recently, intelligent systems, specifically ANNs, have been applied as a novel approach for modeling of complex, highly nonlinear, dynamical systems. The main advantage of ANNs is in cases where intrinsic nonlinearities in the dynamics prevent the development of exactly solvable models. In meteorology, all of these criteria are present in the sense that the dynamics are inherently nonlinear, and prediction is one of the main goals. Capturing the radar signature and classifying this signature using ANNs has been accomplished by Skapura (1996). Trafalis et al. (1997), and Jeyabalan et al. (1998) have introduced ANN based models to diagnose mesocyclones for tornado prediction from radar observations. These investigations have shown that, ANN based models typically recognize tornados and nearly always outperform the existing detection algorithms for prediction of tornados. Of particular interest to the paper is the study of Xiao and Chandrasekar (1997) who have developed an ANN based algorithm for rainfall estimation from reflectivity of radar observation. A three-layer perceptron ANN was developed. The ANN is trained using the radar measurements as the input and the rain gauge measurements as the output. The rainfall estimates obtained from the ANN are shown to be better than those obtained from empirical based existing techniques despite the fact that only reflectivity was utilized. All of these studies report an improvement in performance by using ANNs. The main objective of the paper is developing an intelligent pattern recognizer to improve the WSR-88D rainfall estimation.

PROBLEM STATEMENT

For over forty years, rainfall estimation from radar has been an active area of research. For the most part, the issue has been addressed through reflectivity-rainfall relations (known as Z-R). The Z-R relation was pioneered by Marshall and Palmer (1948). Currently, radar rainfall estimates are computed from a parametric Z-R relation that can be demonstrated in various ways. The most common form of this relationship can be written as follows:

$$Z = aR^b \qquad (1)$$

The current WSR-88D default values for a and b are *300* and *1.4*, respectively (Fulton et al. 1998). However, the values of a and b vary from place to place, season-to-season, and time to time (Wilson and Brandes 1979). It is obvious that an estimation technique based on this relation will not be very successful since no single values for a and b will give a good estimate of the rainfall over a broad range of conditions. What does occur is a large uncertainty in estimating rainfall from reflectivity. The bias is not a constant and cannot be corrected with existing algorithms. Therefore, it is essential to develop a new technique that reduces this variability of the Z-R relation.

WSR-88D records digital base data containing three variables: velocity (V), reflectivity (Z), and spectrum width (W). Current rainfall detection algorithms use only Z data (Fulton et al., 1998). The other two variables V and W are native output of the WSR-88D containing additional information and, remarkably, have not been explored. The primary focus of the paper is to capitalize on these additional variables instead of using only Z. Our paper proposes to (1) use a multiparametric approach by using Z, V, and W for precipitation prediction from the Norman, Oklahoma WSR-88D, (2) apply data mining techniques to Z, V and W in order to understand the naturally occurring inter-relationships and signatures of these data when rain is detected and (3) use ANNs for precipitation prediction.

DATA AND ANALYSIS

The rainfall data is collected for 1998 calendar year from the Chandler, Oklahoma Mesonet station (Basara et al., 1999). Rainfall is measured at 5-minute intervals. The WSR-88D radar has an effective range of approximately 230 km. The digital base data containing the three variables to be used herein is limited to a range of 180 km. The radar performs approximately 10 elevation scans that comprise of a single volume scan. For each elevation scan, the radar revolves a full 360 degrees about the vertical and makes about 366 azimuthal scans. For the azimuthal on the five lowest elevation scans, we will utilize V, W, and Z. W and V are measured once in every 0.25 kilometers or at 920 points and Z is measured once in every kilometer or at 230 points for every azimuthal scan. The radar used in this study (KTLX) is approximately 54 Km WSW of the Chandler raingauge. The radar data are divided approximately into a 5 by 5 array of 1 sq. km horizontally spaced grids centered on the Mesonet raingauge. There were five elevation angles retained which yields 25 grid boxes per level times 5 elevations for a total of 125 boxes serving to describe the atmosphere above the gauge.

Given the 5-minute time resolution of both observing and measurement platforms, it is clear that there will be a large amount of data. Therefore it is necessary to preprocess and do some feature extraction from the raw data for the following two reasons. First of all, raw data may suffer from problems such as noise, missing data points, or probability distribution errors (Rosenfeld et al., 1993). Secondly, using the raw data may lead our search procedure to suffer from the curse of dimensionality and yield poor results. Therefore, it is important to transform the raw data into some new representation by using data mining techniques. The technique used to analyze patterns of inter-related three-dimensional volume scans in this study uses principal component analysis. The analysis is exploratory with a goal of determining which grids at five elevations cluster together.

METHODOLOGY

WSR-88D and Oklahoma Mesonet station data are extracted, preprocessed and applied to the multilayer feedforward ANN model as shown in Fig. 1. Our objective is to find the set of optimal connection weights that achieve the correct mapping between observed precipitation data $RF_g(k)$ by rain gauges k and calculated rainfall data $RF_n(k)$ by the ANN for a set of given input data Z and V for k^{th} pattern by WSR-88D. This problem is commonly known as supervised ANN training. The training phase of ANNs (Haykin, 1999) can be seen as a nonlinear optimization problem where the training error is to be minimized. This problem can be formulated as follows:

$$\text{minimize} \sum_{i=1}^{N_j} [RF_g(i) - RF_n(i)]^2 \qquad (2)$$

The minimization is with respect to the connection weights of the ANN. Several nonlinear optimization techniques can be used (Bazaraa et al. 1993). In our experiments, the ANN will be trained to estimate rainfall precipitation at one Mesonet station (Chandler, OK) by using rain gauge data at this station and WSR-88D data. Note that, for WSR-88D data, only the part whose partitioning coincides directly with the location of an underlying rainfall station will be used. This problem can be viewed as time series analysis or nonlinear dynamical system approximation.

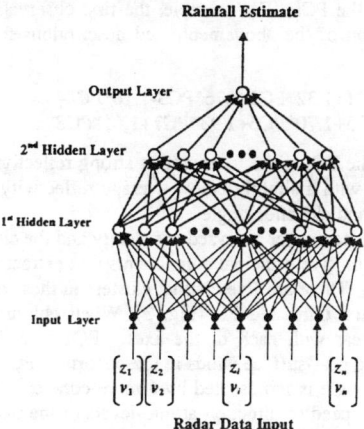

Figure 1. ANN architecture.

RESULTS OF DATA MINING

The correlation structure of the radar reflectivity projected on the aforementioned 125 variables is investigated by projecting the structure on an uncorrelated set of basis vectors using PCA. A plot of the leading two PC loading vectors (Fig. 2) clearly indicates that there is a major locus of points in a large diffuse cluster. The first PC extracts 88 percent of the reflectivity variability. However, by scrutinizing Fig. 2, it is apparent that there are five isolated clusters, each of which exhibits low within group variability, residing near PC1. This merging of multiple clusters is not surprising as the first PC seeks to explain the maximum variation in the data (Richman, 1986).

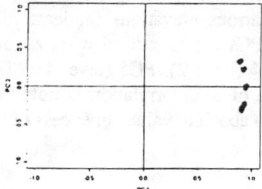

Figure 2. Unrotated PC1 vs. PC2 Reflectivity.

In order to determine if (1) the clusters can be separated onto individual PCs and (2) if those clusters have any physical interpretation, a simple structure rotation is applied. A Promax k=2 oblique transformation is selected to isolate this cluster structure so that each cluster is associated with one PC. After some experimentation to achieve the best signal-to-noise separation and decomposition of individual clusters, it was decided that retention of eight PCs accounting for 96.6 percent of the variance in reflectivity led to a compact description of the natural clustering in the dataset. The description of each PC is as follows: PC1 clusters the lowest elevation angle; PC2 clusters the fourth and fifth elevation angles; PC3 clusters the third elevation angle; PC4 clusters the second elevation; PC5 clusters the western portion of the domain for the lower two elevation angles; PC6 clusters the fifth elevation angle; PC7 clusters the fourth elevation angle and PC8 clusters the southeast portion of the domain in levels two through 5. The advantage of the oblique rotation is that it allows this highly correlated structure to be separated into individual clusters. The correlation between each of the seven PCs indicates every pair of PCs is correlated at an absolute value in excess of 0.5 with 7 of the 28 pairs in excess of 0.7. There are 1009 individual 5-minute observations in the data set. The reflectivity pattern for each observation can be reconstructed with 95 percent overall accuracy with a linear combination of the PCs. For example, the first observation (Jan. 05, 1998 0611 GMT) is a combination of the abovementioned descriptions for the PC loadings with these multipliers:

$$0.24*PC1+1.32*PC2-0.76*PC3-1.16*PC4-0.01*PC5+1.70*PC6+1.93*PC7+1.61*PC8$$

The interpretation of the PCs is the following: a strong reflectivity signal on the highest three elevation angles with a slightly below average reflectivity in the northeast part of the domain for the second elevation angle.

The second native parameter analyzed is velocity and the clustering is accomplished by projecting the data on five principal components that extract 94.6 percent of the total variation of these data. There are five distinct clusters in these data in a highly collinear arc surrounding the first unrotated PC (Fig. 3). When obliquely rotated, five distinct clusters emerge collinear with each of the axes. PC1 is a diffuse cluster, which is consistent with variable near-surface winds in these storms. PC2 clusters the radar boxes on the fifth elevation angle is represented by a more coherent cluster that indicates less variability in the wind speed or direction at mid-levels of the troposphere. Interestingly, levels 3 and 4 cluster on the third PC, with more saturation on level three. This suggests that those two elevation angles are related in velocity. PC4 isolates level four more than any other; therefore, there are two modes of variation involving level 4. one in isolation and one where it covaries somewhat with level 3. PC5 clusters level 3 strongly with level 4 less coherently. The absolute value of the correlations accounting for the angular separation between the PC axes confirm the correlated nature of the clusters seen in Fig. 3 with 8 of the 10 pairs having values in excess of 0.5 and 4 of 10 exceeding 0.7

The third radar parameter analyzed is spectrum width. There are five eigenvalues above the noise spectrum and the eigenvectors associated with these are retained for analysis and account for 53.8 percent of the spectrum width variance. Note this is much lower than for the analysis of either reflectivity or velocity, indicating a more chaotic behavior in the spectrum width data. Investigation of the first pair of unrotated PCs (Fig. 4) indicates two distinct clusters and one elongated one. However, it is noted that, by obliquely rotating the unrotated reference frame, five clusters become distinct. The spectrum width on each level maps to a single PC that indicates the turbulence is small scale and not connecting various elevations (at least in any degree out of the noise spectrum). The following PCs are identified with various levels: PC1 (level 5), PC2 (level 1), PC3 (level 3), PC4 (level 2), PC5 (level 4). The oblique PCs are correlated, where a much lower degree of intercorrelation is noted among the axes on which the clusters are projected with no absolute values in excess of 0.5 but 3 of 10 values in excess of 0.3.

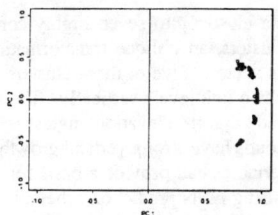 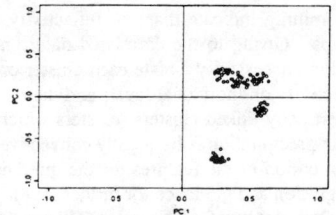

Figure 3. Unrotated PC1 vs. PC2 Velocity. Figure 4. Unrotated PC1 vs. PC2 Spectrum Width.

RESULTS OF NEURAL NETWORKS

We apply ANN to facilitate rainfall prediction. Reflectivity (Z) and spectrum width (W) data obtained from WSR-88D were used as inputs to ANN. There are 550 observations for the training and 26 observations for testing. The dimension of each data point is 1 by 250 column vector. During the experiments examination of some architectures of ANNs is performed in order to find the best one which produced the smallest error. Four different architectures in term of the number of hidden layers and the number of nodes for each layer are investigated. Each architecture is run ten times and the average MSE of these runs is presented in Table 1.

Table 1. MSE of training and testing for different architectures.

Architectures	Training Avg MSE	Testing Avg MSE	Hrly rainfall error (mm/hr)
250-5-1	0.0029	0.3117	6.70
250-5-5-1	0.0014	0.6752	9.86
250-10-1	0.0029	0.5092	8.56
250-10-10-1	0.0028	0.5027	8.51
250-25-1	0.0098	0.8202	10.87
250-25-25-1	0.0013	0.5330	8.76

From Table 1 it can be observed that a small error on training does not guarantee to produce a small error on the testing. This might occur because the values that we are trying to predict (rainfall) do not have a certain recurrent patterns or are very random. Most likely, the single year of rainfall data are too small to capture these necessary features. Therefore it is difficult to produce prediction values that are close to the actual data The MSEs of testing are relatively high compared to MSEs of the training. Increasing the complexity of neural network architecture from 1-hidden layer to 2-hidden layers does not have much influence for the MSE values. Despite this, the lowest testing error was 6.70 mm/hr ($\frac{1}{4}$ inch/hr) which is reasonable considering the coarse resolution of the radar and small size of the raingauge. To investigate if ANN over-training was a problem, ANN with a lower number of epochs during the training is applied. Additionally, SVM is applied to the data. In both cases the results do not improve, suggesting that the source of error is not over-training.

SUMMARY AND CONCLUSIONS

Analysis of the three native variables from the WSR-88D radar in Norman, OK and experiments for predicting rainfall rates are undertaken with data mining and ANNs. The importance of using Mesonet data is that they are recorded in 5 minute intervals that nearly match the volume scan time of the radar. Additionally, the lowest five elevation angles are used to provide a three-dimensional snapshot of the atmosphere above the raingauge. The goals of the research are twofold: data mining and prediction. The former attempts to describe how the 125 grid boxes interrelate for the three native variables during periods of rainfall. By using principal component analysis, results of the data mining indicate that, for reflectivity, the data cluster into seven highly correlated groups. Owing to the correlated nature of the clusters, an oblique transformation was used to successfully isolate each cluster on a basis vector. Five of these clusters are the reflectivity on individual levels and the remaining two link levels vertically. The finding of vertically linked clusters (clusters which spanned adjacent elevation angles) is logical as the precipitation is frequently convective (the clouds have strong vertical growth). The combination of the features for the three native variables can provide a basis for pattern recognition and guidance for further investigation using ANN prediction schemes.

The results of ANNs using ZW are relatively accurate with an average hourly error of 6.7 mm for a 250-5-1 model. The generalization properties of the used architectures might be improved further if additional rainfall data are extracted. In the near future, we plan to extract the 1999 rainfall data. We expect that by increasing the number of data will provide better estimation results. Experimentation with more data points and validation techniques will be investigated in future research. Our objective will be to find optimal architectures with good generalization properties.

Acknowledgements: This research was supported by DOC/NOAA Grant NA67RJ0150. We extend thanks to Tim O'Bannon for assistance in interpreting the radar tapes and providing program to interpret Level 2 radar data and to Tim Kwiatkowski for programming assistance. Additional gratitude is extended to Suat Kasap for data processing assistance and Jean Shingledecker for administrative assistance. This research was performed on The University of Oklahoma ECAS Cray-J90

REFERENCES
Basara, J.B., Brotzge, J.A., and Crawford, K.C., 1999, "Investigating land-atmosphere interactions using the Oklahoma Mesonet", *Third Symposium on Integrated Observing Systems*, January 10-15, Dallas, TX, American Meteorological Society, pp. 120-123.
Bazaraa, M.S., Sherali, H.D., and Shetty, C.M., 1993, *Nonlinear Programming: Theory and Algorithm,*. John Wiley & Sons, Inc., New York.
Fulton, R.A., Breidenbach, J.P., Seo, D.-J., and Miller, D.A., 1998, "The WSR-88D rainfall algorithm", *Weather and Forecasting*, Vol. 13, pp. 377-395.
Haykin, S., 1999, *Neural Networks: A Comprehensive Foundation*, Second Ed., Prentice Hall, New York.
Jeyabalan C., White, A., and Trafalis, T.B., 1998, "Statistical analysis of base data from the WSR-88D weather radar and description of mesocyclone features using neural networks", *Proceedings of the 14th Conference on Probability and Statistics in at Atmospheric Sciences*, January 11-16, Phoenix, AZ, American Meteorological Society, pp. J93-J96.
Marshall, J.S., and Palmer, W.M., 1948, "The distribution of raindrops with size", *Journal of Meteorology*, Vol. 5, pp. 165-166.
OFCM, 1991, *Doppler Radar Meteorological Observations, Part C, WSR-88D Products and Algorithm,* Federal Meteorological Handbook 11, FCM-H11C-1991, Office of the Federal Coordinator for Meteorological Services and Supporting Research, Washington, D. C.
Richman, M.B., 1986, Review paper: "Rotation of principal components", *International Journal of Climatology*, Vol. 6, pp. 293-335.
Rosenfeld, D., Wolff, D.B., and Atlas, D., 1993, "General probability-matched relations between radar reflectivity and rain rate", *Journal of Applied Meteorology*, Vol. 32, pp. 50-72.
Skapura, D.M., 1996, *Building Neural Networks*, ACM Press Books, New York.
Trafalis, T.B., Couellan, N.P., Li, P.-I., Stumpf, G., and White, A., 1997, "An affine scaling neural network training algorithm for prediction of tornados", In *Smart Engineering System Design: Neural Networks, Fuzzy Logic, Data Mining, and Evolutionary Programming*, Vol. 7, Eds. C. H. Dagli, M. Akay, O. Ersay, B. R. Fernandez and A. Smith, ASME Press, New York, pp. 213-218.
Wilson, J.W., and Brandes, E.A., 1979, "Radar measurement of rainfall- a summary", *Bulletin of the American Meteorological Society*, Vol. 60, pp. 1048-1058.
Xiao, R., and Chandrasekar, V., 1997, "Development of a neural network based algorithm for rainfall estimation from radar observations', *IEEE Transactions on Geoscience and Remote Sensing*, Vol.35, pp. 160-171.

USING NEURAL NETWORKS TO PREDICT RAINFALL PATTERNS IN SOUTH TEXAS

GARY R. WECKMAN
Department of Mechanical and
Industrial Engineering
Texas A&M University-Kingsville
Kingsville, Texas 78363

ROBERT A. MCLAUCHLAN
Department of Mechanical and
Industrial Engineering
Texas A&M University-Kingsville
Kingsville, Texas 78363

ABSTRACT

The objective of this paper is to help explain water availability behavior caused by rainfall patterns. A three-layer neural network with backpropagation learning has been used to predict rainfall patterns in South Texas. NeuroSoulutions V3.0 was used to build, train and operate the neural network. Water management in South Texas is becoming more critical for human consumption, agriculture, ranching and industry as local populations and demand grow. In 1996 the south Texas region suffered one of the most severe droughts in many years resulting in water restrictions and rationing. Data was collected for the cities of Kingsville, Austin and Corpus Christi from various water resources centers in Texas. The collected information was divided into a training dataset and a test dataset. The test results indicate that the three-layer backpropagation neural network can be used to predict rainfall patterns in South Texas. This neural network system shows promise as a predictive modeling tool that could be transportable for use in other regions.

INTRODUCTION

Neural networks (NN) are massively parallel, distributed processing systems that can continuously improve their performance via dynamic learning (Hagan et al.; 1996, Lin and Lee, 1996; Zurada, 1992). They can provide successful mechanisms for (a) associative memory, (b) pattern recognition, and (c) abstraction. NN architectures can provide a feasible approach to the design and implementation of pattern recognition based systems for intelligent control and model identification. In multilayer perceptron form, these architectures typically use error based dynamic learning laws such as backpropagation (Lin & Lee, 1996; Zurada, 1992).

NN have more recently begun to emerge as an entirely new approach for the modeling of complex, nonlinear, dynamical phenomena (Marzban and Stumpf, 1996). Neural Computers have opened the door to many applications which are difficult – or nearly impossible - for conventional computers to carry out. NN are suited for applications involving pattern recognition, forecasting and complex data analysis. Furthermore, as neural computers require no programming in the conventional sense, they can reduce the burden of software development. The range of applications vary from predictions in the Stock Market (Gately, 1996) to the study of jet engine removal behavior (Weckman and McLauchlan, 1999). The NN is tolerant of some imprecision and is especially useful for approximating functions, where hard and fast rules cannot be applied. A NN model is good at capturing complex nonlinear relationships but unlike other statistical methods, a NN makes no prior assumption concerning the data distribution. Highly accurate predictions can be made, eventhough qualitative physical understanding may be lacking (Marzban and Stumpf, 1996).

Consider that the relationship between meteorology and rainfall prediction is complex and nonlinear, then this system is well suited for modeling by artificial NNs. In a field such as meteorology, as Marzban suggests, all of these criteria are present and the system dynamics are inherently nonlinear (Marzban and Stumpf, 1996). Recent studies have utilized NN models using meteorological data as input to forecast ground level ozone for metropolitan areas (Cobourn et al, 2000), as well as employed in lightning, tornado and wind prediction (Marzban and Stumpf, 1996; Marzban and Stumpf, 1998). In Italy, NN have recently been used to predict river levels based on the average basin rainfall measurement, water level and hydropower production data (Campolo et al., 1999). Based on these and other studies, NNs have demonstrated that they are a good alternative to traditional statistical approaches. With this thought in mind, it was decided to use NN as a tool for rainfall pattern recognition in South Texas. This approach was applied to the Texan cities of Kingsville, Austin, and Corpus Christi.

MODEL DEVELOPMENT - KINGSVILLE

The rainfall estimation problem was set up as a supervised learning, input-to-output mapping problem. With supervised learning, the network is able to learn from the input and the error (the difference between the output and the desired response). Determining what is to be forecast along with choosing the correct inputs is very critical in a successful process. Collecting data sets can typically be a very extensive undertaking, because of the fact that the data must be available, and have a significant relationship to what is being forecast. In the case of the ozone prediction (Cobourn et al., 2000) four key predictors where used – maximum temperature, average dew-point temperature, average wind speed, and daily total sunshine. For this study actual meteorological data depicting total rainfall, average daily max temperature (high) and average daily min temperature (low) grouped by month as the desired system input during training was used to predict a total rainfall pattern.

The raw data was collected from The Texas Water Development Board for a time period of 5 years for the City of Kingsville. Figure 1 is an example of this data for two of the years collected. The maximum high temperatures would range from 68 to 109 degrees F, while the minimum low temperatures would range from 45 to 74. Rainfall levels ranged from a maximum of 4.27 inches to a minumum of 0.69 inches. When the preprocessing was complete the spreadsheet was imported into the NN software. The data was then divided in two datasets. The first 5 years of data was used for the training set and the final year was used for the test set. The training set is data that the NN uses to find patterns and interrelationships. The test dataset is reserved to validate the trained NN using information the NN has not seen during training.

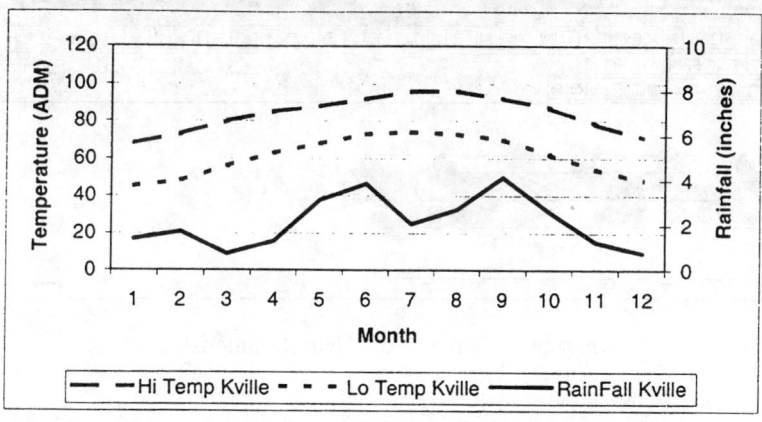

Figure 1: Sample Climatic Historical Data for Kingsville

Though NN can have any number of hidden layers, typically three-layer networks are used. The specific number of input, hidden, and output nodes depends on the characteristics of the particular model. For this model we used a back-propagation, three-layer Multilayer Perceptron (MLP) was used as the architecture. A back-propagation Neural Network is one which compares the forecast to the actual results and uses the difference to adjust the strength of the interconnections between neurons. The Neural Network design for estimating rainfall, using the Multilayer Perceptron, can be summarized as follows:

Multilayer Perceptron: Input Processing Elements = 3
Output Processing Elements = 1
Exemplars = 60

Hidden Layer: Processing Elements (PEs) = 6
PE Transfer Function = TanhAxon
Learning rule = Momentum with
Parameter = 0.7 and Step Size = 1.0

Output Layer: PE Transfer Function = Axon
Learning rule = Momentum with
Parameter = 0.7 and Step Size = 1.0

Maximum Number of Epochs = 1000

The Training data is selected according to NeuralWizard File Format. Theory of Cross Validation was not checked since only 1000 Epochs were needed to train the network. In this work, the NeuroSolutions V3.0 Educator Edition software package (NeuroSolutions, 1998) was used to implement this design. Figure 2 is an example of the NeuroSolutions illustration of this network. The NN was trained utilizing the input vector (Hi Temp., Low Temp., and Month) and a target vector (Rainfall) containing the available data for the location under study. At this point, the NN model predicted a rainfall pattern similar to the training dataset.

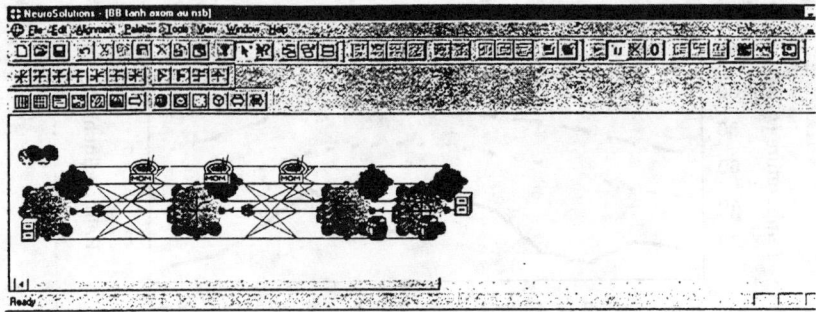

Figure 2: Screen from NeuroSolutions

KINGSVILLE RESULTS AND TRANSPORTABILITY TO OTHER CITIES

The NN is trained from the training data (5 years of historical records) and tested on testing data with 1000 epochs including the momentum rule. In the training phase for Kingsville, the mean square error (MSE) measure went from approximately 0.24 to 0.01 as shown in the Figure 3. Once the network was properly trained its reproducibility was tested with data from the other cities in South Texas. The same previously described procedure was performed on data collected from the cities of Austin and Corpus Christi. The source of data for these cities was collected from the Water Resources Institute website from College Station Texas. The training for Corpus Christi and Austin resulted in final MSE of approximately 0.005 and 0.009 respectively. The validation from the testing phase was performed with the remaining data and the NN generated the prediction of rainfall (output) as listed in Table 1 and Figure 4. The MSE results for the testing phase for Kingsville, Corpus Christi and Austin were 0.07, 0.15 and 0.07. The NN demonstrated a very good fit to the rainfall pattern for all three cities with Kingsville having the largest annual error.

A comparison was made to a more classical approach - multiple regression. The results from the neural network model outperformed the regression model. The regression model produced consistently higher MSE of approximately 0.58 to 0.79.

CONCLUSIONS AND RECOMMENDATIONS

These results indicate that the neural network can be used to predict rainfall patterns in South Texas reasonably well. The accuracy of the neural network for the three cities was at least 93% on an annual basis with a mean error less than 0.293 inches on a monthly basis (see Table 2). In this case, the rainfall pattern was recognized during the training of the network and predictions were fairly accurate. However, the rainfall forecasted for the next period was a typical climatic year for South Texas. The next step in modeling the rainfall pattern might be to collect many more years, perhaps one hundred or so, which would include occurances of atypical climatic seasons, such as drought cycles, toevaluate if the NN could predict these unusual rainfall patterns. In addition,

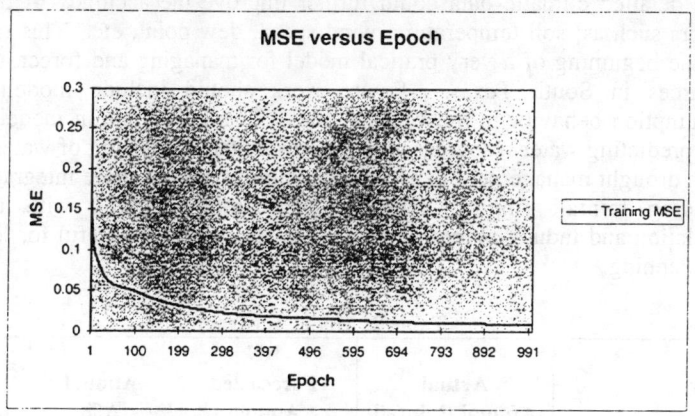

Figure 3: Kingsville – Training Neural Network.

Table 1. Test Data for Kingsville

Month	Hi Temp	Lo Temp	Rainfall	NN Output
January	68	45	1.50	1.527
February	72	48	1.83	1.376
March	79	55	0.89	1.030
April	85	63	1.56	1.237
May	88	68	3.35	3.116
June	92	72	4.02	3.964
July	95	74	2.15	2.472
August	96	73	2.91	2.529
September	92	70	4.31	4.209
October	86	62	2.71	2.303
November	78	54	1.36	1.195
December	71	47	1.01	0.906

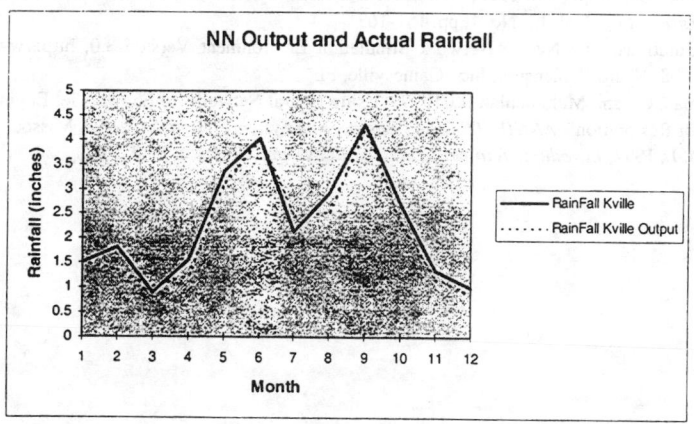

Figure 4: Kingsville -NN Prediction and Actual Rainfall

more detailed climatic data could further improve the accuracy of the model. Factors such as; soil temperature, wind speed, dew point, etc. This research is just the beginning of a very pratical model for managing and forecasting water resources in South Texas. Future work should include modeling water consumption behavior in the various cities. This consumption modeling along with predicting water resources would allow better forecasting of water levels to aid in drought management to avoid related problems. A more integrated model including rainfall (input), consumption (output) coupled with increasing population and industrial requirements (growth) would be useful for long range city planning.

Table 2: Model Accuracy

City	Actual Annual Rainfall	NN Forecasted Annual Rainfall	Annual Δ%	Mean Absolute Error
Kingsville	27.6	25.9	6.28%	.263 in.
Corpus Christi	28.5	28.2	0.92%	.293 in.
Austin	30.7	31.1	1.42%	.203 in.

REFERENCES

Campolo, M., Soldati, A., and Andreussi P., 1999, "Forecasting River Flow Rate During Low-flow Periods Using Neural Networks", *Water Resources Research*, Vol. 35, No. 11, pp. 3547-52.

Cobourn, L. D., French, M., and Hubbard M.C., 2000, "A Comparison of Nonlinear Regression and Neural Network Models for Ground-Level Ozone Forecasting", *Journal of Air and Waste Management Association*, Vol. 50, pp. 174-185.

Gately, E., 1996, *Neural Networks for Financial Forecasting*, John Wiley & Sons, Inc., New York.

Hagan, M. T., Demuth, H. B., and Beale, M., 1996, *Neural Network Design*, PWS, Boston.

Lin, C., Lee, C., 1996, *Neural Fuzzy Systems*, (Chapters 9, 10) Prentice-Hall, Upper Saddle River, New Jersey.

Marzban, C., Stumpf G. J., 1996, "A Neural Network for Tornado Prediction Based on Doppler Radar-Derived Attributes", *Journal of Applied Meteorology*, Vol. 35, pp. 617.

Marzban, C., Stumpf G.J., 1998, "A Neural Network for Damaging Wind Prediction", *Weather and Forecasting*, Vol. 13, No. 1, pp. 151-163.

NeuroSolutions: The Neural Network Simulation Environment Version 3.0, http://www.nd.com/, 1998, Neuro-Dimension, Inc., Gainesville, FL.

Weckman, G., and McLauchlan R., 1999, "Using Neural Networks to Forecast Jet Engine Removals for Restoration", *ANNIE '99 Smart Engineering System Design*, St. Louis, Missouri.

Zurada, J., 1992, *Introduction to Artificial Neural Systems*, West, St. Paul, MN.

COMPARISON OF HOURLY AND DAILY NEURAL NETWORK MODELS FOR FORECASTING HOURLY ELECTRIC LOAD

Swaminathan Vaithianathasamy
Center for Intelligent Systems
Department of System Science
and Industrial Engineering
Binghamton University

David Enke
Smart Engineering Systems Lab
Intelligent Systems Center
Engineering Management Department
University of Missouri – Rolla

ABSTRACT
As a result of deregulation, many electric companies are finding that they can no longer rely only on long-term load forecast for capital planning. Many of these companies now require short-term hourly load forecasts that allow them to predict their daily load requirements, as well as meet the information requirements of the deregulatory agency. This paper discusses the design and development of two feed forward neural network models that are used to predict short-term electric load. Both of the models predict the load for each hour. The primary difference between the two models involves the type of weather data provided as inputs to the networks. One model uses the hour of the day as a continuous variable, while the other model recognizes each hour as belonging to a separate group and therefore builds 24 different models to address each individual hourly load profile. After describing the models, the results obtained using these network models are then compared. The paper concludes with a discussion of the advantages and disadvantages of each model.

INTRODUCTION

Electricity markets are undergoing changes that involve the creation of stock-exchange-like markets for the sale of electricity. This move has made the purchase and sale of electricity very competitive. Since electricity has been made into a commodity within a deregulated environment, decisions on whether to produce or to purchase electricity becomes important. Short-term load forecasts can assist in predicting the energy needs of customers every hour of the day. Accurate load forecasting models are indispensable to plan the production and purchasing of power/electricity since errors in predicting the energy demand could have a significant economic impact for a power company. A conservative estimate of a 1% reduction in forecasting error for a 10,000 MW utility could save a company up to $1.6 million annually [Hobbs et al., 1999].

Literature provides a number of papers and articles that discuss different approaches and techniques that can be used to build short-term load forecasting models. Some of the prominent techniques include the use of time-series-based load forecasting models [Yang et al., 1996], regression-based models [Papalexopoulos et al, 1990], and expert systems [Rahman et al., 1993]. There have also been hybrid models that use the benefits of the different techniques [Kim et al., 1997]. The non-linear nature of the load-forecasting problem makes neural networks a very attractive tool for forecasting electric load. Within neural networks, numerous paradigms have also been tried and suggested, including multi-layered perceptrons [Enke and Vaithianathasamy, 2000], radial basis function networks [Kodogiannis, 2000], Kohonen maps [Yoo et al, 1999], and functional link networks [Dash et al., 1999]. However, the multi-layer perceptron seems to be the most popular due to its success in achieving the intended task of forecasting in general, and load forecasting in particular [Khotanzad et al., 1998]. Feed forward multi-layer perceptrons essentially utilize supervised training, which requires a set of inputs and a set of outputs based on the specific problem at hand. This paper deals with the design, development, and validation of two such network models. The primary difference between the two network models lies in the format of the input data.

The following begins with an overview of the design and development of the neural network models. Validation and testing of the network models is then highlighted. A discussion of the advantages and disadvantages of the two models will then conclude.

INPUT AND OUTPUT VARIABLES FOR THE NEURAL NETWORK MODELS

For electric load forecasting with neural networks, the output can be based on either hourly or daily data. Models based on daily data will typically consist of a single network with 24 different output nodes (one for each hour) or a single output for hourly load along with an extra input data point to specify the current hour being forecast. On the other hand, hourly models simply have one output node, although 24 individual models are necessary to represent the entire day. Fortunately, although additional models must be developed, the individual models are often easier to train than the single network with 24 output nodes due to the vast number of training parameters that must be determined for a single, yet larger network [Hippert et al., 2001]. With regard to the specific inputs that should be used in either type of model, there seems to be a broad consensus that inputs to the models can be divided into the three general categories of weather data, seasonal data, and trend data.

Weather related information is a critical part of the input set. The comfort of the customer, which is affected by the weather, is important to energy demand. Weather data includes air temperature, humidity, cloud cover, and wind speed, among others. Air temperature, in particular, is among the most important factors that should be considered for load forecasting [Drezga et al., 1998]. For this study, the trend in the air temperature was used as an input. Humidity was also considered as an input for the daily weather model. During this research it was observed that cloud cover did not significantly influence the load values. In fact, it actually resulted in confusing the network models and increasing the prediction error.

With regard to the use of seasonal data, it is also fairly intuitive that electricity demand is dependent on the season of the year. In order to represent the date of the observations, two variables, namely the month of the year and the daytype of the week, were used. The daily weather model used an additional variable in the form of the hour for which the observation was made. Daytype was classified into five different categories: Monday, Tuesday/Wednesday/Thursday, Friday, Saturday, and Sunday. All holidays were considered as Sundays. With regard to the trend data, past research also suggests the use of the previous load data as one of the important inputs. In the current research, trend data does not involve the use of previous day or previous hourly load. Instead, modeling concentrated on the trend in the weather data.

ARCHITECTURE DESIGN CONSIDERATIONS

There were several differences between the network models that arose from the data and data format. However, some design considerations, such as the training algorithm, transfer function, and stopping criteria, were similar. The training algorithm used was resilient backpropagation. The transfer function used was the hyperbolic tangent. The stopping criterion was a pre-fixed number of epochs/training cycles. A formal statistical design of experiment approach was also employed for determining the appropriate architecture size [Enke et al., 2000]. The chosen architecture had a reduced number of training parameters and was shown to perform better than networks with more parameters.

DESIGN AND DEVELOPMENT OF THE DAILY AND HOURLY WEATHER MODELS

The daily weather neural network model was built utilizing a dataset with 7489 data points, representing about one year of weather and load data. The weather data used for this model included current high and low forecasted air temperatures and humidity measurements for each hour of the day. Additional inputs in the form of month of the year, daytype, and hour of the day were also used to capture the seasonal data. As mentioned before, trend data is essential in forecasting applications, although there are many ways in which trend data can be represented. Different representations of the trend data were formulated and tested. Among the successful representations were the use of Simple Moving Averages, Exponential Smoothing, and Mean Average Convergence and Divergence [Enke and Vaithianathasamy, 2000]. For the current modeling, a five-hour Exponential Moving Average (EMA) of the high and low forecasted temperatures was used to represent the trend data. Table 1 lists the inputs that were used for the daily model.

Table 1: Inputs for the Daily Model

	Daily Model
1	Hour of the day
2	Day of the week
3	Month of the year
4	Hourly percent humidity
5	Current hour minimum temperature
6	Current hour maximum temperature
7	EMA of minimum temperature
8	EMA of maximum temperature

Since the focus of the research was short-term load forecasting, it was essential to predict load for each hour separately. For the daily model, this involved running the model every hour with the new forecasted weather values. For the hourly models, 24 different models were developed to individually forecast the next 24 hourly loads using one weather data set. For the network modeling, it was not required for each of the models to have the same network architecture.

The models were built using 980 data points representing about two and one-half years of weather and load data. During modeling, it was also verified that the load data did not show a definite trend for the period under consideration. The weather data for each of the 24 models contain only the daily high and low forecasts of the air temperature. Data corresponding to the daily humidity values were not utilized. Similar to the case with the previous daily model, seasonal data was captured in terms of month of the year and daytype of the week. The trend data used was simply the previous day's high and low air temperatures. Table 2 lists the inputs that were used for the hourly model.

Table 2: Inputs for the Hourly Models

	Hourly Models
1	Day of the week
2	Month of the year
3	Current hour minimum temperature
4	Current hour maximum temperature
5	Previous day's minimum temperature
6	Previous day's maximum temperature

In essence, while the daily weather model used hour as a continuous variable, the hourly weather model builds individual models corresponding to each hour of the day. The following briefly summarizes the main differences between the daily and hourly models:

1. There was only one daily model developed, as opposed to 24 different hourly networks.
2. The daily model contained forecasted weather data that is updated hourly, whereas the hourly models each used the same daily weather forecast values, collected once each day.
3. The input set for the hourly models did not contain data related to humidity.
4. The input set for the daily models used an exponential moving average of the high and low hourly temperatures, whereas the hourly models only used the previous day's high and low temperature to represent the weather trend.
5. The daily model had to be run 24 times a day to generate daily load forecasts based on new hourly forecasted weather values, while the 24 hourly network models were each run once a day on the same weather data set.

The key neural network characteristics for the two models are given in Table 3.

Table 3: Characteristics of the Two Network Models

Parameters	Daily Weather Model	Hourly Weather Model
Number of Input Nodes	8	6
Number of Output Nodes	1	1
Number of Hidden Layers	3	3
Number of Training Data Points	5984	799
Number of Testing Data Points	1496	199
Number of Training Parameters	581	209
Number of Training Cycles	1000	500
Transfer Function Used	Hyperbolic Tangent	Hyperbolic Tangent
Transfer Function Between Output and Hidden Layer	Linear	Linear
Training Algorithm Used	Resilient Backpropagation	Resilient Backpropagation
Number of Models Built for Forecasting Loads for One Day	1	24

VALIDATION OF THE NETWORK MODELS

During validation, the best architecture was chosen for both models. The entire dataset was divided into two parts (80% training and 20% testing with the testing data set randomly chosen from the entire data set). Table 4 shows the comparison of the results. Most of the literature reports load forecasting performance in terms of MAPE (Mean Absolute Percentage Error). Performance measures used for comparing the models should be easy to follow, and should also be easily understood by those who use the system. For this reason, MAE (Mean Absolute Error) would be preferred [Hippert et al., 2001]. However, large deviations from the actual values are not penalized appropriately by the above measures. Therefore, the performance is also measured in terms of RMSE (Root Mean Squared Error), allowing for three methods for measuring the performance of the network models.

Table 4: Performance Comparison for the Two Models

Model	MAPE (in %)	MAE (in Watts)	RMSE (in Watts)
Daily Weather Model	2.94	47.12	63.09
Hourly Weather Model	3.18	46.26	63.78

RESULTS AND DISCUSSION

From the above results it is seen that the two models do not appear to be significantly different in terms of error performance. With regard to the MAPE, the daily weather model averages 2.94% error while the hourly weather model gives an error of 3.18% (0.24% difference). Similarly, for the RMSE measurement the daily weather model performs better by less than one watt. Ironically, when the MAE is considered, the hourly weather model performs better than the daily weather model. Regardless, the differences of each measure are quite small, and it appears that neither model is superior to the other with respect to error.

Since the differences between the two models are small in terms of the error, choosing one method over the other may come down to the availability of data, or simply personal preference. Nonetheless, there are some differences between the two methods, and for practical reasons, the hourly models may be preferred. Advantages of the hourly model include:

1. It costs more to collect weather data every 24 hours from a weather service station as required by the daily model.
2. The amount of data that needs to be handled is less when using the hourly model. The daily model would require 24 times more weather data than the hourly model. This could become a significant factor. There are also less data inputs for the hourly model (six compared to eight), although this may change if a different modeling strategy is used.
3. By building separate models for each hour (each model is independent), the load pattern can be modeled more efficiently.
4. It is easier to implement and maintain a system in which both a neural network and an expert system (or human expert) use the same set of data to make predictions. Current research shows that many human experts make individual hourly electric load forecasts using a similar day approach that uses a single temperature data set updated only once a day.

Disadvantages of the hourly weather model include:

1. Training the model may require considerable time, as it is necessary to find an appropriate architecture for each of the 24 hours.
2. The daily model would have weather data that is more continuous than the hourly weather model data, and therefore has the potential to be more accurate.

With regard to the first disadvantage, it was found that the hourly models seemed to work equally well even when the 24 networks had the same architecture and the same number of training cycles. Given increased computer speed, and the need to update models only once a day (at most), this also should not pose a serious problem. With regard to the second disadvantage, it should be noted that the performance difference was not significant to reject the hourly approach. Nonetheless, more research is needed to determine which networks and modeling strategies are preferred based on the price and availability of weather and historical load data.

REFERENCES

Dash, P.k, Liew, A.C., and Satpathy, H. P., "A functional-link-neural network for short term electric load forecasting," Journal of Intelligent and Fuzzy Systems, Vol.7, No.3, pp. 209-221, 1999.

Drezga I., and Rahman. S., "Input variable selection for ANN-based short term load forecasting", IEEE Transaction on Power Systems, Vol. 13, No.4, pp- 1238-1244, 1998.

Enke D., and Vaithianathasamy S., "Electric Load forecasting using Trend Data and a feed-forward neural network", Intelligent Engineering Systems Through Artificial Neural Networks, Vol. 10, 2000.

Enke D., Vaithianathasamy S., Diwe P., "Factorial Design for Developing Feed Forward Neural Network Architectures", Intelligent Engineering Systems Through Artificial Neural Networks, Vol. 10, 2000.

Hippert, H. S., Pedreira, C. E., and Souza, R. C., "Neural Networks for Short Term Load Forecasting: A Review and Evaluation", IEEE Transactions on Power Systems, Vol. 16. No.1, Feb 2001.

Hobbs B.F., S. Jitprapaikulsarn, S. Konda, V. Chankong, K. A. Loparo, and D.J. Maratukulam, "Analysis of the value for unit commitment of improved load forecasting," IEEE Transactions. Power Systems, Vol. 14, No. 4, pp. 1342-1348, 1999.

Khotanzad A., R. Afkhami-Rohanni, and D. J. Maratukulam, "ANNSTLF - Artificial neural network short term load forecaster - Generation Three," IEEE Transaction on Power Systems, Vol. 13, No.4, pp. 1413 - 1422, 1998.

Kodogiannis, V.S, "Comparison of advanced learning algorithms for short term load forecasting," Journal of Intelligent and Fuzzy Systems, Vol. 8, No. 4, pp - 243-259, 2000.

Kwang-Ho Kim; Jong-Keun Park; Kab-Ju Hwang; and Sung-Hak Kim, "Implementation of hybrid short term load forecasting system using Artificial Neural Networks, and fuzzy expert systems," IEEE Transactions on Power Systems, Vol.10, No. 3, pp 1534-1539, 1995.

Papalexopoulos A. D. and T. C. Hesterberg, "A regression based approach to short term system load forecasting," IEEE Transaction on Power Systems, Vol. 5, No. 4,pp. 1535 -1547, 1990.

Rahman S. and O. Hazim, "A generalized knowledge based short term load forecasting technique," IEEE Transactions on Power Systems, Vol. 8, No.2, pp- 508 -514, 1993.

Yang H. T., C.M. Huang, C. L. Huang, "Identification of ARMAX model for short term load forecasting: An evolutionary programming approach," IEEE Transactions on Power Systems, Vol. 11, No. 1, pp - 403-408, 1996.

Yoo H. and R. L. Pimmel, "Short Term load forecasting using a self supervised adaptive neural network," IEEE Transactions on Power Systems, Vol. 11, No.2, pp -779-784, 1999.

A COMPOUND TECHNIQUE FOR ESTIMATING MISSING DATA OF WIND SPEEDS

PUNNEE SIRIPITAYANANON
Department of Computer Science,
The University of Alabama
Tuscaloosa, Alabama 34587-0290
E-mail: sirip001@bama.ua.edu

HUI-CHUAN CHEN
Department of Computer Science,
The University of Alabama
Tuscaloosa, Alabama 34587-0290
E-mail: chen@cs.ua.edu

KANG-REN JIN
Okeechobee Division
South Florida Water Management District
West Palm Beach, Florida 33416-4680
E-mail: kjin@sfwmd.gov

ABSTRACT

Wind data is important for hydrodynamics and sediment transport in a lake system because wind is the major driving force for most lakes. However, there exist a lot of missing data caused by instrumental failure due to bird, thunderstorm, or other unexpected events. An estimation of these missing data becomes an important task. Lake Okeechobee, the second largest freshwater lake within the United States, is used as a test sample for this study. Measurement of wind speed and wind direction was made at four different stations in the lake. An estimation model, including basic statistics approaches coupled with correlation identification and nearest-neighbor analysis, is used to solve missing data problem by this study. The developed model demonstrates its ability to reproduce accurate wind speed of the year 1996 and 1999 at different stations.

Keywords: Wind speed, time series, estimation, missing data.

INTRODUCTION

Lakes provide habitat for a wide variety of wild animals especially birds and fish. Large lakes sometimes are important for an economical and commercial fishery. Scientists recognize that water quality models are beneficial to the lake's animal, plant, and human being communities. For this reason, there are ongoing efforts to evaluate environmental responses to different management alternatives. Sophisticated water quality and hydrodynamic model have been developed to support engineers and scientists in analyzing the impacts of each scenario. However, accurate modeling results are relied on solid and accurate input data, especially wind data.

Lake Okeechobee, the second largest fresh water lake totally within the USA, is located in South Florida. Lake Okeechobee wind data are routinely collected; however, they often contain missing data caused by instrumental failure due to birds, thunderstorms, or other unexpected phenomena. In order to utilize the data to drive models, a consecutive data set is required and missing data should be filled by an appropriate approach. Therefore, the reliable methodology for estimating missing data is of essential importance for model simulations in Lake Okeechobee.

There have been a number of studies on the use of time-series to model for predicting and estimating wind speed. A variety of statistics have their abilities to make successful predicting and estimating for nonlinear dynamic models. Evidences for accomplishing correlation coefficient and auto regression can be seen in recent researches. For instance, Salmon and Walmsley (1999) modified and tested a two-site correlation model for wind speed, direction and energy estimates. Huang and Chalabi (1995) applied auto regression (AR) to model and forecast wind speed. Alexiadis, et. al, (1999) introduced a technique for forecasting wind speed and power output based on spatial correlation models. This paper introduces the proposed model that fills the missing data based on a compound data mining technique.

APPLICATION DOMAIN

Lake Okeechobee extends from the Kissimmee River in the north to the Everglade Agricultural Area in the south (Fig. 1). A total surface area of 1,750 km^2 makes Lake Okeechobee a principal aquatic resource for wildlife and recreation in South Florida. The natural lake is very shallow, with mean and maximum depths of 2.7 and 5.5 m, respectively. Recent studies show that mud sediments in the central region of the lake have accumulated large amounts of phosphorus (P) from these excessive external loads, and this sediment phosphorus is recycled to the water column (internal load). This internal load of phosphorus has a strong correlation to the wind, sediment, and hydrodynamics of the lake. The study of wind is important because wind is the major driving force for hydrodynamic and sediment transport modeling in the lake (Jin and Wang, 1998). Conventional estimation algorithms for missing data are not adequate because wind speed is a non-stationary variable. This paper illustrates data mining techniques for estimating the missing data of wind speed based on cross correlations at neighboring stations.

The data sets used in this study are Lake Okeechobee wind data. Wind speed (mi/hr) and wind direction (degree) data were collected for every 15 minutes at weather stations L001, L005, L006, and Lz40 (Fig. 1). The wind data sets (November 1996 and 1999) from weather stations in Lake Okeechobee were used to develop and test the model in this study.

DATA PREPROCESSING

Wind data sets were collected by South Florida Water Management District (SFWMD) and resided in databases maintained by the SFWMD. We selected November, 1996 wind data as a training data set because it is a complete data set of wind speed and wind direction. Table 1 shows basic statistics of wind speed in November 1996. There are the arithmetic average (Avg.), the standard deviation of each station (S.D.), and the correlation coefficient of wind speed between stations.

RULES FOR ESTIMATING MISSING DATA

We assume that one station might have missing data ("target station") whereas the other three stations still have collected data. This assumption can lead us to calculate the estimated wind speed from the other three stations ("source stations"). Lz40 is selected as the target station that has incomplete wind speed data in November 1996. The actual wind speed data in November 1996 will be used to compare the estimated data, and the performance of the model will be evaluated. When the wind speed is missing, wind direction information is also lost. In this paper, we will focus on the estimation of wind speed only.

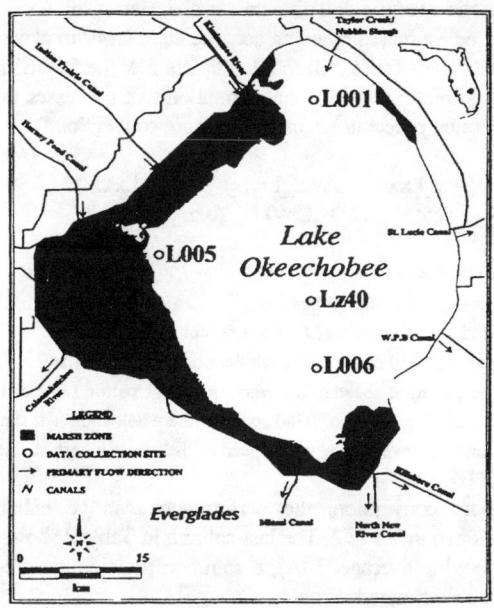

Figure 1. Locations of all stations in Lake Okeechobee.

Table 1. Basic statistics of wind speed in November 1996.

Ws (mi/hr)	L001	L005	L006	Lz40
Avg.	12.1964	12.5504	13.6466	13.5312
S.D.	5.3646	5.1246	6.4946	6.3297
Correlation coefficient		L005	L006	Lz40
	L001	0.83678	0.83725	0.85580
	L005		0.83310	0.86015
	L006			0.96387

Nearest neighbor rule

The estimation of the missing data usually can be obtained by using the actual value of the source stations, which are nearest to the target station (Webb, 1999). This is a very common approach in data mining (Berry and Linoff, 1997). In order to use this nearest neighbor rule, distance between station Lz40 and other stations will be computed. It is found that station L006 has the nearest distance from the station Lz40 and also has the highest correlation. Therefore, the estimated data at Lz40 will be the data collected at station L006. To measure the performances of this approach, the actual value (measured every 15 minutes) at Lz40 is compared with the estimated value, and the absolute difference between these two values is considered as "absolute error." Table 2 shows the mean of these absolute errors, not only estimating Lz40 from L006 but also from other source stations L005 and L001. In order to gain the information regarding how much the estimating value is above the actual value and how much it is below the actual value, "mean of difference over" is used to measure the average amount above the actual value. Similarly, "mean of difference under" is used to measure the average amount below the actual values. The proposed nearest neighbor rule works in this case because the errors from L006 are smaller than other stations (Table 2).

Assuming that the average and the standard deviation of wind speed at Lz40 in November 1996 can be estimated by using actual data of Lz40 in November 1995. Then, instead of using the data of L006 to fill into Lz40 data file, the following equations can be used to convert the estimating value to an estimation that possesses the desired statistics of Lz40. This conversion procedure is called "z-score conversion."

$$z\text{-score} = (Ws\_Lxxx - Avg.\_Lxxx) / S.D.\_Lxxx \qquad (1)$$
$$\text{and } Ws\_Lz40 = z\text{-score} * S.D.\_Lz40 + Avg.\_Lz40 \qquad (2)$$

where z-score = standard score
     Ws_Lz40 = wind speed at station Lz40
     Ws_Lxxx = wind speed at station L001 or L005 or L006
     Avg._Lz40 = average wind speed of the whole month at station Lz40
     Avg._Lxxx = average wind speed of the whole month at station L001 or L005 or L006
     S.D._Lz40 = standard deviation of wind speed of the whole month at station Lz40
     S.D._Lxxx = standard deviation of wind speed of the whole month at station L001 or L005 or L006

By using z-score conversion, the performance can be enhanced by reducing noticeable error as shown in Table 2. The last column in Table 2 shows another estimated result by applying moving average (MA), a common practice in time-series analysis, to the z-scores conversion estimated values.

Table 2. Error of estimating at station Lz40 by other stations.

Station	Error	Actual value	Z-scores	MA
L001	Mean of absolute errors	2.80803	2.60654	2.45211
	Mean of difference over	2.33244	2.78651	2.54437
	Mean of difference under	3.02737	2.44841	2.37359
L005	Mean of absolute errors	2.5413	2.53377	2.41867
	Mean of difference over	1.97877	2.41631	2.26
	Mean of difference under	2.90718	2.66323	2.58593
L006	Mean of absolute errors	1.25217	1.21937	1.23646
	Mean of difference over	1.2771	1.22023	1.13657
	Mean of difference under	1.22349	1.21852	1.33173

Next, we assumed that there are missing data at station L001 that is located the farthest north. The errors of estimating by other source stations after applying moving average technique are shown in Table 3 (time lag = 0). Among three source stations, Lz40 provided the best estimations, but they are not significantly preferable due to the long distance between L001 and Lz40 as shown in Fig.1. Hence, rather than nearest neighbor rule, other methods will be proposed to estimate the target station that located far away from its source stations.

Time-lag rule

In the case of missing data at station L001, the correlation between L001 and L006 is higher than the correlation between L001 and L005 even though the distance between L001 and L006 is longer than L001 and L005. This reflects the fact that the relationships between data correlations and station distances are not significantly related when the distance is beyond 20 kilometers. Since wind takes time to travel from one station to another station and the wind direction is varying from time to time, thus the time lag effect must be considered. We also computed the correlations with time lag up to one hour (1 lag = 15 minutes). The highest correlation between L001 and the other source stations takes place in Lz40 at time lag = -2. This means that the wind speed of L001 has the strongest correlation to the wind speed of Lz40 with 30-minute time lag. Therefore, now the wind speed in Lz40 with lag = -2 is used as the estimating value for missing data at L001. The errors of estimating are shown in Table 3 (time lag = -2). By comparing the results between using data at time lag = 0 and time lag =-2, we concluded that using Lz40 with 30-minute time lag later than station L001 could improve the estimating wind speed.

Table 3. Error of estimating at station L001 by other stations.

Station	Error	Time lag = 0	Time lag = -2
Lz40	Mean of absolute errors	2.20868	2.1348
	Mean of difference over	2.03224	2.01409
	Mean of difference under	2.40233	2.26013
L005	Mean of absolute errors	2.32214	2.22472
	Mean of difference over	2.19687	2.12343
	Mean of difference under	2.45311	2.32814
L006	Mean of absolute errors	2.37717	2.34266
	Mean of difference over	2.23979	2.1845
	Mean of difference under	2.5216	2.5122

Majority-votes rule

Now, instead of using a single station to predict the missing data, another proposed approach is to use all available source stations to predict the missing data. However, the value in one station is different from the other stations. In order to combine the different values from the different stations, only two values that are close to each other are selected. These two close values are considered as the "majority votes" from the source indications. We use these majority votes to compute a potential estimation as given in the following algorithm. The detail of this new approach is described in the following steps:

Step 1: Compute potential estimation by using majority-votes rule.
    - Sort 3 wind speed values from 3 source stations (L005, L006, and Lz40)
    - Compute average between high and middle or between low and middle
    base on which one is majority where majority votes = min[(high-middle),(middle-low)].
Step 2: Separate data into 4 periods of time and compute all basic statistics of each period.
Step 3: Estimate missing wind speed by z-score conversion based on the time period it falls in and the corresponding statistics.

The errors of estimating missing data at station L001 in November 1996 from this rule are demonstrated in Table 4. It is noted that this approach has better performance compared with results shown in Table 3. In addition, we also evaluated the majority-votes rule to the data given in 1999 with the statistics obtained in 1996. The errors of estimating are listed in the last column of Table 4.

RESULTS AND DISCUSSION

The estimated wind speed from the majority-votes rule yields the best solutions. It is surprising to learn that the technique can improve estimating accuracy substantially. We also obtained percentages of absolute errors where the estimated wind speeds deviate from the measured values over particular thresholds as shown in Table 5.

From the computational results, only a few points can be found where the estimated wind speeds are significantly deviated from the observed values (the error over 9.0 mi/hr). The maximum deviation is 11.49 (mi/hr). It is expected that some unusual local events may have happened at that particular time to cause abrupt changes in the wind speed. Thus, in addition to the wind speed information from the source stations, some other local information such as wind direction, or solar radiation, or air temperature can be included for future research.

Table 4. Error of estimating at station L001 by majority-votes rule.

Error	November 1996	November 1999
Mean of absolute errors	2.08306	1.9393
Mean of difference over	2.01546	1.89987
Mean of difference under	2.14858	1.98101

Table 5. Percentages of absolute errors estimated by majority-votes rule.

Percentages of absolute error	November 1996	November 1999
Over 10.0 mi/hr	0.34 %	0.20 %
Over 9.0 mi/hr	0.62 %	0.48 %
Over 8.0 mi/hr	0.97 %	0.72 %
Over 7.0 mi/hr	1.56 %	1.52 %
Over 6.0 mi/hr	2.88 %	2.71 %

CONCLUSIONS

This paper described the use of data mining techniques with time-series analysis to model and estimate missing data of wind speed at the target station from neighboring stations. The analysis has demonstrated that the following approaches can enhance the efficiency of estimating.

(1) Using basic statistics (arithmetic average and standard deviation) to adjust estimated values via z-scores converting.
(2) Considering the best correlation coefficient among the target station and neighboring stations with time lag analysis.
(3) Combining some major effective stations in the majority-votes approach.

REFERENCES

Alexiadis, M.C., Dokopoulos, P.S., and Sahsamanoglou, H.S. (1999). "Wind Speed and Power Forecasting Based on Spatial Correlation Models." *IEEE Trans. Energy*, 14(3), 836-842.

Berry, M. and Linoff, G. (1997). Data Mining Techniques for Margeting, Sales, and Customer Support. John Wiley & Sons, Inc.

Huang, Z. and Chalabi, Z.S. (1995) "Use of Time-series Analysis to Model and Forecast Wind Speed." *J. Wind Eng. Ind.Aerodyn.*,56(1995), 311-322.

Jin, K.R., Hamrick, and J. H.,Tisdale, T. S. (2000). "Application of Three-dimensional Hydrodynamic Model for Lake Okeechobee." *J. Hydraul. Eng.*, ASCE, 126(10), October, 2000, 758-771.

Jin, K.R., and Wang, K.H. (1998). "Wind Generated Waves in Lake Okeechobee." *J. AWRA*, 34(6), 1-12.

Salmon, J.R. and Walmsley (1999). "A Two-site Correlation Model for Wind Speed, Direction and Energy Estimates." *J. Wind Eng.Ind Aerodyn.*, 79(1999), 233-268.

Webb, A. (1999). Statistical Pattern Recognition. New York, Oxford University Press, Inc.

DETECTING REGIMES IN TEMPERATURE TIME SERIES

PATRICK J. CLEMINS
Dept. of Electrical and Computer Eng.
Marquette University
Milwaukee, WI

RICHARD J. POVINELLI
Dept. of Electrical and Computer Eng.
Marquette University
Milwaukee, WI

ABSTRACT

In the field of climate prediction, regimes are used to model long-term cyclic trends. Although air pressure regimes have been discovered, there has been little exploration into the possibility of temperature regimes. This paper develops an approach to finding regimes in a temperature time series. First, the time series is reconstructed in a phase space. Then, a clustering algorithm is used to search the phase space for clusters. Finally, the number of transitions between these clusters is recorded. A low ratio between the number of transitions into the cluster and the number of points in the cluster indicates that a regime structure is present. Results are given for various temperature time series.

INTRODUCTION

The Earth's climate is a complex system with an undetermined number of variables. Many long-term prediction models have been proposed, but most are based on the assumption that the earth's climate is a linear system. The sentiment in the field of meteorology seems to be that linear models are accurate enough for prediction even though evidence has been uncovered that seems to indicate some nonlinear trends in the Earth's climate (Palmer, 1993). However, there has not been much research on these nonlinear climatic trends, and these trends may provide insight into climate prediction. If regimes could be found in the earth's climate, they could be used to help predict future climatic trends.

One technique used to expose patterns in a nonlinear time series is to reconstruct the time series in a phase space (Povinelli, 1999). A phase space is constructed by creating a vector space [s(t) $s^{(1)}(t)$ $s^{(2)}(t)$ ... $s^{(n)}(t)$] where n+1 is the embedding dimension and $s^{(x)}(t)$ is a time delayed s(t) or a time series of a related system parameter. Appropriate time delays can be determined by various statistical methods such as autocorrelation or auto mutual information (Abarbanel, 1996, Kantz, 1997).

An example of a non-linear system is the Lorenz system (Figure 1). The Lorenz system was developed to model atmospheric flow. The system is defined by three differential equations.

$$dX = -aX + aY$$
$$dY = -XZ + rX - Y$$
$$dZ = XY - bZ \qquad (1)$$

Two regimes, or states, are easily discernable in the Lorenz system. Once the system is in one regime, it tends to stay in that regime. The area between the two regimes

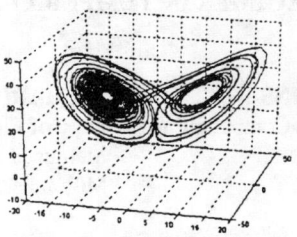

Figure 1 – Lorenz System

is called the unstable region, because once the system enters this region it has a higher likelihood of switching regimes. For the purpose of this paper, a regime is defined as a cluster in phase space where the trajectory of the original time series tends to stay for a period of time. To determine whether a regime exists, the transitions out of each cluster are compared to the number of intra-cluster transitions. If the number of intra-cluster transitions is large compared to the number of transitions out of the cluster, then a regime is present.

Research has shown that regime structures are present in certain parts of the Earth's climate (Palmer, 1999). Some parts of the globe follow the traditional linear model, while other areas, such as the northern hemisphere during the winter, tend to follow a regime structure. The El Nino cycle, which has two quasi-equilibrium states, also exhibits non-linear regime structure. Here, we propose a method for detecting temperature regimes using temperature time series phase spaces.

To detect these regimes and count transitions between them, the phase space must be clustered. The general idea of clustering is to group similar objects together. In the context of this experiment, an object is a point in phase space. Euclidean distance is used to quantify two data point's similarity.

In this paper, the shape, size and number of the clusters is unknown. Density-based and hierarchical clustering techniques such as OPTICS (Ankerst, 1999), and Chameleon (Karypis, 1999) have been shown to work well on such data sets. However, these algorithms are not simple to implement and are unnecessarily complex for this research. Therefore, ideas from both of these algorithms were used to create a simple clustering algorithm that can be used to computationally find clusters given the appropriate parameters.

METHOD

To show that a nonlinear structure is present in the earth's climate, a phase space is constructed using temperature time series. If the temperature time series are nonrandom, a phase space plot should show clusters of points, or the points forming a line as the time series is plotted. However, if the time series are random, the phase space would simply be a random scattering of data points.

A clustering algorithm is formulated from K-means (step 3) and hierarchical (step 4) clustering techniques. A visual summary of the algorithm is in figure 2. The algorithm is as follows:

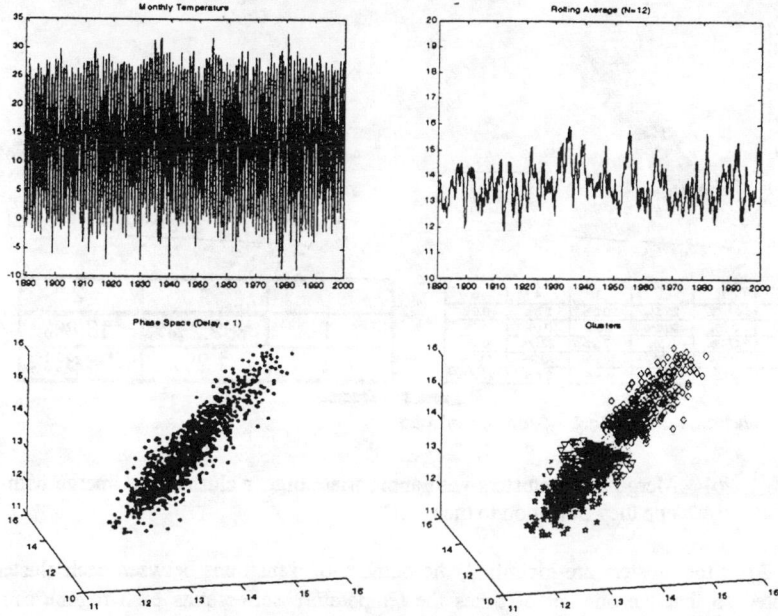

Figure 2 – Methodology
Upper right – Original time series (Wichita, KS); Upper left – Rolling average using 12 months; Lower right – Phase space; Lower left – Discovered clusters (5 clusters found)

1) Preprocess data.
   a) Rolling average, yearly average, etc.
2) Construct phase space.
3) Do an initial K-means binary split clustering of the phase space.
   a) Number of clusters is user-defined (.1*N is a good choice, where N is number of data points)
   b) Remove empty clusters in each iteration (random initialization may give a cluster center in an empty part of the phase space)
4) For each pair of clusters i and j, merge them if there are any two points between the clusters that have a Euclidean distance less than the linear density of cluster i times a scaling factor (values between 3 and 4 worked well). The linear density is defined by:

$$(D_{XY}) / N_i \qquad (2)$$

where D_{XY} is the maximum Euclidean distance between any two points X and Y, and N_i is the number of points in cluster i. This is equivalent to saying merge the clusters if they are close with respect to their densities.
5) Repeat 4 until no more clusters can be merged.
6) Postprocessing of clusters
   a) If a cluster only transitions to one other cluster, merge the clusters

730

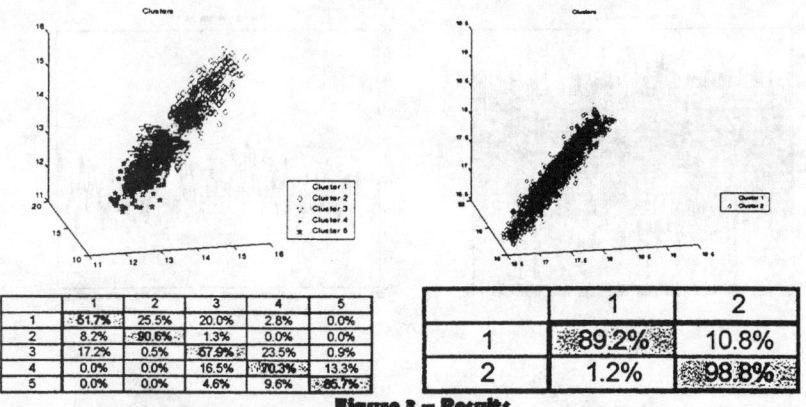

Figure 3 – Results
Left – Wichita, Kansas; Right – Sydney, Australia

b) Merge small clusters with appropriate bigger clusters (i.e. merge with the one they transition to the most)

After the clusters are identified, the number of transitions between each cluster is counted so that the number of times the temperature series exits each regime can be determined.

DATASETS

Two kinds of datasets are analyzed in this paper, modern and paleoclimatic. The modern temperature data consists of average monthly temperature readings for about the last 150 years. These datasets were extracted from the Global Historical Climatology Network (GHCN) database that is available online (2000). The GHCN database contains readings from thousands of weather stations around the globe. The biggest disadvantage with this type of dataset is that it is rather short. Some climatic trends (such as an ice age) last for thousands of years and take hundreds of years to develop. Therefore, this type of dataset may not contain enough points for long-term regime structures to be identified. This dataset does have the advantage that all measurements are accurate.

Paleoclimatic datasets contain estimated temperatures over the last few hundred years. These estimates are calculated from ice core samples, tree rings, and other geologic temperature indicators. The biggest disadvantage with these datasets is that the data points are estimates. The advantage with these data sets is that there are enough data points for longer regimes to become apparent.

RESULTS

The regime structure of many different time series was explored. An examination of the GHCN data from Wichita, KN will be discussed first. This time series has average monthly temperatures from 1989 to 1999. After taking a moving average (N=12), a three-dimensional phase space was constructed using time delays of 1 and 2. These time delays were chosen because upon inspection of most of the temperature time series, the

	1	2	3	4
1	50.9%	28.2%	1.2%	19.6%
2	27.2%	71.8%	0.0%	1.0%
3	1.3%	0.0%	70.7%	28.0%
4	17.4%	8.7%	14.5%	59.4%

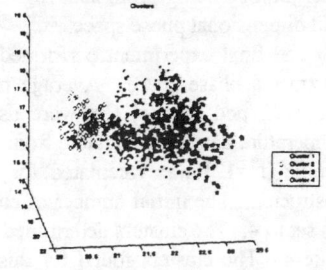

	1	2	3	4
1	30.4%	10.7%	27.7%	31.3%
2	15.3%	34.7%	31.9%	18.1%
3	19.4%	14.2%	41.9%	24.5%
4	16.9%	6.1%	16.8%	59.8%

	1	2	3	4
1	75.4%	1.9%	15.7%	6.9%
2	6.0%	78.9%	2.3%	12.8%
3	10.7%	0.6%	81.3%	7.4%
4	8.3%	5.1%	8.6%	78.0%

Figure 4 – Results
Left – Mann Reconstruction (3D at top, 8D at bottom); Right – Multiple Sites

autocorrelation function decreased linearly with increasing τ. This phase space was then clustered using 256 initial clusters and a combination threshold of 3. The algorithm detected five clusters. The transition matrix and final clustering are shown in figure 3.

The transition matrix is formed by counting the number of transitions between each cluster. The rows represent the clusters the time series is transitioning from and the columns represent the clusters the time series is transitioning to. The percentage of transitions from cluster i to cluster j is given by the value in row i, column j. From the transition matrix for the Wichita time series, it appears that a regime structure is present because once the time series enters a phase space cluster, especially cluster 2, there is a high probability that it will stay in that cluster.

The next GHCN temperature station analyzed is Sydney, Australia. This time series consists of average monthly temperatures from 1859 to 1991. A moving average (N=12) was applied to the time series and time delays of 1 and 2 were chosen for the phase space embedding. The number of initial clusters was set at 256 and a combination threshold of 3.5 was used. The clusters found by the algorithm and transition matrix are in figure 3. This time series, although it has fewer clusters than the Wichita time series, also shows a very prominent regime structure with a very low inter-cluster transition probability.

The next dataset presented is a paleoclimatic reconstruction created by Mann et al (1998). A three-dimensional phase space was constructed using time delays of 1 and 2. The transition matrix is shown in figure 4. The initial number of clusters was set at 64 and a combination threshold of 3 was used.

An eight-dimensional phase space was also considered for this time series. The eight-dimensional phase space was constructed using time delays of 1 through 7. The phase space was clustered using 128 initial clusters and a combination threshold of 3.5. The transition matrix after the clustering is in figure 4.

The Mann reconstructed time series does not show as prominent a regime structure as some of the other time series reviewed here. This might be because a moving average was not applied to the time series before analysis to remove some of the high frequency components of the time series. It is also interesting to note that the algorithm did a much better job of finding prominent regimes in the three-dimensional phase space than in the eight dimensional phase space.

The final experiment performed was using time series from different locations to construct a phase space. Average monthly temperatures from Corpus Christi, TX, El Paso, TX, and Fresno, CA were used to construct a three-dimensional phase space. Temperature data was available from all three sites for the years 1888 – 1998. A moving average (N=12) was calculated for all three time series before the phase space was constructed. The initial number of clusters was set at 256 and the combination threshold was set to 4. The clusters determined by the algorithm and transition matrix are shown in figure 4. The clusters found for this phase space are rather complex and bend around each other. There appears to be a strong regime structure at work in these time series as well because of the low percentages of inter-cluster transitions. This was the expected result for this trial because all three cities are located within the effect of the El Nino cycle.

CONCLUSION

Our analysis shows that regimes can be detected in most temperature time series. However, the strength of the regimes varies greatly. These results are expected however since previous research has shown that only certain parts of the earth have a regime-dominated climate. Regimes were found in various phase space constructions with different types of data preprocessing and different phase space construction techniques.

One thing to explore in the future is to analyze the duration of each visit to a cluster to see if these durations are consistent in anyway. Also, the clustering algorithm could be improved to include some kind of transition minimization instead of relying mostly on the density characteristics of the phase space.

REFERENCES

Abarbanel, Henry D., 1996, *Analysis of Observed Chaotic Data*, Springer-Verlag, New York, New York.

Ankerst, Mihael, Breunig, Markus M., Kriegel, Hans-Peter, Sander, Jörg, 1999, "OPTICS: Ordering Points to Identify the Clustering Structure", *Proc. ACM SIGMOD '99 Int. Conf. On Management of Data*, Philadelphia, PA.

Kantz, Holger, 1997, *Nonlinear Time Series Analysis*, Cambridge University Press, New York, New York.

Karypis, George, Han, Eui-Hong, Kumar, Vipin, 1999, "Chameleon: A Hierarchical Clustering Algorithm Using Dynamic Modeling", *Computer*, Vol. 32, No. 8, pp. 68-75.

National Climatic Data Center, 2000, *Global Historical Climatology Network Version 2 Dataset*, ftp://www.ncdc.noaa.gov/pub/data/ghcn/v2/.

Mann, Michael E., Bradley, Raymond S., Hughes, Malcolm K., 1998, "Global Six Century Temperature Patterns", *IGBP PAGES/World Data Center-A for Paleoclimatology Data Contribution Series # 1998-016*, NOAA/NGDC Paleoclimatology Program, Boulder, CO.

Palmer, T. N., 1993, "A nonlinear dynamical perspective on climate change", *Weather*, Vol. 48, pp. 314-326.

Palmer, T. N., 1999, "A nonlinear dynamical perspective on climate prediction", *Journal of Climate*, Vol. 12, pp. 575-591.

Povinelli, Richard J., 1999, "Time Series Data Mining: Identifying Temporal Patterns for Characterization and Prediction of Time Series Events", Ph.D. Dissertation, Marquette University, Milwaukee, WI.

FORECASTING WARRANTY CLAIMS: A COMPARISON OF SVMS WITH STATISTICAL METHODS AND NEURAL NETWORKS

RATNA BABU CHINNAM, PH.D.
Associate Professor
Industrial & Mfg. Engineering Department
Wayne State University
4815 Fourth Street
Detroit, Michigan 48202, USA
r_chinnam@wayne.edu

VINAY S. KUMAR
Graduate Student
Industrial & Mfg. Engineering Department
Wayne State University

GARY WASSERMAN, PH.D.
Associate Professor
Industrial & Mfg. Engineering Department
Wayne State University

ABSTRACT

The economic necessity for developing accurate forecasts of extended warranty costs on durable goods is well established in the literature. This paper illustrates the benefits of using Support Vector Machines and Multi-layer Perceptron networks. The results are quite promising.

INTRODUCTION

With increasing global competition in the last couple of decades, producers of durable goods are working hard to improve the quality of their products and services. Concurrent with these efforts, companies are extending limits of their product warranties to build increased customer satisfaction. If the impact of offering the extended warranty cannot be fully evaluated, it can result in unexpected lower firm profitability, which is typically true of new vehicle launches in the automobile industry. Thus, the need for more effective methods of predicting warranty life costs is evident. It is our intent herein to provide the reader with a useful overview of a set of basic strategies that will assist the warranty analyst in the development of improved warranty claims forecasts.

MODELING APPROACHES FOR FORECASTING WARRANTY CLAIMS

The strategies for forecasting warranty failures promoted in the literature can be roughly grouped into three categories: static predictive models, dynamic predictive models, and non-parametric methods.

Static Predictive (Multiple Regression or Time Series) Models

Of the three warranty failure forecasting strategy categories listed above, it is our experience that the use of static, linear predictive modeling approaches involving the use of either regression or time series models are the most widely used and best understood by warranty analysts. The common practice in industry is to use time-in-service as a regressor variable in a regression model to forecast warranty claims. The linear regression model is made up of a linear function in the regressor variables, $\mathbf{x}_t = [x_{0,t}, x_{1,t}, \ldots, x_{p,t}]^T$ as follows:

$$y_t = \mathbf{a}^T \cdot \mathbf{x}_t + \varepsilon_t \tag{1}$$

where y_t might either denote repairs per 1000 units ($R/1000$) or in terms of the actual warranty costs on a per unit basis at period $t = 1, 2, \ldots, n$; $\mathbf{a}^T = [\alpha_0, \alpha_1, \ldots, \alpha_p]^T$ denotes a vector of unknown parameters; ε_t = random error terms having zero mean and possessing any arbitrary covariance structure. Note that the predictor variables could also consist of leading indicator variables that reference various customer-usage categories, or warranty information from related processes or phenomena.

The regression approach discussed above is based on the observation that the time series is a deterministic function of time, and actual observations are generated by adding an independent random error component to the mean. Successive observations in many time series are highly dependent. If this is the case, then regression models are inappropriate. Forecasting methods that exploit this dependency are available and generally produce superior results.

Time series models constitute a specific class of linear predictive models characterized by model features that allow for serially correlated response variables, $y_t, t = 1, 2, \ldots$, or error terms, $\varepsilon_t, t = 1, 2, \ldots$. In this case, lagged values of the error terms and/or of y serve as model predictors. In such cases, the Box-Jenkins autoregressive, integrated, moving-average – ARIMA (p, d, q) – time series model comprises a popular approach for capturing such phenomena (Box et al., 1994). This model is given by

$$\phi(B)\nabla^d y_t = \theta(B)\varepsilon_t, \tag{2}$$

where B is the backwards shift operator defined by $y_{t-q} = B^q y_t$; $\phi(B) = I - \phi_1 B - \ldots - \phi_p B^p$; $\theta(B) = I - \theta_1 B - \ldots - \theta_q B^q$; and $\nabla^d = (1-B)^d$.

In the early stages of the product life cycle, however, the forecast error may be excessive owing to dynamic changes in the trend of a process resulting from the institutionalization of product design modifications. In general, static, predictive models are adequate for interpolation, but their use is risky when forecasts are obtained using an extrapolation of a least-squares fit when the underlying process behavior is highly nonlinear, or is perturbed by localized phenomena.

Dynamic Linear Predictive (Kalman filter) Models

While many different kinds of dynamic predictive models are promoted in the literature of forecasting problems, we will focus here on a dynamic linear models (DLMs) for their ability to adapt to local process trends. A DLM provides the needed flexibility to capture local trends while maintaining the simplicity of a linear model. The constitutive linear relationships for a dynamic model are given below:

Observation equation: $y_t = \alpha_t^T \cdot \mathbf{x}_t + \varepsilon_t$; $\varepsilon_t \sim N[0, V]$; (3)

System equation: $\alpha_t = \gamma_t \cdot \alpha_{t-1} + \omega_t$; $\omega_t \sim N[0, \mathbf{W}]$. (4)

To provide for process dynamics, the parameter vector $\alpha_t^T = [\alpha_{0,t}, \alpha_{1,t}, \ldots, \alpha_{p,t}]^T$ is indexed by usage period, t, with time-varying behavior described by the system Eq. (4). The parameter, γ_t, is used to model growth phenomena or to capture dynamic changes in the process mean at time, t, due to a process intervention; otherwise, we will use $\gamma_t = 1$. The system covariance, $\mathbf{W}$, governs the evolution of the state equation parameters α_t. If the elements of $\mathbf{W}$ are small relative to V, then the evolution in α_t is smooth.

The vector of parameters, α_t, may be estimated recursively by using the Kalman filter equations. To initialize the procedure, a prior distribution on the parameter vector, α_0 must be specified – $\alpha_0 \sim N[\mathbf{a}_0, \mathbf{R}_0]$. The estimate of the parameter vector, α_t, will then be updated sequentially using

$$\alpha_t = \gamma_t \cdot \alpha_{t-1} + \mathbf{K}_t \cdot e_t,$$ (5)

$$\mathbf{R}_t = \mathbf{C}_{t-1} + \mathbf{W}_t,$$ (6)

where $\alpha_t \sim N[\mathbf{a}_t, \mathbf{R}_t]$; e_t is the forecast error at time period t; $\mathbf{C}_{t-1}$ is the posterior covariance matrix for α_{t-1}; $\mathbf{K}_t$ is the *Kalman filter gain* $= \mathbf{R}_t \cdot \mathbf{x}_t \cdot Q_{t-1}^{-1}$, where Q_t denotes the prediction variance at time t.

It is evident that the weakness in the use of dynamic linear models approach stems from the inherent complexity in the mathematical relationships and assumptions required in the use of such models. In addition, the forecasts obtained with the use of these models may be quite sensitive to initial choice of the parameters of the prior distribution.

Nonparametric (Neural Network) Modeling Approaches

The use of nonparametric modeling approaches provides the needed flexibility to capture nonlinearities in a data set. These models must be fitted carefully, because an over-fitted model may result in a model fit that is too responsive to local perturbations. The use of a class of feed-forward neural networks, known as multi-layer perceptron networks, for forecasting warranty claims is discussed next.

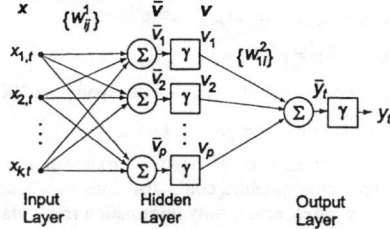

Figure 1: A two-layer MLP neural network.

A typical multi-layer perceptron network (MLP) with an input layer, an output layer, and one hidden layer is shown in Figure 1 (referred to normally as a two-layer network; input layer is not counted). The most popular non-linear nodal function for MLP networks is the sigmoid. The illustrated network carries a single output in the output layer and can be extended to multiple outputs without loss of generality. Each layer of the network can then be represented by the operator

$$\mathbf{N}_l[\mathbf{x}] = \Gamma[\mathbf{W}^{(l)} \mathbf{x}],$$ (7)

and the input-output mapping of this two-layer the MLP network can be represented by

$$y = \mathbf{N}[\mathbf{x}] = \Gamma\left[\mathbf{W}^{(2)} \Gamma\left[\mathbf{W}^{(1)} \mathbf{x}\right]\right] = \mathbf{N}_2 \mathbf{N}_1[\mathbf{x}].$$ (8)

The weights of the network $W^{(1)}$ and $W^{(2)}$ are adjusted to minimize a suitable function of error e between the predicted output y of the network and a desired output y_d, resulting in a mapping function N[x].

Hornik et al. (1989) showed that even an MLP network with just one hidden layer and arbitrarily large number of nodes can approximate any continuous function $f \in C(R^N, R^M)$ over a compact subset of R^N to arbitrary precision, providing the motivation to use MLP networks for forecasting warranty failures.

In training MLP networks, the objective is to determine an adaptive algorithm that adjusts the weights of the network based on a given set of input-output pairs. An error-correction learning algorithm will be discussed here, and readers can see Haykin (1999) for information regarding other training algorithms. If the weights of the networks are considered as elements of a parameter vector θ, the error-correction learning process involves the determination of the vector θ^* that optimizes a performance function J based on the output error. In error-correction learning, the gradient of the performance function with respect to θ is computed and adjusted along the negative gradient as follows:

$$\theta(s+1) = \theta(s) - \eta \frac{\partial J(s)}{\partial \theta(s)} \quad (9)$$

where η is a positive constant that determines the rate of learning and s denotes the iteration step. Typically, J is defined as a function of the mean-square error, where error $e = y - y_d$. A well-known method for determining the gradient in Eq. (9) for MLP networks is the back-propagation method. The analytical method of deriving the gradient is well known in the literature. Numerous modifications and extensions have been proposed in the literature to the gradient descent based back-propagation learning algorithm. For example, Levenberg and Marquart's algorithm (Haykin, 1999), along with conjugate gradient learning procedures have been shown to improve the rate of convergence of the network. For a more in depth treatment of the subject of artificial neural networks, see Haykin (1999).

SUPPORT VECTOR MACHINES

SVMs are a category of universal feed-forward networks pioneered by Vapnik (Vapnik, 1998). Bascially, SVM is a *linear machine* with some very nice properties. It is perhaps easiest to explain the behavior of an SVM by starting with the case of separable patterns that could arise in the context of binary pattern classification. In this context, the SVM in essence constructs a decision surface hyperplane that maximizes the separation *margin* between positive and negative examples. It achieves this desirable property by following a principled approach rooted in statistical learning theory. More precisely, SVM is an approximate implementation of the *method of structural risk minimization* (Haykin, 1999). These concepts are concisely explained below.

Suppose the task of a machine is to learn the mapping $x_i \mapsto y_i$. The machine is actually defined by a set of possible mappings $x_i \mapsto f(x, \alpha)$, where the functions $f(x, \alpha)$ themselves are labeled by the adjustable parameters α. The "expected risk" for a trained machine is therefore:

$$R(\alpha) = \frac{1}{2} \int |y - f(x, \alpha)| p(x, y) dx dy \quad (10)$$

This expression is not very useful unless we have an estimate for $p(x, y)$. In the absence of knowledge of $p(x, y)$, one could alternately work with an "empirical risk" measure $R_{emp}(\alpha)$ expressed as follows:

$$R_{emp}(\alpha) = \frac{1}{2N} \sum_{i=1}^{N} |y_i - f(x_i, \alpha)| \quad (11)$$

where N denotes the number of training patterns.

Now, choose some η such that $0 \le \eta \le 1$. Then, for losses taking these values, with probability $1 - \eta$, the following bound holds (Vapnik, 1998):

$$R(\alpha) \le R_{emp}(\alpha) + \sqrt{\frac{h}{N}\left[\log\left(\frac{2N}{h}\right) + 1\right] - \frac{1}{N}\log\left(\frac{\eta}{4}\right)} \quad (12)$$

Table 1: $R/1000$ data for prior and current model years (Wasserman and Sudjianto, 1996). The first 30 data points are used to train the data; the last six data points are used to validate the model.

Time	Prior year	Current year
1	0.55	1.27
2	4.68	1.58
3	9.88	3.15
4	12.80	7.71
5	15.50	11.40
6	20.00	18.70
7	25.90	26.30
8	33.10	35.40
9	37.50	42.00
10	45.40	46.40
11	56.40	48.20
12	68.90	48.80
13	78.20	53.30
14	84.00	61.80
15	84.70	75.90
16	89.00	90.80
17	97.80	102.00
18	110.00	105.00
19	125.00	113.00
20	139.00	125.00
21	150.00	139.00
22	158.00	151.00
23	163.00	159.00
24	166.00	165.00
25	167.00	172.00
26	173.00	186.00
27	179.00	199.00
28	185.00	211.00
29	187.00	228.00
30	189.00	249.00
31	196.00	271.00
32	205.00	290.00
33	211.00	300.00
34	213.00	328.00
35	215.00	375.00
36	221.00	430.00

where h is a non-negative integer called the Vapnik-Chervonenkis (VC) dimension and is a measure of the capacity of the learning machine. With the knowledge of h, one can easily compute the right hand side. Thus, given several different learning machines and a fixed η, the machine that minimizes the right hand side leads to a machine with the lowest upper bound on the actual risk. In the case of separable patterns, a SVM produces a value of zero for the first term on the right hand side of (12) and minimizes the second term. In the case of nonseparable patterns, one introduces a new set of nonnegative scalar *slack variables* that allow misclassification of data points. The cost function that is to be minimized will now carry an extra term weighted by parameter C that penalizes any SVM that generates too many nonseparable patterns.

A notion that is central to SVM learning algorithms is the inner-product kernel (denoted by $K(\mathbf{x}, \mathbf{x}_i)$) between a "support vector" $\mathbf{x}_i$ and the vector $\mathbf{x}$ drawn from the input space (Haykin, 1999). To keep SVM training algorithms computationally efficient, one has to always choose inner-product kernels that meet the *Mercer's theorem* that arises in functional analysis (Vapnik, 1998). The support vectors consist of a small subset of the training data extracted by the algorithm. Depending on how this inner-product kernel is generated, one may construct different learning machines. In particular, one may use the support vector learning algorithm to construct the following types of learning algorithms (among others): polynomial learning machines, radial-basis function networks, and two-layer perceptron networks.

ε - *Insensitive Loss Function*

For the sake of computational convenience, a loss function that is commonly used for optimizing learning machines is the quadratic loss function. However, a least-squares estimator is sensitive to the presence of outliers, and performs poorly when the underlying distribution of the additive noise has a long tail (Haykin, 1999). To overcome these limitations, in constructing an SVM for approximating a desired response d, we shall use the following ε-insensitive loss function, proposed by Vapnik (1995):

$$L_\varepsilon(d, y) = \begin{cases} |d - y| - \varepsilon_{\text{Insensitivity}} & \text{for } |d - y| \geq \varepsilon_{\text{Insensitivity}} \\ 0 & \text{otherwise} \end{cases} \tag{13}$$

where $\varepsilon_{\text{Insensitivity}}$ is a prescribed parameter.

Support Vector Machines for Nonlinear Regression

Consider a nonlinear regressive model described by

$$d = f(\mathbf{x}) + \upsilon \tag{14}$$

where the scalar-valued nonlinear function $f(\mathbf{x})$ is defined by the conditional expectation $E[D|\mathbf{x}]$, D is a random variable with a realization denoted by d, and υ is an additive noise term that is statistically independent of the input vector $\mathbf{x}$. One can postulate an estimate of d, denoted by y, as an expansion in terms of a set of nonlinear basis functions $\{\varphi(\mathbf{x})\}_{j=0}^{m}$ as follows:

$$y = \sum_{j=0}^{m} w_j \varphi_j(\mathbf{x}) = \mathbf{w}^T \boldsymbol{\varphi}(\mathbf{x}). \tag{15}$$

It is assumed that $\varphi_0(\mathbf{x}) = 1$, so that the weight w_0 represents the bias b. For reasons outlined in Vapnik (1998), the issue to be resolved is to minimize the empirical risk

$$R_{emp} = \frac{1}{N} \sum_{i=1}^{N} L_\varepsilon(d_i, y_i) \tag{16}$$

subject to the inequality

$$\|\mathbf{w}\|^2 \leq c_0 \tag{17}$$

where c_0 is a constant.

One can reformulate this constrained optimization problem by introducing two sets of nonnegative *slack variables* $\{\xi_i\}_{i=1}^{N}$ and $\{\xi_i'\}_{i=1}^{N}$ that are defined as follows:

$$d_i - \mathbf{w}^T \boldsymbol{\varphi}(\mathbf{x}_i) \leq \varepsilon_{\text{Insensitivity}} + \xi_i, \quad i = 1, 2, \ldots, N \tag{18}$$

$$\mathbf{w}^T \boldsymbol{\varphi}(\mathbf{x}_i) - d_i \leq \varepsilon_{\text{Insensitivity}} + \xi_i', \quad i = 1, 2, \ldots, N \tag{19}$$

$$\xi_i \geq 0, \quad i = 1, 2, \ldots, N \tag{20}$$

$$\xi_i' \geq 0, \quad i = 1, 2, \ldots, N \tag{21}$$

The slack variables describe the ε-insensitive loss function defined in Eq. (13). This constrained optimization problem may therefore be viewed as equivalent to that of minimizing the cost functional

$$\Phi(\mathbf{w},\xi,\xi') = C\left(\sum_{i=1}^{N}(\xi_i + \xi_i')\right) + \frac{1}{2}\mathbf{w}^T\mathbf{w} \qquad (22)$$

subject to the constraints of Eqs. (18) to (21). By incorporating the term $\mathbf{w}^T\mathbf{w}/2$ in the functional $\Phi(\mathbf{w},\xi,\xi')$ of Eq. (22), we dispense with the need for the inequality constraint of Eq. (17). The two parameters $\varepsilon_{\text{Insensitivity}}$ and C control the VC dimension of the approximating function and must be tuned simultaneously. For a more in depth treatment of the topic of SVMs, see Vapnik (1998).

FORECASTING WARRANTY CLAIMS USING SUPPORT VECTOR MACHINES AND OTHER STRATEGIES

This section presents the results from using SVMs for forecasting the warranty claim data provided in Wasserman and Sudjianto (1996). We also compare these results against the results from using ARIMA models, Kalman filter models, and MLP networks as provided in Wasserman and Sudjianto (1996).

Data Set Properties

In the marketplace, it is not uncommon for process interventions associated with market events, short-term changes in product quality, or design fixes to result in changes in local process behavior. Additionally, wear-out phenomena, which may arise during the latter stages of the warranty or product life cycle, may manifest themselves as a sudden increase in the observed failure rates. Accordingly, Wasserman and Sudjianto (1996) constructed the 36-point data set in such a way as to favor the use of models that are adaptable to local changes in the trend of a process. They also state that an obvious change in the local trend is introduced in the data set at $t = 30$ usage units. They however did not reveal the true data generating equation. This data set is provided in Table 1.

SVM Design

We chose to work with the simplest SVMs that used a polynomial kernel of order two ($K(\mathbf{x}_1,\mathbf{x}_2) = (\mathbf{x}_1 \cdot \mathbf{x}_2 + 1)^2$) and three ($K(\mathbf{x}_1,\mathbf{x}_2) = (\mathbf{x}_1 \cdot \mathbf{x}_2 + 1)^3$). Better results can be potentially achieved by using more complex kernels, however, the objective here is to demonstrate the power of SVMs even while using simple kernel operators, and hence the above choices. All SVM training runs were carried out by setting the loss insensitivity constant $\varepsilon_{\text{Insensitivity}}$ at 10^{-5}. The maximum number of iterations allowed for convergence is set at 10^5. After preliminary experimentation, the C parameter for the objective function is varied from 5 to 100 in steps of five. Obviously, better results can be potentially achieved by increasing the search resolution for the C parameter.

Predictor variables based upon both prior and current model year information were studied. The use of prior model year information is useful when there is reason to believe that there exists sufficient similarity in the reliability of prior model year product with the current model year.

The partial autocorrelation function plot for the first 30 data points for the current model year warranty claim data suggested a low-order autocorrelation. Similar behavior has been observed in the prior model year data as well. In preparing the input vector for the SVM, we chose to vary the number of lags for the current model year data from $P=2$ to 4 and for the prior model year data from $Q=1$ to 3, resulting in nine distinct combinations. For the case where $P=2$ and $Q=1$, the input vector is four-dimensional, $\mathbf{x}_t = [y_{t-1}^{CMY}, y_{t-2}^{CMY}, y_t^{PMY}, y_{t-1}^{PMY}]$. The superscript differentiates the current model year (CMY) data from previous model year data (PMY). In a similar manner, for the case where $P=4$ and $Q=3$, the input vector is eight-dimensional. Note that the number of training patterns will differ as we change P and Q.

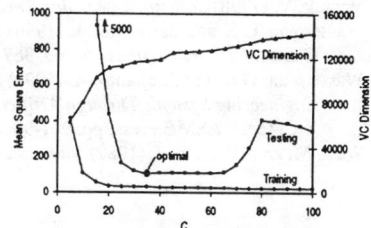

Figure 2: Results from using SVM with $P=2$ and $Q=3$ for forecasting warranty claim data.

SVM Results

All the SVM implementation results reported here are obtained by implementing *mySVM* software.[*] *mySVM* is based on the *SVMlight* optimization algorithm described in Joachims (1999).

Experiments revealed that the third-order polynomial kernel is overfitting the training data. Hence, it was decided to strictly focus on the second-order polynomial kernel. Among the nine cases studied, the best overall model resulted in the case where $P=2$, $Q=3$, and the parameter C is set at 35, as illustrated in Figure 2.

Results from ARIMA, DLM, and MLP Networks

Wasserman and Sudjianto (1996) reported that the best Box and Jenkin's time series model that they could develop for forecasting the current model year data is the ARIMA(1,2,0) model. They also reported that the best Kalman filter based DLM model that they could fit was a first-order model in the leading indicator variable and the autoregressor (i.e., $P=1$, $Q=0$). They also reported that the best MLP network that they developed is a two-layer network with two sigmoid nodes in the hidden layer and a linear output neuron. The results from all these models including the best SVM identified earlier are summarized in Table 2 in the form of Training and Testing MSEs.

It appears that both the ARIMA model and Kalman filter model are overfitting the training data. Numerous attempts have been made to reproduce the MLP Network results reported in Wasserman and Sudjianto (1996) without any success. Our own experiments suggested that an MLP network could not reach the performance of the best SVM reported in Table 2. As stated earlier, one could potentially achieve even better results with SVMs by using other kernel operators, increasing the resolution of parameter C, and changing the input vector properties.

CONCLUSION

The use of several techniques for warranty claims forecasting is demonstrated. While statistics based static and dynamic predictive models are popular in industry for warranty claim forecasting, as an alternative, the warranty analyst should consider the use of nonparametric modeling approaches that also offer the flexibility to model local trends and any general nonlinear phenomena. In particular, MLP networks and SVMs might offer better forecasting results, as is the case in our test study.

Table 2: Summary of training and testing MSEs for four fitted models.

Model	Training	Testing
ARIMA(1,2,0)	11.55	306.34
Kalman Filter Model ($P=1,Q=0$)	20.55	177.17
MLP Network ($P=1,Q=0$)	12.84	78.18
Support Vector Machine ($P=2,Q=3$)	32.67	106.16

(Mean Square Error)

REFERENCES

Box, G.E.P., Jenkins, G.M. and Reinsel, G.C. (1994) *Time Series Analysis: Forecasting and Control*, 3rd edn, Prentice Hall, Engle-wood Cliffs, NJ.

Haykin, S. (1999) *Neural Networks: A Comprehensive Foundation*, 2nd edn, Prentice Hall, NJ: Upper Saddle River.

Hornik, K., Stinchcombe, M., and White, H. (1989) Multi-layer Feed-forward Networks are Universal Approximators. *Neural Networks*. 2.

Joachims, T. (1999) Making large-scale SVM learning practical. In Schölkopf, B., Burges, C. J. C., and Smola, A. J. (eds), *Advances in Kernel Methods–Support Vector Learning*, Cambridge, MA: MIT Press, pp. 41-56.

Vapnik, V. (1998) *Statistical Learning Theory*, John Wiley and Sons, New York.

Wasserman, G.S. and Sudjianto, A. (1996) A Comparison of Three Strategies for Forecasting Warranty Claims, *IIE Transactions. 28.* pp. 967-977.

Wasserman, G.S. and Sudjianto, A. (1992) Neural Networks for Forecasting Warranty Claims, *Intelligent Engineering Systems Through Artificial Neural Networks*, Vol. 2, Dagli, C.H., Burke, L.I., and Shin, Y.C. (eds), ASME Press, pp. 901-906.

West, M. and Harrison, J. (1989) *Bayesian Forecasting and Dynamic Models*, Springer-Verlag, New York.

---

[*] *mySVM* software is developed by Dr. Stefan Rüping of the Department of Computer Science at University of Dortmund, Germany (http://www-ai.cs.uni-dortmund.de).

USING NEURAL NETWORKS AND TECHNICAL ANALYSIS INDICATORS FOR PREDICTING STOCK TRENDS

SURAPHAN THAWORNWONG, DAVID ENKE, & CIHAN DAGLI
Smart Engineering Systems Lab
Intelligent Systems Center
University of Missouri – Rolla, MO, USA

ABSTRACT
Recent studies reflect a growing interest in applying neural networks to answer stock behavior. Most of these studies rely heavily on fundamental analysis factors to determine future stock prices. In fact, there exists another approach, called technical analysis, which attempts to predict the stock trend by using data surrounding past prices and volumes. This paper investigates whether using these indicators as inputs to a neural network will provide more accurate predictions of future stock trends and whether they will yield higher trading profits than the traditional technical indicators. Feed-forward, probabilistic, and learning vector quantization neural networks are then examined to predict the short-term trend signals of several major stocks in different industries. The overall results indicate that the proportion of correct predictions and the profitability of trading exercises guided by these neural networks are consistently higher than those guided by the buy-and-hold strategy and the individual technical indicators.

INTRODUCTION
Stock trend or stock price prediction is an important financial subject that has attracted researchers' attention for many years. This is due to the fact that a successful prediction model could result in substantial monetary rewards. In recent years, technical analysis has been widely used as a viable analytical option among both financial economists and brokerage firms (Achelis, 1995). Technical analysis appears to be a compromising tool since it offers a relative mixture of human, political and economical events. However, the main problem with this technique is that it relies heavily on the discovery of strong empirical regularities in observations of the price and volume movements (Liu and Lee, 1997). In reality, these regularities are not always evident, often masked by noise, and vary from security to security, making it difficult for investors who use this technique to consistently and accurately determine future prices.

This study investigates whether neural networks could be used to better uncover the regularities of these underlying price and volume movements. Accordingly, several popular technical indicators were selected as input variables used to train the neural networks. Three neural network models, including feed-forward, probabilistic, and learning vector quantization neural networks, were then examined for their ability to provide an effective forecast of future stock trend signals. Three major stocks across different market industries were selected to support the robustness of the neural network models. Trading strategies were also developed to account for the best opportunity to take trading positions. Finally, the predictability and profitability of trading practices directed by these models were compared against those of trading guided by a buy-and-hold strategy and other individual technical analysis indicators.

INDICATOR AND VARIABLE SELECTION

It has been long known that neural network training can be made more efficient if certain preprocessing steps have been performed on the network inputs. This approach is particularly applicable to certain aspects of technical indicators. Though the neural networks can independently learn any function, neural network performance could be further improved if such stock related values as highs, lows, and volumes are preprocessed into more meaningful information which is indeed the technical indicators. In this study, three selected stocks include Lockheed Martin Corp. (LMT) representing the aerospace & defense, Caterpillar Inc. (CAT) representing the construction & agricultural machinery, and Delta Air Lines Inc. (DAL) representing the airline group.

This present study focuses on short-term stock prediction. Therefore, the daily data, including the opens, highs, lows, closes, and volumes, were collected to perform the technical indicator calculation for a total of 846 trading days (08/22/1996 – 12/29/1999). The resulting daily indicator values were then divided into two periods. The first period included the indicator values from 08/22/1996 to 06/30/1999 (720 days), while the second period contained the indicator values from 07/1/1999 to 12/29/1999 (126 days). The former period was used for training and validating the neural networks. The latter period was reserved for out-of-sample evaluation among the neural networks and regular technical indicators. All data were downloaded from the finance.yahoo.com website. They also had been adjusted for all applicable stock splits and dividend distributions.

Five technical indicators, including the relative strength index (RSI), money flow index (MFI), moving average (MA and Closing Price), stochastic oscillator (%K and %D), and moving average convergence/divergence (MACD and Signal Line), were selected in this study. Readers can refer to Achelis (1995), Murphy (1999), and Schwager (1996) for a comprehensive explanation of these technical indicators. Since different authors often use different criteria to capture the buy and sell signals based on their accumulated experience, only the most commonly used criteria were considered as a default to remove conflicting recommendations in this paper. Regarding the five technical trading criteria, there are a total of eight variables, including the Closing Price (CP), RSI, MFI, MA, %K, %D, MACD, and Signal Line (SL), required to capture the underlying buy and sell signals of the observed stock. More importantly, recent historical data of indicator and stock price movements are also required to make the stock trend prediction.

In this study, the difference $(CP_t - CP_{t-1})$ of stock closing price was employed so that they can be compared in terms of relative change with the daily stock price movements. Similarly, the MA, SO, and MACD trading guidelines require that line crossing (below or above) must occur before the buy or sell signal can be initiated. Accordingly, the differences of these lines were calculated so that they can be measured in terms of how close the lines will finally be crossed over each other. The crossing can explicitly be captured by sign changes of the resulting values of the line differences. The differences for MA, SO, and MACD indicators were defined as $MA_t - CP_t$, $\%K_t - \%D_t$, and $MACD_t - SL_t$, respectively. Note that no preprocessing steps were done for the RSI_t and MFI_t indicators because the indications of market reversals can be identified by comparing the relative change of stock price movements $(CP_t - CP_{t-1})$ with these two indicators.

Finally, these six selected inputs were included in the base sets with three-day time lags to account for recent movements of the technical indicators. As a result, 18 input variables, including $CP_t - CP_{t-1}$, $CP_{t-1} - CP_{t-2}$, $CP_{t-2} - CP_{t-3}$, RSI_t, RSI_{t-1}, RSI_{t-2}, MFI_t, MFI_{t-1}, MFI_{t-2}, $MA_t - CP_t$, $MA_{t-1} - CP_{t-1}$, $MA_{t-2} - CP_{t-2}$, $\%K_t - \%D_t$, $\%K_{t-1} - \%D_{t-1}$, $\%K_{t-2} - \%D_{t-2}$, $MACD_t - SL_t$, $MACD_{t-1} - SL_{t-1}$, and $MACD_{t-2} - SL_{t-2}$, were employed to predict the direction of the next day stock price (C_{t+1}). These time lags were used throughout the experiment to maintain realistic situations when the technical indicators and stock prices are calculated and gathered.

TRADING PRACTICES AND STRATEGIES

Investors today are allowed to profit from both an increase and decline in stock prices. As commonly known, investors purchasing a share of stock to open a long position profit from price increases. Alternatively, there is another trading practice called a short sale which allows investors to profit from the decline in stock prices. In this paper, a 1% round-trip transaction cost (sell a share to close a long position and sell a borrowed share to create a short position; or purchase a share to cover a short position and purchase a share to open a long position) was charged to the investor when an asset allocation of security positions had been made. To examine if the neural network forecasts could generate higher profits, a trading strategy was devised as trading criteria in connection with the predicted signals of neural network forecasts.

The trading assumes that an investor either maintains a current position or makes an asset allocation of whether to shift from long to short or from short to long when the stock market is opened for each trading day. After the market is closed for each trading day, the investor has to decide whether to maintain the current position or make the asset allocation, depending on the signal generated by the neural network forecasts calling for a long or short position in the next trading day as compared to the current day. The following describes the trading criteria with respect to the predicted signals (directions) of the neural network forecasts:

If $C_{t+1} = +1$, then
Maintain the current long position, or cover a short position and open a long position (pay 1% transaction cost and receive profit/loss of the covered short position)
Else (if $C_{t+1} = -1$), then
Maintain the current short position, or close a long position and create a short position (pay 1% transaction cost and receive profit/loss of the closed long position)

where C is the predicted direction of the next day stock prices given by the neural network forecasting models. The +1 and -1 represent the predicted upward and downward directions of the next day's stock price movement, respectively.

It should be noted that the trading strategies and practices of the technical indicators employed in this study are similar to those of the neural network forecasts for comparable performance evaluations. As such, a sell signal identified by the technical indicators will be used to create a short position in the beginning of the next trading day, while the long position will be created for the next trading day when a buy signal of the technical indicators is found. Similarly, the current investment position will be maintained if the technical indicators signal neither a sell nor buy recommendation. To measure profitability, trading gains or losses were accumulated for the whole trading period.

NEURAL NETWORK MODELING

Feed-Forward Neural Networks (FNN) have been widely used for financial forecasting due to their ability to correctly classify and predict the dependent variable (Vellido et al., 1999). In this study, a single hidden layer and a resilient backpropagation-learning algorithm were chosen. A hyperbolic-tangent function was also selected as the activation function. Two output neurons were employed for the output layer to represent different classes of the predicted direction. The vectors [+1 -1] and [-1 +1] represent the predicted upward and downward directions of the next day stock trend, respectively. The output neuron with the highest value was taken to represent the predicted directions. An early stopping technique was also employed to achieve better generalization (Demuth and Beale, 1998). In this study, the 576 trading days (80%) of the first period were randomly selected as a training set for determining the network specifications. The validation set, for a total of 144 trading days (remaining 20% of the first period), was consequently used to decide when training should be stopped. Note that the values of the input variables were first preprocessed by normalizing them within a range of −1 and +1. The connection weights and biases were initially randomized and then determined during the training process. The number of hidden neurons and appropriate learning rate that minimize the classification errors of the validation set were also determined during network training.

Probabilistic Neural Networks (PNN) learn from sample data instantaneously and use probability density functions to compute the nonlinear decision boundaries between classes in a way that approaches the Bayes optimal (Specht, 1990). The design of the PNN is fast and straightforward. In fact, neither validation nor early stopping is required during its design. Therefore, there would be no need to randomly partition the data into training and validation sets. To take this unique advantage, the first period of the data set was used in network modeling. The +1 and +2 were used as the network outputs to represent the predicted downward and upward directions, respectively. Also, a smoothing parameter equal to 1.00 was selected to entirely consider several nearby design vectors. Again, the PNN design employed the same pre-processing techniques as those implemented for the FNN.

Learning Vector Quantization Neural Networks (LVQ) are two-layer networks that can classify input vectors into target classes. The first hidden layer is normally called a competitive layer, and the second hidden layer is known as a linear layer (Kohonen, 1995). The learning algorithm selected for this study was the LVQ2. The network outputs were similar to those used during the PNN design. The class percentages of 0.5 upward and 0.5 downward were selected to represent an equivalent distribution of future stock movements. The LVQ employed the same training set, validation set, and pre-processing technique as those used for the FNN modeling. Similarly, the training set was used to determine the appropriate learning rate and number of hidden neurons that minimize the classification errors of the validation set.

EMPIRICAL RESULTS

The predictive performance of the developed neural networks and technical analysis indicators was evaluated using the untouched second period. There is some evidence in the finance literature suggesting that traditional measures of forecasting performance may not be strongly related to profits from trading (Pesaran and Timmermann, 1995). An alternative approach is to look at the proportion of time that the signals of stock price changes (SIGN) are correctly predicted. Therefore, the SIGN reported in Table 1 was selected as the performance measure in this study.

Table 1. Predictability Results

		SIGN			
		LMT	CAT	DAL	Average
Technical Indicators	RSI	0.4880	0.4828	0.4800	0.4836
	MFI	0.4884	–	–	–
	MA	0.5439	0.4435	0.4530	0.4801
	SO	0.4960	0.5050	0.5242	0.5084
	MACD	0.4727	0.5210	0.6168	0.5368
Neural Networks	FNN	0.5476	0.6429	0.5794	0.5900
	PNN	0.5873	0.5476	0.5317	0.5555
	LVQ	0.5556	0.5476	0.5794	0.5609
	PortNN	0.5714	0.5873	0.5794	0.5794
Buy-and-Hold			0.3968	0.4444	0.4683	0.4365

The predictability results obtained from always investing in each stock (Buy-and-Hold) are also provided as the benchmark for performance comparisons in the study. To explore further, the outputs of the three neural network models (FNN, PNN, and LVQ) were combined to form a portfolio network model (PortNN). As such, the decisive predicted direction of stock price movement of the PortNN was derived from the majority of the three combined portfolio network outputs. The average of SIGN across all three stocks for each developed model is also provided in the last column of Table 1. Note that the MFI failed to identify the strong indications of market reversals hidden in the CAT's and DAL's securities resulting in two inoperative SIGN calculations.

According to Table 1, the results show that both the technical indicators and neural networks successfully generate higher averaged SIGN than the buy-and-hold account. Nonetheless, it seems that neither the neural networks nor the technical indicators can accurately signal the direction of stock price movement because of the relatively low averaged SIGN, although each of the neural network models is unquestionably better than the model using the individual technical indicators or the Buy-and-Hold. This reveals that the neural networks perform more accurately in predicting the direction of future stock movement. Particularly, the PortNN has achieved a relatively constant predictability performance among all three securities examined in the study.

Simulated Trading

After performing the trading simulation, the total number of transactions (trades) and the total returns on investment obtained from stock trading practices guided by the neural networks and the technical indicators over the one-sliding trading days of the second period presented in Table 2 are calculated.

Table 2. Profitability Results

		LMT		CAT		DAL		Portfolio Returns
		# of Trades	Total Returns	# of Trades	Total Returns	# of Trades	Total Returns	
Technical Indicators	RSI	8	−12.65%	4	−0.19%	2	−10.30%	−7.71%
	MFI	2	12.89%	0	0	0	0	4.30%
	MA	14	38.33%	52	−46.15%	48	−43.89%	−17.24%
	SO	8	−19.19%	6	−8.30%	8	13.10%	−4.80%
	MACD	12	−41.88%	32	20.72%	36	22.02%	0.29%
Neural Networks	FNN	32	16.90%	90	6.04%	100	−13.96%	2.99%
	PNN	18	33.45%	30	14.00%	30	9.74%	19.06%
	LVQ	50	4.67%	56	7.66%	92	12.09%	8.14%
	PortNN	30	13.67%	62	6.32%	86	5.13%	8.37%
Buy-and-Hold		2	−47.52%	2	−25.60%	2	−14.44%	−29.19%

As previously mentioned, the trading exercises of the CAT and DAL securities directed by the MFI never took place since the strong regularities were not detected by the MFI during the trading periods. The last column in Table 2 indicates the rate of return that can be obtained from holding an equally weighted portfolio of the three securities. The reason for forming this stock portfolio in the present study is to limit an investor's risk exposure of always investing in a particular security. In this study, the portfolio return can be derived as $R_P = wR_{LMT} + wR_{CAT} + wR_{DAL}$; where R_p is the rate of return on holding the stock portfolio, $w = 1/3$ is the weight of investment, and R_{LMT}, R_{CAT}, and R_{DAL} are the returns of investing in the LMT, CAT, and DAL securities, respectively.

Similar to the averaged SIGN, the portfolio returns directed by the neural networks and technical indicators are better than that generated by the buy-and-hold strategy. In fact, the buy-and-hold account obtains the lowest portfolio return (–29.19%) over the 126 trading days. The trading results show that the MFI and MACD of the technical analysis, as well as all developed models of the neural networks, generate positive portfolio returns. Particularly, the results show that the PNN, LVQ, and PortNN of the neural networks significantly generate higher portfolio returns than those of the technical analysis. Specifically, the PNN, which yields a total portfolio return of 19.06%, is the best performer among the developed models evaluated in this study. Even though the highest returns of the neural networks are not superior to those of the technical indicators, this empirical finding indicates that the overall trading profits directed by the neural network predictions are consistently better than those guided by the technical analysis recommendations.

CONCLUSIONS

An attempt has been made in this study to improve the predictive ability of technical indicators by adopting neural networks to uncover the underlying nonlinear patterns of these technical indicators for short-term stock trend prediction. Three major stocks across different market industries were tested to support the robustness of the neural network models. To this end, we find that different technical indicators often do not consistently work well for all securities, which in turn generates varied stock signals resulting in diverse trading profits. Neural networks, however, seem to be a perfect tool to manage these diversities since they independently learn the underlying relationships of particular technical indicators on a selected security.

This study covers only technical indicators, while fundamental analysis remains intact. It is far from perfect as the fundamental available information has been proven to provide some predictive factors in stock price and stock return forecasting. In fact, there are numerous studies done by both academics and practitioners in this area. If both technical and fundamental approaches are thoroughly examined and included during neural network modeling, it would no doubt be a major improvement in predicting future stock movements. Furthermore, additional criteria of the trading strategies should be further designed since the success of trading practices is critically dependent on the effective control of both limiting losses and protecting profits. Finally, future research should also consider the trading simulations under different scenarios of transaction costs and individual-tax brackets to replicate realistic investment practices.

REFERENCES

Achelis, S.B. (1995). *Technical analysis from A to Z: covers every trading tool from the Absolute Breadth Index to the Zig Zag*. Chicago, IL: Probus Publisher.
Demuth, H. & Beale, M. (1998). *Neural Network Toolbox: for use with MATLAB*. 5th edition. Natick, MA: The Math Works, Inc.
Kohonen, T. (1995). *Self-Organizing Maps*. New York, NY: Berlin, Springer.
Liu, N.K. & Lee, K.K. (1997). An intelligent business advisor system for stock investment. *Expert Systems*, 14, 129-139.
Murphy, J.J. (1999). *Technical analysis of the financial markets: a comprehensive guide to trading methods and applications*. New York, NY: New York Institute of Finance.
Pesaran, M.H. & Timmermann, A. (1995). Predictability of stock returns: robustness and economic significance. *Journal of Finance*, 50, 1201-1227.
Schwager, J.D. (1996). *Technical analysis*. New York, NY: John Wiley & Sons.
Specht, D.F. (1990). Probabilistic neural networks. *Neural Networks*, 3, 109-118.
Vellido, A., Lisboa, P.J.G., & Vaughan, J. (1999). Neural networks in business: a survey of application (1992-1998), *Expert Systems with Applications*, 17, 51-70.

Predicting Monthly Flour Prices through Neural Networks, RBFs and SVR

THEODORE B. TRAFALIS
School of Industrial Engineering
University of Oklahoma
ttrafalis@ou.edu

BUDI SANTOSA
School of Industrial Engineering
University of Oklahoma
bsant@ou.edu

1. INTRODUCTION

Neural network approaches have been used to analyze multivariate time series by several authors. Chakraborty et al. (1992) applied neural networks to forecast the behavior of time series related to monthly flour prices. Specifically they used real world observations of flour prices in three cities in their work. Remarkable success has been achieved in their study compared to a traditional statistical model. Chakraborty et al. (1992) used the error back propagation algorithm to train the network using mean square error (MSE) over the training sample as the objective function.

In this paper a comparative study of univariate or multivariate models is given, in order to forecast flour prices in three cities. Specifically neural networks, radial basis function, and support vector regression are investigated with the objective of identifying the approach that produces the smallest error.

The paper is organized as follows: in section 2 we give the necessary background of the architectures used. In section 3 we provide the methodology of our experimental setting. Section 4 provides computational results.

2. BACKGROUND

Usually RBF and SVR architectures perform well in the case of nonlinear estimation problems (Haykin, 1999). From preliminary checking of the data plots (not presented here), it is obvious that the data are not linear. Motivated by the above observation we expect that, applying RBFs and SVR should work well in predicting monthly flour prices. The following architectures will be used.

Neural Network: A feedforward network will be used to learn and predict the behavior of multivariate time-series.

Radial Basis Function:Radial: Basis function methods have their origins in techniques for performing exact interpolation of a set of data points in a multi-dimensional space (Haykin, 1999). The exact interpolation problem requires every input vector to be mapped exactly onto the corresponding target vector.

Support Vector Regression (SV)R: Suppose we have been given training data $\{(x_1,y_1),......,(x_m,y_m)\} \subset X \times P$, where X denotes the space of the input patterns and $P \subseteq R$. We want to find a function $f(x)$ that has at most ϵ deviation from the actual target y_i for all training data, and the function is as flat as possible. For the linear case, suppose we have the following function as a regressor:

$$f(x) = <w,x> + b, \qquad (1)$$

where $<.>$ denotes the dot product in X. Flatness in case (1) means that one should seek small w, in terms of length, where b is a threshold. One way to accomplish this objective is by minimizing the Euclidean norm $\|w\|^2$. Then we have the following problem:

$$\text{minimize } \frac{1}{2}\|w\|^2$$

$$\text{subject to } \begin{cases} y_i - \langle w, x_i \rangle - b \leq \varepsilon \\ \langle w, x_i \rangle + b - y_i \leq \varepsilon \end{cases} \qquad (2)$$

We assume that there is a function f that approximates all pairs (x_i, y_i) with precision ϵ. In this case, we assume that the problem is feasible. In the case of infeasibility, one can introduce slack variables ξ, ξ^* to cope with infeasible constraints of the optimization problem. Then the above problem can be formalized as (Vapnik, 1995):

$$\text{minimize } \frac{1}{2}\|w\|^2 + C\sum_{i=1}^{\ell}(\xi_i + \xi_i^*)$$

$$\text{subject to } \begin{cases} y_i - \langle w, x_i \rangle - b \leq \epsilon + \xi_i \\ \langle w, x_i \rangle + b - y_i \leq \epsilon + \xi_i^* \\ \xi_i, \xi_i^* \geq 0 \end{cases} \quad (3)$$

The constant $C > 0$ determines the trade off between the flatness of function f and the amount up to which deviations larger than ϵ are tolerated.

Risk Functional: Here, we have some training data $X = \{(x_1, y_1), \ldots, (x_m, y_m)\}$. We will assume now, that the set of training data has been drawn i.i.d from some probability distribution $P(x, y)$. Our goal is to find a function f that the minimizes a risk functional (Vapnik, 1982).

$$R[f] = \int c(x, y), f(x))dp(x, y), \quad (4)$$

where $c(x, y), f(x))$ denotes a loss function determining how we will penalize estimation errors based on empirical data X. Since we do not know the distribution of $p(x, y)$, then we can only use X for estimating a function f o minimize $R[f]$. The possible approximation to the functional risk is what we call empirical risk.

$$R_{emp}[f] = \frac{1}{\ell}\sum_{i=1}^{\ell} c(x_i, y_i, f(x_i)). \quad (5)$$

A first attempt would be to find the function $fo = \arg\min_{f \in H} R_{emp}[f]$ for some hypothesis class H. However, if H is very rich, i.e its capacity is very high, as for instance when dealing with few data in very high-dimensional spaces, this may not be a good idea, as it will lead to overfitting and thus bad generalization properties. Hence one should add a capacity control term, which in the SV case results to be $\|w\|^2$, which leads to the regularized risk functional (Smola and Scholkopf, 1998)

$$R_{reg}[f] := R_{emp}[f] + \frac{\lambda}{2}\|w\|^2$$

or

$$R_{reg}[f] = R_{emp}[f] + \lambda\|w\|^2 = \sum_{i=1}^{l} C(f(x_i) - y_i) + \lambda\|w\|^2 \quad (6),$$

where ℓ denote the sample size, $C(.)$ is a cost function and λ is a regularization constant. By some manipulation with Lagrange multiplier and dual optimization (Haykin, 1999) we get

$$w = \sum_{i=1}^{l}(\alpha_i - \alpha_i^*)x_i \quad (7)$$

and therefore

$$f(x) = \sum_{i=1}^{l}(\alpha_i - \alpha_i^*) <x_i, x> + b = \sum_{i=1}^{l}(\alpha_i - \alpha_i^*) <x_i, x> + b, \quad (8)$$

In eq (7) w can be completely described as linear combination of training patterns x_i. Kernel function, $k(x,x')$, can be used to substitute the dot product $\langle x, x' \rangle$.

Loss Function

The cost term $C(.)$ in (6) denotes the penalty of the differences between $f(x)$ and the actual a y. Some loss functions that are usually applied in the support vector regression are the ϵ-insensitive loss and quadratic loss functions respectively.

3. METHODOLOGY

The error back propagation algorithm will be used to train the network using the mean square error (MSE) over the training samples as the objective function. The MSE (training) is calculated as $\Sigma (Y_k - t_k)^2 / M$, where t_k and Y_k $1 \leq k \leq M$ are the desired and the actual network output respectively, for each of the M training patterns. The MSE (test) is similarly defined for the test patterns. The same data sets are used for Radial Basis Function training. Here we plug in the input data, target and spread. The same data sets are also applied to the SVR. The configuration of the input data was arranged due to the neural network design. During the training we have to specify what types of kernel function we use, the value of C, the p value for the polynomial function or spread for RBF function, and what type of loss function we are using. Those parameters are very important in the training phase of SVR. From the training we will obtain the optimal weights, parameter alpha and bias values. Then we can use those values to test any other set of data.

4. COMPUTATIONAL RESULTS
4.1 Experimental Setting

In the experiment a trivariate time-series $X_T = \{(x_t, y_t, z_t) | t = 1,2...T\}$ where T ranges up to 100 was used. The data used in this paper are the logarithms of the indices of monthly flour prices for Buffalo(x_t), Minneapolis (y_t), and Kansas City (z_t) over the period from August 1972 to November 1980. These data are obtained from Tiao and Tsay (1989). The data are divided into two sets: $t = 1,..,60$ as a training set and the rest as a test set. Here we applied 0.3 as learning rate (lr) and 0.6 as momentum(mc). Next, we consider the following network architectures that achieve the best results.

Model 2-2 1: In this neural network model for each pattern there are two input units, two units in the hidden layer and one unit output. To forecast the x_{t+1} value, x_{t-1} and x_t are used as inputs. The network is shown in Figure 1.

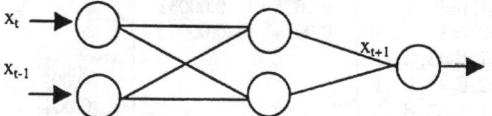

Figure 1. Neural Networks Model 2-2-1.

For each model three architectures were used: neural network, radial basis function and support vector machine respectively.

Model 6-6-1: In this model, there are six input units for each pattern, six hidden nodes, and one output node. To predict the value of x_{t+1}, we used x_{t-1},

x_t, y_{t-1}, y_t, z_{t-1}, and z_t as inputs. The first was used in univariate time series forecasting. This was based on what was suggested by Chakraborty et al. (1992). They concluded that for univariate time series model 2-2-1 is the most appropriate for the data. While for multivariate time series we only use two architecture 6-6-1 and 8-8-1 respectively. These models were selected based on preliminary results by Chakraborty et al.(1992).

4.2 Results

The results of the experiment are given in the following tables.

Table 1. Model 2-2-1 (univariate) testing errors for neural network, RBF and SVR.

	Neural Network epochs = 15000, lr = 0.3, mc = 0.6	Radial Basis Function spread =1	SVR RBF C = 1, p = 2	SVR polynomial C = 0.9 p = 3
Testing error	0.00037	0.00037	0.0171	0.000315

Table 2. Model 6-6-1 (univariate) testing errors for neural network, RBF and SVR.

	Neural network: Epochs= 25000, lr = 0.3, mc = 0.6	Radial Basis Function spread = 20
Testing error	0.000354	0.00083

For this model, we present the results (Table 3) for some different kernel functions, parameters and two different loss functions.

Table 3. SVR Model 6-6-1(univariate) testing errors for some parameter values.

ϵ-Insensitive loss function	SVRRBF	SVR Poly	SVR Sigmoid
C=1, p=2	0.018	0.0145	
C=2, p=2	0.017	0.0152	
C=1, p=3	0.019	0.0162	
C=2 p=3	0.019	0.0164	
C=1, β_0=3, β_1=1			0.0017
C=2, β_0=3, β_1=1			0.0714
C=1, β_0=3, β_1=3			0.0025
C=2, β_0=3, β_1=3			0.0014
Quadratic loss function			
C=1, p=2	0.0017	0.00037	
C=2, p=2	0.00093	0.000314	
C=1, p=3	0.0025	0.00033	
C=2 p=3	0.0014	0.00032	
C=1, β_0=3, β_1=1			
C=2, β_0=3, β_1=1			0.003
C=1, β_0=3, β_1=3			0.0041
C=2, β_0=3, β_1=3			0.0043

The parameter that gives the best results are given by a polynomial SVR, quadratic loss function, C = 0.31, p = 3. Specifically, the testing error is 0.0003115

Model 8-8-1: For these models, special for the SVR, only for the *quadratic* loss function, the results are presented. This loss function produces better results than the ϵ-*Insensitive* loss function.

Table 4. Model 8-8-1(multivariate) testing errors for RBF and SVR.

	Parameter	Testing Error
Neural Network	lr=0.3, mc=0.6, epochs=5000	6.25e-05
Radial Basis Function	Spread =25	2.05e-04
SVR, polynomial kernel, quadratic loss function	C = 2, p = 4	2.206e-05
	C = 1.75, p = 4	2.202e-05 (the best)
	C = 1, p = 3	4.1e-05
	C = 2, p = 3	3.49e-05
	C = 20, p = 3	2.6e-05
	C = 60, p = 3	2.4626e-5
	C = 35, p = 3,	2.5e-5

4.3 Analysis

Based on the results above we can perform some analysis.

Error Comparison: To judge the results intentionally we used MSE of testing. This is done to assure that we use the same calculation method to find the errors. The smaller the MSE value, the better the model is. The smaller error indicates the closeness between the output of the experiment and the actual data. In Table 5, we summarized the testing errors for the three models and only for the best parameter combinations presented in the above table.

Table 5. The best results from each model for Neural Network (NN), RBF and SVR.

	NN	RBF	SVR
Model 2-2-1	0.00037	0.00037	0.000315
Model 6-6-1	0.000354	0.00083	0.0003115
Model 8-8-1	0.0000625	0.0002	0.00002

Here, we see that model 8-8-1 is the best one to predict the flour prices, especially if we use SVR approach. This means that in this case multivariate model forecasts better than univariate. The flour price in one city is affected by the prices in the two other cities. In addition, the price for the next period is affected by the prices in the two previous periods. Chakraborty et al. (1992) showed that *neural network* outperforms *statistical approach*. Here we can conclude that SVR is the best among statistical approach, NN and RBF respectively. From the computational aspect these differences are very significant. In addition, the only available data set is 100 data points. If there were more data for training, the conclusion about the best approach might be different.

Computing time. For small problems with respect to the number of data patterns, RBF and SVR are excellent in computing time. Neural network is time consuming, because of the mechanism of back propagation. For the same architecture, neural network approach needs almost 30 times greater of computing time than RBF and SVR. For a large set of data we might find that RBF and SVR need more time to perform. In RBF we have to save data into the matrix Φ with dimension equal to the square of the input data. Also in SVR we need to work with a kernel matrix with dimension equal to the square of input data.

Parameter values. For SVR if we decrease the degree of polynomial function, to get smaller error we need to increase C. The use of quadratic loss function in SVR can decrease the computing time.

Economic analysis. Now it is important to consider if the MSE difference between NN and SVR for Model 8-8-1 is worth it from the economic point of view. From Table 5 the mean errors (square root of MSE) for both NN and SVR respectively are 0.0079 and 0.0045. The difference is 0.0034. As stated in 4.1, the data used in the experiments is the logarithms of monthly flour price indices. Then, the number 0.0034 is equivalent to $ 1.00. From economic consideration, this value is significantly different.

CONCLUSIONS

From the above study we conclude that (1) to forecast the flour price, multivariate time series is a better model than univariate time series model, (2) Model 8-8-1 is the best model for multivariate time series than 6-6-1, (3) SVR with polynomial kernel function showed the best performance compared to other kernel functions, or to neural network and RBF, (4) choosing an appropriate function for kernel is very important to get the best result in SVR and (5) determining the parameter values of kernel function are also very important in SVR

REFERENCES

Bishop, C.M., (1995), *Neural Networks for Pattern Recognition*, Oxford U. Press.

Chakraborty, K., Mehrotra, K., Mohan, C. K., and Ranka, S., 1992, "Forecasting the Behavior of Multivariate Time Series Using Neural Network", *Neural Networks*, Vol. 5, pp. 961-970.

Haykin, Simon, 1999, *Neural Networks: A comprehensive Foundation*, 2nd edition, Prentice-Hall, Upper saddle River, NJ.

Hines, W. W., and Montgomery, D. C., 1980, *Probability and Statistics in Engineering and Management Science*, 2nd edition, John Wiley and Sons., New York.

Mehrotra, K., Mohan, C. K., and Ranka, S., 1997, *Elements of Artificial Neural Networks*, The MIT Press, Massachusetts.

Müller, K. R., Smola, A.J., Scholkopf, B., Kolmorge, J. Vapnik, V., "Predicting Time Series with Support Vector Machine", *http://svm.first.gmd.de/*

Vapnick, V. N., 1998, *Statistical Learning Theory*, John Wiley and Sons, New York.

ESTIMATION OF THE TEMPERATURE FIELD IN A STEEL PLATE THROUGH SOLUTION OF THE INVERSE HEAT TRANSFER PROBLEM

MARIA M. PETKOVA
Johanson Dielectric Inc., 823 Gateway Center Way, San Diego, CA 92102. Phone: (619) 266-1098, fax: (619) 266-0718

ABSTRACT

This paper deals with the solution of the inverse heat transfer problem, applied to describe the heat transfer in the process of cooling of a thick metal bar during its movement on the intermediate table of a hot strip mill. The optimal dynamical filtration as a method of solution of the inverse heat transfer problem is discussed. The questions of the optimal number of the measurement data and the optimal location of the measurement points are posed. As an example the estimation of the temperature field in a thick metal plate through a robust Kalman filter is discussed.

INTRODUCTION

Mathematical models of distributed parameter plants are used for investigation, design and identification of the parameters of the distributed parameter control systems. Most widespread is the *direct problem*, where the output signal (for example, the temperature field) is determined the from the known input signal and plant characteristics. Less studied is the identification of the input signal from the known output – the so called *inverse problem*. The following classification of the inverse heat transfer problems may be accepted (Petkova, 1998):

- External – from known initial conditions and thermophysical characteristics of the plant to determine the unknown boundary conditions;
- Internal – from known boundary conditions and plant geometry to identify unknown parameters of the model;
- Retrospective – from experimental data for past moments of time it is necessary to define the initial thermal state of the plant;
- Combined – if at the same time is necessary to determine unknown parameters and initial or boundary conditions.

The solution of the inverse problem may be found from the following matrix equation, where X is the state vector and Y – the vector of the experimental data:

$$AX=Y \tag{1}$$

The methods for solution of inverse heat transfer problems may be divided in general to deterministic and stochastic. The stochastic methods are more popular because of the natural random character of the information, used for the solution. One of the most widely spread among them is the Kalman filter (Kraskevitch V. et al, 1988; Haykin, S., et al, 1994; Xie, L. et al, 1994). In this paper the Kalman filter is used for the solution of the inverse heat transfer problem for the reconstruction of the temperature field in a metal plate from experimental data with significant disturbances in the measured temperatures.

MATHEMATICAL MODELING OF HEAT TRANSFER ON THE INTERMEDIATE TABLE

The process of cooling of a steel plate during its movement on the intermediate table of a hot strip mill has been discussed by Laws, W. et al, 1990; Petkova, M., 1998 and others. The technological scheme of this process in the presence of thermal reflecting panels on the intermediate table is presented at Fig. 1. Knowing the thermal state of the metal during this movement is important in order to estimate the effectiveness of the heat retention panels and the actual conservation of heat at this stage. A mathematical model has been created to predict the temperature field in the system metal bar – surrounding panels, under changeable boundary conditions. The effect of different factors, influencing the change in the metal temperature profile, is studied. The major difficulty in investigating the temperature profiles is in the fact that the temperature can only be measured at the surface of the bar and the screens. Besides, these measurements are not quite accurate because of the so called "red spots" on the metal surface and the oxidation layer, introducing big disturbances to the information. In order to restore the actual thermal field inside the body and follow its changes along the intermediate table, a Kalman filter is used to solve the inverse heat transfer problem.

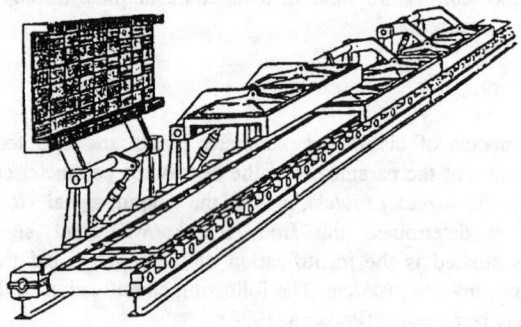

Fig.1. Heat retaining panel system on the intermediate table of the hot strip mill.

The metal cooling process on the intermediate table has two aspects. On the one hand, this is the heat exchange between the surface of the heated metal and the environment and the surfaces of the insulating panels. This is the external problem. The determination of the temperature profiles inside the sheet, as well as inside the surrounding panels, is the so cold internal problem. As in the external heat exchange the major part belongs to radiation because of the high temperatures, for the calculation of the heat fluxes the zonal method is used. Both metal and reflecting panels have been separated to several zones. The resultant fluxes, determined in this way, are set as boundary conditions to the internal heat exchange problem.

To find the temperature distribution inside the metal sheet, the two-dimensional nonlinear transient heat transfer equation is used:

$$c(\theta)\rho(\theta)\frac{\partial \theta(\tau, x, y)}{\partial \tau} = \lambda(\theta)\frac{\partial^2 \theta(\tau, x, y)}{\partial x^2} + \lambda(\theta)\frac{\partial^2 \theta(\tau, x, y)}{\partial y^2} \quad (2)$$

where $\lambda = \lambda_m$ is the heat conduction coefficient of metal, [W/m.K]; $c = c_m$ - specific heat capacity of metal, [J/kg.K]; $\rho = \rho_m$ - density of metal and the panels, [kg/m³]; θ -

temperature of a given node of the investigated cross-section of the plate and the panels, [^{0} C]; x, y - space dimensions, [m]; τ - time , [s].

In this case the initial conditions are unsteady. The boundary conditions are determined by the decision of the external heat transfer problem, written for each zone of the metal plate and the reflecting surfaces, and have the following form :

$$\lambda(\theta)\frac{\partial \theta_m}{\partial x}\bigg|_{x=0} = Q_{3,i}^{res}$$

$$\lambda(\theta)\frac{\partial \theta_m}{\partial x}\bigg|_{x=M} = Q_{4,i}^{res}$$

$$\lambda(\theta)\frac{\partial \theta_m}{\partial y}\bigg|_{Y=0} = Q_{1,i}^{res}$$

$$\lambda(\theta)\frac{\partial \theta_m}{\partial y}\bigg|_{Y=N} = Q_{2,i}^{res}$$

(3)

The right part of these equations represents the resultant heat fluxes for each side of the discussed cross section of the metal bar or insulation screen. M and N are the height and width of the metal bar or the screen. The thermophysical properties are nonlinear functions of temperature. They are determined depending on the mean temperature of the body.

The decision of Eq. (2) is found through the finite difference method, using the explicit numerical technique. It determines the temperature distribution in a given cross-section of the body for each zone of metal or the insulating surfaces.

With the help of this model was traced the behavior of the temperature field in metal under different cooling conditions. The aim of these investigations is to find the optimal geometrical parameters of the insulating chamber, that guarantee preservation of maximum quantity of heat in metal before rolling in the finishing group, as well as to predict dynamically its temperature during its movement on the table.

MULTIDIMENSIONAL TEMPERATURE ESTIMATION IN A METAL BAR

The heat transfer process in the metal plate is described by the non-stationary heat transfer equation (2). Together with the corresponding initial and boundary conditions it can be transformed into a system of linear algebraic equations:

$$\Theta^{\tau+\Delta\tau} = \dot{A}\Theta^{\tau} + Bu \qquad (4)$$

where $\Theta^{\tau+\Delta\tau}$ (1 x n) is the temperature vector at each point of the investigated cross section at the moment $\tau + \Delta\tau$; Θ^{τ} (1 x n) is the temperature vector at the moment τ; A (n x n) - matrix of coefficients in front of the state vector; B (n x m)   - matrix of coefficients in front of the vector of control actions; u (1 x m) is the control actions vector; m – number of the control actions; n – number of the nodes in the discussed cross section.

The choice of a state vector x, control actions u and observable variables y ensures the transition from the model discussed above to a state space description in a discrete form:

$$\begin{cases} x(k+1) = \Phi(k)x(k) + \Gamma(k)u(k) + \eta(k) \\ y(k) = H(k)x(k) + \xi(k) \end{cases} \quad (5)$$

where matrices $\Phi(k)$, $\Gamma(k)$, $H(k)$ are functions of temperature. The temperature vector $\Theta(k)$ is chosen as a state vector $x(k)$, and the vector of the resultant heat fluxes $Q(k)$ is chosen as a control actions vector $u(k)$. The temperatures measured at several points of the body make the vector of observations $y(k)$. $\eta(k)$ and $\xi(k)$ are uncorrelated random disturbances.

The matrix H is determined by the number and position of the measuring points and has dimensions (r x n), where n is the number of nodes in the studied cross-section and the number of elements in the state vector $x(k)$, and r is the number of the measurement points. The matrices Φ and Γ have dimensions (n x n) and (n x m) correspondingly (m is the number of the control actions) and have the following form:

$$\Phi = \begin{bmatrix} 1-2Fo_x-2Fo_y & 2Fo_x & 0 & \cdots & 2Fo_y & 0 \\ Fo_x & 1-2Fo_x-2Fo_y & Fo_x & 0 & \cdots & 0 \\ 0 & Fo_x & 1-2Fo_x-2Fo_y & Fo_x & 0 & 0 \\ \cdots & \cdots & \cdots & \cdots & \cdots & \cdots \\ 0 & \cdots & 0 & Fo_x & 1-2Fo_x-2Fo_y & Fo_x \\ 0 & 2Fo_y & \cdots & 0 & 2Fo_x & 1-2Fo_x-2Fo_y \end{bmatrix}$$

$$\Gamma = \begin{bmatrix} 2Fo_y \frac{\Delta y}{\lambda} & 0 & 2Fo_x \frac{\Delta x}{\lambda} & 0 \\ \cdots & \cdots & \cdots & \cdots \\ 2Fo_y \frac{\Delta y}{\lambda} & 0 & 0 & 0 \\ 0 & 0 & 2Fo_x \frac{\Delta x}{\lambda} & 0 \\ \cdots & \cdots & \cdots & \cdots \\ 0 & 2Fo_y \frac{\Delta y}{\lambda} & 0 & 0 \\ 0 & 2Fo_y \frac{\Delta y}{\lambda} & 0 & 2Fo_x \frac{\Delta x}{\lambda} \end{bmatrix} \quad (6)$$

Here Fo is the number of Fourier.

The state vector $x(k)$ is estimated on the basis of temperature measurements at several points of the body by means of the Kalman Filter:

$$\hat{x}_{k+1/k} = \Phi(k)\hat{x}(k) + \Gamma(k)u(k) \quad (7)$$

$$P_{k+1/k} = \Phi(k)P_{k/k}\Phi^T(k) + Q(k) \quad (8)$$

$$K(k+1) = P_{k+1/k}H^T\left[HP_{k+1//k}H^T + R(k+1)\right]^{-1} \quad (9)$$

$$\hat{x}_{k+1/k+1} = \hat{x}_{k+1/k} + K(k+1)\left[y(k+1) - H\hat{x}_{k+1/k}\right] \quad (10)$$

$$P_{k+1/k+1} = [I-K(k+1)H]P_{k+1/k}[I-K(k+1)H]^T + K(k+1)R(k+1)K^T(k+1) \quad (11)$$

The quality of the filter may be estimated from the following criterion, representing a square form of the measurement error:

$$J(k+1) = \left[Hx_{k+1/k} - y(k+1)\right]^T \left[HP_{k+1/k}H^T + R(k+1)\right]^{-1} \left[Hx_{k+1/k} - y(k+1)\right] \quad (12)$$

An effective improvement of the filter's quality is achieved with the introducing of a scalar coefficient S in Eq. (8):

$$P_{k+1/k} = S(k+1)\left[\Phi(k)P_{k/k}\Phi(k)^T + Q(k)\right] \quad (13)$$

RESULTS AND DISCUSSION

The state vector x(k) is estimated on the basis of temperature measurements at several points of the body by means of the Kalman filter. The object of the investigation is a steel plate with the following dimensions: thickness 0,030 m; width 0,15 m; length 0,30 m. The temperature field is studied in one cross-section, divided to 36 points, i.e. the matrix and vector dimensions are: x(k) - (36x1); u(k) - (4x1); Φ(k) - (36x36); Γ(k) - (36x4). The observation vector y(k) is formed from the temperature measurement data, that comes from thermocouples, placed at several points of the studied cross-section.

Fig. 2 shows the temperature field in this cross-section at the moments of entry and at the end of the zone with the retention panels, reconstructed through the Kalman filter and the described above model. The observation vector includes measurement data from 5 points - y(k+1)=[Θ$_8$ Θ$_{16}$ Θ$_{22}$ Θ$_{28}$ Θ$_{35}$] - and the matrix H(k) has dimensions (5x36).

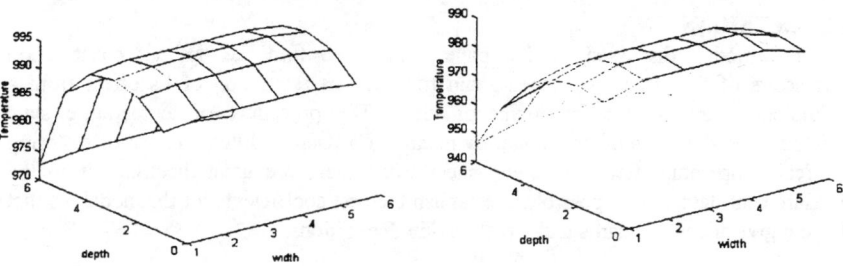

(a) Temperature field at the beginning of the zone   (b) Temperature field at the end of the zone

Fig.2. Two-dimensional fields before and after the heat retention panels zone at one cross-section

Fig. 3 and Fig. 4 show the results from the reconstruction of the temperature field at the centre and at the surface of the metal bar. The experimental data includes measurements with different disturbances (in the first case the mean square deviation of the temperature is σ = 50 °C and in the second case σ = 20 °C). It is important to mention that the accuracy of the filter is different depending on the distance between the measurement point and the estimated one. For nodes located at a greater distance, the information is more unreliable and the estimates are more inaccurate. But as a whole the Kalman filter with the used here modification shows robust performance and is able to give stable solution of the inverse problem

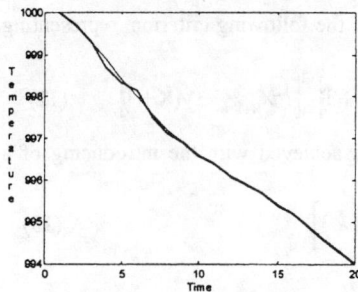
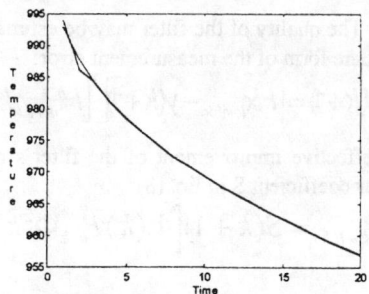

Fig. 3. Real and estimated temperatures at the centre

Fig. 4. Real and estimated temperature at the surface

Another significant side of this discussion is the number and location of the measurement points. The experiments showed that there is an optimal number of sensors, the increasing of which does not improve the accuracy and convergence of the filter. The stable performance of the Kalman filter is additionally improved by the preliminary processing of the experimental data, where by means of diagnostics the false data is eliminated.

The results show that the Kalman filter is capable to cope with noise disturbances of 10%, 20% and even more and restore the temperature field and the unknown boundary conditions with a very good precision.

CONCLUSIONS

The model, discussed in this paper, allows to find the optimal construction parameters of the system of panels and to make an estimation of its effect from a technological and economical point of view. The procedure of temperature state estimation on the basis of this model by means of a Kalman filter, capable of restoring the real temperature field inside the sheet when there are great disturbances in the measurement data, makes possible to establish the best coefficients for the model, so that it could give adequate results under real working conditions.

REFERENCES

Kraskevitch V. E., and . Korbicz, J., 1988, "Suboptimal Kalman filter for distributed parameter systems", *Int. J. Syst. Sci.*, 6, p. 225-234.

Haykin, S., and Liang, L., 1994, "Modified Kalman Filtering", *IEEE Trans. On Signal Process.*, Vol. 5, p. 1239 – 1242.

Laws, W. R., Reed, G. R., Walker, Y. J. and Nickolls, R., 1990, "Heat Conservation During Hot Rolling of Beams and Sections: The Encoscreen", *Iron and Steel Eng.*, , January, p. 37 – 41.

Petkova, M., 1998, "Mathematical Modeling and Control of Thermal Processes in a Hot Strip Mill", Ph.D. Thesis, University of Chemical Technology and Metallurgy, Sofia, Bulgaria.

Xie, L., and Soh, Y. C., 1994, "Robust Kalman Filtering for Uncertain Systems", *Systems and Control Letters*, No 22, p. 123 – 129.

USING SUPPORT VECTOR MACHINES FOR RECOGNIZING SHIFTS IN CORRELATED AND OTHER MANUFACTURING PROCESSES

RATNA BABU CHINNAM, PH.D.
Associate Professor
Industrial & Mfg. Engineering Department
Wayne State University
4815 Fourth Street
Detroit, Michigan 48202, USA
r_chinnam@wayne.edu

VINAY S. KUMAR
Graduate Student
Industrial & Mfg. Engineering Department
Wayne State University
4815 Fourth Street
Detroit, Michigan 48202, USA

ABSTRACT

Traditional statistical process control (SPC) techniques of control charting are not applicable in many process industries due to the fact that data from these facilities are auto-correlated. Several attempts have been made in the literature to extend traditional SPC techniques to deal with auto-correlated parameters. However, these extensions pose several serious limitations. This paper demonstrates that support vector machines can be extremely effective in minimizing both Type-I errors and Type-II errors in both auto-correlated processes and non-correlated processes.

INTRODUCTION

Traditional statistical process control (SPC) techniques of control charting are not applicable in many process industries to identify the presence of assignable causes because data from these facilities are auto-correlated (the value of a parameter is dependent on the previous value of that parameter). When traditional $\bar{X}$ and R or individual and moving-range charts are applied for monitoring correlated process data, they result in a large number of false out-of-control signals [1].

Several attempts have been made in the literature to extend traditional SPC techniques to deal with auto-correlated parameters. Several authors (for example, [2-6]) recommended the use of time series modeling techniques for monitoring correlated process data. This essentially involves fitting the correlated process data to an adequate time series model and applying a traditional control chart to the stream of residuals from the time series model. If the process quality parameter data do not undergo any shifts in the mean, variance, or aurocorrelative structure, the residuals of the time series model will exhibit a mean of approximately zero and non-significant autocorrelations at all lags [7]. The performance of time series control charting techniques was evaluated by several including Wardell, Moskowitz, and Plante [4] and Runger, Willemain, and Prabhu [5]. The results in general show that the performance of these time series modeling techniques is marginal, in particular for detecting small shifts. The literature discusses several machine learning methods based on radial basis function (RBF) networks [1] and multi-layer perceptron (MLP) networks [8-9] to address these problems, with some success. The results are still not impressive. For example, in Cook and Chiu [1], the Type-I error (probability that the method would wrongly declare the process to be out of control or generate a false alarm) for the best model on the Viscosity data set is larger than 50%.

Here, we demonstrate that Support Vector Machines (SVMs) can be extremely effective in minimizing both Type-I errors and Type-II errors (probability that the method will be unable to detect a true shift or trend present in the process) in both auto-correlated processes as well as non-correlated processes.

SUPPORT VECTOR MACHINES

Support Vector Machines are a category of universal feed-forward networks pioneered by Vapnik [10,11]. Bascially, SVM is a *linear machine* with some very nice properties. It is perhaps easiest to explain the behavior of an SVM by starting with the case of separable patterns that could arise in the context of binary pattern classification. In this context, the SVM in essence constructs a decision surface hyperplane that maximizes the separation *margin* between positive and negative examples. It achieves this desirable property by following a principled approach rooted in statistical learning theory. More precisely, SVM is an approximate implementation of the *method of structural risk minimization* [12]. These concepts are concisely explained below.

Suppose the task of a machine is to learn the mapping $x_i \mapsto y_i$. The machine is actually defined by a set of possible mappings $x_i \mapsto f(x, \alpha)$, where the functions $f(x, \alpha)$ themselves are labeled by the adjustable parameters α. The expectation of the test error or "expected risk" for a trained machine is therefore:

$$R(\alpha) = \frac{1}{2}\int |y - f(\mathbf{x},\alpha)| p(\mathbf{x},y) dx dy \qquad (1)$$

This expression is not very useful unless we have an estimate for $p(\mathbf{x},y)$. In the absence of knowledge of $p(\mathbf{x},y)$, one could alternately work with an "empirical risk" measure $R_{emp}(\alpha)$ that can be expressed as:

$$R_{emp}(\alpha) = \frac{1}{2N}\sum_{i=1}^{N}|y_i - f(\mathbf{x}_i,\alpha)| \qquad (2)$$

where N denotes the number of training patterns.

Now choose some η such that $0 \leq \eta \leq 1$. Then, for losses taking these values, with probability $1-\eta$, the following bound holds [11]:

$$R(\alpha) \leq R_{emp}(\alpha) + \sqrt{\frac{h}{N}\left[\log\left(\frac{2N}{h}\right)+1\right] - \frac{1}{N}\log\left(\frac{\eta}{4}\right)} \qquad (3)$$

where h is a non-negative integer called the Vapnik-Chervonenkis (VC) dimension and is a measure of the capacity of the learning machine. In the literature, the right hand side of (3) is referred to as the "risk bound" and the second term on the right hand side is called the "VC confidence". With the knowledge of h, one can easily compute the right hand side. Thus, given several different learning machines and a fixed (sufficiently small) η, selecting the machine that minimizes the right hand side leads to a machine that gives the lowest upper bound on the actual risk. This procedure, in essence, results in a principled method that is essential to the idea of structural risk minimization. In the case of separable patterns, a SVM produces a value of zero for the first term on the right hand side of (3) and minimizes the second term, thus minimizing the overall risk. In the case of nonseparable patterns, we introduce a new set of nonnegative scalar *slack variables* that tolerate misclassification of data points. The cost function that is to be minimized will now carry an extra term weighted by parameter C that penalizes any SVM that generates too many nonseparable patterns.

A notion that is central to the construction of the support vector learning algorithm is the inner-product kernel (denoted by $K(\mathbf{x},\mathbf{x}_i)$) between a "support vector" $\mathbf{x}_i$ and the vector $\mathbf{x}$ drawn from the input space [12]. To keep SVM training algorithms computationally efficient, one has to always choose inner-product kernels that meet the *Mercer's theorem* that arises in functional analysis [11]. The support vectors consist of a small subset of the training data extracted by the algorithm. Depending on how this inner-product kernel is generated, one may construct different learning machines. In particular, one may use the support vector learning algorithm to construct the following types of learning algorithms (among others): Polynomial learning machines, Radial-basis function networks, and Two-layer perceptron networks. Table 1 summarizes the inner-product kernels for these three common types of SVMs.

For the sake of computational convenience, a loss function that is commonly used for optimizing learning machines is the quadratic loss function. However, a least-squares estimator is sensitive to the presence of outliers, and performs poorly when the underlying distribution of the additive noise has a long tail [12]. To overcome these limitations, in constructing a support vector machine for approximating a desired response d, we shall use the following ε-insensitive loss function, originally proposed by [10]:

$$L_\varepsilon(d,y) = \begin{cases} |d-y| - \varepsilon_{\text{Insensitivity}} & \text{for } |d-y| \geq \varepsilon_{\text{Insensitivity}} \\ 0 & \text{otherwise} \end{cases} \qquad (4)$$

where $\varepsilon_{\text{Insensitivity}}$ is a prescribed parameter.

Table 1: Summary of Inner-Product Kernels

Type of Support Vector Machine	Inner-Product Kernel $K(\mathbf{x},\mathbf{x}_i),\ i=1,2,\ldots,N$	Comments
Polynomial learning machine	$(\mathbf{x}^T\mathbf{x}_i + 1)^p$	Power p is specified *a priori* by the user
Radial-basis function network	$\exp\left(-(1/(2\sigma^2))\|\mathbf{x}-\mathbf{x}_i\|^2\right)$	The width σ^2, common to all the kernels, is specified *a priori* by the user
Two-layer perceptron network	$\tanh(\beta_0 \mathbf{x}^T\mathbf{x}_i + \beta_1)$	Mercer's theorem is satisfied only for some values of β_0 and β_1

For a detailed treatment of the subject of SVMs, see [10,11].

SVMS FOR RECOGNIZING SHIFTS IN CORRELATED MANUFACTURING PROCESSES

In this section, we will use the papermaking dataset from Pandit and Wu [13] and the viscosity dataset from Box and Jenkins [14] as representative manufacturing datasets for analysis. These were the same exact datasets studied by Cook and Chiu [1] to evaluate the effectiveness of RBF neural networks for detecting shifts in correlated processes, and hence, can be regarded as benchmark datasets.

Autoregressive processes of lag 1 (i.e., AR(1)) are reported to be quite common in manufacturing processes [15]. Autoregression is a form of regression where the dependent variable is related to past values of itself at varying time lags. An AR(1) process can be represented by the equation

$$X_t = \mu + \phi(X_{t-1} - \mu) + \varepsilon_t \qquad (5)$$

where X_t is the value of the time series at time t, μ is the mean of the data series, X_{t-1} is the value of the time series at time $(t-1)$, ε_t is a normal, independently distributed error term, and ϕ is the autoregressive coefficient restricted to lie between -1 and $+1$.

Cook and Chiu [1] reported that spectral analysis as well as autocorrealation function analysis identified the AR(1) model to be the appropriate model for both the papermaking dataset and the viscosity dataset. They reported the models to be:

Papermaking Data: $X_{P,t} = 32.02 + 0.90(X_{P,t-1} - 32.02) + \varepsilon_{P,t}$; $\varepsilon_{P,t} \sim N(0, \sigma_{P,\varepsilon}^2)$ (6)

Viscosity Data: $X_{V,t} = 9.10 + 0.86(X_{V,t-1} - 9.10) + \varepsilon_{V,t}$; $\varepsilon_{V,t} \sim N(0, \sigma_{V,\varepsilon}^2)$ (7)

Not unlike Cook and Chiu [1], these time series models were used for simulating shifts in the process parameter values for testing the performance of the proposed SVM approach.

Training and Testing Dataset Development

Here, we strictly follow the dataset development procedure discussed in Cook and Chiu [1]. This involves first standardizing the raw datasets to result in datasets with mean 0 and standard deviation ($\tilde{\sigma}_x$) of 1. The resulting time series models are shown below:

Papermaking Data: $\tilde{X}_{P,t} = 0.90\tilde{X}_{P,t-1} + \tilde{\varepsilon}_{P,t}$, $\tilde{\varepsilon}_{P,t} \sim N(0, \tilde{\sigma}_{P,\varepsilon}^2)$ (8)

Viscosity Data: $\tilde{X}_{V,t} = 0.86\tilde{X}_{V,t-1} + \tilde{\varepsilon}_{V,t}$, $\tilde{\varepsilon}_{V,t} \sim N(0, \tilde{\sigma}_{V,\varepsilon}^2)$ (9)

It can be easily shown that

$$\tilde{\sigma}_\varepsilon^2 = \tilde{\sigma}_x^2(1 - \phi^2). \qquad (10)$$

For the papermaking dataset, ϕ of 0.9 with $\tilde{\sigma}_{P,x}$ of 1 will yield a $\tilde{\sigma}_{P,\varepsilon}$ of 0.4359. Similarly, for the viscosity dataset, ϕ of 0.86 with $\tilde{\sigma}_{V,x}$ of 1 will yield a $\tilde{\sigma}_{V,\varepsilon}$ of 0.5103.

Strictly following Cook and Chiu [1], three training datasets were developed for both the papermaking and viscosity series and each contained the original data with no shift in the process mean. In addition, the first training set contained values of the parameter with the process mean shifted by one standard deviation; the second set contained values shifted by one and a half standard deviations; and the third dataset contained values shifted by two standard deviations. The papermaking data training sets included 123 data vectors of nonshifted data and 123 data vectors of shifted data. The test datasets consisted of 35 shifted and 35 nonshifted data vectors. The viscosity training datasets consisted of 227 data vectors of nonshifted data and 227 data vectors of shifted data. The viscosity test datasets consisted of 80 shifted data vectors and 80 nonshifted vectors.

The SVMs were trained to recognize shifts. Cook and Chiu [1] used the following inputs to train their RBF neural networks: the pervious parameter value at time $t-1$ (i.e., $\tilde{X}_{t-1}$), the current parameter value ($\tilde{X}_t$), the time series prediction based on the previous parameter value ($\widehat{\tilde{X}}_t$), the difference between the time series prediction and the parameter value at time t (i.e., $\widehat{\tilde{X}}_t - \tilde{X}_t$), and the differences between the parameter values at time t and time $t-1$ (i.e., $\tilde{X}_t - \tilde{X}_{t-1}$). Obviously, the last two inputs are completely redundant (for all this information is available in the first three inputs). Both RBF networks and SVMs are capable of extracting any such information. Hence, we decided to train the SVMs strictly using the first three inputs alone. Shifts were induced immediately after time $t-1$ so that the parameter values at time t

are potentially shifted. Not unlike Cook and Chiu [1], the SVM output was a single node that signaled the presence or absence of a shift in the process parameter.

SVM Design

We chose to work with the simplest SVMs that used a simple dot product kernel ($K(x_1, x_2) = x_1 \cdot x_2$) and a polynomial kernel of order two ($K(x_1, x_2) = (x_1 \cdot x_2 + 1)^2$) and three ($K(x_1, x_2) = (x_1 \cdot x_2 + 1)^3$). Better results can be potentially achieved by using more complex kernels, however, the objective here is to demonstrate the power of SVMs even while using simple kernel operators, and hence the above choices. All SVM training runs were carried out by setting the loss insensitivity constant $\varepsilon_{Insensitivity}$ at 0.01 (constant that the prediction can deviate from the functional value without being penalized). The maximum number of iterations allowed for convergence is set at 10^6. After preliminary experimentation, the C parameter for the objective function is varied from 0.01 to 167772.16 by selecting 13 levels that follow an approximate geometric series with r set at 4. The actual values are as follows: 0.01, 0.04 (i.e., $0.01r = 0.01 \times 4$), 0.16 (i.e., $0.01r^2 = 0.01 \times 4^2$), ..., 167772.16 (i.e., $0.01r^{12} = 0.01 \times 4^{12}$).

Results

All the SVM implementation results reported here are obtained by implementing *mySVM* software.[1] *mySVM* is based on the *SVMlight* optimization algorithm described in Joachims [16].

Table 2 reports the process state classification accuracy for the papermaking process while using SVMs and compares them to using RBF networks proposed by Cook and Chiu [1]. While the DOT kernel was simply not able deal with this classification problem, the results from the polynomial kernels were superior to the results from the best RBF network. While the Type-I errors (i.e., false alarms) are slightly higher in the case of SVMs (i.e., while dealing with the nonshifted data set), unlike the RBF network, the Type-II errors are zero in the case of SVMs (i.e., while dealing with the shifted data set).

It is also clear that there is strong agreement between the training results and the testing results, indicating no overfitting problems with SVMs. Unfortunately, the training set errors were not reported by Cook and Chiu [1] to discuss the generalization performance of the RBF networks.

Table 3 reports the process state classification accuracy for the viscosity data sets. While the DOT kernel was once again unable to deal with the classification problem, the results from the polynomial kernels were extremely superior to the results from the best RBF networks. This is clear when we examine the case of the Type-I errors associated with the 1 σ case. For this case, the best RBF network resulted in an unacceptable Type-I error of 53.7%, unlike the SVM network with an error of 12.5%. Once again, there are no Type-II errors in the case of SVMs while the error varied from 2.5% to 3.7% in the case of RBF networks. It is also clear that there is strong agreement between the training results and the testing results, indicating no overfitting problems. Once again, the training set errors were not reported by Cook and Chiu [1] to discuss the generalization performance of the RBF networks.

Table 2: Process state classification accuracy using SVMs and RBF networks for papermaking process.

Data Set			Best RBF Network [1]	Best SVM		
				DOT Kernel	Polynomial Kernel	
					2nd Order	3rd Order
Nonshifted	1 σ	Testing	91.4%	45.7%	85.7%	91.4%
	1.5 σ	Testing	100%	57.1%	97.1%	94.3%
	2 σ	Testing	100%	80.0%	97.1%	97.1%
Shifted	1 σ	Testing	91.4%	57.1%	100%	100%
	1.5 σ	Testing	97.1%	57.1%	100%	100%
	2 σ	Testing	91.4%	57.1%	100%	100%

SVMS FOR RECOGNIZING SHIFTS IN NON-CORRELATED MANUFACTURING PROCESSES

In this section, we will use the artificially generated datasets from Smith [9] to study the effectiveness of SVMs for detecting both location and variance shifts. We will compare the results from SVMs with the results of MLP models proposed by Smith [9] and the results from standard Shewhart control charts. The

---

[1] *mySVM* software is developed by Dr. Stefan Rüping of the Department of Computer Science at University of Dortmund, Germany (http://www-ai.cs.uni-dortmund.de).

object is to use a rational subgroup of size 10 generated from a process state to recognize the presence or absence of mean and variance shifts. In particular, Smith [9] developed two data sets with 1500 samples each, labeled A and B, as illustrated in Table 4.

Table 3: Process state classification accuracy using SVMs and RBF networks for viscosity data.

Data Set		Best RBF Network [1]	Best SVM		
			DOT Kernel	Polynomial Kernel	
				2nd Order	3rd Order
Nonshifted	1σ Testing	46.3%	0%	86.3%	87.5%
	1.5σ Testing	88.8%	0%	87.5%	100%
	2σ Testing	98.8%	0%	98.8%	98.8%
Shifted	1σ Testing	96.3%	100%	100%	100%
	1.5σ Testing	97.5%	100%	98.8%	100%
	2σ Testing	97.5%	100%	100%	100%

Table 4. Process states used to generate training and testing data sets described by Smith [9].

Data Set Label	Data Points	Process State	State Label
A	500	$X \sim N(\mu=0; \sigma=1)$	In Control
	500	$X \sim N(\mu=0; \sigma=2)$	Small Variance Shift
	500	$X \sim N(\mu=1; \sigma=1)$	Small Mean Shift
B	500	$X \sim N(\mu=0; \sigma=1)$	In Control
	500	$X \sim N(\mu=0; \sigma=3)$	Large Variance Shift
	500	$X \sim N(\mu=3; \sigma=1)$	Large Mean Shift

Training and Testing Dataset Development

Smith [9] partitioned each of the two data sets by allocating randomly 75% of the data for training and the remaining 25% for testing purposes. Smith reported that efforts to train an MLP network to recognize parameter shifts using raw data alone (i.e., the 10 sample points) were unsuccessful and chose to either augment the 10 sample points with three calculated subgroup sample statistics, the subgroup mean, standard deviation, and range (leading to a 13 dimensional input space) or simply work with the three calculated sample statistics (resulting in a three dimensional input space).

Unlike Smith [9], we chose to strictly work with the three calculated sample statistics for training the SVMs, with the confidence that SVMs will still give better results than those achieved by Smith [9]. Figure 6 reports the properties of the three summary statistics for both the data sets.

Table 5. Process state classification accuracy using SVMs, MLP networks, and Shewhart control charts for data sets from Smith [9].

Data Set		Inputs	Best MLP Network [9]	Shewhart Charts [9]	Best SVM		
					DOT Kernel	Polynomial Kernel	
						2nd Order	3rd Order
A	Testing	Statistics + Data	88.3%	72.5%	N/A	N/A	N/A
		Just Statistics	71.5%	72.5%	84.2%	93.3%	92.7%
B	Testing	Statistics + Data	99.5%	99.5%	N/A	N/A	N/A
		Just Statistics	100%	99.5%	97.1%	99.7%	99.7%

SVM Design

For the reasons explained earlier, we once again chose to work with the simplest SVMs that used a simple dot product kernel ($K(\mathbf{x}_1, \mathbf{x}_2) = \mathbf{x}_1 \cdot \mathbf{x}_2$) and a polynomial kernel of order two ($K(\mathbf{x}_1, \mathbf{x}_2) = (\mathbf{x}_1 \cdot \mathbf{x}_2 + 1)^2$) and three ($K(\mathbf{x}_1, \mathbf{x}_2) = (\mathbf{x}_1 \cdot \mathbf{x}_2 + 1)^3$). All SVM training runs were once again carried out by setting the loss insensitivity $\varepsilon_{Insensitivity}$ at 0.01 and allowing 10^6 iterations for convergence. The C parameter was once again varied as a geometric series with r set at 4 and took on the same values.

Results

The SVM results reported here were also obtained by implementing *mySVM* software. Table 5 reports the process state classification accuracy obtained from using SVMs and compares them to using MLP networks proposed by Smith [9] as well as Shewhart control charts (the results for Shewhart Control chart category came from Smith [9]).

While there exists no real difference between SVMs, MLPs, and Shewhart control charts in the case of large shifts in location and variance (i.e., Data Set *B*), there is significant improvement in classification accuracy while using SVMs in the case of small shifts in location and variance (i.e., Data Set *A*). The results from SVM implementation are significantly better than the results from using an MLP network that utilizes the three summary statistics as well as all the 10 raw data points.

The results demonstrate that a second-order polynomial kernel is adequate to achieve good accuracy (with small Type-I and Type-II errors). While the performance of the polynomial kernel is comparable to the performance of the MLPs and Shewhart control charts (and sometimes even better), overall, the DOT kernel is not able to adequately deal with this non-separable input space.

CONCLUSION

The paper clearly demonstrates that SVMs can be very effective in recognizing shifts in non-correlated as well as correlated manufacturing processes. SVMs deserve serious consideration in developing manufacturing process control and monitoring algorithms. The results achieved are superior to alternate machine learning methods proposed in the literature. In some cases, the errors were reduced by more than 75%.

REFERENCES

[1] Cook, D.F. and Chiu, C. (1998) Using radial basis function neural networks to recognize shifts in correlated manufacturing process parameters. *IIE Transactions*, 30, 227-234.
[2] Alwan, L.C. and Roberts, H.V. (1988) Time-series modeling for statistical process control. *Journal of Business & Economic Statistics*, 6, 87-95.
[3] Harris, T.J. and Ross. W.H. (1991) Statistical process control procedures for correlated observations. *Canadian Journal of Chemical Engineering*, 69, 48-57.
[4] Wardell, D.G., Moskowitz, H., and Plante, R.D. (1994) Run-length distributions of special-cause control charts for correlated processes (with discussion). *Technometrics*, 36, 3-27.
[5] Runger, G.C., Willemain, T.R., and Prabhu, S. (1995) Average run lengths for CUSUM control charts applied to residuals. *Communications in Statistics-Theory and Methods*, 24, 273-282.
[6] Box, G.E.P. and Luceno, A. (1997) *Statistical Control by Monitoring and Feedback Adjustment*. John Wiley and Sons, New York.
[7] Yourstone, S.A. and Montgomery, D.C. (1989) A time-series approach to discrete real-time process quality control. *Quality and Reliability Engineering International*, 5, 309-317.
[8] Pugh, G.A. (1991) A comparison of neural networks to SPC charts. *Computers and Industrial Engineering*, 21, 253-255.
[9] Smith, A.E. (1994) X-bar and R control chart interpretation using neural computing. *International Journal of Production Research*, 32, 309-320.
[10] Vapnik, V. (1995) *The Nature of Statistical Learning Theory*, Springer-Verlag, New York.
[11] Vapnik, V. (1998) *Statistical Learning Theory*, John Wiley and Sons, New York.
[12] Haykin, S. (1999) *Neural Networks: A Comprehensive Foundation*, 2nd edn, Prentice Hall, NJ: Upper Saddle River.
[13] Pandit, S.M. and Wu, S. (1983) *Time Series and Systems Analysis with Applications*. John Wiley and Sons, New York.
[14] Box, G.E.P., and Jenkins, G.M. (1976) *Time Series Analysis: Forecasting and Control*, Holden-Day, Oakland, CA.
[15] Longnecker, M.T. and Ryan, T.P. (1993) Charting correlated process data, Working paper, Department of Statistics, Texas A&M University, College Station, TX 77843.
[16] Joachims, T. (1999) Making large-scale SVM learning practical. In Schölkopf, B., Burges, C. J. C., and Smola, A. J. editors, Advances in Kernel Methods–Support Vector Learning, pages 41-56, Cambridge, MA: MIT Press.

IMPROVING ACCURACY OF NEAREST NEIGHBOR ALGORITHM IN HIGHWAY ACCIDENT PREDICTION

CHAKARIDA NUKOOLKIT
Dept. of Computer Science
The University of Alabama
Tuscaloosa, AL 35487-0290
E-mail: nukoo001@bama.ua.edu

HUI-CHUAN CHEN
Dept. of Computer Science,
The University of Alabama
Tuscaloosa, AL 35487-0290
E-mail:chen@cs.ua.edu

DAVID BROWN
Dept. of Computer Science
The University of Alabama
Tuscaloosa, AL 35487-0290
E-mail: brown@cs.ua.edu

ABSTRACT

A prediction model using nearest neighbor classification is proposed as part of knowledge discovery enhancement of a traffic safety information system. There are more than two hundred variables in this database encoded in categorical formats. The first step of data analysis starts with reducing the number of dimensions of data by identifying the most significant variables. Then two models of the nearest neighbor were studied. The first one used Euclidean distance combined with k-mean clustering. The second one used Value Difference Metric (VDM) distance combined with k-mode clustering. Unfortunately, these two models did not perform as well as we expected. Therefore, an alternative approach is proposed. This new method uses a decision tree to assist in clustering the data. Then a new distance function is developed which uses the digit patterns of variables in each cluster. The computational results show that the proposed method yields higher accuracy in prediction than the models with Euclidean and VDM distance function.

Keywords: nearest neighbor classification, distance function, decision tree, data transformation, knowledge discovery.

1 INTRODUCTION

The objective of this paper is to present an effective and efficient nearest neighbor prediction model that automatically predicts whether a car crash will have either an *injury* or *non-injury* outcome. The accuracy of dangerous crash pattern prediction is crucial in the development of traffic safety controls, preventive regulations, and other crash countermeasures [13, 14, 22]. In this study the data comes from a subset of Alabama interstate alcohol-related crashes of the year 2000. There are 2000 training records and 1000 testing records.

This paper discusses the challenges of improving prediction accuracy of the nearest neighbor model in a categorical database. To overcome the categorical issue, the first model uses a frequency based data transformation to transform categorical data to numerical format. The transformed data is ready for the nearest neighbor model with Euclidean distance function[1, 8, 19], which later is combined with k-mean clustering. The second model uses a VDM distance function[17, 18, 21] to deal with categorical data directly. This model also makes used of k-mode clustering. However, these two models still do not give high accuracy prediction. Thus, an alternative nearest neighbor prediction model is proposed. This model uses a decision tree to aid in clustering. Then the distance function is developed based on the digit patterns of variables in each cluster.

2 METHODOLOGY

2.1 Nearest Neighbor Prediction Model

The nearest neighbor analysis is a classification method [1, 18, 19, 20] in which the class of an unknown record is assigned after comparisons between this unknown record and all known records (or called training data) in data repository are made. The class of the unknown record will be the same as the class of the most similar record in the data repository. The degree of similarity between different records is determined by a function called the *distance function*. That is, the records that are the most similar should have the smallest distance between them. There are many challenges in the nearest neighbor analysis. The first major challenge is the computational burden during the classification phase, due to the large number of comparisons needed, which is proportional to the number of records in the data repository. The second challenge lies in finding the appropriate distance function for the particular problem domain, especially when the records consist of categorical features.

With these challenges in mind, we proposed an approach consisting of four stages: (1) Perform the Cramer's V Coefficient test [23] to identify significant variables that cause injury crashes thereby reducing the dimensions of the data needed for the classification. (2) Preprocess categorical variables by transforming their codes to numerical values suitable for distance calculation. (3) Cluster training data into groups to reduce unnecessary comparisons. (4) Calculate distance between the unknown record and the cluster centers and assign the class for prediction.

2.2 Variable Reductions by Cramer's V Coefficient Test

The number of injuries (V20) is used as the target variable in this study. A value of zero in V20 corresponds to the *non-injury* class, while a positive value corresponds to the *injury* class. Initially, there are 220 variables in the database; however, based on the previous reports on the CARE web pages [2, 3, 4, 5, 6, 7], a total of 105 relevant causal variables were identified. To ensure that the nearest neighbor prediction model will not be confused with irrelevant information during the distance calculation stage, we strive to further reduce the number of causal variables. This is done by performing Cramer's V Coefficient test[23] with a threshold value of 0.4 for each causal variable. As a result, the number of causal variables was reduced from 105 to 20.

2.3 Categorical Data Transformation

Data transformation is one of the most important steps in the knowledge discovery process. It involves transforming data into a suitable format for a particular analysis tool [9]. In our study, the raw data of selected variables were encoded in categorical codes in the range from code 0 to code 99. The objective of the proposed data transformation method is to properly transform each variable with its codes from [0,99] to numerical values between [0,1] for nearest neighbor analysis with a Euclidean distance function.

The basic idea behind the proposed transformation method [12, 14] is to represent the categorical code of a particular variable with a numerical value derived from its relative frequency between *injury* and *non-injury* outcomes.

2.4 Distance Calculation

Distance is a proximity measure between objects or records. The more likeness two records share, the less distance there is between them. On the other hand, the records that share nothing in common have the furthest distance between them. Theoretically, there are several ways to define distance between two objects [1, 19].

2.4.1 Euclidean Distance Function with K-Mean Clustering

One of the most popular distance functions is called Euclidean distance, in which the distance is defined as:

$$D_{XY} = \sqrt{\sum_{i=1}^{n}(X_i - Y_i)^2} \qquad (1)$$

Where D_{XY} = Euclidean distance   n = the number of features in a record
X_i, Y_i = values of the ith-feature of an unknown and a known record respectively

Originally, our nearest neighbor prediction model used Euclidean distance as its distance function because of its popularity. However, this model did not perform well. It yielded prediction errors in the range of 30-40 percent as seen in Table 1. Several variations of class assignment themes [1], such as N-nearest neighbor with majority voting and inverse weight voting (i.e., giving more creditability to the nearer neighbor), were also considered. However, these additional strategies improved the prediction performance only slightly.

Another data mining strategy named k-mean clustering [1, 8, 19] was coupled into the prediction model. K-mean clustering is a clustering method in which each record is assigned to a cluster that has the nearest centroid. The k-mean clustering helped improve the efficiency during distance calculation where there was only one comparison to each cluster center rather than to each record in the training set. However, it did not help improve the prediction accuracy for our problem domain as we expected (Table 1). We need to search for another distance function other than the Euclidean distance because the Euclidean distance assumes that the importances of each feature are independent. However, our highway accident data set does not conform to this assumption.

2.4.2 Value Difference Metric Distance Function with K-Mode Clustering

Another distance function suitable to categorical data is called Value Difference Metric (VDM). VDM originated in 1986 [18] as an alternative distance function to directly handle categorical attributes in memory-based reasoning [17, 21]. VDM defines the distance between two attributes of two patterns (D_{XmYm}) as the sum of the value differences across all attributes as following:

$$D_{XmYm} = \sum_{i=1}^{k}\left|P(C_i \mid X_m) - P(C_i \mid Y_m)\right| \qquad (2)$$

Where k = the number of classes C_i = the ith-class
X_m, Y_m = value of the mth-attribute of two patterns
$P(C_i|X_m)$, $P(C_i|Y_m)$ = probability of class C_i given that the attribute has value X_m and Y_m respectively.

In order to reduce computation efforts, we also coupled k-mode clustering with the VDM distance. The k-mode clustering algorithm [10, 11] is the extension of the k-mean algorithm. The k-mode clustering works with categorical variables and use the mode values as the cluster centers. Again, this VDM with k-mode clustering does not improve prediction accuracy as much as we expected (the error rate is still around 35-40%).

2.4.3 A New Distance Function Based on Decision Tree Clustering

Since a decision tree can be created easily with the categorical data, the records that fall into the same branch usually have the same properties. The records that fall into the same sub-tree are likely to have similar properties. Therefore, the tree structure can be a good indicator of the similarity (or dissimilarity) of records.

To generate a decision tree, we have to divide the numerical codes currently in the range [0, 1] to proper subsets. To do so, we arrange their numerical codes in ascending order. Then we group them into four subsets by locating the cut off points where the numerical code has a big gap from the current numerical code to the next numerical code. With these rules in mind, we are able to group all numerical codes into 4 subsets, named digit i, where $i=1, 4$. The higher the digit value, the more likely this subset of variable codes is toward *injury* crash outcomes. Once the numerical codes are in proper digit i format, a decision tree based on an algorithm named C4.5 [15, 16] is generated. Based on this decision tree, the given training data will be grouped into several clusters. The number of clusters will grow proportional to the number of major branches in the decision tree. Figure 1 shows a sub-tree from the generated decision tree. Within each leaf node, the first number represents the number of cases that resulted in an *injury* crash outcome. The second number represents the number of cases that resulted in a *non-injury* crash outcome.

From Figure 1, all records that fall into this sub-tree are considered in cluster 1, and each record is encoded in 7 digit patterns:[v39-v15-v31-v36-v239-v24-v38]. The value of each digit is taken from the value of a particular variable on the branch of this sub-tree. Although there is another variable, v17, in this cluster, this variable will not be included in the bit patterns because it is not on the main branch of this sub-tree. Here the digit pattern corresponding to the longest branch of the sub-tree will be considered as a cluster center. An unseen record will be encoded relative to the digit pattern of the longest branch of each sub-tree. The leftmost digit, v39, will act as the most significant digit. For example, assume an unseen record is encoded as 1-1-3-3-3-2-1. Since we let the left most digit (v39) act as the most-significant digit, this record can be encoded as 1133321.

In order to catch the detail variation in the cluster, the distances between the unknown record and the known records in the cluster are computed, instead of computing only the distance to the cluster centers. During the training phase, all training records in each cluster are also encoded in this format. Assume there are three records in training data that fall into this cluster, and their digit patterns are 1133321 (*p1*), 113322 (*p2*), and 1133232 (*p3*). Now in the classification phase, we compare distances between the above unseen record with these three records in the cluster. The unseen pattern will lead to the distance zero (absolute value of 1133321-1133321) relative to *p1*, distance of one (absolute value of 1133321-1133322) relative to *p2*, and a distance of 89 (absolute value of 1133321-1133232) relative to *p3*. According to nearest neighbor rule, we should predict the class of the unseen record as *non-injury* similar to the class of *p1*, which has the nearest distance.

In general, once the distance of each pattern is calculated, the class of the unseen record will be assigned according to the class of the record in training data, which results in the nearest distance. In case there is a tie between several training records, for example, one record points to *injury*, the other points to *non-injury*, the final class of the unseen record will be declared based on the value of other variables that are not included in the distance digit patterns of each training record involved in the tie.

3 COMPUTATIONAL RESULT

Let N represent the total number of records in the training data set and n represent the maximum number of records in a particular cluster. It is noted that $N \gg n$. Let m represent the total number of clusters. We can describe accuracy and efficiency of our variations of nearest neighbor prediction models in Table 1. The results show that our proposed nearest neighbor with decision tree clustering performs best among all other general nearest neighbor algorithms mentioned in [1, 18, 19]. In terms of efficiency, the proposed model is much more efficient than the general nearest neighbor prediction methods.

Table 1. Performance of Various Nearest Neighbor Prediction Models

Method	Prediction Error (%)	Number of Comparisons During Classification
Euclidean distance with 1 vote	42	O(N)
Euclidean distance with 3 majority vote	40	O(N)
Euclidean distance with 1 vote with k-mean clustering (k=100)	37	O(m)
Euclidean distance with 3 majority vote with k-mean clustering (k=100)	33	O(m)
VDM distance with 1 vote	40	O(N)
VDM distance with 3 majority vote	38.5	O(N)
VDM distance with 1 vote with k-mode clustering (k=100)	42	O(m)
VDM distance with 3 majority vote with k-mode clustering (k=100)	45	O(m)
Proposed distance function with decision tree clustering	19	O(m+n)

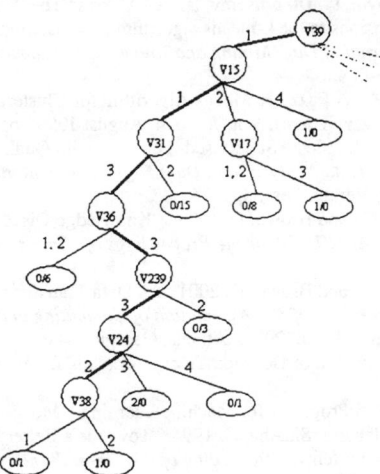

Figure1. A sub-tree depicts cluster 1

4 CONCLUSIONS

The proposed nearest neighbor algorithm relies on decision tree clustering. Each cluster corresponds to one main branch in the tree. The distance function is derived from using the digit sequence of variables in that particular branch. The distance is calculated by taking the absolute difference between the two records under consideration. When a new unknown record comes in, it is compared to each digit pattern of each main branch in a sub-tree to determine the nearest cluster, which the unknown record belongs to. Then that record is compared to all records inside the nearest cluster. This method proves to give more accurate prediction results in our data set than the traditional Euclidean and new VDM distance function. The proposed method also saves a great number of unnecessary comparisons from unrelated clusters. Hence, the proposed method improves both accuracy and computational efficiency for the nearest neighbor classification.

5 REFERENCES

[1] Berry, M. and Linoff, G. 1997. "Data Mining Techniques for Marketing, Sales, and CustomerSupport" John Wiley & Sons, Inc.
[2] Brown, D. 1992. "Comparative analysis of crashes in which causal driver did not have legitimate license", http://care.cs.ua.edu/care/reportImpLic.html.
[3] Brown, D. 1993. "Three-day weekend holiday study: summary of finding", http://care.cs.ua.edu/care/Report3Day.html.
[4] Brown, D. 1994. "Comparative analysis of Mobile County crashes (1994) alcohol/drug proxy: single vehicle nighttime crashes", http://care.cs.ua.edu/care/ReportCityStudy.html.
[5] Brown, D. 1995. "Special study-youth and alcohol", http://care.cs.ua.edu/care/ReportYthAlc.html.
[6] Brown, D. 1997. "Basic principles of problem identification and evaluation using crash records", http://care.cs.ua.edu/care/ReportProbID.html.
[7] Department of Computer Science, University of Alabama. 1999. *The CARE official web site.* http://care.cs.ua.edu
[8] Fasulo, D. 1999. "An Analysis of Recent Work on Clustering Algorithms," Technical Report #01-03-02, Department of Computer Science & Engineering, University of Washington., Seatle, WA
[9] Fayyad, U.M, Piatesky-Shapiro, G., and Smyth P., "From Data Mining to Knowledge Discovery: An Overview." *Advances in Knowledge Discovery and Data Mining*, editors, Fayyad, U. M., Piatesky-Shapiro, G., Smyth, P., Uthurusamy, R. AAAI Press/The MIT Press, pp.1-34.
[10] Huang, Z., 1998. "Extensions to the k-means algorithm for clustering large data sets with categorical values," *Journal of Data Mining and Knowledge Discovery*, Vol. 2, No. 3, September 1998. pp. 283-304.
[11] Huang, Z., Ng, M., 1999. " A Fuzzy K-Modes Algorithm for Clustering Categorical Data," *IEEE Transactions on Fuzzy Systems*, Vol. 7, No. 4, August 1999. pp. 446-452.
[12] Kauderer H., and Mucha H. 1997. "Supervised Learning with Qualitative and Mixed Attributes." *Classification, Data Analysis, and Data Highways*, editors, Balderjahn, I., Mathar, R., Schader, M., Springer-Verlag Press, pp. 374-382.
[13] Nukoolkit, C., Chen, H. C., and Brown, D. 1999. "Knowledge Discovery for Injury Reduction from the CARE system," *ANNIE '99 for the Proceedings of Smart Engineering System Design*, Vol. 9, pp. 477-482.
[14] Nukoolkit, C., Chen, H. C., and Brown, D. 2001. "A Data Transformation Technique for Car Injury Prediction.," *Proceedings of 39th Association of Computing and Machinery--South Eastern Conference*, March 15-16, 2001, Athens, Georgia.
[15] Quinlan, J. R. 1986. "Induction of Decision Trees", *Machine Learning*. Kluwer Academic Publishers. pp. 81-106.
[16] Quinlan, J. R. 1993. "C4.5:Programs for Machine Learning," Morgan Kaufmann Publishers.
[17] Rachlin, J., Kasif, S., Salzbertg, S., Aha, D. 1994. "Towards a Better Understanding of Memory-BasedReasoning Systems," *Proceeding of the Eleventh International Conference on Machine Learning*, pp. 242-250.
[18] Stanfill, C., Waltz, D. 1986. "Toward Memory-Based Reasoning," *Communications of the ACM*, Vol.29, No. 12, December 1986, pp. 1213-1228.
[19] Webb, A. 1999. Statistical Pattern Recognition. Arnold Publishers.
[20] Weiss, S. and Kulikowski, C. 1991. *Computer Systems That Learn: Classification and Prediction Methods from Statistics, Neural Nets, Machine Learning, and Expert Systems.* Morgan Kaufmann Publishers, Inc.
[21] Wilson, D., Martinez, T., 1997. "Improved Heterogeneous Distance Function," *Journal of Artificial Intelligence Research* Vol.6, pp.1-34.
[22] Yang, W., Chen, H. C., and Brown, D. 1999. "Detection Safer Driving Patterns by a Neural Network Approach." *ANNIE '99 for the Proceedings of Smart Engineering System Design*, Vol. 9, pp. 839-844.
[23] Zembowicz, R. and Zytkow, J. M. 1996. "From Contingency Tables to Various Forms of Knowledge in Database.," *Advances in Knowledge Discovery and Data Mining*, editors, Fayyad, U. M., Piatetsky-Shapiro, G., Smyth, P., Uthurusamy, R. AAAI Press/The MIT Press, pp.329-349.

SELECTION OF KEY DATA VARIABLES USING SENSITIVITY ANALYSIS TO PREDICT JET ENGINE MAINTENANCE REMOVALS WITH NEURAL NETWORKS

GARY R. WECKMAN
Department of Mechanical and
Industrial Engineering
Texas A&M University-Kingsville
Kingsville, Texas 78363

ROBERT A. MCLAUCHLAN
Department of Mechanical and
Industrial Engineering
Texas A&M University-Kingsville
Kingsville, Texas 78363

ABSTRACT

In this investigation Neural Networks were used to predict removal patterns of jet engines. A three-layered neural network with backpropagation algorithm was used to predict the jet engine removal pattern for maintenance. Sensitivity Analysis was used to extract the cause and effect relationship between the inputs and outputs of the network to improve the model's efficiency. The removal pattern, a time series measured in monthly periods, was predicted from actual field data based on engine delivery, usage, age, and operating environment. The collected information was divided into a training dataset and a test dataset. A comparison of these results was made to a more traditional time series analysis using ARIMA (autoregressive integrated moving average) model.

INTRODUCTION

Today, reliability is the key to many financial decisions throughout the aviation industry. In fact, reliability considerations are becoming an increasingly popular tool in forecasting resource requirements for various airlines. Because of the financial impact, it is a high priority agenda item for engineering departments and other affected organizations within airlines (Kececioglu, 1993). It is critical for airlines to understand how an engine will age towards its mature time-on-wing (TOW). The ability to forecast this is key in determining budgets, shop capacities, spare engine/module investments and spare parts inventories.

In today's highly competitive market, airlines must find more effective ways of reducing costs, since the expense of engine spare parts and restoration are steadily increasing. Economic pressures are pushing aircraft maintenance work in new directions around the globe. Airlines are desperate to find ways to maintain profitability. In most cases, the focus has shifted to aircraft maintenance, which currently represents about 20% of the total cost of running an airline. The direct maintenance costs (DMC) effect on direct operating costs (DOC) is minimal, about 5%. Although it does have a direct impact on fuel consumption (27%), and is therefore a significant variable in the controllable costs, which should be closely examined for economic advantages. Other categories, such as fees, flight crews, and insurance, tend to be a more fixed operating cost (Kleinert, 1990).

Neural networks are massively parallel, distributed processing systems that can continuously improve their performance via dynamic learning (Hagan et al. 1996; Lin and Lee, 1996 and Zurada, 1992). They can provide successful mechanisms for (a) associative memory, (b) pattern recognition, and (c) abstraction. Neural network architectures can provide a feasible approach to the design and implementation of pattern recognition based systems for intelligent control and model identification. In multilayer perceptron form, these architectures typically use error based dynamic learning laws such as backpropagation (Lin and Lee, 1996 and Zurada, 1992). NN have more recently begun to emerge as an entirely new approach for the modeling of adaptive, distributed, and mostly nonlinear systems. Neural Computers have opened the door to many applications which are difficult for conventional computers to carry out. NN are suited for applications involving complex systems. When applied correctly, a neural or adaptive system can outperform other methods (Principe et al., 2000). With this thought in mind, it was decided to use NN as a tool for forecasting jet engine removals.

Engine removal data was collected from an airline database with a substantial sample size of removals (Note: Due to proprietary reasons the data has been masked). The removal data was analyzed in equal monthly time periods. The jet engine removal estimation problem was set up as a supervised learning, input-to-output mapping problem. With supervised learning, the network is able to learn from the input and the error (the difference between the output and the desired response). The key variables used for predicting the engine removals was the number of engines delivered, cumulative engines delivered, engine flying hours (EFH), delta from previous in EFH, average daily maximum temperature, average precipitation, and shop visit removals with a 12 month lead.

SENSITIVITY

One of the possible ways to preprocess the data for a NN model is by using lags, moving averages, or by subtracting, adding, dividing, and/or multiplication of key variables. In this case, the basic input data was preprocessed using a number of various combinations along with a 12 and 24 month lag and a 12 month moving average. To test the NN model 80% of the data was used to train the model and 20% of the data was saved for testing the effectiveness of this forecasting network.

Sensitivity analysis is a method for extracting the cause and effect relationship between the inputs and outputs of the network. After training a neural network the sensitivity analysis feature of NeuroSolutions can be used to analyze the effect that each of the network inputs has on the network output. Each input channel to the network was varied between its mean +/- 1 standard deviation, while all other inputs were fixed at their respective means. The network output is computed and the corresponding change in the output(s) is reported. The input channels that produce low sensitivity values can be considered insignificant and can most often be removed from the network. This will reduce the size of the network, which in turn reduces the complexity and the training time, typically improving the network performance. This process is repeated for each input variable. A report is generated which summarizes the variation of each output with respect to the variation in each input as shown in Figure 1.

The selection of inputs (Number of Processing Elements) was based on the results of the Sensitivity Analysis. Different models were created reducing the number of inputs by removing the least sensitive variables from the model. The Neural Network design for estimating jet engine removals, using the Multilayer Perceptron (MLP) as the architecture, can be summarized as follows:

Multilayer Perceptron: Input Processing Elements = 7 to 60
                          Output Processing Elements = 1
                          Exemplars = 73 to 85
Hidden Layer: Processing Elements (PEs) = 7 to 43
                          PE Transfer Function = TanhAxon
                          Learning rule = Momentum with
                          Parameter = 0.7 and Step Size = 1.0
Output Layer: PE Transfer Function = TahnAxon
                          Learning rule = Momentum with
                          Parameter = 0.7 and Step Size = 1.0
Maximum Number of Epochs = 1000

In this work, the NeuroSolutions V3.0 Educator Edition software package (1998) was used to implement this design. The design of the NN is an extension of work developed in conjuction with Siddiqui (2000). Theory of Cross Validation was not checked because only 1000 Epochs were enough to train the network. For example, in the training phase for Model D (PEs = 8) the mean square error (MSE) measure went from approximately 0.2000 to 0.0036 (see appendix).

ARIMA MODEL

To test the ability of the neural network model, a comparison was made between it and a more traditional model – ARIMA. A powerful but flexible approach used to analyze complex systems operating in dynamic operational and environmental conditions is the Box - Jenkins models. These methodologies can be used to model the wearout process, actual failure times, times between failures, interactions between failure times and maintenance times of complex systems. Generally, a time series generated by such a complex system will be a non-stationary process made up of a combination of deterministic trends, periodicies, random fluctuations and other interventions. Time series characteristics can be generalized into two states (Box 1994):
- stationary: process remains in equilibrium near the constant mean
- nonstationary: no natural constant mean over time

In most situations, the behavior of the process needs to be analyzed in a stationary state. The technique uses differencing to reduce a nonstationary stochastic time series to a stationary one. This model is based on successive values that are highly dependent, and generated from a series of independent "shocks" a_t, which are a sequence of random variables $a_t, a_{t-1}, a_{t-2}, \ldots$ called a white noise process. The white noise is transformed to z_t by a linear filter which takes the weighted sum of previous random shocks so that $z_t = \mu + a_t + \psi_1 a_{t-1} + \psi_2 a_{t-2} + \ldots = \mu + \psi(B)a_t$ where μ represents the "level" of the process and $\psi(B)$ is the transfer function that transforms a_t into z_t. The ARIMA process of order (p,d,q) is characterized where d is the number of unit roots and the dth difference is required for the process to be stationary. An ARIMA model generally used for a seasonal time series is defined by:

$$\phi(B)\Phi(B^s)\nabla^d\nabla_s^D Z_t = \delta + \theta(B)\Theta(B^s)e_t \qquad [1]$$

where B^s is the backshift operator defined as $Bz_t = z_{t-1}$
$\theta(B)$ and $\Theta(B^s)$ are the seasonal AR and MA operators
D and ∇ are the seasonal and backward difference operator
where $\nabla z_t = z_t - z_{t-1} = (1-B)z_t$ and $[e_t]$ is the white noise process

The abbreviated form for the model is ARIMA (p,d,q) x (P,D,Q)s where s = number of periods (seasonal components). A key step in the ARIMA analysis is model identification. Identification of the models was determined by analyzing the Autocorrelation and Partial Autocorrelation functions.

In analyzing the data, the Autocorrelation function indicated that the data was not stationary. The initial nonstationary process was transformed into a stationary process after one differencing operation. After the data was differenced the Autocorrelation and Partial Autocorrelation functions was analyzed to see what type of ARIMA model best fits the data. Using the characteristics of these two functions, an ARIMA model was chosen based on Box-Jenkins criteria. The model chosen was the ARIMA (0,1,1) based on a differencing of 1 and moving average of 1.

PERFORMANCE PREDICTION RESULTS

The models for both the NN and ARIMA were trained and tested. Mean square error (MSE) was used for both methods in determining the model's ability for both the Training and Testing (See Table 1). The NN models were selected according to the number of inputs (Processing Elements), which have been varied according to the result of Sensitivity Analysis. Overall, the NN models during training resulted in substantial lower MSE values (range of 0.0013 to 0.0193) than the ARIMA model (4.2693). However, the ARIMA model was closer to the NN models in the testing phase. The testing resulted in MSE values in the range of 10.2 to 21.6. Figures 2 and 3 illustrated actual versus forecasted removal results for the NN and ARIMA models.

CONCLUSIONS

These results indicate that the neural network can be used to predict jet engine removals. Based on the above measurements, the NN models with the least number of PEs (Model D) forecasted removals with the least error (MSE). By performing the Sensitivity Analysis on all the inputs it was much easier to determine which of the inputs contributed most to the required output. This reduced the size of the model and the amount of data preparation. In this case, the smaller model was both more efficient and the most accurate. Some modifications could possibly enhance the accuracy of the model by incorporating some additional data variables such as engine workscope level.

The accuracy for the ARIMA models was reasonably capable of forecasting the removals but was higher than the 'best' NN model. This was in part due to the fact that this methodology 'fits' the testing (historical) data without fully understanding the underlying process. In this case the pattern is difficult to identify by the ARIMA model due to hidden phenomenon in the series. This could be an acceptable model, when reasonable accuracy is sufficient, or when the required manpower and capability for more detailed analysis is not available. In spite of this, it is typically better than other methods used today.

For the aviation industry, this level of accuracy would represent significant savings. Present day feelings among airlines agree that current methodologies provide inadequate forecasting of engine removals. If the improvement in forecasting accuracy could make the difference of ordering an additional spare engine, the company could avoid spending somewhere between $5-10 million dollars. This money could be used more effectively on other projects with greater potential earnings.

REFERENCES

Box, G.P., Jenkins, G.M., and Reinsel, G.C., 1994, *Time Series Analysis: Forecasting and Control*, Prentice Hall, Englewood Cliffs, New Jersey.

Hagan, M. T., Demuth, H. B., and Beale, M., 1996, *Neural Network Design*, PWS, Boston.

Kececioglu, D., 1993, *Reliability and Life Testing Handbook: Volume I & II*, Prentice Hall, Inc., New Jersey.

Kleinert, G., 1990, "The Economics of High Thrust Turbofan Maintenance", *Aircraft Maintenance World*, pp. 17-19.

Lin, C., Lee, C., 1996, *Neural Fuzzy Systems*, (Chapters 9, 10) Prentice-Hall, Upper Saddle River, New Jersey.

NeuroSolutions: The Neural Network Simulation Environment, Version 3.0, http://www.nd.com/, 1998, Neuro- Dimension, Inc., Gainesville, FL.

Principe, J.C., Euliano, N.R., and Lefebvre, W.C., 2000, *Neural and Adaptive Systems: Fundamentals through Simulations*, John Wiley & Sons, Inc.

Siddiqui, D. S., 2000, "Predicting Jet Engine Maintenance Removals Using Neural Networks", Graduate Research Project, Texas A&M University-Kingsville, Kingsville, Texas.

Zurada, J., 1992, *Introduction to Artificial Neural Systems*, West, St. Paul, MN.

APPENDIX

Table 1: Actual versus Forecasted Removals

Methodology		MSE (Training)	MSE (Testing)
NEURAL	Model A (PE = 19)	0.0039	14.1360
NETWORKS	Model B (PE = 23)	0.0021	12.5038
	Model C (PE = 47)	0.0013	19.9567
	Model D (PE = 08)	0.0036	10.2043
	Model E (PE = 13)	0.0118	11.7582
	Model F (PE = 15)	0.0193	12.9337
	Model G (PE = 55)	0.0046	21.5819
	Model H (PE = 11)	0.0094	13.5520
ARIMA	ARIMA (0,1,1)	4.2693	15.5057

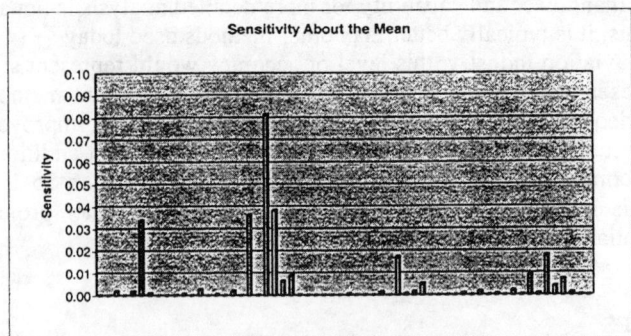

Figure 1: Sensitivity Analysis Report

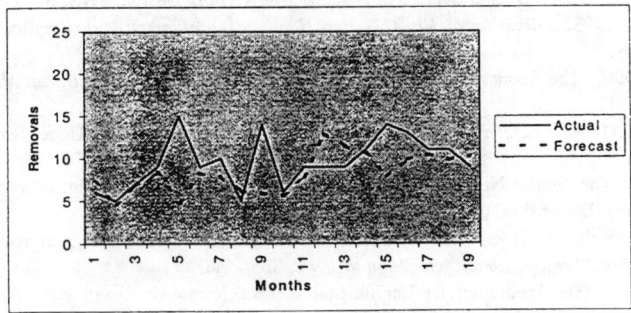

Figure 2: Model D – Actual versus Forecasted NN Removals

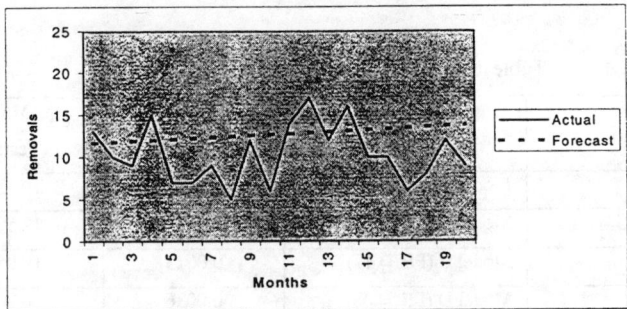

Figure 3: ARIMA (0,1,1) – Actual versus Forecasted Removals

HIERARCHICAL NEURAL NETWORKS FOR REDUCING SYSTEMATIC PREDICTION ERRORS

S. J. HOLL, A. M. FLITMAN, K. A. SMITH
School of Business Systems, Monash University, Victoria 3800, Australia

ABSTRACT
All prediction and forecasting techniques produce results that may contain a combination of systematic and random errors. The challenge for improving the accuracy of the predictions is to isolate the effect of these two types of errors, and to correct them where possible. This paper proposes a hierarchical neural network approach to reducing the systematic errors that may lie in prediction results. Using a simple function approximation problem, we present a series of experiments designed to demonstrate the collective potential of a hierarchy of networks for reducing both systematic and random errors.

INTRODUCTION
Neural network are well known for their ability to approximate arbitrary functions from training data [1]. A trained neural network, however, may still produce errors when applied to both the training data and validation data sets. These errors may be due to conflicts in the data, systematic errors, or random errors. Systematic errors are frequently found in the predictions of neural network models, as well as other forecasting techniques. A common example is the consistent under-estimation of extreme changes in financial forecasting data, or sales forecasting figures. Such systematic errors are prime candidates for identification and removal via dedicated neural networks.

Previous researchers have successfully applied neural networks to reduce systematic errors in a variety of engineering applications [2, 3, 4]. In this paper we propose hierarchical neural networks to isolate and remove the effect of systematic and random errors in prediction outcomes. A number of experiments are conducted to demonstrate that the effect of this hierarchy of neural networks is to smooth random errors and to adjust the predictions for systematic errors.

In Section 2 we demonstrate the ability of neural networks to approximate a simple function – the calculation of the area of a circle. Then we will show in Section 3 how neural networks can remove systematic errors introduced into the data by learning to adjust the predicted values. Random errors are introduced into the data in Section 4, where we show how neural networks can learn to reduce the effect of these errors also. Finally we will show how hierarchical networks can be used to improve prediction accuracy from forecasting systems.

NEURAL NETWORKS AS FUNCTION APPROXIMATORS
Arguably the most approximated value in history is that of the irrational number π. Since Archimedes first approximated its value in the 2nd century BC many variously accurate approximations of π have been used. Archimedes' initial calculation (made

using a 96 sided regular polygon) predicted π to have a value of somewhere between $^{220}/_{71}$ and $^{22}/_{7}$. The latter of these still remains a popular choice as a reasonably accurate estimate of the value of π. In the 2nd century AD the Chinese approximated π as √10. Around this time Ptolemy calculated π to equal 3.1416, which at the time was a considerable achievement. Later in the 16th century a curious fraction 355/113 was discovered, and it turns out to be a surprisingly accurate approximation, good to the 7th decimal place. Recently π has been calculated to the quadrillionth decimal place using a formula expressing π as an infinite sum, and a world-wide network of 1734 computers in 50 different countries [5].

The experiments conducted in this paper are all based on the ability of a neural network to correct the errors in a prediction. Any technique could be used to generate the original prediction or forecast – we are using a neural network for initial prediction as well. We have chosen a simple function approximation problem to generate the initial predictions, and will then deliberately introduce systematic and random errors into these predictions before using hierarchical neural networks to remove such errors.

The first experiment conducted was to see how well a neural network could approximate a value for π. We generated 1000 random circle radius' to use in our testing and calculated the actual values of the area of the corresponding circles using a spreadsheet. This set was used as the training data for a neural network. The radius was used as the single input, and the actual area of the circle as the single output. We constructed a Ward net [6] with the following structure. For all neural networks discussed in this paper, the learning rate is 0.3 and the momentum rate is 0.4:

Input Layer	1 Neuron using Linear Transfer function
1st Hidden Layer	15 Neurons, 8 Logistic, 1 Tanh, 6 Linear
2nd Hidden Layer	3 Neurons, 1 Logistic, 1 Tanh, 1 Linear
Output Layer	1 Neuron using Linear Transfer Function

This network was trained to an R^2 of 0.9999. A second set of 1000 random numbers was generated to validated the network. These were presented to the trained network and also achieved an R^2 of 0.9999 with a MSE of 0.000141826. These results show that the neural network can approximate π with reasonable accuracy.

REDUCING SYSTEMATIC ERRORS WITH NEURAL NETWORKS

In this next section we investigate whether a network can identify a systemic error in an approximation and remove it. To do this we calculated the area of the original 1000 circles using two approximations of π, that is $^{22}/_{7}$ and √10. The known error for the two approximations based on the 1000 randomly generated circles is:

Area = $r^2 \cdot {^{22}/_{7}}$ (MSE = .0000003045)
Area = $r^2 \cdot \sqrt{10}$ (MSE = .0000718221)

We then trained a neural network to learn to correct the systematic error in the area calculation using the $^{22}/_{7}$ approximation. The single input is the predicted area using the approximation for π, while the single output is the actual area of the circle. For this network we used a relatively simple architecture consisting of the following:

Input Layer	1 Neuron using Linear Transfer function
Hidden Layer	5 Neurons using the Linear transfer function
Output Layer	1 Neuron using Linear Transfer Function

This network was trained to an R^2 of 1.0. We then created the validation set using the same approximation for π for the predicted areas and presented it to the network achieving an R^2 of 1.0 as well. This network has identified the systematic error caused by the approximation, and removed it. We repeated the process using the $\sqrt{10}$ approximation and achieved an R^2 of 1.0 on the training and an R^2 of 1.0 on the validation set with the following network. Again the introduced systemic error was completely removed.

Input Layer	1 Neuron using Linear Transfer function
Hidden Layer	12 Neurons using the Linear transfer function
Output Layer	1 Neuron using Linear Transfer Function

REDUCING RANDOM ERRORS WITH NEURAL NETWORKS

We will now add in a random error to the approximations and re-train the neural networks. To do this we generated a random number ε between 0 and 1 and added it to the calculated area.

$$Area = A_1 = r^2 \cdot {}^{22}/_7 + \varepsilon \quad (MSE = 0.331892)$$
$$Area = A_2 = r^2 \cdot \sqrt{10} + \varepsilon \quad (MSE = 0.338517)$$
$$Area = A_a = r^2 \cdot \pi + \varepsilon \quad (MSE = 0.332292)$$

Then we created a network that used the A_1 value calculated from 1000 random values for radius as the only input and the actual areas for those radii as an output. The best performing network was the following:

Input Layer	1 Neuron using Linear Transfer function
Hidden Layer	9 Neurons, 3 Logistic, 6 Tanh transfer functions
Output Layer	1 Neuron using Linear Transfer Function

We then calculated a second set of A_1 values for validation by using a new set of 1000 random radii and passed it through the network achieving an R^2 of 0.92344 and a MSE of 0.06735. This is a substantial improvement given that the original data had a MSE of more then 0.33.

In order to understand why the network achieved such an improvement we plotted the first 50 pairs of radius and A_1 from the validation set and included the actual area (see Figure 1). We also added a line of best fit to our data. If this line were taken as the estimate of the actual area we would see a fairly constant (systematic) error. In our previous set of experiments we showed just how well neural networks could remove these types of systematic errors.

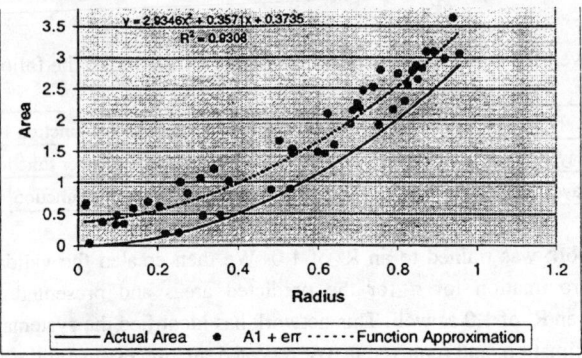

Figure 1: An interpolation function showing the remaining systematic error

We repeated the same experiment using A_2 with the following network structure, achieving an R^2 of 0.92405 and a MSE of 0.06613 on the validation set. Again, this represents a considerable improvement on the original MSE of 0.338517, and demonstrates a reduction in the effect of both systematic and random errors.

Input Layer	1 Neuron using Linear Transfer function
Hidden Layer	7 Neurons, 3 Logistic, 4 Tanh transfer functions
Output Layer	1 Neuron using Linear Transfer Function

We also repeated the experiment using A_π (which uses the exact value of π) to see how well the neural networks could reduce the effect of only a random error in the data using the following network:

Input Layer	1 Neuron using Linear Transfer function
Hidden Layer	4 Neurons, 2 Logistic, 2 Tanh transfer functions
Output Layer	1 Neuron using Linear Transfer Function

We achieved an R^2 of 0.92353 and an MSE of 0.06673, again it was a sizeable improvement from the MSE of the A_π approximation of 0.332292. Thus a simple neural network seems capable of removing random errors from prediction values, as well as systematic errors. The random errors are removed through a smoothing effect, while the systematic errors are removed by learning to adjust the predicted values.

In these three experiments we introduced a random error that was somewhat constrained. One could argue that if the error was randomly distributed between zero and 1 a good approximation for the error is 0.5. Therefore a good approximation of the actual area would be $A_x = A_1 - 0.5$. This however would result in a MSE of 0.083235 for the validation set used in this experiment.

HIERARCHICAL NEURAL NETWORKS FOR ERROR REDUCTION

So far we have shown that neural networks can remove a portion of the errors in our approximations by learning the error in the predictions and then attempting to remove it. This has been shown to be effective in removing both systemic and random errors. Given this we will apply the same principles to the approximations provided by the neural networks constructed thus far and attempt to further refine them by putting these through a second network. We are in effect creating a hierarchy of networks where the output from the first network becomes the input for the second network. The intention is to achieve improved prediction accuracy by smoothing of random errors and adjustment of systematic errors.

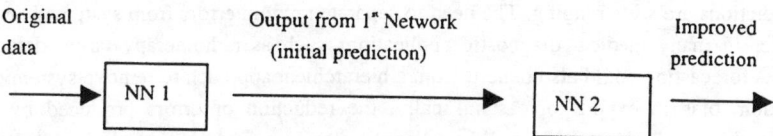

To do this without biasing the results we must create a third set of approximations. We will use the first set to train the first network. We then create a validation set for this network and pass this through the 1ˢᵗ network to generate an output set. This output set is then used to train the second network. The third set of data is used as a validation set for the second network. It is passed through the first network and then through the second network to validate it.

We firstly used this principle on the very first network we created (the one that calculated the area based on the exact value of π). The 1ˢᵗ validation set produced an R^2 of 0.9999 with a MSE of 0.000141826 as before. The results from this network were then used to train a 2ⁿᵈ network with 23 hidden neurons (and linear transfer functions). A 2ⁿᵈ validation set was then run through these two networks. These hierarchical networks achieved an R^2 of .9999 and an MSE of .0000904654. This compares favourably with the MSE of the initial net of 0.000141826.

We repeated the experiment using the A_1 value and network from section 4. We constructed a second network with 33 hidden neurons (11 with logistic, 11 with Tanh and 11 linear transfer functions). We then passed the 2ⁿᵈ validation set through both networks. From doing this we achieved an R^2 of 0.92893 and a MSE of 0.06675. Again this represented an improvement on the initial network.

We repeated the experiment once more using the A_2 value and original network. The second network in this hierarchy consisted of 22 hidden neurons (14 with logistic, 3 with Tanh and 5 linear transfer functions). Passing the second validation set through this hierarchy produced an R^2 of 0.92837 and a MSE of 0.06606. Again this represents a small improvement.

CONCLUSIONS

In each case we have seen that the hierarchical approach to the problem produces an improvement over the single network. One question that comes to mind is whether or not this process can be repeated ad-infinitum until we have a completely eliminated any systematic errors in the predictions. A second questions is how well does the hierarchical network generalise after being so finely tuned? The first could probably be answered by considering the pay-off of better performance with that of the added complexity. A large series of networks certainly takes a greater time to develop and maintain than a simple singular network. Furthermore, each pass of data through the network takes longer to process, and may therefore not be suitable for real-time applications. The second question is not easily answered. Testing and training of multiple networks requires a significantly larger amount of data. In our experiments this was not an issue, but in "real world" applications it may cause a problem.

The applications of the proposed hierarchical approach to error reduction in predictions are wide ranging. The need to separate random errors from systematic errors, arises in many medical diagnostic applications such as radiotherapy target detection. Sales forecasting could also benefit from a hierarchical approach to remove systemic and random bias in expert forecasts. Finally, the reduction of errors produced by such hierarchical neural networks could find application in neural network design. A test for determining if a local minium has been encountered in training could be constructed. Local minima would tend to display larger systematic errors than the global minima. If systematic error can be removed by using a second network then we can conclude that a local minimum must have been encountered. The process can continue until no further error reduction occurs.

REFERENCES

[1] Hornik K, Stinchcombe M, White H, "Multilayer feedforward networks are universal approximators", *Neural Networks*, vol. 2, pp. 359-366, 1989.

[2] Arpaia P, Daponte P, Grimaldi D, Michaeli L, "Systematic error correction for experimentally modeled sensors by using ANNs", *Proceedings of the 16th IEEE Instrumentation and Measurement Technology Conference*, vol.3, pp.1635-1640, 1999.

[3] Yang Q, Butler C, Baird P. "Error compensation of touch trigger probes on CMM", *Measurement*, vol.18, no.1, pp.47-57, May 1996.

[4] Xu S, Crilly PB. "Application of neural networks on color error reductions in television receivers", *IEEE Transactions on Consumer Electronics*, vol.39, no.3, pp.630-635, Aug. 1993.

[5] "Sliced-up pi", New Scientist, no. 2266, page 11, 25th November 2000.

[6] Ward Systems Group, "Neuroshell 2 manual", http://www.wardsystems.com

PART IX: BIOLOGY AND MEDICINE

NEURAL NETWORK CLASSIFICATION OF MALARIA PROTEIN SEQUENCES

KATE A. SMITH[1], TERRY SPITHILL[2], ROSS COPPEL[2]

[1] School of Business Systems, [2] Department of Biochemistry and Molecular Biology, and Department of Microbiology, Monash University, Victoria 3800, Australia

PETER SMOOKER

Department of Biotechnology and Environmental Biology, RMIT University Bundoora, Victoria 3083, Australia

ABSTRACT

This paper describes the use of a neural network to learn to classify two types of protein sequences from the malaria parasite *Plasmodium falciparum*. The first class of protein sequences are those containing an epidermal growth factor (EGF)-like domain, and the second class are sequences that do not contain an EGF-like domain. EGF domains are important in protein-protein interactions on cell surfaces and many potential vaccine candidates exhibit such domains. Therefore, the ability to quickly and automatically identify EGF domains from the large volume of predicted sequences that are emerging as part of the malaria genome DNA sequencing effort is potentially important for future vaccine studies. The sequences used in this study contain blocks of 50 amino acids and are taken from proteins predicted from the DNA sequence of the malaria genome. Different groupings of amino acids based on chemical similarities are explored as a pre-processing technique for reducing the dimensionality of the data. The generalisation of the learning is shown to be highly dependent on the chosen pre-processing approach.

INTRODUCTION

Malaria is one of the most important infectious diseases of humans and represents a major international public health threat, killing 3,000-10,000 children each day [1]. An international consortium was established in 1995 to sequence the 30 million bases of the *Plasmodium* genome, distributed among 14 chromosomes, in order to facilitate drug and vaccine discovery [1]. The sequence of chromosomes 2 and 3 are now complete [2, 3]. The raw sequence of the other chromosomes is about 80% complete, with finalisation of the sequence expected by the end of 2001. Once the complete genome is available, there will be a global effort to apply bioinformatics to study genome organisation and identify the proteome (proteins encoded in the genome) for functional protein analysis.

Data mining techniques play a key role in bioinformatics research due to their ability to discover subtle relationships in large amounts of similar and seemingly repetitive data. A number of significant studies have been conducted over recent years applying data mining techniques to genome analysis problems [4, 5, 6, 7]. In this paper we consider the relationships that may lie in sequences of amino acids. There are 20 different amino acids, and their arrangement in a protein determines the functionality and properties of the protein. Collectively these proteins act to determine all the properties of the organism. In this research we are concerned with identifying sequences of amino acids that might identify potential vaccine candidates for malaria. Several studies have identified that at least three malaria proteins that are promising components for inclusion in a malaria vaccine contain an epidermal growth factor (EGF)-like domain. We hypothesised that it would be worthwhile identifying further proteins with this domain so that they may be assessed for their utility as vaccine components. Therefore, the ability to quickly and automatically identify EGF domains from the large volume of predicted sequences that are emerging as part of the malaria genome DNA sequencing effort would be useful for future vaccine studies.

This paper describes the use of a neural network to learn to classify two types of protein sequences from the malaria parasite Plasmodium falciparum. The first class of protein sequences are those containing an epidermal growth factor (EGF)-like domain, and the second class are sequences that do not contain an EGF-like domain. The sequences used in this study contain blocks of 50 amino acids and are taken from proteins predicted from the DNA sequence of the malaria genome. Different groupings of amino acids, based on chemical similarities and domain knowledge, are explored as a pre-processing technique for reducing the dimensionality of the data.

DIMENSION REDUCTION

For most genome analysis exercises using neural networks, the first step is to convert the sequence of nucleotides (the DNA sequence of A, C, G and T's) into numeric form. This is usually achieved using 1-out-of-4 encoding to preserve the hamming distance between nucleotide representations. Thus the nucleotides are translated into quadruples as:

$$A = 1\ 0\ 0\ 0 \quad C = 0\ 1\ 0\ 0 \quad G = 0\ 0\ 1\ 0 \quad T = 0\ 0\ 0\ 1.$$

In this research, however, we are using protein sequences rather than genome sequences. For each group of three nucleotides, an amino acid is formed, and there are twenty different amino acids possible. A protein sequence comprises these chains of amino acids. In order to use a neural network, we need to convert the sequences of amino acids into numeric form. Adopting a similar 1-out-of-20 encoding approach as above is clearly impractical. For a sequence of 50 amino acids, this would result in a binary input vector of length 1000. Not only would this make training of a neural network on massive amounts of data time consuming, but it also makes rapid evaluation of new sequences more difficult.

One solution to reducing the dimensions of the data it to use functional domain knowledge to group the amino acids based on chemical similarity. We propose two grouping schemes to this end based on common textbook groupings. The first groups the amino acids into 4 distinct classes: hydrophobic, charged, polar, and glycine, and using 1-out-of-4 encoding this results in 200 inputs for a 50 amino acid sequence. The second grouping reduces the number of classes to three: hydrophilic, hydrophobic, and special. This results in 150 inputs for a 50 amino acid sequence. Tables 1 and 2 show the grouping of the twenty amino acids under the two schemes respectively.

Class name	Amino Acids in Class
Hydrophobic	Ala, Val, Phe, Ile, Leu, Pro, Met
Charged	Asp, Glu, Lys, Arg
Polar	Ser, Thr, Tyr, Cys, Asn, Gln, His, Trp
Glycine	Gly

Table 1: Grouping of amino acids into four classes (Grouping A)

Class name	Amino Acids in Class
Hydrophilic	Lys, Arg, His, Asp, Glu, Ser, Thr, Asn, Gln
Hydrophobic	Ala, Val, Phe, Ile, Leu, Met, Trp, Tyr
Special	Cys, Gly, Pro

Table 2: Grouping of amino acids into three classes (Grouping B)

Using this domain knowledge about chemical similarity of amino acids, we have therefore been able to reduce the dimensions of the input vectors for the neural network. One question this raises though, is how does the choice of pre-processing strategy affect the learning ability of the neural network?

DATA SETS

Two data sets have been used for developing our neural network classification model. The first is used for training and testing the neural network, and consists of 18 EGF-domain sequences, and 683 non-EGF-domain sequences. The test set is randomly extracted as 20% of this data set. Each sequence contains 50 amino acids and is converted into numeric form using a grouping scheme described in the previous section. A second data set is then used for validation of the developed model. The validation set contains 13 EGF-domain sequences and 226 non-EGF-domain sequences. Table 3 summarises the data set information.

	training	testing	validation
# of EGF-domain sequences	15	3	13
# of Non-EGF-domain sequences	546	137	226

Table 3: Description of Data Sets

The two different pre-processing strategies are then applied to these data sets, before modelling with a neural network commences.

RESULTS OF NEURAL MODELLING

In order to determine the effect that each pre-processing strategy has on the neural modelling, we have conducted a series of experiments separately as described below:

Pre-processing using Grouping A

A number of different neural networks were experimented with in the Neuroshell2 environment, before arriving at the best one. None of the architectures or parameter combinations were able to obtain satisfactory performance on the test set. The following results in Table 4 represent the best model obtained using a probabilistic neural network. While 100% accuracy was obtained on the training data, the neural network was unwilling to classify most of the test set sequences (although those it did classify were correct). Furthermore, on the validation data set, none of the 239 sequences were classified by the neural network. Clearly, using grouping scheme A, a neural network cannot find a suitable generalisation of the relationships between amino acid sequence and the existence of an EGF domain.

	Classified EGF	Classified non-EGF	Failed to classify
EGF-domain sequences	2	0	1
Non-EGF-domain sequences	0	4	133

Table 4: Test set results using grouping A

Pre-processing using Grouping B

Again, a number of experiments were conducted to arrive at the best performing achitecture, which was (again) a probabilistic neural network. The results in Tables 5 and 6 show the accuracy obtained on the test set and validation set respectively. While 100% accuracy was obtained on the training data as was found when using grouping A, the main improvement is that all sequences in the test and validation sets are now classified into one of the two classes (EGF or not-EGF). Furthermore, the accuracy on the test set is 100%, and on the validation set the overall accuracy is 87.9% (including correctly classifying 11 of the 13 EGF-domain sequences in the data set). Clearly grouping scheme

B, allows the selected neural network to model the relationships between amino acid sequence and the existence of an EGF domain, and to generalise these relationships to unseen sets of sequences.

	Classified EGF	Classified non-EGF	Failed to classify
EGF-domain sequences	3	0	0
Non-EGF-domain sequences	0	137	0

Table 5: Test set results using grouping B

	Classified EGF	Classified non-EGF	Failed to classify
EGF-domain sequences	11	2	0
Non-EGF-domain sequences	27	199	0

Table 6: Validation set results using grouping B

CONCLUSIONS

In this paper we have used neural networks to model an important problem from bioinformatics: recognising EGF-domains within protein sequences. The ability to perform this task effectively and efficiently holds great potential for assisting in the search for vaccine solutions to malaria.

One of the first challenges in protein sequence analysis with data mining techniques is to efficiently encode the sequence in numeric form. We have examined two amino acid grouping schemes that attempt to reduce the dimensionality of the problem, and tested their effect on neural network performance. The results presented have shown that the choice of pre-processing approach has a great effect on the ability of the neural network to learn the relationships between amino acid sequence and the existence of an EGF-domain. In fact, while domain knowledge about chemical similarity of amino acids has been used to propose both grouping schemes, one grouping scheme clearly masked the relationships that obviously exist in the data.

Pre-processing of genome and proteome data is essential in order to achieve the required efficiencies for processing such massive amounts of data generated by the world-wide sequencing effort. Therefore, we expect that much more research needs to be conducted to determine appropriate data reduction schemes that maintain the relationships in the data. Some techniques such as singular value decomposition have been applied from a purely numeric compression perspective [8]. The use of domain knowledge to reduce the dimensions of the task is an alternative approach that appears to warrant further investigation.

REFERENCES

[1] Hoffman, S. L, Rogers, W. O, Carucci, D. J, Venter, J. C. (1998), "From genomics to vaccines: Malaria as a model system", *Nat. Med.* 4:1351-1353.

[2] Gardner, M. J et al. (1998), "Chromosome 2 sequence of the human malaria parasite *Plasmodium falciparum*", *Science* 282 :1126-32.

[3] Bowman S. et al. (1999), "The complete nucleotide sequence of chromosome 3 of *Plasmodium falciparum*", *Nature* 400, 532-538.

[4] Chervitz, S. A. et al. (1996), "Data mining the yeast genome: searching for structure", *Proceedings Yeast Genetics and Molecular Biology*, Madison, Wisconsin.

[5] Hoffman, P. et al. (1997), "DNA visual and analytic data mining", *Proceedings IEEE Visualization Conference*.

[6] Reese, M. G. et al. (1996), "Large scale sequencing specific neural networks for promoter and splice site recognition", *Biocomputing: Proceedings of the 1996 Pacific Symposium*, L. Hunter and T.E. Klein (eds.), World Scientific Publishing Co., Singapore.

[7] Petersen, A. G. and Nielsen, H. (1997), "Neural network prediction of translation initiation sites in eukaryotes: perspectives for EST and genome analysis", *Proceedings Intelligent Systems in Molecular Biology*, vol. 5, pp. 226-233.

[8] Wu, C. and Shivakumar, S. (1994), "Back-propagation and counter-propagation neural networks for phylogenetic classification of ribosomal RNA sequences", *Nucleic Acids Research*, vol. 22, no. 22, pp. 4291-4299.

NEURAL PCA NETWORK FOR LUNG OUTLINE RECONSTRUCTION IN VQ SCAN IMAGES

G. Serpen[1], Ph. D., R. Iyer[1], H. M. Elsamaloty[2], M. D., E. I. Parsai[3], Ph. D.

[1]Electrical Eng. & Computer Science, The University of Toledo, Toledo, OH 43606 USA
[2]Department of Radiology, Medical College of Ohio, Toledo, OH 43614 USA
[3]Department of Radiation Oncology, Medical College of Ohio, Toledo, OH 43614 USA

ABSTRACT

This research focuses on design of a software-based image analysis system towards facilitating automated diagnosis of Pulmonary Embolism using ventilation-perfusion scans and correlated chest x-rays. This proposed system takes the digitized ventilation-perfusion scan images of lungs as input, identify a template according to the size and shape of the lungs and thereby approximate and reconstruct the outline of the lung. The proposed lung outline reconstruction system was designed to facilitate the PIOPED-compliant diagnosis procedure. The system was trained with actual patient lung images, where lung images were compiled to represent the shape and size variation of the population in general. Both adult male and female lung images were utilized. A neural principal component analysis network was used and tested with actual patient lung images obtained through the Medical College of Ohio patient repository, which represented five probability classes as they exist in the PIOPED criteria. Testing results, which were obtained through MATLAB™ simulation, indicate that neural PCA algorithm trained with generalized Hebbian learning performed well although it demonstrated performance degradation for high probability Pulmonary Embolism cases. The work as presented proved in concept that it is feasible to construct the outline of the lungs and is open to enhancement in numerous ways.

INTRODUCTION

Image processing and recognition are extensively used in medicine for the purpose of automating identification of various organs in diagnositic images and diagnosing the diseases related to the same. Lungs being a fundamental part of the human respiratory system, a number of researchers have developed algorithms to identify the human lung and the defects related to them in a variety of imaging modalities. One of the many research projects being carried out is for the diagnosis of Pulmonary Embolism [Gabor F.V., 1994], [Fisher R. E., 1996], [Armato S., 1997]. Pulmonary Embolism is a disease of the lung where the arteries to various anatomical regions of the lung are occluded by emboli, which originate from venous thrombosis, interfering with the normal gas exchange. Anatomically, lungs are divided into two regions: the right lung and the left lung. The right lung is further divided into the upper, middle and lower lobe bronchi; and the left is divided into only the upper and lower lobe bronchi. The lobes are further divided into segments.

In Ventilation-Perfusion (VQ) scanning, the ventilation scans are taken before the perfusion scans. In ventilation imaging, 10 to 20 mCi of Xenon 133 is administered by employing a number of commercially available delivery and rebreathing units. The scan is taken while the patient holds his breath for 15 seconds. Perfusion scans are taken by injecting particulate radiopharmaceuticals during quiet respiration when the patient is in the supine position. The perfusion scans are obtained on a large-field-of-view or a standard-field-of-view camera with a diverging collimator. Images are obtained in the anterior, posterior and both lateral positions. These VQ scans are employed by the radiologists to determine the possibility of the patient suffering from Pulmonary Embolism. The radiologists look for possible defects in the perfusion scans and corresponding mismatch in the ventilation scans to determine the possibility of Pulmonary Embolism.

The following is the mental procedure followed by nuclear radiologists to extract feature data from VQ scans and correlating chest x-rays to aid in diagnosis of Pulmonary Embolism:

- First construct a mental image of the outline of the lungs.
- Mentally superimpose the segmented anatomy over the reconstructed outline of the lungs.
- Mentally locate and list the number and size of the defects in each segment.
- Finally, match the observed defects with the PIOPED criteria to conclude a diagnosis.

Manual interpretation of the ventilation-perfusion scans and the correlated chest x-rays may not be very precise, since these images are of relatively low resolution and can at times be fuzzy. Such interpretations also depend on the skill and experience level of the reader, in this case a nuclear radiologist. Therefore, there may be significant variations in the diagnosis from one reader to another. Automation of the scan-reading procedure that mimics the radiologists' approach to extract data from the VQ scans has the potential to eliminate inter-observer variability.

Even though a considerable amount of research has been carried out in the field of lung imaging, it is still not at a stage to directly benefit the automation of the diagnosis of Pulmonary Embolism (PE). The automated outline reconstruction procedure should yield a smooth curve without eroded surfaces, which can result in false estimation of the probability of Pulmonary Embolism to the desired degree. The lung image also needs to be partitioned into anatomic segments so that the location of a defect in a certain segment in the lung can be pinpointed. Determination of the segment, where the defect is present, is necessary to evaluate the probability of Pulmonary Embolism according to the PIOPED criteria [PIOPED 1990].

A defect in any of the lung segments in the perfusion scan is known as a segmental perfusion defect. A large segmental perfusion defect is one where more than 75% of the lung is affected with PE. A moderate segmental perfusion defect results when 25% to 75% of a lung segment is damaged and, when less than 25% of a lung segment is damaged, it is known as a small segmental perfusion defect.

Reported studies for lung outline reconstruction using VQ scan images are limited in number and mostly appeared during the second half of the last decade. Nevertheless, some of these studies report significant progress towards achieving the goal of a lung outline reconstruction system. Gabor et. al. [1994], in their study, made use of an expert system to diagnose Pulmonary Embolism. The analysis performed by the expert system was on the basis of ventilation-perfusion scans only and excluded correlating chest x-rays. Analyzing the abnormalities in the images required standardizing the images with respect to shape and size: creating a template of lung anatomy for all projections facilitated the standardization and the image from each projection was stretched onto the appropriate anatomic template. A pixel-to-pixel comparison was made between the stretched patient image and the composite normal file to identify the VQ defects within the images. Any pixel with a standard deviation value below 2.2 was considered as abnormal, and to label any area as abnormal required five or more contiguous pixels from that area. The stretching of the images caused mis-registration of the defect into different pulmonary segments on some views. This resulted in the same defect being counted twice and falsely increased the probability of Pulmonary Embolism.

Research carried out by Armato et. al. [1997] presented a distinct method for reconstructing the outline of the lung in the ventilation and perfusion scans with the help of digital chest radiographs. They superimposed ventilation and perfusion images on the chest radiographs to determine the outline of the lung. This was done by identifying lung contours on all the images using an iterative global level thresholding scheme. Once the lung contours had been identified, scaling factors were obtained to appropriately match the dimensions of images from the two modalities. Alignment was achieved by means of vertical translation based on the apex locations and horizontal translations based on the mediastinum locations on the images. The result was a set of three superimposed images. Since this method makes use of the chest x-ray, the lung outline cannot be reconstructed from all views (right lateral, left lateral, right lateral oblique, etc.).

Hasegawa et. al. [1998] made use of a shift invariant neural network to reconstruct the lung outline from the digital chest radiographs. The method of lung segmentation consisted of three stages, preprocessing for the background adjustment, the convolution network to extract the rib cage, and the post processing to smooth the boundaries as well as to reduce noise. The input images were miniaturized and then processed by the histogram equalization algorithm, which adjusts the gray tone distribution of the input image. The images were then processed by the trained neural network to enhance the boundaries of the lung fields. In post processing, the lung boundaries were extracted and the boundaries were smoothed using a boundary smoothing technique. This technique calculates the tangent change at each point on the boundary. The change of tangents at eroded areas is large whereas the change is small in smooth areas. Thus detecting a large change in the absolute value of the tangents can identify eroded areas. This method of lung segmentation leaves many eroded areas on the boundary of the lung image after reconstruction, which can be misinterpreted as a defect. Also, this method cannot be extended to views other than anterior or posterior since chest x-rays required for those views are not ordinarily available.

Although above referenced algorithms appear to be promising to address the problem of lung outline reconstruction, we tend to believe there is merit in exploring an alternate algorithm for this problem. Hence, we propose principal component analysis methods since these algorithms naturally lend themselves to application in image reconstruction. There is vast and stable literature on principal component analysis. Specifics and particulars of this problem makes it possible to identify neural principal component analysis networks as potentially good paradigms to address the problem of outline reconstruction for VQ scan images of lungs.

The ultimate goal for a completely automated image recognition system is to emulate the mental process a radiologist follows while analyzing the VQ scans and correlating chest x-ray to diagnose PE. The required elements of such an image recognition system consists of a subsystem to reconstruct outline of VQ images of lungs possibly distorted by defects due to PE, a subsystem to superimpose the segmented anatomy of lungs on VQ scans of dysfunctional lungs, a subsystem to evaluate the complete list of VQ defects, their sizes and location, and finally, another subsystem to correlate defects with chest x-ray images of the same lungs.

The completely automated image recognition system would accept the set of ventilation-perfusion scan images and correlating chest x-rays as input, and generate diagnostic data as required by the PIOPED criteria for the interpretation/classification subsystem. The input image set includes ventilation-perfusion scan images of the lungs (anterior perfusion, posterior perfusion, right lateral perfusion, left lateral perfusion, right posterior oblique perfusion, left posterior oblique perfusion, equilibrium ventilation) and the correlating chest x-ray.

The scope of this work will be limited to lung outline reconstruction through template recall for anterior perfusion scans. Functionality of the block, which reconstructs lung outline through template recall, involves storing a number of lung templates (anterior perfusion scans), and when presented with the image of a partly dysfunctional lung, recalling the closest matching template, thereby reconstructing the outline of the lung image through a best-fit approximation approach.

We will employ a linear principal component analysis (PCA) neural network that will be trained with the generalized Hebbian learning rule. This network is a deceptively simple and easy to implement yet computationally powerful enough to solve the lung outline reconstruction problem. The algorithm will be tested using cases created using the PIOPED criteria. The recalled images or reconstructed outlines will be validated by a nuclear radiologist with the Medical College of Ohio (MCO).

SIMULATION STUDY

A simulation study was performed to test the performance of neural PCA algorithm on a set of actual human lung images, which were obtained from the Medical College of Ohio (MCO) patient repository. The software implementation of neural PCA algorithm was realized in MATLAB™ version 5.3.0.

The healthy lung templates employed in the simulation, where three images are for males and eight images for females, are perfusion images in the anterior view and were chosen to model the variation and therefore to represent a good initial approximation to the general population as was also confirmed by a radiologist at the MCO. All the lung templates are stored in the bitmap format (.bmp) with a resolution of 120×128 pixels. The pixel values of the lung image template range from 0 to 255: the images are gray level with 8 bit resolution. Pixel values are quantized to binary values -1.0 or $+1.0$ with a threshold value of 190 prior to processing them through the neural PCA algorithm.

Test cases are generated in accordance with the PIOPED criteria: test cases vary according to the number of defects and the size and location of the defects in the lungs. Compliance of test cases with actual medical cases is also validated by a nuclear medicine radiologist. These images are created by adding segmental or sub-segmental defects in accordance with the PIOPED criteria to the original template lung images: all five probability cases are included in the test image set. The procedure for creating the test cases is as follows: a healthy lung image from the existing set of eleven templates is picked at random. The segmented perfusion lung image in the anterior view is mentally superimposed on the template. Next, the number and size of defects consistent with the PIOPED criteria for a particular probability class is created.

Design of the neural network followed the consideration that the first principal component might be sufficient to extract the needed information. Since only one principal component was considered for each lung template image, the neural PCA network was instantiated with one output node [Oja, 1982; Oja and Karhunen, 1985]. The lung template images when converted into a data vector have 15360 elements: The lung template images consist of 120×128 binary pixels, which result in 15360 binary pixels when represented by a one-dimensional array. Thus, the number of input nodes for the neural PCA network was determined as 15360. The neural PCA network was trained with the generalized Hebbian algorithm. Elements of the weight vector were initialized to real numbers in the range [0, 1] using a uniformly random distribution using the MATLAB™ function Rand() available in version 5.3.0. The neural PCA network topology is as shown in Figure 1.

Training the neural network for principal component analysis involves computing the principal components of each of the eleven lung templates. One template is presented to the network at a time and the network is allowed to stabilize; i.e. the elements of the weight vector stop changing following repeated adaptations through the learning rule. The network is guaranteed to stabilize if the learning rate chosen is not very high. The weight vector is updated using Equation 1, where the synaptic weights of this network are denoted by the term w_i, with $i = 1,2,...,15360$. The term $w_i(k)$ represents the value of ith element of weight vector at time instant k and is updated by the following formula:

$$\Delta w_i(k) = \eta [y(k)x_i(k) - y(k)(w_i(k)y(k))] \text{ for } i = 1,2,...,15360, \tag{1}$$

where $\Delta w_i(k)$ is the change applied to the synaptic weight $w_i(k)$ at time instant k, and η is the learning-rate parameter which is taken as 0.001 for this network. Once the network stabilizes i.e.,

$$\Delta w_i(k) = \Delta w_i(k+1), \forall i = 1,2,...,15360 \text{ and } k \geq m \ni m \in Z^+,$$

the output of the network computed as shown in Equation 2 is the first principal component of the input lung template image vector and the weight vector is its corresponding eigenvector:

$$y(k) = \sum_{i=1}^{15360} w_i(k)x_i, \tag{2}$$

where k represents the time instant at which the network weights stabilized. Thus, the first principal components and their corresponding eigenvectors of all the eleven lung templates are independently computed using the neural network in Figure 1. These eleven eigenvectors are then used for lung template recall through the procedure as outlined in Figure 2.

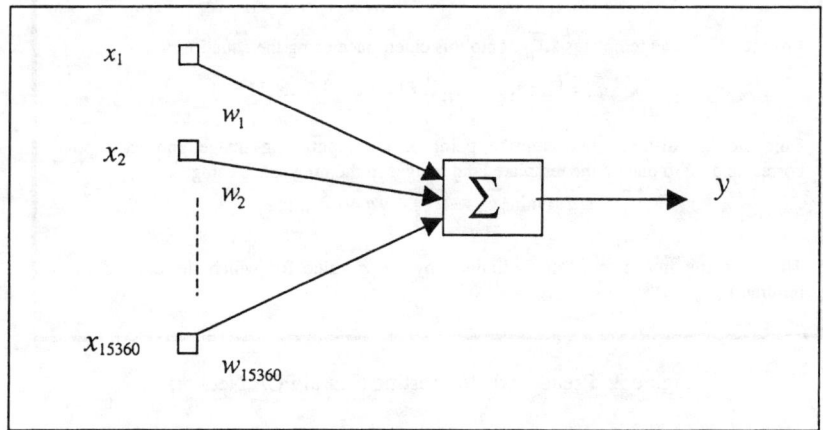

Figure 1. PCA Neural Network Topology

Testing involved images synthetically created using stored templates in accordance with the PIOPED criteria and validated by the MCO radiologists. Figure 3 shows the test cases and recalled templates for all five probability classes under the PIOPED criteria. In order to map the test cases to represent normal probability of Pulmonary Embolism, there should be no segmental or subsegmental defects in the perfusion scans. The template images recalled by the neural PCA algorithm for all five test cases of normal probability of Pulmonary Embolism exactly match the input images. The test cases that belong to very low probability of Pulmonary Embolism have three or less than three small segmental perfusion defects. The neural PCA algorithm recalled all the test images accurately for very low probability of PE. The test cases belonging to the low probability of Pulmonary Embolism should possess three to five small segmental perfusion defects. The neural PCA algorithm accurately identified all five test cases representing low probability of PE. A test case that represents intermediate probability of Pulmonary Embolism should have between one and two large segmental perfusion defects. The neural PCA algorithm when applied to test cases of intermediate probability of PE recalled matching lung templates for all five test cases. In order to map the test cases to represent high probability of Pulmonary Embolism there should be two large segmental perfusion defects or one large and more than two moderate segmental defects. The images recalled by the neural PCA algorithm for four out of the five test cases of high probability Pulmonary Embolism match the expected templates. The template recalled for the fifth case is not the best match for the input image given the stored set of templates. Although a perfect match is not recalled for the fifth test case, the recalled image does resemble the test case to a reasonable degree.

Consider a test image vector, $\bar{x}_{test}$, with dimensions 15360×1.

Compute an inner product of this image data vector, $\bar{x}_{test}$, with each of the 11 eigenvectors $\bar{e}_i$, where each eigenvector has dimensions of 15360×1 and is computed using a single neuron PCA network for a given template image. The resultant is a vector in the eigenspace given by

$$\bar{f}_{test} = [\bar{e}_1 : \bar{e}_2 : \ldots : \bar{e}_{11}]^T \bar{x}_{test}.$$

Project all 11 lung templates, $\bar{x}_q$, onto this eigenspace using the equation:

$$\bar{f}_q = [\bar{e}_1 : \bar{e}_2 : \ldots : \bar{e}_{11}]^T \bar{x}_q, \quad \forall q = 1, 2, \ldots, 11$$

Compute the distance between the point of the input lung image and each point corresponding to one of the template lung images in the lung space using

$$d = \min_q \left\| \bar{f}_{test} - \bar{f}_q \right\|, \quad \forall q = 1, 2, \ldots, 11$$

The matching image template is implied by the q value for which the distance d is minimum.

Figure 2. Pseudocode for Testing Neural PCA Network

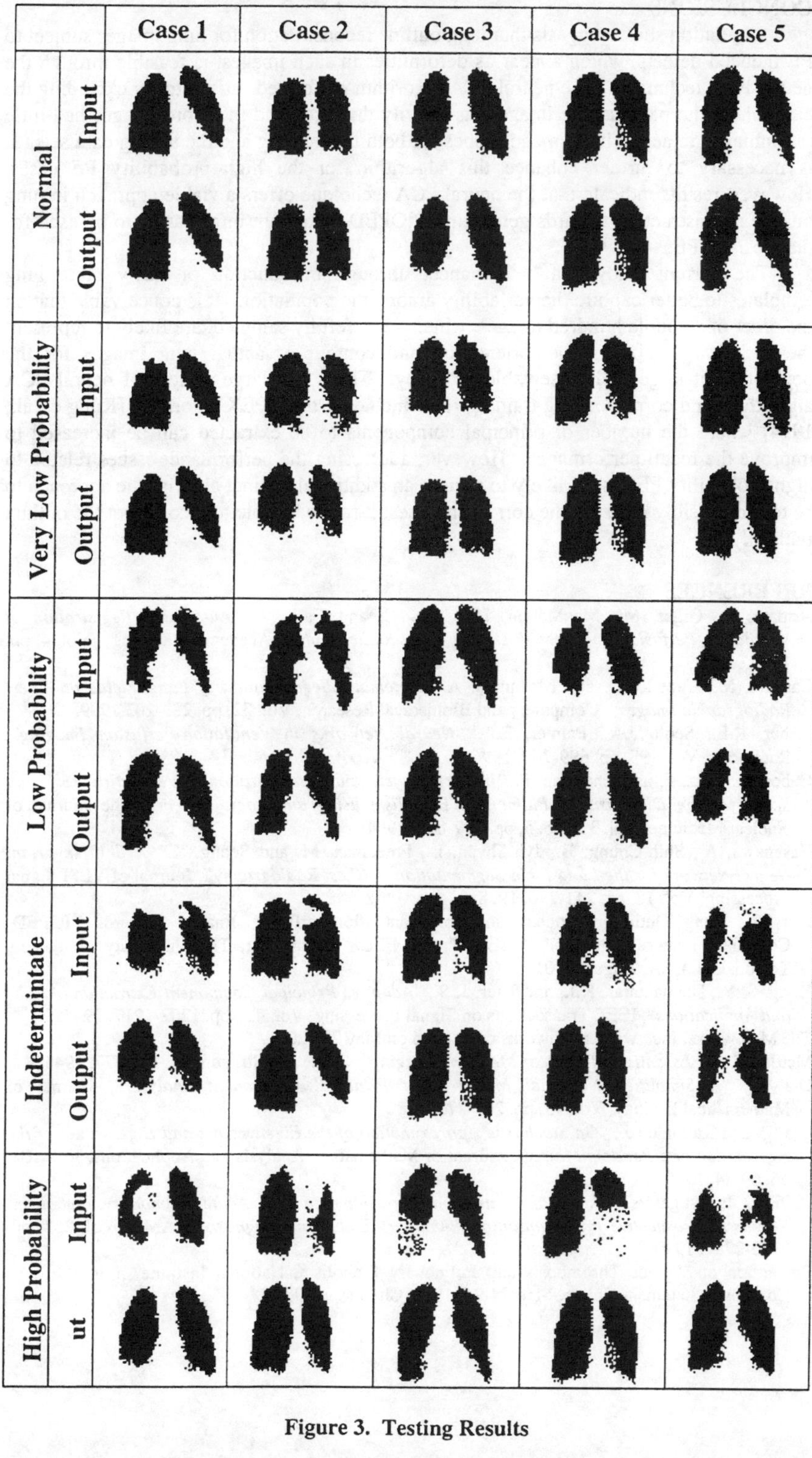

Figure 3. Testing Results

CONCLUSIONS

Simulation study suggests that lung outline reconstruction for lung images subject to PE-induced defects, which appear as deformities in such images, is feasible through the neural PCA technique. The neural PCA algorithm performed satisfactorily excluding the case where the patient lung image was heavily damaged and therefore images had little resemblance to actual lung image shapes for both the training and the testing data sets. It is necessary to further enhance this algorithm for the high probability PE cases. However, results indicate that the neural PCA technique offers a viable approach in lung outline reconstruction towards generating PIOPED-compliant information to be used for diagnosis of PE.

The current study can be enhanced through introduction of many more lung templates to better capture the variability among the population. It is conceivable that on the order of a couple hundred images, which are carefully sampled, are likely to represent the variation in size, shape, orientation and contour of actual lung images for the population at large with acceptable accuracy. There is a large variety of neural PCA algorithms and corresponding training rules including the APEX algorithm [Kung et. al., 1994] where the number of principal components to be extracted can be increased to improve the recall performance. However, addressing the performance issues related to high-probability PE case is likely to require an additional distinct algorithmic approach to be tested possibly utilizing the correlating chest x-ray to be able to reconstruct the outline faithfully.

REFERENCES

Armato, S., Giger, M., MacMahon, H, Chen, C. and Carl, V. *"Automated Registration of Ventilation-Perfusion Images with Digital Chest Radiographs"*, Academic Radiology, Vol. 4, pp. 183-192, 1997.

Carreira, M., Cabello, D. and Mosquera, A. *"Automatic Segmentation of Lung Fields on Chest Radiographic Images"*, Computers and Biomedical Research; Vol. 32, pp. 283-203, 1999.

Fisher, R.E., Scott, J.A., Palmer, E.L. *"Neural Networks in Ventilation-Perfusion Imaging"*, Radiology, Vol. 198, pp. 699-706, 1996.

Gabor, F., Datz, F. and Christian, P. *"Image Analysis and Categorization of Ventilation Perfusion Scans for the Diagnosis of Pulmonary Embolism using an Expert System"*, The Journal of Nuclear Medicine; Vol. 35, No. 5, pp. 797-802, 1994.

Hasegawa, A., Shih-Chung, B., Jyh-Shyan, L., Freedman, M. and Seong, K. *"A Shift Invariant Neural Network for the Lung Field Segmentation in Chest Radiography"*, Journal of VLSI Signal Processing; Vol. 18, pp. 241-250, 1998.

Iyer, R., Lung Outline Reconstruction for Ventilation-Perfusion Images Towards PIOPED-Compliant Feature Extraction, MSEE Thesis, EECS Department, The University of Toledo, Toledo, OH, USA, August 2001.

Kung, S. Y., Diamantaras, K.I., and Taur, J. S., *"Adaptive Principal Component Extraction (APEX) and Applications"*, IEEE Transactions on Signal Processing, Vol. 42, pp. 1202-1217, 1994.

The Mathworks, Inc.; www.mathworks.com, cited on May 7th, 2001.

Mettler, F. A.,*"Essentials of Nuclear Medicine Imaging"*: Crune and Stratton, pp. 140-177, 1945.

Oja, E., *"A Simplified Neuronal Model as a Principal Component Analyzer"*, Journal of Mathematical Biology, Vol. 15, pp. 267-273, 1982

Oja, E. and Karhunen J., *"On Stochastic Approximation of the Eigenvectors and Eigenvalues of the Expectation of a Random Matrix"*, Journal of Mathematical Analysis and Applications, Vol. 104, pp. 69-84, 1985.

PIOPED Investigators. *"Value of Ventilation Perfusion Scans in Acute Pulmonary Embolism: Results of Prospective Investigation of Pulmonary Embolism Diagnosis."* JAMA; Vol. 263, pp. 2753-2759, 1990.

Prevention of Venous Thrombosis and Pulmonary Embolism, National Institute of Health Consensus Statement Online, Mar. 24-26, 1986, Cited May 2001.

A DYNAMIC MODEL FOR NUCLEAR MEMBRANE OF A NANOSCALE BIO-MOLECULAR MOTOR

A.S.C. Sinha, R.M. Pidaparti, P.A. Sarma
Purdue University, Indianapolis

G. Vemuri
Department of Physics, Purdue University, Indianapolis

A.M. Gacy
Department of Pharmacology, Mayo Clinic and Foundation, Rochester, MN

ABSTRACT

We propose to understand the mechanisms of nucleo-cytoplasmic transport and the interactions between elements of the Nuclear Pore Complex (NPC), [see details of NPC structure in figure1(a)] that drive the physical process based on theory of creep strain model. To achieve this we developed a simple idealized model of a nuclear membrane between two nuclear pore complexes. Each pore can facilitate both import and export. The transport through NPCs can be rapid (estimated at several hundred molecules/pore/second) resulting in a nuclear membrane deformation which is elastic in nature. The proposed model studies the creep behavior of the nuclear membranes elastic material and its recovery after the transport of a molecule through the NPC. The model assumes that the upper and lower nuclear membranes behave according to an elastic modulus with creep strain estimated by two exponential functions. The results obtained indicate that there is an improved understanding of the mechanisms of nuclear membrane dynamics of nuclear pore complex structure.

INTRODUCTION

The two-way exchange of proteins and RNA across the double nuclear membrane, between the cytoplasm and the nucleus, is a critical cellular process affecting nearly all-cellular functions. For example, gene expression is controlled by the entry of transcription factors into the nucleus from the cytoplasm, and translation is controlled by the regulation of messenger RNA transport from the nucleus to the cytoplasm. Other functions include DNA chromatin packing effected by the influx of histones into the nucleus, and in the control of cell entry into mitosis have been studied by Martin, et.al., (1995) and Pines, (1999). A cell has many presumably identical NPCs each of which participates in the import and export of nuclear material from the cytoplasm nucleus (Akey, et. al., 1993, Akey, 1989, and, Pante and Aebi 1996). Despite the obvious complexity of a system being involved in so many processes, the basic mechanism of transport can be broken into four components: passive transport which does not require extra energy, active transport which requires the energy from ATP/GTP hydrolysis, import into the nucleus, and export from the nucleus. We will focus our attention on import and export utilizing active transport, since it is a classic example of an energy driven, motor-like process, and most biologically relevant molecules are transported by this mechanism. However, both the mechanisms are briefly described below.

Passive transport is the diffusion driven transport of small macromolecules (<40 kDa) to and from the nucleus. While this is a distinct pathway and has been the subject of much work, we will not study passive diffusion for two reasons. First, very few (less than 1%) biologically relevant molecules are passively transported into the nucleus. Second, passive diffusion apparently does not take advantage of the pumping motor of the NPC, and so is not as regulable as the more complex active transport. Active transport is the energy dependent process responsible for bi-directional movement of the majority of macromolecules (> ~60-kDa), as well as for smaller proteins that possess a nuclear import (NLS) or export (NES) signal. In principle, two major processes in bi-directional nuclear transport are distinguished: a fast transport and binding of the protein to the nuclear membrane and a relatively slower, highly energy-dependent, translocation of the protein through the nuclear membrane. While both import and export require energy to transport large molecules, the primary means of recognition and transport of molecules through the nuclear membrane is quite different.

There has been an increasing interest in nuclear membrane dynamics of the pore complex. The current challenge is to understand the nuclear membrane dynamics when the molecular transport through an NPC takes place, and the mechanical mechanisms that drive the physical processes of translocation through it. The objective of the present study is to develop a mathematical model describing the double nuclear membrane structure between NPCs, and investigate the dynamic behavior through creep strain modeling for the purpose of understanding the forces and constraints the member applies to the NPC during translocation. Preliminary results obtained from the mathematical model are presented and discussed.

MATHEMATICAL MODEL

We begin with a simple model which consists of two distinct NPCs attached to a nuclear membrane with material transport taking place in and out of the NPCs as shown in Fig.1(b). When the transport occurs across an NPC, the nuclear membrane under goes several changes and finally returns to a steady state condition. We model the nuclear membrane dynamics through a creep strain phenomena and explain the implication when transport occurs through NPC.

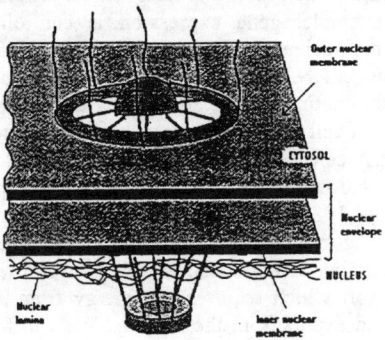

Figure 1(a). A Schematic of an Idealized Nuclear Pore Complex Structure

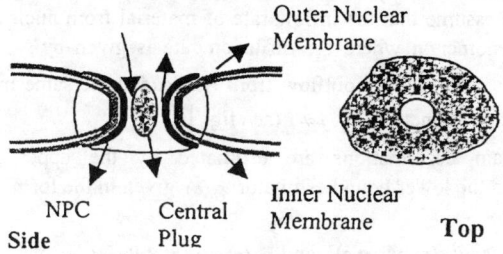

Figure 1(b). Top and Side View of a Single NPC.

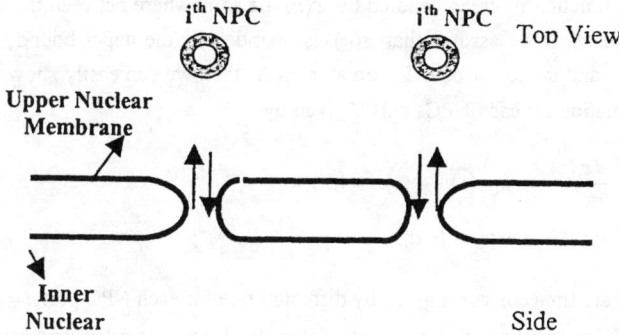

Figure 1(c). A Schematic of an Idealized Nuclear Membrane comprising an Upper and Lower Nuclear Membranes and Two NPCs

The driving force for the nuclear membrane dynamics is the force exerted during molecular transport through NPC. We assume that the nature of the force is similar to impulses regulating neuron cell phenomena. Therefore, we propose to understand the mechanisms of nucleo-cytoplasmic transport and the interactions between the elements of the NPC and nuclear membrane that drive the physical process based on the theory of creep strain. This implies that the change of the amount of total out flux of material through NPC port i, $1 \le i \le n$, in any interval of time equals the change of the amount of total influx of material through NPC port j, $1 \le j \le n$ in the same time interval (see Fig. 1(b)). When the active transport happens through NPCs, the nuclear membrane undergoes creep strain deformation. This creep strain deformation is related to the physical parameters of the nuclear membranes in terms of *activation energy, thermal and membrane material elasticity.* This creep strain deformation in nuclear membrane is similar to engineering material for example, polymers, where the creep parameter μ is related to $\mu = -\dfrac{Q}{RT}$. Here Q is a special physical constant called the *activation energy*. Q is a measure of energy barrier that must be overcome for a molecular motion to occur. T is an absolute temperature in kelvins (K) and the quantity R is the universal gas constant. We assume that the inflow rate of material from nuclear is estimated by the diffusion phenomenon where the diffusion rate is given by $-\mu_{ii}$, for NPC port $i=1,\ldots,n$. The rate of material outflow from *Nucleus* at the same instant is given by the so-called transport function $k_{ji}, j \neq i$ (see fig. 1(c)).

The creep strain deformations are estimated by the upper bound estimator function $\varepsilon_u(c)$ and the lower bound estimator $\varepsilon_l(c)$ given in the form

$$\varepsilon_l(c) = \alpha_l + \beta_l^T c[\exp(-\mu_l c)] \text{ and } \varepsilon_u(c) = \alpha_u + \beta_u c[\exp(-\mu_u c)] \tag{1}$$

where $\alpha_l, \alpha_l, \beta_u$ and β_u are determined from the estimate curve. μ_u and μ_l are positve constants representing the upper and lower membrane diffusion rates. Let the elastic modulus as a function of creep denoted by $\varepsilon(c)$ lie somewhere between the upper and lower membranes. If we assume that $\varepsilon(c)$ is bounded by the upper bound estimator function $\varepsilon_u(c)$ and the lower bound estimator $\varepsilon_l(c)$, then we can easily show that $e(c)$ satisfy the equation for each NPCs port *I* given by

$$\frac{d\varepsilon_i(c)}{dc} = -\mu_i \varepsilon_i(c) + g_i(c), i=1,\ldots,n \tag{2}$$

$$g_i(c) \triangleq \beta_i \exp[\alpha_i c] + \alpha_i \mu_i$$

where $\mu_i > 0$ are time constants given by diffusion rate for each NPC port, $i=1,\ldots,n$, and α_i and β_i are small positive constants, determined from the estimate curve.

We make the following additional mathematical assumptions:
(i)The rate of material outflow from the *Nucleus* to the *cytoplasm* has variation with creep strain. And the coefficients k_{ij} and b_{ij} for $i \neq j$ are coupling terms indicating an interaction between adjacent NPCs through the membrane. The coefficients k_{ij} and b_{ij} for $i \neq j$ may be assumed to be zero if only one port is active; and
(ii) The delay $\gamma_j > 0$ $(j=1,\ldots,N)$ is reaction lag due to effect of the transport function, which is similar to rheology effect (Sinha, 1983).

From these assumptions the regulation mechanism for NPC is similar to neuron cell regulated by summation of impulses as control mechanisms of the diffusion process. Therefore, the rate of change of creep strain for n NPC ports is governed by the functional differential equation:

$$\frac{d\varepsilon_i(c)}{dc} = -\mu_{ii}(c)\varepsilon_i(c) + \sum_{\substack{j=1\\j\neq i}}^{n} k_{ji}(c)\varepsilon_i(c) + \sum_{j=1}^{n} b_{ji}\varepsilon_i(c-\gamma_j)$$

$$+ g_i(c) + u_i(c) \tag{3}$$

$$g_i(c) = \beta \exp[\mu_{ii}c] + \alpha\mu_{ii}; u_i(c) = \sum_{k=0}^{n} \delta(c-c_k)$$

where $\mu_{ii} > 0$, for each $i = 1,...,n$ is diffusion rate for the i th port. Equation (3) is examined to see how the nuclear membrane dynamics change with force mediated by transport through NPC match the known surface and the parameters for NPC.

MAIN RESULTS

Several variations in NPC structure have been reported, generated from a variety of experimental techniques. We have considered the creep strain model to explain these changes in terms of structural changes within the adjoining membrane. A model with known parameters has been developed based on known creep strain theory driven by summation of impulse inputs. We have used impulse inputs similar to neural net inputs. In order to examine our system, we consider the homogeneous part of the system (3) and develop some theoretical basis of analysis from the large-scale theory point of view. We first examine the linear homogeneous system and obtain some conditions and then consider the nucleocytoplasmic transport mechanism. Let us consider the subsystem of equation. (3)

$$\frac{d\varepsilon_i(c)}{dc} = -\mu_{ii}(c)\varepsilon_i(c) \qquad (4)$$

Let the transition matrix the equation (4) be given by $W_i(s,c)$ which is known to satisfy the property

$$\begin{cases} \frac{\partial W_i(s,c)}{\partial c} = -\mu_{ii}(c)W_i(s,c) \\ W_i(s,s) = I_i \end{cases} \qquad (5)$$

To gain further insight into the structures mediating nuclear membrane for the recovery within interconnected tubular ER membranes throughout the cytoplasm, consider the linear equation (3) where the coupling terms coefficients k_{ij} and b_{ij} for $i \neq j$, and the reaction lag due to the effect of the transport function $\gamma_j > 0$ $(j = 1,...,N)$ are included.

$$\frac{d\tilde{\varepsilon}_i(c)}{dc} = -\mu_{ii}(c)\tilde{\varepsilon}_i(c) + \sum_{\substack{j \neq i \\ j=1}}^{n} k_{ji}(c)\tilde{\varepsilon}_i(c) + \sum_{j=1}^{n} b_{ji}\tilde{\varepsilon}_i(c - \gamma_j) \qquad (6)$$

We have the following results:

Theorem 1. If there exist constants $\mu_{ii} > 0$ $(i = 1,...,n)$ such that

$$\|W_i(s,c)\| \leq Me^{-\mu_{ii}(c-s)} \quad i = 1,...,n \qquad (7)$$

The nuclear membranes then recover, i.e. the solutions to equation (6) has converging properties, provided the following conditions hold:

(i) $\mu_{ii} > 0$ $(i = 1,...,n)$

(ii) $\text{Re}\left[\lambda\left(-\mu_{ii} + \|k_{ij}\|_{j \neq i} + \|b_{ij}\|\right)\right]_{n \times n} < 0$ \qquad (8)

The above theorem 1 gives sufficient conditions for the recovery within interconnected tubular ER membranes throughout the cytoplasm. We have given a general framework for modeling cell biological process, to explain nuclear envelope breakdown.

The next theorem considers the case where the nonlinear term is present. The dynamics of Nuclear Pore Complex has also impulse input as the controlling mechanism for the transport through nuclear pores.

Theorem 2. Let the conditions of the Theorem 1. In addition if

$$(i) \sum_{j=1}^{n} e^{-\mu_i(c_j - s)} \leq \rho < \infty, \nu_{ii} > 0 \text{ for each } i = 1, \ldots, n \tag{9}$$

then the creep strain solution of (3) approaches to a constant value.

Remark. The proof of the theorem is based on the application of variation of parameters and Liapunov functional, which has been omitted for sake of brevity.

Using the classical approach, the rate of material from the *Nucleus* to the *Cystoplasm* is modeled allowing for transport delay. The control mechanisms responsible for specific conformal transitions in the NPC are not fully understood. It is proposed that a potential regulation mechanism for the NPC is similar a to neuron cell regulated by the summation of impulses. Currently, the modeled equation is being studied to match the elastic behavior of the nuclear membrane and the NPCs from atomic force microscopy experiments.

Conclusions

Our work has attempted to model the nuclear membrane dynamics based on established results of creep strain. The elastic modulus creep strain model adds to the prospects for a good progress in our understanding the mechanisms of nuclear pore complex structure. It appears that the nuclear membrane of pore complex exhibits the properties of creep strain.

REFERENCES

Akey, C.W., and . Radmacher, M., 1993, *J. Cell. Biol.* 122, pp.1-19.
Akey, C.W., 1989, *J. Cell. Biol.* 109, pp. 955-970.
Martin, D., Vesenka, J.P. and, Henderson, E. and Dobbs, D.L., 1995, *Biochemistry*, 34(14), pp. 4610-4616.
Pante, N., Aebi, U., 1996, *Curr. Opin. Struct. Biol.* 4, pp. 187-196
Pines, J., 1999, "Nature," *Cell Biol.* 1, pp. 73-79.
Sinha, A.S.C., 1983, *Int. J. Systems Science*, 15, 1049

Acknowledgement

The author (RMP) thanks the Office of Professional Development of IUPUI for their support.

ARCHITECTURES FOR EQUINE GAIT ANALYSIS

S. AHMAD
Computing and Info. Science
University of Guelph, ON, Canada

D. CALVERT
Computing and Info. Science
University of Guelph, ON, Canada

D.A. STACEY
Computing and Info. Science
University of Guelph, ON, Canada

J. THOMASON
Biomedical Sciences
University of Guelph, ON, Canada

ABSTRACT

This work described the analysis of equine gait data using several artificial neural network architectures. The analysis attempts to identify characteristics of the horse, including if it is shod and at what pace it is walking. The data consists of a sequence of strain and angle readings taken from hooves of several horses. A comparison is made between different network architectures which are used to classify the data.

KEYWORDS: Bio-Medical Engineering Applications, Data Analysis, Backpropagation, SOM, Fuzzy-ART.

INTRODUCTION

The manner in which a horse walks can give indications as to the health of the animal. Problems with a horse's feet can lead to lameness and to its eventual destruction. There would therefor be considerable economic benefits to developing a system which would allow the early detection of lameness. This would provide an opportunity to treat the problem in its early stages and thereby improve the likelihood that the treatment will be successful. Anecdotal evidence shows that expert horse handlers can identify the early stages of lameness in a horse through observation. The final objective of this work is to develop a system which duplicates the expert's ability.

This work describes techniques and results from an initial examination of the data set and the data collection mechanism. It explores the utility of using strain gauges on horse's hooves to identify gait characteristics. The data is analyzed to identify if the horse is shod and at what pace it is walking.

DATA COLLECTION AND PREPROCESSING

The data collection system consisted of five mechanical sensors placed on the horse's hooves. These recorded the magnitude in two directions and the angle of strain placed on the hoof while the horse was walking. The gauges were placed around the hoof on the anterior (front), lateral side, medial side, mid-anterior medial, and mid-anterior lateral positions. The force which the horse exerts upon the sensors leads to a considerable amount of noise in the data and was occasionally sufficient to disable the device. Each sample was collected over a three minute period. The continuous data stream was divided into twenty-five samples for each stride where the time between each sample was 0.01499 second. Four horses were used to collect this data. Distinct training and testing data sets were used in all of this work, each consisting of the samples from 400 strides. Data was normalized between zero and one in all cases.

CLUSTERING EXPERIMENTS

These tests were all performed to explore the ability of the sensors to discern differences in the horse's gait caused by it being shod. The tests attempt to classify if the horse is shod or unshod given the stride data.

The first experiment uses SOM and Fuzzy ART to cluster a stride information and then apply labels to the clusters which indicate if the it represents shod or unshod patterns. Each input sequence consisted of the 25 samples for a stride and each of the sensors was tested individually. The SOM used for this work consisted of a ten by twelve hexagonal grid, with a radius of five, a learning rate of 0.05, and a training duration of 5000 epochs. As a cluster in the SOM seldomly contained a single class of inputs, a method was developed to identify the clusters which represent either shod or unshod classes. Clusters which contained less than four sequences were not considered as they contain too little information to provide an accurate classification. Experiments were conducted by applying a label to a cluster only if the patterns which mapped to it contained 80%, 90% or 100% of a single class. The results of these tests can be seen in Tables 1-3. The medial strain gauge produced the best results for 100% accuracy and the lateral strain gauge produced the best results for 90% and 80% when used to label clusters. Results using this method for classifying shod or unshod patterns were in generally in the 25%-30% accuracy range and didn't improve with larger hidden layers or different training parameters.

Mapping/Actual	Shod	Unshod	Unknown
Shod	0.234	0.0	0.266
Unshod	0.0	0.005	0.495

Table 1: Truth table when using 100% clusters for classification.

Mapping/Actual	Shod	Unshod	Unknown
Shod	0.259	0.006	0.235
Unshod	0.009	0.067	0.424

Table 2: Truth table when using 90% clusters for classification.

Mapping/Actual	Shod	Unshod	Unknown
Shod	0.308	0.03	0.162
Unshod	0.019	0.14	0.341

Table 3: Truth table when using 80% clusters for classification.

Similar experiments were conducted which use Fuzzy ART to cluster the stride data and apply labels to each cluster. The first element in a cluster created using Fuzzy ART represents the centre of the cluster. The first element was therefore used to label the cluster which it created. The network used in this work consisted of 25 input units, one for each sample within the stride. Tests were made using 30, 60, and 120 internal clustering units. The vigilance setting was 0.5 and fast learning was used. The best results using this method were from the mid-lateral anterior sensor. When using 30 hidden units the system correctly classified 80% of the training set. With 60 hidden units the accuracy increased to 83% and with 120 units accuracy was 86%. It appears that in this instance using a larger number of hidden units doesn't greatly increase the performance of the system.

CLASSIFICATION EXPERIMENTS

This group of tests used a Back-propagation network to classify whether the sample represented a shod or unshod horse. The initial test presented each of the 25 samples from a single sensor for one stride to the network. The network consisted of 25 input, 10 hidden, and 1 output unit with a learning rate of 0.3, a momentum value of 0.8, and training for 3000 epochs. The best results were achieved from the mid-anterior medial strain sensor which were 84.88% correct classification. Linear discriminant analysis was used to identify the 8 most significant samples of a 25 element stride. This led to a Backpropagation network similar to that in the previous test but with only 8 inputs which achieved a marginally better classification accuracy of 87.5%.

All of the previous experiments focussed on categorizing the stride based upon a sequence recorded from a single sensor. A Back-propagation network was also tested using data from ten sensors. Strain and angle information from each sensor was used for this test. Each of the stride sequences was reduced from 25 to 8 elements by sampling every third point. This lead to a network with 80 inputs,

20 hidden, and 1 output unit. The system was trained from 10,000 epochs, with a learning rate of 0.02, and a momentum value of 0.8. This lead to a system with a successful classification rate of 92.75%. This experiment was repeated with fifteen sensors where using every fifth data element from a stride. The network used had 80 input, 40 hidden, and 1 output node and used a learning rate of 0.8. This resulted in the successful classification 90.63% of the testing set. It appears that better classification can be achieved through the use of data from multiple sensors. This appears true even if the temporal resolution of the data is reduces by as much as 75%.

DATA REDUCTION

The final tests classify the shoe characteristics are an attempt to reduce the training set to a smaller size in order to extract a set of features which are most indicative of the classes being examined. Two tests were performed. The first attempted to cluster the stride sequence data using a SOM and use a single representative from the cluster to represent a class. The second test simply sampled random elements from the data set. To act as a baseline and demonstrate whether the SOM was a better technique for choosing representative sequences then random selection. Using the SOM to extract significant sequences produced a subset of the initial training set containing 479 stride sequences. When this subset was used to train a Back-propagation network it produces a 91.86% accuracy when used to classify the testing set. When a random selection of 500 sequences from the training set were used to train a similar network it produced a classification of 90.2% accuracy suggesting that the SOM was not successful in reducing the data set in a useful manner.

GAIT ANALYSIS

This experiment used the same set of data as those previously mentioned but was intended to classify whether the horse was either trotting or cantering. Due to the successful application of Back-propagation to the shoe classifications, the same structure of network was applied to the gait analysis. The network was presented with data from individual sensors during training and testing. It consisted of 25 input, 10 hidden, and 1 output node and was trained with a learning rate of 0.2. The mid-medial anterior strain sensor produced the best results and was capable of correctly classifying the testing set with 99% accuracy. It appears from these results that it is relatively easy to identify the horse's gait from the hoof sensor data.

CONCLUSIONS

No one sensor location or type of reading demonstrated that it was most indicative of a horse's shoe characteristics. The best classification results were achieved using a combination of sensors and both strain and angle readings. The best results for classifying the shoe were 92% and were found using a Back-propagation network. Attempts to use a SOM to identify the elements in the training set which are most representative of the sequence were not overly successful and produced results that were no better than random selection of training elements. The use of a Back-propagation network to identify gait were very successful. This suggests that it is relatively easy to distinguish gait from the hoof sensor data.

Given these results it appears that the sensor data is sufficient to identify at least some characteristics of a horse's walk. Work is continuing with this data in order to identify if the initial signs of lameness can be similarly detected using these techniques.

REFERENCES

Gail A. Carpenter, Stephen Grossberg, and David B. Rosen. Fuzzy Art: An Adaptive Resonance Algorithm for Rapid, Stable, Classification of Analog Patterns, In *Proceedings of the IEEE Conference on Neural Networks*, volume II, pages 411-416, 1994.

Gail A. Carpenter and Stephen Grossberg. Artmap: Supervised real-time learning and classification of nonstationary data by a self-organizing neural network. In Gail. A. Carpenter and Stephen Grossberg, editors, *Pattern Recognition by Self-Organizing Neural Networks*, MIT Press, 1991.

Simon Haykin, *Neural Networks*, second edition, Macmillan College, 1991.

Teuvo Kohonen. *Self-Organization and Associative Memory*, Springer-Verlag, 1989.

FEASIBILITY OF USING RECURRENT NEURAL NETWORKS TO CLASSIFY MENTAL WORKLOAD

KELLY A. GREENE, Ph.D.
USAF Air Force Test Pilot School
Edwards Air Force Base, CA

ABSTRACT

This paper summarizes a feasibility study that was conducted in order to investigate the use of recurrent neural networks for classifying mental workload. One of the more common measures of the brain's electrical activity is ongoing electroencephalography (EEG). Previous to this work, recurrent neural networks (RNNs) had never been used to classify ongoing EEG or mental workload. This study investigated the feasibility of using an Elman RNN to classify mental workload using ongoing EEG activity in the presence of noise. The results indicate that Elman RNNs show promise for classifying mental workload.

INTRODUCTION

The purpose of this feasibility study was to investigate the use of Elman recurrent neural networks (RNNs) (Elman, 1990) for classifying mental activity using electroencephalography (EEG) in the presence of noise. If the US Air Force is to ever classify pilot workload via an Elman RNN using EEG collected during flight, than an Elman RNN classifier must be robust to the effects of noise. There are many sources of potential noise in a cockpit including vibration, movement, talking on the radios, and G forces. For this feasibility study, EEG was collected from a test subject performing three types of mental activity. An Elman RNN was first trained using 10 features derived from the α EEG frequency band to classify the type of mental activity being performed. Ten test sets with varying levels of noise were used to evaluate the Elman RNN's robustness to noise. The measure of effectiveness was the test set classification accuracy CA_{test}. Next, an Elman RNN was trained using 90 features derived from nine EEG frequency bands: Δ, θ, α, α_1, α_2, β_1, β_2, $\mu\beta_1$, and $\mu\beta_2$. Again, 10 test sets with varying levels of noise were used to evaluate the Elman RNN's robustness to noise.

DATA

The test subject used in this feasibility study is a 50-year old male who is in excellent health and takes no medications. EEG was collected from the test subject at the Flight Psychophysiological Laboratory, Wright-Patterson AFB, Ohio. The Workload Assessment Monitor (WAM) (Wilson, 1994) recorded EEG from five electrodes and two reference electrodes at a sampling rate of 128 Hz. The first mental task for the test subject was to read an article from *Science* magazine. He was told that there would be a quiz after data collection so that he concentrated on reading the material (he really was never given a quiz though). The reading task was performed for three minutes. The test subject's next task was to sit quietly but with his eyes open for three minutes. The third and final task was to sit quietly but with his eyes closed for three minutes.

The WAM preprocessed the EEG signals. The preprocessed data from the WAM was then further processed using *MATLAB* code to calculate the log of the power and the variance of the power over a moving 10-second window with 50% overlap for each frequency band for each electrode. There were a total of 34 exemplars for each of the three classes of mental activity. Overall, there were 102 exemplars. Figure 1 shows the log power of the α-band collected at the Cz electrode broken into the three classes of mental activity. A unit on the x-axis of each subfigure represents a 10-second moving window. In addition, the dotted line in each of Figure 1's subfigures shows the mean log power for the corresponding mental activity class. Figure 2 shows the variance of the log power of the α-band collected at the Cz electrode. The α-band log power as shown in Figure 1 and the variance of the α-band log power as shown in Figure 2 is representative of other EEG frequency bands and other EEG electrodes used in this feasibility study.

METHODOLOGY

Ten Input Features

The first Elman RNN trained as shown in Figure 3 had a 10+20/20/3 architecture. Only the α-band features were used as inputs to the first Elman RNN trained. It is expected that the log power of the α-band will increase as the test subject transitions from "reading" to "eyes open" to "eyes closed" as shown in Figure 1. The α-band log power collected at all five electrodes exhibited this behavior. There were a total of 10 input features to the Elman RNN representing the α-band log power and the variance of the α-band log power from five electrodes. Each input feature was normalized between 0.0 and 1.0.

Twenty context nodes were used. There were three output classes: reading, eyes open, and eyes closed. The hidden/context nodes were activated by the sigmoid nonlinear transfer function. The output nodes were activated by a linear transfer function with slope = 1. The Elman RNN was trained using backpropagation with momentum and an

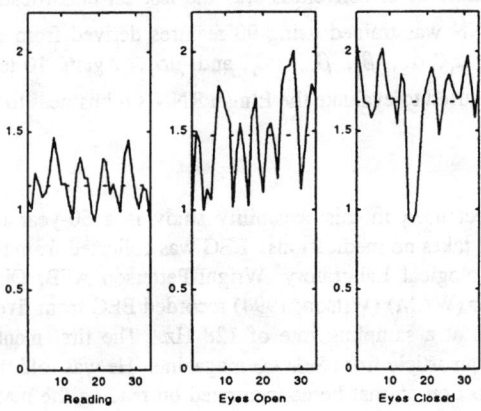

Figure 1. α-Band Log Power Collected at Electrode CZ

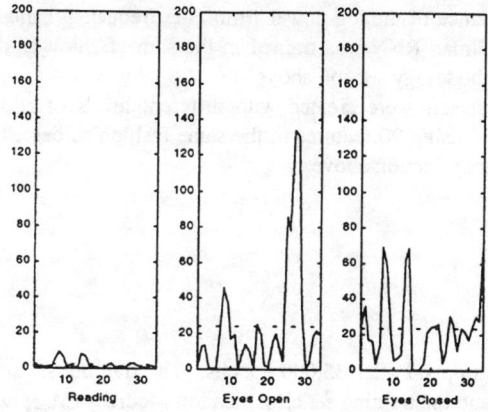

Figure 2. Variance of α-Band Log Power Collected at Electrode Cz

adaptive learning rate. The initial learning rate was set to 0.001.

Ten test sets were created with different levels of added noise to test the Elman RNN trained using 10 features from the α-band. For each of the 10 test sets, noise following a Uniform random distribution was added to each normalized input feature in the training set. For the first test set, noise following a Uniform random distribution between 0.00 and 0.05 denoted as U(0.00,0.05) was added to each input feature. For the second test set, noise following a U(0.00,0.10) distribution was added to each input feature. And so on up to a test set containing noise following a U(0.00,0.50) distribution.

Ninety Input Features

The next Elman RNN trained had a 90+180/180/3 architecture. It used all 90 features available from the collected EEG data. The 90 features represented the log

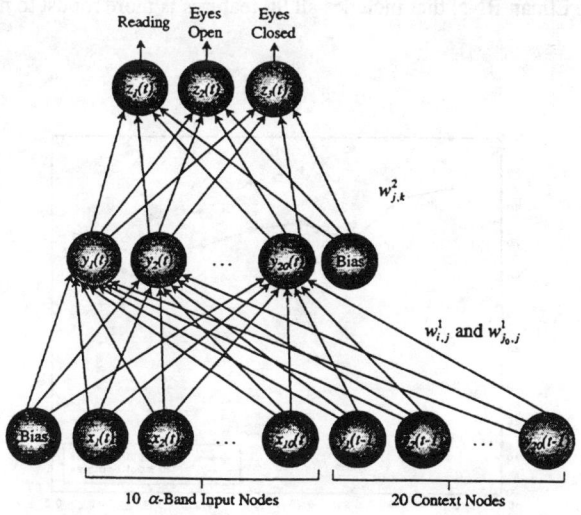

Figure 3. 10+20/20/3 Elman Recurrent Neural Network

power and the variance of the log power from nine frequency bands collected from five electrodes. This Elman RNN was trained in the same fashion as described in the Ten Input Features Methodology section above.

Again, ten test sets were created with different levels of added noise to test the Elman RNN trained using 90 features in the same fashion as described in the Ten Input Features Methodology section above.

RESULTS

Ten Input Features

Training was stopped after 35,000 epochs. The training set sum of squared errors SSE_{train} was 1.70 and the training set classification accuracy CA_{train} was 100%. The solid curve in Figure 4 is the CA_{test} using 10 features for each level of noise added. About 95% of the misclassifications from the test sets were the results of the Elman RNN misclassifying the mental activity as "reading" instead of "eyes open." This implies that the log power and the variance of the log power of the α-band from five electrodes may not be enough to separate "reading" from "eyes open" for this 50-year old male test subject. The other 5% or so of the misclassifications occurred at the transitions from "reading" to "eyes open" or from "eyes open" to "eyes closed."

Ninety Input Features

Training was stopped after 10,500 epochs. The SSE_{train} was 0.90 and the CA_{train} was 100%. The dashed curve in Figure 4 is the CA_{test} using 90 input features for each level of noise added. There does not appear to be any trend associated with the misclassifications using 90 input features. By comparing the two curves in Figure 4, it appears that an Elman RNN that includes all 90 features is more robust to noise.

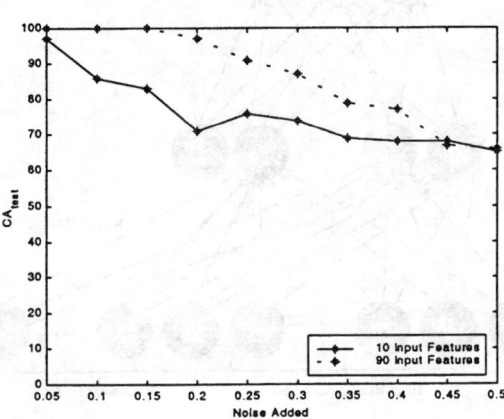

Figure 4. Test Set Classification Accuracy for Differing Levels of Noise Added

CONCLUSIONS

This feasibility study shows that an Elman RNN can adequately classify among three types of mental activity even in the presence of added noise. In both Elman RNNs trained, the CA_{train} with no added noise was 100%. With only 10 input features derived from the α-band, the CA_{test} remained greater than 80% so long as the noise added was no larger than 0.15. With all 90 input features, the CA_{test} remained greater than 80% so long as the noise added was no larger than 0.30. The Elman RNN trained with 90 features appears to be more robust to the effects of added noise.

As with anything else a lot more work could be done on the pattern recognition of EEG for the purpose of classifying mental workload. Future research in this area may improve upon several things. Feature preprocessing, selection, and screening may be used to optimize input features (Bauer et al, 2000; East et al, 2000; Greene, 1998; Greene et al, 1996; Greene et al, 1997; Greene et al, 1998; Greene et al, 2000; Laine et al, 1999).

In conclusion, the Elman RNN shows promise for classifying pilot workload in addition to air traffic controller workload.

ACKNOWLEDGEMENTS

The author would like to thank Ms Corina Monet and the Flight Psychophysiological Laboratory, Wright-Patterson AFB, Ohio, for the data collection effort used in this feasibility study.

NOMENCLATURE

CA_{test}	test set classification accuracy
CA_{train}	training set classification accuracy
SSE_{train}	training set sum of squared errors
$w^2_{j,k}$	second layer weight connecting hidden node j to output node k
$w^1_{i,j}$	first layer weight connecting input feature i to hidden node j
$w^1_{jo,j}$	first layer weight connecting context node j_0 to hidden node j
$x_i(t)$	input feature i at time t
$y_j(t)$	hidden/context node j at time t
$z_k(t)$	output node k at time t
Δ	Delta EEG frequency band, 1.0 – 3.0 Hz
θ	Theta EEG frequency band, 4.0 – 7.0 Hz
α	Alpha EEG frequency band, 8.0 – 11.0 Hz
α_1	Alpha1 EEG frequency band, 8.0 – 9.0 Hz
α_2	Alpha2 EEG frequency band, 10.0 – 11.0 Hz
β_1	Beta1 EEG frequency band, 12.0 – 14.0 Hz
β_2	Beta2 EEG frequency band, 15.0 – 30.0 Hz
$\mu\beta_1$	UltraBeta1 EEG frequency band, 31.0 – 36.0 Hz
$\mu\beta_2$	UltraBeta2 EEG frequency band, 37.0 – 42.0 Hz

REFERENCES

Bauer, K.W., Alsing, S.A., and Greene, K.A. "Feature Screening Using Signal-to-Noise Ratios," *Neurocomputing*, 9 Mar 00, Vol 31, Issue 1-4, pp. 29-44.

East, J.A., Bauer, K.W., and Lanning, J.W. "Feature Selection for Predicted Pilot Mental Workload," *Intelligent Engineering Systems through Artificial Neural Networks*, Proceedings of Artificial Neural Networks in Engineering (ANNIE) International Conference, St. Louis, MO, 5-8 Nov 2000, Vol 10.

Elman, J.L. "Finding Structure in Time," *Cognitive Science*, Vol 14, 1990, pp. 179-211.

Greene, K.A. *Feature Saliency in Artificial Neural Networks with Application to Modeling Workload*. Ph.D. Dissertation, Air Force Institute of Technology, Wright-Patterson Air Force Base, OH, 1998.

Greene, K.A., Bauer, K.W., Kabrisky, M., Rogers, S.K., Russell, C.A., and Wilson, G.F. "A Preliminary Investigation of Selection of EEG and Psychophysiological Features for Classifying Pilot Workload," *Intelligent Engineering Systems through Artificial Neural Networks*, Proceedings of Artificial Neural Networks in Engineering (ANNIE) International Conference, St. Louis, MO, 10-13 Nov 1996, Vol 6, pp. 691-697.

Greene, K.A., Bauer, K.W., Kabrisky, M., Rogers, S.K., and Wilson, G.F. "Estimating Pilot Workload Using Elman Recurrent Neural Networks: A Preliminary Investigation," *Intelligent Engineering Systems through Artificial Neural Networks*, Proceedings of Artificial Neural Networks in Engineering (ANNIE) International Conference, St. Louis, MO, 9-12 Nov 1997, Vol 7, pp. 703-708.

Greene, K.A., Bauer, K.W., Kabrisky, M., Rogers, S.K., and Wilson, G.F. "Determining the Memory Capacity of an Elman Recurrent Neural Network," *Intelligent Engineering Systems through Artificial Neural Networks*, Proceedings of Artificial Neural Networks in Engineering (ANNIE) International Conference, St. Louis, MO, 1-4 Nov 1998, Vol 8, pp. 37-42.

Greene, K.A., Bauer, K.W., Wilson, G.F., Russell, C.A., Rogers, S.K., and Kabrisky, M. "Selection of Psychophysiological Features for Classifying Air Traffic Controller Workload in Neural Networks," *International Journal of Smart Engineering System Design*, Vol 2, 2000, pp. 315-330.

Laine, T.I., Bauer K.W., and Lanning, J.W. "Multiple Crewmember Workload Classification using Neural Networks with Input Feature Selection," *Intelligent Engineering Systems through Artificial Neural Networks*, Proceedings of Artificial Neural Networks in Engineering (ANNIE) International Conference, St. Louis, MO, 7-10 Nov 1999, Vol 9.

Wilson, G.F. "Workload Assessment Monitor," *Proceedings of the Human Factors Society*, 1994, pp. 944.

FEASIBILITY OF USING TIME DELAY NEURAL NETWORKS TO CLASSIFY MENTAL WORKLOAD

KELLY A. GREENE, Ph.D.
US Air Force Test Pilot School
Edwards Air Force Base, CA

ABSTRACT
This paper provides a summary of a feasibility study that was conducted in order to investigate the use of time delay neural networks (TDNNs) for classifying mental workload. A common measure of the brain's electrical activity is the evoked potential (EP) as derived from electroencephalography (EEG). Previous to this work, TDNNs had never been used to classify EPs. This study investigated the feasibility of using a TDNN to detect an EP in an EEG signal. The results indicate that a TDNN unfortunately does not show promise for EP detection.

INTRODUCTION

Whereas ongoing electroencephalography (EEG) focuses on a continuous recording of spontaneous brain electrical activity, the analysis of evoked potentials (EPs) focuses on segments of EEG activity that are "evoked" by specific stimulus events. EPs are the small changes in voltage in EEG that are time locked to a stimulus or cognitive event. Two averaged EPs and several components (N1, P2, N2, and P3) are shown in Figure 1 (Wilson et al, 1994). The EPs in Figure 1 were collected from a USAF F-4 pilot performing the *oddball paradigm* in which two different tones were monaurally delivered to the pilot via a small speaker placed inside his helmet ear cup (Wilson et al, 1994). The trials were repeated 100 times and the pilot was instructed to covertly count the number of times one of the tones was presented and then report this number at the end of the test (Wilson et al, 1994). The averaged EP on the left side of Figure 1 was collected during a baseline condition while the pilot was performing the oddball paradigm only (i.e. not

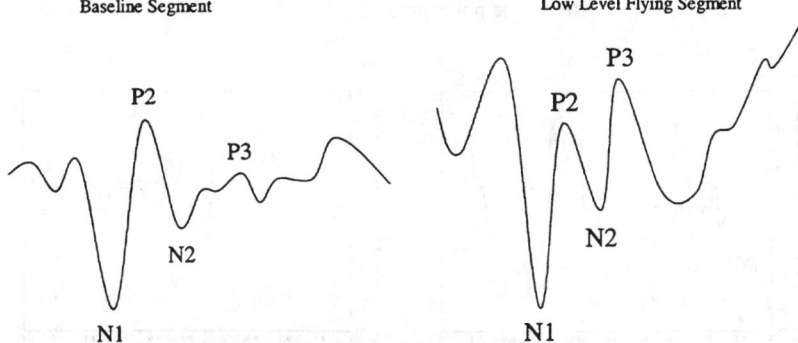

Figure 1. Averaged Evoked Potential Collected from F-4 Pilot Performing Oddball Paradigm (Adapted from Wilson et al, 1994, with Permission from Dr. Wilson)

flying). The other averaged EP was collected while the pilot was flying low level and performing the oddball paradigm.

Analysis of EPs is predominantly accomplished using averaged EPs because ongoing background EEG is much larger in amplitude than the EP (especially at higher EEG frequencies). In fact, the Signal-to-Noise Ratio (SNR) between an EP and EEG is typically around −20 dB.

Most attempts at single EP analysis use a template derived from the averaged EP but have thus far been unreliable (Caldwell et el, 1994). Only recently have investigators begun to focus on single EP responses which typically last only 0.2 seconds. The major problem is determining what portion of the EEG signal is evoked by the response to the stimulus and what portion represents the continuation of ongoing background EEG. Unfortunately, background EEG typically looks like noise. If EPs are ever to be used to classify pilot workload or air traffic controller workload, then the ability to *detect* and then *classify* single EPs must be possible. The purpose of this feasibility study was to investigate the use of TDNNs to *detect* single EP responses in an EEG signal. This feasibility study has two parts. The first part attempted to detect a rectangle pulse in EEG at five varying SNRs. The second part attempted to detect an EP in EEG at five varying SNRs.

DATA

Rectangle Pulse

A simulated EEG signal was generated at a sampling rate of 50 Hz by summing five incommensurate sine waves so that each sine wave represented an EEG band. A total of 1000 samples representing 20 seconds were created such that the maximum peak amplitude was less than 20 microvolts. Next, 17 rectangle pulses lasting 0.2 seconds (10 samples each) were randomly placed throughout the EEG signal. Five time series were created such that the SNR between the rectangle pulse and the EEG were different. An example is shown on the left side of Figure 2. Each sample of the time series was placed into one of four classes:

Class 1.	EEG only
Class 2.	Slight chance that a rectangle pulse is present
Class 3.	Rectangle pulse is more than likely present
Class 4.	Rectangle pulse present

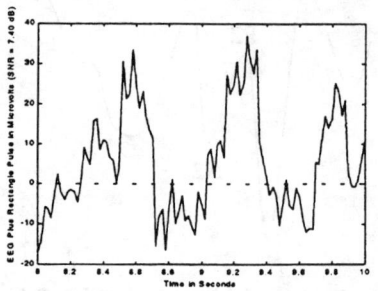
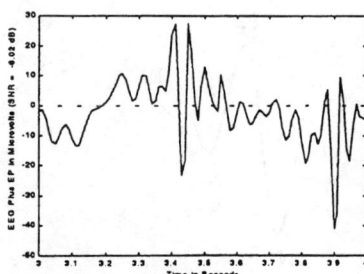

Figure 2. Example Time Series Plots

Class 1 contained all time samples that consisted of only EEG (no rectangle pulse) in addition to the first two time samples of the rectangle pulse. Class 2 contained the next three time samples of the rectangle pulse (samples 3-5). Class 3 contained the next three time samples of the rectangle pulse (samples 6-8). Finally, class 4 contained the last two time samples of the rectangle pulse (samples 9-10).

Evoked Potential (EP)

Next, the EP on the left side of Figure 1 replaced the rectangle pulse and the sampling rate was increased to 100 Hz. With the higher sampling rate, there were now 2000 time samples of the EEG signal in 20 seconds and the 17 randomly placed EPs lasting 0.2 seconds each contained 20 samples. Note in Figure 1 that the EP has two high peaks (P2 and P3) and two low peaks (N1 and N2) thus allowing something for a TDNN to pick up on for classification purposes. Five time series were created such that the SNR between the rectangle pulse and the EEG were different. An example is shown on the right side of Figure 2. Each sample of the time series was placed into one of four classes in a fashion similar to that described in the Rectangle Pulse Data section above.

METHODOLOGY

Rectangle Pulse

A time lag of 10 periods was used in a 11/25/4 TDNN as shown in Figure 3 for rectangle pulse classification to account for the time of the rectangle pulse. The effective number of exemplars became 990 ($1000 - 10 = 990$) due to the lags required. The training set contained 495 exemplars, the test set contained 248 exemplars, and the validation set contained 247 exemplars. All inputs were standardized such that each input feature had zero mean and unit variance. All weights were initialized between -0.5 and 0.5. A separate TDNN was trained for each of the varying SNRs via instantaneous backpropagation using a fixed learning rate of 0.3 and no momentum.

Evoked Potential (EP)

A time lag of 20 periods was used in a 21/50/4 TDNN for EP classification to account for the time of the EP. The effective number of exemplars became 1980 ($2000 - 20 = 1980$) due to the lags required. The training set contained 990 exemplars,

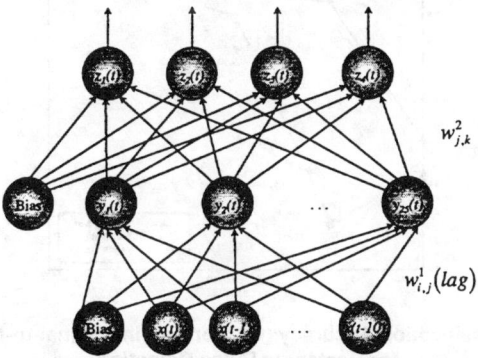

Figure 3. 11/25/4 Time Delay Neural Network

the test set contained 495 exemplars, and the validation set contained 495 exemplars. All inputs were standardized such that each input feature had zero mean and unit variance. All weights were initialized between -0.5 and 0.5. A TDNN was trained for each of the varying SNRs via instantaneous backpropagation using a fixed learning rate of 0.3 and no momentum.

RESULTS

Rectangle Pulse

A total of ten TDNNs were trained. Figure 4 plots the classification accuracy (CA) for the training, test, and validation sets for the trained TDNNs. The TDNNs for rectangle pulse classification performed adequately when the SNR was +27.40 dB, +17.41 dB, or 7.40 dB. When the SNR was -2.59 dB or -12.60 dB, the TDNN for rectangle pulse classification did not perform adequately. At -2.59 dB, the TDNN did not performed adequately on the validation set and is evidenced by only 18.75% of the Class 2 exemplars being correctly classified, 22.22% of the Class 3 exemplars being correctly classified, and 16.67% of the Class 4 exemplars being correctly classified. In the majority of misclassifications at the -2.59 dB level, the exemplar was misclassified as belonging to Class 1.

The TDNN for rectangle pulse classification at -12.60 dB did not perform adequately on its validation set, either. In fact, 0.00% of the Class 2, Class 3, and Class 4 validation exemplars were correctly classified. All but two validation exemplars were classified as belonging to Class 1.

Evoked Potential (EP)

A total of ten TDNNs were trained. Figure 5 plots the CA for the training, test, and validation sets for the trained TDNNs. The TDNNs for EP classification performed adequately when the SNR was -1.59 dB, −3.53 dB, -6.02 dB, and -9.55 dB. The TDNN did surprisingly well when the SNR was -9.55 dB. The TDNN for EP classification at -15.57 dB did not perform adequately but it performed better than the TDNN for rectangle pulse classification at -12.82 dB. At −15.57 dB, the TDNN for EP classification correctly classified 96.89% of the validation Class 1 exemplars, 14.29% of the Class 2 exemplars, 26.67% of the Class 3 exemplars, and 15.79% of the Class 4 exemplars.

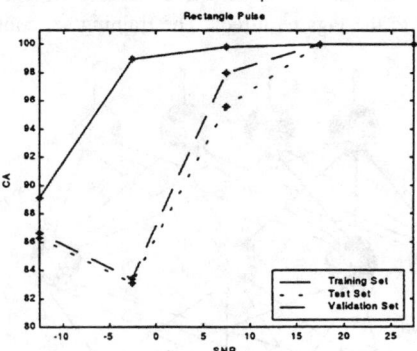

Figure 4. Classification Accuracy (CA) for Varying Signal-to-Noise Ratios for Rectangle Pulse Detection

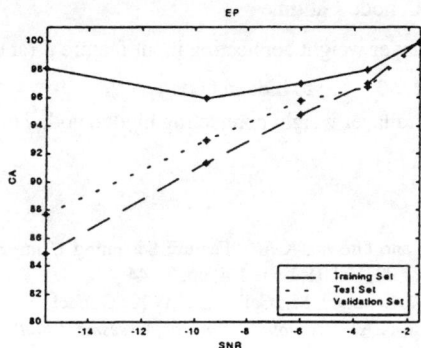

Figure 5. Classification Accuracy (CA) for Varying Signal-to-Noise Ratios for Evoked Potential Detection

CONCLUSIONS

Since the actual SNR between an EP and EEG is -20 dB, single event EP classification via a TDNN does not show promise for modeling of pilot workload or air traffic controller workload.

As with anything, a lot more work could be done on the pattern recognition of EPs in an EEG signal. Future research in this area may improve upon several things. This feasibility study did not consider optimization of the number of time lags. Applying Takens' Theorem can provide an upper and lower bound to the number of time lags using the fractal dimension of the EEG signal, the EP, or the EEG signal with the EP embedded (Takens, 1984). Another idea may be to utilize feature preprocessing, selection, and screening methods to help select the optimal number of time lags (Bauer et al, 2000; East et al, 2000; Greene, 1998; Greene et al, 1996; Greene et al, 1997; Greene et al, 1998; Greene et al, 2000; Laine et al, 1999).

This feasibility study simply used the raw amplitude of the time series. There may be other features that can provide valuable information to a TDNN for classifying an EP. For example, an average of the time samples over a fixed window may be utilized in an attempt to smooth the "noise" of the EEG. The standard deviation of a fixed number of time samples may also provide a measure of the fluctuations in the time series so that a high standard deviation may flag the presence of an EP starting or an EP ending and a low standard deviation would indicate the presence of no EP.

An EEG signal with a rectangle pulse randomly placed throughout could be classified via a 11/25/4 TDNN when the SNR is +27.40 dB, +17.41 dB, or 7.40 dB. An EEG signal with an EP randomly placed throughout could be classified via a 21/50/4 TDNN when the SNR is -1.59 dB, -3.53 dB, −6.02 dB, and -9.55 dB. The actual SNR between a typical EP and EEG is -20 dB. In conclusion, a TDNN will more than likely not be able to classify single event EPs in real EEG data.

NOMENCLATURE

$x(t)$ input feature at time t or at time lag t-1, t-2, t-3, ...

$y_j(t)$ hidden node j at time t

$z_k(t)$ output node k at time t

$w_{i,j}^1(lag)$ first layer weight connecting input feature i (at time t or at time lag t-1, t-2, t-3, ...) to hidden node j

$w_{j,k}^2$ second layer weight connecting hidden node j to output node k

REFERENCES

Bauer, K.W., Alsing, S.A., and Greene, K.A. "Feature Screening Using Signal-to-Noise Ratios," *Neurocomputing*, 9 Mar 00, Vol 31, Issue 1-4, pp. 29-44.

Caldwell, J.A., Wilson, G.F., Centiguc, M., Gaillard, A.W.K., Gundel, A., Lagarde, D., Makeig, S., Myhre, G., and Wright, N.A. *Psychophysiological Assessment Methods*, Advisory Group for Aerospace Research and Development (AGARD) Advisory Report 324. Paris, France: AGARD, 1994.

East, J.A., Bauer, K.W., and Lanning, J.W. "Feature Selection for Predicted Pilot Mental Workload," *Intelligent Engineering Systems through Artificial Neural Networks*, Proceedings of Artificial Neural Networks in Engineering (ANNIE) International Conference, St. Louis, MO, 5-8 Nov 2000, Vol 10.

Greene, K.A. *Feature Saliency in Artificial Neural Networks with Application to Modeling Workload*. Ph.D. Dissertation, Air Force Institute of Technology, Wright-Patterson Air Force Base, OH, 1998.

Greene, K.A., Bauer, K.W., Kabrisky, M., Rogers, S.K., Russell, C.A., and Wilson, G.F. "A Preliminary Investigation of Selection of EEG and Psychophysiological Features for Classifying Pilot Workload," *Intelligent Engineering Systems through Artificial Neural Networks*, Proceedings of Artificial Neural Networks in Engineering (ANNIE) International Conference, St. Louis, MO, 10-13 Nov 1996, Vol 6, pp. 691-697.

Greene, K.A., Bauer, K.W., Kabrisky, M., Rogers, S.K., and Wilson, G.F. "Estimating Pilot Workload Using Elman Recurrent Neural Networks: A Preliminary Investigation," *Intelligent Engineering Systems through Artificial Neural Networks*, Proceedings of Artificial Neural Networks in Engineering (ANNIE) International Conference, St. Louis, MO, 9-12 Nov 1997, Vol 7, pp. 703-708.

Greene, K.A., Bauer, K.W., Kabrisky, M., Rogers, S.K., and Wilson, G.F. "Determining the Memory Capacity of an Elman Recurrent Neural Network," *Intelligent Engineering Systems through Artificial Neural Networks*, Proceedings of Artificial Neural Networks in Engineering (ANNIE) International Conference, St. Louis, MO, 1-4 Nov 1998, Vol 8, pp. 37-42.

Greene, K.A., Bauer, K.W., Wilson, G.F., Russell, C.A., Rogers, S.K., and Kabrisky, M. "Selection of Psychophysiological Features for Classifying Air Traffic Controller Workload in Neural Networks," *International Journal of Smart Engineering System Design*, Vol 2, 2000, pp. 315-330.

Laine, T.I., Bauer K.W., and Lanning, J.W. "Multiple Crewmember Workload Classification using Neural Networks with Input Feature Selection," *Intelligent Engineering Systems through Artificial Neural Networks*, Proceedings of Artificial Neural Networks in Engineering (ANNIE) International Conference, St. Louis, MO, 7-10 Nov 1999, Vol 9.

Takens, F. "On the Numerical Determination of the Dimension of an Attractor," *Lecture Notes in Mathematics, Proceedings of Dynamical Systems and Bifurication Workshop*, Groningen, The Netherlands, Vol 1125, Apr 1984, pp. 99-106.

Wilson, G.F., Fullenkamp, P., and Davis, I. "Evoked Potential, Cardiac, Blink, and Respiration Measures of Pilot Workload in Air-to-Ground Missions," *Aviation, Space, and Environmental Medicine*, Feb 1994, pp. 100-105.

MODELING OF HEART RATE VARIABILITY

RUPA BALAN
University of Oklahoma
Electrical and Computer Engr
Norman, OK 73019
STEPHEN S. HULL, JR
University of Oklahoma HSC
Oklahoma City, OK 73104

SUDHIR H. RAI
University of Oklahoma,
Electrical and Computer Engr
Norman, OK 73019
JOHN Y. CHEUNG
University of Oklahoma
Electrical and Computer Engr
jcheung@ou.edu

ABSTRACT
Neural networks have been modeled to classify the test subjects into various categories viz. Mi Susceptible, Non-Mi Susceptible, Mi Resistant and Mi-dead based on the heart rate variability data in the time and frequency domains. Initially all the available data were used to train and test the various networks designed. Networks trained using Gaussian threshold and nine neurons, gave the same accuracy as networks using large number of neurons. Hence this design was used to train and test the six data sets, four of which used various types of pre-processing and windows (overlapping data points), one used windows and no pre-processing and one did not use both windows and pre-processing. In the case of the six data sets, 60% data was used as train data and 40% data was used as test data. It was found that pre-processing did not give a greater accuracy but using windows did give a greater accuracy.

INTRODUCTION
The goal of the present study is to build a neural network system to model the heart rate variability of test subjects. Heart Rate Variability refers to the precise description of the time interval series of the interbeat time typically resolved to a few milliseconds. Earlier researchers [1] had tested the hypothesis that low values of heart rate variability provided risk assessment both before and after myocardial infarction with use of an established test subject model of sudden cardiac death. In resistant test subjects, myocardial infarction did not affect any measure of heart rate variability. By contrast, after myocardial infarction, susceptible test subjects showed significant decreases in all measures of heart rate variability.

EXPERIMENTAL DATA
More detailed description of the experimental setup and how the data was obtained has been detailed elsewhere [1]. For the present study, the digitized electrocardiogram data was used to obtain the beat-to-beat interval. This constitutes the heart period signal in the time domain data. This data series is analyzed in the time domain data to obtain the mean RR interval, the standard

deviation of mean RR interval, the coefficient of variance, the 3 db width of the histogram and the 6 db width of the histogram. The same data is also analyzed in the frequency domain. The frequency domain data include the spectral power of very low frequencies, low frequencies, and high frequencies.

PRE-PROCESSING

In order to eliminate any inconsistencies in the raw data of each test subject, the raw data was pre-processed in four ways before the data was actually processed into the time domain and frequency domain parameters. Due to the nature of the data representing the heart period between beats, the data can be regarded as uniform sampling by some researchers and non-uniform sampling by others. The main task of preprocessing is to eliminate data points related to premature ventricular contractions that generally caused long periods of inaccurate heart rate data. Four methods were used:

Elimination: In this method, any heartbeat with heart period greater or lesser than a certain preset value are simply removed form the series. While chronologically speaking, there is a whole missing period there. But from the data series is concerned, the gap does not exist.

Replace by single mean: In this method, any heartbeat with heart period greater or lesser than a certain preset value are replaced by the mean value of that heart beats for that individual.

Replace by maximum/minimum: In this method, any heart beat with heart period value greater than maximum value of the test subject's heart beats by that value and any value less than the minimum by the minimum value.

Replace by appropriate number of mean values: Replace any value greater than the maximum value of the test subject's heartbeats by n values of value/n where n=value/mean.

The pre-processed data for each subject are all processed using windows in both time domain and frequency domain. 512 points are processed at a time and the step size is 128. This is shown below.

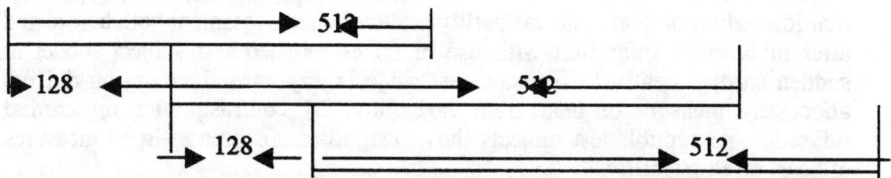

Then the average gives the resultant values in time domain and frequency domain for each subject. Raw data processed directly to obtain HRV data in time domain and frequency domain was obtained both with and without windows. Hence, there were six sets of processed data altogether, out of which four are pre-processed in four different ways and two were not involved in any pre-processing and are processed with and without windows respectively.

NEURAL NETWORK DESIGNS AND IMPLEMENTATIONS

The major difficulty with the present data set is that there is a lack of data for each category and each category does not necessarily have the same number of data points. Since these are real experimental data, it is not possible to control the number of data points for each category. Four different experimental setups have been designed to test the efficacy of neural networks for classifying the obtained data.

First the classifications were carried out using various designs of neural networks. All the data were used as train data and test data i.e. 100% data as test data and 100% data as train data. Here the raw data was processed directly into Heart Rate Variability data in time domain and frequency domain without using windows. Only one test run of each test program was carried out. Only in the case of neural network design using nine neurons and Gaussian threshold, two test runs were carried out to facilitate comparison with the neural network implemented using each of the six data sets. Designs in sections 1 to 5 in observations and results use 100% data as test data and 100% data as train data. They do not use pre-processing and processing by windows.

Then the classifications were carried out using 60% of the data as train data and 40% of the data as test data on each of the six data sets. For all the data sets at least five test runs of the test program was carried out. Only for the data set without pre-processing and not using windows, only two test runs were carried out. The results are given in later sections. In all cases, the transfer function used was the tangent sigmoidal function, the training algorithm used was the trainbfg function provided in MATLAB and a 3 layer feed-forward network was used. Only in the case of classification of all categories in the time domain using 50 hidden neurons, the training algorithm used was trainscg.

RESULTS

The primary test for the efficacy of the neural network classification was by means of the confusion matrix built from the prediction of the network on training and unseen testing data. The confusion matrix gave the number of test subjects classified into the given category i.e. the number of test subjects classified correctly and the number of test subjects classified into another category i.e. the number of test subjects classified incorrectly based on the threshold. For each classification, the number of test subjects whose output did not fall in the specified range of any category, if any, are mentioned. In case of multiple test runs, he confusion matrix gave the mean of all the test runs. Based on the confusion matrix, the error of the network based on the number of incorrect classifications and the number not classified into any category and the accuracy of the network for each classification is obtained.

Comparison With Different Number Of Training Data

In the case of most classifications of various categories of test subjects using all data as train data and all data as test data with various designs, the percentage error is inversely proportional to the number of hidden neurons. But when number of hidden neurons increases, number of epochs required for training and hence time needed is more. But a low error goal is attainable. Less number of neurons means number of epochs needed for training is less. But the error goal attained is high.

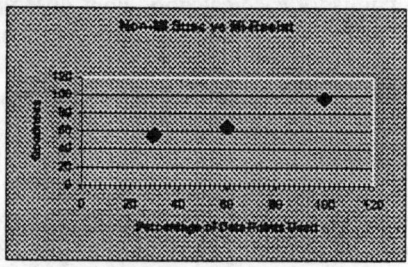

Comparison With Different Number Of Hidden Neurons

The training with less number of neurons was done because it was felt that networks with large number of neurons might be memorizing data. But the accuracy of the networks with less number of neurons is not very less compared to networks with more neurons.

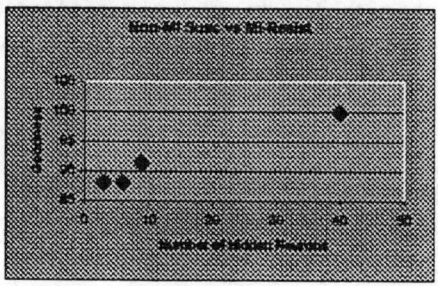

Comparison With The Use Of Gaussian Threshold

With Gaussian threshold and nine neurons, the accuracy of the network is comparable to the accuracy of networks with large number of neurons. Hence this network with nine neurons and Gaussian threshold is used to obtain classifications with each of the six data sets.

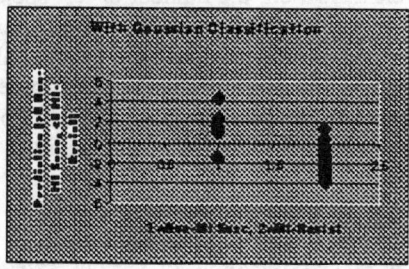

Comparison With The Use Of Preprocessing Strategy

Thus the overall accuracy of each of the six data sets can be found by taking the mean of the accuracies of each of the four classifications i.e. Mi-Resistant vs. Non-Mi Susceptible in time domain, Mi-Resistant vs. Non-Mi Susceptible in frequency domain, Mi Resistant vs. Mi Susceptible in time domain, Mi Resistant vs. Mi Susceptible in frequency domain. It is as follows:

Pre-processing by elimination: Not applicable
Pre-processing by replacing by maximum/minimum: 60.5%
Pre-processing by replacing by single mean: 60.86%
Pre-processing by replacing by appropriate number of means: 60.86%
No Pre-processing and using windows: 61.56%
No Pre-processing and not using windows: 57.5%

Pre-Processing Techniques	Non-MI Susceptible vs. MI-Resistant in the time domain	Non-MI Susceptible vs. MI Resistant in the frequency domain	MI Susceptible vs. MI Resistant in the time domain	MI Susceptible vs. MI Resistant in the frequency domain
Elimination	70.25	75.21	0	47.8
Maximum-Minimum	72.5	60	54.11	55.4
Single Mean	61.3	68.75	54.6	68.75
Appropriate number of means	63.87	66.66	55.88	57.06
No Pre-processing	67.5	71.5	50.85	55.4

OBSERVATIONS

Thus the accuracies of the data sets using pre-processing except the one using elimination are not very different from one another and also not different from the data set using no pre-processing and using windows. But the accuracy of the data set using no pre-processing and no windows is less than that with windows. Thus pre-processing does not improve the results much, but using windows does give a greater accuracy. But the accuracy obtained using all test subjects as train data and all test subjects as test data is much greater than that obtained from the six data sets using 60% as train data and 40% as test data showing that if more number of test subjects of each category are taken into consideration, then with more training and test data available, training with less neurons would give greater accuracy.

Finally, the accuracy of the networks can be enhanced if the weights of the neurons, the number of hidden neurons, and the number of hidden layers are properly adjusted to obtain the desired results. Efficient algorithms other than the ones implemented can be used to make the network more reliable for the classification of various categories of test subjects.

REFERENCES

Demuth H and Beale M, 1994,"Neural Networks Toolbox User's Guide, For Use with Matlab", The MathWorks, Inc., Natick, MA.

Haykin S, 1994,"Neural Networks, A Comprehensive Foundation", Macmillan College Publishing Company, New York.

Hull S.S et al, 1990, "Heart Rate Variability Before and After Myocardial Infarction in Conscious Test subjects at High and Low Risk of Sudden Death ", JACC, Vol.16, No.4, pp. 978-985.

DYNAMIC MODELING OF BEHAVIOR OF CENTRAL OLFACTORY SYSTEM: SIMULATION ANALYSIS AND INFORMATION PROCESSING

NATACHA GUEORGUIEVA
Department of Computer Science
College of Staten Island/CUNY
2800 Victory Blvd.
Staten Island, NY 10314

IREN VALOVA
Dept. of Comp. and Inf. Sciences
Univ. of Massachusetts Dartmouth
285 Old Westport Rd.
North Dartmouth, MA 02747

GEORGE GEORGIEV
Department of Computer Science
University of Wisconsin Oshkosh
800 Algoma Blvd., Oshkosh, WI 54901

ABSTRACT
We discuss the first few stages of olfactory processing in the framework of a layered structure. Our model consists of inhibitory and excitatory formal neurons with dendrodendritic interactions. We show that in a noisy background the network functions as an associative memory, in spite of the fact that the network operates in an oscillatory mode. When receiving a complex input that is composed of several odors, the network segments it into its components.

INTRODUCTION
Although we now know much of the structure and functional attributes of the brain, and can understand some of the its mechanisms of information processing, we lack certainty regarding the overall dynamical properties of the brain. That is, we have little idea how the subcomponents of the brain are integrated into a functional whole.

Being an outgrowth of the forebrain, the olfactory bulb provides us with an open window to the brain (Davis & Eichenbaum 1991) (Shepherd 1979). It receives all of the chemical sensory input from the olfactory epithelium, and projects a processed transformation of it to the cortex. Although thoroughly explored experimentally, the functionality of the synaptic organization, and the computational properties of the bulb are yet unclear.

The main olfactory system can be described as a three-layer system (Masson et al. 1993; Laurent & Davidowitz, 1994): sensory neurons build up the first layer; the second layer (the antennal lobe of insects, the olfactory bulb of mammals) features relay neurons, whose connections with the axons of sensory neurons are located in neuropilar structures called glomeruli; the third layer is built up of the cortical regions where axons of neurons from the second layer project (mushroom bodies in insects, piriform cortex in mammals).

Chemical sensory information enters the bulb via the olfactory nerve that converges onto the glomeruli structures. In the glomeruli, axons of sensory cells terminate on the distal dendrites of excitatory mitral/tufted' cells (Shepherd 1992), that in turn project their output to the olfactory cortex and other processing centers. Building on the models of (Freeman, 1998) and (Hopfield 1991) we construct a model of the glomerular layer that is capable of performing odor separation from odor signals entering the bulb.

The second computational stage in our model is the external plexiform layer. Recent studies (Hendin et al. 1994), we have addressed the form of information processing that takes place in this layer via interactions of projection mitral cells with inhibitory granule cells. These interactions are dendrodendritic in nature, and are mediated by reciprocal synaptic connections.

Figure 1 shows the layered structure of the olfactory system.

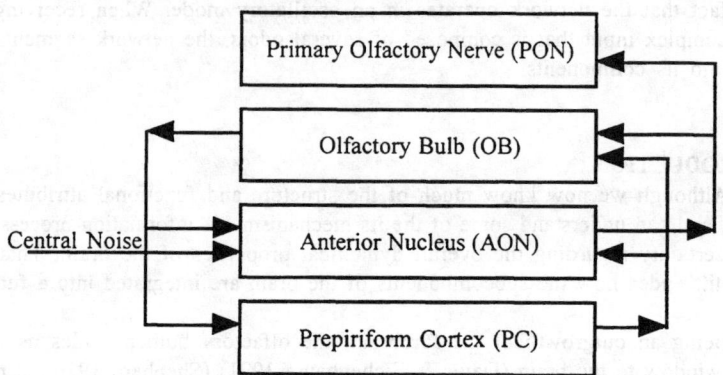

Figure 1 Structure of the olfactory system

MODELS OF THE BULB LAYERS

The olfactory bulb model can be separated into several layers such as glomerular, mitral cells, and granule cell layers. The first synapse of the olfactory sensory neurons is in the glomeruli structures. There, sensory axons terminate on the distal dendrites of mitral/tufted cells. Outputs of sensory neurons in the olfactory epitelium are projected into periglomerular cells and mitral cells. The periglomerular cells and mitral cells are connected to the mitral cells with inhibitory synapses (Shepherd 1979). Outputs of the mitral cells interconnect to each other with excitatory and inhibitory synaptic connections. The granule cells receive feedback input from the olfactory cortex, and the input is transfered to the mitral cells through the inhibitory synaptic connections. Following (Hopfield 1991) we assume that the mitral cells, whose membrane potential will be denoted by u_i, undergo the dynamics

$$\tau_{m,i} \frac{dx_{m,i}(t)}{dt} = -x_{m,i}(t) + \sum_{j=1}^{N} s_{mm,ij} U_{m,j}(t) + \sum_{k=1, k \neq i}^{N} s_{mp,ik} U_{p.k}(t) - \sum_{j=1}^{N} s_{mg,ij} V_{g,j}(t) + s_{mp,ii} V_{p,i}(t) + I_{m,i}(t) \quad (1)$$

where $x_{m,i}(t)$ is the cell potential of the i-th mitral cell at time t; $\tau_{m,i}$ is the decay time of mitral cell; N is the number of mitral cells; $s_{mm,ij}$, $s_{mg,ij}$, and $s_{mp,ii}$ are the strength of the axodendritic synaptic connections from mitral cell j to mitral cell i, and of the dendrodendritic synaptic connection from granule cell j to mitral cell i, and from periglomerular cell to mitral cell. $V_{g,j}(t)$ and $V_{p,i}(t)$ are dendtritic outputs of i-th periglomerular cell, and i-th granule cell respectively. $U_{m,j}(t)$ and $U_{p.k}(t)$ are axonal outputs of i-th mitral cell and i-th periclomerular cell and i-th granule cell respectively. $I_{m,i}(t)$ are the inputs from the olfactory nerves sensitive to a particular odorant molecule to the i-th mitral cell. Figure 2 outlines the olfactory bulb layer model based on the discussed mathematical model.

$$\tau_{p,i} \frac{dx_{p,i}(t)}{dt} = -x_{p,i}(t) + \sum_{j=1}^{N} s_{pm,ij} V_{m,j}(t) + I_{p,i}(t) \quad (2)$$

$$\tau_{g,i} \frac{dx_{g,i}(t)}{dt} = -x_{g,i}(t) + \sum_{j=1}^{N} s_{gm,ij} V_{m,j}(t) + I_{g,i}(t) \quad (3)$$

where $x_{p,i}$ and $x_{g,i}$ are the cell potentials of i-th periglomerular cell, and i-th granule cell at time t. $\tau_{p,i}$ and $\tau_{g,i}$ are the decay times of these cell potentials; $s_{pm,ij}$ and $s_{gm,ij}$ are the strengths of the axodendritic synaptic connections from mitral cell to periglomerular cell, and from mitral cell to granule cell respectively. $V_{m,j}(t)$ is the axonal output from i-th mitral cell; $I_{p,i}(t)$ are the inputs from the olfactory nerves sensitive to a particular odorant molecule to the pareglomerular cell i; $I_{g,i}(t)$ is the feedback input from the olfactory cortex to the granule cell i.

The olfactory layer is connected to the AON layer. AON receives an oscillatory input from the bulb while the bulb receives non-oscillatory input and it has excitatory-to-excitatory connections, while the bulb module does not. This layer is governed by similar system of differential equations as the olfactory layer, however the parameters which are considered characterize the specifics of this layer.

There are two major differences between the interactions of the mitral and granular neurons of olfactory bulb, as observed in the AON and in the cortex. The first is that there is no direct evidence for excitatory-excitatory interactions in the OB. The second is that most interactions are dendrodendritic in contradiction to the axodendritic interactions that are the rule in the AON and in the cortex.

$$dv_i/dt + \mu v_i + \eta h_{granule}(w_i) = \sum_j K_{ij}^0 h_{mitral}(v_j) - \sum_j L_{ij}^0 h_{granule}(w_i) + I_i^{ob}$$

$$dw_i/dt + \delta w_i - \lambda h_{mitral}(v_j) = \sum_j Q_{ij}^0 h_{mitral}(v_j)$$
(4)

where v_i and w_i represent the membrane potentials of the AON excitatory and inhibitory cells, and I_i^{ob} are the net inputs from the bulb via the feedforward inhibitory units.

$$\frac{dw_i}{dt} + \delta w_i - \lambda h_{mitral}(V_j) = \sum_j Q_{ij}^0 h_{mitral}(V_j) + I_i^{oc}(t)$$

where $I_i^{oc}(t)$ are the net inputs from the cortex via feedforward inhibitory units.

The AON layer is connected to the third layer in our system, i.e. the prepiriform cortex. This layer also consists of coupled oscillators with excitatory/ inhibitory cells.

$$\tau_i^p \frac{dp_i(t)}{dt} = -p_i(t) + \sum_{j=1}^N M_{ij}^o h_{mitral}(p_j) - \sum_{j=1}^N N_{ij}^o h_{mitral}(q_j) -$$

$$-\gamma h_{granule}(q_i) + I_i^{ob}(t)$$

$$\tau_i^q \frac{dq_i(t)}{dt} = -\varepsilon q_i + \sigma h_{mitral}(p_i) + \sum_{j=1}^N J_{ij}^o h_{mitral}(p_i)$$

where M^o is a matrix based on excitatory-to-excitatory connections, N^o is inhibitory-to-excitatory connections, and J^o is excitatory-to-inhibitory connections. p_i and q_j are the membrane potentials of the cortex excitatory and inhibitory cells; and $I_i^{ob}(t)$ are the net inputs from the bulb via the feedforward inhibitory connections.

COMPUTER SIMULATIONS

In numerical simulations of the olfactory system, Eqns. (1-4) were integrated using a second order Runge-Kutta method. The integration time step (t) is 0.10 ms. Stability of the results was checked against varying the number of time steps in the simulations and their size. When burst duration and interburst interval were measured, we discarded the first few bursts to eliminate the effects of the initial values and averaged these quantities over 10 to 20 bursts.

Results for one input odor are presented. Figure 2 shows the sensory input over the time. After the inhalation has started the firing frequency of the sensory neurons, which are sensitive to this odor, increases approximately linearly as does the concentration of odor molecules in the nasal mucous membrane. During exhalation the frequency drops exponentially, until it reaches the level of spontaneous activity of the sensory neurons. Figure 3 presents the activation in OB, AON, and PC layers.

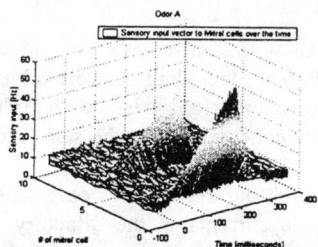

Figure 2 Sensory input over time

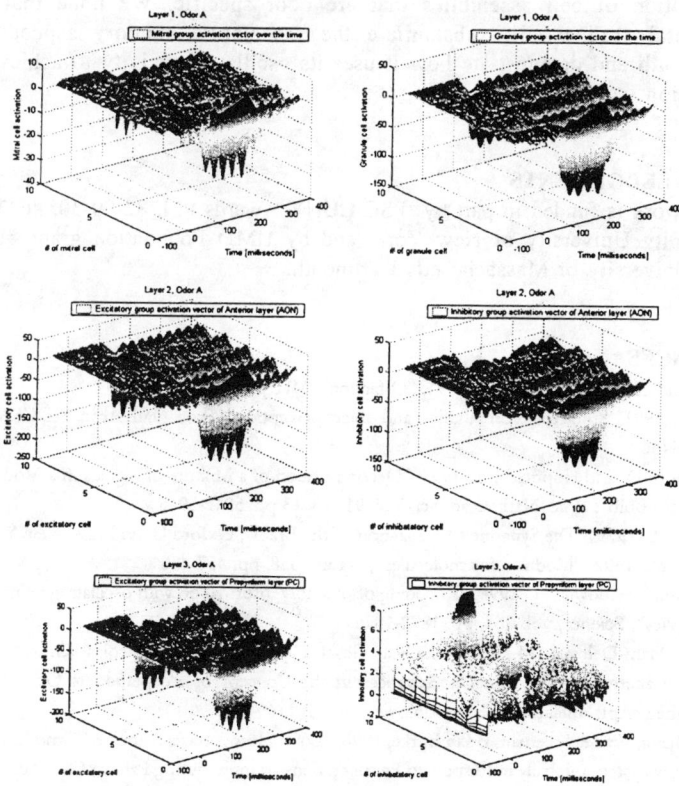

Figure 3 Activation of cells in the olfactory bulb, AON, and PC

We added Gaussian noise to the input. Simulations show that the dynamic behavior of the model is stable under the influence of noise. The model bulb responds to different odor input with spatio-temporal activation patterns, which are unique for each simulated odor. After inhalation has started a burst of oscillatory activity emerges.

CONCLUSIONS

Computer simulations show that different odors induce different amplitude and phase patterns in cells, and some, including the zero-odor input, do not induce any coherent oscillations; the oscillations rise with inhalations and fall with exhalations; all cells oscillate coherently with the same frequency which depend on odor signal.

We conclude that, although our system uses only schematic descriptions of the neural elements and dendritic currents in the olfactory bulb, it may account for cognitive abilities and behavioral patterns. Its activity resembles electrophysiological results (Laurent and Davidowitz 1994), displaying oscillations and formation of cell assemblies that are odor specific. We hope that future experimental studies will substantiate the associative memory aspect of the olfactory bulb and demonstrate how it uses its oscillatory activity to achieve odor segmentation.

ACKNOWLEDGEMENTS

This work is funded in part by PSC-CUNY Awards #61782-00-30, #63374-00-32 from City University of New York, and by UMD Foundation grant #UMDF-525360, University of Massachusetts Dartmouth.

REFERENCES

Davis J. L. and Eichenbaum H. (eds), 1991, "Olfaction", MIT press.
Hopfield J. J., 1991, "Olfactory computation and object perception", Proc. Natl. Acad. Sci. USA, pp. 6462-6466.
Hendin 0. Horn D. and Hopfield J. J., 1994, "Decomposition of a Mixture of Signals in a Model of the Olfactory Bulb", Proc. Natl. Acad. Sci. Vol. 91 No. 13 pp. 5942-5946.
Shepherd G. M., 1979, "The Synaptic Organization of the Brain", Oxford Univ. Press, New York.
Shepherd G. M., 1992, "Modules for molecules", Nature 358, pp. 457-458.
Laurent G. and Davidowitz H., 1994, "Encoding of olfactory information with oscillating neural assemblies", Science, vol. 265, pp. 1872-1875.
Masson, C., Pham-Delegue, M. H., Fonta, C., Gascuel, J., Arnold, G., Nicolas, G. and Kerszberg, M., 1993, "Recent Advances in the Concept of Adaptation to Odours in the Honeybee, Apis mellifera L", Apidologie, 24, (3), 169-194.
Hung-Jen Chang, Walter Freeman, Brian Burke, 1998, "Local Homeostasis Stabilizes A model of the Olfactory System Globally in Respect to Pertuberations by input during Pattern Classification", International Journal of Bifurcation and Chaos", Vol. 8, N0 11, 2107-2123.

VARIANCE ANALYSIS OF A VESTIBULO-OCULOMOTOR SYSTEM STATE IN A SINUSOIDAL STIMULATION ON THE BASIS OF A RECONSTRUCTED COMPONENT METHOD

ANATOLY A. BORISKEVICH AND VADIM O. KUDRYAVTSEV
Institute of Engineering Cybernetics of the National Academy of Sciences of Belarus
Minsk, BELARUS

ABSTRACT

The variance approach based on using principal component analysis for estimating VOS state is proposed. Its essence consists in controlled ENG signal decomposition on a number physiologically easily interpreted reconstructed components by means of a selection of optimal relation among a time window length of the ENG signal analysis, ENG signal length and its basic period. Quantitative estimation of ENG signal structure variability is based on using the proposed new diagnostic parameters obtaining from an eigenvalue spectrum of the initial ENG signals. Result of modeling for the groups of patients with different types of VOS disturbances shows diagnostic significance of proposed parameters for estimating VOS states.

INTRODUCTION

One of the most practical and enough informative characteristics of a vestibulo-oculomotor system (VOS) behavior is electronystagmograms (ENG) representing one-dimensional realization of the complex oscillatory process arising in moving of the eyes (Engelken, et al., 1989), (Sigalev, 1995), (Scollan, et al., 1997). The most simple and spread system providing sinusoidal movement of the visual target in a horizontal plane was used as a VOS stimulator.

In this case VOS behavior is characterized by two types of arbitrary eye movements whose parameters are changed according to disturbances of trunkular-cerebellar structures of cerebrum brain responsible for a regulation of the vestibulo-oculomotor function: smooth and fast tracking movements of eyes (STME and FTME). Variation of STME and FTME is a sensitive factor of VOS disturbances (Sigalev, 1995). The necessity of continuation of searching and analysis of diagnostic parameters characterizing STME and FTME variation is due to the great variability of traditional diagnostic features because of influence of individual singularities of VOS responses.

To solve the given problem uses the differential-integral approach based on using controlled decomposition` of ENG signal in a series of physiologically easily interpreted reconstructed components by optimal choice of a time window of the ENG analysis and estimation of the component variability on the basis of their eigenvalue spectrum parameters.

VARIANCE ANALYSIS OF A VESTIBULAR-OCULOMOTOR SYSTEM STATE

The realization of ENG signal on a finite time interval $t_o \leq t \leq t_o + T_N$, where t_0 is an arbitrary time reference, is represented by discrete time series of one of dynamic VOS variable $\mathbf{x}^T = [x(0), x(\Delta t),...,x((N-1)\Delta t)] = [x_0, x_1,...,x_{N-1}]$, where N is the number of sample values of ENG signal $x(t)$; $T_N = N\Delta t$ is analyzed time interval of VOS state; Δt is a sampling interval.

The proposed approach is based on dynamic VOS behavior describing as M-dimensional state vectors $\mathbf{x}_j^T = [x_j, x_{j+1},...,x_{j+(M-1)}]$, $i = 0,1,...,L-1$, $L = N - (M-1)$ in space R^M (Boriskevich, et al., 1994). The experience shows that the choice of parameter M for the most qualitative ENG signal decomposition on reconstructed components should be carried out by the relationship $M\Delta t = f_{max}^{-1}$, where f_{max} is a basic frequency in a spectrum of ENG signal.

The reconstructed components $\mathbf{x}_i^r$ of ENG signal in the form of N-point sequence $\mathbf{x}$ are computed by the system of linear convolution relationships based on PCA-method and defining operations of filtering and reconstruction of ENG signal with minimal mean square error

$$y_{ij} = \sum_{k=0}^{M-1} u_{ik}^* x_{j-k}, \; j = 0,1,...,M+N-2, \qquad (1)$$

$$x_{ij}^r = M^{-1} \sum_{k=0}^{M-1} u_{ik} y_{j-k}, \; j = 0,1,...,N-1. \qquad (2)$$

Here $\mathbf{u}_i^T = \{u_{ik}, \; k = 0,...,M-1\}$ is i-th eigenvector of a correlation matrix $\mathbf{C}_x$ of the sequence $\mathbf{x}$; superscript * denotes time-reversed order of samples; $\mathbf{y}_i^T = \{y_{ij}, \; j = 0,...,M+N-2\}$ is i-th principal component of ENG signal $\mathbf{x}$; $\mathbf{x}_i^{rT} = \{x_{ij}^r, \; j = 0,...,N-1\}$ is i-th reconstructed component of ENG signal. Note that the coefficient M using for multiplying j-th sample value by M/j in Eq. (2) is designed for correcting tails representing "incomplete convolutions" due to limitation of convolved sequences and consisting of $M-1$ initial and final sample values of the linear convolution. Thus, the reconstructed initial ENG signal is represented by means of the reconstructed component sum $\mathbf{x} = \sum_{i=1}^{M_{min}} \mathbf{x}_i^r$. The choice of M_{min} is carried out by prescribing of the mean square error of the initial ENG signal recovery on the basis of the restricted number of the reconstructed components.

One of the most informative characteristics of reconstructed ENG signal is the spectrum of its reconstructed component eigenvalues. Thereby the different parameters of eigenvalue spectrum and their relations are used as measures of reconstructed ENG signal parameter variability. We propose the following set of new diagnostic parameters for recognizing level and localization of VOS disturbances.

The ratio of m first eigenvalues sum to the sum of all eigenvalues defines the relative contribution of the first reconstructed components characterizing smooth tracking movements of eyes to a total variance describing statistical variability of smooth and saccadic components of tracking movements of eyes and is given as

$$\bar{\lambda}_{slow} = \sum_{i=1}^{m} \lambda_i / \sum_{i=1}^{M_{min}} \lambda_i, \qquad (3)$$

where m is the integer number, its value choice depending on frequency of target displacement and VOS state.

The ratio of last $M-m$ eigenvalues sum to the sum of all eigenvalues defines the relative contribution of the last $M-m$ reconstructed components characterizing fast movements of eyes to a total variance describing statistical variability of smooth and saccadic components of tracking movements of eyes and is written as

$$\bar{\lambda}_{fast} = \sum_{i=m+1}^{M_{min}} \lambda_i / \sum_{i=1}^{M_{min}} \lambda_i \qquad (4)$$

where M_{min} is the selected parameter of ENG signal expansion with the given error of its recovery.

Estimation of a relation describing interrelation between smooth and fast components of tracking eye movements $\bar{\lambda}_{slow} / \bar{\lambda}_{fast}$ can be considered as a measure of spectral shift from the area $\sum_{i=1}^{m} \lambda_i$ (corresponding to slow reconstructed ENG signal component $\mathbf{x}_{slow}$) to the area $\sum_{i=m+1}^{M_{min}} \lambda_i$ (corresponding to fast reconstructed ENG signal component $\mathbf{x}_{fast}$). The equations (3) and (4) can characterize vestibulo-oculomotor disturbances accompanying by a change of smooth or saccadic tracking movements of an eye.

For increasing informative value of the above parameters it is worth carrying out decomposition of the fast reconstructed ENG signal component in a series of the secondary reconstructed components. Thereby the improved values of variance coefficients are given as

$$\bar{\lambda}_{slow} = (\sum_{i=1}^{m} \lambda_i + \lambda_{1\,fast}) / \sum_{i=1}^{M_{min}} \lambda_i \qquad (5)$$

$$\bar{\lambda}_{fast} = (\sum_{i=m+1}^{M_{min}} \lambda_i - \lambda_{1\,fast}) / \sum_{i=1}^{M_{min}} \lambda_i \qquad (6)$$

where $\lambda_{1\,fast}$ is the eigenvalue corresponding to the low-frequency part of the fast reconstructed ENG signal component. The given relations will allow more reliably to evaluate a degree of changing smooth and fast component parameters of tracking eye movement.

The differential coefficient characterizing the ratio of the variance of the reconstructed ENG signal to the variance of the reconstructed differentiated ENG signal can be used as recognition feature of VOS disturbance characterizing changing saccadic component of eye movement

$$\bar{\lambda}_{diff} = \sum_{i=1}^{M_{min}} \lambda_i / \sum_{i=1}^{M_{min}} \lambda_i^{diff} \qquad (7)$$

The above diagnostic parameters can be obtained also for differentiated initial ENG signal. It should be noted that the differentiated signal characterizes a velocity distribution of eye movements and activation of saccadic tracking subsystem of VOS when visual target movement velocity exceeds a critical upper limit of a velocity of smooth tracking subsystem of a VOS.

The number of degree of freedoms of a eigenvalue spectrum is defined as

$$M_{tr} = (M\ S(\lambda))^2 / M\ S^2(\lambda) = (tr\ \mathbf{C}_x)^2 / tr\ \mathbf{C}_x^2, \qquad (8)$$

where $S(\lambda)$ is a eigenvalue spectrum of a correlation matrix $\mathbf{C}_x$; $tr\ \mathbf{C}_x = \sum_{i=1}^{M} \lambda_i$ is a trace of a correlation matrix $\mathbf{C}_x$; $tr\ \mathbf{C}_x^k = \sum_{i=1}^{M} \lambda_i^k$ is k-th moment of correlation matrix eigenvalues.

Statistical parameters directly characterizing singularities of a eigenvalue spectrum structure of a correlation matrix are defined as

$$H_n = H / H_{max} = ID, \qquad (9)$$

where $H_n = M(-\log \lambda_{ni})/H_{max} = -(\sum_{i=1}^{M} \lambda_{ni} \log \lambda_{ni})/\log M$ is a normalized entropy of a normalized eigenvalue spectrum characterizing a measure of a disorder in VOS behavior, $\lambda_{ni} = \lambda_i / \sum_{i=1}^{M} \lambda_i$ is i-th normalized eigenvalue of a correlation matrix of researched ENG signal; $H = M(-\log \lambda_{ni})$ and $H_{max} = \log M$ are entropy and maximal entropy of a structural singularities of normalized eigenvalue spectrum respectively characterizing a VOS state with the uniform and nonuniform statistical contribution of the covariance matrix eigenvalues; $D = H/(H_{max} - H)$ is the measure of structural VOS disorder; $I = (H_{max} - H)/H_{max}$ is the measure of order for VOS state.

EXPERIMENTS AND RESULTS

The experimental research was carried out in the darkened room by presentation of a target as projection of a laser point of 0.5 degrees diameter displaced on the flat screen in a horizontal plane by the sinusoidal law with frequency of 0.25, 0.5, and 1.0 Hz and amplitude of 10 degrees. Eye movements were recorded on a paper by the polygraph with the DC amplifier.

Scannered and digitalled ENG signal was preprocessed in order to remove the trend caused by recording method imperfection. This was done by means of PCA method and removing from initial ENG signal the reconstruction component containing the most lower frequencies (including DC component) of the signal.

Figure 1 shows the real ENG signals and various combination of its reconstructed components for VOS with central disturbance recorded by target motion with frequency equal to 0.5 Hz. It is evident from Fig. 1 b,c,d,e, that the proposed approach allows to decompose reliably the ENG signal into physiologically easily interpreted components: STME and FTME. Note that the degree of freedom influence on the choice of the number of the first components forming STME.

The eigenvalue spectrum of the signal presented in Fig. 1a is shown in Fig. 2. This figure demonstrates the relation of the degree of freedom with the most significant reconstructed components.

Table 1. Diagnostic parameter values for distinguishing VOS state.

VOS disturbance	$\overline{\lambda}_{slow}$ (m=3)	$\overline{\lambda}_{slow}$ (m=5)	$\overline{\lambda}_{diff}$	M_{tr}	H_n D	H_n I
central	9.0855	12.3198	0.01163	2.5321	0.1903	
					0.235	0.8097
peripheral	15.5408	23.3482	0.01533	2.3229	0.1592	
					0.1894	0.8408
none	115.0736	250.7658	0.06619	2.0502	0.1131	
					0.1275	0.8869

From the table 1 it is seen that the proposed parameters are reliably significant for recognizing the level of the VOS disturbance.

Result of modeling for the groups of patients with different types of VOS disturbances shows diagnostic significance of proposed parameters for estimating VOS dynamics. It is of importance for the clinician who uses ENG signal for diagnostic purposes.

CONCLUSION

The variance approach based on using principal component analysis is proposed for estimating VOS state. Its essence consists in controlled ENG signal expansion on a number physiologically easily interpreted reconstructed components by means of selection of optimal relation among a time window length of the ENG signal analysis, ENG signal length and its basic period. Besides quantitative estimation of ENG signal structure variability is based on using the proposed new diagnostic parameters obtaining from an eigenvalue spectrum of reconstructed component of initial and differential ENG signals. Result of modeling for the groups of patients with different types of VOS disturbances shows diagnostic significance of proposed parameters for estimating VOS states.

Further studies are needed for developing the proposed approach to better recognizing the localization degree of various VOS disturbance.

REFERENCES

Boriskevich, A.A., Dayludenko, V.F., Krot, A.M., 1994, "Phase space reconstruction methods on results of experiment for diagnostics and prediction of system state with complex behavior", Preprint, Institute of Engineering Cybernetics, NAN of Belarus, Minsk, No. 24. (in russian).

Engelken, E.J., Stevens, K.W., and Enderle, J.D., 1989, "Computer Analysis of Smooth Pursuit Eye Movements", *Biomedical Sciences Instrumentation*, Vol. 25, pp. 127-134.

Scollan, D.F., Nakamoto, B.K., Shelhamer, M., 1997, "Adaptability and Variability in the Oculomotor System", *Computational Neuroscience*, Trends in Research, JM Bower (ed.), Plenum: New York, pp. 827-831.

Sigalev, E.E., 1995, "Diagnostic possibilities of research of tracking eye movements of in otoneurology," *Vestnik of otorhinolaryngology*, No.1, pp. 44-49 (in Russian).

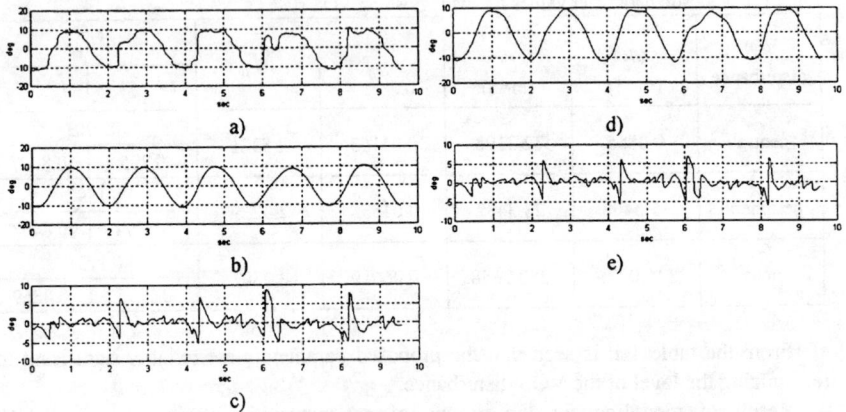

Fig. 1. Reconstructed components of ENG signal for central VOS pathology:
a) initial ENG signal; b) sum of the three first reconstructed components; c) sum of the rest reconstructed components for the case (b); d) sum of the five first reconstructed components; e) sum of the rest reconstructed components for the case (d)

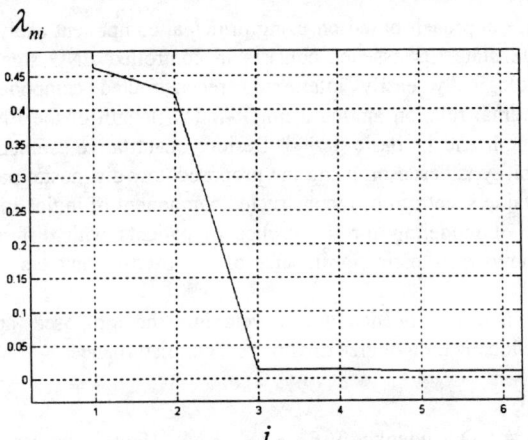

Fig. 2. Eigenvalue spectrum of the initial ENG signal for central VOS pathology

Application of Support Vector Machines To Breast Cancer Classification Using Mammogram and History Data

Anab Akanda
Binghamton University
Binghamton, NY
Animatin@aol.com

Walker H. Land, Jr.
Binghamton University
Binghamton, NY
wland@binghamton.edu

Joseph Y. Lo
Duke University Medical Center
Durham, NC
jyl@deckard.mc.duke.edu

ABSTRACT

Support Vector Machines (SVM) are new and radically different type of classifiers and learning machines that use a hypothesis space of linear functions in a high dimensional feature space, generally trained with learning algorithms from optimization theory that implement a learning bias derived from statistical learning theory. This relatively new paradigm, based on Structural Risk Minimization (SRM), has many advantages over traditional neural networks based on Empirical Risk Minimization (ERM). The most important one being that, unlike neural networks, SVM training always finds a global minimum. The SVM has the inherent ability to solve pattern classification without incorporating any problem-domain knowledge. In this study, the Support Vector Machine was employed as a pattern classifier, operating on mammography data used for breast cancer detection. The main focus of the study was to construct and seek the best SVM configuration for optimum specificity and positive predictive value at very high sensitivities.

Using the DUKE mammogram database of 500 biopsy proven samples, the best performing SVM, on average, was able to achieve (under statistical 5-fold cross-validation) a specificity of 45.0% and a positive predictive value (PPV) of 50.1% while maintaining 100% sensitivity. At 97% sensitivity, a specificity of 55.8% and a PPV of 55.2% were obtained.

1. INTRODUCTION

Cancer of the breast is second only to lung cancer as a tumor-related cause of death in women. In 1993, it was reported that 180,000 new cases and 45,000 deaths occurred just in the US [1]. It has been proposed, however, that the mortality from breast cancer could be decreased by up to 25%, provided that all women in appropriate age groups were regularly screened [2]. Currently, the method of choice for the early detection of breast cancer is mammography, which is non-invasive, widely available, low cost, and speedy. At the same time, while mammography is sensitive to the detection of breast cancer, it has a low positive predictive value (PPV), resulting in costly and invasive biopsies that are only 15%-34% likely to actually show malignancy at histological examination [3].

Computational intelligence has been applied in various forms by the authors and others. Previous applications which serve as a foundation for this work include backpropagation-trained Multiple Layer Feedforward Neural Networks (MLFNs) [4], Evolutionary-Programming (EP) derived neural networks [5], and an Adaptive Boosting (AB)/EP hybrid [6]. This paper describes the extenuation of the research previously done but employing Support Vector Machines (SVM) instead of neural networks as the pattern classifier. The best performing SVM obtained so far produced results that are as good as or better than that of EP-AB hybrid scheme.

In this paper, the application of SVM as a pattern classifier for breast cancer detection from mammography data was investigated. Two different kernel functions were considered in the construction of Support Vector Machines, namely: 1) Simple dot product kernel (linear, non-separable) and 2) Radial Basis Function (RBF) kernel (non-linear, non-separable) [11,16,17]. The quadratic optimization problem associated with SVM was solved using two different algorithms: 1) Stochastic Gradient Ascent (Kernel-Adatron) and the 2) Conjugate Gradient Method [11,14,15,18]. The search for optimum SVM configuration (kernel and its parameters) is still on going. Preliminary results obtained thus far are presented in this paper. The mammogram database structure from Duke University is discussed in the next section. In section 3, a brief summary of the mathematical foundation of SVM is presented. Also, a brief summary of the SVM architectures and their implementation are discussed in section 4. Finally, in section 5, the results obtained from SVM are compared with that of EP-AB hybrid scheme [5] followed by the conclusions.

2. DUKE UNIVERSITY MAMMOGRAM DATABASE

The Duke University mammogram database consists of 500 cases of non-palpable, mammographically-suspicious breast lesions which were selected randomly from patients seen at Duke University Medical Center. Each case underwent needle localization and excisional biopsy, resulting in definitive histopathologic diagnosis. Of these 500 cases, 326 (65%) were benign. The mean age of women was 55.5 years, with a range of 24 to 86 years.

For each case, ten mammographic findings were extracted by one of four experienced mammographers who were blinded to the biopsy outcome. The ten findings included three pertaining to calcifications, four pertaining to masses, and three other miscellaneous findings. The findings were encoded according to the BI-RADS™ lexicon, a standard adopted by the American College of Radiology to improve consistency and accuracy of mammographic interpretation [8]. In addition, six variables pertaining to patient history (such as age, personal history of breast cancer, family history of breast cancer, etc.) also were recorded for each case. The resulting sixteen variables were each linearly scaled into floating point numbers between 0 and +1 (later rescaled to a range of –1 to +1,), with greater values corresponding to a priori increased likelihood of malignancy. The database has successfully been used in several experiments involving the use of neural networks to predict breast cancer malignancy [4,9-10].

3. Brief Summary of SVM Background

There are several texts that provide extensive tutorial on the mathematical foundation of support vector machines [11-13,16,17]. The key concepts of SVM covered in these references are briefly summarized in this section. In the context of breast cancer detection from mammogram data, the main idea of a support vector machine is to construct an "optimal hyperplane" as the decision surface such that the margin of separation between positive (malignant) and negative (benign) cases is maximized. Support Vector Machines are based on some key ideas: i) Structural/Empirical Risk Minimization (SRM/ERM), ii) The VC (Vapnik-Chervonenkis) dimension, iii) The optimal hyperplane and iv) The quadratic optimization procedure for obtaining the optimal hyperplane. The reader should consult [11-13,16,17] for details.

4. Brief Summary of SVM Implementation

In this study, two support vector machines were constructed: i) Linearly Non-separable and ii) Non-linearly Non-separable. In the first case, a linear dot-product kernel function was used with the regularization parameter C, where the optimum value of C was determined experimentally. In the second case, the Gaussian Radial Basis Function (GRBF) $exp(-|x_i - x|^2/(2\sigma^2))$ was used as the kernel function along with the regularization parameter C. In this case, the radius σ and C were also obtained experimentally. The performance of these machines was evaluated via the standard use of training /test set 5-fold cross-validation procedure. In the training phase, the dual formulation optimization problem was solved using two different methods i) The Stochastic Gradient Ascent and ii) The Conjugate Gradient Method. The details of the optimization techniques are discussed in [11,14,15 and 18].

5. Preliminary Results

To evaluate the efficiency of the SVM performance, five-fold cross-validation evaluations where performed [5]. Cross validation is a statistical process whereby a data set of limited size may be partitioned to better evaluate neural network performance against a given set of data. Using the mammogram data set, containing 500 samples, were partitioned into 5 groups of 100 each. The first 100 samples were "held out" as the validation set and the remaining 400 samples used as the training set. The held-out 100 sample set was used as the validation set and the ROC A_z index was computed for the given SVM.

The ROC curve is a plot of the true positive ratio (TPR) as a function of the false positive ratio (FPR); both of which are in the [0,1] closed interval. The area under this curve, called the A_z index, represents an overall performance over all possible (TPR, FPR) operating points. The statistically derived ROC curves are used to assess the effectiveness of a classification procedure and are obtained by evaluating the system under all possible threshold settings. The probability of detecting a malignancy is traded off as a function of the likelihood of obtaining a false positive result. As the threshold value is lowered, the SVM will correctly identify a greater number of malignancies, but this comes at the expense of a higher false positive rate. Conversely, raising the threshold value can lower the false positive rate, but this, in turn, decreases the sensitivity of the procedure. The process of holding out 100 samples and using the remaining 400 for training was continued for the 2nd through 5th set, and ROC indices were computed for each "fold". In addition to the A_z index, the positive predictive value (PPV) and the specificity of the systems were also computed at 100% and 97% sensitivity level for all folds.

As mentioned in the introduction, two different kernel functions were considered in the construction of Support Vector Machines : 1) Dot product (linear, non-separable) and 2) Gaussian Radial Basis Function (RBF) kernel (non-linear, non-separable). For both machines, the 5-fold cross validation procedure was used to determine the optimum kernel parameters and the regularization parameter C. For each set of parameters, a 5-fold cross validation was performed and the corresponding average A_z index, PPV and specificity values were monitored. The process was repeated over the parameter space and the best result was recorded for each machine. These values are compared with that of the EP-AB Hybrid scheme in Tables 1-3.

Table 1. Linearly Non-Separable SVM [C = 5.0; Explicit Bias b = 0]

Sensitivity	Metric	Fold #1	Fold #2	Fold #3	Fold #4	Fold #5	Average	Std
100%	PPV	45.349	59.322	44.737	51.613	49.275	50.059	5.902
	Specificity	22.951	63.077	36.364	55.882	46.970	45.049	15.889
97%	PPV	46.914	65.385	55.000	51.613	56.897	55.162	6.859
	Specificity	29.508	72.308	59.091	55.882	62.121	55.782	15.926
	Az Area	0.7867	0.9125	0.8329	0.8502	0.9057	0.858	0.053

Table 2. GRBF SVM [σ = 8.0; C = 245.0; Explicit Bias b = 0]

Sensitivity	Metric	Fold #1	Fold #2	Fold #3	Fold #4	Fold #5	Average	Std
100%	PPV	45.349	50.725	43.038	45.070	50.000	46.836	3.350
	Specificity	22.951	47.692	31.818	42.647	48.485	38.719	11.041
97%	PPV	45.238	55.738	51.563	45.070	54.098	50.341	4.964
	Specificity	24.590	58.462	53.030	42.647	57.576	47.261	14.146
	Az Area	0.8024	0.8736	0.8405	0.8212	0.8835	0.844	0.034

Table 3. Best EP-AB Hybrid Results [Az stdv = 0.047]

Sensitivity	Metric	Best
100%	PPV	48.500
	Specificity	43.300
98%	PPV	52.000
	Specificity	51.800
	Az Area	0.8510

The best results obtained thus far came from the linearly non-separable dot-product SVM with the regularization parameter C = 5.0 and explicit bias b = 0. The GRBF non-linearly non-separable SVM did not perform as well. The best combination of the radius σ and C for this machine proved to be very difficult to obtain through brute-force numerical experiments.

6. CONCLUSIONS

Results obtained thus far have already shown the potential of SVM for breast cancer detection. There are numerous advantages in using this paradigm over traditional neural networks: i) Unlike neural networks, one does not need to worry about the "best network architecture" ii) SVM training always finds global minimum of the risk functional, iii) For small sizes problems, (e.g., mammogram data) the training and validation process is many orders of magnitude faster than the EP-AB Hybrid scheme (minutes as opposed to days). However, the biggest limitation of SVM lies in the choice of the kernel function and its parameters. In this study, only the 'dot-product' and GRBF kernels were investigated to a limited extent. Other kernel functions such as the polynomial kernel or the two-layer neural network kernel might produce better results [11,16]. Analytical, as opposed to experimental, determination of kernel parameters could potentially help in making that decision. These issues warrant further investigation. As far as the optimization algorithm is concerned, the conjugate gradient method appears to be a good choice for small size problems. Speed of convergence and the ability to enforce linear constraints are the motivating factors for this choice.

REFERENCES

[1] C.C. Boring, T. S. Squires, and T. Tong. "Cancer Statistics." *Cancer Journal for Clinicians*, vol. 43, pp. 7-26, 1993.

[2] P. Strax. Make Sure that You Do Not Have Breast Cancer. St. Martin's, NY, 1989.

[3] J. Y. Lo, J. A. Baker, P. J. Kornguth, J. D. Iglehart, and C. E. Floyd. "Predicting Breast Cancer Invasion with Artificial Neural Networks on the Basis of Mammographic Features." *Radiology*, vol. 203, pp. 159-163, 1997.

[4] J. Y. Lo, J. A. Baker, P. J. Kornguth, and C. E. Floyd, Jr. "Effect of Patient History Data on the Prediction of Breast Cancer from Mammographic Findings with Artificial Neural Networks." *Academic Radiology*, vol. 6, pp. 10-15, 1999.

[5] W. H. Land, T. Masters, J. Y. Lo, and D. McKee. "Using Evolutionary Computation to Develop Neural Network Breast Cancer Benign/Malignant Classification Models." *4th World Conference on Systemics, Cybernetics and Informatics*, vol. 10, pp. 343-7, 2000.

[6] W. H. Land, Jr., T. Masters, and J. Y. Lo. "Application of a New Evolutionary Programming/Adaptive Boosting Hybrid to Breast Cancer Diagnosis." *IEEE Congress on Evolutionary Computation Proceedings*, 2000.

[7] D. B. Fogel, *Evolutionary Computation: Toward a New Philosophy of Machine Intelligence*. IEEE Press, Piscataway, NJ, 1995.

[8] BIRADS, Breast Imaging - Reporting and Data System (BI-RADS). *American College of Radiology*, 1993.

[9] J. A. Baker, P. J. Kornguth, J. Y. Lo, M. E. Williford, and C. E. Floyd, Jr. "Breast Cancer: Prediction with Artificial Neural Network based on BI-RADS standardized lexicon." *Radiology*, vol. 196, pp. 817-22, 1995.

[10] J. Y. Lo, J. A. Baker, P. J. Kornguth, and C. E. Floyd, Jr. "Computer-aided diagnosis of breast cancer: Artificial Neural Network Approach for Optimized Merging of Mammographic Features." *Academic Radiology*, vol. 2, pp. 841-50, 1995.

[11] Cristianini, N., and Shawe-Taylor, J., "An Introduction to Support Vector Machines and other Kernel-Based Learning Methods", Cambridge University Press, 2000.

[12] Haykin, S., "Neural Networks", Prentice Hall, New Jersey, 1999.

[13] Gunn, S., "Support Vector Machines for Classification and Regression", ISIS Technical Report, Image Speech and Intelligent Systems Group, University of Southampton, May, 14, 1998.

[14] Chong, E.K.P., Zak, S.H., "An Introduction To Optimization", John Wiley & Sons, Inc., New York, 1996.

[15] Luenberger, D.G, "Linear and Nonlinear Programming", Addison-Wesley, Massachusettes, 1989.

[16] Vapnik, V.N., "The nature Of Statistical Learning Theory", John Wiley, New York, 1995.

[17] Burges, C.J.C., "A Tutorial on Support Vector Machines for Pattern Recognition", Data Mining and Knowledge Discovery, 2(2):121-167, 1998.

[18] Saunders, C., Stitson, M.O., Watson, J., Bottou, L., Scholkopf, B., Smola, A., "Support Vector Machine – reference manual. Department of Computer Science, Royal Holloway, University of London. http://svm.dcs.rhbnc.ac.uk/.

PART X: SMART ENGINEERING SYSTEMS

D3D - A SOFTWARE FOR GRAPHICAL REPRESENTATION OF OBJECTIVE FUNCTIONS

P.A. SIMIONESCU
Dept. of Mechanical Engineering
and Solidification Design Center
Auburn University, 202 Ross Hall,
Auburn AL, 36849

D. BEALE
Dept. of Mechanical Engineering
Auburn University, 202 Ross Hall,
Auburn AL, 36849

ABSTRACT

The main features of a software named *D3D* specially devised for visualizing objective functions as contour plots and 3D surfaces are presented. These include detailing the minimum areas by logarithmically spacing the level curves, the possibility of truncating the upper parts of the surface (particularly useful in representing penalized objective functions) and representing the gradient of the function as a vector field projected on the bottom plane in combination with 3D surface plots.

INTRODUCTION

Objective functions arising in optimisation problems are single valued functions of one or more variables. Functions of one, two, even three variables allow for graphical representation in the 2 or 3 dimensional Cartesian space aaa1] (Encarnacao et al 1990). Such representations permit attaining a good feeling of the monotonicity, convexity, and the existence of multiple minima of the function, and are therefore usual encounter in most numerical optimisation and operations research publications.

The purpose of this paper is to present a computer software named *D3D* specially devised for representing objective functions of two variables, that has several features not yet at hand in commercially available mathematical and visualisation software. These features will be introduced based on a simple optimisation problem, of finding the equilibrium position of a spring-restrain double pendulum.

EXAMPLE PROBLEM

The dynamic analysis of constrained multibody systems includes, among time response and kinetostatic analysis, the problem of determining the static equilibrium configuration. Several methods for solving the equilibrium problem are known (Simionescu and Fawcett 1997) of which searching for the minimum of total potential energy is the most simple and easy to apply. On the other hand, determining the equilibrium position of spring and weight systems are good illustrative examples of optimisation problems that can arise in practice (Vanderplaats 1984).

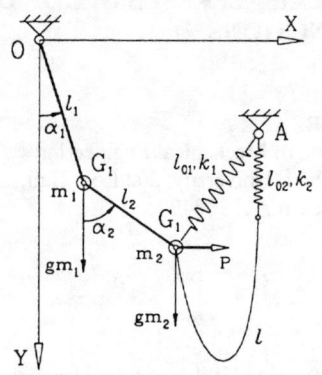

Fig. 1 A double pendulum restrained with elastic springs (g=9.81m/s² is the acceleration due to gravity)

The mass spring system considered for determining the equilibrium position is that of the double pendulum in Fig. 1 amplified with a linear spring (k_1, l_{01}) having one end attached to the outer lumped mass and the other end fixed to the frame at point $A(x_A, y_A)$. Mounted in between the same points G_2 and A, there are an ideal inextensible massless thread of length l, connected in series with a second spring (k_2, l_{02}).

For the system under consideration, the total potential energy is the sum of the gravity potentials of masses m_1 and m_2, that due to the constant force P (which can also be considered as deriving from a potential) and that of the elastic energy of the two springs. Choosing the reference position in deriving the potentials of m_1, m_2 and P is arbitrary. In this example the extreme position of point G_2 along the positive axis Ox for the potential of P, and G_1 and G_2 along Oz for the gravity potentials, have been used as reference. Therefore the total potential energy of the system is:

$$U(\alpha_1, \alpha_2) = (l_1 - z_{G1}) \cdot m_1 g + (l_1 + l_2 - z_{G2}) \cdot m_1 g +$$
$$+ (l_1 + l_2 - x_{G2}) \cdot P + \frac{1}{2} \cdot k_1 \cdot (AG_2 - l_{01})^2 + Q(\alpha_1, \alpha_2) \tag{1}$$

where

$$Q(\alpha_1, \alpha_2) = \begin{cases} 0 & \text{if } AG_2 \leq l + l_{02} \\ \frac{1}{2} k_2 (AG_2 - l - l_{02})^2 & \text{if } AG_2 > l + l_{02} \end{cases} \tag{2}$$

and

$$AG_2 = \sqrt{(x_{G2} - x_A)^2 + (z_{G2} - z_A)^2} \tag{3}$$

$$z_{G1} = l_1 \cdot \cos\alpha_1 \qquad\qquad z_{G1} = l_1 \cdot \cos\alpha_1$$
$$x_{G2} = l_1 \cdot \sin\alpha_1 + l_2 \cdot \sin\alpha_2 \qquad z_{G2} = l_1 \cdot \cos\alpha_1 + l_2 \cdot \cos\alpha_2 \tag{4}$$

It is to observe that the term $Q(\alpha 1, \alpha 2)$ acts as a barrier penalty function (Press et al 1992), with the penalty factor k_2, corresponding to the case when the point G_2 is constrained to remain inside a circle of centre A and radius $l+l_{02}$.

NUMERICAL RESULTS

The parameters of the system in Fig. 1 were considered to be: l_1=0.45m, l_2=0.35m, m_1=0.75kg, m_2=0.5kg, x_A=0.25m, z_A=-0.25m, k_1=9.5N/m, l_{01}=0.2m, k_2=2·10⁴N/m, l_{02}=0.1m and P=1.8N. Initially the thread length l was chosen

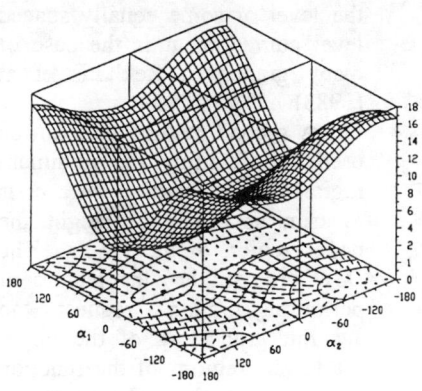

Fig. 2 Graphical representation of the potential function $U(\alpha_1,\alpha_2)$ for $k_2=0$. Also shown as arrows projected on the bottom plane is the gradient of the function.

sufficiently long so that the spring k_2 never becomes effective i.e. the penalty term Q equals zero irrespective of the value of α_1 and α_2 [identical results are obtained considering $k_2=0$ in equation (2)]. For this particular case, the potential function $U(\alpha_1, \alpha_2)$ has the shape in Fig. 2. Two points of extrema can be identified, of which the maximum point corresponds to an unstable equilibrium position.

The static equilibrium position was determined by minimising the function (1) using Fletcher-Reeves algorithm (Press et al 1992), resulting in $\alpha 1=6.0762°$, $\alpha 2=88.0589°$ for which U=3.72088 Nm.

For $l=0.65$m the shape of the potential function $U(\alpha_1, \alpha_2)$ changes to that in Fig. 3, showing that the point of maximum have moved to a different location, however without modifying the location and value of the global minimum.

As one can notice, a whole z-axis range representation of the function U in this case is less suggestive. The solution chosen for detailing the minimum area of the function as shown in Fig. 3, was to map logarithmically spaced level curves on the function surface. This way the level curves are concentrated in the lower region of the function surface and the global minimum highlighted.

The relation used for calculating the height of the horizontal cutting plane that generates the j-th level curve was:

$$z_j = z_{min} + \exp\left[(j-1)\cdot\ln\frac{z_{max}-z_{min}+1}{n-1}\right] - 1 \qquad (5)$$

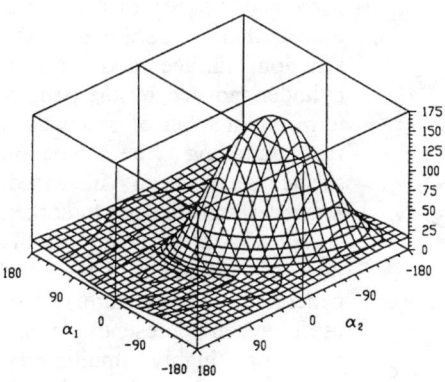

Fig. 3 A whole range representation of the potential function for $l=0.65$m and $k_2=2\cdot10^4$N/m.

where n is the total number of level curves, and z_{min} and z_{max} are the lower and upper range of the z-axis, which can coincide or not with the minimum and maximum function values from among the grid points used for sampling the function surface.

The idea of generating unequally spaced level curve diagrams is not a new concept, but as proposed, there is the certitude that the minimum region is well detailed, without the need of interactively editing

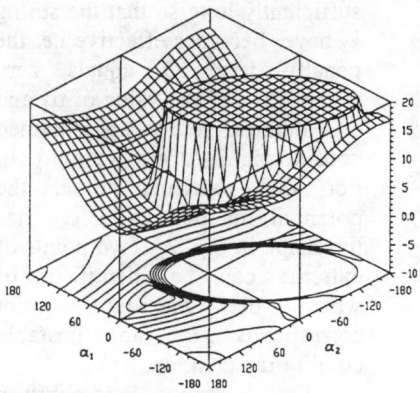

Fig. 4 Truncated representation of the potential function for $l=0.65$m and $k_2=2\cdot10^4$N/m.

the level of some equally spaced level curves, as it is the case of many graphs in Reklaitis et al (1983).

A second method that can be used for detailing the minimum region of the function-surface, is to trim at a certain height the upper region of the graph. The *D3D* program has the feature of permitting upon request an accurate truncating of the upper (or lower) regions of the function surface (Fig. 4). Commercially available software like MS EXCEL™ and MATHEMATICA™ allow truncating the function surface, but the method used (of replacing the z values outside the $[z_{min}..z_{max}]$ domain with exactly z_{min} or z_{max}) is not very accurate (Fig. 5), mainly if the number of points used in sampling the function is small. A different solution in service of MATLAB™ through the *Axis* function is to blind the upper and lower parts of the graph, which is obviously even less satisfactory.

In order to generate an equivalent of the infinite-barrier penalty function (Reklaitis et al 1983), for the same double-pendulum, $l=1.0$m and $k_2=1020$N/m have been considered. The shape of the potential function $U(\alpha_1,\alpha_2)$ looks as shown in Fig. 6, case in which the lower region of the function surface appears completely flatten, hence any details of the minimum region are hidden. Thus a truncated representation of the function surface proves even more useful that in the previous case.

By reducing the upper limit of the z-axis domain from $1.5\cdot10^{17}$ to 20, the diagram in Fig. 7 has been obtained. To increase the accuracy of representing the intersection between the function surface and vertical cylinder induced by the penalty term, the number of points used for sampling the function surface has been increased. However, in order not to darken completely the surface, only the lines of constant a1 have been generated. This feature is also useful when representing noisy data or highly multimodal functions such as Ackley's test objective function (Gen and Cheng 1997):

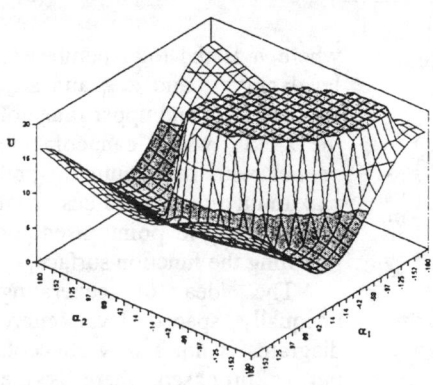

Fig. 5 The same representation as in Fig. 4 produced with MS-EXCEL™. A similar effect has the *Plotrange* option in MATHEMATICA™ software.

$$F(x_1...x_n) = 20\left[1-\exp\left(-0.2\sqrt{\frac{1}{n}\sum_{i=1}^{n}x_i^2}\right)\right] - \exp\left(\frac{1}{n}\sum_{i=1}^{n}\cos(2\pi x_i)\right) + e \qquad (6)$$

shown in Fig. 8 for n=2 as lines of constant x_1 combined with contour lines projected on the bottom plane. shown in Fig. 8 for n=2 as lines of constant x_1 combined with contour lines projected on the bottom plane.

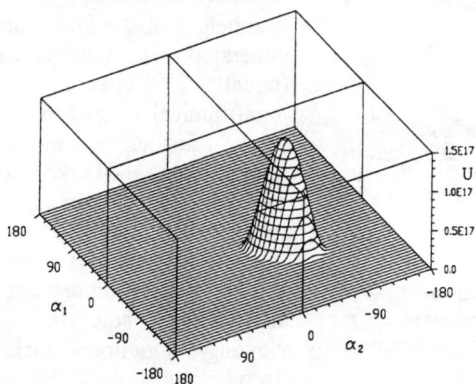

Fig. 6 A whole range representation of $U(\alpha_1,\alpha_2)$ for l=1.0m and k_2=10^{20}N/m.

CONCLUSIONS

The case of the total potential function of a spring restrained double pendulum has been considered to illustrate the features of the **D3D** function representation software, particularly design to facilitate representing graphically objective functions of two variables. These features are summarised briefly in the following:

-plotting single-valued functions of two real variables z=f(u,v) as polylines of constant u, polylines of constant v, crossed-hatch representations or raised level-curves (constant z profiles) with or without the hidden lines removed;
-accurate solving of the intersection problem between horizontal plane(s) and the function surface in truncated plots over z-axes, both in crosshatch and in constant x, y and z profile representations;
-generating contour maps of the same types of functions;
-plotting the gradient of the function as a vector field projected on the bottom plane using the same grid of heights for numerically calculation of its components [resent implementations of this feature in mathematical and visualization software allow the representation of the gradient only in top view projections, and require the user to provide separately the components of the gradient over the x and y axes – see Quiver function in MATLAB 6 or Champ function in SCILAB 2.5 (http://www-rocq.inria.fr/scilab/)]

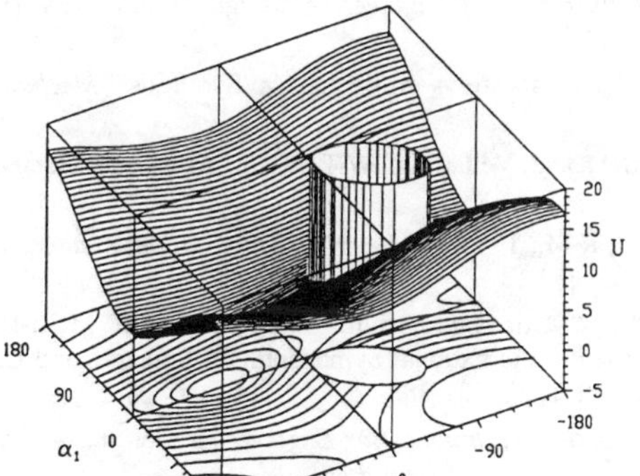

Fig. 7 Truncated representation of the potential function surface $U(\alpha_1,\alpha_2)$ for l=1.0m and k_2=10^{20}N/m.

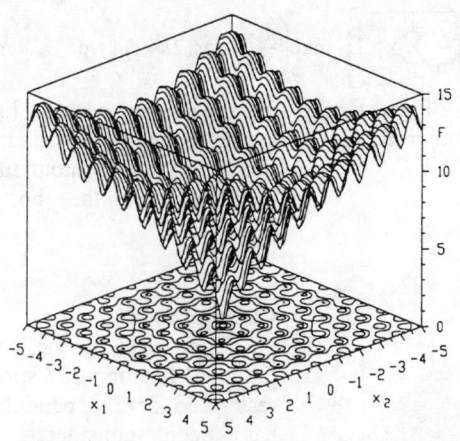

Fig. 8 Ackley's function represented as lines of constant x_2 and 15 contour lines projected on the bottom plane, equally spaced over the [0..15] interval.

- generation of logarithmically spaced level curve diagrams (projected, or mapped on the function surface) for detailing the global minimum area in objective functions;
- the height of the level curves generated automatically (equally spaced or logarithmically spaced), can be further modified interactively and saved to a file from where they can be read later on;
- the orientation of the z-axis can be reversed on request, an alternative solution for viewing function surfaces from below;
- all the graphic images can be exported as PCX or DXF files, and further used in spreadsheets and reports.

More representations produced with D3D software can be found on the Internet at: http://www.auburn.edu/~simiope/fxy.htm

REFERENCES

Encarnacao, J.L., Lindner, R. and Schlechtendahl, E.G., 1990, *Computer Aided Design. Fundamentals and System Architecture*, Springer-Verlag, pp. 335-342.

Gen, M., and Cheng, R., 1997, *Genetic Algorithms and Engineering Design*, Willey, New York, pp. 44.

Kota, S. and Chiou, S.J., 1993, "Use of Orthogonal Arrays in Mechanism Synthesis," *Mechanism and Machine Theory*, Vol. 28, pp. 777-794.

Press, W.H., Flannery, B.P., Teukolsky, S.A. and Vetterling, W.T., 1992, *Numerical Recipes*," Cambridge University Press.

Reklaitis, G.V., Ravindran, A., and Ragsdell, K.M., 1983, *Engineering Optimization, Methods and Applications*, Wiley, New York.

Simionescu, P.A. and Fawcett, J.N., 1997, "Static Equilibrium Determination of Multi-DOF Systems. A Review," Proceedings of the Seventh International Symposium on Linkages and CAD Methods, SYROM'97, Bucharest, Romania, Vol. 1, pp. 281-288.

Vanderplaats G.N., 1984, *Numerical Optimisation Techniques for Engineering Design*, McGraw-Hill.

A NEW APPROACH TO UPDATE PROBABILITY DISTRIBUTIONS ESTIMATES OF AIR TRAVEL DEMAND

IOANA C. BILEGAN
LAAS-CNRS and ENAC
Toulouse, France
KARIM ACHAIBOU
LAAS-CNRS
Toulouse, France

WALID EL MOUDANI
LAAS-CNRS and ENAC
Toulouse, France
FELIX MORA-CAMINO
LAAS-CNRS and ENAC
Toulouse, France

ABSTRACT

The American Airline Deregulation Act, in 1978, opened a new way of maximizing revenues for airliners, by matching the offer to the estimated air travel demand. This innovative solution, called Revenue Management, became crucial for airline companies as market competition exploded. Therefore, the design of new decision support systems, providing optimal seat inventory control processes with real-time capabilities, becomes unavoidable. In this paper a theoretical framework for updating the probability distributions of demand for reservations is presented. This updating process, facing the stochastic nature of demand for travel, is a cornerstone for the design of an efficient on-line decision support system to control the reservation process for a flight by an airline. The considered problem is formulated as a dual geometric problem to which an unconstrained non-convex, primal geometric problem is associated. A genetic algorithm optimization approach is proposed to solve the primal geometric problem, then the classical geometric primal-dual transformations provide the solution to the initial problem.

INTRODUCTION

During the last forty years which followed the first publications about booking control, the passenger booking systems have known a large evolution as a result of the development of computer science and telecommunications technologies on one side and of demand and decision theories on the other side. Today, the huge development of new communication devices, such as the Internet, contributes to make the flight/airline choice process more efficient for potential customers in the context of a quasi-perfect information situation. Nowadays, for airlines as well as for other transportation companies, Revenue Management has become a critical activity for their success in highly competitive markets.

Demand forecasting is a cornerstone of the planning process in all industries. As far as Airlines' Revenue Management is concerned, it is a prerequisite to establish booking limits and overbooking levels. Before the seventies, almost all the research in this field was focused on the subject of overbooking control. This concern stimulated research about detailed demand forecasting techniques concerning the number and the distribution of bookings, booking cancellations, no-shows and go-shows. In the case of air transportation, the forecasting techniques have to take into account a multiplicity of complex factors such as distribution of fares and routes, seasonal effects, time schedules etc. All these factors turn the air transportation demand forecasting into a very difficult task.

In order to take into account the highly stochastic nature of booking requests, forecasts should be updated with the latest information available to improve the efficiency of the Revenue Management decision process. Now, a feedback control loop can be established between a step-by-step Dynamic Programming optimization process and the Demand Forecasting Updating process, leading to a market reactive Revenue Management system. The structure of the proposed system is then such as in the Figure 1.

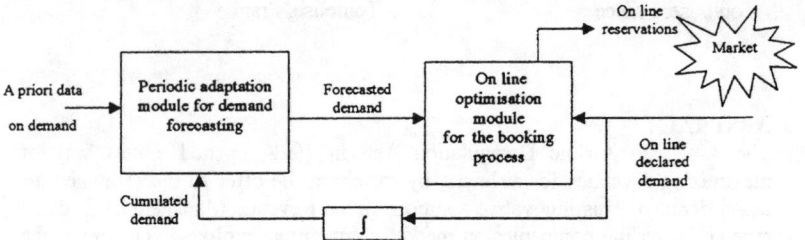

Figure 1: The proposed structure for the RM process

THE DEMAND PROBABILITY DISTRIBUTION UPDATING PROBLEM

Classical probabilistic distributions, such as Poisson distribution, binomial distribution or gamma distribution and variations or combinations of them have been proposed to model the dynamic process of booking, including cancellations, no-shows and go-shows. Empirical studies showed that the normal distribution provides an acceptable continuous approximation for the distribution of the total demand for a flight, but researchers argue that this distribution becomes inappropriate when temporal distributions of demand are needed.

Several stochastic models of booking requests have been proposed in the literature: homogenous and non-homogenous Poisson processes, leading to the cumulative Poisson process and the censored Poisson process. Data available in the records of past reservations which are used to provide demand forecasts are usually biased by the existence of booking limits for each fare class: in general booking requests exceeding the limits of these fare classes, are not recorded.

Once a prior probabilistic distribution of demand along the whole booking horizon is turned available, an on line adaptation process devoted to the updating, according to current data, of the temporal probability distributions of demand, should be started. This need leads to the formulation of a new optimization problem which must be solved at each time step of the booking process.

To state the corresponding optimization problem at time step n, the notations presented in the *Nomenclature* are adopted. In the simple case where demand suffers only temporal shifts, δ_n is given by:

$$\delta_n = D - d_n \tag{1}$$

Otherwise, complex estimation techniques, including qualitative reasoning to take advantage of expert knowledge, must be used to get an updated mean value of the remaining demand δ_n as well as of its variance σ_n^2. Consistency implies that the following relations hold:

$$\delta_n - \sum_{k=n}^{N}\sum_{j=0}^{J} j \cdot p_{jk}^n = 0 \qquad (2)$$

$$\sigma_n^2 + \delta_n^2 - (\sum_{k=n}^{N}\sum_{j=0}^{J} j^2 \cdot p_{jk}^n) = 0 \qquad (3)$$

In the proposed approach, the a priori distribution probabilities have to be corrected at the beginning of each decision period n, over the remaining decision periods $k = n$ to N, taking into account the most recent available information (the expressed demand as well as other newly foreseen effects) obtained during the last decision period. To update the demand probability distribution, its potential changes (between the predicted and the real recorded demand) have to be spread over the remaining decision periods to meet the new consistency conditions (2) and (3). This leads to the minimization of the *information gain* criterion given in the following expression:

$$\min \sum_{k=n}^{N}\sum_{j=0}^{J} p_{jk}^n \cdot \log\left(\frac{p_{jk}^n}{p_{jk}^{n-1}}\right) \qquad (4)$$

which corresponds also to the maximization of a relative entropy between a measure of the distance between the $(n-1)^{th}$ and the nth demand probability distributions estimations.

Here, p_{jk}^{n-1} denotes the previous probabilities computed at the beginning of the $(n-1)^{th}$ decision period, hence available for the nth decision period and the p_{jk}^n are the new probabilities to be obtained from the solution of the optimization problem whose set of constraints (2)-(3) is completed by :

$$\sum_{j=0}^{J} p_{jk}^n = 1, \qquad \text{for all } k=n \text{ to } N \qquad (5)$$

$$\text{and } p_{jk}^n \geq 0, \qquad \text{for } j = 1 \text{ to } J \text{ and } k=n \text{ to } N \qquad (6)$$

AN EQUIVALENT GEOMETRIC PROGRAMMING PROBLEM

The above optimization problem can be reformulated as a dual Geometric Programming problem.

Let $x_{jk}^n = \dfrac{p_{jk}^n}{N-n}$ for $j = 1$ to J and $k=n$ to N, then the objective function (3) can be rewritten as:

$$\min \cdot \sum_{k=n}^{N}\sum_{j=0}^{J} x_{jk}^n \cdot \log\left(\frac{x_{jk}^n}{x_{jk}^{n-1}}\right) \qquad (7)$$

The normality constraints (5) are equivalent to the set of constraints:

$$\sum_{k=n}^{N}\sum_{j=0}^{J} x_{jk}^n = 1 \qquad (8)$$

$$\sum_{j=0}^{J} x_{j\,n+1}^{n} - \sum_{j=0}^{J} x_{jk}^{n} = 0 \qquad \text{for all } k=n \text{ to } N. \tag{9}$$

The first consistency relation (2) can be rewritten as:

$$\sum_{k=n}^{N} \sum_{j=0}^{J} ((N-n)j - \delta_n) \cdot x_{jk}^{n} = 0 \tag{10}$$

while the second consistency relation (3) can be restated as:

$$\sum_{k=n}^{N} \sum_{j=0}^{J} ((N-n)j^2 - (\sigma_n^2 + \delta_n^2)) \cdot x_{jk}^{n} = 0 \tag{11}$$

with

$$x_{jk}^{n} \geq 0 \quad \text{for } j = 1 \text{ to } J \text{ and } k=n \text{ to } N \tag{12}$$

The optimization problem (7) to (12) can be identified as a Dual Geometric Programming problem where the x_{jk}^{n} are the dual geometric variables, relation (8) is the geometric normality condition and relations (9) to (11) represent geometric orthogonality conditions while relation (12) represents the geometric positivity conditions.

To this Dual Geometric Programming problem is associated a Primal Geometric problem given by min $\phi(\theta,t)$ with:

$$\phi(\theta,t) = \sum_{j=0}^{J} p_{j\,n}^{n-1} \cdot \theta_1^{(N-n)j-\delta_n} \cdot \theta_2^{(N-n)j^2 - (\sigma_n^2+\delta_n^2)} \cdot \prod_{k=1}^{N-n-1} t_k +$$
$$\sum_{k=1}^{N-n-1} \sum_{j=0}^{J} p_{j\,n+k}^{n-1} \cdot \theta_1^{(N-n)j-\delta_n} \cdot \theta_2^{(N-n)j^2 - (\sigma_n^2+\delta_n^2)} \cdot t_k^{-1} \tag{13}$$

with $(N-n+1)$ primal variables θ_k and t_k which are such that:

$$\theta_1 > 0, \quad \theta_2 > 0 \text{ and } t_k > 0, \text{ for all } k = 1..N-n-1 \tag{14}$$

and with $(N-n)\cdot(J+1)$ posynomes associated to the $(N-n)\cdot(J+1)$ dual geometric variables.

Observe that the rearrangement of relation (5) into relations (8) and (9) avoids the inclusion of additional constraints into the above Primal Geometric Program. This problem, which is not analytically constrained, is in general non convex and $\phi(\theta,t)$ may present a large number of local minimums, invalidating classical minimization techniques.

SOLUTION OF THE PROBABILITY DISTRIBUTION UPDATING PROBLEM

Since the following conditions are in general satisfied:

$$t_k > 0 \text{ and } (N-n)J^2 > \delta_n^2 + \sigma_n^2 \text{ for } n = 0..N-1. \tag{15}$$

then the solutions of the successive Primal Geometric Programs indexed by n, are contained in a compact set of $(\Re^*)^{N-n+1}$. Writing the first order necessary optimality conditions, we get after some treatment two sets of conditions, one with respect to t:

$$t_k^* = 1 \quad \text{for all } k=1 \text{ to } N-n-1 \tag{16}$$

and the other one with respect to θ_1 and θ_2:

$$\begin{cases} \sum_{j=0}^{J}((N-n)j-\delta_n)\cdot\left(\sum_{k=n}^{N}p_{jk}^{n-1}\right)\cdot(\theta_1)^{(N-n)j}\cdot(\theta_2)^{(N-n)j^2}=0 \\ \sum_{j=0}^{J}((N-n)j^2-(\sigma_n^2+\delta_n^2))\cdot\left(\sum_{k=n}^{N}p_{jk}^{n-1}\right)\cdot(\theta_1)^{(N-n)j}\cdot(\theta_2)^{(N-n)j^2}=0 \end{cases} \quad (17)$$

Then θ_1^* and θ_2^* have to be taken within the set of positive solutions of equations (17). It is possible to reduce the order of these equations by taking:

$$X_1 = (\theta_1)^{(N-n)} \quad \text{and} \quad X_2 = (\theta_2)^{(N-n)} \quad (18)$$

to get:

$$\begin{cases} E_1(X_1,X_2)=\sum_{j=0}^{J}((N-n)j-\delta_n)\cdot\left(\sum_{k=n}^{N}p_{jk}^{n-1}\right)\cdot(X_1)^j\cdot(X_2)^{j^2}=0 \\ E_2(X_1,X_2)=\sum_{j=0}^{J}((N-n)j^2-(\sigma_n^2+\delta_n^2))\cdot\left(\sum_{k=n}^{N}p_{jk}^{n-1}\right)\cdot(X_1)^j\cdot(X_2)^{j^2}=0 \end{cases} \quad (19)$$

The analysis of equations $E_i(X_1,X_2)$, $i = 1,2$, shows that the positive real roots of these equations stay in an open domain Δ_X which is such that:

$$X_1 > 0, \ X_2 > 0 \quad (20)$$

$$X_1(X_2)^{J_0^i+1} < (1+J_0^i)(\max_{j=0\,to\,J_0^i}\{|\alpha_j^i|\})/(\alpha_{J_0^i+1}^i) \quad \text{for } i=1,2 \quad (21)$$

where:

$$\left.\begin{aligned} \alpha_j^1 &= ((N-n)j-\delta_n)\left(\sum_{k=n}^{N}p_{jk}^{n-1}\right) \\ \alpha_j^2 &= ((N-n)j^2-(\delta_n^2+\sigma_n^2))\left(\sum_{k=n}^{N}p_{jk}^{n-1}\right) \end{aligned}\right\} \quad \text{for } j = 0 \text{ to } J \quad (22)$$

and where J_0^i is such that:

$$\alpha_{J_0^i}^i \leq 0 < \alpha_{J_0^i+1}^i \quad \text{for } i = 1,2 \quad (23)$$

Using the non singular transformation over $(\Re^*)^2$ defined by relation (18), an open search domain Δ_θ can be easily defined for θ_1 and θ_2. In general, J_0^1 and J_0^2 are different integers and in this case, Δ_θ is a bounded set. Then Genetic Algorithms techniques can be used to search for the global minimum of the following reduced minimization problem:

$$\min_{\theta_1,\theta_2}\left\{\sum_{k=0}^{N-n-1}\sum_{j=0}^{J}p_{j(n+k)}^{n-1}\cdot\theta_1^{(N-n)j-\delta_n}\cdot\theta_2^{(N-n)j^2-(\sigma_n^2+\delta_n^2)}\right\} \quad (24)$$

with $\quad \{\theta_1,\theta_2\}\in\Delta_\theta \quad (25)$

using as initial seed the neutral point *(1,1)*.

Then the solution of the original Dual Geometric Program can be obtained using the primal-dual relationship relating primal and dual Geometric Program solutions. This leads to the expressions of the dual solution *for k = 0* to *N-n-1* and *j = 0* to *J*:

$$x^{n^*}_{j(n+k)} = (N-n) \cdot$$

$$\frac{p^{n-1}_{j(n+k)} \theta_1^{*(N-n)_j - \delta_n} \theta_2^{*(N-n)_j^2 - (\delta_n^2 + \sigma_n^2)} \cdot \omega_k}{\sum_{j=0}^{J} p^{n-1}_{jn} \cdot \theta_1^{*(N-n)_j - \delta_n} \theta_2^{*(N-n)_j^2 - (\delta_n^2 + \sigma_n^2)} \prod_{l=1}^{N-n-1} t_l^* + \sum_{l=1}^{N-n-1} \sum_{j=0}^{J} p^{n-1}_{j(n+l)} \cdot \theta_1^{*(N-n)_j - \delta_n} \theta_2^{*(N-n)_j^2 - (\delta_n^2 + \sigma_n^2)} t_l^{*-1}} \quad (26)$$

with $\omega_0 = \prod_{l=1}^{N-n-1} t_l^*$ and $\omega_k = t_k^{*-1}$ for $k = 1$ to $N-n-1$.

Finally, we get the probability updating process at the beginning of the nth period:

$$p^{n^*}_{j(n+l)} = (N-n) \frac{p^{n-1}_{j(n+l)} \theta_1^{*(N-n)_j} \theta_2^{*(N-n)_j^2}}{\sum_{k=0}^{N-n-1} \sum_{j=0}^{J} p^{n-1}_{j(n+k)} \cdot \theta_1^{*(N-n)_j} \theta_2^{*(N-n)_j^2}} \quad \text{for } l = 0 \text{ to } N\text{-}n \text{ and } j = 0 \text{ to } J \quad (27)$$

CONCLUSIONS

The present communication presents two main contributions. First, an on-line probability distribution updating process is proposed so that Dynamic Programming approaches of Revenue Management Decision Making can be turned reactive to the market. Observe that this process could be also applied to other fields of application related with perishable asset revenue management.

Second, the proposed solution approach mixes for the first time an advanced Mathematical Programming technique, Geometric Programming, and a stochastic numerical solution technique, Genetic Algorithms, to solve a highly constrained non linear optimization problem. The application of the proposed approach to a set of scheduled flights offered by a regular airline is under way and should provide information about its numerical efficiency.

NOMENCLATURE

D : initial global demand forecast (before starting the booking process), by fare class;
N : total number of decision periods ($n=N$ is the last period, just before departure);
d_n : real demand recorded from $n=1$ until the end of the nth decision period;
δ_n : mean global remaining demand, estimated at the end of the nth decision period;
σ_n^2 : variance of the remaining demand, estimated at the end of the nth decision period.

REFERENCES

Bilegan, I.C., 2001, "Développement de Systèmes d'Aide à la Décision pour les Opérations Aériennes", *2éme Congrès des doctorants, Ecole Doctorale Systèmes*, Toulouse.

Goldberg, D.E., 1989, "Genetic Algorithms in Search, Optimization and Machine Learning", Addison-Wesley.

Lee, A.O., 1990. "Airline Reservations Forecasting: Probabilistic and Statistical Models of the Booking Process", Ph.D. thesis, Flight Transportation Laboratory, M.I.T., Cambridge, MA.

Lee, T.C. and Hersh, M., 1993, "A model for Dynamic Airline Seat Inventory Control with Multiple Seat Bookings", *Transportation Science*, Vol. 27, pp. 252-265.

McGill, J.I. and Van Ryzin, G.J., 1999, "Revenue Management: Research Overview and Prospects" *Transportation Science*, Vol. 33, pp. 233-256.

Mora-Camino, F., 1978, "Introduction à la Programmation Géométrique", COPPE, Rio de Janeiro.

AN HEURISTIC APPROACH FOR THROUGHPUT OPTIMIZATION OF SMT PLACEMENT MACHINES

SREEKRISHNA PALAPARTHI
Department of Systems Science and Industrial Engineering
Binghamton University
Binghamton, New York 13902
passkrishna@hotmail.com

SARAH S. Y. LAM
Department of Systems Science and Industrial Engineering
Binghamton University
Binghamton, New York 13902
sarahlam@binghamton.edu

KRISHNASWAMI SRIHARI
Department of Systems Science and Industrial Engineering
Binghamton University
Binghamton, New York 13902
srihari@binghamton.edu

DENNIS WARHEIT
Universal Instruments Corporation
Binghamton, New York 13902
warheit@uic.com

ABSTRACT

Increased competition among Electronic Manufacturing Service providers has resulted in enhanced interest for improving the efficiency of surface mount placement equipment. Consequently, the suppliers of electronics assembly equipment are making efforts to optimize the performance of their equipment. One performance measure that is frequently considered for improvement, is the throughput obtained through a reduction of component placement time. Component placement is typically the bottleneck operation in a surface mount assembly line and has received increasing attention over the past few years. The placement machine optimization problem is a NP-complete problem with several constraints. This problem consists of three dependent sub-problems: the feeder assignment problem, the pick and place sequencing problem, and the nozzle assignment problem. This paper discusses a heuristic approach to optimize the throughput of surface mount placement machines while considering a large number of constraints. Factors that are considered include the type and location of feeders, the selected components and their dimensions, the component placement location, and the layout of the panel.

INTRODUCTION

Original Equipment Manufacturers (OEMs) are increasingly outsourcing their Printed Circuit Board (PCB) assembly needs to Electronic Manufacturing Service (EMS) providers. Since electronic manufacturing is a low margin industry, EMS providers are seeking increased throughput of the PCB assembly process in order to cope with reduced lead times. High-density PCB production with larger throughput and greater accuracy requires the automation of the PCB assembly process. The Electronic Manufacturing Equipment Suppliers (EMES), who supply component assembly machines to the EMS providers, are faced with the challenges of producing these machines with faster and better optimization routines.

Throughput rate is one of the primary measures of the performance of a component placement machine. It is also referred to as the "Components Per Hour" (CPH) placement metric in the electronic manufacturing industry. The throughput rate is driven by the complexity of the product types, the dimension and density of the component types, and the setup of the feeders. The PCB assembly planning process consists of two primary sub-processes: the feeder arrangement planning and the placement sequence planning. Both of these planning problems belong to the class of NP-hard problems. Therefore, the optimization of throughput of the PCB assembly process is a NP-complete problem, which makes it computationally intractable. Because of the complexity of this problem, the automation of the optimal planning of the PCB assembly process has become one of the major challenges in the arena of electronics manufacturing.

Component placement machines come in different configurations and accordingly they have different environment variables and constraints. The focus of this paper is based on a particular type of Surface Mount Technology (SMT) placement machines. Detailed descriptions of the SMT machine and the problem formulation are discussed in the following sub-sections.

SMT Placement Machine

The SMT placement machine is an X-Y gantry type machine with two moving heads and a stationary table (see Figure 1). The machine has two beams supporting two heads, one at the front end of the machine and another at the rear end of the machine. Each head has four to seven spindles, each of which can take one nozzle. The machine has four banks, two on each side. Slots for feeder racks are present in the banks of the machine. Component feeders are mounted on the machine in the feeder slots. The board on which placements are to be made is placed in the middle of the machine with the help of a conveyer. A nozzle changer panel is located on one side of the machine in front of the banks. Nozzles are mounted on the head and can be changed on the nozzle changer panel. An upward looking camera is mounted on one side of the machine to perform vision processing on larger components. In addition, a camera is mounted on each of the heads to perform vision processing on smaller components. The machine is programmed so that the beams do not collide when placing the components on the board.

Problem Formulation

The throughput optimization problem of the placement machine can be viewed as consisting of three dependent sub-problems: the feeder assignment problem, the placement-sequencing problem, and the nozzle assignment problem. The feeder assignment problem is the problem of assigning a feeder slot for the feeder of a particular component type. The placement-sequencing problem is to find the order in which the components should be picked from their feeder slots and placed at their respective locations on the board, so that the throughput of the machine is optimized. This involves defining the group of components that should be picked in a particular travel of the head from the feeder slots to the board, as well as, defining the sequence in which the placement of components within this group should be performed. The nozzle assignment problem is that of finding the nozzle set that should be mounted on the head during a given set of picking operations. These sub-problems are to be integrated in such a way that the throughput is optimized.

There are several constraints on the problem. The primary constraint is the optimization time. The optimization time in our problem formulation is required to be below two minutes. Sometimes the feeder setup requires that some of the component feeders be assigned only to a particular side of the machine. In the decision for the placement sequence, it is important that the components of greater height are placed after the components of lesser height. Even as they are being placed on the board, care should be taken so that the head travels across the board above the height of the tallest component placed. Components need to be checked either by the camera on the head or the upward looking camera mounted on the machine, depending on the size of the components. Larger components must be assigned to the same side of the upward looking camera, and hence, there is a restriction on the feeder assignment. While one of the heads is placing the components on the board, the other head must wait until the current head has finished with the placements.

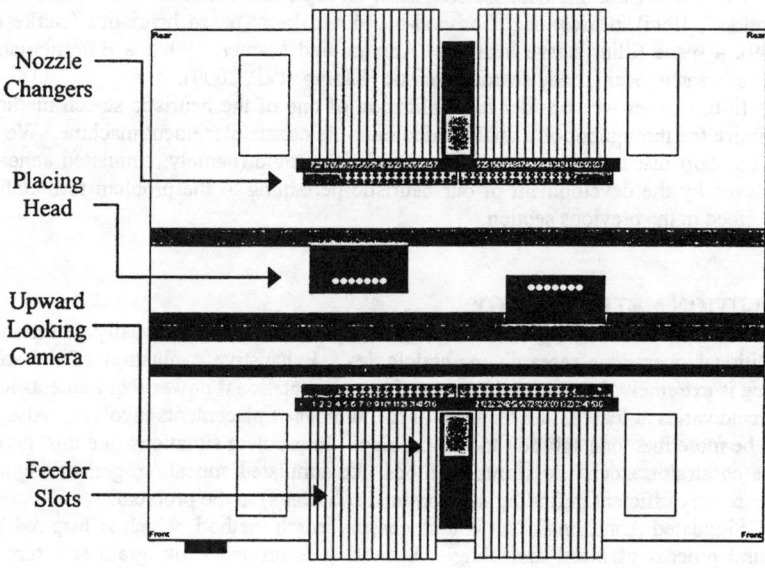

Figure 1. Schematic Drawing of a Placement Machine.

LITERATURE REVIEW
Heuristic search methods such as genetic algorithms, genetic programming, simulated annealing and tabu search, are nature-inspired stochastic search methods. These search methods have been applied to many classes of non-linear combinatorial optimization problems. With proper adjustment and fine-tuning of these heuristic methods near optimal solutions to NP-hard and NP-complete optimization problems can be obtained. Some of the examples include the bin-packing problem, the traveling salesman problem, and the quadratic assignment problem. Aside from the traditional operations research type of problems, the throughput optimization problem of component placement machines has drawn increasing interest in recent years. Following, some of the more recent research papers are summarized.

Ball and Magazine (1988) modeled PCB assembly planning as a rural postman problem. Kumar and Li (1995) modeled the feeder setup problem as a minimum weighted matching problem and the pick and place sequencing problem as a traveling salesman problem.

Wang et al. (1999) applied genetic algorithms to solve a problem case involving multiple placement heads with one nozzle per head. Khoo and Ng (1998) discussed the application of a genetic algorithm to the problem and analyzed the impact of various parameters of the genetic algorithm on the result. Nelson and Wille (1995) discussed the use of simulated annealing and evolutionary computation to obtain near optimal solutions to this problem. They concluded that simulated annealing would give good results in long run time, but evolutionary computation would yield the best results if the time constraints were a bit flexible. Su and Fu (1998) used simulated annealing to solve the problem for a dynamic pick and place machine.

Other heuristics include: Convex Hill, Or-Opt, and rule based heuristics (Or and Demirkol, 1995), nearest-neighborhood-based and k-opt-based heuristics (Burke et al., 1999), a space filling curve heuristic (Ahmadi and Mamer, 1999), a Hamiltonian path heuristic and a local greedy search heuristic (Klomp et al., 2000).

In this paper we consider an application of one of the heuristic search methods to optimize the throughput of a dual headed multiple nozzle placement machine. We begin with a short discussion on our chosen heuristic method (namely, simulated annealing), followed by the development of our heuristic pertaining to the problem formulation as described in the previous section.

SOLUTION METHODOLOGY

The solution to the problem is very complicated and is not easily determined by traditional operations research methodologies. Exhaustive evaluation of the solution space is extremely demanding in terms of the computational power requirements and the demand varies at least as the factorial of the number of placements involved. Also, there can be more than one solution to this problem. In problem situations like this, given the time constraint, stochastic search methods like simulated annealing, genetic algorithms etc. are very efficient in finding near-optimal solution(s) to the problem.

Simulated Annealing (SA) is a stochastic search method, which is inspired by the natural process of metal annealing. A metal gets refined in its grain structure when heated to its characteristic annealing temperature and cooled at a very slow cooling rate. This process of cooling is called annealing. Kirkpatrik et al. (1983) were the first to apply this process of nature to optimization problems. The initial temperature for the heuristic is fixed based on the problem specifications. Given an initial (random or constructive) single solution, the algorithm generates a new solution using a move operator and accepts it if it is better than the previous solution. Otherwise the solution is accepted probabilistically based on the current "temperature" and the badness of the new solution. This is similar to the energy states change in an actual metal annealing process. After a certain number of moves, the system will be cooled to the next temperature based on the cooling rate. This process would continue until a stopping condition is reached.

Simulated Annealing is a relatively faster settling heuristic compared to other soft-computing techniques like genetic algorithms and tabu search. Because of the primary interest to reach a near-optimal solution in a relatively small run-time of the optimization process, this approach was chosen for this research effort.

To estimate whether or not the solution has improved over the previous solution, a pay-off information or the objective function value is necessary for quantifying the amount of improvement or worsening. In effect, this information, drives the optimization. In order to provide this information to the heuristic, a simulator was designed to reflect the approximate functioning of the actual machine. The simulator takes the inputs, simulates the functioning of the machine and returns the total time for populating the printed circuit board.

The Simulated Annealing heuristic for the throughput optimization problem of the placement machine uses a given feeder setup, a placement sequence, and a nozzle assignment as input. It starts at an initial temperature ($T=T_0$) and the temperature cools at a rate of alpha after 'm' moves have been attempted. Hence the annealing function is alpha times the temperature ($\alpha*T$). Annealing takes place after every 'm' moves have been attempted at a temperature. A k-opt algorithm is used as a move operator. The temperature cools when 'm' number of moves at the current temperature have been attempted before cooling the system to the next temperature. After each move, the simulator is called to evaluate the solution quality of the new feeder setup, placement sequence, and the nozzle assignment. CPU run time is used as the stopping criterion for the heuristic, since optimization time is the primary concern.

RESULTS AND DISCUSSION

The Simulated Annealing heuristic was applied to the optimization problem and tested on two printed circuit boards. These boards have 95 and 200 placements respectively. Experiments were conducted with different parameters of the Simulated Annealing algorithm. CPU run time was constrained to less than 100 seconds. The random seeds used for these experiments are time-based. The time-based random seed is obtained by feeding the current system time to the library function provided with the compiler. The results, averaged across 10 replications for each parameter setting for the annealing schedule, show the approximate percentage increase in throughput when comparing the best solution with the initial solution (see Tables 1 and 2). The improvement of the solutions ranged from 18% to 38% over the initial solutions. However, the experimental results do not seem to outperform the throughput rates from the constructive heuristic implemented by the industry. This observation is due to some of the assumptions made in the encoded simulator and modifications on the code are currently underway.

Based on these preliminary results, the percentage of improvement does not seem to be affected significantly by the annealing schedule. As expected, the board type with larger number of placements has a lower percentage of improvement in throughput compared to the one with smaller number of placements. This is most likely due to the increasing density of the components on the board.

Table 1. Results for Experiments based on Product 1 (95 Placements).

Average Run Time (s)	Total # Temperatures Attempted (G)	# Moves Attempted per Temperature	Cooling Rate	% Improvement over Initial Solution
50	100	500	0.9	38
45	100	G*10+50	0.8	35
55	500	50	0.8	32

Table 2. Results for Experiments based on Product 2 (200 Placements).

Average Run Time (s)	Total # Temperatures Attempted (G)	# Moves Attempted per Temperature	Cooling Rate	% Improvement over Initial Solution
70	100	500	0.9	18
96	100	G*10+50	0.9	20
44	500	50	0.7	22

CONCLUSIONS

The Simulated Annealing heuristic was developed and applied to a highly constrained throughput optimization problem of a dual beam placement machine in an Electronic Manufacturing Suppliers' environment. Various parameters of the annealing were experimented with and the results indicated that the percentage of improvement of the throughput time is not sensitive to the parameter setting for the annealing schedule based on two board types. In addition, near-optimal solutions can be obtained within 2 minutes of computer run time on a 600 MHz Intel Pentium III machine. Research work is currently underway to apply the same algorithm to a larger set of PCB assemblies.

The success of the SA heuristic depends heavily on the payoff information provided by the objective function. In most real world applications, the payoff information cannot be put in a simple analytical form. In situations like these, simulation of the situation can be used to provide information in the objective of optimization. In our case, a simulator was designed for simulating the performance of the machine. Better results can be obtained by magnifying the payoff information at places where there is more scope for optimization. Complex optimization problems of greater dimensions and those with large number of constraints can be solved well using these heuristic procedures.

REFERENCES

Ahmadi, R.H. and Mamer, J.W., 1999, "Routing Heuristics for Automated Pick and Place Machines," *European Journal of Operational Research*, 117 (3), 533-552.

Ball, M.O. and Magazine, M.J., 1988, "Sequencing of Insertion in Printed Circuit Board Assembly," *Operations Research*, 36, 192-201.

Burke, E.K., Cowling, P.I. and Keuthen, R., 1999, "New Models and Heuristics for Component Placement in Printed Circuit Board Assembly," *Proceedings 1999 International Conference on Information Intelligence and Systems*, 133-140.

Khoo, L.P. and Ng, T.K., 1998, "A Genetic Algorithm Based Planning System for PCB Component Placement," *International Journal of Production Economics*, 54 (3), 321-332.

Kirkpatrick, S., Gelatt, C. and Vecchi, P., 1983, "Optimization by Simulated Annealing," *Science*, 220, 671-679.

Klomp, C., van de Klundert, J., Spieksma, F.C.R. and Voogt, S., 2000, "The Feeder Rack Assignment Problem in PCB Assembly: A Case Study," *International Journal of Production Economics*, 64 (1-3), 399-440.

Kumar, R. and Li, H., 1995, "Integer Programming Approach to Printed Circuit Board Assembly Time Optimization," *IEEE Transactions on Components, Packaging and Manufacturing Technology – Part B*, 18 (4), 720-727.

Nelson, K.M. and Wille, L.T., 1995, "Comparative Study of Heuristics for Optimal Printed Circuit Board Assembly," *Proceedings of Southcon '95*, 322-327.

Or, I. and Demirkol, E., 1995, "Optimization Issues in Automated Production of Printed Circuit Boards: Operations Sequencing and Feeder Configuration Problems," *Proceedings 1995 INRIA/IEEE Symposium on Emerging Technologies and Factory Automation*, 3, 479-487.

Su, C.T. and Fu, H.P., 1998, "A Simulated Annealing Heuristic for Robotics Assembly Using the Dynamic Pick-and-Place Model," *Production Planning and Control*, 9 (8), 795-802.

Wang, W., Nelson, P.C. and Tirpak, T.M., 1999, "Optimization of High-Speed Multistation SMT Placement Machines Using Evolutionary Algorithms," *IEEE Transactions on Electronics Packaging Manufacturing*, 22 (2), 137-146.

A NEW SELF-ORGANIZING NEURAL NETWORK FOR SOLVING THE TRAVELLING SALESMAN PROBLEM

F. GUERRERO, S. LOZANO, D. CANCA AND J.M. GARCÍA
Department of Industrial Management, University of Seville, Spain

K. A. SMITH
School of Business Systems, Monash University, Australia

ABSTRACT

In this paper, we present a neural approach to the Travelling Salesman Problem. The approach is a Self-Organizing Neural Network that has been previously used to solve scheduling, frequency assignment and cellular manufacturing problems. The neural approach uses neuron normalization as well as a conscience mechanism to consistently find good feasible solutions. Numerical examples of the proposed approach and further research topics are suggested..

INTRODUCTION

There are two main neural approaches that have been used for solving combinatorial optimization problems. The first one is the method of Hopfield & Tank (Hopfield and Tank, 1985), which maps the objective function onto an energy function that depends on the activation of the units in a fully-connected, recurrent neural network. Over the last decade, this approach has received criticisms over its inability to ensure feasibility of the final solution, or its poor solution quality. Modifications have been made which have resulted in new energy minimization variations of this method like Mean Field Annealing (Peterson and Anderson, 1987; Bilbro et al, 1989) and Boltzmann machines (Ackley et al, 1985).

The second main approach is the use of self-organization. Based upon the early work of Kohonen (Kohonen, 1982), these neural networks exploit the principles of neuron organization, and the brain's apparent ability to find structure and order from information. Self-organizing neural approaches are basically clustering techniques, and hence their applicability to solving optimization problems has traditionally been restricted to solving Euclidean problems like the Traveling Salesman Problem.

In previous work (Smith et al, 1996), however, a self-organizing neural network (SONN) which enables solution of combinatorial optimization problems with generalized assignment constraints was proposed. This SONN has been succesfully applied to solving practical optimization problems such as car sequencing problem (Smith et al, 1986), frecuency assignment problem (Smith et al, 1997), and manufacturing cell formation problem (Lozano et al, 1998).

In this paper, a new neural approach, which is based upon the previous SONN is designed to solving the Travelling Assignment Problem (TSP). In this new neural approach, constraints are adaptively enforced using neuron normalization as well as a penalty function that is computed through a conscience mechanism (Burke and Damany, 1992). The paper is organized as follows. In section 2 the TSP is introduced. Section 3 describes the new SONN. In section 4, first computational results are presented. Finally, Section 5 gives the main conclusions as well further research topics.

THE TRAVELLING SALESMAN PROBLEM

The TSP consists of finding the shortest possible tour through a set of n cities, visiting each exactly once. A tour which passes through each city exactly once is known as a feasible tour. A complete review of many exact a aproximate algorithms for solving the TSP can be found in (Laporte, 1992). The TSP can be modelled as:

Let

N   be the number of cities

d_{il} be the distance between cities i and l

The solution space for the TSP can be represented by using a NxN matrix X where the component X_{ij} indicates

$$X_{ij} = \begin{cases} 1 & \text{if city i is assigned to position j in the tour} \\ 0 & \text{otherwise} \end{cases}$$

The TSP can be formulated as

$$\text{minimize } F(\overline{X}) = \sum_{i=1}^{N}\sum_{l=1}^{N}\sum_{k=1}^{N} d_{il} X_{ik} X_{l k+1} \qquad (1)$$

subject to:

$$\sum_{j=1}^{N} X_{ij} = 1 \quad \forall i \qquad (2)$$

$$\sum_{i=1}^{N} X_{ij} = 1 \quad \forall j \qquad (3)$$

$$X_{ij} = 0,1 \quad \forall i,j \qquad (4)$$

The objective function (1) minimizes the length of the tour. Constraints (2) ensure that each city has no more than one position. Constraints (3) imply that no position in the tour is assigned to more than one city. Constraints (4) denote the integrity of the decision variables.

SELF-ORGANIZING NEURAL NETWORK

The SONN we propose is similar to those in (Smith et al, 1996), (Smith et al, 1997) and (Lozano et al, 1998) where a car sequencing problem, a frequency assignment problem, and manufacturing cell formation problem, respectively, have been solved. The architecture of the SONN for this problem is shown in Figure 1.

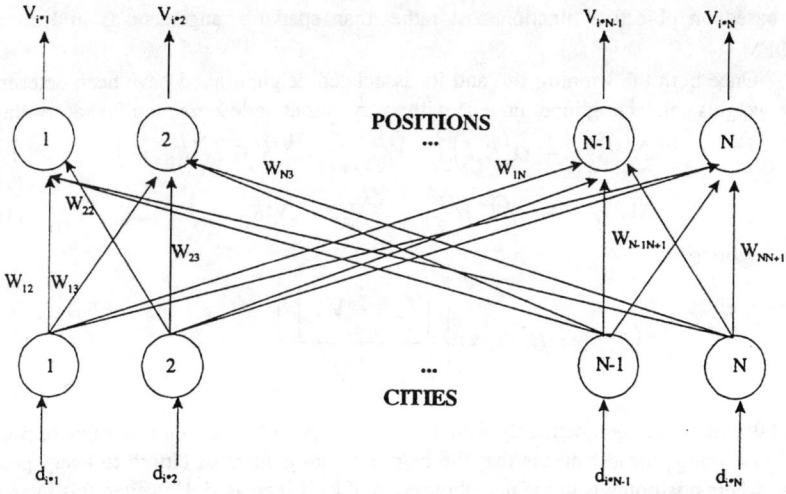

Figure 1. Architecture of the SONN

It is a feedforward neural network with an input layer of N nodes (one for each city), and an output layer of N nodes (one for each position in the tour). The weight between an input node i and an output node j is given by W_{ij+1} and can be interpreted in a fuzzy sense as the degrees of assignment of each city i to each position j+1.

The training set consists of one input pattern for each city. The input pattern corresponding to a city i* is a vector of N components where the l-th component is d_{i^*l}. When the input pattern corresponding to a city i* is presented, the net input to each node k of the output layer is the potential V_{i^*k}, which is a linear combination of the cost of assigning city i* to position k+1 and the degree of violation of constraints (3) (i.e., the same position, k+1, is occupied by different cities):

$$V_{i^*k} = \sum_{l} d_{i^*l} W_{lk+1} + \hat{a} \sum_{l \neq k+1} W_{i^*l} \tag{5}$$

where β is a penalty parameter. Note that the first term of the potential can be alternatively seen as the partial derivate of F(W) with respect to W.

The competition between the N output nodes gives the winning position, i.e. that with the minimum potential

$$k_0 = \arg \min_{k} V_{i^*k} \tag{6}$$

Furthermore, ranking the output nodes according to its potential (closest neighbor to farthest neighbor) as:

$$N(k_0, i^*) = \{k_0, k_1, k_2, \ldots, k_\eta\} \tag{7}$$

where

$$V_{i^*k_0} \leq V_{i^*K_1} \leq V_{i^*K_2} \leq \ldots \leq V_{i^*K_\eta} \leq \ldots \leq V_{i^*K_N} \tag{8}$$

and η ≥ 1 is the neighborhood size. It should be noted that this concept of neighborhood is based on objective function cost rather than spatial arrangement as in Kohonen's SOFM.

Once both the winning cell and its associated neighborhood have been determined, the weights connecting input node i* with every output node k are modified according to

$$\Delta W_{i^*k+1} = \alpha(k,t)\left[1 - W_{i^*k+1}\right] \quad \forall k \in N(k_0, i^*) \qquad (9)$$

$$\Delta W_{i^*k+1} = -\beta(t) W_{i^*k+1} \quad \forall k \notin N(k_0, i^*) \qquad (10)$$

where

$$\alpha(k,t) = \beta(t)\exp\left[\frac{\left|V_{i^*k_0} - V_{i^*k}\right|}{\left|V_{i^*k_0} - V_{i^*k_\eta}\right|}\right] \qquad (11)$$

and $\beta(t)$ decreases geometrically with the training epoch t from an initial value $0 \leq \beta(0) \leq 1$.

Learning rule (9) means that the degree of assignment of city i* to every position k+1 where position k is in the neighbourhood of k_0 is increased. However, this increase is not uniform but depends on the position potential: the higher the potential (i.e. the closer to the potential of the winning position), the higher the increase. Learning rule (10) means that the degree of assignment of city i* to the position k+1 with position k not in the neighborhood of the winning position is decreased. This process is helped by linearly decreasing the neighborhood size σ after the presentation of a fixed number of input patterns so that in the end the neighbourhood contains only the winning position.

After that, and in order to explicitly enforce the constraints (2), the weights for each output node k are normalized as follows:

$$W_{ik+1} = \frac{\exp\left(-(1-W_{ik+1})/T\right)}{\sum_i \exp\left(-(1-W_{ik+1})/T\right)} \qquad (12)$$

where T is a parameter (referred as temperature), which is lowered as the learning process proceeds. The normalization operation guarantees that when convergence is complete, each city is assigned to only one position. A similar normalization procedure was used in (Van den Bout and Miller, 1989).

During the learning process, the neighborhood size, the magnitude of the weight adaptations, and the temperature are gradually decreased. The complete algorithm follows:

1. Randomly initialize the weights:
$$0 \leq W_{ik} \leq 1 \quad \forall i,k \qquad (13)$$
2. Randomly choose a city i^* and present its corresponding input pattern.
3. Compute the potential V_{i^*k} for each output node k according to (5).
4. Determine the winning node, k_0, as well as its neighboring nodes according to (6)-(8).
5. Update weights, W_{i^*k+1}, connecting input node i^* with every output node k according to (9)-(10).

The updated weights are:

$$W_{i^*k+1} \leftarrow W_{i^*k+1} + \Delta W_{i^*k+1} \quad (14)$$

6. Normalize weights to enforce the constraints (2) according to (12).

Repeat from Step 2 until all cities have been selected as input patterns. This is a training epoch t. Repeat from Step 2 until $|\Delta W_{ik+1}| \approx 0 \quad \forall i, k$.
Anneal T to encourage integrality constraints (4). Decrease the neighborhood size η, and $\beta(t)$.

COMPUTATIONAL EXPERIENCES

In this stage of the research, the first objective we have to be tested is the ability of the neural network to generate feasible solutions. Thus, the new neural approach described in the previous section was implemented in C. Computational experiments have been perfomed on a Intel Pentium 166 Mhz PC. Three problem instances named eil51 (51-city problem), eil76 (76-city problem) and rd100 (100-city random problem) which have been extracted from the TSPLIB (Reinelt, 1991) were used. Also, optimal solutions for the three data sets are known and available from the TSPLIB.

Due to the randomness in the order of presentations of the training pattrens, it is appropiate to solve each problem more than once. Thus, each problem has been solved 10 times. The best results using the new neural approach for each one of the 3 problem instances are presented in Tables 1 and 2.

Table 1 shows, for each one of the problem instances (identified by its name), the optimum as well as the best integer objective function achieved from 10 replications of the algorithm are presented. Furthermore, deviations from best objectives to optimum solutions are presented.

Table 1. Results of experiments: objective function

Problem instance		Best results	
Name	Optimum	Best Integer Obj. Funct. Value achieved	Deviation (%)
Eil51	426	426	0
Eil76	538	538	0
Rd100	7910	8425	6.5

Table 2 presents the CPU requirements (in seconds) for obtaining the best results shown in the Table 1.

Table 2. Results of experiments: CPU requirements

Name	CPU time (in seconds)
Eil51	251
Eil76	420
Rd100	1241

In terms of solution quality it can be said that:
- For each one of the 30 runs of the algorithm (10 runs x 3 problem instances), feasible solutions were always obtained.
- The worst deviation from its corresponding optimum solution was less than 6.5%.

CONCLUSIONS

In this paper, we have proposed a new neural approach for the Travelling Salesman Problem. The net is an unsupervised Self-Organizing Neural Network, which uses neuron normalization and a penalty term to enforce feasibility in a adaptative way. In order to compute the penalty term, a conscience mechanism has been introduced.

We believe this neural approach presents attractive characteristics:
- Constraints are enforced withing the learning process, i.e., without any kind of aditional tuning or second stage (Smith et al, 1996).
- The net gives feasible solutions reliably.
- This neural approach has been designed as an extension of an existing hybrid neural approach which has been conveniently modified and is opened to extension to other neural features.

Extensions of this research should follow two ways. The first one, to benchmark this neural approach with other existing neural approaches for the TSP on the same problem instances. Furthermore, extensive computational experiments are required to evaluate this new neural approach. The second line to explore will be to extent the neural approach with other neural features.

REFERENCES

Ackley, D. H., Hinton, G. E. and Sejnowski, T. J., 1985, A learning algorithm for Boltzmann machines. *Cognitive Science*, 9, 147-169.

Bilbro, G., Mann, R., Miller, T.K., Snyder, W.E., Van den Bout, D.E. and White, M., 1989, Optimization by Mean Field Annealing. In: *Advances in Neural Information Processing Systems I*, D.S. Touretzky (ed.), Morgan Kaufmann, 91-98.

Burke, L.I., and Damany, P., 1992, The Guilty Net for the Travelling Salesman Problem, *Computers and Operations Research* 19, 255-265.

Hopfield, J.J. and Tank, D.W., 1985, "Neural" Computation of Decisions in Optimization Problems. *Biol. Cybern.* 52, 141-152.

Kohonen, T., 1982, Self-organized formation of topologically correct feature maps. *Biological Cybernetics*, 43, 59-69.

Laporte, G., 1992, The travelling salesman problem: an overwiew of exact and aproximate algorithms, *European Journal of Operational Research* 59, 231-247.

Lozano, S., Guerrero, F., Eguía, I., Canca, D., and Smith, K.A., 1998, Cell formation using two neural networks in series, *Intelligent Engineering Through Artificial Neural Networks, vol. 8*, C.H. Dagli et al (eds.), ASME Press, 341-346.

Peterson, C. and Anderson, J.R., 1987, A Mean Field Theory Learning Algorithm for Neural Networks. *Complex Systems*, 1, 995-1019.

Reinelt, G., 1991, TSPLIB. A Travelling Salesman Problem Library, *ORSA Journal on Computing* 3, 376-389.

Smith, K., Palaniswami, M. and Krishnamoorthy, M., 1996, A hybrid neural approach to combinatorial optimization. *Computers Ops. Res.* 23, 597-610.

Smith, K. and Palaniswami, M., 1997, Static and Dynamic Channel Assignment Using Neural Networks. *IEEE Journal on Selected Areas in Communications*, 15, 238-249.

Van den Bout, D.E., and Miller, T.K., 1989, Improving the Performance of the Hopfield-Tank Neural Network Through Normalization and Annealing, *Biol. Cybern.* 62, 129-139.

OPTIMIZATION OF THE MOLD ORIENTATION ON AN INVESTMENT CASTING CENTRIFUGE

P.A. SIMIONESCU
Dept. of Mechanical Engineering
and Solidification Design Center
Auburn University, 202 Ross Hall,
Auburn AL, 36849

D. BEALE
Dept. of Mechanical Engineering
Auburn University, 202 Ross Hall,
Auburn AL, 36849

R.A. OVERFELT
Dept. of Mechanical Engineering
and Solidification Design Center
Auburn University, 202 Ross Hall,
Auburn AL, 36849

ABSTRACT

An optimum orientation of the mold inside the spinning drum of a casting centrifuge is sought, at which the resulting g-force has a maximum overall value inside the part's volume. An optimization problem is defined and solved with the aid of a hybrid algorithm that combines an evolutionary algorithm named 3p in series with Nelder and Mead's simplex algorithm. The 3p algorithm, employs an elitist, rank-based selection and introduces a new concept, that of *consanguination-avoidance crossover*.

INTRODUCTION

The Solidification Design Center of Auburn University (http://metalcasting.auburn.edu) was established by NASA in 1997, and together with its affiliates is working to develop advanced manufacturing products through improved understanding and control of gravitational and related effects in industrial processes. The R&D conducted in the field of gravitational effects upon metallurgical processes includes investment, centrifugal and semisolid casting as well as microgravity and high-g directional solidification studies.

Centrifugal-investment casting is a new process, where the tree-shaped mold as used in the classical investment casting process (Davis, 1998) is spun relative to a vertical axis on a centrifuge while pouring the molted metal inside (Fig. 1). The high g-forces thus created provide a better filling of the cavities with molted metals and, after solidification, an improved grain structure by separating the nonmetallic impurities in the tree shaft, close to the axis of rotation of the centrifuge.

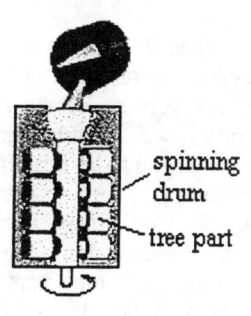

Fig. 1 Centrifugal-investment casting process (the parts are shown as short cylinders forming a tree).

Fig. 2 The Digibot II laser digitizing system.

In order to maximize the output per casting, a compact arrangement must be chosen to the parts inside the centrifuge drum. Secondly, if the resultant g-force is constantly large throughout the part's volume, the metal structure will be more homogenous after solidification. Such a goal can be achieved by locating the cavities corresponding to the parts as close as possible to the circumference of the centrifuge drum, and avoiding any thin protrusions to be oriented towards the axis of rotation of the centrifuge. The problem of optimum orientation of a single part only will be considered in this paper and solved for the actual case of a complex-shape part.

PROBLEM FORMULATION

For a given diameter of the centrifuge drum spinning at a constant RPM, an optimum disposition of the mold requires finding the position and orientation for which the distances between the infinitesimal volumes of material that will form the parts and the centrifuge's axis of rotation is on the average maximum.

One issue in defining an appropriate objective function to be maximized for achieving this goal was the availability of a suitable 3D representation of the part to be cast. The solution considered was to digitize the outside surface of an existing part using a 3D laser scanner available at Solidification Design Center of Auburn University (Fig. 2). The output file used from the laser-digitizing system is a DXF encoding of a number of equally spaced polylines, the vertices of which are acquired using a laser beam in a process described on http://www.digibotics.com/3adaptive.htm. These polylines can be viewed as a collection of level-curves equally spaced over the vertical axis as shown in Fig. 3.

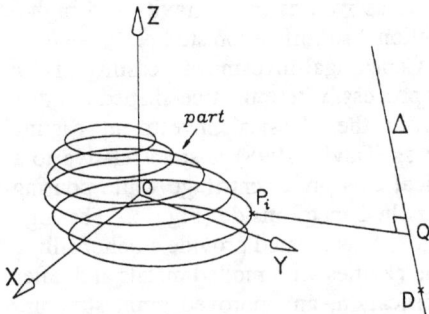

Fig. 4 Schematic used for defining the objective function (3). P_iQ_i is the distance segment between a current vertex P_i and the centrifuge's

Although this is not a "volume representation" of the part, the vertices of the polylines provide a useful numerical representation to be further used in formulating an optimization problem. Conversely, because the vertex concentration is higher in the areas of the part having increased detailed contours, the respective zones will favorably gain importance inside the objective function.

In order to avoid applying repeated translation and rotation transformations to a large number of points, an inverse-motion approach was adopted i.e. the part was considered fixed while the position and orientation of the axis of rotation (noted Δ in Fig. 4) was changed during the optimization process. The axis of the centrifuge has the equation:

$$(x - x_D)/l = (y - y_D)/m = (z - z_D)/n \tag{1}$$

and is repositioned by modifying the x_D and y_D coordinates of the through point D (Fig. 4), while its orientation can be changed by modifying two of its direction cosines, say l, m. The z_D coordinate of the through point is maintained at a constant value (possibly at $z_D=0$), while the third direction cosine n can be calculated using the equation:

$$l^2 + m^2 + n^2 = 1 \tag{2}$$

The vertices read from the DXF file (summing a total of np points) were used in defining the following objective function:

$$F(x_D, y_D, l, m) = \sum_{i=1}^{np} \text{dist}(P_i, \Delta) \tag{3}$$

subjected to the constraints:

$$\text{dist}(P_i, \Delta) \leq R$$
$$l^2 + m^2 \leq 1 \tag{4}$$
$$0 \leq l \leq 1$$
$$0 \leq m \leq 1$$

The first of these inequalities restricts the location of the part inside the centrifuge drum of radius R, while the remaining three constraints ensure that the direction cosines l, m and n satisfy equation (2).

In the above relations dist(P_i,D) is the distance between a current vertex $P_i(x_i, y_i, z_i)$, as read from the DXF file, and the axis of rotation of the centrifuge Δ, and is given by the formula:

$$\text{dist}(P_i, \Delta) = \sqrt{\frac{(m\,dx - l\,dy)^2 + (n\,dx - l\,dz)^2 + (n\,dy - m\,dz)^2}{l^2 + m^2 + n^2}} \tag{5}$$

where dx, dy and dz are:

$$dx = x_i - x_D, \quad dy = y_i - y_D \quad \text{and} \quad dz = z_i - z_D \tag{6}$$

THE SEARCHING SUBROUTINE

An evolutionary algorithm named 3p in series with a standard Simplex algorithm (Nelder and Mead 1965) has been used to maximize the objective function (3). The constraints (4) were introduced in the optimization problem as ramp-type penalty functions (Michalewicz 1996) that are very easy to program.

The 3p algorithm uses vectors of real numbers as individuals and has the following structure:
1. randomly generate an initial population of p members where p is a given number satisfying the inequality $3p>n^*$ (n^* is the number of variables of the objective function);
2. select the most fit individual (the dominant male) from among the total population of $3p$ members;
3. sort from among the remaining $3p-1$ members of the population the most distant to the dominant male $p-1$ individuals i.e. the Euclidean norm between the 2nd ranked individual and the dominant male will be the largest; the Euclidean norm between the 3rd ranked individual and the dominant male will be the second largest etc., up to the pth rank (the remaining $2p$ individuals will remain unsorted);
4. crossover the dominant male with the individuals ranked between 2 and $p-1$ and replace the individuals ranking between $p+1$ and $2p$;
5. replace by random generation the individuals ranked between $2p+1$ and $3p$ (the mutation step);
6. go to step 2 until for q successive iterations (q a given number); the dominant male remains unchanged (the stop test should be done after completing step 3);

It can be seen that the 3p algorithm employs an elitist, rank-based selection (Bäck at al 1997, Poloni and Pediroda 1998, Yao 1999). This type of selection forces the best member of the population (the dominant male) to recombine with the most distant p individuals in the total population which avoids the search to settle in the vicinity of the dominant male. This inverse-distance ranking is equivalent to a *consanguination-avoidance* encountered in nature (like in African lions or modern humans).

The crossover operator used was:

$$x_{p+j,i} = \alpha_i \cdot x_{1,i} + \beta_i \cdot x_{j,i} \quad (i=1..n, \ j=1..p) \tag{7}$$

where $x_{\square,\square}$ represents an entry in the $3p \times n^*$ matrix storing the members of the population and α_i and β_i are uniform random numbers between 0 and 1. This type of crossover in turn facilitates a wider exploration of the design space of the objective function.

After the 3p algorithm reaches its stop condition, the search is continued by a Nelder-Mead simplex subroutine. The purpose of applying a simplex search subsequent to the evolutionary algorithm was to refine the optimum solution. The simplex step is less time consuming and was inspired by the similarities that exist between this particular zero-order searching algorithm and evolutionary algorithms in general.

Nelder and Mead's simplex algorithm has received constant attention ever since it was published due to its robustness and the capability of coping well with nonsmooth, nondifferentiable objective functions (Humphrey and Wilson 2000, Parkinson and Hutchinson 1969, Trabia 2000). The vertices of the simplex in Nelder-Mead algorithm can be considered individuals of a population that experience successive directional crossovers and selection operations. If a mutation operator is introduced (like for example to replace every iteration some

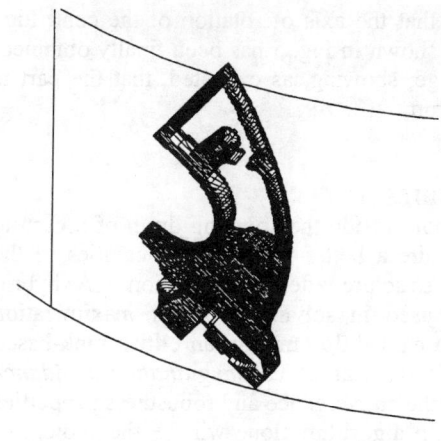

Fig. 5: Isometric view of the part in its optimum position inside the drum

of the vertices of the simplex with other vertices randomly generated), it can be converted into an evolutionary algorithm.

The hybrid algorithm used by the authors for maximizing the objective function (3) employs a classic variant of simplex algorithm, as it was originally described in Nelder and Mead (1965). After the $3p$ algorithm reaches its stop condition, the dominant male together with other n^* individuals are transferred to Nelder-Mead subroutine. These n^*+1 individuals will form the initial simplex and the search will continue until the convergence criteria or other predefined stop condition is achieved.

Regarding the transfer of data to the simplex algorithm, it was found more advantageous to exit the $3p$ subroutine after passing through step 3 (not immediately after step 2) and consider for the vertices of the initial simplex the first n^*+1 ranked members of the last population. In this way the exploratory contribution of the Nelder-Mead algorithm is increased to some extent.

NUMERICAL RESULTS

The actual case of a complex-shape part was considered as a numerical example. Every time the objective function is evaluated the (x_i, y_i, z_i) coordinates of 15224 points (corresponding to 138 polyline levels) have to be read from a

Fig. 6: Top view of the part in its optimum position inside the drum

DXF file. This file sums a total of 185 513 lines of ASCII symbols, which must be read sequentially making the evaluation of the objective function notably time consuming.

After 8436 function evaluations, of which 7258 were consumed in the $3p$ algorithm (for a stop parameter $q=75$), the following results were obtained corresponding to an inner radius of the centrifuge drum R=6": $F_{max}=88492.016$ (objective function value), $x_D=0.1447$", $y_D=-5.7521$", $z_D=1.72597$", $l=0.577919$, $m=0.001415$, $n=0.816093$. The parameters positioning the axis of the centrifuge are expressed relative to the reference system attached to the part (the same used by the laser digitizing system).

By changing the reference frame, so that the axis of rotation of the centrifuge becomes the z axis, the configuration shown in Fig. 5 has been finally obtained. Fig. 6 is a top view of the same image, showing, as expected, that the part is aligned with the inner surface of the drum.

CONCLUSIONS AND FURTHER STUDIES

An optimum orientation of the mold inside the spinning drum of a casting centrifuge was sought in order to ensure a better filling of the cavities of the mold and improve the metal grain structure after solidification. A hybrid evolutionary-simplex algorithm was used in solving a proper maximization problem. The evolutionary algorithm named $3p$ employs an elitist, rank-based selection and introduces a new concept, that of *consanguination-avoidance crossover*. Thorough investigation of the convergence and robustness properties of this hybrid algorithm, and of the $3p$ algorithm alone will be the subject of future investigations the authors are considering to pursue.

REFERENCES

Bäck, T., Hammel U., Schwefel, H. P., 1997, "Evolutionary Computation, Comments on the History and Current State," *IEEE Transactions on Evolutionary Computation*, Vol. 1, pp. 3-16.

Davis, J.R., 1998, *Metals Handbook*, ASM International.

Humphrey, D.G., Wilson, J.R., 2000, "A revised Simplex Search Procedure for Stochastic Simulation Response Surface Optimization," *INFORMS Journal on Computing*, Vol. 12, pp. 272-283.

Michalewicz, Z., 1996, *Genetic Algorithms + Data Structures = Evolution Programs*, Springer.

Nelder J., Mead, R., 1965, "A Simplex Method for Function Minimization," *Computer Journal*, Vol. 7, pp. 308-313.

Parkinson J. M., Hutchinson D., A. (1969) An Investigation Into the Efficiency of Variants on the Simplex Method, (F.A. Lootsma, ed.) Numerical Methods for Non-linear Optimization, Academic Press, London, pp. 115-135.

Poloni, C., Pediroda, V., 1998 "GA Coupled With Computationally Expensive Simulation: Tools to Improve Efficiency," in *Genetic Algorithms and Evolutionary Strategies in Engineering and Computer Science*, (D. Quagliarella, J., Périaux, C. Poloni, G. Winter, editors), Willey.

Trabia, M.B., 2000, "A Hybrid Fuzzy Simplex Genetic Algorithm," *Proc. of the ASME 2000 Design Engineering Technical Conference*, Baltimore MD, (DETC2000/DAC-14231).

Yao, X., (ed.), 1999, *Evolutionary Computation, Theory and Application*, World Scientific.

TRADITIONAL CALCULUS AND MODERN FUZZY APPROACH TO OPTIMIZATION

GRACE WOO
Illinois Math and Science Acad.
1500 W. Sullivan Rd.
Aurora, IL 60506

PENG-YUNG WOO
Electrical Engineering Dept.
Northern Illinois University
DeKalb, IL 60115

ABSTRACT

Traditional Calculus has often emerged as the most effective approach to solving optimization problems in science and engineering. Whether that be in context of complex manufacturing systems, aeronautics, hydraulics or others, researchers have always turned to calculus. This paper presents an alternate approach to these predicaments by using fuzzy logic. Specifically, this new approach offers simpler solutions to multi-constraint issues.

INTRODUCTION

Traditional Mathematics has allowed us to believe that the more closely a phenomenon is modeled with equations the better we are able to solve the problem. This is true as long as we are able to do that. If we feed in a crisp input, we get in return a crisp output. The study of probability, probably the closest classical mathematics to fuzzy logic, deals with the concept of "chance" as opposed to "ambiguity". Often times, a system is too complex to be modeled with Aristotelian or Cartesian functions. In these situations, fuzzy logic, a coarser, less refined approach proves to be an appropriate solution.

Fuzziness describes the ambiguity of an event. The ambiguity of a set can be described by using a membership function [1]. If a man is five feet and two inches, he could be considered "short" because he has a membership value of 0.8 in the "short" set. On the other hand, he could also be considered slightly "tall" because he has a membership value of .5 in the "tall" set.

Traditional Calculus has always emerged as the most effective approach to solving optimization problems in science and engineering. This paper presents an alternate approach to these predicaments by using fuzzy logic. An adapted version of a classical applied mathematics problem is used hereunder to discuss optimization [2].

> Soup cans are to be manufactured to contain a given volume V. No waste is involved in cutting the material for the vertical sides, but the top and bottom (circles of radius r) are cut from squares that measure 2r units on a side. Find the ratio of height to diameter for the most economical can (i.e., the can requiring the least material).

TRADITIONAL CALCULUS APPROACH

Modeling the phenomenon with crisp equations and using a little bit of calculus solves the problem. Below, A represents the total surface area of the can and V (a constant) represents the volume of the can. r and h represent the radius and height of the can, respectively.

Model the phenomenon: $\quad A = 2\pi rh + 8r^2 \qquad V = \pi r^2 h \qquad (1)$

Solve for h in terms of V and r: $\quad V/\pi r^2 = h \qquad (2)$

Substitute in for h: $\quad A = 2\pi r \left[V/\pi r^2 \right] + 8r^2 \qquad (3)$

Simplify: $\quad A = \dfrac{2v + 8r^3}{r} \qquad (4)$

Take the derivative with respect to r: $\quad \dfrac{dA}{dr} = \dfrac{r(24r^2) - (2V + 8r^3)}{r^2} \qquad (5)$

Solve for V when $\dfrac{dA}{dr}$ equals zero (assuming a minimum): $\quad V = 8r^3 \qquad (6)$

Substitute back into the original equation for V and solve for h: $\quad h = 8r/\pi \qquad (7)$

Reduce the ratio for r/h: $\quad r/h = \dfrac{\pi}{8} \qquad (8)$

Through strict differentiation and algebraic calculation, the ideal ratio of r to h to save material is found. If V is assumed to be 100 units, r and h should be 2.32 and 5.91 units, respectively. Thus, a unique real output is returned in exchange for crisp inputs. Although, for this simple problem, the conventional calculus approach is acceptable, it might be impractical to utilize this approach for a temporal complex system, e.g., a multi-constraint system with some of the constraints unable to be modeled.

MODERN FUZZY APPROACH

V is still assumed to be 100 units. Using this value, (4) is reduced to:

$$A = \dfrac{200 + 8r^3}{r} \qquad (9)$$

For experimental purposes, different groups of membership functions for r are designed in terms of different domains and different graphical configuration. Each membership function is modeled by using simple algebraic equations. The domain of r is defined generally in the interval $[a, b]$.

One possible membership graphical configuration is shown in Fig. 1, where [0.5, 5.0] is defined as the domain of r.

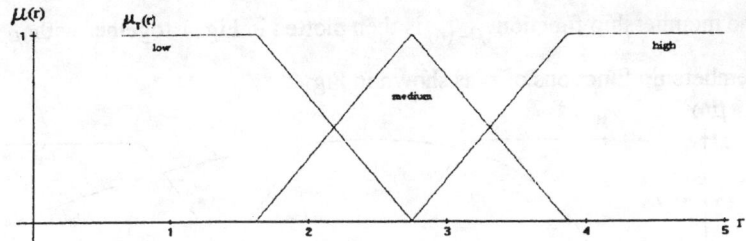

Fig. 1 The membership functions of r (configuration 1)

This configuration depicts the membership functions of radius r in three distinct linguistic sets. The left one represents the set identifiable as "low", the middle one represents the set identifiable as "medium", and the right one represents the set identifiable as "high". Note that the "low" function and the "medium" function overlap (the same do the "high" and the "medium"), but never do the "low" function and the "high" function overlap. Each function is distributed somewhat evenly over the defined domain. Also, the peak of the "medium" (the most "medium") is placed exactly in the middle of the defined domain. These membership functions can be expressed by the equations below.

$$y_l = \begin{cases} 1 & for \quad a \leq r \leq (b+3a)/4 \\ \left(-4/(b-a)\right)r + \left((4b+a)/(a-b)\right) & for \quad (b+3a)/4 < r \leq (b+a)/2 \end{cases} \tag{10}$$

$$y_m = \begin{cases} \left(4/(b-a)\right)r - (b+3a)/(b-a) & for \quad (b+3a)/4 < r \leq (b+a)/2 \\ \left(-4/(b-a)\right)r - (2b+2a)/(b-a) & for \quad (b+a)/2 < r \leq (3b+a)/4 \end{cases} \tag{11}$$

$$y_h = \begin{cases} \left(4/(b-a)\right)r - \left((2b+2a)/(b-a)\right) & for \quad (b+a)/2 < r \leq (3b+a)/4 \\ 1 & for \quad (3b+a)/4 < r \leq b \end{cases} \tag{12}$$

In 1972, Zadeh [3] proposed a conversion of the objective function $y = f(x)$, which is to be *maximized*, into a pseudogoal $\underset{\sim}{G}$ with membership function

$$\mu_{\underset{\sim}{G}}(x) = \frac{f(x) - m}{M - m} \tag{13}$$

where $M = \underset{x \in X}{\sup} f(x)$ and $m = \underset{x \in X}{\inf} f(x)$. Without saying, M and m differ for different domains. For [0.5, 5.0], the membership function of $\underset{\sim}{G}$ corresponding to the objective function (9) in our problem

$$\mu_{\underset{\sim}{G}}(r) = \frac{\dfrac{(-200 - 8r^3 + 402r)}{r}}{(-129 + 402)} = \frac{(-200 - 8r^3 + 402r)}{273r} \tag{14}$$

The membership function $\mu_G(r)$ is then plotted in Fig. 1 together with $\mu_r(r)$, the membership functions of r, as shown in Fig. 2.

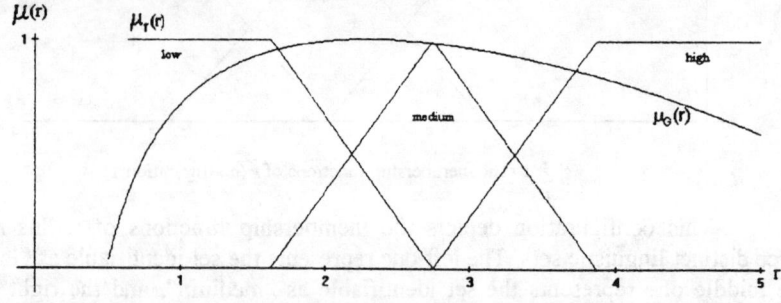

Fig. 2 Find the optimal r^*s (Configuration 1)

The intersections of $\mu_G(r)$ and $\mu_r(r)$ are in fact the optimal values denoted as r^*. In this case, there are four of these values and they, together with their associated membership values, are displayed in the first row of Table 1. The shaded data in the first low has the largest membership value 0.9886 and thus is the r^*_{global} we seek.

The same procedure is repeated with different values of a and b. The resulted optimal r^*s are demonstrated in Table 1. In this case, there are four r^*s at a time. Each of these domains provides a slightly different value for r^*_{global}.

a	b	value of r*	value of r* (2nd)	value of r* (3rd)	value of r* (4rth)
0.5	5	1.6745, .9560	2.7372, .9886	2.7649, .9868	3.7302, .8713
1	5	2.0195, .9835	2.9350, .9340	3.1060, .8940	3.6958, .6958
1.5	5	2.1621, 1.0000	3.1475, .8829	3.4286, .7959	3.8153, .6460
0.5	5.5	1.7851, .9719	2.9639, .9711	3.0459, .9633	4.0235, .8188
1	5.5	2.1274, .9979	3.1514, .9123	3.4170, .8516	4.0076, .6734
1.5	5.5	2.5007, .9992	3.3646, .8646	3.7385, .7615	4.1298, .6298
0.5	6	1.8968, .9841	3.1798, .9499	3.3487, .9282	4.2992, .7630
1	6	2.1621, 1.000	3.3679, .8943	3.7281, .8174	4.3207, .6566
1.5	6	2.634, .9920	3.582, .8507	4.0471, .7359	4.4450, .6178

Table 1 Resulted optimal r^*s for configuration 1 (Fig. 2).

Using the returned answer $r = 2.32$ from the conventional calculus approach as a premise, percent errors are calculated. For configuration 1, the smallest percent error is approximately 7.8%, which occurs at $r^*_{global} = 2.5007$ in the domain [1.5, 5.5]. Antithetically, the largest percent error is approximately 23.1%, which occurs at $r^*_{global} = 1.7851$ in the domain [0.5, 5.5].

Membership functions $\mu_r(r)$ with different graphical configurations are used to investigate the membership function's impact on the result. For example, Fig. 3 demonstrates another possible membership graphical configurations of $\mu_r(r)$ for the same domain [0.5, 5.0] and consequently the new intersections of $\mu_G(r)$ and $\mu_r(r)$.

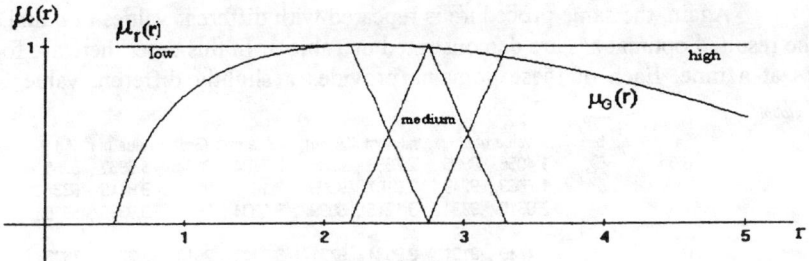

Fig. 3 Find the optimal r^*s (Configuration 2)

The membership functions $\mu_r(r)$ in Fig. 3 may be expressed by similar equations as in the previous example but are left out in veer of redundancy. Once more, the same procedure is repeated with different values of a and b. The resulted optimal r^*s are demonstrated in Table 2. In this case, there are up to six r^*s at a time. Each of these domains provides a slightly different value for r^*_{global}.

a	b	value of r^*	value of r^* (2nd)	value of r^* (3rd)	value of r^* (4rth)	value of r^* (5th)	value of r^* (6th)
0.5	5	2.1621 , 1	2.7434 , .9882	2.7571 , .9873	3.2773 , .9373		
1	5	2.1621 , 1	2.4871 , 1	2.5005 , .9991	2.9643 , .9266	3.0453 , .9095	3.4024 , .8048
1.5	5	2.1621 , 1	2.4871 , 1	3.1932 , .8701	3.3242 , .8305	3.5751 , .7430	
0.5	5.5	2.1621 , 1	2.981 , .9696	3.0214 , .9657	3.5613 , .8981		
1	5.5	2.1621 , 1	2.4871 , 1	2.6968 , .9835	3.1955 , .9032	3.3202 , .8752	3.6871 , .7770
1.5	5.5	2.1621 , 1	2.4871 , 1	3.0323 , .9352	3.4248 , .8496	3.5986 , .8028	3.8613 , .7227
0.5	6	2.0361 , .9945	3.2123 , .9452	3.8365 , .8531	3.788 , .86412		
1	6	2.4871 , 1	2.8954 , .9674	3.4268 , .8829	3.595 , .8478	3.9725 , .7561	
1.5	6	2.1621 , 1	3.2333 , .9186	3.6567 , .8341	3.8726 , .7821	4.181 , .7076	

Table 2 Resulted optimal r^*s for configuration 2 (Fig. 3).

For configuration 2, the smallest percent error is approximately 6.8%, which occurs in several domains where the r^*_{global} is equal to 2.1621. Antithetically, the largest percent error is 7.2%, which occurs in several domains also where r^*_{global} is equal to 2.4871.

A third configuration is considered where the membership functions for "low" and "high" are restricted further, which in turn increases the possible values of the membership function for "medium". In other words, the membership function for "medium" becomes "fatter". Fig. 4 demonstrates this variation for the membership functions of r.

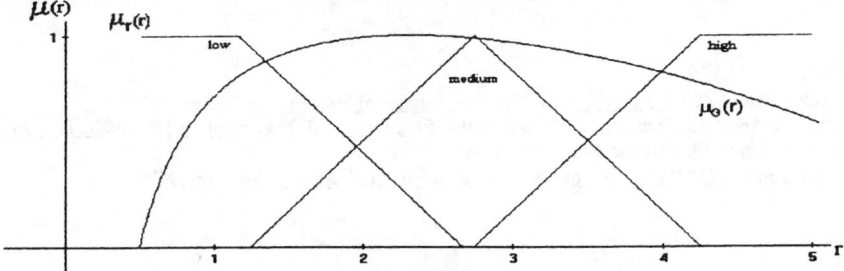

Fig. 4 Find the optimal r^*s (Configuration 3)

Again, the same procedure is repeated with different values of a and b. The resulted optimal $r*$s are demonstrated in Table 3. In this case, there are four $r*$s at a time. Each of these domains provides a slightly different value for $r*_{global}$.

a	b	value of r*	value of r* (2nd)	value of r* (3rd)	value of r* (4rth)
0.5	5	1.4055 , .8963	2.7333 , .9888	2.7704 , .9864	3.9982 , .8255
1	5	1.7669 , .9249	2.9181 , .9386	3.1612 , .8791	3.8445 , .6334
1.5	5	2.0910 , .9934	3.1215 , .8899	3.5304 , .7597	3.9395 , .5910
0.5	5.5	1.4749 , .9151	2.9534 , .9721	3.0642 , .9615	4.2789 , .7673
1	5.5	1.8166 , .9556	3.1261 , .9174	3.5084 , .8277	4.1715 , .61434
1.5	5.5	2.1621 , 1.000	3.3304 , .8928	3.8763 , .7178	4.2696 , .5772
0.5	6	1.5431 , .9311	3.1604 , .9512	3.3923 , .9224	4.5475 , .7078
1	6	1.8757 , .9746	3.3343 , .9006	3.8564 , .7862	4.5002 , .6001
1.5	6	2.1621 , 1.000	3.5397 , .8598	4.2197 , .6869	4.6005 , .5670

Table 3 Resulted optimal $r*$s for configuration 3 (Fig. 4).

For configuration 3, the smallest percent error is approximately 6.8%, which occurs in two domains, [1.5, 5.5] and [1.5, 6.0], where $r*_{global}$ is 2.1621. Antithetically, the largest percent error is approximately 36.2%, which occurs in the domain [0.5, 6.0], where $r*_{global}$ is 3.1604.

The significance of these percent errors can be explained through domain restrictions and graphical configuration of the membership functions. The two largest percent errors that are over 10% both occur in a domain starting with $a = 0.5$. Physically, this means that a radius of 0.5, in practical units, is unreasonable. In fact, the $r*_{global}$s in all of the domains starting with $a = 0.5$ have a relatively high percent error. Furthermore, enough evidence exists to show that the graphical configuration of the membership functions does not play a significant role in producing large percent errors.

CONCLUSIONS

Although for the simple case discussed above there is not enough evidence to say that the fuzzy approach is more effective, it is demonstrated that the fuzzy approach does yield similar results as the conventional calculus approach. We thus conclude that the fuzzy approach is valid in solving optimization problems. Currently, multi-constraint optimization problems, where the conventional calculus approach fails to work, are under investigation by using the modern fuzzy approach.

REFERENCES

Kosko, Bart (1993), Fuzzy Thinking: The New Science of Fuzzy Logic, Hyperion, NY.
Ostebee, Arnold & Zorn, Paul (1997), Calculus from Graphical, Numerical, and Symbolic Points of View, Orlando, Fl: Harcourt Brace & Company.
Ross, Timothy (1995), Fuzzy Logic with Engineering Applications, New York, NY.

A RECURSIVE ASYNCHRONOUS NEURAL NETWORK THAT FINDS THE LARGEST INPUT

THOMAS S. DRANGER and **ROLAND PRIEMER**
Signalysis   University of Illinois at Chicago
22300 Nichols Drive   ECE Department
Chicago Heights, IL 60411   Chicago, IL 60607

ABSTRACT: A new recursive, asynchronous, and discrete time neural network is given. It identifies that input among m inputs that is inputting the largest value, and it outputs this largest value. All parameters are known *a-priori*. Its physical complexity is $O(m)$. We prove convergence in time inversely proportional to the difference between the largest input value and the next largest input value.

INTRODUCTION

Given here is a new recursive neural network that finds the largest number of the input set of numbers $\{V_1, V_2, ...V_m\}$. As the network operates, a subset of the input set emerges in ascending value order. The network is shown to be stable by proving convergence within time proportional to $O(m/\eta)$, where η is the difference between the largest and next to largest number in the input set.

Finding the largest number is needed in fuzzy logic, to determine the winner in Kohonen's [4] competitive learning algorithm, and to implement the recognition layer in adaptive resonance theory [1] [3]. Indeed, finding the largest number, and the source of the largest number is a fundamental operation useful in many computations. Neural networks to find the largest number have been realized as feed forward networks [5] and in the form of recursive neural networks called MAXNET [5]. To preserve the largest value, other networks and various modifications of MAXNET were developed by Winters and Rose [7], Suter and Kabrisky [6], and Yadid-Pecht and Gur [8]. Previous networks are synchronous systems that require more complex processing elements, careful selection of connection weights, or dynamic modification of connection weights. The new network is recursive and asynchronous using simple nonlinear processing elements.

THE NETWORK

The new neural network is represented by a digraph having m neurons and m input ports, one for each neuron. The neurons are fully connected by edges among themselves. Input is by means of an edge directed from each port to its corresponding neuron. Also, each neuron feeds back to itself by an edge having connection weight z^{-1} denoting one increment of delay and unity gain. Port-to-neuron edges also have connection weight z^{-1}. Edges among the neurons have connection weight $-z^{-1}$, denoting a change of sign as well.

Each input port has a single output having value V_j. Let the m-dimension vector V be composed of components V_j that are the outputs of the m input ports.

Let k be the discrete time index. The initial state occurs at time $k=0$. Let the m-dimensional vector $X(k)$ represent the state of the m neurons at time k. Let $\sigma(k)$ be the sum of the elements in $X(k)$, so that

$$\sigma(k) = \Sigma_j X_j(k) \qquad (1)$$

Let $S_j(k)$ be the sum of the signals arriving at the jth neuron at time k from sources other than itself, and we can write

$$S_j(k) = V_j + X_j(k-1) - \sigma(k-1) \qquad (2)$$

Let the m dimensional vector $S(k)$ comprise the elements $S_j(k)$.

Each neuron functions in one of two modes. In one mode, a neuron updates by changing state according to the function $Q(S_j(k))$, a half-rectify function so that

$$Q(S_j(k)) = \begin{cases} S_j(k), & \text{if } S_j(k) > 0 \\ 0, & \text{otherwise} \end{cases} \quad (3)$$

In the other mode, a neuron does nothing but maintain a constant state in X_j. The effect on X_j is given by

$$X_j(k) = \begin{cases} Q(S_j(k)), & \text{in the update mode} \\ X_j(k-1), & \text{otherwise} \end{cases} \quad (4)$$

At every instant $k > 0$, one and only one neuron shall update its state. Define a round to be m steps in time, with the first round starting at time $k = 1$, the second round when $k = 1+m$, the third round when $k = 1+2m$, and so on. Let every neuron update exactly once in each and every round in arbitrary sequence. Let the update index selector function $\gamma(k)$ denote the index of the neuron that updates at time k. Then $X_{\gamma(k)}$ is the only component of X that changes value at time k. For example, suppose $X_5(k)$ is initially zero, remains constant until time $k = 2$, and then updates to a value of one. Here then, $\gamma(2) = 5$, the index of the neuron that updates at time $k=2$. The time series giving the value of $X_5(k)$, beginning from time $k=0$, is $\{0, 0, 1, 1, 1, 1, ...\}$. Evidently, $X_{\gamma(2)}(1) = X_5(1) = 0$, and $X_{\gamma(2)}(2) = X_5(2) = 1$. In general, $X_{\gamma(i)}(k)$ is the value, at time k, of the only element $X_{\gamma(i)}$ of X that updates at time i.

RESULTS

Let each port hold a distinct constant real valued signal in the interval $[-\alpha, 1]$ for time $k \geq 0$, where α is non-negative. Since the network is symmetrical and homogeneous, and $\gamma(k)$ is arbitrary, without loss of generality the constant vector V can be arranged in ascending value order so that $V_1 < V_2 < ... < V_m$.

Theorem 1. If the initial state is $X(0) = \mathbf{0}$, then the distinct positive values of $\sigma(1)$, $\sigma(2), ... \sigma(m)$ in the order they appear in the first round are an ascending sorted subset of V, ending with the largest number in V.

Theorem 2. If all components of the initial state $X(0)$ are non-negative, then the neural network converges to the stable state $X = [0 \; ... \; 0 \; Q(V_m)]$ in not more than $2 + 2Q(V_m)/(Q(V_m) - Q(V_{m-1}))$ rounds.

The proof of these two theorems is simplified by employing some characteristics of the network, as stated by the following lemmas.

Lemma 1. If a neuron updates via Q to a positive state, then the value of σ will be the input of that neuron. That is, if $X_{\gamma(k)}(k) > 0$ then $\sigma(k) = V_{\gamma(k)}$.

Proof: Since $\sigma(k)$ differs from $\sigma(k-1)$ only by the change in $X_{\gamma(k)}$, we have
$$\sigma(k) = \sigma(k-1) - X_{\gamma(k)}(k-1) + X_{\gamma(k)}(k)$$
Since V is constant, we get
$$X_{\gamma(k)}(k) = V_{\gamma(k)} + X_{\gamma(k)}(k-1) - \sigma(k-1)$$
resulting in
$$\begin{aligned}\sigma(k) &= \sigma(k-1) - X_{\gamma(k)}(k-1) + V_{\gamma(k)} + X_{\gamma(k)}(k-1) - \sigma(k-1) \\ &= V_{\gamma(k)}\end{aligned} \quad (5)$$

Lemma 2. If $X_{\gamma(k-1)}(k-1) > 0$, then
$$X_{\gamma(k)}(k) - X_{\gamma(k)}(k-1) = V_{\gamma(k)} - V_{\gamma(k-1)} + c \quad (6)$$
where
i. $\quad c = 0 \quad$ for $X_{\gamma(k)}(k) > 0 \quad (7)$
ii. $\quad c \geq 0 \quad$ for $X_{\gamma(k)}(k) = 0 \quad (8)$

Proof: For Part i, where $c = 0$, note that $\sigma(k)$ differs from $\sigma(k-1)$ only by the change in

one element of X, that is,
$$\sigma(k) - \sigma(k-1) = X_{\gamma(k)}(k) - X_{\gamma(k)}(k-1)$$
By Lemma 1, $\sigma(k) = V_{\gamma(k)}$ and $\sigma(k-1) = V_{\gamma(k-1)}$, so that Equation (6) becomes
$$X_{\gamma(k)}(k) - X_{\gamma(k)}(k-1) = V_{\gamma(k)} - V_{\gamma(k-1)}$$

For Part ii, where $c \geq 0$, and in light of Equation (3), the condition $X_{\gamma(k)}(k) = 0$ requires that
$$S_{\gamma(k)}(k) = V_{\gamma(k)} + X_{\gamma(k)}(k-1) - \sigma(k-1) \leq 0$$
By Lemma 1, $\sigma(k-1) = V_{\gamma(k-1)}$, resulting in
$$S_{\gamma(k)}(k) = V_{\gamma(k)} + X_{\gamma(k)}(k-1) - V_{\gamma(k-1)} \leq 0$$
Rearranging, and adding the condition $X_{\gamma(k)}(k) = 0$ to the greater side yields
$$X_{\gamma(k)}(k) - X_{\gamma(k)}(k-1) \geq V_{\gamma(k)} - V_{\gamma(k-1)}$$
giving $c \geq 0$.

Lemma 3. Suppose $0 < i < j < k$, $X_{\gamma(i)}(i) > 0$, $X_{\gamma(k)}(k) > 0$, and $X_{\gamma(j)}(j) = 0$ for $j = i+1$, $i+2, ..., k-1$. Then,

i. $\quad \sigma(j) = V_{\gamma(i)} - \Sigma_{q=i+1,...,j} X_{\gamma(q)}(q-1)$ (9)

Proof: Since $X_{\gamma(i+1)}(i+1) = 0$ is given, then
$$\sigma(i+1) = \sigma(i) - X_{\gamma(i+1)}(i)$$
Likewise we have
$$\sigma(i+2) = \sigma(i+1) - X_{\gamma(i+2)}(i+1) = \sigma(i) - X_{\gamma(i+1)}(i) - X_{\gamma(i+2)}(i+1)$$
and so on, until
$$\sigma(j) = \sigma(i) - X_{\gamma(i+1)}(i) - X_{\gamma(i+2)}(i+1) - ... - X_{\gamma(j)}(j-1)$$
By Lemma 1, $\sigma(i) = V_{\gamma(i)}$. By substitution for σ, and condensing the sum we get Equation (9).

ii. $\quad X_{\gamma(k)}(k) = V_{\gamma(k)} - V_{\gamma(i)} + \Sigma_{q=i+1,...,k} X_{\gamma(q)}(q-1)$ (10)

Proof: By Equations (3) and (4), and recalling that V is constant, we have
$$X_{\gamma(k)}(k) = V_{\gamma(k)} + X_{\gamma(k)}(k-1) - \sigma(k-1)$$
Substituting the result of part i results in
$$X_{\gamma(k)}(k) = V_{\gamma(k)} + X_{\gamma(k)}(k-1) - V_{\gamma(i)} + \Sigma_{q=i+1,...,k-1} X_{\gamma(q)}(q-1)$$
which gives Equation (10).

iii. $\quad X_{\gamma(k)}(k) - X_{\gamma(k)}(k-1) = V_{\gamma(k)} - V_{\gamma(i)} + \Sigma_{q=i+1,...,k-1} X_{\gamma(q)}(q-1)$ (11)

Proof: Subtract $X_{\gamma(k)}(k-1)$ from both sides of Equation (10).

Lemma 4. Suppose $0 < i < j < k$, $X_{\gamma(i)}(i) > 0$, $X_{\gamma(k)}(k) > 0$, and $X_{\gamma(j)}(j) = 0$ for $j = i+1$, $i+2, ... k-1$, as in Lemma 3, but here, $\gamma(i)$, $\gamma(j)$, and $\gamma(k)$ are all distinct. Then

i. $\quad \sigma(j) = V_{\gamma(i)} - \Sigma_{q=i+1,...,j} X_{\gamma(q)}(i)$ (12)
ii. $\quad X_{\gamma(k)}(k) = V_{\gamma(k)} - V_{\gamma(i)} + \Sigma_{q=i+1,...,k} X_{\gamma(q)}(i)$ (13)
iii. $\quad X_{\gamma(k)}(k) - X_{\gamma(k)}(i) = V_{\gamma(k)} - V_{\gamma(i)} + \Sigma_{q=i+1,...,k-1} X_{\gamma(q)}(i)$ (14)

Proof: Recall Lemma 3 and observe that since the indices of the changing elements of X are all distinct, the state of each element changes only once. The state of each element just prior to change is the same as at time i, allowing the substitution $X_{\gamma(q)}(q-1) = X_{\gamma(q)}(i)$ giving the stated results.

Proof of Theorem 1: If the initial state $X(0) = \mathbf{0}$, then $\sigma(0) = 0$, and subsequently, at any time $k>0$, whenever $X_{\gamma(k)}(k) > 0$, we will have $\sigma(k) = V_{\gamma(k)}$ by Lemma 1. In the first round, $X_{\gamma(k)}(k) > 0$ if and only if $V_{\gamma(k)} > \sigma(k)$. Thus, in the first round σ is zero until a neuron with positive input updates. After that, σ is always equal to some element of V, and any change in σ produces $\sigma(k) > \sigma(k-1)$. At some time in the first round we have $\sigma(k) = V_m$ and then σ remains constant until the end of the first round. The distinct positive values of σ are a subset of V, and since each is larger than the one before, we have a subset of V in ascending value order, ending with the largest.

Proof of Theorem 2: If $V_m < 0$ then $S_i = V_i + X_i - \sigma \leq V_i \leq V_m \leq 0$ for all i, and the proof is trivial. In the case where $V_m > 0$, there are two cases at the end of the first round.

a) No neuron has updated to a positive state, and
$$\sigma(m) = \sigma(0) - \sum_{q=1,\ldots,m} X_{\gamma(q)}(0) = 0$$
b) One or more neurons have updated to a positive state, and
$$\sigma(m) = V_{\gamma(h)} - \sum_{q=h+1,\ldots,m} X_{\gamma(q)}(0) \le V_{\gamma(h)}$$
where h is the time of the last update to a positive state. If $h = m$ then $\sigma(m) = V_{\gamma(h)} = V_{\gamma(m)}$. In either case, $0 \le \sigma(m) \le V_{\gamma(h)} \le V_{\gamma(m)}$ proving that:

i. $\quad 0 \le \sigma(m) \le V_{\gamma(m)}$

Now consider the situation after the first round, so that $\sigma(k)$ is confined to the range $[0, V_m]$ for all $k > m$. Sometime in the second round $\gamma(k) = m$ so that the mth neuron updates its output, resulting in
$$X_m(k) = V_m - (\sigma(k-1) - X_m(k-1))$$
and $\quad X_m(k) \le V_m$, since $\sigma(k-1) - X_m(k-1) = \sum_{q \ne m} X_{\gamma(q)}(k-1) \ge 0$. Since $\sigma(k-1) \le V_m$ and $\sigma(k-1) = V_m$ only if $\gamma(k-1) = m$ so that $X_m(k-1) > 0$, it follows that $X_m(k) > 0$, proving that:

ii. $\quad 0 < X_m(2m) \le V_m$

That is, at the end of the second round, X_m is greater than zero.

Next consider the changes in X_m after the first round. Let h occur in the round beginning at time $rm+1$, where r is a positive integer. Let k occur in the following round. Then $h < k$. Let h and k indicate two instances when X_m updates. Then $\gamma(h) = \gamma(k) = m$. By Lemma 3, Part iii,
$$X_m(h) - X_m(h-1) = V_m - V_{\gamma(i)} + \sum_{q=i+1,\ldots,h-1} X_{\gamma(q)}(q-1)$$
where i is the time of the most recent positive update giving $X_{\gamma(i)}(i) > 0$, for $i < h$. Likewise, we have
$$X_m(k) - X_m(k-1) = V_m - V_{\gamma(j)} + \sum_{q=j+1,\ldots,k-1} X_{\gamma(q)}(q-1)$$
where j is the time of the most recent positive update giving $X_{\gamma(j)}(j) > 0$, for $j < k$. Note that $i < h \le j < k$. Combining these two equations gives
$$X_m(k) - X_m(h-1) = X_m(k-1) - X_m(h) + 2V_m - V_{\gamma(i)} - V_{\gamma(j)}$$
$$+ \sum_{q=i+1,\ldots,h-1} X_{\gamma(q)}(q-1) + \sum_{q=j+1,\ldots,k-1} X_{\gamma(q)}(q-1)$$
Observe that the first summation is zero if $i > h-2$, and the second summation is zero if $j > k-2$. Since h and k occur in consecutive rounds, then
$$X_m(k-1) = X_m(h)$$
leading to
$$X_m(k) - X_m(h-1) = 2V_m - V_{\gamma(i)} - V_{\gamma(j)}$$
$$+ \sum_{q=i+1,\ldots,h-1} X_{\gamma(q)}(q-1) + \sum_{q=j+1,\ldots,k-1} X_{\gamma(q)}(q-1)$$
and since V_m is the largest element in V we have
$$X_m(k) - X_m(h-1) \ge 0$$
Now consider the two possibilities, case (a) where $\gamma(i) = \gamma(j) = m$, and case (b) where $\gamma(i) \ne m$ or $\gamma(j) \ne m$.

a) To be zero, we must have $\gamma(i) = \gamma(j) = m$. Then i occurs in the previous round just prior to the pair being considered since X_m updates only once per round. Also we have $j = h$, since h is the last time X_m updates before time k. Thus, $i \le rm < h = j \le (r+1)m < k$.
Then the summations, which also must be zero, include every other element of X, and $X_{\gamma(q)}(k-1) = 0$ for all $q \ne m$. Then $\sigma(k-1) = X_m(h)$ and $X_m(k) = V_m$.

b) $\quad$ Suppose $\gamma(i) \ne m$ or $\gamma(j) \ne m$. Then
$$X_m(k) - X_m(h-1) \ge V_m - V_{\gamma(i)}, \text{ if } \gamma(j) = \gamma(k) = m$$
$$X_m(k) - X_m(h-1) \ge V_m - V_{\gamma(j)}, \text{ if } \gamma(i) = m$$
$$X_m(k) - X_m(h-1) \ge 2V_m - V_{\gamma(i)} - V_{\gamma(j)}, \text{ if } \gamma(i) \ne m \ne \gamma(j)$$
and in all these cases we have
$$X_m(k) - X_m(h-1) \ge V_m - V_{m-1}$$
proving that:

iii. After round two, X_m updates to V_m, or increases by not less than $V_m - V_{m-1}$ in every pair of consecutive rounds.

Unlikely but still possible, convergence could be slowed by the occurrence of $h = k - 1$ in every pair of rounds. Even so, it would take not more than $V_m/(V_m - V_{m-1})$ pairs to increase X_m from zero to V_m. Totaling up, and counting the first two rounds needed to establish the conditions in parts (i) and (ii), it is proven that:

iv. The state vector X converges to $X = [0 \ldots 0 \ Q(V_m)]$ in not more than $2 + 2V_m/(V_m - V_{m-1})$ rounds if $V_m > 0$.

EXAMPLE : Suppose $X(0) = \mathbf{0}$, $V = [0.1, 0.3, 0.8]^T$. Compare the results of updating by the sequences $\{\gamma(k)\} = \{3, 1, 2; \ldots \}$ and $\{\gamma(k)\} = \{2, 1, 3; 3, 1, 2; \ldots$ (repeating)$\}$. Theorem 2 states that convergence occurs in not more than $2 + 2*0.8/(0.8 - 0.3) = 5.2$, say 6 rounds. Since the initial state is zero, time can be reduced by two rounds, to only four rounds (see part iii in proof of Theorem 2).

Let underscore indicate the newly updated element in X. For the first sequence, $X(1) = [0 \ 0 \ \underline{0.8}]^T$ is the final state, occurring in the first round. For the second sequence, the trajectory of states is $\{X\} = \{[0 \ \underline{0.3} \ 0]^T, [\underline{0} \ 0.3 \ 0]^T, [0 \ 0.3 \ \underline{0.5}]^T; [0 \ 0.3 \ \underline{0.5}]^T, [\underline{0} \ 0.3 \ 0.5]^T, [0 \ \underline{0} \ 0.5]^T; [0 \ \underline{0} \ 0.5]^T, [\underline{0} \ 0 \ 0.5]^T, [0 \ 0 \ \underline{0.8}]^T; \ldots \}$, converging to the final state in the third round.

DISCUSSION

The new network given here is significantly simpler than the networks of Lippman et al. [5], Winters and Rose [7], Suter and Kabrisky [6], and Yadid-Pecht and Gur [8] that are synchronous and require connection weight modification involving mathematical operations that are not biologically plausible for single neurons. Unlike some of these [5] [7], our new network outputs the largest input and also identifies which input is the largest number.

Time τ for convergence to a stable state having the largest input number as the only non-zero output is given by $\tau \leq (2 + 2V_m/(V_m - V_{m-1}))mT$, where m is the number of inputs and T is the discrete time interval. Since $V_m \leq 1$, and η is the difference between the largest and next to largest value in V, we get $\tau \leq (2 + 2/\eta)mT$, so that convergence to the final state occurs in time $O(m/\eta)$. In practical applications, a non-zero η cannot be less than the precision by which inputs are known. Note that absolute stability for all η can be proven by Lyapunov's method [2], in support of the results stated here.

Actual convergence can be faster than given by Theorem 2. Time to convergence depends on the value of σ when the mth neuron updates, especially in the first round. Smaller values of σ lead to faster convergence.

Complexity and speed of algorithms simulating neural networks are often compared, but our asynchronous network is not an algorithms. In synchronous systems, every neuron updates in each time interval. A MAXNET said to converge in time $O(m)$ has computational complexity $O(m^2)$. The networks of Winters and Rose [7], Suter and Kabrisky [6], and Yadid-Pecht and Gur [8] all converge in time $O(\log(m))$, and their computational complexity is $O(m\log(m))$. Note however, operations such as multiplication and division are more complex than a neural activation function of the sum of weighted inputs. For the network given here, both convergence and computational complexity is also $O(m/\eta)$.

We stipulated that the positive elements of V must be distinct. However, the essential requirement for convergence to a final state is that the largest element of V must be distinct. The other elements need not be distinct. Furthermore, even if the largest input is not distinct, so that η is zero or infinitesimal, then the network will reach a locked state within time τ. Here, all neuron outputs are zero except for the neurons having in common the largest input. These neurons will have stable outputs. Their sum will be V_m. In a locked state, σ will be constant and equal to V_m. If only one element of X is positive for two rounds, a final state has been achieved. If there is more than one positive element in X, σ is constant for two rounds and either a final state or a locked state has been achieved.

We may add one neuron to produce σ at its output by summing the outputs of all the other neurons, and transmit σ to all the other neurons. This requires only m pairs of connections. Connection complexity is reduced from $O(m^2)$ to $O(m)$ in this modification.

Now suppose that the neural network is asynchronous in continuous time, and that each neuron updates periodically, but only one neuron updates at any given instant. Then a round becomes the period p of the slowest neuron, resulting in

$$\tau \leq (2 + 2V_m/(V_m - V_{m-1}))p$$

Assuming that transmission is fast enough, convergence time is now independent of the number of neurons, or the size of the problem. Then τ is of $O(1/\eta) < O(m/\eta)$.

CONCLUSION

An asynchronous discrete time neural network has been shown to identify the largest of its inputs, and to preserve and transmit the value of that largest input to each neuron in the network. Moreover, it is shown that from an initial state wherein the output of every neuron is zero, a sorted subset of the inputs emerges from the signals transmitted between the neurons in the first round of updates.

The location and value of the largest input are found in the final state, when the neuron having the largest input has output equal to the largest input and the other neurons have output equal to zero. Time of convergence τ to the final state is finite if m and η are finite. An upper bound on τ is given. If η is infinitesimal or zero, the network enters a locked state in finite time. Thus, the network is absolutely stable for all values of η. These locked states are detected by observing σ. The value of V_m is available to an observer at the end of the first round of updates, and at any time in a locked state.

An upper bound on computational complexity is $O(m/\eta)$. The number of neurons is $O(m)$. The number of connections is $O(m^2)$ in the configuration least susceptible to component failure, but reduced to $O(m)$ in the more economical configuration.

The advantages of the network given here are that the parameters of the network are independent of the number of nodes and known *a-priori*. Connections are all unit weight. The network is simpler than previous networks. There is no limit on the number of nodes or the size of the problem. Nodes can be added or deleted at any time, even when a calculation is under way. There is a plausible resemblance to biological neural networks.

REFERENCES

[1] G. Carpenter and S. Grossberg (1987), "A Massively Parallel Architecture for a Self-Organizing Neural Pattern Recognition Machine", Computer Vision, Graphics and Image Processing 37, pp 54-115.

[2] M.A. Cohen and S. Grossberg (1983), "Absolute Stability of Global Pattern Formation and Parallel Memory Storage by Competitive Neural Networks", IEEE Transactions on System, Man, and Cybernetics SMC-13, pp815-826

[3] S. Grossberg, (1987), "Competitive learning: From Interactive Activation to Adaptive Resonance", Cognitive Science 11, p 23-63.

[4] T. Kohonen (1988), *Self-Organization and Associative Memory, 2nd Edition*, Springer-Verlag, Berlin.

[5] R.P. Lippman, B. Gold and M.L. Malpass (May, 1987), "A Comparison of Hamming and Hopfield Neural Nets for Pattern Classification", Technical Report 769, Lincoln Laboratory, MIT.

[6] B.W. Suter and M. Kabrisky (1992), "On a Magnitude preserving Iterative Maxnet Algorithm", Neural Computation, Vol.4, pp 224-233.

[7] J.H. Winters and C. Rose (1989), "Minimum Distance automata in Parallel Networks for Optimum Classification", Neural Networks, Vol.2, pp 127-132.

[8] O. Yadid-Pecht and M. Gur (May 1995), "A Biologically Inspired Improved MAXNET," IEEE Transactions on Neural Networks, Vol.6, No.3, pp 757-759.

AN IMPROVED PARALLEL ALGORITHM FOR FINDING THE MAXIMUM CLIQUE IN AN ARBITRARY GRAPH[1]

SONGNIAN YU
Shanghai University
Department of Computer
Engineering and Science
Shanghai, China, 200072

XIAOFENG QIAN
Shanghai University
Department of Computer
Engineering and Science
Shanghai, China, 200072

WEIMIN XU
Shanghai University
Department of Computer
Engineering and Science
Shanghai, China, 200072

ABSTRACT

The maximum clique problem is an important NP-Complete problem in graph theory. We describe an improved algorithm of branch and bound for this problem. Two main procedures in the algorithm, both of $O(|E|)$ time complexity, are presented: (1) for finding a maximal clique in an arbitrary graph; and (2) for finding a maximal k-chromatic induced subgraph in an arbitrary graph. Our algorithm was implemented and tested in branch-parallel way with lower communication-spending on clusters.

There are some definitions about *clique* on graph theory as below. Given a simple graph $G = (V, E)$ with vertex set V and edge set E, $S \subseteq V$ is called a clique in graph G if $\forall u, v \in S, u \neq v$ that $(u, v) \in E$. Any clique S in graph G is called a maximal clique in graph G if $\forall v \in (V - S)$, $\exists u \in S, (u, v) \notin E$. Any maximal clique S in graph G is called a maximum clique in graph G if $|S| \geq |S'|$ holds for any other maximal clique S' in graph G.

It is a key NPC problem in graph theory to finding a maximum clique in an arbitrary graph. A lot of famous problems in graph theory, such as information retrieval, experimental design and some lower bounds of Ramsey numbers, have related to it. Many problems can also be transformed to clique problem in polynomial time. Plenty of papers had discussed about clique problem. In this paper we improve the algorithm introduced by Egon Balas and Chang Sung Yu[1], and develop a parallel algorithm to finding a maximum clique in an arbitrary graph in an effective way.

A typical branch and bound procedure is described in this paper. Some definitions should be introduced here: problem set P is a set made up of elements (I, E); every element is consists of $I \subseteq V$ and $E \subseteq V$. Problem set P

---

[1] The research was supported by the Shanghai Development Foundation of Technology under contract No. 00JC14052 and No. 99515036.

is initialized and modified in our algorithm. Set $N(v) = \{w \in (V - \{v\}) | (v,w) \in E\}$ is called the adjacent set of vertex $v \in V$ and set $N(S) = \left[\bigcup_{w \in S} N(w)\right] - S$ is called the adjacent set of subset $S \subseteq V$ if simple graph $G = (V,E)$ is specified.

The paper consists of three parts. Part 1 gives the necessary definitions and some theorems as background information on graph theory. Part 2 outlines our general approach for finding a maximum clique in arbitrary graph, then shows the way to be paralleled. The last part, i.e. Part 3, presents computational experience through some example.

1. FOUNDATION THEORY

Given simple graph $G = (V,E)$, function $h(v)$ is called a coloring if $\forall v_i \in V$, $h(v_i) = c \in C$, that $\forall (v_1, v_2) \in E$, $h(v_1) \neq h(v_2)$.

Definition 1: Given graph $G = (V,E)$, $W \subseteq V$, a coloring $h(v)$ of graph $G(W)$, subgraph $G(W)$ is called a k-**chromatic induced subgraph** of graph G if $|C| = k$. Subgraph $G(W)$ is called a **maximal k-chromatic induced subgraph** of graph G if $\forall v \in V - W$, and $\forall c \in C$, $\exists w \in W$, $(v,w) \in E$ and $h(w) = c$.

Definition 2: Given graph $G = (V,E)$, a maximal clique K of graph G, a maximal $|K|$-chromatic induced subgraph $G(W)$ of graph G, subgraph $G(W)$ is called a **maximal clique K - extended chromatic induced subgraph** of graph G if $K \subseteq W$.

It is obvious that no clique $K \subseteq V$, $|K| > k$ could be found in graph $G = (V,E)$ if graph G was a k-chromatic graph. Then K must be a maximum clique of graph $G(W)$ if $G(W)$ is a maximal clique K - extended induced subgraph of graph G. On the other hand, according to the definition of maximal clique and maximum clique, a maximum clique K of graph G always is a maximal clique of graph G, but a maximal clique K of graph G may not a maximum clique of graph G. Theorem 1 is base on it.

Theorem 1[1]: Given graph $G = (V,E)$, a maximal k-chromatic induced subgraph $G(W)$ of graph G, there is no clique K that $|K| > k$ exists in graph G if $W = V$. Otherwise ($W \subset V$), let $(v_1, v_2, ..., v_m)$ be an arbitrary ordering of the set $V - W$, if G has a clique K^* such that $|K^*| > k$, then K^* is contained in one of the following m sets:
$V_i = \{v_i\} \cup N(v_i) - \{v_1, v_2, ..., v_{i-1}\}, i = 1, ..., m$,
where for $i = 1$ we define $\{v_1, v_2, ..., v_{i-1}\} = \emptyset$.

Proof: If $W = V$, then graph G is k-chromatic. So there is no clique K that $|K| > k$ exists in graph G.
Otherwise $W \subset V$, since $|K^*| > k$, and $G(W)$ is a maximal k-chromatic induced subgraph, that $K^* \not\subset W$. Then $\exists v_i \in V - W, i \in \{1, ..., m\}$, $m = |V - W|$ that $v_i \in K^*$, and $K^* \cap (V - W) \neq \emptyset$. According to the definition of clique, then $\forall v \in L$, $K^* \subseteq (\{v\} \cup N(v))$ if $K^* \cap (V - W) = L$.
Let $L = \{v_{j_1}, v_{j_2}, ..., v_{j_p}, ..., v_{j_q}\}$, $t = \min\{j_1, j_2, ..., j_q\}$, $j_p \in \{1, 2, ..., m\}$, $p = 1, 2, ..., q$, $\exists V_t = \{v_t\} \cup N(v_t) - \{v_1, v_2, ..., v_t\}$ that $K^* \subseteq V_t$. ∎

Corollary 1: Given graph $G = (V,E)$, a maximal clique K of graph G, a maximal clique K-extended chromatic induced subgraph $G(W)$ of graph G, K is a maximum clique of graph G if $W = V$. Otherwise ($W \subset V$),

let $(v_1, v_2, ..., v_m)$ be an arbitrary ordering of the set $V - W$, if G has a clique K^* such that $|K^*| > |K|$, then K^* is contained in one of the following m sets:
$V_i = \{v_i\} \cup N(v_i) - \{v_1, v_2, ..., v_{i-1}\}, i = 1, ..., m$,
where for $i = 1$ we define $\{v_1, v_2, ..., v_{i-1}\} = \emptyset$.

A maximum clique could be found in arbitrary graph G on the basis of Theorem 1 and Corollary 1. It is obvious that according to Theorem 1 or corollary 1, the larger k or K found on $G(W)$, the faster to get the maximum clique on G. Meanwhile, it is also very important to make the procedure finding k or maximal clique K in the algorithm as simple as possible. The following Theorem 2 is concerning maximal clique.

Theorem 2: K is a maximal clique of graph $G = (V, E)$ if and only if $\bigcap_{v \in K}(\{v\} \cup N(v)) = K$.

Proof:
(1) Let K be a maximal clique of graph G. Then $K \subseteq \bigcap_{v \in K}(\{v\} \cup N(v))$ according to the definition of clique. If $\exists w \in V$ and $w \notin K$, that $(K + \{w\}) \subseteq \bigcap_{v \in K}(\{v\} \cup N(v))$, then $w \in \bigcap_{v \in K}(\{v\} \cup N(v))$, as well as $\bigcup_{v \in K}\{(w, v)\} \subseteq E$. Set $K + \{w\}$ is a clique of graph G according to the definition of clique. This conflicts with the definition of maximal clique. So there is no $w \in V$ and $w \notin K$ exists in graph G, that $(K + \{w\}) \subseteq \bigcap_{v \in K}(\{v\} \cup N(v))$. Then $\bigcap_{v \in K}(\{v\} \cup N(v)) = K$ if K is a maximal clique of graph G.

(2) Let $\bigcap_{v \in K}(\{v\} \cup N(v)) = K$. So $\forall v_i \in K$ that $v_i \in \bigcap_{v \in K}(\{v\} \cup N(v))$, as well as $\forall v \in K$, $v_i \neq v$, $v_i \in N(v)$. Then K is a clique of graph G. If $\exists w \in V$ and $w \notin K$, that $K + \{w\}$ is also a clique of graph G, then $w \notin \bigcap_{v \in K}(\{v\} \cup N(v))$. Thus it must be $v \in K, w \notin N(v)$, so set $K + \{w\}$ is not a clique of graph G. This conflicts to the supposition before. Then K is a maximal clique of graph G.

2. DESCRIPTION OF ALGORITHM AND PARALLEL WAY
2.1 Algorithm of Finding A Maximal Clique

According to theorem 2, a prototype algorithm can be concluded. Larger maximal clique could be found by the algorithm of $O(|E|)$ time complexity described below.

Input: simple graph $G = (V, E)$
Output: a maximal clique K of graph G
Steps:
0. Initialization: Set $W = V$, $K = V$;
1. Select $v \in W$ that $|(\{v\} \cup N(v)) \cap K| = \max_{w \in W}\{|(\{w\} \cup N(w)) \cap K|\}$;

   Set $W = W - \{v\} - (V - N(v))$, $K = (\{v\} \cup N(v)) \cap K$;
2. If $W = \emptyset$, then the algorithm is terminated and K is the result; otherwise go to step 1.

The value of expression $|(\{w\} \cup N(w)) \cap K|, w \in W$ for select $v \in W$ may be computed several times in this algorithm to finding a maximum clique before it satisfies the condition $|(\{v\} \cup N(v)) \cap K| = \max_{w \in W}\{|(\{w\} \cup N(w)) \cap K|\}$. The vertex $v \in W$ satisfied with condition $|(\{v\} \cup N(v)) \cap K| = \max_{w \in W}\{|(\{w\} \cup N(w)) \cap K|\}$ is also the vertex v has the maximum degree in graph $G(W)$. It is not necessary to re-compute values of vertex degree in graph $G(W)$ at all because set W reduced one element for every loop. The values of vertex degree could be altered when W had been changed so as to reduce computing time. The expression $W = W - \{v\} - (V - N(v))$ can be simplified as below:

$W - \{v\} - (V - N(v)) = W \cap N(v)$

Another assignment expression $K = (\{v\} \cup N(v)) \cap K$ can be altered to $K = K \cup \{v\}$ with different initialization codes and the same result will be kept. Then for the same computing result a more practical and faster algorithm CNEW is produced below.

Algorithm CNEW
Input: simple graph $G = (V, E)$
Output: a maximal clique K of graph G
Steps:
0. Initialization: Compute the degree $d(v_i) = |N(v_i)|$ of every vertex v_i. Set $W = V$, $K = \emptyset$;
1. Select $v \in W, d(v) = \max_{v_i \in W}\{d(v_i)\}$;

   Modify $d(v_i) = d(v_i) - |(W \cap N(v_i)) - N(v)|$ for every $v_i \in (W \cap N(v))$;

   Set $W = W \cap N(v)$, $K = K \cup \{v\}$;
2. If $W = \emptyset$, then the algorithm is terminated and K is the result; otherwise go to step 1.

2.2 Color Algorithm

The color algorithm may reduce the number of branches produced by the algorithm for finding maximum clique. It speeds up the whole algorithm. So any color algorithm may be used in our algorithm if it satisfied conditions below: the color algorithm should be simple enough; the chromatic induced graph found by the color algorithm should be large enough. Algorithm COLOR1 and COLOR2 are used in different part in algorithm for finding a maximum clique. They are all satisfied with these conditions.

Algorithm COLOR1[1]
Input: simple graph $G = (V, E)$, a maximal clique $K = \{v_1, v_2, ..., v_k\}$ of graph G
Output: a maximal clique $|K|$-extended chromatic induced graph $G(W)$ of graph G, $K \subseteq W$
Steps:
1. Initialize k color classes $C_1 = \{v_1\}, C_2 = \{v_2\}, ..., C_k = \{v_k\}$;
2. If $V = \emptyset$, then set $W = \bigcup_{i=1}^{k} C_i$ and the algorithm is terminated; otherwise select $v \in V$, $V = V - \{v\}$ and check each color class C_i refer to the

ordering $i = 1,2,...,k$ if $C_i \cap N(v) = \varnothing$. If there exists such color class C_i then go to step 3, else go to step 4;
3. Set $C_i = C_i \cup \{v\}$ and go to step 2;
4. Select the color class C_i that has the minimum value of expression
$$|C_i \cap N(v)| \quad . \quad \text{If} \quad C_i \cap N(v) \cap \left[\bigcap_{j=i+1}^{k} N(C_j)\right] = \varnothing \quad \text{then set}$$
$C_i = [C_i - (C_i \cap N(v))] \cup \{v\}$ and check color classes C_j for each $w \in C_i \cap N(v)$ refer to the ordering $j = i+1, i+2,...,k$. If $C_j \cap N(w) = \varnothing$ then set $C_j = C_j \cup \{w\}$ and check next $w \in C_i \cap N(v)$. No matter if the assignment execute or not, go to step 2.

Algorithm COLOR2[1]
Input: simple graph $G = (V,E)$, the chromatic number k of graph G
Output: a maximal k-chromatic induced graph $G(W)$ of graph G
Steps:
Replace step 1 in algorithm COLOR1 to initialize k color classes $C_1 = C_2 = \cdots = C_k = \varnothing$.

2.3 General Algorithm for Finding A Maximum Clique

According to theorem 1 and inference 1, the basic idea of the general algorithm for finding a maximum clique in an arbitrary graph G is to find a maximal clique K then the algorithm check if there exists a clique larger than K. If it is true, then the maximal clique K is such the maximum clique of graph G; otherwise the clique larger than K must be contained in other vertex sets which have not any vertices in K. The algorithm is described as below:
Input: simple graph $G = (V,E)$
Output: a maximum clique K of graph G
Steps:
0   Initialization: $P = \{(I = \varnothing, E = \varnothing)\}$, $k = 0$;
1   If $P = \varnothing$, then the algorithm is terminated and K is the result; otherwise select arbitrary $p = (I,E) \in P$ and set $P = P - \{p\}$. If $|V - E| \leq k$ then go to step 1; else if $|V - E| > k$ and $|I| = k$ then set $S = V - (E \cup I)$ and go to step 2; otherwise set $S = V - (E \cup I)$ and go to step 4;
2   An algorithm for finding a maximal clique K_p of graph $G(S)$: Set $K = |K_p \cup I|$. If $K = S$ then go to step 1; otherwise go to step 3;
3   An algorithm for finding a maximal clique $|K_p|$-extended chromatic induced graph $G(W)$ of graph $G(S)$ such as algorithm COLOR1 ($K_p \subseteq W$). If $W = S$ then go to step 1; otherwise go to step 5;
4   An algorithm for finding a maximal $(k - |I|)$-chromatic induced graph $G(W)$ of graph $G(S)$. If $W = S$ then go to step 1; otherwise go to step 5;
5   Branch algorithm: Select an arbitrary ordering $(v_1, v_2,...,v_m)$ of set $S - W$, set
$P = P \cup \{p_i \mid p_i = (I_i = (I \cup \{v_i\}), E_i = (E \cup (S - (\{v_i\} \cup N(v_i))) \cup \{v_1, v_2,...,v_{i-1}\}))$, $i = 1,2,...,m\}$

2.4 The Parallel Way

There exist three parallel ways for the algorithm introduced in this paper:
1. Use parallel algorithm for finding a maximal clique or parallel algorithm for coloring (see reference [2]);
2. Do loops in the algorithm in a parallel way;
3. Do different branches in a parallel way.

The first parallel way is complicated and not the fastest one. As for the second way since there is very few loops can be done with parallel procedures in our algorithm and forget it. So, the third one was selected in this paper. It is simpler and more effective.

A search tree could be got in our general algorithm for finding a maximum clique. The parallel algorithm will get several maximal cliques by finding maximum cliques of different sub-problems (nodes) in the xth layer of the tree on different processors. The maximum one among them is the answer of the root-problem of the tree. Our parallel algorithm CPAR on n processors is described below:

Algorithm CPAR
Input: simple graph $G = (V, E)$
Output: a maximum clique K of graph G ;
Steps:
1 Set $P = \{p_1, p_2, \ldots, p_m\}$ contains all sub-problems in the xth layer of the search tree;
2 Divide P into n subsets $P_1, P_2, \ldots, P_n$, $P_i = \{p_i, p_{n+1}, p_{2n+1}, \ldots\}, i = 1, 2, \ldots, n$;
3 Finding maximum cliques $K_1, K_2, \ldots, K_n$ for $P_1, P_2, \ldots, P_n$ on n processors by the algorithm introduced in this paper;
4 Exchange the result data on n processors then get $K, |K| = \max_{j=1,2,\ldots,n} \{|K_j|\}$ as the result.

3. EXAMPLE AND COMPUTATIONAL EXPERIENCE

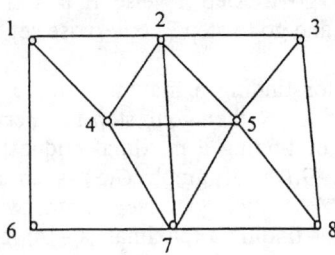

0. $P = \{(I = \emptyset, E = \emptyset)\}$, $k = 0$;
1. Select $p = (I = \emptyset, E = \emptyset) \in P$ and set $P = P - \{p\} = \emptyset$; $|V - E| = 8 > 0 = k$ and $|I| = 0 = k$, so set $S = V - (E \cup I) = \{1,2,3,4,5,6,7,8\}$, go to step 2;
2. Get maximal clique $K_p = \{2,4,5,7\}$ of graph $G(S)$ by the algorithm for finding a maximal clique described in this paper. Set $K = (K_p \cup I) = \{2,4,5,7\}$. $K \neq S$, then go to step 3;
3. Get maximal clique $|K_p|$ -extended chromatic induced graph $G(W)$ of graph $G(S)$

that $W = \{1,2,3,4,5,6,7,8\}$. $W = S$, then go to step 1;
4. $P = \emptyset$, then the algorithm is terminated and $K = \{2,4,5,7\}$ is a maximum clique of graph $G(V, E)$.

Algorithm CNEW and CPAR were implemented on 32×500MHz Pentium III SMP+128MB RAM and connected by 100M ether net with Redhat LINUX 6.2 (Kernal 2.2.14-5.0). Simple graphs consist of 50, 100, 200, 300, 400, 500, 600, 700 vertices on several densities had been tested. We tested our parallel algorithm CPAR on 30 random graphs of each scheme. Parts of the results are listed below. The data such as $a \pm b$ mean that a represents the average computing time and the standard deviation is b. There are two kinds of data for each scheme, the first one corresponding to algorithm CPAR with $x = 1$ and the second one corresponding to algorithm CPAR with $x = 2$.

Graph characteristics			Computing time (seconds)						
Verti ces	Density (%)	Size of maximum clique	Number of CPUs						
^^	^^	^^	1	2	4	8	16	32	64
50	95.0395 ±0.6255	28.4000 ±1.8903	0.0165 ±0.0174	0.0123 ±0.0119	0.0209 ±0.0198	0.0248 ±0.0189	0.0368 ±0.0288	0.0601 ±0.0394	0.1979 ±0.0396
^^	^^	^^	0.0184 ±0.0208	0.0127 ±0.0118	0.0182 ±0.0239	0.0382 ±0.0321	0.0356 ±0.0253	0.0531 ±0.0334	0.1909 ±0.0328
100	80.1488 ±0.5640	19.9000 ±0.7461	8.1233 ±2.3429	4.9459 ±1.4242	3.1595 ±0.9630	2.1769 ±0.6836	1.7216 ±0.6529	1.5701 ±0.5517	1.6702 ±0.5213
^^	^^	^^	8.5940 ±2.4405	5.1036 ±1.3715	3.1418 ±0.7872	1.9823 ±0.6282	1.3633 ±0.4440	1.0161 ±0.3953	0.9216 ±0.2757
200	69.9635 ±0.3159	18.1000 ±0.3958	368.8373 ±71.9049	204.2536 ±34.2999	120.4251 ±20.3713	73.6427 ±12.5509	47.3112 ±8.4174	34.9579 ±7.3391	31.0325 ±6.8202
^^	^^	^^	375.5735 ±73.1501	209.5563 ±38.2704	117.5754 ±17.3999	65.9422 ±10.6009	38.7863 ±7.7118	23.9229 ±4.0810	17.0562 ±3.0060
300	60.0509 ±0.2078	15.1333 ±0.3399	846.2993 ±90.8257	453.1989 ±53.3041	253.2852 ±31.4121	144.0072 ±19.4161	84.8183 ±11.1735	55.2442 ±8.0670	42.5247 ±5.9557
^^	^^	^^	856.2831 ±92.0389	450.5366 ±48.9168	250.4844 ±28.1613	139.7041 ±14.9912	76.9160 ±9.2464	44.6257 ±5.7451	26.5154 ±3.4549
400	49.9682 ±0.2201	12.7667 ±0.4230	590.1129 ±74.1708	320.9487 ±27.3470	170.8864 ±16.0906	92.4592 ±8.1715	52.0594 ±4.9223	32.7136 ±3.6722	22.5807 ±2.5202
^^	^^	^^	599.9048 ±73.8605	325.5285 ±31.1354	172.7141 ±12.3865	90.5205 ±7.2392	50.2535 ±5.1966	30.1910 ±3.1945	22.7870 ±3.8305
500	40.0318 ±0.1463	10.6333 ±0.4819	265.2605 ±29.7777	141.0915 ±7.7261	74.0330 ±5.1299	39.4976 ±2.3895	21.9762 ±1.6517	13.1009 ±1.0143	8.6824 ±0.9188
^^	^^	^^	272.8823 ±28.8698	146.5075 ±10.8810	79.0135 ±4.3815	42.9338 ±2.1299	24.5267 ±1.2298	15.8451 ±0.8865	22.9900 ±2.8148
600	29.9883 ±0.1081	8.9333 ±0.2494	84.6116 ±8.4371	46.3422 ±4.4827	24.4148 ±1.6085	12.7208 ±0.7254	6.9811 ±0.4568	4.0265 ±0.2725	2.9561 ±0.3563
^^	^^	^^	93.7519 ±8.4822	51.8228 ±4.7859	30.0209 ±2.1659	18.2367 ±1.1834	12.218 ±0.6916	9.3349 ±0.4075	21.0079 ±4.7153
700	20.0074 ±0.0719	7.0333 ±0.1795	26.0120 ±1.1490	13.4289 ±0.7185	7.0189 ±0.3409	3.8017 ±0.2449	2.1264 ±0.1369	1.2588 ±0.0854	1.3592 ±0.2156

Graph characteristics			Computing time (seconds)						
Verti ces	Density (%)	Size of maximum clique	Number of CPUs						
			1	2	4	8	16	32	64
			33.1833 ±3.5680	19.3412 ±1.8384	12.4198 ±0.9843	8.9516 ±0.5550	7.2212 ±0.3434	7.0333 ±0.2283	7.6310 ±2.8550

Each curve in the following two figures shows that the computing time is decreasing as the number of CPU is increasing. With the algorithm CPAR a maximum clique was found on a random graph with 300 vertices and 60% dense distribution of edges.

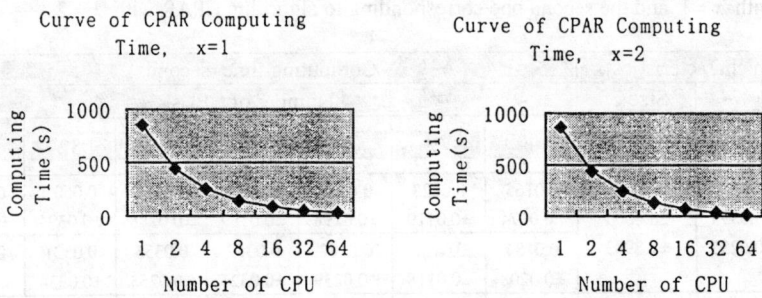

Sum up, it is clear to see that the ratio of the parallel approach is much better if the size of the problem is large enough.

4. REFERENCES

1. Egon Balas and Chang Sung Yu, Finding a Maximum Clique in an Arbitrary Graph, SIAM J. COMPUT. Vol. 15, No. 4, November 1986, P1054-1067.
2. Ceshan Tang, Weifa Liang, Parallel Algorithms for Graph Theory, University Press of Science and Technology of China, 1991.10; P195-220
3. I. Bomze and M. Budinich and P. Pardalos and M. Pelillo. *The maximum clique problem*. In D.-Z. Du and P. M. Pardalos, editors, Handbook of Combinatorial Optimization, volume 4. Kluwer Academic Publishers, Boston, MA, 1999

A MODIFIED HOPFIELD MODEL FOR SOLVING SEVERAL TYPES OF OPTIMIZATION PROBLEMS

IVAN NUNES DA SILVA
UNESP/FE/DEE, CP 473
CEP 17033-360, Bauru-SP, Brazil

ANDRE NUNES DE SOUZA
UNESP/FE/DEE, CP 473
CEP 17033-360, Bauru-SP, Brazil

JOSE ALFREDO C. ULSON
UNESP/FE/DEE, CP 473
CEP 17033-360, Bauru-SP, Brazil

ABSTRACT

Artificial neural networks are dynamic systems consisting of highly interconnected and parallel nonlinear processing elements that are shown to be extremely effective in computation. This paper presents an architecture of artificial neural networks that can be used to solve several classes of optimization problems. More specifically, a modified Hopfield network is developed and its internal parameters are computed using the valid-subspace technique. Among the problems that can be treated by the proposed approach include combinatorial optimization problems, dynamic programming problems and nonlinear optimization problems.

INTRODUCTION

A neural network is a massively parallel distributed processor that has a natural propensity for storing experiential knowledge and making it available for use (Haykin, 1999). Artificial neural networks have been applied to several classes of optimization problems and have shown promise for solving such problems efficiently. The approach developed here uses a modified Hopfield network with equilibrium points representing the solution of the optimization problems. The internal parameters of the network have been computed using the valid-subspace technique (Aiyer et al., 1990; Silva, 1997).

Hopfield networks are single-layer networks with feedback connections between nodes. In the standard case, the nodes are fully connected, i.e., every node is connected to all others nodes, including itself. The node equation for the continuous-time network with N neurons is given by:

$$\dot{u}_i(t) = -\eta . u_i(t) + \sum_{j=1}^{N} T_{ij} . v_j(t) + i_i^b \qquad (1)$$

$$v_i(t) = g(u_i(t)) \qquad (2)$$

where: $u_i(t)$ is the state of the i-th neuron; $v_i(t)$ is the output of the i-th neuron; i_i^b is the offset bias of the i-th neuron; $\eta . u_i(t)$ is a passive decay term; T_{ij} is the weight connecting the j-th to i-th neuron; and $g(.)$ is a activation function.

It is shown in Hopfield (1984) that if T is symmetric and $\eta=0$, the equilibrium points of the network correspond to values $v(t)$ for which the energy function associated with the network is minimized:

$$E(t) = -\frac{1}{2}v(t)^T .T.v(t) - v(t)^T i^b \qquad (3)$$

Therefore, the mapping of optimization problems using the Hopfield network consists of determining the weight matrix T and the bias vector i^b to compute equilibrium points.

THE MODIFIED HOPFIELD NETWORK

In this paper, we have used a modified energy function $E^m(t)$, composed by two energy terms, defined as follows:

$$E^m(t) = E^{conf}(t) + E^{op}(t) \qquad (4)$$

where: $E^{conf}(t)$ is a confinement term that groups the structural constraints associated with the respective optimization problem, and $E^{op}(t)$ is an optimization term that conducts the network output to the equilibrium points corresponding to a cost constraint. Thus, the minimization of $E^m(t)$ of the modified Hopfield network is conducted in two stages:

i-) minimization of the term $E^{conf}(t)$:

$$E^{conf}(t) = -\frac{1}{2}v(t)^T T^{conf}.v(t) - v(t)^T i^{conf} \qquad (5)$$

where: $v(t)$ is the network output, T^{conf} is a weight matrix and i^{conf} is a bias vector belonging to E^{conf}. This corresponds to confine $v(t)$ into a valid subspace generated from the structural constraints imposed by the problem.

ii-) minimization of the term $E^{op}(t)$:

$$E^{op}(t) = -\frac{1}{2}v(t)^T T^{op}.v(t) - v(t)^T i^{op} \qquad (6)$$

where: T^{op} is weight matrix and i^{op} is bias vector belonging to E^{op}. This corresponds to move $v(t)$ towards an optimal solution (the equilibrium points). Thus, the operation of the modified Hopfield network consists of three steps:

Step (I): Minimization of E^{conf}, corresponding to the projection of $v(t)$ in the valid subspace defined by:

$$v = T^{val}.v + s \qquad (7)$$

where: T^{val} is a projection matrix ($T^{val}.T^{val} = T^{val}$) and the vector s is orthogonal to the subspace ($T^{val}.s = 0$). This operation corresponds to an indirect minimization of $E^{conf}(t)$, i.e., $T^{conf} = T^{val}$ and $i^{conf} = s$. An analysis of the valid-subspace technique is presented in Aiyer (1990).

Step (II): Application of a nonlinear 'symmetric ramp' activation function constraining $v(t)$ in a hypercube:

$$g^r(v_i) = \begin{cases} 1 & \text{if } v_i > 1 \\ v_i & \text{if } 0 \leq v_i \leq 1 \\ 0 & \text{if } v_i < 0 \end{cases} \; ; \; \text{where } v_i \in [0,1]. \qquad (8)$$

Step (III): Minimization of E^{op}, which involves updating of $v(t)$ in direction to an optimal solution (defined by T^{op} and i^{op}) corresponding to network equilibrium points, which are the solutions for the optimization problem considered in a specific application. Using the 'symmetric ramp' activation function and given $\eta=0$, Eq. (2) subsequently becomes:

$$v(t) = g^r(u(t)) = u(t)$$

By comparison with Eq. (1) and Eq. (6), we have:
$$\frac{dv(t)}{dt} = \dot{v} = -\frac{\partial E^{op}(t)}{\partial v}$$
$$\Delta v = -\Delta t . \nabla E^{op}(v) = \Delta t . (T^{op}.v + i^{op}) \quad (9)$$
Therefore, minimization of E^{op} consists of updating $v(t)$ in the opposite direction to the gradient of E^{op}. These results are also valid when a 'hyperbolic tangent' activation function is used.

Each iteration has two distinct stages. First, as described in Step (III), v is updated using the gradient of the term E^{op} alone. Second, after each updating, v is directly projected in the valid subspace. This is an iterative process, in which v is first orthogonally projected in the valid subspace, and then thresholded so that its elements lie in the range [0,1].

MAPPING OPTIMIZATION PROBLEMS BY THE MODIFIED HOPFIELD NETWORK

In this section, it is presented the formulation of three types of optimization problems, which are defined by combinatorial optimization problems, dynamic programming problems and nonlinear optimization problems. The notation employed for vectors and matrices, which are used for mapping combinatorial optimization problems and dynamic programming problems, can be found in Graham (1981).

Formulation of Combinatorial Optimization Problems

The combinatorial optimization problem considered in this paper is the matching problem in bipartite graphs. A graph G is a pair $G = (V,E)$, where V is a finite set of nodes or vertices and E has as elements subsets of V of cardinality two called edges. A matching M of a graph $G = (V,E)$ is a subset of the edges with the property that no two edges of M share the same node. The graph $G = (V,E)$ is called bipartite if the set of vertices V can be partitioned into two sets, U and W, and each edge in E has one vertex in U and one vertex in W. For each edge $[u_i, w_j] \in E$ is given a number $P_{ij} \geq 0$ called the connection weight of $[u_i, w_j]$. The goal of the matching problem in bipartite graphs is to find a matching of G with the minimum total sum of weights. As an example, for a bipartite graph with four nodes, the sets V, E, U and W are given by:

$V = \{u_1, u_2, w_1, w_2\}$
$E = \{[u_1,w_1], [u_1,w_2], [u_2,w_1], [u_2,w_2]\}$
$U = \{u_1, u_2\}$
$W = \{w_1, w_2\}$

The equations of T^{conf} and i^{conf} are developed to force the validity of the structural constraints. These constraints mean that each edge in E has one activated node in U and one activated node in W. Thus, the parameters T^{conf} and i^{conf} are given by:
$$T^{conf} = (I^n \otimes R^n) \quad (10)$$
$$i^{conf} = \frac{1}{n}(o^n \otimes o^n) \quad (11)$$
where I^n is the $n \times n$ identity matrix; o^n is an n-element vector of ones; the symbol "$\otimes$" denotes Kronecher product; and R^n is an $n \times n$ projection matrix defined by:

$$R^n = I^n - \frac{1}{n}O^n$$

where O^n is an $n \times n$ matrix of ones. The sum of the elements of each row of a matrix is transformed to zero by post-multiplication with R^n, while pre-multiplication by R^n has the effect of setting the sum of the elements of each column to zero. It is observed that Eq. (10) and (11) satisfy the properties of the valid subspace, i.e., $T^{conf} \cdot T^{conf} = T^{conf}$ and $T^{conf} \cdot i^{conf} = 0$.

The parameters T^{op} and i^{op} are obtained from the corresponding cost constraint given by:

$$T^{op} = 0 \qquad (12)$$
$$i^{op} = -\text{vec}(P) \qquad (13)$$

where vec(P) is a funciton which maps the $n \times n$ matrix U to an $n.m$-element vector, i.e, vec(P) = $[P_{11} P_{21} ... P_{m1} \quad P_{12} P_{22} ... P_{m2} \quad P_{1n} P_{2n} ... P_{mn}]^T$.

Formulation of Dynamic Programming Problems

A typical dynamic programming problem can be modeled as a set of source and destination nodes with n intermediate stages, m states in each stage, and metric data $d_{xi,(x+1)j}$, where x is the index of the stages, and i and j are the indices of the states in each stage. The goal of the dynamic programming problem considered in this paper is to find a valid path which starts at the source node, visits one and only one state node in each stage, reaches the destination node, and has a minimum total length (cost) among all possible paths.

The equations of T^{conf} and i^{conf} are developed to force the validity of the structural constraints. These constraints, for dynamic programming problems, mean that one and only one state in each stage can be actived. Thus, the parameters T^{conf} and i^{conf} are given by:

$$T^{conf} = (I^n \otimes R^m) \qquad (14)$$
$$i^{conf} = \frac{1}{m}(o^n \otimes o^m) \qquad (15)$$

The energy function E^{op} of the modified Hopfield network for the dynamic programming problem is projected to find a minimum path among all possible paths. When E^{op} is minimized, the optimal solution corresponds to the minimum energy state of the network. The energy function E^{op} is defined by:

$$E^{op} = \frac{1}{4}[\underbrace{\sum_{x=1}^{n-1}\sum_{i=1}^{m}\sum_{j=1}^{m} d_{xi,(x+1)j} \cdot v_{xi} \cdot v_{(x+1)j}}_{1st.\,Term} + \underbrace{\sum_{x=2}^{n}\sum_{i=1}^{m}\sum_{j=1}^{m} d_{(x-1)j,xi} \cdot v_{xi} \cdot v_{(x-1)j}}_{2nd.\,Term}] + \qquad (16)$$

$$+ [\underbrace{\sum_{x=1}^{1}\sum_{i=1}^{m} d_{source,xi} \cdot v_{xi}}_{3rd.\,Term} + \underbrace{\sum_{x=n}^{n}\sum_{i=1}^{m} d_{xi,destination} \cdot v_{xi}}_{4th.\,Term}]$$

In this equation, the first term defines the weight (metric cost) of the connection linking the ith neuron of stage x to the jth neuron of the following stage (x+1). The second term defines the weight of the connection linking the ith neuron of stage x to the jth neuron of the previous stage (x-1). The third term provides the weight of the connection linking the source node to all others nodes of the first stage, while the fourth term provides the weight of the connection linking the destination to all other nodes of the last stage. Therefore, optimization of E^{op} corresponds to minimize each term given by Eq. (16) in relation to v_{xi}. From Eq. (16), the matrix T^{op} and vector i^{op} can be given by:

$$[T^{op}]_{pq} = -[P]_{xi,yj} \cdot [Q]_{xy} \quad \begin{cases} [P]_{xi,yj} = \frac{1}{2} d_{xi,yj} \\ [Q]_{xy} = \delta_{(x+1)y} + \delta_{(x-1)y} \end{cases} \quad (17)$$

$$i^{op} = -[\underbrace{d_{source,11} \; d_{source,12} \cdots d_{source,1m}}_{m} \; \underbrace{0 \; 0 \ldots 0 \; 0}_{m.(n-2)}$$
$$\underbrace{d_{n1,destination} \; d_{n2,destination} \cdots d_{nm,destination}}_{m}] \quad (18)$$

where: $T^{op} \in \Re^{nm \times nm}$ and $i^{op} \in \Re^{nm}$
$p = m.(x-1) + i$
$q = m.(y-1) + j$
$x, y \in \{2..n-1\}$
$i, j \in \{1..m\}$

In the next subsection, it is presented the formulation of nonlinear optimization problems by the modified Hopfield network.

Formulation of Nonlinear Optimization Problems

Consider the following general nonlinear optimization problem, with m-constraints and n-variables, given by the following equations:

Minimize $E^{op}(v) = f(v)$ (19)
subject to $E^{conf}(v)$: $h_i(v) \leq 0$, $i \in \{1..m\}$ (20)
$z^{min} \leq v \leq z^{max}$ (21)

where v, z^{min}, $z^{max} \in \Re^n$; $f(v)$ and $h_i(v)$ are continuous, and all first and second order partial derivatives of $f(v)$ and $h_i(v)$ exist and are continuous. The vectors z^{min} and z^{max} define the bounds on the variables belonging to the vector v. The conditions in Eq. (20) and Eq. (21) define a bounded convex polyhedron. The vector v must remain within this polyhedron if it is to represent a valid solution for the optimization problem given by Eq. (19). A solution can be obtained by a modified Hopfield network, whose valid subspace guarantees the satisfaction of condition given by Eq. (20). Moreover, the initial hypercube represented by the inequality constraints in Eq. (21) is directly defined by the 'symmetric ramp' function given in Eq. (8), i.e. $v \in [z^{min}, z^{max}]$.

The parameters T^{conf} and i^{conf} are calculated by transforming the inequality constraints in Eq. (20) into equality constraints by introducing a slack variable $w \in \Re^n$ for each inequality constraint:

$$g_i(v) + \sum_{j=1}^{m} \delta_{ij} . w_j = 0 \quad (22)$$

where w_j are slack variables, treated as the variables v_i, and δ_{ij} is defined by the Kronecker impulse function:

$$\delta_{ij} = \begin{cases} 1, & \text{if } i = j \\ 0, & \text{if } i \neq j \end{cases} \quad (23)$$

After this transformation, the problem defined by Eq. (19), (20) and (21) can be rewritten as:

Minimize $E^{op}(v^+) = f(v^+)$ (24)
subject to $E^{conf}(v)$: $h^+(v^+) = 0$ (25)

$$z^{min} \leq v^+ \leq z^{max}, i \in \{1..n\} \tag{26}$$
$$0 \leq v^+ \leq z^{max}, i \in \{n+1..N^+\} \tag{27}$$

where $N^+=n+m$, and $v^{+T}=[v^T \quad w^T] \in \Re^{N+}$ is a vector of extended variables. Finally, Eq. (7) representing the valid subspace is replaced by the following equation:

$$v = v - \nabla h(v)^T.(\nabla h(v).\nabla h(v)^T)^{-1}.h(v) \tag{28}$$

where $\nabla h(v)$ is the Jacobean matrix in relation to v.

The parameters T^{op} and i^{op} in this case are such that the vector v^+ is updated in the opposite gradient direction that of the energy function E^{op}. Since conditions given by Eq. (20) and (21) define a bounded convex polyhedron, the objective function has a global minimum. Thus, the equilibrium points of the network can be calculated by assuming the following values to T^{op} and i^{op}:

$$i^{op} = -\left[\frac{\partial f(v)}{\partial v_1} \quad \frac{\partial f(v)}{\partial v_2} ... \frac{\partial f(v)}{\partial v_N}\right]^T \tag{29}$$

$$T^{op} = 0 \tag{30}$$

CONCLUSIONS

In this paper, the proposed approach has been applied to solve combinatorial optimization problems, dynamic programming problems and nonlinear optimization problems. The main advantages of using the modified Hopfield network proposed in this paper are the following: i) the internal parameters of the network are explicitly obtained by the valid-subspace technique of solutions, ii) the valid-subspace technique groups all feasible solutions associated with the problem, iii) lack of need for adjustment of weighting constants for initialization, and iv) for all classes of optimization problems, a same methodology is adopted to derive the internal parameters of the network.

ACKNOWLEDGMENTS

The authors express thanks to FAPESP (Process No. 98/08480-0) and CNPq (Process No. 300446/98-5) for providing financial support.

REFERENCES

Aiyer, S. V. B., Niranjan M., and Fallside F., 1990, "A Theoretical Investigation Into the Performance of the Hopfield Network", *IEEE Trans. on Neural Networks*, Vol. 1, pp. 53-60.

Graham, A., 1981, "Kronecher Products and Matrix Calculus," Ellis Horwood Ltd., Chichester, UK.

Haykin, S., 1999, "Neural Networks – A Comprehensive Foundation," Prentice-Hall Inc., Upper Saddle River, NJ.

Hopfield, J. J., 1984, "Neurons With a Graded Response Have Collective Computational Properties Like Those of Two-State Neurons," *Proc. of the National Academy of Science*, Vol. 81, pp. 3088-3092.

Luenberger, D. G., 1984, "Linear and Nonlinear Programming," Addison-Wesley, Reading, MA.

Silva, I. N., Arruda, L. V. R., and Amaral, W. C., 1997, "Robust Estimation of Parametric Membership Regions Using Artificial Neural Networks", *International Journal of Systems Science*, Vol. 28, No. 5, pp. 447-455.

SEGMENTATION OF PLANT FROM BACKGROUND USING NEURAL NETWORK APPROACH

DEV S SHRESTHA
Agricultural and Biosystems Engineering
Iowa State University
Ames-IA 50010
USA

BRIAN L STEWARD
Agricultural and Biosystems Engineering
Iowa State University
Ames-IA 50010
USA

ERIC BARTLETT
Electrical and Computer Engineering
Iowa State University
Ames-IA 50010
USA

ABSTRACT
This paper presents an artificial neural network approach to plant-background segmentation of agricultural field images. A truncated ellipsoidal (TE) surface was found to be most effective in defining the green region in RGB space. However, parameters defining the boundaries were subjective to the lighting condition and non-linear. A feed forward neural network was trained for segmenting early stage corn plant from background in various lighting conditions. Segmentation performance of the neural network approach was compared with normalize difference index (NDI) and Bayes classifier approaches. The neural network effectively estimated the parameters of the TE decision surface from first order image statistics and provided better segmentation performance than the other methods.

INTRODUCTION

Image segmentation is the task of sub-dividing an image into constituent parts or objects. (Gonzalez and Woods, 1993). After segmentation, each image pixel is assigned to one of several specific classes. Segmentation is a basic preprocessing task in many image processing applications. In agricultural machine vision applications, segmentation is essential to separate plant from the background, i.e., soil and residue. Two general classes of methods have been used to separate a plant from its background in color images of agricultural crop fields.

One class of methods transforms the pixels data into a one-dimensional index. Segmentation is then accomplished by threshholding this index's histogram. Meyer et al. (1998) segmented the plant and background by thresholding the excess green color index. Andreasen et al. (1997) segmented images by thresholding the median filtered histogram of the green chromaticity coordinates. Pérez et al. (2000) used a normalized density index (NDI) along with morphological operations for plant segmentation. The motivation behind the NDI is that it is similar to a vegetative index commonly used in agricultural remote sensing to estimate the amount of vegetation represented by a pixel (Rees, 1999). NDI is given by the equation:

$$NDI = \frac{G - R}{G + R} \qquad (1)$$

where G is green channel value and R is red channel value.

In field conditions under daylight, variability in lighting occurs (Steward et al., 1999), and hence practical segmentation algorithms must have the ability to adjust to such changes. Thus, a second class of methods treats the segmentation problem as a pattern recognition problem with the RGB values individually treated as class features. A Bayes classifier is then trained to accomplish segmentation by dividing up the color space with a decision surface. The use of such a classifier allows the training to be accomplished for individual images that represent various lighting conditions typically encountered in outdoor conditions. In addition, because of its general quadratic form, the decision surface produced by the Bayes classifier can take on many different shapes based on class statistics.

Tian and Slaughter (1998) used a Bayes classifier to do plant and weed segmentation with robustness to lighting variations. In order to train the classifier, individual pixels were first classified in a partially-supervised fashion through cluster analysis. Then a Bayes classifier was trained so that a decision surface was defined to segment images with lighting conditions which are similar to those represented by the training image. Further refinements of this approach were documented by Steward and Tian (1998) who analyzed several different classification schemes to divide up the color space. Previously, algorithms were developed to specify Bayes classifier decision surfaces given field images taken under various lighting conditions. The next step in this research is to develop segmentation algorithms that can easily adapt to lighting conditions based on image-derived information.

Although the index approach is not typically thought of in terms of a decision surface, it does, however, functionally define a decision surface (Fig. 1). The only flexibility in the index approach is where the decision surface is positioned. While the classifier approach allows more flexibility in defining the decision surface, it does, however, require initial labelling of pixel classes and estimation of the class statistics. These tasks are both computationally demanding and thus not well suited for real-time adaptation. A neural network approach of segmenting plant from background is presented in this paper. This approach has advantages over other approaches because it offers a flexible decision surface that can be adapted for lighting changes.

OBJECTIVE

The objective of this research was to develop a segmentation method that could easily adapt to changes in lighting conditions. To validate this method, its segmentation performance was compared with that of a NDI index approach and a Bayes classifier approach.

NEURAL NETWORK METHOD

This method accomplished segmentation by using an truncated ellipsoidal (TE) surface in RGB color space as a discrimination boundary between vegetation and non-vegetation regions. This surface was originally developed by plotting curves on constant B value planes to separate regions perceived as green from those perceived as non-green. The motivation behind this method

$$\frac{R^2}{D^2} + \frac{(1-G)^2}{(E \times B + F)^2} = 1 \qquad (2)$$

was to roughly separate green and non-green regions in RGB color space and later fine tune this separation using a neural network. After observing this family of curves, it was determined that the decision surface could be functionally represented by a truncated ellipsoidal (TE) surface given by:

where R, G, and B values were the red, green and blue values of a particular pixel and D, E, and F were the parameters describing the shape of the ellipsoid. For a given set of parameters, the left-hand side of Eq. (2) was used to classify pixels as vegetation if ≤ 1 or background if > 1. A TE decision surface for typical values of D, E and F is shown in Fig. 2.

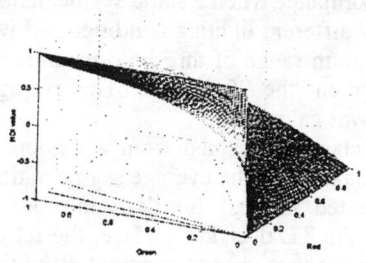 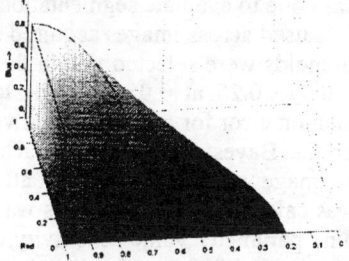

Figure 1. NDI index decision surface for different red and green values with contour lines of NDI values.

Figure 2. TE decision surface values with parameters D = 0.9, E = -0.57 and F = 0.81.

Pixel values are affected by the specific configuration of a color vision system, including factors related to intensity and spectral distribution of illumination, the lens and lens aperture, properties of RGB color filters, the image sensor response and the digitizer (Chang and Reid, 1996). Thus, the optimal values of the decision surface parameters will vary from image to image as lighting condition changes. Lighting will affect relative RGB values and their distribution. Hence, the first order statistics of the image RGB values were used to determine the optimal parameter values.

A fully connected 3-layer feed-forward neural network was developed to estimate the TE surface parameters as a function of the mean and standard deviation of an image's RGB values. Hyperbolic tangent sigmoidal activation functions were used in 2 hidden layers with 5 and 4 nodes respectively, and linear activation functions were used in the output layer. Resilient back propagation was used to find the minimum of error surface (Demuth and Beale, 1998). Initial weights of the neural network were assigned randomly.

SEGMENTATION PERFORMANCE ANALYSIS

A set of 20 images was manually segmented by using visual inspection and the MATLAB AOI selection feature. These images were used for comparing the segmentation performance of the different methods. Segmentation performance was measured using segmentation error, which was defined as the number of pixels segmented into a different class then in the manually segmented image.

The goal of the first set of analyses was to ascertain if the decision surface associated with the three methods produced different performance when optimized for the image being segmented. For the NDI index segmentation approach, the individual NDI threshold for each of the 20 images which minimized segmentation error was found. To evaluate the performance of the Bayes classifier, the multivariate pixel statistics were estimated from each of the 20 manually segmented images. Then the estimated statistics for each image defined the classifier used to segment that image. For the TE decision surface, parameters values were found which minimized the segmentation error of each image. Average segmentation error for the optimized decision surfaces was compared to determine if statistical differences existed between them.

The second analyses evaluated segmentation performance of the three methods using constant parameter values for the entire set of twenty images. This was done to evaluate segmentation performance when a static segmentation scheme is used across images acquired under different lighting conditions. Five NDI thresholds were selected within the optimum range of threshold parameters from 0.05 to 0.25 at 0.05 intervals to segment the 20 images. The average segmentation error for each threshold value was calculated.

For the Bayes classifier approach, a classifier trained from a randomly selected image was used to segment all 20 images, and the average segmentation error was calculated. This process was repeated 5 times. Similarly in order to determine parameter value sensitivity using the TE decision surface, the set of values optimized for a randomly selected image was used to segment all of the 20 images. This process was also repeated 5 times. Average segmentation error for each of the static methods was compared.

For the final analysis, optimum TE parameters were estimated manually for 190 images by a computer-aided method that gave the visually best segmentation of plant. 70% of the images were used to train the neural network described above. The trained NN was applied to the remaining 30% of the 190 images as a validation set. The training and validation process was repeated 10 times by randomly selecting the images with the 7:3 ratio for training and validation. The NN output was compared with manually optimized values. A trained NN was also used to estimate the TE parameters for the 20 manually segmented images. Estimated TE parameters were then used modify the TE surface and then segment that image. Segmentation error was calculated for each segmented image.

RESULTS AND DISCUSSION

Statistically, the TE surface produced the minimum segmentation error of the three methods when each was optimized for individual images. The error for each method was significantly different than the others. Average segmentation error using NDI, Bayes and TE optimized decision surfaces were 6.43, 1.78 and 1.59 % respectively. These results were consistent with qualitative assessment of example image segmentations (Fig. 3). The optimum NDI threshold varied ranged from 0.05 to 0.25 across the 20 images.

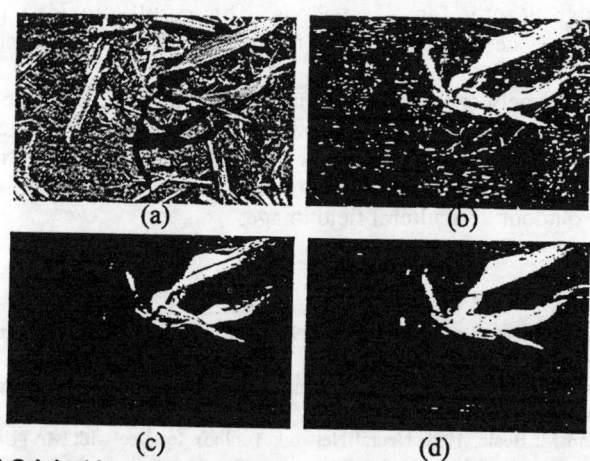

Figure 3. (a) Original image of a corn plant acquired with Sony DCR-TRV900 3CCD Mini DV camcorder. (b) Segmented Image using NDI method. (c) Segmented image using Bayes classifier. (d) Segmented image using neural network method. Optimized decision surfaces were used for these segmentations.

When a constant NDI threshold value was used for segmentation, average segmentation errors using threshold values from 0.05 to 0.25 were 12.3, 7.2, 7.7, 8.0 and 8.0% respectively. At the lowest threshold value, the number of background pixels classified as plant pixels increased. Since there were more background pixels than plant pixels, the average error percentage was higher at the lower threshold value. The 0.10 threshold value resulted in a significantly lower segmentation error than the other threshold values. Segmentation using individually optimized threshold values resulted in significantly lower segmentation error (6.43%) than the best case static threshold value, 0.01.

Similar results were found for the Bayesian and TE methods. The minimum segmentation error obtained across the 5 trials using a static decision surface was 13.9% for the Bayes method and 3% for the TE method. These results confirm that dynamic decision surface parameters are required in order to optimally segment the vegetation irrespective of the method employed.

NN-estimated TE surface parameters were not significantly different than the manually estimated values. When the NN was used to dynamically adjust the TE decision surface parameters, the average segmentation error was 2.07% compared to 1.59% with manually optimized parameters. However, average segmentation error with dynamically adjusted parameters was significantly less than that associated with static decision surfaces.

CONCLUSIONS

The TE decision surface provided the best segmentation performance when each decision surface was optimized for individual images. When a static decision surfaces were to segment a whole set of images, segmentation error increased significantly for all methods. This result provided evidence that decision surfaces need to be adjusted for lighting conditions to achieve optimal segmentation.

An neural network which dynamically estimated TE decision surface parameters for individual images resulted in an average segmentation error which was significantly lower than the error obtained using a static parameters. Thus the neural network approach shows promise as a means of adaptively segmenting outdoor agricultural field images.

REFERENCES

Andreasen, C., M. Rudemo, and S. Sevestre, 1997, "Assessment of weed density at an early stage by use of image processing," Weed Research, Vol 37, pp. 5-18.

Chang Y.C. and J.F. Reid, 1996, "Characterization of a color vision system,"Transactions of the ASAE, Vol 39, pp. 263-273.

Demuth, H. and M. Beale, 1998, Neural Network Toolbox for Use with MATLAB User's Guide Version 3, The MathWorks, Natick, MA, pp. 5.18 -5.20.

Gonzalez, R.C. and R.E. Woods, 1992, Digital Image Processing, Addison-Wesley Publishing Company, Reading, MA.

Meyer, G. E., T. Mehta, M. F. Kocher, D. A. Mortensen, and A. Samal, 1998, "Textural imaging and discriminant analysis for distinguishing weeds for spot spraying,"Transactions of ASAE, Vol. 41, pp. 1189-1197.

Pérez A.J., F. López, J.V. Benlloch, and S. Christensen, 2000, "Colour and shape analysis techniques for weed detection in cereal fields," Computers and Electronics in Agriculture, Vol 25, pp. 197-212.

Rees, G., 1999, The Remote Sensing Data Book, Cambridge University Press, Cambridge, UK,pp. 245-6.

Steward, B. L. and L. F. Tian, 1998, "Real-time weed detection in outdoor field conditions," Proceedings, Precision Agriculture and Biological Quality, G. E. Meyer and J. A DeShazer, eds., SPIE., Bellingham, WA, pp. 266-278.

Steward, B. L., L. F. Tian, and L. Tang, 1999, "Detection of outdoor lighting variability for machine vision-based precision agriculture," ASAE Paper No. 99-3032, ASAE, St. Joseph, MI.

Tian, L. F. and D. C. Slaughter, 1998, "Environmentally adaptive segmentation algorithm for outdoor image segmentation," Computers and Electronics in Agriculture, Vol. 21, pp. 153-168.

HYBRID NEUROCHIP WITH LEARNING ON-CHIP

FRANK STÜPMANN
STEFFEN RODE
NORBERT SCHMIDT
WOLFGANG FREDRICH

Institute of Electronic Appliances
and Circuits, Department of
Electrical Engineering and
Information Technology
University of Rostock, Germany

ABSTRACT

The chip described below is a self-learning classifier. Its evident target is the use in analytical chemistry and lab automation for evaluation of multi-sensor arrays. The decision-making function is a trainable integrated analog neural network structure. The circuit not only contains the reproduction path but also the learning on-chip. Learning patterns for the neural chip are provided in a memory unit. These patterns are automatically presented to the network. A microcontroller core also integrated on the chip supervises the learning process. The process of weight change (i.e. learning) is fully integrated. The information processing speed from the input to the output of the chip is 2 μs in the reproduction process. The number of neurons integrated in the whole chip is 100 in the input layer, 100 in the hidden layer and 20 in the output layer. The back propagation algorithm is implemented in an analog circuit.

INTRODUCTION

The chip is meant to be used for the evaluation of signals of sensor arrays. A sensor array and an external pattern memory work as input for the classifier and an evaluation unit as its output. The sensor array delivers permanently the data to the classifier. These are the patterns to be recognized and be classified by the chip. The classifier consists of the units *switch*, *classification* and *control*. The *switch* unit carries out the switching between learning vectors and input vectors requested from the unit *classification* in correspondence to the learning process or the working process respectively. The *switch* unit takes the data to the controller where they are converted into digital signals and stored in the external pattern memory. In order to provide learning vectors in pattern memory, all required pattern samples are read from the input first. The *control* unit is carried out by the controller core which delivers the control signals for the initialization of the weights, the learning state signal, the learn rate and the refresh signal. It also supervises the learning process. The patterns are presented to the unit *classification's* input via *control* and *switch* in the learning process. During the reproduction process, different patterns from the sensor array are presented to the network. In the working process the classifier recognizes the patterns and sends the decicion to an evaluation unit connected to the output. The net's topology integrated in the chip is the multi-layer perceptron. The learning algorithm used is the back propagation algorithm (BPA), this algorithm is well-known with all its advantages and disadvantages. A lot of users will apply this chip and more than fifty percent of current applications of neural networks use the BPA. The chip uses a SIMD-architecture. In the final version it will have 100 input neurons, 100 hidden neurons and 20 output neurons. The time the data take from input to output in the working process is 2μs. The internal resolution is 6 bit and the resolution of input/output is 10 bit. The chip has analog input/output buffers. The technology used is the 0,6 μm CMOS-technology CUP from Austria Mikrosysteme (AMS). The controller core ARM7TDMI will be integrated on the chip later. It occupies approximately 3 mm^2 of the core area of the chip in that technology.

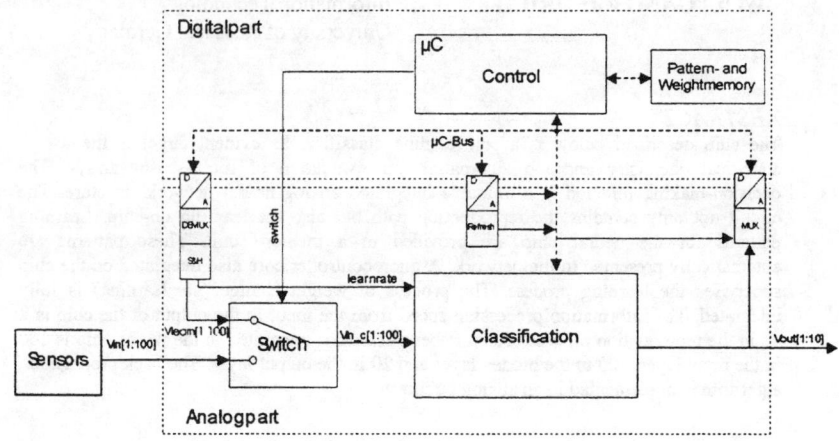

Fig.1 Block diagram of the chip

In Fig.1 the whole neural classifier is shown with its units *control*, *switch* and *classification*. It is necessary to split the project into an analog and a digital part with well-defined interfaces between both parts. The function *switch* is realized by an analog switch.

Switch and *control* manage the input from both the outside of the chip and from the external pattern memory. It is presented to *classification* via these units. The pattern memory is an EEPROM. But the heart of the classifier is the unit *classification*. Due to four reasons it is completely analog designed:
- the transfer rate,
- the size of recources to be used,
- the power consumption of the chip and
- the possibility of the direct use of the parameters of the transistors.

The neurons and synapses are arranged in rows and columns at the core area of the chip. So the synapse in the matrix between two layers is the connection between the neuron in the corresponding row and column.

LEARNING AUTOMATION

The steps to find the optimal learning parameters are automated. The network has to learn the parameters within a certain number of steps starting with a high learn rate. The learn rate is reduced when the network is not converging. The weights are initialized again and than a number of steps will be learned. To observe the success of learning the error of the network will be calculated on the basis of the test-patterns.

The topology of the net is not changeable. Learning does not exclusively take place in the unit *control*. The unit *control* also proceeds the test of the net. It reads the test-patterns from the pattern memory and presents it to the unit *classification*. From the difference between the actual output (of the unit *classification*) and the expected output (stored in the pattern memory) the error is calculated.

FORWARD-NEURON, ERROR-NEURON AND SYNAPSES

When the back propagation algorithm is implemented some calculations are the same for all synapses which are connected to an output neuron. They are computed in the unit *error-neuron*. Following signals are calculated:
- calculation of the error as subtraction of expected and actual value,
- derivation of the activation function is (1 - Vout) • Vout,
- multiplication of the two values to form δ - error signal of the preceding layer
- multiplication of this intermediate result with the learn rate to form a *com*-signal for the calculation of a weight update value of the input synapses of the present neuron.

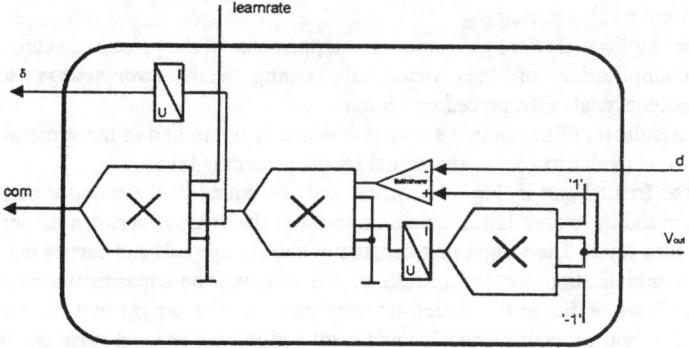

Fig.2 Unit *error-neuron* of the output layer (Stüpmann, 2001)

Error-neurons are only used during the learning process. In the working process the unit *error-neuron* is out of task. The unit *error-neuron* of the hidden layer is not similar to the output layer. In the output layer it calculates the error signal as a difference between the expected (d) and the actual output (Vout). That error signal δ is multiplied in the synapses between hidden and output layer with the weight and forms the back propagated signal $δw_{jk}$. The unit *error-neuron* in the hidden layer adds these backpropagated signals. No error is needed to be calculated for the perceding layer since there is no perceding layer. In this case the input comes from the following synapses and the output is the neuron belonging to it.

The unit *forward-neuron* only plays a role in the reproduction process but not in the back propagation process. In back propagation state the *error-neuron* is used to make calculations as mentioned above. The unit *forward-neuron* itself calculates the sum of its inputs and uses them with an activation function. The *forward-neurons* of the hidden layer do not differ from those of the output layer.

Tasks of the unit *forward-neuron*:
- addition of the input signals
- calculation of the activity

- realization of the non-linear output function of the neuron

The addition of the input signals is done in a current voltage converter using Kirchhoff's current law. The other tasks are carried out by a multiplier and an output current voltage converter.

SYNAPSE WITH WEIGHT-UNIT

The *synapse* multiplies the input values with the stored weight in the reproduction process. In the learning process it calculates the alteration of the weight. During the initialization phase the weights are initialized with different small values. In the learning process the *synapse* in the hidden layer only multiplies the value *com* of the unit *error-neuron* with the input activation (output of preceding neuron). The old value of weight is changed to this new value.

The *synapse* in the output layer calculates the feedback of the error signal δw_{ij} (delta*weight) using the already described operations by multiplication of δ (it comes from *error-neuron* of the following neuron) with the stored weight. This signal is offered to *error-neuron* of the preceding layer. The tasks of the *synapse* are:
- weight value storing
- multiplication of weight value with output signal of the preceding neuron
- multiplication of the *com*-signal coming from *error-neuron* with the outputsignal of the preceding neuron
- calculation of the weight's update δ which is multiplied in the synapse with its weight value to get an error signal for the preceding layer

The red framed part in Fig.3 below can only be found in the *synapse* between the hidden layer and the output layer. It does not exist in the synapse between the input layer and the hidden layer. The weight unit consists of the storage cell and carries out also the update, the initialization and the refresh of the weight. The capacitive storage of the weights are favorite due to the analog learning process. The weight unit has the analog update voltage and the control signals *update*, *ini* and *refresh* as input variables. The only output variable is the analog weight voltage. The tasks of the weight unit are:
- weight value storing
- continuous refreshing of the weight capacity against leakage current
- weight updating with 2 capacities
- initialization of the weights

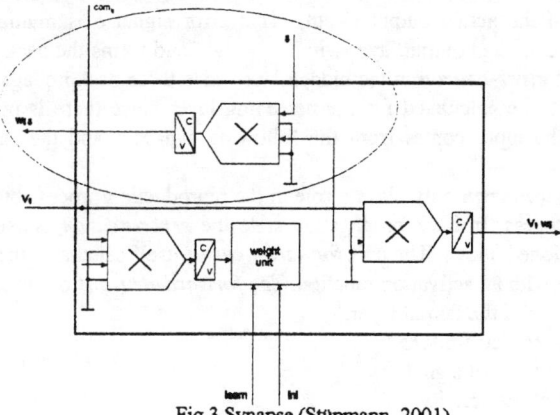

Fig.3 Synapse (Stüpmann, 2001)

REFRESH AND FIRST REALIZATION

The requirement of a long storage time of over 30 seconds leads to the centralization of the refresh. The refresh of one storage cell requires 192μs. Everyone of the 64 staircase steps needs 3μs to execute the comparison between the step voltage and the voltage in the storage cell. This leads to a refresh of all storage devices within the 30 seconds. A bus concept was developed for the centralization of the refreshs and also for the initialization function. It enables the microcontroller and the refresh unit to access every storage cell.

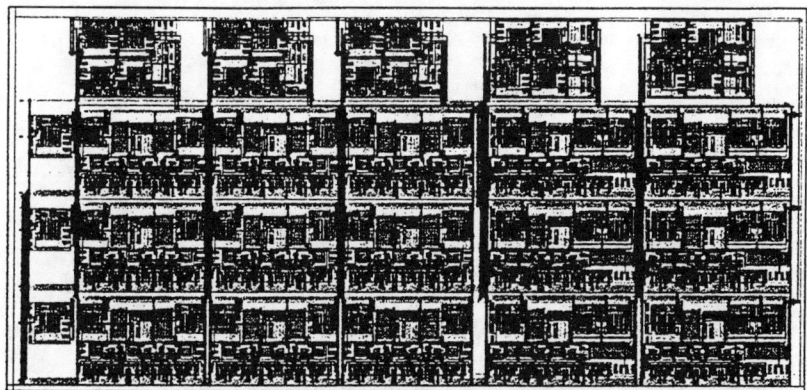

Fig.4 First realization of the core (Rode, 2001)

In Fig. 4 a first realization of the core of the chip with 3 input neurons on the left is shown, 3 hidden neurons on the upper side left and 2 output neurons on the upper side right. The area of this net is 0.8 mm^2. The synapse matrix between input and hidden layer is on left (9 synapses) and the synapse matrix between hidden and output layer is on the right (6 synapses).

APPLICATIONS

The chip can be used to solve a wide variety of problems. One typical field of application is the analysis of signals of sensor arrays, in particular for image processing applications.
Because of its speed the chip is able to solve real-time image recognition problems. (e.g. locates or classifies any kind of images)
The pattern adaptation is limited by the number of available input neurons (100), but clustering and grouping of data sets is one way to solve this problem and make 3D image analysis possible.
Simulations with a standard particle image have shown good results not only for the 2D case, but also for the 3D case. Particle positions are extracted from image series first. In a second step, noisy particles were identified from images.
Due to the results, the chip should be a superior solution for extremely complex and difficult problems (variable content, impurities) in a running system, which demand fastest decisions to increase in efficiency.

CONCLUSIONS

The chip is a self-learning classifier. Since the chip is very fast and stable it could be used for vision processing. At the moment it is tested if the topology of the chip could be used in the 3D-vision processing.

REFERENCES

Rode, S.: Realisierung eines Multilayerperzeptrons in analoger VLSI-Technik, Universität Rostock FB ETIT, Master's thesis, 2001

Stüpmann, F.:Selbständig lernende neuronale Struktur–ein Beitrag zur analogen Hardwarerealisierung neuronaler Netze, Universität Rostock FB ETIT, PhD thesis, 2001

AUTOMATED SEPARATION OF CLODS FROM AGRICULTURAL PRODUCE USING MECHANICAL SEPARATOR WITH INTEGRATED NEURAL NETWORK MACHINE VISION

DEMIAN MORQUIN
Manufacturing Engineering Dept.

MOUNIR BEN GHALIA
Electrical Engineering Dept.

SUBHASH BOSE
Manufacturing Engineering Dept.
The University of Texas – Pan American
Edinburg, Texas, U.S.A.

ABSTRACT

Mechanical harvesting of onion and other agricultural produce from below ground has not been quite successful due to a large number of factors affecting the performance of harvesters. This paper discusses the integration of a neural network-based vision system with mechanical harvesters for separation of onion from soil clod to improve the efficiency of mechanical separator system. The vision system consists of a multi-layer neural network classifier that maps textural features computed from gray-scale images of onions and clods into the right object. Texture features were computed from co-occurrence matrices that specify the spatial relationship between gray-levels in the image. The network was trained using the back-propagation algorithm. Based on this textural feature classification, the effect of changing the network configuration on separation effectiveness was also characterized. It has been demonstrated that the neural network vision system significantly improves the separation effectiveness.

INTRODUCTION

Neural network based vision system has lot of potential for application in agricultural fields that could revolutionize today's technology and bring it to the state-of-the-art technology. This paper focuses on the application of neural networks for mechanical harvesting systems, in particular separation of onions from clods during harvesting. Application of this concept will improve efficiency and reduce damage and wear to trimming process. Techniques used by Harlick et al. (1973) for image classification using textural features is applied to this research. Harlick et al. (1973) have used texture to identify objects or regions of interest in an image. Computable textural features based on gray tone spatial dependencies were used on photomicrograph of sandstone. The textural features are based on statistics, summarizing the relative frequency distribution of one gray tone to another on the image.

Feller et al. (1984) have developed a system that separates onions from soil clods based on the coefficient of restitution property of onion and clod. Since onion has a greater coefficient of restitution, it rebounds to a greater distance than clods when dropped onto a bouncing surface. Efficiency of separation of onions and clods was found to be 93%. However, damaging of onions due to the blades used to cut the roots of the onion is still an important issue. Preventing clods from falling on the blades improve the life of the blades and will increase the speed and quality of the harvesting process (Cuellar et al., 2000). Integration of vision system with mechanical harvesters for separation of onion from clod will significantly improve the efficiency of the system. This paper presents a neural network-based vision system for separating onion from clod and discusses its integration with a mechanical separator.

TEXTURE ANALYSIS

Texture is an important attribute that can be used to distinguish onions from clods using a machine vision system. Jain (1988) defines texture as the spatial repetition of basic patterns. Each pattern is formed by an ensemble of pixels. Figure 1 shows an image of an onion and an image of a clod. In an onion image, there are quasi-regular gray level transitions resulting in a distinct pattern texture that exhibits a visible quasi-regularity (Figure 1(a)). There is also a variation in gray level in a clod image. However, such variation exhibits a random texture, that is no regular pattern texture (Figure 1(b)). Difference in surface texture forms the basis for separation of onion from clod.

While random textures can be characterized by statistical properties such as standard deviation and auto-correlation of gray level in the image, pattern textures require additional measurements to quantitatively characterize the nature and directionality of the pattern. These measurements, called *texture features* (Haralick et al., 1973), are computed from the image of the object (onion or clod).

The Co-Occurrence Matrix

Based on the assumption that a pattern texture information in an image I is contained in the spatial relationship between the gray levels of the image, this spatial relationship can be specified by a matrix P, called the *co-occurrence matrix* (Tamura et al., 1978). The co-occurrence matrix is a square matrix of dimension $N \times N$, where N is the number of different gray levels in the image I. An element P_{ij} of a co-occurrence matrix P of an image I represents the relative frequency with which two pixels, one with gray level i and the other with gray level j, separated by a distance d along a direction θ, occurs on an image. That is, an element P_{ij} is computed as the number of times the gray levels i and j occur in two pixels separated by the distance d (d=1 pixel, or 2 pixels, etc.) along the direction θ ($\theta = 0°$, or $45°$, or $90°$, or $135°$ etc.) in an image I. For one same image, different co-occurrence matrices can be formed for each combination of distance d and directional angle θ. Once the co-occurrence matrices are formed, texture features can be computed.

Selecting Texture Features

The texture feature measures considered in this study are homogeneity, contrast, energy, and variance (Haralick et al., 1973). Some texture features can be perceived by the human eye, such as contrast and homogeneity. However, many other texture features are insensitive to a human eye. It has been reported that a human eye cannot perceive texture differences higher than second order (Tamura et al., 1978). For this application, the selection of the texture features is based on some experimental investigation, the results of which are reported in this paper. The obtained results have shown the usefulness of these features for effectively discriminating between an image of an onion and that of a clod.

Process of Computing Texture Features

Since the texture context of an object is independent of an object's orientation, texture features computed from co-occurrence matrices must produce the same separation results independently of the orientation of the object. To achieve this, four different co-occurrence matrices corresponding to four different values of orientation angle θ (0°, 45°, 90°, and 135°) are established for each image and for a distance $d = 1$. Each feature measure is computed from each co-occurrence matrix. For each textural feature the average value is used as one of the inputs to the neural network separator.

TEXTURE-BASED SEPARATION: EXPERIMENTAL RESULTS
Experimental Data Sets

A set of onions and clods were collected after they have been picked up by an onion harvester in the Rio Grande Valley of Texas. One hundred 64 x 64 array images of onion and one hundred 64 x 64 array images of clod were taken. To obtain these large numbers of images, several images were taken of a same set of onions and clods but with different orientations. Average values of each texture feature were then obtained. The 800 values (400 different texture features for onion and 400 different texture features for clod) constitute the pool of data for the neural network separator. Half of the 800 values were used for training the neural network, and the other half was used for testing.

Separation Algorithm
Neural Network Structure. For this type of problem in which extracting higher-order statistics is of primary importance, a fully-connected multi-layer feed-forward neural network becomes an appropriate candidate for the separation task. Although in a multi-layer neural network structure, several hidden layers might be included between the input layer and the output layer, for this particular application only one hidden layer was used. However, we considered several network topologies differing by the number of input nodes, the combination of textural features used as inputs to the network, and the number of neurons in the hidden layer. In total, we examined 33 different neural network configurations which are summarized in Table 1.

Neural Network Training. The single most significant aspect in the implementation of neural network is mapping of the training set of data to the output layer. Its significance directly influences the learning behavior and performance of the network.

The training consists in (i) presenting the network with textural features measures to its input nodes, (ii) computing the corresponding network output, (iii) computing the error between the desired output and the actual output of the network, and (iv) back-propagating the error and adjusting the network weights.

Separation Results

The objective of the separation process is to recover onions and reject clods. We define the *recovery* variable, R_c, and the *rejection* variable, R_j, for onions and clods, respectively, as the ratios given by

$$R_c = \frac{O_p}{O_p + O_r} = 1 - \frac{O_r}{O_p + O_r}$$

and

$$R_j = \frac{C_r}{C_r + C_p} = 1 - \frac{C_p}{C_r + C_p}$$

where, O_p : onions in the product exits; O_r : onions rejected with clods; C_r : clods rejected; C_p : clods remaining with onions. The *separation effectiveness* (*SE*) is calculated as follows (Brown et al., 1951):

$$SE = 100 \times R_c \times R_j$$

The performance of the neural network as a separator is evaluated by computing its separation effectiveness *SE*. The latter is evaluated for all the 33 neural network configurations. When the neural networks were presented with test data, the highest separation effectiveness was obtained for the following structure, 3-2-1 with inputs En-Co-Ho (Figure 2). The combination of textural features consisting of energy, contrast and homogeneity (En-Co-Ho) used as inputs, provided consistent high separation effectiveness for different network topologies; 3-2-1, 3-3-1, and 3-5-1, with separation effectiveness values of 96%, 90%, and 92%, respectively (Figures 2-4).

If the separation effectiveness (SE) is evaluated based on the number of input textural features, we see from Figures 2-4 that in most cases using two textural features lead to low separation effectiveness, especially for the following combinations: En-Va, En-Ho, and Ho-Va. The simulation results also show that the combination En-Co-Va does not lead in any case to a separation effectiveness greater than 60%.

If the separation effectiveness (SE) is evaluated based on the number of hidden neurons of the neural network, we see that the highest values of separation effectiveness are obtained with two hidden neurons. In addition, the number of hidden neurons has a significant effect on the network performance. From all the cases examined, the network topology (3-2-1 with inputs En-Co-Ho) gives the highest value of separation effectiveness reaching 96%.

PROPOSED INTEGRATION OF VISION SYSTEM WITH THE MECHANICAL HARVESTER

In this project, a mechanical separator consisting of a rotating drum has been designed (Cuellar et el., 2000). Onions are dug out and lifted from the soil by a link type conveyer to a mechanical separator, which has a solid surface to break the clods due to impact and rolls over a screen to a rotating drum. Smaller pieces of clods fall through the screen in the initial stage. The rotating drum has slots along its length and along the surface. The inner surface of the cylinder is glued with a material called poron which prevents damage of onions, but hard enough to break the clods (Cuellar et al., 2000).

A final stage of separation is accomplished using a neural network-based vision system mounted on a cup type conveyer. The cup type conveyer consists of several rows of cups that will be filled with either onion or left over clod. As the onion and clod move along the conveyer, the neural network-based vision system will identify onion from clod and signal to open the cup which holds a clod to fall. The cup with an onion empties at the end to an inclined channel type collector, which sends the onions to the trimming process (Cuellar et al., 2000). The neural networks based vision system eliminates any left over clods from the mechanical separator, preventing damage and wear of the blades in the trimming process.

CONCLUSIONS

It has been demonstrated successfully in this paper that a neural network-based vision system could be used to separate onion from clod based on textural features. The neural network with topology 3-2-1 with inputs energy, contrast, and homogeneity gave the best separation effectiveness of 96%. The proposed integration of the neural network-based vision system with a mechanical separator has been discussed.

REFERENCES

Brown, G. G., Katz, D., Foust, A. S., and Schneidewind, R., 1951, *Unit Operations*, John Wiley and Sons, New York, p.15.

Cuellar, E., Fielder, S., Guidici, E., Bose, S., and Freeman, R., 2000, "Design of an Onion-Clod Separator," *Senior Design Report*, University of Texas-Pan American, Edinburg, Texas.

Feller, R., Nahir, D., and C.G. Coble, 1984, "Separation of soil clods from onions using impact," *Transactions of ASAE*, Vol. 27, No. 2, pp. 353-357.

Haralick, R. M., Shanmugam, K. and Dinstein, I., 1973, "Textural Features for Image Classification," *IEEE Trans. on Systems, Man and Cybernetics*, Vol. 3, pp. 610-621.

Jain, A. K., 1988, *Fundamentals of Image Processing*, Prentice-Hall, New York.

Tamura, H., Mori, S., and Yamawaki, T., 1978, "Texture Features Corresponding to Visual Perception," *IEEE Trans. Systems, Man, and Cybernetics*, Vol. 8, pp. 460-473.

Table 1- Neural network topologies examined in this study

Possible combinations of textural features used as inputs to the network Contrast (Co), Energy (En), Homogeneity (Ho), Variance (Va)	Network Topology
Co-En-Ho-Va	4-5-1
Co-En-Ho-Va	4-3-1
Co-En-Ho-Va	4-2-1
Co-En-Ho, Co-En-Va, Co-Ho-Va, En-Ho-Va	3-5-1
Co-En-Ho, Co-En-Va, Co-Ho-Va, En-Ho-Va	3-3-1
Co-En-Ho, Co-En-Va, Co-Ho-Va, En-Ho-Va	3-2-1
Co-En, Co-Ho, Co-Va, En-Ho, En-Va, Ho-Va	2-5-1
Co-En, Co-Ho, Co-Va, En-Ho, En-Va, Ho-Va	2-3-1
Co-En, Co-Ho, Co-Va, En-Ho, En-Va, Ho-Va	2-2-1

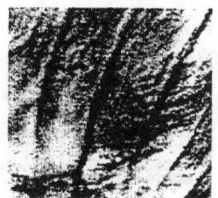
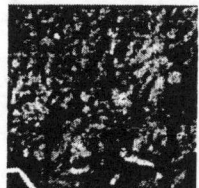

(a) Onion Texture   (b) Clod Texture

Figure 1- Visual perception of the difference in surface texture between

(a) an onion and (b) a clod.

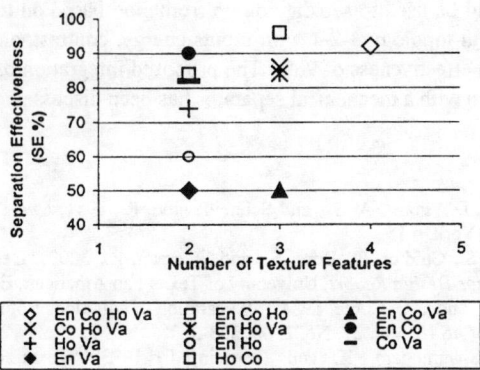

Figure 2- Separation effectiveness with 2 hidden neurons.

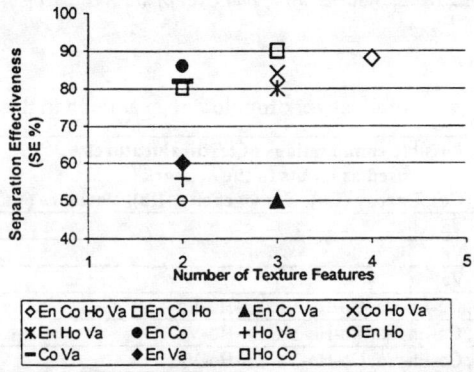

Figure 3- Separation effectiveness with 3 hidden neurons.

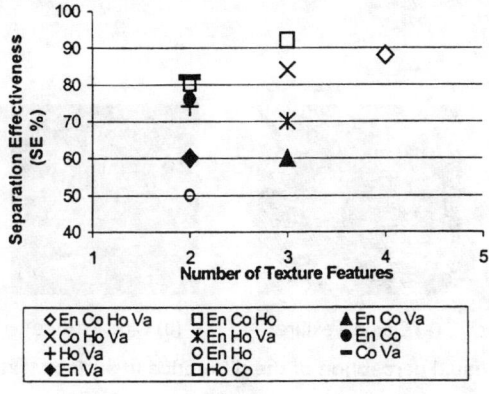

Figure 4- Separation effectiveness with 5 hidden neurons.

NEURAL NETWORK BASED SPEED BOAT EMULATOR

I.N. TANSEL
Mechanical Engineering Dept.
Florida International University,
Center for Engineering & Applied Science
10555 West Flagler Street, Miami, FL 33174

R. SELTZER, W. YUEN
Naval Air Warfare Center
Training Systems Division
12350 Research Parkway
Orlando, FL 32826-3224

ABSTRACT

Complex hull designs are used for speedboats to allow them to come to plane to reduce resistance. Hydrodynamic model of the system is complicated and requires extensive tests to obtain all the necessary parameters of the analytical model. Numerical models are computationally intensive and cannot be executed in real time if the hull shape is closely represented. In this study, use of basic physical model of motion and four neural networks are proposed to simulate response of a speedboat to operator inputs. Neural networks represent the boat's acceleration, rise, rotation and roll. The neural networks of the proposed system can be trained with experimental data one by one. The system is fast enough to operate in the real time on the microcomputer platforms. Each network represented the characteristics of the test functions with less than 5% error (respect to the output range) in more than 95% of the tested cases.

INTRODUCTION

Speedboats are designed to reach high speeds without encountering high resistance forces after the hull speed. These boats come to plane between 6 and 15 knots and operate very efficiently at this state. The nose of the boat rises depending on the speed, and boat turns depending on the direction of rudder or water jet. During the maneuvers, their V or semi V shaped hulls inclines to the sides (rolls) and boat skids sidewise depending on the speed and turning radius. A realistic numerical simulation of this complex system by considering the intricate hull shape is very computer intensive. An analytical model requires many costly experiments and necessary assumptions to compromise the accuracy. Since the boats are cheaper and easier to handle then the airplanes, boat simulators were not needed in the past. However, with the change of the characteristics of the military equipment and use of sophisticated small crafts such as Mark V in military operations, the need for the boat simulators will increase. In this paper, use of multiple neural networks and basic equations of motion is proposed to simulate the characteristics of speedboats from experimental data.

Steering of ships has been carefully studied (Tzeng, 1998; Tiano and Blanke, 1997; Roberts et al., 1997; Derong et al., 1999; Shen et al., 2000, Mo et al., 1999; Nejim, 2000; Enab, 1996; Zhang et al., 1996; Zirilli et al., 2000) and neural networks have been widely used (Koushan and Mesbahi, 1998) with that purpose. Studies are very limited on the modeling of planning hulls (Hicks and et al., 1993) and real time simulation of a boat response on a motion base. Compared to analytical or numerical approaches, it is very easy to observe the characteristics of an existing boat experimentally and to emulate its responses to the operator input. In this paper, use of four neural networks is proposed to emulate a speedboat. Basic motion equations were used to calculate the position and interim variables to obtain the necessary information for the neural networks. Although, the concept is mainly developed for military boats since their characteristics are classified, a small speedboat with 50 HP engine is considered to create and test the accuracy of the neural networks.

PROPOSED SYSTEM

The proposed boat simulator uses four neural networks as presented in Figure 1. One neural network estimates the boat acceleration according to current speed and the position of the throttle, the second network estimates the derivative of turning rate (angular acceleration) according to current velocity and rudder (or jet) direction change. The velocity and position of the boat are calculated by using the basic motion equations. The third neural network estimates the raise of the nose of the boat according to velocity and rotation radius. The fourth neural network estimates the roll using the same inputs.

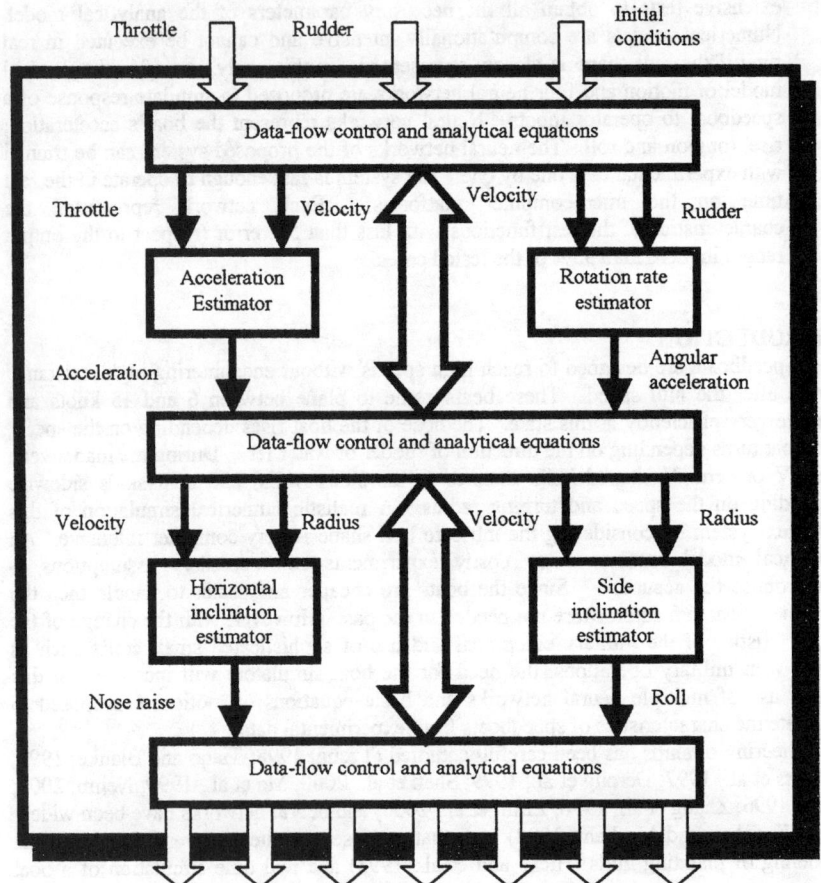

Position, Velocity, acceleration, direction, lateral acceleration, inclination, lateral inclination

Figure 1. The proposed model of a speedboat with neural networks.

TRAINING OF NEURAL NETWORKS

Neural networks were trained by using four simple analytical functions to evaluate the feasibility of the system (Rumelhart et al., 1986). Since we could not find simplified functions in literature for the task, we prepared them to represent the general characteristics of the system. Each function had two inputs and one output. A reasonable operating range was selected for each input. These simple functions were used to simulate the acceleration, angular acceleration, nose raise and roll characteristics of a small boat with 17 ft length and 50 HP engine. No real measurements were done during the work. When the proposed system is used in real life, these characteristic plots will be obtained experimentally in the acceptable range. Some of the regions of the plots could be too dangerous to test or may cause serious accidents. These ranges should be estimated.

RESULTS AND DISCUSSION

Since all the neural networks have two inputs and one output, the characteristics of the boat are represented with four surfaces. Backpropagation type neural networks were trained by using the data obtained from the simple functions. After the training neural networks were tested on the data they never experienced before. In these tests, the midpoints of the inputs, which were used during the training, were used. For example, if the training data were generated at the 1, 2, 3 m/sec velocities, the test data were generated at the 1.5, 2.5, 3.5 m/sec velocities. In more than 95% of the studied cases, backpropagation type neural networks represented the characteristics of the system with less than 5% error respect to the range of the output. For the acceleration, the test data neural network estimations are presented in Figures 2a and 2b, respectively. The theoretical test data and neural network estimations for the angular acceleration, nose raises and roll are presented in Figures 3, 4 and 5, respectively.

The validity of the models can be easily visualized by considering the characteristics of a boat. For each throttle position, the boat reaches a constant speed and acceleration becomes zero. As long as boat moves at that speed, the acceleration will not change unless the environment changes. If the throttle position is changed, the acceleration of the boat increases or decreases depending on if the throttle is opened or closed.

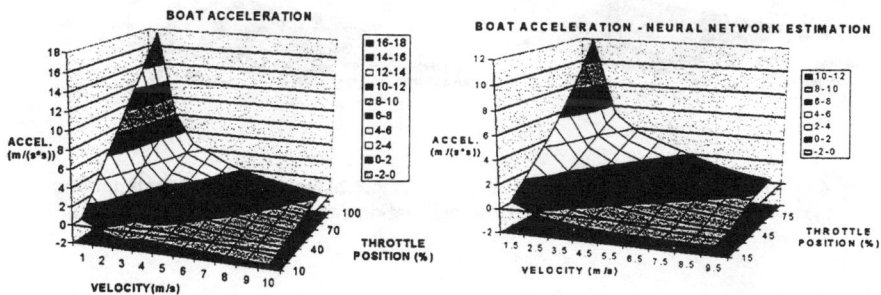

Fig.2.a) Test cases   Fig.2.b) Neural networks estimations
Figure 2. Evaluation of the acceleration estimation capability of neural networks.

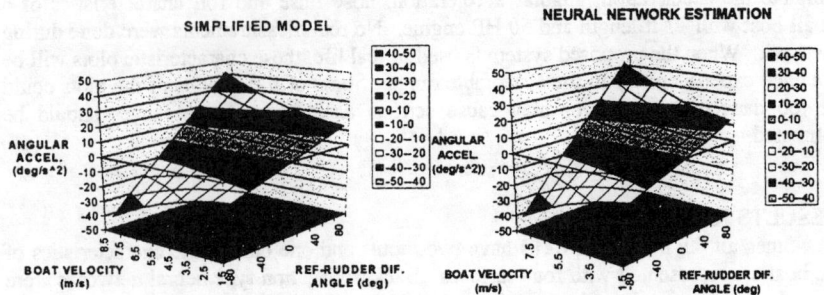

Fig.3.a) Test cases   Fig.3.b) Neural networks estimations
Figure 3. Angular acceleration estimation accuracy of neural networks.

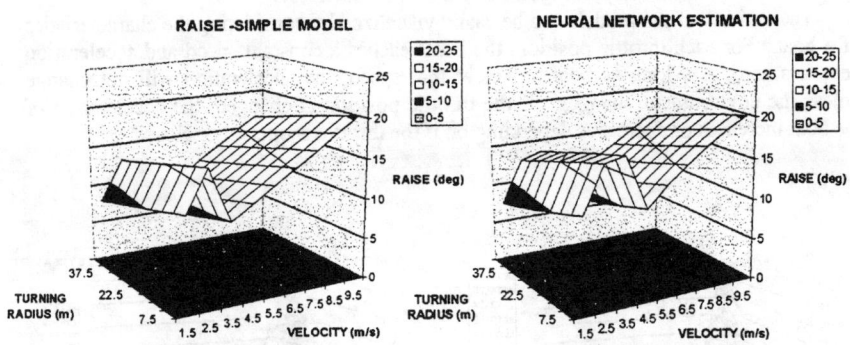

Fig.4.a) Test cases   Fig.4.b) Neural network estimations
Figure 4. Evaluation of the nose raise estimation capability of neural networks.

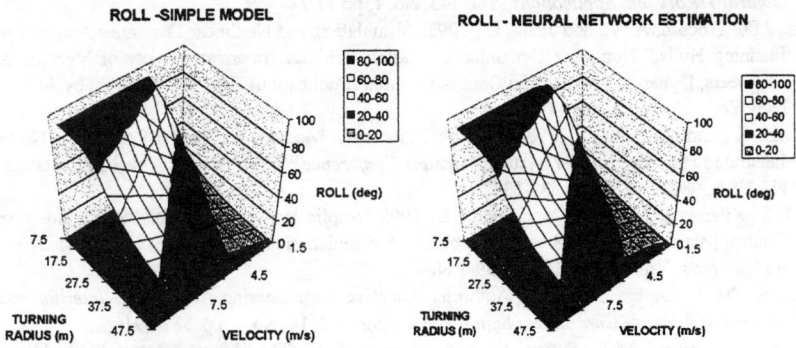

Fig.5.a) Test cases   Fig.5.b) Neural network estimations
Figure 5. Evaluation of the roll angle estimation capability of neural networks.

CONCLUSIONS

A multi-neural network system is proposed to emulate speedboat response to operator inputs. The accuracy of the neural networks was better than 5% of the output range in more than the 95% of the test cases. Such system should provide reasonably accurate guidance to the operator to emulate the response of small boats.

The neural networks do not need the training data to be obtained by following a perfectly distributed grid. Proposed system can be directly trained by using the collected experimental data. In addition, the neural networks may be used to generate a grid, which can be easily modified by experts to improve the accuracy of the system or to impose the response of the system to possible inputs, which cannot be safely tested in the field.

Boat and land vehicle simulators will find wide interest in future because of the increasing safety concerns and declining hardware prices. Also, small but costly vehicles with highly sophisticated control systems will be widely used in military operations since terrorism has become an important threat while the probability of large scale wars have been diminishing. Neural networks will allow development of emulators for these vehicles from the experimental data. The development time and cost will be much lower than the conventional approaches which use analytical or numerical models.

ACKNOWLEDGMENT

The U.S. NAVY – ASEE Summer Faculty Research Program supported this study. Authors appreciate the support and Mr. Bob Cannaday and Mr. Ken Geis' help during the preparation of the paper.

REFERENCES

Derong, L., Patino, P.H., and Daniel, H., 1999, "Adaptive Critic Designs for Self-Learning Ship Steering Control," *IEEE International Symposium on Intelligent Control – Proceedings*, pp. 46-50.

Enab, Y.M., 1996, "Intelligent Controller Design for the Ship Steering Problem," *IEE Proceedings: Control Theory and Applications*, Vol. 143, No. 1, pp 17-24.

Hicks, J.D., Troesch, A.W., and Jiang, C., 1993, "Simulation and Nonlinear Dynamics Analysis of Planning Hulls," Nonlinear Dynamics of Marine Vehicles American Society of Mechanical Engineers, Dynamic Systems and Control Division (Publication) DSC Vol. 51, Publ by ASME, New York, NY, USA, pp. 41-55.

Koushan, K., and Mesbahi, E., 1998, "Empirical Prediction Methods for Rudder Forces of a Novel Integrated Propeller-Rudder System," *Oceans Conference Record (IEEE)*, Vol. 1., Piscataway, NJ, USA, 98CB36259, pp. 532-537.

Mo, Y.S., Kitamura, T., Zhu, R., and Zhang, L., 1999 "Application of a Combined Fuzzy Adaptive Control Method to a Ship Steering Problem - A Simulation Research," *Advances in Modeling and Analysis*, Vol. 53, No. 1, pp. 1.59-1.68.

Nejim, S., 2000, "Design of Limited Authority Adaptive Ship Steering Autopilots," *International Journal of Adaptive Control and Signal Processing*, Vol. 14, No. 4, pp. 381-391.

Roberts, G.N., Sharif, M.T., Sutton, R., and Agarwal, A., 1997, "Robust Control Methodology Applied to the Design of a Combined Steering/Stabilizer System for Warships," *IEE Proceedings: Control Theory and Applications*, Vol. 144, No. 2, pp. 128-136.

Rumelhart, D.E., Hilton, G. and Williams, R.J., 1986, "Learning Internal Representations by Error Propagation," Parallel Distributed Processing: Explorations in the Microstructure of Cognition, Vol. 1, Ed. E. Rumelhart, J.L. McClelland, MIT Press.

Shen, D., Ma, X., Sun, L., and Luo, R., 2000, "Prediction on Ship Maneuvering Performance in Wave," *Journal of Ship Mechanics*, Vol. 4, No. 4, pp. 15-27.

Tiano, A., and Blanke, M., 1997, "Multivariable Identification of Ship Steering and Roll Motions," *Transactions of the Institute of Measurement and Control*, Vol. 19, No. 2, pp. 63-77.

Tzeng, C.Y., 1998, "Analysis of the Pivot Point for a Turning Ship," *Journal of Marine Science and Technology*, Vol. 6, No. 1, pp. 39-44.

Zirilli, A., Roberts, G.N., Tiano, A., and Sutton, R., Dec 2000, "Adaptive Steering of a Containership Based on Neural Networks," *International Journal of Adaptive Control and Signal Processing*, Vol. 14, No. 8, pp. 849-873.

Vukic, Z., Pavlekovic, D., and Ozbolt, H., 1998, "Rudder Servo-System Fault Diagnosis Using Neural Network Fault Modeling," *Oceans Conference Record (IEEE)*, Vol. 1, Piscataway, NJ, USA, 98CB36259, pp 538-543.

Zhang, Y., Hearn, G.E., and Sen, P., 1996, "Neural Network Approach to Ship Track-Keeping Control," *IEEE Journal of Oceanic Engineering*, Vol. 21, No. 4, pp. 513-527.

SYNCHRONIZATION OF HYPERCHAOS IN SYSTEMS WITH DELAY

YECHIEL CRISPIN
Department of Aerospace Engineering
Embry-Riddle University
Daytona Beach, FL 32114
crispinj@db.erau.edu

ABSTRACT

A generalized method for the synchronization of chaos in systems governed by ordinary differential equations and hyperchaos in systems governed by delay-differential equations is developed. The method is applicable to the field of secure communications. The parameters of the driver and the response systems vary as a function of time and carry the encrypted messages. The method is based on an adaptive parameter identification algorithm. In order to validate the method, the synchronization algorithm is applied to a cryptosystem based on two hyperchaotic systems described by nonlinear first order delay differential equations of the Mackey-Glass type.

INTRODUCTION

The possibility of synchronizing two chaotic physical systems has attracted considerable attention in recent years [Boccaletti, 1997]. A major motivation is the potential of applying the synchronization methods to the field of secure communications by chaotic scrambling of messages [Goedgebuer, 1998]. Other important applications are in the field of nonstationary time series analysis and system identification [Parlitz, 1996]. So far, most of the attention, however, has been directed towards the study of stationary chaotic systems, that is systems with constant independent parameters. To date, the possibility of using systems with nonstationary parameters for secure communications has received little attention.

This paper presents a method for synchronizing both chaotic and hyperchaotic systems with nonstationary parameters. The variable parameters of the response system are controlled such as to follow the variable parameters of the driving system. A system of differential equations governing the evolution of the controlled nonstationary parameters are derived systematically using a hydrodynamic analogy. The synchronization error between the two chaotic systems is viewed as a fluid property advected by a marker particle evolving along a trajectory in the vector field of the response system. The controlled variable parameters of the response system are varied continuously such as to minimize this error. The method is applied to hyperchaotic systems described by delay-differential equations of the Mackey-Glass type [Losson, 1993] with variable parameters.

SYNCHRONIZATION BY HYDRODYNAMIC ANALOGY

Consider two similar dynamical systems described by ordinary delay-differential equations:

(1) $\quad \frac{dx}{dt} = f(x(t-T), p(t))$

(2) $\quad \frac{dy}{dt} = f(y(t-T), q(t))$

where t is time, T is the time delay, $x \in \mathbf{R}^n$ are the state variables of the driver, and $y \in \mathbf{R}^n$ are the state variables of the response system. $p(t) \in \mathbf{R}^k$ and $q(t) \in \mathbf{R}^k$ are independent time varying parameters of the respective systems and $f : \mathbf{R}^n \times \mathbf{R}^k \mapsto \mathbf{R}^n$ is a nonlinear vector function of the state variables. It is assumed that the initial functions of the state variables $x(t) = \varphi_1(t)$ and $y(t) = \varphi_2(t)$ for $t \in (-T, 0)$ are not necessarily the same. Similarly the initial values of the parameters $p(0) = p_0$ and $q(0) = q_0$ of the driver and response systems are different. Since chaotic systems are sensitive to initial conditions, the driver and response systems will not synchronize, unless the response system is controlled and forced to synchronize with the driver system using a transmitted signal. For instance, a single scalar signal $s(t)$, which is a function of the state $x(t)$, can be transmitted by the driver and used to enslave the response system [Tamasevisius, 1997], [Peng, 1996]:

(3) $\quad s(t) = h(x(t))$

The purpose of this paper is to propose a generalized method of synchronization of chaos and hyperchaos when the parameters $p(t)$ of the driver system vary as a function of time. In the context of secure communications, it would be possible to encode a message in the parameters of the driver system rather then in a state variable. Since the message is encoded in a time variable parameter, synchronization of the state variables of an eavesdropping response system with the driver will not ensure recovery of the message, unless the variable parameter is identified. Therefore, the proposed method improves the security of communications by chaos or hyperchaos masking. Also, synchronization can be achieved even when the parameters p and q are initially substantially different. This is accomplished by controlling the response system y such that the parameters $p(t)$ of the driver system x are eventually identified.

In order to achieve synchronization, the differential equations governing the evolution of the response parameters $q(t)$ need to be derived systematically. A hydrodynamic analogy based on the Lagrangian approach for describing the evolution of a scalar quantity convected in a flow field [Lamb, 1995, Milne-Thomson, 1968] is used for this purpose. According to this approach, the equations of motion of two marker particles advected in the fluid flows described by the vector fields $w(x, p) = f(x, p)$ and $w(y, q) = f(y, q)$, are given by Eq.(1-2), where the right hand sides are to be interpreted as the local velocity vectors $w(x, p)$ and $w(y, q)$ at any given point $x \in \mathbf{R}^n$ and $y \in \mathbf{R}^n$ of the two flow fields, respectively, i.e,

(4) $$\frac{dx}{dt} = w(x,p,T) = f(x(t-T),p)$$

$$\frac{dy}{dt} = w(y,q,T) = f(y(t-T),q)$$

Unlike in hydrodynamics, where the vector fields $w(x,p), w(y,q) \in \mathbf{R}^n$ and phase spaces $x, y \in \mathbf{R}^n$ have a dimension $n \leq 3$, here the analogy is extended to vector fields of higher dimensions. Consider the time variation of a scalar property $J(x,y)$ of the flow along a trajectory of the response system as it evolves in the phase space $y \in \mathbf{R}^n$. The rate of change of this scalar property is due to two contributions: a local contribution due to its time variation plus a contribution due to the rate of change of the property due to its advection along the trajectory in the phase space. The total rate of change is then given by the substantial derivative of the scalar property, the derivative following the flow:

(5) $$\frac{DJ}{Dt} = \frac{\partial J}{\partial t} + f(y(t-T),q)\nabla_y J$$

where ∇_y is the gradient with respect to y and the product $f(y,q)\nabla_y J$ is a scalar or dot product. Consider a scalar property J based on the distance $y - x$.

(6) $$J = \tfrac{1}{2}|y-x|^2 = \tfrac{1}{2}\sum_{i=1}^{n}(y_i - x_i)^2 = \tfrac{1}{2}\sum_{i=1}^{n}e_i^2$$

where $e_i = y_i - x_i$ are components of the error vector function $e = y - x$. The local component of the derivative of J with respect to time is given by $\frac{\partial J}{\partial t} = \sum_{i=1}^{n} e_i \frac{de_i}{dt}$, whereas the components of the gradient $\nabla_y J$ are $\frac{\partial J}{\partial y_i} = e_i$. Using these results together with Eqs.(4), (5), the substantial derivative of J is written as:

(7) $$\frac{DJ}{Dt} = \sum_{i=1}^{n} e_i [\frac{de_i}{dt} + f_i(y(t-T),q)]$$

Since $e = y - x$ and $de/dt = dy/dt - dx/dt = f(y,q) - f(x,p)$, Eq.(7) reduces to:

(8) $$\frac{DJ}{Dt} = \sum_{i=1}^{n} e_i[2f_i(y(t-T),q) - f_i(x(t-T),p)]$$

Eq.(8) defines the rate of change of the positive scalar property J in terms of the state variables x and y of the driver and response systems, the independent driver parameters p and the controllable parameters q of the response system. The question now is how should the control vector q be varied such as to continuously minimize J?

The substantial derivative should be continuously decreased in order to achieve synchronization. A possible control law is to vary the control vector q such as to decrease the derivative DJ/Dt. This can be achieved by varying the control q in a direction opposite to the gradient $\nabla_q(DJ/Dt)$:

$$(9) \quad \frac{dq}{dt} = -G' \nabla_q \{\sum_{i=1}^{n} e_i [2f_i(y(t-T),q) - f_i(x(t-T),p)]\}$$

where ∇_q is the gradient with respect to q. The matrix G' is a (kxk) matrix of control gains. Since $f_i(x,p)$ does not depend on q, it follows that $\nabla_q f_i(x(t-T),p) = 0$. Eq.(9) then reduces to:

$$(10) \quad \frac{dq}{dt} = -2G' \nabla_q [\sum_{i=1}^{n} e_i f_i(y(t-T),q)]$$

SYNCHRONIZING HYPERCHAOTIC SYSTEMS WITH DELAY

As an example, consider the class of dynamical systems described by nonstationary first order delay-differential equations of the form [Kolmanovskii, 1992], [Losson, 1993], [Mansour, 1998], [Hegger, 1998] and [Just, 1998]:

$$(11) \quad \frac{dx}{dt} = f(x,p) = p_1(t) F[x(t-T)] - p_2(t) x$$

$$(12) \quad \frac{dy}{dt} = f(y,q) = q_1(t) F[y(t-T)] - q_2(t) y$$

where the driver system is governed by Eq.(11) and the response system by Eq.(12). The transmitted signal is the scalar state variable $s(t) = h(x) = x(t)$. The dynamics of the identification parameters are governed by Eq.(10), which in the case of the systems of Eqs.(11,12), can be written as:

$$(13) \quad \frac{dq_1}{dt} = -G_{11}(y-x) F[s(t-T)] - G_{12}(x-y) y$$

$$(14) \quad \frac{dq_2}{dt} = -G_{21}(y-x) F[s(t-T)] - G_{22}(x-y) y$$

As a numerical example, consider a cryptosystem [Goedgebuer, 1998] based on the Mackey-Glass equation [Losson, 1993, Mansour, 1998], where the form of the function appearing on the right hand side of Eq.(11) is given by:

$$(15) \quad F[x(t-T)] = \frac{x(t-T)}{1+x^m(t-T)}$$

where m is a constant. A value $m = 11$ and a delay $T = 5$ were used in this example. The case with one identification parameter $q_1(t)$ is treated in the numerical simulation. In this case it is assumed that the second identification parameter in Eq.(12) is a known constant, that is, $p_2 = q_2 = 1$. Then Eqs.(13-14) become:

(16) $\dfrac{dq_1}{dt} = -G_{11}(y-x)F[x(t-T)]$

(17) $\dfrac{dq_2}{dt} = 0$

In the simulations, a value of $G_{11} = 200$ was used. It was found that the systems synchronize for a wide range of the gain G_{11}. With the assumption $p_2 = q_2 = 1$, the driver and response systems, Eqs.(11-12) reduce to:

(18) $\dfrac{dx}{dt} = p_1(t)F[x(t-T)] - x$

(19) $\dfrac{dy}{dt} = q_1(t)F[y(t-T)] - y$

In order to solve the system of delay-differential equations described by Eqs.(16-19), the following initial functions were chosen, with the delay $T = 5$:

(20) $x(t) = \varphi_1(t) = 0.1$, $y(t) = \varphi_2(t) = 0.2$, $q_1(t) = \varphi_3(t) = 4$

in the time interval $t \in [-T, 0]$. The signal is chosen as a sinusoidal wave

(21) $p_1(t) = \lambda_1 + d\lambda \sin(\dfrac{2\pi t}{T_p})$

with $\lambda_1 = 2$, $d\lambda = 0.2$ and $T_p = 20$. Note that the initial value of the identification parameter $q_1(t) = 2\lambda_1 = 4$ is greater than the average value of the signal to be identified $<p_1(t)> = \lambda_1 = 2$. The systems synchronize despite this large difference in the values of the initial functions. The results show that after synchronization has been achieved, the difference $|y(t) - x(t)|$ is about two orders of magnitude less than the state variables, which in this case is of order one. Also, the transient period before synchronization is very short. The identified parameter $q_1(t)$ is shown in Fig.(1). Ideally q_1-λ_1 should be a sinusoidal wave q_1-$\lambda_1 = \delta\lambda \sin(2\pi t/T_p)$ as described by Eq.(21) for the hidden signal, however there is some noise superimposed because of the hyperchaos in the delay system.

In conclusion, a generalized method for the synchronization of chaos and hyperchaos based on a physical argument of a hydrodynamic analogy for the transport of fluid properties along streamlines in a flow field has been developed. It has been successfully applied to chaotic systems described by ordinary differential equations as well as to hyperchaotic systems described by delay-differential equations. Some implications of using the method in the field of secure communications where the transmitted information is masked by chaos or hyperchaos have also been demonstrated.

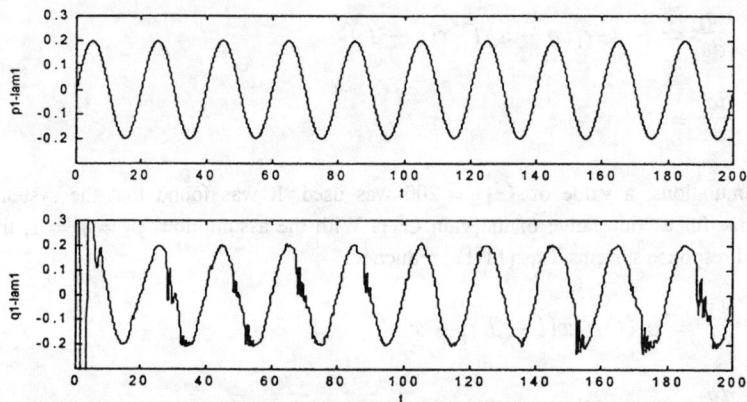

Fig.1: The transmitted signal $p_1-\lambda_1$ and the identified signal $q_1-\lambda_1$. Ideally $q_1-\lambda_1$ should be a sinusoidal wave $q_1-\lambda_1=\delta\lambda \sin(2\pi t/T_p)$. However there is some noise superimposed because of the hyperchaos in the delay system.

REFERENCES

S. Boccaletti, A. Farini and F.T. Arecchi, "Adaptive Synchronization of Chaos for Secure Communication", Physical Review E, 55, 5, 4979-4981, 1997.

J-P. Goedgebuer, L. Larger and H. Porte, "Optical Cryptosystem Based on Synchronization of Hyperchaos Generated by a Delayed Feedback Tunable Laser Diode", Physical Review Letters, 80, 10, pp.2249-2252, March 1998.

R. Hegger, M.J. Bunner H. Kantz and A. Giaquinta, "Identifying and Modeling Delay Feedback Systems", Physical Review Letters, 81, 3, pp.558-561, July 1998.

W. Just, D. Reckwerth, J. Mockel, E. Reibold and H. Benner, "Delayed Feedback Control of Periodic Orbits in Autonomous Systems", Physical Review Letters, 81, 3, pp.562-565, July 1998.

V. Kolmanovskii and A. Myshkis, Applied Theory of Functional Differential Equations, Mathematics and its Applications, Vol. 85, Kluwer Academic Publishers, 1992.

H. Lamb, Hydrodynamics, Sixth Edition, Cambridge University Press, NY, 1995.

J. Losson, M.C. Mackey and A. Longtin, "Solution Multistability in First-order Nonlinear Differential Delay Equations", Chaos, 3, 2, pp.167-176, 1993.

B. Mansour and A. Longtin, "Chaos Control in Multistable Delay Differential Equations and Their Singular Limit Maps", Physical Review E, 58, 1, pp.410-422, July 1998.

B. Mansour and A. Longtin, "Power Spectra and Dynamical Invariants for Delay Differential and Difference Equations", Physica D 113, 1, pp. 1-25, February 1998.

L.M. Milne-Thomson, Theoretical Hydrodynamics, MacMillan, NY, 1968.

U. Parlitz, "Estimating Model Parameters from Time Series by Autosynchronization", Physical Review Letters, 76, 8, pp.1232-1235, 1996.

J.H. Peng, E.J. Ding, M. Ding and W. Yang, "Synchronizing Hyperchaos with a Scalar Transmitted Signal", Physical Review Letters, 76, 6, 904-907, 1996.

A. Tamasevicius and A. Cenys, "Synchronizing Hyperchaos with a single Variable", Physical Review E, 55,1, 297-299, 1997.

VIRTUAL SENSORS FOR PREVENTIVE HEALTH MAINTENANCE OF AIRCRAFT ENGINES

ANNA L. BUCZAK[*A], ONDER ULUYOL[B], EMMANUEL NWADIOGBU[C]

[a] *Philips Research, 345 Scarborough Rd., Briarcliff Manor, NY 10510-2099*
[b] *HTC, Honeywell, Inc., 101 Columbia Road, Morristown, NJ 07962-1021*
[c] *Engines & Systems, Honeywell Inc., 111 S. 34th Street; Phoenix, AZ 85034*

ABSTRACT:

This paper describes a Virtual Sensor developed for an aircraft engine application. The Virtual Sensor mimics the operation of a physical sensor that is not mounted on the engine for physical and/or economical reasons. It infers the value of a physical quantity from the readings of other sensors that provide relevant information. The Virtual Sensor was implemented as a feedforward neural network and trained with Levenberg-Marquardt algorithm. The measurement accuracy obtained on actual data is very high even though the data set contains a large amount of noise.

1. INTRODUCTION

Virtual sensors (sometimes also called *soft sensors*) are devices that measure a physical quantity not through a direct physical sensor but by inferring its value based on the readings of other sensors that provide relevant information. Virtual sensors are very useful in applications in which the number of physical sensors is limited due to such factors as the weight, reliability and the cost of the overall system, and its complexity. When developing a virtual sensor, the selection of appropriate base sensors, the calibration of the virtual sensor, and the choice of method for mapping the available information into the virtual sensor readings are issues to be addressed.

In this paper, neural networks are chosen for the mapping method for their proven ability to learn and represent nonlinear relationships, fault and noise tolerant characteristics. Since the mapping is from many-to-one, a multi-layered feedforward neural network is used. Calibration is not an issue since we have actual physical data collected for the virtual sensor being developed.

2. SYSTEM DESCRIPTION

The virtual sensors developed are part of a full system for validation and recovery of sensors in aircraft engines. The full system, with special emphasis on sensor validation, was described in detail in [Ulu-01]. In the overall system (see Figure 1) data coming from sensors is processed by the Covariance Analysis and Noise Analysis modules. The Covariance Analysis categorizes sensors into groups. They constitute a set of inputs/outputs of a neural network for sensor validation. The Noise Analysis module

---

[*] Research done while with HTC, Honeywell, Inc., Morristown, NJ

performs computation of statistical measures of noise in data such as mean, standard deviation, Z-score, signal to noise ratio, and percent range deviation. The noise measures calculated are used for simulation of faulty sensor signals.

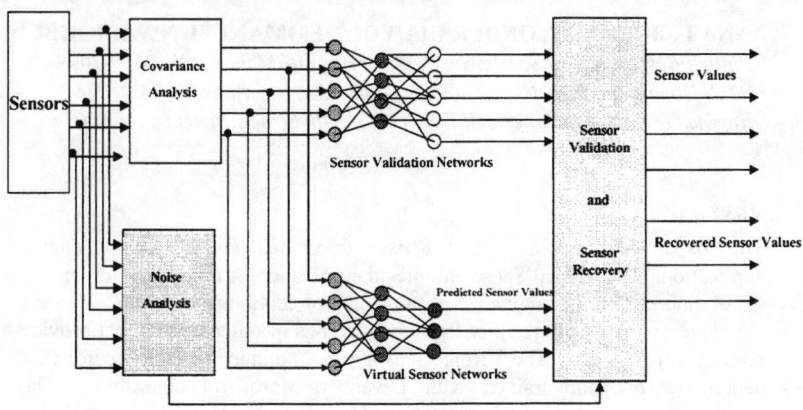

Figure 1. System Architecture.

Data from sensors that are classified by Covariance Analysis as belonging to one group are used as inputs/outputs to auto-associative and/or hetero-associative Sensor Validation neural networks. These networks together with the residual analysis developed are used for validation and recovery of engine sensors. Sensor recovery predicts the values of sensors that were detected as faulty by the sensor validation process.

In addition to sensor validation and sensor recovery, Virtual Sensor networks are an intricate part of the overall predictive health maintenance system for the engine. The Virtual Sensor networks are used to predict values of those sensors that are present in development but absent in production. Virtual sensor networks are also used to predict, in real time, values of parameters, such as turbine inlet temperatures, whose measurement location and environment are extremely harsh and cannot physically support reliable and durable sensors.

3. CROSS-COVARIANCE ANALYSIS

There are approximately 200 to 400 sensors, engine pressure taps and temperature monitors, used during engine development and only about 15 will actually be used on production units, for the Auxiliary Power Unit (APU) under study. Conventional approaches to sensor validation require redundant sensors. In our case however, the absence of redundancy and the small number of production sensors necessitate novel approaches that would exploit nonlinear relationships among sensors to perform sensor validation. As a first step, we categorize sensors into relevant groups. Sensors that belong to the same group must be related, i.e. one is needed to predict the value of the other. Mathematically speaking, it means that they should have high values of cross-correlation and cross-covariance.

The cross-correlation sequence is a statistical quantity defined as:

$$\gamma_{xy}(m) = E\{x_n y^*_{n+m}\} \quad (1)$$

where x_n and y_n are stationary random processes, $-\infty < n < \infty$, and $E\{\}$ is the expected value operator. The cross-covariance sequence is calculated by first removing the mean and then estimating the cross-correlation sequence:

$$C_{xy}(m) = E\{(x_n - \mu_x)(y^*_{n+m} - \mu^*_y)\} \quad (2)$$

However, since we have only a finite-length record of random process in real-life applications, we use the following relation to estimate the deterministic cross-correlation sequence (also called the time-ambiguity function):

$$R_{xy}(m) = \begin{cases} \sum_{n=0}^{N-|m|-1} x_n y^*_{n+m} & m \geq 0 \\ R^*_{xy}(-m) & m < 0 \end{cases} \quad (3)$$

where $n=0,...,N-1$.

The estimated cross-covariances are normalized to be 1 at the zero-lag. The correlations are investigated with a maximum lag envelope of [-50,50]. Hence for K channels, a cross-covariance matrix of the size (50+50+1) by (K^2) is generated.

4. FEEDFORWARD NEURAL NETWORKS WITH LEVENBERG-MARQUARDT LEARNING

The virtual sensor network is implemented using a feedforward neural network with one hidden layer. The network weights are adjusted during a training procedure. Out of many learning algorithms, the Levenberg-Marquardt (LM) algorithm is found to be the most suitable for our task. By combining the speed of the Newton algorithm with the stability of the steepest descent method, the LM algorithm provides an efficient training procedure [Hag-94].

For objective functions with continuous second derivatives, the Levenberg-Marquardt learning algorithm approximates the error surface by a parabolic approximation, and the minimum of the paraboloid constitutes the solution at each step during the iteration.

The LM algorithm finds an optimum solution to the objective function by solving at each step the following equation:

$$\Delta w = (J^T J + \gamma I)^{-1} JE \quad (4)$$

where J is the Jacobian matrix, E is the cumulative error vector, and w is the weight matrix. The parameter γ is automatically adjusted during the training for convergence. Besides the fast convergence property of the LM algorithm, it has been shown to have a regularization effect [Cha-96]. The regularization by constraining the search space of the weight vector ensures a smooth network output.

5. RESULTS

Pressure sensor PS1, measuring turbine inlet pressure, was selected for proof of concept demonstration of usage of virtual sensors in an operating engine environment. PS1 sensor readings are usually measured during engine development, to ascertain the turbine efficiency. However when the engine goes to production this particular sensor shares the fate of about 300 other sensors that will not be mounted on the operating engine.

The correlation analysis of the pressure sensor PS1 determined that there are 16 sensors highly correlated with PS1. The cross-covariance values for sensor PS1 are presented on Figure 2. The light-colored bars indicate production sensors. PS1 (Sensor #14) and the rest of the development sensors are shown in dark-colored bars.

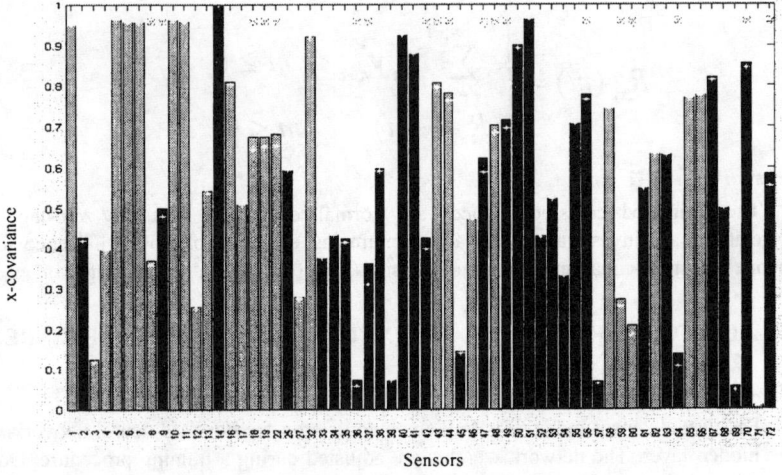

Figure 2. Cross-correlation for sensor PS1.

Using the 16 correlated base sensor data as input and PS1 data as output, a feedforward neural network is trained. For training, 5% of the whole data set is used, while the rest is kept for validation and testing. The training set comprises only19,000 time frames of data, while the complete set is 380,000. The network uses a very small hidden layer in order to achieve good generalization, which determines how well a network keeps its output close to the desired values when the input is partially corrupted. This is a feature that is definitely desired in a virtual sensor application since the base sensors are prone to failure.

The virtual sensor network is trained for 100 epochs using the Levenberg-Marquardt learning algorithm. The top plot on Figure 3 shows the actual and the predicted values. Despite its small size, the network has learned to generate a virtual sensor output that is very close to the actual one. Moreover, because of the generalization ability of the neural network, the virtual sensor seems to filter out the noise. The prediction Root Mean Squared Error (RMSE) when no input sensors are failed is 1.8% of the sensor range. The second plot on Figure 3 shows the residual between the actual and the predicted values. This prediction error stays bounded even at regions of steep change.

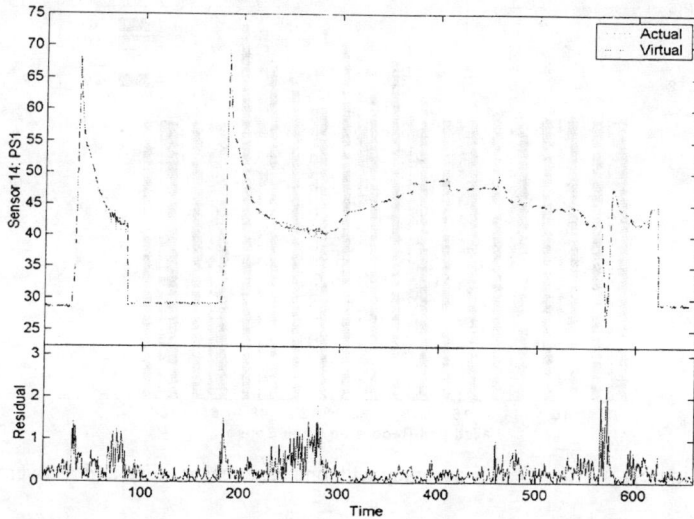

Figure 3. Actual values and values predicted by virtual sensor neural network for sensor PS1.

In order to test the generalization ability of the developed virtual sensor network further, the failed and then recovered base sensors are used as input. Ten of the 16 sensors, for which we have already designed failure detection, identification and recovery networks (see [Ulu-01]), are assumed to fail ±10 and ±20 standard deviations. These sensors with offset values are first processed through their respective sensor validation and recovery networks and then the resulting values are used as input to the virtual sensor network. Figure 4 summarizes the results. The RMSE is in between 1.8% and 2.4% of range (2.5 and 3.5 times standard deviation) for all cases. We should note here that this data set, namely the failed and recovered sensor data, was not included in the training set at all. The small error value and the consistency of it despite a failure of a different sensor in each case, show that when base sensors are chosen appropriately, a feedforward neural network can be trained to reliably act as a virtual sensor.

6. CONCLUSIONS

The work presented in this paper is a part of a system developed for validation and recovery of sensors in aircraft engines. The special difficulty encountered in such environments is the lack of any sensor redundancy. The only sensors that are mounted on production engines are the ones strictly necessary for control.

The methodology involves the covariance and the noise analyses of sensor data, auto-associative and hetero-associative backpropagation neural networks for sensor validation, detection logic for sensor faults, and virtual sensors. In this paper we are especially emphasizing the Virtual Sensor developed.

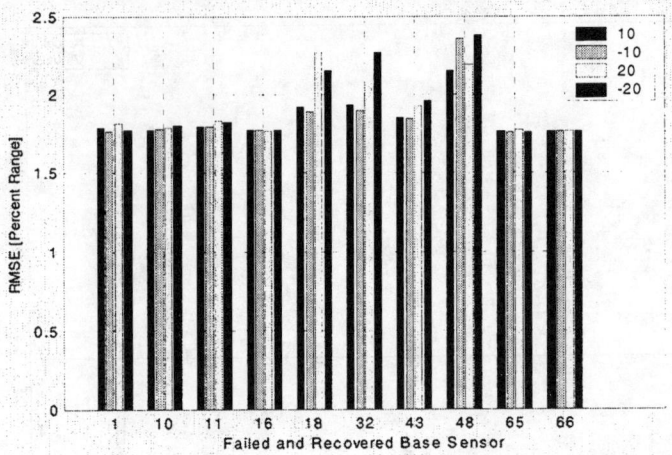

Figure 4. Virtual Sensor accuracy with one base sensor failed and recovered.

The Virtual Sensor mimics the operation of a physical sensor by inferring the value of a physical quantity from the readings of other sensors that provide relevant information. The Virtual Sensor was implemented as a feedforward neural network and trained with Levenberg-Marquardt algorithm. The inputs to the network were determined by using covariance analysis. The prediction accuracy of the Virtual Sensor was tested on actual engine data not only when the basis sensors were operating properly, but also when they were failed by ±10 and ±20 standard deviations. When no basis sensors were failed the prediction RMSE was very low (1.8% of sensor range). When one of the basis sensors failed, and was recovered through the recovery mechanism developed, the prediction RMSE continued to be low (1.8% to 2.4% of range). Our results clearly demonstrate that virtual sensors implemented as feedforward neural networks are a viable alternative for providing additional sensory information in aircraft engines.

7. REFRERENCES

[Ulu-01] O. Uluyol, A.L. Buczak, E. Nwadiogbu, "Neural Networks Based Sensor Validation and Recovery Methodology for Advanced Aircraft Engines", *Proceedings of AeroSense 2001*, SPIE Vol. 4389, Orlando, Florida, 16-20 April, 2001.

[Hag-94] M. T. Hagan, and M. Menhaj, "Training Feedforward Networks with the Marquardt Algorithm", *IEEE Transactions on Neural Networks,* Vol. 5, No. 6, pp. 989-993, 1994.

[Cha-96] L. Chan, "Levenberg-Marquardt Learning and Regularization," in S. Amari, L. Xu, L. W. Chan, I. King, and K. S. Leung, editors, *Progress in Neural Information Processing*, pp. 139-144, Springer Verlag, 1996.

NEURAL NETWORK MODEL OF HARDENING COEFFICIENT OF AIRENGINE DETAILS

VALERIY I. DUBROVIN
Dep. of Design and Production of
Radio equipment, Zaporozhye State
Technical University, Zhukovskiy
str., 64, Zaporozhye, 69063, Ukraine
e-mail: vdubrov@zstu.edu.ua

SERGEY A. SUBBOTIN
Dep. of Design and Production of
Radio equipment, Zaporozhye State
Technical University, Zhukovskiy
str., 64, Zaporozhye, 69063, Ukraine
e-mail: subbotin@zstu.edu.ua

VIKTOR K. YATZENKO
Dep. of Technology of Airengines
and Machine-building, Zaporozhye
State Technical University,
Zhukovskiy str., 64, Zaporozhye,
69063, Ukraine

ABSTRACT

The solution of a problem of hardening coefficient simulation on the basis of neural networks is considered. The algorithms allowing to execute valuation of selfdescriptiveness and selection of informative features are developed. The offered algorithms allow essentially to increase the accuracy of construction of mathematical models of complex objects and processes.

INTRODUCTION

One from the important problems in calculation of a safety factor of gasturbine airengine details is the tentative estimation of hardening coefficient $\cdot^y = \cdot y_{.1} / \cdot_{.1}$, where $\cdot y_{.1}$ - limit of endurance of a hardened detail, $\cdot_{.1}$ - limit of endurance of a detail finally treated on a serial process engineering by grinding or polishing.

For calculation of hardening coefficient it is necessary to conduct fatigue tests of defined number of details. It is expensive and hardly executive task at the designing stage. At the present time calculation of safety factor of details is executed on results of a fatigue test of standard exemplars with various concentrators of powers. In this case the similarity of the intense state in a contact zone at deformation hardening and the change of hardening coefficient $\cdot^y$ at transition from a hardened sample to a detail is not always observed (Boguslaev et al., 1993). The efficiency of a diamond smoothness, which used in an airengine building depends on selected modes, physical-mechanical and geometrical characteristics of hardened details and deforming tool.

Research problem (Yatzenko et al., 1985) included obtaining of a mathematical model of hardening coefficient with the help of theories of a similarity and analysis of dimensionalities considering parameters of the process of a diamond smoothness, physical and mechanical characteristics of a detail and tool materials with allowance for changes of efficiency of hardening (if there is concentration of powers) and a scale factor of details on the stage of design and implementation of the technological process. For the hardening coefficient modelling we offer to use the factors described in the table 1 because these factors most completely reflecting the process of a diamond smoothness of details.

Table 1 - Factors for modelling of hardening coefficient

№	Factor desig-nation	Factor measu-rement unit	Factor description	№	Factor desig-nation	Factor measu-rement unit	Factor description
1	HB	MPa	Hardness of a material	9	R_{a1}	micron	Parameter of a source grain of a detail
2	q_{max}	MPa	Average contact pressure in a zone of deformation	10	P_y	N	Smoothing force
3	a	mm	Semi axis of the ellipse in a zone of elastic contact	11	R_{sp}	mm	Sphere radius of diamond tool
4	S	mm per revolution	Feed at a smoothness	12	R_{a2}	micron	Parameter of a grain after detail smoothness
5	σ_B	MPa	Structural limit	13			Theoretical coefficient of powers concentration of a full-scale detail
6	$\sigma_{0.2}$	MPa	Yield point of a material	14	D	mm	Diameter of a sample in dangerous section
7	N		Metric deformation of a hardening.	15	R	mm	Radius of rounding galtel or cut
8	σ_{techn}		Theoretical coefficient of powers concentration from processing tracks	16	$\bar{G}$	mm^{-1}	Relative gradient of first main straining

In the work (Yatzenko et al., 1985) the statistical models of hardening coefficient were constructed on the basis of obtained experimental data. These models have an error more than 10 % at σ^y calculation.

In the present work for the construction of a hardening coefficient model we offer to use the artificial neural networks (NN) because they have such properties as the high adaptive ability and also as the ability to approximating many-dimensional dependencies. After the neural network model construction it is possible to realize the analysis of an informativety of the factors, which is carried out an evaluation of value of hardening coefficient, with the purpose of simplification of a model and rise of its reliability. For this purpose it is offered to use a combination of the correlation analysis with a neural network evaluation of features informativety.

THE MULTILAYER NEURAL NETWORK

The standard multilayer feed-forward NN (MNN) consists of M layers of neurons. The µ-th layer has N_μ neurons. The last layer is the network output. The j-th neuron of µ -th layer has $N_{\mu-1}$ inputs $x_i^{(\mu,j)}$, where i = 1,2.., $N_{\mu-1}$. Each i-th input of the j-th neuron of µ-th layer is weighted with an appropriate weight $w_i^{(\mu,j)}$.

The sum $\sum_{i=1}^{N_\mu} x_i^{\mu,j} \cdot w_i^{\mu,j}$ of the weighted inputs of j-th neuron and the bias $w_0^{(\mu,j)}$ form the input of the transfer function $\psi^{\mu,j} = w_0^{\mu,j} \cdot \dfrac{1}{1 + e^{-\sum x \cdot w_0^{\mu,j}}}$, µ = 1,2.., M, j = 1,2.., N_μ

(Dubrovin et al., 2000).

The network can be trained for function approximation, pattern association, or pattern classification. During training weights and biases of the network are iteratively adjusted to minimize the network performance function - the sum-squared error between the network outputs y_i and the target outputs y^*_i. For the NN training it is expedient to use a standard Levenberg-Marquardt algorithm because it is capable to ensure in practice much faster convergence, than majority of other methods (Nguyen et al, 1990, Himmelblau, 1972).

NEURAL NETWORK INFORMATIVE FEATURES SELECTION

For the MNN that has one output and contained on the last layer only one neuron the degree of influence of i-th feature x_i to an output signal y, in general, are determined by weights values and MNN entry signals, which values go to inputs of neurons of the first layer.

The quantitative evaluation of a significance of i-th feature will be determined as a sum of significances of i-th inputs of neurons of the first layer. For an evaluation of a significance of MNN inputs it is necessary to evaluate not only their significance concerning a neuron, but also significance of an output of a neuron as input of neurons of the next layer and so on up to neurons of the last layer. Obviously, what to realize the similar procedure it is easier in the back order, moving from the last layer to first.

For implementation of such approach we propose to use the procedure similar to wide known error backpropagation algorithm. Algorithm of a back evaluation of significance of features of MNN has the following form.

Step 0. Set the counter of samples L = 1.
Step 1. Set the counter of layers q = M.
Step 2. For all neurons of q-th layer calculate quotients of a significance of their inputs concerning their outputs (these significances we shall name as quotients, as they do not take into account an informativety of an output of the given neuron as input of neurons of the next layer):

$$z^* x_i^{*q,j*L} \bullet = \frac{|w_i^{*q,j} x_i^{*q,j*L}|}{\sum_{d=1}^{N_q} |w_d^{*q,j} x_d^{*q,j*L}|}.$$

Step 3. For all neurons of q-th layer determine significances of their inputs concerning outputs with allowance for significances of an output of a neuron for neurons of the following layer:

$$z^* x_i^{*q,j*L} \bullet = z^* x_i^{*q,j*L} \bullet \sum_{p=1}^{N_{q+1}} z^* x_j^{*q+1,p*L} \bullet.$$

For neurons of M-th layer set all $\sum_{p=1}^{N_{q+1}} z^* x_j^{*q+1,p*L} \bullet$ equal to 1.

Step 4. If q>1 to set q = q-1 and go to step 2, else go to step 5.
Step 5. For all features x_i, i = 1,2..., N calculate evaluations of their significance:

$$z^* x_i^{*L} \bullet = \sum_{j=1}^{N_1} z^* x_i^{*1,j*L} \bullet.$$

Step 6. If L < s, where s - size of learning sample, set L = L + 1 and go to step 1.
Step 7. Calculate average evaluations of a significance of features:

$$z^* x_i \bullet = \frac{1}{s} \sum_{L=1}^{s} z^* x_i^{*L} \bullet, i = 1,2..., N.$$

However, such computing procedure will be rather complex, as will contain 4 nested cycles and will be rather slow. Therefore, in practice for acceleration of calculations as well as in case of a single-layer perceptron it is possible to limit by a crude estimate of a significance of features in the first approximation by accepting all $x_i^{(q,j)}$ equal to 1.

The evaluations $z(x_i^*)$ in common will contain a defined error and there is no warranty, that these evaluations always will be close to $z(x_i)$. However, it is possible to assume, that in practice of an evaluation $z(x_i^*)$ will allow correctly to determine a qualitative evaluation of a significance of features, that quite can be rather for many applications.

EXPERIMENTS AND RESULTS

The results of fatigue tests of 57 batches were used for the modelling of hardening coefficient. The exemplars have diameter varying from 7.5 up to 60 mm. They were made from the high alloyed steels of different brands (Boguslaev et al/, 1993).

The diamond smoothness of details was produced by the tools with sphere radiuses varying from 0.8 up to 3 mm. The hardness of researched materials was in bounds HB = 2350-3300 MPa, strength $\bullet_B$ = 950 - 1150 MPa, yield point $\bullet_{0.2}$ = 600 - 1000 MPa, metric of deformation hardening n = 0.103 – 0.131, force of a smoothness P_y = 100 - 500 H, feed S = 0,03 - 0,16 mm / revolution, relative gradient of the first main power $\overline{G}$ = 0.43 - 2.51 mm^{-1}. For the smooth samples $\overline{G}_0$ = 0.5 mm^{-1} (d = 7,5 mm and r = 10 mm).

The fatigue tests were carried out on electromagnetic installation in a mode of resonance oscillations (v = 310 - 320 Hz) at plane with alternating signs curving of the console fixed sample and at pure curving with rotation on the machine MVP-10000 (v = 50 Hz). The bounds of endurance hardened exemplars $\bullet_{-1}^y$ and the bounds of source exemplars $\bullet_{-1}$ for probability of destruction P = 50 % were determined for each batch (10 - 12 samples) (Boguslayev et al., 1993, Yatzenko et al., 1985).

The modelling of hardening coefficient was realized on the basis of two-layer perceptron (Nguyen et al., 1990, Dubrovin et al., 2000). The first layer of perceptron was contained 4 neurons, and the second layer was contained 1 neuron. All neurons had sigmoidal transfer functions: $\psi(x)=1/(1+e^{-x})$.

The values of the factors were moved to the NN inputs. The value of hardening coefficient for an appropriate sample was moved to the NN output.

The NN training was produced on the basis of the Levenberg-Marquardt algorithm. At NN learning the the training step was equal to 0.00001, the maximum number of training cycles (epochs) was equal to 500.

The matrix of weight coefficients obtained in result of NN learning is represented in the table 2.

The time of NN learning was total the 106.7 sec., the number of learning cycles was equal to 339, and the sum-squared error was equal to 9.94262 $\cdot 10^{-7}$.

After obtaining a neural network model of hardening coefficient the neural network evaluation of the factors informativety was made, and also the correlation coefficients of the factors and the hardening coefficient were calculated. The results of an evaluation of informativeties of the factors are represented in the table 3.

The solutions on sharing the factors to the two groups (informative and small informative) were provided on the basis of obtained values of correlation coefficients and estimations of informativeties of the factors. The factors with informativeties and correlation coefficients exceeded particular defined threshold values were shared to informative features group. Other factors were shared to small informative features group. The following thresholds of a significance were used at decision making on an informativety of features: for correlation coefficients - 0.1979, for neural network estimations of informativeties of features - 0.0625.

After decision making on an informativety of features the small informative features (1,4,9,13,15) were eliminated from the learning set. Then the repeated modelling of hardening coefficient on the basis of NN was made. Thus all parameters of NN and learning process were the same, as well as in the previous case, except for maximum number of learning cycles, which was enlarged epochs = 1000.

Table 2 - Matrix of NN weight coefficients

μ	j	i				μ	j	i
		1	2	3	4			1
1	0	14,1894	-5,1831	-10,8277	-5,3176	2	0	-34,3184
	1	-21,1447	4,6313	-0,7001	0,9456		1	37,3173
	2	-7,8715	-3,2767	0,4041	3,4354		2	-32,6482
	3	-11,5879	35,883	6,7379	5,2636		3	29,7206
	4	-2,8764	15,19	-1,1862	2,0939		4	63,3830
	5	34,8108	-4,2685	0,3106	-2,1849			
	6	-14,0461	11,9728	-0,4762	2,2898			
	7	-1,3948	28,2757	-36,5509	19,7894			
	8	-4,4415	-0,8422	-1,7184	3,4117			
	9	-5,6667	-8,1141	13,0482	1,4346			
	10	6,1819	-7,5646	-2,5819	-1,7134			
	11	-1,5746	-10,854	-3,4903	2,2829			
	12	10,664	4,4055	5,3613	-8,4166			
	13	-0,7028	0,5237	-1,0671	1,2218			
	14	-21,3477	15,2453	-8,8919	5,6718			
	15	-0,9858	7,8709	15,4614	-2,6364			
	16	0,4963	-7,2036	30,9704	0,4219			

Table 3 - Results of an evaluation of the factors informativeties

fac-tor №	Coefficients of correlation between the features and predicted factor	NN evaluation of the feature informativety	Solution on a level of an feature informativety	fac-tor №	Coefficients of correlation between the features and predicted factor	NN evaluation of the feature informativety	Solution on a level of an feature informativety
1	-0.088	0.0456	low informative	9	0.1665	0.0459	low informative
2	0.288	0.0380	informative	10	0.2139	0.0330	informative
3	0.3076	0.1033	informative	11	-0.4946	0.0345	informative
4	-0.0769	0.0374	low informative	12	0.2442	0.0814	informative
5	-0.199	0.0736	informative	13	0.0428	0.0108	low informative
6	-0.2947	0.0512	informative	14	0.1308	0.0993	informative
7	-0.042	0.2096	informative	15	-0.0919	0.0491	low informative
8	0.1979	0.0314	informative	16	0.2875	0.0558	informative

The results of NN simulation of hardening coefficient are shown in the table 4.

Table 4 - Fragment of results of neural network simulation after exception of the small informative factors

№	$\beta^y_{exp.}$	$\beta^y_{calc.}$	№	$\beta^y_{exp.}$	$\beta^y_{calc.}$	№	$\beta^y_{exp.}$	$\beta^y_{calc.}$	№	$\beta^y_{exp.}$	$\beta^y_{calc.}$
1	1.16	1.1594	9	1.21	1.2103	25	1.44	1.4398	41	1.58	1.5853
2	1.27	1.2706	10	1.51	1.5098	26	1.13	1.1343	42	1.6	1.6004
3	1.38	1.38	11	1.51	1.51	27	1.29	1.2901	43	1.56	1.5593
4	1.54	1.5396	12	1.38	1.38	28	1.29	1.29	44	1.66	1.6572
5	1.46	1.4602	21	1.45	1.4503	29	1.32	1.3196	45	1.6	1.6006
6	1.35	1.3499	22	1.5	1.5006	30	1.42	1.42	46	1.59	1.5912
7	1.39	1.3907	23	1.61	1.61	31	1.57	1.57	47	1.55	1.5497
8	1.35	1.3492	24	1.64	1.6404	32	1.47	1.47	48	1.56	1.5651

Here $\beta^y_{exp.}$ is the value of hardening coefficient obtained experimentally, $\beta^y_{calc.}$ is the value of hardening coefficient calculated on the basis of NN.

The time of NN learning was total 190.87 sec. for 1000 learning cycles (95.44 sec for 500 cycles), the sum-squared error was equal to $3.98 \cdot 10^{-4}$. The error of calculation of hardening coefficient in comparison with the previous case was increased, because the NN memory was decreased at the expense of decreasing of weights number of deleted features, and also deleting of the information containing in deleted features from the learning sample. On the other hand, the obtained result is quite acceptable. Let's mark also, that in the latter case the speed of NN learning (at fixed amount of learning cycles and also at speed of NN work) was increased in comparison with the previous case.

The coefficient of multiple correlation for data from the table 4 was equal to 0.99(9) for described methods and 0.95 for statistical models obtained in (Boguslaev et al., 1993). It testifies that the obtained neural network model is more exact in comparison with statistical models (Boguslaev et al., 1993).

CONCLUSIONS

The high accuracy ensured at hardening coefficient modelling based on NN allows to calculate a bound of endurance of details at the stage of development of the technological process. The results of modelling of hardening coefficient of airengine details based on the NN are acceptable for application in practice. The combined analysis of an informativety of the factors based on NN and correlation coefficients allows to simplify and to optimize a hardening coefficient model, and also to increase reliability of obtained results.

NOMENCLATURE

M: number of neural network layers
N_μ: number of neurons in μ-th layer of neural network
$w_i^{(\mu,j)}$:   weight of i-th input of j-th neuron of μ-th layer of neural network
$x_i^{(\mu,j)[L]}$:   i-th input of j-th neuron of μ-th layer of neural network for L-th sample
z(x):   informativety of factor x

REFERENCES

Boguslaev, V.A., Yatzenko, V.K., and Pritchenko, V.F., 1993, "Technological supporting and prediction of sufferability of gas turbine engine details", Kiev: Manuscript Pub., 333 p.

Dubrovin, V.I., Subbotin, S.A., Morshchavka, S.V., and Piza, D.M., 2000, "The plant recognition on remote sensing results by the feed-forward neural networks", Smart Engineering Systems Design: Neural Networks, Fuzzy Logic, Evolutionary Programming, Data Mining, and Complex Systems, ANNIE 2000: the 10-th Anniversary edition, ed. C. H. Dagli et al., ASME Press, vol. 10, P. 697-701.

Himmelblau, D.M., 1972, "Applied Nonlinear Programming", New York: McGraw-Hill.

Nguyen, D., and Widrow, B., 1990, "Improving the learning speed of 2-layer neural networks by choosing initial values of the adaptive weights", Proceedings of the International Joint Conference on Neural Networks, Vol. 3, pp. 21-26.

Yatzenko, V.K., Zaytsev, G.Z., Pritchenko, V.F. et al., 1985, "Increase of sufferability of machine details by the diamond smoothness", Moskow: Mashinostroyeniye Pub., 232 p.

INTELLIGENT TRIANGULATIONS WHICH APPROXIMATE 3D POINTS FOR PRODUCTION PURPOSES

STEFAN PITTNER[1], KISHORE POCHAMPALLY,
SRIKANTH VADDE AND SAGAR V. KAMARTHI
Northeastern University, Dept. of Mechanical, Industrial and Manufacturing Engineering,

ABSTRACT

One of the most important problems in reverse engineering is an economic approximation of an object surface from a point cloud in $\mathbb{R}^3$, which is the output of a measurement process. The construction of a triangular mesh for a free-form surface of an object is often the first step when manufacturing or recreating such an object through reverse engineering. A triangular mesh allows a simple graphical representation of an object and obviates a difficult surface segmentation step. However, there is no satisfactory solution known for this problem. This paper demonstrates that the marching cubes algorithm, which has been designed for 3D arrays, can be employed to create a triangular mesh for discrete points and to eliminate noise. Using a practical example, this paper shows a possibility for improving the output of the marching cubes algorithm to obtain a decimated mesh with a small number of well-shaped triangles, which fits the geometry of the original surface. In addition, applications of surface triangulations such as in CAD/CAM and rapid prototyping are pointed out.

INTRODUCTION

The term "reverse engineering" is borrowed from hardware and software development (Chikofsky and Cross, 1990; Pressman, 2001; Rekoff, 1985) and has been used in the manufacturing literature since the beginning of the last decade (Sarkar and Menq, 1991). Reverse engineering is a methodology for manufacturing physical parts for which no drawings are available; it generally starts with digitizing an existing prototype, creating an accurate three-dimensional computer-aided design (CAD) model (i.e. a representation of the part in a commercial CAD system), and then using it to manufacture the component (Puntambekar et al., 1994). In some cases, one needs to acquire the geometrical information partially or completely through reverse engineering due to inadequate documentation of the part (Ma and Kruth, 1998).

In order to construct a smooth and accurate representation of parts with free-form shapes, reverse engineering uses two stages. In a segmentation stage, the surface of the part is partitioned into regions using sharp edges and surface locations with high curvature as the boundaries. During the second stage a parametric surface is fitted to each region and the resulting surface patches are directly imported into the CAD system to get a CAD model of the part. Bézier, B-spline and non-uniform rational B-spline (NURBS) surfaces are the most popular surfaces used for these approximations (Ma and Kruth, 1998). After a possible analysis and modification by the user, the constructed CAD model is used to manufacture the object.

The advantages of an accurate approximation of a point cloud, which has been obtained by scanning a part, by a set of triangular facets for reverse engineering are twofold. Firstly, such triangulations can serve as the initial step for building the parametric surfaces used by CAD systems (Bernard, 1999). Approaches have

---

[1] Corresponding author

been proposed for this purpose by Eck and Hoppe (1996), Hoppe (1995) as well as Krishnamurthy and Levoy (1996). Secondly, the manufacturing of the part is often based on triangular surface approximations derived from a CAD model. This is especially true for rapid prototyping technologies, where the stereolithography (STL) file format has become a de facto standard (Chen and Wang, 1999), but also for many computer-aided manufacturing (CAM) software systems, e.g. for CNC milling (Friedhoff, 1997; Bradley et al., 1999; Sun et al., 2001). Consequently, it has been suggested to use only a triangulation process in reverse engineering for these applications before the actual manufacturing stage, which would avoid the design of a CAD model (Bernard, 1999; Chen and Wang, 1999).

In the next section the requirements for an appropriate triangular approximation of scanned free-form surfaces are summarized and a well-known algorithm, which has been proposed in the context of medical imaging, is adapted for this problem.

USING THE MARCHING CUBES ALGORITHM FOR 3D POINT CLOUDS

Objectives and Existing Methods

In reverse engineering applications, an intelligent approximation of a given set of points in $\mathbb{R}^3$ by a triangular mesh, i.e. by a finite set T of triangles, has to satisfy the following major requirements:

- Each egde of the triangles in T has exactly two adjacent triangles in T.
- Two different triangles of T may intersect each other only along a common edge.
- The approximating mesh of triangles T should reflect the geometric and topological information of the true surface using a small number of triangles.
- The computed mesh of triangles T should be robust with respect to noise in the point cloud.
- All triangles in T should have a moderate aspect ratio. The aspect ratio of a triangle Δ is usually defined by c/h_c, where c stands for a longest side of Δ and h_c for its altitude from c (Bern et al., 1994). It is bounded below by the optimum value $2/\sqrt{3} \approx 1.15$, which is reached just for all equilateral triangles.

The first and the second property mean that the mesh T represents the surface of a solid object which has a nonzero thickness everywhere. The third requirement demands that flat areas of the object surface be represented by fewer triangles than those required for representing surface areas with high curvature. The condition that almost all triangles of the mesh have a small aspect ratio is a desirable property to reach a homogenous mesh of triangles of almost equal shape. This aesthetic consideration becomes very important when the user interacts with the graphical model to improve a constructed mesh T or to evaluate the approximation quality of T. The reason is that for irregular part surfaces it is difficult to recognize the actual shape and size of inhomogenous triangles.

The current scanning devices measure 10^4 to 10^5 points per second (Bernardini et al., 1999), which means they give densely sampled surfaces. Therefore, the third property indicates that the triangles constructed for T need to give a parsimonious representation of a densely sampled surface, i.e. the number of vertices of T should be only a small fraction of the number of data points in the point cloud. Some of the main limitations of the existing algorithms (about a dozen or so) for representing scanned surfaces by triangulations are that they can handle only a small set of data points (e.g. Hoppe et al., 1993), they can process only a surface shell (corresponding to a function) instead of a closed surface (which can be described by a parametric equation, e.g. Margaliot and Gotsman, 1994), or their implementation is complicated (e.g. Bernardini et al., 1999). But most importantly, they completely ignore the aspect ratios of the triangles constructed.

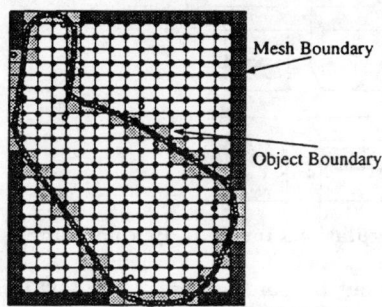 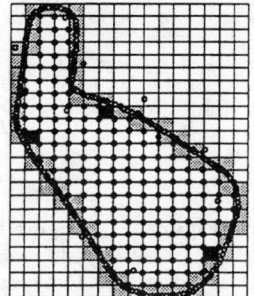

(a) The boundary cubes of the scanned obect and of the mesh

(b) The result of the first three steps of the triangulation method

Figure 1: Two-dimensional illustration of the triangulation method. The inner vertices are represented by bold dots.

The only exception is the approach proposed by Sun et al. (2001). They designed a decimation algorithm for point clouds which takes the inherent geometric and topological information into account. All of the points of the original point cloud which remain after this filtering are then triangulated utilizing a number of heuristic rules. It becomes clear from this method that even the requirement that each edge in the mesh belongs to just two triangles is difficult to satisfy.

The Principal Steps of the Triangulation Algorithm

The marching cubes algorithm (Lorensen and Cline, 1987) has been introduced as a simple method for producing surface models for 3D images (Russ, 1999), which are also called 3D arrays, as they are acquired in certain medical applications, and has become a standard in this area. In the remainder of this section we briefly summarize a novel method which uses the idea of the marching cubes algorithm to triangulate scanned surfaces.

Step 1. The method starts by covering the point cloud by a regular mesh M of cubes whose edge length is chosen considering the density of the given points and the desired level of accuracy. Both M and the vertices of the cubes in M are stored as binary arrays.

Step 2. The first objective of the method is to identify all cubes of M that lie entirely outside the point cloud. The boundary cubes of the scanned object are marked in the mesh M. To make the algorithm robust with respect to noise and outliers among the data points, these boundary cubes are defined to contain at least two points. In addition, the cubes that bound M are stored in a list L (see Fig. 1a).

Step 3. Take the first available cube C of the list L and examine each of its 6 neighbors along the faces of C. If such a neighbor is not marked in M, it is added to L. Delete C from the list, mark each of its 8 vertices as being outer vertices. Perform this step for the next available cube of L until L is empty.

As shown in Fig. 1b, this leaves only cubes at the boundary and inside the object. Furthermore, just the vertices inside the object remain unmarked. For the remaining step only the grid of the vertices of the cubes is considered. It is assumed that the surface is located just between the inner (unmarked) and the outer (marked) vertices.

Step 4. For each cube which contains both inner and outer vertices, take the center of each edge which connects an inner and an outer vertex as vertex for triangles. Create a triangulation of one to four triangles from the constructed vertices as shown in Fig. 2.

Given that the vertices of a cube can be either inner or outer vertices, there exist $2^8 = 256$ different kinds of cubes. Using an exchange of the vertex types and

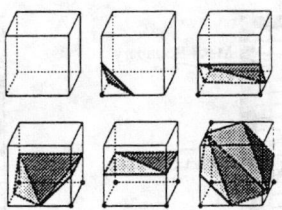

Figure 2: Scheme for triangulations inside individual cubes.

rotations leaves the 15 types pointed out by Lorensen and Cline (1987). From a practical point of view, one gets just the 6 cases shown in Fig. 2 by considering only surfaces of a simple type.

Properties of the Triangulation Algorithm

From the description in the previous section, it is immediately obvious that intersections of two triangles of a triangulation constructed by the proposed algorithm can occur only at an edge and that each edge always connects just two triangles. Moreover, step 2 of the algorithm has been designed to deal with noise and outliers, and a simple calculation based on Fig. 2 shows that the maximum aspect ratio of the triangles computed by the new algorithm is at an acceptable level ($2\sqrt{3} \approx 3.46$).

One disadvantage of the proposed triangulation algorithm is that in general smooth parts of a surface will be represented by approximations which are locally nonsmooth because of the relatively limited number of possible orientations of the triangles constructed with Fig. 2. On the one hand the algorithm can construct surface approximations that require considerably less memory than the originial point cloud and for fixed edge length of the mesh its computational complexity grows only linearly with the number of points in the point cloud. On the other hand the algorithm allows surfaces to be approximated with arbitrary accuracy as long as the point cloud consists of a sufficient number of noise free sampling points. However, as with the original marching cubes algorithm, decimation or refinement of the ouput triangulation can be achieved only evenly distributed over the whole surface (by modifying the edge length of the cubes). This means that for an accurate approximation, one has to accept a large number of triangles (Shu et al., 1995). Accordingly, a methodology for transforming an arbitrary triangular mesh to a new one that has significantly fewer triangles but retains the geometric structure and accuracy of the original is proposed in the next section.

INTELLIGENT DECIMATION OF TRIANGULATIONS

A variety of decimation methods for triangulations have been proposed in the literature, some of which are reviewed by Heckbert and Garland (1997). The simplest principle for decimating triangulations is called edge decimation (Lee et al., 2000). It consists of moving one vertex V_0 of an edge $\overline{V_0V_1}$ along $\overline{V_0V_1}$ to the other vertex V_1 (see Fig. 3). Along with V_0 also all edges adjacent to V_0 are moved to V_1.

We constructed an intelligent methodology for iteratively coarsening arbitrary triangulations T. The key idea of the methodology is derived from the discussion in the second section of this paper: The less the curvature of the surface the higher the number of edges that can be eliminated. A sketch of the edge decimation procedure is as follows: For each vertex V_0 in the mesh an "average plane" is fitted to V_0 and the cycle of vertices $V_1, V_2, \ldots, V_n$ of the triangles surrounding V_0. This average plane is used to decide if the mesh T is almost flat around V_0 and it is applied for the implementation of a special necessary condition for avoiding invalid triangles which would result if the shortest edge from the vertex V_0 is decimated. The mesh decimation algorithm allows the user to set an upper bound for the aspect ratio of the triangles. However, the algorithm in its present state

949

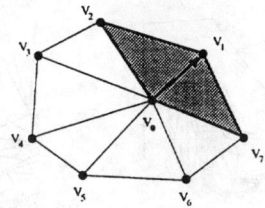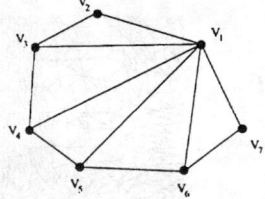

Figure 3: Decimation of an edge from a vertex V_0 to a vertex V_1.

does not guard against possible self-intersections in the triangle-based surface representation (de Cougny, 1998).

AN EXAMPLE PROBLEM

We implemented the triangulation algorithm presented in this paper in the C programming language and the mesh decimation algorithm in MATLAB 5.3. The triangulation algorithm took several seconds to approximate one million random sampling points of the surface of an object similar to a table tennis bat when using an edge length of 0.3 (on a Sun Blade 100 (UltraSPARC-IIe) workstation). This triangulation, which is shown in Fig. 4a, consists of 37628 triangles with an average aspect ratio of 2.05, and the points of the original point cloud are located at an average distance of 0.05 and at a maximum distance of 0.17 from this surface representation. After white noise with a standard deviation of 0.1, i.e. one third of the side length, was added to the sampling points an increase of the ede length was required for the triangulation procedure to run properly. Compared to the computation with the original data using this same new edge length, the only difference was a slight increase in the number of triangles in the mesh. When the maximum aspect ratio was set to 10, the mesh coarsening algorithm reduced the number of triangles from 37628 to 4038 in about 7 minutes (see Fig. 4b). The triangles have the average aspect ratio of 2.79. The average error remained almost constant while the maximum error increased to 1.3.

CONCLUSIONS AND FUTURE WORK

In this paper two new methods for efficiently computing and improving piecewise linear approximations for free-form surfaces in reverse engineering were presented. The advantages and drawbacks of both methods, especially regarding the aspect ratios of the computed triangles and noise in the data, were discussed and demonstrated with a test example. The triangulation procedure presented provides a simple means for approximating densely sampled surfaces by triangular meshes. The mesh coarsening algorithm will be described in more detail in forthcoming papers. Future work will also include a study of ideas for fairing triangular surfaces as proposed by Welch and Witkin (1994).

REFERENCES

Bern, M., Eppstein, D., and Gilbert, J., 1994, "Provably Good Mesh Generation," *Journal of Computer Systems Science*, Vol. 48, pp. 384–409.

Bernard, A., 1999, "Reverse Engineering for Rapid Product Development: A State of the Art," *Proceedings of the SPIE - The International Society for Optical Engineering*, Vol. 3835, pp. 50–63.

Bernardini, F., Bajaj, C. L., Chen, J., and Schikore, D. R., 1999, "Automatic Reconstruction of 3D CAD Models from Digital Scans," *International Journal of Computational Geometry and Applications*, Vol. 9, pp. 327–370.

Bradley, C., Wei, S., Zhang, Y. F., and Loh, H. T., 1999, "Generation of Polyhedral Models from Machine Vision Data," *Proceedings of the SPIE - The International Society for Optical Engineering*, Vol. 3832, pp. 26–37.

Chen, Y. H., and Wang, Y. Z., 1999, "Genetic Algorithms for Optimized Re-Triangulation

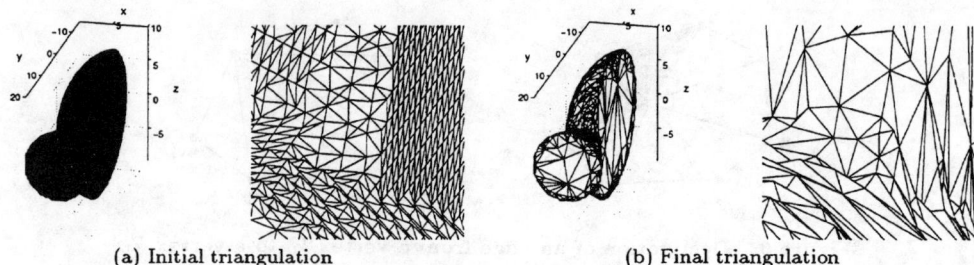

(a) Initial triangulation (b) Final triangulation

Figure 4: Example problem

in the Context of Reverse Engineering," *Computer-Aided Design*, Vol. 31, pp. 261–271.
Chikofsky, E. J., and Cross, J. H., 1990, "Reverse Engineering and Design Recovery: A Taxonomy," *IEEE Software*, Vol. 7, pp. 13–17.
de Cougny, H. L., 1998, "Refinement and Coarsening of Surface Meshes," *Engineering with Computers*, Vol. 14, pp. 214–222.
Eck, M., and Hoppe, H., 1996, "Automatic Reconstruction of B-Spline Surfaces of Arbitrary Topological Type," *Computer Graphics Proceedings (SIGGRPAH 96)*, pp. 325–334.
Friedhoff, J., 1997, *Aufbereitung von 3D-Digitalisierungsdaten für den Werkzeug-, Formen- und Modellbau*, Ph.D. Thesis, Vulkan Verlag, Essen, GER.
Heckbert, P. S., and Garland, M., 1997, "Survey of Polygonal Surface Simplifaction Algorithms," Technical Report, Computer Science Department, Carnegie Mellon University, Pittsburgh, PA.
Hoppe, H., 1995, "Generation of 3D Geometric Models from Unstructured 3D Points," *Proceedings of the SPIE - The International Society for Optical Engineering*, Vol. 2410, pp. 424–431.
Hoppe, H., DeRose, T., Duchamp, T., McDonald, J., and Stuetzle, W., 1993, "Mesh Optimization," *Computer Graphics Proceedings (SIGGRPAH 93)*, pp. 19–26.
Krishnamurthy, V., and Levoy, M., 1996, "Fitting Smooth Surfaces to Dense Polygon Meshes," *Computer Graphics Proceedings (SIGGRPAH 96)*, pp. 313–324.
Lee, J. K., Kim, J. S., Kim, D. W., and Kim, S. I., 2000, "Fast and Memory Efficient Generation of Simplified Meshes," *Proceedings of the SPIE - The International Society for Optical Engineering*, Vol. 3976, pp. 490–495.
Lorensen, W. E., and Cline, H. E., 1987, "Marching Cubes: A High Resolution 3D Surface Construction Algorithm," *Computer Graphics*, Vol. 21, pp. 163–169.
Ma, W., and Kruth, J.-P., 1998, "NURBS Curve and Surface Fitting for Reverse Engineering," *International Journal of Advanced Manufacturing Technology*, Vol. 14, pp. 918–927.
Margaliot, M., and Gotsman, C., 1994, "Piecewise-Linear Surface Approximation from Noisy Scattered Samples," *Proceedings, IEEE Conference on Visualization (Visualization '94)*, Washington, DC, pp. 61–68.
Pressman, R. S., 2001, *Software Engineering: A Practitioner's Approach*, 5th ed., McGraw-Hill, Boston, MA.
Puntambekar, N. V., Jablokow, A. G., and Sommer, H. J. III, 1994, "Unified Review of 3D Model Generation for Reverse Engineering," *Computer Integrated Manufacturing Systems*, Vol. 7, pp. 259–268.
Rekoff, M. G., Jr., 1985, "On Reverse Engineering," *IEEE Transactions on Systems, Man, and Cybernetics*, Vol. 15, pp. 244–252.
Russ, J. R., 1999, *The Image Processing Handbook*, 3rd ed., CRC Press, Boca Raton, FL.
Sarkar, B., and Menq, C.-H., 1991, "Smooth Surface Approximation and Reverse Engineering," *Computer Aided Design*, Vol. 23, pp. 623–628.
Shu, R., Zhou, C., and Kankanhalli, M. S., 1995, "Adaptive Marching Cubes," *The Visual Computer*, Vol. 11, pp. 202–217.
Sun, W., Bradley, C., Zhang, Y. F., and Loh, H. T., 2001, "Cloud Data Modelling Employing a Unified, Non-Redundant Triangular Mesh," *Computer Aided Design*, Vol. 33, pp. 183–193.
Welch, W., and Witkin, A, 1994, "Free-form Shape Design Using Triangulated Surfaces," *Computer Graphics Proceedings (SIGGRPAH 94)*, pp. 247–256.

ABSTRACTION BASED SOFTWARE SYSTEM DESIGN

CHRISTOPHER LANDAUER, KIRSTIE BELLMAN
cal@aero.org, bellman@aero.org
Aerospace Integration Science Center
The Aerospace Corporation
Los Angeles, California 90009-2957, USA

ABSTRACT

In building a complex software system, we almost never know enough to do "top-down" design. Instead, we have to do some kind of "knowledge first, sideways" design, in which we only know some of the requirements, and have to infer others later. In this paper, we describe "abstraction-based" design, which makes fundamental use of abstractions to delay design decisions until they can be made properly, and rescind them when new information makes them unsuitable (retaining all information and decisions for retrospective analysis). Throughout the design process, the decision points, the relevant considerations, the applicable problem specification, and the decision that is made, are all made explicit and computable.

INTRODUCTION

This paper is about a design process for what we have called *Constructed Complex Systems*, that is, complex systems monitored or mediated by computing systems (Landauer and Bellman, 2001). We describe some characteristics of environments that can support the design and development processes necessary for these complex systems, and an approach to design that makes fundamental use of a computational notion of abstraction. We call our approach "abstraction-based design". The idea is to use proper abstractions, and be able to explain them, change them, elaborate them, and compute with them.

First, we recommend that the design process be supported by a model development environment that includes many different modeling methods, analytic tools for examining and comparing models, and processes for searching through complex design spaces. We advocate using many models and relationships among the models, and note that executable models allow more powerful analytical tools to be used early in the process (so the design process is not restricted to verbal discussions about pictures). Since NO one model or modeling method is adequate for describing a complex system (Bellman and Brock, 1960), we expect there to be many alternative models of the system and its components, until they can be properly evaluated in their expected context of use.

Second, we recommend that the design system keep the decision points, the relevant considerations, the applicable problem specification, and the decision that is made, for all design decisions, so that new information can cause decisions to be rescinded. This will not help for those design decisions that are not recognized as such, but we also recommend that the design support system allow identification of and changes to decisions.

Third, we recommend that all information be made explicit and machine interpretable, or at least as much as possible, since no other way allows the design support system to help. This includes information about the design system itself, so it can help its users use it.

In general, abstraction is difficult, and modeling is difficult, so the best we can expect to do is to provide the tools that can make these processes easy to discover, record, explain, and expand.

The most important property of such a design system is "Computational Reflection" (Bellman, 2000) (Landauer and Bellman, 2001), since it allows the system to use explicit models of its own behavior to adjust that behavior. To this end, there have to be active coordination processes that use the models to coordinate among the other system functions. They also provide overview and navigation tools, context maintenance functions, monitors, and other explicit infrastructure activities. It is this ability of the system to analyze its own behavior that provides some of the power and flexibility of resource use. The self-model supports explicit software engineering functions from within the same system (Landauer and Bellman, 1998), and provides a basis for self-monitoring, explanation, and failure diagnosis. Another way to view this property is that the system can make recursive calls in the "meta-direction", i.e., to switch from studying some application problem to studying the problems of its own internal workings.

ABSTRACTION

In this section, we describe our definition of abstraction. We make a distinction between the definition of *an abstraction*, which is a relationship between configurations of models, and the *process of abstraction*, which is a way to construct abstraction relationships. We should note here that we are describing criteria and algorithms for one kind of abstraction, but we do not believe that it is possible for one algorithm to implement all kinds of abstraction. While the computations are easy, the choices of which computations to use are not.

Abstraction and Modeling

In the sense in which we use them, abstracting and modeling are almost synonymous: they are both about inventing a more explicit and more formal description of some phenomenon, and relating its structures and behaviors to those of the phenomenon. The difference is that we usually take modeling as the first step, of making a computable or formal object that is as like the external phenomenon as possible in the ways in which we are interested, but which does not necessarily allow any formal analysis of the relationship (only experiments), whereas abstraction mostly takes place in the computable or formal world, as a relationship between two models, which is therefore subject to analysis (since both sides are accessible).

For our purposes in this paper, models are computational or formal objects. Here we use formal exclusively to mean mathematically formal, and distinguish formal from systematic or computational objects, which are those that can be simulated by computer files or programs. There is a large overlap in these categories, but they are not co-extensive. Descriptions of formal objects are adequate only if they are computational (since we want to compute with them), and semantics for computational objects are adequate only if they are formal or computational (since we want to prove and observe properties about them).

There is a well-known distinction in Mathematics, which appears under many guises, between the *extensional* description of a system, which exhibits its structure, components, or behavior, and the *intensional* description of the same system, which provides properties of its structure, components, or behavior. It is most useful when the two kinds of description meet in the middle: that is, when the conditions specified uniquely determine the elements. The connection to our modeling approach is that extensional models are usually computational, because they exhibit model behavior, and intensional ones are usually formal, since they are assertions about properties of the model.

Modeling Process

For any phenomenon, whether natural, systematic, computational, or formal, effective modeling is a five-fold process:

1. the modeler decides what properties of the phenomenon are important,
2. the modeler chooses a formal or computational space in which those properties can be represented,
3. the modeler maps the phenomenon into the formal or computational space,
4. the modeler manipulates the elements of the representation space, according to the rules of that space, and
5. the modeler remaps the results back into the phenomenon.

This process succeeds when the formal or computational manipulations provide new information about the phenomenon, which happens surprisingly often (Wigner, 1960). The first two steps above are among the most important, and they are also the ones usually left out of the descriptions. In any case, this *modeling relationship* is essential to understanding of modeling, because it makes explicit the difference between the model and the phenomenon, and because it emphasizes the modeler's choices. The basic purpose of modeling is to find a formal space and map the phenomena into it so that inference (operations in the formal space), when mapped back into the phenomena, correspond to entailment (the natural processes of the phenomena) (Russell, 1931).

Since we know that NO one model is adequate for describing a complex system (Bellman and Brock, 1960), we expect several different kinds of model spaces and modeling maps, and it then becomes interesting to try to combine the information they provide, using various different kinds of integration methods (Bellman and Landauer, 2000).

Model Definitions

In this subsection, we describe the computational notions of model and abstraction that we use in the design environment. These notions are our models, for the purposes of the design environment, of what models, modeling, and abstraction are. The definitions provide a relatively large class of model spaces, but of course, they cannot define all modeling approaches, and one of the necessary flexibilities in any such modeling system is to allow user specified extensions and alternatives to the model types and processes.

A *model* (as an object) is a set of concepts, structures, processes, causes (consequences), and relationships among all of these (connections and constraints). Since these models are built for computer processing, we define the concepts by terms and constraints, and the structures using a few standard construction mechanisms (enumerated types, tuples, lists, and sets, etc.). The processes are resources that exhibit behaviors associated with the models, or that compute changes in the models. The causes are implications that map certain behaviors into others (for example, if this happens, then that happens). The relationships are implications that map certain constraints into other properties of the models (for example, if this property holds in the model, then that one must also hold in these other models).

An *abstraction* (as a relationship between two models) is a set of relationships of concepts, choices, and conflations. So model A is an abstraction of model B if we can map all of the concepts and structures of model A into corresponding concepts and structures of model B. Model B may have other parts not defined by model A, and it may have different parts of model A map to the same parts of model B. This relationship is computable, since model A only has a finite number of parts, but it is sometimes difficult to derive.

Abstraction (as a process) is a set of identifying assumptions and conflations (especially implicit ones), defining classes in which the assumptions make constraints, and defining generators out of which existing structures and behaviors can be built (or will happen). Given a model, computing an abstraction of it takes several steps, each of which has some choices to make. Any of the spaces in which the concepts lie can be replaced by larger spaces. Any of the spaces in which the structures lie can be replaced by more general spaces (for each structure, there are simple ways to find more general ones). Any constraint on the model can be replaced by a relaxed or eliminated constraint. Any implicit assumptions or unstated properties can be made explicit, once they are found, by defining a space in which the assumption or property can hold. Finally, any process can be replaced by a simpler one, with some steps coalesced (though the most effective ones to coalesce is difficult to determine).

The most important process in this approach to modeling is abstraction. There are several kinds of abstraction operations, all of which change an element X, in a possibly unspecified space, into an element Y that refines to X in a space Z. The element X is in some space as soon as it is defined, unless it is the original phenomenon being modeled. For example, we can change an instance to a collection and a set of selection rules or criteria for the instance, or we can change a constraint into a constraint space and selection rules or criteria for the constraint, or we can change an observation into a description and explanation of the observation. Changing from a set or sequence of random values to the corresponding probability distribution is an example of the first kind of abstraction. Another kind of abstraction is to change the result of a decision to the corresponding problem and decision criteria. This is a special case of the first kind of abstraction described earlier, since the result of a decision is some choices, and the problem specifies those choices.

All modeling begins with some phenomenon or behavior, which is the most concrete version of the phenomenon, and infers successively more abstract models (these are all empirical constructs). The observation model contains the data and the collection criteria, and then various kinds of empirical processes identify such properties as distributions, commonalities, prevalent features, apparent constraints, time behavior, trends and fluctuations in the observations. Each of these is a model for the observations, and each is more abstract than the observations. If the observations are also from a model, then the relationship is an example of the abstraction relationship. This means that modeling and the abstraction process are essentially the same thing, one applied to the external world, and one to the models themselves. This is why Computational Reflection is so useful: it allows the system to apply the same processes to the models as it applies to the environment.

System Design

A system design based on abstraction contains models, abstractions, and design processes, and needs many facilities to manage their relationships (Landauer and Bellman, 2001). A design process is about elaboration and evaluation of designs, and selection of alternatives based on those evaluations. Elaboration of designs includes making choices, and making some things more specific or more implicit. This point should be emphasized: design does not make components explicit, it makes them more specific, and also makes more explicit the choices of what is to remain implicit.

Refinement and elaboration are not the opposite of abstraction exactly, but they are special cases (Fishwick, 1989a, 1989b), since they make models more concrete. The simplest kinds of abstraction are based on sets, and the best descriptions of them make them simple category theory notions (see Chapter XV of (MacLane and Birkhoff, 1967)): "induction" is the same as quotient objects, and "reduction" is the same as subobjects.

Our push for explicity is directly opposed to the increasing reliance on emergence (i.e., magic) as explanation. We want to go from "it just does this sometimes" to "this is what it does under these circumstances", and even more, to "this is why it does this under these circumstances". We want to go from behavior observation to behavior description, explanation, and prediction. These successive properties require successively more detailed models, of what happens, when it happens, and even why it happens.

Finally, it is important to know when we are finished modeling, that is, when we can stop constructing models and start analyzing them. We require our models to be analyzable, which means executable or interpretable for process models, and interpretable or interrogatable for data models. These criteria mean that we need end-to-end coverage in our models, so that we have an entire system design, at some level of detail, before we can study the important model interactions. These are not necessarily deep models, but they must be broad.

CONCLUSIONS AND PROSPECTS

We have argued here that an effective design system must contain many models, relationships among those models, and processes that create, interpret, change, relate, and combine those models. We have shown elsewhere (Landauer and Bellman, 1998, 2001), that we can build such complex design support systems using Wrappings, our Computationally Reflective integration infrastructure, and that it is important and useful to do so, because they allow great flexibility in the use and organization of complex collections of analytical and computational resources. We have argued that, using abstraction as the fundamental operation in organizing the large and diverse set of models required in a design process, we can build systems that retain the design decisions at their appropriate level of detail, so that new information or requirements can be used to change the design at an appropriate level of abstraction, and the necessary changes that ripple down from that original one can be managed more easily.

We have not provided methods for automatically computing abstractions, since we do not think that is possible in general, but we have provided methods that support users' choices in creating abstractions, and that help assess their utility. Design is a difficult creative process, and we think that organizing the support systems using Wrappings and abstraction mappings allows the system to do the important part of design support: staying out of the creative designer's way.

REFERENCES

Kirstie L. Bellman, "Mathematical Fictions and Physiological Realities: The Challenges of Brains Reasoning about Themselves", *Proc. IMACS'2000 (CD), Invited Session on Bioinformatics*, 21-25 Aug 2000, Lausanne (2000)

Kirstie L. Bellman, Christopher Landauer, "Towards an Integration Science: The Influence of Richard Bellman on our Research", *J. Math. Anal. and Appls.*, Vol. 249, No. 1, pp. 3-31 (2000)

Richard Bellman, P. Brock, "On the concepts of a problem and problem-solving", *Amer. Math. Monthly*, Vol. 67, pp. 119-134 (1960)

Paul A. Fishwick, "Process Abstraction in Simulation Modeling", Chapter 4, pp. 92-131 in Lawrence E. Widman, Kenneth A. Loparo, Norman R. Neilson (eds.), *Artificial Intelligence, Simulation and Modeling*, Wiley (1989a)

Paul A. Fishwick, "Abstraction Level Traversal in Hierarchical Modeling", Chapter V.2, pp. 393-429 in Maurice S. Elzas, Tuncer I. Ören, Bernard P. Zeigler (eds.), *Modelling and Simulation Methodology: Knowledge Systems' Paradigms*, North-Holland (1989b)

Christopher Landauer, Kirstie L. Bellman, "Wrappings for Software Development", pp. 420-429 in *Proc. HICSS'98*, 6-9 Jan 1998, Kona, Hawaii (1998)

Christopher Landauer, Kirstie L. Bellman, "Flexible Design Support Systems", These proceedings (2001)

Saunders MacLane, Garrett Birkhoff, *Algebra*, Macmillan (1967)

Bertrand Russell, *The Scientific Outlook*, George Allen and Unwin, London; W. W. Norton, New York (1931)

Eugene Wigner, "The unreasonable effectiveness of mathematics in the natural sciences", *Comm. Pure and Appl. Math.*, Vol. XIII, pp. 1-14 (1960)

A SEARCH ENGINE FOR TWO-DIMENSIONAL VECTOR GRAPHICS

MICHAEL A. FISHER
Object Computing Inc.
St. Louis, Missouri

DANIEL C. ST. CLAIR
Computer Science Department
University of Missouri-Rolla
Rolla, Missouri

ABSTRACT

Searching for vector based graphics objects such as those found in CAD drawings has, to date, been largely a manual process. Present approaches to searching require developers to provide "clues" to graphic content by using esoteric file naming conventions or other types of written communication. While these approaches provide a file level guide to searching, users are required to conduct manual searches within the files. In efforts to better automate such searches, the present work extends the work of St. Clair et al (1998) in which a graphics search engine was developed that could take a vector graphic as input and search a series of graphics files for occurrences of that graphic. In addition to improving the former search engine's performance, the new search engine supports user-specified search criteria as well as an algorithm for ranking the similarity between graphics objects. A prototype software system was developed to evaluate the new approach.

INTRODUCTION

While a large volume of graphics information exists, few search engines have been developed for performing graphics searches. The difficulty of this problem is exacerbated by the fact that there are a large variety of graphics types including vector graphics, computer aided design (CAD) graphics, and digital pictures (raster graphics). Vector graphics are characterized as being a collection of lines, circles, and arcs. CAD graphics objects include vector graphics objects as well as text and properties of the CAD graphics object. The graphics search problem is further complicated by the fact that there is no single standard file format (Brown and Shepherd, 1995).

Much of the work in graphics search has focused on finding similarities between digital pictures. Many approaches used for searching digital pictures are "pattern matching" approaches. One common approach is based on using wavelets to "filter" out certain properties while retaining others (Jacobs et al, 1995).

Searching for graphics objects in vector based graphics, although extremely important, has received much less attention. The concept is to identify a source object, such as a house, and then to search candidate target graphics objects for the same or similar objects. This type of search requires that graphics objects be identified according to specific characteristics so that searches can be performed on these characteristics. St. Clair et al (1998) developed some basic techniques for identifying graphics object characteristics and for searching for these characteristics. The approach was slow, not always accurate, and was unable to provide a ranking of search results.

The present work, based on St. Clair et al's work, provides a more robust search algorithm. The new approach supports user-specified search criteria as well as an algorithm for ranking the similarity between the source graphics object and the matches that are found. A prototype software system was developed to evaluate the new approach.

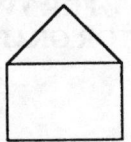

Figure 1. A Single 2-D Graphics Object

THE PROBLEM OF SEARCHING FOR GRAPHICS OBJECTS

Vector and CAD graphics are used to represent a set of objects that are of interest to the designer as well as to users of the graphic. Such a graphic consists of a set of graphics objects, each of which is represented as a collection of graphics entities such as points, lines, arcs, and circles (Foley, Van Dam, et al 1994). Associated with each graphics entity is information for drawing the entity that includes points, color, width, etc.

A graphics object is defined as any object of interest to the user of the graphic. Such designation is usually made when the graphic is drawn. For example, the graphic shown in Figure 1 could be thought of as a single graphics object that represents a house. In this context, the object could only be broken down into graphics entities (i.e. 6 lines with 5 nodes). It would not be thought of as a triangle that rests on top of a rectangle (i.e. 7 lines with 7 nodes).

The present approach for developing a method for searching graphics objects requires four basic components:

1. A knowledge structure (KS) that contains properties important in a search,
2. A learning algorithm for learning knowledge about graphics objects,
3. An algorithm for comparing and ranking knowledge structures, and
4. A search engine.

The knowledge structure contains attributes of importance about the graphics objects. The learning algorithm is used to extract this knowledge from each of the objects under consideration. Once the knowledge has been "learned," the comparison algorithm uses the knowledge obtained to make comparisons during the search process. The search engine provides a technique for performing the active search. In order to make the graphics search engine as fast as possible, both the learning and comparison algorithms must be efficient. The search engine must also be able to provide some type of ranking as to how "close" two objects match.

THE 2-D GRAPHICS COMPARISON APPROACH (2-DGCA)

The 2-DGCA approach differs from St. Clair et al's approach in several ways. First, the 2-DGCA has a different set of graphics object attributes. This leads to a new KS that provides information useful in speeding-up and improving the accuracy of the comparison algorithm. The new approach also supports user-configurable search specifications. This feature provides users with the ability to select specific search attributes with user-specified tolerances and user-specified weight values.

The knowledge structure created for the 2-DGCA (see Figure 2) contains information about each two-dimensional graphics object. In the 2-DGCA attribute information is learned from each graphics object and stored in a knowledge structure. The "Object ID" in Figure 2 is simply a unique identification number assigned to each

959

1)	Object ID	4)	Number of nodes in the object (*)
2)	Line properties	5)	Degree of each node in the object (*)
	a) Endpoints	6)	Connectivity attributes: (*)
	b) Slope of non-vertical line (*)		a) Connectivity matrix
	c) Line length		b) Angle matrix
3)	Number of lines in the object	7)	Intersection points within object (*)

Figure 2. A Knowledge Structure for 2-D Line Based Graphics Objects
[(*) denotes quantities not present in St. Clair et al 1998)

graphics object. The "Number of lines in the object" can be used as a quick filter for comparing graphics objects. In counting the "Number of nodes in the object," shared end points are counted only once. The "Degree of each node in the object" refers to the number of lines that share an end point with the given node. The Connectivity matrix is an n x n matrix that tells which nodes are connected by a line segment, where n equals the number of nodes in the graphics object. A non-zero entry at row i and column j is the length of the line segment connecting nodes i and j. The angle matrix is m x m, where m is the number of lines in a graphics object. Each non-zero entry at row i and column j in the angle matrix represents the angle between the connected lines i and j. The attribute "Intersection points within object" refers to the points where lines intersect.

The learning algorithm extracts information needed in the KS by breaking a graphics object down into its fundamental entities and then learning their associated attributes and relations. The learning algorithm works by traversing the data structure representation of the graphics object and gathering information about the object as the traversal proceeds. The learning and other algorithms are described in Fisher (2000).

The comparison algorithm takes a graphics object as input and searchs a collection of graphics objects to find all that "match" the input. The output of the comparison algorithm is a list of graphics objects that match the input graphic based upon user-provided search parameters.

The user-provided search parameters include a list of attributes from the KS, an associated tolerance and weight for each attribute, and a check for isomorphism between the source object and each candidate target object. (The tolerance and weight parameters are part of the ranking algorithm and will be discussed in the next section.) The user can select which attributes are to be compared during the search and can use Boolean expressions to specify how the search attributes are used in the comparison. For example, assume that the user has chosen to search based on the following attributes: number of lines, number of nodes, and isomorphism. Once the search parameters are selected, the user might choose one of the following search conditions: number of lines AND number of nodes AND isomorphism, (number of lines AND number of nodes) OR isomorphism, number of lines AND number of nodes AND NOT isomorphism, etc. The interface for entering user defined search parameters is shown in Figure 3. In particular, Figure 3 illustrates the user defined Boolean search condition "(C AND A) OR E."

The ranking algorithm sorts graphics search results in order of best to worst match. The ranking of search results is based on matched attributes, an associated attribute tolerance, and an associated attribute weight.

The attribute tolerance is a user-specified (+/-) range of values between 0 and 10 in which an attribute's values may fall and still be considered an acceptable match. For instance, a tolerance for the number of lines might be defined to be +/- 3. Given a +/- 3 tolerance, a square might match a triangle or a septagon. Since a triangle is closer to a square in terms of the number of lines than is a septagon, comparing a square with a triangle should provide a higher rank than comparing a septagon with a triangle.

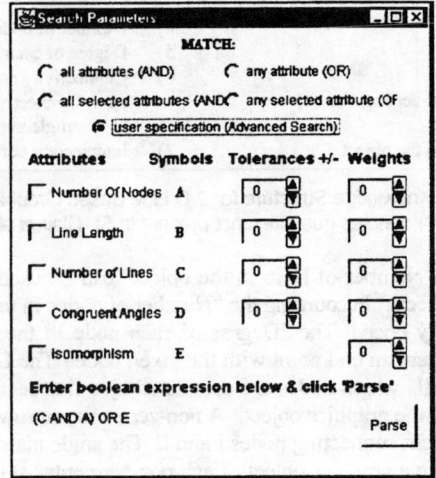

Figure 3. 2-DGCA Search Parameter Screen

The attribute weight is a user-specified value between 0 and 10. The weight is used to represent the importance of an attribute in the consideration of a possible match. A user-specified weight value of 10 represents an attribute of most importance. Conversely, a value of 0 will represent an attribute with no weight association.

Using the 2-DGCA adjusted weight calculation; the rank of a matched object is calculated by taking the sum of each attribute's adjusted weight. An attribute's adjusted weight is calculated as ($W_i - |\Delta V_i|$); where W_i is the attribute i's weight and ΔV_i is the difference between the source and the match object's attribute values.[1] For example, if a search is based on the search parameters specified in Table 1, the attribute and adjusted weight values associated with Match Object 1 are shown in Table 2. Equation (1) shows the resulting rank calculations. Higher ranking values indicate closer similarity.

Table 1. Source Object's Attribute Values, Tolerances, and Weights

Attribute	Actual Attrib. Val (V1)	Tolerance	Weight
Number of Lines	4	+/-2	6
Number of Nodes	4	+/-3	2

Table 2. Target Object 1's Attribute and Adjusted Weight Values

Attribute	Actual Attrib. Val (V2)	Adjusted Rank = ($W_i - |\Delta V_i|$)
Number of Lines	3	(6 - |4 − 3|) = 6 − 1 = 5
Number of Nodes	3	(2 - |4 − 3|) = 2 − 1 = 1

$$\text{Target Object 1's Rank} = \sum (W_i - |\Delta V_i|) = 5 + 1 = 6 \qquad (1)$$

---
[1] The variable i represents attributes A, B, C, D, or E as shown in Figure 3.

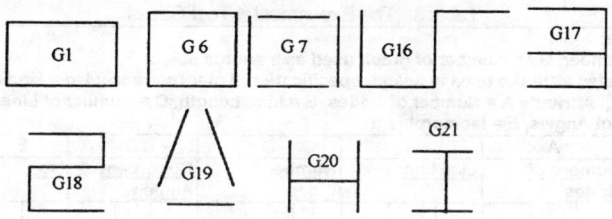

Figure 4. Figures Matched in Tests 58 and 59

Note that the weights in the current ranking equation only influence the overall level of ranking since $\sum(W_i - |\Delta V_i|) = \sum W_i - \sum |\Delta V_i|$. Other ranking schemes could use the weight values to modify the difference in attribute values or in other ways.

A prototype software system was developed to demonstrate the 2-DGCA concepts described earlier. The system consists of two applications, a Graphics Editor/Viewer and a Graphics Search Engine. The Graphics Editor/Viewer Application provides several functions. First of all, the editor/viewer allows the user to easily create 2-D graphics objects through a graphics user interface (Leather). Secondly, it provides a way to save graphics objects created to a file by storing the graphics objects' KS data. Lastly, the editor/viewer provides the capability to open, read, and edit a graphics data file.

The Graphics Search Engine Application reads KS data from two graphics data files. The two files are a source data file and a target data file. The source graphics data file is used for selecting a query graphics object. The target graphics data file is used for searching for a graphics match. Once a source file has been opened, the user can select a graphics object to use in performing a search. Then, after the search has been completed, a list of results is returned which displays all possible matches.

PROOF-OF-CONCEPT EVALUATION

Experiments were conducted to evaluate the two-dimensional graphic comparison approach. Of specific interest was the performance of the approach on exact matches of graphics objects as well as on how the ranking algorithm performed when various tolerances and weights were included in the selection criteria.

The test data for this experiment consisted of a set of 2-D objects. The complete set of twenty-two objects is described by Fisher (2000). Due to limited space, only two of the test results will be reported here. Figure 4 shows the objects discussed below.

While a large number of experiments were performed using a test layout similar to that shown in Table 3, only tests 58 and 59 from Fisher (2000) will be discussed due to the limited space available. The test #'s and object #'s (G#'s) match those reported by Fisher (2000).

Test # 59 illustrates the use of a tolerance specification on a single attribute. Test #59 compares G6 (a rectangle) with all objects in the target file whose number of nodes is between 6 and 8. Since G7 has 8 nodes, it matches G17, G18, G19, G20, and G21. Table 4 shows the results from these two experiments as well as from the other experiments.

Test # 58 illustrates a comparison in which (A AND B) OR C must be satisfied by the search results. The source graphic object is G6. Table 4 shows the search results from highest (G1) to lowest (G16, G17) ranking.

Table 3. The Experimental Test Cases

Abbreviations:
Test = test number, G# = Number of graph used as a source object
use = Y indicates attribute used in search specification, Tol = Tolerance; Iso = Isomorphism;
Weight = Wgt, Attribute A = Number of Nodes, B = Line Length, C = Number of Lines,
D = Congruent Angles, E = Isomorphism

Test	G#	A – Number of Nodes			B – Line Length			C – Number of Lines			D – Congruent Angles			E – Iso		Boolean Exp.
		use	Tol	Wgt	use	Tol	Wgt	use	Tol	Wgt	use	Tol	Wgt	use	Wgt	
58	6	Y	0	6	Y	0	2	Y	0	10	N	0	0	N	0	(A AND B) OR C
59	7	Y	2	0	N	0	0	N	0	0	N	0	0	N	0	A

Table 4. Experimental Test Results

Test #	Actual Results "[ ]" groups objects of the same rank.
58	[1][7][16,17]
59	7, 17, 18, 19, 20, 21

CONCLUSIONS

The importance of providing search engines for vector-based graphics is clear to those involved with the storage and retrieval of CAD drawings such as schematics. The research reported in this paper has focused on providing a prototype environment in which graphics searchs can be performed on vector based graphics objects.

The 2-DGCA provides a fast search engine for two-dimensional graphics objects composed of lines and points. The approach extends the work of St. Clair et al (1998) by improving the speed and accuracy of their basic search engine and by adding user-specified criteria to the search engine. While the present focus has been on line based two-dimensional graphics, it is easily extendable to all two-dimensional vector graphics, including circles and arcs. Preliminary evaluations suggest that it can be extended to three-dimensional vector objects.

User-specified search criteria which produces a ranking of matched objects is a new concept in vector graphics search engines. While the ranking algorithm developed in this research is a start, other, more descriptive algorithms will be developed as this research continues. The present approach provides a structure on which to test additional ranking approaches.

REFERENCES

Brown, Wayne C., and Barry J. Shepard, 1995, "Graphics file formats: Reference and Guide," Manning Publications Co.

Fisher, M.A., 2000, "A Two Dimensional Graphics Comparison System," MS Thesis, University of Missouri – Rolla.

Foley, James D., Andres Van Dam, Steven K. Feiner, John F. Hughes, and Richard L. Phillips, 1994, "Introduction to Computer Graphics," Reading, MA: Addison-Wesley.

Jacobs, Charles E., Adam Finkelstein, and David H. Salesin, 1995, "Fast multiresolution image querying." *Proceedings of SIGGRAPH '95*, New York, ACM, pp. 277-286.

Leather, Mark. "Simple Object Based Draw Program," *Java Gallery Page*. Silicon Graphics, Inc. < http://www.sgi.com/fun/java>

St. Clair, D.C., G. Hagedorn, C.M. Miller III, and E. Boyle, 1998, "An Intelligent Search Engine For Finding Graphics Objects," *Artificial Neural Networks in Engineering '98*, St. Louis.

MULTI- AGENT TRANSACTIONAL NEGOTIATION FOR E-COMMERCE

V.K.MURTHY
Research School of Information Sciences and Engineering,
Australian National University,Canberra, ACT, 0200, Australia
email:murthy@discus.anu.edu.au

ABSTRACT
This paper proposes a multi-agent transactional paradigm using object-based rule systems for realising agent negotiation protocols in E-Commerce. The construction of the protocol is carried out in two stages: first expressing a program into an object-based rule system and then converting the rule applications into a set of transactions.We also describe an algorithm to prove termination of the negotiation among the agents.

INTRODUCTION
In a recent paper Fisher [3] describes a philosophical approach to the representation and execution of agent-based systems.This paper develops Fisher's philosophical approach into a concurrent multi-agent programming paradigm based on the transactional logic model [2].The multi-agent system consists of the following subsystems, Dignum and Sierra [2], Genesereth and Nilsson [4], Ishida [5], Fig .1.

(1) Worldly states or environment U:
Those states which completely describe the universe containing all the agents

(2)Percept: Depending upon the sensory capabilities (input interface to the universe or environment) an agent can partition U into a standard set of messages T , using a sensory function Perception (PERCEPT):

PERCEPT :U $\to$ T. PERCEPT can involve various types of perception: see, read, hear, smell . The messages are assumed to be of standard types based on an interaction language that is interpreted identically by all agents.

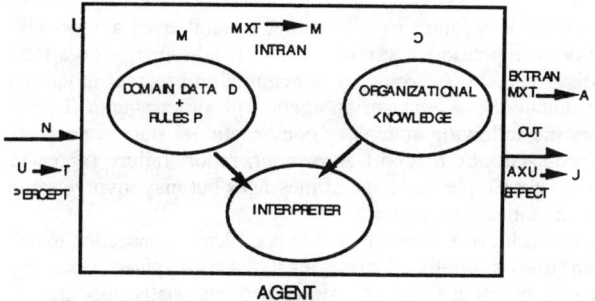

Figure 1

(3) Epistemic states or Mind M:
We assume that the agent has a mind M (that is essentially a problem domain knowledge consisting of an internal database for the problem domain data and a set of problem domain rules) that can be clearly understood by the agent without involving any sensory function. The database D sentences are in first order predicate calculus (also known as extensional database) and agents mental actions are viewed as inferences arising from the associated rules that result in an intensional database, that changes (revises or updates) D.

The agent's state of belief, or a representation of an agent's state of belief at a certain time is represented by an ordered pair of elements (D, P). D is a set of beliefs about objects, their attributes and relationships stored as an internal database and P is a set of rules expressed as preconditions and consequences(conditions and actions). When T is input, if the conditions given in the lefthand side of P match T the elements from D that correspond to the rightside are taken from D and suitable actions are carried out locally (in M) as well as on the environment.

(4) Organizational Knowledge (O): Since each agent needs to communicate with the external world or other agents, we assume that O contains all the information about the relationships among the different agents. For example, the connectivity relationship for communication, the data dependencies between agents, interference among agents with respect to rules, information about the location of different domain rules are in O.

(5) Intran: M is suitably revised or updated by the function called Internal transaction (INTRAN). Revision means acquisition of new information about the world state, while update means change of the agent's view of the world. Revision of M corresponds to a transformation of U due to occurrence of events and transforming an agent's view due to acquisition of new information that modifies rules in P or their mode of application (deterministic, nondeterministic or probabilistic) and corresponding changes in database D. Updates to M correspond to changes in U due to the occurrence of events that changes D but not P. That is: INTRAN: M X T $\rightarrow$ M

(6) Extran: External action is defined through a function called global or external transaction (EXTRAN) that maps an epistemic state and a partition from an external state into an action performed by the agent. That is : EXTRAN: M X T $\rightarrow$ A ; that is current state of mind and new input activates an external action from A.

(7) Effect: The agent also has an effectory capability on U by performing an action from a set of actions A (ask,tell, hear, read,write, speak, send,smell,taste, receive, silent), or more complex actions. Such actions are carried out according to a particular agent's role and governed by an etiquette called protocols. The effect of these actions is defined by a function EFFECT, that modifies the world states through the actions of an agent:

EFFECT: A X U $\rightarrow$ U; EFFECT can involve additions, deletions and modifications to U.

Thus an agent is defined by a 9-tuple:
(U,T,M(P,D),O,A,PERCEPT,INTRAN,EXTRAN,EFFECT).

The interpreter repeatedly executes selected rules in P, until no rule can be fired [5-12].

NEGOTIATION

In E-commerce, " negotiation" is a joint process among a number of agents with varying degrees of cooperation, competition, commitment that results finally in a total agreement, consensus or a disagreement. Accordingly, a negotiation protocol is viewed as a set of public rules that dictate the conduct of an agent with other agents. Thus a negotiation protocol involves the following actions or conversational states : propose, accept, refuse, modify, no-proposal, abort, report agreement, report failure (agree to disagree)[2]. Note that these are not simple exchange of messages but may involve some intelligent or smart computation- called transactions.

Multiagents can cooperate to achieve a common goal to complete a transaction to aid the customer. The negotiation follows rule-based strategies that are computed locally by its host server. Here competing offers are to be considered; occasionally cooperation may be required. Special rules may be needed to take care of risk factors, domain knowledge dependencies between attributes, positive and negative end conditions. When making a transaction several agents have to negotiate and converge to some final set of values that satisfies their common goal. Such a goal should also be cost effective so that it is in an agreed state at the minimum cost or a utility function . To choose an optimal strategy each agent must build a plan of action and communicate with other agents. We illustrate this by converting a distributed greedy algorithm that finds a minimal cost path in a graph into a negotiation problem.

PLANNING , REASONING AND NEGOTIATION

To solve a problem through negotiation, we need to look for a precondition that is a negation of the goal state and look for actions that can achieve the goal. This strategy is used widely in AI [4,13] and forms the basis to plan a negotiation. Such a planning is possible for clear-cut algorithmic problems. To systematically derive a multi-agent negotiation protocol (MAN) we use the specification approach [1,9]

DESIGN OF NEGOTIATION PROTOCOL

We now describe how to carry out distributed multi-agent negotiation by sending, receiving, handshaking and acknowledging messages and performing some local computations. messages. A multi-agent negotiation has the following features [2,3,14-18] :

1. There is a seeding agent who initiates the negotiation.
2. Each agent can be active or inactive.
3. Initially all agents are inactive except for a specified seeding agent which initiates the computation.
4. An active agent can do local computation, send and receive messages and can spontaneously become inactive.
5. An inactive agent becomes active if and only if it receives a message.
6. Each agent may retain its current belief or revise its belief as a result of receiving a new message by performing a local computation. If it revises its belief, it communicates its revised state of belief to other concerned agents; else it does not revise its solution and remains silent.

To explain the negotiation protocol we consider the lowest-cost path problem in a graph.

EXAMPLE

Consider the problem of finding a lowest cost path between any two vertices in a directed graph whose edges have a certain assigned positive costs (Figure 2). This problem requires the entity set of vertices, the relationship set of ordered pairs of vertices (x,y) representing edges, and the attribute of cost c for each member of the relationship set, denoted by (x,y,c). Given a graph G the program should give for each pair of vertices (x,y) the smallest sum of costs path from x to y. The vertex from which the lowest cost paths to other vertices are required is called the root vertex r (vertex 1 in this example). Let c denote the cost between any two adjacent vertices x and y, and let s denote the sum of costs along the path from the root to y; we assume that c (and hence s) is positive. This information is described by the ordered 4-tuple(x,y,c,s): (vertex label, vertex label, cost, sum of costs from root). The fourth member of the 4-tuple, namely the sum of costs from a specified root remains initially undefined and we set this to a large number *.We then use the production rules to modify these tuples or to remove them.To find the lowest cost path to all vertices from a specified root r, we use the multi-agent negotiation for tuple processing and let the 4-tuples interact; this interaction results in either the generation of modified 4-tuples or the removal of some 4-tuples of the representation .

Specification to find the shortest path

Let C(i,j) be the cost of path (i,j). A better path is one that can pass through some vertex k such that: C(i,k) +C(k,j) < C(i,j).That is our production rule is:
If C(i,k) +C(k,j) < C(i,j) then delete C(i,j) and set C(i,j) = C(i,k) +C(k,j). The invariant is: if C(i,j) is the initial cost then all the costs are always less than or equal to C(i,j) .We refine this by using the rule : If C(i,k) < C(p,k) , delete C(p,k) and retain C(i,k)
Thus the following three production rules result:

Rule 1: If there are tuples of the form (r,r,0,0) and (r,y,c,*), replace (r,y,c,*) by (r,y,c,c) and retain (r,r,0,0).Rule 1 defines the sum of costs for vertices adjacent to the root, by deleting * and defining the values.

Rule 2: If there are tuples of the form (x,y,c1,s1) and (y,z,c2,s2), where s2 > s1+c2 then replace (y,z,c2,s2) by (y,z,c2,s1+c2); else do nothing. Rule 2 states that if s2 > s1+c2 we can find a lower cost path to z through y.

Rule 3:If there are tuples of the form (x,y,c1,s1) and (z,y,c2,s2) and if s1< s2 , then remove (z,y,c2,s2) from the tuple set; else do nothing. Rule 3 states that for a given vertex y which has two paths, one from x and another from z, we can eliminate that 4-tuple that has a higher sum of costs from the root.

The above **three rules** provide for local computation by many agents and we are left with those tuples that describe precisely the lowest cost path from the root. We assume that there are n agents with names identical to the nodes in the graph and each agent is connected to other agents in an isomorphic manner to the given graph. Such an assumption on the topology of the network simplifies the organizational knowledge O. Using O, each agent knows the identity of its neighbours, the direction and cost of connection of the outgoing edges. Thus for a given directed graph the outdegree of each node is the number of out-channels and the indegree is the number of in-channels.

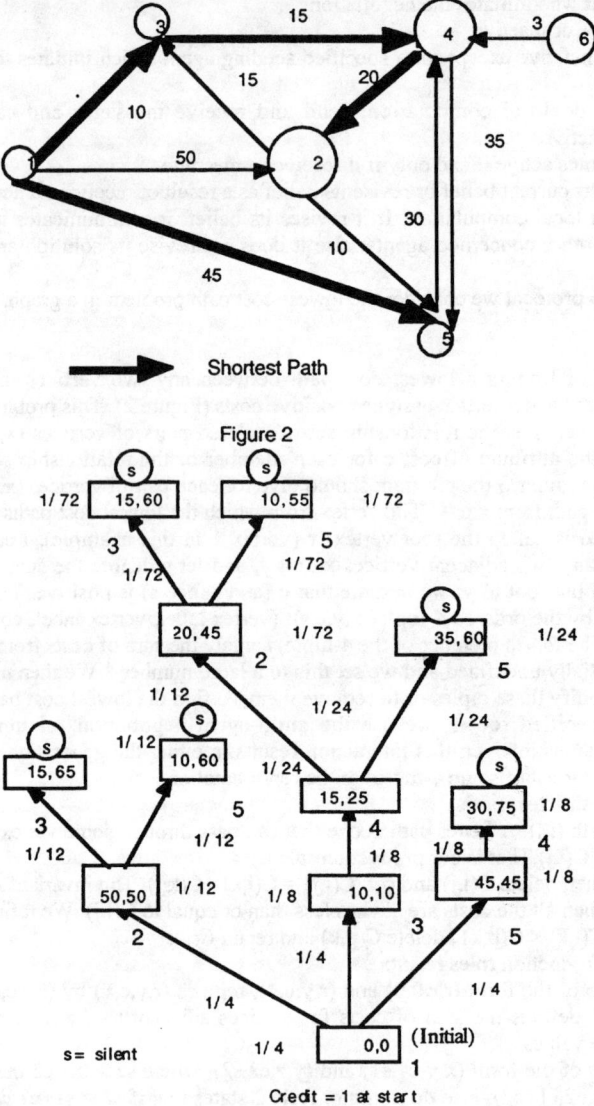

Figure 2

MULTI- AGENT NEGOTIATION

Figure 3

The revised production rules for multi-agent computation are as follows (Figure 3):
a. Agent 1 (root) sends to all its neighbours x the tuple (1,x,c,c) describing the name of the root, and the distance of x from the root (c); all the neighbours of the root handshake, receive, and store it. This corresponds to the initialization of beliefs.
b. Each agent x sends its neighbour y at a distance c1 from it, the tuple (x,y,c1,c+c1) describing its name, its distance to y and the distance of y from the root through x using its distance to the root c. This is the initial set of beliefs of the agents.
c. Each agent y compares an earlier tuple (x,y,c1,s1) got from a neighbour x, or the root, with the new tuple (z,y,c1',s1') from another neighbour z. If s1< s1', then y retains (x,y,c1,s1) and remains silent; else it stores (z,y,c1',s1') and sends out the tuple (y,w,c2,s1'+c2) to its neighbour w at a distance c2, advising w to revise its distance from the root.;or each agent revises its beliefs and communicates its beliefs to concerned agents.
d. An agent does not send messages if it receives a message from another agent that tells a higher value for its distance from the root and ignores the message, i.e., it does not revise its beliefs. Thus it contains only the lowest distance from the root. All the agents halt when no more messages are in circulation and the system stabilizes. An algorithm to detect the termination of negotiation is described in the next section.

NEGOTIATION TERMINATION DETECTION

We now describe an algorithm "Commission-Savings-Tally Algorithm (COSTA)" for global termination detection of a negotiation protocol. This is applied to the above example.

Let us assume that the N agents are connected through a communication network represented by a directed graph G with N nodes and M directed arcs. Let us also denote the outdegree of each node i by Oud(i) and indegree by Ind(i). Also we assume that an initiator or a seeding agent exists to initiate the transactions. The seeding agent (SA) holds an initial amount of money C. When the SA sends a data message to other agents, it pays a commission:$C/(Oud(SA) + 1)$ to each of its agents and retains the same amount for itself. When an agent receives a credit it does the following:
a. Let agent j receive a credit $C(M(i))$ due to some data message M(i) sent from agent i. If j passes on data messages to other agents j retains $C((M(i))/(Oud(j)+1)$ for its credit and distributes the remaining amount to other Oud(j) agents. If there is no data message from agent j to others, then j credits $C(M(i))$ for that message in its own savings account; but this savings will not be passed on to any other agent, even if some other message is received eventually from another agent.
b. When no messages are received and no messages are sent out by every agent, it waits for a time-out and sends or broadcasts or writes on a transactional blackboard its savings account balance to the initiator.
c. The initiator on receiving the message broadcast adds up all the agents' savings account and its own and verifies whether the total tallies to C.
d. In order to store savings and transmit commission we use an ordered pair of integers to denote a rational number and assume that each agent has a provision to handle exact rational arithmetic. If we assume C=1, we only need to carry out multiplication and store the denominator of the rational number.

We prove the following theorems to describe the validity of the above algorithm:

Theorem 1: If there are cycles present among the agents (including the initiator itself) then the initiator cannot tally its sum to C.

Proof: Assume that there are two agents i and j are engaged in a rule dependent argument cycle. This means i and j are revising their beliefs forever without coming to an agreement, and wasting the common resource C. Let the initial credit of i be x. If i passes a message to j, then i holds x/2 and j gets x/2. If eventually j passes a message to i, then its credit is x/4 and i has a credit x.3/4 ; if there is continuous exchange of messages for ever then their total credit remains $(x - x/2^k)$ with $x/2^k$ being carried away by the message at k th exchange. Hence the total sum will never tally in a finite time.

Theorem 2: COSTA terminates if and only if the Initiator tallies the sum of all the agents savings to C.
Proof: If part: If the initiator tallies the sum to C this implies that all the agents have sent their savings and no message is in transit carrying some credit and there is no chattering among agents.
Only if part: The credit assigned can be only distributed in the following manner:
a. An agent has received a message and credit in a buffer; if it has sent a message then a part of the credit is lost; else it holds the credit in savings.
b. Each message carries a credit; so, if a message is lost in transit or communication fails then total credit cannot be recovered.
Thus termination can happen only if the total sum tallies to C, i.e., the common resource is not wasted and all the agents have reached an agreement on their beliefs.

Figure 3 illustrates the above algorithm for the example of Section 6.1; the credits are indicated on the nodes and arcs . Note that at the termination of the algorithm the nodes have the following savings summing up to 1:
Node 1: 18/72; Node 2: 7/72; Node 3: 16/72; Node 4: 12/72; Node 5: 19/72.

CONCLUDING REMARKS

The multi-agent negotiation paradigm described here has the following features:
1.The production system provides a programming methodology free from control management.The transactional implementation provides for propose-verify-revise strategy.
2. It provides for the application of locality principle in protocol construction.
3. It will have applications in programming as well as design of multi-agent architectures [14-18] consisting of agents that serve as processes, functions, relations or constraints.

REFERENCES
[1]Bauer,F.L and Brauer,W.,1993,*Logic and Algebra of Specification*, Springer Verlag, New York.
[2]Dignum, F. and.Sierra, C.,2001, *Agent Mediated E-Commerce*, Lecture notes in Artificial Intelligence,Vol.1991; Vol..2003,Springer Verlag,New York.
[3]Fisher,M., 1995,*Representing and executing agent-based systems, in Intelligent Agents*, in M.Woolridge and N.R.Jennings (Eds.), Lecture Notes in Computer Science,Vol. 890, pp. 307-323, Springer-Verlag, New York.
[4]Genesereth,M.J., and Nilsson,N.J.,1987, *Logical Foundations of AI*,Morgan Kaufmann, New York.
[5]Ishida,T., 1994, *Parallel, Distributed and multiagent Production Systems*,Lecture Notes in Computer Science,Vol .878,Springer Verlag, New York..
[6]Krishnamurthy, E.VandMurthy,V.K.,1991,*Transaction Processing Systems*, Prentice Hall, Sydney.
[7]Kuo,S., and Moldovan, D.,1992, The state of the art in parallel production systems, *J. Parallel and distributed computing*, 15,pp. 1-26.
[8]Murthy,V.K. and Krishnamurthy,E.V.,1995, Probabilistic Parallel Programming based on multiset transformation, *Future Generation Computer Systems*,11,pp.283-293..
[9]V.K. Murthy,V.K., and E.V. Krishnamurthy, 1996,Gamma programming paradigm and heterogeneous computing, *Proc. Hawaii Intl. Conf. on System Sciences (Software Technology Track), HICSS-29* I, IEEE Computer Society Press, New York,pp. 273-281
[10]Murthy and Krishnamurthy, E.V.,1995, Automating Problem Solving using transactional paradigm, *Proc. Intl. Conf. on AI & Expert Systems*, Gordon Breach ,New York,721-729.
[11]Murthy, V.K.,1996,Transactional programming for distributed agent systems, *Proc. IEEE Intl. Conf. on Parallel and Distributed Systems, ICPADS'96*, Japan, IEEE Computer Society Press, New York,pp.64-71.
[12] Paton,N.W., and Williams,M.H.,1994, *Rules in database Systemss*, Springer Verlag,New York..
[13] Rich,E and Knight, K.,1991,*Artificial Intelligence*, McGraw Hill,New York.
[14]Singh,M.P.,1994, *Multiagent Systems*, Lecture Notes in Computer Science,Vol.799,Springer-Verlag, New York.
[15]Van de Velde,w., and Perram,J..W., (Eds.), 1996,*Agents Breaking Away*, Lecture Notes in Computer Science, Vol. 1038, Springer-Verlag, New York.
[16] Wittig, T.,1992,*ARCHON: An architecture for multiagent systems*,Ellis Horwood, New York.
[17]Woolridge,M., and Jennings, N.R., 1995,*Agent theories, Architectures and Languages, A survey*, Lecture Notes in Computer Science,Vol. 890,Springer-Verlag, New York,pp.1-39.
[18]Wooldridge, M.,Muller, J.P.,Tambe, M., (Eds.)1996, *Intelligent Agents II*, Lecture Notes in Computer Science, Vol. 1037, Springer-Verlag, New York.

FLEXIBLE DESIGN SUPPORT SYSTEM

CHRISTOPHER LANDAUER, KIRSTIE BELLMAN
cal@aero.org, bellman@aero.org
Aerospace Integration Science Center
The Aerospace Corporation
Los Angeles, California 90009-2957, USA

ABSTRACT

In this paper, we describe an approach to the infrastructure of a conceptual design support environment for space systems, which can involve thousands of organizations and millions of components, and must live for decades. To design such a system, we need a great deal of computational assistance, especially in the early design stages, when the most costly mistakes are made, and before we have enough information to use detailed engineering support tools. A designer needs to start wherever something is known, and look for design drivers. Many models and analytic tools are needed to help construct and evaluate partial design alternatives and their implications. Source data can come from models, tables in files, programs or simulations, or simple educated guesses, and the system needs to retain the source "pedigree" of all information and models used, so they can be examined and justified or replaced later on.

INTRODUCTION

Space systems are huge: they involve hundreds of organizations, thousands of people, sometimes millions of components, and are expected to last for decades (Bellman et al., 1993) (Landauer et al., 1993). They include the satellites, the launch systems, the ground stations and communication networks, and even the specialized manufacturing facilities that build some of the unique components. There are no reliable methods of developing such systems, and no methods at all for keeping track of all of the requirements and design decisions. In particular, that means that the different phases of system development are often isolated from each other, with little or no information passed along the development process.

We concentrate in this paper on conceptual design, that is, on the very early stage in which some of the most expensive design mistakes are made, and on computing systems that support the design process in a flexible way. We describe an approach to conceptual design that attempts to reduce the complexity of design, extend the coverage of the design analysis, and keep track of all the relationships among different artifacts of the process, so that any decision can be examined, reconsidered, and possibly rescinded. One of the most important properties of these design environments is that they not restrict the creativity of the designer. They must therefore be very flexible in their reactions to designer's activity, and they must provide many facilities for changing, examining, and augmenting a design.

A design environment for a complex application requires many models and even many modeling methods, since NO one model or language or method can suffice to model a complex system (Bellman and Brock, 1960). It must allow designers to make their reasoning processes and modeling decisions explicit, so the reasoning and justifications can be examined later, in the context of other design decisions, new capabilities or technologies, or new system requirements.

INTEGRATION INFRASTRUCTURE

In this section, we briefly describe the Wrapping integration infrastructure that allows a system to manage all of the disparate kinds of models needed. We provide only a few details here; the rest can be found elsewhere (Landauer and Bellman, 1999a, 1999b), and references therein.

Wrapping

In a long series of papers originally aimed at the software and systems engineering community (Landauer and Bellman, 1998, 2000), we defined and developed a Computationally Reflective integration infrastructure for Constructed Complex Systems that we called *Wrapping*. It is based on processing explicit qualitative information about all of the system components and their interconnection architecture. These systems are complex collections and interactions of components. They often contain heterogeneous processes, difficult and possibly unknown requirements, and function in complex environments. Designing and building such systems requires explicit models of the system, its architecture, and the environment in which it is expected to operate.

The Wrapping approach to integration has four essential features that underlie its flexibility and power of expression:

1. ALL parts of a system architecture are *resources* that provide some kind of *information service*, including programs (computational tools and services), data, user interfaces, human users, architecture and interconnection models, meta-tools (such as genetic algorithm and neural net constructors, knowledge-bases and other resources that use resources), and everything else that provides any information service.

2. ALL activities in the system are *problem study*, (i.e., all activities *apply* a resource to a *posed problem* in a *problem context*), including user interactions, information requests and announcements within the system, service or processing requests, and all other processing behavior. We therefore specifically separate the problem to be studied from the resources that might study it.

3. *Wrapping Knowledge Bases* contain *Wrappings*, which are explicit machine-processable descriptions of all of the resources and how they can be applied to problems to support the five *Intelligent User Support* functions (Bellman, 1991):

    - *Selection* (which resources can be applied to a particular problem),

- *Assembly* (how to let them work together),
- *Integration* (when and why they should work together),
- *Adaptation* (how to adjust them to work on certain kinds of problems), and
- *Explanation* (why certain resources were or will be used or not used).

That is, the Wrappings describe not only "how", but also "why", "when", and "whether" the use of a resource is appropriate for a given problem in a given context.

4. *Problem Managers (PMs)*, including the *Study Managers (SMs)* and the *Coordination Manager (CM)*, are active integration processes that interpret the Wrapping descriptions to collect and select resources to apply to problems. They use Knowledge-Based Polymorphism, which is a kind of implicit invocation, both context- and problem-dependent, to choose and organize resources. We emphasize these processes of integration as well as the data, since the processes define the semantics of the data. The PMs are also resources, and they are also Wrapped (and therefore selected, applied, etc.).

The most important conceptual simplifications that the Wrapping approach brings to integration are the uniformities of the first two features, which together provide a kind of consistency of expression at the meta-level to support heterogeneity of content and behavior in the system: the uniformity of treating everything in the system as resources, and the uniformity of treating everything that happens in the system as problem study. The most important algorithmic simplification is the reflection provided by treating the PMs as resources themselves: by considering these programs that process the Wrappings to be resources also, and Wrapping them, all of our integration support processes apply to themselves, too. The entire system is therefore Computationally Reflective (Landauer and Bellman, 1998, 2000). In particular, the same flexibility of resource use occurs in the system internals, and, in fact, there are NO privileged processes or data sources at all: every part of the system is selectable at run time (provided of course that the resources that do the work have been made available to the system). Therefore, the system has a complete model of its own behavior (to some level of detail), so it can retain design decisions and their rationales, to revisit and examine or evaluate them again. It is this ability of the system to analyze its own behavior that provides some of the power and flexibility of resource use.

Wrapping Processes

The heartbeat of the system is in the Coordination Managers (CMs), which perform a sequence of steps that manage all of the system activity (see Figure 1). In particular, they repeatedly cycle through a loop of getting a problem posed and using a Study Manager (SM) to study it.

These steps are interpreted as follows: to "Find context" means to establish a context for problem study, possibly by requesting a selection from a user, but more often getting it explicitly or implicitly from the system invocation. It is our placeholder for conversions from that part of the system's invocation environment that is necessary for the system to represent to whatever internal context

```
define
CM [ <usr> ]:
    [
    <cxt> = Find context [ <usr> ],
    for ever :
        [
        <prob> = Pose problem [ <usr> ],
        <res> = Study problem [ <usr>, <prob>, <cxt> ],
        Present results [ <usr>, <res> ]
        ]
    ],
```

Figure 1: Default Coordination Manager (CM) Step Sequence, in *wrex*

structures are used by the system. To "Pose problem" means to get a problem to study from the problem poser (a user or the system), which includes a problem name and some problem data, and to convert it into whatever kind of problem structure is used by the system (we expect this is mainly by parsing of some kind). To "Study problem" means to use an SM and the wrappings to study the given problem in the given context, and to "Present results" means to tell the poser what happened. Each step is a problem posed to the system by the CM, which then uses the default SM to manage the system's response to the problem. The first problem, "Find context", is posed by the CM in the initial context of "no context yet", or in some default context determined by the invocation style of the program. This Problem Manager is written in *wrex*, our generic programming notation (Landauer and Bellman, 1999a, 1999b).

The default Study Manager (SM) is the main problem solving algorithm (see Figure 2). The SM process begins with a problem poser, a problem defined

> **Interpret problem** :
>> **Match resources** : get candidate list
>> **Resolve resources** : reduce list, make some bindings
>> **Select resource** : choose the one to apply
>> **Adapt resource** : finish the bindings
>> **Advise poser** : describe resource and bindings chosen
>
> **Apply resource** : go do it
>
> **Assess results** : evaluate the result

Figure 2: Default Study Manager (SM) Step Sequence

by its name and associated data, and the context in which the problem was originally posed. This default SM step sequence represents a basic inline planner (Landauer and Bellman, 1998). Computational Reflection now comes from a "meta-recursion", by taking each step above for the CM and SM also to be a posed problem, and using the same kind of selection process to choose resources to apply to those problems. In each case, there is a simplest resource that applies (for example, the default SM is the simplest resource that applies to the problem "Study Problem" posed by the CM).

Wrapping Knowledge Bases

Since the process steps that interact with the Wrapping Knowledge Base (WKB) are themselves posed problems, we can use completely different syntax and semantics for different parts of the WKB in different contexts, and select the appropriate processing algorithms according to context. Even though the WKBs have a variable syntax, there are some common features in the semantics, which we describe next. The collected Wrapping Knowledge Bases (WKBs) contain an entry for each style of use of each resource. It describes how to use a resource, when and whether to use it and in which particular contexts it can be used, but does not necessarily define what the resource is or does.

Each Wrapping is a list of "problem interpretation" entries, each of which describes one way in which this resource can be used to deal with a problem. There may be several problem interpretation entries for the same problem if the resource has many different ways of dealing with it. Each of these problem interpretation entries has lists of context conditions that must hold for the resource to be considered or applied. All of these conditions are checked using the current local context. These sets of conditions are important at different times; one at resource consideration or planning time, and the other at resource application time. These act as pre-conditions for application of the resource. The corresponding post-condition is the "product" list of context component assignments, which describes what information or services this resource makes available when it is applied. This means that ANY Knowledge Representation mechanism that: (1) can express the above semantic features, (2) can be queried by problem (in order to match resources to problems), and (3) can be filtered by problem and context information (in order to resolve resource selections), can be used for Wrappings.

INTEGRATED DESIGN ENVIRONMENTS

In this section, we briefly describe an integrated design environment based on Wrappings. It is loosely descended from the VEHICLES system, which was a conceptual design environment for space systems (Bellman et al., 1993) (Landauer et al., 1993). This application of Wrapping supports a very large set of models and resources for the design support system, selected and integrated by the Wrapping processes, according to the information in the Wrapping Knowledge Bases.

First, the system contains a large number of computational utilities, such as optimizers, equation solvers, visualization programs, inductive inference and pattern recognition algorithms, etc., that are independent of the problem domain. Second, the system contains a collection of tools specific to an application domain, such as design rationale capture, design space study methods, including alternative comparison and similarity computations, analysis planning algorithms, etc.. Along with these resources are the corresponding Wrappings, which define what kinds of problems they are intended to address, how to set up and use the resources for that kind of problem, and how to interpret the results.

Third, the system contains many resources for a given design project, organized into an abstraction hierarchy, of models, of partial design problems, of partial design solutions, and of connections among the models. Finally, the Computational Reflection provided by the Wrapping approach allows the system to contain and reason about the design processes, and purposes within the design process of using various resources. These facilities allow the system to help the designer, by making suggestions or noting comparisons with other activities, models, or partial designs.

CONCLUSIONS

Modeling complex systems requires many kinds of models, modeling methods, and even modeling approaches, which means that there must be sufficient descriptive information about all of those models, and sufficient computational processes to interpret those descriptions, to make very flexible use of those models. Organizing this large and diverse a set of models and interpreters requires a powerful and flexible infrastructure to support it. We have shown how Wrappings can provide such an infrastructure, using Computational Reflection to monitor its own actions. This approach does not make the design problem easy. Modeling is still hard, and model integration is hard, but we can provide computer support for different kinds of model creation, analysis, and integration.

REFERENCES

Kirstie L. Bellman, "An Approach to Integrating and Creating Flexible Software Environments Supporting the Design of Complex Systems", pp. 1101-1105 in *Proc. WSC'91*, 8-11 Dec 1991, Phoenix, Arizona (1991)

Kirstie L. Bellman, April Gillam, and Christopher Landauer, "Challenges for Conceptual Design Environments: The VEHICLES Experience", *Revue Internationale de CFAO et d'Infographie*, Hermes, Paris (Sep 1993)

Richard Bellman, P. Brock, "On the concepts of a problem and problem-solving", *American Mathematical Monthly*, Vol. 67, pp. 119-134 (1960)

Christopher Landauer, Kirstie L. Bellman, "Wrappings for Software Development", pp. 420-429 in *Proc. HICSS'98, Software Process Improvement Mini-Track*, 6-9 Jan 1998, Kona, Hawaii (1998)

Christopher Landauer, Kirstie L. Bellman, "Generic Programming, Partial Evaluation, and a New Programming Paradigm", Chapter 8, pp. 108-154 in Gene McGuire (ed.), *Software Process Improvement*, Idea Group (1999a)

Christopher Landauer, Kirstie L. Bellman, "Problem Posing Interpretation of Programming Languages", in *Proc. HICSS'99, Engineering Complex Computing Systems Mini-Track*, 5-8 Jan 1999, Maui, Hawaii (1999b)

Christopher Landauer, Kirstie L. Bellman, "Reflective Infrastructure for Autonomous Systems", in *Proc. EMCSR'2000, Symp. Autonomy Control: Lessons from the Emotional*, 25-28 Apr 2000, Vienna (2000)

Christopher Landauer, Kirstie L. Bellman, April Gillam, "Software Infrastructure for System Engineering Support", *Proc. AAAI'93 Workshop on Artif. Intell. for Software Engr.*, 12 Jul 1993, Washington, D.C. (1993)

A NEW DECISION MAKING STRATEGY FOR DISTRIBUTED CONTROL OF CELLULAR MANUFACTURING SYSTEMS

PAOLO RENNA
University of Basilicata, DIFA,
Potenza, Italy
GIOVANNI PERRONE
University of Basilicata, DIFA,
Potenza, Italy

MICHELE AMICO
University of Palermo, DTPM,
Palermo, Italy
MANFREDI BRUCCOLERI
University of Palermo, DTPM,
Palermo, Italy

ABSTRACT

In order to play in the environment of high market shifting, rapid introduction of new technologies and customer needs focalizations, enterprises need to build reactive and agile manufacturing systems. Due to their reconfiguration features, agent - based manufacturing have demonstrated their ability to be very agile. System efficiency is one of the most important issue in agile manufacturing system management. This paper proposes a new decision making strategy for autonomous agents in cellular manufacturing system; such a strategy is based on a fuzzy rule which dynamically takes into account the differences among the cells' efficiencies during the negotiation process.

INTRODUCTION

The nowadays competition is played in an environment characterized by high market shifting, rapid new technologies development and introduction, global competition and customer need focalization. In order to play in such an environment, enterprises need to build reactive or agile manufacturing systems. In the field of manufacturing this objective can be achieved by operating both at system and control level. At the system level the most common solution is the decomposition of the manufacturing system into smaller units, e.g. manufacturing cells. At the control level a way to guide the complexity of operation management problems to simplicity, reactiveness, scalability and fault tolerance is to take advantages from distributed control by the implementation of Multiple Agent Systems (MAS) controls (Shen and Norrie, 1999).

Two types of distributed real time controller for making scheduling decision in distributed manufacturing systems can be distinguished in literature: (1) those where agents are responsible for local scheduling, while the global schedule is obtained through a merge of such local schedules; such an approach is very similar to centralized scheduling (McEleney et al., 1998); (2) those where an agent manages a single resource and the overall scheduling is obtained by means of a negotiation with other resource agents. This approach is similar to dispatching rules application in real time scheduling.

Most of the research efforts within the second approach concern the way how autonomous agents negotiate with each other. Among these studies, those related to the negotiation protocol mechanism include the Contract Net Protocol

(Smith, 1980), and the Market-Driven Contract Net (Baker, 1991).

Nevertheless, very few studies concern the way how autonomous agents make dispatching decisions in order to get better global system performances. This problem is indeed very important, as stressed in the huge literature concerning dispatching rules application in manufacturing systems (Xiang and O'Brien, 1995). This paper stresses that autonomous agent decision making strategies have to be selected in order to improve global systems performances; such strategies need to be flexible and system status dependent. In particular, this paper proposes a new decision making strategy for autonomous agents in cellular manufacturing systems; this strategy is based on the evaluation of the cell efficiency during the time flowing. A proper simulation experiment plan will show a comparison between two different approaches of the proposed strategy: the fixed weight approach and the fuzzy weight approach.

THE DECISION MAKING STRATEGY

The manufacturing environment considered here consists of M production cells, that are able to perform any kind of manufacturing operation. In such a manufacturing system, scheduling problems are mere dispatching decisions. The control system has only to decide in which cell the part will perform the next operation. The dispatching problem is solved through a negotiation between autonomous agents representing resources and parts in a real time fashion.

A resource agent is associated to each cell; it is an intelligent entity whose main aim is to schedule the resource tasks in order to improve the resource efficiency. Moreover, when a new part enters the system the corresponding part agent is created; whenever the part requires a technological operation, the part agent locates the cell where that operation should be performed. Dispatching problems for each required operation are faced through a negotiation between the agents of the part and of the resources; this negotiation is based on a contract net protocol. For detailed description of the negotiation process steps, see Renna et al. (2001).

The contract net protocol defines the environmental relations of the autonomous agents involved in the network, but no indication about the agent decision making behavior (productive function) is given. The productive function of the resource agent consists in evaluating and providing to the part agent the following three parameters:

Expected Waiting Time (EWT): it expresses the time that the part should wait in queue before being processed by the k-th resource (EWT_k). It is computed by summing up the processing times, PT_k^i, of the $NQ_k(t)$ parts waiting in the resource queue, plus the residual processing time of the part that is being processed at the negotiation time t, rPT_k:

$$EWT_k(t) = \sum_{i=1}^{NQ_k(t)} PT_k^i + rPT_k; \qquad (1)$$

Resource Failure Index (RFI): the k-th resource failure index is the ratio between the cumulated resource failure time in the interval [0, t], $FT_k(t)$, and the interval length t:

$$RFI_k(t) = FT_k(t)/t; \qquad (2)$$

Resource Processing Index (RPI): the *k*-th resource processing index expresses the ability of a cell to perform a technological operation more or less fast; it depends on the technical characteristics of the cell machines, on the status of its tools, on the specific technological operation under consideration. The actual processing time of the part over the *k*-th resource is obtained by multiplying the RPI_k by the expected processing time of the generic part i, PT^i, i.e.:

$$PT_k^i = RPI_k(t) \cdot PT^i . \tag{3}$$

It has to be noticed that the first of the above measures, i.e. the *EWT*, is a system dependent performance measure, while the *RFI* and the *RPI* are exclusively resource dependent. This means that the *EWT* can be modified by the part assignment policy, while the *RFI* and the *RPI* values can not.

The last two parameters concur in determining the *Internal Resource Index (IRI)*, while the first one, *EWT*, sets the *External Resource Index (ERI)*. Finally the *Resource Efficiency Index (REI)* is the weighted sum of *IRI* and *ERI*.

In particular the part agent productive function operates through the following steps:

(1) Resource performance measures normalization; after having collected the three performance measures described above, the part agent normalizes them by computing the following indexes for each resource:

$$\eta_k^I(t) = [\max_h \{I_h(t)\} - I_k(t)] / [\max_h \{I_h(t)\} - \min_h \{I_h(t)\}] \tag{4}$$

where *I* is *EWT*, *RFI* or *RPI*, and $k, h = 1, \ldots M$.

(2) *IRI* computation; this index is computed as it follows:

$$IRI_k(t) = [J_k(t) - \min_h \{J_h(t)\}] / [\max_h \{J_h(t)\} - \min_h \{J_h(t)\}] \tag{5}$$

where $J_k(t) = \eta_k^{RPI}(t) + \eta_k^{RFI}(t)$.

(3) *ERI* computation: $ERI_k(t) = \eta_k^{EWT}(t)$;

(4) *REI* computation; this index is computed as it follows:

$$REI_k(t) = w \cdot ERI_k(t) + (1-w) \cdot IRI_k(t) \tag{6}$$

where *w* varies within the interval [0, 1].

(5) Part assignment; the part agent assigns the part to the most efficient resource, k^*, at time t, i.e.:

$$k^* \mid REI_{k^*}(t) = \max_k \{REI_k(t)\} \tag{7}$$

In the above procedure the part agent has to assign a value to the weight *w* of relation (6). According to Amico et al. (2000), in order to accomplish this task two different procedures can be considered: the fixed weight and the fuzzy dynamic weight approaches. In the former approach, the weight is fixed a priori and it doesn't change throughout the production process; in the latter one, the weight is computed by adopting a fuzzy rule.

THE FUZZY RULE FOR THE WEIGHT COMPUTATION

The weight *w* is inferred through a proper fuzzy system, whose knowledge base is the statement: "If the cells have a very similar *IRI* values, then the weight of *ERI* in the *REI* computation is high". Therefore, the fuzzy system consists of a single rule whose input is the standard deviation of the Internal Resource Indexes and the output is the weight *w*.

The *IRI* standard deviation is shown in the following relation:

$$\sigma = \sqrt{\frac{1}{M} \cdot \sum_{k=1}^{M} \left(IRI_k - \overline{IRI} \right)} \qquad (8)$$

where $\overline{IRI}$ is the average of IRI_k; σ varies from 0, when the cells have all the same *IRI*, to 0.5, when cells' *IRI* are very different from each other. It could be demonstrated that the maximum value of σ is 0.5, when *M* is even. In fact, if *M* is even, the maximum standard deviation is obtained when the IRI_k values are half 0 and half 1. Thus the fuzzy rule can be expressed as: "If σ is *low* then *w* is *high*". The membership function of the linguistic terms *low* and *high* are reported in Fig. 1. Because of the crispness of the input, the output fuzzy set is obtained applying a fuzzy implication to the antecedent membership grade and the consequent linguistic term. In particular the Wu fuzzy implication (Klir et al., 1995) has been chosen, and finally, to obtain the weight crisp value, the center of gravity (COG) of the output fuzzy set is taken (Amico et al., 2000).

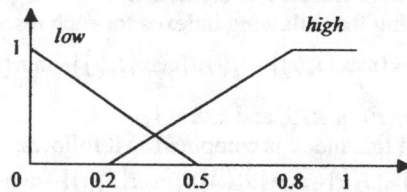

Fig. 1: Fuzzy linguistic terms

THE EXPERIMENTAL ENVIRONMENT

The proposed decision making strategy should be able to improve the control system reaction to the internal changes of the manufacturing system. In order to demonstrate the effectiveness of this strategy a dynamic environment has been considered. Renna et al. (2001), demonstrated this effectiveness by adopting the fixed weight approach. The objective of this paper is to verify the efficacy of the fuzzy weight approach respect to the fixed weight one: for this reason the same simulation experiment plan of Renna et al. (2001) has been considered. Four manufacturing systems have been considered, containing two, four, six, or eight general purpose cells. Each manufacturing system consists of only two kinds of cell; moreover, it has been assumed that each system has an equal number of the two cell typologies. The manufacturing systems are requested to produce a set of four different parts. Each part needs several operations to be performed by whichever cell. The number of operations for each part is reported in Table 1, where the production mix is also provided.

Table 1: Production run external attributes

	Part 1	*Part 2*	*Part 3*	*Part 4*
N. of operations	3	3	3	2
Mix	30%	40%	15%	15%

Parts enter the system following an exponential arrival stream whose mean inter-arrival times are reported in Table 2. Furthermore, parts are characterized by a processing time (*PT*), which is uniformly distributed within the ranges reported in Table 2.

Table 2: Inter-arrival times and *PT* (minutes)

Number of cells	Inter-arrival time	PT
Two cells	23	Un[14; 18]
Four cells	21	Un[29; 33]
Six cells	18	Un[36; 40]
Eight cells	14.5	Un[40; 44]

In order to emulate a dynamic environment the four manufacturing system configurations have been tested through a production run consisting of several alternating stages; each stage is characterized by different internal attributes, that refer to cells operative condition, i.e. the *RPI* and the *TBF* (Time Between Failures). The *TBF* is exponentially distributed, with the mean computed multiplying the mean PT ($\overline{PT}$) by a numerical coefficient; this coefficient depends on the cell type and the production run stage, as it is shown in Table 3.

Table 3: Production run internal attributes

	RPI		TBF	
	Type 1 cell	Type 2 cell	Type 1 cell	Type 2 cell
Stage 1	0.9	0.6	5	8
Stage 2	0.81	0.66	5.75	6.4
Stage 3	0.73	0.72	6.61	5.12
Stage 4	0.65	0.8	7.6	4.1
Stage 5	0.59	0.88	8.74	3.27

Moreover, the two kinds of cell not only have a different efficiency in performing manufacturing operations, but this efficiency varies over the time according with the production stages. In particular it has been assumed that cells of type one have an increasing efficiency, vice versa cells of type two.

Finally, the repairing time is exponentially distributed, with mean $1.5 \cdot \overline{PT}$.

EXPERIMENTAL RESULTS AND CONCLUSIONS

Four different stage lengths have been tested; specifically, the stage length, *sl*, has been obtained by the expression $sl = k \cdot \overline{PT}$, where *k* is a stage length factor that could be respectively 20, 10, 5, 2.5. By reducing the stage length, a more dynamic and changing production environment is obtained. The simulation run length has been fixed in order to avoid transitory influence on the results of each run and manufacturing system configuration. Of course, this means that the simulation run length is longer than just 5 production stages; for this reason, within the same simulation run, data of Table 3 are replicated several times. By combining the manufacturing system size and the different stage lengths, 16 experiment classes are obtained. The system has been modeled by using the Arena® environment (Kelton et al., 1998). Results have been compared in term of the work in process, the throughput time and the average queue length of the cells. For each experiment class a number of replication able to assure a 5% confidence interval for each performance measure has been conducted.

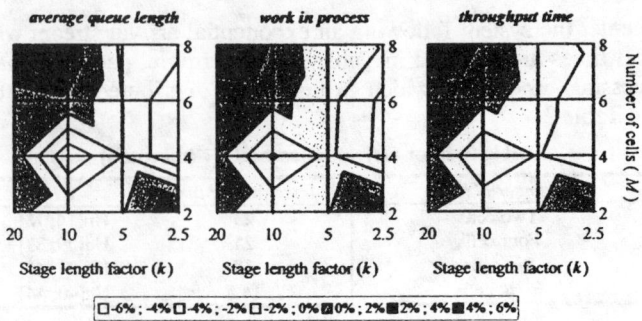

Fig. 2: Experimental results

Figure 2 shows the experimental results in terms of per cent difference of the three performance measures obtained by using the fuzzy weight approach in front of the fixed weight approach. All the performances clearly show an improvement, by using the fuzzy weight approach, in the following cases: low stage length factor ($k \leq 5$) and high or medium number of cells ($M \geq 4$); medium stage length factor and medium number of cells (around the point $k = 10$, $M = 4$). In other words, the fuzzy weight approach seems to perform better for high manufacturing system sizes in very dynamic environments. Several other issues should be investigated within this research framework; the most relevant concerns the influence of external changes in the choice between fuzzy dynamic weight and fixed one; this research has, indeed, investigated the effect of internal changes on the two methods; but, of course, manufacturing systems are also interested to external changes such mix variation, congestion increasing and so forth. These issues are object of further developments.

REFERENCES

1. Amico, M., Bruccoleri, M., Perrone, G., Noto La Diega, S., 2000, "Designing an Intelligent Controller for Agent-based Manufacturing through Fuzzy Systems", *Intelligent Engineering Through Artificial Neural Networks, Volume 10*, ASME Press: 399-404;
2. Baker, A.D., 1991, "Manufacturing Control with a Market-Driven Contract Net", *PhD thesis, Rensselaer Polytechnic Institute*, NY;
3. Kelton W. D., Sadowski R. P., Sadowski D. A., 1998, Simulation with Arena, McGraw Hill;
4. Klir, G. J., Yuan, B., 1995, *Fuzzy sets and fuzzy logic, theory and applications*, Prentice Hall;
5. McEleney, B., O'Hare, G.M.P., Sampson, J., 1998, "An Agent Based System for Reducing Changeover Delays in a Job-Shop Factory Environment", *Proceedings of PAAM'98*, London;
6. Renna, P., Perrone, G., Amico, M., Bruccoleri, M., Noto La Diega, S., 2001, "A performance comparison between market like and efficiency based approaches in Agent Based Manufacturing environment", *Proceedings of the 34th CIRP International Seminar on Manufacturing Systems*, Athens, May 2001: 93-98;
7. Shen, W., and Norrie D., H., 1999, "Agent-Based Systems for Intelligent Manufacturing: A State-of-the-Art Survey", *Knowledge and Information System, an Int. J.*, 1(2), 129-156;
8. Smith R. G., 1980, "The contract net protocol: high level communication and control in a distributed problem solver", *IEEE Transactions on computers*, Vol. 12, 1104-1113;
9. Xiang, D. and O'Brien, C., 1995, "Cell Control research – current status and development trends", *International Journal of Production Research*, Vol.33, No.8, 2325-2352.

NEURAL NETWORK SYSTEM FOR WHITE COLOR BALANCE ADJUSTMENT IN TELEVISION RECEIVERS

FRANCISCO DEL PUERTO
Electrical Design Department
Television Div., L.G. Electronics
Reynosa, Tamaulipas, Mexico

MOUNIR BEN GHALIA
Electrical Engineering Dept.
The University of Texas-Pan American
Edinburg, Texas, U.S.A.

ABSTRACT

This paper discusses the application of neural networks to the white tracking adjustment of television receivers during production. High quality levels of tracking for the color temperature 8,000 °K were obtained with four-layer (7-10-10-6) network. The network input set consists of brightness level, high and low luminance levels, and "x" and "y" coordinates on the chromaticity diagram for both high and low luminance. The network output set consists of recommended adjustments for brightness, red, green, and blue cutoffs, and green and blue gains. The network was trained using the back-propagation algorithm. The experimental study has shown that the application of neural networks has reduced the testing time which has led to an increase in production rate.

INTRODUCTION

Companies that manufacture television (TV) receivers need an automated process to make white color adjustment during production. In order to ensure a desired high production rate, the adjustment process should not take more than twenty seconds. Moreover, the adjustment process should be repeatable in order to obtain similar responses of thousands of TV sets for one same model and CRT size. A TV system combines the basic additive colors, namely red, blue, and green in order to obtain the white color and any other possible secondary color. The information carried in a video signal is assembled to display the original image on the screen. A TV set has three cathodes that produce electron beams each targeting a particular basic color (red, blue, and green) of the phosphor-coated internal screen (Webster, 1999), (Keller, 1997).

A color TV set is considered balanced for white when it displays a picture of a neutral white card as "pure white," and not some identifiable color. Neutral or pure white has no color: the color saturation is zero (Grob, 1984). Before any adjustment, a typical response of a CRT is characterized by uneven current intensities in the three cathodes resulting in an unbalanced white color, which results in an unfaithful reconstruction of color images on the screen. Because any image is reconstructed by mixing three colors and theirs intensities, it is mandatory to have the white color balance in order to display the original image with the correct chrominance and amount of detail.

This paper presents the results of design and implementation of a neural network-based white balance adjustment system. The use of the neural network adjustment system helps to eliminate the existing iterative adjustment approach used to compute the optimal values of the control variables to be downloaded into the TV circuit board during testing.

CONTROL VARIABLES RELATED TO VIDEO SIGNAL

The CRT response can be adjusted by modifying the values of six control variables accessible on the TV video processor unit. These control variables are related to the video signal (see Figure 1) and are defined next.
- *RF Brightness Control*: adjusts the brightness level for the three cathodes.
- *Red, Green, & Blue Cutoffs*: adjust the black level (DC) for each cathode separately.
- *Green & Blue Gains:* adjust the amplitude of AC video signal for the two cathodes.
- *Luminance:* "Y" (luma) indicates the amount of light intensity.
- *Chrominance:* color defined by a pair of coordinates "x" and "y" on the chromaticty diagram (Keller, 1997).
- *Color Temperature:* an object heated to any temperature above 650-900 °K will produce a broad-spectrum emission of light with its color directly related to its temperature. This is termed blackbody radiation. The color progresses from a very deep red through orange, yellow, white, and finally bluish-white as the temperature increases. From marketing considerations, TV sets must display a specific color temperature corresponding to some specific "x" and "y" coordinates on the chromaticity diagram for low luminance (1 Ft/L) and high luminance (20-50 Ft/L).

NEURAL NETWORK-BASED WHITE BALANCE ADJUSTMENT
Neural Network Structure

In this study a fully connected multi-layer feed-forward neural network structure has been chosen for the white balance adjustment. Although, in a multi-layer neural network structure, several hidden layers might be included between the input layer and the output layer, for this particular application two hidden layers were used. Several network topologies differing by the number of hidden layers and the number of neurons in each hidden layer have been considered. In this paper, we present the neural network topology (7-10-10-6) that has provided us with the best results. The structure of the network is given in Figure 2. The following subsections give more insight into the selected topology.

Input Layer. There are seven input variables, each targets one node of the input layer of the neural network. These variables are summarized in Table 1. The first input variable (RF-B1) corresponds to the initial integer value of the black level. The other six input variables Y_1, x_1, y_1, Y_2, x_2, y_2 are directly measured from the CRT response using a color analyzer equipment. Luma 1 (Y_1) and Luma 2 (Y_2) are measured in footlambert (Ft/L) (1 Ft/L = $1/\pi$ cd/ft^2). Note that luminance is the measurement of light from a surface while brightness is the subjective appearance of the surface.

Output Layer. The neural network has six output control variables which are summarized in Table 2. Once the optimal values of the control variables are computed by the neural network, they are downloaded into the PC board of the TV chassis, resulting in a CRT response that corresponds to the desired coordinates of the white color on the chromaticity diagram.

Objective. The objective of the white color balance adjustment is to obtain the correct coordinates of the white color on the chromaticity diagram corresponding to a desired color temperature. For each TV model, a color temperature is requested. Low-end TV sets have only one color temperature. Middle-end TV sets have two color temperatures. High-end TV sets have three color temperatures. The training objectives are given in Table 3.

Network Training

A 27 inch CRT model that has only one color temperature, set at 8,000 °K, was selected for this study. One of the reasons for selecting this TV model is that it has been produced at high volume. The first task was to collect data to be used for the neural network training. The existing iterative adjustment program was modified to allow for collecting adjustment data during production. In one production day, 1,405 TV sets were adjusted and the data were recorded. For each TV set, the variables associated with the original CRT responses were measured and the values of the control variables were saved after they have been loaded into the TV memory. In addition, the values (x,y) at low luminance and high luminance were recorded after adjustment was completed. Hence, a matrix of data of size 1,405x17 was obtained. The 17 values from each TV set are: 7 inputs; 6 Outputs, 2 final "x" and "y" coordinates at low luma and 2 final "x" and "y" coordinates at high luma.

The single most significant aspect of the implementation of neural network is mapping of the training set of data to the output layer (Haykin, 1997). Its significance directly influences the learning behavior and performance of the network. Initial weights have also a crucial role on the performance of the neural network training. For our study, the initial weight values were randomly assigned, and they were adjusted during training via the back-propagation algorithm (Haykin, 1997).

Data corresponding to 1,335 TV sets were used for neural network training. The training consists in (i) presenting the network with values of the variables to its input nodes, (ii) calculating the network output vector, (iii) computing the error between the desired output and the actual output of the network, and (iv) back-propagating the error and adjusting the network weights. After the neural network has been trained and final values of the weights have been obtained, its performance in white balance adjustment was tested using the data from the remaining 70 TV sets.

Network Testing and Simulation Results

The objective of the neural network-based white balance adjustment is to eliminate the existing iterative adjustment approach used in the computation of the optimal values of the control variables. The simulation results are presented in Figures 3-8. It can be seen that overall the values of the control variables computed by the neural network are very close to the target or desired values. The trained network is then implemented for experimental testing during on-line production.

Test System Set-Up. A special program has been developed to implement an on-line neural network-based white balance adjustment system. The test system set-up is shown in Figure 9.

Results. The neural network was experimentally tested on 368 new 27 inch TV sets. Very promising experimental results have been obtained using the neural network-based white balance adjustment system. It took the neural network only 17 seconds to adjust a TV set compared to the lengthier traditional process (taking between 26 and 35 seconds). In addition, the target x-y coordinates are reached with +/- 0.005 accuracy using the neural network approach, whereas with the traditional approach, the accuracy is about +/-0.013. Therefore, the neural network approach not only has helped to meet the 20 seconds adjustment constraint, but also has provided better adjustment results than the traditional approach. The results are summarized in Table 4.

CONCLUSIONS

Companies that manufacture television (TV) receivers need an automated process to make white color adjustment during production. In order to ensure a desired high production rate, the adjustment process should not take more than twenty seconds. In addition, the adjustment process should be repeatable in order to obtain similar responses of thousands of TV sets for one same model and CRT size.

This paper has presented the results of design and implementation of a neural network-based white balance adjustment system. The use of the neural network adjustment system has helped to eliminate the existing iterative adjustment approach used to compute the values of the control variables that need to be downloaded into the TV circuit board. An important benefit resulting from the implementation of the neural network is the increase in the production rate by reducing the time required for TV set adjustments.

In this study, a 27 inch CRT model with only one color temperature, set at 8,000 °K, was selected. Future work is needed to design a neural network capable of adjusting TV models that require more than one color temperature.

REFERENCES

Grob, B., 1984, *Basic Television and Video System*, Fifth Edition, McGraw-Hill, New York.
Haykin, S., 1997, *Neural Networks: A Comprehensive Foundation*, Prentice Hall, New Jersey.
Keller, P. A., 1997, *Electronic Display Measurement*, Wiley, New York.
Webster, J. G., 1999,"Cathode Ray Tube Displays," *Wiley Encyclopedia of Electrical and Electronics Engineering*, Wiley, New York.

Table 1- Inputs to the neural network

#	Input	Symbol	Unit	Description
1	RF Brightness 1	RF-B1	-	Initial Black level
2	Luma 1	Y1	Ft/L	Low Luma
3	X Coordinate 1	x1	-	"x" at Low Luma
4	Y Coordinate 1	y1	-	"y" at Low Luma
5	Luma 2	Y2	Ft/L	High Luma
6	X Coordinate 2	x2	-	"x" at High Luma
7	Y Coordinate 2	y2	-	"y" at High Luma

Table 2- Outputs of the neural network

#	Output	Symbol	Unit	Description
1	RF Brightness 2	RF-B2	-	Final Black Level
2	Red Cutoff 1	RC1	-	Red Cutoff
3	Green Cutoff 1	GC1	-	Green Cutoff
4	Blue Cutoff 1	BC1	-	Blue Cutoff
5	Green Gain 1	GG1	-	Green Gain
6	Blue Gain 1	BG1	-	Blue Gain

Table 3 - Training objectives

Variable	Symbol	Description	8,000 °K
x Coordinate 1	x_1	"x" at Low Luma	$x_1 = 0.295$
y Coordinate 1	y_1	"y" at Low Luma	$y_1 = 0.305$
x Coordinate 2	x_2	"x" at High Luma	$x_2 = 0.295$
y Coordinate 2	y_2	"y" at High Luma	$y_2 = 0.305$

Table 4- Experimental testing results

Method	Processing Time	x-y Accuracy at Low Luma	x-y Accuracy at High Luma
Original	25 to 36 seconds	+/-0.007	+/-0.013
Neural Network	17 seconds	+/-0.005	+/-0.005

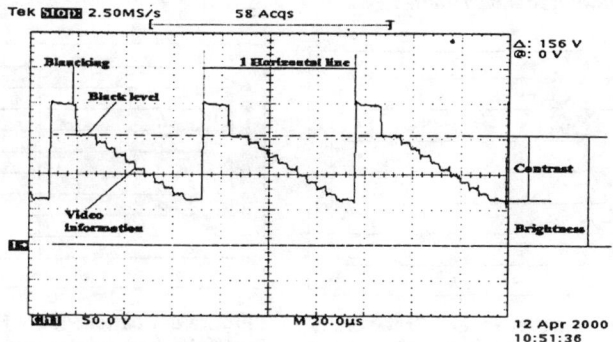

Figure 1- Video signal wave.

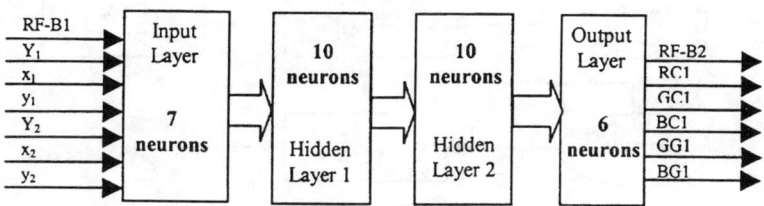

Figure 2- Structure of the neural network.

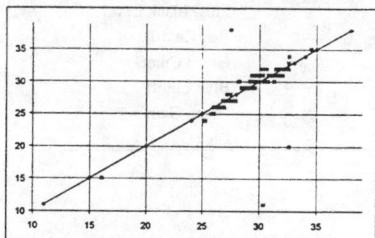

Figure 3- Scatter Plot of RF Brightness- Desired values versus neural network generated values.

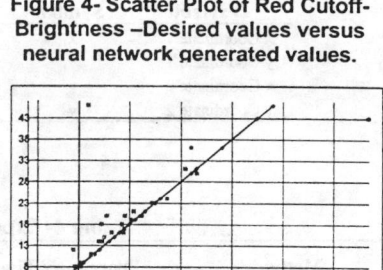

Figure 4- Scatter Plot of Red Cutoff- Brightness –Desired values versus neural network generated values.

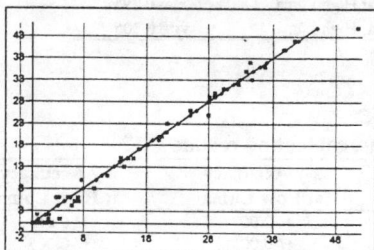

Figure 5- Scatter Plot of Green Cutoff- Desired values versus neural network generated values.

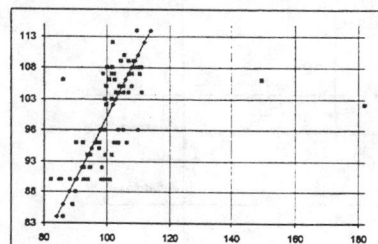

Figure 6- Scatter Plot of Blue Cutoff- Desired values versus neural network generated values.

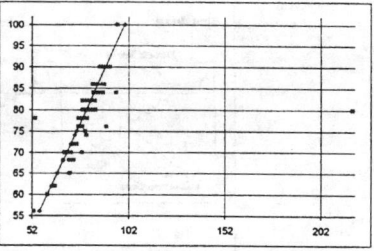

Figure 7- Scatter Plot of Green Gain- Desired values versus neural network generated values.

Figure 8- Scatter Plot of Blue Gain- Desired values versus neural network generated values.

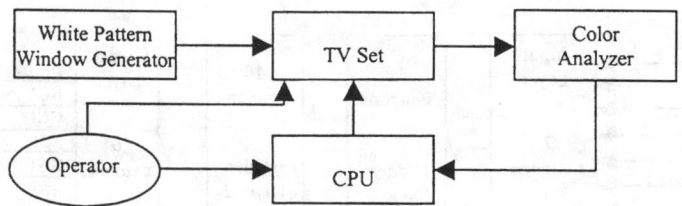

Figure 9- Hardware system set-up.

SMART SYSTEMS, METRIC ENTROPY, PHASE TRANSITIONS AND TIME-ARROW

E.V.KRISHNAMURTHY,
Computer Sciences Laboratory,
Australian National University,
Canberra, ACT 0200, Australia.
email: abk@discus.anu.edu.au

VIKRAM KRISHNAMURTHY,
Department of Electrical Engineering,
University of Melbourne,
Victoria 3010, Australia.
email: vikram@ee.mu.oz.au

ABSTRACT

A smart system has the four important properties:(i)Collective, coordinated and parallel operation(ii)Adaptive and Flexible(iii)Seamless- scale free(iv) Self -organization through emergence. We propose a hierarchy among the control and computational systems- classical or quantum - that transcend from the unsmart to the smart (or from being to becoming in the sense of Prigogine) based on metric entropy. When faced with a nondeterministic choice among several possibilities, a smart system undergoes a spontaneous breaking of symmetry so that there is a maximal thermodynamic profitability by minimizing the consumption of energy or other resources. The smart systems lack structural programmability and are non- unitary leading to the arrow of time from past to the future. Thus the price we pay for a system to be smart is "ageing " that would degrade the system ,and inability to obey a predetermined structural program.

INTRODUCTION

Smart systems have no formal definitions. For our purpose we define the smart systems as those systems having the following important properties.

1. Collective, Coordinated and Highly efficient Parallel Operation: They collectively and cooperatively perform actions, coordinating their actions when there is competition, to obtain maximal efficiency.

2. Adaptive and flexible Operation:
Smart systems such as biological systems are always flexible to change-they can modify their past behaviour and adapt to environmental changes.

3. Seamless Scale-free Operation:
The smart systems- as in biological systems operate on a scale that is very wide from molecular to the macroscopic. The artificial scaling laws encountered in physics - such as microscopic, mesoscopic and macroscopic-seem to work totally in a seamless manner.

4. Self organization through emergence
The system's behaviour cannot be interpreted as a sum of behaviour of its components and new properties emerge abruptly and the whole is not a sum of parts.

Obviously the biological information processing systems are smart and they satisfy the above properties. Our purpose in this paper is to consider the following issues:
1. Whether control and computational systems can exhibit such a smart behaviour?
2. If so whether such systems are easy to design and what are the requirements for realising such a smart behaviour systems?
3. Is there a hierarchy in the degree of smartness in computational and control systems?
4. What about quantum computational and control systems? Can one show the Quantum computational systems will result in a smart system?
5. What are the current open problems and what are the limitations in our knowledge?

Prigogine [16] pioneered in understanding the theory of open systems and its applications to biology. In his work Prigogine emphasises why the irreversible (non-unitary) process play a fundamental role in biological Systems which are open systems.

According to Prigogine, since future and past play the same role in time- reversible, unitary systems they cannot explain the emergence of new dynamical patterns involved in the biological systems. We will see later that for quantum systems too time reversal symmetry breaks down resulting in chaos and emergence of new properties.

CONRAD'S PRINCIPLE

According to Conrad [2], a general computing system cannot have all the three following properties:
1. Structural Programmability
2. High Computational Efficiency
3. High evolutionary adaptability and flexibility

Properties 1 and 2 are mutually exclusive. This means we cannot have high computational efficiency and a recursively structured programmable system. Properties 1 and 3 are mutually exclusive in the region where maximum computational efficiency exists. This means that whenever maximal computational efficiency exists, the system should be highly adaptive rather than structurally programmable . Programmable machines and the evolutionary system operate in radically different domains. These evolutionary systems turn out to be much more efficient in coupling material resources to problem solving. Efficient coupling makes enormous difference for problems such as pattern recognition, survival etc. For such problems structural programming systems cannot cope up. Viewed another way the coupling of material resources to problem solving can result in symmetry breaking transformations which are not structurally programmable, Conrad [2]. Based on this we can say that self organization is not possible in a structurally programmable system. Thus it appears smartness and structural programmability do not coexist! .Is there a further support for this view? We will consider these below.

INTEGRABILTY, COMPLEXITY AND METRIC ENTROPY

In a recent paper [10] we studied the relationship between the notion of computability and complexity in computer science, and the notion of integrability in mathematics and physics to understand the complexity aspects of quantum computing. Similar to the notions of computability and algorithmic complexity that play important roles in logic and computing, the notion of integrability plays an important role in Physics. We now explain how the notion of integrability affects the measurement of properties of systems. In an interesting paper, Eckhardt, Ford and Vivaldi [3] describe in detail the relationship between integrability, measurement and Metric entropy.

Metric entropy is the average information per measurement obtained from an infinite (denumerable) sequence of finite precision - identical measurements made on a time evolving system from a time minus infinity to plus infinity. If the metric entropy is zero it implies that the measurement sequence, begun in the remote past, ceases to provide new additional information after a finite time T. Metric entropy can be related to algorithmic complexity or algorithmic entropy. This is done by measuring the bit-length of the minimal program that describes a functional rule to generate a required output sequence from a given input sequence. For example, consider an infinite sequence that is totally erratic. This cannot be described as an output sequence of some input sequence by using a functional rule whose program length is smaller than the desired output length.

Whenever, the predictions of a physical theory can be obtained from noninfinite or terminating algorithms , such a theory is representable by a Turing machine or recursive function theory. Thus, there is a close analogy between measurement (or evaluation) procedure and algorithm. In both cases there is a well-defined input and we want a well-defined output after a finite time that can be described by finitary means. This analogy permits us to use the algorithmic complexity measure to measure the complexity of the measurement procedure. That is the bit-length of the minimal program that describes the functional rule to generate a required output from a given input can be used to define the complexity of the measurement procedure as well as algorithms.

We can define the algorithmic entropy measure thus: Let K(n) define the bit length of a minimal program that can output any n bit finite subsequence using a functional rule. Also let K = Lim n → ∞ [K(n)/n]. We note that 0 ≤ K ≤ 1. Thus when K > 0 the length of the program required to generate a desired output sequence goes on increasing or the rule is not finite and turns out to be more and more complex. However, for K = 0 the program length is much shorter than the required output thus defining a deterministic, orderly and predictable output according to an algorithmic rule. Thus K can be used to measure metric entropy.

In general, systems with zero entropy are tractable (predictable) and those with positive entropy are not predictable and hence truly nonintegrable. Nonintegrable systems cannot be described in any simpler way than they are and there exist no formal grammatical rules that can be used to parse them and understand the exact meaning. Well structured objects e.g., Context free grammars, regular grammars and serial-parallel orders, provide for easy description through functional rules and hence have zero metric entropy. The systems with zero metric entropy are classified "algorithmically integrable". Such systems are amenable for accurate measurements in a finite time and their orbital sequences are algorithmically predictable. In nonintegrable dynamic systems, the rate of evolution can be exponentially faster than that of the evaluation or measurement scheme. Such systems cannot be measured reliably as the measurement complexity grows exponentially; we can only hope to optimise a cost function by minimising rather than annihilating the entropy.

Quantum and Algorithmic Entropy

Classical motion is chaotic if the flow of trajectories in a given region of phase space has a positive definite K-entropy: K(E)= [<Σ L> Average}, where L>0, called the Lyapunov characteristic numbers or L-Numbers. The L- numbers measure the rate of exponential separation between two initially neighbouring points in the phase space and the average is taken in some particular neighbourhood. Thus qualitatively K(E) reflects the hypersensitivity on the initial conditions. Unlike classical trajectories, quantum trajectories are not single valued space -time trajectories. This was a major concern in undestanding quantum chaos [1,4,5,6,9-13]. However, recently Faisal [5] has suggested that by using De Broglie-Bohm formulation of Quantum mechanics, quantum trajectories are describable in configuration space in a consistent way so that the standard definition of L- numbers can be defined for Quantum mechanics thus:

L= Lim (t → ∞ ; d(0) → 0) [t^{-1} log (d(t)/d(0)] where d(t) is the Euclidean distance in phase space at a time t (evolving from two initially adjacent trajectories). Thus quantum motion is chaotic if the flow of de-Broglie-Bohm trajectories has positive definite K-entropy: K(QE)= K(E)= [<Σ L> Average}, where L>0; and average is taken over some particular neighbourhood. Physically L measures the rate of exponential divergence of two initially neighbouring trajectories or K(QE) is the rate of loss of information during the quantum evolution. Thus we can observe quantum chaos whenever K(QE) turns positive.

Algorithmic Information and K entropy

Consider K = Lim (t → ∞ ; K(t)/t); where K(t) is the algorithmic information needed to record a piece of the trajectory in the interval of time t. Thus if for positive K-entropy , in the long run the recording of information for evolution increases unbounded and the evolution cannot be followed deterministically, unless a disproportionately long or infinite time is devoted to this task beyond a critical time t = T(c). As we approach T(c), the recording is critically slowed down -much like in phase transition phenomena. At this time there is a spontaneous breakdown of time -reversal symmetry. Note that K is also the bit length of a minimal program that can output any t bit finite subsequence using a functional rule and is the Chaitin- Kolmogorov entropy. Thus quantum chaos results in

time-reversal symmetry breaking. The critical length of time T(c) over which the evolution is deterministic and reversible is estimated by the Chirikov ratio
C= L|T(c)|/|Log e| ≤ 1 ; where e is the error or accuracy in recording.
.For example if e= O(2^{-100}) then LT(c)= 100 so that C=1.If higher and higher accuracy are used we can only postpone the occurrence of chaos, but not avoid it.

Hierarchy of Smart Machines

Thus we can establish a hierarchy among control and computational system based on metric entropy independent of whether the system is classical or quantum . That is the metric entropy seems to play an important scale-free role in deciding whether the sytem is smart. We infer that only positive entropy machines (whether classical or quantum) can become smart! The price we pay for this is the arrow of time that leads to ageing and structural non-programmability.

The two classes are :

I. Completely structured, Deterministic, Exact behaviour, Zero entropy, Algorithmic or Unsmart machines with well-defined trajectories in space and time.
This class contains: Finite State machines, Push down-stack machines,
Turing Machines (Deterministic, Nondeterministic Probabilistic,Alternating) that halt.
Exactly integrable Hamilton flow machines, Quantum mechanical Machines governed by unitary transformations.

II. Partially Structured,Non-deterministic,Probabilistic,Average-behaviour,Positive Entropy Non-Algorithmic or Smart machines
This class contains: [9-12]: Ergodic machines in which the phase space flow preserves shape and voume, and more general K-flow machines, Equilibrium and non-equilibrium Statistical mechanical machines , Quantum - field - theoretic machines, Chemical and Biological machines. For these more general machines (beyond ergodic), space-time trajectories disappear and the shape in phase space is distorted. The macro behaviour of the system is no longer a sum of the micro behaviour of its subsystems.

The positive entropy machines do not obey the unitary transformation law. Their actions are irreversible. That is the time-reversal symmetry is spontaneously broken. Such machines- whether quantum or classical- can become smart ; but they will have to experience an arrow of time and age. Prigogine [16] suggests the use of non-unitary transformation, called star-Hermetian operators. Thus to extend the capabilities of computational systems to reflect average behaviour we require the tools of both equilibrium and non equilibrium quantum statistical mechanics .

ALGORITHMIC AND DYNAMICAL SYSTEMS: PHASE TRANSITION

The classical computational or algorithmic paradigm is described by using a syntax that defines the legal sentences that are admissible, e,g., successor, predecessor, recursion operators in primitive recursive functions . In Recursive function theory, the meaning of such sentences are then interpreted through a well-defined semantics that assigns the meaning in the respective domains. A particular form of semantics- called denotational semantics- uses a functional approach to describe the relation between the input and output states of any computational procedure (program). The denotational semantics is specified as a mapping from the program to the function or relation it computes. In other words, denotational semantics defines each language construct in terms of certain mathematical entities (numbers, truth values , functions, operators) that model their meaning. The total meaning of a program can then be reconstructed as a composition of the meanings of the individual basic constructs ; that is "a whole "is described in terms of its "parts". That is, if the program is recursively defined , its meaning is the input -output function corresponding to the least or minimal fixed point of a transformation associated with the program. This ensures that the program has a well-defined functional relationship between the input and output states and the program halts, Manna [15] .

Unlike in recursive function, which may have a fixed point, many fixed points or no fixed points (nonhalting),the evolution of a living system (dynamical system) can belong to four classes, called Wolfram classes , Langton et al [13].
1. Evolution to a fixed homogeneous state in living systems (limit points in dynamical systems) corresponding to fixed points in programming.
2. Evolution to simple separated periodic structures in living systems (or limit cycles in dynamical systems) corresponding to cycles of deadlock or livelock in concurrent computation.
3. Evolution to a chaotic behaviour yielding aperiodic patternsin living systems ((strange attractors in dynamical systems) that has no correspondence in computer science.
4. Evolution to complex patterns of localized structures in living systems - has no analog in dynamical systems or in computer science .

Also between periodic and chaotic behaviour there is a phase transition. While the periodic and chaotic regions are governed by rules the transition region is not governed by any rules . The phase transition therefore separates the space of computation into an ordered and a disordered regime- which can be thought to be in correspondence to halting and nonhalting computations. Phase transition means symmetry breaking caused by a change in external parameters that lead to a new macroscopic spatio-temporal pattern of the system and an emegence of a new order- such as growth of an organism. Phase transition arises in nature whenever there is a degenerate situation in which there are several equivalent possibilities. In such a case ,the system has to make a nondeterministic choice among the possibilities. Combinatorial explosion paralyses such a system if it is truly algorithmic and domain-general. In an adaptive problem the use of domain general program has to evaluate all alternatives- sometimes they may be unknown. Hence a *"while-loop "* in an algorithmic paradigm may involve exponential growth of complexity or may never terminate. By the time the system analyses a biological problem of standard complexity, if it has no domain specific rules of relevance, procedural knowledge or privileged hypothesis, it cannot solve the problem in the amount of time it needs to solve that problem for its survival. Hence the system undergoes a spontaneous breaking of symmetry so that there is a maximal thermodynamic profitability in terms of minimizing the consumption of energy or other resources. Dynamics in the vicinity of the phase transition gives rise to a critical slowing down and the various complexity classes (from constant , linear , polynomial, exponential) are encountered. Critical slowing down of a system corresponds to computation of an intractable problem. Phase transition like situation arises in a wide variety of algorithms and heuristic used for search problems in NP class or beyond, Gent and Walsh[.7] , Walsh [17], Zhang and Korf [19]. The phenomenon of critical slowing down also occurs in quantum systems,Faisal[5],resulting in the breakdown of time reversal -symmetry-similar to phase transition .

QUANTUM SYSTEMS

We mentioned that if a program is recursively defined, its meaning is the input -output function corresponding to the least or minimal fixed point of a transformation associated with the program . This assumption also holds true in classical mechanics where the real states are always local and do not contain statistical correlations, among the states of their local subsystems. However, such an assumption breaks down in the quantum field theory (QFT).The quantum system can exhibit both kinematic and measurement-induced nonlocality, Zeh [18]. The many particle hamiltonian expression may not obey the basic assumption used in denotational semantics, namely, a composite system is composed of parts and the state of whole is definable in terms of the states of its parts. There are quantum correlation between different subsystems due to the superposition principle. In fact, the whole is more than its parts,even if we assume there is no interaction among the subsystems. So,we may not be able to analyse a quantum mechanical system into its parts and synthesize the system using its components; this is also called non-separability. In other words, the concept of analytico-synthetic approach which forms the basis of entire computer science and logic fails here. Therefore, what we intended to achieve through a program will not be the outcome of that program. In fact, a new macroscopic system may emerge due to phase transitions satisfying new conservation laws.That is, nature will find a means to break the symmetry that satisfies a new conservation law when there are restrictions on available energy for a physical system or food for living systems. Thus we may end up with a new set of rules or laws leading to a nonmonotonic open logical system. Accordingly, the structural programmability of the system is lost and the system evolves as a whole, on its own. This is Nature's own way of solving problems with a high efficiency and leads to the Conrad's [2] trade-off principle described earlier. Hence, if we attempt to solve intractable mathematical problems using nature's way ,we may not succeed in minimizing the resources needed, unless nature has its own necessity to solve that problem. The fact that such systems can become nonprogrammable due to the absence of a recursive function support means that algorithmic approaches to solve the problems through programmed instructions may not be available.

Much work remains to be done to understand quantum-to-classical asymptotics (time tending to infinity and Plank's constant tending to zero), Kolomogorov- Sinai entropy, quantum-nonintegrability, near-integrability, complexity ,time-arrow [4,8,14,18] and chaos and star-unitary evolutionary aspects. Also further studies on the entropy operator introduced by Prigogine [16] will be useful to unify deterministic -reversible and nondeterministic- irreversible quantum computations. Studies on these aspects can help sharpen the results on quantum computability theory, integrability theory and complexity theory.

REFERENCES
[1]Casati,G., and Chirikov,B.V.,1996, *Quantum chaos*, Cambridge University Press, Cambridge.
[2] Conrad,M.,1990, Molecular Computing, *Advances in Computers*, Vol 31,pp. 235 -325, Academic Press, New York.
[3]Eckhardt.B,Ford,J., and Vivaldi,F.,1984, Analytically Solvable Dynamical Systems which are not Integrable, *Physica*,Vol.13 D, pp.339-356.
[4] Ezawa,H., and Murayama,Y.,1993, *Quantum control and Measurement*, North Holland, Amsterdam.
[5] Faisal,F.H.M.,1997, Quantum Chaos, Algorithmic Paradigm and Irreversibility,in, Ph.Blanchard and A.Jadczyk (Editors),*Quantum Future*,Lecture Notes In Physics, Springer Verlag, New York,pp.47-57.
[6]Feng ,D.H.and Yuan .J,1993, *Quantum Non-integrability*, World Scientific, Singapore.
[7] Gent,I., and Walsh,T.,1996, Phase transitions from real computational problems, *Proc.8 th Symposium on AI* ,pp.356-364.
[8]Halliwell,J.J.,Mercader,J.P., and Zurek W.H.,1994, *Physical Origins of Time Asymmetry*, Cambridge University Press, Cambridge, U.K.
[9] Krishnamurthy,E.V.,1998,"Computational power of quantum machines, quantum grammars and feasible computation",*International J. Modern Physics*, Vol.9, pp.213-241.
[10] Krishnamurthy,E.V.,1999, "Integrability,Entropy and quantum computation", *International J. Modern Physics*, Vol.C10, No. 7,pp.1205-1228.
[11] Krishnamurthy,E.V., and Krishnamurthy,V.,2001,Quantum field theory and computational paradigms,*International Journal of Modern Physics*, to appear .
[12] Krishnamurthy,V., and Krishnamurthy,E.V., "Rule-Based programming paradigm: A formal Basis for Biological, Chemical and Physical Computation', BioSystems, Vol.49, 205-228, 1999.
[13]Langton,C.E., et al.,1992, Life at the Edge of chaos, pp. 41-91, in *Artificial Life II*, Addison Wesley, Reading, Mass., see also, Computation at the edge of chaos,1990, *Physica D* Vol. 42 , pp.12-37.
[14]Mackey,Michael. C.,1992, *Time's Arrow: the origins of ThermodynamicBehaviour*, Springer Verlag, New York.
[15] Manna,Z.,1974 *Mathematical Theory of Computation*, McGraw Hill, New York.
[16] Prigogine,I.,1980, *From being to becoming*, W.H.Freeman and Co, San Fransisco.
[17]Walsh,T.,1996, The constrainedness knife edge, *Proc.AAAI-98*,pp..406-411.
[18]Zeh,H.D.,1989,*The Physical Basis of the Direction of Time*, Springer, New York.
[19]Zhang,W.,and Korf,R.,A study of complexity transitions on the asymmetric travelling salesman problem,1996, *Artificial Intelligence,* Vol 81,(2),pp.223-239.

NATURAL LANGUAGE TRANSLATION SYSTEM USING NEURAL NETWORKS

SAROJ KAUSHIK
Department of Computer Science & Engineering,
Indian Institute of Technology,
New Delhi - 110016
INDIA

MANOJ KUMAR
Department of Electrical Engineering,
Engineering College,
Kota, Rajasthan,
INDIA

ABSTRACT

The aim of this paper is to test the ability of neural networks to perform natural language translation. Simple feed forward three layered neural network is used for performing such task. Error back propagation training algorithm is used for training the neural networks. Such translation neural networks can be trained to perform translation between any pair of natural languages provided dictionaries for these languages are carefully designed. We have considered translation from English to Hindi (Indian Language). The sentences in both the languages are represented by series of 20-bits binary codes. The system simply learns mapping from one language to another language. The scope of translation is simple sentences (all forms) of three tenses (past, present and future). The performance of the system relies upon the size and types of sentences used for training. It is implemented in Visual C++ on pentium platform.

INTRODUCTION

Machine translation of natural languages have been tackled in many ways in the past. To name some of them, direct approach between pair of languages, interlingua approach by Dorr(1993a), lexical and corpus based translation by Dorr(1993b) and example based machine translation by Nirenburg(1992). Neural Net has also been used for machine translation from English to Spanish by Allen (1987) and more recently from English to Serbo Croatian (European Language) by Konkar and Guthrie(1994). We have explored the usefulness of feed forward back propagation neural network to perform translation work from English to Hindi. Most of the conventional systems for translation have hard coded rules (with varying depth of meaning) that can be modified only by a knowledge engineer who is an expert in expressing such rules in the language. Neural networks, having ability to learn from examples, are a possible solution to this problem. If such a neural network translation system wrongly translates a sentence, it can be corrected and taught the proper translation by user without any expert knowledge of how computer stores and represents rules. This work demonstrates the ability of neural networks in precisely this area.

We have proposed a Translation Neural Network (TNN) which is a three-layer feed forward neural network containing one hidden layer along with input and output layers. Nodes at input layer of the proposed system accept encoded form of a sentence in source language (English), which are binary codes corresponding to the words of a sentence. A word in both the languages is assigned 20 bit code.

The system generates binary codes corresponding to the words of a sentence at output layer. This code is further decoded to give rise a sentence in the target language. A training set contains various types of translation examples with varying size of sentences. These translation examples are classified according to their English sentence size and encoded forms are stored into separate files. During training phase of TNN, connection weights are adjusted until it produces output with acceptable level of bit errors for all the sentences used for training. Connection weights thus generated between layers of neural network are represented as matrices and stored in data files. These weights are retrieved from files when TNN is used for translation. TNN makes use of dictionaries of both the languages which are intelligently designed. The performance of TNN is calculated in terms of errors at word and sentence levels. It is noted that performance is directly proportional to the size of training set.

OVERALL DESIGN OF TNN

Fig.1(appendix) gives overall design of the proposed system. It consists of decoder, encoder, Trained Translation Neural Network and dictionaries of both the languages.

Encoder

An encoder encodes source sentence to binary form by replacing each word with its corresponding 20-bit binary code using English dictionary. Input sentence size varies from 2 to 9 words.

Translation Neural Network

It is a three layered feed forward neural network. Input layer contains maximum of 20 * N neurons, where N is number of words in English sentence. The middle layer (hidden layer) contains suitable number of neurons (generally between number of input neurons and output neurons). Output layer contains 20 * (N+2) neurons because Hindi sentence may have maximum N+2 words corresponding to N-word English sentence.

Decoder

Decoder decodes the binary codes obtained at output layer word by word using Hindi dictionary and displays sentence in Hindi. Fig.2 (appendix) shows a flow diagram for translation process using 2-word input sentence and 3-word output sentence. Script for Hindi is Devanagri which contains vowels and consonents. For the sake of simplicity and non availability of software for Hindi script, we have used Roman script. The coding conventions are given as follows:.

Hindi Vowels: a -> a, aa -> A, i -> i, ii ->I, u -> u uu -> U, e -> e, ee-> E, o -> o, oo -> O
Hindi consonants:

k	kh	g	gh	n	| c	ch	j	jh	n
|T	Th	D	Dh	N	| t	th	d	dh	n
|p	ph	b	bh	m	| y	r	l	v	
|sh	S	s	h						

CODING SCHEME FOR WORDS

Words in both the languages have been divided into several categories such as {nouns, verbs, adjectives, adverbs, prepositions etc.}. Each of these categories are further divided into sub-categories depending upon the target language. Each word is coded into 20 bits. The code is divided into two parts. Part one consists of high order 8 bits which are reserved for indicating sub-category and part two consists of lower 12 bits used for assigning a unique sequence number of the word in that category. The code size 20 bits are sufficient for our system that can handle maximum 256 categories each having maximum of 4096 words within each category.

Encoding of nouns

Noun can be broadly divided into singular and plural each of which is either masculine or feminine. In English, there is only one direct form of plural noun whereas in Hindi there are two forms (direct and oblique) for plural. For example, *laDake* and *laDakon* (plural masculine in Hindi) corresponding to *boys* (plural masculine in English) and *laDakiyAn* and *laDakiyon* (plural feminine in Hindi) corresponding to *girls* (plural feminine in English). We have 6 categories in Hindi and 4 categories in English.

Some of masculine and feminine nouns have only one direct form in Hindi corresponding to singular and plural nouns whereas they have different singular and plural forms in English. The verb in Hindi sentence decides whether the noun is singular or plural. For examples: *AdamI* is used for both singular and plural (direct) in Hindi whereas there are two forms {*man, men*} in English. However such nouns have different oblique plural forms in Hindi (e.g., *Adamiyon*). So we need 4 separate categories for such noun words in both the languages. Thus in English we have 8 categories whereas in Hindi 10 categories. Classification of noun categories is based on the target language. We have used the following notation: EH_Nx - category valid for English Hindi both.
                H_Nx - category valid for Hindi and not available in English.

- Nouns having different singular-plural form in Hindi corresponding to noun in English:
    EH_N1 - Singular Masculine Nouns,   EH_N2 - Plural Masculine Nouns,
    EH_N3 - Singular Feminine Nouns,   EH_N4 - Plural Feminine Nouns.
- Nouns having same singular-plural form in Hindi corresponding to noun in English:
    EH_N5 - Singular Masculine Nouns,   EH_N6 - Plural Masculine Nouns,
    EH_N7 - Singular Feminine Nouns,   EH_N8 - Plural Feminine Nouns.
- Oblique plural in Hindi (not available in English):
    H_N9 - Plural Masculine Nouns,   H_N10 - Plural Feminine Nouns

Encoding of nouns in English and Hindi are shown in Table below:

Cat. No.	English Word	Hindi Word	Code	Decimal range
EH_N1	boy	laDakA	1000 0010 xxxx xxxx xxxx	532480 - 536575
EH_N2	boys	laDake	1000 0011 xxxx xxxx xxxx	536576 - 540671
EH_N3	girl	laDakI	1000 1010 xxxx xxxx xxxx	565248 - 569343
EH_N4	girls	laDakiyan	1000 1011 xxxx xxxx xxxx	569344 - 573439
EH_N5	man	AdamI	1000 0000 xxxx xxxx xxxx	524288 - 528383
EH_N6	men	AdamI	1000 0001 xxxx xxxx xxxx	528384 - 532479
EH_N7	sparrow	ciDiA	1000 1000 xxxx xxxx xxxx	557056 - 561151
EH_N8	sparrows	ciDiA	1000 1001 xxxx xxxx xxxx	561152 - 565247
H_N9	men	Admiyon	1000 0111 xxxx xxxx xxxx	552960 - 557055
H_N10	sparrows	ciDiyon	1000 1111 xxxx xxxx xxxx	585728 - 589823

Encoding of verbs

In English, verbs can have 5 forms: *go, goes* {Present}, *went* {Past}, *gone* {Past perfect} and *going* {Continuous}. All verbs need not have different present, past, and past perfect forms. So the verbs have been classified according to their present, past and past perfect forms and are shown in the following Table with example.

	Category Type	Category Examples	Code Range
1.	Similar present, past and past perfect (e.g., cut)	cuts	0001 0000 xxxx xxxx xxxx
		cut	0001 0001 xxxx xxxx xxxx
		cutting	0001 0101 xxxx xxxx xxxx
2.	Similar past and past perfect. (e.g., think)	thinks	0011 0000 xxxx xxxx xxxx
		think	0011 0001 xxxx xxxx xxxx
		thought	0011 0010 xxxx xxxx xxxx
		thinking	0011 0101 xxxx xxxx xxxx
3.	Similar present and past perfect (e.g., run)	runs	0101 0000 xxxx xxxx xxxx
		run	0101 0001 xxxx xxxx xxxx
		ran	0101 0010 xxxx xxxx xxxx
		running	0101 0101 xxxx xxxx xxxx
4.	Different present-past and past perfect (e.g., go)	goes	0111 0000 xxxx xxxx xxxx
		go	0111 0001 xxxx xxxx xxxx
		went	0111 0010 xxxx xxxx xxxx
		gone	0111 0011 xxxx xxxx xxxx

A verb in Hindi may take 16 different forms. The following Table contains all 16 categories of Hindi verbs. For example, verb *go* in English has five forms as: *go-goes-went-gone-going*. In Hindi corresponding verb *jA* has 16 forms as *jAnA-jAtA-jAtI-jAte-jAyegA-jAyegI-jAyenge* etc. 'PPPP' in the code will be replaced by most significant 4-bits of corresponding English verb code.

	Verb Examples	Code Range
1.	'jAnA', 'khAnA'	PPPP 0000 xxxx xxxx xxxx
2.	'jAtA', 'khAtA'	PPPP 0001 xxxx xxxx xxxx
3.	'jAtI', 'khAtI'	PPPP 0010 xxxx xxxx xxxx
4.	'jAte', 'khAte'	PPPP 0011 xxxx xxxx xxxx
5.	'jAyegA', 'khAyegA'	PPPP 0100 xxxx xxxx xxxx
6.	'jA', 'khA'	PPPP 0101 xxxx xxxx xxxx
7.	'jAyegI', 'khAyegI'	PPPP 0110 xxxx xxxx xxxx
8.	'jAyenge', 'khAyenge'	PPPP 0111 xxxx xxxx xxxx
9.	'gayA', 'khAyA'	PPPP 1000 xxxx xxxx xxxx
10.	'gayi','khAyi'	PPPP 1001 xxxx xxxx xxxx
11.	'jAoge', 'khAoge'	PPPP 1010 xxxx xxxx xxxx
12.	'gaye', 'khAye'	PPPP 1011 xxxx xxxx xxxx
13.	'jAogI', 'khAogI'	PPPP 1100 xxxx xxxx xxxx
14.	'jAiye', 'khAiye'	PPPP 1101 xxxx xxxx xxxx
15.	'Jane', 'khAne'	PPPP 1110 xxxx xxxx xxxx
16.	'jAo',,'khAo'	PPPP 1111 xxxx xxxx xxxx

Encoding of adjectives

Adjectives in English language can appear in three forms viz., simple, comparative and superlative (e.g., *fast, faster, fastest*). Encoding for adjectives in English is given the following table.

Adjective type	Code syntax
Simple	0000 0001 00xx xxxx xxxx
Comparative	0000 0001 01xx xxxx xxxx
Superlative	0000 0001 10xx xxxx xxxx

The adjectives in Hindi can appear in three forms depending upon the type of noun being used with the adjective. These forms are used with singular masculine nouns{ *'baDA'*, *'chotA'*, *'AcchA'*}, plural nouns { *'baDe'*, *'chote'*, *'Acche'*} and singular feminine nouns{ *'baDI'*, *'chotI'*, *'AcchI'*}. The following Table shows coding scheme for adjectives in Hindi.

Adjective type Examples	Code syntax
'baDA', 'chotA'	0000 0001 00xx xxxx xxxx
'baDe', 'chote'	0000 0001 01xx xxxx xxxx
'baDI', 'chotI'	0000 0001 10xx xxxx xxxx

Encoding of adverbs and prepositions

Encoding for adverbs and prepositions are fairly simple. Similar codes are assigned for English and Hindi words in this categories. Encoding range for adverbs is from 0001 1010 0000 0000 0000 to 0001 1010 1111 1111 1111 and for prepositions is from 1100 0000 0000 0000 0000 to 1100 0000 1111 1111 1111.

TRAINING OF TNN

A training set consisting of translation examples is prepared for TNN. A file stores these translation examples (English - Hindi sentence pair) with varying size of sentences. These examples are classified according to their English sentence size and encoded forms are stored into separate files. Here encoder encodes each word into corresponding decimal code. Thus encoded sentence is a sequence of decimal codes. Each of these codes are later converted into 20-bit binary codes at the time when these sentences are used for training the network. For each encoded input sentence, TNN is trained one-by-one until the desired output sentence is generated at output layer with acceptable level of bit error. During training phase, iterations are performed by adjusting the weights until the TNN produces acceptable outputs for all the input sentences. The trained network weights are then stored in data files which are loaded later while performing the actual translation. It is a feed forward neural network where each neuron in a layer is connected to every neuron in the next higher layer. Back-propagation training algorithm is used for training this network. Weight $W_0(i,j)$ connects i^{th} neuron of input layer to j^{th} neuron of hidden layer. Weight $W_1(i,j)$ connects i^{th} neuron of hidden layer to j^{th} neuron of output layer.

DICTIONARY DESIGN

Dictionary design plays an important role in the translation. An intelligent encoding of words in dictionaries helps the neural network to learn the concepts easily. Each word in a language is assigned 20-bit binary code. Two words in English having same meaning in Hindi are assigned same binary code. Similarly one word in English can have multiple equivalent words in Hindi, so there will be multiple entries for such words in Hindi with different codes. For example the word 'place' can be used either as a noun (Hindi translation 'jagah') or as a verb (Hindi translation 'rakhanA').

Two separate dictionaries for English and Hindi are used. Each dictionary contains fixed length records having two fields viz., word and corresponding code. Records are entered randomly in the file with no specific ordering. For faster searching and insertion, *Binary Search Trees (BSTs)* are used as an index tree to the dictionaries. The dictionaries are indexed on word and code separately and these indexes are created automatically in the form of Binary Search Trees (some other indexing technique may be used). The various operations required on dictionaries are search, insert new word and code, delete an existing word and code etc. Dictionary consists of three files: dictionary record files (containing word-code pair), code index file and word index file (both implemented as binary search trees). When a new word in inserted, word-code pair is appended at the end of dictionary record file as fixed length record and new index entries (code-word pair) are inserted in the binary search trees.

PERFORMANCE MEASURE

The performance of the system and its accuracy of translation relies upon the size and selection of translation examples used in the training. The system is capable of translating only those types of sentences correctly for which sufficient number of translation examples are available in training sets. The TNN is trained using about 1800 sentences. The results of test performed after training the network is given in the following table which clearly indicates that accuracy of translation is not good for long sentences as the training set is small in size.

Sentence Size	Number of sentences in Training set	Test result in % Sentence error	Word error
2	270	1.5%	1.1%
3	545	2.3%	1.5%
4	380	4.2%	2.4%
5	210	5%	2.2%
6	80	11.6%	3.3%
7	45	12.5%	4.1%
8	30	12.2%	3.8%
9	30	13.5%	4.3%

The proposed system learns rules from the examples after training. Some of these are listed below:
1. form of verb changes with feminine and masculine nouns.
    boy **eats** -> laDakA **khAtA** hai | girl **eats** -> laDakI **khAtI** hai
2. use of direct or oblique plural form of noun
    boys ate food ->**laDakyon** ne khAnA khAyA
 and **boys** went to school->**laDake** school gaye
3. change in the form of adjectives

boy is having big car -> laDakA ke pAs baDI car hai
and   boy is having big house -> laDakA ke pAs baDA ghar hai

There are limitations of the proposed system. Phrases can not be translated because a group of words represent some meaning not directly related to word by word translation. Further, the system can not understand semantics of a given sentence. For example 'child drinks milk' and 'stone drinks milk' both will be translated to *bacchA dUdh pItA hai and patthar dUdh pItA hai* respectively, but second sentence is semantically wrong. Punctuation can not be used in the sentences. If we want to use punctuation, then preprocessing of sentences will be required before feeding them for translation. Another problem with punctuation is that they sometimes change the meaning of a given sentence. Such system can not be used for performing translation of bulk texts. It can be used for translating sentences one by one. Main application of this system can be for teaching translation of simple sentences to English speaking children. The same frame work can be used to design other translation systems to perform translation between any pair of languages provided the codes for words are carefully designed.

CONCLUSIONS

Neural Network has been used for performing machine translation of simple sentences of all forms of three tenses (past, present and future) from English to Hindi (Indian language). Since Neural Network learns through examples after training, this capability has been exploited for machine translation. Simple feed forward neural networks are used. Error back propagation training algorithm is used for training the neural networks. Such translation neural networks (TNNs) can be trained to perform translation between any two natural languages provided dictionaries for these natural languages are available. The system is implemented in Visual C++ on pentium platform.

REFERENCES

Allen, R. B. (1987), Several studies on Natural Language and Back Propagation: Proc. First International conference on Neural Networks, vol.2.

Bonnie J. Dorr (1993a), Interlingual Machine Translation: a parameterized approach, Journal of Artificial Intelligence 63, pp 429 -492.

Bonnie J Dorr (1993b), Machine Translation: A View from the Lexicon, the MIT press, USA.

Raman, S. and Alwar, N.(1990), An AI based Approach to Machine Translation in Indian Languages, Communication of ACM, volume 33, Number 5, PP 521 -527.

Nirenburg, S. et. Al. (1992), Two approaches of Matching in Example based Machine Translation, Proceedings of TMI - 93, Kyoto, Japan.

N. K. Bose and P. Liang, (1998), Neural network fundamentals with graphs algorithms, and applications, Tata Mcgraw-Hill.

Nenad Konkar and Gregory Guthrie, (1994), A natural language translation neural network, Proceedings of the Conference on New Methods in Language Processing (NeMLaP), pp. 71-77, Manchester (UK).

D.E. Rumelhart, G.E. Hinton, and R.J. Williams, (1986), Learning internal representations by error propagation. In E.E. Rumelhart and J.L.McClelland, editors, *Parallel Distributed Processing*, vol. 1, pp.318-362. MIT Press, Cambridge, MA.

APPENDIX

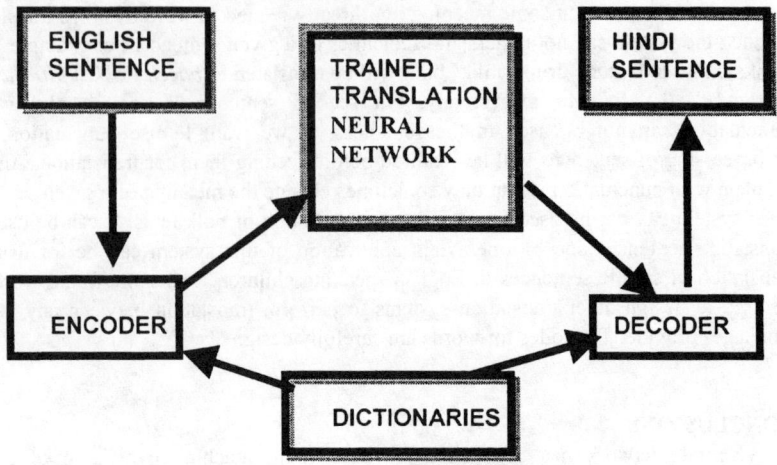

Fig 1. Overall Design of TNN

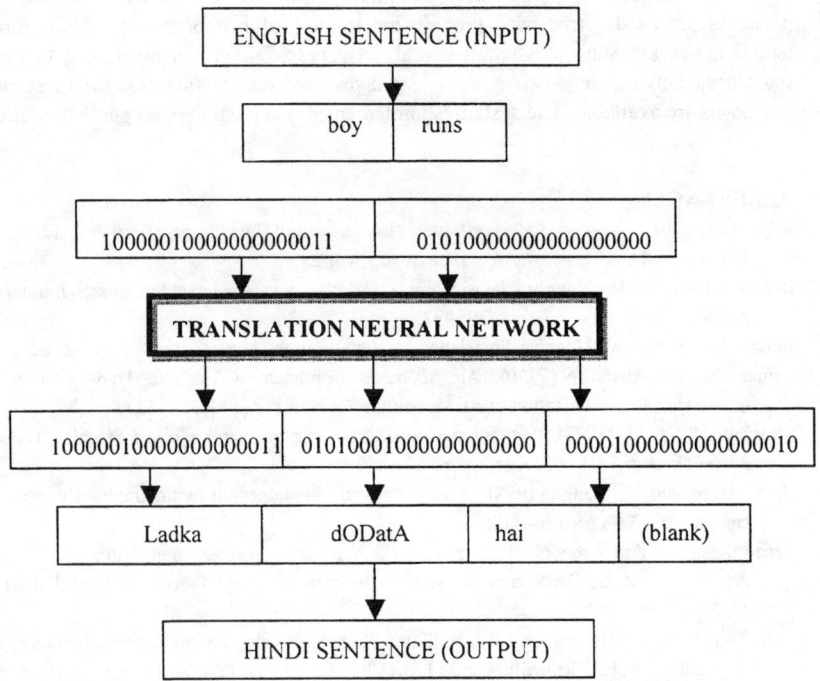

Fig. 2: Flow diagram of TNN

COMPARATIVE PERFORMANCE ANALYSIS OF STRUCTURED NEURAL NETWORKS AND OBSERVERS FOR CNC FEED DRIVE DISTURBANCE FORCE ESTIMATION

MELIK DOLEN
Dept. of Mech. Engr.
Middle East Technical University
Ankara, Turkey

EKREM KAYIKCI   ROBERT LORENZ
Dept. of Electrical & Comp. Engr.
University of Wisconsin – Madison
Wisconsin, USA

ABSTRACT

This paper presents a structured neural network for feed drives of CNC vertical machining centers to accurately estimate machining forces. The estimation performance of the network is evaluated through a detailed simulation study. The accuracy of a Luenberger-type disturbance observer, which is a feasible alternative to the network, is compared to that of the network. With similar system parameter estimates for both methods, the network is found to be considerably more accurate than the observer.

I. INTRODUCTION

Cutting force sensors in CNC technology can potentially provide the most crucial information for adaptive control, tool condition monitoring, and detection of chatter vibrations etc. Although force sensors are currently available in the market for a wide variety of CNC machines, these products have not gained recognition in industry due to their significant drawbacks. Not surprisingly, the technological trend in the industry is towards the employment of estimation schemes and the elimination of (advanced) sensors wherever the estimation methods are feasible alternatives.

To estimate the machining forces accurately, the CNC machine tool feed drives, which are directly subjected to the machining forces, must be thoroughly modeled. The research in this field has generally concentrated on the development of advanced digital control algorithms utilizing rather simple linear models of the controlled <u>feed drive</u> (FD) system (Van Brussel and Vastmans, 1981; Koren, 1984 and Kulkarni et al., 1984). Only a handful of studies concentrate on the estimation of machining forces by utilizing accurate physical FD models. For instance, Stein et al. (1986) conducted a sensitivity analysis on the currents drawn by the FD motor of a CNC lathe. Their analysis reveals the feasibility of designing a disturbance force estimator that employs current measurements. Similarly, Altintas (1992) developed a simple electromechanical FD model to estimate the cutting force components. Although the estimation results are in good agreement with the steady-state experimental data; the method, which ignores inertial forces, produces considerable deviations when estimating transient forces.

As a result, the main motivation of this paper is to develop an accurate force estimator for CNC machine tools with the utilization of advanced FD physical models. Since nonlinearities are dominant in such systems, <u>structured neural networks</u> (SNNs) are the major estimation and modeling tool used in this study. The organization of this paper is as follows. It begins with a brief model of a typical FD system. Based on this background, the next section develops an efficient SNN to estimate the fundamental machining forces. The following section evaluates the performance of the network through a detailed simulation study. The accuracy of the network is compared to that of a (Luenberger-type) disturbance force observer. The differences in estimation performance between the techniques are highlighted in this section. Finally, Section V. summarizes the key conclusions from this paper.

II. FEED DRIVE MECHANICAL SYSTEM MODELING

To estimate the disturbance on a FD system, one needs to develop the electromechanical model of a typical FD as shown in Fig. 1. The differential equations governing the motion for the FD table can be written as

$$(m_x + m_w)\frac{d^2x}{dt^2} + F_x + F_{f1}(F_x, F_y, F_z, x, y, z, dx/dt, d^2y/dt^2) = F_{s1} \quad (1)$$

Here F_{s1} is the force applied by the nut to the table; F_{f1} is the friction force on the table. With respect to the motor side,

$$J_1\frac{d\omega_1}{dt} + T_{f1} + T_{s1} = T_{em1} \quad (2)$$

where T_{f1}, T_{s1}, and T_{em1} are friction torque, load torque, and electromagnetic torque respectively.

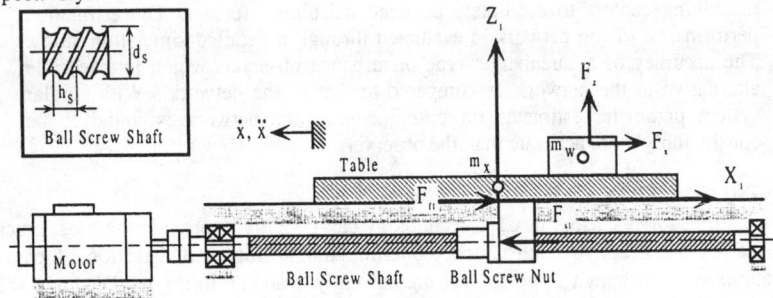

Figure 1 – Typical, mechanical feed drive system.

The motor side contains several preloaded ball bearings causing frictional losses. Employing the (empirical) bearing friction model (Harris, 1991) yields the motor friction torque:

$$T_{f1} = (b_1|\omega_1|^{2/3} + T_{01}) \cdot sgn(\omega_1) \quad (3)$$

where b_1 and T_{01} are positive constants; while ω_1 refers to the speed of the motor. Utilizing all three equations yields the differential equation of the FD system:

$$J_{e1}\frac{d\omega_1}{dt} + \frac{h_s}{2\pi\eta_s}(F_x + F_{f1}) + (b_1|\omega_1|^{2/3} + T_{01}) \cdot sgn(\omega_1) = T_{em1} \quad (4)$$

where J_{e1} is the equivalent inertia; h_s is the lead of feed screw [m] while η_s refers to the efficiency of the screw. Assuming that the same type of screw and the motor are utilized in the Y-axis FD (a.k.a. "saddle"), one can derive a similar differential equation. Note that the friction force F_f (F_{f1} and F_{f2}) in (4) can be expressed as

$$F_f = \mu(v)F_N \quad (5)$$

where F_N is the equivalent normal force induced by the dynamic load on the table. Armstrong and Dupont (1993) gives the following empirical expression for kinematic friction coefficient μ:

$$\mu(v) = sgn(v) \cdot (c_1 + c_2 e^{-c_3 v}) + c_4 v \quad (6)$$

where table speed $v = h_s\omega/2\pi$ ($v >> 0^+$) ; $c_1, ..., c_4$ are (positive) empiric friction coefficients. It is clear from (4) that the friction force (F_{f1} or F_{f2}) needs to be determined accurately to estimate the process forces themselves. Dolen et al. (2000b) presented a detailed friction (normal force) model for hydrodynamic guideways which are commonly used in the machine tool industry. The normal force is heavily affected by the magnitude of the cutting force components. However, Dolen (2000) shows that the model boils down to a simple Coulomb (dry) friction model for light machining conditions where the magnitudes of average machining forces are less than 1kN.

III. STRUCTURED NEURAL NETWORKS FOR FORCE ESTIMATION

In the previous section, the electromechanical FD model was presented. Employing the neural network programming technique of Dolen and Lorenz (1999), this section develops a SNN representing the given model.

Disturbance force estimation has three important components: **i.** acceleration estimation; **ii.** current filtering and/or electromagnetic torque estimation; **iii.** friction force estimation and decoupling. It is critical to note that the force components in milling are periodic in nature. Therefore, all relevant states of the FD system (e.g., velocity, electromagnetic torque), which are affected by these forces, are periodic as well and can be conveniently represented by a Fourier series. For instance, the motor speed can be expressed as

$$\omega(t) = \sum_{n=0}^{N} A_n \cos(n\Omega t + \varphi_n) \quad (7)$$

where A_n and φ_n are amplitude and phase of the nth harmonic. Here, Ω is the angular speed of the cutting tool (a.k.a. spindle speed) and can be assumed constant for all practical purposes.

Figure 2 illustrates the network design incorporating these important ingredients. By taking advantage of the Fourier series representation, the network first generates the basis functions in (4). A <u>Recursive Discrete Fourier Transform</u> (RDFT) network, which is introduced by Dolen et al. (2000a), is used as an inertial torque estimator. Figures 3 and 4 illustrate such a network and its general-purpose (stationary frame) RDFT unit respectively. Dolen et al. (2000a) show that by modifying the weights of the unit, one can implement spectral decomposition (filtering) and differentiation of harmonic components for periodic waveforms. Here, the RDFT network with differentiating (a.k.a. D^1) units calculates the inertial torque (acceleration) by differentiating the harmonic components of the (measured) motor speed. Similarly, the second RDFT network (with filtering (D^0) network units) serves as a band-pass filter to eliminate the switching noise in the armature current (i_a) of a conventional permanent magnet DC servomotor ($\propto T_{em}$).

The remaining network elements simply calculate the viscous and the Coulomb friction torques in (4). The feedforward neural network labeled BVF ("bearing viscous friction"), whose properties are given by Dolen (2000), approximates $|\omega|^{2/3}\text{sgn}(\omega)$ while $\text{sgn}(\omega)$ of the Coulomb friction model $[T_0\text{sgn}(\omega)]$ is generated by saturating a bipolar sigmoid using a very high gain (γ_1). Likewise, the "friction coefficient estimator" essentially calculates the friction coefficient in (6). Similarly, all individual inputs of the summing neurons (#1 and #2) have to be properly scaled to utilize the linear portion of the bipolar sigmoid. For instance, the output of neuron #1 corresponds to the scaled version of the disturbance force estimate on the tableside. This quantity initially includes the friction force on the table. Therefore, a feedback from the "Friction Force Estimator" is required to separate out the cutting force component (F_x or F_y). The recurrent network ("Friction Decoupling Network"), shown in Fig. 2, carries out this decoupling process using the friction force estimate. The accuracy, convergence rate, and stability of this recurrent network are further discussed in Dolen et al. (2000b).

IV. PERFORMANCE EVALUATION AND DISCUSSION

A well-designed disturbance observer such as the one in Fig. 5 (Lorenz, 1996) could be a viable alternative to the presented neural network in terms of accuracy, bandwidth, hardware cost, etc. Therefore, a detailed simulation study has been conducted to study the performance of these two competing techniques.

First, the accuracy of each methods is computed (at steady state) as the function of spindle speed (Ω) under different simulated machining conditions. A disturbance force in the form of (8) can be applied to the simulated drive system for a given set of harmonic frequencies:

$$F_d(t) = \sum_{n \in \{0,1,4\}} F_n \cos(n\Omega t + \varphi_n) \qquad (8)$$

where φ_n are random numbers with $U(0,2\pi)$ (i.e. uniform probability density distribution between 0 and 2π); F_n are random variables with $U(-F,F)$ such that $|F| \in \{1000, 3000, 5000\}$ (in [N]) represents light, normal, and heavy machining conditions respectively.

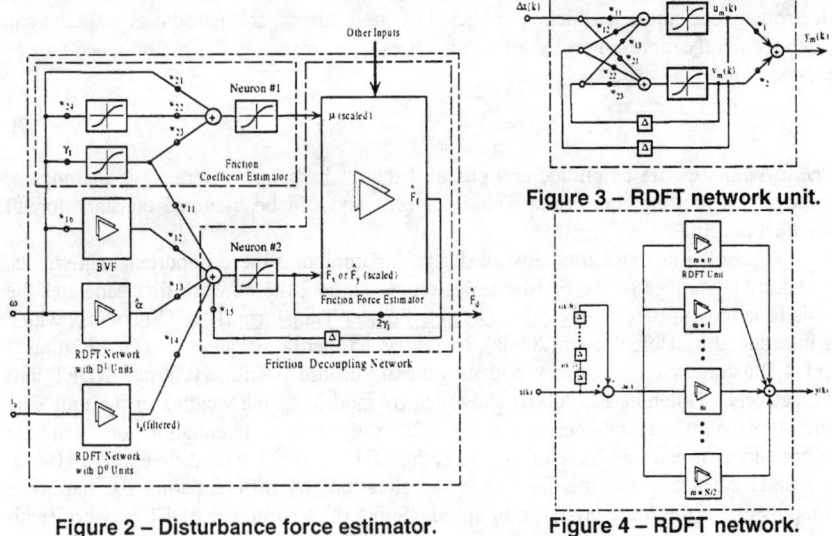

Figure 2 – Disturbance force estimator.   Figure 3 – RDFT network unit.

Figure 4 – RDFT network.

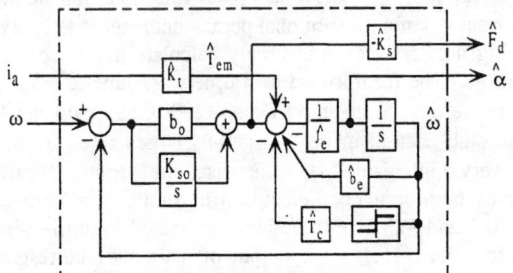

Figure 5 – Continuous-time, disturbance force observer.

The accuracy $P(\Omega)$ can be calculated for a number of spindle frequencies:

$$P(\Omega) = \frac{1}{T} \int_{t=0}^{T} |\hat{F}_d(t)/F_d(t)| dt \qquad (9)$$

The physical parameter estimates employed by both methods must be equivalent to make a fair comparison. To accomplish that, various tests are conducted on the simulated drive, including acceleration tests, constant speed (with no load) tests, a variety of light machining tests. The observer parameters (J_e, b_e, T_c, K_t, K_s) are indirectly computed using the data obtained from these tests. It has been observed that the resulting parameter estimation errors are generally within 3%-7% error range. The observer is then tuned to a bandwidth frequency of 100Hz. With respect to the network, similar tests, which additionally include medium and heavy machining conditions, are performed to form various training sets. Using the genetic algorithm presented by Dolen et al. (1999), the free weights of the network have been adjusted to minimize the disturbance estimation errors. After the training, the average training error is observed to be slightly than 2%.

Figure 6 illustrates the accuracy of both methods. Omitting the characteristics of acceleration estimation for both methods, the only major difference between them is that the network incorporates a detailed friction model. As can be seen, the SNN is significantly more accurate than the observer at almost every machining condition. The mean relative-errors can be given as 8.14% and 3.29% for the observer and the SNN respectively. Hence, the resulting accuracy improvement is a factor of 2.5.

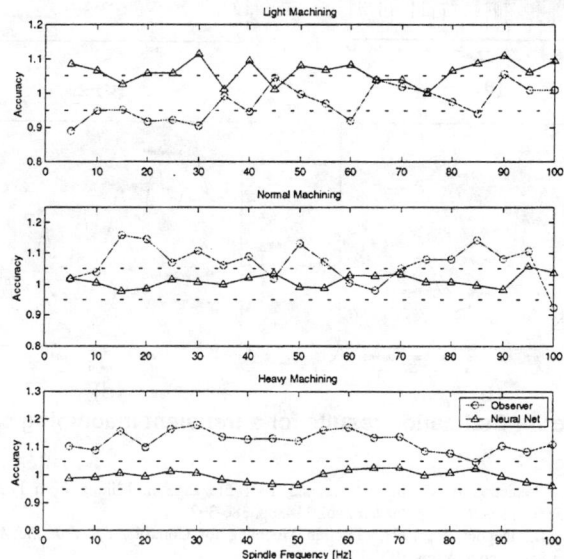

Figure 6 – Accuracy plots under various machining conditions.

To evaluate the transient estimation capabilities of these competing methods, a transient machining case shown in Figs. 7a (e.g. a step change in the axial depth of cut) is studied. The (averaged) estimation errors are shown in Figs. 7b. As can be seen from the figure, the network's error is much smaller than its counterpart. Finally, a speed transient of the drive as demonstrated in Fig. 7c is considered. In this case, the velocity signals of the drive motors have low frequency components that have not been modeled by the RDFT network. Since the low frequency components of the speed appears as a constant over a moving average period, the RDFT network fails to estimate the low frequency acceleration components. As can be seen from Figs. 7d, this causes intolerable estimation error. On the other hand, the observer produces very good results for this case. Hence, there are certain practical (i.e. hardware) limitations to model a signal with low frequency components.

V. CONCLUSION

For cutting force estimation, this paper has focused on developing FD electromechanical model to design a SNN. The performance of the network is compared to that of a Luenberger disturbance observer through a detailed simulation study. The investigation has revealed that the accuracy of the net is 2.5 times better than that of the observer for steady-state and transient machining cases. However, the neural network has also demonstrated poor estimation results when the drive speed has un-modeled low frequency components.

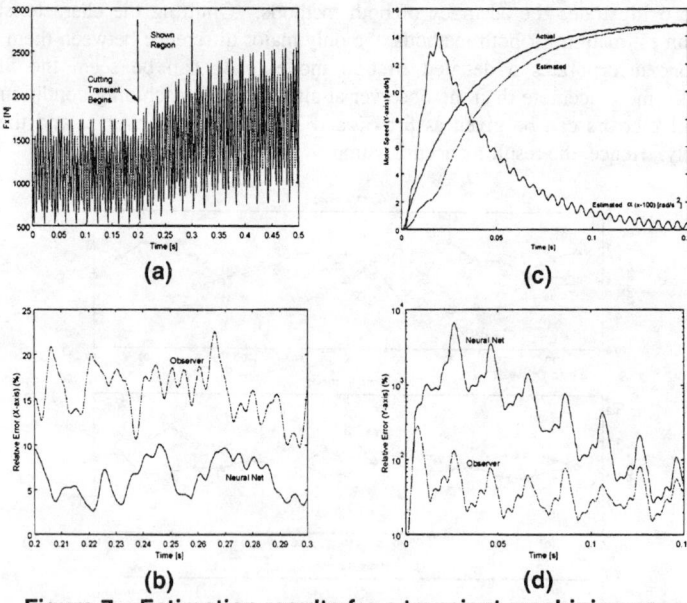

Figure 7 – Estimation results for a transient machining case.

REFERENCES

Altintas, Y., 1992, "Prediction of Cutting Forces and Tool Breakage in Milling from Feed Drive Current Monitoring," *ASME Journal of Eng. for Ind.*, vol. 114, pp. 386-392.

Armstrong-Helouvry, B., Dupont, P., 1993, "Friction Modeling for Control," *Proc. of the American Control Conference*, San Fransisco, CA, pp. 1905-1909.

Dolen, M., 2000, "Modeling and Estimation by Structured Neural Networks for CNC Machine Tools," *PhD Dissertation*, University of Wisconsin – Madison, 2000.

Dolen, M., Kayikci, E., and Lorenz, R. D., 2000a, "Recurrent Neural Network Topologies for Spectral State Estimation and Differentiation," in *Intelligent Engineering System Design through Artificial Neural Networks*, C. H. Dagli et al. (Eds.), vol. 10, pp. 159-164, ASME Press, NY.

Dolen, M., Kayikci, E, Chung, P. Y., and Lorenz, R. D., 2000b, "Disturbance Force Estimation for CNC Machine Tool Feed Drives by Structured Neural Network Topologies," in *Intelligent Engineering System Design through Artificial Neural Networks*, C. H. Dagli et al. (Eds.), vol. 10, pp. 851-856, ASME Press, NY.

Dolen, M. and Lorenz, R. D., 1999, "General Methodologies for Neural Network Programming," in *Intelligent Engineering System Design through Artificial Neural Networks*, C. H. Dagli et al. (Eds.), vol. 9, pp. 155-160, ASME Press, NY, 1999.

Dolen, M., Lorenz, R. D., and Kayikci, E., 1999, "Genetic Algorithms for Continuous Design Domain," in *Intelligent Engineering System Design through Artificial Neural Networks*, C. H. Dagli et al. (Eds.), vol. 9, pp. 337-342, ASME Press, NY.

Harris, T. A., 1991, *Rolling Bearing Analysis*, 3/e, John Wiley & Sons, Inc., NY.

Koren, Y., 1984, *Computer Control of Manufacturing Systems*, McGraw Hill, NY.

Kulkarni, P. K., Srinivisan, K., and Johnson, W. C., 1984, "Simulation and Discrete Time Modeling of Machine Tool Feed Drive Dynamics," *Proc. of the 1984 American Control Conference*, San Diego, CA, pp. 474-481.

Lorenz, R. D., 1996, "New Drive Control Algorithms (State Control, Observers, Self-sensing, Fuzzy Logic, and Neural Nets)," *in Proc. of PCIM Conf.*, Las Vegas, pp. 275-289.

Stein, J. L., Colvin, D., Clever, G., and Wang, C. H., 1986, "Evaluation of DC Servo Machine Tool Feed Drives as Force Sensors," *ASME Journal of Dynamic Systems, Measurement and Control*, vol. 108, pp. 279-288.

Van Brussel, H., and Vastmans, L., 1981, "Direct Digital Control of Feed Drives," 13th *CIRP Int'l Seminar on Manufacturing Systems, Microprocessors in Manufacturing Systems*, Leuven, Belgium.

LOAD TORQUE ESTIMATION FOR SPINDLE DRIVES USING STRUCTURED NEURAL NETWORK TOPOLOGIES

MELIK DOLEN	EKREM KAYIKCI	ROBERT LORENZ
Dept. of Mech. Engr.	Dept. of Electrical & Comp. Engr.	
Middle East Technical University	University of Wisconsin – Madison	
Ankara, Turkey	Wisconsin, USA	

ABSTRACT

This paper presents a structured neural network for the spindle drives of CNC machine tools to estimate the load torque accurately. Since preloaded bearings play an important role in the dynamics of such systems, major emphasis is placed on bearing friction modeling. The accuracy of the proposed topology is compared to that of a Luenberger type of disturbance torque observer. The performance of the network is found to be superior to that of the observer, which is quite sensitive to quantization noise.

I. INTRODUCTION

Designing reliable, sensitive, flexible, and accurate sensors and estimation schemes plays a critical role in realizing the productive and highly sophisticated machine tools of the future. Of many sensory devices being used in computer numerical control (CNC) technology, cutting torque/force sensors can provide the most crucial information for adaptive control of machine tools, detection of chatter, tool wear/breakage, and machine condition monitoring. Not surprisingly, torque transducers, which are currently available in the market for a wide variety of CNC machines, have not gained recognition in industry due to their numerous drawbacks. The technological trend in the industry is towards the employment of non-invasive estimation schemes and the elimination of (additional) sensors wherever the estimation methods are feasible alternatives.

One of the pioneering analyses in this field has been done by Matsushima et al (1982) to monitor the disturbance torque on a spindle drive (SD) of a lathe for the purpose of tool failure detection. Since the spindle system has a very large inertia and inertial torque has not been estimated in their study, the resulting estimation scheme has a very low bandwidth (<0.5Hz). Similarly, Stein et al. (1986) further investigate the armature current of a field controlled DC spindle motor of a CNC lathe through perturbation analysis. Their experimental results are, not surprisingly, a confirmation of Matsushima et al (1982). Due to large spindle inertia of their lathe (as well as neglect of inertial torque in their analysis), the bandwidth of the estimator was very low. In summary, most estimation methods in the CNC research literature have neglected the important effects of inertial torque. The resulting estimation scheme is of very low bandwidth, which seriously limits its practical use.

Consequently, the main motivation of this paper is to develop an advanced torque estimator with improved accuracy and adequate bandwidth. With the utilization of detailed spindle drive (mechanical system) model, which can accommodate all major nonlinearities such as bearing friction, non-ideal gear behavior etc., this paper intends to evaluate the potential for implementing such a disturbance torque estimator.

II. SPINDLE DRIVE MECHANICAL SYSTEM MODELING

To develop a SD mechanical model, let us consider the system (Cincinnati-Milacron 5VC-750) shown in Fig. 1. The differential equation for the first gearing stage of this system can be written as

$$J_{e1}\frac{d^2\theta}{dt^2} + T_{f1}(F_{t1}, d\theta/dt) = T_{em} - F_{t1}\frac{D_1}{2} \quad (1)$$

where T_{f1} is total friction torque [Nm]; T_{em} refers to electromagnetic (motor) torque [Nm]; F_{t1} denotes tangential (spur gear) force [N]; J_{e1} is total equivalent inertia of the input shaft (including motor inertia) [kg·m^2] while D_1 is pitch circle diameter of pinion [m]. Assuming a gear efficiency of 100%, the following equation for the second stage can be obtained:

$$J_{e2}N_1\frac{d^2\theta}{dt^2} + T_{f2}(F_{t1}, F_{t2}, d\theta/dt) = F_{t1}\frac{D_1}{2} - F_{t2}\frac{D_{2i}}{2} \quad (2)$$

where $N_1 \equiv D_1/D_{2i}$; $i \in \{1,2\}$ is transmission-mode index. With respect to the output stage of the spindle system, one can the dynamic equation as follows:

$$J_{e3}N_1 N_{2i}\frac{d^2\theta}{dt^2} + T_{f3}(F_{t2}, F_x, F_y, F_z, T_c, d\theta/dt) = F_{t2}\frac{D_{3i}}{2} - T_c \quad (3)$$

where $N_{2i} \equiv D_{2i}/D_{3i}$; T_c is disturbance (load) torque while F_x, F_y, and F_z are the principal milling force components. To calculate the friction torques (T_{f1}, T_{f2}, T_{f3}) on each individual shaft, one needs to study the friction process in pre-loaded ball/roller bearings. The friction torque on such bearings (T_{fb}) is mainly affected by external loads, bearing design, and lubrication technique. Harris (1991) gives the following empirical expression for friction torque (T_f):

$$|T_f| = aF_0(F_e)^d + b|\omega|^{2/3} + c \quad (4)$$

where ω is (mean) angular velocity of the bearing [rad/s]; F_e and F_0 are equivalent (static) load and parametric force respectively. Here a, b, c, and d are (positive) empiric factors that can be obtained through the tables given by Harris (1991). Both F_0 and F_e are (nonlinear) functions of radial, tangential, and axial components of bearing load. Once these forces for each bearing are determined, the friction torque can be calculated using (4). The summation of individual bearing torques yields the total friction torque on each shaft.

III. STRUCTURED NEURAL NETWORKS FOR LOAD TORQUE ESTIMATION

In the previous section, the mechanical SD model, which form the basis for designing a structured neural network (SNN), were introduced. Employing neural network programming technique (Dolen and Lorenz, 1999), a recurrent neural network (RNN) can be developed to estimate disturbance torque accurately. Disturbance torque estimation has three basic components: **i.** acceleration estimation; **ii.** electromagnetic (motor) torque estimation; **iii.** friction force estimation and decoupling. Furthermore, the milling torque is periodic in nature. All relevant SD states (e.g., velocity, motor torque), which are affected by this torque, are periodic as well and can be conveniently represented by a Fourier series. For instance, the motor speed can be expressed as

$$\omega(t) = \sum_{n=0}^{N} A_n \cos(n\Omega t + \varphi_n) \quad (5)$$

where A_n and φ_n are amplitude and phase of the nth harmonic. Here, Ω is the steady-state cutting tool (angular) speed. Figure 2 illustrates a SNN incorporating these important features (Dolen, 2000). By taking advantage of Fourier series representation, this network first generates the basis functions in (1–3). Here, a Recursive Discrete Fourier Transform (RDFT) network, which is introduced by Dolen et al. (2000a), is used as an inertial torque estimator. Figures 3 and 4 illustrate such a network and its general-purpose (stationary frame) RDFT unit respectively. Dolen et al. (2000a) show that by modifying the weights of the unit, one can implement spectral decomposition (filtering) and differentiation of harmonic components for periodic waveforms. In Fig. 2, the RDFT network with differentiating (a.k.a. D^1) units calculates the inertial torque (acceleration) by simply differentiating the harmonic components of the sample average motor speed (ω_m). Since the induction-machine rotor position (θ) (i.e. input shaft position) is to be measured, the sample average speed is calculated through first differences method. As illustrated in Fig. 2, the filtering (a.k.a. D^0) network units are needed to filter out the quantization noise in ω_m.

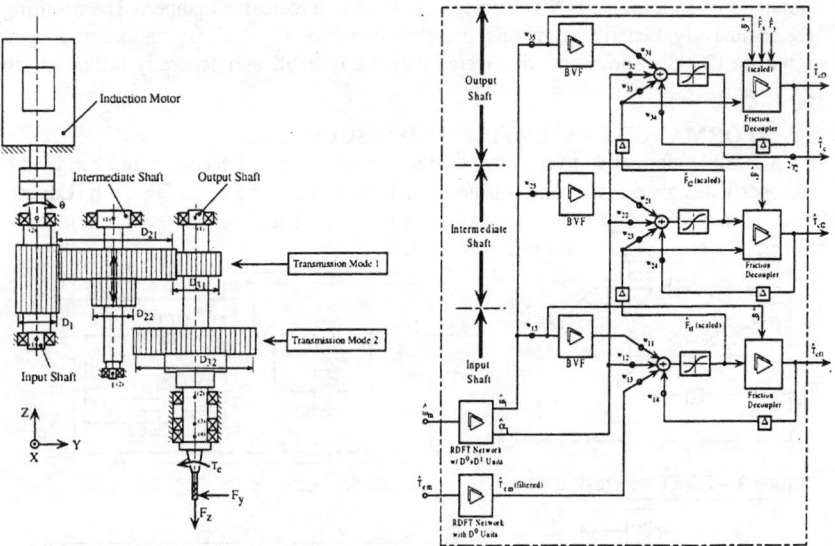

Figure 1 – Spindle drive system.   Figure 2 – Disturbance torque estimator.

Electromagnetic (airgap) torque developed by the induction motor (T_{em}) is the cross product of stator flux vector and stator current vector in the stator reference frame and can be calculated in a straightforward fashion when the stator currents as well as motor terminal voltages are measured (Novotny and Lipo, 1995). Due to inverter switching noise, the electromagnetic torque (estimate) is extremely noisy. Therefore, another RDFT network (with filtering (D^0) network units) is specifically employed in this topology as a band-pass filter to eliminate this noise.

The presented network implements the remaining terms in the differential equations in a piecewise fashion. For instance, a pre-trained feedforward neural network (FNN) labeled BVF is used to approximate the total bearing viscous friction torque [i.e. $f(x) = |x|^{2/3} \text{sgn}(x)$ in (4)]. The characteristics of this network are further discussed in Dolen (2000). Similarly, the "friction decouplers" of Fig. 2 are FNNs designed to compute the Coulomb friction torque (CFT) in (4):

$$T_{cf,j} = \sum_{k=1}^{K} a_k F_{0,k}(F_{e,k})^{d_k} \text{sgn}(\omega_j) \qquad (6)$$

where K is the number of bearings on shaft j. Since $F_{0,k}$ and $F_{e,k}$ do require the loads on bearing k under the action of external forces on the shaft, the computation of (6) poses major difficulty in this estimation scheme. To solve for the unknown tangential gear (external) forces, the CFT terms are neglected in the first pass and then the forces are solved utilizing (1-4):

$$\hat{F}_{t1} \approx \frac{2}{D_1}[\hat{T}_{em} - \hat{J}_{e1}\hat{\alpha}_1 - \hat{b}_{e1}|\omega_1|^{2/3}\text{sgn}(\omega_1)] \qquad (7a)$$

$$\hat{F}_{t2} \approx \frac{2}{D_{2i}}[\hat{F}_{t1}\frac{D_{2l}}{2} - \hat{J}_{e2}N_1\alpha_1 - \hat{b}_{e2}|N_1\omega_1|^{2/3}\text{sgn}(N_1\omega_1)] \qquad (7b)$$

$$\hat{T}_c \approx \hat{F}_{t2}\frac{D_{3i}}{2} - \hat{J}_{e3}N_1N_{2i}\alpha_1 - \hat{b}_{e3}|N_1N_{2i}\omega_1|^{2/3}\text{sgn}(N_1N_{2i}\omega_1) \qquad (7c)$$

With the friction model in (4), more accurate external force components (F_{t1}, F_{t2}) could be obtained when improved CFT estimates are used in a recursive manner. The resulting RNN exclusively benefits from this predictor/corrector type of estimation paradigm. Dolen et al. (2000b) shows that the overall network is stable and generally settles within a few sweeps.

IV. PERFORMANCE EVALUATION AND DISCUSSION

A feasible alternative to the presented network in terms of accuracy and bandwidth is a properly designed disturbance observer such as the ones shown in Fig. 5a-b (Lorenz, 1996). Therefore, a detailed simulation study has been conducted to study the performances of these two competing techniques.

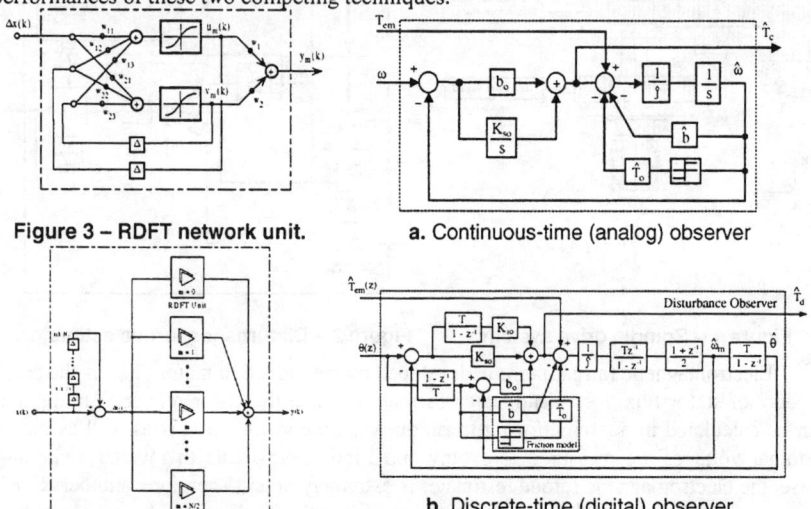

Figure 3 – RDFT network unit.   a. Continuous-time (analog) observer

Figure 4 – RDFT network.   b. Discrete-time (digital) observer

    Figure 5 – Disturbance torque observers.

The accuracies of both methods (at the steady-state) are to be computed as a function of spindle frequency (Ω) under different simulated machining conditions. For this purpose, a disturbance torque in the form of (6) could be applied to the simulated SD system for a given set of frequencies $n \in \{0, 1, 4\}$. In (6), φ_n are taken as the random numbers with uniform probability density distribution between 0 and 2π while the cutting torques magnitudes (A_n), which corresponds to light – normal machining conditions, uniformly lie in between 2 and 8 [Nm]. The accuracy $P(\Omega)$ can then be calculated for a number of spindle frequencies:

$$P(\Omega) = \frac{1}{T}\int_{t=0}^{T}|\hat{T}_d(t)/T_d(t)|dt \qquad (8)$$

The physical parameter estimates employed by both methods must be equivalent to make a fair comparison. First, the observer parameters (J, b, T_0) were computed through various "tests" on the simulated drive. The resulting parameter estimation errors were generally within 10% error range. Then, the observer was tuned to a bandwidth of 30

Hz. With respect to the network, similar tests, which additionally included light and medium machining conditions, were performed to form a variety of training sets. Using the genetic algorithm of Dolen et al. (1999), the free weights of the network were adjusted to minimize the estimation errors on these training sets. After the training phase, the average estimation error for all training sets was 1.82%.

Fig. 6 illustrates the accuracies of both methods. Since the resolution of position sensor plays a key role in the estimator dynamics, the accuracy is calculated for two different encoder resolutions: 12 bit and 16-bit. As can be seen, the network is significantly more accurate than the observer itself at almost every resolution range. Note that the accuracy plot for 16-bit resolution range also includes the characteristics of an "analog" observer with a bandwidth of 30 Hz. This observer, which is shown in Fig. 5a, employs noise-free velocity measurements and thus it is hypothetically equivalent to a "digital" observer using extremely high-resolution position measurements. In summary, the mean relative-errors for all resolution ranges are 1.59% (SNN) and 15.73% (digital observer). The resulting accuracy improvement is remarkably on the order of 10. Similarly, the corresponding error of the analog observer is 3.11% indicating two times accuracy-improvement.

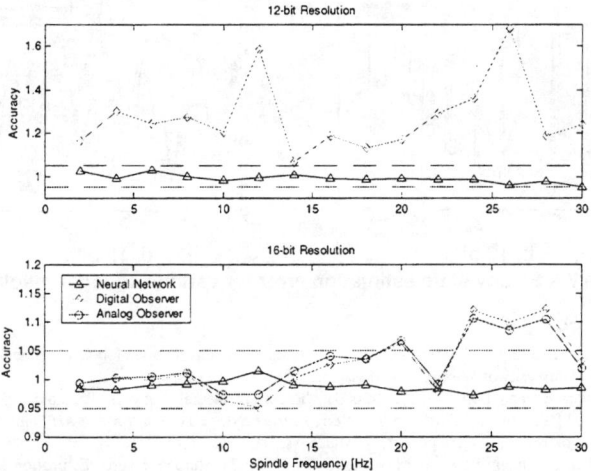

Figure 6 – Accuracy plots under various machining condition.

Finally, Fig. 7 illustrates the steady-state responses of both methods when large disturbance torque is applied. As can be seen from the figures, the quantization level of the position measurements seriously affects the estimation performance of the digital observer. On the other hand, the network exhibits an exceptional performance independent of the quantization noise level because the RDFT network, which is used to estimate acceleration, responds only to the predefined harmonics of the measurement and filters out the rest. Also notice that in Figs. 7a,c; the electromagnetic torque estimate lacks the high frequency disturbance components that are essentially contained in the acceleration. Therefore, the conventional techniques, which neglect acceleration, yield very low bandwidth including the references cited in Section I.

V. CONCLUSION

This paper has focused on various SD mechanical components and their physical models for the purpose of disturbance torque estimation. The mechanical SD model of a specific CNC machine is developed. Based on this model, a SNN is designed to estimate the cutting process torque. The accuracy of the proposed topology is compared to that of a discrete-time Luenberger type of disturbance torque observer. The performance of the neural network is found to be superior to that of the observer, which is quite vulnerable to quantization noise. Contrary to the common conjecture in the industry, a relatively high bandwidth (80-120Hz.) can be easily attained using this method.

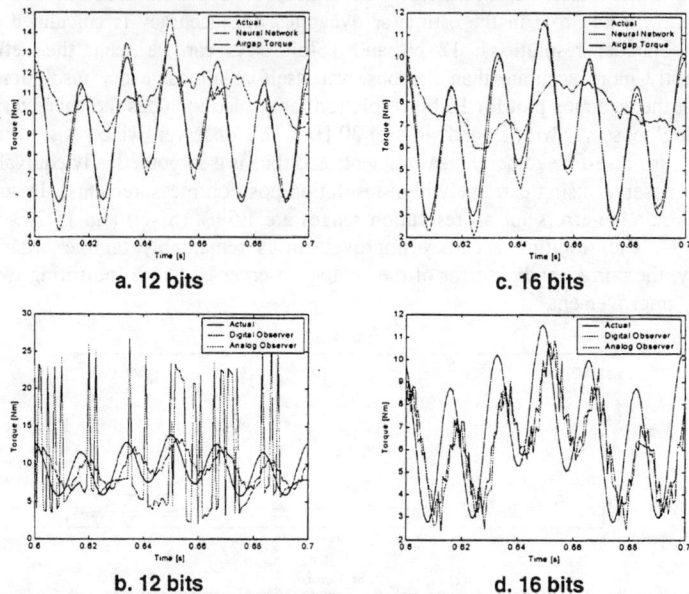

a. 12 bits   c. 16 bits

b. 12 bits   d. 16 bits

Figure 7 – Steady-state estimation error for various encoder resolutions.

REFERENCES

Dolen, M., 2000, Modeling and Estimation by Structured Neural Networks for CNC Machine Tools, *Ph.D. Dissertation*, University of Wisconsin – Madison.

Dolen, M., Kayikci E., and Lorenz, R. D., 2000a, "Recurrent Neural Network Topologies for Spectral State Estimation and Differentiation," in *Intelligent Engineering System Design through Artificial Neural Networks*, C. H. Dagli et al. (Eds.), vol. 10, pp. *-*, ASME Press, NY.

Dolen, M., Kayikci, E, Chung, P. Y., and Lorenz, R. D., 2000b, "Disturbance Force Estimation for CNC Machine Tool Feed Drives by Structured Neural Network Topologies," in *Intelligent Engineering System Design through Artificial Neural Networks*, C. H. Dagli et al. (Eds.), vol. 10, pp. 851-856, ASME Press, NY.

Dolen, M. and Lorenz, R. D., 1999, "General Methodologies for Neural Network Programming," in *Intelligent Engineering System Design through Artificial Neural Networks*, C. H. Dagli et al. (Eds.), vol. 9, pp. 155-160, ASME Press, NY.

Dolen, M., Kayikci, E., and Lorenz, R. D., 1999, "Genetic Algorithms for Continuous Design Domain," in *Intelligent Engineering System Design through Artificial Neural Networks*, C. H. Dagli et al. (Eds.), vol. 9, pp. 337-342, ASME Press, NY.

Harris, T. A., 1991, *Rolling Bearing Analysis*, 3/e, John Wiley & Sons, Inc., NY.

Lorenz, R. D., 1996, "New Drive Control Algorithms (State Control, Observers, Self-sensing, Fuzzy Logic, and Neural Nets)," *in Proc. of PCIM Conf.*, Las Vegas, pp. 275-289.

Matsushima L., Bertok, P., and Sata, T., 1982, "In-process Detection of Tool Breakage by Monitoring the Spindle Motor Current of a Machine Tool," *ASME Meas. and Cont. for Batch Manufacturing*, ASME Book, pp. 14-19.

Novotny, D. W., and Lipo, 1995, T. A., *Dynamics and Control of Induction Motor Drives*, Clarendon Press, Oxford.

Stein, J. L., and Shin, K. C., 1986, "Current Monitoring on Field Controlled DC Spindle Drives," *ASME Journal of Dynamic Systems, Measurement and Control*, vol. 108, pp. 289-295.

MODEL IDENTIFICATION FOR RESTRUCTURABLE CONTROL SYSTEM CONSIDERING FAULT TOLERANCE

YUJI KUWASHIMA
Autonomous System Engineering,
Complex System Engineering,
Hokkaido University, Japan

HIROSHI YOKOI
Autonomous System Engineering,
Complex System Engineering,
Hokkaido University, Japan

YUKINORI KAKAZU
Autonomous System Engineering,
Complex System Engineering,
Hokkaido University, Japan

ABSTRACT

Fault tolerance is very important for spacecraft, which can achieve tasks even got into unexpected troubles. One of the methods to realize fault tolerance is to build a restructurable control system. In such a restructurable control system, after dynamical changes of spacecraft caused by troubles, the model of an airframe is identified, and the control system is rebuilt using the present airframe, based on the identified model. In this study, a method of a restructurable control system that consists of model identification by parameter search using genetic algorithms (GAs), model construction by artificial neural network (ANN) and a restructurable controller by genetic algorithms is proposed. The effectiveness of the fault tolerance system was verified using a rotary motion simulation of spacecraft.

INTRODUCTION

As space technologies have accomplished great developments recent years, missions in the space by spacecraft tend to be more complicated and protracted. In order to execute such missions, control systems with fault tolerance such as self-repair function are necessary, since the missions should be succeeded not only in the external environment in which they were considered in design stage, but also in unpredictable troublesome environments, under bad equipment conditions. Though the reliability of components have been raised, since a space machine is constituted intricately and built from a great number of parts, it is difficult to prevent the occurrence of failures beyond the presumption. Also because the body should be considered working in the ultimate environment of the universe, it is next to impossible to restore the body by man's hand.

Structural redundancy by reserving spare components can be as one method of realizing the fault tolerance. However, as for a spacecraft, the weight of loading energy or the body is restricted, and the method would cause a high cost. Another method is to prepare a repertoire of considerable failure cases. But, this method cannot deal with the situation beyond presumption. And it is also impossible to predict all troublesome situations, in the case of spacecraft missions. Robust control is also a candidate, however, the method raises the robustness to faults by cutting down the performance in the normal state.

One method to realize fault tolerance control while restraining the redundancy of hardware, is to build a restructurable control system. By such restructurable control systems, when failures occur, a model of the body after dynamical characteristic changes is identified first. After that, a control system that can accomplish tasks is restructured on the basis of the dynamical model of the body identified, using the present remaining system.

However, in order to control using the identified model, a restructurable control system should first detect failure and to perform model identification after failures occur.

In this study, a method of failure detection, model identification by incorporating genetic algorithms (GAs) and artificial neural networks (ANN) and restructurable controller by genetic algorithms was proposed, and it was shown that the proposed method is effective using a rotary motion simulation of spacecraft.

THE MODEL OF THE SPACECRAFT

The rotary motion of spacecraft used is expressed by equation (1), deduced from the Eulerian's equation and the definition of angular momentum.

$$M = \dot{I} \cdot \omega + I \cdot \dot{\omega} + \dot{\omega} \times (I \cdot \omega) \tag{1}$$

$$M = f(\hat{M})$$

M : Torque $\hat{M}$: Control Signal
ω : Angular Velocity $\dot{\omega}$: Angular Acceleration
I : Inertia Matrix (3 × 3 matrix) $\dot{I}$: Variation of Inertia Matrix

Torque M is generated by injection of thruster around the directions of 3 axes, i.e., x-axis (roll), y-axis (pitch), and z-axis (yaw). And the spacecraft has 6 thrusters, as there are 2 thrusters around each axis (+/-). Normally, inertia parameters are continuous, and do not change, so that $\dot{I}$ is equal to 0. But inertia parameters continuously change when the airframe is not rigid body. For example, if a part of body is in a movable state, moves in response to acceleration, the inertia matrix will change continuously as the body accelerates.

The $\hat{M}$ is a control variable for thrusters. Normally, thrusters operate according to this variable, so that the relation between torque M and the control variable $\hat{M}$ is equal. But, when several of thrusters break down, the relation between torque M and the control variable $\hat{M}$ is not necessarily equal.

Following equation (2) is deduced from education (1).

$$\dot{\omega} = I(t)^{-1} \cdot (M - \dot{I}(t) \cdot \omega + \omega \times (I(t) \cdot \omega)) \tag{2}$$

The angular acceleration is dependent on not only torque but also angular velocity in order to prove from equation (2). Therefore, it is possible that the attitude is controlled by getting angular acceleration by controlling angular velocity for the adequate value, even if any one thruster becomes impossible to generate the torque by breaking down. And, a feedback gain is used as a controller of the thrusters.

THE PROPOSED METHOD

Normally, the condition of the spacecraft is always observed to in order to detect trouble immediately.

If the failure is detected, identification of the dynamic model after the trouble and restructurization of the controller are carried out while sensing.

The algorithm of the proposed method is shown in Fig.1.

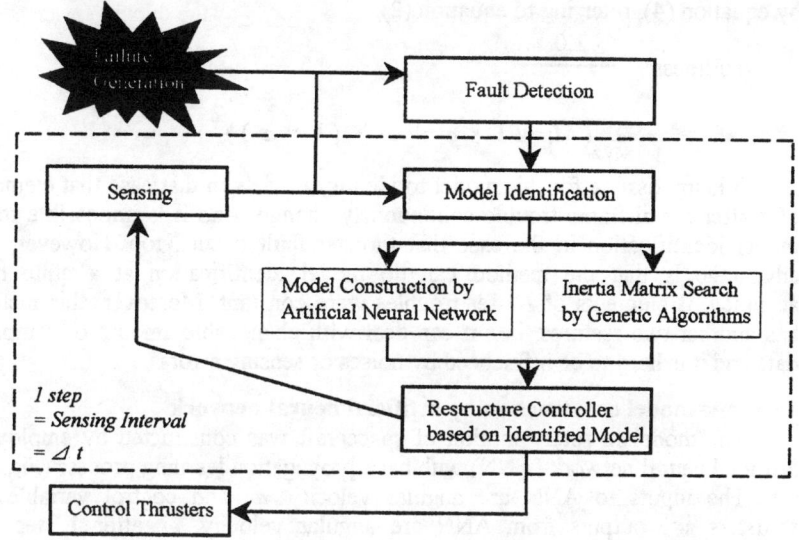

Fig.1 Algorithm of the Proposed Method

The fault detection

To begin with, the following method is used for fault detection.

Each element of inertia matrix I in normal state is a constant. So the following equation (3) is obtained, referring to equation (1) in the last section.

$$M = I \cdot \dot{\omega} + \omega \times (I \cdot \omega) \qquad (3)$$

By substituting data (M, ω, $\dot{\omega}$) sensed from 3 data collecting processes into equation (3) and solving, inertia matrix I is obtained.

When a spacecraft has broken down, changes appear in I, which could be deducted by solving this equation. Therefore whether changes in I occurred or not could be confirmed by calculating I every step. Then, if changes in I were detected for a quite long period, the control should be shifted to the stage of dynamic model identification for troubled airframe.

The dynamic model identification after the failure

We employed a method incorporating genetic algorithms (GAs) and artificial neural network (ANN) approaches to identify the dynamic model after the trouble occurred. Then, spacecraft orbits estimated by two identified models were compared with the real orbit, and the model resulting in smaller error would be selected.

Inertia matrix search by genetic algorithms

By using genetic algorithms (GAs), inertia matrix I of the troubled airframe was searched. The locus of gene of GAs used by this method is real number, and matrix I was searched by substituting each element of matrix I into the locus of gene. At every sensing interval individuals are generate. And it is assumed that elements of I after troubles are constant, so fitness of gene would be expressed by equation (4), referring to equation (2).

$$\text{Fitness} = \frac{1.0}{E + 1.0} \quad (4)$$

$$E = \sum_{p} \sum_{j=x,y,z} (M_j - (I \cdot \dot{\omega} + \omega \times (I \cdot \omega)))^2$$

It is impossible for this model to identify models in that case that elements of I after the airframe trouble continuously change. And it is impossible to do model identification in the case that thruster fault occurs, too. However, one advantage is that the method can do model identification at a quite high accuracy, if elements of I after troubles were constant. Moreover, this method has another two features, i.e., it can deal with changeable amount of sampling data and it is hard to be influenced by noises or sensing errors.

Dynamic model construction by artificial neural network

The model in rotary motion of spacecraft was constructed by employing artificial neural network (ANN) with back propagation learning process (Fig.3).

The inputs to ANN are angular velocity ω and control variable for thrusters $\hat{M}$, outputs from ANN are angular velocity ω after 1 step and variations of angle $\triangle\theta$. Teacher signals are angular velocity ω got by sensing and variations of angle $\triangle\theta$. At every sensing interval the weights of artificial neural network are recalculated.

The model identification of spacecraft can be achieved by using a minimum set of input/output data, when torque M is different from the control variable $\hat{M}$, possibly caused by thruster breaking down or I changes due to the shape change of airframe.

Restructure Controller by Genetic Algorithms

Using the identified model, the controller was restructured.

The locus of gene of GAs used by this method is real number, and controller F was restructured by substituting each element of matrix F into the locus of gene. Fitness of gene would be expressed by equation (5).

$$\text{Fitness} = \frac{1.0}{T + 1.0 - \triangle t} \quad (5)$$

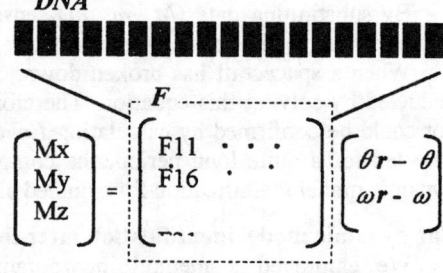

Fig.2 Feedback Gain Controller

T : Time until the model reach target attitude
$\triangle$ t : Sensing Interval

F is a 3×6 matrix that expresses the feedback gain. (Fig.2)

EXPERIMENTS

Experiments of the spacecraft using motion simulation were carried out.

While the data was senses after the trouble occurred, the dynamic model identification of the airframe after the trouble was carried out based on the data, and the controller was restructured based on the identified model, and control. Target attitude is set the attitude of which the angles and the angular velocities are equal to 0.0. Trouble generation is set at 0.0 second, and sensing interval was set 0.1 second.

The experiments were carried out for 3 kinds troubles, described as follows.
(1) The case in which only I changes by the deformation of airframe, and elements of I after the trouble is a constant.
(2) The case in which only I changes by the deformation of airframe, but elements of I after the trouble is variable.
(3) The case in which one thruster breaks down and cannot generate torque.

On case (3) the thruster which could generate – torque around x-axis broke down and became unable to generate torque.

Fig.3 showed the results of the experiment on case (1). In Fig.3-Left the angles were plotted against time, and so the angular velocities in Fig.3-Right. That of case (2) was shown in fig.4, and that of case (3) was shown in fig.5.

For case (1), as we discussed before, the model after the case (1) failure could be quickly and effectively identified by using GAs, so spacecraft reached to the target attitude in the short time.

while the models after the case (2) failures could not be effectively identified using GAs. Consequently, fig.3 showed a more accurate prediction than that in fig.4. When the method using GAs can be used, it can identify the model quite quickly and accurately.

On case (3), even when a thruster broke down, model identification and restructuring controller was successful so spacecraft reached so spacecraft reached to the target attitude.

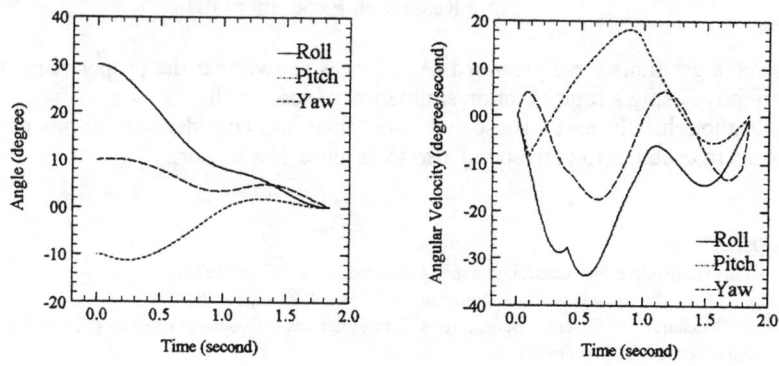

Fig.3 Results on Experiment (1)

CONCLUSIONS

The method of failure detection, model identification by incorporating genetic algorithms and artificial neural networks and restructurable controller by

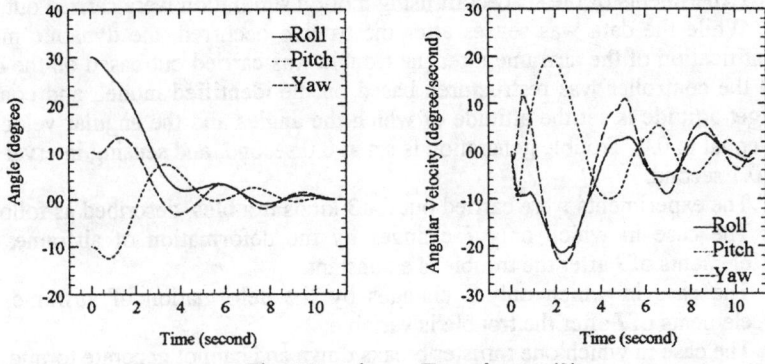

Fig.4 Results on Experiment (2)

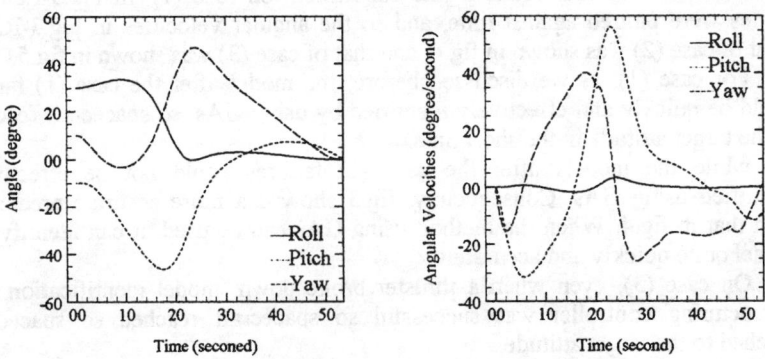

Fig.5 Results on Experiment (3)

genetic algorithms were proposed. And it was shown that the proposed method is effective using a rotary motion simulation of spacecraft.

Although failure of a sensor was not taken into consideration in this paper, it plans to enable it to correspond also to fa ilure of a sensor.

REFERENCES

Kaplan, M.H.; Modern Spacecraft Dynamics & Control, John Wiley (1976)

Chobotov, V.A.; Spacecraft Attitude Dynamics and Control, Krieger (1991)

Zvigniew Michalewicz; Genetic Algorithms + Data Structures = Evolution Programs, Third Revised and Extended Edition (1996)

R.Beale and T.Jackson; Neural Computing: An Introduction, Adam Hilger, Bristol, Philadelphia and New York (1990)

J.E.Dayhoff; Neural Network Architectures: An Introduction, Van Nostrand Reinhold, New York (1990)

M.Kurotaki, M.Kohno, S.Ishida, H.Yoneda, H.Yokoi, and Y.Kakazu; "An Evolutionary Reconfiguration for A Space Flight Controller", Artificial Neural Networks in Engineering Conference (ANNIE '99), pp.745-750, St. Louis, Missouri, U.S.A., November, 1999

DEEPSIA - FROM SUPPLY CHAINS TO SUPPLY WEBS

PEDRO A. C. SOUSA
Universidade Nova de Lisboa
Faculdade de Ciências e Tecnologia
P-2825-114 Monte de Caparica
Portugal

ADOLFO STEIGER-GARÇÃO
Universidade Nova de Lisboa
Faculdade de Ciências e Tecnologia
P-2825-114 Monte de Caparica
Portugal

JOÃO P. PIMENTÃO
Universidade Nova de Lisboa
Faculdade de Ciências e Tecnologia
P-2825-114 Monte de Caparica
Portugal

ABSTRACT

The strategic aim of the project DEEPSIA "Dynamic on-linE IntErnet Purchasing System, based on Intelligent Agents", the IST project Nr. 1999-20 283, funded by the European Union, is to create a state-of-the-art system to support companies' usual day to day purchasing requirements via the Internet. DEEPSIA's approach, based on a purchaser-centered solution and tailored to individual requirements, will assist SME – Small and Medium Enterprises, into the e-commerce process of finding the most suitable offer of goods and/or services for their needs. This solution will support the development of supply webs rather than supply chains, since the purchaser, in a step-by-step approach, will be able to avoid the traditional procedures of acquiring, substituting them, by an electronic process based on an intelligent multi-agent system. The system, based on a 'Framework for Internet data collection based on intelligent agents", will offer a user-friendly interface customized through a purchaser's-personalized catalogue, which is automatically updated with information gathered from available e-business portals. The framework is sustained on a multi-agent system using intelligent agents, (Bond, 1988) which are responsible for gathering web data and processing it (Mitchell, 2000), in order to collect relevant business information (costs, description of particular characteristics, etc.). The major technical hitches to be overcome by the proposed framework are: standard definition in order to create an Internet meta-data description; data format conversion in order to integrate differentiated data formats; creation of an ontology for data classification; and knowledge data storage for relevant Internet data recognition; and integrity validation of Internet links. Intelligent agents, human agents, catalogue management, database technology and up to date communication technologies will be used towards the integration of the flow of business information to and from buyers and suppliers. This paper outlines the objectives, the strategy and architecture, designed for the support of the DEEPSIA system.

INTRODUCTION

In the near future, huge network infrastructures and new information and telecommunication technologies will be one of the most basic and important tools for all business processes in organizations from the industrial sector, through services and government sectors to financial institutions. However, the proliferation of information, already evident in recent times, is starting to cause considerable problems both for individuals and organizations. As a consequence there is both the need to effectively manage existing collections of information and to access and import new relevant information. (Bokma, 2000) The procurement process is no exception and it is also affected by these problems magnified by the myriad of commercial information available. In this context, Business-to-Business (B2B) e-commerce models usually focus on SME as suppliers, often within virtual shops or marketplaces. Nevertheless, all SMEs are purchasers of goods and services. Depending on the size of the SME, this function may be performed by one person – the owner/manager, or the purchasing manager – or may be open to a number of staff.

DEEPSIA's purpose is to provide a purchaser centered solution, available on the desktop and tailored to individual requirements, rather than a supplier centered marketplace.

Therefore, the final aim is to introduce SMEs to e-commerce as purchasers, rather than just suppliers, enabling traditional supply chains' substitution by Internet supply webs. It will support/help the end user processing more cost and time effective purchases, providing an interface with catalogues containing offers from various suppliers.

The reduction of time in collecting data will be achieved by presenting the user with a catalogue of the set of entries that he/she is looking for, avoiding having to browse through a wide range of web pages and links that are not related to his/her needs. The catalogue will store organized pre-selected information, collected from different web sites, avoiding useless visits to several sites. The process of presenting the catalogue-collected data will bring all the similar items to one page, therefore simplifying the task of comparing them.

The Multi-Agent System (MAS) is responsible for an autonomous data collection, and semi-automatic catalogue updating process. This multi-agent system is transparent to the end-user, meaning that the "dirty" work (interfacing the web and the private databases) is running in the background, generating and obtaining information for the catalogue. The Multi-Agent System increases its performance, taking advance of learning methods dedicated to Internet information retrieval based on a predefined ontology and Rocchio relevance feedback algorithm (Salton, 1991).

Using DEEPSIA approach, and due to the low cost of the channel being used, SMEs and can easily join the e-commerce society. The main business innovation is its capacity to enable "old data collection process" companies to address "new electronic data collection process" issues through the adaptation to an "e-access" of existing processes.

2 THE ADOPTED FRAMEWORK

The adopted framework is composed by three subsystems, responsible for performing specific tasks, and interacting with each other through data and information flow. The three basic modules are a Dynamic Catalogue (DC), responsible for data maintenance and presentation; a Multi-Agent System (MAS), responsible for autonomous data collection, and semi-automatic catalogue process update; and an optional Portal Interface Agent (PIA), responsible for an optional and privileged interface with the web data suppliers.

2.1 DYNAMIC CATALOGUE AND THE PORTAL INTERFACE AGENT

The Dynamic Catalogue is the user interface, and it is responsible for presenting the multi-agents' collected information based on users' best preferences. The dynamic on-line catalogue will communicate with intelligent agents in order to provide end-users with all required information. The catalogue will not only hold information about the data selected by the agents on the web but, it will also hold information about the sites contacted, the in-use ontology (configurable by the user) and all user requirements towards the data gathered and stored. The electronic catalogue's data will be stored in a database (DB) and will be made available through a web-browsing interface.

The portal interface agent is responsible for creating a privileged interface between the web portals and the multi-agent system (MAS). This agent works as facilitator agent in the web data supplier's site. With this agent the web data supplier will create a dedicated interface to interact with the MAS, thus enabling querying to their site's private DB. It is not mandatory for this agent to be adopted by the web data suppliers, since the MAS will be able to collect information based on a web-crawling strategy. Nevertheless, its inclusion will provide the web data suppliers with a much more integrated and flexible interface and will enable the possibility for direct querying to web data suppliers' DB, i.e. the MAS will be able to query the web data supplier about specific data.

2.3 MAS - MULTI-AGENT SYSTEM

The Multi-Agent System (MAS) is responsible for autonomous data collection, and semi-automatic catalogue process update. Going into a deeper detail, one will find that the system is composed by a set of agents that do a lot of subsidiary tasks that contribute to the ultimate goals of the system.

All the agents will need a starting knowledge base upon which their behavior will be based. This knowledge base will have to be tailored to localized end-users attitudes and to specific SME's strategic objectives.

In deeper view of the multi-agent system we can identify the Web Crawler Agent (WCA), the Miner Agent (MA), the Human Agent (HA), the Facilitator Agent (FA) and the Collector Agent (CA), that are presented in the following sections.

2.3.1 THE WEB CRAWLER AGENT

The Web Crawler Agent (WCA) will automatically fetch web pages looking for interesting pre-selected data defined by the user. The search starts by fetching a seed page, and then all the pages referenced by the seed page, in a recursive approach. Therefore, each time the end-user finds a new site containing interesting data, he can send its URL to the WCA in order to start a deep search for useful information starting on this location.

The WCA is responsible to perform the first selection and classification of the contacted web pages, for posterior complete processing by the Miner Agent. Before the WCA starts to perform classification, it must be trained in an off-line process in order to be able to correctly decide about new pages found.

The algorithm in use represents a page as vector $\vec{p} = (p_1, ..., p_{|M|})$, so that similar pages have similar vector representation, according to a similarity metric (the nearest neighbor). Each p_i element represents a word Wi and |N| is the number of words considered.

p_i element is calculated as a combination of word frequency, WF(Wi,p), (the number of times the word Wi occurs in document p) and page frequency PF(Wi) (the number of documents in which the word Wi occurs at least once). All the words included in predefined stop list are immediately eliminated.

The off-line agents' learning process is achieved using a set of pre-classified pages (training set) representative of each class that will be recognized (e.g. web-pages of product promotion, web-pages that present artificial intelligence themes). For each class, the agent will automatically create a prototype vector, using an average of the positive sample pages vectors. These prototype vectors will be used to univocally represent each Class, therefore the Crawler classification is done based on the similarity between the page's vector representation and the prototype vector that represents each class. Using the learned prototypes and the similarity metric, the WCA agent makes the first classification and assigns a confidence factor to each URL found, it then sends the pages to the miner agent for posterior classification. Applying the Multi-agent systems to product promotions identification scenario will result in the following tasks division. WCA agent is the responsible for deciding if a page is a product promotion page or not. The MA is responsible for the product identification and the promotion associated condition (price, quality, guaranties, etc).

Experiments are being performed in augmenting the agent's accuracy with a large pool of unlabeled web pages. The strategy focuses on making a first training just with the labeled pages, and than use the achieved prototypes to classify the unlabeled pages. Then, in an iterative process until convergence is achieved, re-create the prototypes with all pages and re-classify the unlabeled pages. (NIGAM, 2000)

2.3.2 THE MINER AGENT

The miner agent (MA) is responsible for the analysis of the web pages found by the WCA. Once found and classified by the WCA, the MA analyzes each page and the relevant information presented on the page will be selected for storage on the miner's database.

The main challenge for this agent will be the selection of the really useful data to be introduced on the catalogue. In fact, it is expectable to find a huge amount of non-significant data that must be avoided in order to produce an easy-to-read and directly comparable catalogues. There are two fundamental hitches to be overcome, concept identification and concept relations.

To this end, this agent uses a knowledge base that allows the identification of the relevant data. The relevant data is described using an ontology of concepts, enabling the agent with the capability to identify the concepts included in the web pages, and a set of rules for concept relations. The usage of ontology and conceptual graph approaches for concept storage is based on the need to create a user-friendly interface. Philosophers and psychologists recognize that the human memory is associative, and that humans tend to understand and classify reality by devising abstract models and classifications, in order to categorize observations and to interpret them. (Tan, 1993)

Some of the important problems to be addressed are the multitude of variants, synonymous, multiple identifications, abbreviations, etc.

Despite the classification given to each URL by the WCA, the MA will overwrite this value according to the quality of the information found during the analysis process. The facilitator agent will use this classification to automatically add the new pages to catalogue or to interrogate the user about its validity. However, the inclusion of the source URL on each reference will always allow the user to easily confirm the information presented on the catalogue.

The MA will also determine the temporal validity of the information gathered in the catalogue. When it is possible to define the expiration date for each item, this date will be stored in order to allow the automatic update of the information. When the expiration date is not present (or detected) the system assigns a default validity that will be used for information updating purposes.

2.3.3 THE HUMAN AGENT

This agent will be the user interface that will allow the validation of the belief of the identified resources and sites. Through the Human Agent it is possible to add a specific site to the Miner's input, as a site that is identified by the human user as a potential site that may not yet have been found by the Crawler.

2.3.4 THE FACILITATOR AGENT

The Facilitator will work as an interface between the catalogue and the knowledge gathered by the Miner and/or the Collector Agent. The production of data for the catalogues is a task that occurs at configurable time intervals. The Facilitator maintains levels of belief for the resource hierarchy that allows it to decide when to request the validation of a given resource from the Human Agent. When a request is issued from the Catalogue, to gather "on the spot" information concerning a given resource, the Facilitator forwards this request to the Collector Agent to search for it on the known sites.

2.3.5 THE COLLECTOR AGENT

This agent works as a direct interface with sites that have joined this system through the PIA (Portal Interface Agent). It maintains a database of all the sites that have joined the service. A methodology (similar to yellow pages service, DNS or NIS) will have to be devised in order to maintain this database. When accessed by the Facilitator it selectively forwards the requests to the PIA and gathers the information that is then conveyed back to the Facilitator.

CONCLUSIONS

Besides the framework development, DEEPSIA project plans to perform a deep analysis on business process changes involving organization sciences, economics, law and the social sciences. (Filos, 2000)

DEEPSIA is distinctive in adopting a demand led approach (rather than supply side) to the role of SMEs in e-commerce. It addresses the situation of SMEs as purchasers within e-commerce business models. This does not ignore or exclude their participation as suppliers, and in fact may assist SME suppliers to access a wider customer base, but this is an add-on to the main focus of the project.

"Framework for Internet data collection based on intelligent agents" is a generic approach; its context definition is done using the training process of the Crawler agents and the ontology and rules definition on Miner Agent level. Depending on the class, the ontology and rules definition, the system will be adapted to distinctive objectives and we hope time and experiment will continue to support our expectancies.

DEEPSIA's approach will contribute to support new organizational schemata to enable businesses and organizations to take advantages of their new environment: e.g. enhance the efficiency, create new working environments, and establish new supplier/customer relationships.

In short, the project's main business innovation is its capacity to enable "old economy" companies to address "new economy" issues through the adaptation or "e-commercialization" of existing business processes. This will provide immediate economic benefits (streamlined processes and cost savings) and will also provide evidence to existing and potential investors of the company embracing the new economy.

REFERENCES

Bokma, Albert, 2000, "CogNet: Integrated Information and knowledge management and its use in virtual organizations", in E-business and Virtual Enterprises, Managing Business-to-Business cooperation, Luis Camarinha-Matos, Hamideh Afsarmanesh, Ricardo Rabelo, Kluwer Academic publishers, ISBN 0-7923-7205-0

Filos E., 2000, "Moving construction towards the digital economy, in Product and Process Modeling in Building and Construction", Balkema, Rotterdam, ISBN 90 5809 179 1

Mitchell, Tom M, 1997 "Machine Learning", ISBN 0-07-042807-7, McGraw-Hill.

Bond A., 1988, "Readings in Distributed Artificial Intelligence", L.Gasser (eds.), Morgan Kaufman.

Salton G, 1991, "Developments in Automatic Text Retrieval", Science, Vol. 253, pages 974-979

Tan TC, 1993, "The development of Intelligent Conceptual storage and retrieval system", Phd Thesis, University of Sunderland, UK

Nigam, K., Maccallum K et al., 2000, "Text Classification from labeled and Unlabeled documents using EM", Machine Learning, Kluwer Academic Publishers, Boston. Netherlands

COMPUTERIZED ARRANGEMENT OF VOCAL MUSIC

MATT D. JOHNSON
University of Missouri-Rolla
Computer Science Department
Rolla, MO 65409

RALPH W. WILKERSON
University of Missouri-Rolla
Computer Science Department
Rolla, MO 65409

ABSTRACT

This paper explores the ability of a computer to arrange music which conforms to the rules of a particular music genre. The motivation for this research was to implement a software package which could be utilized by musicians to reduce the drudgery associated with music arrangement. To realize this goal, the decision was made to focus on four part vocal music. A C++ program was written which would read a melody from a file and produce four part harmony in the style of a Protestant church hymn. Roughly 50% of all the chord names generated by the Computerized Arranger of Vocal Music (hereafter referred to as CAVM) were almost identical to the original chords. Eighty-three percent of the chord names generated by CAVM were either identical or very similar to the original chords. Seventeen percent of the chords were significantly different from the original chord. However, in all of the tested songs, the new arrangements sounded good, despite the noticeable differences. The resulting arrangements were not perfect, but they do illustrate that computers can be programmed to arrange music. From this humble beginning, it seems reasonable to predict that in the future computers will be able to arrange music on a much larger and complex scale.

INTRODUCTION

Music arrangement is the process of assigning notes for each voice in a music performance group. The performing group or ensemble could be one person playing the piano, or an entire symphony orchestra. The melody, which is the lead part of a song, is already known by the arranger and a version of that melody will be used in the new arrangement. The goal of music arrangement is to create a new setting of a given melody. Different settings of a piece of music are intended for different performing groups. For example, someone might arrange "Amazing Grace" for a large choir, and another person might arrange the same song for a men's quartet. Stated another way, to arrange a piece of music is to interpret a song and write out that interpretation for a certain type of musical ensemble (Guralink, 1977) (Ottman, 1989).

Music arranging is done by many types of people all over the world. Many individuals who arrange music are professionals who have studied music most of their lives. Arranging is not limited to the professional musician though. Arranging is also done by the local garage band that plays an occasional "gig" in a nearby club. There is no single correct way to arrange music. What is important about arranging music is that the resulting score is playable or singable, and it is enjoyable by the targeted audience. The music produced by the average garage band is a far cry from the masterpiece of a professional musician. However, they both share a common thread; they both organize sound in a manner which is pleasing to someone's ear.

Every style of music has its own rules of arrangement which accompany that style. In some cases, the rules are unofficial rules of thumb. This is the case of pop music, which usually consists of a very simple chord progression with a couple of solos thrown in to make the music more interesting. The church hymn, which is the focus of this research, has a more interesting chord progression, and the rule set that governs the part movement(how the voices change from chord to chord) is more structured and generally well-known. It is important that an arranger follow the accepted rules of arranging that govern the particular music genre. Changing the rules of arrangement at random in music can lead to very unpleasant results. If different rules of arrangement are used within one piece, the change should be implemented to create a specific musical effect.

COMPUTERIZED ARRANGEMENT OF VOCAL MUSIC

This section steps through CAVM in the order in which each step is executed in the actual program. It also discusses the development of some of the algorithms used in the code. The author assumes the reader will have a basic knowledge of general music.

The Melody

Since the general concept of CAVM is to arrange music for a given melody, the first step taken by the software is to get the melody. A melody of any length must be converted into a text format, and read in by the software. The melody can have no accidentals, and no key changes are allowed. The melody must also be in the range of a soprano singer.

The Chord Progression

While reading in a melody from a file is a relatively straight forward task, determining what chords should go with that melody is a completely different matter. The assumption that every note would be in the melody makes the job of assigning chords simpler. For any given note in a key, there are three triads in the key that can be formed, but deciding which triad to use can be quite challenging. To answer this question, experimentation with heuristics ensued. No heuristic was found which produced satisfactory results. As a result, the decision was made to use C4.5, a classifier system which can analyze data and discover patterns which are useful for the classification of future cases (Quinlin, 1992). In order to produce a chord progression from strictly a melody, "well behaved" hymns were chosen as examples. Six Methodist hymns were typed one note at a time into a data file suitable for C4.5 to read. Each note was recorded in the file with its time signature, duration, octave, scale degree, and chord. C4.5 analyzed the data and produced a decision tree which would suggest a chord based on characteristics of the note.

While C4.5 provided logic for generating the chord progression, it also provided another key piece of information - the predicted error rate. C4.5 predicted 45.7% of the chords generated would be differ with the original version. This prediction is actually quite accurate, because 50% of the notes in the test cases were assigned the same chord that was in the original arrangement of the song. While 50% may seem a bit low, it is actually quite remarkable, because in the training data, there are a number of cases where multiple occurrence of a note have different chords to go with it.

The Bass Line

Once the chord progression was generated, getting the bass note was quite simple, because the chord number provides the bass note. If the chord is in root inversion, the degree of the bass note is assigned to the chord value. If the chord is in first inversion, the degree will be the scale degree the chord is built on plus 2, with wrap around. If the chord is in second inversion, the degree will be the scale degree plus 4 with wrap around. After the scale degree is established, then the octave must be selected. The octave selection depends on a number of criteria. First, a jump in the bass line greater than an octave is generally hard to sing, so this should be avoided. Secondly, the bass line should be at least an octave away from the soprano, so there is room for the tenor and alto in between. If the note happens to be an E, then it has to be in octave 0. If the note is a C or D, then it could go in octave 0 or octave 1. Anything else can be octave -1 or 0. A random number is generated so that half the time, it will chose octave 0. The other times, it will try to go to the extreme octave, assuming that no conditions will be violated by doing so. This mechanism was chosen as a means of assigning the octave because there does not seem to be a good rule of thumb dictating which octave to use. The randomness combined with the qualifications mentioned above generally picks an octave for the bass note which sounds fine.

Alto and Tenor

After the bass and soprano notes have been determined, the alto and tenor notes must be filled in. Since a triad has three notes and CAVM is arranging for four voice parts, one note must be doubled. A doubling decision will be made based on the type of chord at hand and the arrangement of the existing soprano and bass notes. After the note to be doubled is selected, all possible combinations for alto and tenor are enumerated; assignments which exhibit the following conditions are eliminated.

1. Tenor is greater than alto
2. Tenor is less than bass
3. Alto is greater than soprano
4. Alto is equivalent to Tenor and it should not be
5. Tenor and Alto are more than an octave apart
6. Alto and Soprano are more than an octave apart.

It is possible that all the possibilities could be eliminated. This will happen only if the bass and soprano are too close together. If this situation occurs, then the alto will be assigned a note above the soprano. The next step is to determine which assignment of alto and tenor is best. In many cases, there is only one option, so that option is used. However, if there is more than one available note assignment to pick from, the one that moves the least from the previous chord is selected.

TEST RESULTS

The results of CAVM are very promising for the area of computer arranged music. The goal of this research was to create a program which arranges music that is pleasing to the ear, easy to sing, and generally follows part writing rules. The music produced by CAVM achieves all of these goals.

The major strengths of the arrangements include a good chord progression, a smooth bass line, and fluid inner voices. Weaknesses include occassional cross voices and occassional parallel motion. The strengths of the arrangements far outweigh the minor problems.

Table 1 shows the tallied results of comparisons of a few original songs to the CAVM versions of the song. The comparisons are based strictly on the chord progression. The chord progression is the structure of the song, and can be used to illustrate similarites between various tunes. The row labeled "Differences" indicates the number of chords in the CAVM version which are significantly different from the original version. Two chords are significantly different if they have no notes or only one note in common. For example, a 1 chord and a 5 chord share only the fifth note of the scale. Two chords which are "Exact Matches" must have the exact same chord name. Two chords which are "Close Matches" must be built on the same scale tone, or share all but one note. A 5 chord is built on the same scale tone as a 57 chord. The only difference between these two chords is one note. While a 1 chord and a 6 chord are not as similar to each other as a 5 and a 57, they are still considered similar. A 1 chord consists of a 1, 3, and 5. A 6 chord consists of a 6, 1, and 3 chord. Both chords have both a 1 and a 3 tone, so these are considered similar. The "Close Matches" category includes exact and similar matches. The "Total Chords" row is simply the number of chords in the song. The "Totals" column is the sum of the preceding numbers in that row. The "Percentages" column is the total of that row over the total number of chords.

	Beauty	King	Grace	Morning	Totals	Percentages
Differences	6	8	4	9	27	16.8
Exact	24	21	16	20	81	50.6
Close	38	33	31	31	133	83.1
Total	44	41	35	40	160	

Table 1. Test Results

If a chord in the CAVM version of a song is exactly the same as the original, then it will sound almost the same. The octaves could be different, and the assignment of the respective parts of that chord could be different, but the sound will be quite similar. Fifty percent of all the chords produced by CAVM in these test cases were the same as in the original version. Eighty-three percent were either exactly the same or similar to the original version. Most of these sound very much like the original as well. Seventeen percent of the chords are significantly different from their original counterparts. These differences are easily noticed by the casual listener who hears both songs.

RELATED WORK

NetNeg is a system utilizing both neural networks and distributed artificial intelligence to implement computer music composition of polyphonic music (Goldman, 1995). The main thrust of NetNeg is to determine whether this type of approach has any advantages in the music field.

Polyphonic music consists of two voices. Each voice should be a melody unto itself, and also work well with the other melody. Polyphonic music was chosen because of its relative simplicity, and because it is a good way to learn the basics of music composition.

In order to produce software which will compose music, the NetNeg team first observed that a human musician composing a polyphonic piece of music will typically start with one part of one voice, and then add the other part. The musician must consider the beauty of each individual voice, while at the same time considering the sound of both parts together. Both parts will have to meet certain criteria, such as the highest note appearing only once. The melodies also must be flexible and non-redundant. Rules which dictate consonant and dissonant sounds must also be followed. The process of applying all these rules and still getting both melodies to sound good individually and together is something of a negotiation process. In this negotiation process, it is inevitable that tradeoffs will occur. In some cases, one part may not be quite as good as it could be, but the resulting work will be better as a whole.

Goldman points out that most people who listen to music and find it aesthetically pleasing have learned to like certain aspects of music. However, there is typically not a rule set that dictates what a person likes. While this may be true, it does not mean a rule-based system could not be found to formulate new music.

The first step in NetNeg is the neural network. The neural network is trained using two part counter melodies. At each time unit, the neural network will predict the expectation that each possible note will occur next. After the possibilities are generated and ranked, two agents, one for each voice, decide on the best combination. The agents must consider all the rules which dictate their own voice, and must also consider the sound of the two melodies together. This is accomplished with a ranking system, and an elimination system. Any combination which is "illegal" is eliminated. The remaining combination with the highest rating is the winner.

The NetNeg system produces good results which follow most of the rules for this style of music. By using a neural network to learn the character of the melodies, and using agents to hash out the best combination of notes, the system seems to mimic one particular method of composition fairly well, even though it does not do so perfectly, and is not the most creative system. The concept of replicating the natural process may certainly lead to more interesting computer compositions in the future.

CONCLUSIONS

The results of this research clearly indicate that music arrangement is a task feasible for a computer to perform. All the music arranged sounds good, even though there are usually a number of differences between the new arrangement and the original. This is actually a good feature, because if the new arrangement were identical to the original, then it would not be a new arrangement. The important thing is that the results sound good.

This research also illustrates the versatility of the computer. Besides the potential for computers in music, it also seems reasonable that a computer could perform other tasks which are traditionally thought of as being only possible by humans. A rule set, a decision making process, and appropriate representation of data are all critical for a computer doing an art-like task. The rule set and the decision tree may be derived only after extensive analysis of works in the field, or through intimate knowledge of the field.

Computers excel in performing many calculations and making decisions rapidly. They are especially well adapted to performing large numbers of calculations with relative speed and ease. It is this ability to calculate and make decisions based on data that allows CAVM to arrange music. The concept that dominated the development of CAVM is one of learning rules from input music and combining those rules with general music knowledge to form a program to arrange music. Most of music arrangement is a task which depends on rules and actions. Any task which can be defined strictly in terms of rules and actions is within the reach of AI.

While the computer excels in areas of calculating and decision making, it is inherently lacking in one area - emotion. The ability of human beings to feel, express, and create emotions is perhaps the greatest obstacle that computer scientists face in creating true artificial intelligence. It is this barrier that makes exceptionally beautiful computer generated music so hard to attain. If some method can accurately produce the emotion and creativity associated with almost every aspect of our lives, then high level artificial intelligence is right around the corner. If computers cannot duplicate human emotion and creativity, then computers will never be truly intelligent.

REFERENCES
Goldman, C. V., Gang, D., and Rosenchein, J. S. "Net-Neg: A hybrid system architecture for composing polyphonic music." *Workshop on Artificial Intelligence and Music at the Fourteenth International Joint Conference on Artificial Intelligence*, August 1995.
Guralink, D. B., Ed. *Webster's New World Dictionary*. New York: William Collins + World Publishing Company Inc., 1977.
Ottman, R. W. *Elementary Harmony. 4th ed*. Englewood Cliffs: Prentice Hall, 1989.
Quinlan, J. R. *Programs For Machine Learning*. Morgan Kaufmann Publishers, 1992.

BEST VALUE PROCUREMENT USING ARTIFICIAL INTELLIGENT 'MODIFIED DISPLACED IDEAL MODEL'

DEAN T. KASHIWAGI
Director, Performance Based Studies Research Group,
Del E. Webb School of Construction, Arizona State University

ABSTRACT

Many of the old rules for project delivery that held sway in the construction industry no longer have the same potency as they once did. Design-bid-build still has its many followers, but it does not always ensure performance due to the pressures of low cost. There is a growing movement in the construction industry towards Performance Based and Best Value procurement.
Performance based procurement advocates that contractors who can optimize both price and performance can achieve best value. The major difficulty is finding an unbiased way to evaluate value vs. performance. The paper explores the development of a modified Displaced Ideal Model (Artificial Intelligent Decision Maker) to minimize subjectivity in "Best Value" procurements.

INTRODUCTION

Facility owners in the worldwide competitive marketplace have been forced to minimize cost and increase performance. It has led to:
1. Restructuring management.
2. Outsourcing as much as possible.
3. Identify and increase the "value" of all services.
4. Use the latest information technology to minimize risk.

DESIGN OF "BEST VALUE" SYSTEM

The objective of this research is to develop a "Best Value" modeling process that achieves the above requirements. The most difficult task is to identify value of construction services. The minimization of first cost or price has taken precedence over the identification of value. Another problem caused by the lack of performance information, is the subjective decision-making of facility owner representatives. The contractors have been motivated to influence the owner's representatives' by "marketing" rather than increased performance.

With the current lack of performance information, which is required to identify "value," the best value system must accomplish the following:
1. Identify performance criteria.
2. Identify construction performance data.
3. Identify construction value.

The following theorems from Information Measurement Theory (IMT) are used to design the "best value" procurement system:
1. Every event's final condition is predictable if information is known.
2. No two events can have the same initial and final conditions.
3. Factors of an alternative's environment will identify the alternative.
4. The more information that is known, the more accurate the prediction.
5. The closer the source, the more accurate the information.
6. All alternatives are related and have relative measurements.
7. Every person has different experiences, thought patterns, and biases.
8. Decisions are due to a lack of perception or information.
9. Mathematical models are a biased interpretation of reality. The bias is the coefficient of error. Error is the inaccuracy of the prediction.
10. All entities control their future environments.
11. A full information system optimizes "self improvement" by minimizing the expectations of others, and passing information back to the entity.

PERFORMANCE INFORMATION PROCUREMENT SYSTEM (PIPS)
PIPS was first designed in 1991. The process has the following steps:
1. The participating contractors set performance criteria, submit references, and have performance data collected.
2. Contractors attend a pre-bid conference and submit bids, management plans, which identify risk to the owner and the minimization of risk.
3. Contractor's management plans are reviewed and rated.
4. Contractor's key personnel are interviewed and rated.
5. The past performance of the general and critical subcontractors, key personnel (past performance), management plan rating and interview of key personnel (current capability and ability to effectively plan out the future) are put into the AI model.
6. The AI model then prioritizes by measuring differences.
7. The top prioritized alternative is then requested to re-review the project in detail for construct-ability, answer all technical questions that were brought up in the process, and sign a contract to construct the project according to the intent of the facility owner's designer.
8. After award and construction, the contractor is rated and the rating becomes 25% of their future performance rating.

SELECTION OF AN ARTIFICIAL INTELLIGENCE DECISION-MAKING MODEL FOR PIPS
Two models were considered for the procurement decision making model:
1. The Analytical Hierarchical Process (AHP).
2. The Displaced Ideal Model (DIM).

The DIM was selected due to its simplicity and ability to take a "biased" requirement regardless of "level of consistency" of the relationship of the weighting between criteria. Its theoretical basis has been discussed extensively by Kosko (1995) and Zadeh (1985). Its use of the entropy

equation and the natural log function provides a minimization of risk that is required for the procurement model. In early testing, one of the requirements of facility owners was to quickly set the weights. The greatest obstacle for using the AHP was the requirement for pair wise comparison of all the variables. A typical general construction project may have 500 criteria. Pair wise comparison would make setting the weights a lengthy process. The strengths of the DIM model were:
1. Simplicity. It considered the distance from the best number from each criterion, the weight given the criteria, and an information factor, which was the reciprocal of the entropy equation (the natural log function.)
2. Congruent to Information Measurement Theory (measurement of differences and logic processes).
3. Ability to take "biased" set of weights.
4. Measure the relationship of objective measured criteria, subjective performance criteria, and risk related factors, without translation. This includes risk minimization factors such as number of references and jobs, customer satisfaction ratings, and measurements such as size of job, complexity of job, and number of subcontractors on the job.

DISPLACED IDEAL MODEL

The DIM has the following general steps:
1. Identifies the optimal value of each attribute from all the alternatives.
2. Divides each value of each attribute by the optimal attribute value to make all values relational to each other and divides each alternative's relational value by the sum of all the alternatives' relational values for that attribute. This normalizes all the model's attributes, giving a relation of values within and between attribute values.
3. The model then uses the entropy equation to identify the entropy of each value of each attribute and then the entropy is summed for each attribute.
4. In order to normalize the entropy between attributes each sum of attribute entropy is then divided by the maximum possible entropy for each attribute, which is the natural log of the number of alternatives.
5. The information factor for the attribute is then calculated as the reciprocal of the entropy of each attribute.
6. The model then multiplies the normalized distance of each attribute value by the attribute's information factor and the weight factor.

A more detailed explanation using mathematical equations and symbols is presented below. The process can be divided into two modules:
   a) Identifying fuzzy membership functions.
   b) The identification of the importance of attributes that describe the alternatives.

For the following set of pairs: $\{x^k_i, d^k_i\}$ i = 1,..........,m, k= 1,..........,n

Where d^k_i is a membership function mapping the scores of the ith attribute into the interval [0,1]. Hence the degree of closeness to x^*_i for individual alternatives could be computed as:
1. If x^*_i is a max. Then $d^k_i = x^k_i / x^*_i$
2. If x^*_i is a min. Then $d^k_i = x^*_i / x^k_i$
3. If x^*_i is a feasible goal value or Coomb's ideal value, for example, x^*_i is preferred to all x^k_i smaller and larger than x^k_i, then
$d^k_i = [½ \{ (x^k_i / x^*_i) + (x^*_i / x^k_i) \}]^{-1}$
4. If, for example, the most distant feasible score is to be labeled by zero regardless of its actual closeness to x^*_i, we can define: $x_{i\bullet} = \text{Min } x^k_i$
And $d^k_i = [(x^k_i - x_{i\bullet})/(x^*_i - x_{i\bullet})]$
The above four functions d^k_i indicate that x^j is preferred to x^k when $d^k_i < d^j_i$

To measure the attribute importance, a weight of attribute importance λ_i, is assigned to the ith attribute as a measure of its relative importance in a given decision situation is directly related to the average intrinsic information generated by the given set of feasible alternatives through the ith attribute and to the subjective assessment of its importance. This reflects the decision maker's cultural, psychological, and environmental background.

The more distinct and differentiated are the scores, i.e., the larger is the contrast intensity of the ith attribute, the greater is the amount of decision information contained in and transmitted by the attribute.

For the vector $d_i = (d^1_i ... d^m_i)$ characterizes the set D in terms of the ith attribute and let $D_i = \sum^m_{k=1} d^k_i$ i=1...n.

Then the entropy measure of the ith attribute contrast intensity is
$e(d_i) = -K \sum^m_{k=1} (d^k_i/D_i) \ln(d^k_i/D_i)$.

If all d^k_i became identical for a given i, then $d^k_i/d_i = 1/m$, and $e(d_i)$ assumes its maximum value, that is, $e_{max} = \ln m$. Thus by setting $K = 1/e_{max}$ we achieve $0 \leq e(d_i) \leq 1$ for all d_i's. Such normalization is needed for comparative purposes.

We shall also define total entropy as: $E = \sum^n_{i=1} e(d_i)$. Because weights $\tilde{\lambda}_i$ are inversely related to $e(d_i)$, we shall use $1-e(d_i)$ rather than $e(d_i)$ and normalize to assure that $0 \leq \tilde{\lambda}_i \leq 1$ and $\sum^n_{i=1} \tilde{\lambda}_i = 1$: $\tilde{\lambda}_i = [1/(n-E)][1-e(d_i)]$
n = number of criteria.

Both w_i and $\tilde{\lambda}_i$ are determinants of importance in parallel fashion. The most important attribute is always the one having both w_i and $\tilde{\lambda}_i$ at their highest possible levels. The overall importance weight (information and weight) λ_i can be formulated as follows: $\lambda_i = \tilde{\lambda}_i . w_i$
Or after normalization: $\lambda_i = [\tilde{\lambda}_i . w_i] / \sum^n_{i=1} [\tilde{\lambda}_i . w_i]$ i=1,.....,n.
Calculation of Relative distance R_i of each variable would then be:
$R_i = \lambda_i [1 - d^k_i]$ i=1,.....,n

MODIFICATION OF THE DIM

Modifications to the DIM were required to improve the effectiveness of PIPS. The following modifications were made to the DIM:
1. The ability for the model to handle minimums and maximums at the same time was a modification to the model.
2. Further increase of differentials of performance.
3. Making the value of an alternative more understandable to the construction industry.

An inconsistency was noticed in the treatment of distances from the best minimal value and the best maximum value (as described above.) The distance from the best minimal value has a greater impact than distances from the maximum value. This is caused by the distance from the best value calculated as the absolute value of one minus the percentage of the best value. The model was giving the advantage to alternatives with best numbers in the "smallest number is better" criteria. To nullify this advantage, the author changed the equation:

(IF) x (WF) x |(1-alternative value/best minimal value)| to
(IF) x (WF) x |(1 −1/(alternative value/best value)|

Where:
Information factor = IF
Weight factor = WF

The new equation resulted in the following:
1. All performance numbers in the range of 0 to 1.
2. Equal advantage to alternatives with "minimal and maximum" best values.

In criteria where the owners rated the contractors (1-10), owners were staying within a range of (7-10). To identify the differences in these criteria, and to stress the importance of the criteria, the lowest score was subtracted from every alternative's score. This creates a range from the lowest score to the highest score, instead of a percentage of the highest score, which had very little impact. It creates a differential, which gives a contractor an advantage for performance.

1. To assist the contractors understand "value," a new method was designed to allow the price and performance to be considered. Price was moved from the performance criteria and was considered once performance was identified.

TEST RESULTS

The modified DIM has been tested out on construction in various locations in the United States. The State of Hawaii Public Works Group has been trained to use the PIPS modified DIM. The following are results of the tests:

1. The State of Hawaii Public Works Group has run 90 tests. They have been able to create, run, and explain the model results to the contractors.
2. An environment has been created that has improved contractor performance and value as identified by the end users. Contractors have been motivated to increase their performance through understanding their relative strengths and weaknesses. There has been 98% customer satisfaction, no contractor generated change orders, 98% of the projects finished on-time, and 100% of the projects finished within budget.

CONCLUSIONS

This research has resulted in the following conclusions:
1. Produced the first documentation of a facility management group procuring construction on a repeated basis using an artificial intelligent system that identifies "best value."
2. Successfully trained a state agency to use an artificial intelligent system to replace human decision-making.
3. Modified a multiple criteria decision-making tool that mimics the human mind to make it more usable, understandable, and more efficient in increasing performance.
4. Tested out the modified model to ensure usability, consistency, and understandable results that motivate continuous improvement.

REFERENCES

Kashiwagi, Dean T. and Al-Sharmani, Ziad (1997) "A Performance-Based Procurement System Used by the State of Wyoming" *Cost Engineering*, Vol. 39. No. 12, pp. 37 – 41.

Kashiwagi, Dean T. (1999) "The Development of the Performance Based Procurement System (PIPS)" *Journal of Construction Education*, Associated Schools of Construction, Vol. 3, No. 2, pp. 204 – 214.

Kosko, Bart (1993) *Fuzzy Thinking, The New Science of Fuzzy Logic*, Hyperion, NY.

Saaty, Thomas L. and Joyce M. Alexander (1989) *Conflict Resolution*, Praeger, NY.

Zeleny, Milan (1984), *Multiple Criteria Decision-Making*, McGraw-Hill Publishing, NY.

DOCUMENT CONTROL SIMULATION FOR CUSTOMS CLEARANCE IN A SEA PORT

SHAMSUDDIN AHMED
Department of Business Administration
College of Business and Economics
United Arab Emirates University
P O Box 17555, Al Ain
United Arab Emirates
Email: Dr_SUA@Yahoo.Com

ABSTRACT

A simulation study is undertaken to analyze document-processing time in a seaport of a Gulf Cooperation Council (GCC) country and recommend restructuring of staffing at the customs department. The document processing time is constrained by the number of employees in different departments and the type of documents requiring clearance. The distribution times of the documents processed are identified by a system study. A discrete event simulation model is developed to study the effect of a number of document dispatch and processing rules. The study identifies the optimal number of staffing and it results in a 33.68% reduction in document processing time. It also suggests a 7.8% reduction in labor at the customs department is required.

Key-Words: Simulation, GPSS, Sea-Port, Manpower-planning, Document processing,

INTRODUCTION AND SYSTEM DEFINATION

Import related document-processing operation in a GCC country is considered in this study. In particular the customs related documents and their processing by the customs authority at the seaport is evaluated. To identify the sequence of document processing operations, a study is carried out to identify the number of documents processed. The probability distribution of the document processing time is determined. According to the sequence of operations and the order of preference, the departments are identified and named according to the nature of work done. They are namely numbering, verifying, reviewing, and stamping departments. There is also a supervisory department manned by a manager. He decides how the documents are to be processed, sequenced and prioritized. A fee collection counter is allocated to collect customs related document processing cost. There are thirty-nine employees deployed for this work. The documents are first presented to the fees collection counter, to pay storage charges. The documents are then successively processed at a number of document control departments. A data flow analysis study (Ahmed, 1998) at the customs clearance center show that the documents arrive at the document control department following a normal distribution with a mean of 28.60 minutes and standard deviation of 12 minutes. The documents first enter into the document-screening department to control the black listed or default customers. If a customer is black listed, the documents of that customer are refused for additional processing. The department manager distributes the documents; which are to be processed further; to the numbering employees for allocating reference number and other policy related information. The verification agent validates the completeness and authenticity of the documents before being passed to the review department which checks to see if the information is in order. The stamping agent completes final phase of operation. The documents are stamped and preparations are done for dispatch. Operational time and the type of probability distribution to process documents are summarized in Table 1.

The computer simulation model evaluates and controls the document processing operation. Two scenarios are evaluated. They are defined as Scenario I and Scenario II. Scenario I is based on first in first out (FIFO) dispatch rule. In scenario II the documents are processed sequentially from one department to another when a predetermined number of documents have been collected. This creates a bulk of documents before being transferred to the next department. The current document management policy is recommended to bulk a maximum of 8 documents. Once the maximum number of document is collected in a department, they are passed to the next department for further processing.

The simulation study evaluates the present document processing policy and identifies the optimum number of employees at the customs department. The mean resident time is the critical parameter in the system. The mean resident time is defined as the total time spent by the documents inside the customs department. The aim is to minimize the document processing time using simulation, thus determining the optimum number of personnel.

Efforts are also made to reduce the variance in the experiment to validate the model (Law and Kelton, 1982; Pritsker et al., 1998). Statistics that are of interest are the average number of documents processed in a day or in a specified time, and the average utilization of service facilities. These statistics are used in evaluating different scenarios. Delays in processing the specified number of documents that are allowed for bulking may result in lost business and customer dissatisfaction.

MODEL DESCRIPTION

A discrete event simulation model based on GPSS (General Purpose Simulation System) computer simulation language (O'Donovan, 1979; Schriber, 1974) is developed to analyze the system behavior. The parameters of the simulation model are derived from a systems study (Ahmed, 1998). The simulation experiment is carried out over a period of 80 hours, which is equivalent to ten days of operation. The numbers of employees in different departments are taken according to the suggestions of the customs authority. The standard GPSS statistical outputs are collected from the simulation study and are shown in Table 2.

Table 1 The Distribution Time for the Document Processing

Document operations	Probability distributions (minutes)	Mean time (Minutes)	Standard deviation (Minutes)
Arrival of Documents	Normal	28.60	12
Scrutinizing of documents	Normal	21.74	8
Numbering on Documents	Exponential	62.74	
Verification of Documents	Exponential	53.82	
Review of Documents	Exponential	13.88	
Document distribution by manager	Uniform	10.00	5

The numbers of employees in different departments are modeled with the help of *Storage block*. Using GPSS system random numbers, two distribution functions are defined: the normal and exponential distributions, to schedule the arrival of document at different sections. This is done using a GPSS *Generate block*. The *Test block* is used to model bulking operations. A *Transfer block* with predetermined probability distribution is used to reject black listed customers. The *Advance block* symbolizes the document processing time. *Gate block* in *test mode* is used to control the number of document processed in a section. An *Assign block* is applied for computations of the system statistics. *Tabulate blocks* are introduced into the model to monitor system statistics during the entire length of simulation time and a *Report* block collects the total report of the simulation study.

Table 2 Summary statistics of scenario I and scenario II

Departments	Scenario							
	I	II	I	II	I	II	I	II
	Average utilization		Average content in queue		Average delay in queue		Total time in system (Minutes)	
RECEIVE	0.259	0.321	0.97	0	279.57	0	191	207
NUMBER	0.309	0.285	0	0	0	0		
VERIFY	0.326	0.349	0	0	0.16	0.16		
REVIEW	0.086	0.084	0	0	0.13	0.13		
STAMP	0.066	0.064	0	0	0	0		
PORTER	0.317	0.321	0	0	-	-		
MANAGER	0.468	0.321	0	0	-	-		

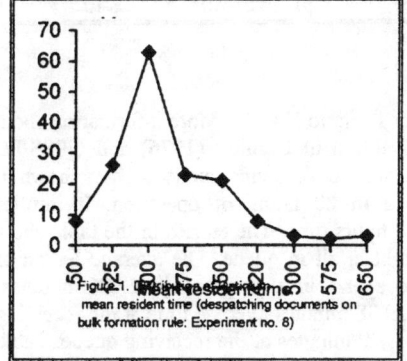

Figure 1. Distribution of mean resident time (despatching documents on bulk formation rule: Experiment no. 8)

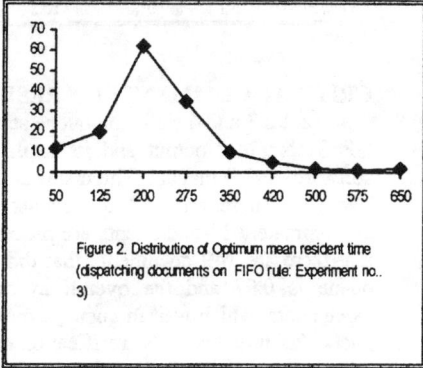

Figure 2. Distribution of Optimum mean resident time (dispatching documents on FIFO rule: Experiment no. 3)

MODEL VERIFICATION AND VALIDATION

The FIFO document processing rule or scenario I, is evaluated over a period of 4800 minutes. The numbers of employees are varied to minimize the variance and mean resident time of the document processing. Based on the distribution time in Table 1, one can compare scenario II or any other processing rule against scenario I. The results for the two scenarios with relevant statistics are given in Table 2, Table 5 and Table 6. The mean document processing time is used as a point estimate and the simulated time is counted to record the relative frequencies to estimate the distributions. The distributions of the mean resident times in minutes are summarized in Figures 1 and 2. It corresponds to the experiment number 3 and 8 as shown in Tables 5 and 6 respectively. Scenario I can be considered as the benchmark case.

Table 3 Queue statistics with FIFO rule to dispatch documents

Queue name	Max length (Nos.)	Content in queue	Entries: total	Entries (0): Zero entry	Average content in queue (nos.)	Average time in queue (Minutes)	Average zero entry (Nos)
RECEIVEQ	18	18	167	149	0.97	27.96	259.38
NUMBERQ	1	0	137	137	0	0	0
VERIFYQ	1	0	137	137	0	0.016	2.5
REVIEWQ	1	0	137	137	0	0.013	1.9
STMPINGQ	1	0	149	149	0	0	0

Table 4 Queue statistics with bulk formation rule for dispatch

Queue name	Max length (Nos.)	Content in queue	Entries: total (Nos.)	Entries (0): Zero entry	Average queue content	Average queue time (Minutes)	Average zero entry (Nos.)
RECEIVEQ	1	0	165	165	0	0	0
NUMBERQ	1	0	156	156	0	0	0
VERIFYQ	1	0	154	153	0	0.016	2.5
REVIEWQ	1	0	150	149	0	0.013	1.9
STMPINGQ	1	0	157	157	0	0	0

CRITICAL COMMENTS ON RESULTS

Tables 3 and 4 show the queue statistics for scenario II and I. More information about GPSS computer output and its explanation is found in Bobllier (1976) and GPSS/PC software (1986) manual. The maximum queue length in receiving queue is 18 as shown in Table 3. There are 167 documents processed in 80 hours of operation. It implies approximately 17 documents are processed in 8 hours time. The entries in the table show that there are 149 documents that did not weight at all in queue. The average length of queue is 0.97 and the overall average time spend in the queue is 27.96 minutes. Documents, which wait in queue, experience 259.38 minutes average time at the receiving queue, 2.5 minutes at the verification queue and 1.9 minutes at the receiving queue. Table 4 depicts almost similar results and represents scenario II, except that there was no queue at the receiving queue.

Tables 5 and 6 show the effect of the number of employees on the mean resident document processing time. The experiment numbers 3, 4 and 5 show that the minimum mean resident time is 191 minutes for the FIFO document-processing rule. The standard deviation in experiment number 3 is 84 minutes and is the lowest; hence the optimum number of total number of employees is 36 against the initial total number 39. The simulation study suggests to employ 13 receiving, 7 numbering, 5 verifying, 5 reviewing, 2 stamping, 1 porter and 3 supervisory personnel. It is clear from experiment numbers 1, 2, 3 and 9 that changing the number of the stamping personnel and the manager improves the mean document processing time and standard deviation in the system. The study shows that the number of managers and stamping personnel are among the critical factors for bottleneck operations. Table 6 shows similar results for scenario II. The experiment numbers 7 and 8 show that the minimum mean resident time is 207 minutes. The standard deviation in experiment number 8 is 113 minutes, which is the lowest value. The experiment numbers 1, 4, 7, 8 and 9 indicate that the number of stamping personnel and managers also affecting the mean resident time. The variation in the staffing also improves the standard deviation of the document processing time in the system. As a result the simulation study suggests that the optimum number of employees is 36 to process the documents in a customs department at the seaport.

Table 5. Optimization of mean resident time (FIFO dispatch rule)

Run no.		Receive	Number	Verify	Review	Stamp	Porter	Supervisor	Total Server (Nos.)	Mean (Minutes)	Standard Deviation (Minutes)
1	Number	14	8	6	6	2	2	1	39	288	134
	Utilization	.419	.249	.367	.065	.123	.145	.931			
2	Number	12	7	6	4	1	1	3	34	195	90
	Utilization	.412	.317	.289	.121	.137	.325	.331			
3	Number	13	7	5	5	2	1	3	36	191	84
	Utilization	.259	.309	.326	.086	.066	.317	.325			
4	Number	14	6	5	5	2	1	2	35	191	103
	Utilization	.451	.329	.360	.102	.067	.310	.468			
5	Number	12	6	6	4	2	1	2	33	191	104
	Utilization	.522	.287	.284	.120	.062	.285	.425			
6	Number	12	6	6	4	1	1	2	32	192	99
	Utilization	.280	.363	.249	.101	.128	.315	.477			
7	Number	12	5	5	4	1	1	2	30	195	104
	Utilization	.394	.436	.316	.091	.129	.311	.473			
8	Number	11	7	6	5	2	1	2	34	192	103
	Utilization	.559	.246	.284	.095	.062	.284	.426			
9	Number	6	3	2	2	1	1	1	16	537	225
	Utilization	.694	.483	.712	.153	.096	.228	.684			

Table 6. Optimization of mean resident time (bulk formation)

Run No		Receive	Number	Verify	Review	Stamp	Porter	Supervisor	Total server (Nos.)	Mean (minutes)	Standard Deviation (Minutes)
1	Number	14	8	6	6	2	2	1	39	445	140
	Utilization	.307	.249	.370	.072	.126	.142	.939			
2	Number	10	6	4	5	1	1	1	28	358	153
	Utilization	.473	.267	.359	.087	.131	.301	.910			
3	Number	8	4	3	3	1	1	1	21	407	130
	Utilization	.395	.496	.560	.138	.126	.303	.935			
4	Number	6	3	2	2	1	1	1	16	580	166
	Utilization	.542	.619	.778	.161	.122	.287	.874			
5	Number	12	5	5	4	1	1	2	30	216	117
	Utilization	.293	.444	.342	.113	.131	.313	.479			
6	Number	11	6	4	4	1	1	3	30	209	110
	Utilization	.547	.338	.457	.103	.133	.314	.319			
7	Number	12	7	6	4	1	1	3	34	207	115
	Utilization	.429	.304	.267	.088	.131	.316	.309			
8	Number	13	7	5	5	2	1	3	36	207	113
	Utilization	.321	.285	.349	.084	.064	.312	.312			
9	Number	13	7	5	5	2	1	2	35	218	102
	Utilization	.162	.300	.370	.092	.065	.315	.490			

CONCLUSIONS

The proposed simulation model evaluates the document processing time and document management policy at the seaport to accomplish the shortest mean processing time. It is recommended that there should be 36 employees, composed of 13 in receiving, 7 in numbering, 5 in verifying, 5 in reviewing, 2 in stamping, 1 in porter and 3 in supervisory positions. The mean document processing time is of the order 207 minutes with a standard deviation of 113 minutes. It suggests that there are a variety of documents. This result is obtained for scenario II; which allows bulking of 8 documents before forwarding them to any other department. The FIFO document processing rule, which is scenario I results 191 minutes mean document processing time and a standard deviation of 84 minutes. It uses the same combination of employees. The bottleneck operation is identified in stamping section and the manager's workstation. Finally, the study suggests 7.7% reduction in labor and 28.12% decrease in document processing time against Scenario II. While in Scenario I the reductions are 7.8% and 33.68% respectively. The document processing policy should follow Scenario I as a result of this study.

REFERENCES

Ahmed, Shamsuddin, "Analysis of document in a sea port", A feasibility report, KAAU, SA, 1998.

Bobllier, P.A., Kahan, B.C., Probst, A.R., "Simulation with GPSS and GPSS V." Prentice Hall Inc. Englewood Cliffs, NJ, 1976.

Law, Averill M. and Kelton, W. David., "Simulation study and analysis." McGraw Hill Book Co., NY, U.S.A., 1982.

Minuteman, "GPSS/PC, General purpose Minuteman Software." P.O.Box # 171, Stow, Massachusetts-01775, U.S.A., 1986.

O'Donovan, Thomas M., "GPSS; Simulation Made Simple." John Wiley and sons Ltd., Chichester, 1979.

Pritsker, A., O'Reilly, J., and LaVal D., "Simulation with Visual SLAM and AweSim, John Wiley, 1998.

Schriber, Thomas J., "An Introduction to simulation using GPSS/H." John Wiley and Sons, NY, 1991.

SUBJECT INDEX

A

Acceleration, 15
Accuracy, 1019
Adaptive Architecture, 363, 471
Adaptive Behavior, 135, 423
Adaptive Control, 223, 235, 477, 485, 485, 491, 491, 497, 509, 853, 927
Adaptive Control Dynamics, 235
Adaptive Fuzzy Control, 223, 235
Adaptive Learning, 261, 325, 405, 417, 449, 485
Adaptive Prediction, 853
Adaptive Systems, 449
Aerospace, 57, 969
Aircraft, 595, 673, 853, 939
Algorithm, 423, 545, 1031
Algorithms and Training, 9, 15, 33, 113, 363, 497, 509
Algorithms Learning, 75, 449, 485, 497, 509
ANFIS, 539
ANN, 81
Application Neural Networks, 939
Application of Neural Networks, 821
Applications, 209, 261, 539, 673, 859
Applications and Algorithms, 847, 1025
Approximation, 33
Architecture, 375
Architecture Optimization, 497, 509, 939, 969
Architectures, 47, 471, 703, 951, 969
Architectures Neural Networks, 509
Artificial Intelligence, 95, 101, 113, 141, 417, 429, 435, 461, 471, 569, 827, 951, 1025, 1031
Artificial Life, 399, 405, 461
Artificial Neural Network, 15, 33, 57, 75, 95, 101, 183, 417, 521, 589, 665, 673, 745, 789, 897, 981
Artificial Neural Network Applications, 75, 551, 789
Artificial Neural Network Architectures, 75, 81
Artificial Neural Networks, 113
Associative Memory, 69, 601
Attention, 471
Automated Inspection, 981
Automatic Feature Extraction, 405, 645, 939
Autonomous Agent, 63, 399, 471, 569
Autonomous Learning, 405, 569

B

Backpropagation, 107, 183, 261, 521, 589, 665, 703, 745, 803, 915, 981, 1013
Back-Propagation Algorithm, 821
Backpropagation Learning, 715
Backpropagation learning algorithm, 551
Backpropagation Neural Network, 685, 745, 775
Basis Function Networks, 41, 417
Behavior, 63, 399, 449, 569
Benchmark Problems, 357
Biological Inspired Networks, 75, 101, 107, 441, 461, 619, 809, 827
Biological Nervous System, 101, 827
Biological Systems, 101, 135, 141, 423, 455, 461, 619, 833
Biologically Inspired Networks, 69, 101, 107, 113, 441, 569, 809
Biology, 441, 455, 455, 833
Biomedical Applications, 267, 295, 461, 783, 827, 833
Biomedical Engineering, 797
Biomedical Engineering Applications, 789, 797, 803
Bio-Medical Engineering Applications, 295, 441, 783, 797, 803, 815, 833
Biomedical Image Processing, 789
Breast Cancer Diagnosis, 839
Building Blocks, 129

C

C++, 1025
CAD, 123, 285, 583
Case based reasoning, 969
Changing Environment, 63
Chaos, 927, 927
Character Recognition, 651
Chemistry, 345, 695
Civil Engineering, 123, 939
Classification, 9, 295, 375, 381, 387, 577, 639, 783, 815, 915, 1019
Classification Recognition, 15
Classifiers, 3, 9, 113, 363, 417, 545, 589, 607, 645, 665, 1025
Cluster, 9
Clustering, 113, 351, 363, 369, 387, 545, 583, 645, 703, 865, 1019
CMAC, 229, 477
Combinatorial Optimization, 423, 859, 865, 897
Comparison of Networks, 107, 183, 497, 509
Competitive Environment, 135
Complex Systems, 255, 399, 399, 405, 411, 411, 417, 417, 423, 423, 429, 429, 435, 461, 461, 491, 703, 927
Complexity, 399, 461
Complexity Reduction, 63, 455
Composite Material, 533
Computational Intelligence, 141, 827
Computational Neuroscience, 101, 113, 569
Computer Engineering Applications, 325, 685, 909, 939, 945, 945
Computer Topography, 307
Computer Vision, 107, 633, 847
Conceptual Design, 969
Conjugate Gradiant, 839
Connectionist NLP, 993
Constrained Optimization, 141, 423, 853, 859, 871, 897
Constrained Satisfaction Programming, 135
Constraint Satisfaction, 123
Constraints, 141, 853, 859
Construction, 1031
Control, 209, 247, 261
Control Application, 285
Control Systems, 235, 497
Controller, 261, 497, 509
Convergence, 883
Convergence Analysis, 129
Convergence Time, 129
Correlation, 939
Correlation Analysis, 645, 703
Critic Based Design, 75
Cross Validation Error, 183
Crossover, 129, 183
Curse of Dimensionality, 63
Curve Fitting, 295

D

Data Acquisition, 235, 533, 1019
Data Analysis, 95, 313, 703, 709, 715, 769, 803, 1031
Data Mining, 339, 345, 351, 381, 703, 783, 975
Data Mining Applications, 333, 333, 345, 375, 387, 387, 685, 727, 763, 1019
Data Mining Methods and Algorithms, 95, 113, 171, 215, 215, 333, 357, 357, 363, 369, 375, 375, 471, 727, 951
Data Mining Tools, 333, 357, 375, 957
Data Reduction, 345, 1031
Data Visualization, 351
Database, 387
Decision Analysis, 969
Decision Making, 267, 313, 375, 449, 539, 821, 853, 903, 939, 1025
Decision Making Application, 915
Decision Making Applications, 325, 903, 909, 915, 1025, 1031
Defuzzification, 235, 267, 285, 307
Design, 261
Design Automation, 969
Design of Experiments, 533
Detection, 625, 639, 665
Diagnosis Systems, 625, 821
Dictionaries, 993
Dimensionality Reduction, 113, 387, 545, 703
Directional Derivative, 81
Discrete Fourier Transform, 589
Discrete Wavelet Transform, 477
Disease, 267, 295
Distance Measures, 9, 369, 569, 1031
Distributed Autonomous Agent, 399, 423
Distributed Data Mining, 975
Dynamic Genetic Algorithms, 135
Dynamic Neural Network Control, 551
Dynamical Systems, 135, 235, 625, 853, 897
Dynamics, 129, 625
Dynamics Critic, 75
Dynamics, Critic Learning, 75

E

Edge Detection, 619
Efficient Computation, 123, 165
Electroencephalography, 815
Embedded Systems, 477, 981
Emergent Computing, 461
Energy Forecasting, 715
Energy Function, 897
Engineering, 261, 1013
Engineering Application, 123, 545, 551, 625, 673, 695, 709, 769, 871, 921, 1013, 1019, 1031
Engineering Applications, 533, 563
Engineering Optimization, 123, 165, 859, 981
Enhanced Genetic Algorithms, 209, 369
Entropy, 147, 833, 1031
Environmental Engineering, 695
Equations, 533, 927
Error Estimation, 165, 673
Error Minimization, 107, 165, 485, 497, 589, 1031
Estimation, 33, 165, 369, 703, 721, 853, 1001, 1007
Euclidean Distance, 9
Evaluation, 165, 357, 435
Evolution, 147
Evolution Learning, 423, 449
Evolutionary Algorithm, 123, 129, 135, 141, 165, 183, 273, 369, 423, 871
Evolutionary Computation, 135, 141, 165

Evolutionary Programming, 159, 159, 171, 171, 189, 189, 197, 197, 215, 369, 951
Experimental Design, 533
Expert Systems, 95, 285
Exploration, 563, 951
Exploratory Data Analysis, 369, 703, 809

F

Face Recognition, 57
Factorial Design, 533
Failure Analysis, 387
Fast Fourier Transform, 589
Fault Detection, 545, 595, 625, 1013
Fault Diagnostics, 1013
Feature Classification, 95, 589, 703
Feature Detection, 471, 703
Feature Extracting, 295
Feature Extraction, 57, 95, 645
Feature Identification, 589
Feature Identification and Classification, 357, 545, 589, 633, 633, 645, 645, 685, 783, 945
Feature Ranking, 589
Feature Selection, 273, 345, 471, 589, 939
Feedforward Networks, 75, 665
Feedforward Neural Network, 497, 551, 589, 673, 1001, 1007
Feedforward Neural Networks, 75, 739, 939
Financial and Economic Applications, 739
Financial Applications, 171, 215, 313, 739
Finite Automata, 633
Fitness Function, 129, 135, 859
Flight Psychophysiological Laboratory, 589
Floating-Point Coding, 209
Forecasting, 709, 739, 769, 775, 853
Fractal, 539, 695
Function Approximation, 33, 41, 563
Fundamentals of Learning, 33, 47, 485, 601, 827
Fuzzy, 229
Fuzzy Aggregation, 267, 325
Fuzzy Applications, 235, 267, 295
Fuzzy Basis Function, 241
Fuzzy CMAC, 229
Fuzzy Concept, 9, 267
Fuzzy Control, 209, 223, 223, 235, 235, 255, 255, 261, 267, 285, 301, 319
Fuzzy Control Applications, 285
Fuzzy Controller, 267
Fuzzy Expert Systems, 1031
Fuzzy Identification, 235, 295
Fuzzy Logic, 235, 247, 247, 261, 261, 267, 273, 273, 279, 279, 307, 307, 539, 877, 1031
Fuzzy Logic and Systems, 295
Fuzzy Logic Control, 267
Fuzzy Membership, 273, 307
Fuzzy Memberships, 9
Fuzzy Modeling, 301
Fuzzy Neural Nets, 375, 381, 679
Fuzzy Reasoning, 877, 1031
Fuzzy Rules, 267, 307
Fuzzy Set Memberships, 267
Fuzzy Sets, 1031
Fuzzy System, 267, 273, 285
Fuzzy Systems, 113, 209, 235, 241, 241, 285, 301, 301, 313, 313, 877

G

General Engineering Applications, 159, 503, 503, 613, 613, 657, 733, 733, 757, 757, 809, 877, 889, 889, 933, 933, 957, 957, 963, 963, 987, 987, 1013, 1013
Generalization Error, 3
Genetic Algorithm, 135, 183, 273, 375, 539, 569, 853, 1013
Genetic Algorithms, 123, 129, 129, 135, 141, 141, 147, 147, 153, 153, 165, 165, 177, 189, 203, 203, 209, 369
Genetic Operators, 129, 147, 165
Genetic Programming, 171, 215, 405
Geoscience, 307, 563
Global Optimization, 33, 129, 141, 171, 215, 369, 497, 509, 601, 853, 859, 871, 877
Grade of Membership, 267, 285
Gradient Descent Algorithm, 325, 369, 551
Gray Level, 915
Grouping, 399, 865

H

Hardware Implementation, 491
Health Monitoring, 295, 625
Heuristic Knowledge, 285, 313, 969, 981
Hierarchical Neural Networks, 3, 775
Highly Coupled Systems, 255
Hill-Climbing, 123
Hopfield Network, 865, 897
Human Machine Interface, 461
Hybrid Algorithm, 141, 209, 369, 871
Hybrid Architecture, 569
Hybrid Artificial Neural Systems, 69, 95, 95
Hybrid Neural Network, 865
Hybrid Systems, 539

I

Identification, 295, 639, 1013, 1019
Image Interpretation, 307
Image Processing, 57, 107, 295, 307, 619, 633, 639, 645, 695, 915
Image Processing Interpretation, 417
Image Reconstruction, 107, 789
Implementation, 981
Implicit Redundant Representation, 405
Inference, 307
Intelligent Control, 209
Interaction, 63, 135, 325, 423
Interactive Data Exploration, 351, 847
Interactive Simulations, 135, 563
Iterative Learning Control, 485
Iterative Learning Process, 75, 325, 1019

K

Kalman Filters, 751
Kernel Functions, 9, 15, 703
K-means Algorithm, 369
Knowledge Base Systems, 255, 1019
Knowledge Based System, 267
Knowledge Extraction, 95, 405, 545, 645, 969
Knowledge Representation, 449
Kohonen Network, 865

L

Language, 449
Learning, 69, 107, 601
Learning Accuracy, 171, 215
Learning Algorithm, 33, 485, 601, 915
Learning Algorithm and Training, 509
Learning Algorithms, 247, 485, 1019
Learning Algorithms and Training, 3, 9, 15, 21, 21, 27, 27, 33, 41, 41, 63, 75, 107, 113, 325, 417, 449, 449, 497, 889, 921, 921, 981, 1025
Learning Network, 107
Learning Pattern, 449
Learning Speed, 171
Learning Vector Quantization, 645, 739
Least Square, 33, 1001
Life, 461
Linear Activation Functions, 673, 1001, 1007
Linear Programming, 897
Logic, 461, 515
Lyapunov Function, 255, 485

M

Machine Intelligence, 839
Machine Learning, 3, 69, 75, 107, 273, 545
Machine Monitoring, 387, 625
Machine Translation, 993
Machine Vision, 619
Machine Vision and Image Processing, 695, 695
Manipulator Control, 509
Manufacturing, 209
Manufacturing Application, 545, 613, 665, 733, 757, 859, 877, 963, 981, 987
Manufacturing Applications, 203, 235, 945, 957, 981
Markov Chain Model, 645
Mathematical Model, 129, 295, 461
Mathematical Modeling, 129, 235, 563, 751, 939
Mathematical Programming, 33, 853, 865, 897
MATLAB, 3, 123, 209, 255, 261, 539, 563, 673, 751, 789, 827
Mean-Square Error, 815
Mean-Square Estimation Error, 325
Mechanical Design, 235, 325
Mechatronics, 197, 235, 491, 515
Medical Diagnosis, 295, 789
Medical Imaging, 295, 477, 477, 789, 789
Medicine, 783
Membership, 267
Membership Function, 285, 295, 313, 1019
Memory-based Neural Network, 41, 57, 69, 113
Minimum Distance, 569
Minimum Distance Classifier, 645, 1019
Mining Methods and Algorithms, 375, 703, 1019
Min-Max Method, 69, 183
Missing Data, 721
Mixed linear and nonlinear systems, 1001, 1007
MLP, 95, 939
Mobile Robots, 63, 569
Model Estimation, 3, 171, 215, 461, 471, 951, 1013
Model Generation, 123, 471, 951
Model Reference Adaptive System(MRAS), 235
Modeling, 241, 471, 595, 601, 625, 709, 769, 951, 1013
Modeling of Social Behavior, 285, 399
Models System Modeling, 569, 821
Momentum, 509
Monitoring, 209, 533
Motion Analysis, 921
Multi Layered Feed Forward Networks, 95

Multi-Agent System, 399, 423
Multi-Layer Feed Forward Neural Networks, 75, 183, 665
Multi-Layer Perceptron, 183, 563, 577, 685, 703, 709, 769, 821
Multi-Objective Optimization, 423
Multi-Sensor Fusion, 63
Multivariable Control, 255

N

NARMA, 241
Nearest Neighbor Classifier, 9, 57, 763
Network Architecture, 745
Network Topology, 821, 915
Neural Control, 261
Neural Controller, 509
Neural Network, 15, 107, 261, 375, 569, 651, 745, 839, 883, 915, 993, 1013
Neural Network Algorithm, 15, 417
Neural Network Applications, 107, 325, 563, 625, 673, 703, 709, 769, 915, 1001, 1007
Neural Network Architecture, 41, 107, 497, 703, 715, 745, 827
Neural Network Architectures, 993
Neural Network Chips, 57, 57
Neural Network Classifiers, 95, 589
Neural Network Control, 477, 497
Neural Network Dynamic, 551
Neural Networks, 539, 551, 595, 625, 709, 769
Neurally Based Networks, 47, 89, 101, 921, 981
Neurally-Based Network Architectures, 89, 625, 745, 821, 827, 1001, 1007
Neuro Control, 503
Neuro-control, 527, 527
Neuroscience, 101, 449, 827
Noisy Input, 9, 703
Nonlinear Dynamic System, 235, 399, 551, 563
Non-Linear Dynamics, 129, 235, 727, 751, 927
Nonlinear Function, 847
Nonlinear Function Approximation, 171, 183, 215, 325
Nonlinear Optimization, 123, 171, 215, 853, 859, 877
Nonlinear System, 241
Nonlinear System Modeling, 171, 215, 551, 563, 927, 1001, 1007
Nonlinear Systems, 619, 897
Non-linear Systems, 235, 255
Non-Linear Systems and Modeling, 129, 235, 515, 551, 557, 557, 563, 751
Nonstationary Data, 751
Normalization, 9, 307
Numerical Approximation, 165

O

Object Oriented Design, 1025
Object Recognition, 633, 639
Obstacle Avoidance, 63
On-Line Adaptation, 325, 1001, 1007
Operations Research, 853, 897
Optimal Trajectory, 497, 509
Optimization, 3, 33, 123, 129, 135, 141, 159, 165, 177, 177, 417, 423, 847, 847, 853, 853, 859, 859, 865, 871, 871, 877, 877, 883, 897, 897

P

Paradigms, 455, 969
Parallel Computing, 273, 1001, 1007
Parallel Data Mining Algorithms, 273
Parallel Implementations, 3, 273, 1001, 1007
Parallel Processing, 273
Parallel Techniques for Search in Data Mining, 47
Parameter Estimation, 171, 203, 307, 325, 833, 897, 927, 1001, 1007
Path Planning, 859
Pattern Categorization, 113, 471
Pattern Classification, 15, 113, 417, 645
Pattern Recognition, 3, 69, 113, 189, 369, 387, 417, 521, 557, 563, 583, 601, 601, 639, 639, 645, 651, 651, 665, 665, 679, 685, 685, 709, 769
Perception, 449, 569, 685
Performance Analysis, 15, 47, 47, 357, 533, 625, 945
Performance Estimation, 165
Performance Evaluation, 533, 639, 1001, 1007
Pharmaceuticals, 345
Piezoelectric Sensors, 533
Pilot Workload, 577, 589
Planning, 471, 569, 969
Population Size, 129
Power Engineering Applications, 715
Prediction, 183, 307, 357, 709, 709, 715, 745, 769, 769, 775, 775, 783, 921
Prediction Error, 171, 215, 775, 1031
Principal Component Analysis (PCA), 183, 789, 833
Probability Density Estimation, 853
Process Control, 203, 209, 209, 235, 477, 539, 539, 545, 551, 665, 933, 1037
Process Modeling, 203, 471, 551, 951, 1001, 1007
Process Monitoring, 203, 471, 521, 521, 533, 545, 545, 551, 951
Psychophysiological Data, 399, 449

Q

QSAR, 345
Quality Control, 539, 939, 981

R

Radial Basis Function, 9
Random Process, 853
RBF, 57
Recognition, 915
Recurrent Neural Network, 815, 897, 1001, 1007
Regression Analysis, 307, 533
Regression Models, 171, 215, 533
Reinforcement Learning, 57, 63, 75, 503
Reliability, 63
Reverse Engineering, 583
Robot Control, 63, 101, 261, 285, 509, 569
Robustness, 33, 255, 369, 471
Rough Sets, 279, 339, 339
Rule Extraction, 381, 545
Rule Formulation, 285
Rule Formulations, 267

S

Saliency Measure, 589, 769
Saliency Metrics, 589
Scheduling, 423, 477

Search Heuristics, 141, 859, 865, 969
Self Organizing Feature Map, 351, 583
Self-Adaptive Training, 81
Self-organizing feature maps, 351
Self-Organizing Neural Network, 865
Self-Tuning Control, 485
Semiconductor Manufacturing, 183
Sensitivity Analysis, 183, 769, 969
Sensor, 569
Sensor Fusion, 69, 673
Separating Boundaries, 3
Sequential Minimal Optimization, 839
Set Membership, 273
Shape Recognition, 633
Sigmoid Activation, 497, 509
Sigmoid Activation Function, 417, 521, 589, 1001, 1007
Signal Interpretation, 833, 1001, 1007
Signal Processing, 539, 557, 607, 607, 619, 619
Signal-To-Noise Ratio (SNR), 607, 815
Similarity, 319, 387, 423, 545
Similarity Measures, 113, 369, 1019
Simulated Annealing, 859
Simulation, 235, 399, 595, 915, 1037, 1037
Simulation Approximation, 307
Sliding Mode Control, 255
Slow Convergence, 485
SNR Saliency measure, 577, 589
SNR Screening Method, 577, 589
Software, 847, 1025
Space-Time, 703
Spatial-Temporal Processing, 645, 673
Speakers, 539
Speed, 533, 939
Stability, 255
State Estimation, 751, 833
Statistics, 33, 387, 639, 673, 703
Stochastic Learning, 75
Supervised Learning, 273, 325, 485, 521, 639, 709, 769, 1019
Supervised Training, 3, 645, 665
Support Vector Machines, 3, 9, 15, 639, 839
Surface Reconstruction, 583
System Control, 569
System Identification, 33, 247, 551, 563, 897
System Modeling, 171, 215, 551, 625, 709, 769, 939
Systems, 261, 625, 709, 769

T

Target Recognition, 57
Televisions, 209, 981
Temporal Pattern Identification, 595, 595, 645, 673, 679, 679, 727
Testing, 533, 709, 769, 981
Texture, 915
Texture Classification, 695
Theory, 33, 147, 279, 455, 577
Three-Dimensional Imaging, 847
Time Series Analysis, 171, 215, 471, 539, 721, 721, 727, 727, 739, 815, 833
Time Series Forecasting, 171, 215, 313, 471, 715, 739
Topology, 307
Training, 33, 625, 639, 709, 769
Trajectory Prediction, 497, 509, 545
Transfer Function, 417, 563
Transformation, 417
Traveling Salesman Problem (TSP), 147, 859
Tuning Process, 209

U

Uncertainty, 307, 471, 539, 853
Unsupervised Classification, 113, 369
Unsupervised Learning, 69, 865

V

Validation and Verification, 563, 969, 1001, 1007
Variable Learning Rate, 81
Vehicle Control, 921
Vibration Analysis, 521
Visualization, 345, 363, 449, 583, 703

W

Wavelet Transform, 645
Weight Analysis, 939

AUTHOR INDEX

A

Aboul-Hassan, Sawsan, 685
Abu-Mahfouz, Issam, 521
Acar, Levent, 601, 619
Achaibou, K., 853
Agogino, Alice, 197
Ahmad, S., 803
Ahmed, Shamsuddin, 81, 1037
Aiello, Gaetano, 441
Akamine, Yuhei, 411
Akanda, Anab, 839
Aksenova, Tetyana, 557
Albert, Laura, 165
Alexander, Jr., John, 101
Al-Hindi, K., 41
Allamehzadeh, Hamid, 255
Alleyne, Andrew, 515
Al-Mashouq, Khalid A., 89
Amico, A., 975
Amirat, Yacine, 247
Ancona, Nicola, 639
Anderson, Steve, 307
Arciniegas, Fabio, 345
Ashlock, Dan, 405
Assawasantakul, Warawat, 295
Au, James, 645
Aur, Dorian, 569

B

Bach-y-Rita, Paul, 441
Balan, Rupa, 821
Banicescu, Ioana, 273
Bao, W. Y., 203
Bartlett, Eric, 903
Bauer Jr., Kenneth W., 577, 589
Beale, D., 847, 871
Beh, Chris, 551
Bellman, Kirstie, 471, 951, 969
Bennett, Kristin, 345
Bilegan, Ioana, 853
Biswas, Sudip, 607
Bogullu, Vamsi, 313
Boriskevich, Anatoly, 833
Bose, Subhash, 915
Bouchaffra, Djamel, 685
Bouzerdoum, Abdesselam, 81
Branca, A., 639
Breneman, Curt, 345
Bress, Robert, 153
Bridges, Susan, 273
Brown, David, 763
Brown-VanHoozer, Stefania, 449
Bruccoleri, M, 975
Bryden, Kenneth, 405
Buczak, Anna, 933
Buehrer, Daniel J., 279
Bulut, Mehmet, 527
Burns, Scott, 123

C

Calvert, David, 607, 803
Campa, Giampiero, 595
Canca, D., 865
Cansever, Galip, 527
Castillo, Oscar, 209, 539
Caulfield, H. John, 435
Challoo, Rajab, 261
Chen, C.L. Philip, 223
Chen, Hui-Chuan, 721, 763
Cheng, Z., 307
Cheung, John Y., 255, 821
Chi, Hoi-Ming, 3, 333
Chibirova, Olga, 557
Chih-Ming, Hsieh, 279
Chinnam, Ratna Babu, 733, 757
Chu, Y., 307
Chueh, Hao-En, 319
Cicchello, Orlando, 607
Cicirelli, Grazia, 639
Ciesielski, Vic, 357
Clemins, Patrick, 727
Coello, Carlos, 141, 177
Constantin, Joseph, 497, 509
Cook, George, 235
Coppel, Ross, 783
Coppock, Sarah, 339
Cordes, Gail, 613
Crispin, Yechiel, 927
Cross, Jim, 81
Cruz-Cortes, Nareli, 141

D

da Silva, Ivan Nunes, 897
Dagli, Cihan, 9, 313, 375, 739
Dalinghaus, Klaus, 679
de Souza, Andre Nunes, 897
Distante, Arcangelo, 639
Djouani, Karim, 247
Dolen, Melik, 1001, 1007
Dranger, Thomas, 883
Drew, Mark S., 645
Dryga, Oleksandr, 557
Duan, Minglei, 171, 215
Dubrovin, Valeriy, 939
Duenas, Felipe, 209

E

El Moudani, W., 853
Elqaq, Eyad, 301
Elsamaloty, H., 789
Embrechts, Mark, 153, 345
Endo, Satoshi, 135, 411, 423
Enke, David, 313, 375, 715, 739
Ersoy, Okan, 3, 333
Ewald, Hartmut, 665

F

Ferrier, Nicola, 107
Fisher, Mike, 957
Flanagan, C., 665
Flitman, Andrew, 775
Fotouhi, Farshad, 69
Fravolini, M., 595
Fredrich, Wolfgang, 909

G

Gacy, A., 797
Gantzer, C., 307
Garcia, J., 865
George, Laurent, 477
Georgiev, George, 827
Ghalia, Mounir, 915, 981

Ghioca, Teodora, 569
Goldberg, David, 123, 129, 165
Golosinki, Tad, 387
Gorban, Alexander N., 363, 657
Gray, Robert, 673
Greene, Kelly, 577, 809, 815
Gueorguieva, Natacha, 827
Guerrero, Fernando, 865

H

Hamad, Denis, 497, 509
Haskell, Kevin, 613
He, Aijing, 339
Hemminger, Thomas, 673
Hempoonsert, Jiranun, 695
Hemsathapat, Korakot, 375
Hernandez, Karen, 545
Hideg, Laszlo, 485
Holl, S., 775
Hou, Maxwell, 279
Hu, Haibo, 515
Hu, Hui, 387
Hu, W., 15
Hu, Y.H., 107
Hull, Jr., Stephen, 821

I

Ibrahim, Remzi, 357
Ishida, Takashi, 399
Iyer, R., 789

J

Jin, Kang-Ren, 721
Johnson, Matt, 1025
Johnson, Peter, 405
Jones, James, 613
Jones, N., 563

K

Kakazu, Yukinori, 63, 399, 491, 1013
Kamarthi, Sagar, 945
Kashiwagi, Dean, 1031
Kaushik, Saroj, 993
Kayikci, Ekrem, 1001, 1007
Kercel, Stephen, 461
Kesavan, Divya, 267
Knopf, George, 351, 583
Kojima, Fumihiro, 503
Kong, Xiangdong, 515
Koseeyaporn, P., 235
Kreesuradej, Worapoj, 241
Kremer, Stefan C., 607
Krishnamurthy, E. V., 987
Krishnamurthy, Vikram, 987
Krishnapuram, Raghu, 369
Kropas-Hughes, C., 203
Kudryavtsev, Vadim, 833
Kulkarni, Arun, 381
Kumar, Manoj, 993
Kumar, Vinay, 733, 757
Kuwashima, Yuji, 1013

L

Lam, Sarah, 183, 859
Lamm, Ross, 57

Land Jr, Walker, 839
Landauer, Christopher, 455, 471, 951, 969
Lanning, Jeffrey, 589
Leep, Herman, 533
Lewis, E., 665
Li, Hongxing, 223
Li, L., 203
Li, Shuhui, 261
Lin, Chun-Shin, 41
Lin, Nancy P., 319
Liu, Qitao, 273
Lo, James Ting-Ho, 33
Lo, Joseph, 839
Lopez-Dominguez, Jose, 285
Lorenz, Robert, 1001, 1007
Lozano, Sebastian, 865
Lu, Cheng, 645
Lu, Qiang, 633
Lursinsap, Chidchanok, 95
Lyons, W., 665

M

Marar, Joao Fernando, 417
Matheus, Justo, 545
Matsumura, Yoshiyuki, 159
Mazlack, Lawrence J., 339
McLauchlan, Robert, 261, 625, 709, 769
Melin, Patricia, 209, 539
Mercier, Gilles, 477
Mezura-Montes, Efren, 177
Mikhailov, Alexei, 113
Miller, David, 27
Miyagi, Hayao, 135, 423
Mo, Zhiwei, 381
Mora-Camino, F., 853
Morquin, Demian, 915
Mrazova, Iveta, 21
Murthy, V.K., 963

N

Nagai, Takashi, 491
Naito, Dai, 63
Nakamura, Sayaka, 399
Napolitano, M, 595
Nasr, Chaiban, 497, 509
Nasraoui, Olfa, 369
Nawaz, Zaygham, 89
Nerome, Moeko, 135
Neubert, Jeremiah, 107
Noel, Jeremy, 589
Norris, W., 285
Nukoolkit, Chakarida, 763
Nwadiogbu, Emmanuel, 933

O

Ohkura, Kazuhiro, 159, 503
Omar, S. Iqbal, 261
Overfelt, R., 871
Ozdemir, Muhsin, 345

P

Palaparthi, Sreekrishna, 859
Pallerla, Suresh, 625
Parsai, E. I., 789
Parthasarathy, Prasanna, 123
Perrone, Giovanni, 975

Petkova, Maria, 751
Pidaparti, R.M., 797
Pimentao, Joao, 1019
Pister, Kris, 197
Pitenko, A., 363
Pittner, Stefan, 945
Plett, Gregory, 75
Pochampally, Kishore, 945
Pok, Yang-Ming, 113
Popova, T., 657
Povinelli, Richard, 171, 215, 727
Priemer, Roland, 301, 883
Puerto, Francisco Del, 981

Q

Qian, Jixin, 229
Qian, Xiaofeng, 889

R

Rai, Sudhir, 821
Ranson, Aaron, 545
Raviwongse, Rawin, 295
Reeder, Florence, 429, 435
Reen, N., 203
Renna, P., 975
Richman, Michael, 703
Rizki, Mateen, 69
Rode, Steffen, 909
Ruxpakawong, Phongthep, 651

S

Sadovsky, M., 657
Sangole, Archana, 351, 583
Sankaravadivoo, S., 267
Santosa, Budi, 703, 745
Sarma, P., 797
Sastry, Kumara, 129
Schleis, George, 69
Schlindwein, F., 563
Schmidt, Norbert, 909
Scott, Peter, 633
Seltzer, R., 921
Serpen, Gursel, 789
Shrestha, Dev, 903
Simionescu, Petru-Aurelian, 847, 871
Sinha, A, 797
Siripitayananon, Punnee, 721
Smith, Kate, 357, 551, 775, 783, 865
Smithmaitrie, P., 235
Smooker, Peter, 783
Sohn, Sunghwan, 9
Song, Qing, 15
Sotelo, Fernando, 539
Sousa, Pedro, 1019
Specht, Donald, 57
Spithill, Terry, 783
Sreenivas, R., 285
Srihari, Krishnaswami, 183, 859
St. Clair, Dan, 957
Stacey, Deborah, 147, 803
Steiger-Garcao, Adolfo, 1019
Steward, Brian, 903
Strauss, A., 235
Stupmann, Frank, 909
Subbotin, Sergey, 939
Sugiyama, Tetsuya, 47
Suzuki, Takayuki, 63

T

Takahashi, Nobuyuki, 491
Tansel, Berrin, 695
Tansel, Ibrahim, 203, 695, 921
Terpenny, Janis, 325
Tetko, Igor, 557
Thammano, Arit, 651
Thanwornwong, Suraphan, 739
Thomason, J, 803
Toma, Naruaki, 423
Touati, Youcef, 247
Trafalis, Theodore, 703, 745
Tse-Win, Lo, 279

U

Ueda, Kanji, 159, 503
Ulson, Jose, 897
Uluyol, O., 933
Ustun, Seydi Vakkas, 527

V

Vadde, Srikanth, 945
Vaidyanathan, Ganesh, 267
Vaithianathasamy, Swaminathan, 183
Vaitianathasamy, Swaminathan, 715
Valova, Iren, 827
Van Ausdeln, Leo, 613
VanHoozer, W., 449
Vasquez, Edmundo, 405
Vemuri, G., 797
Vilcu, Dana, 477
Villa, A.E.P., 557
Vivas, Angel, 545

W

Wang, Jiachuan, 325
Wang, Ligong, 261
Warheit, Dennis, 859
Wasserman, Gary S., 733
Watanapa, Wattana, 295
Webley, Paul, 551
Weckman, Gary, 709, 769
Wettayaprasit, Wiphada, 95
Weyde, Tillman, 679
White, Anderson, 703
Wilkerson, Ralph, 1025
Wiwattanakantang, Chokchai, 241
Woo, Frederick, 357
Woo, Grace, 877
Woo, Peng-Yung, 877
Woodley, Robert, 601, 619
Wunsch, Donald, 363, 657

X

Xiang, Jianping, 563
Xu, Weimin, 889
Xue, Yuncan, 229

Y

Yamada, Kazuaki, 503
Yamada, Koji, 135, 411, 423
Yan, Lian, 27
Yang, L. X., 147
Yatsuzuka, Yohtaro, 47

Yatzenko, Viktor, 939
Yokoi, Hiroshi, 63, 399, 491, 1013
Yu, Songnian, 889
Yuen, W., 921

Z

Zhang, Qin, 285, 515
Zhou, Ningning, 197
Zhu, Bo, 197
Zhu, Yaoyao, 339
Zinovyev, A., 363
Zmuda, Michael, 189

533
drilling, no ANN.